STRUCTURAL MODELING AND EXPERIMENTAL TECHNIQUES

SECOND EDITION

Harry G. Harris
Drexel University
Philadelphia, Pennsylvania

and

Gajanan M. Sabnis
Howard University
Washington, D.C.

CRC Press
Boca Raton   London   New York   Washington, D.C.

Acquiring Editor: Tim Pletscher
Project Editor: Carol Whitehead
Marketing Managers: Barbara Glunn, Jane Lewis, Arline Massey, Jane Stark
Cover design: Jonathan Pennell
PrePress: Kevin Luong, Gary Bennett
Manufacturing: Carol Slatter

Library of Congress Cataloging-in-Publication Data

Harris, Harry G.
    Structural modeling and experimental techniques / Harry G. Harris
and Gajanan M. Sabnis. -- 2nd ed.
        p.   cm.
    First ed. pub. under title: Structural modeling and experimental
techniques.
    Includes bibliographical references and index.
    ISBN (invalid) 0-8493-2469-6 (alk. paper)
    1. Structural frames--Models--Testing.   2. Structural design.
I. Sabnis, Gajanan M. II. Title.
TA643.H37   1999
624.1′773--dc21   98-46557
                                                                           CIP

This book contains information obtained from authentic and highly regarded sources. Reprinted material is quoted with permission, and sources are indicated. A wide variety of references are listed. Reasonable efforts have been made to publish reliable data and information, but the author and the publisher cannot assume responsibility for the validity of all materials or for the consequences of their use.

Neither this book nor any part may be reproduced or transmitted in any form or by any means, electronic or mechanical, including photocopying, microfilming, and recording, or by any information storage or retrieval system, without prior permission in writing from the publisher.

The consent of CRC Press LLC does not extend to copying for general distribution, for promotion, for creating new works, or for resale. Specific permission must be obtained in writing from CRC Press LLC for such copying.

Direct all inquiries to CRC Press LLC, 2000 Corporate Blvd., N.W., Boca Raton, Florida 33431.

Trademark Notice: Product or corporate names may be trademarks or registered trademarks, and are only used for identification and explanation, without intent to infringe.

© 1999 by CRC Press LLC

No claim to original U.S. Government works
International Standard Book Number 0-8493-2469-6
Library of Congress Card Number 98-46557
Printed in the United States of America 1 2 3 4 5 6 7 8 9 0
Printed on acid-free paper

Preface

Since the first edition of this book, many developments in the use of structural modeling have taken place. Significant among these is the extension of the modeling technique to masonry structures and the popular use of models on shaking tables to study the earthquake resistance of structures. This second edition attempts to document these and other developments and presents to the student an up-to-date text suitable for use in advanced undergraduate or graduate level courses in structural modeling or in the behavior of civil and architectural engineering structures using experimental techniques.

Structural models have always played a significant role in structural engineering research, design, and education. More recently, the modeling technique has been used in structural product and structural concept development and is illustrated in the second edition with several detailed examples. Spurred by the extensive developments in microminiaturization of instrumentation and electronics and renewed interest in using experimentation in the new engineering curriculum, the use of structural modeling techniques has gained new importance in civil and architectural engineering education. The purpose of this textbook is to present a current up-to-date treatment of structural modeling for applications in design, research, education, and product development.

This extensively illustrated textbook treats equally the techniques of modeling of reinforced and prestressed concrete, masonry, steel, and wooden structures under static and dynamic loads. New chapters on model fabrication methods and use of structural models in civil and architectural engineering education have been added to the second edition. Many worked out examples and case studies are used to guide the student or practicing engineer through the necessary steps of using the structural modeling technique for themselves in understanding structural behavior. Numerous problems of varying degrees of difficulty after each chapter test the understanding of the reader.

The authors view the structural model as a complement to the mathematical model and not as a competitor nor a replacement for analysis. A rule of thumb is that if an appropriate, well-tested analytical approach exists for a given situation, it will usually be less expensive and quicker than an experimental approach. If analysis is not feasible, or if boundary conditions are poorly defined or highly variable, the model test may be the only solution to the problem. Models must be used selectively, and their range of application is constantly changing as analytical methods get increasingly more powerful. This book should prepare the reader to form the proper perspective as to when and where models should be utilized.

Models for determining the elastic response of structures have been used for many years, and considerable information is available in the form of research reports and books. Many elastic modeling techniques have been replaced by computer-based analysis methods, and therefore they are treated here in a relatively brief manner.

The major emphasis is on the modeling of the true inelastic behavior of structures. Compared with elastic models, the problems in inelastic modeling are considerably greater — starting from the selection of materials to be used in the models, the techniques of fabrication, instrumentation, and testing to the interpretation of model results to predict the behavior of the simulated prototype structure.

Applications of the modeling techniques to real structures help one to better understand the actual process of model analysis. They also assist in forming a perspective on the types of structures for which physical modeling is important. There are some types of special structures where the models approach plays an essential role in design. These topics are given detailed treatment in the form of case histories.

Chapter 1 discusses the historical background of model analysis and similitude principles that govern the design, testing, and interpretation of models. Eight well-chosen case studies illustrate the wide use of model studies covering application to a variety of structural forms such as buildings, bridges, and special structures and a variety of construction materials including reinforced concrete, masonry, and wood. Chapter 2 presents the theory of structural modeling under static and dynamic loading. Chapter 3 deals with the various aspects of elastic models with additional material added to cover the properties of balsa and other wooden models. Chapter 4 treats materials for concrete and concrete masonry models. Extensive material has been added in this chapter to cover the properties of block masonry structures. Chapter 5 presents the material modeling requirements for structural steel and reinforcements for reinforced and prestressed concrete structures. Accurate modeling of the properties of both concrete and steel is absolutely essential and is one of the more difficult parts of the modeling process. Chapter 6 is a new chapter that covers the various model fabrication techniques.

Instrumentation techniques are treated in Chapter 7 with emphasis on strain measurement and interpretation. Chapter 8 presents selected laboratory techniques and loading methods. Developing a sound familiarity with these techniques is an integral part of model analysis. A new Chapter 9 deals with size effects, accuracy, and reliability of materials and models. An understanding of the nature and source of scale effects is crucial in understanding the capabilities and limitations of the modeling technique. Chapter 10 covers actual applications of structural modeling under static or quasi-static loading. Case studies of a number of important uses of modeling in design, research, and product development are presented. These are drawn from the experience of the authors and their colleagues and cover a wide spectrum of applications.

Dynamic load effects such as wind, blast, impact, and earthquakes are treated in Chapter 11. Case studies of important model structures under dynamic loading are also presented. The problems discussed in earlier chapters dealing with static loading become more difficult in dynamic studies. Additional similitude requirements and new experimental techniques are necessary when dealing with dynamic structural models.

A new Chapter 12 dealing with educational models for civil and architectural engineers has been added to the second edition. A large number of applications, mainly from the teaching experience of the senior author at Drexel University, illustrates the usefulness of the structural modeling technique in teaching structural behavior to undergraduate and graduate students alike.

We believe that this book will be of substantial assistance not only to students of civil and architectural engineering curricula but also to other engineering students with interest in experimental techniques. It should be helpful to practicing engineers, designers, and architects who are dealing with structures. It should also be useful to those engaged in testing large or full-scale structures since the instrumentation techniques and overall approaches used in testing large structures are very similar to those utilized in small-scale modeling work.

Information on typical structural models courses and how the material presented in this second edition can be introduced into the classroom can be found on the senior author's web page: www.pages.drexel.edu/faculty/harrishg/

We express many thanks to Dr. Richard N. White, Cornell University, advisor, mentor, and colleague who contributed greatly to the first edition and without whose collaboration this book would not have been possible. Also, thanks go to Dr. M. Saeed Mirza, McGill University, for his contributions to the first edition. The authors are indebted to Dr. Stuart Swartz, Kansas State University, and Dr. Philip Perdikaris, Case Western Reserve University, for reviewing the entire manuscript and to Drs. Mohamed Elgaaly, Aspa Zerva, Mahmoud El-Sherif, and Prof. James Mitchell, Drexel University, for reviewing Chapters 5, 9, and 12, respectively. Former graduate students: Dr. Vincent Cacese, University of Maine, Dr. Bechara Abboud, Temple University, Ivan J. Becica, Dr. Ahcene Larbi, Dr. Win Somboonsong, and Basem Dow reviewed parts of the manuscript for which we are very grateful. Thanks are also due to our many colleagues who provided photographs and other material for this book. Among those, we would like to thank Drs. Helmut Krawinkler, Daniel Abrams, Philip Perdikaris, Miha Tomazevic, Nabil Grace, Ahmad Hamid, Theodor Krauthammer, Emin Aktan, Vincent Caccece, Bechara Abboud, Ahcene Larbi, Salah Sagur, Kang-Ho Oh, and Win Somboonsong and Mr. Norman Hoffman. A heartfelt thanks to George Papayiannis, who labored through the equations and tables of the draft. Finally, we would like to thank our many students over the years who persevered through our structural models courses and whose needs were the inspiration for this book.

Harry G. Harris
Gajanan M. Sabnis
Philadelphia, Pennsylvania

Foreword

This revised edition of an earlier edition entitled *Structural Modeling and Experimental Analysis* by Sabnis, Harris, White, and Mirza is a very welcome addition to the structural engineering literature. It provides an up-to-date exposition of the bases, principles, and practice of experimental study of the actual physical behavior of a broad class of structures, ranging through all the common materials of construction with particularly strong treatment of concrete and masonry structures. It is written from the premise that carefully done experimental studies are essential in developing a clear understanding of how a structure behaves at high load levels (up through the failure stage). The resulting insight also is required in the formulation of meaningful mathematical models for further study through analytical means.

While the basic thrust of the book is directed to reduced-scale models, these same principles are just as applicable to full-scale programs. One of the strongest features of the book is its very solid engineering approach to all phases of experimental study of structures, including initial planning, materials selection, instrumentation, loading design and application, error analysis and size effects, and conducting the test and reporting on results.

The expanded coverage in this new edition includes detailed information on loading systems — static and dynamic, including wind, earthquake, blast, and impact. Dynamic experimental structures studies are now perhaps the most important of all, given the fact that nonlinear static response can be increasingly well treated analytically.

Perhaps the best way to truly appreciate this book is to study the many excellent applications and five detailed case studies covered in Chapter 10. Each case study provides a wealth of information on experimental investigations of complex structural behavior.

A new feature of the book is a comprehensive chapter (Chapter 12) on educational models for civil and architectural engineering students. The use of physical models in teaching structural engineering provides students with a level of understanding not achievable through theory alone. They are also invaluable in motivating and encouraging students to appreciate the complexities and subtleties of structural behavior. I heartily encourage my colleagues in academia to incorporate these kinds of experiments in their courses at both the undergraduate and graduate levels.

Richard N. White
James A. Friend Family Professor of Engineering
Cornell University
Ithaca, New York

The Authors

Harry G. Harris has been a professor of Civil and Architectural Engineering at Drexel University since joining the faculty in 1974. Prior to this, he worked in several positions for Grumman Aerospace Corporation (Northrop-Grumman). He earned his B.E. (civil) degree from the American University of Beirut, S.M. and C.E. degrees from MIT, and his Ph.D. from Cornell University where he was an Instructor. He has authored or co-authored over 130 publications and several books dealing with structural engineering problems — many of them related to structural modeling and experimental techniques of concrete, masonry, steel, and wooden structures. He has conducted research for the Air Force, NASA, DOD, HUD, NRC, NSF, and many industrial concerns. Some of his areas of investigation include progressive collapse and earthquake resistance of precast concrete buildings; tasks under the reinforced concrete and masonry portions of the U.S.–Japan Cooperative Program on Earthquake Engineering Research; earthquake qualification of masonry walls in nuclear power plants, and the development of ductile hybrid FRP (D-H-FRP) reinforcement for concrete structures. He is active in several professional societies including ASCE, ACI, TMS, and EERI.

Gajanan M. Sabnis has been Professor of Civil Engineering at Howard University since 1973. Prior to this, he worked for Bechtel Corporation as well as the American Cement Corporation. He took his undergraduate studies in Mumbai, India and obtained his Ph.D. from Cornell University in 1967. He has authored over 100 publications and several books related to concrete, including rehabilitation of concrete structures. Dr. Sabnis is active in various technical committees of professional societies, which include ASCE, ACI, ASEE, and PCI. He has traveled and lectured extensively internationally. In addition to his academic achievements, Dr. Sabnis heads two design and research related firms and is a registered professional engineer in many states, including California, Maryland, and New York.

Table of Contents

Chapter 1 Introduction to Physical Modeling in Structural Engineering
1.1 Introduction ...1
1.2 Structural Models — Definitions and Classifications ...2
1.3 A Brief Historical Perspective on Modeling ...6
1.4 Structural Models and Codes of Practice ...7
1.5 Physical Modeling and the New Engineering Curriculum ...8
1.6 Choice of Geometric Scale ...9
1.7 The Modeling Process ...10
1.8 Advantages and Limitations of Model Analysis ...11
1.9 Accuracy of Structural Models ...12
1.10 Model Laboratories ...13
1.11 Modeling Case Studies ...13
1.12 Summary ...34
Problems ...36
References ...37

Chapter 2 The Theory of Structural Models
2.1 Introduction ...42
2.2 Dimensions and Dimensional Homogeneity ...42
2.3 Dimensional Analysis ...45
2.4 Structural Models ...56
2.5 Similitude Requirements ...62
2.6 Summary ...76
Problems ...77
References ...81

Chapter 3 Elastic Models – Materials and Techniques
3.1 Introduction ...86
3.2 Materials for Elastic Models ...87
3.3 Plastics ...88
3.4 Time Effects in Plastics — Evaluation and Compensation ...96
3.5 Effects of Loading Rate, Temperature, and the Environment ...100
3.6 Special Problems Related to Plastic Models ...103
3.7 Wood and Paper Products ...104
3.8 Elastic Models — Design and Research Applications ...121
3.9 Determination of Influence Lines and Influence Surfaces Using Indirect Models — Müller-Breslau Principle ...121
3.10 Summary ...123
Problems ...124
References ...127

Chapter 4 Inelastic Models: Materials for Concrete and Concrete Masonry Structures
4.1 General 130
4.2 Prototype and Model Concretes 130
4.3 Engineering Properties of Concrete...131
4.4 Unconfined Compressive Strength and Stress–Strain Relationship............133
4.5 Tensile Strength of Concrete 142
4.6 Flexural Behavior of Prototype and Model Concrete 146
4.7 Behavior in Indirect Tension and Shear 148
4.8 Design Mixes for Model Concrete 153
4.9 Summary of Model Concrete Mixes Used by Various Investigators..........159
4.10 Gypsum Mortars 165
4.11 Modeling of Concrete Masonry Structures 170
4.12 Strength of Model Block Masonry Assemblages..188
4.13 Summary 202
Problems...202
References 205

Chapter 5 Inelastic Models: Structural Steel and Reinforcing Bars
5.1 Introduction 210
5.2 Steel..210
5.3 Structural Steel Models..214
5.4 Reinforcement for Small-Scale Concrete Models.......................................230
5.5 Model Prestressing Reinforcement and Techniques....................................252
5.6 FRP Reinforcement for Concrete Models...255
5.7 Bond Characteristics of Model Steel...259
5.8 Bond Similitude 266
5.9 Cracking Similitude and General Deformation Similitude in Reinforced Concrete Elements 267
5.10 Summary 272
Problems...272
References 274

Chapter 6 Model Fabrication Techniques
6.1 Introduction 280
6.2 Basic Cutting, Shaping, and Machining Operations...................................281
6.3 Basic Fastening and Gluing Techniques..283
6.4 Construction of Structural Steel Models 287
6.5 Construction of Plastic Models 288
6.6 Construction of Wood and Paper Models 296
6.7 Fabrication of Concrete Models 299
6.8 Fabrication of Concrete Masonry Models...309
6.9 Summary 312
Problems...312
References 315

Chapter 7 Instrumentation — Principles and Applications
- 7.1 General 320
- 7.2 Quantities to be Measured 320
- 7.3 Strain Measurements 322
- 7.4 Displacement Measurements 348
- 7.5 Full-Field Strain Measurements and Crack Detection Methods 351
- 7.6 Stress and Force Measurement 353
- 7.7 Temperature Measurements 361
- 7.8 Creep and Shrinkage Characteristics and Moisture Measurements 362
- 7.9 Data Acquisition and Reduction 364
- 7.10 Fiber Optics and Smart Structures 367
- 7.11 Summary 377
- Problems 377
- References 381

Chapter 8 Loading Systems and Laboratory Techniques
- 8.1 Introduction 383
- 8.2 Types of Loads and Loading Systems 384
- 8.3 Discrete vs. Distributed Loads 389
- 8.4 Loadings for Shell and Other Models 390
- 8.5 Loading Techniques for Buckling Studies and For Structures Subject to Sway 400
- 8.6 Miscellaneous Loading Devices 404
- 8.7 Summary 407
- Problems 407
- References 409

Chapter 9 Size Effects, Accuracy, and Reliability in Materials Systems and Models
- 9.1 General 412
- 9.2 What Is a Size Effect? 414
- 9.3 Factors Influencing Size Effects 414
- 9.4 Theoretical Studies of Size Effects 415
- 9.5 Size Effects in Plain Concrete—Experimental Work 420
- 9.6 Size Effects in Reinforced and Prestressed Concrete 431
- 9.7 Size Effects in Metals and Reinforcements 433
- 9.8 Size Effects in Masonry Mortars 434
- 9.9 Size Effects and Design Codes 435
- 9.10 Errors in Structural Model Studies 437
- 9.11 Types of Errors 439
- 9.12 Statistics of Measurements 441
- 9.13 Propagation of Random Errors 444
- 9.14 Accuracy in (Concrete) Models 450
- 9.15 Overall Reliability of Model Results 457
- 9.16 Influence of Cost and Time on Accuracy of Models 458
- 9.17 Summary 458
- Problems 459
- References 460

Chapter 10 Model Applications and Case Studies
10.1 Introduction 466
10.2 Modeling Applications 466
10.3 Case Studies 529
10.4 Summary 572
Problems 573
References 579

Chapter 11 Structural Models for Dynamic Loads
11.1 Introduction 586
11.2 Similitude Requirements 587
11.3 Materials for Dynamic Models 588
11.4 Loading Systems for Dynamic Model Testing 593
11.5 Examples of Dynamic Models 604
11.6 Case Studies 649
11.7 Summary 673
Problems 673
References 675

Chapter 12 Educational Models for Civil and Architectural Engineering
12.1 Introduction 680
12.2 Historical Perspective 681
12.3 Linearly Elastic Structural Behavior 681
12.4 Nonlinear and Inelastic Structural Behavior 694
12.5 Structural Dynamics Concepts 712
12.6 Experimentation and the New Engineering Curriculum 725
12.7 Case Studies and Student Projects 729
12.8 Summary 751
Problems 752
References 753

Appendix A Dimensional Dependence and Independence
A.1 The Form of Dimensions 757
A.2 Method I: The Numeric Method 759
A.3 Method II: The Functional Method 761
A.4 Illustrative Examples 763
References 766

Appendix B A Note on the Use of SI Units in Structural Engineering
B.1 Geometry 768
B.2 Densities, Gravity Loads, Weights 768
B.3 Force, Moment, Stress, and Other Stress Resultants 768
B.4 Miscellaneous (Angles, Temperature, Energy, Power) 768
B.5 SI System Standard Practice 769

Index 771

To Helen and Sharda

CHAPTER 1

Introduction to Physical Modeling in Structural Engineering

CONTENTS

1.1	Introduction	1
1.2	Structural Models — Definitions and Classifications	2
	1.2.1 Models Classification	3
	1.2.2 Physical Models in Other Engineering Disciplines	5
1.3	A Brief Historical Perspective on Modeling	6
1.4	Structural Models and Codes of Practice	7
1.5	Physical Modeling and the New Engineering Curriculum	8
1.6	Choice of Geometric Scale	9
1.7	The Modeling Process	10
1.8	Advantages and Limitations of Model Analysis	11
1.9	Accuracy of Structural Models	12
1.10	Model Laboratories	13
1.11	Modeling Case Studies	13
	1.11.1 Case Study A, TWA Hangar Structures	14
	1.11.2 Case Study B, R/C Frame-Wall Structures	15
	1.11.3 Case Study C, Reinforced Concrete Bridge Decks	20
	1.11.4 Case Study D, Lightly Reinforced Concrete Buildings	21
	1.11.5 Case Study E, Prestressed Wooden Bridges	24
	1.11.6 Case Study F, Interlocking Mortarless Block Masonry	27
	1.11.7 Case Study G, Pile Foundations	30
	1.11.8 Case Study H, Externally/Internally Prestressed Concrete Composite Bridge System	31
1.12	Summary	34
Problems		36
References		37

1.1 INTRODUCTION

A perspective on physical modeling of structures is presented in this chapter, including classification of the various types of physical models that have evolved over the years and some comments on the general role of these categories of models in education, design, research, and product and concept development in North America and elsewhere.

Structural models (and reduced-scale structures sometimes called replica models) have always played a significant role in structural engineering education, research, and design. Experiments on reduced-scale structures and specimens have always been important in teaching structural mechanics and structural engineering concepts. A wide range of problems is met in planning, conducting, and interpreting an experimental study of structural behavior. Each of these areas, which range from theoretical similitude requirements to the rather extensive discipline of experimental stress analysis, will be treated in detail in subsequent chapters. This chapter gives the reader an overview and, it is hoped an appreciation of structural modeling in its broadest physical sense and of how structural modeling is used today in the engineering profession. This new expanded second edition of the book has added new material on the teaching of structural concepts to civil and architectural engineering students and on the incorporation of structural modeling techniques in teaching structural design concepts in the new engineering curriculum. Purely architectural models are also important in planning new construction and correlating spaces, but this type of model will not be considered here because its role is completely different from the structural model. Structural models used in architectural engineering education will be covered extensively, however, in several chapters of the book since these become the vehicle by which the student can be introduced to important structural behavior concepts in both analysis and design.

1.2 STRUCTURAL MODELS — DEFINITIONS AND CLASSIFICATIONS

A structural model is defined as "any **physical representation** of a structure or a portion of a structure. Most commonly, the model will be constructed at a **reduced scale**." This definition has been evolved by ACI Committee 444, Experimental Analysis for Concrete Structures. It applies equally well to models of structures made of any material, of course. A second definition given by Janney et al. (1970) is

> A structural model is any structural element or assembly of structural elements built to a *reduced scale* (in comparison with full size structures) which is to be tested, and for which *laws of similitude* must be employed to interpret test results.

Both definitions encompass a broad class of modeling studies on prototype (full-size) structures such as buildings, bridges, dams, towers, reactor vessels, shells, aerospace and mechanical engineering structures, undersea structures, etc. Loadings include static, simulated seismic, thermal, and wind effects.

Many reduced-size structural elements are customarily used in research studies; some investigators also class these structures as models even though similitude conditions are not normally applied to the large-scale research models. Instead, design methods and equations are based directly upon the observed behavior of these research models and are accordingly given full acceptance by the design profession. It should be noted that in the above studies prototype materials can still be used. An important distinction is usually made when prototype materials cannot be used for the reduced-size structure. In that case, appropriate model materials must be substituted and the reduced-size structure is then properly designated a "model."

This book contains extensive material on models of reinforced, prestressed, or precast concrete structures with a geometric scale factor of about 10, and block masonry structures with a geometric scale factor of 3 to 5. Material is also provided for steel and timber structures with a geometric scale factor of 5 to 15. Much of the material to be covered is applicable to models of other types of structures and to different scale factors.

1.2.1 Models Classification

Structural models can be defined and classified in a variety of ways. The definitions adopted here relate to the intended function of the model. That is, what do we expect to achieve from the tests? Do we want only *elastic response,* or do we expect to load the model up to failure to observe its complete behavior, including the *failure mode and capacity?* Are we content to work with *influence lines* determined from the models, or do we need actual *strain measurements* for prescribed loadings? The models needed in each of these applications have been given well-accepted names.

Elastic Model — This type of model has a direct geometric resemblance to the prototype but is made of a homogeneous, elastic material that does not necessarily resemble the prototype material. The elastic model is restricted to the elastic range of behavior of the prototype and cannot predict postcracking behavior of concrete or masonry, postyield behavior of steel, nor the many other inelastic behavior modes that develop in actual structures when they are loaded. Chapter 3 treats elastic models in detail, including selection of materials. Plastics such as methyl methacrylate (Plexiglas, Lucite, Perspex) and polyvinylchloride (PVC) are most widely used in constructing elastic models, even though their time-dependent properties present difficulties. Wood of many commercial species and balsa wood are also covered in Chapter 3 since these find many applications in modeling timber structures and are easy to work with. In many applications of elastic models, especially in the demonstration of structural behavior to students, materials with low modulus of elasticity such as rubber, paper products, and plastics are used to accentuate deformations making it easy to observe the behavior. These materials are also discussed in Chapter 3.

Indirect Model — An indirect model is a special form of the elastic model that is used to obtain *influence diagrams* for reactions and for internal stress resultants such as shearing forces, bending moments, and axial forces. The loading applied to indirect models has no correspondence to the actual loads expected on the prototype structure since load effects are obtained from superposition of the influence values. An indirect model often does not have a direct physical resemblance to the prototype; for example, a frame whose behavior is controlled by its flexural stiffness properties (EI) can be modeled with an indirect model that correctly reproduces the relative stiffness values. The latter can be done without precise scaling of the cross-sectional shape (circular shapes in the indirect model can represent prototype wide-flange sections), and the element areas may be grossly distorted without affecting the results.

In the past, most applications of the indirect model have been for nonuniform members in indeterminate frames, but now this type of model finds very little use because these purely elastic calculations are better done by computer.

Direct Model — A direct model is geometrically similar to the prototype in all respects, and the loads are applied to it in the same manner as to the prototype. Strains, deformations, and stresses in the model for each loading condition are representative of similar quantities in the prototype for the corresponding loading condition. Thus, an elastic model can also be a direct model. Examples of indirect and direct models are shown in Figure 1.1.

Strength Model — This type of model is also called ultimate strength or realistic model or *replica model* and is a direct model that is made of materials that are similar to the prototype materials such that the model will predict prototype behavior for all loads up to failure. A strength model of a reinforced concrete element or structure must be made from model concrete and model-reinforcing elements where each of the materials satisfies the similitude conditions for the prototype materials. The latter represents the most difficult problem met in strength models for concrete structures; Chapters 4 and 5 cover these topics in depth. Strength models can also be made for steel structures,

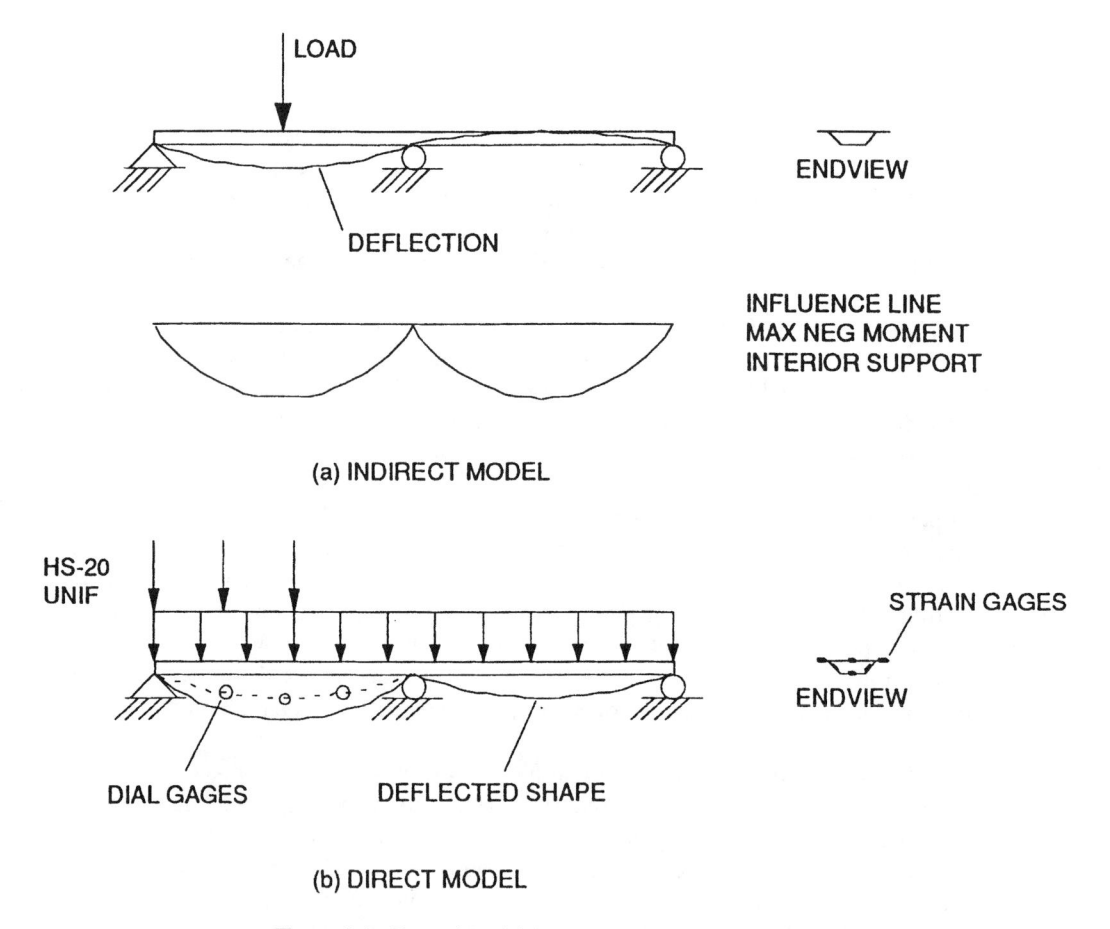

Figure 1.1 Examples of (a) indirect and (b) direct models.

timber structures, etc., and in each case the major problem is in finding the proper materials and fabrication techniques for the models.

A strength model must be a direct model, by definition. To use the results of indirect models, one must rely on superposition of results, and the superposition principle is not valid for the postlinear response found in all strength models. It is not economical to build strength models and use them only in the elastic range of behavior.

Wind Effects Model — There are various ways of classifying wind effects modeling. We can utilize *shape* or *rigid models,* where either total forces or the wind pressures on the structure may be measured, and *aeroelastic models,* where both the shape and stiffness properties of the prototype structure are modeled in order to measure the wind-induced stresses and deformations and the dynamic interaction of the structure with the wind.

Dynamic Models — These models are used to study vibration or dynamic loading effects on structures. They may be tested on a shaking table for studying earthquake loading effects or in a wind tunnel for studying aeroelastic effects. Dynamic models can also be used to study internal or external blast effects or impact effects on structures. These form a very important group of models and are extensively covered in Chapter 11.

Instructional, Research, and Design Models — One often sees the classification of instructional models, research models, and design models. While the use of each is obvious, it is worthwhile to point out that the degree of sophistication needed in each may be markedly different. Instructional models should be made as simple as possible to demonstrate the concepts under study, and similitude distortion that does not markedly influence the desired behavior is permissible. Research models, from which theories may be substantiated and generalizations made for a class of structures, usually must be made with as much accuracy as the laboratory technicians can master. Design models may range in accuracy requirements from the instructional model to the research model, depending upon the desired results. Some design models may be used only as a conceptual tool to get a better idea of how a proposed structure deforms under load; others may be expected to predict the true load capacity of the structure. The use of structural models as direct aids in design is one of the most powerful applications of structural models. An engineer is often called upon to design structures such as the Trans World Airlines maintenance hangar facility located in Kansas City, MO and described in Chapter 10. In this instance a series of plastic models and a strength model were used as the main approach in the final design.

Another major physical modeling application in design is to help verify calculations for very large and monumental structures where failure consequences could be extremely serious (such as heavy loss of life or capital investment, or disruption of essential lifeline services). A nuclear reactor structure is a good example of this application.

Perhaps the major disadvantages of using a model for design purposes, from the standpoint of the consulting engineer, are the time and money involved in the modeling study. This topic is given further attention in Section 1.8.

Other Model Classifications — Other classifications of models include thermal models, where effects of temperature gradients are studied. Thermally loaded models are usually elastic, direct models, although some attempts have been made to combine mechanical loads and thermal loads for strength models. There is also a group of *photomechanical* models that utilize optical effects, such as the *photoelastic effect* for stress intensities and directions and *interference effects* from grids to measure plate displacements, internal strain fields, and deflections of framed structures. *Construction procedure* models are used to help plan the building of very complex structures, such as in reinforcement placement in nuclear reactor containments, and in cantilever bridges.

1.2.2 Physical Models in Other Engineering Disciplines

Hydraulic engineering models have been studied as early as the late 1800s with considerable success. Studies of fluid motion in pipes, pumps, and open channels, wave action, beach erosion, silting due to tides and tital currents, and the extent of contaminants due to pollution in important ecstuaries and river basins have all been successfuly carried out by means of physical models. Today, all important hydraulic structures are designed and built after certain preliminary model studies have been completed.

Naval architects have for a long time relied on the use of physical models in the design of ships. All the important features of ship design, such as ship maneuverability in smooth and rough seas, ship bending and vibrations, frictional and wave-making characteristics, etc., have been studied by means of scale models in model testing tanks or water basins.

The automotive and aerospace industries have relied on the use of physical models for studying a number of phenomena ranging from the wind flow over aerodynamic bodies to vibration testing and thermal stress effects. Important contributions to the advancement of automotive and aeronautical sciences and space exploration would not have been made possible were it not for model testing. Important examples in these areas are covered by Schuring (1977).

The geological, geomechanical, and soil mechanics sciences have also used physical models to study soil and rock phenomena. More recently, model studies dealing with geosynthetics and geomechanics have also been carried out. Modeling of soil–structure interaction is briefly treated in the modeling of structural systems using a centrifuge in Chapter 2 and in an example in Chapter 11.

1.3 A BRIEF HISTORICAL PERSPECTIVE ON MODELING

The use of small-scale models by engineers and builders dates back many hundreds and even thousands of years. However, these early models were primarily aids for planning and constructing structures and were not useful for predicting deformations and strengths of prototypes. They more nearly resemble the modern architectural model and should not be thought of in the same context as structural models.

Most models used to predict structural behavior require measurement of strains, displacements, and forces. Thus, the development of modeling as a practical tool has been sharply influenced by abilities in experimental stress analysis. The most-used techniques in experimental stress analysis have been established only since the turn of the century. They include:

1. Photoelasticity for elastic stress analysis of complex geometries;
2. Deformeters developed by Beggs, Eney, Gottschalk, and others for introducing deformations into indirect models and then determining influence lines by use of the Müller–Breslau principle;
3. Mechanical and optical strain gages for measurement of surface strain;
4. Electrical resistance strain gages;
5. Linear variable differential transformers (LVDT), linear potentiometers, and similar devices for electrical recording of displacements;
6. Brittle coatings, moiré and interference fringe methods, and photoelastic coatings for "full-field" strain measurements on the surface of a structure or model;
7. Automated data acquisition systems that use a minicomputer to control and process many channels of data.

Item 2, deformeters in indirect models, has been used to study influence lines and to determine deflections in skeletal structures since the early 1920s. Beggs (1932) and Eney (1939) were the first to use specially designed deformeters to obtain influence lines for skeletal structures made of plastic and other materials. The simplest application is the use of a long, flexible strip of wood, brass, or steel and to measure the model deformations directly. The Gottschalk (1926) continostat is an improvement on this technique. However, large deformations must be imposed on the model, which causes other kinds of errors (Kinney, 1957). Bull (1930) used the brass spring model for indirect analysis of articulated structures. Development of other deformeters is reported by Ruge and Schmidt (1939) at MIT (moment deformeter) and by Moakler and Hatfield (1953) at RPI (RPI deformeter).

Item 4, the electrical resistance strain gage, is perhaps the single most important development in terms of providing an easily used method for determining either static or dynamic strains in a structure. The same gage forms the sensing device in commonly used load cells and transducers. Thus, its introduction in the 1940s can be considered to be the basis for modern experimental stress analysis and structural model analysis (Ruge, 1943/4; Perry and Lissner, 1955).

Relatively little model analysis other than photoelastic studies and indirect models was done prior to 1940. The $1/240$-scale Hoover Dam model built in 1930 and models of other great dams of that era built by the Bureau of Reclamation, Denver, CO were a notable exception (Savage, 1934). Another notable exception of the early 1930s was the work of George Beggs and his co-workers at Princeton on suspension bridge models (Beggs et al., 1932). Since that time, the technology needed for rapid construction, instrumentation, and testing of structural models has continued to develop. The current use of structural modeling is introduced in this chapter, and the full range of applicability will become apparent as the reader progresses through the book.

1.4 STRUCTURAL MODELS AND CODES OF PRACTICE

Modeling has received relatively little attention in most North American building codes and specifications. However, most codes do contain special provisions that permit the engineer to make rather substantial use of models in the design process. For example, the 1969 City of New York Building Code contains the following paragraph:

(5) MODEL TESTS. Tests on models less than full size may be used to determine the relative intensity, direction, and distribution of stresses and applied loads, but shall not be considered as a proper method for evaluating stresses in, nor the strength of, individual members unless approved by the commissioner for this purpose. Where model analysis is proposed as a means of establishing the structural design, the following conditions shall be met:
  a. Analysis shall be made by a firm or corporation satisfactory to the commissioner.
  b. The similitude, scaling, and validity of the analysis shall be attested to by an officer or principal of the firm or corporation making the analysis.
  c. A report on the analysis shall be submitted showing test set-ups, equipment, and readings.

ACl 318-95, *Building Code Requirements for Structural Concrete*, permits model analysis for shell structures in Section 19.2.4:

19.2.4 — Experimental or numerical analysis procedures shall be permitted where it can be shown that such procedures provide a safe basis for design.

This is further elaborated on in the Code Commentary R19.2.4:

Experimental analysis of elastic models[19.14] has been used as a substitute for an analytical solution of a complex shell structure. Experimental analysis of reinforced micro-concrete models through the elastic, cracking, inelastic, and ultimate ranges should be considered for important shells of unusual size, shape, and/or complexity.
  For model analysis, only those portions of the structure which affect significantly the items under study need be simulated. Every attempt should be made to ensure that the experiments reveal the quantitative behavior of the prototype structure.

Some countries, such as Australia, permit complete designs of certain types of structures by model analysis alone. Thus, there is a relatively healthy potential usage of modeling in design codes. Many engineers would be even more receptive of modeling for design if they would only realize that many of the code provisions that they apply analytically every day are in fact derived mainly from tests on reduced-scale models. The engineer who wishes to use models should not hesitate to contact the responsible building official to seek approval and should also seek proper assistance from an expert in model analysis.

There are numerous situations in which these code provisions might be applied in practice; in most cases it is where the analytical approach is not fully adequate. Basic doubts may arise in applying existing analytical techniques to new and complex structural forms. Analytical methods are not yet developed to handle the extremely complicated behavior of reinforced concrete structures loaded to near-failure or certain other limit-state conditions. This is why modeling is often used by engineers studying the failure of structures.

Types of structures suitable for possible structural model studies during the design phase include:

1. Shell roof forms of complex configuration and boundary conditions;
2. Tall structures and other wind-sensitive structures for which wind tunnel modeling is indicated;

3. New building structural systems involving the interaction of many components;
4. Complex bridge configurations such as multicell prestressed concrete box girder highway bridges;
5. Nuclear reactor vessels and other reinforced and prestressed concrete pressure vessels;
6. Ordinary framed structures subjected to complex loads and load histories, such as wind and earthquake forces;
7. Structural slabs with unusual boundary or loading conditions, or with irregular geometry produced by cutouts and thickness changes;
8. Dams;
9. Undersea and offshore structures;
10. Detailing.

Item 10 points out an important use of models: for studying problems that arise in only a limited region of a structure, such as an involved connection detail or localized stresses due to large prestressing forces. Carefully designed and tested, the partial model can be extremely important in clarifying these situations and in leading to an improved design. The major difficulty with the partial model is in providing proper boundary conditions: an inadequate boundary condition in a physical model may produce even worse results than a poor boundary condition in an analytical solution.

1.5 PHYSICAL MODELING AND THE NEW ENGINEERING CURRICULUM

The new engineering curriculum, under development and implementation at Drexel and other universities with strong undergraduate engineering programs, relies on teaching basic engineering concepts by giving the student hands-on experiences in the laboratory (Quinn, 1994). Mastery of experimental techniques is expected in the first and second years. The students become familiar with how data are acquired, processed, and analyzed as well as how basic experimental techniques, devices, and methods are used in a wide variety of engineering disciplines. The experimental experience is not lost in upper-level engineering courses but is integrated with the theory. This means that experimental techniques must become more widespread to reach a much wider audience of the engineering community. For undergraduate students studying the fundamentals of structures, it means that laboratory facilities must exist where the design, fabrication, testing, and observation of the structural behavior can be carried out. This kind of experience can be invaluable to the new student who cannot afford to wait to be a practicing engineer with several years of experience to obtain such information. Students' understanding is greatly enhanced by observation and correlation with the theory that they are learning.

It should be pointed out that in the new engineering curriculum experimentation is an integral part of the engineering student's professional development. The incoming freshmen start engineering courses from day one. Emphasis is given to experimental methods in engineering because of their wide use in analysis, design, development, and manufacturing. Special attention is given to the interpretation and effective presentation of experimental results in written and oral forms. The computer is used effectively as a research and design tool. The students engage in professional design projects solving real-world problems. They learn interactively, through teamwork, gaining life-long learning skills.

Regardless of discipline, practicing engineers perform a wide variety of experiments throughout their careers. Mastering the techniques of experimentation is therefore very essential. During the freshman year, students conduct a three-hour laboratory each week in the *Engineering Test, Design, and Simulation Laboratory.* This state-of-the-art facility provides students with opportunities to exercise their imagination, satisfy their curiosity, and experience the joy of engineering. Through the use of computers and computer-controlled laboratory instruments, students are able to explore how experimentation is used in engineering applications. They become familiar with how data are acquired, processed, and analyzed as well as how basic experimental techniques, devices, and methods are used in a wide variety of engineering disciplines.

During the sophomore and junior years, the laboratory component is shared among several courses rather than being an integrated course as was the case in the freshman year. These courses teach the fundamentals of structural engineering and form the backbone of the young structural engineer's academic training. The laboratory work is completed by the student majoring in structures in courses such as *Construction Materials*, *Statics*, *Mechanics of Materials*, and *Introduction to Structural Analysis*. These courses continue the student's laboratory training in a variety of hands-on experiments and demonstrations using the techniques developed in the first year.

During the senior year, the structural engineering major takes courses such as *Project Design*, *Structural Design of Steel, Concrete, Masonry and Timber Structures,* and *Senior Seminar*. In addition, advanced seniors are allowed to take introductory graduate-level courses including *Model Analysis of Structures* and *Experimental Analysis of Nonlinear Structures I* and *II*. Examples of structures, drawn from these courses and demonstrating different levels of behavior, are described in Chapter 12 to illustrate how the laboratory component of the new engineering curriculum is implemented.

1.6 CHOICE OF GEOMETRIC SCALE

Any given model being built in a given laboratory has an optimum geometric scale factor. Very small models require light loads but can present great difficulties in fabrication and instrumentation. Large models are easier to build but require much heavier loading equipment. The latter requirement is not serious in a laboratory that is fully equipped to conduct tests on large structures, but it is a severe handicap in a smaller laboratory. Typical scale factors for several classes of structures are as follows:

Type of Structure	Elastic Models	Strength Models
Shell roof	$\frac{1}{200}$ to $\frac{1}{50}$	$\frac{1}{30}$ to $\frac{1}{10}$
Highway bridge	$\frac{1}{25}$	$\frac{1}{20}$ to $\frac{1}{4}$
Reactor vessel	$\frac{1}{100}$ to $\frac{1}{50}$	$\frac{1}{20}$ to $\frac{1}{4}$
Beam/slab structures	$\frac{1}{25}$	$\frac{1}{10}$ to $\frac{1}{4}$
Dams	$\frac{1}{400}$	$\frac{1}{75}$
Wind effects	$\frac{1}{300}$ to $\frac{1}{50}$	Not applicable

The rationale behind this table should become more apparent as we progress through the chapters of this book. Strength models of concrete structures have many practical dimensional limitations, such as minimum feasible thickness, bar spacing, cover, etc. Maintaining materials similitude requirements is a crucial problem in this category of models. As will become apparent later in the text, it is desirable always to try to use the largest physical size model for a particular application. An example of the various factors that influence the choice of model size is given in the following example.

Example 1.1

It is required to test two reinforced concrete strength models to demonstrate the principle of "over" and "under" reinforced concrete beam behavior. The available testing machine has a capacity of 45 kN and the work area to accommodate the model is approximately 0.5 × 2 m. Based on these physical limitations, a simply supported beam of up to 50 × 100 mm cross section made of sand and fine aggregate ("pea" gravel) can be fabricated using up to a No. 2 bar (6.35 mm) for the steel reinforcement. A model of the hypothetical prototype beam experiencing this behavior would be at ¼ to ⅙ scale. If only wire reinforcement of less than 6.35 mm in diameter is available, then a beam of appropriately smaller size will have to be used with a scale as small as ¹⁄₁₀ or ¹⁄₁₂. Obviously, a strength model of a larger portion of a reinforced concrete structure to be studied, using the same equipment, will have to be of a smaller cross section, hence scale.

1.7 THE MODELING PROCESS

The successful modeling study is one that is characterized by careful *planning* of the many diverse steps in the physical modeling process. An experimental study of an engineering structure is a small engineering project in itself, and as in any engineering venture, a logical and careful sequencing of events is an absolute necessity.

Detailed planning of an experiment is even more essential than planning of an analytical approach because refinement of a structural model halfway through the modeling process is usually impossible. A major aspect of planning is deciding what is expected from the model. Do we need only elastic stresses and displacements, or do we want to see how the structure behaves at overloads leading up to failure? Is instability a possible failure mode? Are thermal stresses involved? Do we have to simulate dynamic effects? The time required to complete the model study can range from perhaps a week or two for a very limited elastic model of a portion of a structure to 6 months or more for a detailed, ultimate-strength reinforced mortar model for predicting failure behavior of a complete structure. We obviously must guard against "overdoing" the model study just as we have to avoid excessive analysis of a structure. The engineer who bears final responsibility for the project must be the key person in prescribing precisely what the model is supposed to accomplish.

A typical modeling study can be broken into the following multistep process:

1. Define the *scope* of the problem, deciding what is needed from the model and what is not needed.
2. Specify *similitude* requirements for geometry, materials, loading, and interpretation of results. Pay particular attention to those similitude requirements that cannot be met, such as the desired equality of Poisson's ratio for concrete and plastics when doing elastic modeling of shell and slab structures (Chapter 2).
3. Decide on the size of model and required level of *reliability* or *accuracy*. What size model should be used consistent with the accuracy? If ±30% is adequate for design purposes, then an attempt to achieve ±10% accuracy is wasted effort and time (see Chapter 9).
4. Select model *materials* with proper attention to steps 1, 2, and 3 above (Chapters 3 through 5).
5. Plan the *fabrication* phase in consultation with the technicians who will be constructing the model, and follow the fabrication activities closely. This can be a frustrating part of modeling because it is often quite time-consuming (Chapter 6).
6. Select *instrumentation* and recording equipment for strains, displacements, forces, and other quantities. This step must be closely coordinated with steps 5 and 6, particularly if embedded strain gages are to be used in concrete models. Special strain gages and other equipment must be ordered well in advance of the actual time of usage (Chapter 7).

7. Design and prepare the *loading equipment;* new systems should be thoroughly checked out and calibrated before use on a model (Chapter 8).
8. Observe the *response* of the model during loading, taking complete notes and photographic records of the behavior. Do not rush through a test, and never leave anything to memory. Some investigators use tape or video recorders in this phase to record detailed comments on load history, cracking development, instability modes, and other information that may be difficult to describe numerically. Approximate calculations should be done before the experiment to estimate expected levels of response. Equilibrium checks should be done on results obtained early in the test.

    Because of the great importance in properly recording data when it must be done manually, a few specific comments are in order here.
    a. Prepare a ruled sheet with columns; put the date, names of test personnel, and the model designation on the first sheet.
    b. Record the readings directly, and do not attempt to reduce the data in one's head.
    c. Record the zero readings, allowing at least two lines of space since zero readings often must be taken more than once.
    d. Allow adjacent columns for reduction of results.
    e. Take readings at lower load increments as failure is approached.
    f. Take readings as yielding or failure actually occurs, even if the level of accuracy achieved is not high — approximate readings can give a better idea of behavior.
    g. Take final readings when the load is removed.
9. *Analyze the data and write the report* as soon as possible, while the entire test is still fresh in the mind. In addition to reporting the results, suggestions for improvement in techniques should be recorded to facilitate better modeling results in subsequent experiments.

Most of these steps are merely statements of common sense, but it is surprising how often common sense is ignored or left out of a crucial step in an experimental study. Several "laws" should be kept in mind when thinking about the difficulties of experimental work:

Murphy's Law: If something can go wrong, it will.
O'Toole's Law: Murphy is wildly optimistic.
Harris–Sabnis Law: Things are never as bad as they turn out to be.

There is a moral in these tongue-in-cheek laws, and that is the fact that experiments must be carefully planned, controlled, and interpreted if they are to succeed.

1.8 ADVANTAGES AND LIMITATIONS OF MODEL ANALYSIS

The main advantage of a physical model over an analytical model is that it portrays behavior of a complete structure loaded to the collapse stage. Although substantial progress is continually made in computer-based procedures for analysis of structures, we still cannot predict analytically the failure capacity of many three-dimensional structural systems, especially under complex loadings.

The prime motivation to conduct experiments on structures at reduced scales is to reduce the cost. Cost reductions come about from two areas: reduction of loading equipment and associated restraint frames, etc. and a reduction in cost of test-structure fabrication, preparation, and disposal after testing. The load-reduction factor is most dramatic since the concentrated load on a prototype is reduced in proportion to the square of the geometric scale factor of the model (a 100-kN prototype load is 0.25 kN on a ¹⁄₂₀-scale model). This reduction is even more dramatic when a low-modulus material such as plastic is used in the model.

The major limitations of using structural models in a design environment are those of time and expense. In comparing physical models with analytical models, one finds that the latter are normally less expensive and faster, and one cannot expect physical models to supplant or replace analytical modeling of structures when the latter procedure leads to acceptable definition of behavior of the prototype structure. Thus, physical models are almost always confined to situations where the mathematical analysis is not adequate or not feasible. Another limiting factor is that changes in the prototype design resulting from the results of a model study may require a second model to check the design. Practical considerations therefore often dictate that the model will be used to verify a "nearly finalized" design.

The time involved in modeling is often subjected to further pressures because the decision to go to a physical study is often made at the last minute, after more conventional approaches are proved inadequate. An engineer who is accustomed to getting all answers by analytical means is naturally hesitant to admit that the analysis is insufficient and that a physical model is needed. Suitable efforts must be made to predict earlier in the design process that a test is needed. This would enable earlier planning and a smoother, less hectic approach to the model study.

Design applications of structural models have been outlined earlier in this chapter. Structural models are also widely employed in research programs in such applications as the following:

1. Development of experimental data for verification of the adequacy of proposed analytical methods.
2. Study of basic behavior of complex structural forms such as shells.
3. Parametric studies on member behavior. Much of our basic research on reinforced concrete flexural members has been done on large-scale models.
4. Behavior of complete structural systems subjected to complex loading histories, such as coupled shear walls and connecting beams.
5. Development of new structural systems. The "dry stack" interlocking block masonry units described in Chapter 10 are an example.

Many of these areas of research modeling will be explored through examples in subsequent chapters. It is well recognized that research models play an invaluable role in improving knowledge of structural behavior and thereby pave the way for new and improved design methods. This role will always be important in structural engineering because it is a discipline founded strongly on physical behavior of real systems made of ordinary materials of construction.

1.9 ACCURACY OF STRUCTURAL MODELS

The reliability of the results from a given physical modeling study is perhaps the single most important factor to the user of the modeling approach. This topic is explored in depth in Chapter 9, and only a few general comments will be given here to stimulate the reader into thinking about this important topic. Adequate definitions of reliability and accuracy are difficult to formulate. One obvious measure is the degree to which a model can duplicate the response of a prototype structure. The problem met in such a comparison is the inherent variability in the prototype itself, particularly if it is a reinforced concrete structure. Two supposedly identical reinforced concrete elements or structures will normally show differences, sometimes as high as 20% or more, and when one must compare a model to a single prototype, the difficulty in making a firm conclusion on accuracy becomes rather apparent. Multiple prototypes and multiple models are needed to treat the results statistically, but the expense of even a single test structure is usually high, and the availability of sufficient data for application of statistical tests of significance is severely limited.

The factors affecting model accuracy include model material properties, fabrication accuracy, loading techniques, measurement methods, and interpretation of results. Elastic models can be built to give extremely high correlation with detailed computer-based results. The only limitation is in the cost of properly fabricating and loading the model. Elastic models of reinforced concrete structures can only predict elastic response and thus will have high accuracy (errors on the order of less than 5 to 10%) for structures with minimal cracking, such as shells.

Carefully designed and tested strength models of reinforced concrete beams, frames, shells, and other structures normally have maximum errors on the order of less than 15% for prediction of postcracking displacements and ultimate load capacity of the structure, provided that bond between steel reinforcement and model concrete is not the governing factor in behavior.

A better perspective on the degree of reliability to be expected in any particular model testing program can be achieved only by careful study of a large number of individual cases. The material presented in Chapter 9 as well as in other chapters and in the cited references will provide the reader with much of the material needed in studying model reliability.

1.10 MODEL LABORATORIES

There are a number of outstanding laboratories in Europe that have developed excellent reputations in physical modeling. There are no similar commercial laboratories in North America, but there are excellent structural modeling facilities at a number of private and institutional laboratories, including Wiss Janney Elstner and Associates in Northbrook, IL, the Portland Cement Association in Skokie, IL, and educational/research laboratories at Cornell University, McGill University, the University of Texas, Drexel University, and elsewhere. The Boundary Layer Wind Tunnel at the University of Western Ontario and a similar facility at Colorado State University are widely used for both research and investigation of wind effects on actual structures. Dynamic tests on shaking tables may be done at laboratories located at the University of California (UC), in Berkeley and in Los Angeles, the State University of New York (SUNY) Buffalo, the University of Illinois, Stanford University, the University of Calgary, Cornell University, Drexel University, and elsewhere.

The many diverse problems associated with structural modeling make it evident that high-quality structural modeling is best done by skilled engineers and technicians at established laboratories. This statement is not made to discourage newcomers; instead, it is merely a realistic comment on the difficulties of good experimental work. Considerable time and patience are required in establishing a diverse structural testing laboratory, whether it be at full scale or at the greatly reduced small model scale. One particularly important point to be made is that considerable amounts of trial-and-error approaches to materials development are now in the literature, and the careful individual can take full advantage of this material in setting up a laboratory. Thus, advances in structural modeling since 1960, coupled with the simplicity and reliability of modern instrumentation, have at least partially eased many of the difficulties to be faced in beginning a new laboratory operation.

1.11 MODELING CASE STUDIES

Several structural modeling studies will be used in this text to help illustrate the many facets of modeling. These have been selected because they represent a variety of structural systems, materials of construction, and types of loading. Two design model studies (designated A and E), four research model studies (B, C, D, and G), and two product development model studies (F and H) are utilized as case studies. General descriptions are given in this chapter, and the rest of the material is given in Chapters 10 and 11. Additional case studies of educational models are discussed in Chaper 12. Other modeling studies related to more specialized topics such as prestressed concrete

Figure 1.2 Completed Trans World Airlines maintenance facilities at Kansas City International Airport. (Courtesy of Wiss, Janney, Elstner & Associates, Inc., Northbrook, IL.)

pressure vessels, long-span bridges, dynamic response, reinforced concrete slabs with penetrations, shell roof stability, concrete masonry structures, and fiber-reinforced polymer (FRP) reinforced composite beams will also be given in Chapter 10.

1.11.1 Case Study A, TWA Hangar Structures

The Trans World Airlines Maintenance Hangar Facility located in Kansas City, MO, was designed with the aid of a series of structural models. The total facility consists of four long-span shell structures, each located at the corner of a cruciform-shaped framed building. Two completed shells are shown in Figure 1.2. Each shell spans about 96 m and consists of hyperbolic paraboloid (abbreviated here as *hypar*) surfaces spanning between hollow triangular edge members and an arch formed at the intersection of the two hypar surfaces. Each shell varies in thickness from 75 to 150 mm. Four separate model studies were used for the TWA hangar to resolve a series of design questions. They are referred to as models A1, A2, A3, and A4.

Model A1 was a $1/300$-scale shape model of the entire complex of four shells. It was placed in a wind tunnel and loaded to investigate the general character of wind pressures and wind flow over the site. Only the shape of the model is needed in this investigation; the model can be built of any convenient solid material such as softwood.

Model A2 (Figure 1.3) was another wind tunnel shape model of a single hangar shell structure, built at a scale of $1/100$ and tested to obtain a more-detailed and accurate representation of local wind pressures and flow patterns with doors opened, closed, etc.

The wind tunnel model results were utilized in the analytical design phase to help generate the proposed final design. The third model was then constructed to determine elastic behavior under many different load conditions. Model A3 (Figure 1.4) was $1/50$-scale structure made from plastic. It was loaded with discrete concentrated loads and with a vacuum loading. The evaluation of bending effects produced by the heavy maintenance equipment suspended from the shell was of particular importance in this model study. The stress resultants in the shell near the stiff edge members were also given considerable attention.

INTRODUCTION TO PHYSICAL MODELING IN STRUCTURAL ENGINEERING 15

Figure 1.3 Detailed wind model of hangar with doors closed. (Courtesy of Wiss, Janney, Elstner & Associates, Inc., Northbrook, IL.)

Figure 1.4 Elastic model enclosed in vacuum frame. (Courtesy of Wiss, Janney, Elstner & Associates, Inc., Northbrook, IL.)

Model A4 (Figure 1.5) was a realistic or strength model. Built at a scale of $1/10$, the model was designed to simulate the true behavior of a reinforced concrete shell structure loaded to failure. This strength model required duplication of both concrete and reinforcement at small scale. The load–deflection behavior of the shell was studied for several loadings before the model was loaded to failure to determine its failure mode and load capacity and factor of safety against collapse.

The TWA hangar models were done by the firm of Wiss, Janney, Elstner and Associates of Northbrook, IL.

1.11.2 Case Study B, R/C Frame-Wall Structures

As part of the first joint research effort by the planning group of the U.S.–Japan Cooperative Program on Large Scale Testing, a seven-story reinforced concrete frame-wall earthquake-resistant building was designed, constructed, and tested at the Large-Size Structures Laboratory, Building Research Institute, Tsukuba, Japan. The plan of a typical floor and the elevation of the test building prototype is shown in Figure 1.6. As can be seen from this figure, this test building consists of a seven-story space frame with walls. In the direction of loading and longitudinal direction there are

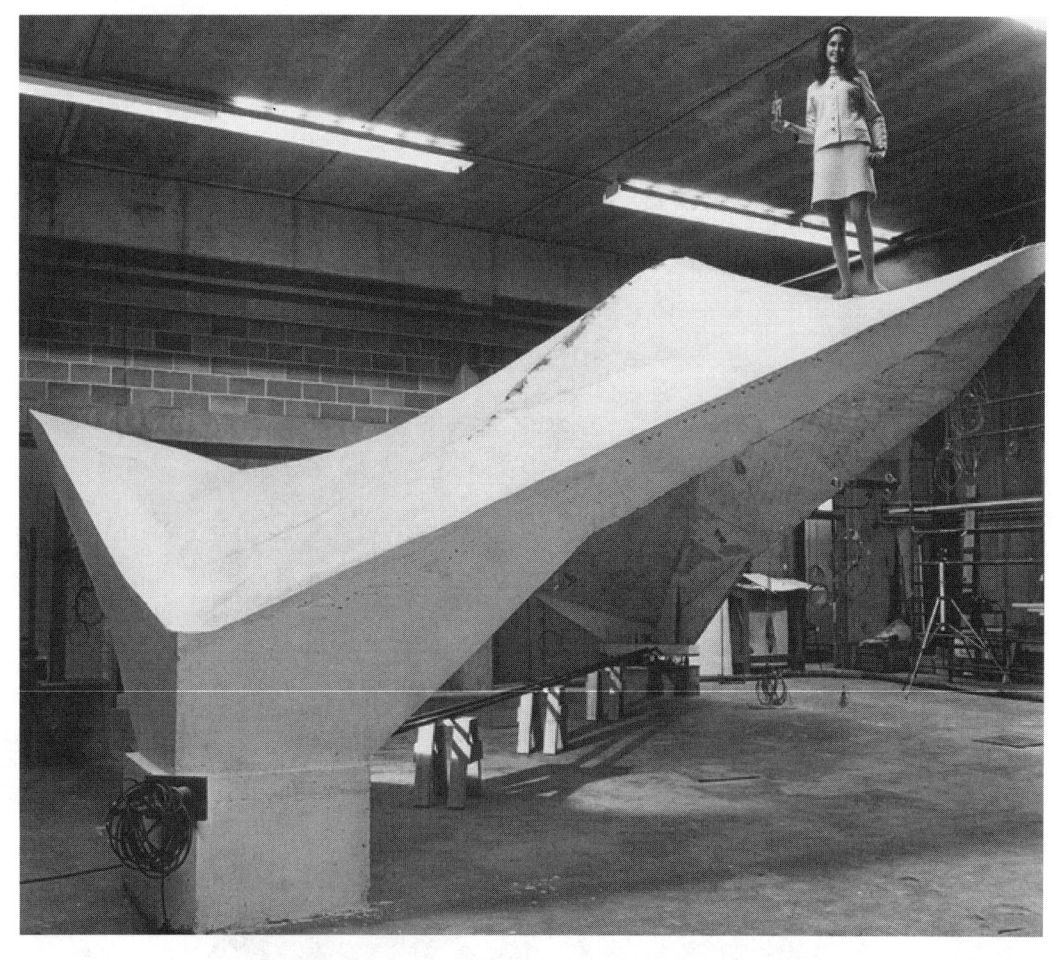

Figure 1.5 Completed microconcrete model and hostess. (Courtesy of Wiss, Janney, Elstner & Associates, Inc., Northbrook, IL.)

three frames, each one of three bays. A full-height shear wall is located at the midbay of the center frame B. In its transverse direction the test building consists of four frames, each one with two bays of 6 m each and cantilevering 2 m at each end. In the extreme transverse frames (lines 1 and 4 in Figure 1.6a), there are full-height shear walls which, as their main purpose, provide torsional resistance for the building. The chosen seven-story reinforced concrete test structure represents a portion of a building having dimensions common to earthquake-resistant construction in both the U.S. and Japan. The lateral load–resistant construction is provided by interacting structural walls and frames. Input from both Japanese and U.S. researchers has gone into the development of the design details, the planning of the instrumentation, and the conduct of the testing of the prototype.

In buildings of this type, it is generally recognized that the properties of the shear wall will greatly influence the aseismic performance of the building. From analysis of the results obtained in the analytical prediction of the responses of the prototype, and considering the limitations of the University of California, Berkeley shaking table, a $\frac{1}{5}$-scale model of the prototype structure was chosen. A detailed discussion of this selection and the final design of the model structure are given by Harris and Bertero (1981), Harris et al. (1981), and Aktan et al. (1983). The maximum capacity of the 6.1 × 6.1 m University of California, Berkeley shaking table is 490 kN. Based on these limitations and the desire to test the model under very large deformations and thus enable the study of the failure mechanisms of the building, a $\frac{1}{5}$-scale reinforced microconcrete model of the prototype structure was chosen; see Figure 1.7.

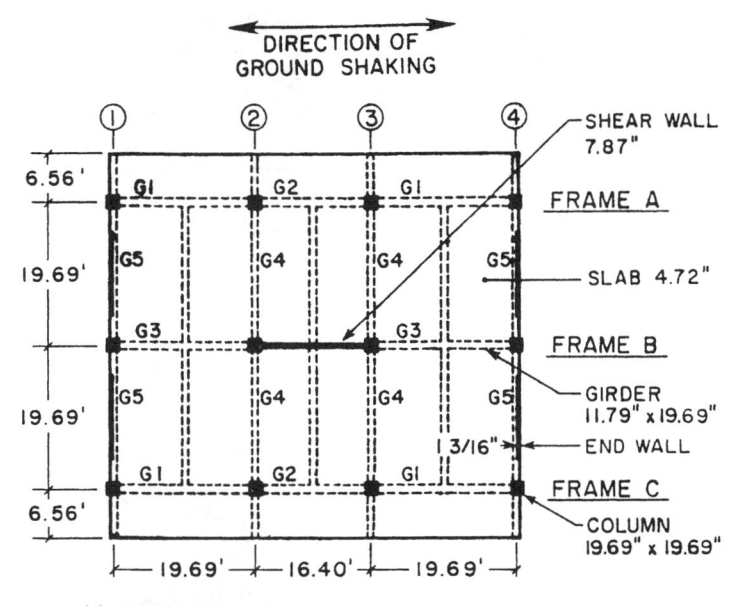

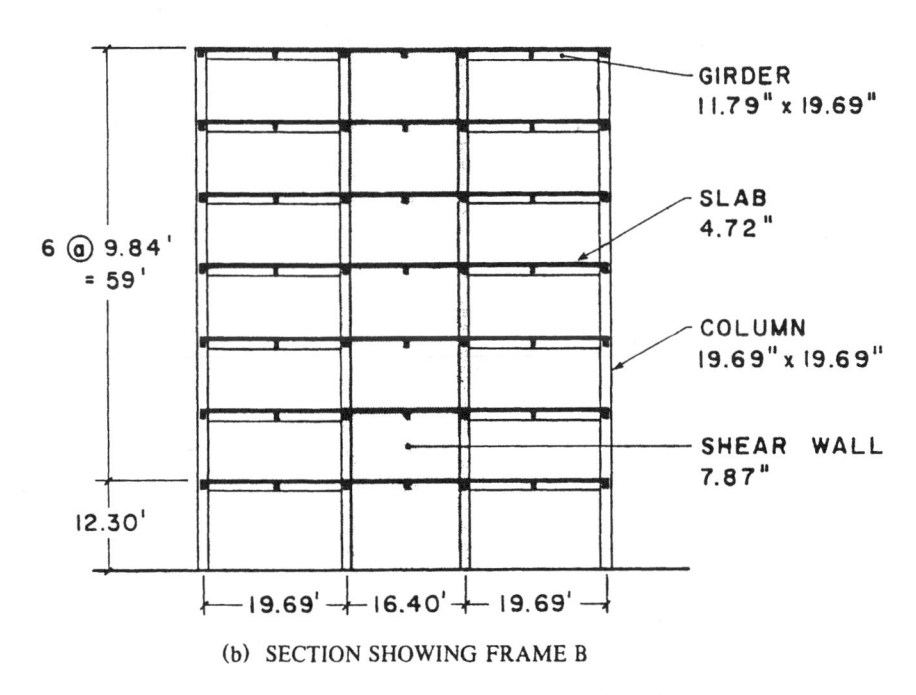

INTRODUCTION TO PHYSICAL MODELING IN STRUCTURAL ENGINEERING   17

Figure 1.6 Plan and section of full-scale frame-wall reinforced concrete structure.

The design of the model was chosen to comply with similitude requirements, in addition to shaking table limitations (Monkarz and Krawinkler (1981)). The ⅕-scale model chosen for the U.S.–Japan program satisfies similitude with regard to geometric and loading parameters and also complies with all material requirements except for the mass density. The detailed mechanical characteristics of model materials and any discrepancies are documented by Bertero et al. (1983; 1984). The mass density is augmented by means of lead ballast weights attached to the roof and

Figure 1.7 View of ⅕-scale model on the U.C. Berkeley shaking table.

all floor slabs in such a manner so as to cause little or no influence to the structural stiffnesses, as well as no significant increase in the mass moments of inertia of the slabs (Figure 1.8). In addition, it added no major changes in the damping characteristics of the structure.

Besides the internal force transducers at midheight of the first-story columns, the model was instrumented to record overall response and local behavior of the critical regions of the most severely strained members. Internal and external gages to measure average strain along the main reinforcing bars and at the surface of the microconcrete were placed at the critical regions of structural members. Displacement transducers were located at each floor of the model and connected to a steel reference frame placed outside the shaking table to record overall lateral response of the model.

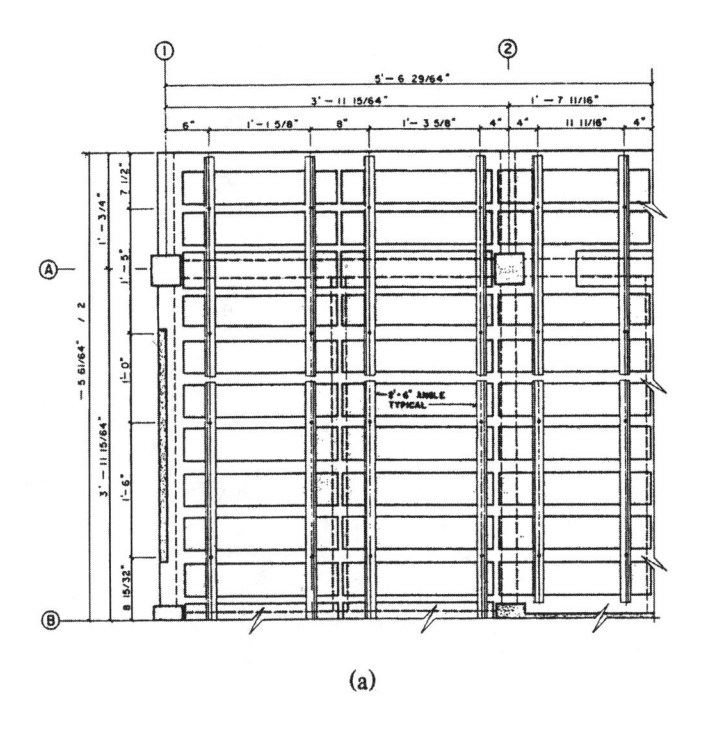

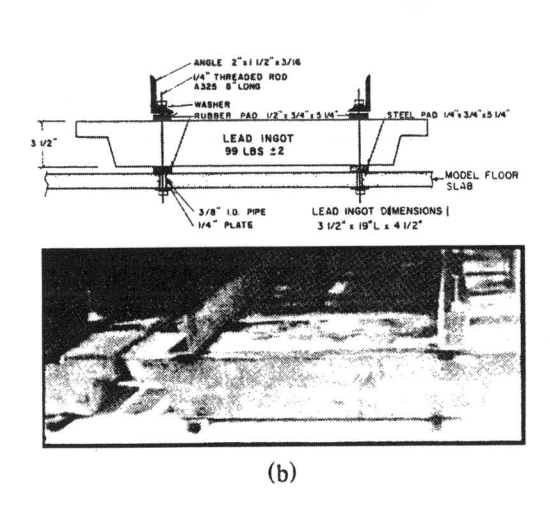

Figure 1.8 One-fifth-scale model loaded with required lead ballast. (a) Distribution of ballast on floor slabs; (b) attachment of lead ingot (used as ballast) to the model slab; (c) one-fifth-scale model loaded with ballast.

The ⅕-scale model of the seven-story R/C test structure tested on the University of California, Berkeley earthquake simulator is one of the largest and most-detailed studies undertaken thus far to study the earthquake resistance of structures under dynamic loading conditions. Perhaps one of the most important objectives of the research conducted on the ⅕-scale model of the structure was to evaluate the reliability of experimental analysis at all the seismic response limit states, utilizing such a small-scaled true replica model of the test structure. The most important difficulty in the attainment of a true replica model, with the same strain response history as the full-scale structure when subjected to similar seismic effects, was satisfying similitude requirements for the response characteristics of the constituent materials of the model structure. Details are given in Chapter 11 and a discussion of some of the consequences of the limitations in attaining perfect similitude in the constitutive relationships of both steel and concrete, as well as their composite action, i.e., the bond characteristics, as reflected on the serviceability, damageability, and collapse limit state responses of the structure.

The research described in Case Study B was conducted at the Earthquake Engineering Research Center, University of California, Berkeley under grant CEE-80-09478 from the National Science Foundation.

1.11.3 Case Study C, Reinforced Concrete Bridge Decks

A research program to study the ultimate and fatigue strength of noncomposite reinforced concrete bridge decks under pulsating and moving loads was conducted at Case Western Reserve University by Perdikaris and his associates (Perdikaris and Beim, 1988; Perdikaris et al., 1989; Perdikaris and Petrou, 1991; Petrou et al., 1994, 1996). The direct modeling approach was used to study a typical 15.24-m-long prototype highway bridge with a 216-mm-thick reinforced concrete deck supported on four W36 × 150 steel girders spaced at 2.13 or 3.05 m. A series of 17 full bridge deck ¹⁄6.6 scale direct models (Figure 1.9) and six individual panels at ⅓ scale were tested in this program.

The objectives of the research were to obtain a better understanding of the fatigue response of noncomposite reinforced concrete bridge decks and to present a description and rational explanation of the various load transfer mechanisms present in the deck during its life expectancy. Small-scale model concrete deck slabs reinforced according to the current *American Association of State Highway and Transportation Officials Code* (AASHTO) (1989) (orthotropic reinforcement) and the *Ontario Highway Bridge Design Code* (OHBDC) (1983) provisions (isotropic reinforcement) were subjected to static and moving constant wheel-load in the direction of traffic. The test setup shown in Figure 1.10 was developed for these studies. A close-up of the moving constant wheel-load testing a ⅓-scale panel is shown in Figure 1.11. Also, fatigue tests under a pulsating load applied at a fixed point were conducted to determine the effect of the load movement on the fatigue strength and level of deterioration of the deck slabs. The determination of the endurance limit of a concrete deck slab reinforced with either "isotropic" steel reinforcement according to the OHBDC Code or "orthotropic" steel according to the *AASHTO Code* and subjected to a moving constant wheel-load would give a better estimate of the actual safety factors against first cracking, steel yielding, or deck punching failure and the true level of conservatism for the two bridge design philosophies.

Several important parameters were studied in this research. The effects of the flexural steel reinforcement ratio, steel girder spacing, rotational and lateral restraint of the deck, and the type and level of applied load on the cracking pattern, stiffness degradation, fatigue life, and failure mode of a concrete bridge deck were investigated. The critical factors contributing to the fatigue failure of concrete bridge slabs were determined. The experimental load–deflection static response, fatigue strength under different levels of moving constant wheel-loads, measurements of strains (steel reinforcement, concrete deck surface, and steel girders), cracking patterns, and the failure mode were all studied. The presence of any size effect was investigated by comparing the test results for the ¹⁄6.6- and ⅓-scale models. These results are summarized in Chapter 10.

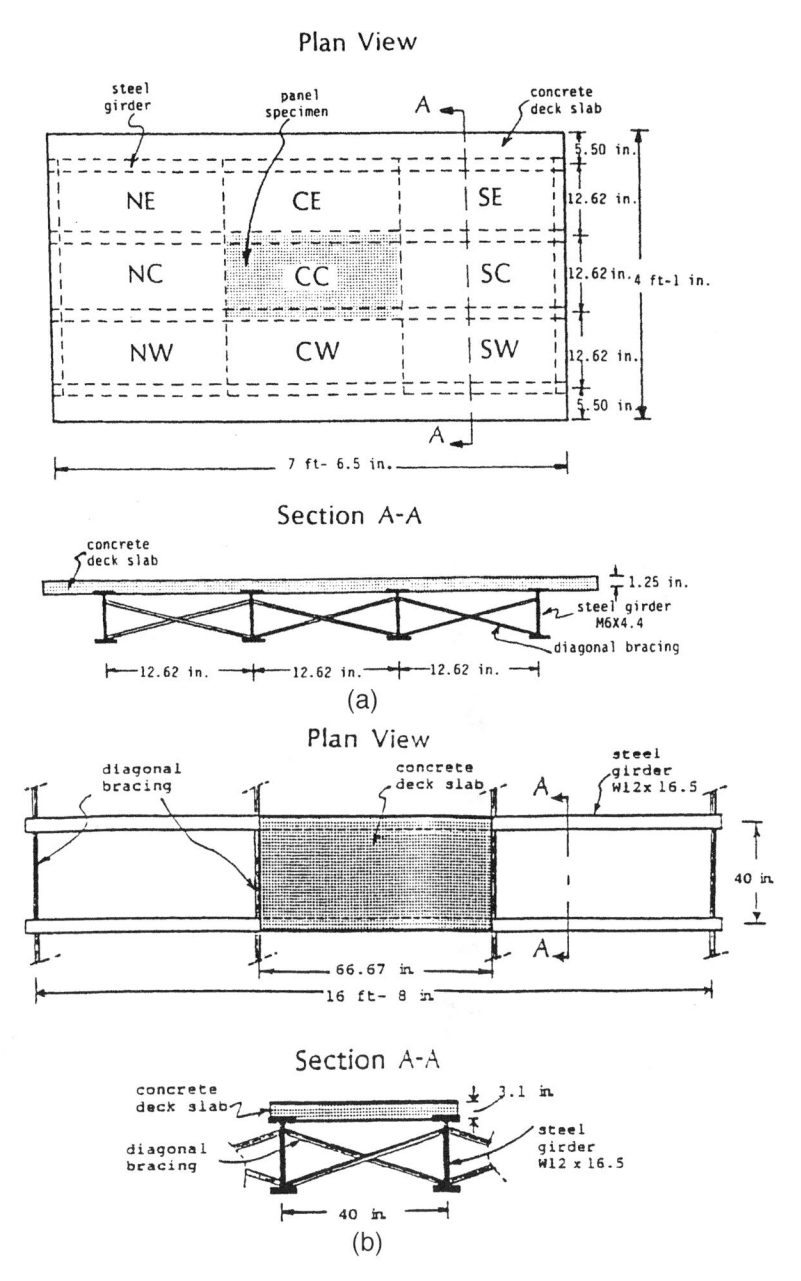

INTRODUCTION TO PHYSICAL MODELING IN STRUCTURAL ENGINEERING 21

Figure 1.9 Models at scales of $1/6.6$ and $1/3$ of noncomposite decks on steel girder bridges. (a) Type B ($1/6.6$ scale) bridge deck model specimen (1 in. = 2.54 cm); (b) dimensions of the $1/3$-scale deck "panel" model (PI-10S and PO-10S), with a prototype girder spacing of 10 ft.

The punching shear failure was found to be interrelated with the instability of a two-dimensional three-hinge compressive rigid-strut model. The compressive arching action mechanism in the concrete deck transferring the load to the steel girders is modeled by the compressible strut members of the truss, while the lateral restraint in the deck is depicted by elastic horizontal springs.

1.11.4 Case Study D, Lightly Reinforced Concrete Buildings

The case study presented is part of a comprehensive research effort conducted by the National Center for Earthquake Engineering (NCEER) on the damage assessment and performance evaluation

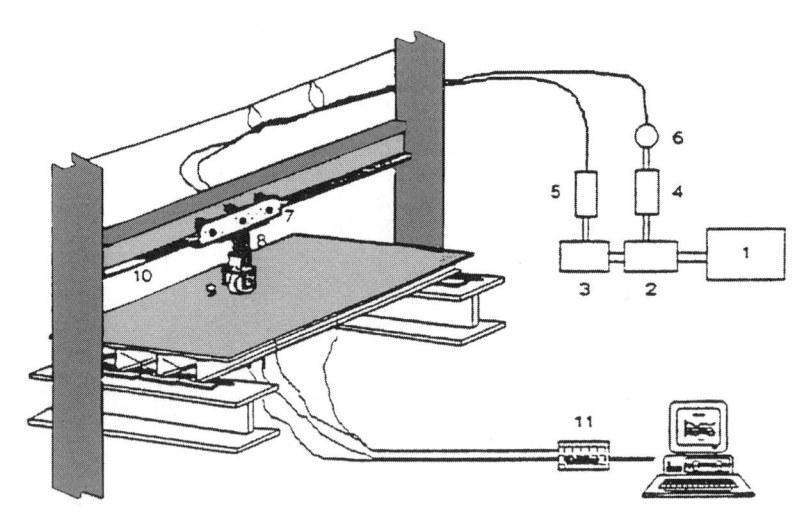

Figure 1.10 Illustration of the moving constant wheel-load setup. (1) Electric motor and control panel; (2) hydraulic cylinder pump; (3) hydraulic motor pump; (4) pressure reducer; (5) wheel speed control; (6) pressure dial gage; (7) moving trailer; (8) hydraulic jack; (9) polyurethane-bonded steel wheel; (10) reaction steel frame; (11) computer data acquisition system.

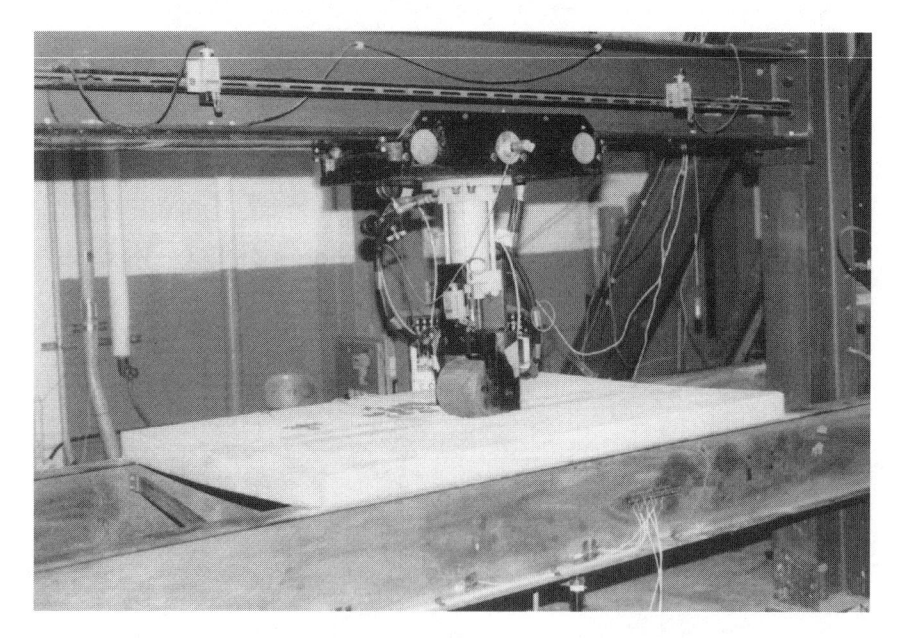

Figure 1.11 Moving constant wheel-load setup (⅓-scale wheel size).

of nonseismically detailed buildings during earthquakes. A ⅛-scale model of a lightly reinforced three-story office building tested on the shaking table at Cornell University (El-Attar et al., 1991) and a ⅓-scale model of the same building tested on the shaking table at SUNY Buffalo (Bracci et al., 1992). The building was designed and detailed to reflect the common design and practice features of the Central and Eastern U.S. during the period 1950 to 1970. Test results from these studies are evaluated and compared with numerical results to investigate the reliability of available analytical tools in predicting the response of this type of building. A general layout of the prototype building is shown in Figure 1.12. The relative dimensions adapted for the frame members of the idealized prototype structure were based on a survey of typical construction practices and the limitations of the NCEER shaking table.

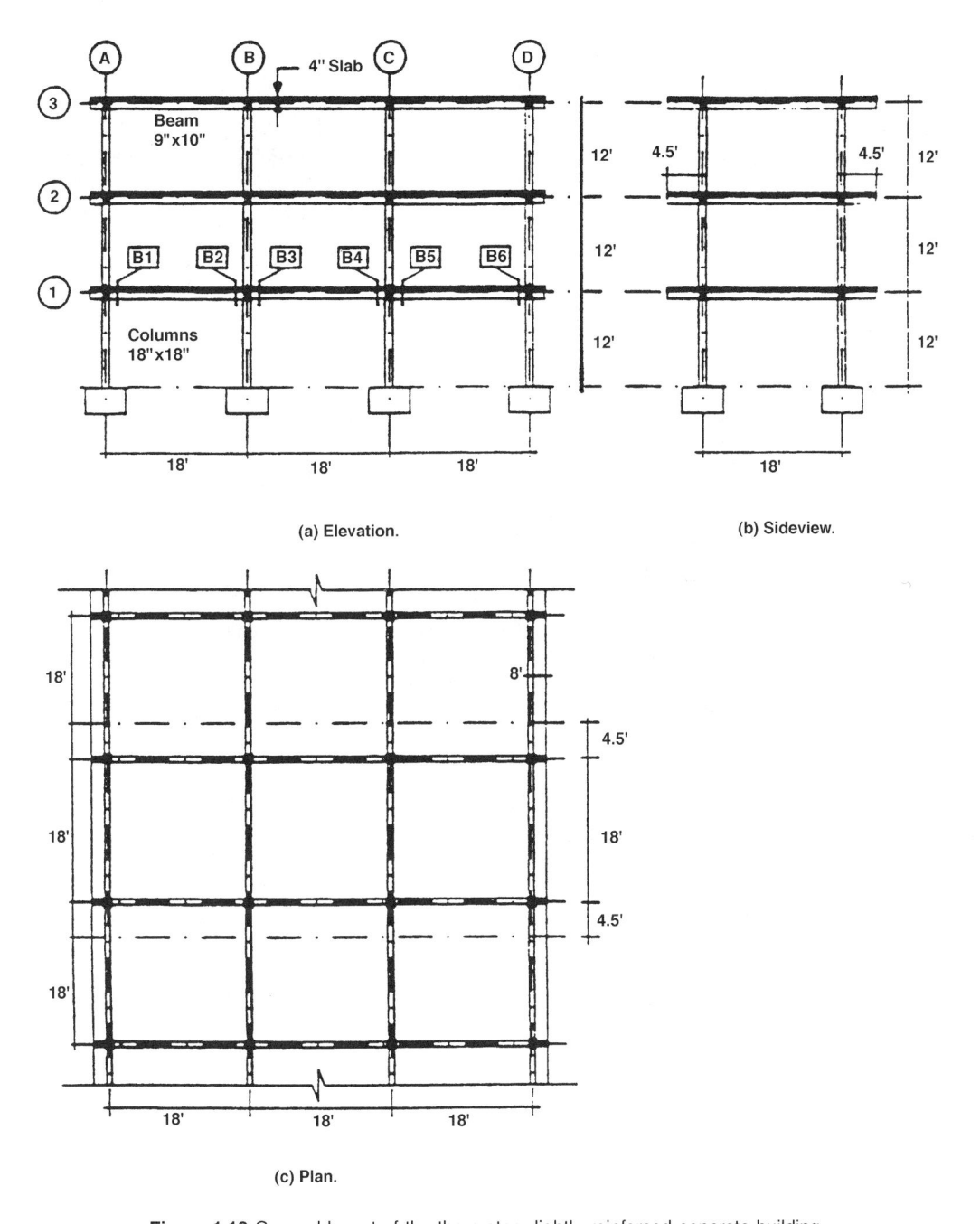

Figure 1.12 General layout of the three-story lightly reinforced concrete building.

Realistic reinforcement details that reflected construction practices during the 1950 to 1970 period were adopted for the models tested. Special attention was paid to critical details such as the beam–column joints and column lap splices. Figure 1.13 shows an interior and an exterior non-seismically detailed joint. What characterizes these joints are the following details: (1) discontinuous bottom beam reinforcement with a very short embedment length (150 mm), (2) no beam or joint confinement except for several stirrups at a spacing of 75 mm (usually) located 75 mm below the beam bottom face, (3) no joint reinforcement (column stirrups discontinuous through joint length),

(a) Interior Joint.   (b) Exterior Joint.

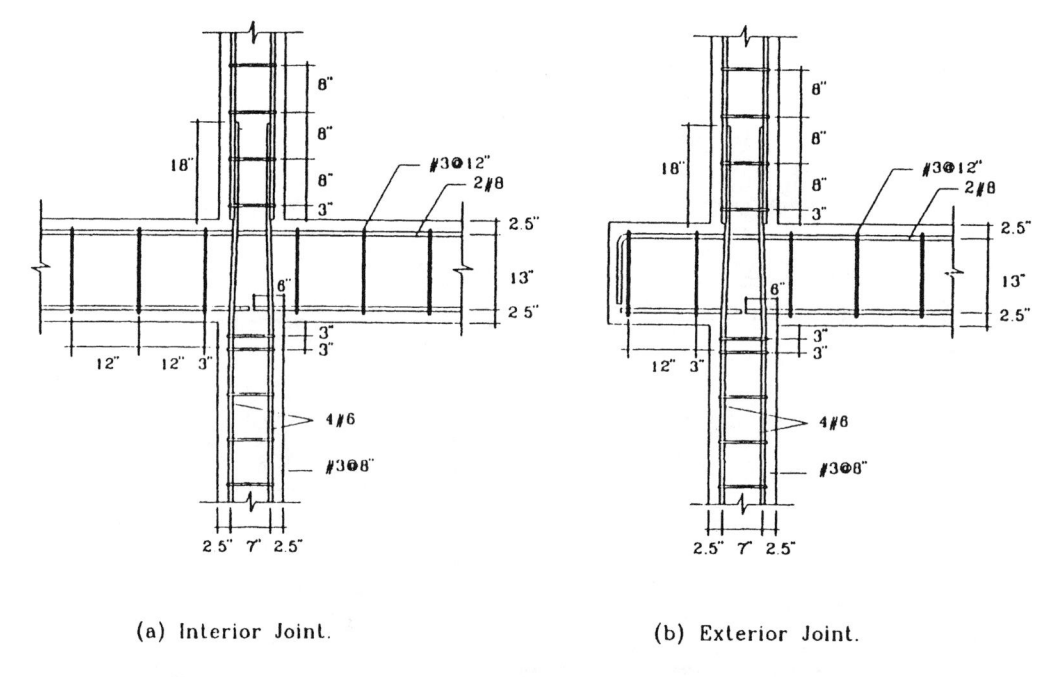

Figure 1.13 Beam–column joint details of the three-story prototype building.

(4) a compressive lap splice that is very short and located immediately above beam top face, and (5) no special confinement required for the lap splice. All the above features are shown in Figure 1.13.

The ⅛-scale model required additional ballast loads in the form of lead bricks symmetrically placed on the floor and roof slabs as shown in Figure 1.14. The mounting technique was designed so that the center of mass of the lead bricks coincided with the slab midthickness and minimized their eccentricity. Bolts 13 mm in diameter passing through 19-mm holes held the lead bricks to the slab. A set of steel washers was installed between the slab and the lead bricks to minimize slab stiffening (insert, Figure 1.14a). The ⅛-scale model is shown on the Cornell shaking table in Figure 1.14b.

A ⅓-scale model of the same critical portion of the lightly reinforced concrete building (Figure 1.12) was built and tested on the NCEER shaking table at the SUNY Buffalo Earthquake Simulation Laboratory. Three views of the ⅓-scale model are shown in Figure 1.15. A series of varying intensity simulated ground motion tests were performed on the ⅓-scale building model using scaled accelerograms on the shaking table to represent minor, moderate, and severe earthquakes. More details of the Cornell and SUNY Buffalo model studies are given in Chapter 11.

1.11.5 Case Study E, Prestressed Wooden Bridges

Recent concern regarding the decay of the infrastructure of the nation has focused attention on the need for replacement of many of the nation's bridges. Of particular concern, because of the vast quantity, are bridges with spans of less than 50 ft. These make up a majority of America's bridges (Oliva and Dimakis, 1987). Many of these bridges are made of steel and reinforced concrete and were constructed in the early and mid part of this century. They have performed for the most part satisfactorily during their lifetime; however, they are approaching old age and are in need of repair or replacement. Repair and replacement can be quite expensive considering the large number of bridges, and labor and material costs. It is obvious that a rapidly constructible, inexpensive replacement bridge would be desirable. Wood is becoming an increasingly attractive alternative to

INTRODUCTION TO PHYSICAL MODELING IN STRUCTURAL ENGINEERING

(a)

(b)

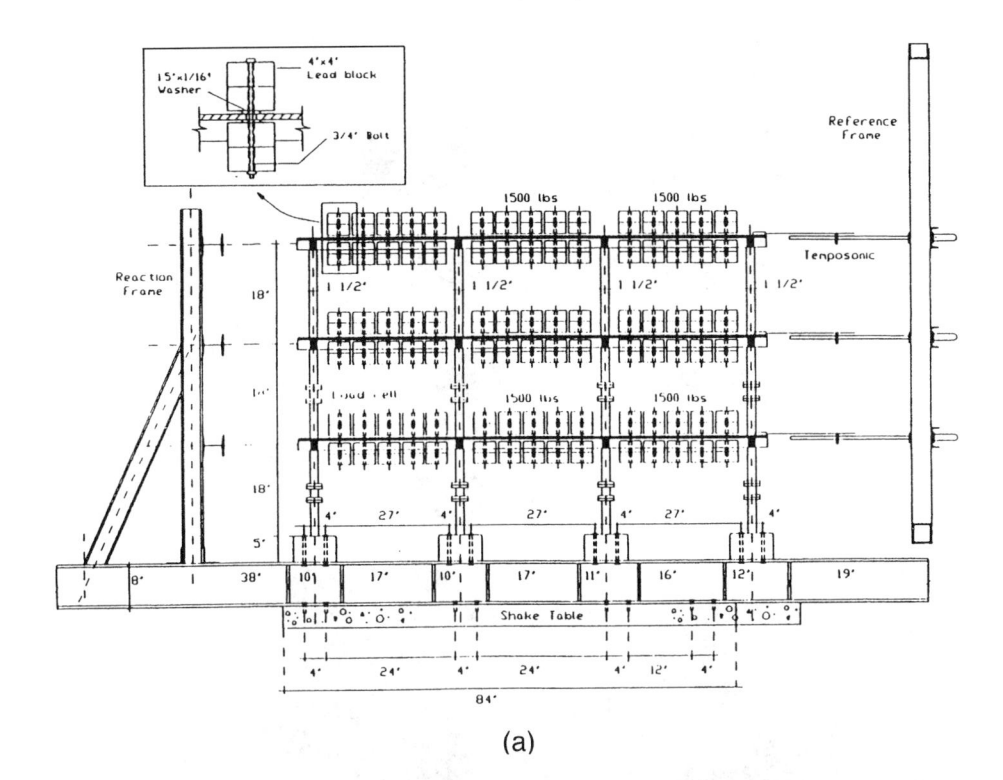

Figure 1.14 One-eighth scale model tested on the Cornell University shaking table. (a) Ballast load setup; (b) ready for test.

Figure 1.15 One-third scale model on the SUNY Buffalo NCEER shaking table.

reinforced concrete and steel replacement bridges. Some unique advantages of wood are low price, availability of short- and medium-length timbers, resistance to salts used in snow removal, ease of use, and the fact that it is a renewable resource. Wooden bridges do have several drawbacks, however, such as potential decay, large long-term creep deflections, low strength and stiffness when compared with reinforced concrete and steel, and limited availability of very large timbers.

A promising new transverse post-tensioning technique attempts to utilize wood to its maximum potential while taking advantage of smaller-sized members. Originated in Canada in the late 1970s the technique has been used successfully in several short-span bridges in both Canada and the U.S. (*Ontario Highway Bridge Design Code,* 1983). Most of the research and applications in the U.S. and Canada of transverse post-tensioned wooden bridge decks have focused on softwood species. An interesting design concept uses North American hardwood species. Specifically, abundantly

INTRODUCTION TO PHYSICAL MODELING IN STRUCTURAL ENGINEERING 27

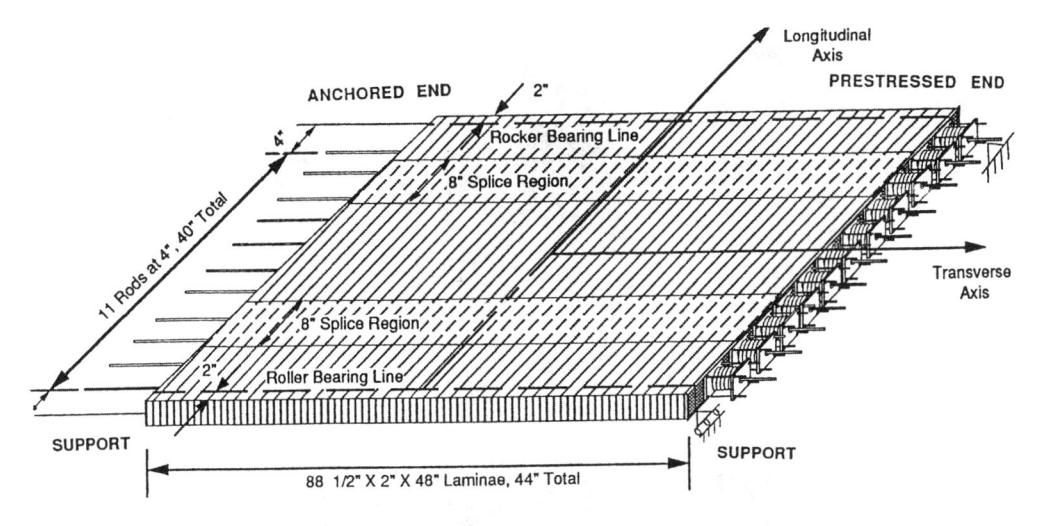

Figure 1.16 Instrumentation details for ⅙-scale model.

available Pennsylvania red oak was used in the replacement design of the Samuel Road bridge. This is located in West Whiteland Township, Chester County, PA and involves the replacement of a severely deteriorated 6.71-m skew span steel girder and reinforced concrete deck bridge. The design was carried out in accordance to the *Ontario Highway Bridge Design Code* and AASHTO requirements. The primary exception is that the anchorage bulkhead channel has been replaced by individual square steel plates at the anchor points.

Two identical ⅙-scale models of the Samuel Road bridge design (Figure 1.16) with the skew angle eliminated were tested at Drexel University in order to verify the design using local creosote impregnated red oak (Hoffman, 1990). Model E1, shown in Figure 1.17, was used for dead load testing and for stress relaxation measurements. Live load simulation using the AASHTO HS-20 truck loading was used to study the behavior of test Model E2. Model E2, shown in Figure 1.18, was tested elastically and then loaded to ultimate failure. The elastic behavior was compared with an FEM analysis. Results of these tests, conducted in the Structural Models Laboratory, Drexel University, are presented in Chapter 10.

1.11.6 Case Study F, Interlocking Mortarless Block Masonry

An interesting and novel application of the direct modeling technique that has far-reaching implications in industrial structural products is illustrated in this example. It involves the use of faithfully reproduced exemplar models in a product development study of new block masonry. Product development involves a large number of engineers from a variety of disciplines including structural engineering, materials, and construction. The search for building procedures that are more rapid and less dependent on workmanship has led to the need to develop "dry stackable" block masonry units which can be laid without mortar. Introduction of interlocking or dry stack mortarless masonry systems in reinforced block masonry construction requires the development of efficient, easy-to-handle, and yet versatile blocks. Two promising interlocking block types were developed at Drexel University (Harris et al., 1992, 1993; Oh et al., 1993; Oh, 1994) for application to reinforced masonry construction including earthquake-resistant structures using ⅓-scale direct models. The main aim in this development was, first, to equal or exceed the structural performance of conventional masonry systems and, second, to provide a more economical and rational solution for the masonry system thus leading to more-competitive designs. The two interlocking block units developed are designated the modified H-Block and the WHD Block. Production of the units (facilitated by ⅓-scale reduction) was followed by strength and stiffness evaluation under compressive, bending, and

Figure 1.17 Model E1: dead load effects and prestress losses.

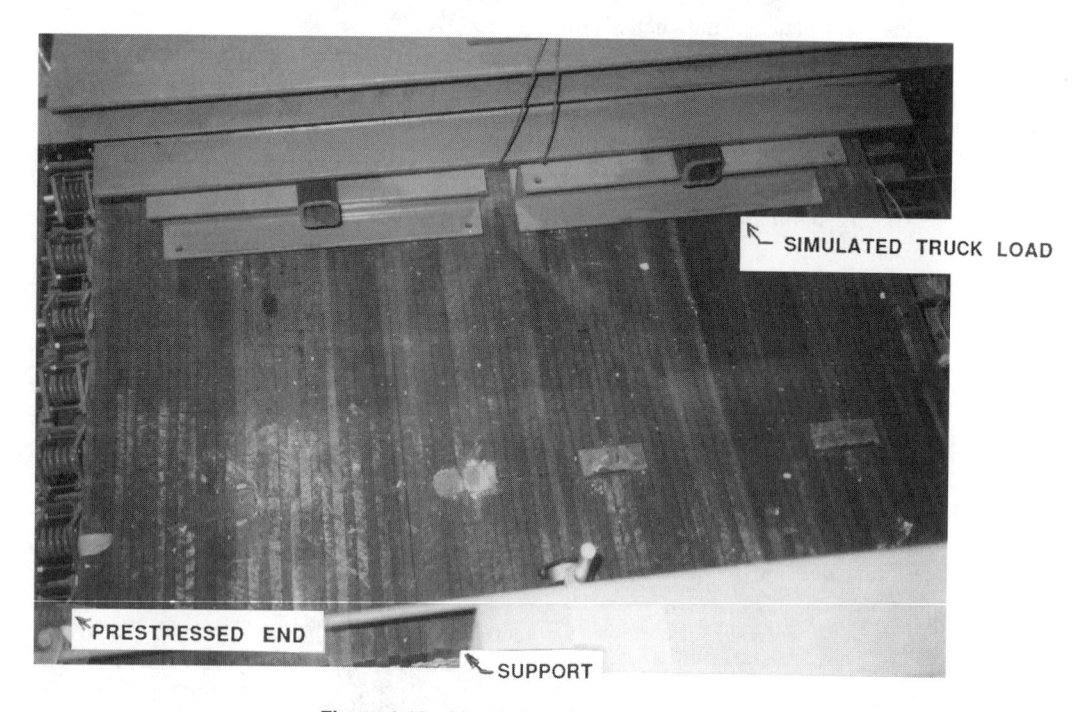

Figure 1.18 Model E2 during live load testing.

shearing loads. All comparisons were baselined to conventional mortared masonry construction. Analytical models were developed to predict behavior under load.

The first of the two interlocking masonry systems investigated was the modified H-Block shown in Figure 1.19. This is a simple open-ended block unit with tongue-and-groove interlock on both the bed and head joints. It can be reinforced in the vertical and horizontal directions (Figure 1.19c)

INTRODUCTION TO PHYSICAL MODELING IN STRUCTURAL ENGINEERING

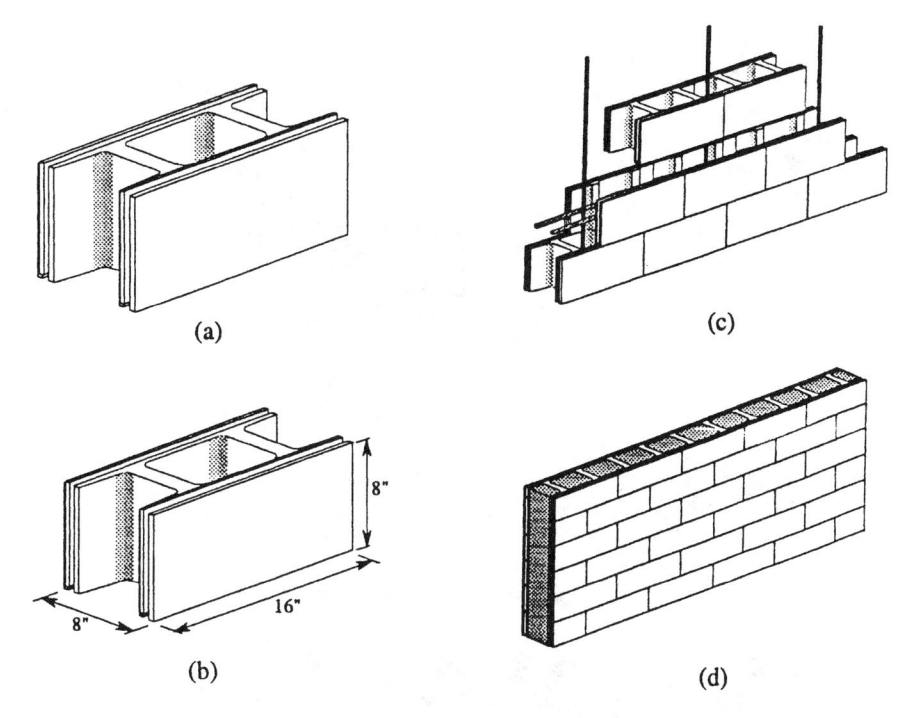

Figure 1.19 Modified H-Block masonry system. (a) Stretcher; (b) nominal dimensions at full scale; (c) reinforcement; (d) stacking in running bond; (e) top view showing alignment of cells; (f) wall dry stacked in running bond.

and is particularly suited to earthquake-resistant construction because it can be easily placed around vertical bars although vertical threading is required. Figure 1.19b shows the dimensions of the full-size prototype. The main advantages of the modified H-Block unit are ease of vertical and horizontal reinforcement, ease of unit cell vertical alignment, minimum tapering of cells, main area of block concentrated in the desirable face shell, optimum cell shape for ease of grouting, and high resistance to water penetration due to raised tongue-and-groove bed joints. These features are illustrated in Figure 1-19. The problem of controlling the height of units to achieve accurate running bond can be achieved by means of specially designed block molds. The grooves can be strengthened by proper mix design and vibration during production. During transportation and handling, simple stacking techniques can be developed in packaging to minimize breakage. To solve the problem of stability during construction prior to full grouting, either preventive partial grouting or extended bracing can be used. The stacking of units in running bond pattern and the reinforcement placing is shown in a hollow wall (Figure 1.19c) and when the wall is fully grouted in Figure 1.19d. The running bond dry-stacking technique is illustrated in Figure 1.19e and f.

The WHD Block was developed independently at Drexel University and was studied using a ⅓-scale direct modeling technique. Units of the WHD Block system are shown in Figure 1.20a. The prototype dimensions are shown in Figure 1.20b. Reinforcement in the vertical and horizontal directions and stacking of the hollow units are illustrated in Figure 1.20c. Significant horizontal and vertical bending stiffness is obtained in the staggered joints of the dry-stacked system (Figure 1.20c). Full grouting of a wall is illustrated in Figure 1.20d. The freshly finished model blocks are shown in Figure 1.20f. A specially designed block-making machine (Figure 1.20e) is described in Chapter 6. Details of the model block strength and stiffness characteristics are given in Chapter 10.

The research described in this case study was conducted in the Structural Models Laboratory, Drexel University under grant No. MSM-9102769 from the National Science Foundation. Additional cases of direct models used in product development are given in Chapters 5 and 10.

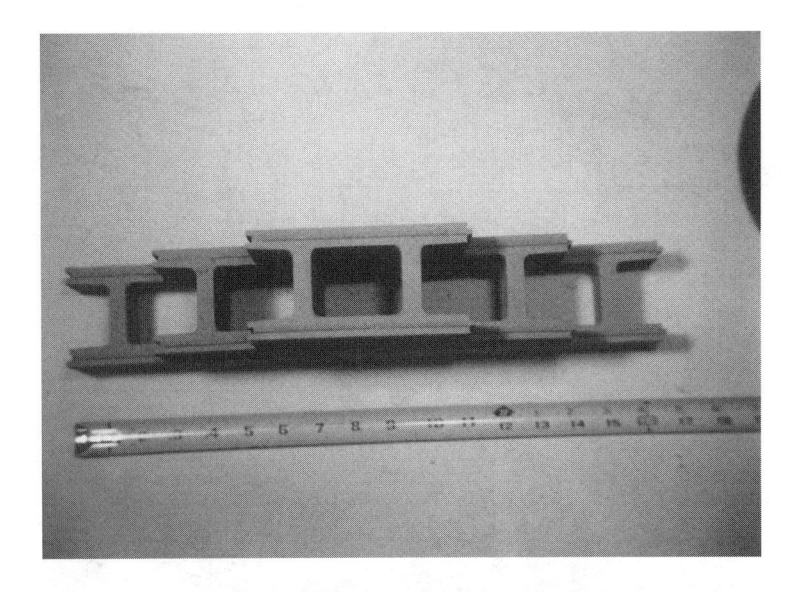

(e) Top view showing alignment of cells

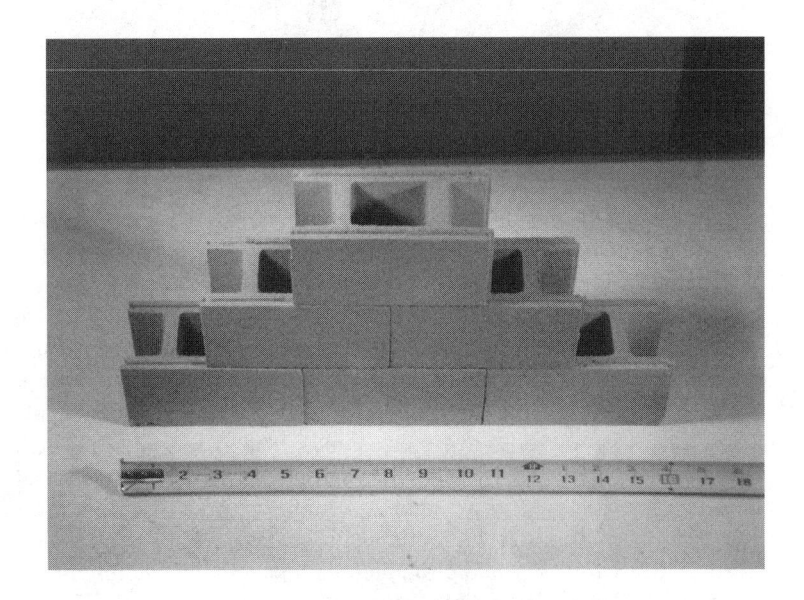

(f) Wall " dry stacked" in running bond

Figure 1.19 (continued)

1.11.7 Case Study G, Pile Foundations

Pile caps form an important element between the column and the piles, through which the load is transmitted to the soil. Often, it is overlooked both by the structural engineer and the soils engineer and is designed somewhat empirically, since the code provisions in many cases cannot be used directly. The problem is one of a deep slab (two deep beams) and has to be handled differently than the code provisions. The pile cap study presented here is based on research carried out at Howard University (Idowu, 1979; Ndukwe, 1982; Dagher, 1988) and at the University of Puerto Rico (Jimenez-Perez et al., 1986) and reported subsequently by Sabnis and Gogate (1980),

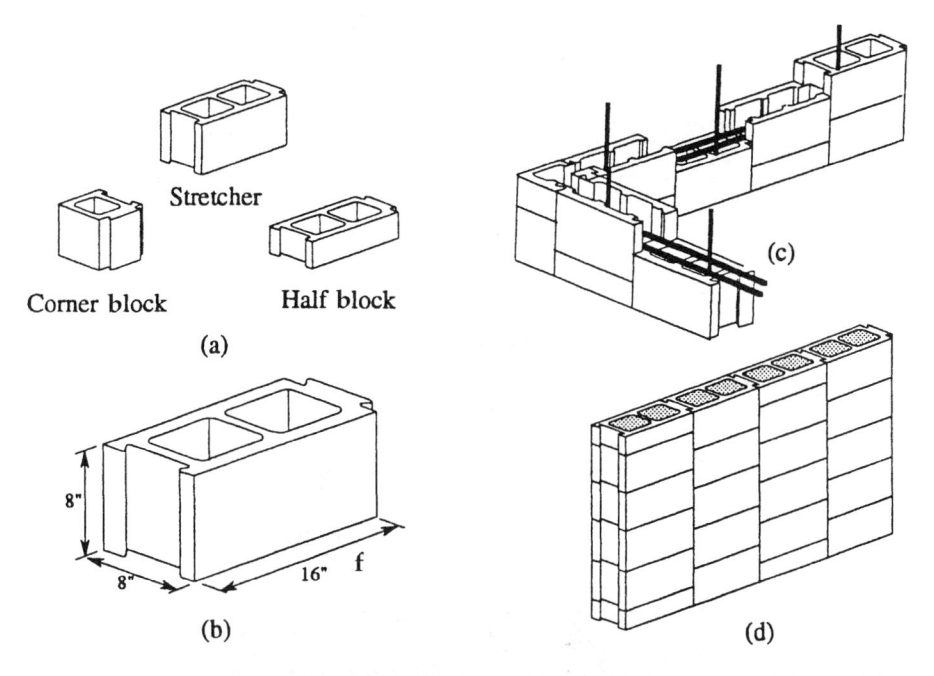

INTRODUCTION TO PHYSICAL MODELING IN STRUCTURAL ENGINEERING 31

Figure 1.20 WHD block masonry system. (a) Type of units; (b) nominal dimensions at full scale; (c) reinforcement; (d) stacking; (e) the mold extracting completed model block; (f) completed model blocks.

and Sabnis and Dagher (1989). The overall problem of structural strength assessment was tackled through various topics; these include the present practice of pile cap design and the experimental testing program using structural models designed to cover a wide range of parameters, with particular emphasis on the variation of reinforcing steel. Two scales for testing were used at two universities and correlated to demonstrate the universal approach of model studies.

The behavior of pile caps was studied by varying the amount of reinforcement, while other parameters, such as concrete strength and steel depth, were considered secondary based on the earlier research work. Furthermore, due to the difficulty of making large specimens and handling them with the limited available resources, it was decided to use scaled models (Figure 1.21) throughout this test program at Howard University. A ⅕ scale was used to suit the facility at Howard University and the available reinforcement. The scale was large enough for one person to handle. Larger specimens at ½.₅ scale were tested at the University of Puerto Rico (Jimenez-Perez et al., 1986).

The results of the testing program and their implications on code revisions are discussed in Chapter 10.

1.11.8 Case Study H, Externally/Internally Prestressed Concrete Composite Bridge System

This case study involves the development of a new structural system for highway bridges by eliminating corrosion and the need of shoring and formwork during construction, thus lengthening the bridge life and cutting down on expenses. The system of the main feature is the use of precast modified double-Tee (DT) panels which are prestressed with internal carbon fiber–reinforced polymer (CFRP) strands. A cast-in-place deck slab reinforced with CFRP bars is connected to the DT girders through shear connectors and an epoxy bonding agent. Externally draped post-tensioned CFRP strands are used for the longitudinal post-tensioning. The various components of the new bridge system are shown in Figure 1.22. The new bridge system consists of:

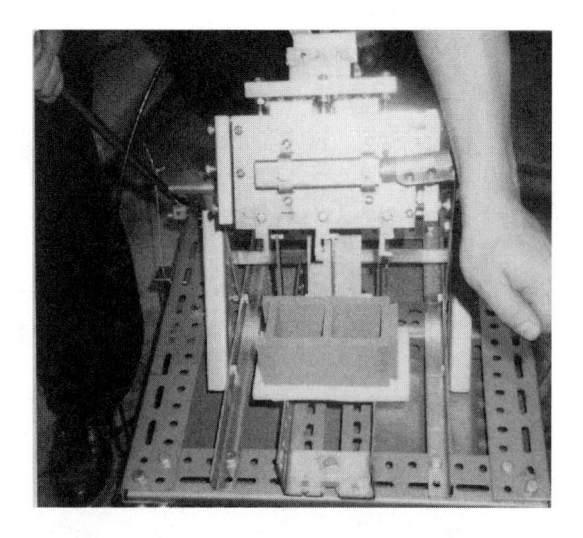

(e) The mold extracting completed model block.

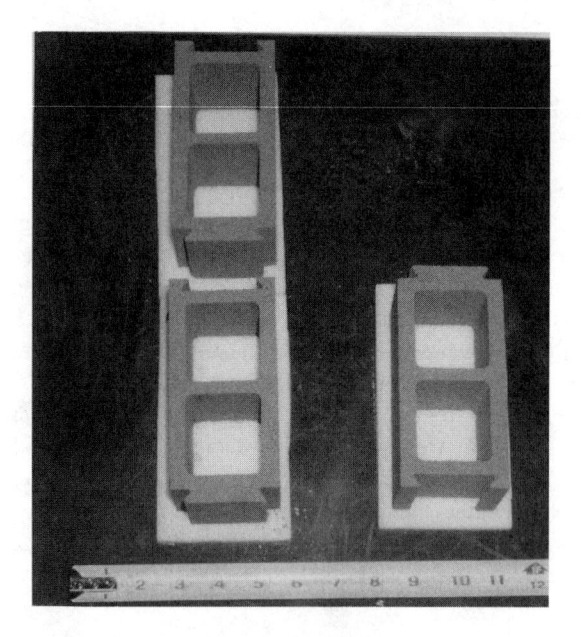

(f) Completed block.

Figure 1.20 (continued)

1. Precast modified DT girders, prestressed internally with CFRP strands.
2. Post-tensioned cross-beams cast monolithically into the DT girder.
3. Cast-in-place CFRP-reinforced deck, connected to the DT girder with CFRP shear connectors.
4. External post-tensioned draped CFRP prestressing strands.

The standard DT girder is modified with the cross beams to accommodate the draped external strands and to enable transverse prestressing of the DT girders. This allows several DT girders to

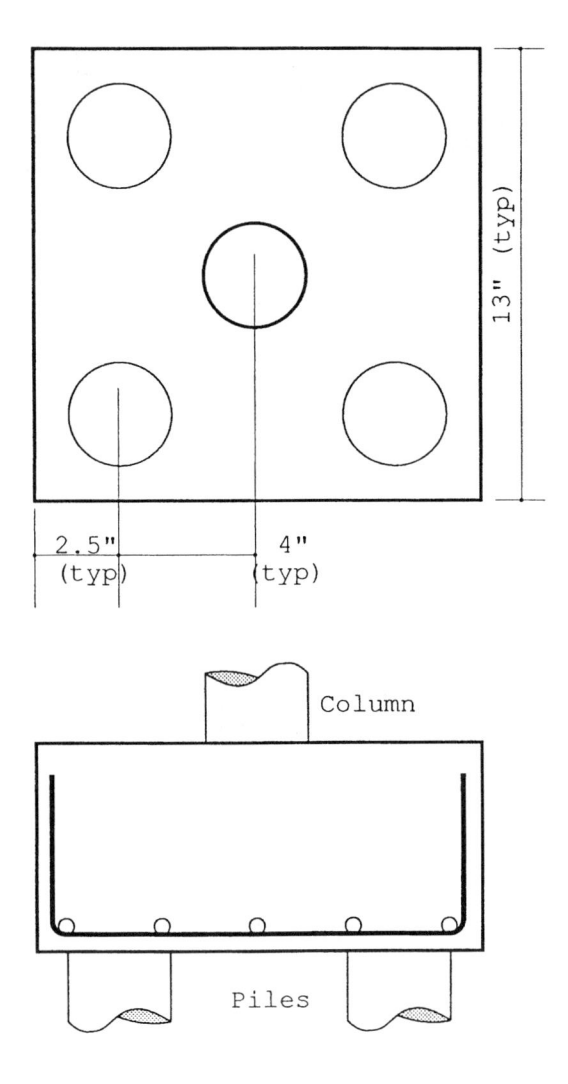

Figure 1.21 Dimensions of ⅓-scale pile cap model.

be placed adjacent to each other and tied together with CFRP prestressing strands. The unbonded post-tensioning in the transverse direction facilitates future widening of the bridge system.

Four ⅓-scale experimental models of the modified DT concrete girder bridge system utilizing carbon fiber reinforcing bars and prestressing strands were constructed and tested at Lawrence Technological University, Southfield, MI to determine their performance under various loading conditions (Grace and Sayed, 1996a, b; 1997a, b; 1998). The large size of the models chosen allowed the use of prototype materials throughout, thus greatly simplifying the modeling process. The test loadings for each model consisted of static/dynamic fatigue (7 million cycles) and ultimate strength. Two of the test structures were right-angle or straight bridges and two were skewed axes. The dimensions of the four model bridges are shown in Figure 1.23. Bridge model DT-30 (with 30° skew ends) is shown in the laboratory undergoing fatigue testing (Figure 1.24). The widest of the four models, Model DT-15 (15° skew) is shown in Figure 1.25 undergoing external post-tensioning.

Results of the testing of all four models are discussed in Chapter 10.

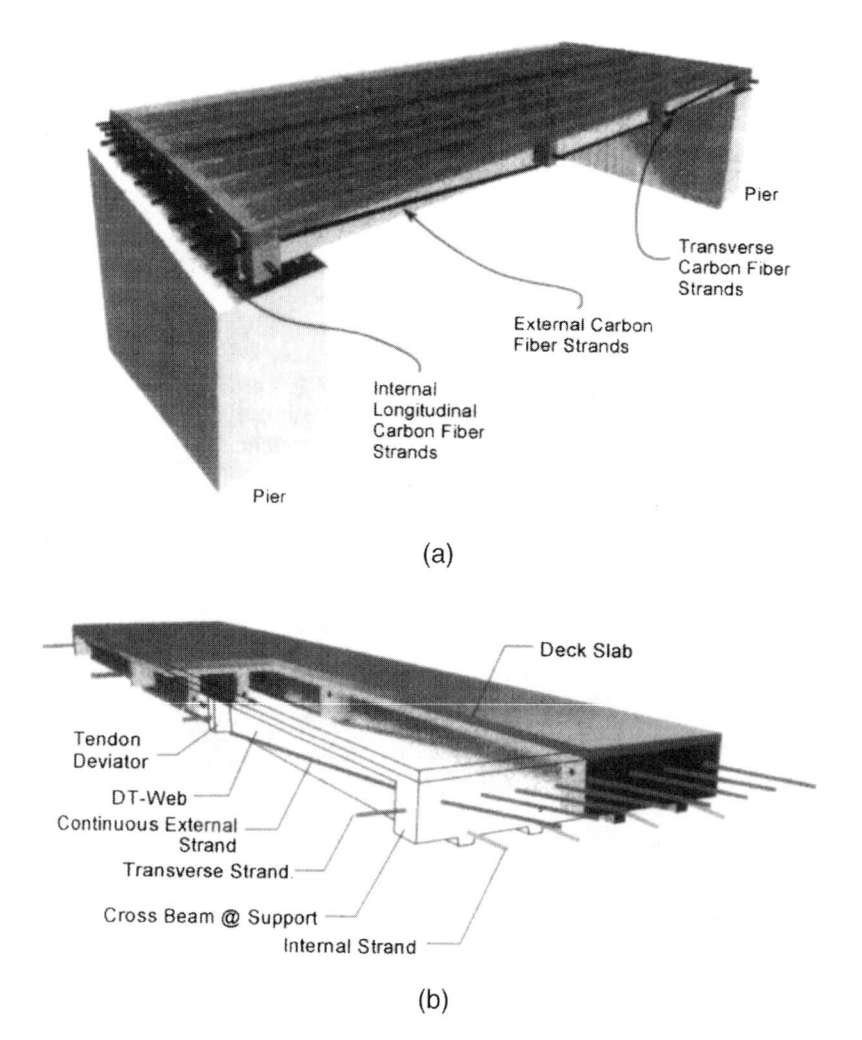

Figure 1.22 Noncorrosive DT bridge system. (a) Three-dimensional view of DT bridge system; (b) components of the DT bridge system.

1.12 SUMMARY

Physical models of structures are used in education, in research, in design, and in product and concept development. A model can be built and tested at a small fraction of the cost of a prototype (full-scale) structure because of the great reductions in loading magnitudes and in construction costs. Many different types of models are used, and the cost and time requirements vary widely for each type.

The applicability of models in design applications changes almost continuously as improved analytical capabilities permit the engineer to model mathematically increasingly complex structures, but the development of new combinations of structural forms and materials will most likely always be one step ahead of analysis. The need of experimental verification will, therefore, always be necessary to help us understand the behavior of complex structural systems.

Experimental stress analysis and the general principles of experimentation form integral parts of the modeling process. The execution of the many steps involved in a structural model study is an engineering project in itself; hence a considerable degree of "art" is involved along with the rather well-developed technology given in this book.

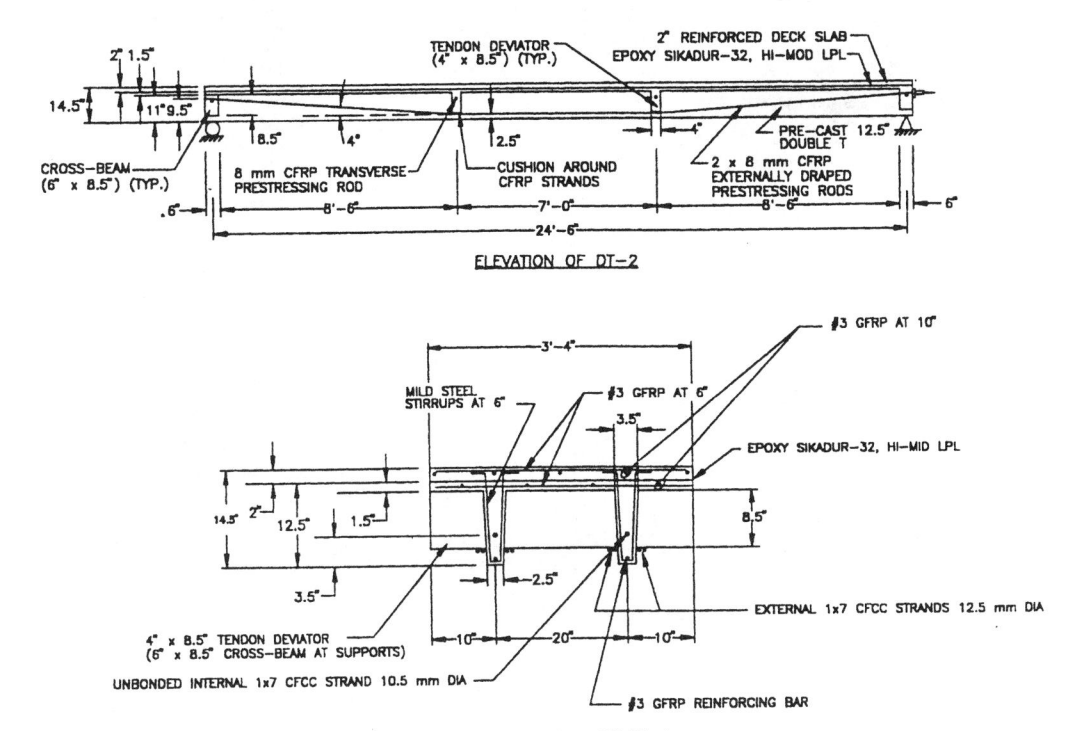

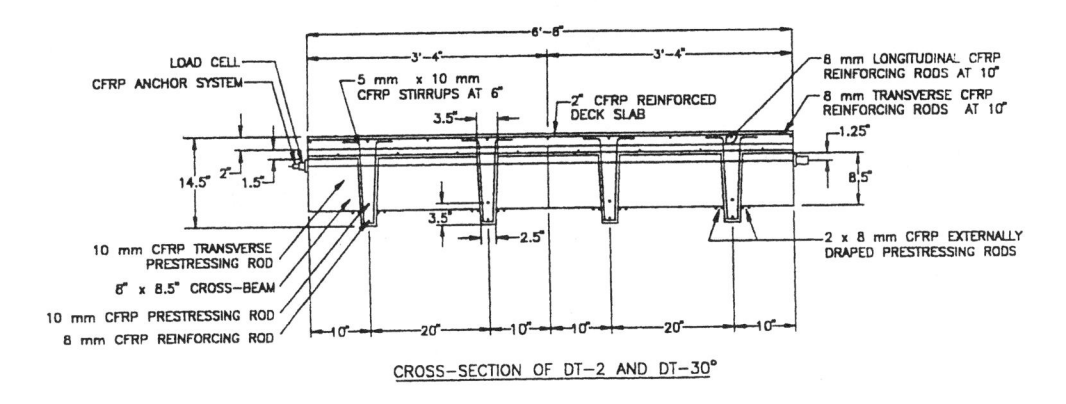

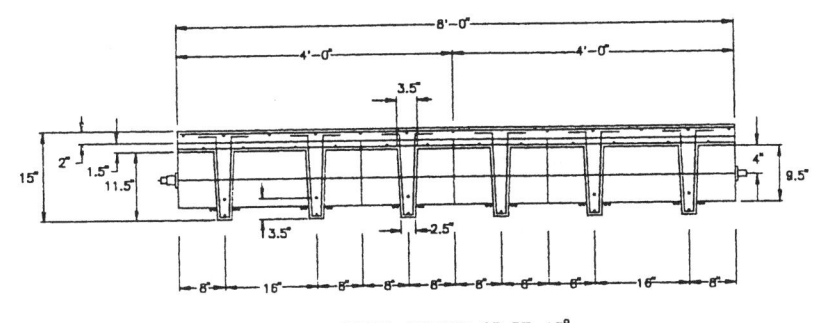

Figure 1.23 Details of tested bridge models.

Figure 1.24 Fatigue load test of bridge model DT-30 (30° skew). (Courtesy Prof. Nabil Grace, Lawrence Technological University, Southfield, MI).

Figure 1.25 External post-tensioning of bridge model DT-15 (15° skew). (Courtesy Prof. Nabil Grace, Lawrence Technological University, Southfield, MI).

PROBLEMS

1.1 Refer to Case Study A. If the shape of this roof was completely flat, how do you think the wind loads would differ if the wind model was tested in a wind tunnel?

1.2 What are the main factors for using small-scale models to conduct research on structural systems at a small university with very limited resources?

1.3 Imagine that you had at your disposal several straight pieces of steel, aluminum, wood, and Plexiglas rectangular sections of small size and various lengths. Buckling of a pin-ended model column can be demonstrated by holding the column vertically on a smooth horizontal surface and pushing at the top with your hand, increasing the force gradually until the column bows out.

Suppose you did not know the Euler buckling equation that governs the behavior of the initially straight, pin-ended axially loaded column but you had at your disposal a balance of appropriate accuracy that you can introduce between your hand and the bottom of the column to measure the load. How would you set up an experiment to demonstrate that the buckling load varies inversely as the length of the column squared and directly as its Young's modulus and its moment of inertia?

1.4 What is the most difficult problem that must be overcome in executing an accurate small-scale model study on a shaking table?

1.5 The old City Hall Building in your town needs to be upgraded to comply with newer, more stringent requirements concerning earthquake loadings. Being a very important historical building constructed of unreinforced masonry, discuss how a possible program, using small-scale models, could help the engineer in charge of the project to strengthen the building and make it more earthquake resistant.

1.6 As new construction materials are developed for stronger and lighter structures, their laboratory evaluation must precede any field applications. Think of situations where the testing of new construction materials on small-scale structural models can save time and money and speed their practical introduction into construction practice.

1.7 One of the current urgent structural needs in the U.S. and other industrialized countries is the repair of the deteriorated urban infrastructure (highways, bridges, buildings, tunnels, etc.). To accomplish this in an economical manner, the use of many novel repair techniques (structural concepts and/or materials, etc.) will be necessary. Think of ways in which the small-scale modeling technique can be used to determine the structural *safety, reliability,* and *durability* of such repair methods in a timely and economical manner.

REFERENCES

AASHTO (1989). *Specifications of Highway Bridges,* 14th ed., American Association of State Highway and Transportation Officials, Washington, D.C.

Aktan, A. E., Bertero, V. V. Chowdhury, A. A., and Nagashima, T. (1983). Experimental and Analytical Predictions of the Mechanical Characteristics of a 7-Story ⅕-Scale Model R/C Frame-Wall Building Structure, Earthquake Engineering Research Center, Report No. UCB/EERC-83/13, University of California, Berkeley.

Banerjee, A. C. (1973). Design of Pile Caps, Central Building Research Institute, Roorkee (India), March.

Beggs, G. E. (1932). An accurate mechanical solution of statically indeterminate structures by use of paper models and special gages, *Proc. Am. Concr. Inst.,* 18, 58–82.

Beggs, G. E., Timby, K. E., and Birdsall, B. (1932). Suspension bridge stresses determined by model, *Eng. News-Record,* June 9.

Bertero, V. V., Aktan, A. E., Harris, H. G., and Chowdhury, A. A. (1983). Mechanical Characteristics of Materials Used in a ⅕-Scale Model of a 7-Story Reinforced Concrete Test Structure, Earthquake Engineering Research Center, Report No. UCB/EERC-83/21, University of California, Berkeley, October.

Bertero, V. V., Aktan, A. E., Charney, F. A., and Sause, R. (1984). U.S.–Japan Cooperative Earthquake Research Program: Earthquake Simulation Tests and Associated Studies of a ⅕-Scale Model of a 7-Story Reinforced Concrete Test Structure, Earthquake Engineering Research Center, Report No. UCB/EERC-84/05, University of California, Berkeley, June.

Blevot, J. and Fremy, R. (1967). Sur pieus, *Ann. Inst. Tech. Batim. Trav. Public* (Paris), 20, February.

Bracci, J. M., Reinhorn, A. M., and Mander, J. B. (1992). Seismic Resistance of Reinforced Concrete Frame Structures Designed Only for Gravity Loads: Part I — Design and Properties of a One-Third Scale Model Structure, Technical Report NCEER-92-0027, National Center for Earthquake Engineering Research, SUNY/Buffalo, December 1.

Bull, A. H. (1930). A new method for the mechanical analysis of trusses, *Civil Eng.* 1, (3), 181–183.

Clark J. L. (1973). Behavior and Design of Pile Caps with Four Piles, Technical Report No. 42.489, Cement and Concrete Association, Wexham Springs, Slough.

CRSI (1986). Pile caps for individual columns, in *CRSI Handbook*, Concrete Reinforcing Steel Institute, Chicago, IL, Chap. 13.

Dagher, R. (1988). Investigation of Thick Pile Caps, M.E. thesis, Howard University, Washington, D.C.

Drexel University (1998). *The Drexel Engineering Curriculum,* College of Engineering, LeBow Engineering Building, Philadelphia.

El-Attar, A. G., White, R. N., and Gergely, P. (1991). Shaking Table Test of a ⅛-Scale Three-Story Lightly Reinforced Concrete Building, Technical Report NCEER-91-0018, National Center for Earthquake Research, SUNY/Buffalo, February 28.

El-Attar, A. G., White, R. N., and Gergely, P. (1997). Behavior of gravity load designed reinforced concrete buildings subjected to earthquakes, *ACI Struct. J.*, 94(2), 133–145.

Eney, W. J. (1939). New deformeter apparatus, *Eng. News-Record*, 122(7), 16, 221.

Gottschalk, O. (1926). Mechanical calculation of elastic systems, *J. Franklin Inst.,* July.

Grace, N. F. and Sayed, G. A. (1996a). Double tee and CFRP/GFRP bridge system, *Concrete Int.,* American Concrete Institute, 18(2), 39–44.

Grace, N. F. and Sayed, G. A. (1996b). Feasibility of CFRP/GFRP prestressed concrete demonstration bridge in the USA, paper presented as Fourth National Workshop on Bridge Research in Progress, National Center for Earthquake Engineering Research, Buffalo, NY.

Grace, N. F. and Sayed, G. A. (1997a) Ductility of prestressed concrete bridges using internal/external CFRP strands, paper presented at Seventh International Conference & Exhibition, Structural Faults + Repair 97, Edinburgh, Scotland.

Grace, N. F. and Sayed, G. A. (1997b). Behavior of externally/internally prestressed concrete composite bridge system, paper presented at Third International Symposium on Non-Metallic (FRP) Reinforcement for Concrete Structures, Sapporo, Japan, Oct. 14–16.

Grace, N. F. and Sayed, G. A. (1998). Ductility of prestressed concrete bridges usinginternal/external CFRP strands, *Concr. Int., ACI,* June.

Harris, H. G. and Bertero, V. V. (1981). One-Fifth Scale Model of a 7-Story R/C Frame-Wall Building under Earthquake Loading, in-house report, University of California, Berkeley, August.

Harris, H. G. and Sabnis, G. M. (1996). Physical models for concrete structures — their role in the new engineering curriculum, paper presented at the ACI Fall Convention, New Orleans, Nov. 3–8.

Harris, H. G., Bertero, V. V., and Clough, R. W. (1981). One-fifth scale model of a seven-story reinforced concrete frame-wall building under earthquake loading, *Proceedings of the Joint I. Struct. E./B.R.E. International Seminar on Dynamic Modeling of Structures,* Building Research Station, Garston, Watford, England, Nov. 19–20.

Harris, H. G., Oh, K., and Hamid, A. A. (1992). Development of new interlocking and mortarless block masonry units for efficient building systems, in *Proceedings of the Sixth Canadian Masonry Symposium,* Saskatoon, Canada, June 15–17.

Harris, H. G., Oh, K., and Hamid, A. A. (1993). Development of New Interlocking and Mortarless Block Masonry Units to Improve the Earthquake Resistance of Masonry Construction, Final Report to the National Science Foundation under Grant No. MSM-9102769, Department of Civil and Architectural Engineering, Drexel University, Philadelphia.

Hobbs, N. B. and Stein, P. (1957). An investigation into the stress distribution in pile caps with some notes on design, in *Proceedings of the Institution of Civil Engineers,* Vol, 7, July, 599–628.

Hoffman, N. S. (1990). Behavior of Transversely Prestressed Wood Bridge Decks under Simulated Short-Term and Long-Term Traffic Loading, M.Sc. thesis, Department of Civil and Architectural Engineering, Drexel University, Philadelphia, March.

Idowu, J. (1979). Investigation of Thick Pile Caps, Research Project Report, Howard University, Washington, D.C.

Janney, J. R., Breen, J. E., and Geymayer, H. (1970). Use of models in structural engineering, in *Models for Concrete Structures,* ACI SP-24, American Concrete Institute, Detroit, MI, pp. 1–18.

Jimenez-Perez, R., Sabnis, G. M., and Gogate, A. B. (1986). Experimental behavior of thick pile caps, paper presented at the Structural Assessment Seminar at London.

Kinney, J. S. (1957). *Indeterminate Structural Analysis,* Addison-Wesley, Reading, MA, 655 pp.

The Ministry, (1983). Ontario Ministry of Transportation and Communication, Ontario Highway Bridge Design Code, 2nd ed., Downsview, Ontario, Canada.

Moakler, M. W. and Hatfield, L. P. (1953). The design and construction of a deformeter for use in model analysis, M.Sc. thesis, Rensselaer Polytechnic Institute, Troy, NY.

Monkarz, P. D. and Krawinkler, H. (1981). Theory and Application of Experimental Model Analysis in Earthquake Engineering, Report No. 50, the John A. Blume Earthquake Engineering Center, Department of Civil Engineering, Stanford University, CA.

Ndukwe, A. (1982). Comparison of Methods of Analysis and Design of Thick Reinforced Concrete Pile Caps, Research Project Report, Howard University, Washington, D.C.

Oh, K.-H. (1994). Development and Investigation of Failure Mechanism of Interlocking Mortarless Block Masonry Systems, Ph.D. thesis, Department of Civil and Architectural Engineering, Drexel University, Philadelphia, June.

Oh, K., Harris, H. G., and Hamid, A. A. (1993). New interlocking and mortarless block masonry units for earthquake resistant construction, in *Proceedings of the Sixth North American Masonry Conference,* June 6–9, Philadelphia.

Ohio DOT (1947). Investigation of the Strength of the Connection between a Concrete Cap and the Embedded End of a Steel H-Pile, Research Report No. 1, Ohio State Department of Highways.

Oliva, M. G. and Dimakis, A. G. (1987). Behavior of Post-tensioned Wood Bridge Decks: Full-Scale Testing, Analytical Correlation, Design Guidelines, Structures and Materials Test Laboratory, 87-1, University of Wisconsin, May, 150 pp.

Perdikaris, P. C. and Beim, S. R. (1988). RC bridge decks under pulsating and moving load, *ASCE, J. Struct. Eng.,* 114(3), March, 591–607.

Perdikaris, P. C. and Petrou, M. (1991). Code predictions vs. small-scale bridge deck model test measurements, Transportation Research Record No. 1290, TRB, Vol. 1, *Bridges & Structures,* March 10–13, 179–187.

Perdikaris, P. C., Beim, S. R., and Bousias, S. (1989). Slab continuity effect of ultimate and fatigue strength of R/C bridge deck models, *ACI Struct. J.,* 86(4), July-August, 483–491.

Perry, C. C. and Lissner, H. R. (1955). *The Strain Gage Primer,* McGraw-Hill, New York.

Petrou, M. F. (1991). Behavior of Concrete Bridge Deck Models Subjected to Concentrated Load — Ontario vs. AASHTO, M.Sc. thesis, Case Western Reserve University, Cleveland, OH, May.

Petrou, M. F., Perdikaris, P. C., and Wang, A. (1994). Fatigue behavior of non-composite reinforced concrete bridge deck models, Transportation Research Record No. 1460, *Bridges and Structures,* December.

Petrou, M. F., Perdikaris, P. C., and Duan, M. (1996). Static behavior of non-composite concrete bridge decks under concentrated loads, *J. Bridge Eng.,* ASCE.

Quinn, R. (1994). The mathematical and scientific foundations for an interactive engineering curriculum, *J. Eng. Educ.,* August.

Ruge, A. C. (1943/4). The Bonded Wire Gage Torque Meter, in *SESA Proceedings,* Vol. I, No. 2.

Ruge, A. C. and Schmidt, E. O. (1939). Mechanical structural analysis by the moment indicator, *Trans. ASCE,* 104; also *Proc. ASCE,* 65, No. 1, January, 161–170 and No. 6, June, 1037–1040.

Sabnis, G. M. and Dagher, R. (1989). Investigation of reinforced concrete pile caps, in *Proceedings of One-Day Conference on Life Structures,* Brighton, England.

Sabnis, G. M. and Gogate, A. B. (1980). Investigation of thick pile caps, *ACI J.,* 77(1), Jan.–Feb., 18–24.

Savage, J. L. (1934). Dam stresses and strains studied by slice models, *Eng. News-Record,* Dec. 6, 720–723.

Schuring, D. J. (1977). *Scale Models in Engineering Fundamentals and Applications,* Pergamon Press, New York.

Whittle, R. T. and Beattie, D. (1972). Standard pile caps I and II, *Concrete,* 6(1), January, 34–36; 6(2), February, 29–30.

Yan, H. T. (1954). The design of pile caps, *Civil Eng. Public Works Rev.,* 49, May and June.

CHAPTER 2

The Theory of Structural Models

CONTENTS

2.1 Introduction 42
2.2 Dimensions and Dimensional Homogeneity 42
2.3 Dimensional Analysis 45
    2.3.1 Buckingham's Pi Theorem 49
    2.3.2 Dimensional Independence and Formation of Pi Terms 50
    2.3.3 Uses of Dimensional Analysis 52
    2.3.4 Additional Considerations in Using Dimensional Analysis 53
2.4 Structural Models 56
    2.4.1 Models with Complete Similarity 56
    2.4.2 Technological Difficulties Associated with Complete Similarity 58
        2.4.2.1 Other Types of Distortion 59
        2.4.2.2 Relaxation of Design Requirements 60
    2.4.3 Models with First-Order Similarity 60
    2.4.4 Distorted Models 61
2.5 Similitude Requirements 62
    2.5.1 Reinforced Concrete Models 63
    2.5.2 Models of Masonry Structures 67
    2.5.3 Structures Subjected to Thermal Loadings 67
    2.5.4 Structures Subjected to Dynamic Loadings 70
        2.5.4.1 Introduction 70
        2.5.4.2 Vibrations of Elastic Structures 70
        2.5.4.3 Fluidelastic Models 73
        2.5.4.4 Blast and Impact, Load Modeling 74
            1. External Blast on Structures 74
            2. Internal Blast Effects 75
        2.5.4.5 Earthquake Modeling of Structures 76
        2.5.4.6 Modeling for Centrifuge Testing 76
2.6 Summary 76
Problems 77
References 81

2.1 INTRODUCTION

Any structural model must be designed, loaded, and interpreted according to a set of *similitude requirements* that relate the model to the prototype structure. These similitude requirements are based upon the theory of modeling, which can be derived from a dimensional analysis of the physical phenomena involved in the behavior of the structure. Accordingly, this chapter examines two distinct topics:

1. Dimensional analysis and similitude theory.
2. Actual similitude requirements for different types of structural models, aimed at studying their response under elastic and ultimate load conditions as well as under dynamic and thermal loadings. Particular emphasis is given to models of structures undergoing dynamic loading effects because this is an area where physical scale model testing can be of significant help to the structural engineer.

It must be emphasized that a strictly formal application of modeling theory to a structural problem, without at least some understanding of the expected structural behavior, can lead to an inadequate and even incorrect modeling program. Similitude theory must be viewed simply as one aspect of the total modeling problem.

2.2 DIMENSIONS AND DIMENSIONAL HOMOGENEITY

The use of dimensions dates from early history when human beings first attempted to define and measure physical quantities. It was essential for these descriptions to have two general characteristics: qualitative and quantitative.

The *qualitative* characteristic enables physical phenomena to be expressed in certain *fundamental measures* of nature. The three general classes of physical problems, namely, mechanical (static and dynamic), thermodynamic, and electrical, are conveniently described qualitatively in terms of the following fundamental measures:

1. Length
2. Force (or mass)
3. Time
4. Temperature
5. Electric charge

These fundamental measures are commonly referred to as *dimensions*. Full chapters of books (such as those by Ipsen, 1960, and Bridgman, 1922) are devoted to establishing and categorizing the fundamental measures.

Most structural modeling problems are mechanical; thus, the measures of length, force, and time are most important in structural work. Thermal problems require the additional measure of temperature.

The *quantitative* characteristic is made up of both a *number* and a standard of *comparison*. The standard of comparison, also called the standard *unit,* was often established rather arbitrarily from traditional usage (such as the inch). Each of the fundamental measures, or dimensions, thus has its associated standard units in the several different unit systems in use today (U.S. Customary, SI, metric, etc.). Dimensions and units are such logical quantities that they are now taken completely for granted. It is difficult to realize that the present state of physical description of occurrences did not always exist.

THE THEORY OF STRUCTURAL MODELS

Keeping the above definitions of dimensions and units in mind, the theory of dimensions can be summarized in two essential facts:

1. Any mathematical description (i.e., equation) that describes some aspect of nature must be in a dimensionally homogeneous form. That is, the governing equation must be valid regardless of the choice of dimensional units in which the physical variables are measured. As an example, the equation for bending stress, $\sigma = Mc/I$, is correct regardless of whether force and length are measured in Newtons and meters, pounds and inches, or other consistent units.
2. As a consequence of the fact that all governing equations must be dimensionally homogeneous, it can be shown that any equation of the form

$$F(X_1, X_2, ..., X_n) = 0 \quad (2.1)$$

can be expressed in the form

$$G(\pi_1, \pi_2, ..., \pi_m) = 0 \quad (2.2)$$

where the π (pi) terms are dimensionless products of the n physical variables $(X_1, X_2,...,X_n)$, and $m = n - r$, where r is the number of fundamental dimensions that are involved in the physical variables.

This second fact, that any equation of the form $F(X_1, X_2,...,X_n) = 0$ is expressible as $G(\pi_1, \pi_2,...,\pi_m) = 0$, has two very important implications:

1. The form of a physical occurrence may be partially deduced by proper consideration of the dimensions of the n physical quantities X_i involved. The deductions are made by *dimensional analysis,* which is discussed at length in Section 2.3.
2. Physical systems that differ only in the magnitudes of the units used to measure the n quantities X_i, such as the quantities for a prototype structure and its reduced-scale model, will have identical functionals G. *Similitude requirements* for modeling result from forcing the pi terms $(\pi_1, \pi_2,..., \pi_m)$ to be equal in model and prototype, which is a necessary condition for the full functional relationships to be equal. Section 2.4 expands upon this concept.

Example 2.1

Hooke's law furnishes a good example of a dimensionally homogeneous relation. For a stretched bar of a perfectly elastic material, the equation which describes the stress–strain relation is

$$\sigma - E\varepsilon = 0$$

It is important to notice that the fundamental measures that describe the physical quantities in this equation combine in a certain fashion. Thus:

$$k_1 \frac{\text{force unit}}{(\text{length unit})^2} - \left[k_2 \frac{\text{force unit}}{(\text{length unit})^2} \cdot k_3 \frac{\text{length unit}}{\text{length unit}} \right] = 0$$

or

$$k_1 \frac{\text{force unit}}{(\text{length unit})^2} - k_2 k_3 \frac{\text{force unit}}{(\text{length unit})^2} = 0$$

and the equation holds for any system of force and length units. Such an equation is said to be *dimensionally homogeneous*.

Consider the special case of a bar made of steel with modulus of $E = 200$ N/mm². Then the equation is

$$\sigma - 200\varepsilon = 0$$

The equation is no longer dimensionally homogeneous. It is obviously correct only when σ and E are measured in terms of newtons and millimeters.

The following examples further verify the fact that physical phenomena are always expressible in dimensionally homogenwous form.

Example 2.2

An algebraic equation describing the deflection of a simple prismatic elastic beam with a span l that is subjected to the triangularly distributed total load W (Figure 2.1a) is

$$y(x) - \frac{Wx}{180EIl^2}(3x^4 - 10l^2x^2 + 7l^4) = 0$$

All quantities may be expressed in the fundamental measures of length (L) and force (F); thus, the dimensions of the left side of the equation are

$$L - \frac{FL}{(F/L^2)L^4L^2}(L^4 - L^2L^2 + L^4)$$

Example 2.3

A general nonlinear ordinary differential equation describing the undamped motion of the simple pendulum (Figure 2.1b) is

$$l\frac{d^2\theta}{dt^2} + g\sin\theta = 0$$

Dimensionally, the left side of the equation is

$$L\frac{1}{T^2} + \frac{L}{T^2}$$

Example 2.4

A partial differential equation describing small deflections of a flat plate subjected to lateral and edge loads N_x, N_y, and N_{xy} (Figure 2.1c) is

THE THEORY OF STRUCTURAL MODELS   45

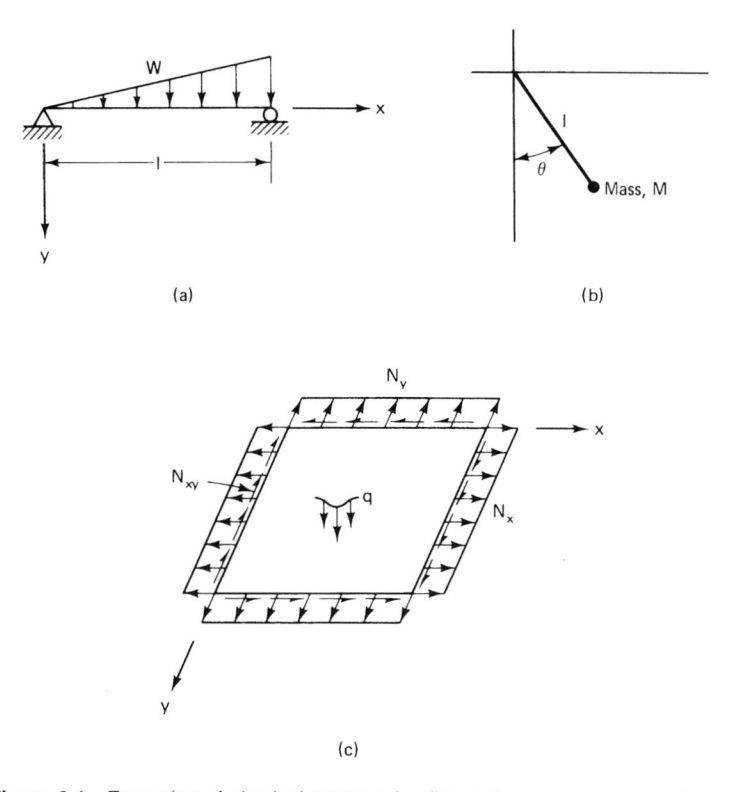

Figure 2.1 Examples of physical systems leading to homogeneous equations.

$$\frac{\partial^4 w}{\partial x^4} + 2\frac{\partial^4 w}{\partial x^2 \partial y^2} + \frac{\partial^4 w}{\partial y^4} - \frac{12(1-v^2)}{Eh^3}\left(q + N_x\frac{\partial^2 w}{\partial x^2} + N_y\frac{\partial^2 w}{\partial y^2} + 2N_{xy}\frac{\partial^2 w}{\partial x \partial y}\right) = 0$$

and the dimensions of the left side become

$$\frac{L}{L^4} + \frac{L}{L^2 L^2} + \frac{L}{L^4} - \frac{1}{(F/L^2)L^3}\left(\frac{F}{L^2} + \frac{F}{L}\frac{L}{L^2} + \frac{F}{L}\frac{L}{L^2} + \frac{F}{L}\frac{L}{LL}\right)$$

Note in each example that the right-hand side of the governing equation is zero. The fact that zero is a constant and thus has no dimensions would seem to indicate a contradiction. Of course, there is no contradiction. Zero is a very special constant as far as dimensional considerations are concerned, and one could either transpose one quantity from the left to the right or nondimensionalize the equation.

2.3 DIMENSIONAL ANALYSIS

Dimensional analysis is of substantial benefit in any investigation of physical behavior because it permits the experimenter to combine the variables into convenient groupings (pi terms) with a subsequent reduction of unknown quantities.

Example 2.5

As an introduction, consider the problem of experimentally determining the maximum stress at a section of a multispan girder subjected to a known uniformly distributed loading q per unit length. The analytical equation for stress σ has the form of Equation 2.1. Assuming that one had a good insight into the nature of this problem, it would be apparent that the stress σ is a function of loading q and a representative length l, or

$$F(q,l,\sigma) = 0 \tag{2.3}$$

Buckingham (1914) proves that any equation of this form can be represented as a product of powers in the form below:

$$\sigma = K q^a l^b \tag{2.4}$$

where K is a dimensionless parameter that may itself be a function of dimensionless groupings of the pertinent physical quantities, but is more often simply a constant.

In dimensional terms, Equation 2.4 takes the form*

$$\frac{F}{L^2} \doteq \left(\frac{F}{L}\right)^a L^b$$

or

$$FL^{-2} \doteq F^a L^{-a+b} \tag{2.5}$$

The dimensional homogeneity requirement forces equal dimensions on each side of the equation, or the exponents on each of the fundamental measures must be equal for the two sides of the equation. Thus, one writes exponential equalities for both F and L, or

$$\begin{aligned} F: &\quad 1 = a \\ L: &\quad -2 = -a + b \end{aligned} \tag{2.6}$$

from which

$$a = 1$$
$$b = -1$$

and thus

$$\sigma = K\left(\frac{q}{l}\right) \tag{2.7a}$$

in which K can be determined experimentally. Note that dimensional analysis alone has shown σ to be a linear function of (q/l).

---

* The symbol $\doteq$ will be used where dimensional equivalence is meant, and the symbol $=$ will be left for those equations where numerical equivalence is also maintained.

THE THEORY OF STRUCTURAL MODELS

Equation 2.7a can also be cast in the form of Equation 2.2:

$$G\left(\frac{l\sigma}{q}\right) = 0 \tag{2.7b}$$

which tells us that the problem is formulated in terms of a single dimensionless ratio and an unknown functional G.

Recourse to a mathematical solution of the problem would lead to

$$\sigma = \frac{Mc}{I} = \frac{a_1 q l^2 (a_2 l)}{a_3 l^4} = \frac{a_1 a_2}{a_3} \frac{q}{l} \tag{2.8}$$

where a_1, a_2, and a_3 are constants that depend upon the geometry of the girder and $a_1 a_2/a_3 = K$ in Equation 2.7a. Of course, the dimensional analysis of the problem could not have determined the magnitude of the constant $a_1 a_2/a_3$.

Example 2.6

Now suppose that the load in the previous example is of a dynamic nature and, therefore, has a variation with time. If one is interested in determining the maximum elastic displacement, then a logical set of pertinent physical parameters would include modulus of elasticity E, geometric length l, time t, loading q, specific weight of steel γ, and the acceleration of gravity g. The last two quantities are needed to account for inertia forces of the girder. The functional relationship implied by Equation 2.1 becomes

$$F(u, E, l, t, q, \gamma, g) = 0 \tag{2.9}$$

or writing the displacement u as a function of the other variables,

$$u = F'(E, l, t, q, \gamma, g)$$

which can also be expressed in the continued product form

$$u = K E^a l^b t^c q^d \gamma^e g^f \tag{2.10}$$

The dimensional equation for this expression is

$$L \doteq (FL^{-2})^a L^b T^c (FL^{-1})^d (FL^{-3})^e (LT^{-2})^f$$

Forcing this expression to be dimensionally homogeneous, one then has three equations for the three fundamental measures of force, length, and time, or

$$\begin{aligned} F: & \quad 0 = a + d + e \\ L: & \quad 1 = -2a + b - d - 3e + f \\ T: & \quad 0 = c - 2f \end{aligned} \tag{2.11}$$

The three equations (Equations 2.11) have six unknowns and thus allow a threefold infinity of solutions. Selecting d, e, and f "arbitrarily,"* the equations can be solved for a, b, and c in terms of d, e, and f. The solutions are

$$c = 2f$$
$$a = -d - e$$
$$b = -d + e - f + 1$$

Then

$$u = K\left(E^{-d-e} l^{-d+e-f+1} t^{2f} q^d \gamma^e g^f\right)$$

or

$$\left(\frac{u}{l}\right) = K\left[\left(\frac{q}{El}\right)^d \left(\frac{\gamma l}{E}\right)^e \left(\frac{t^2 g}{l}\right)^f\right] \tag{2.12}$$

Equation 2.12 can also be written in the alternate form

$$G\left(\frac{u}{l}, \frac{q}{El}, \frac{\gamma l}{E}, \frac{t^2 g}{l}\right) = 0 \tag{2.13}$$

The important fact that can be drawn from this analysis is that the dimensionless ratio involving the desired displacement u can be expressed as a function of a set of dimensionless ratios, or

$$\frac{u}{l} = \phi\left(\frac{q}{El}, \frac{\gamma l}{E}, \frac{t^2 g}{l}\right) \tag{2.14}$$

Thus, we have been able to use dimensional considerations alone to deduce a substantial insight into the problem and to eliminate a number of the exponents involved in the original formulation. The form of the functional relationship would have to be determined by experiments in which the several dimensionless ratios were systematically varied. Murphy (1950) prescribes tests that can be made to determine if the parameter K in Equation 2.12 is merely a constant or if it is a function of the dimensionless parameters.

A fully equivalent formulation of this problem that involves one fewer variable can be obtained by realizing that the acceleration of gravity g enters the problem only indirectly in converting the specific weight to a mass density, and that if mass density ρ had been used instead of specific weight γ, then g would not have been needed. This alternate formulation leads to

$$G\left(\frac{u}{l}, \frac{q}{El}, \frac{\rho l^2}{Et^2}\right) = 0 \tag{2.15}$$

and

$$\left(\frac{u}{l}\right) = \phi\left(\frac{q}{El}, \frac{\rho l^2}{Et^2}\right) \tag{2.16}$$

---

* Later in the chapter it will become evident that this choice is really not arbitrary; the three chosen exponents must embrace all three fundamental measures.

2.3.1 Buckingham's Pi Theorem

The discerning reader will have noticed that an analysis of dimensions has led in the first example from

$$\left. \begin{array}{c} F(q,l,\sigma) = 0 \\ \\ G\left(\dfrac{l\sigma}{q}\right) = 0 \end{array} \right\} \quad (2.17)$$

to

and in the second example from

$$\left. \begin{array}{c} F(u,E,l,t,q,\rho) = 0 \\ \\ G\left(\dfrac{u}{l}, \dfrac{q}{El}, \dfrac{\rho l^2}{Et^2}\right) = 0 \end{array} \right\} \quad (2.18)$$

to

These two examples are illustrations of a general theorem stated by Buckingham (1914). This theorem states that *any dimensionally homogeneous equation involving certain physical quantities can be reduced to an equivalent equation involving a complete set of dimensionless products*. For the structural models engineer, this theorem states that the solution equation for some physical quantity of interest, i.e.,

$$F(X_1, X_2, ..., X_n) = 0 \quad (2.19)$$

can equivalently be expressed in the form

$$G(\pi_1, \pi_2, ..., \pi_m) = 0 \quad (2.20)$$

The pi terms are dimensionless products of the physical quantities $X_1, X_2, ..., X_n$. A complete set of dimensionless products are the $m = n - r$ independent products that can be formed from the physical quantities $X_1, X_2, ..., X_n$. In the previous two illustrations it turned out that three and six physical variables reduced to one and three dimensionless products, respectively. Generally, it can be stated that the number of dimensionless products (m) is equal to the difference between the number of physical variables (n) and the number of fundamental measures (r) that are involved. The first example was a static mechanical problem; the fundamental dimensions were force and length, or $m = n - r = 3 - 2 = 1$. The second problem was a dynamic mechanical problem; the fundamental dimensions were force, length, and time, and $m = 6 - 3 = 3$.

Buckingham's pi theorem occupies a very important place in the theory of dimensional analysis. Before proceeding to the applications of dimensional analysis, the next sections consider in some detail the procedures used in obtaining the dimensionless products that go into Buckingham's pi equations.

Table 2.1 Typical List of Physical Quantities

	Quantity	Units
l	Length	L
Q	Force	F
M	Mass	$FL^{-1}T^2$
σ	Stress	FL^{-2}
ε	Strain	—
a	Acceleration	LT^{-2}
δ	Displacement	L
ν	Poisson's ratio	—
E	Modulus of elasticity	FL^{-2}

2.3.2 Dimensional Independence and Formation of Pi Terms

Table 2.1 presents a sample of physical quantities that might be involved in structural problems, and the dimensional measures required to describe them. Some structural problems will involve temperature and heat considerations; the extension of Table 2.1 to cover these quantities is done in Section 2.5.3.

An examination of the dimensional measures in Table 2.1 makes it obvious that the dimensions that are required to describe any physical quantity occur in the form of a single product. The fact that all such dimensional descriptions occur in the product form is a direct result of the nature of dimensions and the basic foundations upon which scientific measurements were first established.

Now, from any set of physical quantities, such as that listed in Table 2.1, it is possible *dimensionally* to form certain quantities in the set by combining others in the form of products. Thus, the dimensions of length (l) divided by the dimensions of time (t) yield the dimensions of velocity (v). Also, the dimensions of stress (σ) divided by the product of the dimensions of acceleration (a) and of time (t) squared yield the dimensions of specific weight (γ). Consequently, it is seen that the products vt/l and $\gamma at^2/\sigma$ are dimensionless. In any set of physical quantities there is a limited number of quantities that cannot themselves be combined with other quantities in the set to yield a dimensionless product. The quantities involved in the limited set are said to be dimensionally independent, while the other quantities are dimensionally dependent upon the special limited set. The main question met in applying the Buckingham pi theorem pertains to the formation of appropriate pi terms. The following simple points are the only guidelines needed in the formation process:

1. All variables must be included.
2. The m terms must be independent.
3. In general, there is no unique set of pi terms for a given problem; alternate formulations are possible either by forming the pi terms in several different ways or by suitable transformations of one set of pi terms. Thus, it is not possible to state that a set of complete, independent pi terms is either "right" or "wrong" for a given problem.

The best method for arriving at the groupings of pi terms is open to personal preference; there are a number of rather formal methods which involve setting up the appropriate dimensional equations. One less formal approach involves the following steps:

1. Choose r variables that embrace the r dimensions (fundamental measures) required in expressing all variables of the problem, and that are dimensionally independent. This means that if a problem involves the dimensions of force F, length L, and time T, then the three variables chosen must collectively have dimensions which include F, L, and T, but no two variables can have the same dimensions. Variables that are in themselves dimensionless (strain, Poisson's ratio, angles) cannot be chosen in the set of r variables.

THE THEORY OF STRUCTURAL MODELS

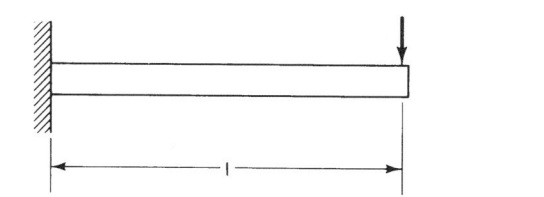

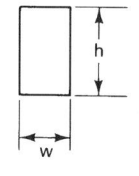

Figure 2.2 Beam stiffness example.

2. Form the m pi terms by taking the remaining $(n - r)$ variables and grouping them with the r variables in such a fashion that all groups are dimensionless. This procedure will guarantee a set of independent, dimensionless terms. It should be noted that the r variables chosen in step 1 above will in general appear more than once in the total set of pi terms while the remaining $(n - r)$ variables will each appear only once.

Example 2.7

How does the elastic stiffness of a rectangular cantilever beam depend on its properties? Assume the beam is loaded at its end, as shown in Figure 2.2, and that stiffness is defined as the force per unit displacement measured at the location of the load. With no knowledge of the actual relationship between beam properties and stiffness, one might choose the stiffness S to be a function of beam length l, depth h, width w, elastic modulus E, and Poisson's ratio ν.

The variables and their dimensions are conveniently represented in tabular form:

	S	l	w	h	E	ν
F	1	0	0	0	1	0
L	−1	1	1	1	−2	0

All six variables are expressible in two dimensions: force and length. Thus, there will be $6 - 2 = 4$ pi terms. Since the desired relationship is for the stiffness S, it is best not to include it in the two variables that may appear more than once. A convenient choice for the two multiple variables is span length l and modulus of elasticity E. These variables embrace both dimensions (F and L) and do not have identical dimensions.

The pi terms may be formed by inspection by appropriate grouping of l and E with the other variables:

$$\pi_1 = \frac{S}{El}, \quad \pi_2 = \frac{w}{l}$$
$$\pi_3 = \frac{h}{l}, \quad \pi_4 = \nu \tag{2.21}$$

The behavior of the beam can then be summarized as

$$G\left(\frac{S}{El}, \frac{w}{l}, \frac{h}{l}, \nu\right) = 0$$

or, alternatively,

$$\frac{S}{El} = G'\left(\frac{w}{l}, \frac{h}{l}, \nu\right) \tag{2.22}$$

Example 2.8

Form a complete, independent set of dimensionless pi terms from the quantities listed in Table 2.1.

The quantities and their dimensions are listed in array form as

	l	Q	M	σ	ε	a	δ	v	E
F	0	1	1	1	0	0	0	0	1
L	1	0	−1	−2	0	1	1	0	−2
T	0	0	2	0	0	−2	0	0	0

In selecting the three independent quantities that will appear at least once in the $9 - 3 = 6$ pi terms, it is evident that either mass M or acceleration a must be included, as these are the only two quantities which possess the dimension of time. The three quantities chosen here are length l, modulus E, and acceleration a. The six pi terms are then formed by inspection:

$$\pi_1 = \frac{Q}{El^2}, \qquad \pi_2 = \frac{Ma}{El^2}$$
$$\pi_3 = \frac{\sigma}{E}, \qquad \pi_4 = \varepsilon \qquad (2.23)$$
$$\pi_5 = \frac{\delta}{l}, \qquad \pi_6 = v$$

A more rigorous treatment of dimensional independence is given in Appendix A.

2.3.3 Uses of Dimensional Analysis

Dimensional analysis can be used by the engineer in two separate ways. First, it can be useful in deducing, from experimental observations, certain theoretical results regarding the behavior of a physical phenomenon. Such a situation could arise if one knew the relevant physical variables that affected the state of some other physical variable but did not know the mathematical relationship that connected these variables. For example, if only three or four physical quantities are involved, dimensional analysis may reveal the solution to within some constant value or some unknown function of one or two variables. Rather simple experiments can then be performed to determine the constant value or the functional relationship. Of course, if there are 10 or 20 physical variables to begin with, there will still be so many dimensionless products remaining as to make experimental analysis difficult or impractical.

The second use in the area of structural design work has been stated very succinctly by Bridgman (1922):

> There are in engineering practice a large number of problems so complicated that the exact solution is not obtainable. Under these conditions dimensional analysis enables us to obtain certain information about the form of the result which could be obtained in practice only by experiments with an impossibly wide variation of the arguments of the unknown function. In order to apply dimensional analysis we merely have to know what kind of a physical system it is that we are dealing with, and what the variables are which enter into the equation; we do not even have to write the equations down explicitly, much less solve them.
>
> Suppose that the variables of the problem are denoted by X_1, X_2, etc., and that the dimensionless products are found, and that the result is thrown into the form

$$X_1 = X_2^{a_1} X_3^{b_1} \ldots \phi\left(X_2^{a_2} X_3^{b_2} \ldots, X_2^{a_3} X_3^{b_3}, \ldots\right) \qquad (2.24)$$

THE THEORY OF STRUCTURAL MODELS

where the arguments of the function and the factor outside embrace all the dimensionless products, so that the result as shown is general. Now in passing from one physical system to another, the arbitrary function ϕ will in general change in an unknown way, so that little if any useful information could be obtained by indiscriminate model experiments. But if the models are chosen in such a restricted way that all the arguments of the unknown function have the same value for the model as for the full-scale example, then the only variable in passing from model to full scale is in the factors outside the functional sign, and the manner of variation of these factors is known from the dimensional analysis.

Stated in this way it would appear that the structural design engineer's model problems are solved, and, in fact, they are if the dimensionless products that are the arguments of the unknown functions have the same value for the model as for the prototype. As will be seen later, technological problems may make it impossible to strictly satisfy this condition, particularly for models that are intended to reproduce the inelastic response of reinforced concrete structures.

2.3.4 Additional Considerations in Using Dimensional Analysis

It is essential that the experimenter who wishes to use dimensional analysis have sufficient insight into the problem to choose the proper physical quantities in the formulation. The multispan beam problem of Example 2.5 is a convenient example for discussion here. To the person who does not have a keen insight into beam bending, it might not be immediately apparent that the modulus of elasticity E does not enter into the stress problem. If E was added to the problem, then the governing relationship would be

$$F(\sigma, q, l, E) = 0 \tag{2.25}$$

which is expressible in the two fundamental measures F and L. By following the procedures set down in Section 2.3.2, it is seen that q and l, q and σ, q and E, l and σ, and l and E are all dimensionally independent. Taking q and l as the dimensionally independent physical quantities, the complete set of dimensionless products is

$$\pi_1 = \frac{\sigma l}{q}, \quad \pi_2 = \frac{El}{q} \tag{2.26}$$

According to Buckingham's pi theorem, Equation 2.26 can be reduced to

$$G\left(\frac{\sigma l}{q}, \frac{El}{q}\right) = 0 \tag{2.27}$$

or

$$\sigma = \left(\frac{q}{l}\right)\phi_1\left(\frac{El}{q}\right) \tag{2.28}$$

There was nothing unique about the selection of q and l as the independent physical quantities. It is seen that q and E could just as easily have been chosen, and a procedure identical to the above would have led to

$$\sigma = E\phi_2\left(\frac{El}{q}\right) \tag{2.29}$$

Equation 2.27 should be compared with Equation 2.7a. The unnecessary inclusion of E as a relevant physical variable does not make Equation 2.27 incorrect. However, it is needlessly complicated, and it would require additional experimental work to find out that the beam stress was in fact independent of E. The experimental design engineer thus is faced with the following dilemma:

1. If a relevant physical quantity is omitted from a dimensional analysis of a problem, the investigation either meets an impasse or leads to erroneous results. This statement is not rigorously proved here, but reason alone indicates its validity.
2. If not only the relevant physical quantities but also some irrelevant ones are included, the dimensional analysis will lead to a result that will make the experimental investigation much more difficult than need be. In fact, it may eliminate a model study as a practical means of obtaining the desired information.

In this light, Langhaar (1951) has stated,

Frequently the question arises: How do we know that a certain variable affects a phenomenon? To answer this question, one must understand enough about the problem to explain why and how the variable influences the phenomenon. Before one undertakes the dimensional analysis of a problem, he should try to form a theory of the mechanism of the phenomenon. Even a crude theory usually discloses the actions of the more important variables. If the differential equations that govern the phenomenon are available, they show directly which variables are significant.

If they are not available, then the engineer must have some other insight into the phenomenon, because it is clear that dimensional analysis can be of no use unless one can identify the relevant physical variables.

Example 2.9

Free transverse vibrations of a flat elastic plate are known to be governed by the partial differential equation

$$\frac{\partial^4 w}{\partial x^4} + 2\frac{\partial^4 w}{\partial x^2 \partial y^2} + \frac{\partial^4 w}{\partial y^4} + \frac{\rho h}{Eh^3/12(1-v^2)}\frac{\partial^2 w}{\partial t^2} = 0 \tag{2.30}$$

subject to certain prescribed boundary and initial conditions, with the pertinent variables being:

x, y = coordinates of points on the surface of the plate
h = plate thickness
E = modulus of elasticity
v = Poisson's ratio
t = time
ρ = mass density
w = out-of-plane displacement of plate middle surface

It is desired to make a model study of a large irregularly shaped plate in order to determine the lowest natural frequency of vibration. What are the relevant physical variables? In the light of a dimensional analysis of the problem, how should a model study be conducted, and is it likely that useful results will be obtained from a model investigation?

Assume that the solution equation for the natural frequency f is of the form

$$F(f, l, \rho, E, v, h) = 0 \tag{2.31}$$

where l is a characteristic plate dimension for either plan or thickness dimensions. Taking l, ρ, and E to be dimensionally independent, Equation 2.31 can be reduced to

$$G\left(\frac{fl\rho^{1/2}}{E^{1/2}}, v\right) = 0 \tag{2.32}$$

or in the solved form

$$f = \frac{1}{l}\sqrt{\frac{E}{\rho}}\phi(v) \tag{2.33}$$

Equation 2.32, which is the most that can be obtained from a dimensional analysis, can be compared with the mathematical solution of the simply supported rectangular plate. Such a solution yields for the fundamental frequencies

$$f = \left(\frac{k^2}{a^2} + \frac{j^2}{b^2}\right)\pi\sqrt{\frac{Eh^2}{12\rho(1-v^2)}}$$

where a and b are the side lengths and m and n are integers indicating the number of half sine waves in the deflected shape of the plate. For a square plate with j and k equal to 1,

$$f = \frac{h}{a^2}\sqrt{\frac{E}{\rho}}\sqrt{\frac{\pi^2}{3(1-v^2)}}$$

Returning to the model problem, Equation 2.33 can be written once for the prototype and once for the model. Dividing the prototype equation by the model equation, one obtains

$$\frac{f_{\text{prototype}}}{f_{\text{model}}} = \frac{(1/l_p)\sqrt{(E_p/\rho_p)}}{(1/l_m)\sqrt{(E_m/\rho_m)}} \frac{\phi(v)_p}{\phi(v)_m}$$

If Poisson's ratio in the model and the prototype are equal, then the magnitude of the function is an identical constant for each. In that case

$$f_{\text{prototype}} = f_{\text{model}} \frac{l_m}{l_p}\sqrt{\frac{E_p\rho_m}{E_m\rho_p}} \tag{2.34}$$

In fact, it might be reasoned that Poisson's ratio does not greatly influence this physical phenomenon, and the experimenter might then allow the model material to have a Poisson's ratio different from that of the prototype. The experimenter must be aware, of course, that the final result would be subject to the error associated with the departure from the dictates of the dimensional analysis.

The boundary conditions also must be modeled. Clearly, if the edges of the prototype plate are rigidly fixed or simply supported, then the edges of the model plate must also be rigidly fixed or simply supported. Similarly, elastically restrained edges in the prototype must be modeled. As a matter of interest, it should be expected that the technological problems associated with providing similar boundary conditions in this problem would be of considerably more concern than the effect of Poisson's ratio.

2.4 STRUCTURAL MODELS

It is a relatively simple matter to apply dimensional analysis principles to the structural model. As the discussion is developed, three types of structural models will be described. These are

1. The *true model*, which maintains *complete similarity*. Any model that satisfies each and every stipulation set forth by a proper dimensional analysis would be said to have complete similarity.
2. The *adequate model*, which maintains *"first-order" similarity*. If an engineer has a special insight into a problem, then it may be possible to reason that some of the stipulations set forth by proper dimensional analysis are of "second-order" importance. For example, in rigid-frame problems it is known that axial and shearing forces are of second-order importance relative to bending moments insofar as deformations are concerned. Thus, it may be adequate to model the moment of inertia but not the cross-sectional areas of members. Thus, any model which satisfies each and every first-order stipulation which is set forth by a proper dimensional analysis but which may not satisfy certain second-order stipulations would be said to have first-order similarity.
3. The *distorted model*, which *fails to satisfy* one or more of the first-order stipulations as set forth by proper dimensional analysis.

Of course, complete similarity is desirable in all structural models, but usually the economic and technological conditions preclude a model study that maintains complete similarity with the prototype. By neglecting certain second-order effects, it is usually possible to make an *adequate model* study to obtain results to predict the behavior of a prototype structure accurately.

2.4.1 Models with Complete Similarity

It has been seen from Buckingham's theorem that the mathematical formulation of any physical phenomenon can be reduced to an equation involving a complete set of dimensionless products,

$$\pi_1 = \phi(\pi_2, \pi_3, \ldots, \pi_n) \tag{2.35}$$

If Equation 2.35 is written once for the prototype and once for the model, the following quotient can be formed:

$$\frac{\pi_{1p}}{\pi_{1m}} = \frac{\phi(\pi_{2p}, \pi_{3p}, \ldots, \pi_{np})}{\phi(\pi_{2m}, \pi_{3m}, \ldots, \pi_{nm})} \tag{2.36}$$

where π_{1m} refers to π_1 in the model and π_{1p} refers to π_1 in the prototype, etc. Complete similarity is defined to be that condition in which all of the dimensionless products are the same in both model and prototype. When complete similarity is maintained,

$$\begin{aligned}\pi_{2m} &= \pi_{2p} \\ \pi_{3m} &= \pi_{3p} \\ &\ldots \\ \pi_{nm} &= \pi_{np}\end{aligned} \tag{2.37}$$

so that Equation 2.36 may be written

$$\frac{\pi_{1p}}{\pi_{1m}} = \frac{\phi(\pi_{2p}, \pi_{3p}, \ldots, \pi_{np})}{\phi(\pi_{2m}, \pi_{3m}, \ldots, \pi_{nm})} = 1 \qquad (2.38)$$

or

$$\pi_{1p} = \pi_{1m} \qquad (2.39)$$

Equations 2.37 and 2.39 are the basis for the model method. The relations between model and prototype quantities as implied in Equation 2.37 are called the *design and operating conditions*; the single Equation 2.39 is the *prediction equation* for the dependent variable of the problem. Thus, in Example 2.9,

$$\pi_1 = \frac{fl\rho^{1/2}}{E^{1/2}}$$

$$\pi_2 = v$$

Equation 2.39 took the form

$$f_p l_p \sqrt{\frac{\rho_p}{E_p}} = f_m l_m \sqrt{\frac{\rho_m}{E_m}}$$

which yielded Equation 2.34.

The similitude relations corresponding to the pi terms of Example 2.8 can be derived by equating $\pi_m = \pi_p$ for each of the six terms and solving for the scale factor $s_i = i_p/i_m$. s_i is defined as the scale factor for the quantity i, and the subscripts p and m denote prototype and model, respectively.

There are three dimensionally independent quantities (l, E, and a) appearing in the six pi terms of Example 2.8. Accordingly, scale factors may be arbitrarily chosen for only three quantities. It is logical to choose S_l, S_E, S_a. The remaining scale factors are then

$$S_Q = S_l^2 S_E, \qquad S_\varepsilon = 1$$

$$S_M = \frac{S_l^2 S_E}{S_a}, \qquad S_\delta = S_l \qquad (2.40)$$

$$S_\sigma = S_E, \qquad S_v = 1$$

It should be noted that model and prototype strains must be identical for true models. Stresses will be identical *only* if $S_E = 1$. This results automatically when the same material is used in model and prototype.

The formulation of scaling relations for any true modeling problem can be established quite easily by the simple translation of π terms into required scale factors. The reader is encouraged to select a complete set of quantities for some more complicated problem, derive a set of π terms, establish the similitude relations, and examine them carefully.

2.4.2 Technological Difficulties Associated with Complete Similarity

Several types of departures from complete similarity can occur, including

1. Accidental overlooking of a pertinent variable;
2. Deliberate violation of a similitude requirement that is considered to be not critical, such as using a model material with Poisson's ratio different from that of the prototype material;
3. Necessary deviations from true modeling, such as using a discrete load system to replace a continuous load.

Lack of complete similarity in the model and prototype dimensionless ratios means that the ratio ϕ_p/ϕ_m in Equation 2.38 is no longer unity. While it is not feasible in most instances to make an evaluation of the true value of the ratio, it is essential to realize that this departure from true similarity affects the model result to the degree that the ratio departs from unity. Lack of similarity, either known or unknown, often leads to differences that are misleadingly called "size effects." Size effects as such do not exist if complete similarity is maintained in geometry, materials properties, and loading (Chapters 3 to 5 and Chapter 8 deal with the latter two aspects in depth).

Example 2.10 Gravity Load Simulation

This example illustrates a departure from similitude of the second type mentioned above. Corresponding to the fact that the maximum number of dimensionally independent quantities equals the number of fundamental dimensions involved in the quantities, the model engineer can select only a restricted number of model quantities without regard for the prototype. Thus, in static and dynamic problems only two and three model quantities, respectively, can be arbitrarily selected. Practical considerations generally demand that the model geometric scale and certain model material properties be selected to be compatible with the available equipment and materials.

If the problem involves only static response and the deadweight of the structure exerts an important influence, then one of the dimensionless products in the problem will be $\gamma l/E$ where γ is the specific weight of the material, E is the modulus of elasticity of the material (or some equivalent quantity which represents the stress–strain characteristics of the material), and l is a representative length. Now only two model quantities can be selected arbitrarily, and thereafter the preceding dimensionless product must have the same magnitude in model and in prototype. Thus

$$\left(\frac{\gamma l}{E}\right)_{model} = \left(\frac{\gamma l}{E}\right)_{prototype}$$

or

$$\gamma_{model} = \gamma_{prototype} \frac{l_p E_m}{l_m E_p} \qquad (2.41)$$

As a typical illustration, consider a reinforced concrete prototype structure simulated with a ¹⁄₁₀-scale polyvinylchloride plastic model. Then

$$\gamma_{model} = 90 \text{ lb/ft}^3 \; (1440 \text{ kg/m}^3)$$

but the dead-load similitude requirement says that the density of the model material should be

$$150(10)\left(\frac{450,000}{3,000,000}\right) = 225 \text{ lb/ft}^3 \; (3600 \text{ kg/m}^3)$$

THE THEORY OF STRUCTURAL MODELS

Figure 2.3 Model for continuous bridge under dynamic loading. (Courtesy of ISMES, Bergamo, Italy.)

Thus, Equation 2.41 is not satisfied. A departure from complete similarity is necessary. If the structure is "slender," the difference between the deadweight dictated by similitude considerations and that furnished by the model material may be added at the surface of the structure. On the other hand, a "massive" structure may require that the additional mass be dispersed throughout the volume of the structure. Thus, many of the model studies of large concrete dams have required such an approach.

Three model quantities can be arbitrarily selected in dynamic problems. This fact should not lead one to believe that the specific weight could be that additional quantity. Length, stress, and specific weight are not dimensionally independent quantities. Thus, deadweight stresses constitute a static phenomenon, and therefore artificial means must ordinarily be used to provide the simulation. Of course, if the structural response that is of interest does not depend significantly upon the deadweight, then the most appropriate procedure may be simply to neglect the discrepancy completely.

Figure 2.3 shows how additional artificial mass was attached to a bridge vibration model to simulate prototype deadweight effects.

2.4.2.1 Other Types of Distortion

There are a number of distortions that are met frequently in modeling, including

1. Discrete loading in place of distributed loading;
2. Lack of similitude in bond strength of reinforcing wires for reinforced concrete models;
3. Model concrete with failure criteria different from that of prototype concrete;
4. Time scaling in dynamic loadings;
5. Violation of $v_m = v_p$ when using elastic models for concrete structures.

Several of these situations will be discussed subsequently.

2.4.2.2 Relaxation of Design Requirements

Another way of looking at the problem of incomplete similarity is to say that the laws of similitude governing a particular physical phenomenon of interest are being "relaxed" so that a scaled model study can be performed. Several methods of relaxation are discussed by Schuring (1977).

2.4.3 Models with First-Order Similarity

Some of the practical difficulties of obtaining complete similarity in the model were discussed in the preceding section. If the experimental method is to be used for both analysis and design, it may be necessary to relax the restriction that the ratio of the ϕ functions in Equation 2.38 be identically 1. First-order similarity is here defined to be that degree of model-to-prototype similarity such that the engineer is willing to neglect the difference between the actual value of the ratio ϕ_p/ϕ_m and unity, or to neglect an error introduced by incomplete similarity of quantities outside the ϕ function. In other words, first-order similarity is when ϕ_p/ϕ_m is approximately equal to 1, or such that, for example, the behavioral difference between a uniformly applied load and a discrete pattern of loads can be neglected.

In line with the examples that were discussed in the preceding section, the bridge vibration model that incorporated a series of additional masses along the length of the bridge and a thin-shell buckling model that is loaded by a grid of concentrated loads instead of a distributed load could be considered to be models with first-order similarity.

In another vein, there are certain types of structural problems that are special situations. The nature of these problems can best be understood by considering first the nature of all structural response. In this regard, the deformations in any structure are dependent upon (1) the force and displacement boundary and initial conditions imposed on the structure, (2) the geometry of the structure, and (3) the materials of which the structure is composed. In the determination of these deformations (and hence stresses, etc.) from a mathematical point of view, certain special types of structural behavior have been categorized. Thus, from an analytical point of view, any structural response is said to arise from either (1) axial, (2) shearing, (3) bending, or (4) torsional deformation or *any* combination of the four. These categories have been invented in an attempt to surmount certain of the difficulties associated with analytical stress analysis. Now, the greatest asset of the experimental method of stress analysis lies in the fact that the model, like the prototype, is unaware of these four categories, and hence simply yields the complete result. Nevertheless, certain special problems may arise in which our knowledge of these four types of response can be usefully applied in a model problem.

For example, consider a planar rigid-frame structure that is very highly statically indeterminate. The component members, even with a variable moment of inertia, are all relatively long in comparison with their cross-sectional configuration. If the external loads are all applied in the plane of the frame and the connections of the members do not involve any lateral eccentricities, then it is known that the resulting deformations are primarily the result of bending. Axial and shearing deformations are of second order, and torsional deformations are nonexistent. If the problem is to determine the bending moment at a certain cross section and the behavior is known to be elastic, it would suffice to maintain similarity with respect to the moment of inertia of any particular cross section rather than to maintain similarity with respect to shape as well. At the price of not being able to account properly for axial and shear force effects, the fabrication problems associated with the model study may have been reduced significantly. However, suitable care must be exercised when taking such liberties. Thus, for example, the fact that bending moments are similar does not ensure that stresses are similar. Further, one might ask whether model and prototype bending moments would be similar if the material stress–strain characteristics were nonlinear.

THE THEORY OF STRUCTURAL MODELS

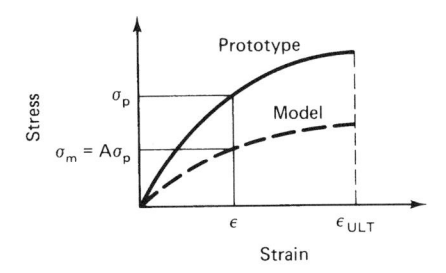

Figure 2.4 Completely similar model material.

2.4.4 Distorted Models

The question now arises whether or not it is possible that deviations from complete similarity can be permitted that will lead to gross dissimilarities between model and prototype behavior. The only answer that can be given is that any deviations can be permitted as long as it is somehow possible to determine the influence of such deviations. It has been said that complete similarity demands that the ϕ_p/ϕ_m ratio in Equation 2.38 be identically 1. When certain second-order deviations from complete similarity are permitted, the ratio might only be approximately equal to 1; if first-order deviations from complete similarity are permitted, the resulting ϕ_p/ϕ_m ratio will, in general, be unknown. Accordingly, the model-to-prototype extrapolation equation given by Equation 2.39 would no longer be correct. Models with such first-order deviations are said to be *distorted*.

Distortion can arise through a dissimilarity in boundary and initial conditions, geometry, or material properties. In structural problems a distortion in boundary and initial conditions or in geometry is seldom necessary or, for that matter, practically advantageous. Murphy (1950) discusses structural models of eccentrically loaded compression blocks that involve geometric distortion, but in these cases the analytical solution is available to enable one to evaluate the ϕ_p/ϕ_m ratio. Of course, if one knows this, the physical model may be less useful than the analytical method.

Geometric distortion has been used in hydraulic models of tidal basins. In these cases the effect of surface tension on the water, which can be neglected in real life, may become significant if the geometric scale is faithfully reduced in the model. If the height dimension of the model is distorted, however, the water is deeper and surface tension again can be neglected. Means exist to account for this distortion.

Of greater significance to the structural model engineer is the possibility of permitting distortion in the reproduction of the prototype material stress–strain characteristics. For complete similarity with regard to a uniaxial stress state, the model material will have to behave according to Figure 2.4. Suppose that such a model material is simply not available but materials are available which follow the stress–strain laws shown in Figure 2.5. Certainly, if a model material conforming to Figure 2.5b were to be used, the strains in the model would be larger than those in the prototype. As a result the model displacements, which are a function of strain times length, would not be similar to the displacements of the prototype. If the structural behavior is dependent upon the displacements (e.g., the stresses in a beam–column or the critical buckling pressure in a thin-shell roof), such a distortion cannot be permitted. However, if the displacements are sufficiently small so as not to disturb the conditions of equilibrium, then such a distortion can be permitted. The corresponding strains, displacements, velocities, accelerations, etc., which would be set up in the model would not be similar to those of the prototype, but would be known to err by the factor that indicates the magnitude of the strain distortion (a_1 and a_2 in Figure 2.5). Beaujoint (1960) gives an extensive discussion of strain distortion.

Another type of material property distortion may arise when the Poisson's ratio of the model material does not equal that of the prototype material. If the structural behavior is known to be

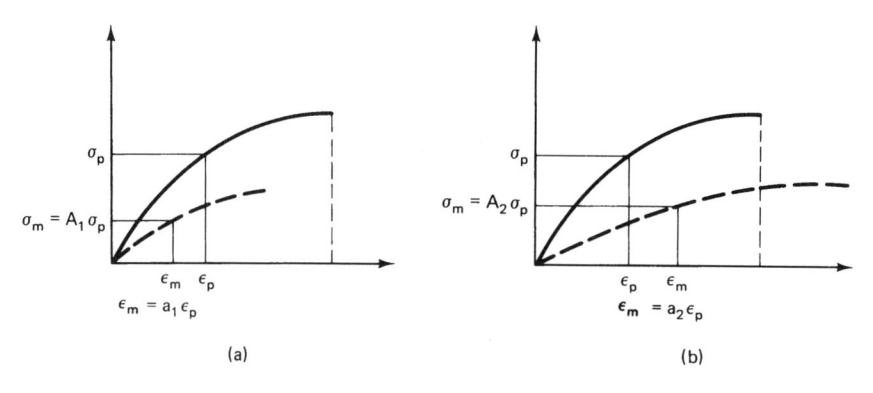

Figure 2.5 Distorted model materials.

Table 2.2 Similitude Requirements, Static Elastic Modeling

Quantities	Dimensions	Scale Factor
Material-related properties		
Stress	FL^{-2}	S_E
Modulus of elasticity	FL^{-2}	S_E
Poisson's ratio	—	1
Specific weight	FL^{-3}	S_E/S_l
Strain	—	1
Geometry		
Linear dimension	L	S_l
Linear displacement	L	S_l
Angular displacement	—	1
Area	L^2	S_l^2
Moment of inertia	L^4	S_l^4
Loading		
Concentrated load, Q	F	$S_E S_l^2$
Uniformly distributed line load, w	FL^{-1}	$S_E S_l$
Pressure or uniformly distributed surface load, q	FL^{-2}	S_E
Moment, M or torque, T	FL	$S_E S_l^3$
Shear force, V	F	$S_E S_l^2$

characterized by plane stress (e.g., the bending of a common beam), a Poisson's ratio discrepancy may very well distort the model strains but not the model stresses, reactions, bending moments, etc. In such cases proper interpretation of the model results can preserve the integrity of the model study. In other instances, analytical or experimental results from a related problem may substantiate that Poisson's ratio v is not of predominant importance. Many of our analytical expressions contain terms like $\sqrt{1-v^2}$. Such knowledge leads one to allow certain models to be fabricated from a material having a dissimilar Poisson's ratio with only a small percentage of error. On the other hand, if the structural behavior is not well understood, then a Poisson's ratio distortion can lead to incorrect results.

2.5 SIMILITUDE REQUIREMENTS

Similitude requirements for static, elastic modeling are summarized in Table 2.2. This is followed by requirements for static inelastic modeling of reinforced and prestressed concrete structures. Special cases of thermal structural modeling are treated, and a detailed treatment of dynamic modeling is presented.

THE THEORY OF STRUCTURAL MODELS

In Table 2.2 the independent scale factors chosen are those for modulus of elasticity and length; all remaining scale factors are either unity or functions of s_E and s_l. The material for an elastic model of a prototype need only satisfy the requirement that it must remain elastic within the model loading range, and that it have the same Poisson's ratio as the prototype material. Assuming that loadings are then scaled by the factors given in Table 2.2, which can easily be derived by forming the appropriate dimensionless ratios, the model stresses are s_E times as small as those in the prototype, while model strains are identical to prototype strains. This type of similitude preserves the correct state of strain for beam–column effects and other geometry-dependent phenomena. If the latter type of behavior is not present, then in many modeling applications the loading is increased by some "slicing" factor that has the advantage of increasing the deformations and displacements of the model. This permits more accurate measurement of the response of the model. The same slicing factor must be taken out again in projecting the model results to the prototype.

The load similitude requirements demonstrate one of the major advantages of reduced-scale elastic modeling in structures. Loads are reduced from prototype loads by the factor $s_E s_l^2$, which is a very large number for a small-scale plastic model of a concrete or steel structure. For typical plastics, E is approximately 2720 MPa; thus s_E is about 8 for a concrete prototype and about 75 for a steel prototype. The above product of $s_E s_l^2$ automatically becomes large, thereby resulting in extremely small loads in comparison with those of the prototype.

2.5.1 Reinforced Concrete Models

It is not easy to model the complete inelastic behavior of a reinforced or prestressed concrete structure, including the proper failure mode and capacity. The highly inelastic nature of concrete under both tensile and compressive stress states is a substantial problem in itself. The other major difficulty is in the reinforcing phase of this two-component material. The strength properties and surface roughness characteristics (bond capacity) of ordinary reinforcement must be given very careful attention if successful models are to be realized. In prestressed models the stress–strain characteristics of the prestressing steel are crucial, as is the anchorage detail.

Because the modeling of reinforced and prestressed concrete structures normally includes loading to failure, the failure criteria for model concrete subjected to multiaxial stresses should be identical with that of the prototype concrete. The lack of a well-defined failure criterion normally leads one to relax this requirement, as outlined below, even for true models:

1. Stress–strain curves must be geometrically similar in model and prototype concrete for both uniaxial tension and compression.
2. $\varepsilon_m = \varepsilon_p$ at failure under uniaxial tension and compression.

These requirements are summarized in Figure 2.6. The corresponding similitude requirements are given in column 4 of Table 2.3.

The use of a nonunity stress scale factor (s_σ) in true models is justified from dimensional analysis. If $s_\sigma \neq 1$ is used for the model concrete, the same nonunity s_σ must be applied to the reinforcing steel, or $s_\sigma = s'_\sigma$, where the primed quantity refers to the reinforcing. The problem of using this type of scaling arises from the fact that s_E must equal s_σ, and the model reinforcing must have a modulus differing from steel by the factor $s'_E = s_E = s_\sigma$. The extensive discussion on materials in Chapters 4 and 5 will show that steel is the only feasible material for reinforced concrete models. Thus, one is led to the conclusion that the only practical way to conduct true modeling of reinforced concrete structures is to use $s_\sigma = s_E = 1$; the corresponding similitude requirements are given in column (5) of Table 2.3. The rather stringent requirement of forcing model concrete to have a stress–strain curve identical to the prototype concrete may be impossible to meet. Finally, it should be noted that the density requirement cannot be satisfied.

(a) Concrete   (b) Reinforcement

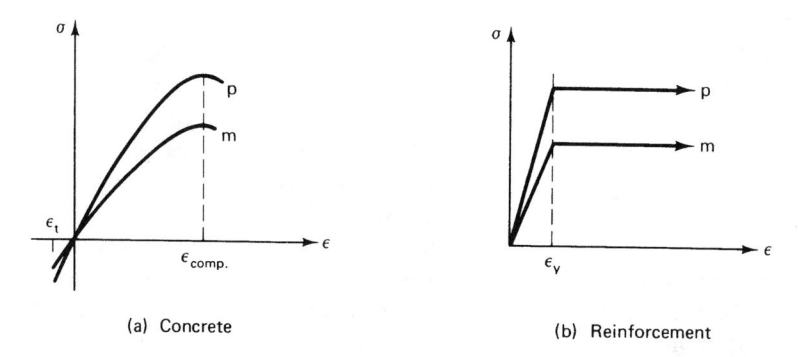

Figure 2.6 Similitude requirements.

Table 2.3 Summary of Scale Factors for Reinforced Concrete Models

(1)	Quantity (2)	Dimension (3)	True Model (4)	Practical True Model (5)	Distorted Model, Case 1 (Figure 2.7) (6)	Distorted Model, Case 3 (Figure 2.8) (7)
Material-Related Property	Concrete stress, σ_c	FL^{-2}	S_σ	1	S_σ	S_σ
	Concrete strain, ε_c	—	1	1	S_ε	S_ε
	Modulus of concrete, E_c	FL^{-2}	S_σ	1	S_σ/S_ε	S_σ/S_ε
	Poisson's ratio, ν_c	—	1	1	1	1
	Specific weight, γ_c	FL^{-3}	S_σ/S_l	$1/S_l$	S_σ/S_l	S_σ/S_l
	Reinforcing stress, σ_r	FL^{-2}	S_σ	1	S_σ	S_σ
	Reinforcing strain, ε_r	—	1	1	S_ε	S_ε
	Modulus of reinforcing, E_r	FL^{-2}	S_σ	1	1	1
	Bond stress, u	FL^{-2}	S_σ	1	S_σ	a
Geometry	Linear dimension, l	L	S_l	S_l	S_l	S_l
	Displacement, δ	L	S_l	S_l	$S_\varepsilon S_l$	$S_\varepsilon S_l$
	Angular displacement, β	—	1	1	S_ε	S_ε
	Area of reinforcement, A_r	L^2	S_l^2	S_l^2	S_l^2	$S_\sigma S_l^2/S_\varepsilon$
Loading	Concentrated load, Q	F	$S_\sigma S_l^2$	S_l^2	$S_\sigma S_l^2$	$S_\sigma S_l^2$
	Line load, w	FL^{-1}	$S_\sigma S_l$	S_l	$S_\sigma S_l$	$S_\sigma S_l$
	Pressure, q	FL^{-2}	S_σ	1	S_σ	S_σ
	Moment, M	FL	$S_\sigma S_l^3$	S_l^3	$S_\sigma S_l^3$	$S_\sigma S_l^3$

[a] Function of choice of distorted reinforcing area.

Table 2.4 Possible Distortions in Reinforced Concrete Models

	Concrete			Reinforcement		
Case	S_ε	S_σ	S_E	S'_ε	S'_σ	S'_E
1	$\neq 1$	S_ε	1	S_ε	S_ε	1
2	$\neq 1$	1	$1/S_\varepsilon$	S_ε	1	$1/S_\varepsilon$
3	$\neq 1$	$\neq 1$	$\neq 1$	S_ε	S_ε	1
4	$\neq 1$	$\neq 1$	$\neq 1$	S_ε	S_σ	$\neq S_E$

It is necessary to utilize a distorted model approach when the available model concrete does not have $s_\varepsilon = s_\sigma = 1$. A number of possible distortions are discussed by Zia et al. (1970) and are summarized in Table 2.4.

THE THEORY OF STRUCTURAL MODELS

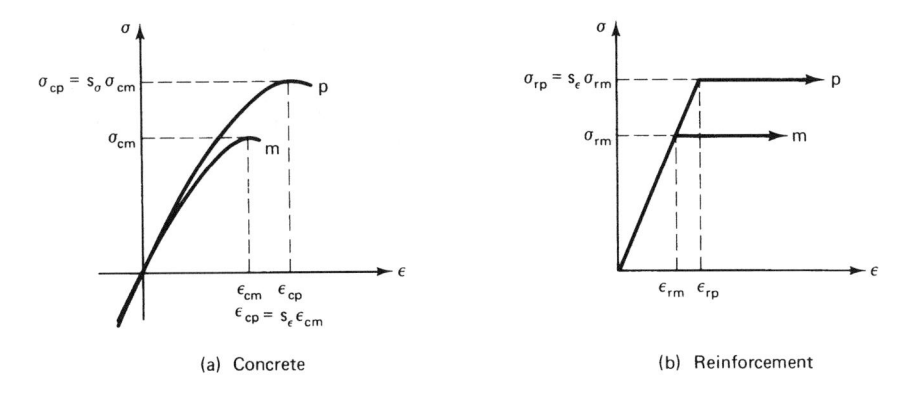

Figure 2.7 Case 1: distortion similitude requirements.

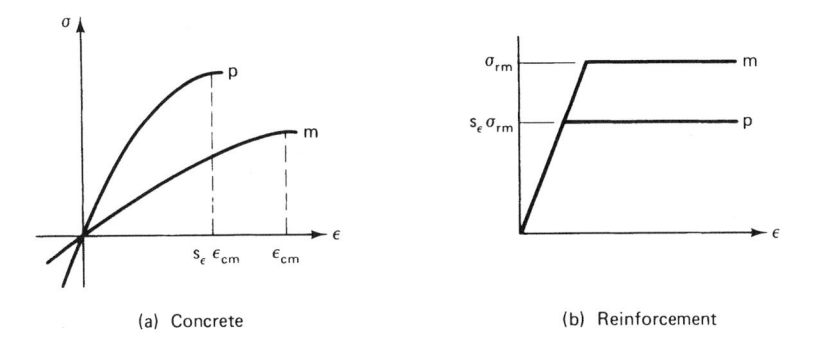

Figure 2.8 Cast 3: distorted models.

Only cases 1 and 3 are of interest because the others require reinforcement made of a material other than steel. Note that both cases utilize a distortion in strain, and thus they should not be used when the structural response is sensitive to absolute magnitude of strain (such as in a beam–column). The case 1 distortion would have material properties as shown in Figure 2.7.

The effects of the strain distortion on the scale factors are given in column 6 of Table 2.3. Loading similitude is not affected by strain distortion; this statement is true for any type of model with a strain distortion. This type of similitude still requires geometrically similar concrete stress–strain curves but has the flexibility of not requiring $s_\sigma = s_\varepsilon = 1$.

Case 3 distortion treats the combination of $s'_E = 1$ and $s_E \neq 1$ (Figure 2.8). In order to have the steel strains distorted in the same fashion as the concrete strains, $s'_\varepsilon = s_\varepsilon$ and with $s'_E = 1$, s'_σ must equal s'_ε and the model steel yield strength must satisfy the requirement ($\sigma_{rp} = s_\varepsilon \sigma_{rm}$). By equating the scaled steel force in the model to that in the prototype ($F_p = s_\sigma s_l^2 F_m$) and using the relation $F = A\sigma$ for model and prototype steels, it follows that

$$\frac{A_p}{A_m} = s_A = \frac{s_\sigma s_l^2}{s_\varepsilon}$$

is the appropriate scaling factor for model reinforcing areas.

In conclusion, it must be emphasized that the tensile strength properties of the model concrete must be properly scaled if such phenomena as shear (diagonal tension) strength and cracking, and even deflections, are to be modeled. Such effects as softening and localization are more difficult to address. Size effects are considered in Chapter 9. Detailed physical properties of appropriate model concretes and reinforcements are given in Chapters 4 and 5, respectively.

Concrete stresses are normally transmitted to reinforcing elements through the actions at the steel–concrete interface (pullout and dowel action). The character of the bond (pullout) action is very important in many structural elements, and cannot be ignored in modeling. The following discussion from Zia et al. (1970) is intended to cast light on bond similitude requirements in reinforced concrete models.

> The basic similitude requirement for bond between concrete and its reinforcement, for true models, is that bond stresses developed by the model reinforcing be identical to those of the prototype reinforcing. It also requires that ultimate bond strength of model and prototype reinforcement be identical. This would be possible only when $s_\sigma = 1$.

The scale of the model is an important factor in the bond similitude problem. For large models, No. 2 and No. 3 deformed bars which satisfy ASTM requirements may be within the desired range of size as model steel. For smaller models, one is forced to utilize either wire (smooth, rusted, or deformed), threaded rod, or stranded cables (for prestressed models). It should be noted, however, that even the use of standard deformed bars does not necessarily ensure true bond modeling because of the difference in bond strength of small and large size bars.

Modeling of bond is seriously complicated by our limited knowledge of the bond mechanism in prototype concrete. According to the current notion of the bond behavior of prototype reinforcing (see Lutz and Gergely, 1967) one must attribute bond strength mainly to the mechanical wedging action afforded by the protruding deformations of the bars. Based upon the ACI Code, the ultimate bond strength per inch of bar length, for normal size bars, is

$$U_\mu = 30\sqrt{f'_c} < 2500D \qquad (2.42)$$

where f'_c = concrete compressive strength in psi and D = bar diameter in inches. It would be grossly misleading to attempt to model this relationship with the types of small reinforcement mentioned previously. In the first place, the above expression is dimensionally inhomogeneous. Second, it was not intended to be applied to the small size reinforcement as it was derived experimentally from tests on ordinary prototype bars. If the above expression were valid for all size bars, the ultimate bond strength of small size reinforcement, 0.25 in. (6 mm) diameter or below, would always be dictated by the $2500D$ limitation. However, from the work of Lutz and others, it can be stated that the ultimate bond strength is not independent of bar size for any range of diameters. Third, the geometric dissimilarity between the wire, threaded rod, stranded cable, and the prototype deformed bars is so great that their bonding action is undoubtedly different. For wires and stranded cables, the lack of protruding lugs greatly reduces the tendency to produce a splitting failure in concrete. On the other hand, pullout tests on threaded rods often produce a complete shearing of the surrounding concrete at the top of the threads, also with little or no splitting accompanying the failure. The effect of concrete cover upon bond strength, and the increased bond strength afforded by stirrups and other steel which tends to prevent splitting from developing adjacent to bars under high bond stress, are also difficult to predict and to model.

In spite of these difficulties, it should be noted that it may be possible to provide model steel which will duplicate local bond behavior of the prototype steel even though its ultimate bond strength is not identical to that of the latter. Cracks normal to the direction of the reinforcement can open through the mechanism of localized slipping of reinforcing without necessarily having longitudinal splitting cracks. Therefore, partial satisfaction of similitude would be provided by this type of model steel.

Another method to improve modeling is strictly dependent upon a detailed knowledge of the bond behavior of model reinforcements. If this information is available, then the bond strength of a certain combination of prototype bars may be modeled by using the proper number and size of model bars. The philosophy of this type of modeling is to provide the correct amount of total bond strength instead of being concerned with scaling unit bond stress. It is felt that this approach has considerable merit and will provide the best answer once a detailed knowledge of both prototype and model reinforcing bond behavior is at hand. True bond stress distribution bears little resemblance to the computed average unit bond stress; it will always be difficult to attempt to model such a poorly defined quantity.

The bond characteristics of several model reinforcements will be presented in Chapter 5. Another problem often associated with concrete model similitude, that of increasing scaled strength with decreasing size of structural element, will also be dealt with in Chapter 9.

2.5.2 Models of Masonry Structures

The most general and useful modeling techniques used in the design and analysis of masonry structures subjected to static and dynamic loads are those which can predict inelastic as well as elastic behavior and have the ability to study with confidence the mode of failure of the structure (Harris and Becica, 1977, 1978; Becica and Harris, 1977; Abboud, 1987). These techniques are, however, very restrictive on the choice of model materials and their methods of fabrication. Let us consider first the case of static loading on the structure, which usually consists of the dead and live loads but could also consist of equivalent static loads for the dynamic effects of wind or earthquake and even abnormal loads such as explosions or impact. Under the assumption that there are no significant time-dependent effects that influence the structural behavior, the pertinent parameters that enter the modeling process are listed in Table 2.5, column 2. For complete similarity of the structural behavior including the inelastic effects of cracking and yielding, a dimensional analysis will give the scale factors shown in Table 2.5, column 4. If it is assumed that the stresses caused by the self-weight of the structure are not significant, as is usually the case in most masonry buildings, the scale factors given in Table 2.5, column 5, will be adequate for modeling masonry structures. For this latter, "practically true" modeling approach, the stress–strain curves of both model and prototype masonry must be the same, presenting a very difficult challenge to the model analyst. The reason for this is that since masonry is a composite material, one has to model all of its constituents: block or brick, mortar, grout, and reinforcements. In addition, fabrication difficulties arise because of the small size of the individual units. This work on the modeling of concrete masonry structures is reported in some detail in Chapters 4, 10, and 11.

2.5.3 Structures Subjected to Thermal Loadings

Modeling of thermal effects has proved useful for several classes of structural problems. Examples include studies of thermal stress in arch dams (Rocha, 1961) and in nuclear reactor vessels, and temperature distribution and temperature stresses in spacecraft structures (Katzoff, 1963). Issen (1966) has studied the problem of modeling the effects of fire on concrete structures. Thermal model analysis should also be considered for other thermostructural problems for which analytical methods are very difficult.

The elastic response of a structure built of a homogeneous, isotropic material will be treated in the initial discussion. Transient heat conduction with no internal (to the material) heat generation will be assumed. With these limitations in mind, the only thermal properties needed are the coefficient of linear expansion, α, and the thermal diffusivity, D. It is further assumed that these thermal properties and the elastic properties of the model material are independent of temperature.

The ten quantities involved in the analysis are defined in Table 2.6. The thermal diffusivity D is equal to $k/(c\gamma)$ where k = thermal conductivity, c = specific heat per unit weight, and γ = specific weight of the material. Four fundamental measures are needed to express the ten quantities: force F or mass M, length L, time T, and temperature θ. Choosing the four independent quantities as l, E, D, and θ, a suitable set of pi terms is:

$$\pi_1 = \frac{\sigma}{E}, \quad \pi_2 = \varepsilon$$

$$\pi_3 = \nu, \quad \pi_4 = \frac{\delta}{l}$$

$$\pi_5 = \alpha\theta, \quad \pi_6 = \frac{tD}{l^2}$$

Table 2.5 Summary of Scale Factors for Masonry

			Static Loading	
Group (1)	Quantity (2)	Dimension (3)	True Model (4)	Practical True Model (5)
Loading	Concentrated load, Q	F	$S_\sigma S_l^2$	S_l^2
	Line load, w	FL^{-1}	$S_\sigma S_l$	S_l
	Pressure, q	FL^{-2}	S_σ	1
	Moment, M	FL	$S_\sigma S_l^3$	S_l^3
Geometry	Linear dimension, l	L	S_l	S_l
	Displacement, δ	L	S_l	S_l
	Angular displacement, β	1	1	1
	Area, A	L^2	S_l^2	S_l^2
Material properties	Masonry unit stress, σ_m	FL^{-2}	S_σ	1
	Masonry unit strain, ε_m	1	1	1
	Modulus of masonry unit, E_m	FL^{-2}	S_σ	1
	Masonry unit Poisson's ratio, ν_m	1	1	1
	Specific weight, γ_m	FL^{-3}	S_σ/S_l	$1/S_l$
	Mortar stress, σ'_m	FL^{-2}	S_σ	1
	Mortar strain, ε'_m	1	1	1
	Modulus of mortar, E'_m	FL^{-2}	S_σ	1
	Mortar Poisson's ratio, ν'_m	1	1	1
	Reinforcement stress, σ_{rm}	FL^{-2}	S_σ	1
	Reinforcement strain, ε_{rm}	1	1	1
	Modulus of reinforcement, E_{rm}	FL^{-2}	S_σ	1

Table 2.6 Summary of Scale Factors for Thermal Modeling

Quantity (1)	Dimension (2)	True Model (3)	Same Materials in Model and Prototype, and Same Temperatures (4)	Distorted Strain Scaling (5)
Stress, σ	FL^{-2}	S_E	1	$S_\alpha S_\theta S_E$
Strain, ε	—	1	1	$S_\alpha S_\theta$
Elastic modulus, E	FL^{-2}	S_E	1	S_E
Poisson's ratio, ν	—	1	1	1
Coefficient of linear expansion, α	θ^{-1}	S_α	1	S_α
Thermal diffusivity, D	L^2T^{-1}	S_D	1	S_D
Linear dimension, l	L	S_l	S_l	S_l
Displacement, δ	L	S_l	S_l	$S_\alpha S_\theta S_l$
Temperature, θ	θ	$1/S_\alpha$	1	S_θ
Time, t	T	S_l^2/S_D	S_l^2	S_l^2/S_D

THE THEORY OF STRUCTURAL MODELS

The scale factors for a true thermal model are summarized in column 3 of Table 2.6. The term π_6, known as *Fourier's number* in heat transfer, leads to the time scaling factor of $s_t = s_l^2/s_D$. With model time inversely proportional to the square of the geometric scale factor, it is possible to model long-time thermal effects in a greatly reduced time.

Whenever identical materials are used in model and prototype, the scale factors shown in column 4 of Table 2.6 should be followed. Taking this approach on a true modeling basis compels one to use the same temperature in model and prototype and eliminates the possible problem of having a distortion introduced because of dependence of material properties on temperature level.

Distortion of strain in thermal modeling is often possible; the criteria for deciding on acceptability of strain distortion is the same as for mechanically loaded models. If thermal actions and mechanical actions are being modeled simultaneously, the same degree of strain distortion must be followed for both actions.

Using a distorted strain scale and different materials in model and prototype leads to the scaling shown in column 5 of Table 2.6. The strain is distorted because the temperature scaling is distorted from its desired true value of $s_\theta = 1/s_\alpha$. It can be reasoned that a distortion of temperature will have a linear effect on strains, stresses, and displacements in the model. Accordingly, a temperature distortion factor d_θ is defined as follows. In the true model, $s_\theta = 1/s_\alpha$; in the distorted model, $s_\theta = d_\theta(1/s_\alpha)$. Solving for the distortion factor, $d_\theta = s_\theta s_\alpha$, the distorted scaling relations for strain, stress, and displacement are formed by multiplying the distortion factor d_θ by the true scaling factors for the affected quantities, or

$$s_\varepsilon = 1 \cdot d_\theta = s_\alpha s_\theta$$

$$s_\sigma = s_E \cdot d_\theta = s_E s_\alpha s_\theta$$

$$s_\delta = s_l \cdot d_\theta = s_l s_\alpha s_\theta$$

Properties of materials suitable for studying thermal stresses in dams at a scale of $1/100$ to $1/500$ will be given in the next chapter. The time scaling factor allows annual temperature waves on the prototype to be modeled in a matter of minutes. Models of this type must be planned very carefully to ensure temperatures and temperature strains large enough to be measured accurately and yet be within the temperature ranges allowed by the model material.

Transfer of heat at a boundary is a function of the surface coefficient of heat transfer, h. The corresponding dimensionless parameter is hl/k, where k = coefficient of thermal conductivity and l = characteristic length. This parameter, called *Nusselt's number* (N), should be equal in model and prototype in order to have properly scaled surface temperature gradients, but a direct contradiction arises when the same material is used in model and prototype. With $h_m = h_p$ and $k_m = k_p$ Nusselt's number can be satisfied only when $l_m = l_p$, or a full-sized model is used. Lack of similitude in temperature gradients at and near the boundary layer in a model has not been a serious factor in either high-temperature fire tests or long-time dam studies (Rocha, 1961; Issen 1966).

Another type of thermal loading is where the heat transfer is by radiation, such as in a spacecraft subjected to the heat of the sun. No additional similitude requirements are placed on the model itself, but the intensity of radiation H (dimensions of Q/L^2T, where Q is a measure of heat) must be scaled in accordance with the parameter $s\theta^4/H$, where s is the Stefan–Boltzmann constant and θ is the absolute temperature. Radiation scaling may lead to substantial difficulties; the reader is referred to Katzoff (1963) for further discussion of thermal radiation modeling.

Thermal effects leading to extensive inelastic actions and structural failures can be modeled, but only with extreme care being given to all material properties. Full discussion of the inherent problems is beyond the scope of this text.

2.5.4 Structures Subjected to Dynamic Loadings

2.5.4.1 Introduction

Physical modeling of structures with dynamic loadings has steadily developed since the end of World War II (Hudson, 1967; Baker, 1973; Castoldi and Casirati, 1976; Schuring, 1977; Krawinkler et al., 1978; Harris, 1982). Time-dependent loadings, because of their complex nature and effect on structures, have placed small-scale structural model techniques on a par with analytical techniques. The dynamic loadings of interest to the structural engineer range from wind- or traffic-induced elastic vibrations to blast and impact loadings that can cause considerable structural damage. Of special interest is the problem of earthquake loading, which, because of its widespread nature and potentially devastating effects, has assumed a greater importance in our highly urbanized society.

Dynamic modeling of structures is important in education, research, and design. In education, simple laboratory experiments demonstrate basic concepts of vibrations to undergraduate and graduate students. In the area of structural research, the small-scale dynamic model has proved to be a powerful tool in extending knowledge and understanding of structural behavior in many complex situations where analytical techniques are inadequate. Also, a carefully constructed model aids the design of many dynamically loaded structures. In recent years, the quantity and the quality of the information obtained from the model test has increased as a result of improved instrumentation and data-processing systems.

The dynamics of any structure is governed by an equilibrium balance of the time-dependent forces acting on the structure. These forces are the inertia forces that are the product of the local mass and acceleration, the resisting forces that are a function of the stiffness of the structure in the particular direction in which motion is occurring, and the energy dissipation of damping forces, whether material or construction related. In addition to these forces that produce dynamic stresses and deformation in the structure, there are certain types of massive structures in which gravity-induced stresses play an important role in dynamic situations and affect modeling. The similitude requirements that govern the dynamic relationships between the model and prototype structure depend on the geometric and material properties of the structure and on the type of loading (Figure 2.9). These relationships can be derived using the pi theorem. A summary of these relationships for the most commonly encountered dynamic loads affecting civil engineering structures follows.

2.5.4.2 Vibrations of Elastic Structures

Vibration problems of elastic structures (Figure 2.9a) are very common in civil engineering practice. Traffic-induced vibrations of bridges, wind-induced vibrations of tall buildings, towers, and chimneys, rotating machinery induced vibrations in buildings, and water flow–induced vibrations of pilings and submerged structures are but a few of many such examples. These problems can be very conveniently studied by means of small-scale models. Consideration of the variables that govern the behavior of vibrating structures reveals that in addition to length (L) and force (F), which we considered in static loading situations, we must now include time (T) as one of the fundamental quantities before we proceed with dimensional analysis.

As an example, consider an elastic structure of a homogeneous isotropic material whose vibration conditions are to be determined. A typical length in the structure is designated by l and a typical force by Q. The materials of both the model and prototype can be characterized by the material constants: the modulus of elasticity E, the Poisson's ratio ν, and the mass density ρ. The important parameters to be determined from the structural vibration are the deflected shapes δ, the natural frequency f, and the dynamic stresses σ. The dimensions of the governing variables in both absolute and common engineering units are shown in Table 2.7. The acceleration due to gravity g

THE THEORY OF STRUCTURAL MODELS

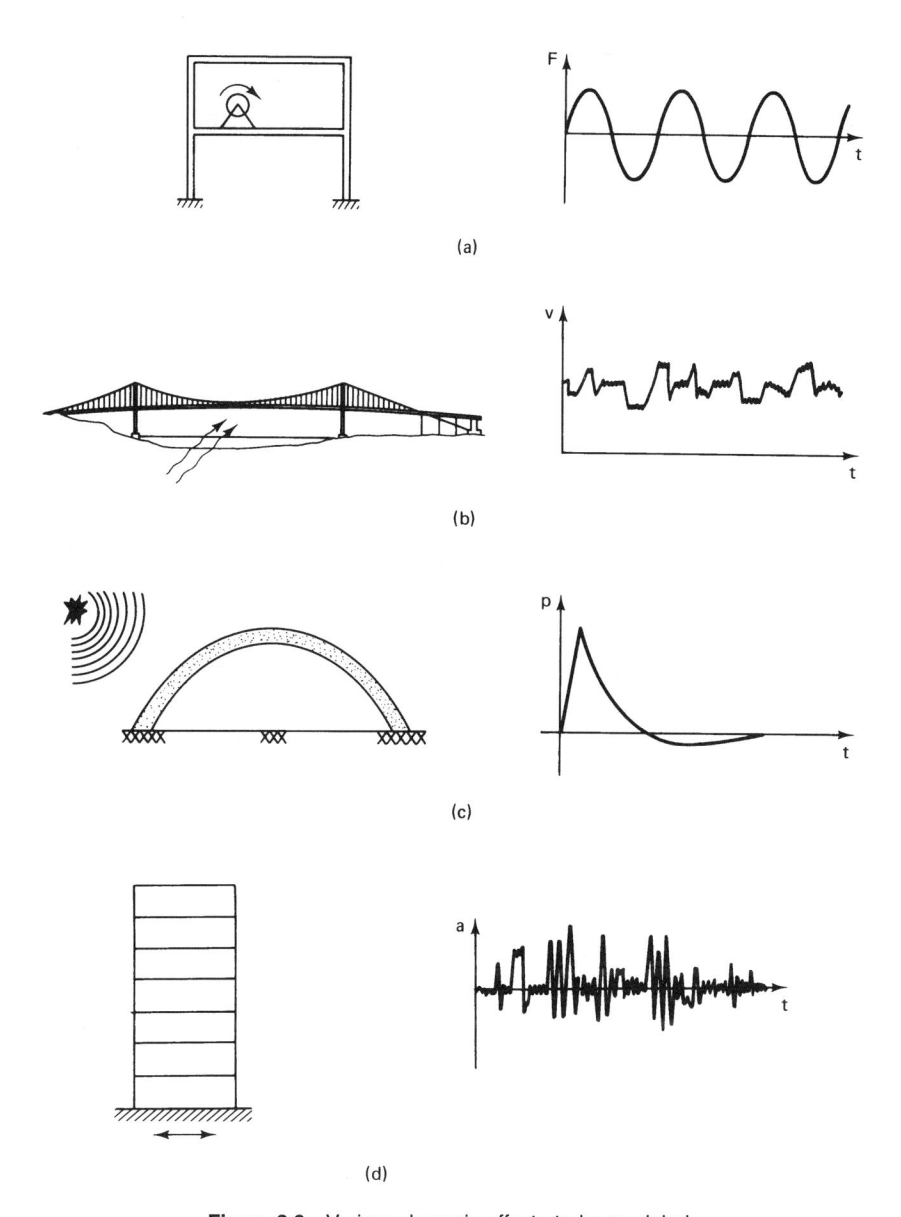

Figure 2.9 Various dynamic effects to be modeled.

must also be included in Table 2.7 since it is common to both model and prototype structures. The relationships that govern both model and prototype can be determined with the help of the pi theorem. For a true model, dimensionless parameters that govern the behavior can be shown to be

$$\phi\left(\frac{\delta}{l}, \frac{\sigma}{E}, \frac{f^2 l}{g}, \frac{\rho g l}{E}, \frac{Q}{E l^2}, \nu\right) = 0 \qquad (2.43)$$

If the deflections are to be of primary interest, the functional relationship becomes

$$\frac{\delta}{l} = \phi'\left(\frac{\sigma}{E}, \frac{f^2 l}{g}, \frac{\rho g l}{E}, \frac{Q}{E l^2}, \nu\right) \qquad (2.44)$$

Table 2.7 Dimensions of Governing Variables

Quantity	Absolute System	Engineering System
Length, l	L	L
Force, Q	MLT^{-2}	F
Modulus of elasticity, E	$ML^{-1}T^{-2}$	FL^{-2}
Poisson's ratio, ν	—	—
Mass density, ρ	ML^{-3}	FT^2L^{-4}
Deflection, δ	L	L
Stress, σ	$ML^{-1}T^{-2}$	FL^{-2}
Frequency, f	T^{-1}	T^{-1}
Acceleration, g	LT^{-2}	LT^{-2}

Table 2.8 Similitude Requirements for Elastic Vibrations

Group (1)	Quantity (2)	Dimension (engineering units) (3)	Exact Scaling (4)	Gravity Forces Neglected (5)
Loading	Force, Q	F	$S_E S_l^2$	$S_E S_L^2$
	Gravitational acceleration, g	LT^{-2}	1	1
	Time, t	T	$S_l^{1/2}$	S_l
Geometry	Linear dimension, l	L	S_l	S_l
	Displacement, δ	L	S_l	S_l
	Frequency, f	T^{-1}	$S_l^{-1/2}$	S_l^{-1}
Material properties	Modulus, E	FL^{-2}	S_E	S_E
	Stress, σ	FL^{-2}	S_E	S_E
	Poisson's ratio, ν	—	1	1
	Specific weight, γ	FL^{-3}	S_E/S_l	Neglected

and similarly if the stresses are of primary interest

$$\frac{\sigma}{E} = \phi''\left(\frac{\delta}{l}, \frac{f^2 l}{g}, \frac{\rho g l}{E}, \frac{Q}{El^2}, \nu\right) \tag{2.45}$$

From Equations 2.44 and 2.45 we can determine the dynamic characteristics of the structure by means of a model test by forcing the dimensionless terms on the right-hand side of Equations 2.44 and 2.45 to be identical in model and prototype. The deflections and stresses then become

$$\left(\frac{\delta}{l}\right)_m = \left(\frac{\delta}{l}\right)_p \quad \text{or} \quad \delta_p = \delta_m S_l \tag{2.46}$$

and

$$\left(\frac{\sigma}{E}\right)_m = \left(\frac{\sigma}{E}\right)_p \quad \text{or} \quad \sigma_p = \sigma_m S_E \tag{2.47}$$

In order to accomplish this, the second, third, and fourth terms in parentheses (Equations 2.44 and 2.45) impose certain restrictions on the model design. The implied scale factors that govern these relationships are summarized in Table 2.8.

As can be seen from Table 2.8, the time scale is equal to $S_l^{1/2}$ for a true elastic model and it is equal to S_l, in the case of a model where gravity loading effects can be neglected. The frequency of vibration of the model, which is inversely proportional to the period, will be $S_l^{-1/2}$ and S_l^{-1} in the above two cases, respectively. This means that the model will have higher frequencies. The shaking table facility or the vibrator, which is exciting the model, must have a frequency and force output capability equivalent to that required for the highest important natural frequency of the model to be studied.

2.5.4.3 Fluidelastic Models

Wind loading forms another major design criterion for many civil engineering structures. Wind effects on buildings and structures of complex shapes are usually studied by means of wind tunnel testing of small-scale models. Two important modeling considerations must be addressed by the model analyst in studying wind effects on structures. The first consideration is a proper modeling of the wind loading through the scaling of the roughness, pressure, and velocity parameters of the flow, and the second is the scaling of the model structure itself. Tests on determining wind effects on buildings date to at least 1893, when Irminger made some experiments in Copenhagen. A report of this early work was presented by Irminger and Nokkentved (1930). Early work in the U.S. includes the tests on the Empire State Building rigid model that were conducted at the National Bureau of Standards by Dryden and Hill (1933) with no attempt to scale the velocity or the turbulence of the model flow. Tests on small-scale rigid building models were first subjected to boundary layer flow by Bailey and Vincent (1943). At about the same time, Jensen (1958) and Jensen and Franck (1965) established on a firm basis the need to use a thick turbulent boundary layer with upwind surface roughness scaled to the same scale as the building.

The modeling requirements for boundary layer type of winds can be easily obtained from dimensional considerations. Detailed derivations of the various similarity requirements are given by Cermak (1975) and Snyder (1972). Exact similarity requirements of the atmospheric boundary layer may be achieved as follows:

1. Topographic relief, terrain roughness, and surface temperature must be simulated.
2. The mean and fluctuating velocity and temperature fields of the approaching flow must be simulated.
3. Equality of the Reynolds, Rossby, Richardson, Prandl, and Eckert numbers needs to be maintained in model and prototype atmospheres.

Because of these restrictions, exact similarity at a scale ratio other than 1 is not possible; however, if requirements 1 and 2 above are satisfied, complete similitude of all of the nondimensional parameters listed in 3 is not necessary. A discussion of the effects of relaxing some of the above requirements can be found in Cermak (1975), Snyder (1972), and Surry and Isyumov (1975).

The field of fluidelasticity covers a wide range of structural dynamics problems that result from the interaction of the structure and the fluid medium that is exciting it. Wind-induced problems range from the classical suspension bridge oscillations (Figure 2.9b) and overhead power line galloping to buffeting of tall buildings and oscillations of tanks, stacks, and towers. Water- and wave-induced problems occur in the oscillations of harbor and offshore structures. A summary of scale factors for fluidelastic models (Scanlan, 1973, 1974; Isyumov, 1976; Cermak, 1977) are given in Table 2.9 for the case of negligible Reynolds number effect (column 4) and negligible effect due to gravity stresses (or neglect of Froude's number) (column 5).

Recent use of improved higher-strength materials and construction techniques has resulted in the development of generally lighter and more flexible structures. The present widespread use of welded steel and prestressed concrete structures are some examples where not only the generally

Table 2.9 Similitude Requirements for Fluidelastic Models

Group (1)	Quantity (2)	Dimension (3)	Scale Factors — Reynolds No. Neglected (4)	Scale Factors — Froude No. Neglected (5)
Loading	Force, Q	MLT^{-2}	$S_\rho S_l^3$	$S_\rho S_l^{-1}$
	Pressure, q	$ML^{-1}T^{-2}$	$S_\rho S_l$	$S_\rho S_l$
	Gravitational acceleration, g	LT^{-2}	1	1
	Velocity, v	LT^{-1}	$S_l^{1/2}$	S_v
	Time, t	T	$S_l^{1/2}$	$S_l S_v^{-1}$
Geometry	Linear dimension, l	L	S_l	S_l
	Displacement, δ	L	S_l	S_l
	Frequency, ω	T^{-1}	$S_l^{-1/2}$	$S_v S_l^{-1}$
Material properties	Modulus, E	$ML^{-1}T^{-2}$	$S_\rho S_l$	$S_\rho S_l$
	Stress, σ	$ML^{-1}T^{-2}$	$S_\rho S_l$	$S_\rho S_l$
	Poisson's ratio, ν	—	1	1
	Mass density, ρ	ML^{-3}	S_ρ	S_ρ

more flexible construction has led to increased fluidelastic action but also the lower damping characteristics of these materials has added to fluid structure interaction. The resulting greater sensitivity to dynamic excitation by wind action has prompted an increased use of wind tunnel simulations for providing design information.

2.5.4.4 Blast and Impact, Load Modeling

Impulsive type loading due to external or internal blast effects, or impact of moving objects on structures, is considered an abnormal type of loading that has a relatively low risk of occurrence. However, the fact that it may occur at all in certain types of structures requires knowledge of the ability to treat its effects on their dynamic response. For these reasons this chapter will consider the modeling of blast effects either external or internal to the structure.

1. External Blast on Structures — Because of the difficulty and expense of conducting studies dealing with blast loading (Figure 2.9c) at full scale and the enormous difficulties in performing analytical studies, the need for model experiments in this particular area has developed from the very beginning (Baker, 1973). The first scaling law to be developed for the blast phenomenon itself has been the *Hopkinson* or *cube-root scaling law*. This law states that self-similar blast (shock) waves are produced at identical scaled distances when two explosive charges of similar geometry and the same explosives but of different size are detonated in the same atmosphere; thus

$$L_m = \frac{L_p}{\overline{E}^{1/3}} = \frac{L_p}{W^{1/3}}$$

where

L_m = length in model

L_p = length in prototype

$\overline{E}$ (or W) = the energy of the prototype explosion

THE THEORY OF STRUCTURAL MODELS

Table 2.10 Summary of Scale Factors for Blast Loading

Group (1)	Quantity (2)	Dimension (3)	True Replica Model (4)	Gravity Forces Neglected, Prototype Material (5)
Loading	Force, Q	F	$S_E S_l^2$	S_l^2
	Pressure, q	FL^{-2}	S_E	1
	Time, t	T	S_l	S_l
	Gravitational acceleration, g	LT^{-2}	1	Neglected
	Velocity, v	LT^{-1}	1	1
Geometry	Linear dimension, l	L	S_l	S_l
	Displacement, δ	L	S_l	S_l
	Strain, ε	—	1	1
Material	Modulus of elasticity, E	FL^{-2}	S_E	1
	Stress, σ	FL^{-2}	S_E	1
	Poisson's ratio, ν	—	1	1
	Mass density, ρ	FT^2L^{-4}	S_ρ	1

This type of modeling is sometimes referred to as *replica* modeling, and the pressure amplitude and loading in such modeling will be similar in form as in the prototype. The pressure amplitude and velocity of the pressure wave are the same. The duration of the pulses will be scaled by the length scale.

The most general and useful modeling techniques used in the design and analysis of structures to predict the blast loads are those associated with the ability to predict the elastoplastic as well as the elastic behavior and the ability to study with confidence the mode of failure of the structure. These techniques are, however, very restrictive on the choice of model materials and methods of fabrication. Extensive research has been conducted at MIT (Antebi et al., 1960; 1962; Harris et al., 1962; 1963; Smith et al., 1963) and several government facilities to develop and utilize such techniques in the study of nuclear blast effects on various structures. Although these studies were conducted for external blast effects on structures, the same methodology can be applied to the case of other gaseous explosions.

For a geometric scale factor of S_l, when the energy of the blast source is scaled according to the cube-root law and the model is made from components of the same strength properties with the same manufacturing details and supports as the prototype, then the scaling laws that must relate the model to the prototype structure are given in Table 2.10. Since specific weight scales the same as the length scale, one cannot, in general, provide a model material with the same strength properties and yet S_l times denser than the prototype. Thus, the effects of gravity stresses cannot be modeled through a choice of a denser material in most direct model studies, unless one resorts to preloading techniques.

2. Internal Blast Effects — The nature of gaseous explosions in domestic surroundings has been studied extensively after the progressive collapse at Ronan Point (Rashbash, 1969; Stretch, 1969; Alexander and Hambly, 1970; Slack, 1971). The 100-psi theoretical maximum pressure that can be reached in a completely confined explosion in a maximum concentration of most gaseous types is rarely realized (Rashbash, 1969). Most gaseous explosions are less concentrated, and venting of the burning gases drastically reduces the theoretical peak pressures. Most experimentally induced explosions have peak pressures in the range of ¹⁄₂₀ to ¹⁄₁₀ of the theoretical maximum.

Venting area and covering can be simulated in dynamic models using scaled-down window and door areas and coverings found in actual construction. An expression that shows clearly the effect of venting, which was developed in the Netherlands by Dragosavic (1973), is also sometimes used in analytical work. Venting can also occur through weak partitions between rooms, although more tests are needed to quantify these effects. The effects of turbulence created as the gas expands from one room into another may create higher pressures in the second room. In the particular problem of interior gaseous explosions, the computed energy release of typical prototype explosions can be scaled, and since the length scale will be dictated by the size of model to be used, the model explosion energy is defined. The model testing can be carried out in a small strong chamber, and the model charges will follow the cube-root scaling law for the prototype gaseous explosion. Extensive verification of Hopkinson's law has been reported (Baker, 1973). More-sophisticated blast scaling laws such as Sach's (Baker, 1973) can be employed in the event that Hopkinson's law proves inadequate for the particular problem at hand.

2.5.4.5 Earthquake Modeling of Structures

Earthquake loading (Figure 2.9d) is an important design consideration in many civil engineering structures because of its potential catastrophic nature. In practical structures it requires that the inelastic behavior of the structure be mobilized in order for the design to be economical. Such considerations, however, impose severe restrictions on the possible choice of materials for model testing. A summary of the scale factors obtained from similitude considerations (Krawinkler et al., 1978) for earthquake loading is given in Table 2.11. True replica models (column 4) satisfy the Froude and Cauchy scaling requirements, which imply simultaneous duplication of inertial, gravitational, and restoring forces. Unfortunately, however, such models are practically impossible to build and test because of the severe restrictions imposed on the model material properties, especially the mass density. Alternate scaling laws shown in columns 5 and 6 of Table 2.11 have been shown to simulate the behavior of the structure adequately. Considerable success has been achieved in the testing of reduced-scale structures and structural components (Mirza et al., 1979) on shaking tables where additional material of a nonstructural nature has been added to simulate the required scaled density of the model. The similitude laws for artificial mass simulation are shown in column 5. Both lumped mass and distributed mass simulation has been reported. A third type of scaling law, shown in column 6, applies to the case where gravity stresses can be neglected in the structural behavior and where the same materials are used in both model and prototype to enable the testing to proceed to failure.

2.5.4.6 Modeling for Centrifuge Testing

Modeling of structures that are founded in soils and the interaction between soil and structure that it entails form a very important class of problems. One of the unique ways to study flow, stability, and dynamic problems in such systems is by using centrifuge loading. Typical centrifuges and results from dynamic testing in such facilities are given in Chapter 11. The principles of dimensional analysis presented at the beginning of this chapter apply equally well to problems of soil–structure interaction. Roscoe (1968) and Nielsen (1977) have presented the scaling relations for soil systems. Scaling factors required in centrifuge testing have been presented by Fuglsang and Krebs Ovesen (1988) and are reproduced here as Table 2.12. It should be noted that the scale factors given in Table 2.12, i.e., the N-values are equal to the inverse of the S_i scale factors defined in Equation 2.40.

2.6 SUMMARY

The theory of structural modeling is presented. Similitude requirements are given for elastic and ultimate-strength models in a variety of structural materials and loading situation. Ultimate-strength model scale factors are given for reinforced and prestressed concrete and masonry structures.

THE THEORY OF STRUCTURAL MODELS

Table 2.11 Summary of Scale Factors for Earthquake Response of Structures

			Scale Factors		
		Dimension	True Replica Model	Artificial Mass Simulation	Gravity Forces Neglected Prototype Material
(1)	(2)	(3)	(4)	(5)	(6)
Loading	Force, Q	F	$S_E S_l^2$	$S_E S_l^2$	S_l^2
	Pressure, q	FL^{-2}	S_E	S_E	1
	Acceleration, a	LT^{-2}	1	1	S_l^{-1}
	Gravitational acceleration, g	LT^{-2}	1	1	Neglected
	Velocity, v	LT^{-1}	$S_l^{1/2}$	$S_l^{1/2}$	1
	Time, t	T	$S_l^{1/2}$	$S_l^{1/2}$	S_l
Geometry	Linear dimension, l	L	S_l	S_l	S_l
	Displacement, δ	L	S_l	S_l	S_l
	Frequency, ω	T^{-1}	$S_l^{-1/2}$	$S_l^{-1/2}$	S_l^{-1}
Material properties	Modulus, E	FL^{-2}	S_E	S_E	1
	Stress, σ	FL^{-2}	S_E	S_E	1
	Strain, ε	—	1	1	1
	Poisson's ratio, ν	—	1	1	1
	Mass density, ρ	$FL^{-4}T^2$	S_E/S_l	a	1
	Energy, EN	FL	$S_E S_l^3$	$S_E S_l^3$	S_l^3

a $(g\rho l/E)_m = (g\rho l/E)_p$.

Static, thermal, and dynamic loadings consisting of wind, blast, impact, and earthquake loadings are discussed and scale factors for each of these situations are presented. Scale factors for centrifuge testing of soil–structure systems are presented. The concept of complete similarity and the technological restrictions that govern its implementation are discussed. Alternatives to complete similarity, where some of the similitude requirements are relaxed, are also presented.

PROBLEMS

2.1 In studying the effect of the size of a submarine on the depth to which it can safely dive, a single design is considered and then the size effect is studied by scaling all linear dimensions (including plate thicknesses) in the same proportion. The critical depth of dive is one where the structure fails in elastic buckling. What can be said about the critical depth as related to the size of the hull and its material? (Courtesy, W. Godden, University of California, Berkeley)

2.2 Derive the similitude relation for displacements in a flexural member,

$$u_m = \frac{1}{S_l S_\varepsilon} u_p$$

from the fact that the displacement can be expressed as

$$\delta = k \frac{ML^2}{EI}$$

where k is a geometric constant, and the model and prototype values of M, L, E, and I are related by scale factors that are readily derived (or recognized by inspection).

Table 2.12 Scaling Factors in Centrifuge Testing

	Parameter	Symbol	Dim. Less Number	Similarity Requirement	Scaling Factor
1	Acceleration	a		$N_a =$	n
2	Model length	l		$N_l =$	$\frac{1}{n}$
3	Soil density	ρ		$N_\rho =$	1
4	Particle size	d	$\frac{d}{l}$	$N_d =$	1
5	Void ratio	e	e	$N_e =$	1
6	Saturation	S_r	S_r	$N_S =$	1
7	Liquid density	ρ_l	$\frac{\rho_l}{\rho}$	$N_{\rho l} = N_\rho =$	1
8a	Surface tension	σ_t	$\frac{\sigma_t}{\rho_l a d l}$	$N_\sigma = N_\rho N_a N_d N_l =$	1
8b	Capillarity	h_c	$\frac{h_c \rho_l a d}{\sigma_t}$	$N_h = N_\sigma N_\rho^{-1} N_a^{-1} N_d^{-1} =$	$\frac{1}{n}$
9a	Viscosity	η	$\frac{\eta}{\rho_l d \sqrt{al}}$	$N_\eta = N_\rho N_d N_a^{\frac{1}{2}} N_l^{\frac{1}{2}} =$	1
9b	Permeability	k	$\frac{k\eta}{d^2 \rho_l a}$	$N_k = N_d^2 N_\rho N_a N_\eta^{-1} =$	n
10	Particle friction	ϕ	f	$N_\phi =$	1
11	Particle strength	σ_c	$\frac{\sigma_c}{\rho a l}$	$N_\sigma = N_\rho N_a N_l =$	1
12	Cohesion	c	$\frac{c}{\rho a l}$	$N_c = N_\rho N_a N_l =$	1
13	Compressibility time	E	$\frac{E}{\rho a l}$	$N_E = N_\rho N_a N_l =$	1
14	Inertia	t_1	$t\sqrt{\frac{a}{l}}$	$N_t = N_l^{\frac{1}{2}} N_a^{-\frac{1}{2}} =$	$\frac{1}{n}$
15	Laminar flow	t_2	$t\frac{k}{l}$	$N_t = N_l N_k^{-1} =$	$\frac{1}{n^2}$
16	Creep	t_3			1

Note: $N_i = 1/S_i$ in the text.

Source: Fuglsang, L. D. and Krebs Ovesen, N., in *Centrifuges in Soil Mechanics*, W. H. Craig et al., Eds., A. A. Balkema Publishers, Brookfield, VT, 1988. With permission.

2.3 For the case of $S_\varepsilon \neq 1$, $S_\sigma \neq 1$, and $E_{rm} \neq E_{rp}$, derive expressions for a reinforced concrete beam for
  a. Required area of model reinforcing A_{rm};
  b. Yield strength of model reinforcing as a function of yield strength of prototype steel. (*Hint*: This can be accomplished by considering the basic flexural mechanics of a simple reinforced concrete beam.)

2.4 a. Using a dimensional analysis, develop a general expression for the distance that a freely falling object will drop in time t, neglecting resistance afforded by the air.
  b. What happens in the analysis if we erroneously assume that the weight of the object is a variable?
  c. Assume the dropped object starts with an initial velocity v_0. Develop an expression for distance traveled in time t.

THE THEORY OF STRUCTURAL MODELS

2.5 Determine the π terms for the beam-stiffness problem discussed in this chapter, expressing the parameters in terms of the fundamental quantities of mass, length, and time. What is the basic difficulty met in this formulation of the problem?

2.6 The following equations are to be checked for dimensional homogeneity. Give values for the dimensions attached to any constants in the nonhomogeneous equations.

a. Column strength formula:

$$P = A_g \left(0.25 f'_c + f_s p_g\right)$$

b. Bending of a transversely loaded plate:

$$\frac{\partial^4 w}{\partial x^4} + 2 \frac{\partial^4 w}{\partial x^2 \partial y^2} + \frac{\partial^4 w}{\partial y^4} - \frac{q}{D} = 0$$

c. AISC column stress formula:

$$F_a = \left[1 - \frac{(kl/r)^2}{2C_c^2}\right] f_y / \text{factor of safety}$$

d. Manning's formula for open-channel flow:

$$V = \frac{1.486}{n} R^{2/3} S^{1/2}$$

e. Buckling load of a pin-ended column:

$$P_{cr} = \pi^2 EI/L^2$$

f. Deflection δ, at end of cantilever of length L, with a tip load P.

$$\delta = PL^3/3EI$$

g. Rankine or total active pressure p_a, on each foot length of wall of height H, caused by cohesive soil:

$$p_a = 1/2 \gamma H^2 - 2cH$$

2.7 You are working in a design office and have just completed the calculation to determine the maximum bending moment in a three-span continuous girder carrying a uniform load of 3.0 kips/ft. Your supervisor then informs you that the spans have been increased from 60, 80, and 60 ft to 72, 96, and 72 ft. Can you, without going through the calculations again, quickly find the new maximum bending moment?

2.8 Given

$$\sigma_{yp} = s_\varepsilon \sigma_{ym}$$

$$\text{Steel } A_p = \frac{s_\sigma s_l^2}{s_\varepsilon} A_m$$

$$f'_{cp} = s_\sigma f'_{cm}$$

$$d_p = s_l d_m$$

$$b_p = s_l b_m$$

Show that $M_p = s_\sigma s_l^3 M_m$ (ultimate moment capacities) by applying conventional ultimate strength analysis

$$M = A_s f_y \left(d - \frac{a}{2} \right)$$

and forming the ratio of M_p to M_m.

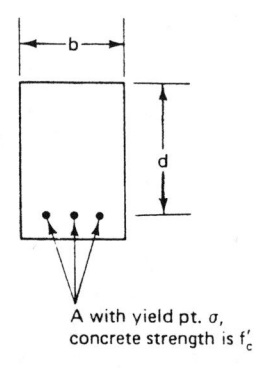

A with yield pt. σ, concrete strength is f'_c

2.9 Prove or disprove the validity of Galileo's statement (given below) by applying similitude theory to the problem of deadweight stresses in a structure of constantly increasing size.

> It would be impossible to build up the bony structures of men, horses, or other animals so as to hold together and perform their normal functions if these animals were to be increased enormously in height; for this increase in height can be accomplished only by employing a material which is harder and stronger than usual, or by enlarging the size of the bones, thus changing their shape until the form and appearance of the animals suggest a monstrosity. If the size of a body be diminished, the strength of that body is not diminished in the same proportion; indeed the smaller the body the greater its relative strength. Thus, a small dog could probably carry on his back two or three dogs of his own size, but I believe a horse could not carry even one of his own size.

2.10 A rectangular tubular steel section is subject to the loading shown below. Use the pi theorem to develop a general expression for the deflection of point a with respect to point b.

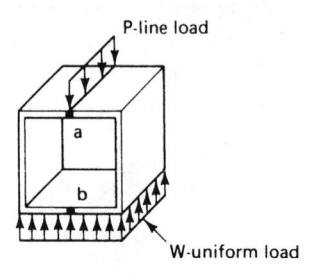

2.11 Use the pi theorem to develop an expression for the natural frequency of a freely vibrating fixed-end beam of prismatic cross section.

$$f_{\text{prototype}} = f_{\text{model}} \, l_m/l_p \sqrt{E_p \rho_m / E_m \rho_p}$$

THE THEORY OF STRUCTURAL MODELS

2.12 It is desired to build and test a Plexiglas model of a large cast steel flywheel having a heavy rim and radial spokes. Establish the similitude conditions. For a 1/20-scale model, what are the stress and velocity scales for a prototype that has an angular velocity of 50 rad/s?

2.13 The natural frequency of a steel tuning fork is 200 vibrations per second. What is the natural frequency of a 1/3-scale aluminum model of the tuning fork? The unit weight of aluminum is 36% that of steel, and its E value is 1/3 that of steel. (A tuning fork is a freely vibrating elastic system.)

2.14 The idealized dynamic loading on a bridge is given by a forcing function that varies periodically with time. The exciting forces are fully specified by the frequency f and the maximum value of force $F_0 = 10,000$ lb (44.5 kN). Determine similitude and scaling relations for a model bridge with a geometric scale factor $S_L = 30$. Both bridges are made of steel. Is it necessary to add additional mass (weights) to the bridge to satisfy similitude? Explain carefully.

2.15 For the beam of Problem 2.14, develop an expression for the maximum stress due to a concentrated load P at midspan, the beam self-weight, and a uniform temperature rise of ΔT. Assume elastic action only.

2.16 The torque on an airplane propeller depends only on the diameter d of the propeller, its angular velocity ω, the velocity of advance v, the mass density ρ, and viscosity μ of the air. Using dimensional analysis, find an expression for the torque, and show that if the effect of viscosity can be neglected then the torque is proportional to air density. (Courtesy of W. Godden, University of California, Berkeley.)

2.17 To study the performance of a high-speed train on its track, a 1/10-scale model is made of the complete system. The track is both curved and banked (tranverse slope) in places. The model train is true scale, and its effective density is found to be twice that of the prototype. The model is to be used to measure the forces exerted by the train on the track and to study the tendency for the train to overturn on corners. Specify the following model ratios required to simulate prototype behavior:

Track:
    a. Horizontal radius of curvature, r
    b. Transverse slope, ϕ
    c. Coefficient of friction, μ

Train:
    d. Velocity, v
    e. Acceleration, a

Forces:
    f. Centrifugal force on the track, F_c at points of curvature
    g. Axial forces on track due to accleration, F_a

Will this model correctly simulate the tendency for the train to overturn? (Courtesy of W. Godden, University of California, Berkeley.)

REFERENCES

Abboud, B. E. (1987). The use of small-scale direct models for concrete block masonry assemblages and slender reinforced walls under out-of-plane loads, Ph.D. thesis, Department of Civil and Architectural Engineering, Drexel University, Philadelphia.

Alexander, S. J. and Hambly, C. E. (1970). Design of structures to withstand gaseous explosions, Parts 1 and 2, *Concrete*, 4 (2), February; 4(3), March.

Antebi, J., Smith, H. D., and Hansen, R. J. (1960). Study of the Applicability of Models for Investigation of Air Blast Effects on Structures, Technical Report to the Defense Atomic Support Agency, Department of Civil Engineering, Massachusetts Institute of Technology, Cambridge, October.

Antebi, J., Smith, H. D., Sharma, H. D., and Harris, H. G. (1962). Evaluation of Techniques for Constructing Model Structural Elements, Research Report R62-15, Department of Civil Engineering, Massachusetts Institute of Technology, Cambridge, May.

Bailey, A. and Vincent, N. D. G. (1943). Wind pressure on buildings including the effects of adjacent buildings, *J. Inst. Civ. Eng.,* London, 20, 243–275.

Baker, W. E. (1973). *Explosions in Air,* University of Texas Press, Austin.

Beaujoint, N. (1960). Similitude and theory of models, *RILEM Bull.,* Paris, No. 7, June.

Becica, I. J. and Harris, H. G. (1977). Evaluation of Techniques in the Direct Modeling of Concrete Structures, Structural Models Laboratory Report No. M77-1, Department of Civil Engineering, Drexel University, Philadelphia, June.

Bridgman, P. W. (1922). *Dimensional Analysis,* Yale University Press, New Haven, CT.

Buckingham, E. (1914). On physically similar systems, *Phys. Rev.,* London, 4(345).

Castoldi, A. and Casirati, M. (1976). Experimental Techniques for the Dynamic Analysis of Complex Structures, Report No. 74, ISMES-Istituto Sperimentale Modelli e Strutture, Bergamo, Italy, February.

Cermak, J. E. (1975). Applications of fluid mechanics to wind engineering, A Freeman Scholar Lecture, *J. Fluids Eng.* ASME, 97(1), March, 9–38.

Cermak, J. E. (1977). Wind-tunnel testing of structures, *J. Eng. Mech. Div., ASCE,* 103(EM6), December, 1125–1140.

Dragosavic, I. M. (1973). Structural measures against natural gas explosions in high-rise blocks of flats,*Heron*, 19(4), Department of Civil Engineering, Technological University, Delft, the Netherlands, 3–51.

Dryden, H. L. and Hill, G. C. (1933). Wind pressure on model of the empire state building,*J. Res. Natl. Bur. Stand.,* 10, 493–523.

Fuglsang, L. D. and Krebs Ovesen, N. (1988). The application of the theory of modeling to centrifuge studies, in *Centrifuges in Soil Mechanics*, W. H. Craig, R. G. James, and A. N. Schofield, Eds., A. A. Balkema Publishers, Brookfield, VT.

Harris, H. G., Ed., (1982). *Dynamic Modeling of Concrete Structures,* Publication SP-73, American Concrete Institute, Detroit, 242 pp.

Harris, H. G. and Becica, I. J. (1977). Direct small-scale models of concrete masonry structures, paper presented at the 2nd Annual ASCE Engineering Mechanics Division Specialty Conference, North Carolina State University, Raleigh, May 23–25, 1977, published in *Advances in Civil Engineering through Engineering Mechanics,* ASCE, New York, 101–104.

Harris, H. G. and Becica, I. J. (1978). The behavior of concrete masonry structures and joint details using small-scale direct models, in *Proceedings of the North American Masonry Conference,* University of Colorado, Boulder, August 14–16.

Harris, H. G., Pahl, P. J., and Sharma, S. D. (1962). Dynamic Studies of Structures by Means of Models, Report R63-23, Department of Civil Engineering, Massachusetts Institute of Technology, Cambridge.

Harris, H. G., Schwindt, R., Taher, I., and Werner, S. (1963). Techniques and Materials in the Modeling of Reinforced Concrete Structures under Dynamic Loads, Report R63-54, Department of Civil Engineering, Massachusetts Institute of Technology, Cambridge, December; also NCEL-NBY-3228, U.S. Naval Civil Engineering Laboratory, Port Hueneme, CA.

Hudson, D. E. (1967). Scale model principles, in *Shock and Vibration Handbook,* Harris, C. M. and Crede, C. E., Eds., McGraw-Hill, New York, Chap. 27.

Ipsen, D. C. (1960). *Units, Dimensions, and Dimensionless Numbers,* McGraw-Hill, New York.

Irminger, J. O. V. and Nokkentved (1930). Wind pressure on buildings, *Ingenioervidensk. Skr.*

Issen, L. (1966). Scaled models in fire research on concrete structures,*J. PCA Res. Dev. Lab.,* 8(3), September, 10–26.

Isyumov, N. (1976). Modeling of Wind Effects on Structures and Buildings, lecture notes.

Jensen, M. (1958). The model-law for phenomena in natural wind, *Ingenioreen* (international edition), 4, 121–128.

Jensen, M. and Franck, N. (1965). *Model-Scale Tests in Turbulent Wind,* Parts I and II, The Danish Technical Press, Copenhagen.

Katzoff, S. (1963). Similitude in Thermal Models of Spacecraft, NASA Technical Note D-1631, April.

Krawinkler, H., Mills, R. S., Moncarz, P. D., et al. (1978). Scale Modeling and Testing of Structures for Reproducing Response to Earthquake Excitation, the John A. Blume Earthquake Engineering Center, Department of Civil Engineering, Stanford University, Stanford, May.

Langhaar, H. L. (1951). *Dimensioal Analysis and Theory of Models,* John Wiley & Sons, New York.

Lutz, L. A. and Gergely, P. (1967). Mechanics of bond and slip of deformed bars in concrete, *Proc. Am. Concr. Inst.,* 64(11), November, 711–721.

Mirza, M. S., Harris, H. G., and Sabnis, G. M. (1979). Structural models in earthquake engineering, in *Proceedings of the Third Canadian Conference of Earthquake Engineering,* June 4–6, Montreal, Quebec, Vol. 1, 511–549.

Murphy, G. (1950). *Similitude in Engineering,* the Ronald Press Company, New York.

Nielsen, J. (1977). Model laws for granular media and powders with special view to silo models, *Arch. Mech.,* Vol. 29, Warsaw.

Rasbash, D. J. (1969). Explosions in domestic structures; part i: the relief of gas and vapour explosions in domestic structures, *Struct. Eng.* (London), 47(10), October.

Rocha, M. (1961). Determination of thermal stresses in arch dams by means of models, *RILEM Bull.* (Paris), (10), 65.

Roscoe, K. H. (1968). Soils and model tests, *J. Strain Anal.,* 3(1), 57–64.

Scanlan, R. H. (1973). Dynamic similitude in models, in *Proceednigs, EDF Conference sur l'Aero-Hydo-elasticite,* CEA-EDF, Chatou, France, September 9, 1972, Eyrolles, Paris, 787–828.

Scanlan, R. H. (1974). Scale Models and Modeling Laws in Fluid-Elasticity, Preprint No. 2247, ASCE National Structural Engineering Meeting, Cincinnati, OH, April 22–26.

Schuring, D. J. (1977). *Scale Models in Engineering — Fundamentals and Applications,* Pergamon Press, New York.

Slack, J. H. (1971). Explosions in buildings — the behavior of reinforced concrete frames, *Concrete,* 5, April, 109–114.

Smith, H. D., Clark, R. W., and Mayor, R. P. (1963). Evaluation of Model Techniques for the Investigation of Structural Response to Blast Loads, Report R63-16, Department of Civil Engineering, Massachusetts Institute of Technology, Cambridge, February.

Snyder, W. H. (1972). Similarity criteria for the application of fluid models to the study of air pollution meteorology, *Boundary Layer Meteorol.,* 3(1), September.

Stretch, K. L.(1969). Explosions in domestic structures; Part 2: the relationship between containment characteristics and gaseous reactions, *Struct. Eng.* (London), 47(10), October.

Surrey, D. and Isyumov, N. (1975). Model studies of wind effects — a perspective on the problems of experimental technique and instrumentation, paper presented at Int. Congress on Instrumentation in Aerospace Simulation Facilities, Ottawa.

Zia, P., White, R. N., and Van Horn, D. A. (1970). Principles of model analysis, in *Models for Concrete Structures,* ACI SP-24, American Concrete Institute, Detroit, MI., 19–39.

CHAPTER 3

Elastic Models – Materials and Techniques

CONTENTS

3.1	Introduction	86
3.2	Materials for Elastic Models	87
3.3	Plastics	88
	3.3.1 Thermoplastics and Thermosetting Plastics	88
	3.3.2 Tension, Compression, and Flexural Characteristics of Plastics	89
	3.3.3 Viscoelastic Behavior of Plastics	92
	3.3.4 Mechanical Properties of Polyester Resin Combined with Calcite Filler	94
3.4	Time Effects in Plastics — Evaluation and Compensation	96
	3.4.1 Determination of the Time-Dependent Modulus of Elasticity and Poisson's Ratio	97
	3.4.2 Loading Techniques to Account for Time-Dependent Effects	98
3.5	Effects of Loading Rate, Temperature, and the Environment	100
	3.5.1 Influence of Strain Rate on Mechanical Properties of Plastics	100
	3.5.2 Effects of Temperature and Related Thermal Problems	101
	3.5.2.1 Temperature Effects on Elastic Constants	101
	3.5.2.2 Temperature Effects on Strength	101
	3.5.3 Coefficients of Thermal Expansion	101
	3.5.4 Thermal Conductivity	102
	3.5.5 Softening and Demolding Temperatures	103
	3.5.6 Influence of Relative Humidity on Elastic Properties	103
3.6	Special Problems Related to Plastic Models	103
	3.6.1 Modeling of Creep in Prototype Systems	103
	3.6.2 Poisson's Ratio Considerations	104
	3.6.3 Thickness Variations in Commercial Shapes	104
	3.6.4 Influence of the Calendering Process on the Modulus of Elasticity	104
3.7	Wood and Paper Products	104
	3.7.1 Balsa Wood	105
	3.7.1.1 Balsa Wood Shapes Available	105
	3.7.1.2 Strength Properties	107
	3.7.1.3 Applications of Balsa Wood to Model Studies	110
	1. Two- and Three-Dimensional Trusses	110
	2. Elastic Buckling Studies	112

		3.7.2	Modeling of Structural Lumber..114
			3.7.2.1 Strength Properties of Wood Used for Structural Models..................114
			3.7.2.2 Examples of Wooden Models..115
		3.7.3	Small-Scale Modeling of Glue-Laminated Structures 115
			3.7.3.1 Paper Products Used for Structural Models.......................................119
			3.7.3.2 Examples of Paper Models..120
3.8		Elastic Models — Design and Research Applications..121
3.9		Determination of Influence Lines and Influence Surfaces Using Indirect Models — Müller-Breslau Principle..121
3.10		Summary 123
Problems..124
References 127

3.1 INTRODUCTION

An elastic model can be used to study the behavior of the prototype only in the linearly elastic range. It cannot be used to predict any inelastic behavior of a loaded structure resulting from material nonlinearities such as the postcracking behavior of concrete, postyielding behavior of steel, or the postbuckling behavior of a plate or column. Historically, elastic models were used to study the linearly elastic behavior of complex redundant structures prior to the development of the digital computer. Elastic models have been used extensively to study the response of multistory buildings, bridges, nuclear reactor pressure vessels, dams, and other types of structures subjected to static, dynamic (including earthquake, blast, and wind loads), and thermal loadings. Elastic models of structural components such as columns, frames, slabs, and shells have been used in elastic stability studies. Many such applications can now be performed by a myriad of computer codes that are readily available to the engineer. Usually the computer solution is more economical and efficient. However, there exist many structural situations where the use of elastic models can be very beneficial. In the case of educational models and architectural engineering models the use of elastic materials is still essential. A greater emphasis on laboratory and structural concept demonstration in the new engineering curriculum also requires physical model testing. A majority of these model applications can be satisfied with linearly elastic model materials. It is with these applications in mind that this chapter is retained in the second edition and is somewhat expanded to satisfy the needs of the civil and architectural engineering student.

Similitude requirements for static elastic modeling were presented in Table 2.2. Independent scaling factors were chosen for modulus of elasticity and length, and all remaining scale factors were established as functions of S_E and S_l. It follows from Table 2.2 that the material for an elastic model of a prototype structure need only satisfy the requirements that it must not be loaded beyond the linear elastic range and that it must have the same Poisson's ratio as the prototype material, although this latter requirement can be waived in one-dimensional structures. However, even in structural situations with more complex states of stress, elastic model materials having a Poisson's ratio close to that of the prototype can be found because the range in this material constant is small (0.15 to 0.35 for most materials of interest). Elastic models can be used as direct models (the pattern of model loading is similar to that of the prototype) or as indirect models (to derive influence lines and influence surfaces, where the pattern of prototype loading does not have to be reproduced). In indirect models, it is not necessary to satisfy the condition of equality of strains at corresponding (homologous) points in the prototype and the model ($\varepsilon_p = \varepsilon_m$).

It follows that the intensity of applied loading can be varied at will; the resulting deformations will be proportional to the applied loads provided the linear elastic range of the material is not exceeded. Conceptually, such experimental models are very similar to the available procedures for elastic analysis of structures and account for the stiffnesses of members and joints; in some ways, especially with respect to boundary conditions, the physical model represents a better idealization of the structure than the mathematical model. The reason for this is that analytical modeling of actual support conditions is over simplified in most, if not all, computer codes of structural analysis.

3.2 MATERIALS FOR ELASTIC MODELS

The choice of materials available for the construction of any structural model is perhaps even greater than that for prototype structures. The advantages, disadvantages, and limitations of commonly used model materials have been examined by several investigators. The selected material must satisfy the laws of similitude presented in Chapter 2, and besides having reproducible mechanical properties and geometric stability, it should be readily available, easily fabricated, and above all *inexpensive*. A detailed discussion of the physical and chemical properties of all materials suitable for the construction of models is beyond the scope of this text, and only a few relevant properties of the more commonly used model materials are presented to assist with the selection process. The most significant model material properties are

1. Proportional (limit) stress
2. Stiffness
3. Failure mechanism
4. Influence of temperature and humidity on material properties
5. Creep characteristics
6. Load rate and strain rate effects
7. Effect of size and shape on material properties

All material properties should be confirmed by appropriate tests, as data given in the various handbooks or manufacturer's catalogs represent average or "obtainable" values and are often unreliable. The properties of some materials such as plastics, wood, and concrete not only show large variability from one sample to another, but are also significantly dependent on the type, shape, and size of the specimen and the rate of loading. Three major types of materials suitable for the construction of elastic models are presented in this text: plastics, wood, and paper (Chapter 3), cementitious materials (Chapter 4), and metals (Chapter 5).

The only condition required to be satisfied for *indirect models* is that the material exhibit a linear elastic stress–strain relationship; in general, plastics or metals are used for constructing indirect models. Plastics have low elastic moduli, and this leads to small load requirements and measurable large deformations in small-scale models. By using structural mechanics, it can be shown that the principles of superposition and reciprocity are valid within the linear elastic range of behavior for the prototype or model structure (this also implies that there are no stability or catenary effects, the deformations are small, and the limit of proportionality of the material is not exceeded). These two principles are useful for determining deformations and stresses in direct models as a result of combinations of a variety of loading conditions such as dead, live, and wind loads. For a general case, to simulate a given state of deformation in the prototype, the following similitude relationship for Poisson's ratio must be satisfied (see Table 2.2):

$$\nu_p = \nu_m \tag{3.1}$$

In planar skeletal structures, such as frames, trusses, arches, and cables, torsion is nonexistent and shearing effects are not as predominant as the flexural and axial effects; therefore, the similitude requirement (Equation 3.1) can be relaxed for these structures. However, if the response of the structure is not independent of Poisson's ratio, as in grids, three-dimensional frames, plates, and shells, then Equation 3.1 should be satisfied.

3.3 PLASTICS

The term *plastic* is normally used for a material derived by chemical synthesis and containing carbon compounds (Preece and Davies, 1964). A wide variety of plastic materials with varying chemical compositions and mechanical properties is available under specific trade names from manufacturers and local suppliers. Only those properties of commonly used plastics that are relevant to the construction and testing of elastic models are presented in this section.

Beggs (1932) is perhaps the first known investigator to use plastic models to solve statically indeterminate structural problems. Since the early 1960s, plastics have been used effectively for the construction of direct and indirect models designed to simulate the linear elastic or linear viscoelastic response of the prototype. Comprehensive studies of structural models constructed from plastics have been reported by Fialho (1962), Litle and Hansen (1963; 1965), and Carpenter et al. (1964). Mechanical properties of plastics directly relevant to structural modeling were reviewed by Rowe (1960), Preece and Davies (1964), and Roll (1968).

3.3.1 Thermoplastics and Thermosetting Plastics

Plastics can be classified into two categories: thermoplastics and thermosetting plastics. *Thermoplastics* become progressively softer at temperatures between 93 and 149°C and can be formed into complex shapes such as shells of complex geometry with little or no pressure and yet retain their shape upon cooling. If a thermoplastic material is reheated, it can be remolded into another shape, although some types have a "memory." This characteristic of thermoplastics is extremely useful for the vacuum-forming process, which will be described later. The commonly used thermoplastics (namely, acrylic plastics and polyvinylchloride, PVC) are generally available in the form of sheets, rods, and tubes. It must be noted that several types and grades are commercially available under a single trade name; for example, there are many different grades of Plexiglas. Plexiglas G, commonly used for many applications, will shrink about 2.2% during the heating process, with a corresponding increase in thickness. By contrast, Plexiglas II UVA, used extensively for research work (e.g., at Cornell University), is preshrunk and is manufactured to more exacting standards of optical quality, surface quality, and thickness tolerances. Other preshrunk types include I-A UVA, 55, and 5009.

The acrylic plastics (Plexiglas, Perspex, Lucite) can be easily machined and cemented, and accurate models can be rapidly assembled. Sheets can be softened by heating and formed into shells of single or double curvature using the vacuum-forming process. Comprehensive studies of acrylic plastics used for models have been reported by Fialho (1962), by Carpenter et al. (1964), and later summarized by Roll (1968). Fumagalli (1973) has reported on the use of Lucite for several elastic model studies at the Istituto Sperimentale Modelli e Strutture (ISMES) in Italy.

PVC is normally available in thinner sheets and sheets of more uniform thickness than acrylic plastics and are specially suited for model studies of various kinds of shells. A series of vacuum-formed shells of PVC have been tested at Cornell University and the Massachusetts Institute of Technology (MIT). Litle (1964) has reviewed the use of PVC for model testing and has reported a considerable amount of strength and stiffness data.

Thermosetting plastics differ from thermoplastics in that they cannot be remolded by heating once they have been cast into their original shapes. Thermosetting plastics such as the epoxies (Araldite) or polyesters (Marco, Palatal) can be used for casting models at room temperature with upper and lower molds without the use of pressure or ovens (Roll, 1968). Since these casting resins are in liquid form prior to polymerization, they are frequently used for casting very intricate models;

however, considerable skill and experience are prerequisites to achieve successful results. Thermosetting plastics are preferred to thermoplastics in the manufacture of shell models with varying thickness. Any complex curved surface with any desired thickness variation can be cast conveniently using thermosetting plastics. One must remember that the thickness of such shells is affected by the variations in thickness due to the forming process. This variation is governed by dissipation of the heat of polymerization, and therefore care must be exercised in using thermosetting plastics.

The advantage of using epoxy resins compared with thermoplastics is that the limited development of the heat of polymerization assures a more homogeneous hardening process, which results in a constant elastic modulus throughout the mass (Fumagalli, 1973). Also, the relatively lower shrinkage that occurs in epoxy resins after casting results in a significant decrease in the internal stresses. These are particularly useful in models of varying thickness where the internal stresses can even lead to fracture of the model. Properties of some plastics commonly used for structural models are described in Table 3.1. It must be noted that *only* typical values of the properties of plastics used by some investigators are listed. There are other available plastics that may be suitable for model work.

Epoxy resins also offer the possibility of modifying their physical properties by adjusting the quantity of hardener or by adding an inert material such as a filler dispersed homogeneously throughout the mass and/or reinforcements consisting of inorganic or organic fibers. Silica sand, powdered metal (aluminum or iron), cork, lead shot, polystyrene granules, and other ingredients have been successfully used as fillers. The addition of fillers alters the material density and modifies the modulus of elasticity within wide limits. It also decreases the temperature rise due to the generated heat of polymerization and reduces shrinkage and the associated internal stresses. Powdered cork, sand, and polystyrene help reduce the values of the Poisson's ratio, while the use of an aluminum powder increases the thermal conductivity, which helps disperse the heat generated from the electrical resistance strain gages later applied to the surface. Properties of some epoxy resin mixes with varying amounts of selected fillers are shown in Table 3.2.

3.3.2 Tension, Compression, and Flexural Characteristics of Plastics

The strength and stress–strain characteristics of plastics are dependent on a number of factors, such as the type of test (tension, compression, or flexure), the specimen size, the rate of loading, and the previous stress history in terms of creep and relaxation. The mechanical properties of plastics are also significantly influenced by temperature and relative humidity, which are discussed in more detail in Section 3.5. The measured properties vary not only from batch to batch but also from one sheet thicknesss to another within the same batch. Reasonable care must therefore be exercised to determine the properties in the laboratory under conditions of temperature and relative humidity similar to those in which the model will be both cast and tested. Also, it is important that if the model is subjected principally to direct stresses, the modulus of elasticity in direct tension or compression should be determined from tension or compression tests on suitable specimens. Similarly, if the model is subjected principally to bending, the modulus of elasticity in flexure should be determined using a cantilever beam test or a similar flexure test.

It is recommended that the tension and flexural specimens should be at least 8 in. (200 mm) long, randomly selected from the material for model construction. A brief description of the specimens recommended by the American Society for Testing and Materials (ASTM) Standards and of specimens used by some investigators is presented in Table 3.3.

Details of the tension specimen Type 1 recommended by ASTM Standard D638 *Standard Test Method for Tensile Properties of Plastics* are shown in Figure 3.1. The specimen has a uniform cross section (12.5 mm × sheet thickness) over a length of 57 mm and a gradual transition to the two enlarged ends to prevent failure at the grips. It is recommended that this specimen be used in model material evaluations.

Table 3.1 Properties of Some Plastics Suitable for Structural Models

Plastic	Thermal Characteristic	Available Shapes	Tensile Strength, psi	Compressive Strength, psi	Flexural Strength, psi	Modulus of Elasticity, psi
Cellulose nitrates (celluloid)	Thermoplastic	Sheets, rods, and tubes	3000–7000	3000–30,000	3000–17,000	$65–400 \times 10^3$
Cellulose acetates (plasticele)	Thermoplastic	Sheets, rods, and tubes	2250–11,000	2200–10,900	2200–11,500	$65–260 \times 10^3$
Methyl methacryenlates[a] (Plexiglas, Lucite Perspex)	Thermoplastic	Sheets, rods, and tubes	7000–11,000	12,000–20,000	3000–17,000	$420–500 \times 10^3$
PVC (Boltaron)	Thermoplastic	Sheets, rods, and tubes	5000–10,000	8000–13,000	3500–13,500	$350–600 \times 10^3$
Polyethylenes (Alkathene)	Thermoplastic	Sheets, rods, tubes, molding powders	1000–5000	—	2000–7000	$17–80 \times 10^3$
Natural or syntheentic rubber	Thermoplastic	Sheets and extruded shapes	1000–4000	—	—	200–350
Polyester resins (Marco, Palatal)	Thermosetting	Casting resins	5000–6000	12,000–20,000	—	$300–400 \times 10^3$
Epoxy resins (Epon, Araldite)	Thermosetting	Casting resins	5000–12,000	15,000–30,000	—	$430–600 \times 10^3$

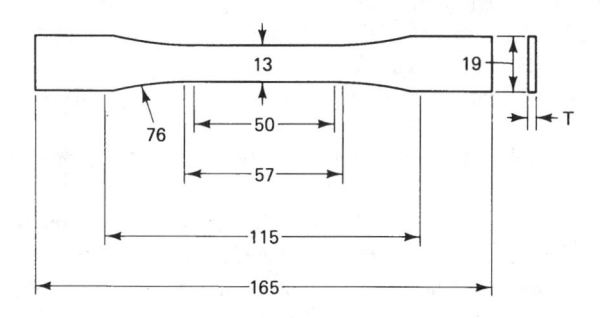

Figure 3.1 Tension test specimen dimensions in millimeters (for sheet, plate, and molded plastics.) (Courtesy of ASTM, Philadelphia.)

The ASTM Standard 695 *Standard for Compressive Properties of Rigid Plastics* recommends that a right cylinder or prism whose length is twice its diameter or its principal width be used to determine the compressive properties of plastics; the preferred specimen size is $12 \times 12 \times 25$ mm for prisms and 12 mm diameter $\times$ 25 mm high for cylinders. For determination of elastic modulus and the yield stress, the test specimen should be of such dimensions to avoid the slenderness or instability problems. The slenderness ratio generally used is between 11 and 15. The preferred specimen size for prisms is $12 \times 12 \times 50$ mm and for cylinders the recommended dimensions are 12 mm diameter and 50 mm height. It is not necessary to machine the test specimen cross sections to the preferred sizes; compression specimens of suitable height can be cut from the material used for the model construction according to the ASTM Standard 695.

Several investigators have used ASTM standard specimen to determine the flexural properties of different plastics. According to the ASTM Standard D790 *Standard Test Methods for Flexural Properties of Plastics and Electrical Insulating Materials* for plastic materials 1.5 mm or greater in thickness, the depth of the specimen for flatwise tests shall be the material thickness and for edgewide tests the depth shall not exceed the width; for all tests the support span shall be 16 times the depth with sufficient overhangs to prevent the specimen from slipping through the supports.

Table 3.1 (continued) Properties of Some Plastics Suitable for Structural Models

Poisson's Ratio	Elongation at Rupture Percent	Specific Gravity	Softening Temperature,°C	Coefficient of Expansion, in./in./°C	Machinability	Jointing Characteristics
0.40–0.42	40–90	1.35–1.70	70–100	11–17×10^{-5}	Excellent	Can be cemented with solvent cements and ethyl acetate (acetone)
0.40		1.24–1.32	250–350	8–16×10^{-5}		Can be cemented with solvent cements and ethyl acetate (acetone)
0.35–0.38	3–10	1.17–1.20	80–160	5–9×10^{-5}	Excellent	Can be cemented with commercial adhesives or a solution of the plastic and chloroform
0.38–0.40	85–100	1.38–1.40	80–105	5–8×10^{-5}	Excellent	Can be welded with PVC rods or cemented with epoxy cements
0.45–0.50		0.91–0.96	85–127	9–18×10^{-5}		Can be welded but difficult to cement
0.50	300–800	0.95–1.20	70–75	9–12×10^{-5}		Can be cemented with rubber cements
0.35–0.45	2	1.20–1.35	80–90	3–6×10^{-5}	Fair	(Solvents are acetone, and cellosolve)
0.33–0.45	5–10	1.20	—	3–9×10^{-5}	Good	Can be cemented with epoxy cements

Note: All plastics with specific gravity of less than 1.0 will float in water.

[a] Acrylic resins.

Table 3.2 Composition and Properties of Some Epoxy Resin Mixes with Added Fillers

Resin Base, % by weight		Filler Materials, % by weight				Physicomechanical Properties	
Epoxy Resin (Alraldite M)	Hardener	Cork Dust	Aluminum Powder	Polystyrene Granules	Silica Sand 0.5–1 mm	Elastic Modulus E, kg/cm^2	Poisson's Ratio
83.3	16.7	—	—	—	—	26,000–32,000	0.38
71.5	14.3	—	14.2	—	—	25,000–30,000	0.28
38.5	7.7	38.4	15.4	—	—	5,000–7,000	0.27
23.7	4.8	—	—	71.5	—	18,000–25,000	0.34
15.4	3.0	—	—	4.6	77.0	80,000–90,000	0.27
9.8	2.0	—	—	—	88.2	140,000–160,000	0.26

After Fumagalli, 1973.

Table 3.3 Test Specimens Used by Various Researchers for Tests on Plastics

Researcher Ref.	Test	Specimen Size	Comments
Struminsky (1971)	Tension	ASTM Standard Specimen	Tensile properties of fiberglass-reinforced polyester resin
Carpenter et al. (1964)	Tension	1.5 × 0.25 × 16 in. (37 × 6 × 400 mm)	Tensile properties of Plexiglas; specimen ends reinforced
Balint and Shaw (1965)	Tension	⅝ in. diam. × 8 in. (16 mm diam. × 200 mm)	Plastrene 97 with varying amounts of calcite filler used for the Australia Square models; enlarged specimen ends
Balint and Shaw (1975)	Compression	2 in. diam. × 4 in. (50 mm diam. × 100 mm)	Compressive properties of the above mixes
Struminsky (1971)	Flexure	ASTM Standard Specimen	Flexural properties of fiberglass-reinforced polyester resin
Tabba (1972), Fam (1973)	Flexure	1 in. × sheet thickness × 8 in. (25 mm × sheet thickness × 200 mm)	Flexural properties of Plexiglas

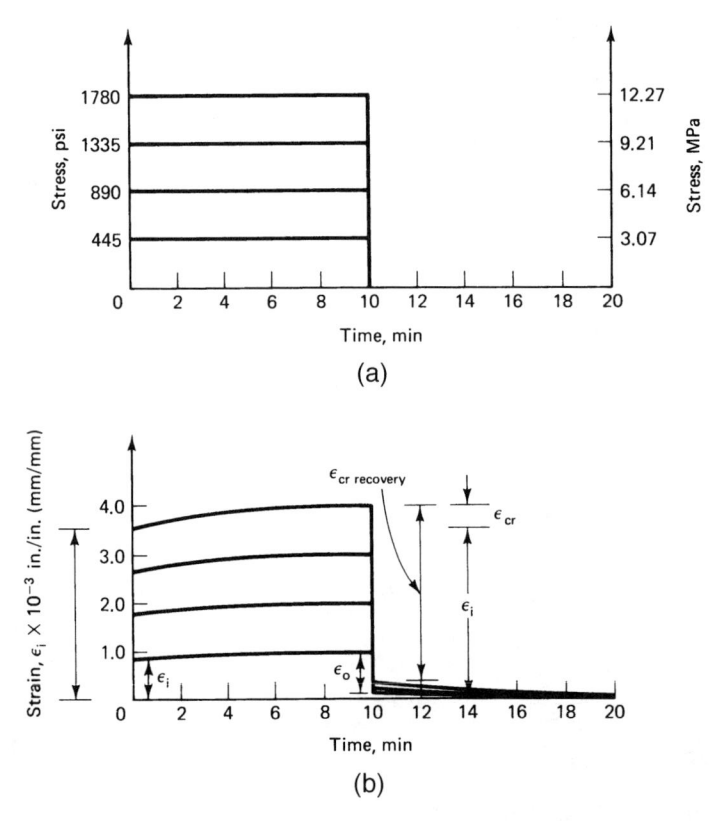

Figure 3.2 Stress–strain–time characteristics of a linear "viscoeleastic" material.

Dimensions of the roller supports and the rounded loading nose, along with other details, are also specified.

3.3.3 Viscoelastic Behavior of Plastics

Most of the commonly used plastics exhibit linear viscoelastic properties; that is, the stress–strain relationship at any particular time after loading is linear although the modulus of elasticity varies with time. In other words, the stress–strain relationship does not conform to Hooke's law, and the strain is a function of time after loading, the loading history, and the stress level. Thus, if a plastic specimen is subjected to stress σ that is maintained constant for a duration of time t (Figure 3.2a), the strain response of the specimen will be as shown in Figure 3.2b. There will be an instantaneous strain ε_i in the plastic immediately upon the application of stress. Under constant stress the strain increases further with time, and this rate of strain increase ($d\varepsilon/dt$) is dependent on the stress intensity. According to Preece and Davies (1964), in most model studies the maximum stress intensity must be controlled so that ($d^2\varepsilon/dt^2$) is negative. If the stress intensity is above the creep strength, the value of ($d^2\varepsilon/dt^2$) will be positive and the strain will proceed to increase at an increasing rate until the specimen fractures. This problem can normally be avoided by limiting the value of the maximum stress in the plastic well below the creep strength of the plastic.

Under a constant stress σ, the total strain ε_t at any time interval t (Figure 3.2b) consists of two parts: the instantaneous elastic strain, ε_i and the time-dependent creep strain ε_{cr}. When the stress is removed, there will be an instantaneous recovery of strain approximately equal to ε_0 followed by a slower strain recovery that gradually disappears with time. If the stress intensity is doubled to 2σ, the corresponding total strain at the same time t will be $2\varepsilon_t$, with the total instantaneous elastic

ELASTIC MODELS – MATERIALS AND TECHNIQUES

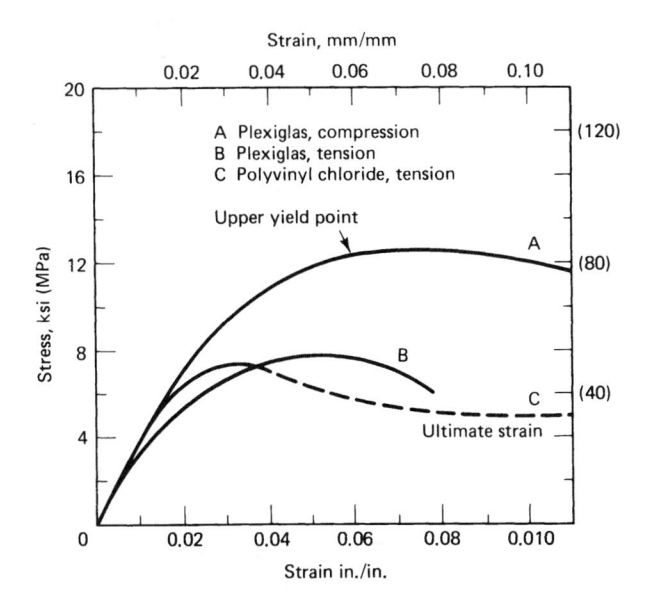

Figure 3.3 Stress–strain curves for PVC and Plexiglas. (After Pahl and Soosaar, 1964.)

and the creep strains being $2\varepsilon_i$ and $2\varepsilon_{cr}$, respectively, if the plastic exhibits linear viscoelasticity. It follows that at any given time interval, the ratios (σ/ε_t), $(2\sigma/2\varepsilon_t)$, ... will be constant, and thus the *effective* elastic modulus

$$E_t = \frac{\sigma}{\varepsilon_t} \qquad (3.2)$$

is instantaneously constant with respect to time.

The typical mechanical properties of commonly used plastics are summarized in Table 3.1. The moduli can vary from about 42 to 55 N/mm² for plastic foams to values between 6900 and 13,800 MPa for some glass-reinforced plastics. For common acrylics, polyesters, celluloids, epoxies, and other plastics, the moduli values range between 2070 and 4140 MPa. These relatively low moduli values (compared with steel and concrete) result in measurable strains and deflections in plastic models without requiring large loads. Typical tension and compression stress–strain curves for Plexiglas (grade G and grade II UVA used at MIT) and a typical stress–strain curve for PVC (normal-impact grade Type I used at MIT); sheet thickness of 0.8 mm and 1.6 mm are shown in Figure 3.3. A series of Cornell University tests for tension on instrumented Plexiglas coupons showed that after about 10 min of loading, the modulus of elasticity decreases to about 90% of the instantaneous value. The specimens ceased to remain linear viscoelastic after a strain of about 1700 μin./in., which is in agreement with the range of 1500 to 2000 μin./in. observed by other investigators. Although these Plexiglas specimens underwent large total deformations, they were not ductile and exhibited little warning of failure. However, some of the softer PVCs are more ductile. Unlike other plastics, ethyl cellulose and polycarbonate (thermoplastic kind) have stress–strain curves in tension similar in shape to that of steel (Figure 3.4) but without strain hardening and have been used by Harris et al. (1962) for model studies of welded steel frame structures subjected to blast-type dynamic loadings. It should be noted that at small strains, the stress–strain curves for most plastics in tension and compression and the corresponding moduli of elasticity are almost identical. Thus, the modulus of elasticity in flexure will be equal to the modulus of elasticity in tension or compression. However, there is a variation between the compressive and the tensile yield strength, the former being greater.

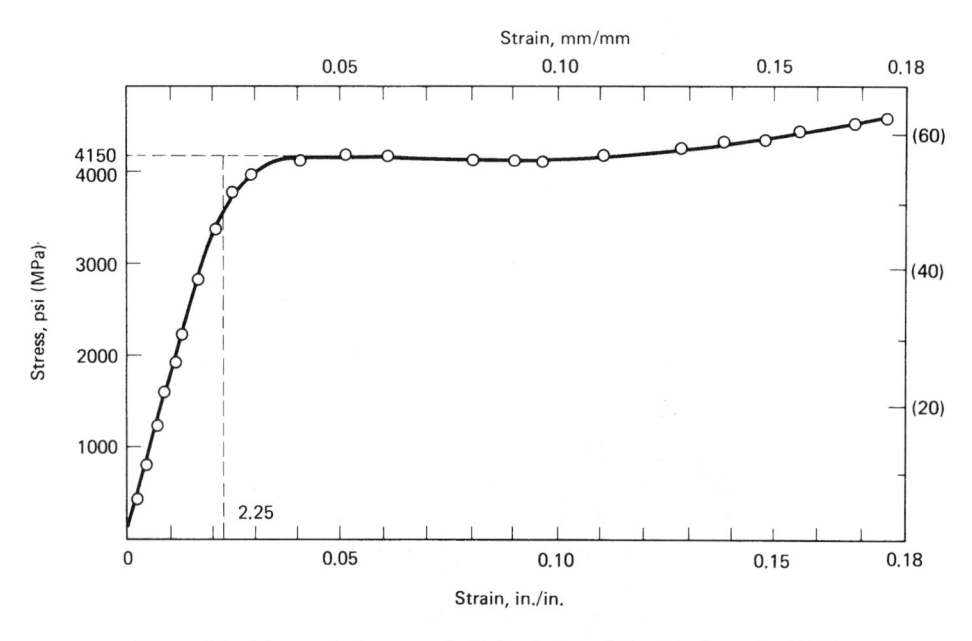

Figure 3.4 Stress–strain curve of ethyl cellulose. (After Harris et al., 1962.)

3.3.4 Mechanical Properties of Polyester Resin Combined with Calcite Filler

Polyester casting resin with the trade name of Plastrene 97 (Balint and Shaw, 1965) is another material used for elastic models. The properties of the resin are varied by using different amounts of calcite filler; thus it is possible to match the moduli of elasticity of the dense and lightweight concretes of the prototype structure with that of the model in order to fulfill the similitude requirements (see Figure 3.5). The influence of variation of the calcite filler on material shrinkage, coefficient of linear expansion, strength, creep characteristics, and glueability have also been investigated. This material was used in the model studies of the circular, 183-m-high Australia Square, Sydney, which is constructed with lightweight concrete in the upper stories (Gero and Cowan, 1970). Plastrene 97 is a moldable material and was used for this model because of the intricate shapes involved, such as recesses around doors, tapered columns and beams, and varied slab thicknesses; the change in the modulus of elasticity of the concrete through the building height; and the need for a low Poisson's ratio. At room temperature, complete polymerization of Plastrene 97 normally took place 90 days after the addition of the catalyst and the accelerator. However, this process was accelerated at elevated temperatures, and curing for about 3 h at 65°C resulted in a fully polymerized product (Balint and Shaw, 1965). Best results were obtained by slow and even cooling of the heat-cured components, guarding them against warping. This slow cooling process resulted in a practically stress-free material. Reheating the cured product to 81°C softened the material and easily eliminated any unwanted warps.

Uniaxial tests on various mixes indicated linear viscoelastic behavior and equal moduli of elasticity in tension and compression. A 5-min test vs. a 3-h test showed the presence of creep strains and a varying modulus of elasticity. Typical creep curves for Mix 23 (23% calcite filler by weight) showing variation of strain with time and temperature at two stress levels 1.7 and 5 MPa are shown in Figure 3.6. The tensile strength of Mix 23 and Mix 130 were 36 and 23 MPa, respectively. Variation of the modulus of elasticity of Mix 23 with time and temperature is shown in Figure 3.7. Testing these fully cured specimens at elevated temperatures resulted in a loss of tensile strength.

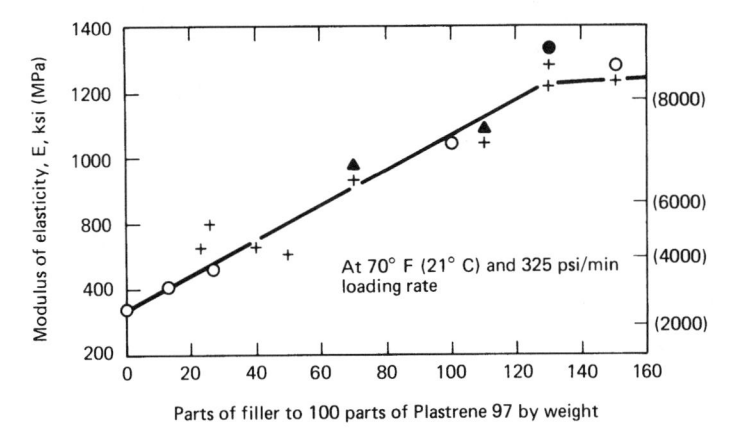

Figure 3.5 Variation of modulus of elasticity of Plastrene 97 with calcite filler. (After Balint and Shaw, 1965.)

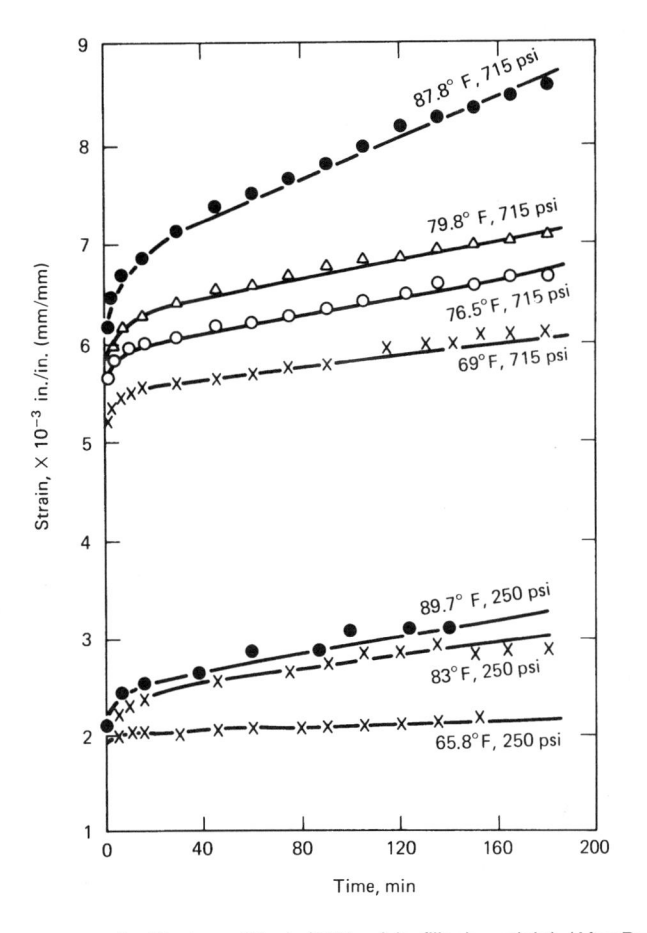

Figure 3.6 Typical creep curves for Plastrene 97 mix (23% calcite filler by weight). (After Balint and Shaw, 1965.)

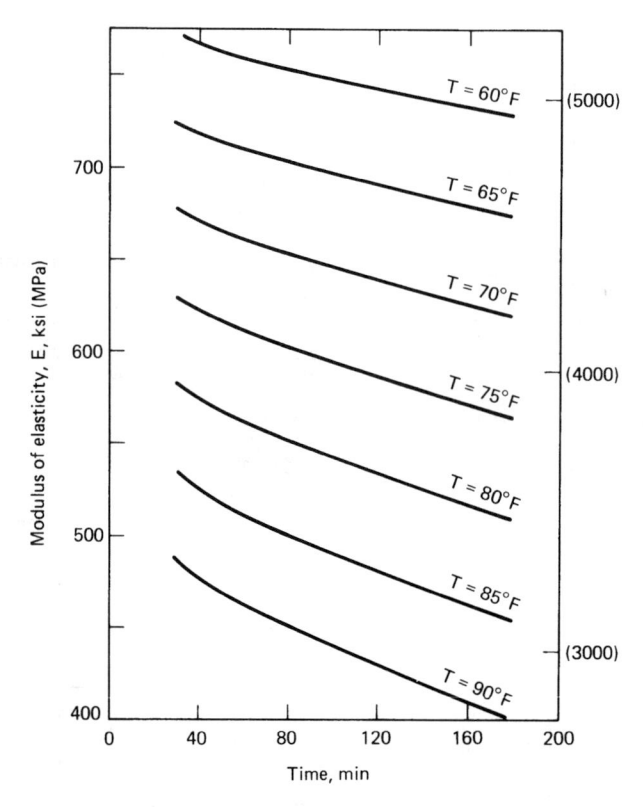

Figure 3.7 Variation of modulus of elasticity of Plastrene 97 mix (23% calcite by weight) with time and temperature. (After Balint and Shaw, 1965.)

3.4 TIME EFFECTS IN PLASTICS — EVALUATION AND COMPENSATION

The time-dependent strain or creep should be accounted for in interpreting the experimental data, especially if several strain gages are to be read and there is a time delay in readings. Because of creep, strain readings under a given applied load are changing with time; therefore, no unique value of stress is associated with a measured strain value, and the time-dependent modulus of elasticity E_t must be used to evaluate the stress values from the corresponding strain values ε_t at a specified time t. In an indirect model test, one derives the influence lines or surfaces from the distorted shape of the model, and time-dependent properties do not influence the results. In these cases, the change in the stress with time at any point, that is, relaxation, is inconsequential. However, in a direct model one cannot ignore the creep of the plastic, and proper care must be used in interpreting the test data.

If the stress values are low (about 20% of the material yield strength), then creep and creep recovery are linear functions of stress and all creep is eventually recovered upon the removal of stress. Similarly, if the material is subjected to a strain, there is a stress corresponding to the initial strain, and this stress decreases with time under a sustained strain. This relaxation is also a linear function of strain if the imposed strain is small. Basic creep and relaxation behavior of plastics are normally determined by simple uniaxial or flexural tests, and it has been observed that in general the creep modulus of elasticity and the relaxation modulus are about equal for a specified time t. Hence, only one time-dependent modulus, called the *apparent modulus*, is necessary.

ELASTIC MODELS – MATERIALS AND TECHNIQUES

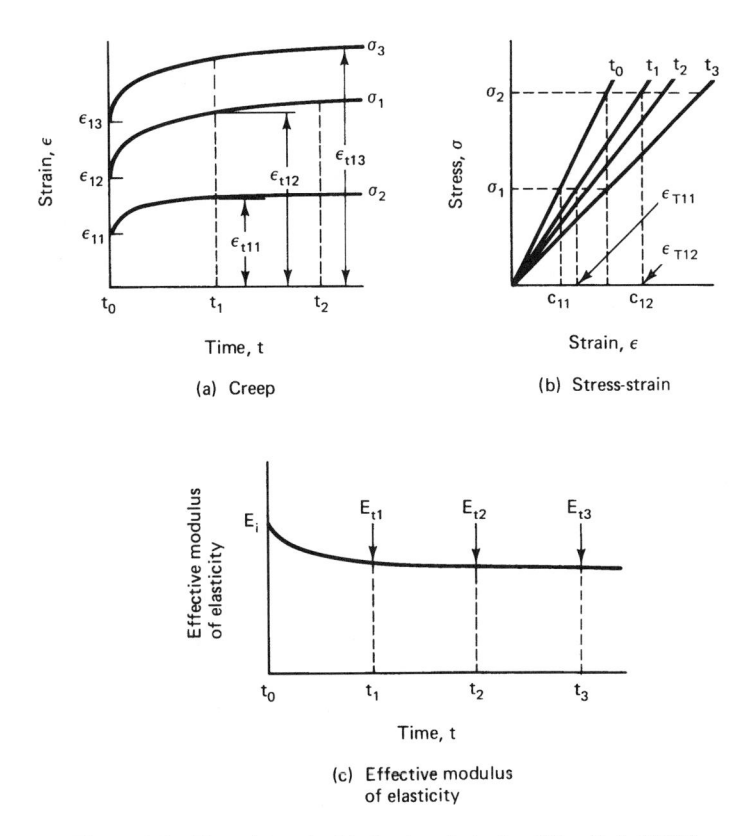

Figure 3.8 Time-dependent behavior of plastic. (After Roll, 1968.)

3.4.1 Determination of the Time-Dependent Modulus of Elasticity and Poisson's Ratio

Roll (1968) suggests that the time-dependent stress–strain curve or the effective modulus of elasticity can be easily determined by conducting creep tests in tension or flexure for as few as two values of stress. It is recommended that for better accuracy these tests be conducted for three different stress values σ_1, σ_2, and σ_3. As shown in Figure 3.8a, at the time t_0 when the stresses are first applied the corresponding initial strains are ε_{i1}, ε_{i2}, and ε_{i3}, respectively. For stress values below the proportional limit, these points $(\sigma_1, \varepsilon_{i1})$, $(\sigma_2, \varepsilon_{i2})$, and $(\sigma_3, \varepsilon_{i3})$ lie on a straight line corresponding to time t_0 (Figure 3.8b), the slope of which is the initial tangent modulus of elasticity E_i (Figure 3.8c). At some time t_1, the corresponding total strain values are ε_{t11}, ε_{t12}, and ε_{t13}, respectively (Figure 3.8a), which result in the stress–strain curve (Figure 3.8b) and the modulus of elasticity E_{t1} (Figure 3.8c). In a similar manner, we can determine the stress–strain curves for times t_2 and t_3 and the corresponding moduli of elasticity E_{t2} and E_{t3}, respectively. Thus, the variation of E_t with time can be determined (Figure 3.8c).

The curve in Figure 3.8c may be represented by the equation:

$$E_t = \frac{E_i}{1 + c(t)E_i} \qquad (3.3)$$

where $c(t)$ is the unit creep function (i.e., the creep due to a unit stress). It must be remembered that Equation 3.3 is based on the assumption that the material is strain-free when the stresses are applied. However, like all viscoelastic materials, plastics exhibit a "memory" effect that is a function of the stress history, including the currently applied stress. As for elastic materials, the principle of superposition is applicable for viscoelastic behavior provided that the governing differential equations are linear and that the resulting deformations do not significantly change the boundary conditions. In such a case, it can be shown that for a linear viscoelastic material, the total strain ε_t at any time is given by

$$\varepsilon_t = \frac{\sigma_t}{E_i} + \sum_i \int_{t_i}^{t} \sigma_i \frac{dc(t)}{dt} dt \qquad (3.4)$$

where: σ_t = the currently applied stress
σ_i = the stress previously applied or removed at time t_i
$c(t)$ = the unit creep function
E_i = the initial modulus of elasticity

It can be noted that if the unit creep function $c(t)$ is independent of stress at low stress levels and if the material is strain-free when stress is applied, Equation 3.4 reduces to Equation 3.3. It is clearly advantageous to eliminate the complications resulting from the memory effects by loading the model in a *strain-free* condition. This can be easily achieved by removing the load after the strain readings have been taken and then allowing enough time for creep recovery before reloading the model. Because of the principle of superposition, one can expect the model to recover completely if it is left unloaded for as long a duration as for which it was loaded.

The Poisson's ratio v_t can be easily determined by using additional strain gages installed perpendicular to the longitudinal gages in a uniaxial tension or flexural test. These gages are used to measure the lateral strains. The value of the Poisson's ratio at time t is given by

$$v_t = \frac{\text{lateral strain}}{\text{longitudinal strain}} \qquad (3.5)$$

In general, the Poisson's ratio is not as sensitive to time effects as is E_t.

Another indirect method to determine the Poisson's ratio is to determine the value of the time-dependent shear modulus of elasticity G_t from a torsion test (Litle, 1964). The Poisson's ratio can then be calculated from the relationship

$$v_t = \frac{E_t}{2G_t} - 1 \qquad (3.5a)$$

This test is cumbersome compared with the flexural test; however, it is a useful and independent check when accurate values of the elastic constants are required. For models made from plastics stressed in the range 0 to 14 MPa, the Poisson's ratio normally varies from 0.3 for the thermosetting plastics and some of the more brittle thermoplastics to about 0.35 to 0.45 for most of the thermoplastics.

3.4.2 Loading Techniques to Account for Time-Dependent Effects

The difficulties caused by the time-dependent effects in plastics can be overcome by several techniques, including deferring measurements until creep has terminated, that is, until the additional

ELASTIC MODELS – MATERIALS AND TECHNIQUES

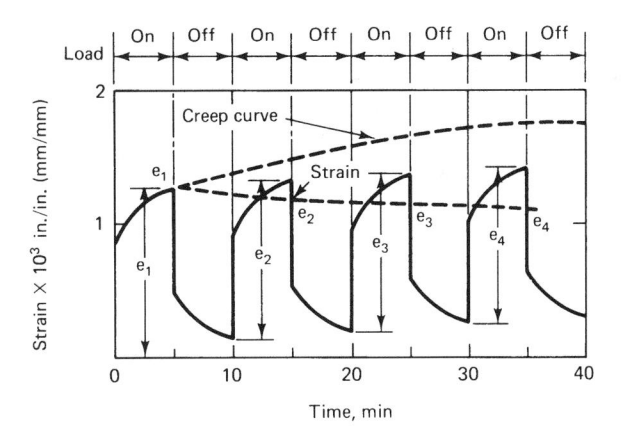

Figure 3.9 Repeated loading of a viscoeleastic material showing constant-strain increment after first loading cycle. (After Rocha, 1961.)

increase of strain with time has become negligible. This time interval depends on the creep characteristics of the material, but in almost all cases it is also dependent on the magnitude of the applied stress. Low stress levels result in a lower creep rate and a smaller required time interval for creep termination. Preece and Davies (1964) suggest a waiting time of 6 h; however, Roll (1968) recommends a far shorter period of 20 to 30 min for low stresses. In any case, creep tests must be conducted at the highest stress expected in the model, so that the asymptotic time for creep termination can be determined for the model material. The stress at any point in the model can then be determined using the known value of the effective modulus of elasticity E_t.

Rocha (1961) has suggested a method of obtaining a sufficient number of readings from a viscoelastic model by cycling the load until the observed strain (or deflection) difference between the loading and unloading cycles becomes constant (Figure 3.9). Under these conditions, the material behaves as if it were elastic, and the effective elastic modulus E_t is dependent upon the strain difference between the commencement of loading and the start of the following unloading cycle.

Another technique to handle the time-dependent effects in plastic models is the *constant strain method*, described by Carpenter (1963). This method has previously been used primarily for indirect tests, such as with the Beggs deformeter system. For tests on direct models, the method consists of imposing and maintaining a displacement at some point on the model and reading the resulting strains and deflections that do not vary with time. Thus, the difficulties of obtaining time-dependent data from a deforming model are avoided. However, although the strains and deformations are not changing with time, the corresponding loads and stresses decrease with time, and some auxiliary method must be used to convert the measured strains into stresses. This is the major disadvantage in the use of the constant-strain method.

Using the "spring balance" concept of Wilbur and Norris (1950) to circumvent this difficulty, Carpenter (1963) used a calibrated "strain cell" as the auxiliary structure in his tests. The function of the strain cell is illustrated in Figure 3.10. The strain cell is made of the same material as the model, and its axis is oriented along the chosen direction of the applied deflection Δ. The strain cell can be instrumented as a load cell with a suitable number of strain gages and a circuitry to achieve maximum sensitivity. The fixed displacement Δ imposed on the load cell results in a force Q_t that is a function of time. Since the displacement Δ is maintained constant, the strain in the load cell, ε_{cell}, and the strain at any point in the structure, ε_p, remain constant. Hence, the strain in the load cell is given by

$$\varepsilon_{cell} = D \frac{Q_t}{E_t} \tag{3.6}$$

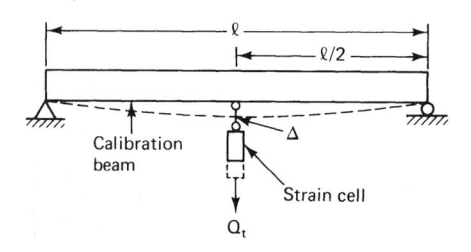

Figure 3.10 Calibration beam and strain cell. (After Carpenter, 1963.)

where D is the calibration constant for the cell. For the model structure,

$$\varepsilon_p = \frac{\sigma_p}{E_t} \qquad (3.7)$$

where σ_p and ε_p are the stress and strain, respectively, at some selected point, and for convenience they are assumed to be uniaxial.

Since the model material and the cell material are the same, by eliminating E_t between Equations 3.6 and 3.7,

$$\frac{\sigma_p}{Q_t} = D\frac{\varepsilon_p}{\varepsilon_{\text{cell}}} \qquad (3.8)$$

is obtained.

Although the stress σ_p and the force Q_t are changing with time, they are both proportional to the same relaxation function of time, and, therefore, the stress per unit load remains unchanged with time. Thus, the response of the structure is the same as that of a model constructed from a creep-free elastic material whose properties are *not* time dependent.

The principal advantage of the constant-strain method is that the strain readings do not change with time, and this eliminates the need to conduct additional creep tests to determine the effective modulus of elasticity. It is not really necessary to know the time-dependent force Q_t corresponding to the imposed displacement Δ. The construction and calibration of the strain cell is relatively simple and can be easily handled by an individual with some laboratory experience.

3.5 EFFECTS OF LOADING RATE, TEMPERATURE, AND THE ENVIRONMENT

Direct tension, compression, and flexure tests of materials at ordinary room temperatures have shown that the stress–strain relationships of most of the common metals are characterized by a typical stress–strain relationship. This is not the case for plastics that are more sensitive to the rate of testing (strain rate), temperature, and environmental effects. These effects are discussed in the following sections.

3.5.1 Influence of Strain Rate on Mechanical Properties of Plastics

The mechanical properties of plastics are influenced by the rate at which stresses or strains are applied. These effects are in addition to those due to time-dependent deformations, such as creep. ASTM D638 and ASTM D695 specify certain rates of crosshead movement for standard tension and compression tests, respectively. For tension tests, four speeds are listed, namely, 1, 5 to 6, 50, and 500 mm/min with preference given to the 5 to 6 mm/min rate. It must be noted that these speeds relate to a 114 mm grip length and may not correspond to strain rates over the gage length. For compression tests a speed of 1 mm/min is specified between the bearing blocks supporting a 25-mm-long specimen.

In a model test, different parts of the structure and, for that matter, different points on the same cross section are subjected to varying rates of straining. This causes serious difficulties in assigning different values of E_t corresponding to the different strain rates for calculating stresses at various points in the structure. The influence of the different rates of straining can be minimized by applying the loads as slowly as possible and by leaving the model in a loaded state for some time before deformations are measured. Under such conditions, the creep strains will be large compared with the strain rate effects, and no significant error will be introduced by using the same value of E_t throughout the structure. It is obvious that serious difficulties can arise in interpreting test results for dynamically loaded models.

3.5.2 Effects of Temperature and Related Thermal Problems

The mechanical properties of plastics are significantly influenced by environmental changes such as changes in temperature and in some cases relative humidity. Therefore, the testing environment must be carefully controlled; otherwise, considerable errors can exist in correlating prototype and model behavior (Roll, 1968). It is also important to know the effects of temperature changes on the mechanical properties of plastics. Of course, a knowledge of thermal properties of plastics is necessary in the model-molding process and in the use of plastic models for thermal studies.

3.5.2.1 Temperature Effects on Elastic Constants

The values of the elastic constants are sensitive to temperature changes. For example, the tensile and shear moduli of elasticity of the acrylic Perspex, if assumed to be "standard" at 20°C, increase by about 1%/°C fall in temperature (Preece and Davies, 1968). These variations are practically linear at room temperature; however, there is a marked decrease in the moduli values at temperatures greater than 100°C. The value of Poisson's ratio for Perspex remains practically unchanged (0.35 to 0.37) at room temperature, but this value increases to 0.5 if the temperature exceeds 100°C.

3.5.2.2 Temperature Effects on Strength

It should be noted that plastic models are normally used to simulate elastic behavior of a prototype; therefore it is essential to know the compressive and tensile strengths in order to limit the working stresses in the model. The maximum stress in a model should be related to its type and duration. Normally, the maximum working stress should not exceed one third of the ultimate tensile strength to ignore creep effects.

3.5.3 Coefficients of Thermal Expansion

Plastics have very high coefficients of thermal expansion relative to metals. According to Preece and Davies (1964), the coefficient of linear expansion for Perspex is 9×10^{-5}/°C over the temperature range −20 to 60°C. This is about nine times the value for steel. The coefficient of linear expansion is assumed to be one third of the coefficient of volumetric expansion, which can be determined experimentally. The high thermal expansion coefficient causes some problems in molding of the specimens. Because plastics either cure or are formed at high temperatures, the dimensions of any piece will be smaller than the mold dimensions after the piece has cooled to room temperature. In practice, it is difficult to separate thermal contraction and shrinkage of the resin. For example, unshrunk methyl methacrylate sheet material will shrink about 21 mm/m (or 2%) in the first heating cycle as a result of further polymerization; however, mold shrinkage is generally less than 1%.

Some plastics exhibit larger values. Fialho (1960) reported a 20% shrinkage in the molding of a polyethylene model. Another direct consequence of the large thermal contractions experienced by molded parts as they cool to room temperatures is manifested as residual stresses in the molded part. This can lead to warping in the final specimen. It must be noted that careful temperature control during fabrication can reduce these effects. However, they cannot be totally eliminated.

3.5.4 Thermal Conductivity

Plastics have a very low thermal conductivity approximating 0.1 Btu/(h) (ft) (°F), compared with 222 for copper, 28 for iron, 0.4 for asbestos, 0.2 for wood, and 0.03 for cork. Thermal conductivity is an important factor in determining stresses in thermal model studies. Also, because of the low thermal conductivity, when vacuum forming it is necessary to heat both sides of plastic sheets that are more than 1.25 mm thick. Otherwise, the temperature differential through the sheet may cause problems.

The low thermal conductivity of plastics causes difficulties when strains are measured with electrical resistance strain gages (see Section 7.3.2). As the strain gage circuit is switched, the low thermal conductivity causes heating of the gage and the material under it, which in turn influences its mechanical properties. The heating of the gage itself causes a drift in the measuring circuit, resulting in a change of the output with time. Moreover, it is not possible to separate the output due to structural response and that due to drifting caused by heating of the gage. Litle et al. (1965) caution that very little quantitative data is available on the gage-heating effects and that one should be aware of this serious problem and should not "just willy-nilly" use strain gages on plastic materials. They also present some experimental data to illustrate the variation of measured strains with time as a result of gage-heating effects in a methyl methacrylate and a PVC specimen.

The following methods have been successfully used by some investigators to handle the problems resulting from gage heating:

1. Investigators at the Portland Cement Association (Carpenter et al., 1964) simultaneously switch into the measuring circuit a cold active gage and a cold dummy gage. If both gages are in the same environment, the drift due to heating will be minimal. It is adequate to have only three or four dummy gages that can be used in turn, giving each gage sufficient time to cool off from the preceding cycle. This method has also been used at the Laboratorio Nacional de Engenharia Civil (LNEC) in Lisbon and at the University of Pennsylvania for strain measurements in tests on Perspex (Plexiglas) models of box-beam highway bridges (Roll and Aneja, 1966).
2. Johnson and Homewood (1961) have handled the gage-heating problem by having the gages heated at all times by means of a separate heating circuit that passes current through all gages not in the measuring circuit. This method has been used generally by the Cement and Concrete Association in England, and also recently by the Portland Cement Association.
3. Yates et al. (1953) used pulse excitation of the strain gage, with the measuring current passing through the gage in short pulses in the millisecond range. Since the time between two consecutive pulses is significantly longer than the pulse itself, the temperature of the gage remains low enough to prevent any significant heating effects.

The bridge voltage of the strain-indicating equipment is an important consideration in the gage-heating problem. The heating effect is proportional to the square of the bridge voltage, and therefore it is important to keep the bridge voltage as low as possible (Roll, 1968). Carpenter et al. (1964) observed that an applied gage voltage of 3.5 V, corresponding to a bridge voltage of 7.0 V, was enough to damage a 1.5 mm gage-length foil gage on Plexiglas. This is seldom a problem at present because of the availability of low-voltage (1.5 V) strain indicators. The use of gages that are

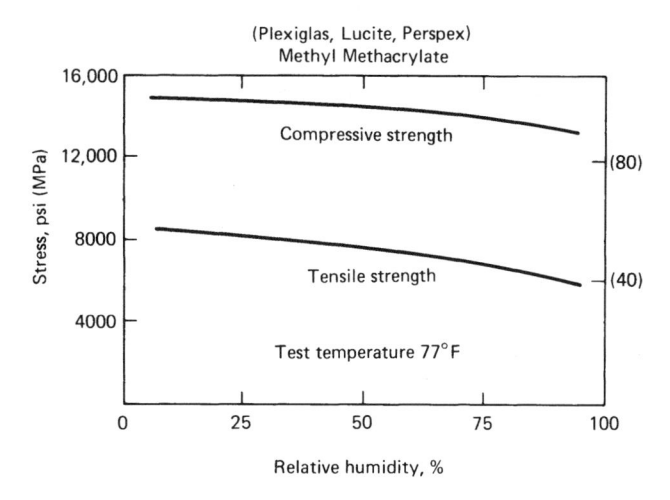

Figure 3.11 Influence of humidity on compressive and tensile strengths of methyl methacrylates. (After Manufacturing Chemists' Association, 1957.)

compensated for plastics does not completely solve the problems; however, in general, the use of temperature-compensated foil gages along with a low-voltage strain indicator will minimize the gage-heating effects. Regardless of the method used to handle the gage-heating problem, the mechanical properties of the plastic should be determined with the same equipment, techniques, and environment used for the model.

3.5.5 Softening and Demolding Temperatures

At the softening temperature, the shape of a plastic sheet can be easily molded to any curved profile. For acrylic plastics, such as Plexiglas, Perspex, and Lucite, this temperature is approximately 110°C (Preece and Davies, 1964). According to Litle (1964), the vacuum-forming temperature for Boltaron 6200 PVC is in the range of 85 to 127°C.

3.5.6 Influence of Relative Humidity on Elastic Properties

Like temperature, relative humidity affects the tensile strength, compressive strength, proportional limit, and yield strength of plastics. However, the elastic properties are not significantly influenced, and therefore the influence of humidity changes on the response of an elastic model can be neglected. The effects of relative humidity on the tensile and compressive strengths of methyl methacrylate (Plexiglas, Perspex, and Lucite) are shown in Figure 3.11.

3.6 SPECIAL PROBLEMS RELATED TO PLASTIC MODELS

There are additional characteristics peculiar to plastics; knowledge of these characteristics can be helpful in the construction and testing of the model and interpretation of the test data.

3.6.1 Modeling of Creep in Prototype Systems

It has been shown that the creep behavior of plastics can be troublesome in interpreting strain data; however, it can be quite useful if the model is used to study the effects of creep on the behavior of the prototype. Ross (1946) simulated the creep behavior of reinforced concrete structures using a fabric-reinforced phenolic plastic. The unit creep of both materials was approximately equal for a timescale factor of 240; that is, the unit creep of plastic in 50 h was equal to the unit creep of the concrete at 500 days. Using this material to construct models of reinforced concrete elements, he conducted experiments in short periods of time to determine creep buckling in columns and to study the redistribution of reactions due to settling supports on continuous beams. Using unplasticized acrylic resin models, he also studied the effect of creep on the horizontal components of reactions in two-hinged arches. An excellent study of creep buckling of cylindrical shell roofs under gravity loading, using Plexiglas models, has been conducted by Kordina (1964).

3.6.2 Poisson's Ratio Considerations

Section 3.2 noted that for skeletal-type structures the similitude requirement of equality of Poisson's ratio can be relaxed. However, in grids, plates, and shell-type structures where the structural response is sensitive to Poisson's ratio, this similitude requirement must be satisfied. If the prototype structure is steel ($v_p = 0.30$), then models constructed from acrylic plastics, with Poisson's ratio of approximately 0.35, will be reasonably satisfactory, with little error in satisfying the similitude requirement. However, if it is required to model a prototype concrete structure ($v = 0.15$ to 0.20), then it may cause some discrepancy between the prototype and the model response, depending on the sensitivity of the structural response to Poisson's ratio.

3.6.3 Thickness Variations in Commercial Shapes

The forming process can cause significant thickness variations in commercial shapes of plastics that can be as large as ±15% of the nominal thickness. If not considered in analysis, this can seriously affect the interpretation of the test data. Therefore, if commercial shapes are used, care must be taken to use the best available grade to minimize this variation.

3.6.4 Influence of the Calendering Process on the Modulus of Elasticity

Thermoplastics such as PVC are commonly manufactured as continuous sheets, up to 2.5 m in width, on rolls called *calenders*. The components are usually mixed and blended on heated rolls called *mills* or milled under pressure in a mixer and subsequently extruded through one or more sets of calenders or rolls at speeds of up to about 60 m/min. This process can produce continuous sheets ranging in thickness from 1 mil or less to about 3 mm or more.

Litle (1964) examined the variability of modulus of elasticity and its sensitivity to some factors in an elastic buckling study of spherical shells, where the models were vacuum-formed from flat sheets of Boltaron 6200 PVC. A summary of the test values from four specimens from each of the 20 shells tested is shown in Figure 3.12. These data establish a sense of variability of the material modulus of elasticity. Also, it is important to note that although the plastic is manufactured by a calendering process, the anisotropy is not large.

Since the fabrication process for the shells (Litle, 1964) involved heat forming, the effect of annealing on the modulus of elasticity was studied. The results presented in Figure 3.13 show the sensitivity of the modulus to annealing. It must also be noted that the vacuum-forming process causes the material to stretch in going from the initial flat sheet to the contour of the model mold. The influence of this stretching on the modulus of the material is shown in Figure 3.14.

3.7 WOOD AND PAPER PRODUCTS

Wood and paper materials are very easy to work with, even for inexperienced model makers. For this reason, they are extensively used for educational models which are needed to demonstrate a variety of structural behavior situations in the elastic range.

ELASTIC MODELS – MATERIALS AND TECHNIQUES

1. Four samples (two in each direction) taken for each model.
2. Modulus determination by cantilever beam test.
3. x implies parallel to calendering. Average = 454,000 psi
4. o implies perpendicular to calendering. Average = 447,000 psi

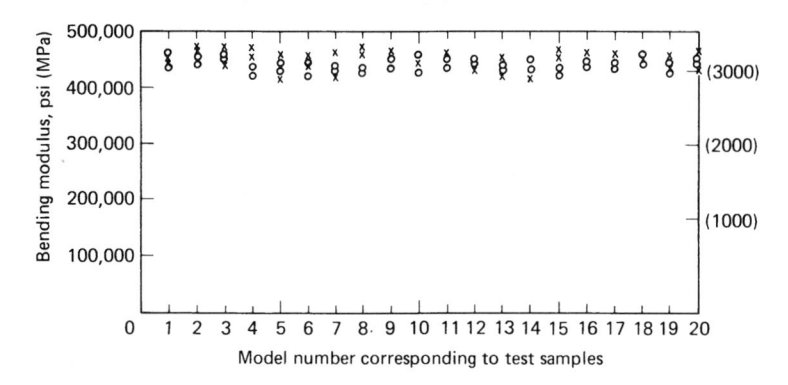

Figure 3.12 Modulus of elasticity of Boltaron 6200 PVC. (After Litle, 1964.)

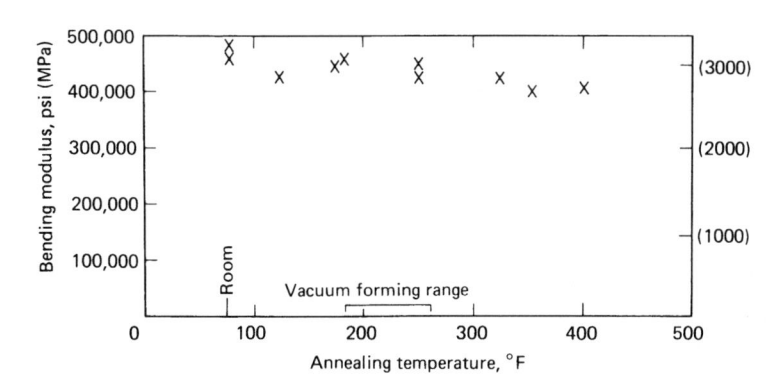

Figure 3.13 Effect of annealing temperature on bending modulus of Boltaron 6200 PVC. (After Litle, 1964.)

3.7.1 Balsa Wood

Balsa wood is a widely used modeling material with very low density (Table 3.4) and a uniform cellular structure giving it excellent insulating properties (Carpenter, 1917). It is inexpensive and easy to shape into structural members without the need of special tools. Balsa wood is readily available in a variety of different shapes that can be easily assembed into model structures. Because of these desirable characteristics, models made of balsa wood can be sized and assembled even by inexperienced students. Examples of models that can be made of balsa wood are wooden trusses, beams and frames, stiffened plates, and plate girders. Elastic behavior, including elastic buckling, can be very effectively demonstrated using carefully made balsa wood models. Examples of its many and diverse model applications are given in Chapter 12.

3.7.1.1 *Balsa Wood Shapes Available*

A large variety of balsa wood shapes are readily available to the model user. These and bass wood come in the form of rectangular and square shapes, and thin sheets as shown in Figure 3.15. The model user need only determine the scale of the model since almost any cross section can be easily assembled by cutting to size and gluing together various available components.

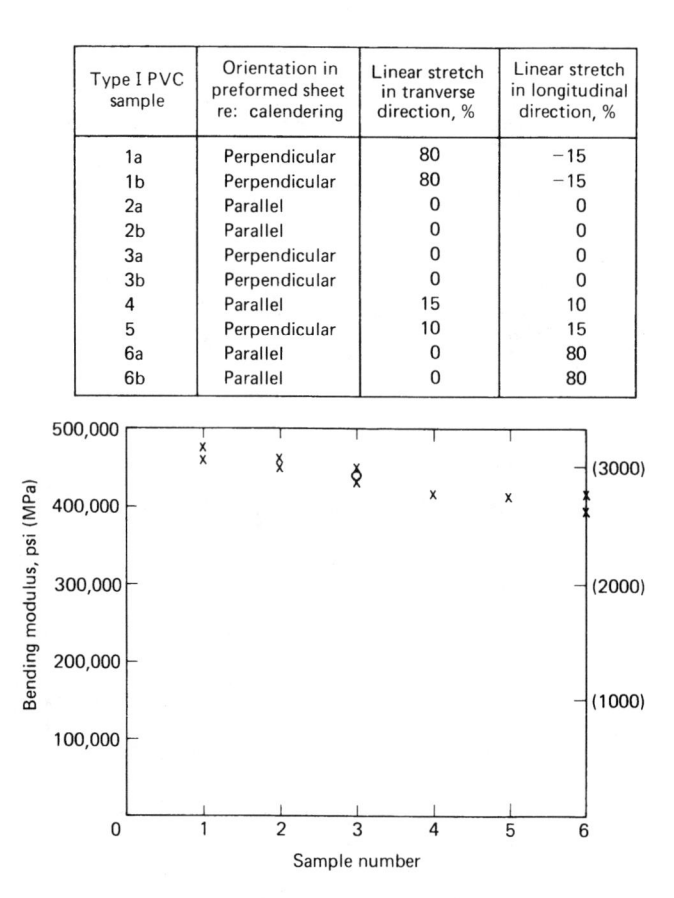

Type I PVC sample	Orientation in preformed sheet re: calendering	Linear stretch in tranverse direction, %	Linear stretch in longitudinal direction, %
1a	Perpendicular	80	−15
1b	Perpendicular	80	−15
2a	Parallel	0	0
2b	Parallel	0	0
3a	Perpendicular	0	0
3b	Perpendicular	0	0
4	Parallel	15	10
5	Perpendicular	10	15
6a	Parallel	0	80
6b	Parallel	0	80

Figure 3.14 Effect of vacuum stretch on bending modulus of Boltaron 6200 PVC. (After Litle, 1964.)

Table 3.4 Weights of Common Wood Species

Common Name	Scientific Name	Specific Weight, lb/ft³
Balsa	*Ochroma lagopus*	7.3
Cork	(Bark from cork oak, *Quercus suber*)	13.7
Missouri corkwood	*Leitneria floridana*	18.1
White pine	*Pinus strobus*	23.7
Catalpa	*Catalpa speciosa*	26.2
Cypress	*Taxodium distichum*	28.0
Douglas fir	*Pseudotsuga mucronata*	32.4
Sycamore	*Platanus occidentalis*	35.5
Red oak	*Quercus rubra*	40.5
Maple	*Acer saccharum*	43.0
Long-leaf pine	*Pinus palustris*	43.6
Mahogany	*Swietenia mahogonia*	45.0
Locust	*Robinia pseudo acacia*	45.5
White oak	*Quercus alba*	46.8
Hickory	*Carya alba*	54.2
Live oak	*Quercus virginiana*	60.5
Ironbark	*Eucalyptus leucoxylon*	70.5
Lignum-vitae	*Guaiacum sanctum*	71.0
Ebony	*Diospyrus*	73.6
Black ironwood	*Krugiodendron ferreum*	81.0

ELASTIC MODELS – MATERIALS AND TECHNIQUES

Balsa Wood: 914.4 mm Lengths

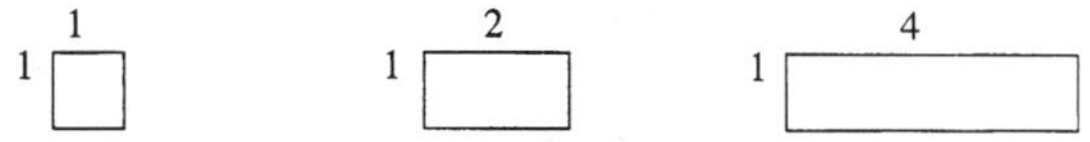

1	2	4
1.59 x 1.59	3.18 x 6.35	1.59 x 6.35
2.38 x 2.38	4.76 x 9.53	2.38 x 9.53
3.18 x 3.18	6.35 x 12.7	3.18 x 12.7
4.76 x 4.76		
6.35 x 6.35		
9.53 x 9.53		
12.7 x 12.7		

In addition, flats ranging in size from 0.79 mm x 25.4 mm to 9.53 mm x 101.6 mm are available.

Bass Wood: 609.6 mm Lengths

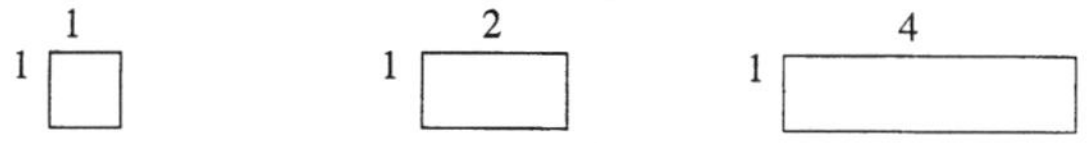

1	2	4
1.59 x 1.59	3.18 x 6.35	1.59 x 6.35
2.38 x 2.38	4.76 x 9.53	2.38 x 9.53
3.18 x 3.18	6.35 x 12.7	3.18 x 12.7
4.76 x 4.76		
6.35 x 6.35		
9.53 x 9.53		
12.7 x 12.7		

In addition, flats ranging in size from 0.79 mm x 76.2 mm to 6.35 mm x 101.6 mm are available.

Figure 3.15 Readily available balsa and bass wood shapes.

3.7.1.2 Strength Properties

The strength of wood and especially balsa wood is very heavily dependent on its density. Extensive studies (Soden and McLeish, 1976) indicate that the two most important variables that influence the strength characteristics of balsa wood are the density and the grain direction relative to the load. Tensile strength tests clearly show the effect on strength of these two variables. The effect of direction of loading to the grain is shown in Figure 3.16 where the tensile specimens were cut from the same sheet; hence, they have the same density. Determination of tensile strength properties is usually made using dog bone–shaped specimens if the material is in the form of a sheet. If the material is in the form of rectangular or square shapes, the strength can be determined by testing equal-density specimens in flexure. This type of test can also be used to determine the Young's modulus, E, of the material by using the linearly elastic load–deflection curve and the elastic deflection equation as illustrated in Figure 3.17. For truss-and-frame structures made of balsa wood, the simply supported beam method has been found to give consistent results with the least effort. Direct tension tests of square or rectangular balsa wood pieces involve considerable specimen preparation to make gripping the ends possible.

The effect of the variation of different density of balsa wood sheets is shown in Figure 3.18. As can be seen from Figure 3.18, large variations in strength can be expected from different sheets of balsa wood. It becomes therefore imperative to check the density of the balsa wood prior to fabrication to ensure that all members of the model have similar strengths. To obtain maximum benefit of the properties of the material it is usually necessary to weigh the material from which the model will be made and select the pieces of suitable density. Compressive strength along and perpendicular to the grain as well as shear strength and tensile strength across the grain of balsa wood sheets as a function of the specific gravity were studied by Soden and McLeish (1976). A

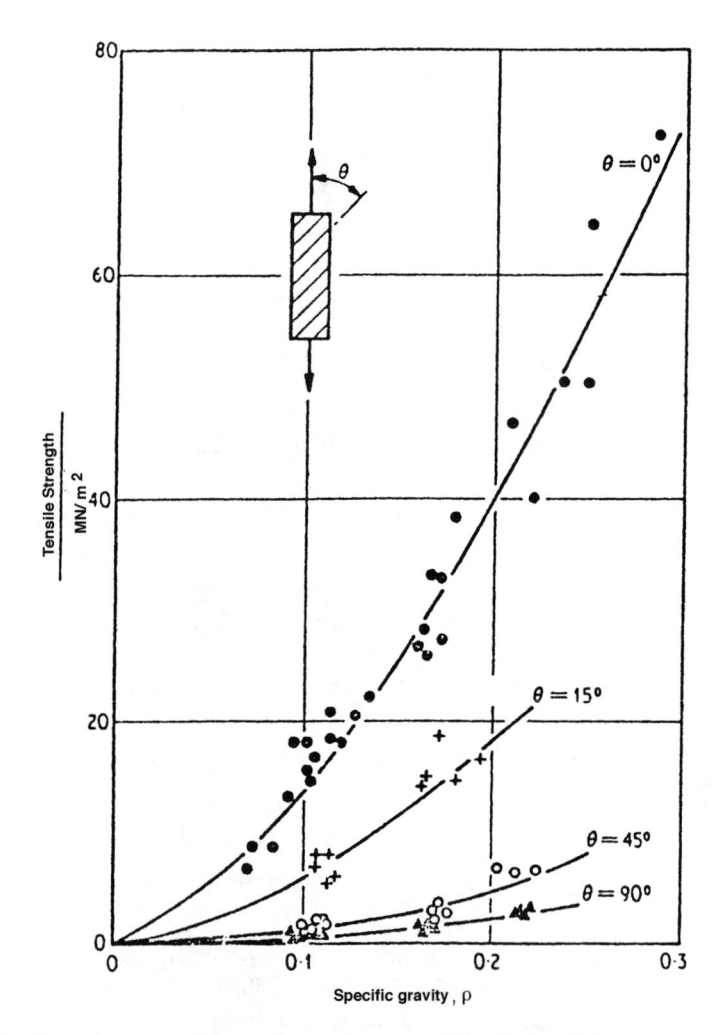

Figure 3.16 Tensile strength at various angles to the grain. (From Soden, P.D. and McLeish, R.D., *J. Strain Anal.*, 11(4), 225–234, 1976. With permission.)

summary of their research on strength properties of balsa wood is given in Table 3.5. Strength properties given in Table 3.5 can be adjusted for any density of balsa wood using the empirically determined equation:

$$P_\rho = P_{0.17}\left(\frac{\rho}{0.17}\right)^n \tag{3.9}$$

where P_ρ is the strength property at the desired specific gravity, ρ, $P_{0.17}$ is the strength property at $\rho = 0.17$, and n is an experimentally determined coefficient as shown in the table.

A significant property of any elastic model is the elastic or Young's modulus of the material from which it is made. In the case of balsa wood, the modulus of elasticity varies with specific gravity as shown in Figure 3.19. The empirically determined equations of tension and compression elastic modulus, E, as a function of specific gravity are also given in Figure 3.19. A comparison of the mechanical properties of balsa wood with some of the more common structural materials is shown in Table 3.6. The specific stiffness and specific tensile and compressive strength is also shown in Table 3.6. The advantage of using balsa wood to study the elastic behavior of steel or reinforced concrete structures is illustrated in the example below.

ELASTIC MODELS – MATERIALS AND TECHNIQUES

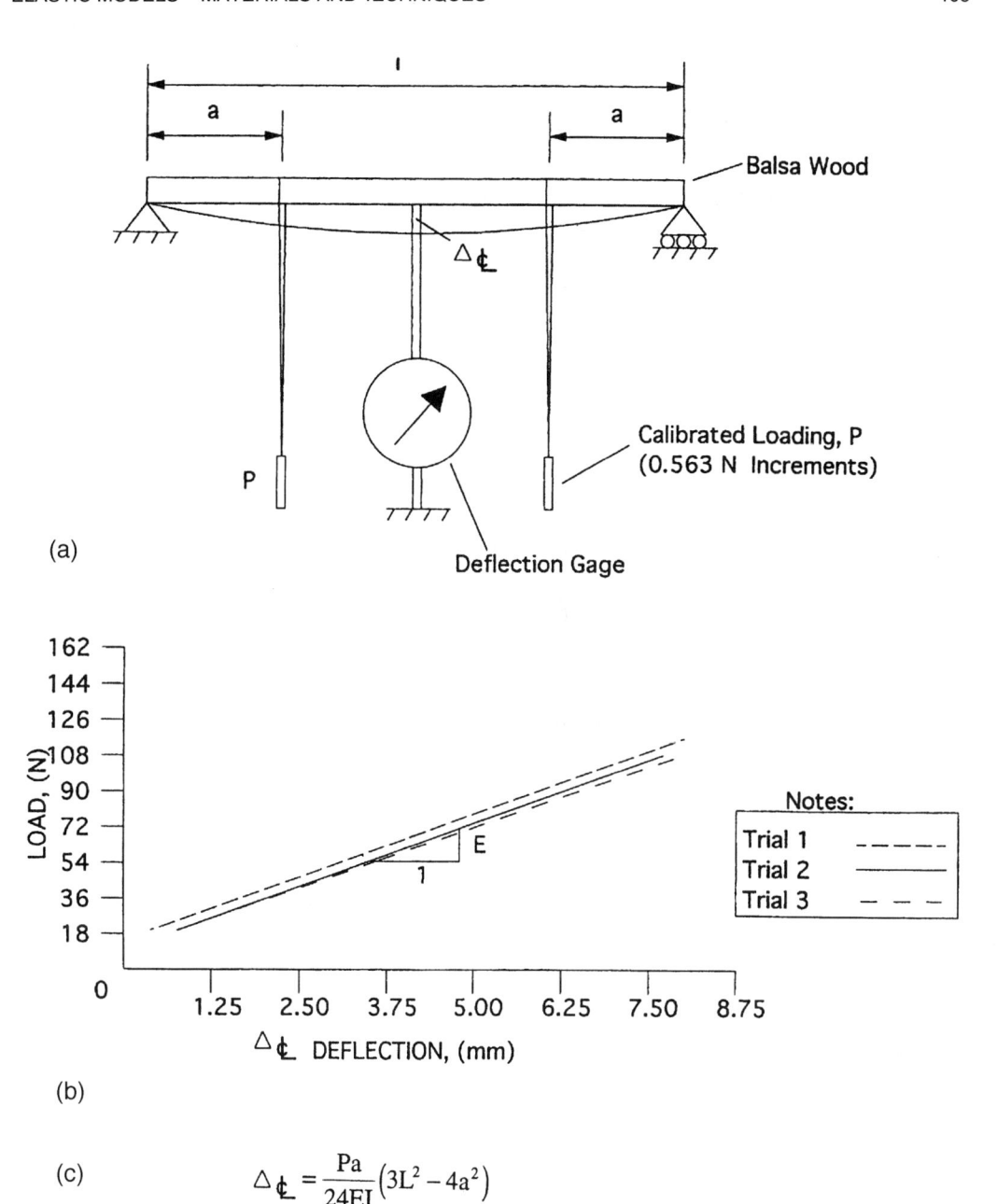

Figure 3.17 Determination of Young's modulus of linearly elastic material from simple beam test.

(c) $$\Delta_{\mathcal{C}} = \frac{Pa}{24EI}(3L^2 - 4a^2)$$

Example 3.1

It is required to demonstrate the deflection characteristics of an irregular steel framework using small-scale, balsa wood model. Young's modulus of steel is 207 GPa and that of balsa wood is taken as 10.4 GPa. With these choices of materials, the deflections in the balsa wood replica model will be approximately 20 times larger than if the model were made of steel. A magnification of the deformations in the model will therefore result in deflections that are very easy to see. The unique low modulus of elasticity, but relatively high specific strength, of balsa wood makes it an excellent material for studying the deformations of complex structures in the elastic range. As will be explained in Chapter 6, balsa wood models are easy to fabricate.

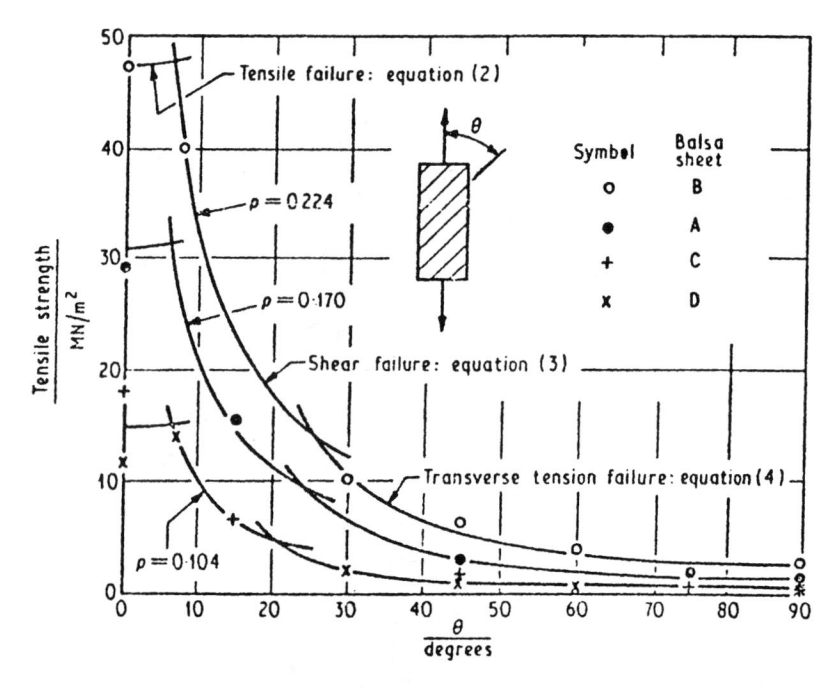

Figure 3.18 Variation of mean tensile strength with direction of loading. (From Soden, P.D. and McLeish, R.D., *J. Strain Anal.*, 11(4), 225–234, 1976. With permission.)

Table 3.5 Summary of Balsa Wood Properties

Property	Mean Value at $\rho = 0.17$ $P_{0.17}$, MN/m²	n[a]	Percentage Error of Estimate	Number of Tests
Tensile strength parallel to grain	31.5	1.52	15.9	27
Tensile strength across grain				
All specimens	1.53	1.93	33.8	65
Specimens loaded across grain	1.51	2.07	31.9	19
Shear strength	3.44	1.45	19.8	20
Compressive strength parallel to grain	14.7	1.67	10.4	39
Compressive strength across grain	0.88	1.58	28.3	14
Tension modulus parallel to grain	5170	1.55	13.4	13
Compression modulus parallel to grain	3770	1.59	10.4	10

[a] $P_\rho = P_{0.17} \left(\dfrac{\rho}{0.17} \right)^n$.

3.7.1.3 Applications of Balsa Wood to Model Studies

1. Two- and Three-Dimensional Trusses — Model studies of wooden trusses made from balsa wood are especially easy to fabricate and test. Figure 3.20a shows a through truss balsa wood bridge model with glued gusset plates used as a teaching aid to demonstrate structural correlation between analysis and experiment. More complex trusses can also be studied using balsa wood models. An example of such applications is the ⅓-scale elastic model study (Kopatz, 1982) of a ³⁰⁄₂₀ GHz microwave Cassegrainian antenna shown in Figure 3.20b. This type of antenna has two smaller reflectors to focus the microwave instead of one larger one. Very stringent deflection requirements of not more than 0.025 in. (0.625 mm) differential displacement between the two reflectors necessitated a very stiff three-dimensional truss. The prototype test truss was itself a ⅓-scale version of the design antenna and was made of square steel tubing welded together to form a rigid structure. The ⅑-scale

ELASTIC MODELS – MATERIALS AND TECHNIQUES

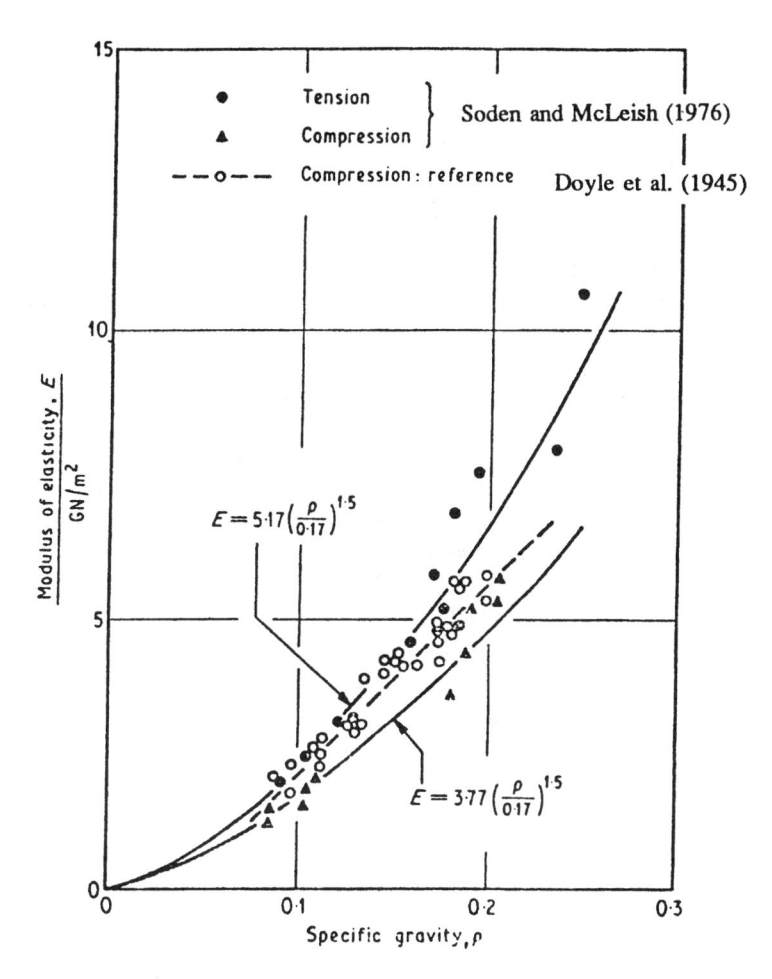

Figure 3.19 Modulus of elasticity of balsa wood along the grain. (From Soden, P.D. and McLeish, R.D., *J. Strain Anal.*, 11(4), 225–234, 1976. With permission.)

Table 3.6 Mechanical Properties of Various Materials

Material	Tensile Strength, σ_u, MPa	Modulus of Elasticity, E, GPa	Specific Gravity, ρ	Specific Stiffness, E/ρ, GPa	Specific Tensile Strength, σ_u/ρ, MPa	Specific Buckling Strength, $\frac{\sqrt{E}}{\rho}$
Mild steel	415	207	7.8	27	53	1.84
High-tensile steel	2100	207	7.8	27	266	1.84
Aluminum (L65)	460	73	2.8	22	166	3.05
Titanium (DTD 5173)	930	110	4.5	24	207	2.33
72% (by volume) S glass fiber in epoxy resin	1900	66	2.12	31	900	3.83
70% (by weight) carbon fibers in epoxy	550	180	1.6	111	350	8.4
Balsa wood	**64**	**10.4**	**0.27**	**38**	**236**	**11.9**

Adapted from Soden and McLeish, 1976

model (Figure 3.20c used square balsa wood sections glued together with 5-min epoxy. The choice of model scale is explained below.

Example 3.2

The structural properties of the prototype space truss members made from five different size steel tubes are listed in Table a below. Select an appropriate size scale model using the available square balsa wood lengths ranging from 1.6 × 1.6 mm to 12.7 × 12.7 mm. In order to answer this question, which of the properties of the prototype truss are the most important must be determined. In truss action, where the load is carried by direct stresses (tension or compression), the most important property of the members to be modeled is the area. The moment of inertia plays a secondary role and therefore can be relaxed. The tendency of members of the truss carrying compressive forces to buckle will be controlled in the model by the higher radius of gyration of the solid model sections and the low load level of the model since its behavior is linearly elastic. Based on this reasoning, a table is constructed showing the equivalent scaled area of all the available balsa wood sections for different model scales. Only the boldfaced areas in Table b appear to match the scaled prototype member areas. As can be seen from Table b, the best match is at a scale of $S_l = 3$.

Table a   Antenna Structural Members

No.	Member	Wall Thickness, mm	Wt., kg/m	Area, mm²	I, mm⁴	r, mm
1	3 × 3	4.763	0.948	1302.6	1081369	28.804
2	3 × 3	3.937	0.799	1098.1	936937	29.210
3	2 × 2	4.763	0.596	818.7	277501	18.412
4	2 × 2	3.175	0.420	576.4	211404	19.152
5	1 × 1	2.413	0.151	206.8	17482	9.190

Note: All members are square hollow tubing A36 steel.

Table b   Areas Modeled by Available Shapes, mm

Scale L_m/L_p	1.588	3.175	4.766	6.35	7.938	9.525	11.113	11.906	12.7
1/10	0.39	1.56	3.52	—	—	—	—	—	—
1/9	0.31	1.28	2.86	—	—	—	—	—	—
1/8	0.25	1.00	2.25	—	—	—	—	—	—
1/7	0.19	0.77	1.72	3.06	—	—	—	—	—
1/6	0.14	0.56	1.27	2.25	—	—	—	—	—
1/5	0.10	**0.40**	**0.88**	1.56	2.43	—	—	—	—
1/4	0.06	0.25	0.56	1.00	1.56	2.25	—	—	—
1/3	0.04	0.14	**0.31**	0.56	**0.88**	1.27	1.72	1.98	2.25

The boldfaced numbers correspond to the properly scaled values of available balsa wood shapes. The majority of the shapes match the truss member requirements at 1/3 scale.

Deflection and strain measurements for the model antenna were carried out and compared with analytical results. The testing procedure, test results, and correlation are presented in Chapter 12.

2. Elastic Buckling Studies — Another example of a balsa wood model used to check the design of a steel beam walkway in which elastic buckling is of prime concern (Simonetti, 1982) is shown in Figure 3.21. The prototype beams (Figure 3.21a) were two W36 × 135 11-m-long which were modeled at 1/12 scale using glued balsa wood strips: 1.5 × 25 × 915 mm for the flanges and 0.75 × 20.25 × 915 mm for the webs. Elmer's wood glue was used for the joining of all balsa wood components. Two separate models (Figure 3.21b) were tested under concentrated and simulated

ELASTIC MODELS – MATERIALS AND TECHNIQUES

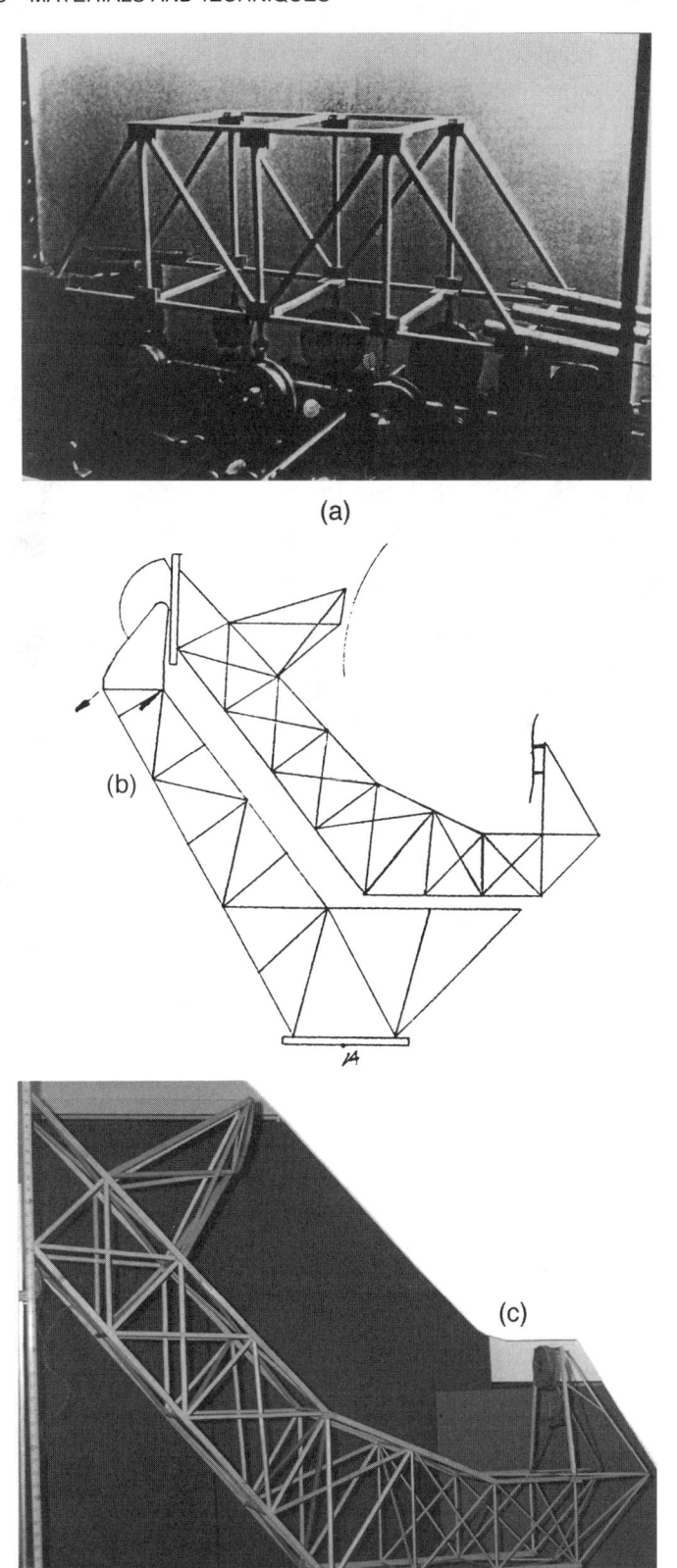

Figure 3.20 Examples of balsa wood models. (a) Through truss bridge; (b) $^{30}/_{20}$ GHz Cassegrainian microwave antenna configuration; (c) three-dimensional 1/9-scale model. (Courtesy of Drexel University, Philadelphia.)

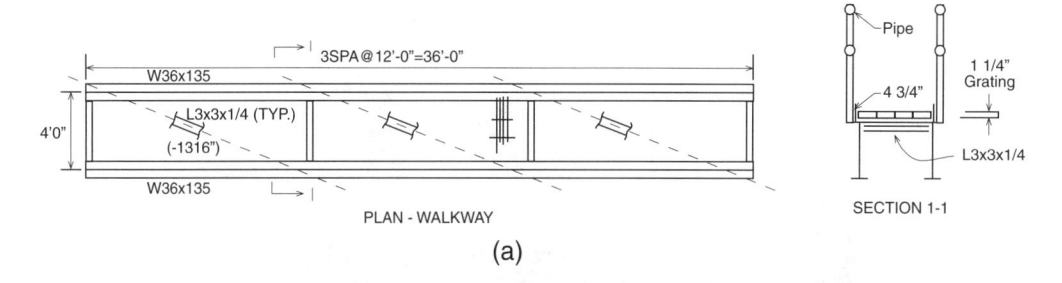

Figure 3.21 One-twelfth scale model of a steel plate girder walkway made of balsa wood. (a) Dimensions of prototype; (b) model during testing. (Courtesy of Drexel University, Philadelphia.)

uniform loads measuring the vertical and lateral deformations of both members until buckling occurred. Elastic web buckling was easily observable. The test results and the correlation with elastic lateral buckling predictions are given in Chapter 12.

3.7.2 Modeling of Structural Lumber

Models of wooden structures are usually constructed from the same species and grade of lumber as the prototype structure. However, if only the elastic behavior is of interest, other types of wood that are more readily available or are easier to finish can be substituted. The easiest woods to work with are basswood and white pine. Basswood, in particular, can be purchased in a variety of small-sized uniform sections just like balsa wood and thus is very easy to use. (See Figure 3.15).

3.7.2.1 Strength Properties of Wood Used for Structural Models

The ASTM Standard D 143-52, *Standard Methods of Testing Small Clear Specimens of Timber,* is used exclusively to obtain the strength properties of visually stress-graded lumber and glue-laminated timber. The most imortant strength properties used in design are static bending, compression parallel

ELASTIC MODELS – MATERIALS AND TECHNIQUES 115

to grain, compression perpendicular to grain, shear parallel to grain, and tension perpendicular to grain. These properties are listed in Table 3.7 for the most common lumber used in models.

3.7.2.2 Examples of Wooden Models

Small-scale wooden models can be easily fabricated using the same species of wood as the prototype structure. Some of the soft woods such as pine and hemlock fir are easier to cut and shape and these can be used in case the wood species is not important in the model study. One such example is a ¹⁄₁₀-scale model of a wooden roof truss made from white pine (Heckler, 1980). The dimensions and layout of the prototype truss used in the roof of a garage is shown in Figure 3.22a and b. The ¹⁄₁₀-scale model was constructed from the following materials:

1. Clear white pine — appearance stud grade, i.e., a white pine that would be used in furniture making, moldings, and the like;
2. Birch veneer plywood — a three-ply, 1.5-mm-thick plied laminate for the gusset plates;
3. White, phenolic resin glue (similar to Elmer's glue);
4. Sanding sealer manufactured by Sherwin-Williams.

The clear white pine was milled using a 3 mm upright table saw. The members were cut, squared, and surfaced by means of a hollow-ground planer blade from 25×50 mm stock. The dimensions of the 50×100 mm and the 50×150 mm members used in the prototype truss at ¹⁄₁₀ scale were 5×10 mm and 5×15 mm, respectively. When the model members were all milled, the the layout of the truss was done. For the layout surface, a piece of plywood approximately 200×950 mm was used. On the plywood surface, member edge lines were drawn according to the geometric specifications. Blocks of wood were then glued on the outside edges of the truss outline to act as stops when gluing commenced. This acted as a jig on which the truss could be assembled. The chord and web members of the model truss were then cut at the appropriate intersecting angles and set aside. The gusset plates were cut from the birch veneer plywood in accordance with the specifications. At this time, the chords and webs were placed on the jig and pinned in place. Glue was applied sparingly to the plywood gusset plates and members. Then they were clamped or weighted in position until dry.

As accuracy of the joint intersections was very important, only one or possibly two joints were set at a time. After the first side was complete, the form of the truss was established and more than one or two joints could be glued at once on the second side. Usually three to five joints were gusseted at once on the second side. On completion of the gusseting, the model truss was sanded lightly with an aluminum oxide sandpaper, 220 grade. Finally, a prime coat of Sherwin-Williams sanding sealer was applied to the truss. This coat was allowed to dry overnight; then it was sanded lightly, and a finish coat was applied.

The model is shown in Figure 3.22c during testing. Load was applied through small holes drilled through the gusseted joints as shown in Figure 3.22d. Test results and comparison of the behavior with two modifications of the basic Fink truss arrangement are discused in Chapter 12.

3.7.3 Small-Scale Modeling of Glue-Laminated Structures

Glue-laminated wood construction is becoming more important in building and other structural applications. Models of glue-laminated structures can be easily fabricated by gluing strips of wood together to simulate the prototype construction. An example of this procedure is illustrated in the testing of two ¹⁄₂₄-scale models of prototype glue-laminated beams 127 mm wide by 610 mm deep and 12.2 m long studied by Johnson (1973). The model goemetry and test set up are shown in Figure 3.23a. The ¹⁄₂₄-scale model beams were 5 mm wide by 25 mm deep and 0.5 m long. The prototype laminations were 38 mm thick and the model laminations were 1.56 mm thick. A commercially available adhesive

Table 3.7 Strength Properties of Some Commercially Important Woods Grown in the United States

Common and Botanical Names of Species	Moisture Content, %	Specific Gravity	Static Bending Fiber Stress at Proportional Limit, MPa	Modulus of Rupture, MPa	Modulus of Elasticity, MPa	Compression ∥ to Grain Fiber Stress at Proportional Limit, MPa	Max. Crushing Strength, MPa	Compression ⊥ to Grain Fiber Stress at Proportional Limit, MPa	Shear ∥ to Grain– Max. Shear Strength, MPa	Tension ⊥ to Grain– Max. Tensile Strength, MPa
Basswood, American (Tilia americana)	105	0.32	18.62	34.48	7171	11.65	15.31	1.45	4.14	1.93
	12	0.37	40.68	59.99	10963	26.20	32.61	3.10	6.83	2.41
Pine, Eastern white (Pinus strobus)	73	0.34	20.69	33.79	6826	14.07	16.82	1.79	4.69	1.72
	12	0.35	39.30	59.30	8550	25.30	33.10	3.52	6.21	2.14
Douglas-fir, Coast type (Pseudotsuga menziesii)	38	0.45	31.03	52.40	10825	21.58	26.61	3.03	6.41	2.07
	12	0.48	53.78	84.12	13445	40.34	51.23	6.00	8.00	2.34
Fir, Balsam (Abies balsamea)	117	0.34	20.69	33.79	6619	14.34	16.55	1.45	4.21	1.24
	12	0.36	35.85	52.40	8481	27.37	31.23	2.62	4.90	1.24
Hemlock, Eastern (Tsuga canadensis)	111	0.38	26.20	44.13	7378	17.93	21.24	3.03	5.86	1.59
	12	0.40	42.06	61.37	8274	27.72	37.30	5.52	7.31	—

Adapted from *Forest Products Laboratory Wood Handbook* (results of tests on small, clear specimens in the green and air-dry condition). U.S. Department of Agriculture Handbook No. 72, U.S. Government Printing Office, Washington, D.C., 1955.

ELASTIC MODELS – MATERIALS AND TECHNIQUES 117

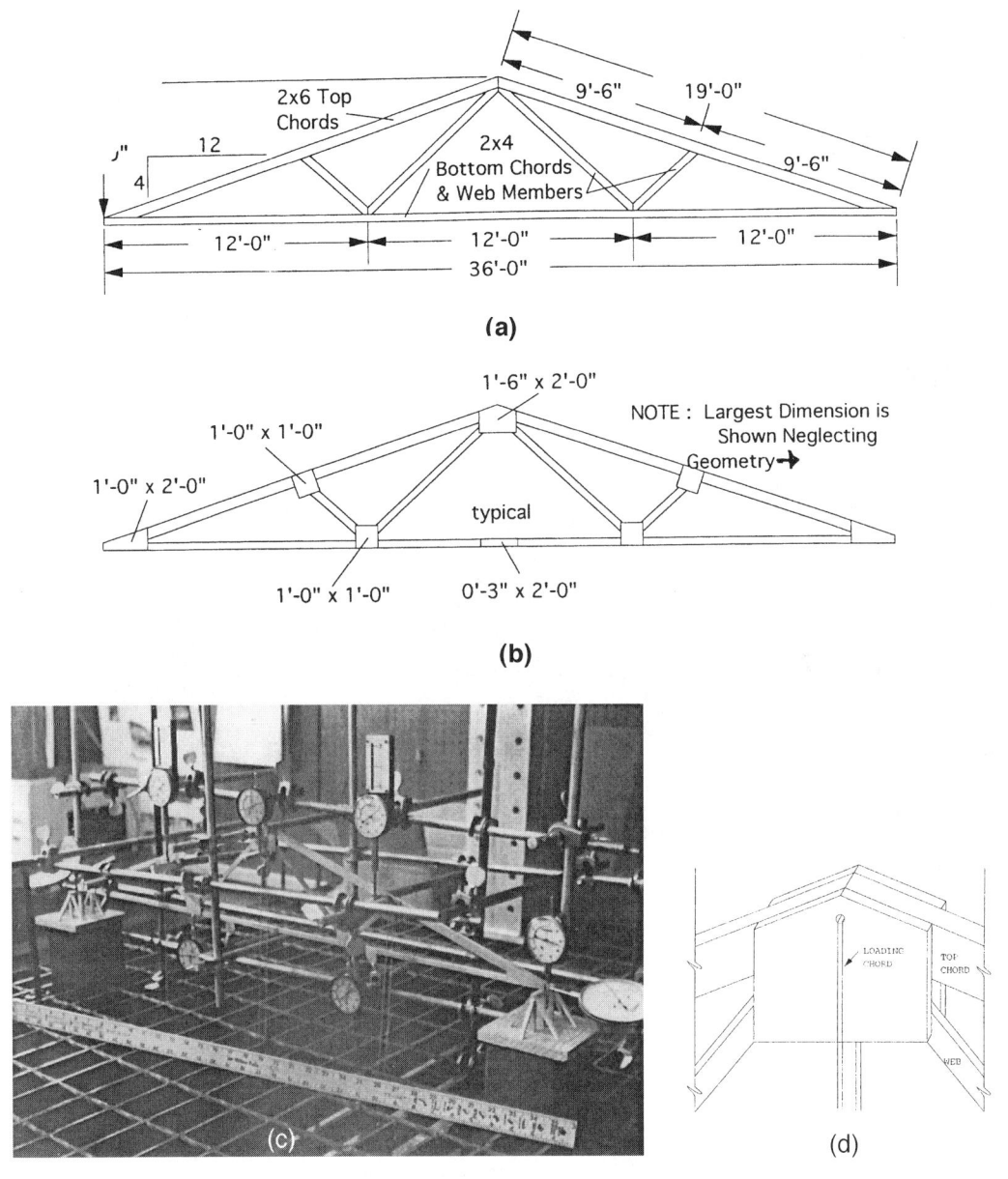

Figure 3.22 Fink roff truss geometry. (a) Prototype member sizes; (b) ptototype gusset plates; (c) one-tenth scale model during testing; (d) node loading detail. (Courtesy of Drexel University, Philadelphia.)

"super glue" was used to construct the two models. As this was an exploratory study, no attempt was made to vary the strength of the laminations with respect to the depth of the beam. Testing to ultimate strength was accomplished by preventing lateral torsional buckling, as shown in Figure 3.23b.

A more complex example of modeling of glue-laminated structural members is that of double-tapered glulam beams (Kopatz, (1984). One form used frequently in roof design is the double-tapered shape, which allows for slope of the roof to either end. The center of the beam becomes very deep, causing the transverse stresses to become large. When this type of beam is fabricated from Douglas fir, which has a lower-tension perpendicular-to-grain strength than other types of wood used in these applications, the transverse stresses must be taken into consideration. It has

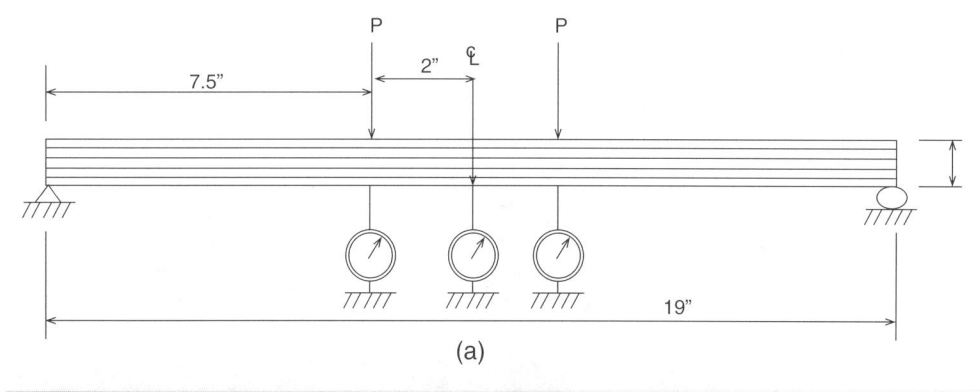

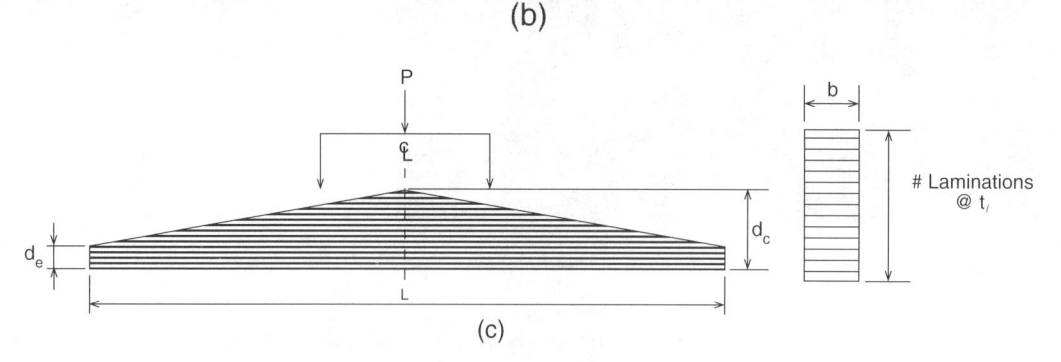

Figure 3.23 Examples of glulam beam models. (a) Straight beam configuration; (b) glulam model during testing; (c) double-tapered glulam beam configuration; (d) one-twelfth scale model during testing. (Courtesy of Drexel University, Philadelphia.)

been observed that an arching effect occurs in such members, the double-tapered beam acting as a deep beam (Gopu and Goodman, 1975).

The geometry of the prototype structure selected (Gutkowski et al., 1982) is shown in Figure 3.23c. The width of the prototype beam was 128 mm, the depth at midspan and the end depth were 1.07 m and 0.30 m, respectively, and the length was 9.14 m. Individual laminae of the prototype were 37.5 mm thick and in the 1/12-scale model 3 mm thick. Both model and prototype

ELASTIC MODELS – MATERIALS AND TECHNIQUES

(d)

Figure 3.23 (continued)

Table 3.8 Mechanical Properties of Manila Folder Paper at 8% Moisture Content

Test Direction	Thickness, mm	Density, g/cm^3	Tension σ_u, MPa	γ_u, $\times$ 100 mm/mm	E, MPa
Machine direction	0.279	0.786	43.44	1.6	4275
Cross machine direction			20.69	3.2	2206

Source: Bodig, J. and Jayne, B. A., *Mechanics of Wood and Wood Composites,* Van Nostrand Reinhold, New York, 1982. With permission.

were fabricated from Douglas fir and a resorcinol glue was used in fabricating the model. Instrumentation consisted of deflection and strain measurements at the critical sections. The ¹⁄₁₂-scale model during testing is shown in Figure 3.23d. The model test results are compared to the prototype and to a finite element analysis. These are discussed in Chapter 12.

3.7.3.1 Paper Products Used for Structural Models

Paper products of different thickness and quality have been successfully used to fabricate elastic structural models. One type of thin (0.28-mm) cardboard of uniform properties called "Manila" or "cartridge" paper is a very useful modeling material that can be combined with other elastic materials such as balsa wood to study the behavior of plate girders, stiffened plates, composite construction, and several types of shells. It is easily cut and glued and has very uniform properties.

Because paper is manufactured from a slurry of fibers, there is a preferential orientation of the fibers as they are deposited. The preferential orientation of the long axis of the fibers is in the machine direction (MD), The direction perpendicular to MD in the plane of the sheets called the cross machine direction (CMD). The strength properties of the Manila folder–type cardboard are given in Table 3.8.

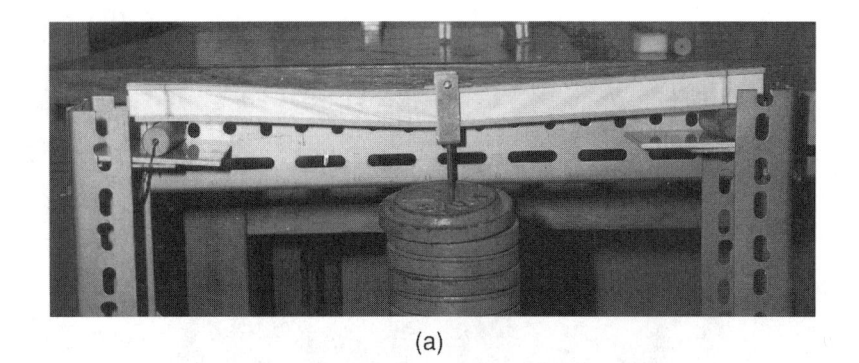

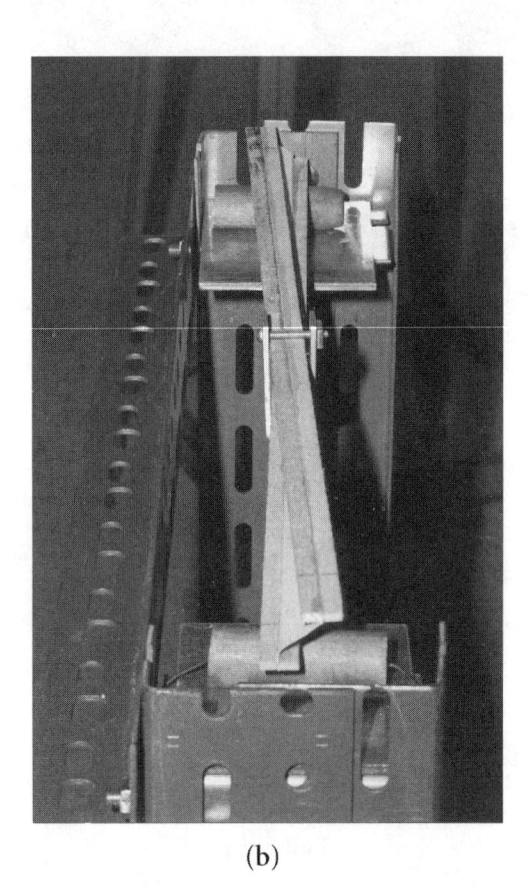

Figure 3.24 Paper and wood plate girder demonstrating buckling behavior. (a) Elevation — note buckled web; (b) lateral torsional buckling. (Courtesy of Drexel University, Philadelphia.)

3.7.3.2 Examples of Paper Models

An example of a mixed or hybrid elastic model made from cartidge paper and balsa wood is that of a plate girder study shown in Figure 3.24. The effect of relative stiffness of flange to web and the effects of end supports and depth to span can be studied using such models even if the lateral buckling theory for such structures is unknown to the student. Note the lateral torsional buckling of the plate girder model when the ends are not prevented from moving laterally in Figure 3.24b. Web buckling of an unrestrained plate girder is shown in Figure 3.24a. Educational models of this type can become powerful tools for demonstrating structural behavior in the classroom.

3.8 ELASTIC MODELS — DESIGN AND RESEARCH APPLICATIONS

For the past several decades, elastic models made from plastics have been used as a complement to a mathematical model in the design process and to obtain solutions for problems with poorly defined or very highly variable boundary conditions for which analysis is not feasible. Leading models laboratories have conducted model studies on many buildings, bridges, dams, and other structures. Development of facilities in North America and elsewhere should encourage more engineers to use models not only as an important complement to the analytical procedures but also as an integral part of research and development programs aimed at improving current design procedures.

Elastic models have been used both to verify the designs prepared using the available analytical tools and as a partial or complete "design aid" to design several important buildings, bridges, and dams and some special-purpose structures such as aircraft hangars, prestressed concrete reactor vessels, reservoirs, and complex shell-and-dome roof structures. Another application of elastic models is the wind effects model that has come into prominence recently and is being used in the design of many large, exposed structures. Elastic models have also been used for studying the dynamic behavior of several important structures. These are discussed in Chapter 11.

3.9 DETERMINATION OF INFLUENCE LINES AND INFLUENCE SURFACES USING INDIRECT MODELS — MÜLLER–BRESLAU PRINCIPLE

An *influence line* is a curve the ordinate of which at any point equals the value of some particular function as a result of a unit load acting at that point. An *influence surface* can be defined similarly. The ordinates for an influence line for a particular structural action (force or moment reaction, or an internal force or moment) at a selected section in a linear elastic structure are evaluated analytically by placing a unit load successively at several possible load positions and calculating the value of the structural action at the selected station. It must be noted that the condition that the materials of the structure behave elastically is not sufficient to ensure that deformation is proportional to load. In addition, it must be ensured that there are no cable or stability effects, that is, the deformations do not influence the actions of the loads, and that the structure is sufficiently stiff so that the deformations under applied loads are small and do not change the geometry of the structure appreciably. Under such conditions, the Maxwell–Betti reciprocal theorem is valid and can be used to derive the Müller–Breslau principle, which provides a very convenient method of computing influence lines and forms the basis for certain indirect methods of model analysis (Norris and Wilbur, 1976). This principle can be defined as follows:

> For any linear elastic structure if the restraint corresponding to any internal stress resultant (axial or shearing force, bending or twisting moment at a section) or a reaction is removed and a corresponding deformation (translation or rotation) is introduced, the ordinates of the deflected shape of the structure represent (to some scale) the influence characteristics for the stress resultant considered.

This principle is applicable to any type of structure — skeletal, surface, or solid structures — statically determinate and indeterminate alike. For statically determinate structures, the material need not be elastic for the principle to be valid; however, for statically indeterminate structures, this principle is limited to structures made of linear elastic materials.

Indirect models have been used to obtain influence lines and influence surfaces for linear elastic structures, especially for skeletal structures. The experimental procedure is quite simple. If influence characteristics are required for force, moment, or reactions at any support, the corresponding restraint (translation or rotation) is released and a prescribed deformation (translation or rotation) is introduced by a mechanical device. The results are interpreted in accordance with the Müller–Breslau principle. Similarly, if the influence characteristics are required for an internal stress resultant (axial or shearing force, bending or twist moment) at a particular section, the indirect model is cut

Figure 3.25 Beggs deformeter equipment (deformation devices, plugs, and microscope). See also Figure 3.16 for similar devices. (Courtesy of Soiltest, Inc., Evanston, IL.)

at this section, and a prescribed deformation (translation or rotation) is imposed on this section by a mechanical device. The results are again interpreted in accordance with the Müller–Breslau principle. The normal practice is to impose a unit deformation; then the deflected shape of the indirect model becomes to some scale an influence line or an influence surface for the given reaction or internal stress resultant. For this reason, an indirect model is often also referred to as a *displacement model*. Therefore, the use of an indirect model to obtain influence characteristics for reactions or internal stress resultants requires cutting of the model and a suitable apparatus to displace the model. This technique has a significant advantage in that only deformations need be measured.

To plot the influence line for a generalized deformation (translation or rotation), a prescribed force is applied to deform the model, and the results are interpreted using the Maxwell–Betti reciprocal theorem. In this case, in addition to measuring the model deformations, it is also necessary to measure the force magnitude.

The first known application of an indirect model is due to Beggs, who developed a small apparatus to impose small displacements to plastic models using plugs and wedges (Beggs deformeter), as shown in Figure 3.25. Models of planar structures are normally mounted on a sheet of smooth paper on a drawing board in a horizontal plane. An appropriate type of cut (representing a generalized force release) is introduced at the section at which the reactive force is required. A specially designed deformeter is used to introduce relative axial, shear, or rotational displacements at the cut section. It is useful to drill holes along the axis of the structure, and a pin is used to mark the positions of the initial and deformed axis on the paper. These deformations are usually very small and must necessarily be measured by a micrometer microscope or by dial gages. To reduce friction, the model is often supported on ball bearings at a selected number of points. The errors arising from the deformation of the structure can be minimized, and in general the accuracy of measurements can be considerably improved by applying equal deformations about both sides of the undeformed axis of the structure and by measuring the displacement between the two deflected curves (Hendry, 1964).

ELASTIC MODELS – MATERIALS AND TECHNIQUES

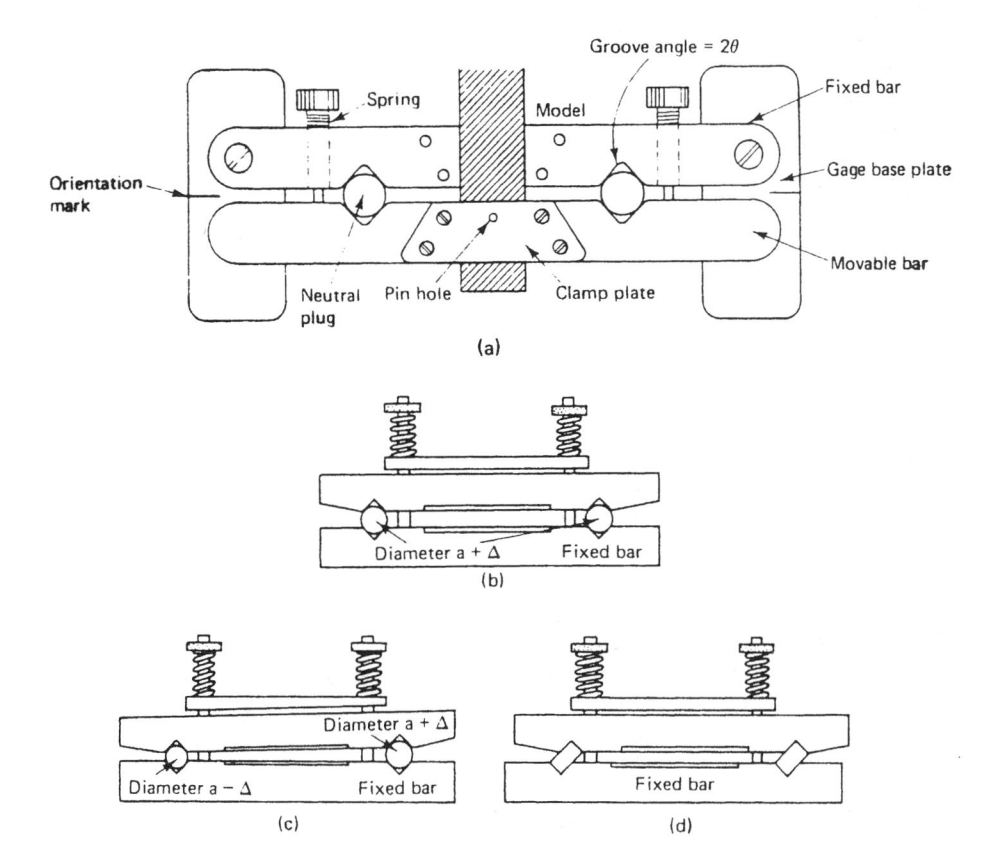

Figure 3.26 (a) Pippard's displacement deformeter for use with the Beggs deformeter: neutral or no displacement position. (After Kinney, G.F., 1957.) Pippard's deformeter in position for determination of influence lines (b,c,d).

Pippard (1947) developed a small displacement deformeter (Figure 3.26) for use with the Beggs deformeter. The ends of the member at the cut section are attached to the two plates, which are separated by two plugs of equal diameter and located in V-shaped grooves with an angle of 2θ (Figure 3.26a). To determine the influence characteristics for axial force, an axial displacement is introduced by replacing the original plugs by plugs of diameter $(a + \Delta)$, thus introducing an axial displacement (elongation) $\Delta \sec \theta$ along the member axis. Similarly, an equal and opposite displacement (shortening) $\Delta \sec \theta$ can be introduced by using plugs of diameter $(a - \Delta)$ (Figure 3.26b). For determining moment influence characteristics, a relative rotation $\Delta \sec \theta / L$ is introduced by using plugs of diameter $(a + \Delta)$ and $(a - \Delta)$ in the two notches a distance $2L$ apart (Figure 3.26c). It is simple to cause a shear displacement by using rectangular plugs, as shown in (Figure 3.26d).

Rocha and Borges (1961) developed a useful procedure for use with the Begg's deformeter, which consists of photographing the model before and after the application of deformations on the same photographic plate, by "double exposure." The displacements can be measured at leisure and more accurately compared with direct measurements on the model using the permanent photographic record.

3.10 SUMMARY

Similitude relationships and materials suitable for elastic models are discussed in this chapter. Since plastics, wood, and paper are the most commonly used materials, their physical properties and uses are presented in detail. The properties and versatility of balsa wood as a material used in elastic scale models are fully discussed. A variety of wood structure modeling studies can be more easily performed if the same species of wood is used for the model and prototype. Examples of such applications are discussed especially as they pertain to glue-laminated timber construction. Examples of elastic direct models are given to illustrate the many applications of elastic models in research and education. Techniques are presented for the use of elastic models as indirect models to obtain influence lines and influence surfaces for any type of structure.

PROBLEMS

3.1 The purpose of this exercise (to be done in the laboratory) is to obtain experience in the use of Plexiglas as a material for structural models. The time-dependent behavior of the material will soon become evident, and the problems of using a plastic with constant-stress loading techniques will be demonstrated.

Specimen: One flexure specimen

Equipment Available: Dial gages, strain gages, strain indicator, loading device with load cell, dead-load weight system.

Questions to be answered:
   a. Is Plexiglas linear viscoelastic in flexure? To what stress level?
   b. How does E vary with time?
   c. How does the behavior of the tensile side of the beam compare with that of the compressive side? Is the neutral axis of bending at mid-depth?
   d. How does response as measured by the dial gage compare with that measured with the strain gage?
   e. Does creep recovery appear to be the same phenomenon as creep itself?
   f. Is there a problem connected with the heating of the gages as the strain-measuring circuitry is activated?
   g. With the given equipment, what problems would you encounter with a relaxation test?
   h. For your specimen, what is the stress level at the peak deflection that you would still consider to be a "small deflection"?
   i. What is the ultimate strength of the material in flexure?
   j. Is the material ductile or brittle in its failure mode?

3.2 The lateral buckling capacity of a shallow circular timber arch is to be modeled because it may be unstable under a *dead* loading of 700 lb/ft of length until the roof is fully completed and can provide full lateral support. Using a plastic model with $E = 400{,}000$ psi, establish the design and operating conditions for stresses, deflections, and buckling load in a 1/40-scale model. How will the model be loaded? How will buckling be detected?

3.3 For the Plexiglas specimens tested in the laboratory:
   a. Plot strain (displacement) values as a function of time for each of the three loadings, following the format of Figure 3.2 in the text.
   b. Plot strain (displacement) values as a function of stress and time. Compute modulus values.
   c. Is the plastic linear viscoelastic? Discuss the validity of this assumption for the two tests conducted.
   d. Suggest any improvements that might be made in the experimental methods used to get the data.

3.4 It is proposed that the Plexiglas model built for a shell structure be loaded at double the true model loading intensity in order to double the strains and make their measurement more accurate. Is this permissible? Discuss.

3.5 A Plexiglas model beam was tested in the laboratory and the following data were obtained at the critical cross section:

ELASTIC MODELS – MATERIALS AND TECHNIQUES   125

Load Step	Midspan Stress, MPa	Midspan Deflection, mm	Midspan Compression Fiber Strain, microstrain	Midspan Tension Fiber Strain, microstrain	Time, min
1	2	3	4	5	6
0	0.000	0.000	0	0	
1	0.378	0.330	95	95	
2	0.756	0.965	262	250	
3	1.134	1.524	412	392	
4	1.512	2.083	562	542	
5	1.890	2.692	710	682	
6	2.268	3.302	878	848	
7	2.646	3.886	1018	992	0
7	2.646	4.064	1028	1004	10
7	2.646	4.115	1052	1025	20
7	2.646	4.191	1120	1068	30
7		4.293	1209	1155	40
8	5.292	7.874	2046	2025	0
8		8.052	2058	2038	10
8		8.153	2083	2058	20
8		8.230	2115	2090	30
8		8.357	2175	2145	40

Questions to be answered
a. Plot the stress vs. deflection and the stress vs. strain curves from 0 to 840 psi.
b. Determine the modulus of elasticity of the model beam as a function of time and plot this. Does this compare to the appropriate figure in the text? Expain similarities and differences between the two.
c. If the accuracy of the instrumentation used to collect the data is not better than ±1% of the range of the values recorded, what can be concluded about the location of the neutral axis of the beam?
d. Estimate the creep strain at the two stress levels tested.

3.6 From the table of available balsa wood sections, select the sizes that one would use to design a ¹⁄₁₀-scale model of the original roof truss for the Academy of Music, Philadelphia, PA, shown below, made from Eastern white pine timbers. Describe how one would fasten and connect the pieces forming the various truss members together. If the wood used in the prototype has a modulus of elasticity of 10 GPa, what would be the value of a measured midspan deflection of 15 mm in the model under the scaled loads shown in the prototype truss?

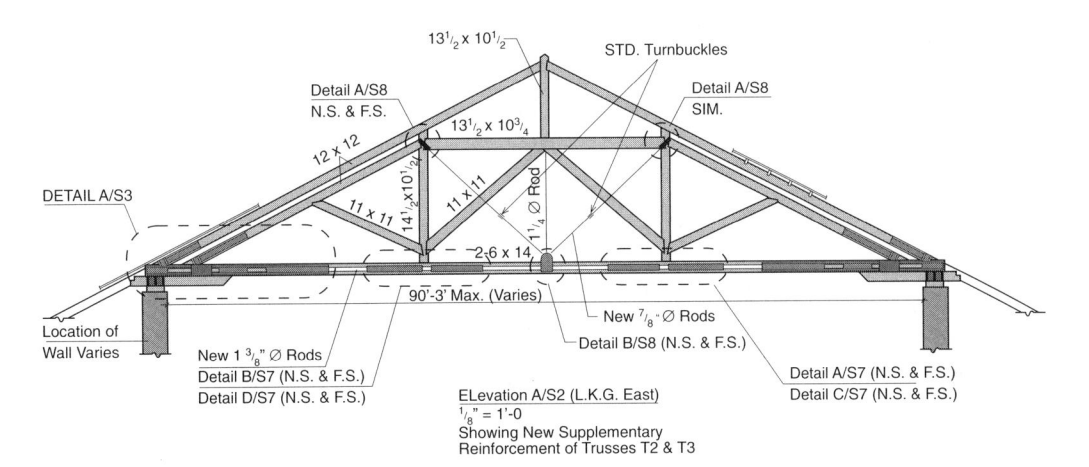

DETAIL A/S3

13½ x 10½
STD. Turnbuckles
Detail A/S8 N.S. & F.S.
Detail A/S8 SIM.
13½ x 10¾
12 x 12
14½ x 10½
11 x 11
11 x 11
1¼ Ø Rod
11 x 11
2-6 x 14
90'-3" Max. (Varies)
New ⅞" Ø Rods
Detail B/S8 (N.S. & F.S.)
Location of Wall Varies
New 1 ⅜" Ø Rods
Detail B/S7 (N.S. & F.S.)
Detail D/S7 (N.S. & F.S.)
ELevation A/S2 (L.K.G. East)
⅛" = 1'-0
Showing New Supplementary Reinforcement of Trusses T2 & T3
Detail A/S7 (N.S. & F.S.)
Detail C/S7 (N.S. & F.S.)

3.7 Footbridges for the Appalachian Trail were designed at Drexel University (Barbato et al., 1985) in a senior design project. This "how to" manual is a guide to inexperienced volunteer groups who can build 5 to 18 m footbridges such as the model shown below over the many streams which the trail crosses. The design needs to be verified and a factor of safety obtained. A $1/12$-scale ultimate-strength model of the 3.7-m (24') truss (see figure) is proposed to be built and tested in the laboratory. The prototype lumber is pressure treated. All prototype joints are formed with marine plywood gusset plates which are glued with a Weldwood® resorcinol glue and nailed to the truss members. Design the model to be tested to ultimate by a uniform load representing a group of scouts watching a canoe race.

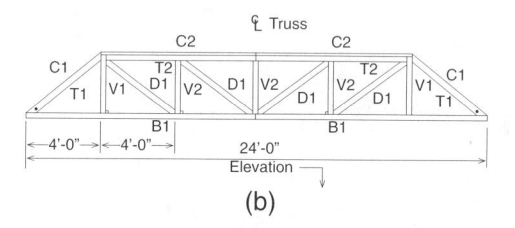

(a)

(b)

3.8 A red oak timber beam of rectangular cross Section 150×300 mm carries a concentrated load of 20 kN at 1.8 m from the left end. If the simply supported span is 5 m, establish the design conditions and write the prediction equation for the deflection at any point if a $1/10$-scale balsa wood model is to be tested.

3.9 A series of buckling tests were performed on long pin-ended balsa wood model columns. The buckling load, Pcr, in such columns is elastic and can be determined by the Euler formula:

$$Pcr = \frac{\Pi^2 EI}{L^2}$$

where E is the Young's modulus, I is the weakest moment of inertia of the section, and L is the length between pin supports. Note that in order to compare the test results with the Euler equation, E of the balsa wood is required.

Answer the following:
   a. Is Euler's formula dimensionally homogeneous?
   b. Set up an experiment to determine E from a statistical sample of extra model columns identical to those used in the buckling tests. Show sketches and describe how you would go about obtaining E by at least two independent methods. Which of the methods is the more reliable?

3.10 A timber beam of rectangular cross section has actual dimensions of 150 mm width and 300 mm depth. It carries a concentrated load of 25 kN at 1.5 m from the left end of its 4-m span. Establish the design conditions and write the prediction equation for the deflection at any point along its length if a plastic beam of an elastic modulus of 3 GPa and 200-mm span is to be used as a model. Assume that the elastic modulus of the wood is 12 GPa.

REFERENCES

American Society for Testing and Materials (1987). Standard Methods of Testing Small Clear Specimens of Timber, D-143, in *Annual Book of ASTM Standards,* Vol. 4.09, Philadelphia.

Balint, P. S. and Shaw, F. S. (1965). Structural model of Australia Square Tower in Sydney, *Archit. Sci. Rev.,* Sydney, 8, 136–149.

Barbato, J. L., Diewald, T. L., and Muir, J. S. (1985). Appalachian Trail Conference Foot Bridge Design Manual, Senior Design Project, Department of Civil and Architectural Engineering, Drexel University, Philadelphia, June.

Beggs, G. E. (1932). An accurate mechanical solution of statically indeterminate structures by use of paper models and special gages, *Proc. Am. Concr. Inst.,* 18, 58–82.

Bodig, J. and Jayne, B. A. (1982). *Mechanics of Wood and Wood Composites,* Van Nostrand Reinhold, New York.

Carpenter, J. E. (1963). Structural model testing — compensation for time effect in plastic, *J. PCA Res. Dev. Lab.* Skokie, IL., 5(7), 47–61.

Carpenter, J. E., Magura, D. D., and Hanson N. W. (1964). Structural Model Testing–Load Distribution in Concrete I-Beam Bridges, Development Department Bulletin D94, Portland Cement Association, Skokie, IL.

Carpenter, R. C. (1917). The properties of balsa wood (*Ochroma Lagopus*), *Trans. Am. Soc. Civil Eng.,* 81, 125–160.

Doyle, D. V., Drow, J. T., and McBurney, R. S. (1945). Elastic properties of wood — the Young's modulus and Poisson's ratio of balsa and quipo, U.S. Department of Agriculture, Forest Products Laboratory Report No. 1528.

Fam, A. R. M. (1973). Static and Free Vibration Analysis of Curved Box Bridges, Ph.D. thesis, McGill University, Montreal.

Fialho, J. F. L. (1960 and 1962). The use of plastics for making structural models, *RILEM Bull.,* New Series No. 8 (September 1960) and Technical Paper No. 185, Laboratorio Nacional de Engenharia Civil, Lisbon, Portugal, 1962.

Forest Products Laboratory: (1955). *Wood Handbook,* U.S. Department of Agriculture Handbook No. 72, U.S. Government Printing Office, Washington, D.C.

Fumagalli, E. (1973). *Statical and Geomechanical Models,* Springer-Verlag, New York, 182 pp.

Gero, J. E. and Cowan, H. J. (1970). Structural concrete models in Australia, in *Models of Concrete Structures,* ACI, SP-24, American Concrete Institute, Detroit, MI, 353–386.

Gopu, V. K. A. and Goodman, J. R. (1975). Full-scale tests on tapered and curved glulam beams, *J. Struct. Div.,* American Society of Civil Engineers, 101(ST12), December.

Gutkowski, R. M., Dewey, G. R., and Goodman, J. R. (1982). Full-scale tests on single-tapered glulam beams, *J. Struct. Div., ASCE,* Vol. 108(ST10), October.

Harris, H. G., Pahl, P. J., and Sharma, S. D. (1962). Dynamic Studies of Structures by Means of Models, Report R63-23, Department of Civil Engineering, Massachusetts Institute of Technology, Cambridge.

Heckler, G. F. (1980). The Analysis, Fabrication and Testing of a Single Fink Truss, a Modified Single-Fink Truss and a Raised-Collar Truss and Similitude Study, term project, Course G218, *Model Analysis of Structures,* Drexel University, Philadelphia, January.

Hendry, A. W. (1964). *Elements of Experimental Stress Analysis,* Pergamon Press, London, 193 pp.

Johnson, A. E. Jr. and Homewood, R. H. (1961). Stress and deformation analysis from reduced scale plastic model testing, *Proc. Soc. Exp. Stress Anal.,* 18(2).

Johnson, J. W. (1973). Efficient fabrication of glue-laminated timbers, *J. Struct. Div., ASCE,* 99(ST3), March.

Kinney, G. F. (1957). *Engineering Properties and Applications of Plastics,* John Wiley & Sons, New York, 278 pp.

Kopatz, K. W. (1982). Deflection Investigation of the $^{30}/_{20}$ GHz Antenna Using a Wooden Model. term project, Course G218, *Model Analysis of Structures,* Drexel University, Philadelphia, May 5.

Kopatz, K. W. (1984). Model Analysis of Double Tapered Glulam Wooden Beams, term project, Course G280, *Experimental Analysis of Nonlinear Structures,* Drexel University, Philadelphia, Sept. 26.

Kordina, K. (1964). The influence of creep on the buckling load of shallow cylindrical shells — preliminary tests, in *Non-Classical Shell Problems,* North-Holland, Amsterdam, 602–608.

Litle, W. A. (1964). *Reliability of Shell Buckling Predictions,* Research Monograph No. 25, Massachusetts Institute of Technology, Cambridge, 149.

Litle, W. A. and Hansen, R. J. (1963). The use of models in structural design, *J. Boston Soc. Civil Eng.,* 50(2), 59–94.

Litle, W. A., Hansen, R. J., et al. (1965). Notes of the Special Summer Course on Structural Models, Massachusetts Institute of Technology, Cambridge.

McLeish, R. D. and Soden, P. D. (1968). A design exercise using balsa wood, *Bull. Mech. Eng. Educ.,* 7, 63–68.

Norris, C. H. and Wilbur, J. B. (1976). *Elementary Structural Analysis,* McGraw-Hill, New York.

Pahl, P. J. and Soosaar, K. (1964). Structural Models for Architectural and Engineering Education, Report No. R64-3, Department of Civil Engineering, Massachusetts Institute of Technology, Cambridge, 269 pp.

Pippard, A. J. (1947). *The Experimental Study of Structures,* Edward Arnold, London, 29.

Preece, B. W. and Davies, J. D. (1964). *Models for Structural Concrete, C. R.* Books, London, England, 252 pp.

Rocha, M. (1961). Determination of thermal stresses in arch dams by means of models, RILEM *Bull.* (Paris), (10), 65.

Rocha, M. and Borges, J. F. (1961). Photographic method for model analysis of structures, *Proc. Soc. Exp. Stress Anal.,* 8(2), 129–142.

Roll. F. (1968). Materials for structural models, ASCE *J. Struct. Div.,* 94(ST6), 1353–1382.

Roll, F. and Aneja, I. K. (1966). Model tests of box-beam highway bridges with cantilevered deck slabs, paper presented at the October 17–21, ASCE Transportation Engineerng Conference held at Philadelphia (Preprint No. 3905).

Ross, A. D., (1946). The effect of creep on instability and indeterminacy investigated by plastic models, *Struct. Eng.* (London), 24(8), August, 413–428.

Rowe, R. E. (1960). Works on models in the Cement and Concrete Association, *J. PCA Res. Dev. Lab.,* 2(1), 4–10.

Simonetti, R. K. (1982). Model Analysis of a Steel Walkway System Using Balsa Wood Models, term project, Course G218, *Model Analysis of Structures,* Drexel University, Philadelphia, April 25.

Soden, P. D. and McLeish, R. D. (1976). Variables affecting the strength of balsa wood, *J. Strain Anal.,* 11(4), 225–234.

Struminski. E. S. (1971). Low Cycle Fatigue Study of Fiberglass Reinforced Plastic Laminates, M. Eng. thesis, McGill University, Montreal.

Tabba, M. M. (1972). Free Vibrations of Curved Box Girders, M. Eng. thesis, McGill University, Montreal.

Wilbur, J. B. and Norris, C. H. (1950). Structural model analysis, in *Handbook of Experimemtal Stress Analysis,* M. Hetenyi, Ed., John Wiley & Sons, New York, chap. 15, 663–699.

Yates, J. G., Lucas, D. H., and Johnston, D. L. (1953). Pulse-excitation of resistance strain gages for dynamic multi-channel observation, *Proc. Soc. Exp. Stress Anal.,* 11(1).

CHAPTER 4

Inelastic Models: Materials for Concrete and Concrete Masonry Structures

CONTENTS

4.1	General	130
4.2	Prototype and Model Concretes	130
4.3	Engineering Properties of Concrete	131
	4.3.1 Prototype and Model Concretes — Effect of the Microstructure	132
4.4	Unconfined Compressive Strength and Stress–Strain Relationship	133
	4.4.1 Prototype Concrete	133
	4.4.2 Model Concrete	134
	4.4.3 Comparison of Prototype and Model Concrete Stress–Strain Characteristics	135
	4.4.4 Creep and Creep Recovery of Concrete	136
	4.4.5 Effect of Aggregate Content	138
	4.4.6 Effect of Strain Rate	139
	4.4.7 Moisture Loss Effects	139
	4.4.8 Strength–Age Relations and Curing	140
	4.4.9 Statistical Variability in Compressive Strength	142
4.5	Tensile Strength of Concrete	142
4.6	Flexural Behavior of Prototype and Model Concrete	146
	4.6.1 Specimen Dimensions and Properties	147
	4.6.2 Stress–Strain Curves	147
	4.6.3 Observed Variations in Modulus of Rupture with Changes in Dimensions	147
	4.6.4 Rate of Loading	148
	4.6.5 Influence of Strain Gradient	148
4.7	Behavior in Indirect Tension and Shear	148
	4.7.1 Tensile Splitting Strength	149
	4.7.2 Results of Model Split Cylinder Tests	151
	4.7.3 Tensile Splitting Strength vs. Age	152
	4.7.4 Correlation of Tensile Splitting Strength to Flexural Strength	152
4.8	Design Mixes for Model Concrete	153
	4.8.1 Introduction	153
	4.8.2 Choice of Model Material Scale	153
	4.8.3 Properties of the Prototype to Be Modeled	155
	4.8.4 Important Parameters Influencing the Mechanical Properties of Concrete	155
4.9	Summary of Model Concrete Mixes Used by Various Investigators	159

4.10	Gypsum Mortars	165
	4.10.1 Curing and Sealing Procedures	168
	4.10.2 Mechanical Properties	169
4.11	Modeling of Concrete Masonry Structures	170
	4.11.1 Introduction	170
	4.11.2 Material Properties	170
	4.11.2.1 Prototype Masonry Units	170
	4.11.2.2 Model Masonry Units	171
	1. One-Fourth-Scale Blocks	171
	2. One-Third-Scale Blocks	173
	3. One-Third-Scale Interlocking Mortarless Blocks	175
	4.11.2.3 Model Mortars	181
	4.11.2.4 Model Grout	185
4.12	Strength of Model Block Masonry Assemblages	188
	4.12.1 Axial Compression	189
	4.12.2 Flexural Bond	192
	4.12.3 Bed Joint Shear	194
	4.12.4 In-Plane Tensile Strength	196
	4.12.5 Out-of-Plane Flexural Tensile Strength	199
	4.12.6 Diagonal Tension (Shear) Strength	202
4.13	Summary	202
Problems		202
References		205

4.1 GENERAL

Direct models of reinforced and prestressed concrete structures require the use of materials to simulate the reinforcement or prestressing tendons, the prototype concrete, and the bond strength and dowel action at the reinforcement–concrete interface. Similarly, in modeling of block masonry structures, it is important to reproduce the tensile strength of the mortar along with the size effects of the stressed volume. It must be noted that these direct models are miniature prototypes and require construction techniques that should be as close as possible to those used in the construction of the prototypes. Therefore, the success of direct models of structural concrete and masonry structures depends upon the degree of success (or accuracy) with which the models engineer can simulate the relevant prototype material properties and loading and environmental conditions.

This chapter emphasizes the modeling of prototype materials used in the construction of reinforced and prestressed concrete and block masonry structures. Modeling of steel and the new fiber-reinforced polymer (FRP) reinforcement along with the associated phenomena such as bond strength are discussed in Chapter 5. Principal characteristics of both prototype and model materials and their various strengths are presented along with a discussion of the influence of different parameters on these strengths.

4.2 PROTOTYPE AND MODEL CONCRETES

Prototype and model concretes consist of a mixture of inert granular substances held together by a cementing agent. More specifically, prototype concrete is a combination of cement, water, coarse and fine aggregates, and possibly admixtures, while model concrete normally consists of fine aggregates (and sometimes fine gravel or crushed stone), cement, water, and possibly admixtures.

Aggregates are hard, chemically inert materials in the form of graded fragments. The most common aggregates are crushed stone, gravel, and sand, although other materials are frequently used for both prototype and model concretes. For example, expanded shales, slates, slags, and clays have been used in making lightweight prototype concrete. Gravels and sands have specific gravities of about 2.7 as compared with a specific gravity of about 1.7 or less for expanded shales. Any aggregate passing the U.S. No. 4 sieve (0.187-in. or 4-mm mesh) is designated as fine *aggregate,* and that which is retained on the sieve is known as *coarse aggregate*. The aggregate grading can influence the mechanical behavior of the resulting concrete. Well-graded aggregates yield the minimum void space and hence require the least amount of cement paste. Well-graded aggregates not only lead to economy in the use of cement, but also yield concrete having maximum strength and minimum volume change due to drying shrinkage.

The maximum aggregate size for prototype concrete is dependent upon the type of construction in which the concrete is to be used. The American Concrete Institute (ACI) recommends maximum aggregate sizes from 10 to 150 mm; 20 to 40 mm are the most commonly used in building construction. There are no universally accepted rules to determine the maximum aggregate size for model concrete, but it is normally established from the model geometric scale, the minimum member thickness in the model, and the reinforcement spacing. A survey of the literature shows that the maximum aggregate size used in model concrete ranges from 10 mm or 6 mm for ½- to ⅓-scale models to the U.S. No. 4 sieve (0.187-in. or 4-mm mesh) for ⅙- to ¹⁄₁₀-scale models. It should be kept in mind that the largest possible size aggregate should always be utilized in order to minimize the relatively high tensile strength of the model concrete.

Portland cement is used as a binder in most prototype and model concretes. It consists of four principal compounds: tricalcium silicate ($3CaO \cdot SiO_2$), dicalcium silicate ($2CaO \cdot SiO_2$), tricalcium aluminate ($3CaO \cdot Al_2O_3$), and tetracalcium aluminoferrite ($4CaO \cdot Al_2O_3 \cdot Fe_2O_3$) with a specific gravity of about 3.15. When Portland cement is mixed with increasing amounts of water, it gradually becomes pasty and then a rather viscous liquid with considerable adhesion. After about 1 h, the liquid begins to stiffen, and after 6 to 10 h, it is hard or fully set. This hydration process can take place under water, and for this reason Portland cement is also known as a *hydraulic material*. No special consideration needs to be given to the water used for mixing concrete; a part of this water is used for the hydration of cement particles, while the remainder assists in generating a mix with reasonable workability for easy placing. Generally, ordinary water that is fit for drinking purposes is entirely satisfactory.

Often it is desired to improve the workability, to increase or decrease the setting time, to increase the strength, or to decrease the porosity of the cement–aggregate mixture over that which would naturally be obtained. Small quantities of materials called admixtures, such as calcium chloride, acetic acid, or an air-entraining agent, may be added in controlled quantities at the time of mixing or included in the cement itself, with the express purpose of increasing workability, etc. or bringing about other desired changes. The air-entraining agents are the most commonly used admixtures.

4.3 ENGINEERING PROPERTIES OF CONCRETE

Concrete is a unique construction material and possesses properties that are not common to other materials; for example, the tensile strength of concrete is less than its shear strength, which in turn is less than its compressive strength. Consequently, prototype concrete cannot normally be replaced by any other material in ultimate-strength models. The engineering properties of the hardened cement–aggregate mass that comprises prototype and model concretes are discussed extensively in the literature Popovics, 1980; Neville, 1987; these are dependent on several factors, including:

1. Water–cement ratio.
2. Cement–aggregate ratio.
3. The nature of the aggregates, i.e., size, hardness, gradation, porosity, surface texture, etc.
4. Type of cement. Ordinary Portland cement Type I and rapid-hardening Portland cement Type III (high early-strength cement) are commonly used in the construction of both prototype and model structures. Rapid-hardening cement develops strength more rapidly; for the same water–cement ratio, the strength developed at the age of 3 days is of the same order as the 7-day strength of ordinary Portland cement. The increased rate of strength gain with rapid-hardening Portland cement is due to a higher tricalcium silicate content and the grinding of the cement clinker (Neville, 1987). Use of rapid-hardening Portland cement is particularly important for model concrete as it decreases the laboratory time requirements.
5. Time history of moisture available for reaction with the cement, and time history of the temperature during this curing period.
6. Moisture content and temperature during testing.
7. Age at testing.
8. Type of stress caused by the applied loading: tension, bending, uniaxial compression, biaxial compression, triaxial compression, etc.
9. Duration of loading.
10. Strain rate of loading.

The effect of these factors and their interaction is not yet completely understood. In spite of significant research efforts since the early part of the century, the complete load–time–deformation behavior of concrete is one of the least well understood of all the common construction materials. These technical shortcomings are compounded by the fact that structural concrete is mixed, placed, and cured under a wide variety of conditions. Consequently, it is difficult to obtain a model material to simulate the prototype concrete by scaling the individual components according to the laws of similitude. Moreover, the physical and chemical processes at the molecular level cannot be scaled down. However, these limitations are not serious, as long as the physical properties of the model concrete, including its stress–strain curve and the failure criteria, are compatible with those of the prototype concrete according to the laws of similitude discussed in Chapter 2. Model concrete and gypsum mortar are commonly used in ultimate-strength model studies basically because of their rheological similarity to the prototype concrete. A knowledge of the relationship between the water–cement or water–plaster ratio and strength is important in the selection of a trial mix for model concrete and is discussed in Section 4.8.

4.3.1 Prototype and Model Concretes — Effect of the Microstructure

A close examination of the structure of concrete, mortar, and cement paste reveals an interesting analogy (Newman, 1965). Model concrete is basically concrete, but on a reduced scale of at least one order of magnitude. The stiff, coarse aggregate in a softer matrix of mortar in concrete is analogous to the hard sand particles in a softer matrix of cement paste of model concrete. A second-order magnification of the hardened cement paste structure reveals the same qualitative composition that is found in concrete and mortar of relatively hard unhydrated cement particles in a matrix of cement gel. These similarities can lead to theories of two-phase models of the behavior of these systems as a first approximation by excluding the void phase. Yet, marked differences in behavior under load are observed when cement paste, mortar, and concrete of the *same consistency* and normal compositions are compared:

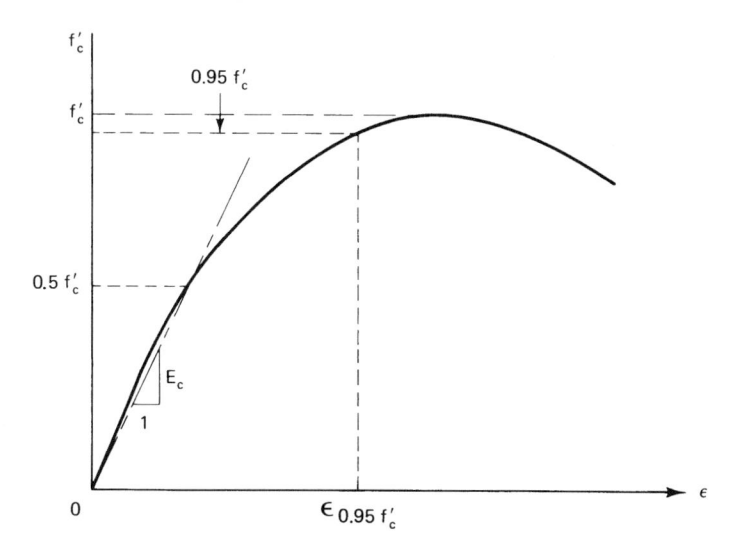

INELASTIC MODELS: MATERIALS FOR CONCRETE AND CONCRETE MASONRY STRUCTURES

Figure 4.1 Typical stress–strain behavior of concrete in uniaxial compression.

1. Cement, mortar, and concrete show the same elastic portion on their stress–strain diagram.
2. The stress–strain curve for concrete starts to deviate from a straight line at a lower stress than either mortar or paste.
3. The paste shows a linear stress–strain curve up to a very sudden and brittle failure.
4. Both mortar and concrete show gradual curving or nonlinear stress–strain curves with some ductility and warning of failure.

The addition of aggregate to the cement paste creates a heterogeneous system of a complex nature that behaves very differently from the paste under load (Hsu and Slate, 1963). Consistency is not a good basis for comparing these systems, however, and, as will be shown in subsequent sections, the effect of the volume of aggregate in the system plays an important role (Ruiz, 1966). In fact, when the volume of aggregates is less than a certain critical value, both mortars and concretes are stiffer than a paste made with the same water–cement ratio (Gilkey, 1961). When this critical volume is exceeded, as is the case with most structural concretes, there is a very sharp decrease in strength and stiffness, with the net result that the paste is stronger than the mortar, which is in turn stronger than the concrete.

4.4 UNCONFINED COMPRESSIVE STRENGTH AND STRESS–STRAIN RELATIONSHIP

4.4.1 Prototype Concrete

The unconfined compressive strength is considered to be the most important property of structural concrete. The stress–strain curve is reasonably linear at low stress levels and the modulus of elasticity E_c is usually taken as the slope of the tangent through the origin at the initial portion of the stress–strain curve (Figure 4.1).

Several different expressions are available for the modulus of elasticity. The ACI Building Code 318-95 uses the expression:

$$E_c = 33w^{1.5}\sqrt{f'_c} \qquad (4.1a)$$

where w is the unit weight of concrete in pounds per cubic foot, f'_c is the compressive strength in pounds per square inch (psi), and E is the modulus of elasticity in pounds per square inch. In SI units

$$E_c = 0.043 w^{1.5} \sqrt{f'_c} \qquad (4.1b)$$

where w is the concrete mass density in kilograms per cubic meter (kg/m³), f'_c is the compressive strength in megapascals (MPa), and E_c is the modulus of elasticity in megapascals.

4.4.2 Model Concrete

The response of any model concrete to external compressive load is of primary importance, because concrete as a structural material is used mainly to carry compressive loads. The behavior of the prototype concrete under compression is normally all that is available when a model substitute is desired. For these reasons the stress–strain behavior in uniaxial compression becomes of primary importance in correlating material similitude between the prototype and the model structure. Model concrete mixes must be designed carefully to simulate not only the strength and the stress–strain curve in compression but also other important properties such as the tensile strength, time-dependent behavior, etc. Design of model concrete mixes is presented in Section 4.8. The compressive strength of model concrete is determined using cylindrical specimens with a length to diameter (L/D) ratio of 2, similar to that used for prototype cylindrical specimens, although Harris et al. (1966) showed that in order to determine the true uniaxial compressive strength, an L/D ratio of at least 2.5 is needed.

The reader is cautioned that all material properties described and illustrated in the various figures herein should be considered to give general trends, and therefore this information can be used as a guide for initiating work on local materials to establish their properties and suitability for use as model materials. Such a study of materials must always be undertaken before initiating any model investigation.

A study of the literature on model concretes reveals that different sizes of cylinders ranging from 12.5 × 25 mm to 150 × 300 mm have been used by various investigators. It will be shown in Chapter 9 that for the same concrete mix, the observed concrete strength increases as the test specimen size is decreased. A typical question that must be answered is: What is for example the strength of concrete in a 12.5-mm-thick reinforced concrete shell model? One should probably attempt to obtain this strength from uniaxial compression tests on 12.5 × 25 mm cylinders; however, because of the very small specimen size, such tests are difficult to perform. The ACI Committee 444, *Experimental Analysis for Concrete Structures*, recommends a 50 × 100 mm cylinder as a standard size along with any other size cylinder that the models engineer selects. This would assist with correlation of strength data on model concretes.

Typical stress–strain curves obtained with bonded electrical resistance strain gages on 25 × 50 mm microconcrete cylinders (Harris and White, 1972) are shown in Figure 4.2a. Also, stress–strain curves from carefully instrumented 25 × 50 mm microconcrete cylinders were studied by McDonald and Swartz (1981) as part of a study on mechanical properties. A mix of 0.5:1:2.2 water:cement:sand by weight was used with curing for 21 days in a moist room (21°C, 100% relative humidity). The cylinders were then removed and strain gages applied, a process taking 6 days. The cylinders were then placed in a free-water tank at 620.6 kPa for one day. They were then placed in the moist room for 4 days before removing and testing. Their investigation showed (Figure 4.2b) that very consistent results can be obtained within each sample.

Stress–strain curves obtained from tests on 50 × 100 mm cylinders of three model concrete mixes used at Cornell University (Harris et al., 1966) with compressive strength, f'_c ranging from 22 to 47 MPa are shown in Figure 4.3. The secant moduli, E_c, ranged from 14.5 to 23.5 GPa. Several other model concretes used at Cornell University followed the same general trends.

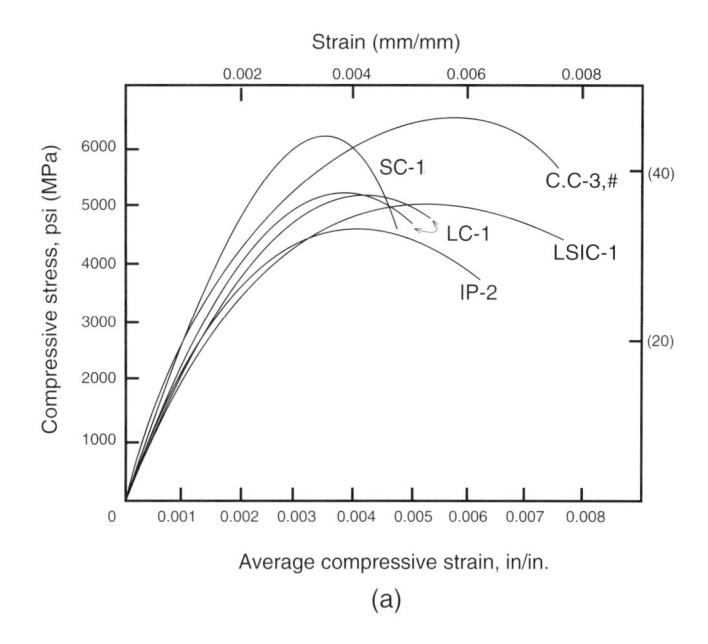

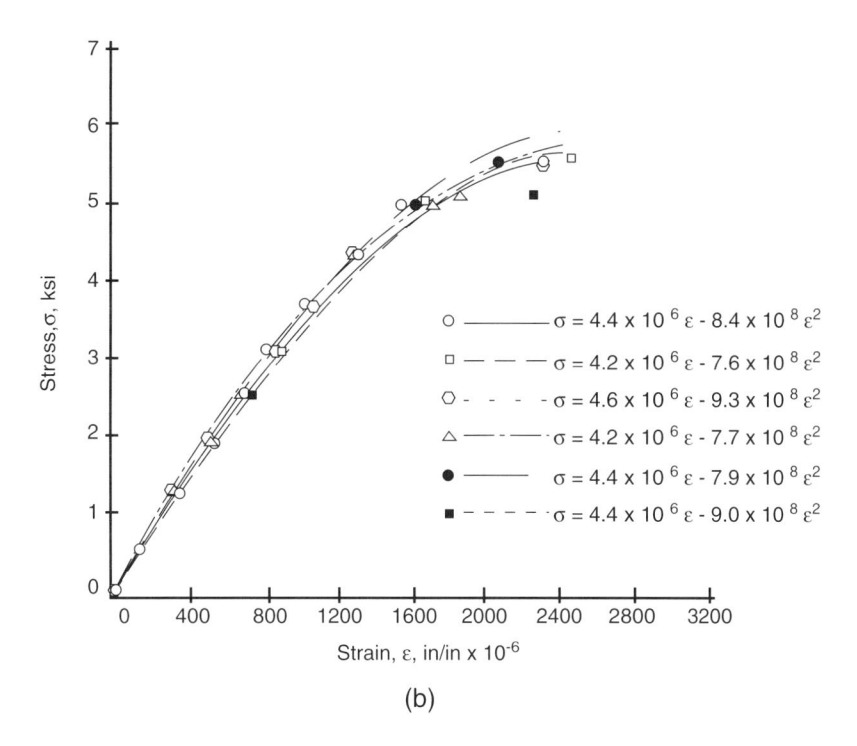

INELASTIC MODELS: MATERIALS FOR CONCRETE AND CONCRETE MASONRY STRUCTURES 135

Figure 4.2 Stress–strain curves of model concrete in compression from 25 × 50 mm cylinders. (Courtesy of (a) Cornell University, Ithaca, NY and (b) Kansas State University, Manhattan.)

4.4.3 Comparison of Prototype and Model Concrete Stress–Strain Characteristics

Several investigators have noted that the typical stress–strain behavior of model concrete cylinders with the same height–diameter ratio of 2 but of smaller dimensions than the standard 150

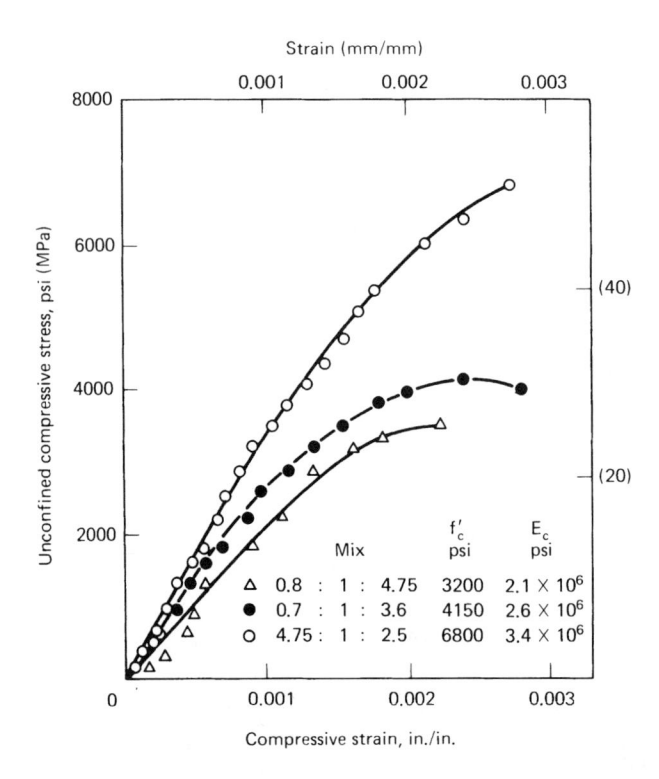

Figure 4.3 Compressive stress–strain curves for 2 × 4 in. microconcrete cylinders. (From Harris, H. G. et al., Rep. No. 326, Department of Structural Engineering, Cornell University, Ithaca, NY, September, 1966. With permission.)

× 300 mm cylinders is very similar to ordinary concrete of the same ultimate strength (Sabnis and Mirza, 1979). Figure 4.4 compares the stress–strain curve of a model concrete mix used at the Massachusetts Institute of Technology and a prototype concrete mix used at the University of Illinois (Johnson, 1962). Both mixes had an ultimate strength of 32.2 MPa, and the type of correlation generally expected is shown in Figure 4.4. There are usually some minor variations in the modulus and the ultimate strain with the variations in the compressive strength, but these are generally within an acceptable range. Sabnis and White (1967) compared the stress–strain curves of prototype and model concrete and gypsum mortar mixes (see Section 4.10) using a non- dimensional basis, as shown in Figure 4.5. Excellent correlation was noted, and the nondimensional stress–strain curves can be considered to be similar for the prototype and model concretes and the gypsum mortar mix. Syamal (1969) tested prototype cylinders 150 × 300 mm and their ½-, ¼-, and ⅙-scale models in compression. The nondimensional stress–strain curves obtained are shown in Figure 4.6, and it can be noted that these stress–strain curves are similar for the various sizes of cylinders, although there are slight variations in the moduli and the ultimate strains. For more details, see the paper by Mirza et al. (1972).

4.4.4 Creep and Creep Recovery of Concrete

When a concrete structure is loaded, the deformations continue to increase with time. Thus, although the initial deformations may be nearly elastic (Figure 4.1), the strains continue to increase even under constant stress. Such additional strains are referred to as *creep strains*. Some of the major factors that influence the creep characteristics of concrete are

1. Age of concrete at the time of loading
2. Water–cement ratio

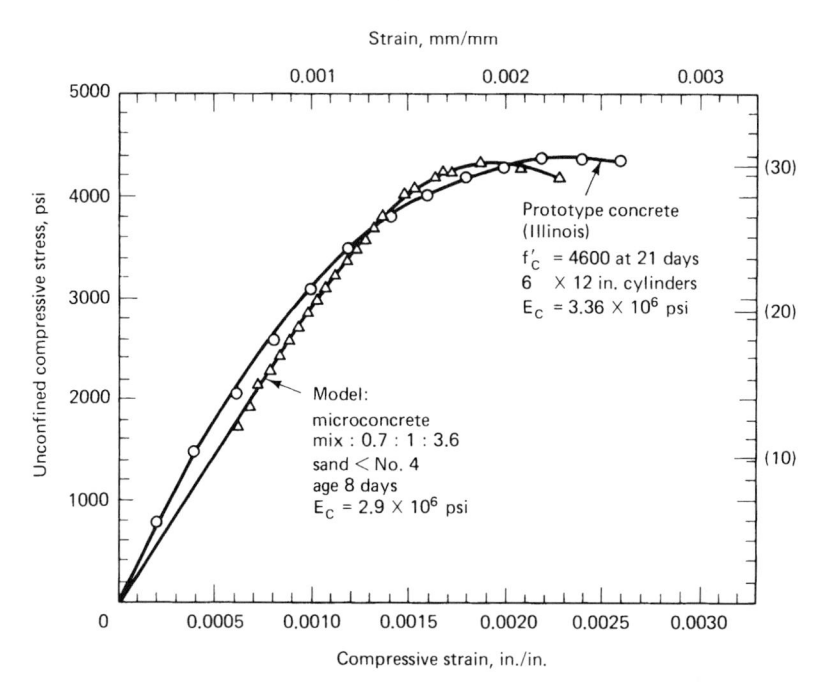

INELASTIC MODELS: MATERIALS FOR CONCRETE AND CONCRETE MASONRY STRUCTURES 137

Figure 4.4 Comparison of stress–strain behavior of actual concrete and microconcrete at about 1/10 scale.

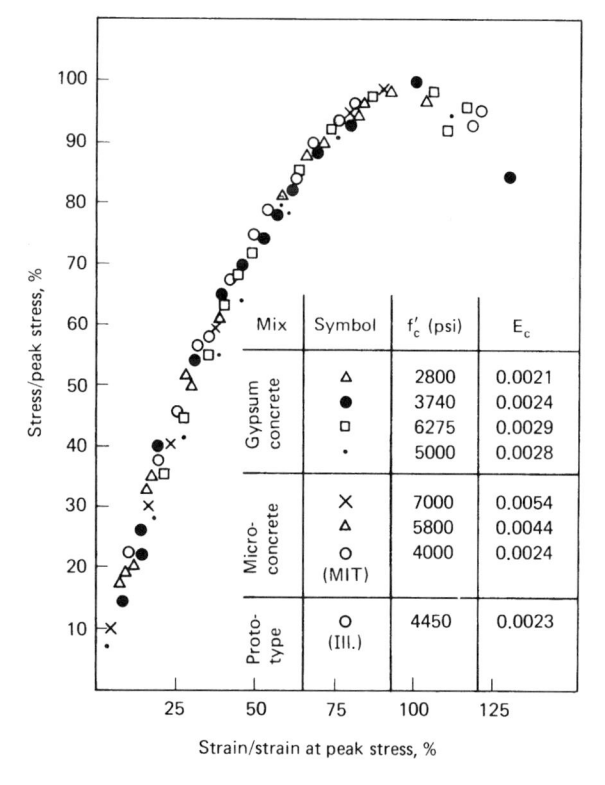

Figure 4.5 Comparison of compressive stress–strain curves for various concretes.

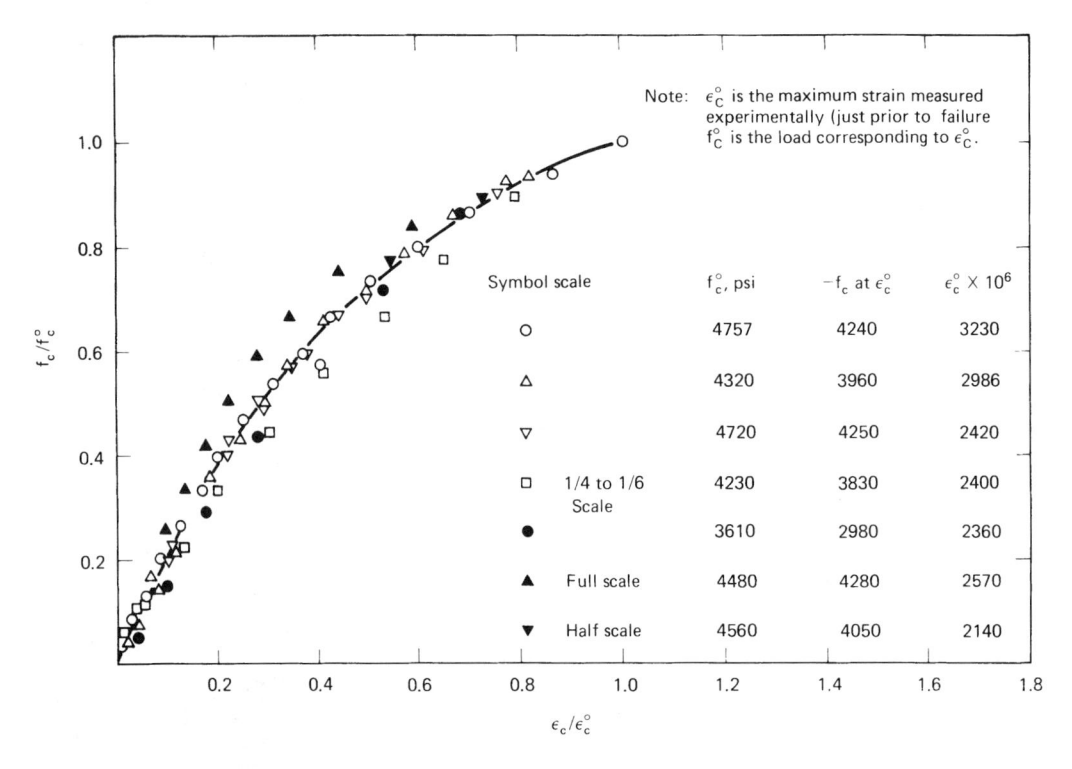

Figure 4.6 Nondimensional stress–strain curve for concrete in compression.

Symbol	scale	f_c^o, psi	$-f_c$ at ϵ_c^o	$\epsilon_c^o \times 10^6$
○		4757	4240	3230
△		4320	3960	2986
▽		4720	4250	2420
□	1/4 to 1/6 Scale	4230	3830	2400
●		3610	2980	2360
▲	Full scale	4480	4280	2570
▼	Half scale	4560	4050	2140

Note: ϵ_c^o is the maximum strain measured experimentally (just prior to failure f_c^o is the load corresponding to ϵ_c^o.

3. Relative humidity
4. Magnitude and time history of the applied stress system
5. Compressive strength f_c'

Results reported by Rüsch (1959) are shown in Figure 4.7, which indicates the effect of time on the stress–strain characteristics of concrete. This shows the significance of modeling the prototype behavior for corresponding loading durations. No systematic studies have been undertaken on model concretes; however, since moisture exchange with the environment has a strong influence on creep and since model concretes (in these small sections) can exchange moisture more readily, creep behavior of model concretes is expected to be different from that of the prototype concrete. Therefore, the shape of the modeled stress–strain curve from a cylinder test must be approached by the same type of loading history that was used on the prototype.

4.4.5 Effect of Aggregate Content

Experiments at Cornell University (Harris et al., 1966) showed that increasing the sand content in model concrete mixes caused the compressive strength to decrease, but the modulus of elasticity increased with an increase in the aggregate–cement ratio. The Cornell University results are presented in Figure 4.8 along with Newman's test data. This variation is due to the higher modulus of elasticity of the aggregate than that of the paste. However, the ultimate strength decreases with the increase of aggregate particles because they produce more stress concentrations and hence a greater probability of starting cracks at a given load level. The data from the Cornell tests using 25×50 mm model cylinders do not fall exactly on the curves given by Newman because his tests were on much larger specimens, $100 \times 100 \times 500$ mm prisms, and at a different age. The trend, however, is the same.

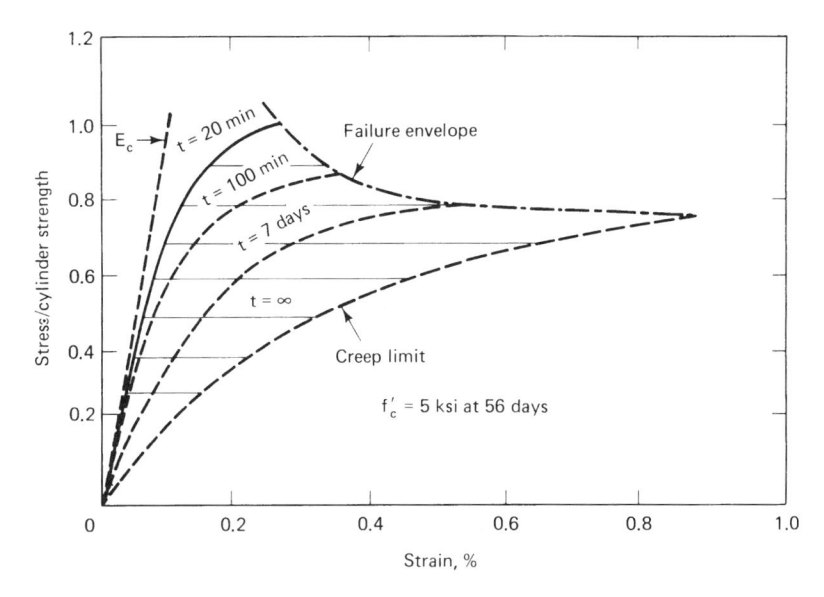

Figure 4.7 Stress–strain curves for various intensities and durations of sustained axial compressive loadings. (After Rüsch, H., 1959.)

The effect of sand concentration on the modulus of elasticity and flexural strength of mortar beams has been studied by Ishai (1961). He found that a critical volume concentration of sand exists above which the material loses strength and stiffness very abruptly. A study of the effect of sand concentration on the modulus of elasticity, the ultimate strength, and the strain at 95 on ultimate strength using square prismatic compressive specimens by Ruiz (1966) confirms Ishai's findings. The results, together with comparisons with some tests on small model cylinders, are shown in Figure 4.9. Ruiz investigated mixes with water–cement ratios of 0.4 and 0.6. The rest of the curves are extrapolated from these findings.

4.4.6 Effect of Strain Rate

The effect of increasing strain rate is shown in Figure 4.10 for ordinary concrete and model concrete using 50×100 mm cylinders. The rate of increase of compressive strength with increasing strain rate seems to be greater for the model material. However, the smaller cylinder used for the model material does have a higher apparent strength than the larger cylinder at strain rates in the range of what is generally considered a "short-time" strength. A constitutive model for the strain rate effect has been developed by Suaris and Shah (1985). More discussion on dynamic strain rate effects on model concretes is given in Chapter 11.

4.4.7 Moisture Loss Effects

It has been observed by various investigators that when wet, cured concrete cylinders are allowed to lose part of their moisture, their compressive strength increases. Increases of up to 25% have been observed when wet, cured specimens were allowed to remain at room conditions for a few days. In the case of model concrete structural elements, this partial drying during instrumentation and testing can be a cause of variations in strength. A method of coating the model that does not retard surface cracking and at the same time retains transparency is highly desirable. Shellac lacquer has reasonably good sealing qualities. It is very brittle in nature and therefore does not inhibit detection of crack formation.

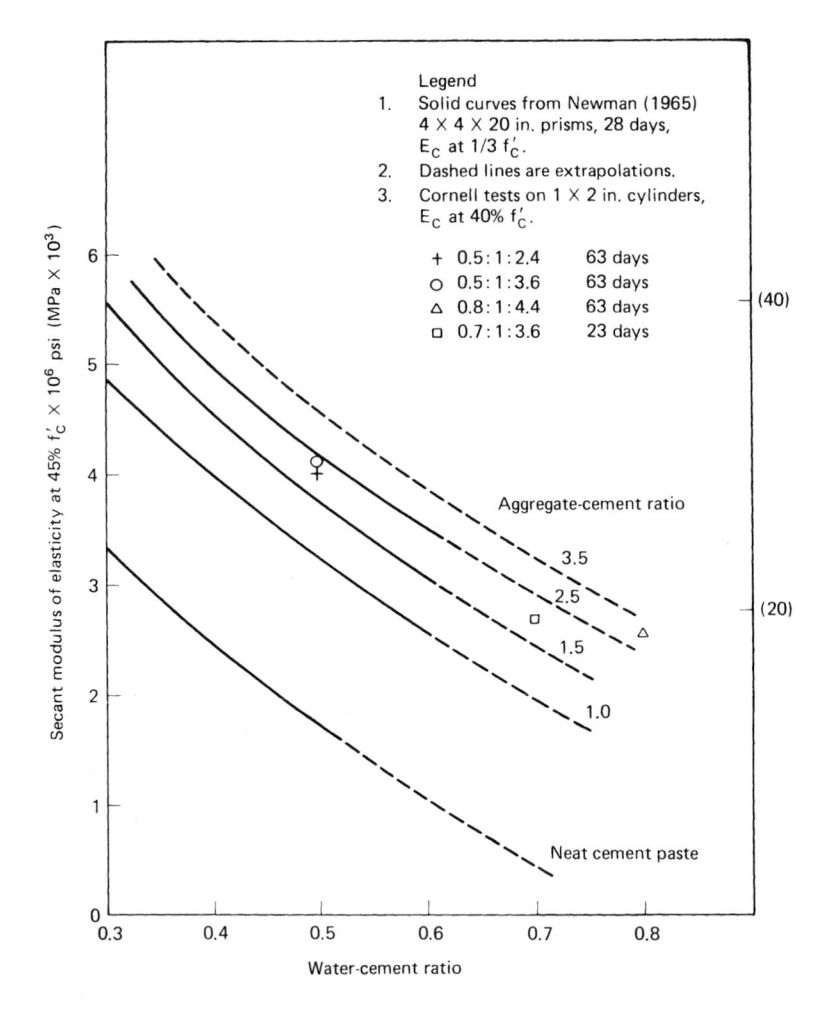

Figure 4.8 Effect of water–cement (W/C) and aggregate–cement (A/C) ratios on elastic properties of mortars. (After Newman, K., 1965.)

An attempt to relate the amount of moisture loss with time in a controlled environment (21°C and approximately 70% relative humidity) was made by Harris et al. (1966). Two groups of 25×50 mm compression cylinders, and one group of three $25 \times 38 \times 89$ mm prisms were used for this purpose. The results of these experiments are shown in Figure 4.11. The behavior of cylinders and prisms was essentially the same. Of the cylinders, the densest mix had the lowest moisture loss, and the mixes that were cured for longer periods of time also had lower moisture loss.

4.4.8 Strength–Age Relations and Curing

Temperature and moisture have pronounced effects on the strength development of model and prototype concretes. The development of strength stops at an early age when the concrete specimen is exposed to dry air with no previous moist curing. This can have special significance in model making, where strength variations during testing need to be minimized. This effect on 150×300 mm cylinders is shown in Figure 4.12; in the first 28 days the moist-cured specimen shows considerable increase in strength, and the air-cured specimen shows no effective increase in strength after 14 days. Figure 4.13 shows that for smaller elements used in model work the increase in strength is

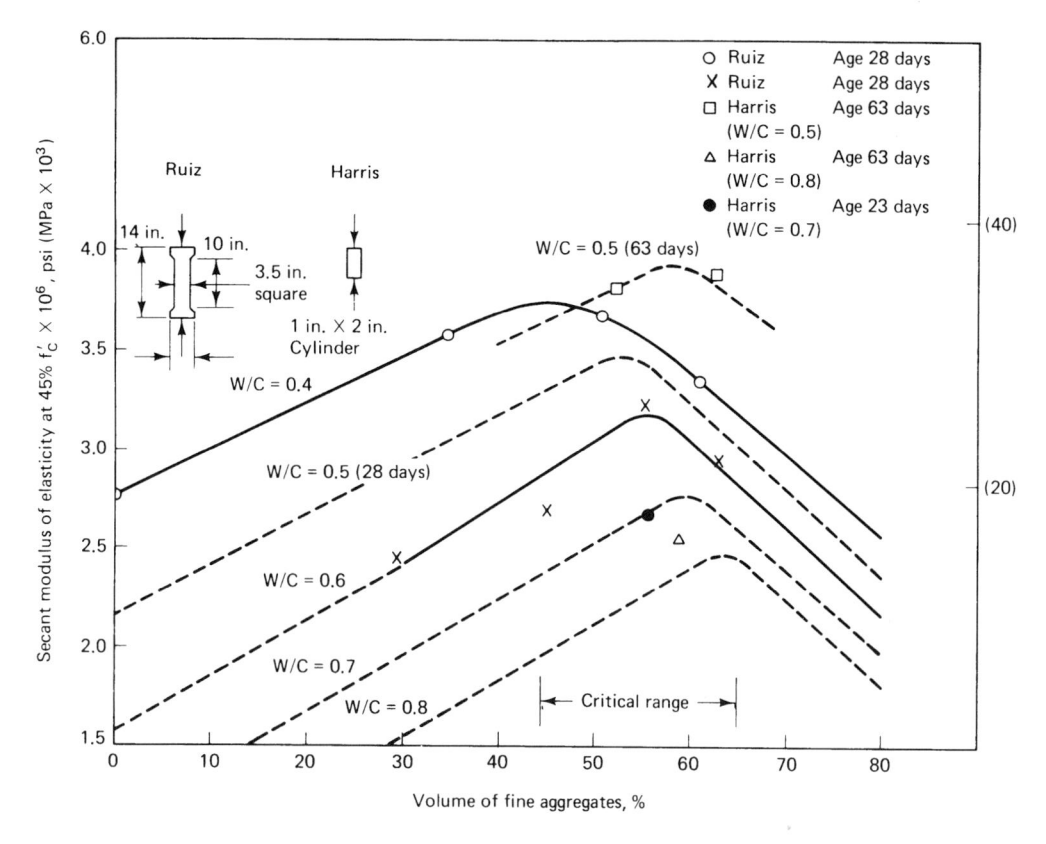

INELASTIC MODELS: MATERIALS FOR CONCRETE AND CONCRETE MASONRY STRUCTURES 141

Figure 4.9 Effect of aggregate on the modulus of elasticity.

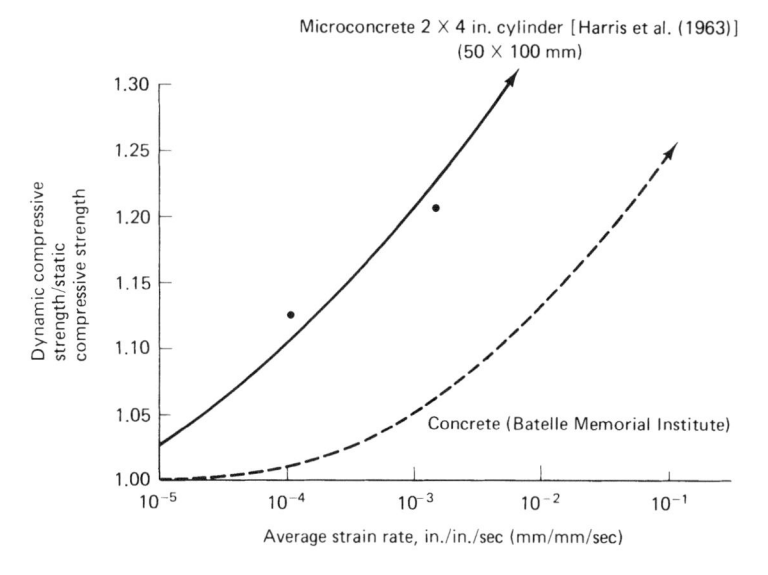

Figure 4.10 Effect of strain rate on compressive strength.

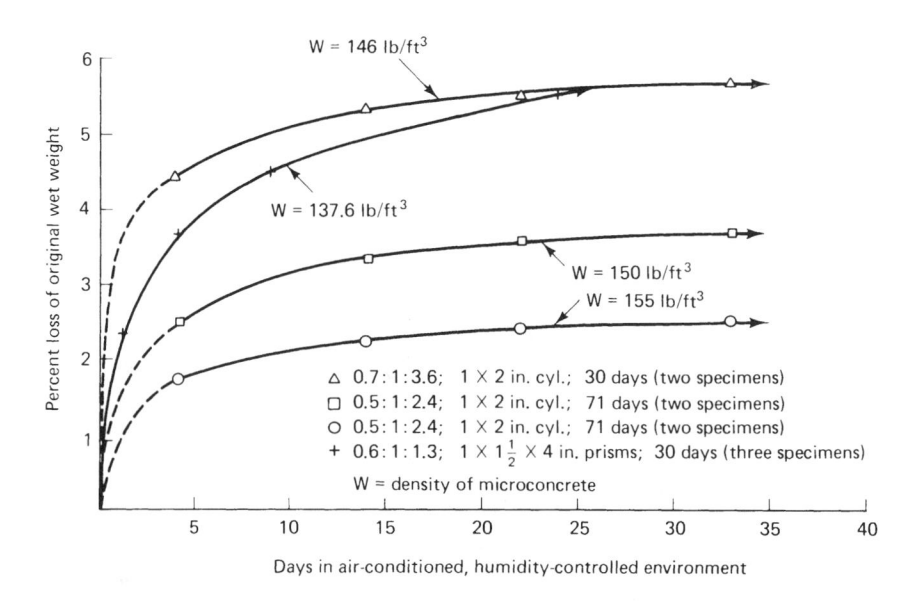

Figure 4.11 Rate of moisture loss in microconcrete in a controlled environment.

achieved in relatively shorter curing times, a desirable feature of the modeling technique. Curing temperatures have a marked effect on the strength development of concrete, as shown in Figure 4.14.

4.4.9 Statistical Variability in Compressive Strength

In order to study the statistical variation of the compressive strength of small cylinders, test series with large numbers of specimens were cast from the same mix and cured and tested in the same way at Cornell University. The standard deviation and coefficient of variation of each series were then computed. Standard deviations obtained from these model tests along with those reported by Rüsch (1964) are shown in Figures 4.15 and 4.16. For a large variety of concrete structures (Figure 4.15a and b) the expected standard deviations show that models of very small scales are feasible (Figure 4.16). The statistical variations of strength obtained from small model cylinders were not greater than 3.5 to 4.8 MPa. The variation of standard deviation with mean compressive strength of 25×50 mm model concrete cylinders (Figure 4.16) shows a trend similar to that for actual prototype structures.

It is known that concrete cast under laboratory conditions has lower values of standard deviation than concrete on actual jobs. Therefore, it can be argued that in modeling, one should use the small-size specimens to duplicate the actual prototype structure rather than those in the laboratory tests. As pointed out by Neville (1959), both the standard deviation and the mean strength decrease with increase in size of specimen, so that for the very small sizes it is to be expected that the spread in the measured strength as indicated by the standard deviation will be rather high. However, it has been shown by several investigators that with suitable care, the statistical variations of model concrete strengths can be made comparable to the variations observed on actual prototype structures.

4.5 TENSILE STRENGTH OF CONCRETE

The tensile strength of concrete is a fundamental property. The tensile stress or strain that can be developed in a prototype structure influences the behavior of the structure in many ways, including:

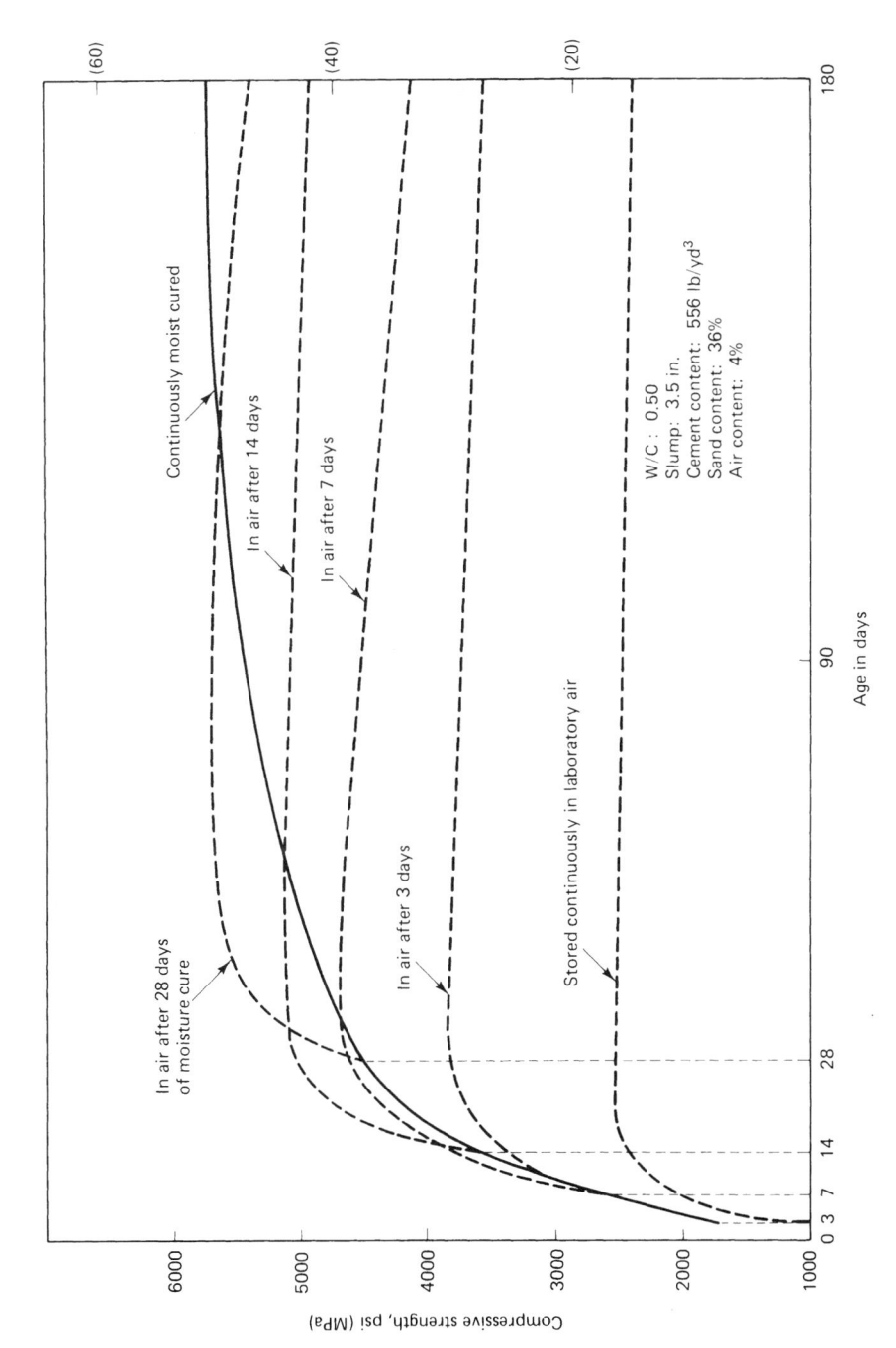

INELASTIC MODELS: MATERIALS FOR CONCRETE AND CONCRETE MASONRY STRUCTURES 143

Figure 4.12 Compressive strength of concrete dried in laboratory air after preliminary moist curing. (From Price, 1951).

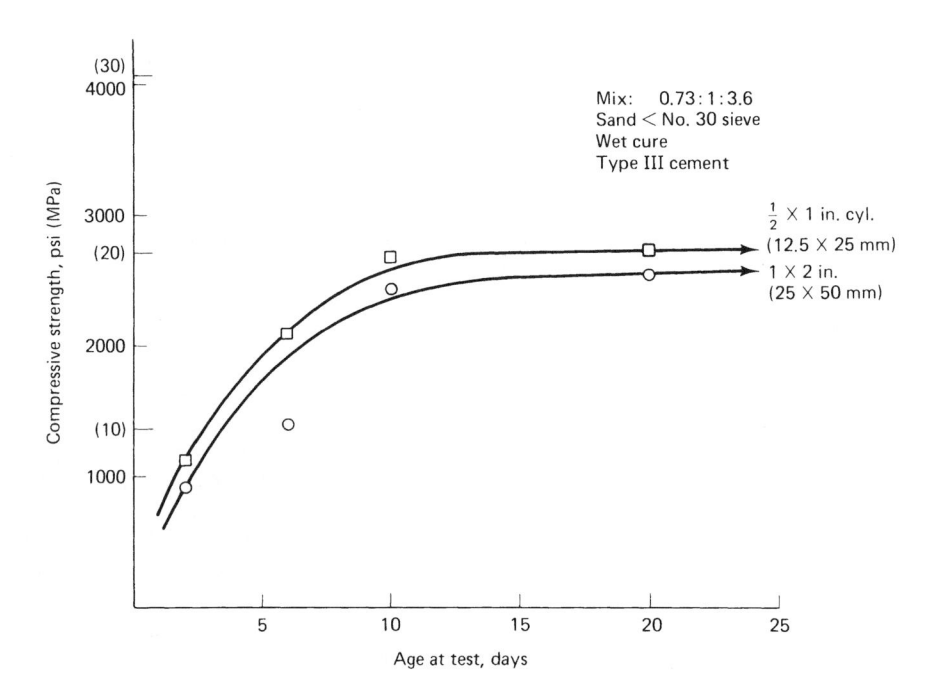

Figure 4.13 Strength–age curves of model cylinders.

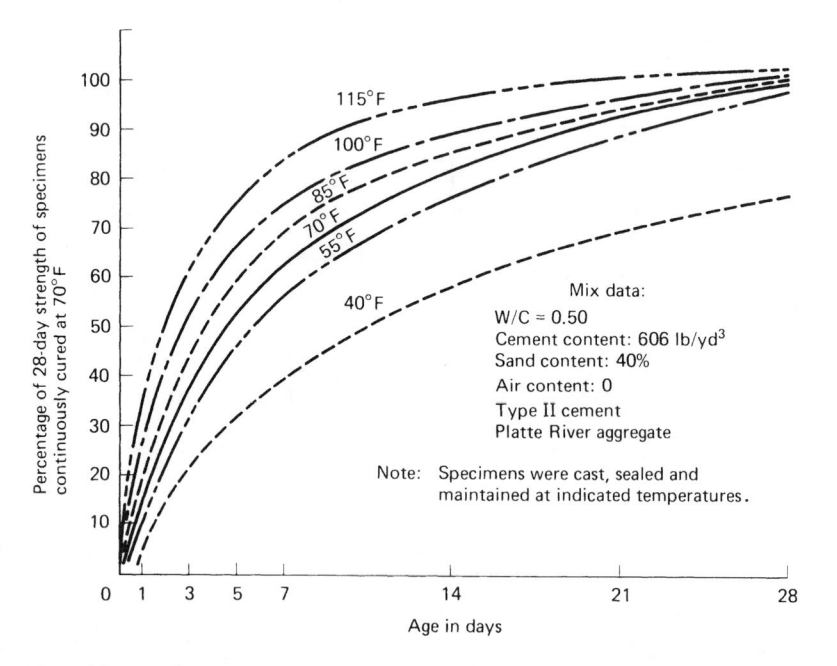

Figure 4.14 Effect of curing temperature on the compressive strength of concrete.

1. Strength in diagonal tension and resistance to shear,
2. Bond strength of deformed bars,
3. Cracking load levels and crack patterns,
4. Effective stiffness of structure and degree of nonlinearity in response to load,
5. Buckling behavior of thin shells.

INELASTIC MODELS: MATERIALS FOR CONCRETE AND CONCRETE MASONRY STRUCTURES 145

A = mass concrete
B = pavements and runways
C = bridges
D = housing, offices, schools
E = multistory structures
F, G = industrial buildings
H = tunnels
J = prefabricated members
L = ready-mixed concrete
Note: Results of an International questionnaire conducted by H. Rüsch.

(a)

(b)

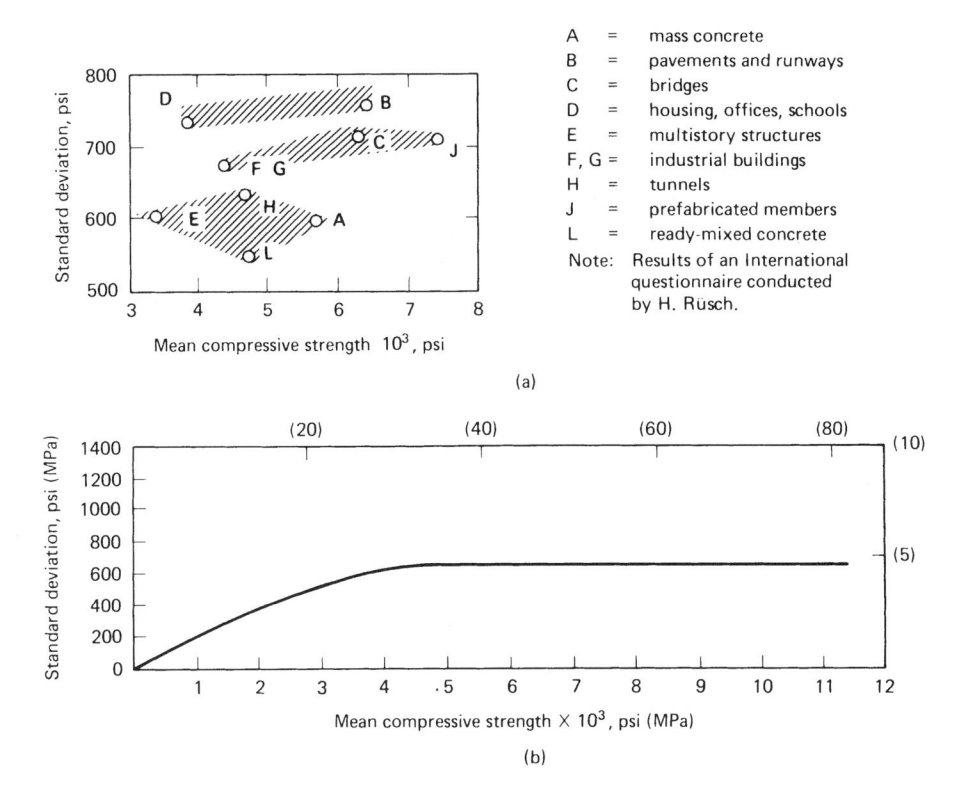

Figure 4.15 Variation of standard deviation with mean compressive strength for various types of concrete structures. (From Rüsch, 1964.)

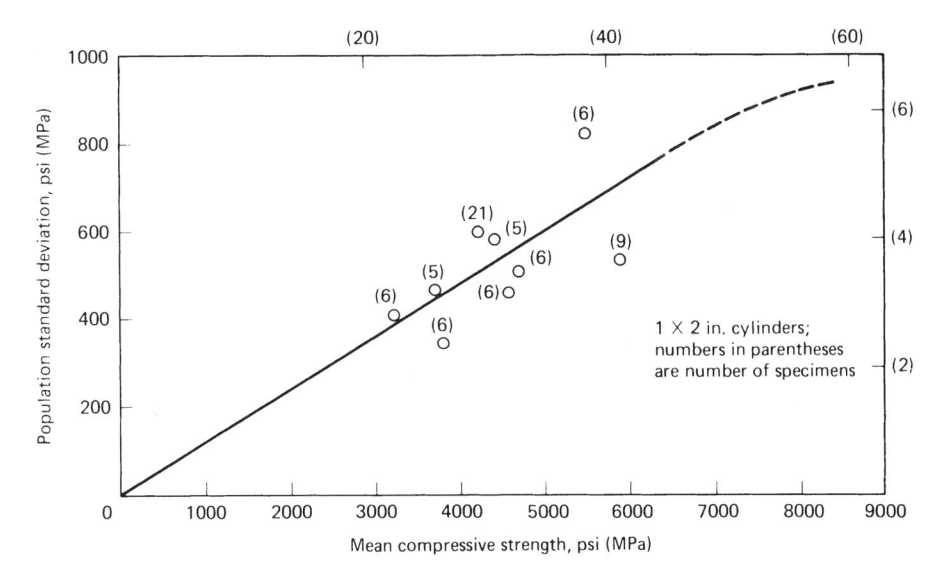

Figure 4.16 Variation of standard deviation with mean compressive strength of model specimens.

In addition, concrete tensile strength also influences the design process for prestressed concrete, pavements, and other structures where tensile strength is sometimes utilized in carrying load or the tensile stresses developed must be kept within certain limits. It is evident that modeling of concrete tensile strength is an important consideration in the overall modeling process and that the tensile properties of model concrete must be determined and understood.

There are several tension tests available, and the final choice of the tests depends on the strain distribution existing in the member. For example, uniform tension across a member requires that the tensile strength be determined using direct tension tests that are difficult to perform and are not commonly used. In a direct tension test, a concrete cylinder is glued with epoxy to a set of coaxial platens that are in turn gripped in the jaws of the testing machine. Indirect tension test (split cylinder test) and torsion tests are used for sections with stress distributions in which the principal compressive and tensile stresses are of the same order. Compressive loads are applied to the cylindrical surface of a concrete cylinder at diametrically opposite ends; the cylinder is laid flat, and the load is distributed through two strips. The flexure test consists of testing a plain concrete beam with a square or a rectangular section under a central or third-point loading. This test is used in connection with reinforced and prestressed concrete flexure members, pavements, etc. Like the compressive strength, the indirect tensile strength and the flexural tensile strength of plain concrete specimens have been noted to be dependent on the following variables (Harris et al., 1966):

1. Type of test
2. Method and rate of loading
3. Specimen size
4. Maximum size of aggregate
5. Water–cement ratio, aggregate–cement ratio, and the method of casting
6. Effect of differential curing
7. Workmanship (quality of specimen)
8. Effects of strain gradient
9. Statistical volume effects
10. Differential temperature

Section 4.6 deals with tensile strength of prototype concrete and model concrete under strain gradient (flexural strength or modulus of rupture). Section 4.7 treats tensile strength under nearly uniform strain as well as the shear strength of model concretes. The various measured tensile strengths are compared in Section 4.7. The tensile properties of gypsum model concrete are presented in Section 4.10. As is the case for the compressive strength of concrete, the tensile strength values obtained from any type of tension test are significantly influenced by the size of the test specimen. This effect of size on the tensile strength of concrete is discussed in Chapter 9.

4.6 FLEXURAL BEHAVIOR OF PROTOTYPE AND MODEL CONCRETE

The strength of unreinforced concrete in flexure is usually obtained by testing prismatic beams under third-point loading in accordance with ASTM recommendations. The usual assumption of a linear distribution of strain across the depth together with a linear stress–strain relation is made. It is generally accepted that the tensile strength of concrete varies as $\sqrt{f'_c}$. The ACI Building Code 318-95 (1995) and the CSA Standard A23.3-M77 (1977) provide the following equations for the flexural tensile strength, of concrete

$$f_r = 7.5\sqrt{f'_c} \tag{4.2a}$$

where f_r and f'_c are in psi and

$$f_r = 0.6\sqrt{f'_c} \text{ in SI units} \tag{4.2b}$$

where f_r and f'_c are in megapascals.

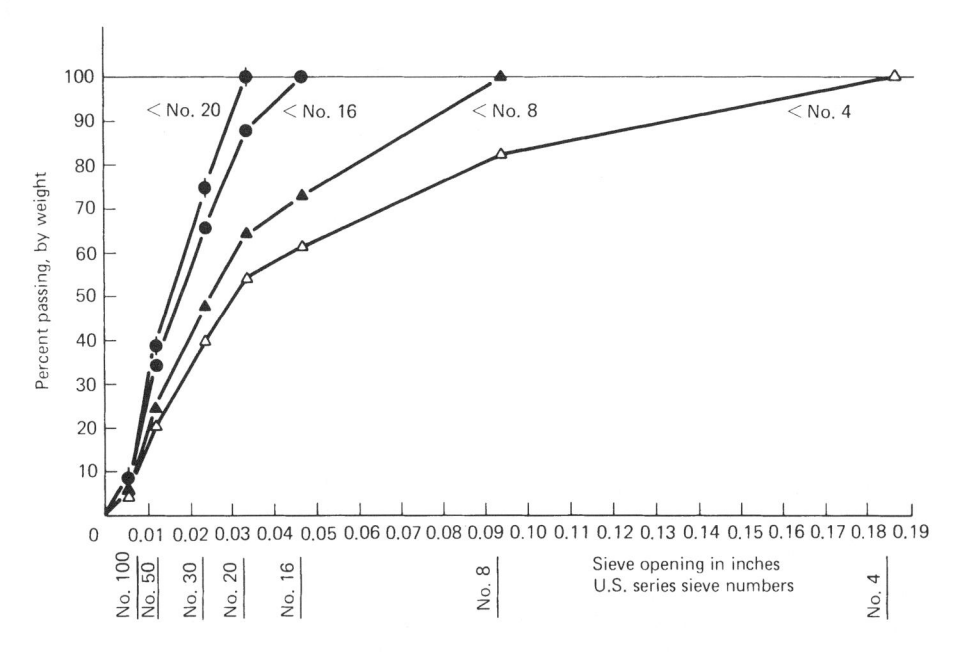

Figure 4.17 Gradation curves of Ithaca, NY glacial deposits sand.

The tensile strength of a model concrete is normally higher than that of a prototype concrete. It is therefore important to understand the flexural tensile properties of model concrete in the overall modeling process.

4.6.1 Specimen Dimensions and Properties

A series of model beams made of plain model concrete with mix proportions of 0.7:1:3.6 and sand passing the No. 8 sieve, and having a typical gradation curve shown in Figure 4.17, were cast from the same batch and cured identically (Harris et al., 1966). The details of sizes and numbers of each are shown in Table 4.1. Each beam had a ratio of depth to width of 1.5 and a clear span between supports of four times the depth. Loading and support were accomplished with steel plates and rollers with dimensions as shown in Table 4.1. Third-point loads were applied to all the beams. Four beams from 25 × 38 × 150 mm series (two cast vertically and two horizontally) and two from the 50 × 76 × 300 mm series (one cast vertically and one horizontally) were instrumented with strain gages at midspan. Vertical casting meant that the larger dimension was vertical during casting. These instrumented beams were tested in a partially dry condition in order to ensure proper bonding of the strain gages, while the remaining beams were tested immediately after removal from the moist room at the age of 21 days.

4.6.2 Stress–Strain Curves

The experimental stress–strain curves were fairly linear except at higher loads, which agreed with Kaplan's findings (1963). Test results also showed that any extra drying of specimens resulted in additional shrinkage, which decreased the measured tensile strength. Hsu and Slate (1963) had made similar observations.

4.6.3 Observed Variations in Modulus of Rupture with Changes in Dimensions

The results of the series of plain beams described in Table 4.1 were analyzed using the assumptions that the strain distribution is linear across the depth and that stress is directly proportional to

Table 4.1 Nonreinforced Beam Series

Section Breadth × Depth	Length Overall	Length Clear Span	Number of Beams Cast Vertical	Number of Beams Cast Horizontal	Rollers Diam.	Pad l	Pad w	Pad t
4×6	29	24	0	3	2	5	4	2
2×3	$14\frac{1}{2}$	12	3	3	1	$2\frac{1}{2}$	2	1
$1 \times 1\frac{1}{2}$	$7\frac{1}{4}$	6	6	6	$\frac{1}{2}$	$1\frac{1}{4}$	1	$\frac{1}{2}$
$\frac{1}{2} \times \frac{3}{4}$	$3\frac{5}{8}$	3	6	6	$\frac{1}{4}$	$\frac{5}{8}$	$\frac{1}{2}$	$\frac{1}{4}$
$\frac{1}{4} \times \frac{3}{8}$	$1\frac{13}{16}$	$1\frac{1}{2}$	6	6	$\frac{1}{8}$	$\frac{5}{16}$	$\frac{1}{4}$	$\frac{1}{8}$

Note: All dimensions are in inches.

strain. The extreme fiber tensile stresses were computed and averaged for each size group. The variation in the modulus of rupture with depth of beam is shown in Figure 4.18. Horizontally and vertically cast beams show the same steep increase in strength with decreasing dimensions. The horizontally cast beams were on the whole 10 to 15% stronger because of their more homogeneous structure as cast. During casting it was observed that a gradual migration of water to the top layers takes place, resulting in partial segregation of the heavier sand particles to the bottom of the mold. The resulting structure of the vertically cast beams is thus less uniform than that of those cast horizontally.

4.6.4 Rate of Loading

Effects due to rate of loading of beams on the modulus of rupture have been investigated by Wright and Garwood (1952) for concrete beams with an L/D ratio of 3 and an age of 28 days. Wright and Garwood found that a linear relation exists between the modulus of rupture and the logarithm of the rate of increase of extreme fiber stress. Stress rates from 0.14 to 7.9 MPa/min were investigated in this study, the results of which are shown in Figure 4.19.

The rates of loading used in the model beam series were determined by the minimum rate of crosshead motion that could be read on the loading scale of the Tinius–Olsen machine. This minimum is 0.05 mm/min and was used for the smallest beams. The rates for the other sizes were increased in the ratio of the scale in order to keep the rate of stress increase constant. It was found, however, that the two larger-size beams could not be loaded at the scaled rates, so smaller rates of loading were used.

It can be seen from the result plotted in Figure 4.19 that the rate of increase of the extreme fiber stress did not cause very large differences in the observed modulus of rupture values in these model beams.

More discussion on dynamic strain rate effects on model concretes is given in Chapter 11.

4.6.5 Influence of Strain Gradient

Experimental data from studies of the effect of strain distribution in rectangular-section prismatic beams (Blackman et al., 1958; Bredsdorff and Kierkegaard-Hansen, 1959; Harris et al., 1966) shows that the ultimate strain at failure can vary considerably with size of beam. Blackman's data and their statistical analysis indicate that the effect of the strain gradient on ultimate strain for a particular shape of specimen can be approximated by a bilinear function.

4.7 BEHAVIOR IN INDIRECT TENSION AND SHEAR

Development of reliable and easily used methods for testing concrete in direct tension has been seriously hindered by the difficulties encountered in applying load to the specimen. Therefore, the

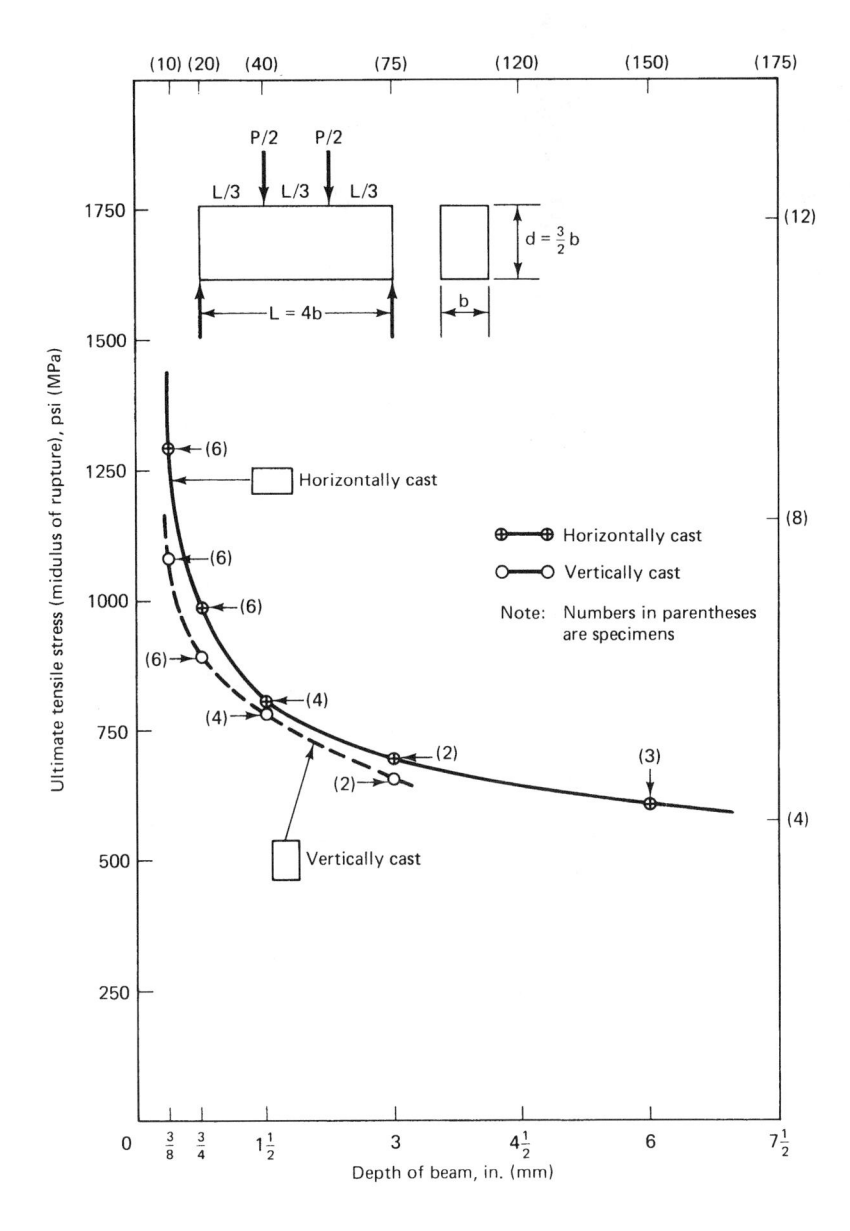

Figure 4.18 Variation of extreme fiber tensile stress with size of specimen.

flexure test was used to measure the tensile strength of concrete even though it does not measure the true tensile strength. However, the tensile split cylinder test and the ring test provide simple methods for determining the tensile strength under nearly uniform strain conditions. The split cylinder test has gained increasing acceptance as a tension test method during the last two decades; the modeling techniques for determining the splitting strength of model concrete and experimental data from some model tests are described in the following sections.

4.7.1 Tensile Splitting Strength

The split cylinder test measures the ultimate load needed to split a cylinder, lying on its side, with compressive line loads applied over a very small width on two opposite generators. This test, invented by Carniero and Barcellos (1953) in Brazil and by Akazawa (1953) in Japan, apparently

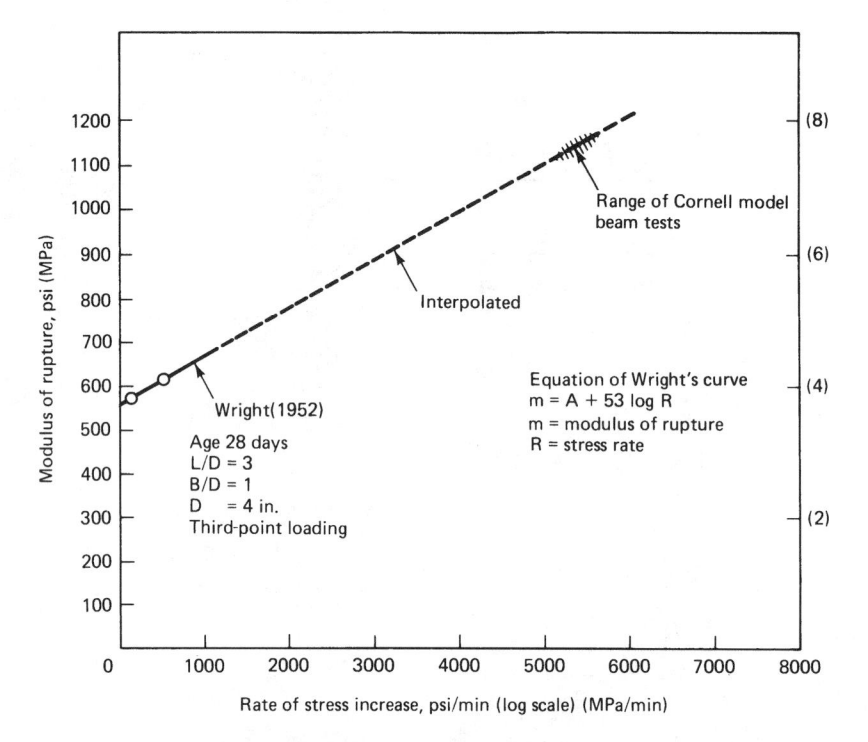

Figure 4.19 Effect of rate of loading on modulus of rupture.

independently, has gained in popularity because of its simplicity and because it uses the same cylindrical specimen needed for compressive strength tests.

A stress state in the cylinder can be computed on the assumption of a linear elastic, isotropic, homogeneous material (Carniero and Barcellos, 1953). The solution gives a uniformly distributed tensile stress over nearly the entire height of the diametral plane, as shown in Figure 4.20, with compression stresses developing in a small region near the loads. The exact distribution of stress is a function of the relative width and stiffness of the loading strip; Figure 4.20 gives principal stresses σ_1 and σ_2 in terms of coefficients C_1 and C_2 for strip widths of $a/D = 0.1$ and $a/D = 0.05$.

The almost uniform horizontal tensile stresses on the center plane, σ_1 (f_{rsp}) is given by

$$f_{rsp} = 2P/\pi LD \tag{4.3}$$

where: f_{rsp} = the tensile splitting strength of the cylinder
P = splitting load at failure
L = length of cylinder
D = cylinder diameter

When using the above relations to compute the stresses at failure, it is tacitly assumed that the stress distribution remains unchanged up to failure. This seems to be a questionable assumption for concrete inasmuch as the stress–strain relation deviates considerably from linearity as failure is approached.

Split cylinder specimens fail by splitting of the tensile zone, although the elastic analysis indicates very high compressive stresses near the load points. However, in the latter areas the two principal stresses are compressive; the apparent strength of concrete under such conditions is increased above its normal uniaxial strength sufficiently to cause failure to initiate in the region where the transverse principal stress σ_1 is tensile.

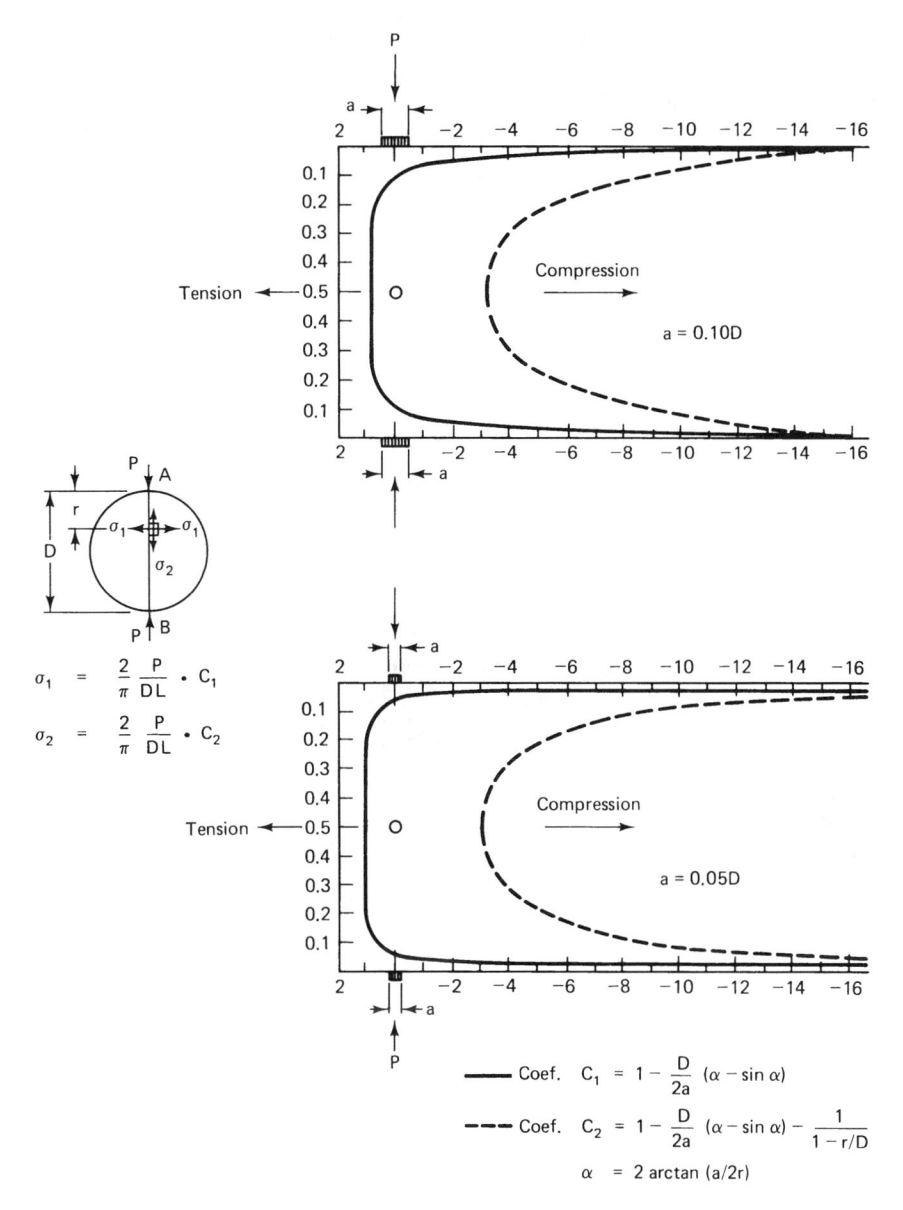

INELASTIC MODELS: MATERIALS FOR CONCRETE AND CONCRETE MASONRY STRUCTURES 151

Figure 4.20 Stress distribution in split cylinder test.

In order to obtain uniform line loads on model cylinders, in practice a soft material such as a strip of thin plywood or hard leather (cowhide) is used. The width of the loading strips are usually taken as one sixth the diameter of the cylinder.

4.7.2 Results of Model Split Cylinder Tests

The results of split cylinder tensile tests on 25×50 mm model concrete cylinders are shown in Table 4.2. A comparison of the splitting tensile strength as a percentage of the compression strength is shown in Figure 4.21 for the data of Table 4.2. A comparison is also made with prototype tests on 150×300 mm cylinders reported by various investigators. The model concrete cylinders have strengths of about $0.12 f'_c$ while the prototype concrete tensile strength range is slightly below this value.

Table 4.2 Split-Cylinder Tensile Strengths of Microconcrete

Size of Specimens, in.	Age at Test, days	Mix	Number of Specimens Tension	Number of Specimens Compression	$f'_{t\,sph}$ psi	f'_c, psi	$\dfrac{f'_{t\,spl}}{f'_c}$
1 × 2	28	0.8:1:3.6 < # 16	8	8	300	2489	0.12
1 × 2	28	0.5:1:2 < # 16	7	6	534	4175	0.128
1 × 2	31	0.6:1:3	6	5	594	4128	0.144
1 × 2	31	0.7:1:3.6	3	3	550	4940	0.112

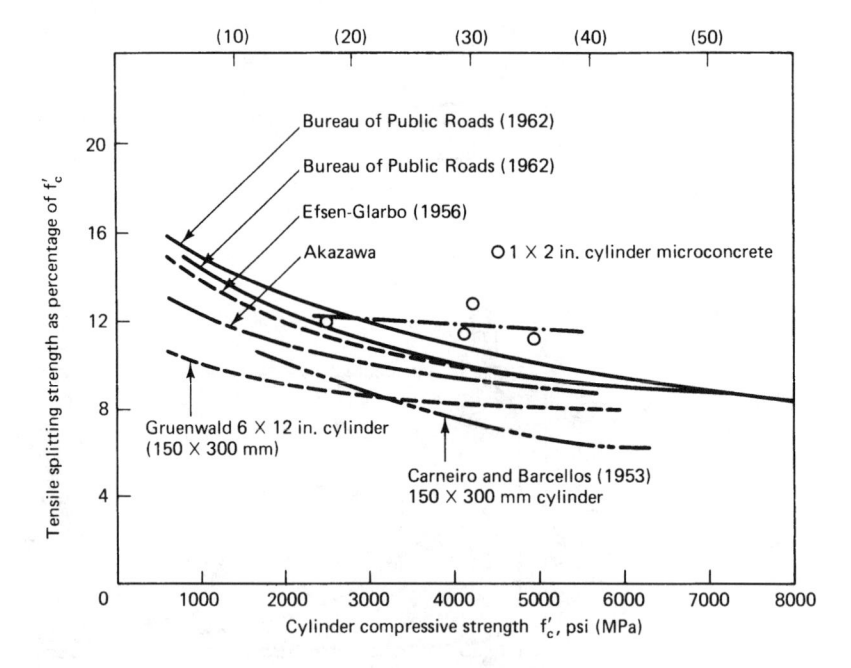

Figure 4.21 Relation between the tensile splitting strength and compressive strength of cylinders.

4.7.3 Tensile Splitting Strength vs. Age

The strength–age relation for cylinders tested by splitting has been investigated by Carniero and Barcellos (1953) using 150 × 300 mm cylinders and three different concrete strengths. Their results are shown in Figure 4.22 and indicate that the tensile splitting strength has a similar dependence on age as the compressive strength.

Although the strength gain in tension is similar to compression strength gain as curing of the material progresses, there are indications that the proportionality between the two is not the same at all ages, and this affords at least partial explanation of why there is so much scatter in combined stress test results performed at various ages. It has also been inferred that the bond strength between the paste and the aggregate deteriorates with age. If this is true, then the tensile strength will be affected more than the compressive strength and will start to decrease when the bond starts to deteriorate. However, up to ages of 3 months the tests of Carniero and Barcellos (Figure 4.22) do not show this trend of reduced strength in tensile splitting.

4.7.4 Correlation of Tensile Splitting Strength to Flexural Strength

The correlation of tensile splitting and flexural strength of microconcrete has not been made. However, comparison of the individual strengths to prototype concretes have been made and are

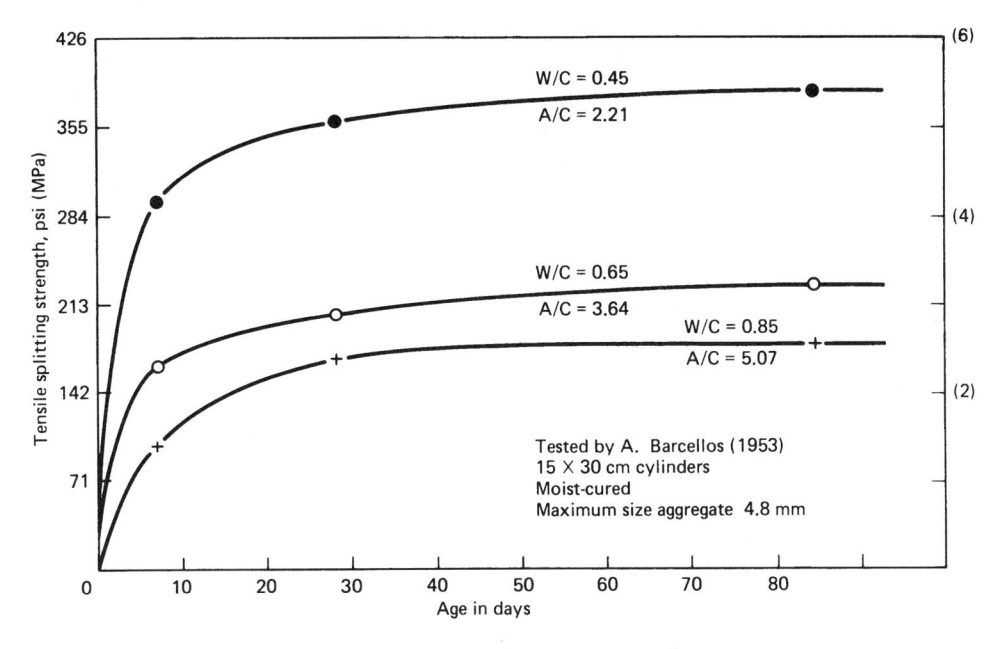

INELASTIC MODELS: MATERIALS FOR CONCRETE AND CONCRETE MASONRY STRUCTURES

Figure 4.22 Splitting strength vs. age curves for mortars.

presented in Figure 4.23. It follows from the trends shown in these comparisons that the relations between the tensile strengths obtained from the three methods indicated will be approximately the same for model concrete as it is for prototype concretes. These relations for the prototype concrete are clearly indicated in Figure 4.24. It is seen from this figure that the splitting tensile strength is closely related to the direct tensile strength over a wide range of compressive strength values, while the flexural strength is considerably higher than the direct tensile strength. At low values of compressive strength there is a greater discrepancy between the three tensile tests as shown in Figure 4.24. The difference between splitting tensile strength and direct tensile strength decreases significantly as compressive strength increases.

4.8 DESIGN MIXES FOR MODEL CONCRETE

4.8.1 Introduction

The choice of a suitable model concrete mix is of considerable importance in direct modeling of concrete structures. After a model scale has been chosen by similitude and other considerations, the material scale is then well defined in terms of the relative size of the coarser particles of the heterophase. Thus, if a mean size of coarse aggregate exists in the prototype concrete, then a corresponding scaled-down mean size of (sand) particles exists in the model concrete. Then mix proportions must be chosen that will ensure mechanical properties similar to those of the prototype within specified limits. This method of design will generally achieve reproducibility within the confidence level common in engineering work and within the statistical variations of testing.

4.8.2 Choice of Model Material Scale

The cement is the same in model and prototype. The difference is in the aggregates. Aggregates used in structural concrete (with the exception of lightweight and high-density concretes) consist of particles whose size ranges from fine sand to coarse particles of a specified maximum average

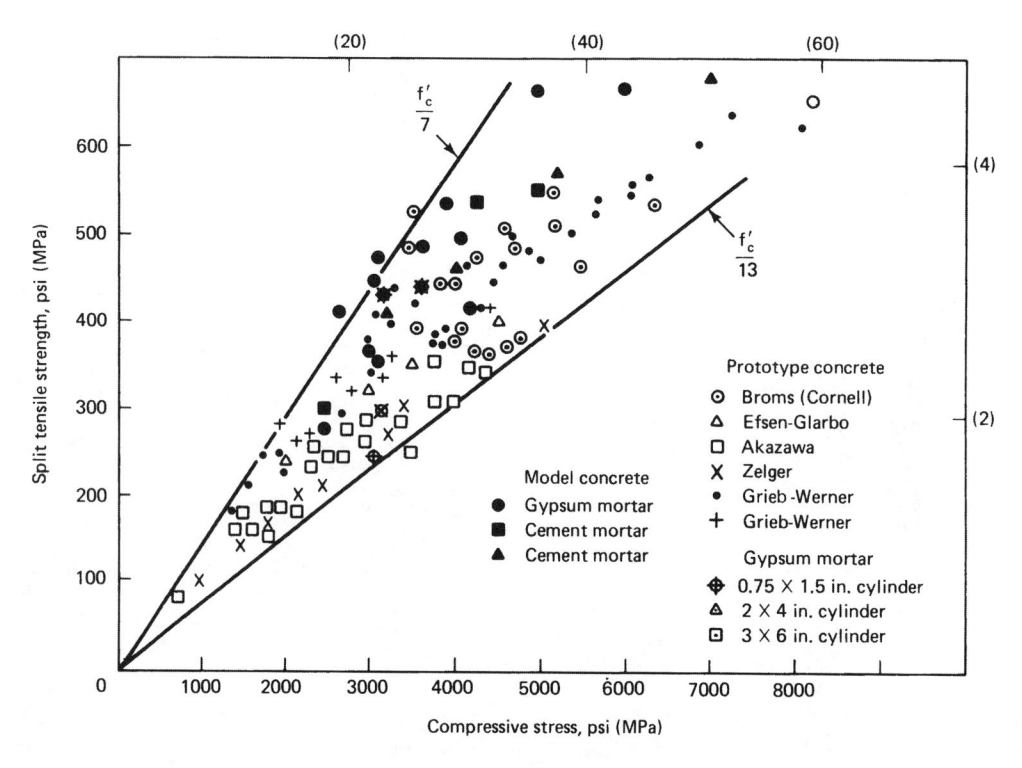

Figure 4.23 Relation between split tensile strength and compressive strength.

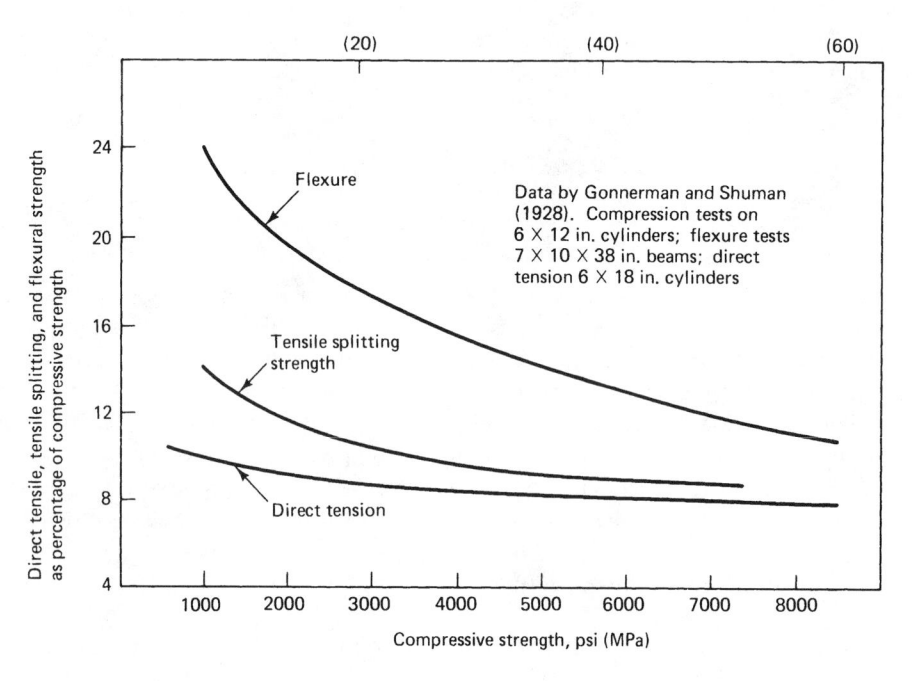

Figure 4.24 Relation of tensile strength to compressive strength.

INELASTIC MODELS: MATERIALS FOR CONCRETE AND CONCRETE MASONRY STRUCTURES

dimension. For the model material, ordinary well-graded concrete sand (and sometimes fine crushed stone or pea gravel for the larger-scale models) is used with scaling of the coarsest particles. It is implicitly assumed that the scale ratio between the maximum-size aggregates of model and prototype materials has the same ratio as the mean sizes of the two. In practice this is usually the case. The finer particles in the model mix are limited to less than 10% passing the U.S. No. 100 sieve (0.149-mm mesh) and is done to prevent the necessity for very high water–cement ratios in order to obtain workable mixes. By using this method of modeling aggregate, it becomes apparent that the gradation curve of the sand becomes steeper and steeper as the size of the model decreases. Generally, the amount of aggregate, and not the gradation, has the greatest effect on the mechanical properties of model concrete.

In some localities some grades of very narrowly graded crushed sands are commercially available. The maximum aggregate size for these crushed sands normally varies from that passing U.S. No. 6 sieve to that passing U.S. No. 100 sieve. In such cases, a desired grading is first established by scaling of the coarsest particles and by limiting the amount of fines passing U.S. No. 100 sieve to less than 10%. Gradings of the various available grades of crushed sands are determined using the standard sieve analysis. Then by trial and error, four or five of the available grades of sands ranging from the maximum aggregate size selected for the model concrete down to the grade with a maximum sand particle size passing U.S. No. 70 sieve are combined to match the selected grading.

4.8.3 Properties of the Prototype to Be Modeled

One of the most challenging tasks met in the small-scale modeling of concrete structures is the selection of the microconcrete mix that can faithfully reproduce the failure criterion of the prototype concrete under a general three-dimensional state of stress. Given that such a criterion is typically not known for a given prototype concrete, this exact similitude requirement is generally relaxed to modeling only four of the most important properties of the prototype concrete. It is generally required that a model concrete material have the following properties under *short-time* load:

1. A specified ultimate compressive strength f'_c;
2. A specified modulus of elasticity such as the secant modulus E_c at 0.45 or 0.5 of the ultimate strength f'_c;
3. A specified ultimate compressive strain ε_{cu}, or a strain at 95% ultimate, $\varepsilon_{0.95cu}$;
4. A specified ultimate tensile strength f'_t

In the case of prototype concrete all these properties are specified for a hypothetical 150 × 300 mm cylindrical specimen to be cast, cured, and tested under prescribed procedures set by the ASTM.

The values of the mechanical properties shown in Figure 4.25 define bounds on the stress–strain relationship but do not define its *shape*.

4.8.4 Important Parameters Influencing the Mechanical Properties of Concrete

The mechanical properties of concrete (particularly those of interest to the models engineer) depend on a large number of parameters. The parameters that have the most influence on short-term behavior of concrete are water–cement ratio, the percentage volume of aggregates in the system, aggregate–cement ratio, and the age of concrete at testing, as shown in Figure 4.26. For clarity, the dependence on time has been assumed to be linear in Figure 4.26. These relations are plotted in Figure 4.27 for an age of 28 days.

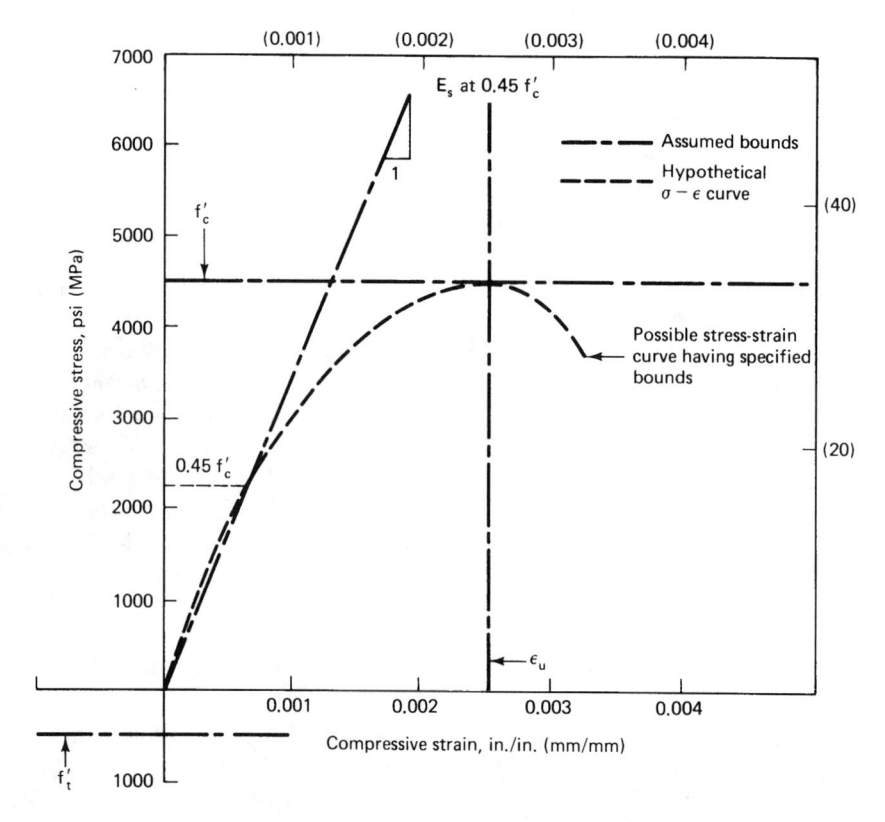

Figure 4.25 Specification of a hypothetical concrete stress–strain curve to be modeled.

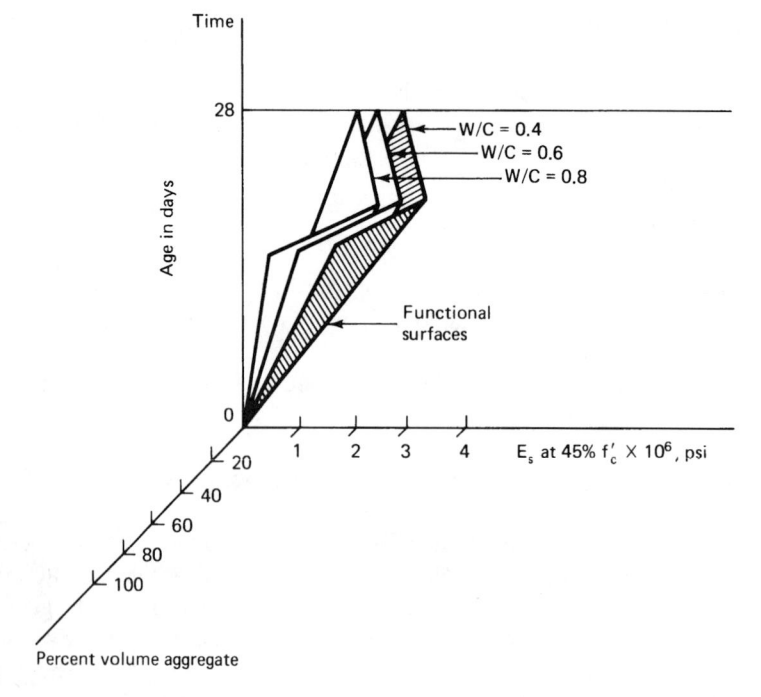

Figure 4.26 Schematic of functional relations existing between the property E_s and parameters W/C, age, and percent volume of aggregate.

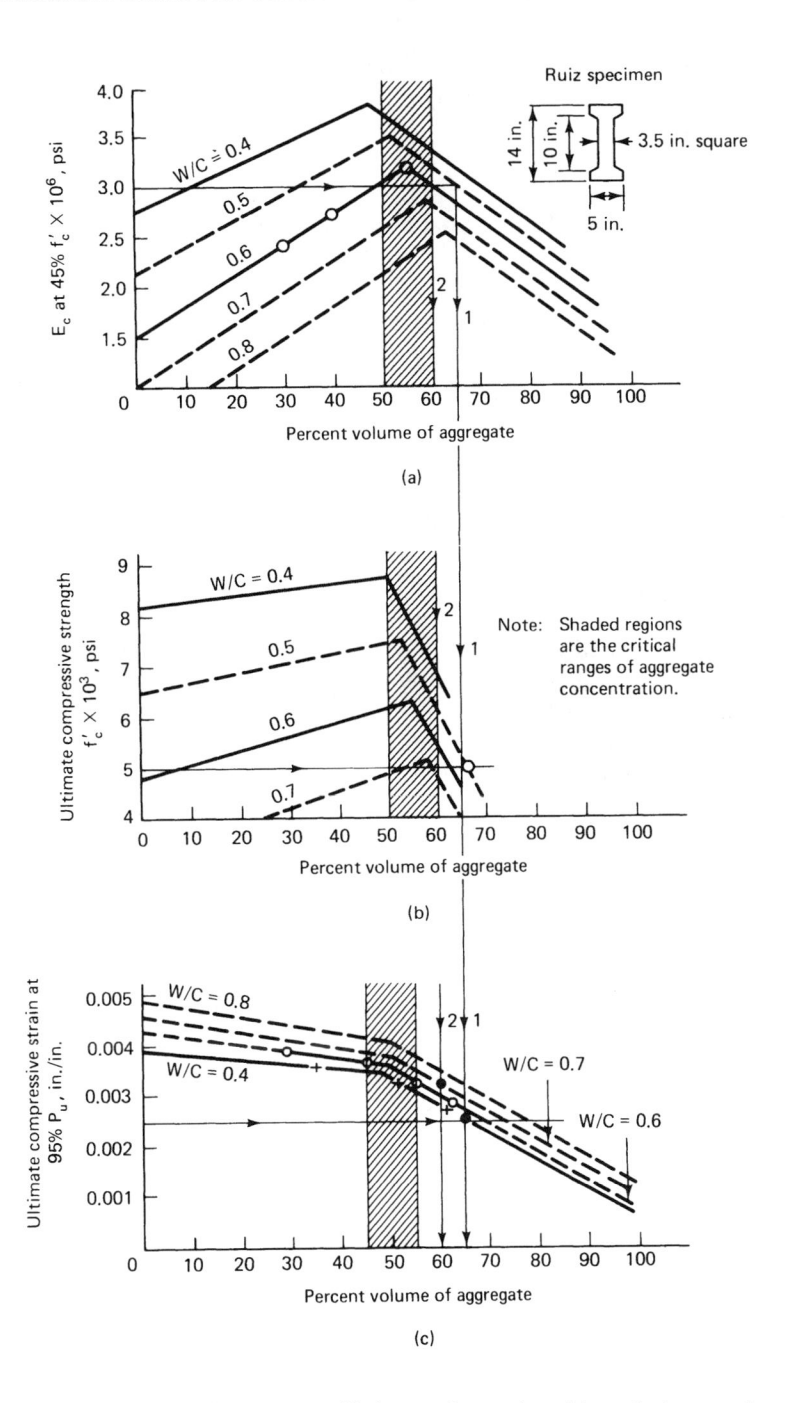

INELASTIC MODELS: MATERIALS FOR CONCRETE AND CONCRETE MASONRY STRUCTURES 157

Figure 4.27 Relations between specified properties and most important parameters.

The choice of an appropriate mix is aided by the curves shown in Figure 4.27, which are based on work done by Ruiz (1966); the extrapolated curves (shown dotted) have similar shapes to those obtained experimentally (full lines). It should be pointed out that in order for the choice to be valid, the model material should be tested at the same age, on the same size specimens, and under similar mixing, casting, curing, and testing conditions as those used to derive such experimental curves as shown in Figure 4.27.

With the accumulation of data and valid extrapolation, sets of curves can be established and used whenever a particular mix is to be designed.

Example 4.1

It is required to design a model mix for a hypothetical prototype concrete with the following characteristics:

$$E_c = 20.6 \times 10^3 \text{ MPa } (3 \times 10^6 \text{ psi})$$
$$f'_c = 34.5 \text{ MPa } (5000 \text{ psi})$$
$$\varepsilon_{0.95cu} = 0.0025$$

using the simplified functional relationships of Figures 4.26 and 4.27. (Note that the prototype concrete tensile strength, f'_t, is neglected in this example for simplicity. If this model concrete were to be used in studies for earthquake loading, for example, this simplification could not be made. More discussion about this point is given in Section 4.9 below and in Kim et al., 1988).

In Figure 4.27a, draw a horizontal line for $E_c = 20.6 \times 10^3$ MPa intersecting the various water–cement ratio curves at different percentages of aggregate volume. Similarly, by drawing horizontal lines for $f'_c = 34.5$ MPa in Figure 4.27b and for compressive strain $\varepsilon_{0.95 f'_c} = 0.0025$ at $0.95 f'_c$, in Figure 4.27c, it is found that a water–cement ratio of 0.5 and a volume of aggregates of 65% (indicated by line 1) satisfy the specified values. The resulting mix is therefore 0.5:1:4 by weight of water, cement, and sand; however, this mix will be too dry to use. A reexamination of Figure 4.27 shows that a higher water–cement ratio of 0.70 and a 60% aggregate volume (indicated by line 2) can also result in a satisfactory mix. However, the ultimate strain will be about 28% higher than desired. Therefore, use the trial mix of 0.7:1:4, which is a workable mix.

In designing the above trial mix, the following observations are made:

1. Often, the specified values cannot be satisfied simultaneously, and compromises must be made.
2. To get low ultimate strains similar to those of prototype concrete, very high percentages of aggregate are needed, as shown in Figure 4.27c. The resulting mixes are unworkable and have low modulus and strength since they are beyond the *critical volume* of 50 to 60%. This means that one must settle for somewhat higher strains (hence higher tensile strength) in the model concrete material.

Having designated a trial mix, one must cast sample cylinders and test them at the required age. If the mechanical properties are not satisfactory, some adjustments may have to be made.

The modeling of the tensile strength of the trial mix can be achieved by any of the techniques described above. It has been found, however, by many investigators, that the tensile strength, f'_t is much more difficult to model in a prototype concrete and the difficulty increases with decreasing scale. In most studies, the tensile strength of the model concrete has been found to be relatively high compared with the compressive strength. Attempts have been reported (Maisel, 1979) to treat the aggregate with mineral oil or silicone resins to reduce the tensile strength to the desired level successfully. Similar results were reported by Kim et al. (1988) who were able to reduce the tensile strength of microconcrete to the same relative value of the compressive strength as the prototype concrete they tested ($f'_t = 0.096 f'_c$) by coating the coarse aggregate with a chemical. Details of the properties of the prototype and four trial model concrete mixes investigated by Kim et al. (1988) are given in the following section.

4.9 SUMMARY OF MODEL CONCRETE MIXES USED BY VARIOUS INVESTIGATORS

Aldridge and Breen (1970) designed a model concrete mix based on an assumed aggregate gradation for the prototype 21-MPa concrete mix (maximum aggregate size 19 mm). The model concrete mix was a ⅛-scale replica of the prototype mixture with the exception of the omission of the very fine particles. The maximum aggregate size used was No. 8 (2.38 mm). Three other model concrete mixes were used for size-effect studies using maximum aggregate sizes of No. 4 (4 mm), 6 mm, and 10 mm and cylinder sizes of 25 × 50 mm, 50 × 100 mm, and 75 × 150 mm, respectively.

A number of gravels and sands, each with a limited distribution over only a few screen sizes, were blended to obtain the desired gradation. The grading of each basic aggregate was expressed by a linear equation indicating the portions retained on certain selected critical sieve sizes. If the basic aggregates are selected reasonably carefully, solution of a set of simultaneous linear gradation equations can indicate the appropriate proportions of each basic aggregate type for blending. The factors for each sieve size of a selected basic aggregate are multiplied by the same proportioning variable X_i, with the sum of all X_i values equal to 1.00. Moreover, the sum of the product X_i and the percentage of the basic aggregates retained on a given critical sieve must equal the desired percentage retained on the sieve in the model concrete mix. Any solution of these equations that results in positive values for all X_i will result in the formulation of a physically attainable model concrete mix. If negative values result for one or more Xi, the solution is not usable and one or more of the basic aggregates must be changed. The workability of the fresh model concrete was used as a visual measure of its consistency, since it was considered impractical to model any of the standard consistency devices. The compressive strengths of both the prototype and the model concrete, measured values of strains at $0.95\ f'_c$, and the values of the secant moduli at $0.45\ f'_c$ showed reasonable agreement. Stress–strain curves were obtained for the prototype and the model concrete from tests on 150 × 300 mm and 25 × 50 mm cylinders, respectively.

Mirza (1967) experimented with several trial mixes aimed at obtaining a compressive strength of 20.7 MPa using high-early-strength cement (Type III) and local sand passing the U.S. No. 4 sieve (4-mm, mesh) and the U.S. No. 8 sieve (2.38-mm, mesh), respectively. Large variations in strength were noted for the same water–cement ratio and aggregate–cement ratio; these were traced back to the inconsistent grading of the sand from batch to batch. A blended mixture of five grades of very narrowly graded crushed quartz sands, namely, No. 10, No. 16, No. 24, No. 35, and No. 70, was used as follows (percentage by weight):

No. 10 crushed quartz sand (2.00-mm, mesh): 20%
No. 16 crushed quartz sand (1.19-mm, mesh): 20%
No. 24 crushed quartz sand (0.707-mm, mesh): 35%
No. 35 crushed quartz sand (0.500-mm, mesh): 25%
No. 70 crushed quartz sand (0.210-mm, mesh): 10%

The water:cement:aggregate ratio used was 0.8:1:3.25. A total of 588 75 × 150 mm control cylinders were tested to determine the compressive strength of model concrete and resulted in an average compressive strength of 21.6 MPa with a standard deviation of 0.94 Mpa. Similarly, 197 75 × 150 mm control cylinders were tested in splitting tension; the average tensile strength was 2.7 MPa with a standard deviation of 0.1 MPa. Thus, excellent quality control was achieved for model concretes used in this investigation. Compressive and splitting tension stress–strain curves for the model concrete were obtained from tests on 20 instrumented 75 × 150 mm cylinders (14 in compression and 6 in splitting tension).

Table 4.3 Microconcrete Mix Details

Mix No.	Water–Cement Ratio	Aggregate–Cement Ratio	Grading Curve Used	Remarks
A-1	0.8	3.50	1	Cement: Type III
A-2	0.775	3.50	1	Aggregate: Sand Nos. 10, 16, 24, 40, and 70
A-3	0.75	3.50	1	
A-4	0.725	3.50	1	
A-5	0.733	3.50	1	
A-6	0.675	3.50	1	
A-7	0.650	3.50	1	
A-8	0.63	3.50	1	
A-9	0.825	4.5	2	Cement: Type III
A-10	0.72	3.75	2	Aggregate: Sand Nos. 10, 16, 24, 40, and 70
A-11	0.635	2.5	2	
A-12	0.575	2.0	2	
A-13	0.535	1.5	2	
A-14	0.45	1.25	2	
B-1	0.72	3.75	2	Cement: Type III
B-2	0.635	2.50	2	Aggregate: Sand Nos. 10, 16, 24, 40, and 70
B-3	0.575	2.0	2	
B-4	0.535	1.5	2	
B-5	0.45	1.25	2	

Tsui and Mirza (1969) developed model concrete mixes with design strength from 17 to 41 MPa. High-early-strength cement (Type III) was used throughout the investigation, and the aggregates consisted of two separate blended mixtures of fine, very narrowly graded crushed quartz sands passing sieve No. 10, No. 16, No. 24, No. 40, and No. 70. The water–cement and aggregate–cement ratios used for 14 trial mixes in these tests are detailed in Table 4.3. The statistical strength variation of these same mixes are given in Table 4.4. Typical stress–strain curves are shown in Figure 4.28. Variation of the direct compressive and the indirect tensile strengths with the water–cement ratio is shown in Figures 4.29. Based on the workability obtained for different mixes, six mix designs have been recommended, as follows:

Water/Cement	Aggregate/Cement	Expected f'_c (MPa)
0.83	4.0	17.2
0.72	3.75	20.7
0.60	3.25	27.6
0.55	2.75	34.5
0.50	2.50	41.4
0.40	2.25	48.3

From his model tests at Queen's University, Batchelor (1972) found that neither a model concrete with the properly scaled aggregates nor a mortar with gypsum was suitable to model concrete. The fraction of a locally available fine limestone passing a 6-mm sieve was used as the aggregate for a new model concrete. The maximum size of the aggregate was selected to match the minimum construction clearances rather than scaling them down, and the grading was then maintained within practical grading limits as shown in Figure 4.30. Typical mix proportions of water:cement (Type III): aggregate were 0.65 to 0.70:1:4 by weight. Typically, the concretes had slumps of 75 to 125 mm, as determined by the standard slump cone, a 14-day compression strength of 35 MPa,

INELASTIC MODELS: MATERIALS FOR CONCRETE AND CONCRETE MASONRY STRUCTURES

Table 4.4 McGill Model Concrete Trial Mix Results

	Direct Compressive Strengths			Indirect Tensile Strengths		
Series No.	Mean Strength, psi	Standard Deviation, psi	Coefficient of Variation	Mean Strength, psi	Standard Deviation, psi	Coefficient of Variation
	Test Results, 3 × 6 in. (75 × 150 mm) Cylinders					
A1	2886	102	0.04	303	15.3	0.05
A2	2753	229	0.08	276	3.54	0.01
A3	2622	149	0.06	311	42.4	0.14
A4	3100	108	0.03	356	32	0.09
A5	2970	561	0.19	330	11.4	0.03
A6	3687	268	0.07	325	32.4	0.10
A7	3848	251	0.07	387	14.3	0.04
A8	3514	214	0.06	400	34.8	0.09
A9	2570	20.5	0.008	278	14.7	0.05
A10	3151	200	0.06	348	9.6	0.03
A11	3817	162	0.04	424	6.2	0.01
A12	4484	88.7	0.02	420	24.8	0.06
A13	6040	51	0.01	482	4.5	0.01
A14	5189	112	0.02	416	16	0.06
	Test Results, 2 × 4 in. (25 × 50 mm) Cylinders					
B1	3533	177	0.05	488	12.1	0.02
B2	5013	242	0.05	631	19.3	0.03
B3	4795	66.3	0.01	597	17.2	0.03
B4	6573	512	0.08	656	13.8	0.02
B5	5942	247	0.04	574	13.6	0.02

and split cylinder tensile strength of about 2.8 MPa. The ratios of the indirect tensile to compressive strengths were about 0.08 and thus were typical of the values to be expected for the prototype. The modulus of elasticity of the model concrete is in general somewhat lower than that for prototype concrete, and this factor has been recognized in research on models.

A study to improve model concrete mixes appropriate for seismic studies was conducted by Kim et al. (1988) at Cornell University. Particular emphasis was placed on the development of a model concrete mix to model the important strength and stiffness properties of full-scale prototype concrete accurately. The gradation of aggregate used in the mix, along with the aggregate-to-cement ratio, was shown to be critical in achieving sufficiently low tensile strength while still maintaining acceptable critical strain levels at compressive failure of the model concrete.

A prototype and four model concrete mixes were used in making the test cylinders, as defined in Table 4.5. Sand having the gradation shown in Figure 4.31 (Curve 1) and passing the U.S. No. 4 standard sieve was commercially obtained. This sand was used in the prototype concrete mixed in the ratio of 3 to 3 by weight with the gravel of Curve 2 to give the gradation of Curve 3 in Figure 4.31.

The microconcretes used only sand and cement without any gravel. In order to get variously graded sands for the microconcretes, the sand was divided into two parts: one had particles larger than No. 8 sieve size and smaller than No. 4 sieve size (called model gravel and denoted by G_m in Table 4.5 and Figure 4.31). The other fraction had particles smaller than No. 8 sieve size (called model sand) and denoted S_m. Sands having different gradation curves were made by recombining the model sand and the model gravel with different mix ratios.

For Microconcretes I and II, the original sand was used with a sand-to-cement ratio of 3 and a sand-to-gravel ratio of 4. In Microconcrete II the coarse particles corresponding to the model gravel size were coated with a plastic. This was done to reduce model concrete tensile strength by

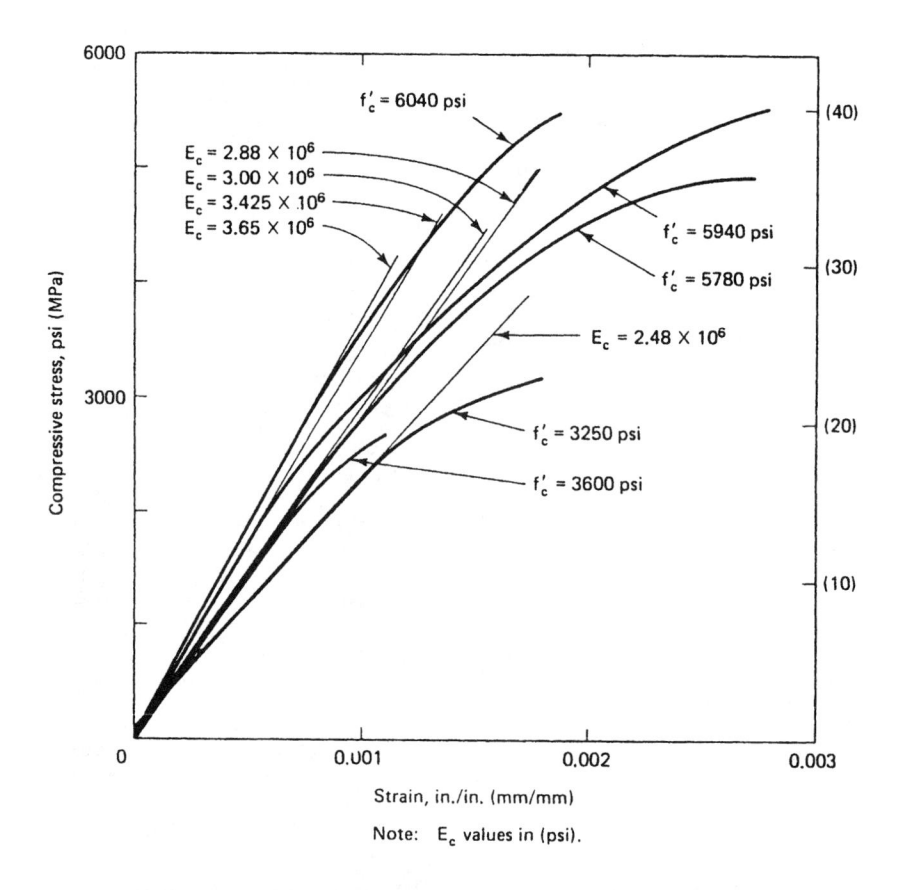

Figure 4.28 Compression stress–strain curves, model concrete.

reducing the bond strength between the cement paste and the coarse aggregate (Darwin, 1968). The coated model gravel was mixed with the model sand to make Microconcrete II. The aggregate for Microconcrete III consisted of the model sand and gravel in a mix ratio of 3 to 3 in order to increase the portion of large particles. To further increase the portion of large particles, the model sand and gravel were mixed in a ratio of 2 to 4, in Microconcrete IV. The gradation curves of the aggregate used for Microconcretes III and IV are represented by Curves 4 and 5, respectively in Figure 4.31.

The observed results from compression and tensile splitting tests on 50 × 100 mm cylinders are summarized in Table 4.6 and in the normalized average compressive stress–strain curves in Figure 4.32. Although the results showed that the microconcretes had less stiffness, and larger compressive strain capacity than the prototype concrete, differences were minimized by appropriate adjustments to the model microconcrete mixes. The results showed that as the ratio of aggregate to cement increases, the compressive stiffness increases, and tensile strength decreases. In addition, the tensile strength decreases with increase of aggregate–cement ratio. Trends in the variation of the stiffness and f'_t of microconcretes with respect to aggregate–cement ratio is shown in Figure 4.33.

White (1976) experimented with 11 series of model concrete mixes detailed in Table 4.7. He used high-early-strength cement (Type III) and varied the water–cement ratio from 0.40 to 0.63. The aggregate–cement ratio ranged from 2.5 to 5.6. River sand passing the U.S. No. 3 sieve (6.73-mm, mesh) and retained on the U.S. No. 100 sieve (0.15-mm, mesh) and two different gradations of locally available Conrock aggregates, maximum aggregates sizes 10 and 6 mm, were

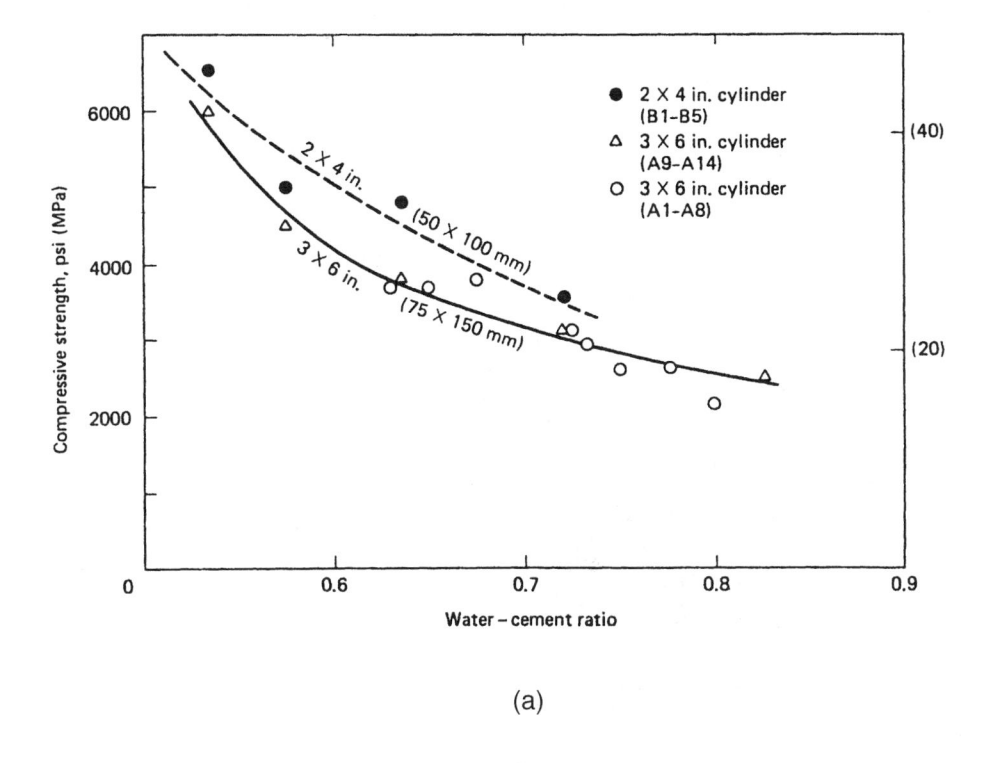

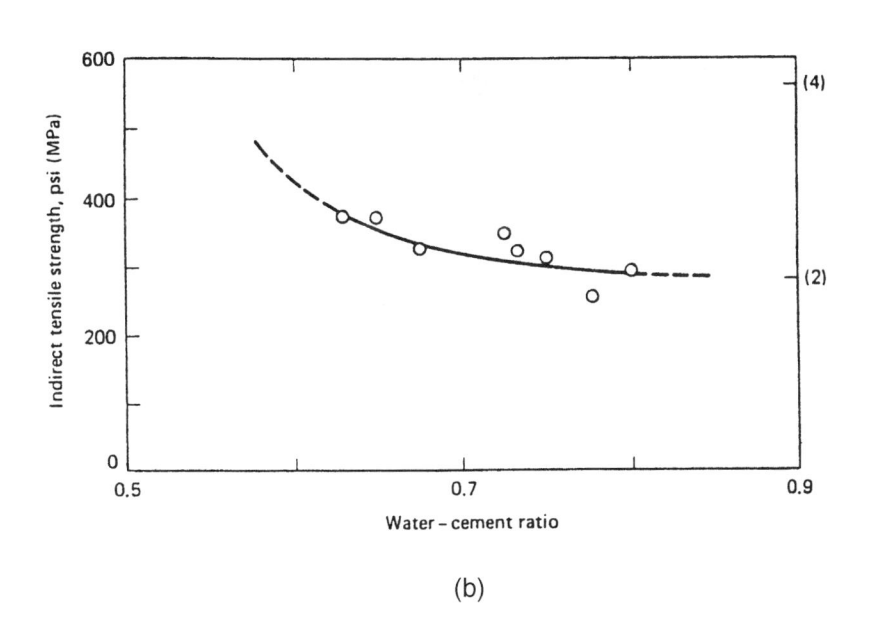

INELASTIC MODELS: MATERIALS FOR CONCRETE AND CONCRETE MASONRY STRUCTURES 163

Figure 4.29 Variation of strength properties with water–cement ratio: (a) compressive strength and (b) tensile strength.

used, with the sand–gravel ratio being 1.0 for nine mixes and 3.0 for two mixes (Table 4.7). The Conrock aggregates were intended to be used as filter sands and were cleaned and kiln-dried by the manufacturer. The compressive strengths, the moduli of elasticity, and the indirect tensile (split cylinder) strengths at 14, 28, and 105 days were obtained using 50×100 mm cylinders. The average

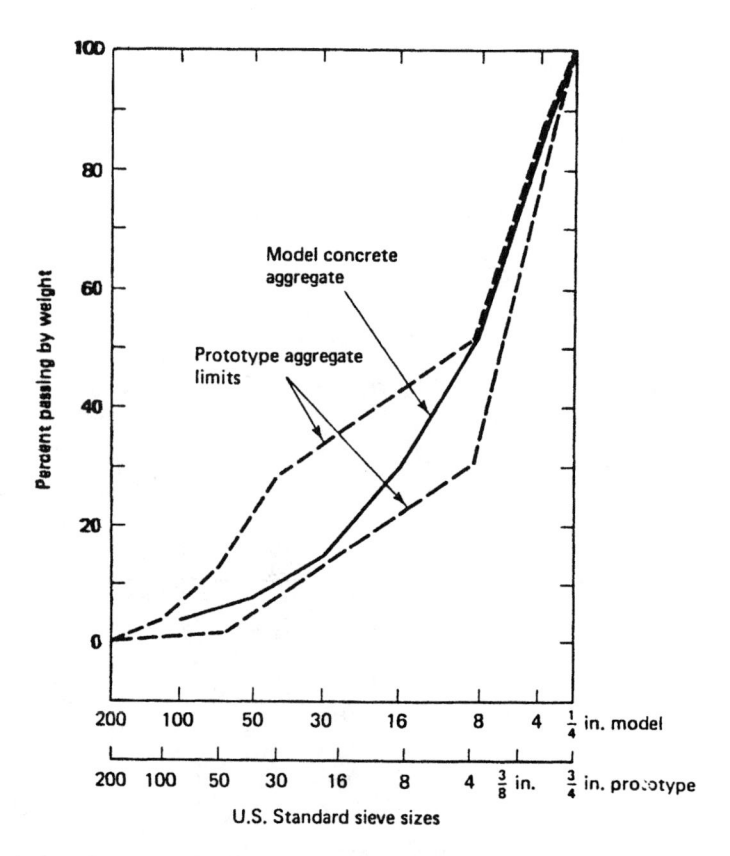

Figure 4.30 Typical grading curves for model and prototype aggregates. (From Batchelor, 1972. Unpublished report.)

Table 4.5 Mix Ratios of Test Specimens

Mix	Water	Cement	Sand ($S_m + G_m$)[a]	Coarse Aggregate
Prototype concrete	0.65	1	3 (2.4 + 0.6)	3
Microconcrete I	0.70	1	3 (2.4 + 0.6)	0
Microconcrete II	0.70	1	3 (2.4 + 0.6)[b]	0
Microconcrete III	0.70	1	6 (3.0 + 3.0)	0
Microconcrete IV	0.70	1	6 (2.0 + 4.0)	0

[a] S_m = model sand defined by particle size smaller than No. 8 sieve.
G_m = Model gravel defined by particle size larger than No. 8 sieve and smaller than No. 4 sieve.
[b] G_m in this mix was coated by a chemical material.
Source: Kim, et al., 1988.

compressive strength at 28 days ranged from 51.4 to 85.7 MPa, while the average tensile strength ranged from about 4.8 to 6.1 MPa. White's work shows that it is possible to design high-strength model concrete mixes; for example, a 83-MPa model concrete mix with reasonable workability can be obtained by using a water–cement ratio of about 0.4 and an aggregate–cement ratio of about 2.5 (with equal sand and gravel of maximum size 6 mm). The extremely high quality Conrock aggregate was the key to success.

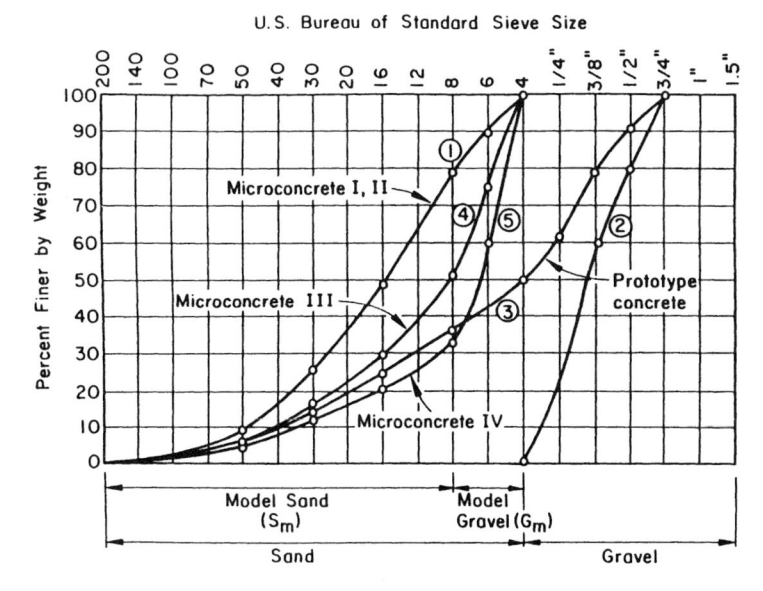

INELASTIC MODELS: MATERIALS FOR CONCRETE AND CONCRETE MASONRY STRUCTURES 165

Figure 4.31 Grading curves of aggregate used. (From Kim et al., 1988.)

Table 4.6 Summary of Concrete Cylinder Tests

Mix	(W:C:S:A)[a]	f'_c, MPa	ε_u, mm/mm	E_{int}, GPa	$E_{o.4f'_c}$, GPa	f'_t, MPa	f'_t/f'_c	$f'_t/\sqrt{f'_c}$
Prototype concrete	0.65:1:3:3	33.58	0.00229	25.41	21.24	3.18	0.095	6.61
Microconcrete I	0.7:1:3:0	31.77	0.00345	13.27	12.18	3.57	0.112	7.63
Microconcrete II	0.7:1:3:0	29.30	0.00424	12.21	10.60	2.64	0.090	5.86
Microconcrete III	0.7:1:6:0	32.29	0.00301	20.66	16.54	3.10	0.096	6.56
Microconcrete IV	0.7:1:6:0	36.92	0.00293	21.62	17.13	2.61	0.071	5.17

Notes: 1. For Prototype concrete, 6 × 12 in. standard cylinders were tested.

2. For all Microconcretes, 2 × 4 in. model cylinders were tested.

3. Various gradations of sand were used in the microconcretes.

[a] Water:cement:sand:aggregate (by weight).

Source: Kim, et al., 1988.

4.10 GYPSUM MORTARS

Gypsum mortars, consisting of gypsum, sand, and water, are used as model concretes in some laboratories. Their main advantage is a fast curing time (1 day or less), while their main disadvantage is the strong influence of moisture content on mechanical properties. Gypsum gains strength and stiffness by drying and tends to become too brittle if allowed to air-dry for periods of several days or longer. Successful use of gypsum mortars in structural modeling involves preservation of a selected moisture content by sealing the surfaces of the models.

Several gypsum products have been used in model concretes, including Hydrocal I-II, Ultracal 30, Ultracal 60, and Hydrostone, typically made in the U.S. Most of the following discussion will concentrate on Ultracal mixes, as their properties have been studied most extensively (Sabnis and White, 1967; White and Sabnis, 1968).

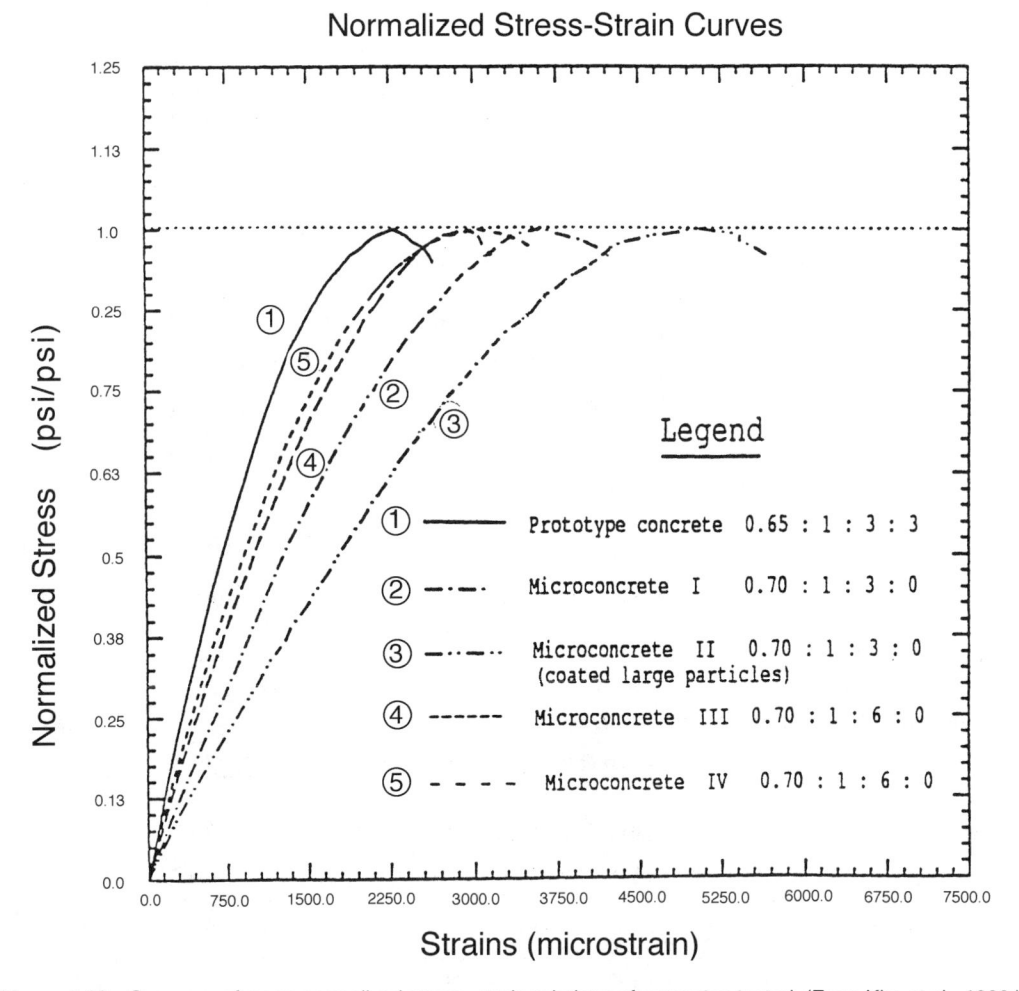

Figure 4.32 Summary of mean normalized stress–strain relations of concretes tested. (From Kim et al., 1988.)

The strength properties of Ultracal, as supplied by the manufacturer and for a water–gypsum ratio of 0.38, are

Compressive strength (wet)	21–35 MPa
Compressive strength (dry)	47–54 MPa
Tensile strength (wet)	3.2–3.8 MPa
Tensile strength (dry)	4.6–5.0 MPa
Modulus of elasticity (dry)	17×10^3 MPa

The wet strengths are measured 3 h after casting, and the dry-strength tests are conducted after specimens are dried to a constant weight at 42°C.

Ultracal 30 and Ultracal 60 differ only in the amount of set retarder added to the product at the mill; the numbers 30 and 60 refer to the time of set in minutes. It is recommended that Ultracal 60 always be used in order to maximize the available time for placing the model concrete in the forms and for making control specimens, although Ultracal 30 can be modified by the addition of a commercially available sodate set retarder (0.1% by weight) to lengthen its setting time.

Ordinary mortar sands and those used for other concretes are satisfactory for gypsum mortars. No special gradations are required; normal practice at Cornell University is to take out the coarser

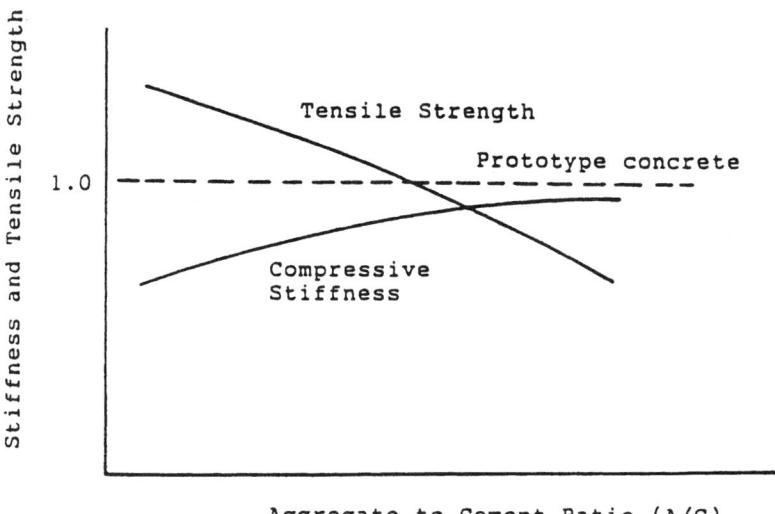

Figure 4.33 Schematic variation of stiffness and tensile strength of microconcrete as function of A/C ratio. (After Kim et al., 1988.)

Table 4.7 Mix Proportions and Summary of Properties of High-Strength Model Concretes

	Mix										
	1	2	3	4	5	6	1-A	2-A	3-A	4-A	5-A
W/C	0.455	0.481	0.50	0.50	0.459	0.40	0.547	0.60	0.547	0.60	0.63
A/C	3.0	3.41	3.8	3.8	3.41	2.5	4.5	5.2	4.5	5.2	5.6
S/G	3.0	1.0	3.0	1.0	1.0	1.0	1.0	1.0	1.0	1.0	1.0
Sand	S3-100	S3-100	S3-100	S3-100	S3-100	S3-100	S3-100	S3-100	S3-100	S3-100	S3-100
Gravel	G-$\frac{1}{4}$	G-$\frac{1}{4}$	G-$\frac{1}{4}$	G-$\frac{1}{4}$	Nelson $\frac{1}{4}$-No. 5	G-$\frac{1}{4}$	G-$\frac{1}{4}$	G-$\frac{1}{4}$	Nelson $\frac{1}{4}$-No. 5	Nelson $\frac{1}{4}$-No. 5	G-$\frac{1}{4}$
Workability	Low	Good+	Low	Good	Good+	Good–	Good	Good–	Good	Good–	Good–
14 day f'_c, psi	9,577	9,037	8,440	8,703	8,587	10,755	8,156	7,003	7,623	7,085	6,121
E, $\times 10^6$ psi	3.92	4.44	4.32	3.79	4.19	4.38	3.79	3.90	3.57	3.81	5.37
f'_t, psi	724	750	744	697	680	739	735	668	664	672	695
28 day f'_c, psi	11,082	10,260	9,634	10,058	10,117	12,425	9,125	8,175	9,204	7,459	—
E, $\times 10^6$ psi	—	3.82	3.85	3.74	4.02	4.36	4.37	4.26	4.50	4.46	—
f'_t, psi	—	893	892	790	807	890	732	704	740	725	—
105 day f'_c, psi	10,816	10,122	9,114	10,090	10,398	12,159	9,634	8,497	9,458	7,676	7,395
E, $\times 10^6$ psi	5.40	4.54	4.23	4.41	6.64	4.32	4.38	4.66	4.04	4.45	4.20
f'_t, psi	918	808	828	755	799	935	831	752	794	728	755

Source: After White, 1976.

particles such that the ratio of the largest particle size to the smallest dimension of a model does not exceed ⅕. The gypsum sand mixes used at North Carolina State University utilize sand that passes through a No. 16 mesh and is retained on a No. 40 mesh.

Suggested proportions for Ultracal mortars with compressive strengths in the range of 17.2 to 27.6 MPa were developed by Sabnis and White (1967) and are given in Table 4.8. Higher-strength mixes are very difficult to control. Experience has shown that mixes that utilized Hydrostone and combinations of Ultracal 60 and Hydrostone are excessively inconsistent from batch to batch, and even within batches. The very early setting time of Hydrostone (about 10 min) may be a factor in these unsatisfactory results. The effect of the water–gypsum ratio and the aggregate–gypsum ratio on the compressive strength of gypsum mortar is shown in Figures 4.34 and 4.35, respectively.

Table 4.8 Properties of Ultracal Model Concrete Mixes

Mix	Sand	Ultracal	Water	Age, h	f'_c, psi (MPa)
1	1.2	1.0	0.35	24	2400 (16.5)
2	1.0	1.0	0.35	24	2600 (18)
3	1.0	1.0	0.3	24	3000 (20.7)
4	0.8	1.0	0.3	24	3400 (23.4)
5	1.0	1.0	0.3	48	4000 (27.6)

Test results:

Specimen Size, in. (mm)	Compression Strength, psi (MPa)	Splitting Tensile Strength, psi (MPa)
3 × 6 (75 × 150)	3015 (20.78)	312 (2.15)
2 × 4 (50 × 100)	3021 (20.83)	312 (2.15)
1.5 × 3 (38 × 75)	3061 (21.11)	306 (2.11)
1 × 2 (25 × 50)	3033 (20.91)	315 (2.17)

Size effect:

Beam Cross Section, in. (mm)	Modulus of Rupture, psi (MPa)
2 × 3 (50 × 75)	531 (3.66)
1 × 1.5 (25 × 38)	574 (3.96)

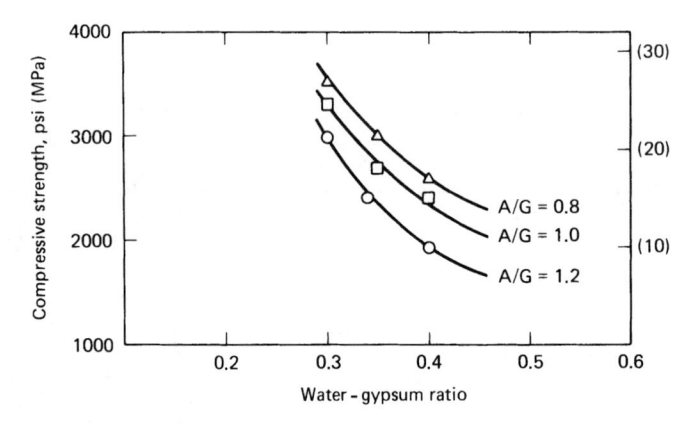

Figure 4.34 Effect of water–gypsum plaster ratio on compressive strength.

4.10.1 Curing and Sealing Procedures

The sealing process will be discussed before mechanical properties are presented inasmuch as the latter are a direct function of the moisture content. The cast material is allowed to air-dry for a specified time, after which all surfaces are sealed with a waterproof material. A seal consisting of two coats of shellac has proved to be better than most commercial sealing compounds.

One problem associated with the sealing process is the selection of the proper age for sealing. The drying rate of a specimen depends upon size and geometry. Data on this important matter are presented in Chapter 9 on size effects.

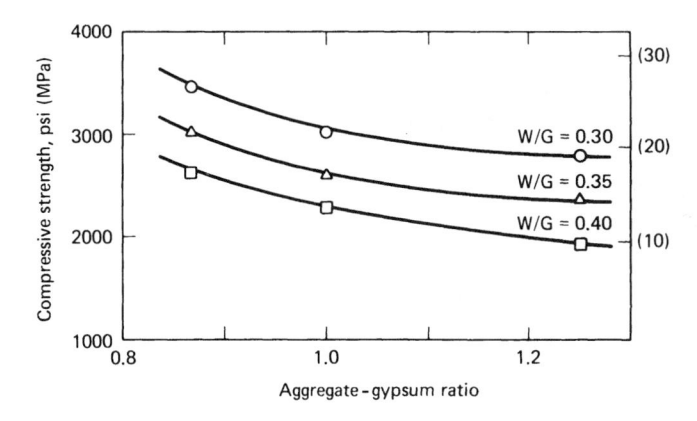

```
                                                                 4000
                                                                        ┌─────────────────────────────┐ (30)
                                                                        │                             │
                                                                        │ ○                           │
                                                              3000      │   ○                W/G=0.30 │ (20)
                                                                        │ △                      ○    │
                                                                        │   △        △       W/G=0.35 │
                                                                        │ □                      △    │
                                                              2000      │   □        □       W/G=0.40 │ (10)
                                                                        │            □           □    │
                                                                        │                             │
                                                              1000 └────┴────────────┴────────────┴───┘
                                                                   0.8              1.0              1.2
                                                                              Aggregate–gypsum ratio
```

Figure 4.35 Effect of aggregate–gypsum plaster ratio on compressive strength.

4.10.2 Mechanical Properties

No significant study of the failure criteria for gypsum mortars has been made; hence, currently one must settle for data on those properties determined from the uniaxial compression tests and the various measures of tensile strength.

The shape of the uniaxial compressive stress–strain curve for several Ultracal gypsum mortars obtained by Sabnis and White (1967) is compared with curves for Portland cement model concrete and prototype concrete in Figure 4.5; the gypsum mortars model this important property very well. Kandasamy (1969) arrived at the same conclusion from the results of his tests on eighty 75 × 150 mm gypsum mortar cylinders.

After some trials, Loh (1969) developed a casting procedure that resulted in uniform density of the test cylinders. Using this procedure, he cast some 100 compression cylinders and 100 splitting tensile cylinders from six separate batches. The cylinder sizes varied from 25 × 50 mm to 75 × 150 mm and were sealed with shellac to maintain uniform moisture content. All cylinders were capped with Hydrostone and were tested using accessories and loading rates that were scaled geometrically to match the specimen size. A summary of the test results follows:

Specimen Size, mm	Compression Strength, MPa	Splitting Tensile Strength, MPa
75 × 150	21.8	2.2
50 × 100	20.8	2.2
38 × 75	21.1	2.1
25 × 50	20.9	2.2

The consistency in strength within each size in a given batch was excellent; coefficients of variation ranged from 0.0005 to 0.064 (with an average of 0.033) for compressive strengths and from 0.004 to 0.102 (average of 0.048) for splitting tensile strengths.

Tests for modulus of rupture were conducted on two beam sizes 25 × 38 mm and 50 × 75 mm in cross section. All beam dimensions and loading accessories were scaled. The compaction procedure and drying times were adjusted to achieve a uniform moisture content and density. Six small beams and three large beams were cast from each of six batches; the coefficients of variation for measured modulus of rupture averaged 0.046 for each of the 12 test series. A size effect was evident from the results:

Beam Cross Section, mm	Modulus of Rupture, MPa
50 × 75	3.66
25 × 38	3.96

It was concluded that there are no discernible size effects in the compressive and splitting tensile strength of Ultracal mortars when the density, moisture content, and loading conditions are identical for various size specimens (Loh, 1969). A small size effect was measured for modulus of rupture; it is felt that this is produced by a combination of strain gradient effects and statistical variation in strength over the uniform moment region in the middle third of the beam. More discussion on size effects is given in Chapter 9. It is evident that proper procedures can reduce size effects in gypsum mortars to negligible levels. These procedures are relatively easy to follow if the models are beams, columns, or frames, but they are difficult to implement in more complex or variable-thickness models.

4.11 MODELING OF CONCRETE MASONRY STRUCTURES

4.11.1 Introduction

Masonry structures represent one of the oldest forms of construction. Many different types of masonry components have evolved over the years both in this country and elsewhere for constructing walls of low-rise and high-rise buildings. The large majority of masonry units fall into two main categories: burnt clay and block. Both of these types find extensive applications in civil engineering structures and have been successfully modeled at reduced geometric scales. A brief historical review of work using the modeling technique in masonry structural studies, together with similitude requirements, has been presented in Chapter 2.

Since the first edition of this book an explosion of research activity has taken place in masonry construction. Prompted by the insistence of national organizations, such as the Nuclear Regulatory Commission, on seismic repair and retrofit of masonry walls in a number of operating nuclear power plants and by the desire of national and local trade organizations to improve the quality of engineered masonry, the National Science Foundation and many other sponsors underwrote a considerable amount of new research. This activity has led to a new level of understanding of masonry behavior and a vast body of new knowledge leading to new codes of practice and design aids for the practitioner. This new body of knowledge is well summarized by Drysdale et al. (1994).

4.11.2 Material Properties

4.11.2.1 Prototype Masonry Units

Concrete or block masonry units are molded of a mixture of sand, aggregate, cement, and water under pressure and/or vibration. Curing is by autoclave and steam curing. There are many types of aggregates used in concrete masonry; these include the normal weights, such as sand, gravel, crushed stone, air-cooled blast-furnace slag, and the light weights, such as expanded shale, clay, slate, expanded blast-furnace slag, sintered fly ash, coal cinders, pumice, and scoria (ACI-TMS-ASCE Joint Committee 531, *Concrete Masonry Structural Design and Construction*).

The terms *normal weight* and *light weight* refer to the density of the aggregates used in the manufacturing process. Generally, local availability determines the use of any one type, and blocks made with any of the aforementioned aggregates are considered concrete blocks. However, the term *concrete block* is used in some locations to describe units made of sand and gravel or crushed stone (Randall and Panarese, 1976).

ASTM C90 *Standard Specifications for Load-Bearing Concrete Masonry Units* classifies concrete masonry units according to type. The type of unit is either Type I, moisture-controlled, or Type II, non-moisture-controlled. From a design standpoint, dry shrinking due to loss of moisture could cause excessive stress buildup and cracking of walls. This specification is designed to eliminate moisture-less effects.

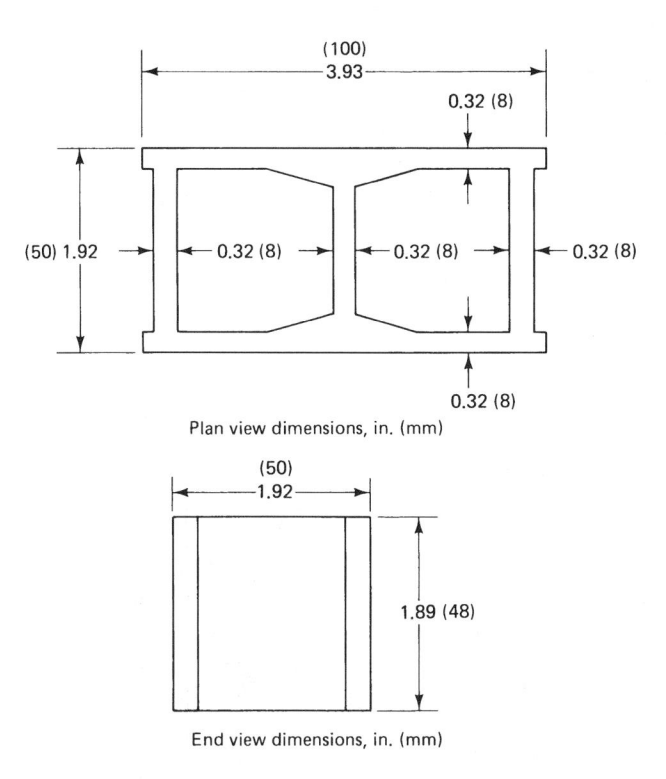

INELASTIC MODELS: MATERIALS FOR CONCRETE AND CONCRETE MASONRY STRUCTURES

Figure 4.36 Model concrete masonry units at ¼ scale.

Minimum face shell and web thicknesses and equivalent thikcness requirements are specified in ASTM C90. Most concrete masonry units have net cross-sectional areas ranging from 50 to 70% (30 to 50% core area) depending on unit width, face shell and web thickness, and core configuration (Randall and Panarese, 1976).

ASTM C140 *Standard Methods of Sampling and Testing Concrete Masonry Units* outlines the procedure for determining compressive strength of block masonry units. Attempts have been made to measure tensile strength, but no accepted technique has been adopted. The compressive strength of hollow concrete blocks is about 3.4 to 20.7 MPa after 28 days, based on gross area. With different types of lightweight aggregates, values from 3.4 to 7.9 MPa are reported (Sahlin, 1971)]. Because curing plays an important role in block masonry strength, no conclusions of compressive strength can be drawn from the knowledge of unit weight, aggregate type, and water–cement ratio. The designer, therefore, must rely on test data of finished blocks.

4.11.2.2 Model Masonry Units

1. One-Fourth-Scale Blocks — The configuration of the first ¼-scale model units reported (Harris and Becica, 1977; Becica and Harris, 1977) resembles the double corner and regular stretcher type of $200 \times 200 \times 400$ mm nominal size concrete masonry blocks. Manufactured by the National Concrete Masonry Association, McLean, VA, these units measured $50 \times 50 \times 100$ mm nominally. Data on the physical properties of both model and prototype units are listed in Table 4.9. The prototype data was collected under the guidelines of the ASTM C140 by Yokel et al. (1971), and modeled as closely as possible. Figure 4.36 shows typical average sections of the units. Note that the outside dimension and face shells were scaled accurately while the webs were oversized by about 30%. The web thickness dimensions shown in Figure 4.36 are the average of 36 measurements taken at six locations on the webs of six regular stretcher units.

Table 4.9 Dimensions and Physical Properties of Concrete Masonry Units

Masonry Unit	Number of Specimens Measured	Actual Dimensions Width, in.	Actual Dimensions Length, in.	Actual Dimensions Height, in.	Minimum Face Shell Thickness, in.	Gross Area, in.²	Net Solid, %	Compressive Strength Gross Area, psi	Dry Weight Density, lb/ft³	Water Absorption, lb/ft³
Prototype: 8-in. (200 mm) hollow expanded slag block (Yokel et al. (1971)	5	$7\tfrac{5}{8}$	$15\tfrac{5}{8}$	$7\tfrac{5}{8}$	$1\tfrac{1}{4}$	119.1	52.2	1100	103.0	14.3
$\tfrac{1}{4}$-scale model: double-corner units	12	1.91	3.91	1.89	0.323[a]	7.48	52.0	901[b]	114.4	13.9
Regular stretcher units	6	1.92	3.93	1.89	0.325[a]	7.54	50.9	1175	110.9	17.1

[a] Average face shell thickness.
[b] Combined average compressive strength of six saturated surface dry (f'_c = 816 psi) and six dry (f'_c = 1086 psi) specimens.

Net solid area and adsorption for 12 units were calculated using the mean values of all dimensions and weights. Adsorption represents the difference of dry-weight density and wet-weight density after a 1-min immersion in water. These model units were then capped on both bearing surfaces with Hydrostone, a gypsum product, and tested in compression, with a loading rate that caused failure in 1 to 2 min. For the double-corner units, Table 4.8 shows a difference in strength between units tested wet and those tested dry, although this may be due only to statistical variation of such a small population.

Since the mid-1970s when the above results were presented, several other model masonry units have been developed at Drexel University (Hamid and Abboud, 1986; Abboud, 1987; Oh, 1988; Labrouki, 1989; Lafis, 1988; Larbi, 1989; Abboud et al., 1990; Abboud and Hamid, 1990; Harris et al., 1990, 1992; Harris and Oh, 1990; Oh and Harris, 1990; Larbi and Harris, 1990; Hamid and Ghanem, 1991; Hamid and Chandrakeerthy, 1992; Ghanem and Hamid, 1992; Oh et al., 1993; Harris et al., 1993; Ghanem et al., 1993; Oh, 1994; Hamid et al., 1994, 1995; El-Shafie et al., 1996) and are described below. An extensive study by Abboud (1987) introduced three new ¼-scale block model units shown in Figures 4.37 through 4.39. The first unit or Type I is a ¼-scale version of a Canadian double-corner kerfed block 150 × 200 × 400 mm nominal dimensions. The second and third units (Type II and Type III) are scaled versions of stretcher units found typically on the East Coast of the U.S., with and without extra mortar bed, respectively.

These model units were made on the same National Concrete Masonry Association (NCMA) block-making machine donated to Drexel University (Figure 4.40a). A new mold box (Figure 4.40b) was made which produced the Type I, II, and III units. The full operation of this model block-making machine and a new more accurate machine for making model interlocking mortarless blocks are described in Chapter 6. The NCMA/Drexel machine is equipped with a mechanical vibrator, and the vibration operation is repeated several times during the process to ensure a good compaction. A dry lubricant is used to ease the release of the units from the molds. After removal, the units are stored in a high humidity room for curing for at least 28 days before they are taken out for testing or storage. Ordinary concrete sand was used having the gradation shown in Figure 4.41 with the maximum size aggregate based on similitude requirements (Chapter 2). A mix proportion of 0.44:1:4 (water:cement:sand) by weight was used in making the ¼-scale model block units. Portland cement Type III was used to accelerate production.

The geometric and physical properties of the new Types I to III ¼-scale units are given in Tables 4.10 and 4.11, respectively. Axial compression tests, following ASTM C140 with properly scaled steel bearing plates, were conducted on all three types of units. Tensile splitting testing was conducted according to ASTM C1006 *Standard Test Method for Splitting Tensile Strength of Masonry Units* as shown in Figure 4.42. The tensile splitting stresses at failure are calculated from the equation

$$f_{bt} = 2P/\pi A \qquad (4.4)$$

where: f_{bt} = the tensile splitting strength of the block
P = splitting load at failure
A = net sectional area of the splitting plane

Figure 4.43 compares the tensile splitting strength of the ¼-scale model blocks and prototype data reported. The results show good agreement. A summary of all the mechanical properties is given in Table 4.12. Extensive characterization of these model block units and their use to study the behavior of slender block walls under cyclic out-of-plane bending was carried out by Abboud (1987). A summary of these results is given in Chapter 10.

2. One-Third-Scale Blocks — A 150 × 200 × 400 mm nominal prototype block unit used throughout the Technical Coordinating Committee for Masonry Research (TCCMAR) was modeled at ⅓ scale, as shown in Figure 4.44 (Harris et al., 1990). This unit has a low net solid ratio (large cell to facilitate placing rebar and grouting), Figures 4.44a and b, and is used in the

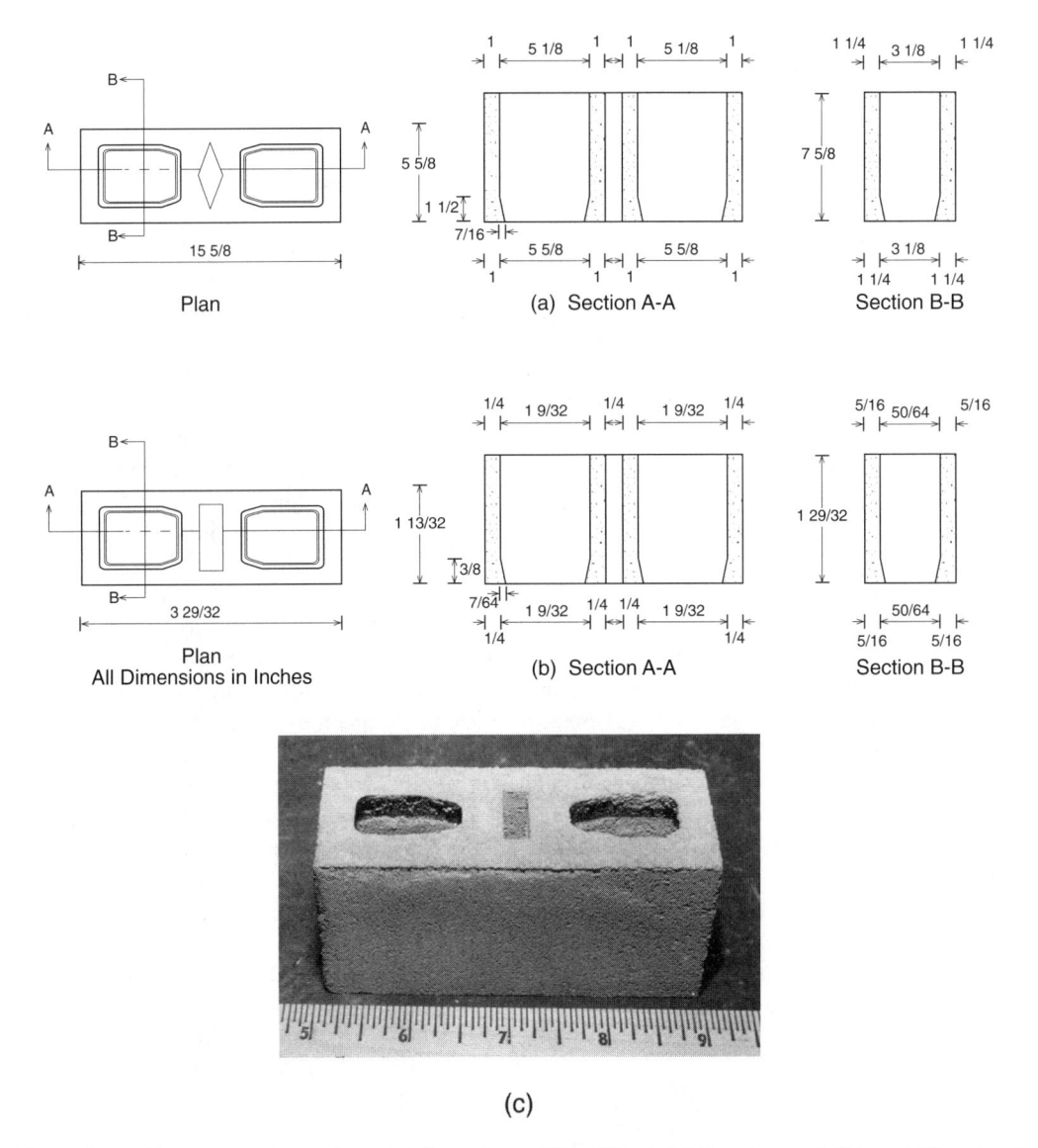

Figure 4.37 Prototype and ¼-scale model dimensions of Type I block. (a) Prototype unit; (b) model Type I unit; (c) model Type I unit. (All dimensions in inches). (After Abboud, 1987.)

western states of the U.S. The ⅓-scale model block is shown in Figure 4.44c. A mold box producing five units per use was made to fit the NCMA/Drexel block-making machine as shown in Figure 4.44d. The geometric properties of the prototype and ⅓-scale model blocks are given in Table 4.13. The constituents used in the units consist of Portland cement Type III, sand aggregates for which the maximum size is determined on the basis of similitude and the minimum thickness of the unit (gradation shown in Figure 4.41), and enough water to achieve complete hydration. The mix proportions by weight were 0.49:1:4.85 (water:cement:aggregate). The mix, obtained after several trials, was designed to achieve a unit compressive strength in the range of 20 to 21 MPa. (on the net area), which agrees with prototype results.

The compressive strength of the model unit was determined according to ASTM C140. The units were capped with Hydrostone gypsum on both bearing surfaces and loaded through steel bearing plates. The tensile splitting strength was obtained according to ASTM C1006. Again the size of the steel rods used for the application of the line loads was scaled appropriately. The physical

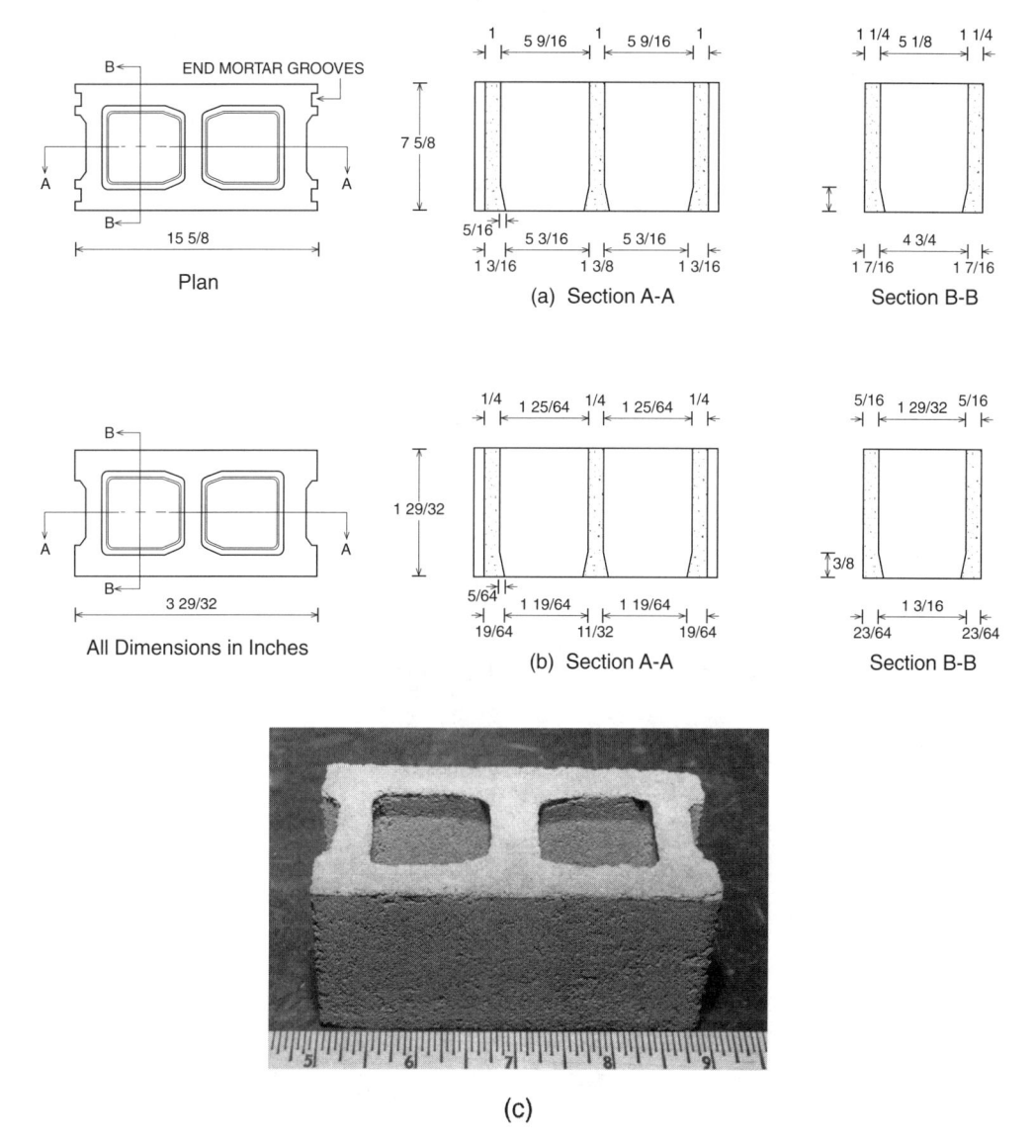

INELASTIC MODELS: MATERIALS FOR CONCRETE AND CONCRETE MASONRY STRUCTURES 175

Figure 4.38 Prototype and ¼-scale model dimensions of Type II block. (a) Prototype unit; (b) model Type II unit; (c) model Type II unit. (All dimensions in inches). (After Abboud, 1987.)

properties of the model block units were obtained in accordance with ASTM C140 for the weight density, the average net and gross volumes, the average net and gross areas, and moisture content. The initial rate of absorption was also determined. Physical and mechanical characteristics of the model and prototype block units are summarized in Table 4.14.

Strength characterizations of masonry assemblies in axial compression, shear, in-plane and out-of-plane tension, and applications to structural components made with the ⅓-scale block units and tested under cyclic reversed loading were carried out. Comparisons of model and prototype results are given below and in Chapter 10, section 10.2.1.2, cases 2 and 3.

3. One-Third-Scale Interlocking Mortarless Blocks — Two interlocking mortarless units developed at Drexel University (Harris et al., 1993) were briefly discussed in Chapter 1 Case Study F. These two units have the prototype dimensions given in Figures 1.19b and 1.20b and are shown at ⅓ scale in Figure 4.45. Their physical properties are listed in Table 4.15. Unlike the other units described above that were scaled from prototype blocks, the interlocking mortarless blocks were

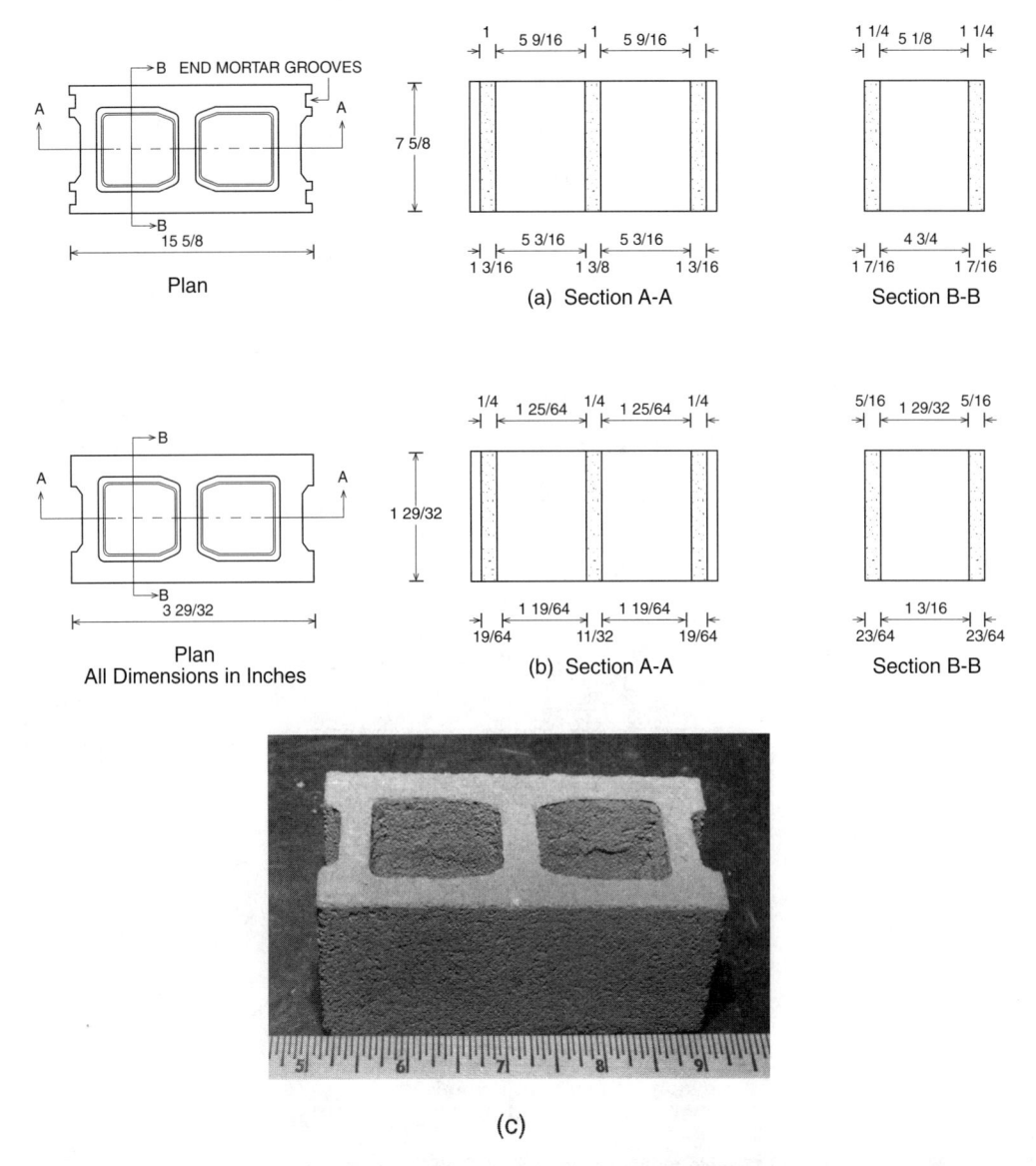

Figure 4.39 Prototype and ¼-scale model dimensions of Type III block. (a) Prototype unit; (b) model Type III unit; (c) model Type III unit. (All dimensions in inches). (After Abboud, 1987.)

part of a product development effort in which the prototype was being designed by means of tests conducted on ⅓-scale direct ultimate-strength models.

The compressive strength of the three types of units studied were determined using six randomly chosen specimens capped with a high-strength gypsum. The top and bottom bearing faces of the modified H-Block units, which have a tongue-and-groove arrangement, were capped without their removal. In the case of the WHD Block specimens, which are unsymmetrical in plan, the load was applied at the centroid of the net area.

The ultimate loads at failure were divided by the unit average net area. Results are shown in Table 4.16. As expected, the higher compressive strengths were obtained in the modified H-Blocks because of their higher densities. The model conventional and the WHD Block units showed compressive strengths of 12.9 and 13.1 MPa, respectively. The modified H-Block units showed conical shear-compression type of failure as shown in Figure 4.46a. The failures were explosive and occurred near the midheight of all of the six units tested. This type of failure mode is typically found in 200-mm-wide

Figure 4.40 Drexel University model block-making machine. (a) Removing model blocks; (b) mold box.

full-scale conventional block units tested at Drexel University (Assis et al., 1989), see Figure 4.46b. The WHD-Block and ⅓-scale model conventional block tests showed asymmetric shear-compression failures through the face shells, as shown in Figure 4.46c. Unlike the modified H-Blocks where failure was sudden and explosive, the failure was gradual, forming long but inclined vertical separations of the face shells. This type of failure mode was similar to the 6-in full-scale conventional blocks tested at Drexel University (Assis et al., 1989), see Figure 4.46d.

The differences in the failure modes between the modified H-Blocks and the other blocks seem to be again related to the different fabrication methods used. The modified H-Block units were fabricated using a brass mold which consolidated the material and resulted in a higher material

Table 4.10 Geometric Characteristics of Model and Prototype Units

Concrete Masonry Unit		Width, mm	Height, mm	Length, mm	Gross Area, mm²	Net Area at Top, mm²	Net Area at Middle, mm²	Net Area at Bottom, mm²	Solid at Top, %	Solid at Bottom, %	Solid at Bottom, %
Kerfed block	Prototype	142.9	193.7	396.9	56704	35897	35897	42989	63.31	63.31	75.81
	Model Type I	35.7	48.4	99.2	3544	2141	2141	2585	60.42	60.42	72.92
Stretcher with extra mortar bed	Prototype	193.7	193.7	396.9	76557	38003	40544	50503	49.64	52.96	65.97
	Model Type II	48.4	48.4	99.2	4785	2398	2557	3179	50.11	53.43	66.44
Stretcher with no extra mortar bed	Prototype	193.7	193.7	396.9	76557	38003	40544	43086	49.64	52.96	56.28
	Model Type III	48.4	48.4	99.2	4785	2398	2557	2715	50.11	53.43	56.75

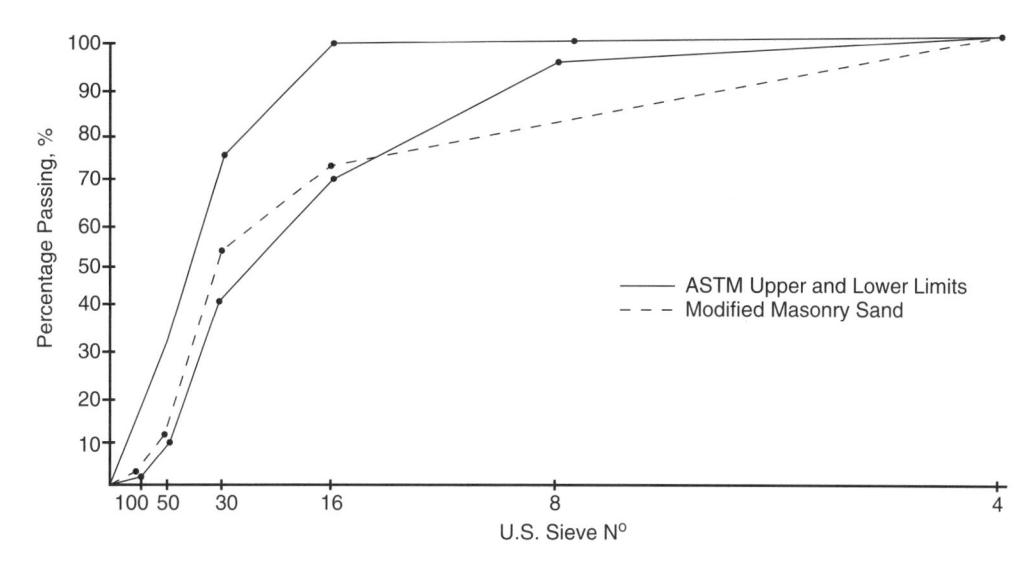

INELASTIC MODELS: MATERIALS FOR CONCRETE AND CONCRETE MASONRY STRUCTURES

Figure 4.41 Aggregate gradation curve for model masonry units.

Table 4.11 Physical Properties of Model Masonry Concrete Blocks

Description	Model Block Type I	Type II	Type III
Weight density, kg/m³ (unit weight)	1762.8	1762.8	1794.9
Average net volume, m³	1.10×10^{-4}	1.27×10^{-4}	1.22×10^{-4}
Average gross volume, m³	1.71×10^{-4}	2.37×10^{-4}	2.34×10^{-4}
Average net area[a], %	64.0	53.6	52.1
Average net area[b], m²	2.26×10^{-3}	2.57×10^{-3}	2.50×10^{-3}
Absorption, kg/m³	242.0	234.0	216.3
Absorption, %	13.8	13.2	12.1
Moisture content, %	9.7	10.2	12.1

[a] Average Net Area, % = $\dfrac{\text{Average Net Volume}}{\text{Average Gross Volume}} \times 100$

[b] Average Net Area, m² = $(\text{Average Gross Area})\left(\dfrac{\text{Average Net Volume}}{\text{Average Gross Volume}}\right)$

Figure 4.42 Model block splitting strength determination.

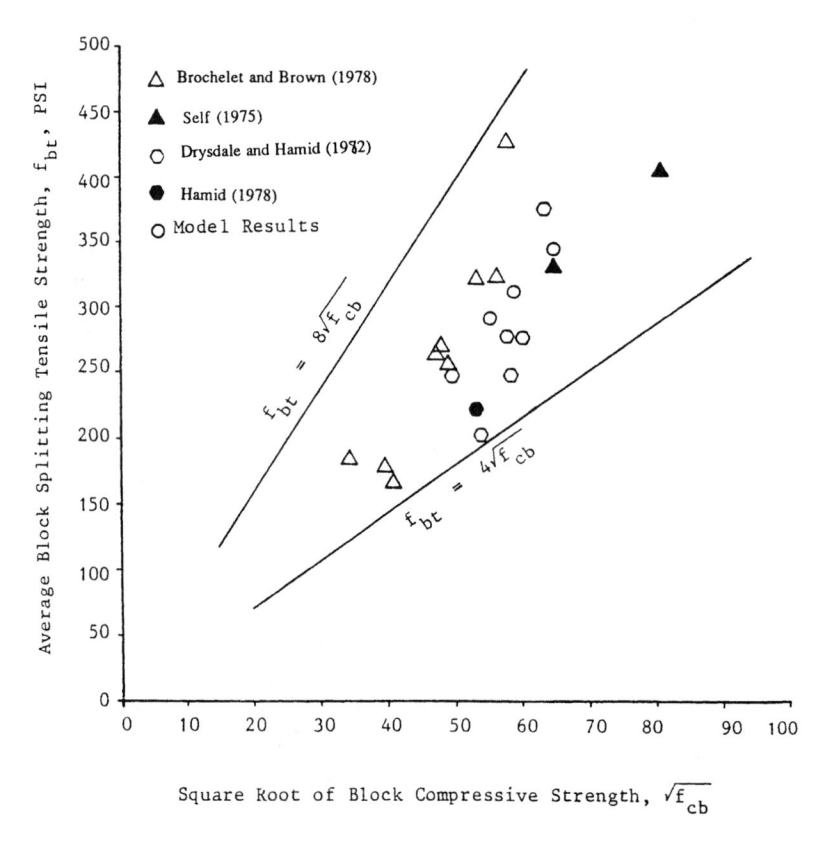

Figure 4.43 Comparison between model and prototype block splitting tensile strength. (After Abboud, 1987.)

Table 4.12 Mechanical Properties of Model Concrete Masonry Blocks[a]

	Model Block					
	Type I		Type II		Type III	
Description	Mean	c.o.v.[(2)]	Mean	c.o.v.	Mean	c.o.v.
Axial compressive strength,[c] f_{cb}, MPa	24.0	0.09	21.4	0.12	29.0	0.07
Modulus of elasticity, GPa						
a. Initial, E_{bi}	18.6	0.04	16.9	0.05	20.7	0.04
b. Secant, 50% ultimate, E_{bs}	16.8	15.7	17.5			
Strain at ultimate stress, ε_{cb}	0.0021	0.002	0.0022			
Tensile splitting strength,[d] f_{bt}, MPa	2.1	0.10	1.9	0.05	2.4	0.11
Modulus of rupture,[d] MPa						
a. In-plane, f_{bir}	4.4	0.07	4.3	0.08	5.1	0.07
b. Out-of-plane, f_{bor}	2.3	0.04	2.2	0.04	2.6	0.03

[a] Average of 20 specimens.
[b] Coefficient of variation.
[c] Based on the average net area.
[d] Based on net area.

density. The WHD Block and the conventional blocks were fabricated using a new machine developed at Drexel and described in Chapter 6 which can consolidate the mix material by means of imposed vibrations of the mold. This causes a less uniform distribution of block density. It seems that, in order to obtain a higher and more uniform density, the mix should be modified with higher water–cement ratio and/or modification to the machine in such a way so that it can produce stronger vibration and pressure. This can be done by adding more weight on the rotating eccentric mass of the vibrator unit and improving the force of the motorized pulling system; refer to Chapter 6.

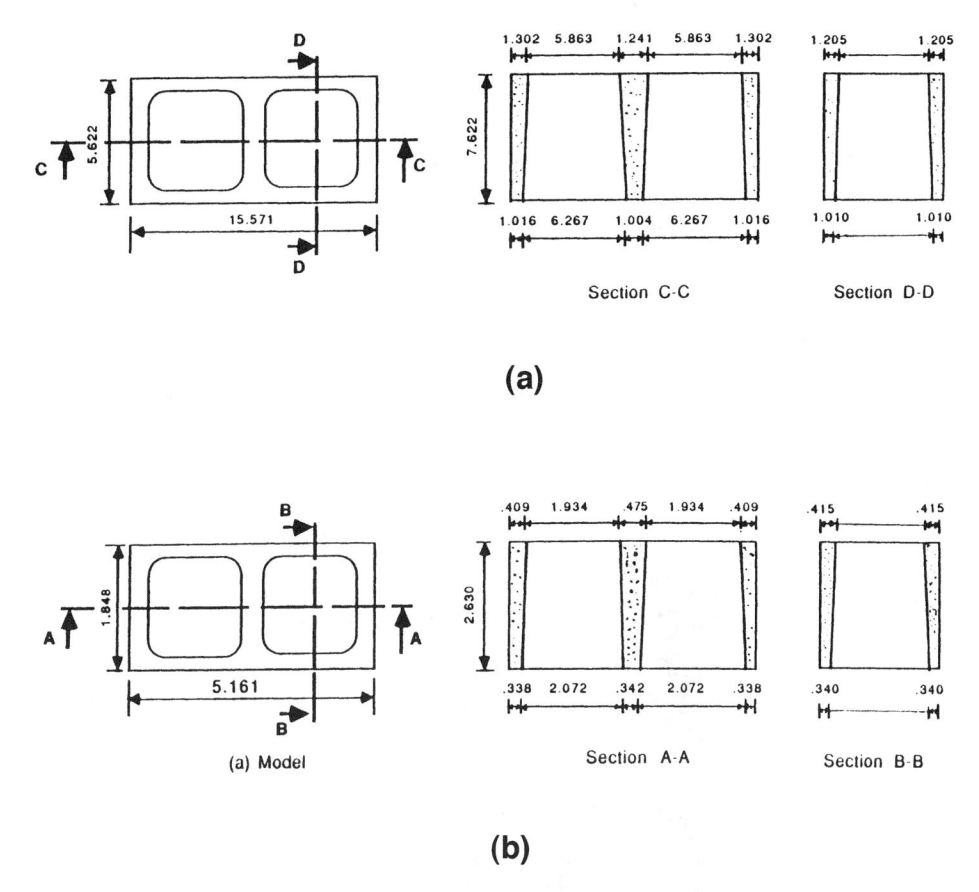

INELASTIC MODELS: MATERIALS FOR CONCRETE AND CONCRETE MASONRY STRUCTURES 181

Figure 4.44 One-third scale model block for the U.S.–Japan TCCMAR program. (a) Prototype dimensions; (b) model dimensions; (c) top and bottom views of unit; (d) mold box (top) and model block units removed.

Four units each of these ⅓-scale blocks were tested for tensile splitting strength according to ASTM C1006 methods. In the case of the modified H-Block, the interlocking tongue and grooves were removed to obtain a more uniform contact with the 6-mm-diameter steel bearing bars. A high-strength gypsum compound was used on the bearing surfaces onto which the loading bars were pressed down until a contact was made. The splitting strength was determined at both cells in the WHD Block and the conventional block units and in the center cell of the modified H-Block unit, as shown in Table 4.17. The tensile splitting strengths reported in Table 4.17 were determined using Equation 4.4. The failure modes are characterized by vertical splitting as shown in Figure 4.42.

4.11.2.3 Model Mortars

ASTM C-270-94 *Standard Specification for Mortar for Unit Masonry* recognizes four mortar types, with acceptance based on either the proportion specification or property specification (Table 4.18). Compressive strength of mortars depends mainly on the type and quantity of cementitious material used (Sahlin, 1971) and the duration of curing (Isberner, 1969). Compressive strength is measured in the laboratory by casting, curing, and testing 50-mm cubes in compression. Because available test data on full-scale hollow-core masonry incorporates the use of ASTM Type N masonry mortar, it was necessary to develop a similar type model masonry mortar. By using the proportion specification as outlined in ASTM C270 as a guide, three mixes were tested in an attempt to match the reported 28-day strength of 5.2 MPa on 50-mm cubes.

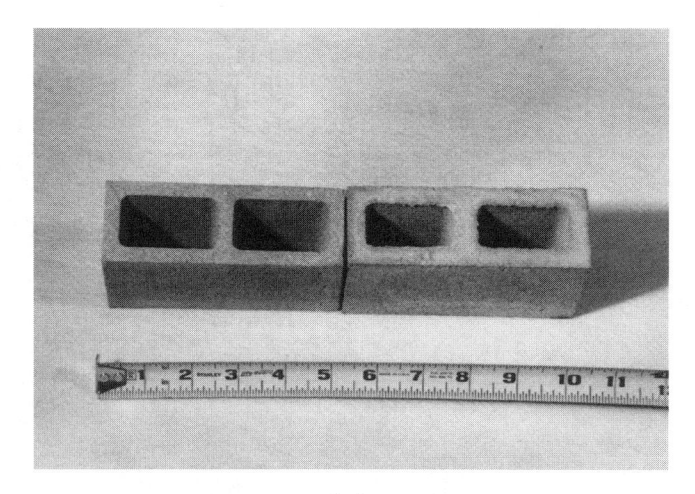

(c)

(d)

Figure 4.44 (continued)

The proportion specification calls for a masonry cement–aggregate ratio of 1:3 by volume. After several trial mixes, which were in general too strong, a mix of 1:1:4 (water:cement:aggregate) proportion by volume was found adequate as a model mortar mix. Generally, the strength of this mix was approximately 5.17 MPa. The workability and retentivity suffered to some degree because of the harshness of the mix, but steps were taken during casting of model masonry components to remedy these conditions. For these reasons, the 1:1:4 mix was chosen as the model of Type N mortar.

The aggregate used for model mortar was a commercially obtained natural masonry sand having the gradation shown in Figure 4.47 and a fineness modulus of 0.8. To properly scale a nominal 10-mm mortar joint as used in practice, a model joint thickness of 2.4 mm was necessary. This

Table 4.13 Geometric Characteristics of Model and Prototype Blocks[a]

Description	ASTM Standards	Prototype[a]	Theoretical ⅓ Scale[b]	Actual Model[c]
Width, mm	C140	142.8	47.6	46.9
Height, mm	C140	193.6	64.5	66.8
Length, mm	C140	395.5	131.8	131.1
Minimum face shell thickness, mm	C140			
At top		30.6	10.2	10.5
At bottom		25.7	8.5	8.6
Minimum central web thickness, mm	C140			
At top		31.5	10.5	12.1
At bottom		25.5	8.5	8.7
Minimum end web thickness, mm	C140			
At top		33.1	11.0	10.4
At bottom		25.8	8.6	8.6
Gross area, mm²	C140	56480	6280	6150
Net area, mm²,				
At Top	C140	32170	3570	3500
At middle		30140	3350	3350
At bottom		27570	3060	3200
Percent solid, %,				
At top	C140	56.97	56.97	56.92
At middle		53.36	53.36	54.51
At bottom		48.82	48.82	52.10

[a] Based on the average of six specimens.
[b] Scaling of prototype.
[c] Average of six model specimens (actual model).

Table 4.14 Properties of Model and Protoype Masonry Concrete Blocks[a]

Description	ASTM Standards	Model Block	Prototype Block
Physical properties:			
Weight density, kg/m³	C140	1885	1635
Average net volume, m³	C140	224×10^{-6}	6.1×10^{-3}
Average gross volume, m³	C140	411×10^{-6}	10.9×10^{-3}
Average net area, m²	C140	3.4×10^{-3}	30.1×10^{-3}
Average net area, %	C140	54.50	54.51
Absorption, kg/m³	C140	237	176
Absorption, %	C140	11.15	10.8
Moisture content, %	C140	6.23	3.83
IRA,[b] gm/min/30 m²	C67	25.7×10^{-3}	28.3×10^{-3}
Mechanical characteristics:			
Axial compressive strength, f_{cb}, MPa	C140	20.6	20.1
Tensile splitting strength, f_{bt}, MPa	C1006	1.6	2.0

[a] Based on the average of six specimens.
[b] Initial rate of absorption.

required the removal of particle sizes greater than a U.S. No. 16 sieve. The result of this reduction is shown in Figure 4.47 and represents the aggregate gradation used to fabricate model masonry components. It has been shown (Harris et al., 1966) that in testing concrete mortars it is the volume and not the gradation that most strongly influences model mortar behavior. As previously indicated, it was necessary to decrease the cement–aggregate ratio from the specified 1:3 by volume to 1:4 in order to achieve the required model mortar strength.

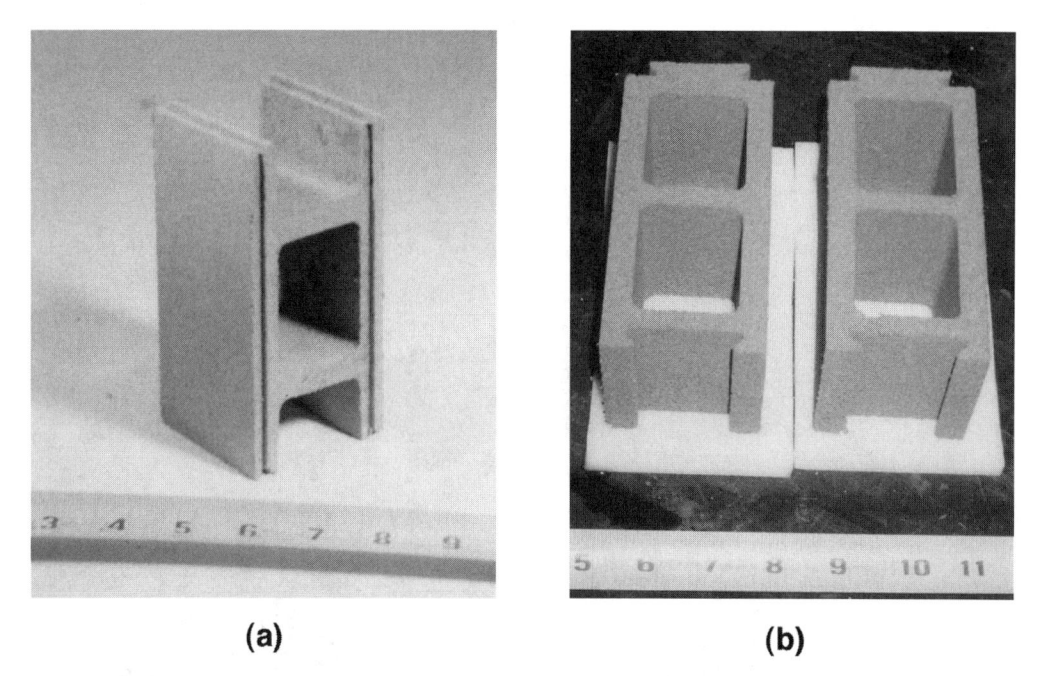

Figure 4.45 One-third scale mortarless interlocking block units developed at Drexel University. (a) Modified H-block; (b) WHD-blocks.

Table 4.15 Density, Volume, and Net Area of Blocks Fabricated

Block Type	Density, kg/m^3	Net Volume, m^3	Gross Volume, m^3	Net Area, m^2	Net Area, %
Modified H	1970	273.2×10^{-3}	566.5×10^{-3}	4.3×10^{-3}	48.2
WHD	1780	307.3×10^{-3}	590.4×10^{-3}	4.5×10^{-3}	52.0
Conventional	1730	269.1×10^{-3}	541.8×10^{-3}	4.1×10^{-3}	49.7

As is typical for cementitious materials, moisture plays a significant role in the strength development of prototype and model mortars. It is reported (Isberner, 1969) that mortar relative humidity must be a minimum of 85% for hydration to continue and that as relative humidity decreases so does the rate of hydration. This can be of significance in small-scale modeling where instrumentation and testing over an extended period of time can cause changes in strength.

Figure 4.48 shows the results of strength–age tests on 25 × 50 mm compression cylinders of 1:1:4 mix. All specimens were moist-cured for 3 days, at which time sufficient specimens were removed from the wet room, shellacked, and left in room air until tested with their counterpart moist-cured cylinders. Each group of tests represents six specimens.

Note in Figure 4.48 that there was an increase in strength of 14% at 7 days for the dry-tested specimens. It has been observed (Harris et al., 1966) that when wet-cured cylinders of normal-weight concrete are allowed to lose some of their moisture, strength increases of up to 25% are obtained. Other studies of curing effects on 150 × 300 mm cylinders show that in the first 28 days the moist-cured specimens showed considerable strength increase, and no increase in strength for the air-cured specimen was observed after 14 days. On the other hand, for small volumes of model material, this effect is accelerated to such a degree that stable strengths are seen as early as 10 days for air-cured specimens.

Four 25 × 50 mm model mortar cylinders were instrumented with two 25 mm SR-4 strain gages on generators 180° apart and tested in compression, as shown in Figure 4.49. Specimens Nos. 1,

Table 4.16 Unit Compressive Strength

Block Type	Sample No.	Ultimate Load, N	Compressive Strength (Based on Net Area), MPa
Modified H	H-1	129000	29.9
	H-2	138800	32.2
	H-3	144600	33.5
	H-4	140100	32.5
	H-5	155700	36.1
	H-6	86300	20.0
	Average	132400	30.7
	c.o.v.[b] (%)	18.28	18.28
WHD	WHD-1	55600	12.3
	WHD-2	64300	14.3
	WHD-3	51200	11.4
	WHD-4	61800	13.7
	WHD-5	62300	13.8
	WHD-6	58300	12.9
	Average	58900	13.1
	c.o.v. (%)	8.32	8.32
Conventional	TR-1	52500	12.7
	TR-2	48500	11.7
	TR-3	66300	16.0
	TR-4	49800	12.1
	TR-5	47600	11.5
	TR-6	54400	13.1
	Average	53200	12.9
	c.o.v. (%)	12.94	12.94

[a] Based on net area.
[b] Coefficient of variation.

2, and 3 are of the mix 1:1:3 and specimen No. 4 is of a mix 1:1:4 by volume. Specimen No. 4 is typical of a batch used in casting shear or diagonal compression specimens. Values for the secant modulus at 50% of the ultimate compressive strength are also shown in Figure 4.49.

In reinforced grouted block masonry construction, ASTM C270 Type S mortar is sometimes specified. A model mortar mix was developed for the ⅓- and ¼-scale block masonry models. To comply with ASTM C144 *Aggregate for Masonry Mortar*, coarse particles with a size greater than a U.S. No. 16 (1.10 mm) were removed. The sand gradation is shown in Figure 4.50. The proportions of the model mortar were 1:3.83:0.212 (cement:sand:lime) by weight with enough water added to achieve an initial flow of 110 to 120% (increase in diameter of a standard cone after 25 drops on a standard flow table, Figure 4.51). Thus, the water-to-cement ratio varied from 0.85 to 0.95 approximately. Control specimens were taken according to ASTM C1019 *Standard Method of Sampling and Testing Grout*. Cylinders 50 × 100 mm and 50-mm cubes were used. These are air-cured and capped with a gypsum plaster prior to testing at the time of the masonry component. Typical compressive strength results of mortar specimens cast with ⅓-scale block masonry shear walls by Larbi (1989) are shown in Table 4.19.

4.11.2.4 Model Grout

Prototype grout is a high slump (200 to 250 mm) concrete made with fine aggregate and pea gravel. It must be fluid enough to fill voids in the grout space and to encase the reinforcement completely. It has been shown that even excess water will be absorbed by the masonry leaving the grout with a low enough water–cement ratio to have adequate strength. A grout used in the construction of model block shear walls (Larbi, 1989) represented a ⅓-scale replica of the coarse grout specified in ASTM C476 *Standard Specification for Grout for Masonry*. The mix proportion

(a)

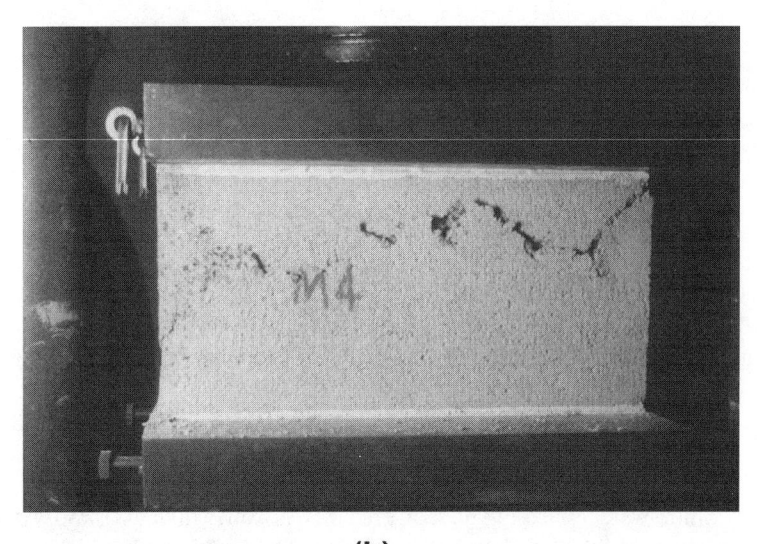

(b)

Figure 4.46 Compressive modes of failure of model and prototype blocks. (a) Typical failure mode of modified H-block; (b) typical failure mode of full-scale block Type I; (c) typical failure mode of WHD-block and conventional block; (d) typical failure mode of full-scale block Type II. (Courtesy of Drexel University, Philadelphia.)

used was 1:2.71:1.71:0.45:1/94 (Portland cement Type III:sand:gravel:water:grout aid) by weight. Regular concrete sand was used for the proportion of sand and gravel as it was seen that it reproduces at ⅓ scale the aggregate of the prototype grout. A grout aid (Sika mix 119/120) was premixed with a small amount of water and then added during the mixing of the grout to provide the necessary workability and is used to increase flow and compensate for shrinkage. The water-to-cement ratio was established to yield a mix that would flow readily without segregation. A minimum slump of 250 mm, was hence, achieved.

For quality control of the model grout, two types of specimens were used: 50 × 100 mm nonabsorbent cylinders and block-molded prisms (Figure 4.52) having a height-to-length ratio of 2 with a height being equal to that of the masonry unit (ASTM C1019). All grout control specimens are air-cured in the same atmospheric conditions as the companion walls and are capped and tested

(c)

(d)

Figure 4.46 (continued)

at the same age as the test components. The cylinders yielded a low grout strength as expected due to the higher water-to-cement ratio; therefore, for better accuracy in reproducing the actual strength of the grout in the wall and for the purpose of correlation to the prototype results, the block-molded

Table 4.17 Unit Tensile Splitting Strength

Block Type	Sample No.	Loading at Splitting, N Left	Center	Right	Average	Tensile Splitting Strength, MPa Left	Center	Right	Average
Modified H	SH-1		5780				2.5		
	SH-2		4000				1.7		
	SH-3		6670				2.9		
	SH-4		3890				1.7		
	SH-5		3560				1.5		
	Average		4780		4780		2.1		2.1
	c.o.v.[a] %		28.53		28.53				28.53
WHD	SW-1	2540		3020		1.1		1.3	
	SW-2	3140		3650		1.3		1.6	
	SW-3	2250		2690		1.0		1.2	
	SW-4	3960		3160		1.7		1.4	
	Average	2970		3130	3050	1.3		1.4	1.3
	c.o.v. %	25.49		12.67	18.52	25.4		12.73	18.52
Conventional	ST-1	1910		2310		0.8		1.0	
	ST-2	1820		1670		0.8		0.7	
	ST-3	2690		2450		1.1		1.1	
	ST-4	2270		2690		1.0		1.1	
	Average	2180		2280	2230	0.9		1.0	0.9
	c.o.v. %	18.15		19.16	17.45	18.01		19.21	15.78

[a] Coefficient of variation.

Table 4.18 Mortar Proportions by Volume (ASTM C-270)

Mortar Type	Minimum Compression Strength at 28 Days, psi (MPa) (2-in. cubes)	Portland Cement	Hydrated Lime Minimum	Maximum	Masonry Cement	Damp Loose Aggregate
M	2500 (17.2)	1	—	$\frac{1}{4}$	—	$2\frac{1}{2}$ to 3 times sum of the cements or the sum of cement plus lime
		1	—	—	1	
S	1800 (12.4)	1	$\frac{1}{4}$	$\frac{1}{2}$	—	
		$\frac{1}{2}$			1	
N	750 (5.2)	1	$\frac{1}{2}$	$1\frac{1}{4}$	—	
		—	—	—	1	
O	350 (2.4)	1	$1\frac{1}{4}$	$2\frac{1}{2}$	—	
		—	—	—	1	

prism strength should be used. Typical compressive failures of grout block-molded prisms are shown in Figure 4.53.

4.12 STRENGTH OF MODEL BLOCK MASONRY ASSEMBLAGES

Masonry is an inelastic, nonhomogeneous, and anisotropic material that consists of masonry units, mortar, steel reinforcement, and grout. Consequently, when subjected to external loads, masonry structural elements do not behave elastically even in the range of small deformations. Modeling of structures in the inelastic range requires a choice of materials that have identical

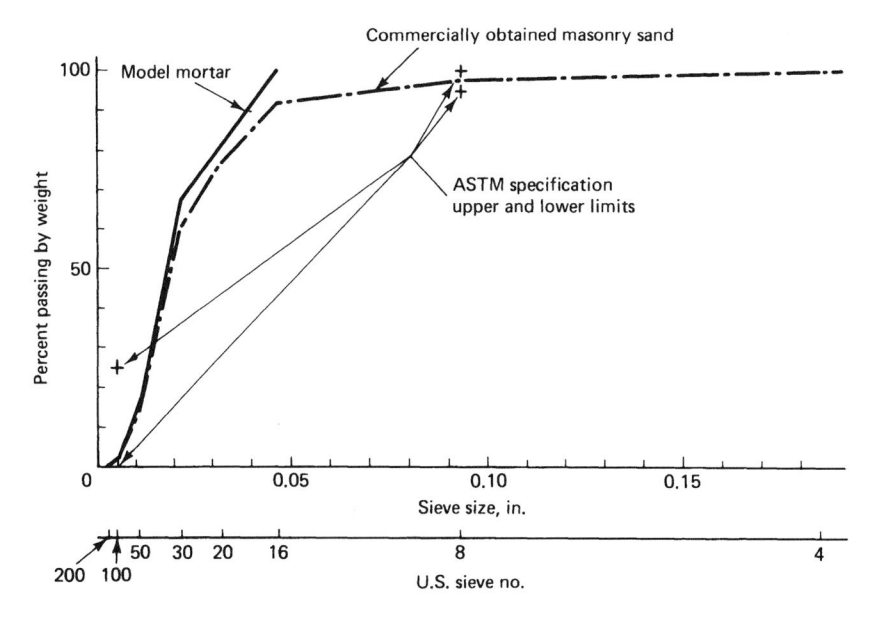

Figure 4.47 Aggregate gradation curves for masonry mortars.

stress–strain characteristics (Chapter 2). In the case of concrete masonry structures, the validity of the direct modeling technique must be demonstrated for simple compression, tension, and shear behavior prior to the testing of structures with complex states of stress. In addition, bed joint shear, in-plane tensile strength, and out-of-plane flexural tensile strength are generally considered to be critical strength characteristics of masonry (Drysdale et al., 1994). For this reason all of these strength characteristics are evaluated by direct comparison with carefully documented prototype tests. Only by this direct comparison of model and prototype assemblage behavior can the validity of the ⅓-scale and ¼-scale direct modeling technique be used with confidence in more complex structures, where prototype results do not exist. For direct modeling of reinforced grouted block masonry structures, the validity of bond strength and development length must also be demonstrated. The results of all of these correlations are compared in the following sections.

4.12.1 Axial Compression

The ASTM E447 *Compressive Strength of Masonry Assemblages* specifies test procedures, test equipment, method of reporting for prisms of height–thickness ratio from 2 to 5 with correction factors for slenderness effects. It is generally felt that end restraints have a large effect on the strength of two-block prisms of full scale. At a reduced scale, such effects are easily amplified and require careful consideration. Although prisms two units high are allowed by ASTM E447, it is recommended that three- and four-unit prisms be used in model testing. The three-block prism was chosen for comparison because prototype data using such specimens were more readily available. A typical prism test set up to record the complete stress–strain behavior is shown in Figure 4.54. Compression stress–strain curves from ⅓-scale grouted masonry prisms are shown in Figure 4.55. Typical modes of failure of hollow and grouted ¼-scale masonry prisms three units high are shown in Figure 4.56. Failure of the three-block hollow prisms was by end splitting, as shown in Figure 4.56a. This is also typical of prototype prisms.

Since masonry strength is directly related to the mortar strength (Fishburn, 1961; Sahlin, 1971; Yokel et al., 1971), the adjusted model strengths are compared with the prototype data in which

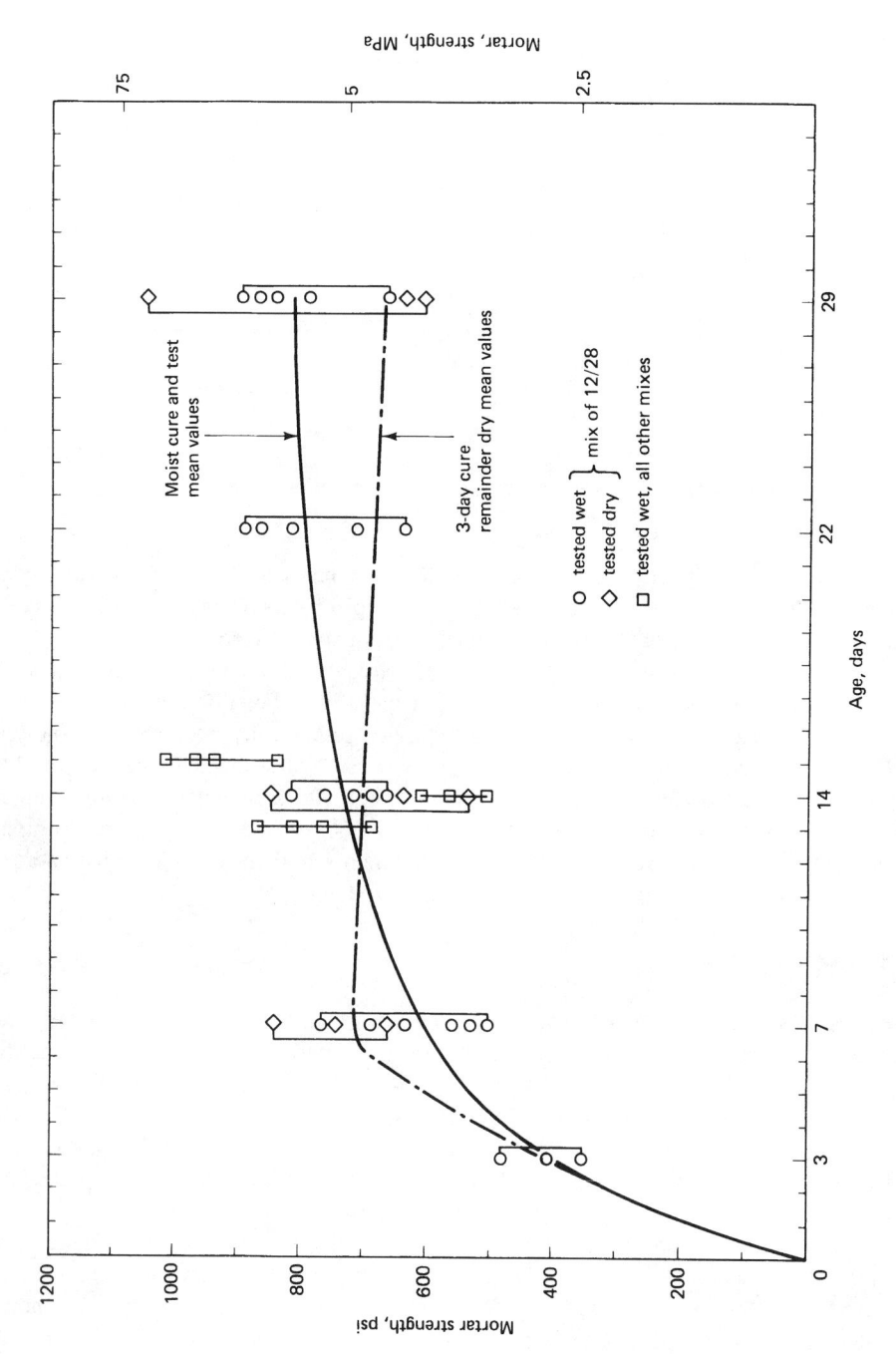

Figure 4.48 Strength–age curves for 1 × 2 in. cylinders of masonry mortar 1:1:4 mix.

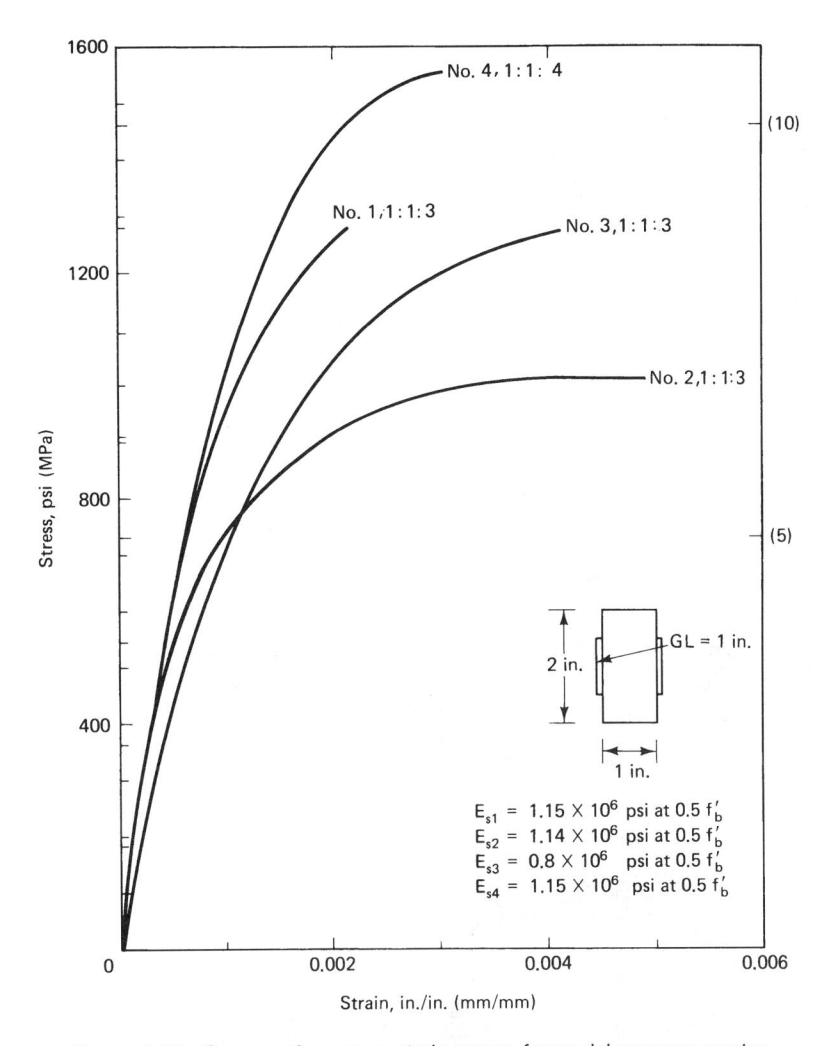

Figure 4.49 Compressive stress–strain curves for model masonry mortar.

the prism compressive strength on the gross area is related to the mortar strength. Comparison of six series of ¼-scale model compression prism data with prototype data (Yokel et al., 1971) is shown in Figure 4.57. Model data are obtained on hollow-core units that had a compressive strength of 7.6 MPa on the gross area. Some imperfections are noted in the scaling of the ¼-scale masonry units used. In particular, the oversized web structures (see Figure 4.36) causes perhaps most of the deviation in the compressive strength of three-course prism test specimens. To compensate for the increased tensile strength that these units could sustain in splitting as a result of their increased web area, a correction to the apparent strength is made. In addition, the effect of volume of mortar in the joints on the compressive strength is taken into account by empirically determining the effect of size of specimen on the unconfined compressive strength (Chapter 9). The model data shown in Figure 4.57 have been corrected for scale and geometry effects. Note the small deviation from the mean curve (shown dashed), which would indicate that the modeling technique developed for prisms in compression is indeed accurate.

The effect of grouting and grout strength on prism compressive strength was studied using ¼-scale masonry units, Type III (see Figure 4.39). The variation of compressive strength of model prisms as a function of the average compressive strength of block molded grout prisms is shown in Figure 4.58 along with the prototype results of Drysdale and Hamid (1979). The model results

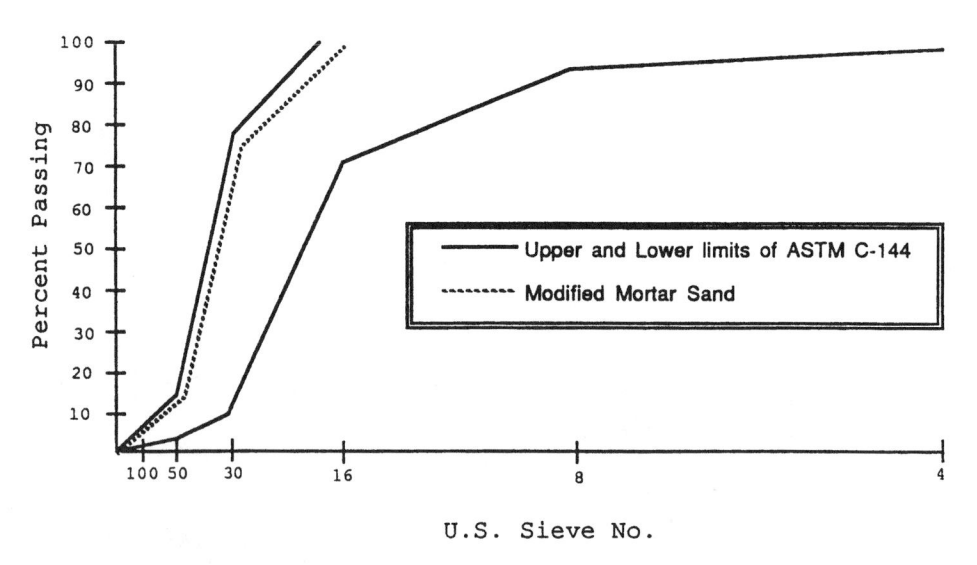

Figure 4.50 Aggregate gradation for model mortar sand.

Figure 4.51 Mortar flow test setup.

agree very well with the prototype results with respect to the observation that prism strength increased moderately with a large increase in grout strength. The Priestly (1986) equation for estimating masonry strength, f'_m, also shown in Figure 4.58, agrees better with the model results.

4.12.2 Flexural Bond

The ASTM C1072 *Measurement of Masonry Flexural Bond Strength* was developed to measure the flexural bond strength of each joint in a prism which eliminates bias in the test results (Figure 4.59). The ASTM bond wrench test employs a relatively short lever arm and is intended to be used in a test machine. The flexural bond strength of model masonry is determined by testing two or more block-high prisms that are clamped in a metal frame (adapted from ASTM C1072) at both the top and bottom of the prism and loaded eccentrically from the centroid of the prism. The prisms are constructed of regular stretcher units with face shell and end-web bonding. The prototype specimens in tests were constructed of similar units with only face shell bonding (Yokel et al., 1971). Because of the relatively

INELASTIC MODELS: MATERIALS FOR CONCRETE AND CONCRETE MASONRY STRUCTURES 193

Table 4.19 Mortar Compressive Strength

Wall Designation	Compressive Strength,[a] MPa Cylinder, 50 × 100 mm	Cube, 50 × 50 mm	Flow, %	W/C Ratio
SW1A	21.9	21.2	119	0.85 to 0.95
SW1B	26.2	24.5	120	0.85 to 0.95
SW1C	26.0	34.8	120	0.85 to 0.95
SW1D	26.9	21.8	115	0.85 to 0.95
SW2A	24.0	33.4	114	0.85 to 0.95
SW2B	20.7	24.0	120	0.85 to 0.95
SW2C	23.2	21.0	115	0.85 to 0.95
SW3A	23.9	28.7	120	0.85 to 0.95
SW3B	25.0	25.9	114	0.85 to 0.95
SW3C	20.8	29.5	119	0.85 to 0.95
Prisms[b]	16.9	—	114.5	0.85 to 0.95

[a] Based on an average of three specimens.
[b] Three-course stack-bond prisms tested under axial compression.

Figure 4.52 Grout control specimens.

Figure 4.53 Typical failure mode of grout prisms.

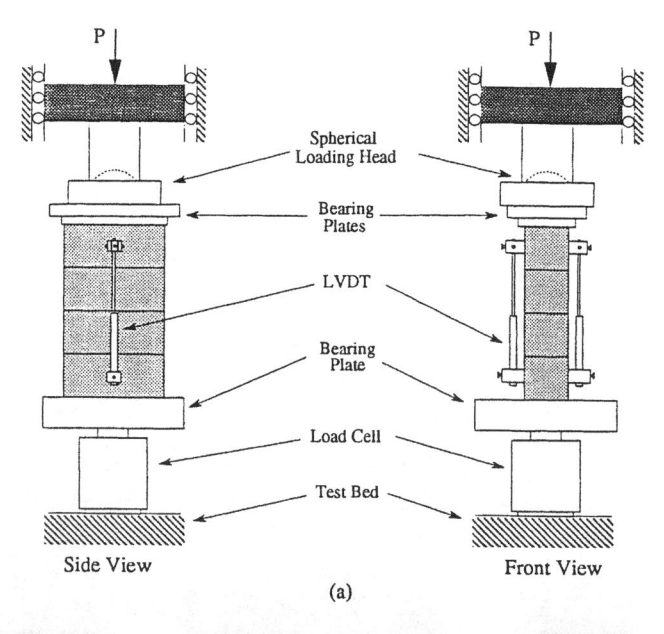

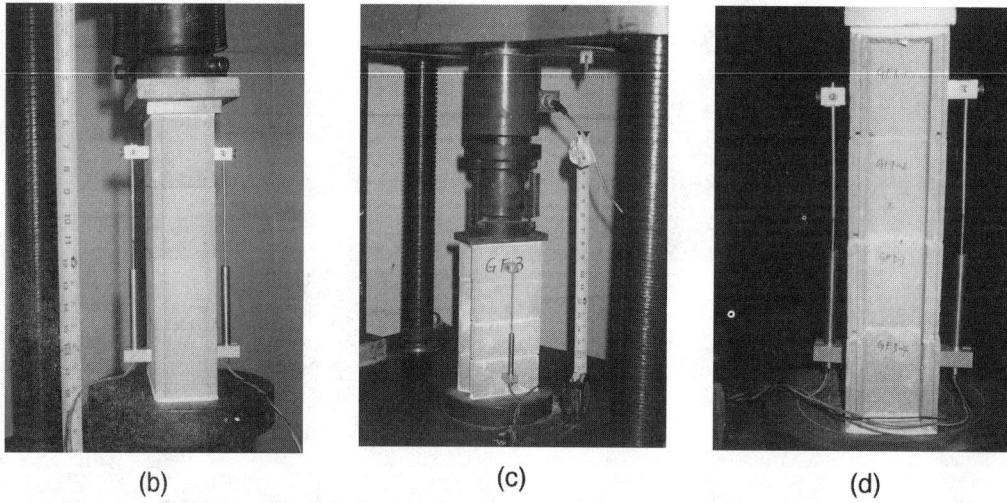

Figure 4.54 Typical test setup of prism compression tests. (a) Schematics; (b, c, d) actual views.

weak bond developed between units, the mortar ultimate loads are small, and thus failure occurs within 1 min of load application. The typical mode of failure is separation at the mortar–unit interface, with one unit remaining free of mortar. The test results of two model series consisting of six specimens each and test results of three prototype specimens were compared by Aboud (1987); series A prisms were cast using dry blocks, whereas series B were cast using saturated surface dry units. The effect of dewatering the mortar via the high block adsorption is a probable contributor to the reported strength difference within the model study. The effects of end-web bonding and high mortar strength have also contributed significantly to the differences between prototype and model results.

4.12.3 Bed Joint Shear

The validity of the model masonry was demonstrated by comparing the bed joint shear strength with prototype data. Since no ASTM testing procedures exist for this strength determination, the modified triplet specimen shown in Figure 4.60 using the ¼-scale blocks illustrated in Figure 4.37.

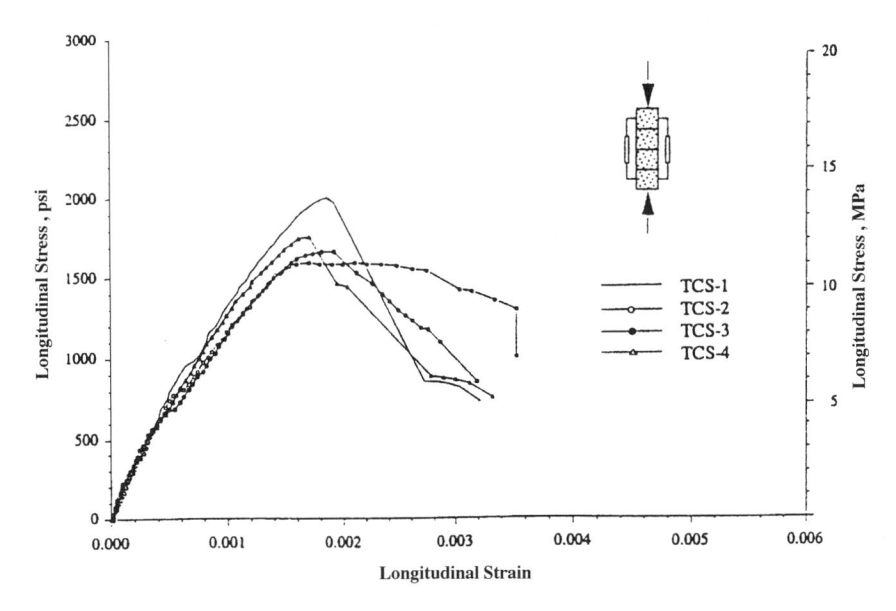

INELASTIC MODELS: MATERIALS FOR CONCRETE AND CONCRETE MASONRY STRUCTURES 195

Figure 4.55 Stress–strain curves of ⅓-scale model grouted conventional block prisms in compression.

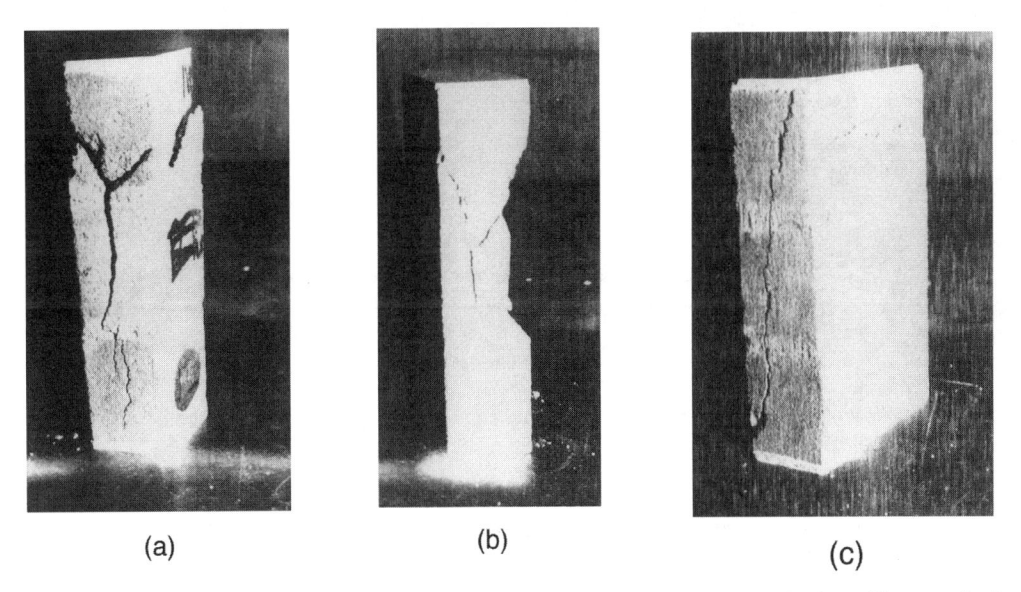

Figure 4.56 Typical modes of failure for three-course model prisms. (a) Ungrouted prism; (b) grouted prism with weak grout (GW); (c) grouted prism with either normal or strong grout (GN or GS).

The advantages of this specimen is that the shear development at the joint is in pure shear without bending and the specimen is easy to construct and test. A total of 12 specimens were fabricated by an experienced mason: 3 hollow and 3 each with three different gout strengths. Failure of the ungrouted specimens was initiated by a debonding of the block–mortar interfaces. With grouted specimens, a sudden failure occurred after the formation of cracks at the block–mortar interfaces, initiated by mortar debonding and followed by diagonal tensile failure of the grout cores. The ¼-scale models had the same mode of failure as those of prototype tests by Hamid et al. (1979). Joint shear values of model and prototype tests are compared in Figure 4.61.

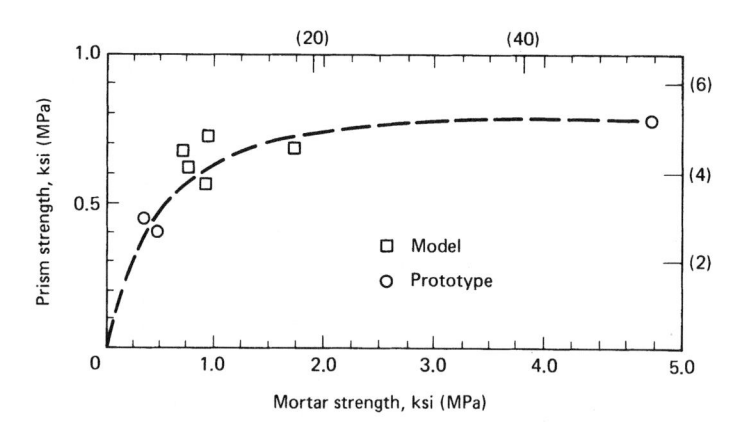

Figure 4.57 Comparison of model and prototype compressive strengths.

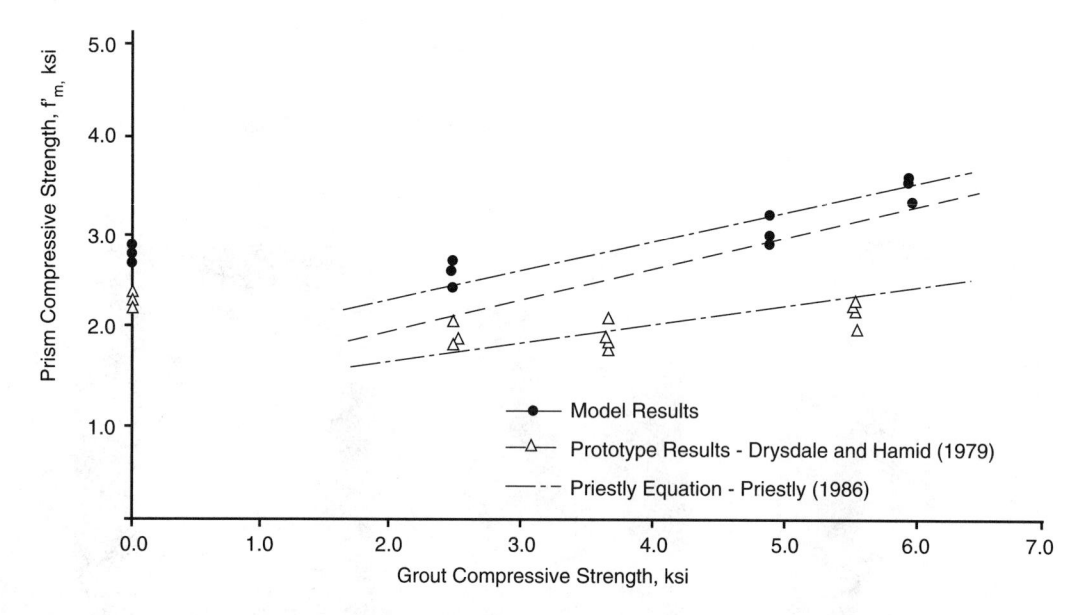

Figure 4.58 Effect of grouting and grout strength on the compressive strength of model and prototype prisms. (After Abboud, 1987.)

The similarity between the model and prototype behavior indicates again the validity of direct modeling in predicting the shear behavior of ungrouted and grouted masonry structures. It has to be noted, however, that the results of the model grouted specimens were approximately 15% higher than those of prototype results. The deviation is mainly attributed to the high tensile strength of model grout and to the stressed volume statistical effect of the model mortar (Chapter 9) that influence the overall model behavior. In general, the deviation is not considered significant and falls within the normal variation of the test results.

4.12.4 In-Plane Tensile Strength

The effect of in-plane loads in a material with low tensile strength such as masonry is to result in cracking. This is the cause of most masonry failures. Splitting tensile tests of disks and square masonry assemblages, such as those shown in Figure 4.62, have been used to study the parameters that effect in-plane tensile strength. This type of test has no ASTM standard yet but is very useful

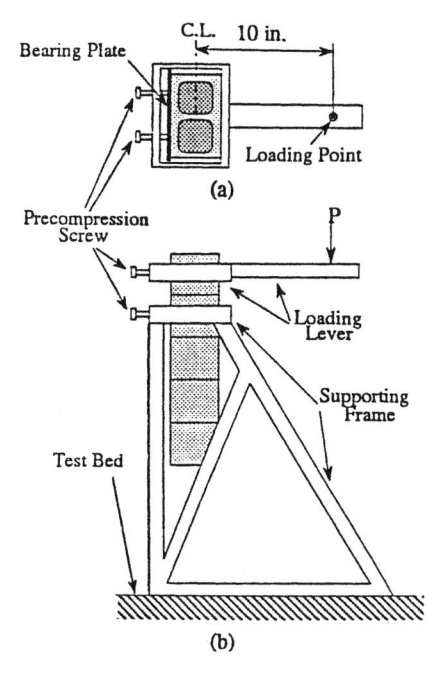

INELASTIC MODELS: MATERIALS FOR CONCRETE AND CONCRETE MASONRY STRUCTURES   197

Figure 4.59 Schematics of test setup. (a) Loading level (top view); (b) side view. (After Oh, 1994.)

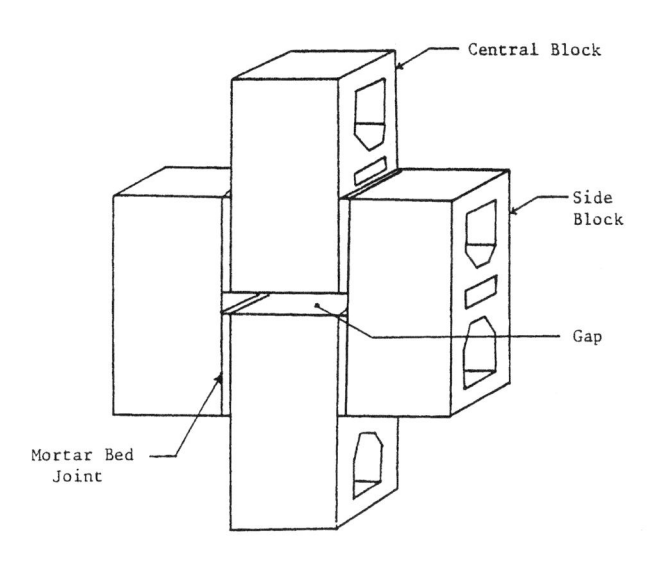

Figure 4.60 Model shear test specimen.

in understanding the factors affecting the in-plane tensile strength of masonry. Prototype test data exist that can be used to measure the accuracy of the modeling technique. The failure in a splitting test follows the load line of action rather than the line of least resistance.

In all, 36 ungrouted and grouted specimens were constructed in running bond with full bed and head joints. All the ¼-scale model specimens were fabricated by the same experienced mason using the same techniques. The specimens were two units long and four units high, as shown in Figure 4.62. Model units of the type shown in Figure 4.38 were used with Type S mortar. Three strengths of grout — weak, normal, and strong — were used. All model specimens, together with

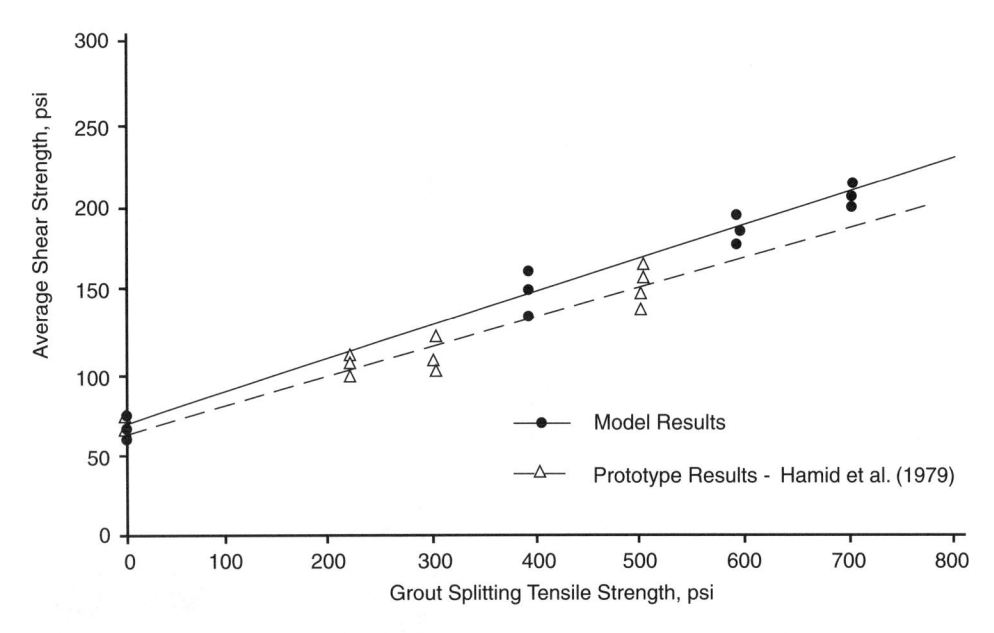

Figure 4.61 Effect of grout and grout strength on the shear capacity of model and prototype block masonry. (After Abboud, 1987.)

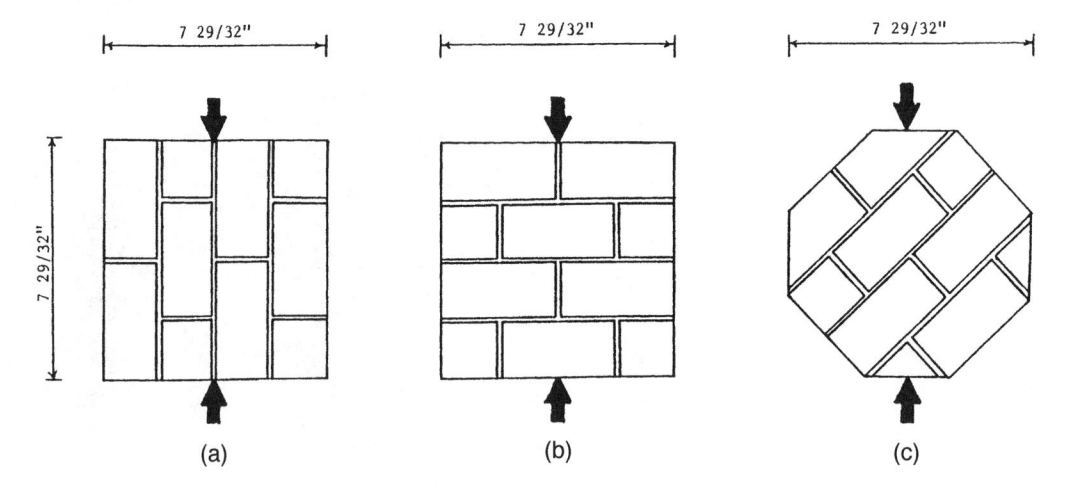

Figure 4.62 Typical dimensions of model splitting test specimens under loads having three different orientations with respect to the bed joints. (a) $\alpha = 0°$; (b) $\alpha = 90°$; (c) $\alpha = 45°$.

their corresponding control mortar and grout specimens, were air-cured in the laboratory under controlled temperature and humidity.

For the specimens that were loaded at 45° to the bed joint, the four corners of the square walls were cut with a diamond-edge blade saw at 26 days. The failure mode of the model specimens was characterized in general by a split along the line of loading. For the specimens loaded along the bed joint, failure took place on a plane along the mortar–block interface in both the ungrouted and grouted walls. For specimens loaded normal to the bed joints, failure in the ungrouted panels was through the face shells and along the mortar–block interfaces of the head joints in a direct line

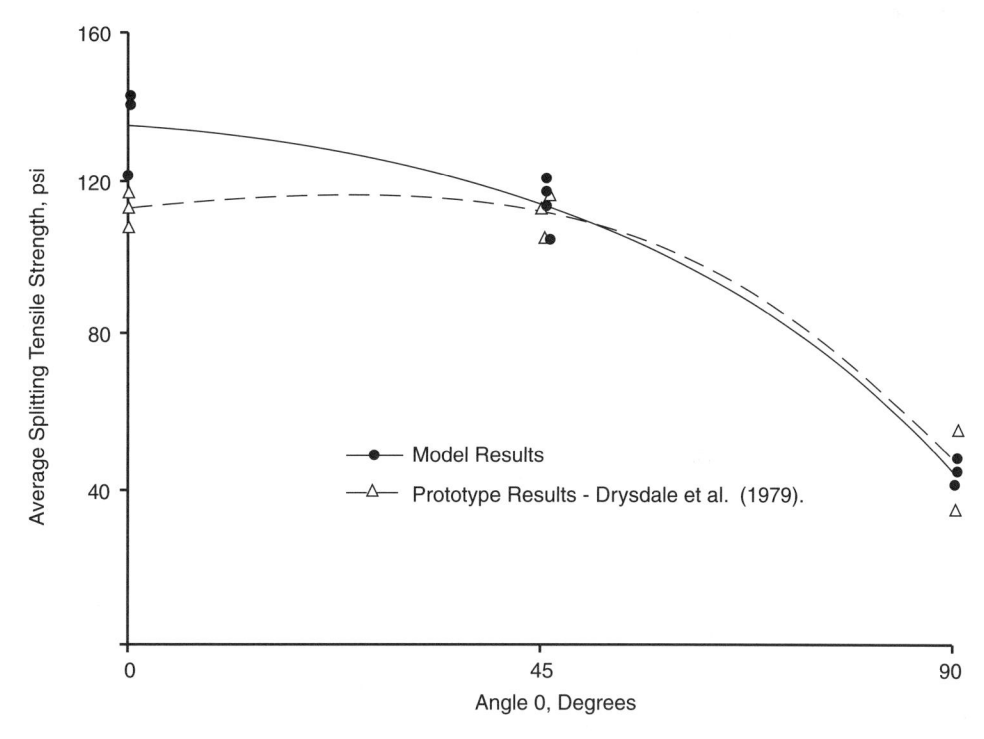

Figure 4.63 Effect of stress orientation on the splitting tensile strength of ungrouted model and prototype block masonry. (After Abboud, 1987.)

between the load points. However, in the grouted panels, the failure plane left both the center web and the grout core intact. For a line load at 45° to the bed joint, the failure plane for ungrouted masonry disks was a fracture crack extending through the blocks and mortar joints in a direct path between load points. For grouted masonry disks, the fracture was a mixed shear (slip at the block–mortar interfaces) and tension (splitting of the block, mortar, and grout) mode of failure.

The tensile failure stress at the center of the specimen was calculated using the following equation:

$$f_{sp} = 2P/\pi A_g \tag{4.5}$$

where: f_{sp} = splitting tensile strength of masonry
P = applied failure load, and
A_g = gross area of the splitting plane

Results of the effect of stress orientation in model masonry specimens are compared with prototype results in Figure 4.63 for the ungrouted specimens. The general variation trend of the tensile strength with different orientations of the applied load shows very good agreement between model and prototype test results.

4.12.5 Out-of-Plane Flexural Tensile Strength

Several ASTM standards exist for reporting the flexural strength of masonry. ASTM E72 *Standard Methods for Conducting Strength Tests on Panels for Building Construction* describes quarter-point or uniform load tests of wall specimens in out-of-plane bending. This type of flexural

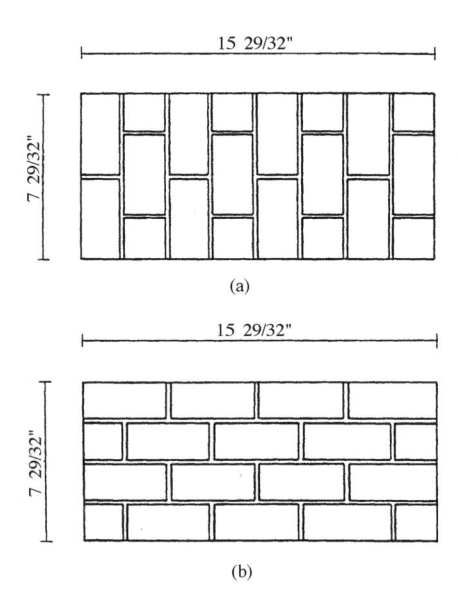

Figure 4.64 Typical dimensions of model flexural wallettes. (a) Flexural tension normal to the bed joints specimens; (b) flexural tension parallel to the bed joints specimens.

test is more realistic for a wall since, unlike the stacked-beam ASTM E518 *Test Method for Flexural Bond Strength of Masonry* or the bond wrench test ASTM C1072, the failure plane is several units long. Wall assemblages or "wallettes" (Figure 4.64) tested in accordance with ASTM E72, with tension normal and parallel to the bed joint, provide results that are similar to full-sized walls. To ensure that simple support conditions exist, testing of the walls as horizontal beams, as shown in Figure 4.65, is preferable.

To correlate the flexural strength of the ¼-scale masonry to prototype test results (Hamid and Drysdale, 1988), a total of 24 ungrouted and grouted flexural tension wallettes were constructed using the ¼-scale units (see Figure 4.39) and the modeling techniques described earlier (Abboud, 1987). The model wallettes tested for tension normal to the bed joints were two blocks wide by eight courses high (Figure 4.64). Model wallettes tested for tension parallel to the bed joints were four blocks wide by four courses high (Figure 4.64). All model specimens were constructed by the same experienced mason in running bond with face shell mortar and partial head joint mortar. No mortar was placed on the webs except for wallettes that were to be grouted later. Mortar was placed on the end webs of these specimens to prevent any grout from flowing out. The joints in the model wallettes were tooled on both sides to compact the mortar joints further. The setup for testing the model wallettes is shown in Figure 4.65.

Correlation of the ¼-scale model and prototype test results is shown in Figure 4.66. As can be seen from Figure 4.66, the general behavior of the models compares very well with that of the prototype tests. Model test results had the following characteristics, similar to those observed for the prototype:

1. The failure mode of ungrouted and grouted specimens was a tensile crack formed along one plane in the constant bending moment region, regardless of the direction of principal stresses. The crack pattern for the model and prototype wallettes was similar in all aspects.
2. The flexural tensile capacity increased substantially with the addition of grout. However, the increase was not proportional to the increase in the grout tensile strength. The maximum contribution of the grout to the flexural tensile capacity occurred when the tensile stresses were normal to the bed joints. There was a very minor contribution when the tensile stresses were parallel to the bed joints.

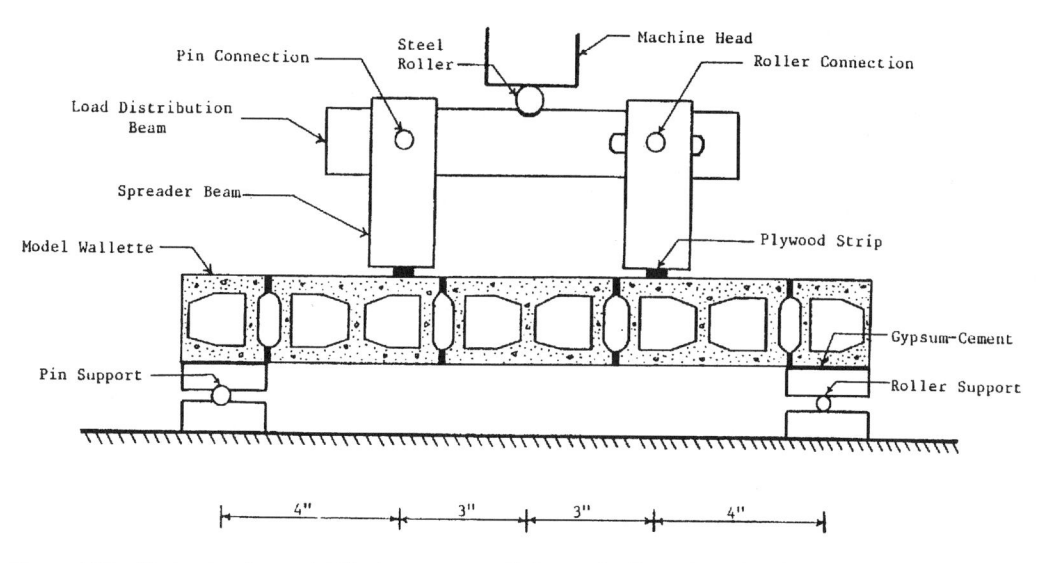

INELASTIC MODELS: MATERIALS FOR CONCRETE AND CONCRETE MASONRY STRUCTURES 201

Figure 4.65 Test setup for model block masonry wallettes under flexural tension (bending parallel to the bed joints).

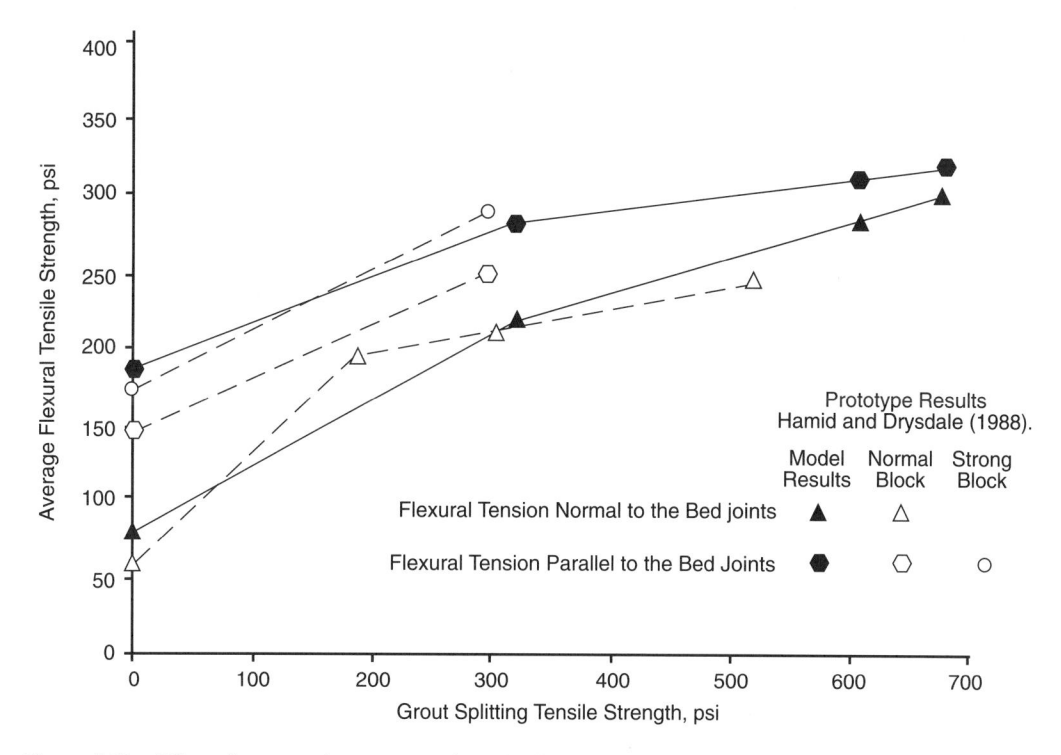

Figure 4.66 Effect of grout and grout strength on the flexural strength of model and prototype block masonry. (After Abboud, 1987.)

3. Anisotropic characteristics of masonry were reduced substantially by the presence of grout.
4. The average orthogonal ratio (ratio of tensile strength parallel to the bed joint to the tensile strength normal to the bed joints) for the ungrouted model specimens was 2.35, whereas for the prototype it was 2.1. For grouted specimens with normal grout, the model specimens and the prototype specimens had an average orthogonal ratio of 1.1 and 1.2, respectively.

4.12.6 Diagonal Tension (Shear) Strength

Determination of masonry shear strength usually consists of testing square prisms by compression along one diagonal, with a resulting failure in diagonal tension (Fattal and Cattaneo, 1974). The ASTM E519 *Standard Test Method for Diagonal Tension (Shear) in Masonry Assemblages* was followed. This procedure calls for the testing of small masonry walls with height-length (H/L) ratio of 1 in diagonal compression. The method was used to determine the diagonal tension (shear) of the ⅓-scale interlocking mortarless masonry units developed at Drexel University (Harris et al., 1993) and shown in Figure 4.45 by direct comparison to companion ⅓-scale conventional masonry with mortar. The model specimen size was chosen to be two units long by four courses high of the running bond pattern. After grouting and curing, the specimens were instrumented with linear variable differential transformers (LVDTs) placed along the compressive and tension diagonals of both faces as shown in Figure 4.67. The loading shoes recommended by the ASTM E519 that fit at opposite ends of the specimen diagonal were fabricated at ⅓ scale. The inside of the loading shoe was lightly oiled and a strong gypsum compound was poured into the device. Figure 4.67 shows a typical shear specimen ready for testing. Shear stress vs. shear strain curves of conventional grouted ⅓-scale masonry panels are shown in Figure 4.68. Failure in all four specimens occurred along the diagonal connecting the line of load.

4.13 SUMMARY

It must be pointed out that, although it is not possible to simulate completely the constituents of concrete at a small scale (aggregates, cement, water, and admixtures), it is adequate to have homologous stress–strain curves and to simulate the tensile strength to compressive strength ratio and the maximum aggregate size. Factors influencing the strength and the behavior of model mixes have been discussed in this chapter. Sufficient data and techniques are available to design model concrete mixes for any strength ranging 17 to 83 MPa for use in model reinforced and prestressed concrete work. It is possible to use either local natural sand if it is suitably graded or a blended mixture of very narrowly graded natural or crushed stones. The sizes of the test cylinder must be selected with due consideration of the minimum effective dimension of the model and the volumes of failure regions. The ACI Committee 444 has recommended that to enable a rational comparative analysis of size effects in concrete strength, 50×100 mm cylinders must be used along with the specimen size selected.

Suitably sealed gypsum plaster mixes can adequately model the behavior of practical concrete mixes in compression and tension. Gypsum mixes can be designed for a given strength and have the advantages of reducing the testing time to 24 h or less.

The methodology for using ⅓- and ¼-scale direct models of concrete masonry structures has been presented. The basic strength evaluation tests for axial compression, flexural bond, bed joint shear, in-plane tensile, out-of-plane flexural tensile, and shear strengths recommended for prototype structures have also been developed, with minor modifications, for the evaluation of model masonry strength. A systematic analysis of the parameters that affect the strength and stiffness of masonry under load has provided the means to compare model and prototype test results. Correlation of the model and prototype results ranged from excellent to good. Excellent correlation of the elastic modulus of concrete masonry in compression, tension, flexure, and shear was obtained from the model tests and the corresponding stress–strain data of prototype masonry reported in the literature.

PROBLEMS

4.1 A demonstration model of a concrete slab that has a large rectangular penetration at its interior is to be built in the laboratory to illustrate the concept of yield lines for an

INELASTIC MODELS: MATERIALS FOR CONCRETE AND CONCRETE MASONRY STRUCTURES 203

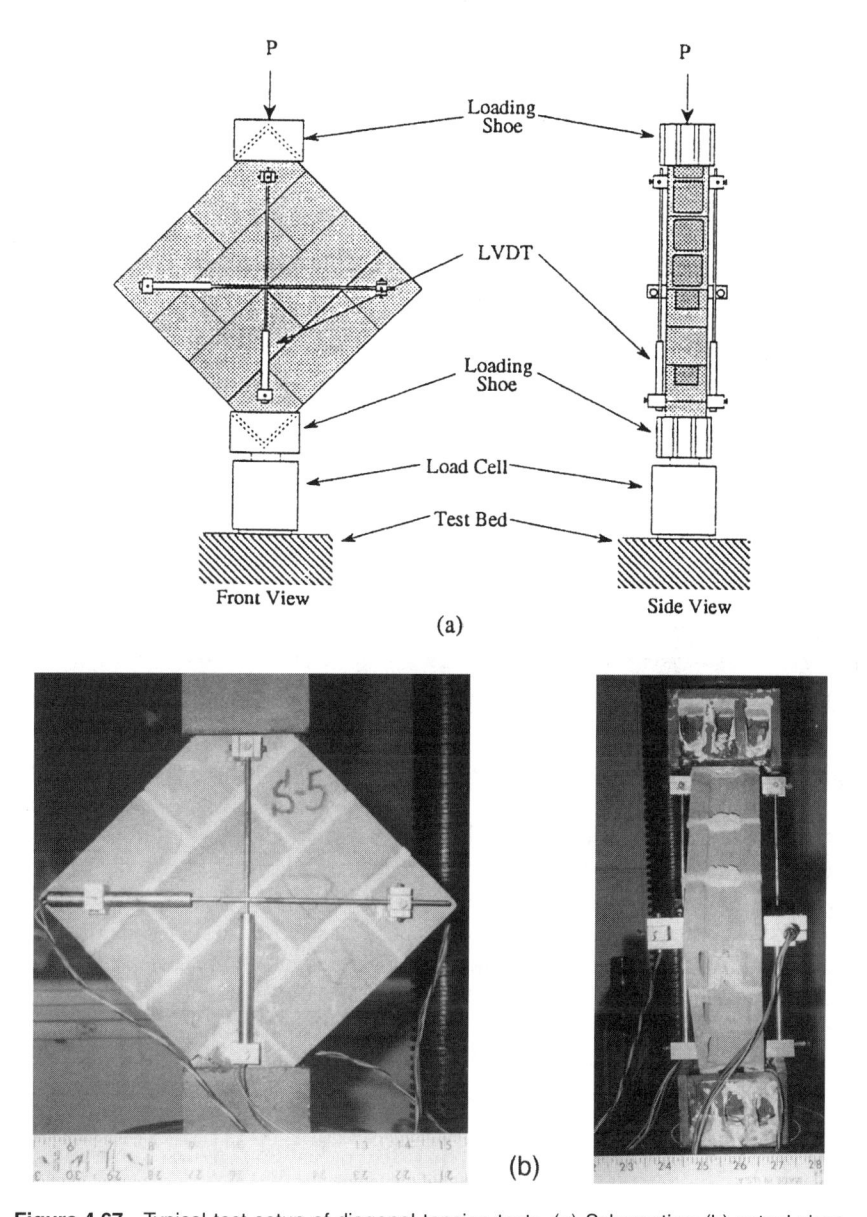

Figure 4.67 Typical test setup of diagonal tension tests. (a) Schematics; (b) actual view.

undergraduate course in reinforced concrete. As the person responsible for this task, a time limit of *1 week* is given to you to construct the model and have the slab ready for instrumentation (minimum is required) and loading which it is anticipated will take only 1 additional day. The demonstration test is to be done in a total of 8 days. Assuming that the proper reinforcing is already fabricated and the casting mold needed is available to you, choose the appropriate model concrete (type, mix proportions, etc.) with a compressive strength of 25 MPa from the data presented in this chapter to complete your assignment. Justify your choice.

4.2 From your understanding of the higher tensile strength of model concrete mixes as compared with prototype concrete, suggest ways to better evaluate this property and ways to overcome this limitation in scale model studies.

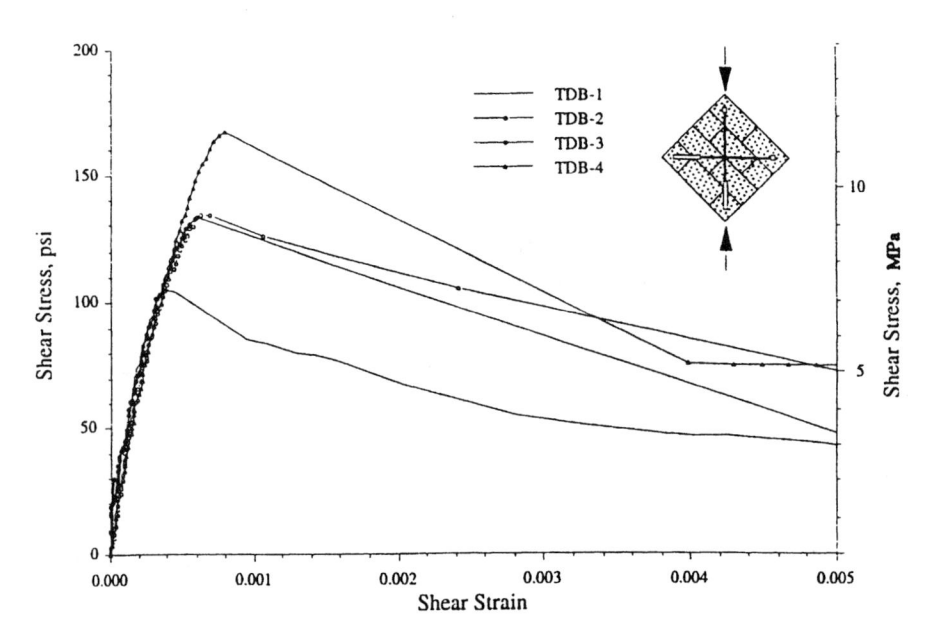

Figure 4.68 Shear stress–strain curves of grouted ⅓-scale conventional block specimens in diagonal compression. (After Oh, 1994.)

4.3 In rigid concrete pavements, it is assumed that the subgrade material to the slab undergoes vertical deflections that are proportional to the pressure imparted by the slab. The ratio k of the pressure to the deflection of the subgrade is considered to be a characteristic constant, called the "subgrade modulus." If a wheel is applied to the slab, the bottom fibers of the slab are in tension, as a result of the bending. If the tensile strength of the concrete slab is exceeded, the slab will crack. It is consequently important to determine the maximum tensile stress f_t that results from a given wheel load. The primary variables that determine the f_t are the subgrade modulus k, the static modulus E of the concrete, the thickness h of the slab, the wheel load Q, and the air pressure p_t in the tire.
   a. Find the governing π (pi) terms of this problem.
   b. Determine an appropriate concrete model material for a 1/16-scale model of a 300-mm airport runway slab under the weight of various jumbo jet wheel loads.
   c. In view of the problem posed in 4.2 above what precautions will you take to minimize any scale effects?

4.4 Describe the steps you would take to design a new solid block masonry unit (100% solid) using ⅓-scale direct models. The new unit must be only ⅓ as dense as conventional block masonry and must be interlocking and mortarless so that it can be quickly formed into shelters in subzero weather conditions if necessary.

4.5 The compressive strength of four unit prisms f'_m of a prototype block masonry is 10 MPa. A ¼-scale model block unit having all the geometric, physical, and mechanical properties of the prototype unit is available to you. Suppose the compressive strength of the block (model and prototype) was 20 MPa. What are the steps that you need to take to ensure that ¼-scale model prisms fall within 10% of the desired prototype values? Will this ensure that the behavior of a ¼-scale "slender" (H/t) block masonry wall (see figure) will behave within 10% of its prototype counterpart? What is needed, if anything, to attain this level of accuracy?

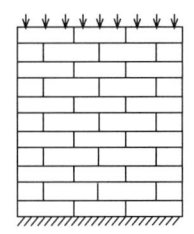

4.6 A high-strength model concrete is desired to be developed for a 1/10-scale model study of a high-rise reinforced concrete frame. Describe the steps that you would need to take to design the mix and characterize its strength properties.

4.7 What are the minimum face shell and web thicknesses specified for 200-mm hollow concrete blocks? What are the practical limitations on scaling the unit at 1/5 and 1/8 scale? Discuss the limitations of building model walls at these scales.

4.8 A grouted model masonry wall is composed of normal-weight 50-mm nominal hollow model concrete blocks having a compressive strength of 20 MPa, Type S model mortar, and 27.5 MPa model grout. If the principle of superposition is expected to hold for the wall, what is the expected compressive strength?

REFERENCES

Abboud, B. E. (1987). The Use of Small Scale Direct Models for Concrete Block Masonry Assemblages and Slender Reinforced Walls under Out-of-Plane Loads, Ph.D. thesis, Civil and Architectural Engineering Department, Drexel University, Philadelphia.

Abboud, B. E. and Hamid, A. A. (1990). Behavior of model masonry slender walls under out-of-plane lateral loads, in *Proc. 5th North American Masonry Conf.,* University of Illinois, Urbana-Champaign, June.

Abboud, B. E., Hamid, A. A., and Harris, H. G. (1990). Small-scale modeling of concrete block masonry structures, *ACI Struct. J.,* 87(2), March–April.

Akazawa, T. (1953). Tension Test Method for Concrete, Bulletin No. 16, International Association of Testing and Research Laboratories for Materials and Structures, Paris, November.

Aldridge, W. W. and Breen, J. E. (1970). Useful techniques in direct modeling of reinforced concrete structures, in *Models for Concrete Structures,* ACI SP-24, American Concrete Institute, Detroit, MI, 125–140.

American Concrete Institute (ACI) Committee 444 (1979). Models of concrete structures — state of the art, *Concr. Int. Des. Constr.,* 1(1), January. 77–95.

American Concrete Institute (ACI) Committee 531 (1970). Concrete masonry structural design and construction, *Proc. Am. Concr. Inst.,* 67(5), May, 380–403.

Assis, G., Hamid, A. A., and Harris, H. G. (1989). Material Models for Grouted Block Masonry, Report No. 1.2 (a)-2, U.S.–Japan Coordinated Program on Masonry Building Research, Department of Civil and Architectural Engineering, Drexel University, Philadelphia, September.

Batchelor, B. (1972). Materials for Model Structures at Queens, University, Department of Civil Engineering Rept. (unpublished), Queen's University, Kingston, Ontario.

Becica, I. J. and Harris, H. G. (1977). Evaluation of Techniques in the Direct Modeling of Concrete Masonry Structures, Structural Models Laboratory Report No. M77-1, Department of Civil Engineering, Drexel University, Philadelphia, June.

Blackman, J. S., Smith, D. M., and Young, M. L. E. (1958). Stress distribution affects ultimate tensile strength, *J. Am. Concr. Inst.,* 55, 675–684.

Bredsdorff, P. K. and Kierkegaard-Hansen, P. (1959). Discussion of paper by J. S. Blackman et al., Stress Distribution Affects Ultimate Tensile Strength, *J. Am. Concr. Inst.,* 55 (June), 1421–1426.

Carniero, F. L. L. B. and Barcellos, A. (1953). Concrete Tensile Strength, Bulletin No. 13, International Association of Testing and Research Laboratories for Materials and Structures, Paris, March.

Darwin, D. (1968). The Effect of Paste-Aggregate Bond Strength upon the Behavior of Portland Cement Concrete under Short-Term Load, M.Sc. thesis, Cornell University, Ithaca, NY.

Drysdale, R. G. and Hamid, A. A. (1979). Behavior of concrete block masonry under axial compression, *ACI J.*, 76(6), June.

Drysdale, R. G., Hamid, A. A., and Heidebrecht, A. C. (1979a). Tensile strength of concrete masonry, *J. Struct. Div., Proc. ASCE*, 105(ST7), July.

Drysdale, R. G., Hamid, A. A., and Heidebrecht, A. C. (1979b). Shear strength of concrete masonry joints, *J. Struct. Div., Proc. ASCE*, 105(ST7), July.

Drysdale, R. G., Hamid, A. A., and Baker, L. R. (1994). *Masonry Structures Behavior and Design,* Prentice-Hall, Englewood Cliffs, NJ.

El-Shafie, H., Hamid, A. A., Okba, S., and Nasr, E. (1996). Masonry shear walls with openings: state-of-the-art report, in *Proc. 7th North Am. Masonry Conf.,* University of Notre Dame, June.

Fattal, S. G. and Cattaneo, L. E. (1974). Evaluation of Structural Properties of Masonry in Existing Buildings, NBSIR 74-520, National Bureau of Standards, Washington, D.C., July.

Fishburn, C. (1961). *Effect of Mortar Properties on Strength of Masonry,* NBS Monograph 36, National Bureau of Standards, Washington, D.C., November 20.

Ghanem, G. and Hamid, A. A. (1992). Effect of steel distribution on the behavior of partially reinforced masonry shear walls, in *Proc. 6th Can. Masonry Symp.,* Saskatoon, Canada, June.

Ghanem, G., Salama, A., Abu Elmagd, S., and Hamid, A. A. (1993). Effect of axial compression on the behavior of partially reinforced masonry shear walls, in *Proc. 6th North Am. Masonry Conf.,* Philadelphia, June.

Gilkey, H. J. (1961). Water-cement ratio vs. strength — another look, *Proc. Am. Concr. Inst.,* 57(4), April, 1287–1312.

Hamid, A. A. and Abboud, B. E. (1986). Direct modeling of concrete masonry under shear and in-plane tension, *J. Testing Eval.,* 14(2), March.

Hamid, A. A. and Chandrakeerthy, S. (1992). Compressive strength of partially grouted concrete masonry using small-scale wall elements, *TMS J.,* 11(1), August.

Hamid, A. A. and Drysdale, R. G. (1988). Flexural tensile strength of concrete block masonry, *Journal of Structural Engineering ASCE, Structural Division,* Vol. 114, No. 1, January.

Hamid, A. A. and Ghanem, G. (1991). Partially reinforced concrete masonry, in *Proc. 9th Int. Brick/Block Masonry Conf.,* Berlin, Germany, October.

Hamid, A. A., Abboud, B. E., and Harris, H. G. (1985). Direct modeling of concrete block masonry under axial compression, in *Masonry: Research, Application and Problems, ASTM STP 871,* American Society for Testing and Materials, Philadelphia, 151–166.

Hamid, A. A., Drysdale, R. G., and Heidebrecht, A. C. (1979). Shear strength of concrete masonry joints, *J. Struct. Div., Proc. ASCE,* 105(ST7), Proc. Paper 14670, July, 1227–1240.

Hamid, A. A., Mahmoud, A., and Abo El Magd, S. (1994). Repair and strengthening of masonry assemblages using fiber glass, in *Proc. 10th Int. Brick and Block Masonry Conf.,* Calgary, Canada, July.

Hamid, A. A., Mahmoud, A., and Abo El Magd, S. (1995). Lateral response of unreinforced solid masonry shear walls: an experimental study, in *Proc. 7th Canadian Masonry Symp.,* McMaster University, Canada, June.

Harris, H. G. and Becica, I. J. (1977). Direct small-scale modeling of concrete masonry, in *Proceedings, Symposium on Advances in Civil Engineering Mechanics,* EMD-ASCE, May 23–25, 101–104.

Harris, H. G. and Becica, I. J. (1978). Behavior of concrete masonry structures and joint details using small-scale direct models, in *Proceedings of the North American Masonry Conference,* Boulder, CO, August, 10-1–10-18.

Harris, H. G. and Oh, K. H. (1990). Seismic behavior of floor-to-wall horizontal joints between block masonry walls and precast concrete hollow core slabs using ⅓-scale direct models, in *Proceedings, Fourth U.S. National Conference on Earthquake Engineering,* 2, 777–786, Palm Springs, CA, May.

Harris, H. G. and White, R. N. (1972). Inelastic behavior of reinforced concrete cylindrical shells, *ASCE J. Struct. Div.,* 98(ST7) July, Proc. Paper 9074, 1633–1653.

Harris, H. G., Labrouki, B., and Lafis, S. G. (1990). Material characterization for direct modeling of reinforced block masonry structures, in *Proceedings of Fifth North American Masonry Conference,* University of Illinois, Urbana-Champaign, June 3–6, Vol. 2, pp. 639–650.

Harris, H. G., Oh, K. H., and Hamid, A. A. (1992). Development of new interlocking and mortarless block masonry units for efficient building systems, *Proceedings of 6th Canadian Masonry Symposium,* June 15–17, Saskatoon, Saskatchewan, Canada.

Harris, H. G., Oh, K. H., and Hamid, A. A. (1993). Development of New Interlocking Blocks to Improve Earthquake Resistance of Masonry Construction, Report to the National Science Foundation, Department of Civil and Architectural Engineering, Drexel University, Philadelphia, 146 pp.

Harris, H. G., Sabnis, G. M., and White, R. N. (1966). Small Scale Direct Models of Reinforced and Prestressed Concrete Structures, Report No. 326, Department of Structural Engineering, Cornell University, Ithaca, NY, September, 362 pp.

Hsu, T. T. C. and Slate, F. O. (1963). Tensile bond strength between aggregate and cement paste or mortar, *Proc. Am., Concr. Inst.,* 60(4), April, 465–486.

Isberner, A. W. (1969). Properties of masonry cement units, in *Design Engineering and Construction with Masonry Products,* Dr. Franklin B. Johnson, Ed., Gulf Publishing Company, Houston, 42–50.

Ishai, O. (1961). Influence of sand concentration on the deformations of beams under low stresses, *Proc. Am. Concr. Inst.,* 58(11), November, 611–623.

Kandasamy, E. G. (1969). Stress–Strain Characteristics of Gypsum Plaster-Sand Mixes under Direct and Flexural Compression, M.S. thesis, North Carolina State University, Raleigh, N.C.

Kaplan, M. F. (1963). Strains and stress of concrete at initiation of cracking and near failure, *Proc. Am. Concr. Inst.,* 60, July, 853–880.

Kim, W. El-Attar, A., and White, R. N. (1988). Small-Scale Modeling Techniques for Reinforced Concrete Structures Subjected to Seismic Loads, Report No. NCEER-88-0041, National Center for Earthquake Engineering Research, State University of New York at Buffalo, November.

Labrouki, B. (1989). Material Characteristics of ⅓-Scale Direct Models of Block Masonry Components, M.Sc. thesis, Department of Civil Engineering, Drexel University, Philadelphia, February.

Lafis, S. G. (1988). Study of Bond and Splices in Reinforced Masonry Using ⅓-Scale Models, M.Sc. thesis, Department of Civil Engineering, Philadelphia, June.

Larbi, A. (1989). Behavior of Block Masonry Shear Walls Under In-Plane Monotonic and Reversed Cyclic Loads Using 1/3-Scale Direct Models, M.Sc. thesis, Department of Civil and Architectural Engineering, Drexel University, Philadelphia, March.

Larbi, A. and Harris, H. G. (1990). Seismic performance of reinforced block masonry shear walls using ⅓-scale direct models, in *Proceedings of Fifth North American Masonry Conference,* University of Illinois, Urbana-Champaign, June 3–6, Vol. 1, pp. 321–332.

Loh, G. (1969). Investigation of the Effective Width of the Slab of Reinforced concrete T-Beams, M.S. thesis, Cornell University, Ithaca, NY.

Maisel, E. (1979). Mikrobeton fur modellstatische Untersuchungen II, Institut fur Modellstatik, Universitat Stuttgart, Germany.

Masonry Standards Joint Committee (1992). Building Code Requirements for Masonry Structures ACI 530-92/ASCE 5-92/TMS 402-92, American Concrete Institute, American Society of Civil Engineers, The Masonry Society, 1992.

McDonald, C. R. and Swartz, S. E. (1981). Use of strain gages on miniature concrete cylinders, *Exp. Tech.,* 5(1), March, 1–3.

Mirza, M. S. (1967). An Investigation of Combined Stresses in Reinforced Concrete Beams, Ph.D. thesis, Department of Civil Engineering and Applied Mechanics, McGill University, Montreal, March.

Mirza, M. S., White, R. N., and Roll, F. (1972). Materials for structural models, in *Proceedings of the ACI Symposium on Models of Concrete Structures,* Dallas, American Concrete Institute, Detroit, MI, March, 19–112.

Neville, A. M. (1959). Some aspects of the strength of concrete, three parts, *Civil Eng. Public Works Rev.,* 54(639), October; 54(640), November; 54(641), December.

Neville, A. M. (1987). *Properties of Concrete,* Vol. I, Pitman and Sons, London.

Newman, K. (1965). The structure and engineering properties of concrete, in *Proceedings of the International Symposium on the Theory of Arch Dams,* Southampton, April, Pergamon Press, New York.

Oh, K. H. (1994). Development and Investigation of Failure Mechanism of Interlocking Mortarless Block Masonry Systems, Ph.D. thesis, Department of Civil and Architectural Engineering, Drexel University, Philadelphia, December.

Oh, K. H. and Harris, H. G. (1990). Seismic behavior of floor-to-wall horizontal joints of masonry buildings using ⅓-scale direct models, in *Proceedings of Fifth North American Masonry Conference,* University of Illionois, Urbana-Champaign, June 3–6, Vol. 1, pp. 81–92.

Oh, K. H., Harris, H. G., and Hamid, A. A. (1993); New interlocking and mortarless block masonry units for earthquake resistant structures, in *Proceedings of Fifth North American Masonry Conference,* Drexel University, Philadelphia, June 6–9.

Price, W. W. (1951). Factors influencig concrete strength, *J. Am. Concr. Inst.,* 47, (February), 417.

Priestly, M. J. N. (1986). Seismic design of concrete masonry shear walls, *J. Am. Concr. Inst., Proc.,* Vol. 83(1), Jan.–Feb.

Randall, F. A. and Panarese, W. C. (1976). *Concrete Masonry Handbook,* Portland Cement Association, Skokie, IL.

Ruiz, W. (1966). Effect of Volume of Aggregate on the Elastic and Inelastic Properties of Concrete, M. S. thesis, Cornell University, Ithaca, NY.

Rüsch, H. (1959). Physikalische Fragen der Betonprufung, [Physical problems in the testing of concrete], *Zem.-Kalk-Gips,* 12(1), Cornell University, Ithaca, NY, (translation by G. M. Sturman).

Rüsch, H. (1964). Zur Statistichen Qualitaskontrolle de Beton (On the Statistical Quality Control of Concrete), *Materialprüssung,* 6, No. 11, November.

Sabnis, G. M. and Mirza, M. S. (1979). Size effects in model concretes? *ASCE J. Struct. Div.,* 105(ST 6), June, 1007–1020.

Sabnis, G. M. and White, R. N. (1967). A gypsum mortar for small-scale models, *Proc. Am. Concr. Inst.,* 64(11), November, 767–774.

Sahlin, S. (1971). *Structural Masonry,* 1st ed., Prentice-Hall, Englewood Cliffs, NJ.

Suaris, W. and Shah, S. P. (1985). Constitutive model for dynamic loading of concrete, *ASCE J. Struct. Eng.,* 111(3), March.

Syamal, P. K. (1969). Direct Models in Combined Stress Investigations, M. Eng. thesis, Structural Concrete Series No. 17, McGill University, Montreal, July.

Tsui, S. H. and Mirza, M. S. (1969). *Model Microconcrete Mixes,* Structural Concrete Series No. 23, McGill University, Montreal, November.

White, R. N. (1976). High Strength Model Concrete Mixes, Unpublished report, Department of Structural Engineering, Cornell University, Ithaca, NY.

White, R. N. and Sabnis, G. M. (1968). Size effects in gypsum mortar, *J. Mater.,* 3(1), March, 163–177, ASTM, Philadelphia.

Wright, P. J. F. and Garwood, F. (1952). Effect of the method of test on the flexural strength of concrete, *Mag. Concr. Res.* (London), 11, 67–76.

Yokel, F. Y., Mathey, R. G., and Dikkers, R. D. (1971). Strength of Masonry Walls under Compressive and Transverse Loads, Building Science Series 34, National Bureau of Standards, Washington, D.C., March.

CHAPTER 5

Inelastic Models: Structural Steel and Reinforcing Bars

CONTENTS

5.1	Introduction	210
5.2	Steel	210
	5.2.1 Reinforcing Steel Bars	211
	5.2.2 Structural Steels	211
	5.2.3 Prestressing Steels	211
	5.2.3.1 General	211
	5.2.3.2 Stress–Strain Characteristics	213
5.3	Structural Steel Models	214
	5.3.1 General	214
	5.3.2 Steel Columns	215
	5.3.3 Steel Beams	218
	5.3.3.1 Fabrication of Wide-Flange Shapes	218
	5.3.3.2 Beam Tests	219
	5.3.4 Steel Frames	221
	5.3.4.1 M.I.T. Ultimate-Strength Model Steel Frameworks	221
	5.3.4.2 Stanford University ⅙-Scale Shaking Table Tests	222
	5.3.4.3 U.S.–Japan Six-Story Steel Building Model Studies	226
5.4	Reinforcement for Small-Scale Concrete Models	230
	5.4.1 General	230
	5.4.2 Model Reinforcement Used by Various Investigators	230
	5.4.3 Wire Reinforcement for Small-Scale Models	232
	5.4.4 Black Annealed Wire as Model Reinforcement	233
	5.4.5 Custom-Ordered Model Wire	233
	5.4.6 Commercially Deformed Wire as Model Reinforcement	234
	5.4.7 Laboratory Wire-Deforming Machines	239
	5.4.7.1 Cornell University Wire Deformer	239
	5.4.7.2 Cement and Concrete Association Wire Deformer	242
	5.4.7.3 Deformeters Used by Various Investigators	245
	5.4.7.4 Drexel University Wire Deformer	246
	5.4.8 Heat Treatment of Model Reinforcement	250
	5.4.8.1 General	250
	5.4.8.2 Furnaces Used for Annealing	250
	5.4.8.3 Annealing Processes	250
	5.4.9 Model Reinforcement Selection	251

5.5	Model Prestressing Reinforcement and Techniques		252
	5.5.1	Model Prestressing Reinforcement	252
	5.5.2	Anchorage Systems	253
5.6	FRP Reinforcement for Concrete Models		255
	5.6.1	General	255
	5.6.2	Nonductile FRP Reinforcement	256
	5.6.3	Ductile Hybrid Fiber-Reinforced Polymer	256
5.7	Bond Characteristics of Model Steel		259
5.8	Bond Similitude		266
5.9	Cracking Similitude and General Deformation Similitude in Reinforced Concrete Elements		267
5.10	Summary		272
Problems			272
References			274

5.1 INTRODUCTION

In this chapter, we shall discuss the modeling of steel and fiber reinforced polymer (FRP) rebar and structural steel for direct scale models. The scaled models are designed to study structural response to applied loads through both the elastic and postelastic ranges of behavior until failure. As most structural concrete elements are underreinforced, that is, the reinforcing provided is not adequate to fully utilize the compressive strength of the concrete (and thus ensure a ductile failure), the stress–strain characteristics of both the prototype and the model reinforcement are critical in determining the structural behavior in the postcracking and postyielding ranges. A brief discussion of steel as a structural material follows.

5.2 STEEL

Most steels have a crystalline structure and consist of a basic iron–carbon system. Relatively small changes in the carbon content and/or other alloys result in significant changes in the mechanical behavior of the resultant steel. The mechanical properties of steel that are of interest to the design engineer are the stress–strain curve; the yield strength, if any; the amount of strain at yield, the percentage elongation at failure, or ductility; the amount and rate of strain hardening; and the ultimate tensile strength. While the mechanical behavior of a particular steel is significantly influenced by its carbon content, other factors that influence its properties are the chemical composition and the method used to shape the molten mass into its final form as steel. The mechanical properties of steel are affected by the following parameters:

1. Chemical Composition
    a. Carbon content
    b. Presence of alloying elements such as nickel, chromium, vanadium, and copper
    c. Presence of other elements such as sulfur, phosphorus, manganese, and silicon
2. Physical Conditions
    a. Slow cooling from the molten state or quenching
    b. Annealing
    c. Hardening characteristics
    d. Shaping operations (e.g., cold working)
    e. Weldability

5.2.1 Reinforcing Steel Bars

The ASTM A615 *Standard Specification for Deformed and Plain Billet-Steel Bars for Concrete Reinforcing* covers billet steel of Grades 40 (minimum yield strength, 276 Mpa), 60 (414 Mpa), and Grade 75 (517 MPa) and with tensile strengths of 483, 621, and 690 MPa, respectively. The billet steel is newly made steel with a carefully controlled chemical composition to obtain the necessary ductility. Rail steel Grade 60 (ASTM A616 *Standard Specification for Rail-Steel Deformed and Plain Bars for Concrete Reinforcement*) and axle steel Grades 40 and 60 (ASTM A617 *Standard Specification for Axle-Steel Deformed and Plain Bars for Concrete Reinforcement*) are also used for manufacturing reinforcing steel bars. Both axle and rail steel bars are made from steel that is rerolled from old axles and rails and are in general less ductile than billet steel (Salmon and Johnson, 1996). Low alloy steel of Grade 60 (ASTM A706 *Standard Specification for Low-Alloy-Steel Deformed Bars for Concrete Reinforcement*) is useful for applications of reinforcing steel bars that involve both welding and bending. The carbon content of these steels is approximately 0.25%. All bar sizes, No. 3 through No. 11, No. 14, and No. 18 are available in Grade 60 and sizes No. 6 to No. 14 and No. 18 are available in Grade 75 billet and low-alloy steels; however, bars of sizes No. 14 and No. 18 are not available either in Grade 40 billet steel or in rail or axle steel.

5.2.2 Structural Steels

The *American Institute of Steel Construction (AISC)*, 9th edition, specification covering the *common* structural steels for buildings includes steels with specified minimum yield strength ranging from 36 ksi (250 MPa) (ASTM A36) to 50 ksi (345 MPa) (ASTM A242 *Standard Specification for High-Strength Low-Alloy Structural Steel* and ASTM A440 *Standard Specification for High Strength Structural Steel*). Since the chemical composition is constant (0.22% carbon for A242 steel and 0.28% for A440 steel), the increased hot working required to produce thin sections increases the yielding strength in thinner sections. The A36 steel has a maximum carbon content varying from 0.25 to 0.29% depending on the thickness. The AISC will replace ASTM A36 as the industry base standard in the future with a 50 ksi (345 MPa) steel having improved performance through better-defined strength and material limits. The structural steelS show a marked yield point, as shown in Figure 5.1. An increase in the carbon content raises the yield strength but decreases the ductility and causes problems with the welding operations if the carbon content is higher than 0.3%. Most of the structural steels have a low carbon content and therefore have good welding characteristics. Also, increased limitations have been placed on the carbon content in recent years as emphasis on good weldability has increased (Salmon and Johnson, 1996).

5.2.3 Prestressing Steels

5.2.3.1 General

Prestressing steels are available in the form of single wires (ASTM A421 *Standard Specification for Uncoated Stress-Relieved Wire for Prestressed Concrete*), wire strands (ASTM A416 *Standard Specification for Uncoated Seven-Wire Stress-Relieved Strands for Prestressed Concrete*), or high-strength bars (AISI 5160 and 9260, 1978). Wire strands are of the seven-wire type with six helically placed outer wires wound tightly around a central wire. Strand diameters vary 6 to 12 mm. Prestressing wires usually have diameters 5 to 7 mm. The high-strength steel bars range in diameter from 19 to 35 mm. The prestressing wires or bars are made by cold-drawing high-carbon (approximately 0.6% carbon content) steel bars. Typical stress–strain curves for prestressing strands and prestress bars are shown in Figure 5.2.

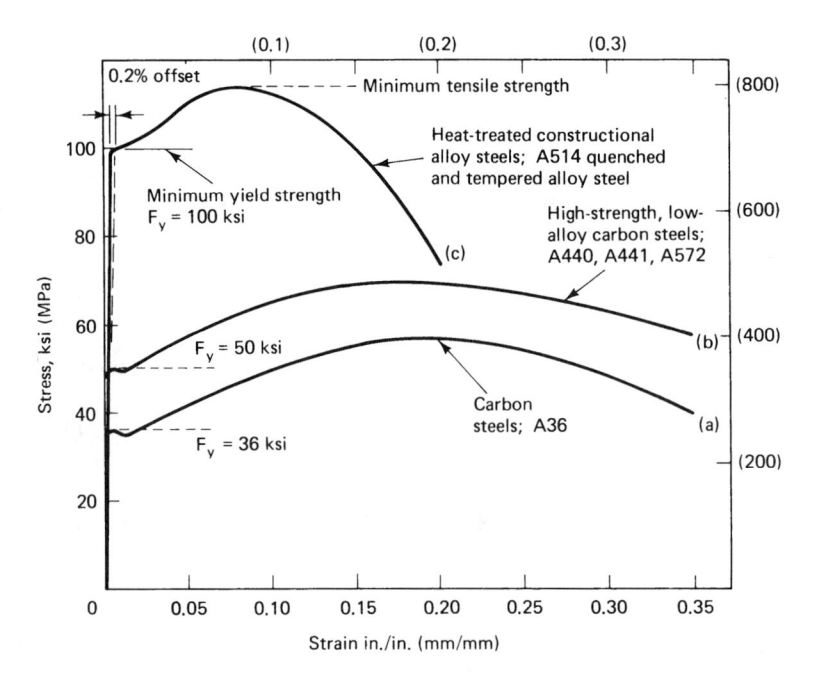

Figure 5.1 Typical stress–strain curves for structural steel.

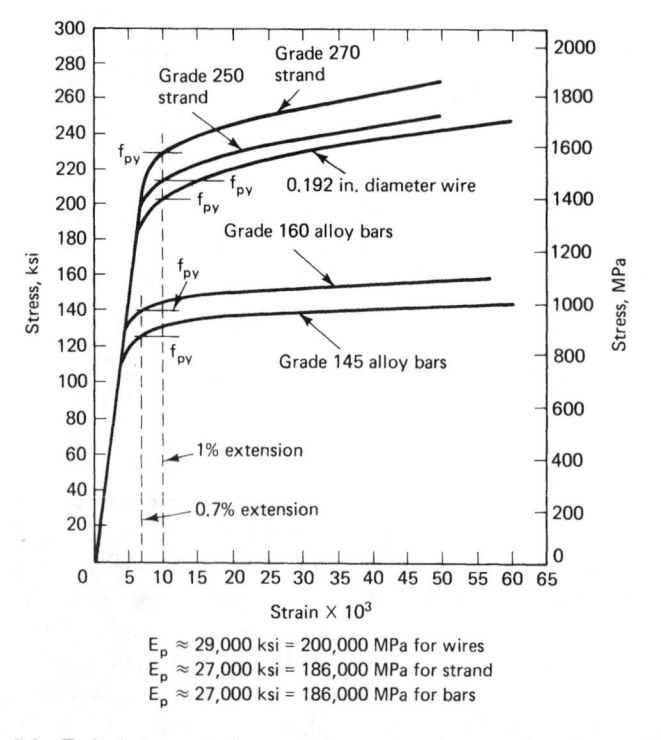

Figure 5.2 Typical stress–strain curves for prestressing strands, wires, and bars.

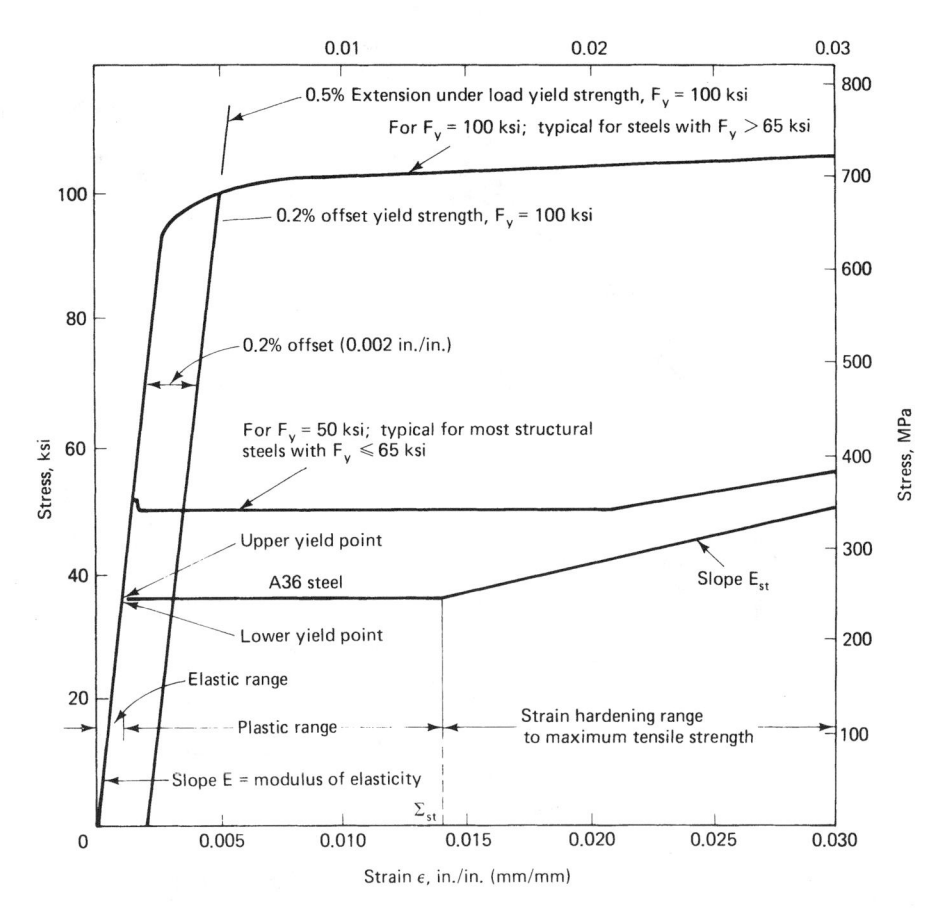

INELASTIC MODELS: STRUCTURAL STEEL AND REINFORCING BARS

Figure 5.3 Enlarged stress–strain curve for A36 and other steels.

5.2.3.2 Stress–Strain Characteristics

An enlarged view of the stress–strain curve for an A36 steel and some other steels for the strain range 0 to 0.030 mm/mm is shown in Figure 5.3. The following points must be emphasized with respect to this stress–strain curve:

1. Even with the high quality control in the manufacture of steel, there will be property variations from structure to structure, and even within a structure. This suggests that the yield strength of steel that is used in design must be carefully determined.
2. Yield strength variations are not normally accompanied by similar changes in the slope of the initial portion of the stress–strain curve. The modulus of elasticity E of the various steels varies between very narrow limits, seldom exceeding 210 GPa. The E value for prestressing steel wire strands is of the order of 190 GPa because of wrapping effects. For most structural steels and reinforcing steel bars, the modulus of elasticity has a typical value of 200 GPa.
3. Steel is a structural material that exhibits the flat yield plateau shown in Figure 5.3. There are many other metals that are ductile, but none possesses this property. Only phosphor–bronze has a definite yield point, as does mild steel; however, it strain-hardens at strains only slightly greater than the yield strain (Antebi et al., 1962; Harris et al., 1962). In seeking a model material that will simulate steel in the elastoplastic range, consideration must be given to the yield plateau, including its extent.

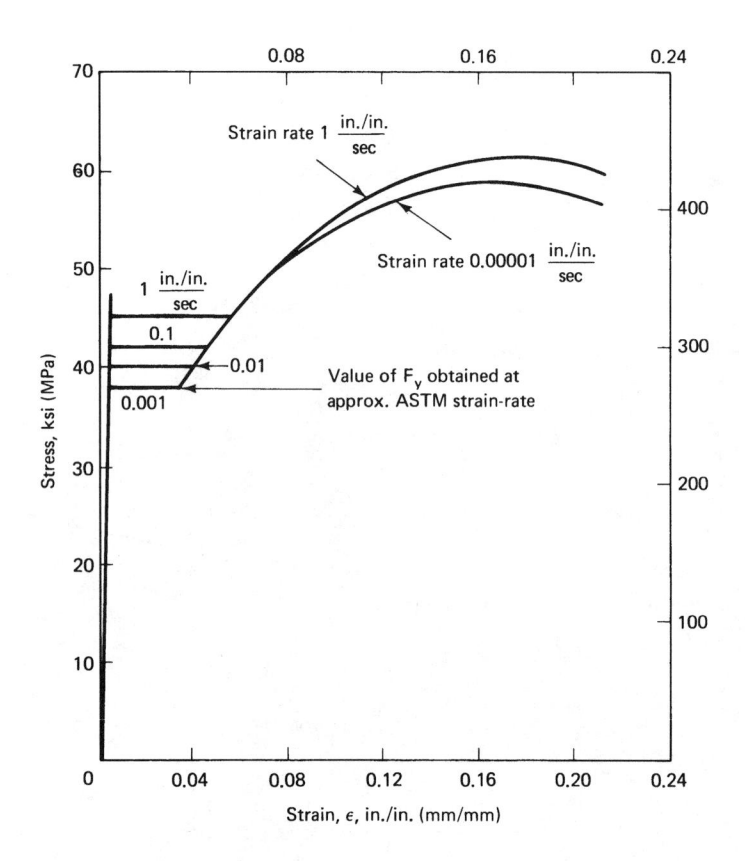

Figure 5.4 Effect of strain rate on stress–strain curve for A7 steel.

4. It must be noted that the stress–strain curves that are ordinarily presented (e.g., Figure 5.1) are normally obtained from small tensile coupons. If one were to test a complete wide-flange section (rolled or welded), the yield plateau might not be easily observed. The reason for the apparent difference in behavior would be due solely to the presence of initial or residual stresses in the wide-flange specimen that result from the differential cooling of hot-rolled or welded shapes. If such initial stresses influence the structural behavior, the initial stresses in the model must simulate those that would exist in the prototype. Finally, the load–deformation characteristics of steel materials are sensitive to the rate at which the strains are induced. A typical example is shown in Figure 5.4.

5.3 STRUCTURAL STEEL MODELS

5.3.1 General

Vastly improved computing facilities and sophisticated design techniques have provided the design engineer with tools that can be used confidently in the design of any conventional structure such as a building or a bridge. However, it must be noted that most of the provisions in the existing codes are empirical in nature and have been derived from interpretation of complex experimental information on member behavior. A rigorous analytical approach may not always be sufficient to obtain a design solution, and it is in the cases of highly unconventional structures that model studies can be useful.

Successful experimental research work on steel structures has led to innovations and improvements in design techniques, e.g., LRFD (Load and Resistance Factor Design) also known as "limit states design" elsewhere, residual stresses, member design under combined loadings, use of corrugated webs, bolted connections, etc. So far, full-scale tests have produced much useful information (Gaylord, et al., 1992; Galambos et al., 1996). However, they are generally restricted to simple members and very simple structures because of the space and laboratory facilities required and the expenditures involved. Use of small-scale models to study the complex behavior of whole structures or their component substructures can overcome these restrictions and can be useful as research and development tools.

In a laboratory of usually limited resources, studies of steel structures at small scale can mean the difference between being able to conduct experimental research and not being able to at all. In addition to the research use of small-scale steel structures, demonstration of the ultimate-strength behavior in the classroom is an important teaching tool which is becoming increasingly more important in the new engineering curriculum with its emphasis on experimental techniques. For this reason, the small-scale model techniques available for studying the behavior of steel structures and in particular structural frameworks is presented in this section.

5.3.2 Steel Columns

Several techniques are considered for fabrication of small-scale wide-flange beams and columns. The most practical of these that give accurate yet economical results are the assembly of separate plates for the flanges and the web and milling of rectangular bar stock. Scaled model steel columns have been studied by Cherry et al. (1952) and Baker et al. (1956) using the former of the two techniques. Litle (1966) and Litle et al. (1968) used the milling of hot-rolled bar stock to fabricate $1/15$-scale model columns as part of steel frameworks. Mills et al. (1979) and Wallace and Krawinkler (1985b) also used the milling of rectangular bar stock to form the columns of their small-scale steel frame models.

Two distinct column problems were studied by Cherry et al. (1952): the effect of residual stresses on ultimate strength and the investigation of combined lateral buckling and flange twist. To study the effect of residual stresses on ultimate strength, the pin-ended models MR-4, MR-5, and MR-6 shown in Figure 5.5 were used. Residual stresses were introduced into the model using a strain transfer–relaxation technique by cutting the web plate oversize, stressing it while the joint between the web and flange was made, and ending with a self-equilibrating system of stresses between the web and the flanges. The simlpified H-type section shown was fabricated from 50×3 mm structural steel bar by milling to the proper width. A 3×1.5 mm slot was milled along the centerline of one of the faces of each flange and the edges of the web plate were subsequently fitted into these grooves during the fabrication operation. In order to reproduce a consistent set of end conditions, knife-edge bearing supports were adopted. These blocks were attached to the base plate of the model column by means of a pair of screws which were tapped into the knife-edge blocks and fitted through holes in the base plates to which they were clamped by means of nuts (see Figure 5.5).

Three column models were tested to investigate the combined lateral buckling and flange twist. Two of the models (MW-1 and MW-2) were $1/12$-scale simplified models of a built-up Section 52-C tested at full scale by Cherry and Mosborg (1950) at the University of Illinois. These specimens were prepared by forming the component sections in a milling machine after which the various elements were assembled as shown in Figures 5.6 and 5.7. A similar method of fabricating model columns has also been used by Baker et al. (1956) at Cambridge University to study column behavior. The material composing the flanges was cut from 19-mm-thick ASTM A-7 structural steel plate. The operational technique used in forming the flange sections in the milling machine is illustrated in Figure 5.8. The web plate of MW-1 was cut from 1.5-mm sheet material and milled to the proper size. Model MW-2 was made with very thin web sheet material of only 0.36 mm in thickness. Solder was used to attach the flanges to the web in these two models.

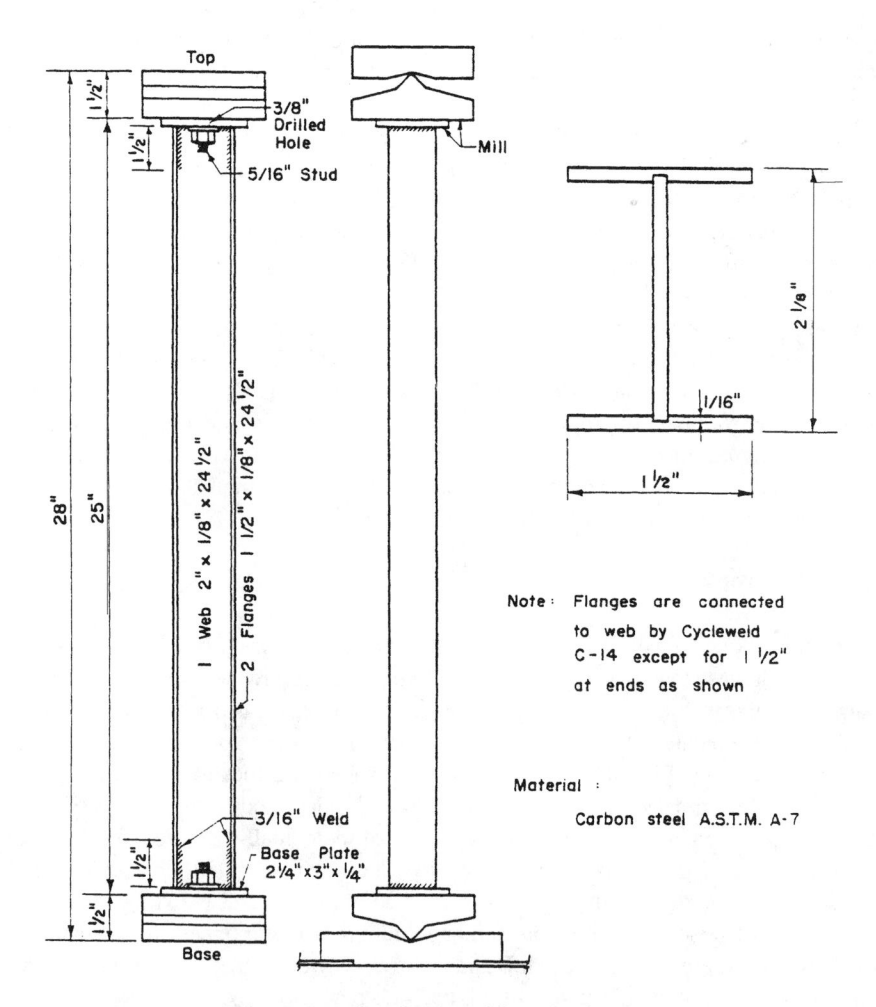

Figure 5.5 Details of model steel columns.

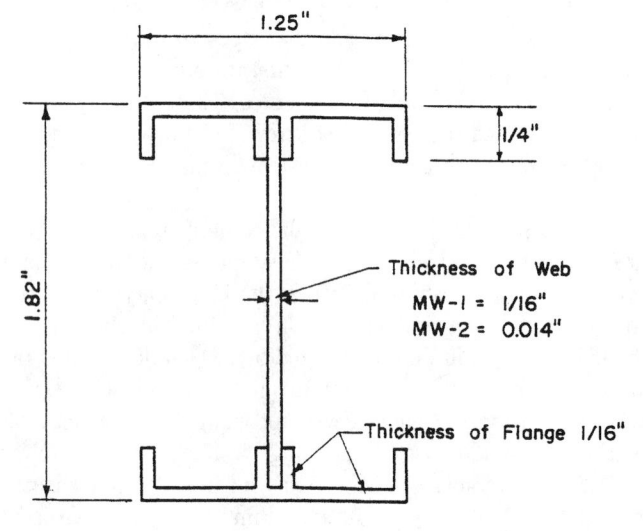

Figure 5.6 Cross-section of model columns MW-1 and MW-2.

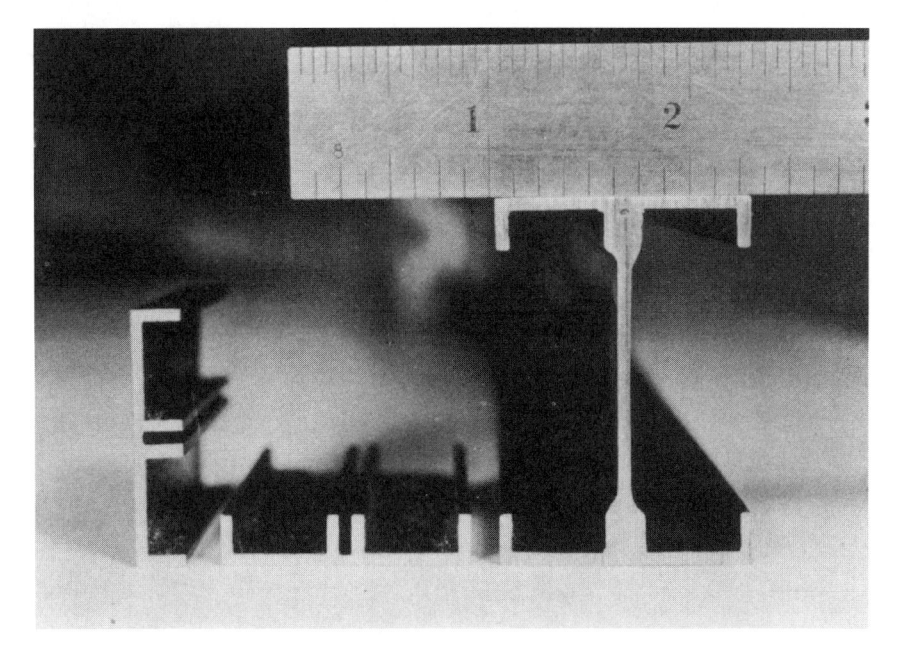

INELASTIC MODELS: STRUCTURAL STEEL AND REINFORCING BARS 217

Figure 5.7 Cross-sectional view of model column MW-1.

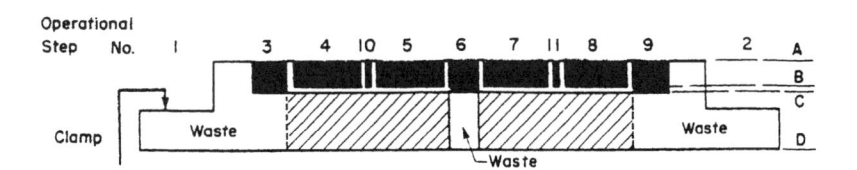

Figure 5.8 Diagram showing milling procedures used in preparing flanges for model columns MW-1 and MW-2.

The third model MW-3 was a 1/6-scale replica model of the same built-up Section 52-C mentioned above. It was fabricated from plates and angles such that the cross-sectional shape was identical with that of the full-scale columns. The dimensions of the 1/6-scale model are shown in Figure 5.9. The angles and and flange plates were cut from 1.5-mm carbon-steel sheet material while the web was cut from 1.9-mm carbon-steel sheet material. Since it was impossible to procure angles of the desired size, the necessary sections were made by bending 25-mm-wide steel strips into the proper shape. This procedure had the disadvantage that a sharp right-angled corner could not be obtained. The plate and angle elements were connected with 3-mm soft iron rivets to form the complete column section shown in Figure 5.9. The mechanical properties of coupons of materials used in fabricating all six column scaled models are summarized in Table 5.1.

The most important conclusions from this exploratory study according to the authors are

1. The behavior of the 52-C series of large columns was fairly accurately reproduced in the test of the 1/6-scale replica model.
2. The ultimate strengths of all three models used in the investigation of lateral buckling and flange twist were not appreciably reduced by the freedom of the flanges to rotate.
3. The strength of prestressed model MR-5 which was fabricated with residual stresses of 49.4 MPa (compression) in the flanges and 71.4 MPa (tension) in the web was approximately 20% less than the strength of the identical stress-free model MR-6 tested in the same manner.

Figure 5.9 Details of model column MW-3

Table 5.1 Mechanical Properties of Coupons of Materials Used for Type MW and Type MR Models

		Compression Tests		Tension Tests				
Model (1)	Location (2)	Yield Point, MPa (3)	Modulus of Elasticity, 10^6 MPa (4)	Yield Point, MPa (5)	Modulus of Elasticity, 10^6 MPa (6)	Ultimate Strength, MPa (7)	Percentage Reduction in Area (8)	Percentage Elongation in 50 mm (9)
MW-1	Flange	279	0.205	265	0.215	464	54.7	29.9
	Web	—	—	260	0.203	348	58.3	34.8
	75 mm section removed from model after failure	286	0.199	—	—	—	—	—
MW-2	Flange	290	0.220	268	0.199	529	41.2	29.0
MW-3	Angle	—	—	313	0.206	376	52.1	26.3
	Flange plate	—	—	190	0.203	302	40.1	31.6
	Web plate	—	—	287	0.202	353	59.6	33.3
MR-4	All from same bar stock	—	—	—	—	374	65.6	36.5
		—	—	260	0.207	377	65.2	35.2
MR-5	Web	—	—	248	0.206	363	60.5	37.0
MR-6	Flange	—	—	252	0.211	363	61.4	37.1

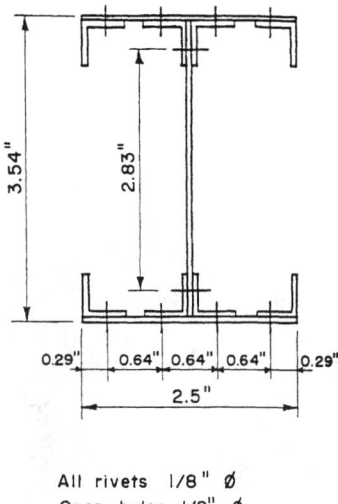

5.3.3 Steel Beams

5.3.3.1 Fabrication of Wide-Flange Shapes

Litle and Foster (1966) undertook a project to fabricate small-scale wide-flange sections, small-scale joints, and a small-scale building frame made of steel. They found that the chemical and mechanical properties of SAE C1020 hot-rolled structural steel (Society of Automotive Engineers,

INELASTIC MODELS: STRUCTURAL STEEL AND REINFORCING BARS

Table 5.2 Mechanical and Chemical Properties of ASTM A36 and SAE B1113, C1010, and C1020

Item	C	Mn	P	S	Minimum Yield Stress, MPa	Minimum Yield Strength, MPa	% Elong. 50 mm Gage
A36	0.25	0.80–1.20	0.040	0.50	248[a]	400[a]	23[a]
B1113 (C.R.)	0.13	0.70–1.00	0.07–0.12	0.24–0.33	413[b]	538[b]	10[b]
C1010 (C.R.)	0.08–0.13	0.30–0.60	0.040	0.050	283[b]	331[b]	20[b]
C1020 (H.R.)	0.18–0.23	0.30–0.60	0.040	0.050	207[b]	379[b]	25[b]

[a] Typical minimum
[b] Specimen minimum

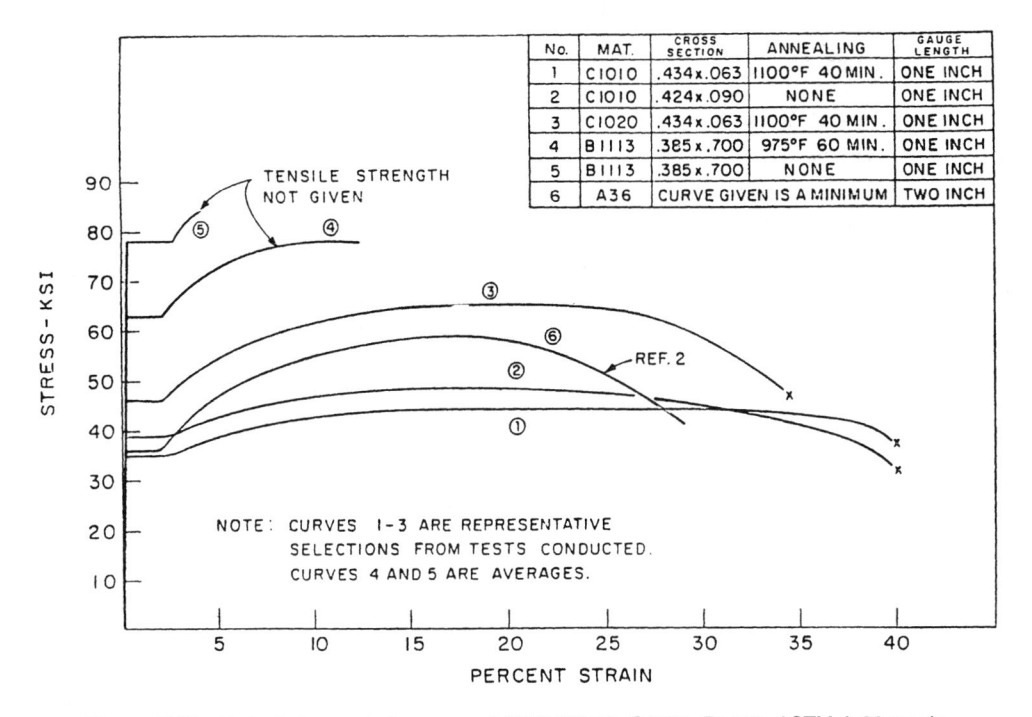

Figure 5.10 Typical stress–strain curves of SAE C1010, C1020, B1113, ASTM A-36 steels.

1964) are such that it may be used satisfactorily in the modeling of steel structures. The chemical and mechanical properties of prototype A36 steel and the three possible model steels they considered are given in Table 5.2. Typical stress–strain curves for the prototype and three model materials they considered are given in Figure 5.10. On the basis of its weldability and mechanical properties, the SAE C1020 hot-rolled steel was selected as the most appropriate material for ultimate-strength models of A36 steel structures. From their investigation, they demonstrated that milling wide-flange sections from hot-rolled bar stock is a reliable and accurate method for fabricating small-scale sections with element thickness as small as 0.6 mm.

5.3.3.2 Beam Tests

Four beams (two W14 × 103 and two W21 × 62) were fabricated and tested with third-point loads to determine the influence of annealing on the behavior of milled wide-flange sections. Comparison of specified and measured dimensions of the 1/15-scale W14 × 103 and W21 × 62 beams are given in Table 5.3. The tensile strength properties of the beams and the stock material

Table 5.3 Comparison of Specified and Actual Dimensions for W14 × 103 and W21 × 62 Sections

Section	Dimension	d, mm	b, mm	t_f, mm	t_w, mm	I, mm[4a]
W14 × 103	Specified	24.13	24.69	1.37	0.91	9573
	Actual average	24.13	24.69	1.32	0.86	9365
W21 × 62	Specified	35.56	13.92	1.04	0.69	10905
	Actual average	35.61	13.92	1.04	0.71	11030

[a] Calculated from dimensions.

Table 5.4 Tension Test for SAE C1020 Hot-Rolled Steel Specimen

Test No.	Description	Area, mm × mm	Annealing	Yield Stress, GPa	Ultimate Strength, GPa	% Elongation 50 mm
1	Plain specimen from stock	1.60 × 11.02	593°C 40 min	0.318	0.441	34
2	Plain specimen from stock	1.60 × 11.02	593°C 40 min	0.321	0.455	32
3	Plain specimen from stock	1.60 × 11.02	593°C 40 min	0.318	0.450	36
4	TIG butt welded	2.87 × 10.92	593°C 40 min	0.340	0.485	14
5	TIG butt welded	2.87 × 10.90	593°C 40 min	0.336	0.490	12
6	Cut from web of $\frac{1}{15}$-scale W21 × 62	0.71 × 9.53	None	0.334[a]	0.441	27
7	Cut from web of $\frac{1}{15}$-scale W21 × 62	0.71 × 10.95	None	0.315[a]	0.450	25
8	Cut from web of $\frac{1}{15}$-scale W21 × 62	0.71 × 11.05	None	0.327[a]	0.445	23
9	Cut from web of $\frac{1}{15}$-scale W21 × 62	0.71 × 11.07	538°C 45 min	0.292	0.424	32
10	Cut from web of $\frac{1}{15}$-scale W21 × 62	0.71 × 11.07	538°C 45 min	0.312	0.424	29
11	Cut from web of $\frac{1}{15}$-scale W21 × 62	0.71 × 11.10	538°C 45 min	0.319	0.427	34
12	Cut from web of $\frac{1}{15}$-scale W14 × 103	0.84 × 11.05	None	0.237[a]	0.413	38
13	Cut from flange of $\frac{1}{15}$-scale W14 × 103	1.35 × 11.07	None	0.247[a]	0.415	23
14	Cut from flange of $\frac{1}{15}$-scale W14 × 103	1.32 × 11.07	538°C 45 min	0.258	0.402	26
15	Cut from flange of $\frac{1}{15}$-scale W14 × 103	1.32 × 11.07	538°C 45 min	0.261	0.407	39

[a] 0.2% offset method.

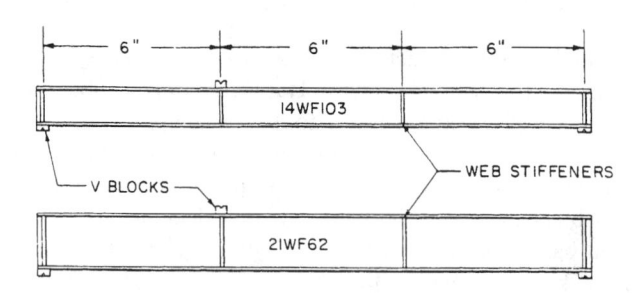

Figure 5.11 Position of web stiffeners and loading blocks.

are given in Table 5.4. Stiffeners and loading blocks were welded to the model beams (Figure 5.11) using a tungsten inert gas (TIG) process discussed in Chapter 6. The test set up is shown in Figure 5.12 with lateral bracing provided just inside the third points of the 450-mm-long beams.

Two beams were cut from a length of the milled model W14 × 103 section and two from a length of the model W21 × 62. One of each was annealed at 520°C for 60 min, while the other

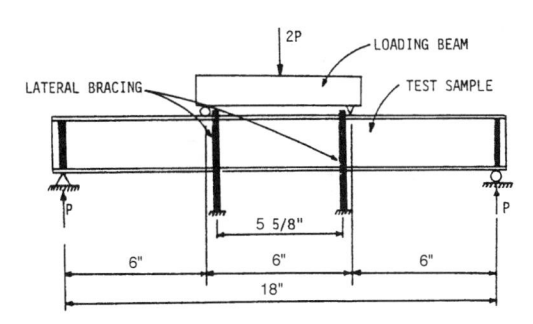

INELASTIC MODELS: STRUCTURAL STEEL AND REINFORCING BARS 221

Figure 5.12 Setup of beam test.

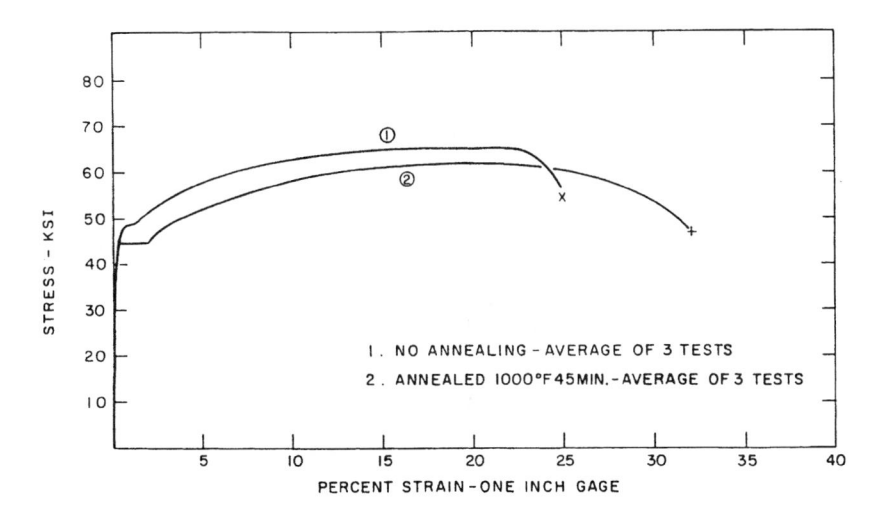

Figure 5.13 Annealing effect on W21 × 62 web tension samples — C1020 H.R. steel.

received no heat treatment. Results of the annealing of the W21 × 62 beam on the tensile properties are shown in Figure 5.13. The midspan moment vs. midspan deflection results shown in Figures 5.14 and 5.15 confirm that the stresses induced by milling do not significantly influence the bending strength of the beam members.

Model beams at ¹⁄₁₂.₅ scale, forming together with model columns interior and exterior welded connections of steel frames, were studied under reversed cyclic loading by Wallace and Krawinkler (1985b). The small-scale model joint results (Lee and Lu, 1984) were compared with available full-scale test results with excellent correlation.

5.3.4 Steel Frames

5.3.4.1 M.I.T. Ultimate Strength Model Steel Frameworks

In order to establish proper techniques, Litle and Foster (1966) fabricated a one- by two-bay three-story space framework using ¹⁄₁₅-scale W14 × 103 members as columns and ¹⁄₁₅-scale W21 × 62 members as beams. TIG welding with 0.8 mm Industrial Stainless 410 filler wire was used throughout. The nominal framework dimensions are shown in Figure 5.16 and the measured dimensions on the finished frame (not annealed) are listed in Table 5.5. The maximum deviations from specified dimensions occur in the bottom story column heights where in one place the column is 2 mm too short. The second- and third-story column heights and all bay widths are everywhere within 1.5 mm.

Table 5.5 Geometry of Completed Space Framework (Refer to Figure 5.16)

Frame	Section	1–2, mm	2–3, mm	3–4, mm
A	1	375.2	269.7	269.0
	2	374.9	269.0	268.5
	3	374.4	269.5	268.7
	4	373.9	269.7	269.0
	5	374.4	269.7	269.5
B	1	374.4	269.7	269.5
	2	376.7	269.7	269.0
	3	376.2	269.7	269.0
	4	376.2	269.5	269.2
	5	374.4	269.7	269.0
C	1	374.1	269.7	268.7
	2	374.1	269.7	269.0
D	1	375.9	269.7	269.0
	2	376.2	269.5	269.0
E	1	374.4	269.7	269.2
	2	374.1	269.7	269.0

Bay Width	2, mm	3, mm	4, mm
DE_0	813.3	813.3	813.6
DE_1	813.8	812.8	813.8
DE_2	812.8	813.1	812.8
DE_3	812.8	813.1	812.8
CD_0	406.4	406.4	406.4
CD_1	406.4	406.1	406.4
CD_2	406.4	405.9	406.4
CD_3	406.4	406.1	406.4
AB_0	383.3	383.3	383.3
AB_1	383.3	383.3	383.3
AB_2	382.8	382.5	382.8
AB_3	382.8	383.3	382.8
AB_4	383.0	382.8	383.0
AB_5	383.3	383.3	383.3

The fabrication of the framework consisted of first cutting and milling the W14 × 103 columns and W21 × 62 beams to size. The girder-to-column web joint detail used is shown in Figure 5.17. Before welding of stiffeners, beam seats, and plates, all parts were cleaned to remove dust, oil, and oxides. A special jig consisting of a flat steel plate with small right angles welded to its surface at each floor level was used in the fabrication of each of the three portal frames in the framework. The columns were clamped to the flat steel plate and then, with the beam seats and top plates snugly against the web of the column, the beams were clamped against the small angles. The beam seats and top plates were first tacked and then welded all around. The filler beams were jigged and then welded. It was found from the study that fabrication of a complete framework is possible, but it is necessary to fix elements during welding and to follow a predetermined sequence of assembly to reduce shrinkage deformations. This sequence may vary with each structure. Until more refinements are made in the welding process, it is necessary to anneal the whole framework to obtain member behavior consistent with the stress–strain characteristics of the material.

5.3.4.2 Stanford University ⅙-Scale Shaking Table Tests

Extensive modeling techniques were developed at Stanford University by Mills et al. (1979) as part of their project to study the problem of testing small-scale models on earthquake simulators.

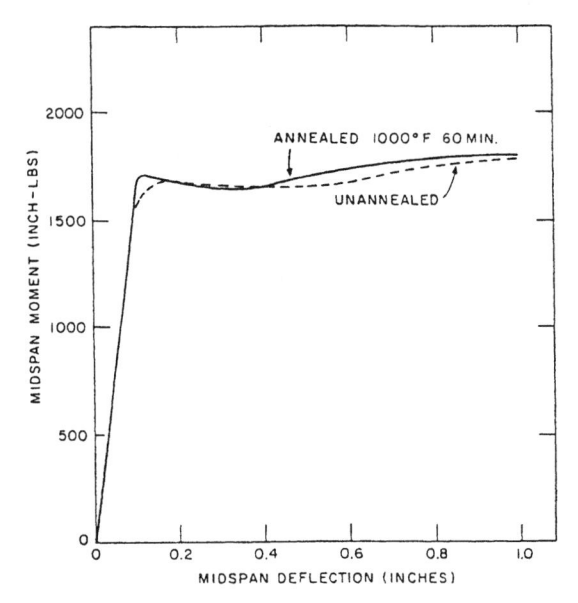

INELASTIC MODELS: STRUCTURAL STEEL AND REINFORCING BARS

Figure 5.14 One-fifteenth scale W14 × 103 beam tests.

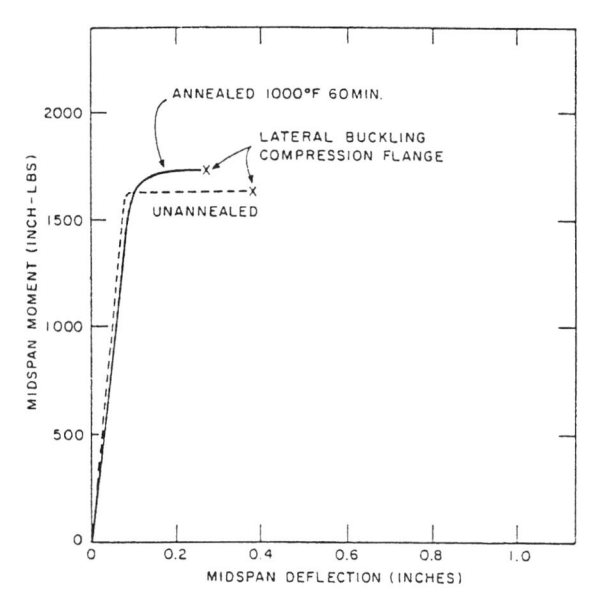

Figure 5.15 One-fifteenth scale W21 × 62 beam tests.

To aid in the development of methodologies for dynamic model fabrication, testing, and analysis, a case study was undertaken involving a small-scale replica of extensive testing of a three-story, single-bay steel frame structure tested at full scale on the University of California, Berkeley shaking table. The prototype steel structure was itself reduced in size (1/2.5) but not in materials or method of fabrication. The prototype consists of two parallel steel frames with moment-resistant, welded connections. These main frames are then joined at floor levels by bolted connections to cross beams and diagonal bracing, thus affecting floor diaphragms rigid in their own plane. End bay bracing in the column weak direction, pretensioned by turnbuckles, resists structural motion transverse to the excitation axis. Artificial mass simulation was used to preserve similitude of the gravitational and inertial effects in both the prototype and 1/6-scale replica model (Figure 5.18).

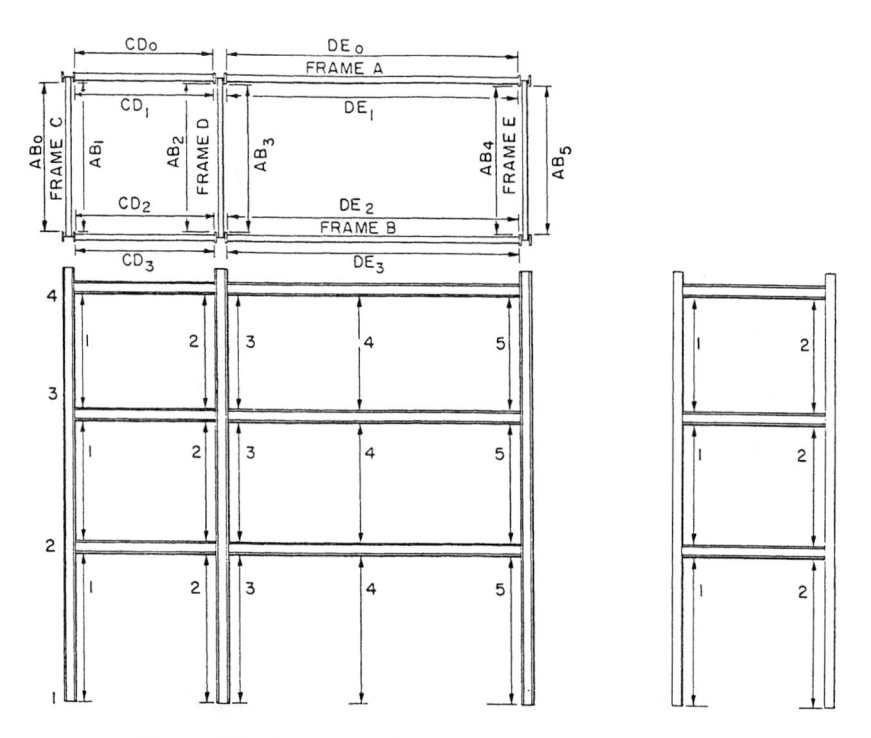

Figure 5.16 Geometry of fabricated 1/15-scale steel framework.

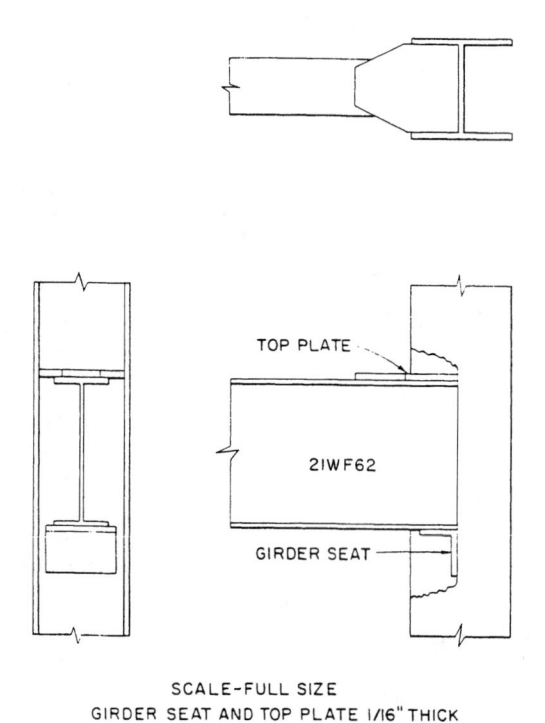

SCALE-FULL SIZE
GIRDER SEAT AND TOP PLATE 1/16" THICK

Figure 5.17 Girder–column web joint detail.

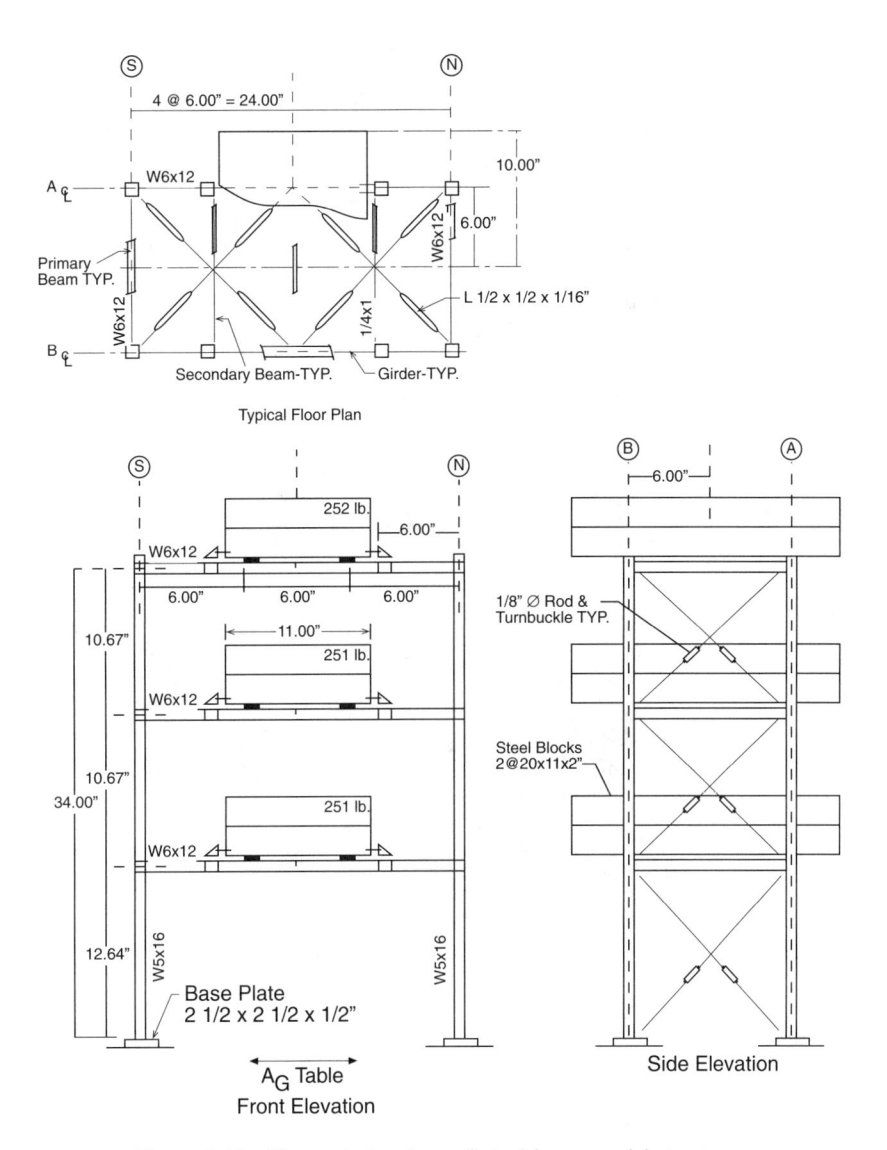

INELASTIC MODELS: STRUCTURAL STEEL AND REINFORCING BARS

Figure 5.18 Plan and elevations of steel frame model structure.

The geometrically scaled prototype cross sections required for the ⅙-scale replica model are shown in Figure 5.19. Fabrication techniques investigated included silver soldering, welding of thin sheet metal, and milling of bar stock. Silver soldering resulted in extreme heat distortions when no preheat was applied and was thus eliminated. The heliarc TIG welding process with argon gas and Stainless 308 0.75-mm-diameter rod was used to produce trial beams for accuracy comparisons with equal-length milled web and flange sections. The tolerances achieved by this technique were unacceptable as compared with those required as shown in Table 5.6. Bar stock of A36 hot-rolled steel material of dimensions 19 × 31.75 mm and 25 × 25 mm were used to fabricate by milling the girder and column members, respectively. The finished sections are shown in Figure 5.20 and the geometric properties of the model elements are given in Table 5.7. The mechanical properties of the prototype and ⅙-scale primary members are listed in Table 5.8. The finished ⅙-scale steel frame model was tested on the John A. Blume Earthquake Engineerring Center shaking table. Results and comparisons with the prototype steel frame tests are given in Chapter 11.

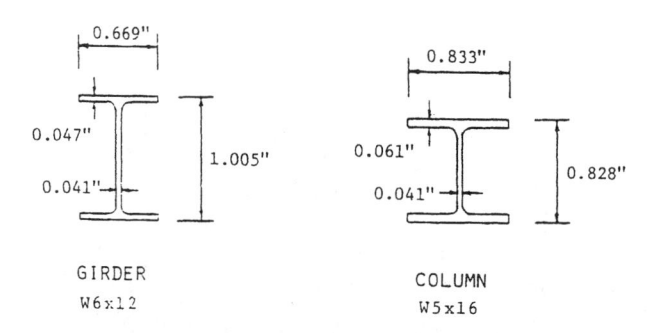

Figure 5.19 Model element specifications.

Table 5.6 Tolerances for Trial Model Elements, $l_r = 1:6$

Item	Scaled Standard Mill Practice	Measured Deviations, mm — Welded Specimen	Measured Deviations, mm — Machined Specimen
Flange out of square	< 1.02 mm	0.76	0.25
Camber	< 0.64 mm	2.54	0.51
Sweep	< 1.27 mm	2.03	0.25

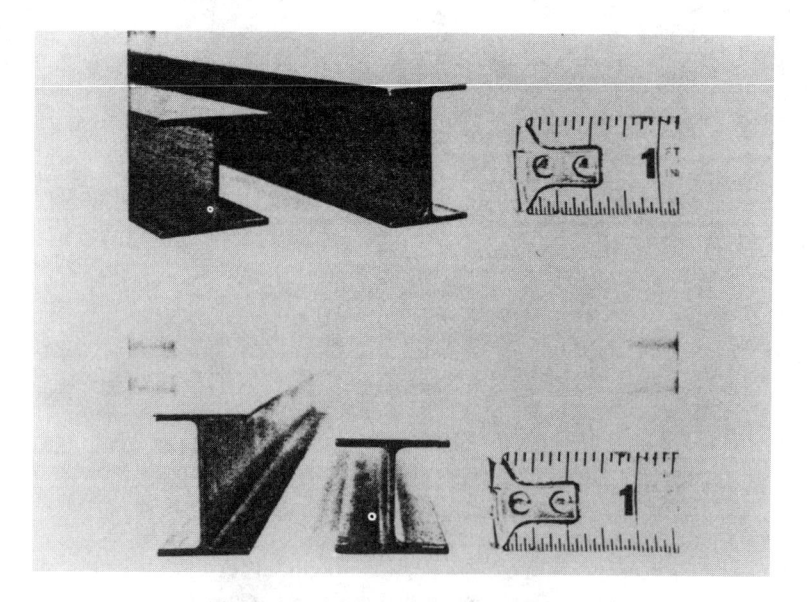

Figure 5.20 Machined model elements.

5.3.4.3 U.S.–Japan Six-Story Steel Building Model Studies

As part of the multifaceted U.S.–Japan Cooperative Program in Earthquake Engineering Research, model studies at small scale of a prototype six-story steel structure were conducted by Wallace and Krawinkler (1985b) at Stanford University. The objectives of the model study were

1. To investigate the feasibility and limitations of small-scale model testing in earthquake engineering,
2. To study the simulation accuracy of specific failure modes in small-scale models,
3. To correlate results at different scales to assess prototype response prediction from experimental studies, and
4. To study the seismic behavior of components and assemblies of braced frame structures.

INELASTIC MODELS: STRUCTURAL STEEL AND REINFORCING BARS

Table 5.7 Machined Sections for Stanford University 1/6-Scale Model

Item	b, mm	d, mm	t_w, mm	t_f, mm	A, mm²	I_x, mm⁴	Camber, mm	Sweep, mm	Flange out of Square, mm
Model girder (W6 × 12 prototype)									
Specified	16.993	25.527	1.041	1.194	64.645	7126	—	—	—
Actual average	17.018	25.552	1.016	1.194	66.322	7359	—	—	—
Tolerance[a]	+1.016/−0.762	±0.508	—	—	±1.935	—	0.635	1.270	1.016
Actual maximum	±0.102	±0.178	±0.102	±0.178	±3.226	—	0.508	0.254	0.254
Model column (W5 × 16 prototype)									
Specified	21.158	21.031	1.041	1.549	84.258	6747	—	—	—
Actual average	21.158	21.057	1.041	1.575	87.484	6997	—	—	—
Tolerance[a]	+1.016/−0.762	±0.508	—	—	±1.935	—	0.965	0.965	1.016
Actual maximum	±0.127	±0.203	±0.102	±0.152	±3.226	—	0.254[b]	1.016	0.254

[a] Tolerances are scaled values from standard mill practice, *AISC Steel Construction Manual*.
[b] Column camber increased to 1.27 mm after welding of stiffeners and heat treatment.

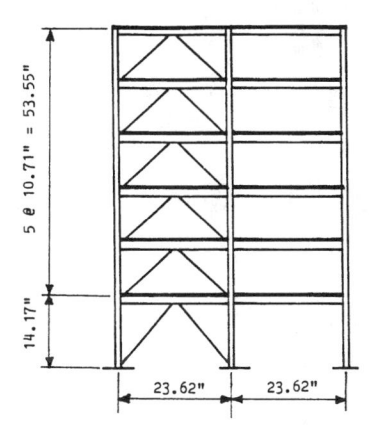

Figure 5.21 Steel-braced frame model dimensions.

The degree to which these objectives were met and the results of these model studies are given in Chapter 10.

The small-scale (1/12.5) model study included three beam–column assemblies and the braced frame (Figure 5.21). This latter model included prototype floor slabs 15.8 m wide but none of the surrounding bracing. The model braced steel frame prior to casting of the microconcrete slabs is shown in Figure 5.22. The three beam–column assemblies included an exterior joint with web connection, an interior joint with flange connection, and an exterior joint with flange connection. Wide flange steel shapes used in the model beam–column assemblies and braced frame specimen were milled from a single piece of A36 steel plate. Two tensile coupons cut from beam webs and four coupons cut from flanges were tested and a typical stress–strain diagram is shown in Figure 5.23.

Fabrication of the steel skeleton of the 1/12.5-scale model braced frame specimen followed these operations: (1) cutting the model beams and columns to length; (2) cutting stiffeners from sheet metal and welding them in place using the TIG welding process; and (3) welding studs to the beams using capacitive discharge stud welding equipment. The model beams and columns were then assembled on a fixture and the joints were welded using the TIG welding process.

Table 5.8 Material Properties – Prototype and Model Coupons: Stanford University ⅙-Scale Model

Item	Location[a]	$\dot{\varepsilon}_{elastic}$, mm/mm/s	$\dot{\varepsilon}_{strain\ hardening}$, mm/mm/s	E, $\times 10^6$ MPa	σ_y^{upper}, MPa	σ_y, MPa	σ_{ult}, MPa	E_{st}, MPa	E_{st}/E, %	ε_{st}, $\times 10^{-3}$ in./in.	$\varepsilon_{st} - \varepsilon_y$, $\times 10^{-3}$ in./in.	$\varepsilon_y/\varepsilon_{st}$, %
Prototype girder	w	3×10^{-5}	2×10^{-3}	0.213	357.2	326.8	466.1	3379	1.6	24.5	22.9	15
	f	2×10^{-5}	8×10^{-4}	0.213	—	261.3	439.2	6068	2.8	14.6	13.3	12
	f	8×10^{-6}	3×10^{-4}	0.212	276.5	272.4	437.8	4068	1.9	15.8	14.4	12
Prototype column	w	3×10^{-5}	2×10^{-3}	0.209	313.7	294.4	448.2	3310	1.6	27.0	25.5	18
	f	2×10^{-5}	8×10^{-4}	0.208	272.4	262.0	—	4482	2.1	18.3	17.0	14
	f	4×10^{-5}	2×10^{-3}	0.203	293.7	268.9	450.9	4482	2.2	21.0	19.6	15
Model girder	w	2×10^{-4}	2×10^{-4}	0.191	—	293.7	467.5	4137	2.2	15.1	13.6	11
	f	2×10^{-4}	2×10^{-4}	0.193	—	279.2	439.9	4206	2.2	12.9	11.4	12
	f	2×10^{-4}	2×10^{-4}	0.165[b]	—	277.8	—	3792	2.3	14.2	12.5	12
Model column	w	2×10^{-4}	2×10^{-4}	—	—	—	460.6	—	—	—	—	—
	w	2×10^{-4}	2×10^{-4}	0.181	335.8	314.4	—	3103	1.7	16.4	14.6	11
	f	2×10^{-4}	2×10^{-4}	0.189	339.9	319.2	—	2620	1.4	19.9	18.1	9
	f	2×10^{-4}	2×10^{-4}	0.185	324.1	320.6	450.2	2758	1.5	19.4	17.4	9
	f	2×10^{-1}	2×10^{-1}	0.207	403.4	353.0	—	2758	1.3	24.3	22.2	8

[a] Location is indicated as "w" for web, "f" for flange coupon.
[b] Measurement error was assumed responsible for the low E value.

Figure 5.22 Steel-braced frame model prior to casting concrete slab.

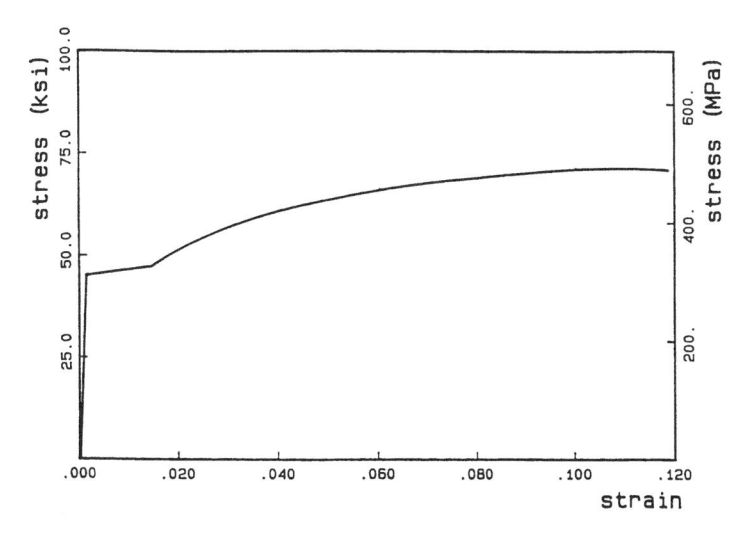

Figure 5.23 Typical stress–strain diagram for beam and column material.

5.4 REINFORCEMENT FOR SMALL-SCALE CONCRETE MODELS

5.4.1 General

Steel bars, rods, and cables are generally used for reinforcing structural concrete, with low- to medium-strength steels used for normal reinforcing and high-strength wires and rods for prestressing tendons. The two types of steel reinforcing commonly used in North America are deformed bars and welded wire fabric. Although the ACI-318-95 *Building Code Requirements for Structural Concrete* permits the use of steel with a yield strength of 517 MPa and without a well-defined yield point, most conventional reinforced concrete members are reinforced with steels that have a well-defined yield point and sufficient ductility to fulfill the requirement of an underreinforced design. The properties of steel that must be considered in the modeling of reinforcement are:

1. Yield and ultimate strength in tension, plus yield strength in compression,
2. Shape of stress–strain curve,
3. Length of the yield plateau (important in modeling seismic dynamic behavior),
4. Rate of strain hardening (important in modeling seismic dynamic behavior),
5. Ductility,
6. Bond characteristics at steel–concrete interface.

The procedure for selecting the correct reinforcement for a concrete scaled model is the single most important step in the whole modeling process. A schematic of the steps necessary for producing model reinforcement with the required characteristics enumerated above is shown in Figure 5.24. The model reinfocement selection process starts with the required prototype stress–strain curve. If the available model wire as purchased is plain, it must be deformed, annealed, and then tested to match the required stress–strain properties as shown. If the model wire is one of the commercially deformed bars available, it may or may not require annealing in order to satisfy the required stress–strain characteristics. Experience has shown that when commercially deformed bars are used some annealing will always be required and a compromise must sometimes be made on the spacing and height of the available surface deformations. For these reasons, the availability of a wire deformeter, with a selection of scaled rollers, and an annealing furnace can be a very useful addition to any structural models laboratory.

5.4.2 Model Reinforcement Used by Various Investigators

Commercially available wires and rods of varying sizes and strengths have been used for model reinforcing steel at different research centers. However, a problem exists with the bond characteristics in conjunction with their use as reinforcing in model concretes, and, generally, deformed wires are necessary for model reinforcing to improve the nature and amount of bond strength and to achieve the best possible cracking similitude. Currently available wires and rods may be grouped as follows:

1. Round steel wire and rod in a variety of sizes and strengths.
2. Cold-rolled threaded steel rods. The highly deformed surface provided by the threads leads to a high bond strength.
3. Commercially available deformed wires as used in manufacturing welded wire fabric.
4. Custom deformed wire in a machine that produces scaled rolled external lugs or deformations.
5. No. 2, No. 3, and 6-mm deformed bars.

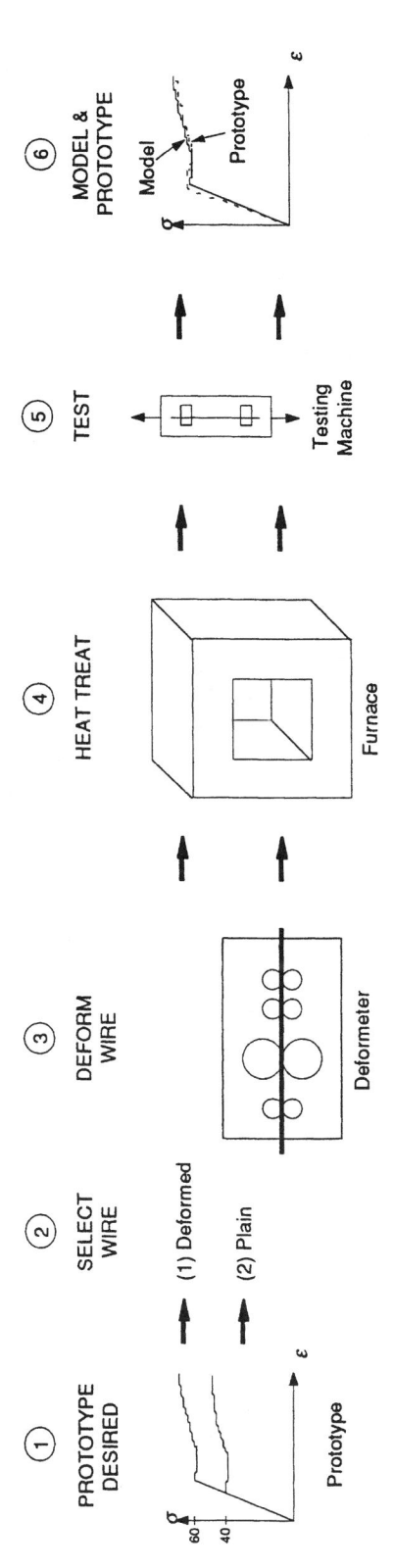

INELASTIC MODELS: STRUCTURAL STEEL AND REINFORCING BARS 231

Figure 5.24 Steps needed for producing scaled model reinforcement.

Careful choice of model reinforcing, combined with the proper annealing processes described later, will result in reinforcing of suitable properties for a given model study. It should be pointed out, however, that the accurate modeling of the steel reinforcement is the single most important step in the successful conclusion of a reinforced or prestressed concrete small-scale model study.

At present, a majority of North American investigators use No. 3 (10 mm) and No. 4 (12 mm) deformed steel bars for reinforcing large-scale models and deformed steel wires (ASTM A496, *Standard Specification for Steel Wire, Deformed for Concrete Reinforcement*) for reinforcing smaller models. However, various other techniques have been attempted by model investigators to simulate the prototype reinforcement. Plain wires with a rusted surface have been used with partial success at MIT, Cornell University, and McGill University. Brock (1959), Lord (1965), Kim et al. (1988), and El-Attar et al. (1991; 1997) have successfully used threaded wires in achieving both a high degree of bond and acceptable cracking similitude. Harris et al. (1966a) developed a simple technique (the Cornell deformer) to cold-deform commercially available plain steel wires by passing them through a special device with two pairs of perpendicular knurling wheels. White and Clark (1978) used a similar technique at the Cement and Concrete Association, U.K., to develop a ⅙-scale model of the 12-mm GK60 deformed bar used in the U.K. Unlike the Cornell deformer that produces internal indentations on the smooth wire, their machine produced external lugs. Maisel (1978a,b) working at the University of Stuttgart reported on a wire-deforming machine that was able to deform model wire with different rib spacing. Noor and Goodman (1979) in a discussion to the paper by White and Clark (1978) reported on an interesting wire deformer that can produce a variety of appropriately sized external deformations for small-scale microconcrete models. Using an identical technique, Subedi and Garas (1978) deformed available ⅙-mm-diameter wires to produce "crimped" wires for use in their model work. They also used plain, threaded, and deformed bars in their investigations. Since the early 1980s the Structural Models Laboratory, Drexel University has been using a wire-deforming machine that puts rolled lugs or external deformations on the wire to scaled spacings and height, thus producing model bars to any desired scale of standard reinforcing bars. A variation of the grooved rolls reported above by Evans and Clark (1978) has been used at Stanford University by Wallace and Krawinkler (1985a) in studying ¹⁄₁₂.₅-scale models of a seven-story reinforced concrete frame-wall structure (Case Study B, Chapter 1). Murayama and Noda (1983) report on a deforming device for deforming 3-mm-diameter wire for their small-scale model work in Japan. The wire deformers are described below.

5.4.3 Wire Reinforcement for Small-Scale Models

The range of wire sizes generally used in small-scale models varies from SWG No. 11, diameter = 3 mm, to SWG No. 21, diameter = 0.8 mm. Although these wires are readily available in the form of rolls of annealed wire, straight lengths of annealed wires can also be specially ordered. Straight wires are generally desirable for ease of working in the laboratory and also for accurate placement of wires in the very small models.

Commercially deformed wires used in producing welded wire fabric are available in sizes above a diameter of 2.7 mm. In addition, deformed wires as small as 1.6 mm are available on special order from some suppliers. Threaded wires in various sizes are readily obtained, but their cost is considerably higher than other types of reinforcement.

Some of the wires do not have a sharp yield plateau and have very limited ductility in tension because of the rather severe deforming process used in making them. The characteristic sharp yield of annealed wires may be significantly altered by the straightening of rolled wire and also by the deformation technique used in the laboratory. Thus, proper strain relieving, or in certain cases annealing, is required before the wire is suitable for use as model reinforcing. The lack of a sharp yield point is sometimes desired when modeling high-strength reinforcement that does not exhibit sharp yielding; this type of model reinforcing may also be produced by careful use of selected heat

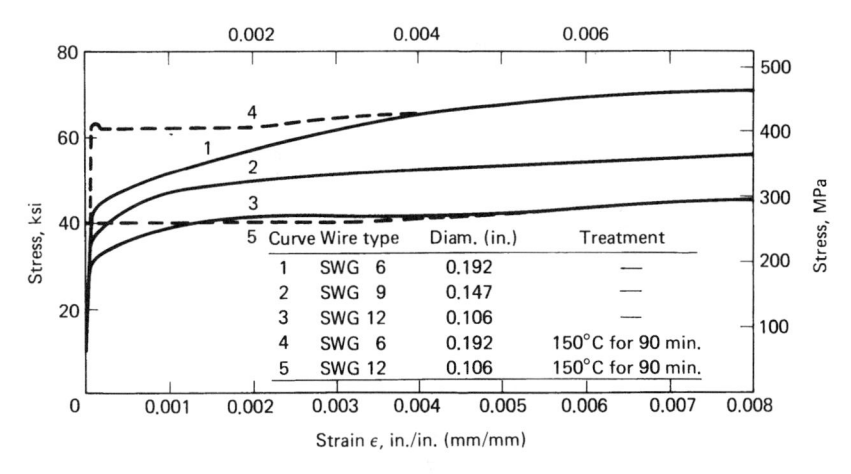

INELASTIC MODELS: STRUCTURAL STEEL AND REINFORCING BARS 233

Figure 5.25 Stress–strain curve of wires.

Curve	Wire type	Diam. (in.)	Treatment
1	SWG 6	0.192	—
2	SWG 9	0.147	—
3	SWG 12	0.106	—
4	SWG 6	0.192	150°C for 90 min.
5	SWG 12	0.106	150°C for 90 min.

treatments. During heat treatment, care must be exercised to check variation of properties along the entire length of wires.

5.4.4 Black Annealed Wire as Model Reinforcement

Microconcrete models at MIT have been reinforced with black annealed wire that is available commercially in the form of rolls. The wire is cold-drawn and annealed in the factory. Before it can be used in models, it must be straightened by pulling, and this process strains the wire sufficiently to destroy the yield point, as shown by Curves 1, 2, and 3 in Figure 5.25. These graphs have been obtained for a black annealed wire that was first straightened, then bent to form part of the reinforcement cage of an arch, and then straightened again before testing. If the cage is allowed to age for 2 months, the wire once more exhibits a definite yield point. The process can be speeded up by keeping the assembled reinforcement cage at 150°C for 90 min. The stress–strain curves of the wire are then given by Curves 4 and 5, which are in good agreement with the stress–strain curves of prototype reinforcing steel. Curve 4 demonstrates, however, that the yield strength of the wire can exceed that of mild steel considerably. The small-diameter wires that are commercially available frequently have too high a strength to be useful in model studies. It must be noted that the yield strength of such wires can be reduced to the desired value by annealing.

Mirza (1967) used closed stirrups made from 11-, 13-, and 16-gage black annealed wires in a study of combined bending and torsion in ¼-scale model beams. Some of these stirrups were instrumented with strain gages to examine the postcracking and ultimate behavior of these model specimens.

5.4.5 Custom-Ordered Model Wire

The large variability in wire stress–strain characteristics that can result when wire for small-scale models is purchased is due to the fact that the chemical composition and in particular the carbon content is not specified. When a large quantity of wire is required or the model study is of such importance, then a special order for one or more ingots with the desired chemical composition can be placed with most wire manufacturers. The wire can then be drawn to the desired diameter, cold-deformed, and heat-treated to the required mechanical properties.

The SWG No. 14 (2 mm) wire used for the shear wall, slabs, ties, and stirrups of the ⅕-scale seven-story frame-wall model tested on the University of California, Berkeley shaking table (Case Study B, Chapter 1) was ordered in billet form and drawn to order from the Davis Wire Company,

BAR	DIAMETER in.	DIAMETER mm	S mm	H mm	$T_1 + T_2$ mm
D10	0.38	9.53	6.7	0.4–0.8	7.5
D19	0.75	1.91	13.4	1.0–2.0	15.0
D22	0.87	22.2	15.5	1.1–2.2	17.5

(a)

(b)

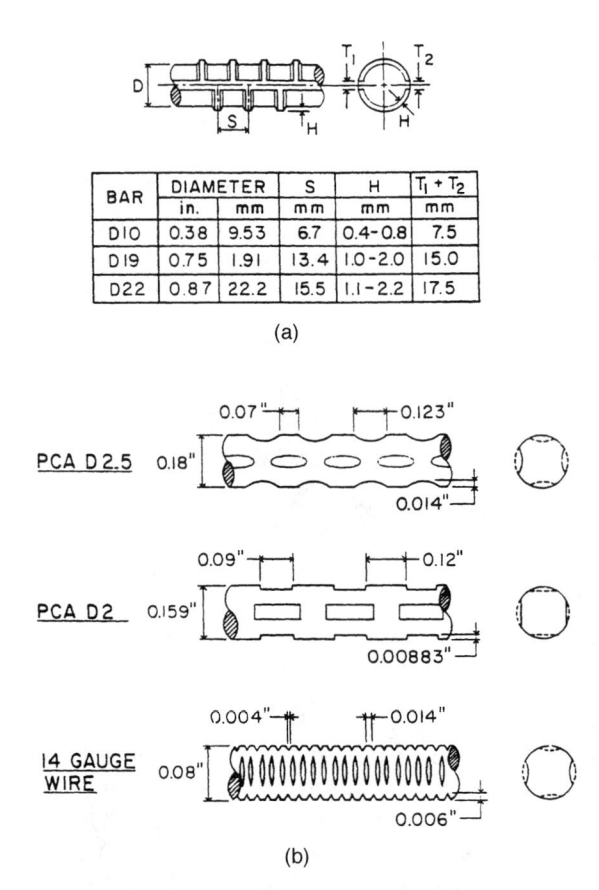

Figure 5.26 Geometric characteristics of the reinforcing bars used in the prototype and model. (a) Prototype reinforcing bars; (b) model reinforcement.

Hayward, CA. The chemical composition (ASTM C1021) allowed heat treatment after knurling with a Cornell-type deformeter borrowed from Prof. H. Krawinkler of Stanford University. The geometric characteristics of the 2-mm model reinforcement are shown in Figure 5.26 together with the prototype and commercially available bars used to model the beam and column reinforcement. After heat treatment in 75-mm-diameter steel tubes (Figure 5.27) at an oven temperature of 568°C for 6 h, after taking 5 h to reach peak, the material was then slowly cooled in the oven with the heat turned off. Stress–strain relations for the 2-mm model bar and D10 prototype are shown in Figure 5.28.

5.4.6 Commercially Deformed Wire as Model Reinforcement

Commercially deformed wires (ASTM A496) normally consist of cold-worked deformed steel wire intended for use in producing welded wire fabric and as reinforcement in concrete construction. These wires normally have a minimum yield strength of 480 MPa and a minimum tensile strength of 550 MPa. These wires can be obtained by special order in sizes D-1 through D-10, with the wire size number indicating the nominal cross-sectional areas of the deformed wire section in 100ths of a square inch. Details of the geometry of deformed steel wire are shown in Table 5.9, which is reproduced in part from the *Manual of Standard Practice of the Reinforcing Steel Institute*. McGill University and Queen's University have used 2.3-mm-diameter wire (approximately 13 gage) in their reinforced concrete model work. This indented wire was supplied by the Steel Company of Canada and was specially annealed to conform to the ASTM specifications for

INELASTIC MODELS: STRUCTURAL STEEL AND REINFORCING BARS 235

Figure 5.27 View of 75-mm steel tubes filled with the reinforcing bars in the furnace.

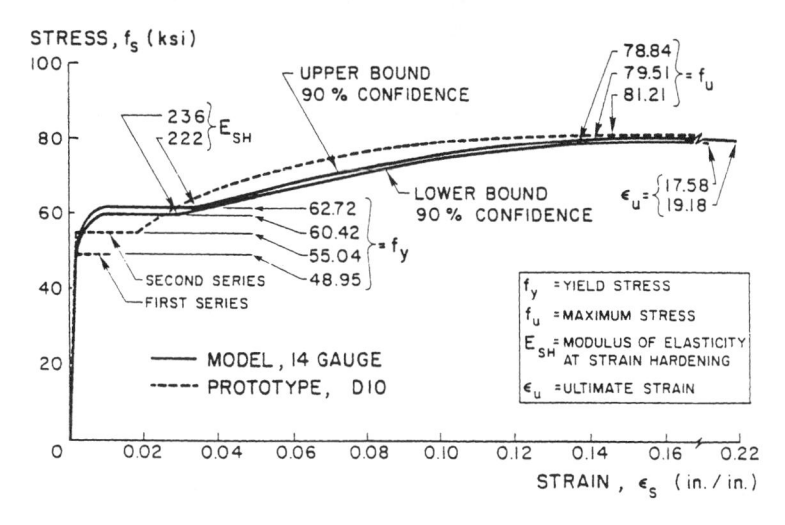

Figure 5.28 Stress–strain relations for wall, slab, tie, and stirrup reinforcement.

intermediate-grade steel. A typical stress–strain curve for the wire is shown in Figure 5.29. The wires used at Queen's University were initially cold-drawn to conform to ASTM Standard A496, and subsequently annealed to give the required stress–strain characteristics. Other McGill work has used deformed steel wires obtained from Lundy Fence Company, Dunnville, Ontario, ranging in sizes from D-2 to D-10. The chemical composition of this wire steel is carbon, 0.13%; phosphorus, 0.007%; sulfur, 0.028%; manganese, 0.68%; and silicon 0.15%.

Labonte (1971) used D-2 wires to model No. 8 (25-mm) bars in a model investigation of the behavior of anchorage zones in prestressed concrete containments. Figure 5.30 shows the stress–strain curve obtained using electrical resistance strain gages for the D-2 wires as obtained from the supplier and after being annealed at 925°C for approximately 30 min, which lowered the yield point and increased ductility. Mirza (1972) also reported his work on heat treatment of steel bars on the deformed wires obtained from the Lundy Fence Company. He heated 16 sets, each set consisting of 20 samples (five bars for each size D-2, D-3, D-4, and D-8), to temperatures ranging from 480 to 870°C for a period of 1 h. Of these sets, 12 were air-cooled, two groups were oil-quenched at temperatures of 750 to 870°C, and the remaining groups were water-quenched at

Table 5.9 Dimensional Requirements for Deformed Steel Wire for Concrete Reinforcement

Deformed Wire Size Number[a]	Unit Weight, lb/ft	Nominal Diameter, in.	Cross-Sectional Area, in.²	Perimeter, in.	Spacing Maximum, in.	Minimum, in.	Minimum Average Height of Deformations, in.[b]
D-1	0.034	0.113	0.01	0.355	0.285	0.182	0.0045
D-2	0.068	0.159	0.02	0.499	0.285	0.182	0.0063
D-3	0.102	0.195	0.03	0.612	0.285	0.182	0.0078
D-4	0.136	0.225	0.04	0.706	0.285	0.182	0.0101
D-5	0.170	0.252	0.05	0.791	0.285	0.182	0.0113
D-6	0.204	0.276	0.06	0.867	0.285	0.182	0.0124
D-7	0.238	0.298	0.07	0.936	0.285	0.182	0.0134
D-8	0.272	0.319	0.08	1.002	0.285	0.182	0.0143
D-9	0.306	0.338	0.09	1.061	0.285	0.182	0.0152
D-10	0.340	0.356	0.10	1.118	0.285	0.182	0.0160

[a] The number following the prefix D identifies the nominal cross-sectional area of the deformed wire in hundredths of a square inch.
[b] The minimum average height of deformations shall be determined by measurements made on not less than two typical deformations from each line of deformations on the wire. Measurements shall be made at the center of indentations.

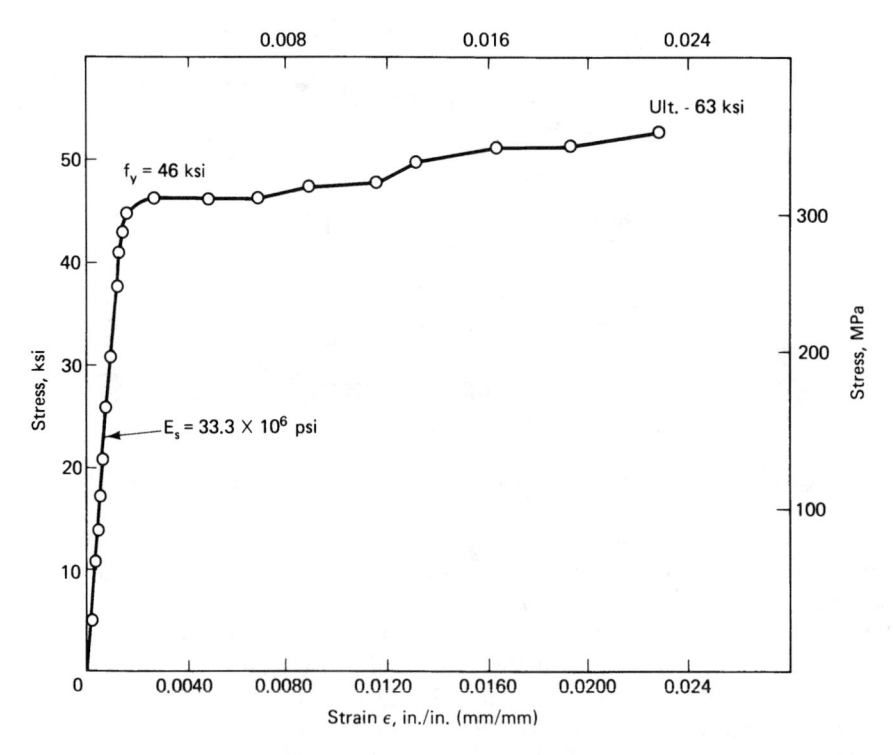

Figure 5.29 Stress–strain curve for wire reinforcement (Stelco — 0.092 in. diameter, indented).

870°C. The stress–strain curves for the bars as obtained from the supplier are shown in Figure 5.31. The effect of annealing followed by air cooling on the mechanical properties of the D-2, D-3, D-4, and D-8 wires is summarized in Figure 5.32. It can be noted that at temperatures between 600 and 650°C the yield strength is lowered to a value between 240 and 275 MPa and the ultimate tensile strength is lowered to approximately 345 MPa. Temperatures above 650°C do not have any additional influence on the yield and the ultimate strengths. For all temperatures above 600°C an

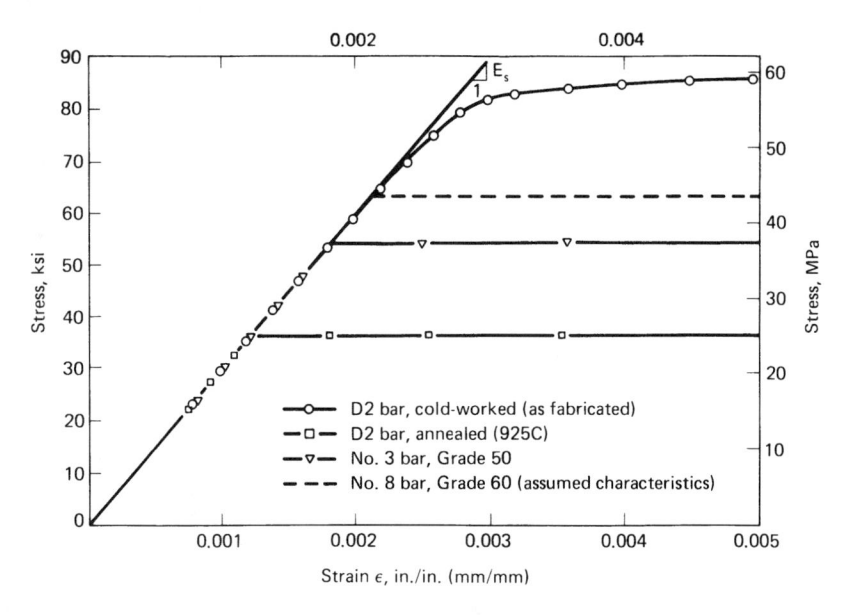

INELASTIC MODELS: STRUCTURAL STEEL AND REINFORCING BARS 237

Figure 5.30 Reinforcement load–deformation characteristics.

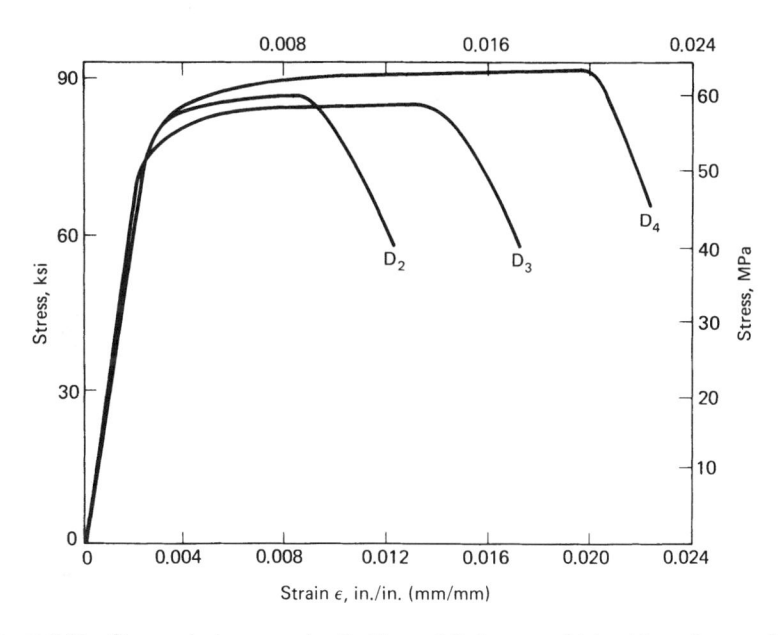

Figure 5.31 Stress–strain curves for D_2, D_3, and D_4 bars as obtained from the supplier.

elongation of more than 20% was observed for all wires over a gage length of 125 mm. It was noted that specimens that were oil-quenched or water-quenched had higher yield and ultimate strengths than the specimens that were air-cooled. The percentage elongation and therefore the overall ductility of oil- or water-quenched steel was much lower than the air-cooled steel.

An extensive study of the annealing characteristics of commercially deformed wires of diameter 2.9 and 4 mm was performed by Chowdhury and White (1971). The wires had four lines of rectangular patterns embossed into the surface at 6-mm spacing. The stress–strain curves for the wires as delivered are shown at the top of Figure 5.33. Because of the cold working involved in the embossing process, the wires did not have a well-defined yield point. The effects of various heat treatments on the wire properties are shown in Figures 5.33 and 5.34. In all, 127 specimens

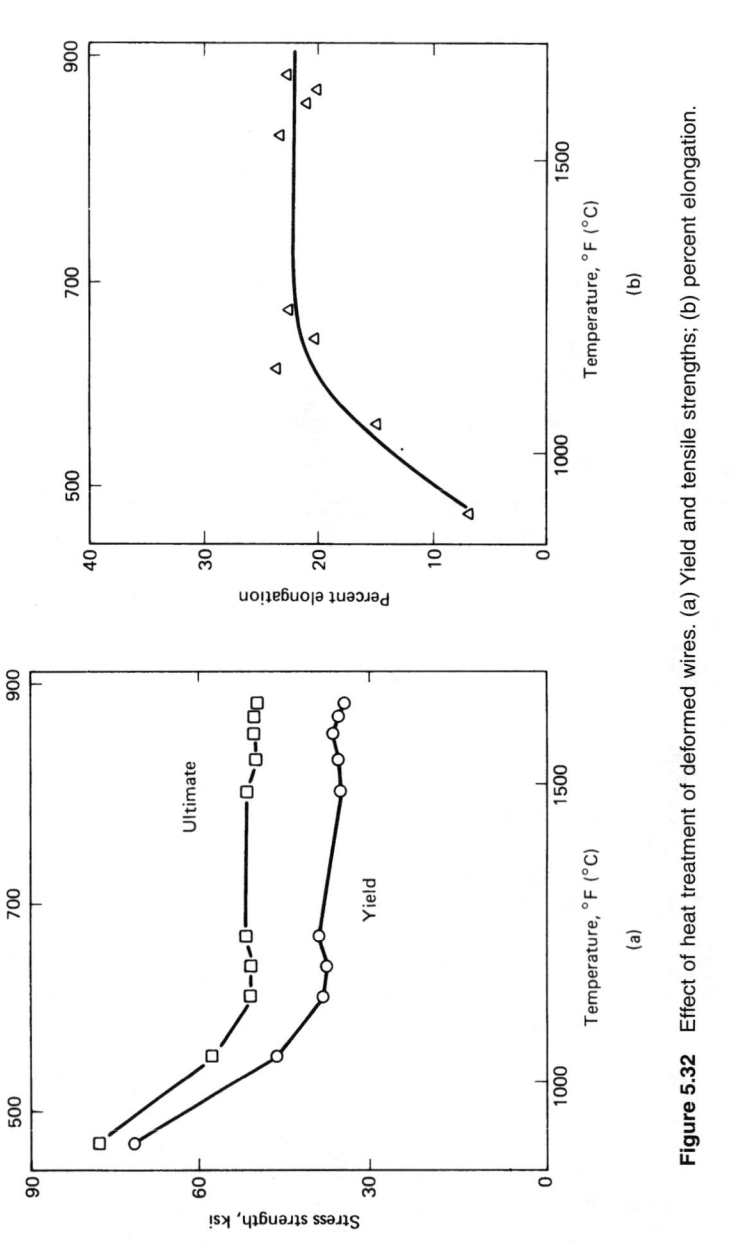

Figure 5.32 Effect of heat treatment of deformed wires. (a) Yield and tensile strengths; (b) percent elongation.

were tested to failure in tension. Full annealing at about 870°C, with slow cooling through the critical range, produced very low yield points. The cooling rate is crucial in full annealing. Normalizing, which consists of heating about 38°C above the critical temperature (910°C in this case) and then cooling in air, gives higher yield strengths and lower ductilities than full annealing. Process annealing was used to achieve the proper model steel yield points in this study. Process annealing is done at 480 to 650°C and does not involve the pearlite–austenite transformations of full annealing. It produces recrystallization of deformed crystals into undeformed or less deformed ones without changing their nature because the temperature is held below the critical point. A more complete description of these heat-treatment processes is given by Chowdhury and White (1971).

Commercially available reinforcement bars (PCA D2.5, 4.4 mm, and D2, 3.8 mm) were used in the ⅕-scale concrete building model (Case Study B, Chapter 1) for the columns and beams, respectively, see Figure 5.26. Note from Figure 5.26 that the Japanese designations of reinforcing bars, i.e., D10 to D22, are the rounded value of the bar diameter in mm and not the Reinforcing Steel Institute (RSI) designation of hundredths of square inch of area. The correlation of the stress–strain relations of the model and prototype reinforcement for the columns and beams after heat treatment is shown in Figures 5.35 and 5.36, respectively.

The variation of properties of individual lengths of model steel is an important quantity, particularly in sections with relatively few reinforcing wires. For wires annealed in a single batch, the maximum coefficient of variation was 0.09. In 81% of the batches the coefficient of variation was below 0.05. Such differences may be produced among others by (1) differences in wire quality and (2) nonidentical surface deformations, which lead to differences in net area. When comparing one batch with another, the differences in heat-treatment temperature and in cooling rates also must be considered. It is recommended that heat treatment of reinforcements be done under closely controlled conditions using a single furnace with temperature control of ±3°C. This precision is necessary to reduce the typical variation of steel strength from the desired value to within ±5%. It is necessary to clean the wires with diluted hydrochloric acid to remove grease and improve the bond to model concrete.

Model stirrups and ties can be made of plain annealed wire, 1.04 mm and 1.4 mm in diameter. Reinforcing steel may be fabricated by tying with 26 SWG wire.

5.4.7 Laboratory Wire-Deforming Machines

5.4.7.1 Cornell University Wire Deformeter

A simple technique was developed in the Cornell Structural Models Laboratory to cold-deform plain wire (Harris et al., 1966a,b), thus making it more suitable for use as scaled deformed bars. The desired deformations are obtained by passing the wire through two pairs of perpendicular knurls, as shown in Figure 5.37. The extent of deformation, which is easily adjusted, is kept proportional to the various wire diameters. In order to get the optimum uniformity, the wire can be continuously deformed for the full length required for a particular model. The Cornell wire deformeter described above cuts small groves internal to the bar diameter, which is unlike the actual external lugs of prototype rebar.

Five different sizes of wire deformed by this technique are shown in Figure 5.38. In addition, three commercially deformed wires are shown. None of the deformation patterns is geometrically similar to the raised lugs used in prototype reinforcing; the threaded rod has a large continuous helical deformation, while the deformations on the laboratory-deformed and commercially deformed wires are rolled into the wire in the shape of internal valleys rather than external protrusions. It is assumed that inward deformations of a given depth have the same mechanical bonding effect as a protrusion of the same height. Although the assumption is open to question, it serves a useful purpose in helping to arrive at a criterion for judging the bond characteristics of the various reinforcements, which are discussed in Section 5.7.

Symbol	Diam., in.	Temp., °F	Annealing time, hr	Type of cooling
●	0.159	As delivered	—	—
△	0.123	As delivered	—	—
○	0.113	1010	$1\frac{1}{2}$	FC (16 hr)
▽	0.159	1075	$1\frac{1}{2}$	FC (36 hr)
▲	0.113	1100	1, 2	FC (16 hr)

Symbol	Diam., in.	Temp., °F	Annealing time, hr	Type of cooling
+	0.159	1675	$\frac{1}{2}$	AC (78°F)
□	0.113	1675	$\frac{1}{2}$	AC (78°F)
▼	0.159	1600	$\frac{1}{2}$	FC (16 hr)
■	0.113	1600	$\frac{1}{2}$	FC (16 hr)

AC - air cooling, FC - furnace cooling

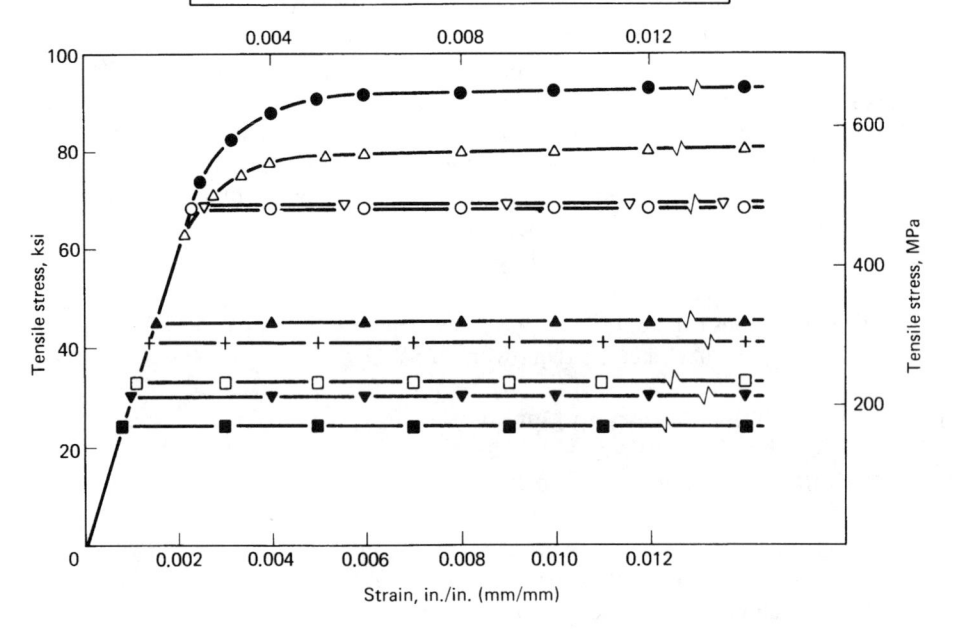

Figure 5.33 Stress–strain curves for commercially deformed wires under various kinds of conditions.

Figure 5.39 shows the effect of the laboratory deformation technique on the measured stress–strain curve of 0.078-in. (2-mm)-diameter annealed wire. The drop in yield strength of wires deformed by the above device with increasing annealing time is shown in Figure 5.40 for three different small-diameter wires. Detailed results of annealing effects are given by Harris et al. (1966a). Suitable annealing temperatures ranged from 540 to 820°C, with the lower portion of the temperature range being most suitable. Experiments will quickly establish the appropriate annealing time for a given wire.

INELASTIC MODELS: STRUCTURAL STEEL AND REINFORCING BARS 241

Symbol	Dia. in.	Annealing time, hr	Type of cooling
△	0.159	1, 2	AC (78° F)
○	0.159	1, 2	FC (16 hr)
●	0.113	1, 2	AC (78° F)
□	0.113	1, 2	FC (16 hr)
■	0.113	1, 1½	FC (36 hr)

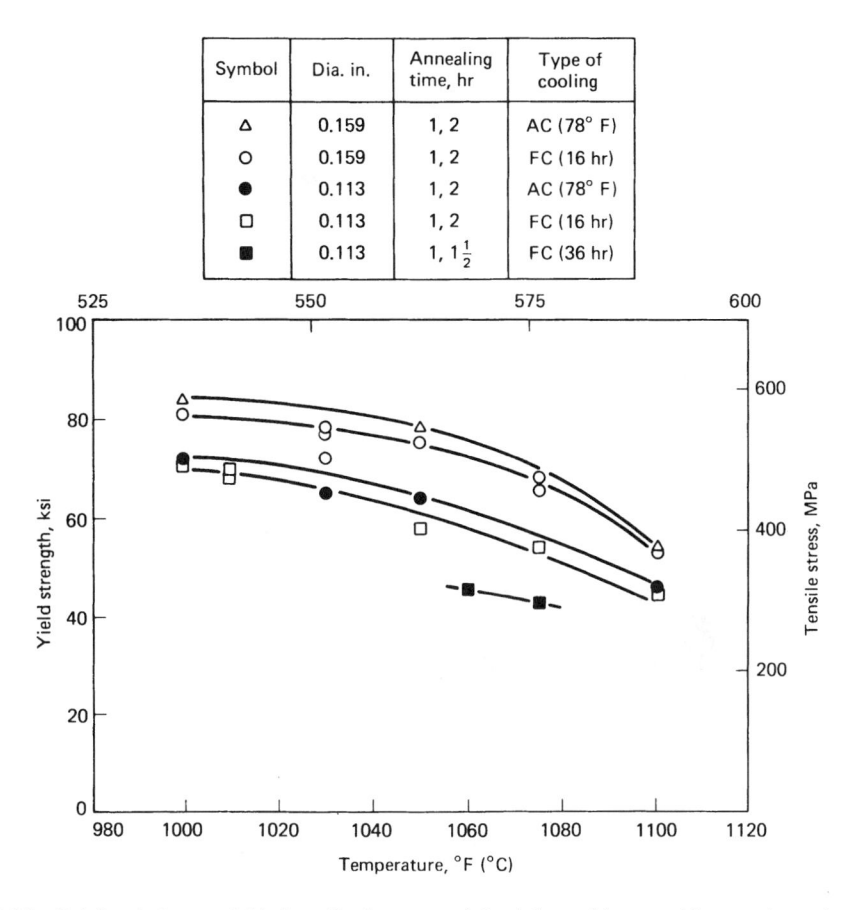

Figure 5.34 Relation between yield strength of commercially deformed bars and temperature of annealing.

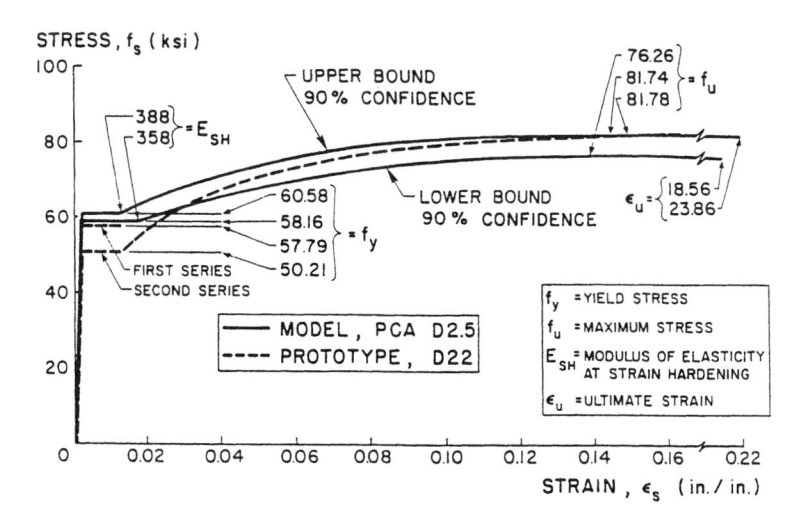

Figure 5.35 Stress–strain relations for column reinforcement.

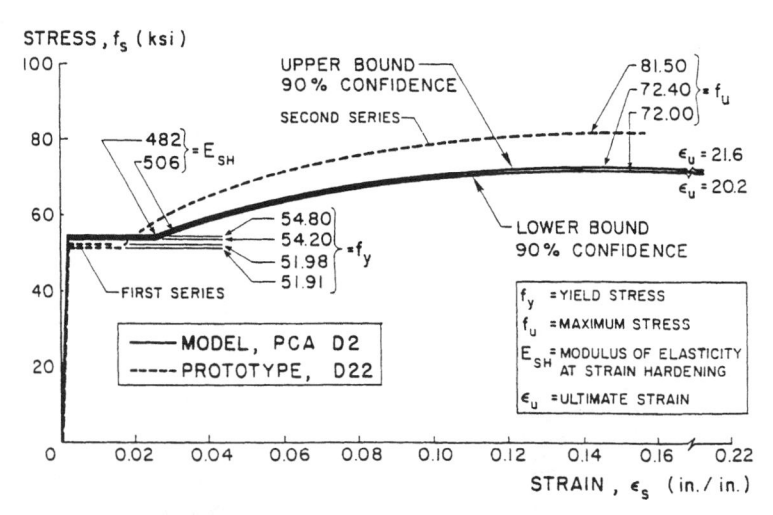

Figure 5.36 Stress–strain relations for beam reinforcement.

Figure 5.37 Cornell University wire deformer.

5.4.7.2 Cement and Concrete Asssociation Wire Deformeter

The Cornell knurling deformeter puts internal indentations onto the surface of a smooth bar. In order to overcome this limitation, wire-deforming machines have been developed that can deform a small-size bar with a scaled set of external deformations similar to those of the prototype. A schematic of the process by which surface deformations are produced on the surface of a smooth bar according to E. Maisel of the Institut of Modellstatik, University of Stuttgart is shown in Figure 5.41. Using this same principle, White and Clark (1978) modeled standard 12.7-mm GK60 deformed bars with a laboratory-deformed 14-gage (2.06-mm-diameter) mild steel wire. The deformations were produced using a knurling machine whose special rolls for deforming the wire are made by cutting grooves which result in wider ribs than desired, as shown in the table below. It was observed that the wire diameter reduced to 1.88 mm after deforming, thus resulting in a linear scale factor of

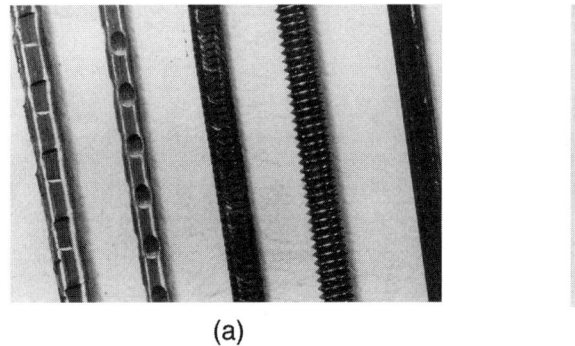

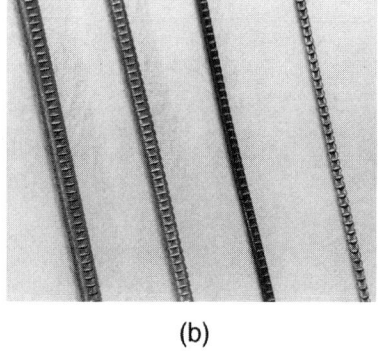

INELASTIC MODELS: STRUCTURAL STEEL AND REINFORCING BARS 243

Figure 5.38 (a) Typical model reinforcements (left to right): Fabribond, U.S. Steel, Cornell deformed, threaded rod, and plain bar. (b) Typical model reinforcements (all deformed in Cornell Laboratory).

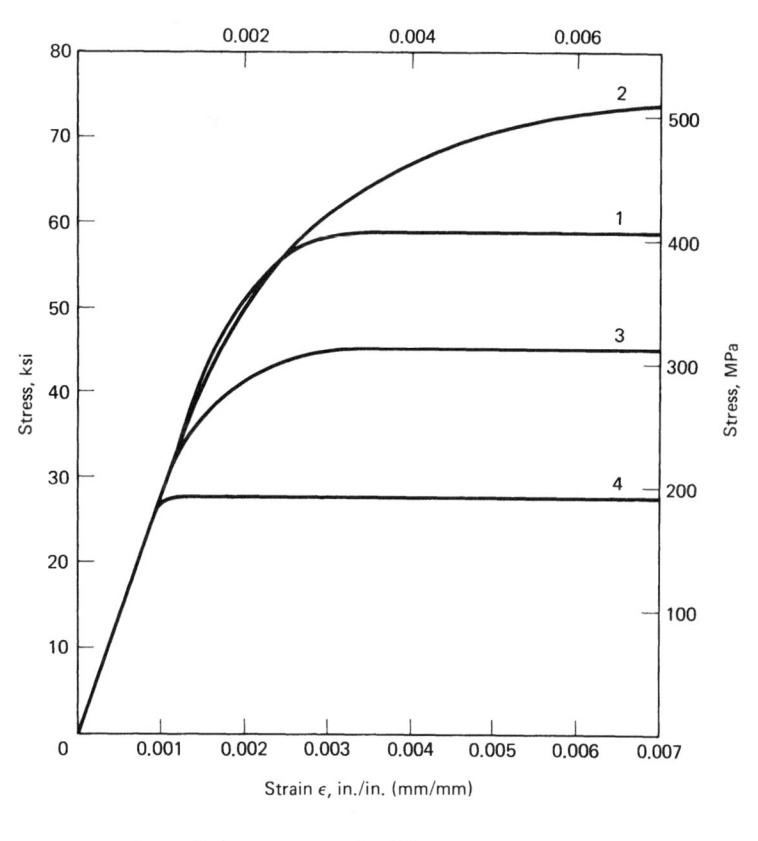

#	σ_y (ksi)	Condition
1	58	As delivered
2	70	$\frac{1}{3}$ turn* deformed — no heat
3	45	$\frac{1}{3}$ turn plus annealed 1000°F-45 min.
4	27	$\frac{1}{3}$ turn plus 1500°F-15 min.

*1 turn = $\frac{1}{28}$ in. motion of knurls developed in the laboratory

Figure 5.39 Effect of deformation and heat treatment on stress–strain behavior or reinforcing wire (size: 14 gage or 0.078 in. diameter).

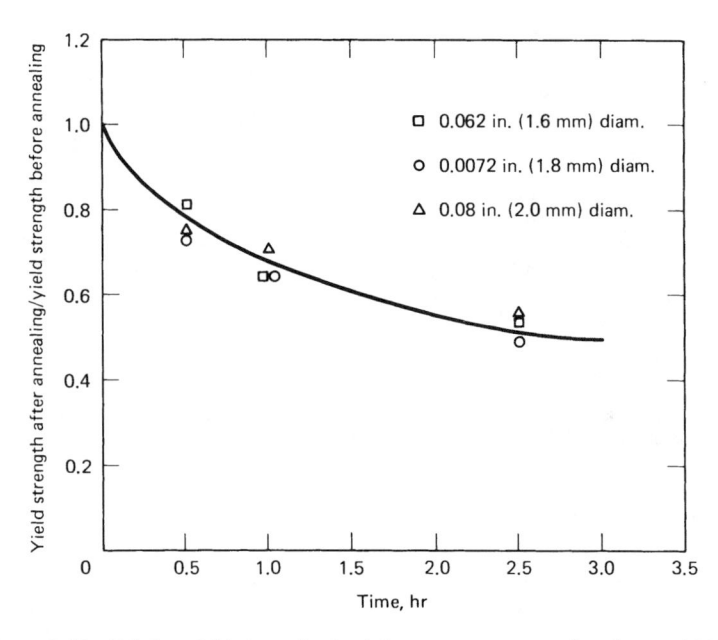

Figure 5.40 Relative yield strength of reinforcement vs. annealing time at 1000°F.

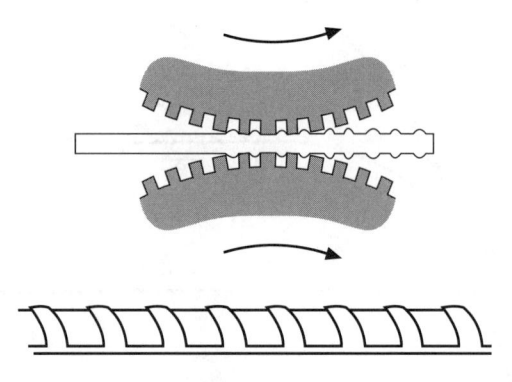

Figure 5.41 Method of deforming model wire. (From Maisel, 1978.)

⅟6.76. The prototype rib dimensions and the required model rib dimensions at a scale of ⅟6.76 and the dimensions obtained after deforming are as follows:

	Prototype Dimension	Model Dimension Required	Model Dimension Achieved
Bar diameter, mm	12.70	1.88	1.88
Rib spacing, mm	8.30	1.23	1.25
Rib width, mm	0.97	0.14	0.30
Rib height, mm	0.53	0.08	0.08
Rib angle, °	30	30	30

Thus, excellent similitude was obtained for all dimensions except the rib width, which was about twice the required width. The model rib width is determined by the width of the transverse saw

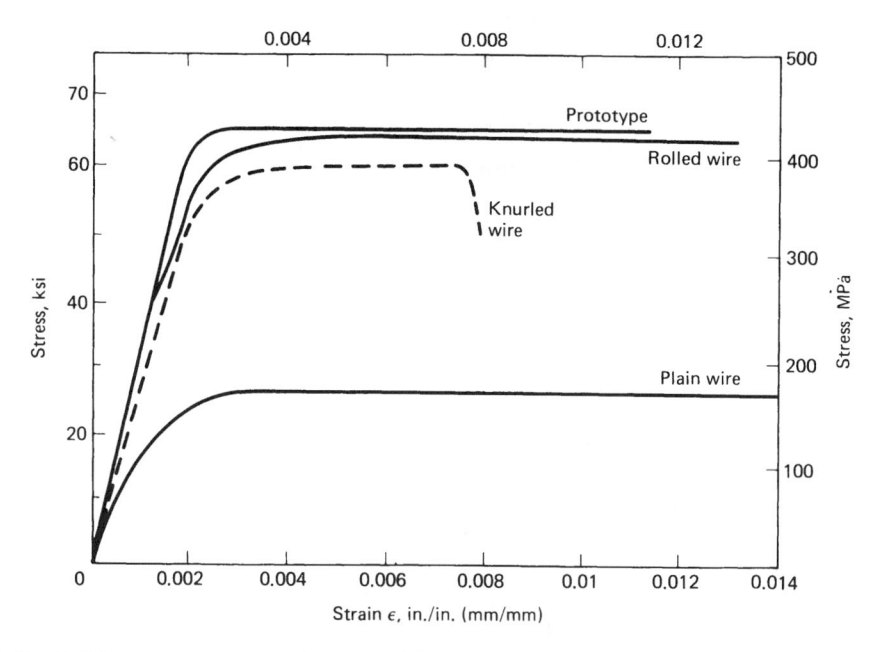

Figure 5.42 Reinforcement stress–strain curves. (From White, I. G. and Clark, L. A., in *Proceedings of the Joint Institution of Structural Engineers/Building Research Establishment Seminar on Reinforced and Prestressed Microconcrete Models*, Garston, England, May, 1978. With permission.)

cuts made in the rolls of the knurling machine, and it is not considered feasible, by conventional workshop practice, to make these any narrower. In addition, if the cuts could be made narrower, the wires would then have to be hot-rolled in order to form the ribs. The stress–strain curves of the prototype bar, the deformed model wire, the Cornell University knurled model wire, and the plain wire are compared in Figure 5.42. Good agreement was obtained between the prototype and models except in the vicinity of the yield point. White and Clark (1978) reasoned that this was because the prototype was hot-rolled whereas the model steels were cold-rolled.

5.4.7.3 Deformeters Used by Various Investigators

Noor and Goodman (1979) in a discussion to White and Clark (1978) reported on an interesting wire deformer that can produce a variety of appropriately sized external deformations for small-scale microconcrete models. They show very narrow rolls for producing deformations of different spacing. These investigators recommend that for best results the wire be annealed and lightly drawn to remove scale before rolling the deformations. Using an identical technique to White and Clark (1978), Subedi and Garas (1978) deformed available 1.6 mm diameter wires to produce "crimped" wires for use in their model work. They also used plain, threaded, and deformed bars in their investigations. A variation of the grooved rolls reported above by Evans and Clark (1978) has been used at Stanford University by Wallace and Krawinkler (1985) in studying $1/12.5$-scale models of a seven-story reinforced concrete frame-wall structure (Case Study B, Chapter 1). A close-up of the Stanford deforming rolls with attached guide sleeve is shown in Figure 5.43a. The model deformed wire has square lugs (Figure 5.43b) unlike the 30° inclined lugs of White and Clark (1978). Typical stress–strain diagrams of the Stanford deformed wire and that of the prototype are shown in Figure 5.44. A machine for producing deformed model reinforcement using the principle shown in Figure 5.41 has been reported by Maisel (1978a,b). Murayama and Noda (1983) report on a deforming device for deforming 3-mm-diameter wire for their small-scale model work in Japan. Dimensions of the three types of model deformed 3-mm bars are given in Table 5.10.

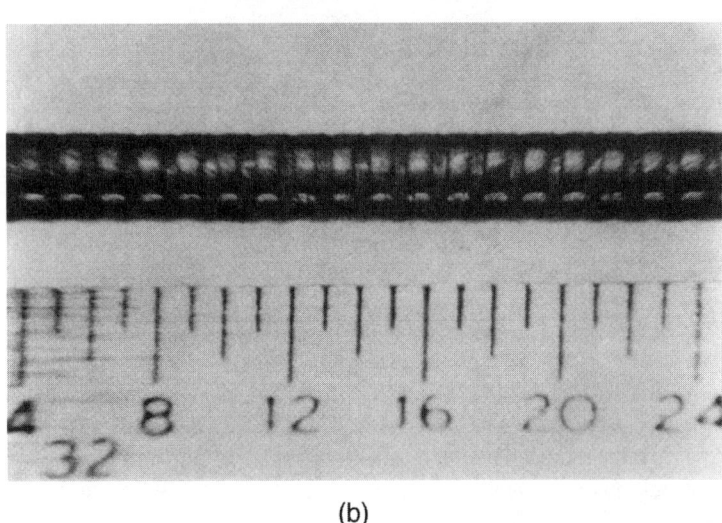

Figure 5.43 Stanford University wire deformeter technique. (a) Wire deforming rolls; (b) model reinforcing bar.

5.4.7.4 Drexel University Wire Deformer

Since the early 1980s the Structural Models Laboratory, Drexel University has been using a wire-deforming machine that puts rolled lugs or external deformations on the wire to scaled spacings and heights, thus producing model bars to any desired scale of standard reinforcing bars. The Drexel model reinforcement deforming machine (Figure 5.45) basically consists of a pair of semicircular gears driven by a constant-speed motor through a gear train. The rollers are of special hardened steel with the indentations produced by a plasma arc process. Rollers for a range of bar diameters and scales are available. A load cell attached to the top gear controls the depth of the protrusion ribs. The wire is guided by sets of grooved rollers placed before and after the rolling operation (Figure 5.45) which tend to straighten it as it is being deformed.

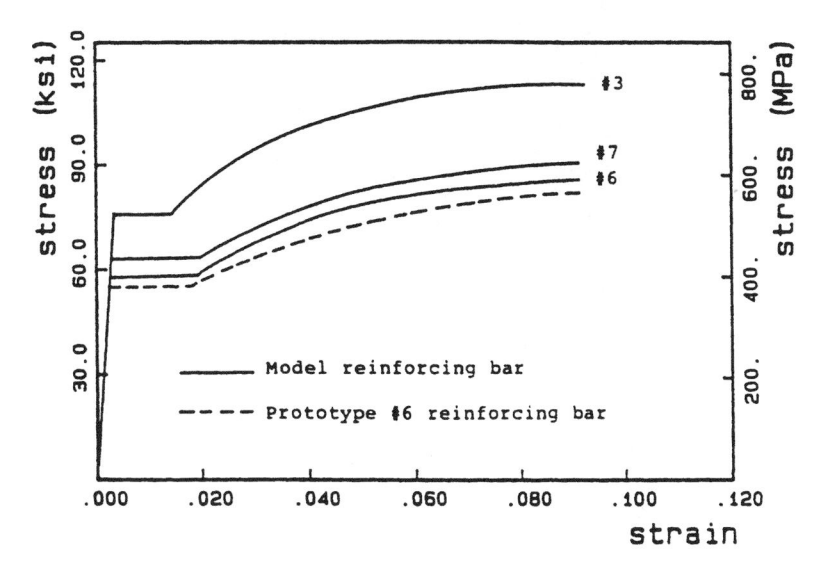

INELASTIC MODELS: STRUCTURAL STEEL AND REINFORCING BARS 247

Figure 5.44 Stress–strain diagrams of model and prototype bars.

Table 5.10 Dimensions of Scaled Rebars, "D3"

Type of D3	r, mm	ln, mm	h, mm	a, mm	b, m	BA, %
D3-1	1.50	3.38	0.095	0.60	0.47	2.4
D3-2	1.45	2.43	0.145	0.50	0.63	4.9
D3-3	1.43	1.38	0.175	0.45	0.53	10.9

Figure 5.45 Drexel University wire-deforming machine.

Desired model bar deformations are scaled from the ASTM A615 *Standard Specification for Deformed and Plain Billet-Steel Bars for Concrete Reinforcement*. Typical prototype bars that have been scaled are shown in Figure 5.46. A wide range of model reinforcing bars have been produced using the machine shown in Figure 5.45. Typical examples of the model bars produced are shown in Figure 4.47.

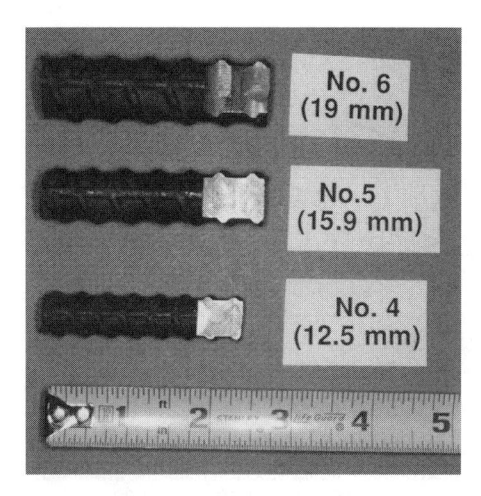

Figure 5.46 Prototype reinforcement models.

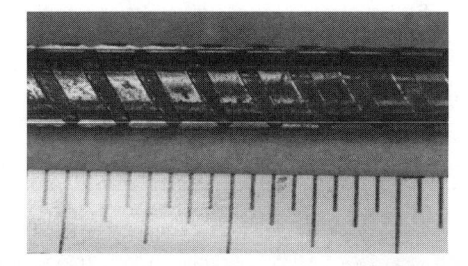

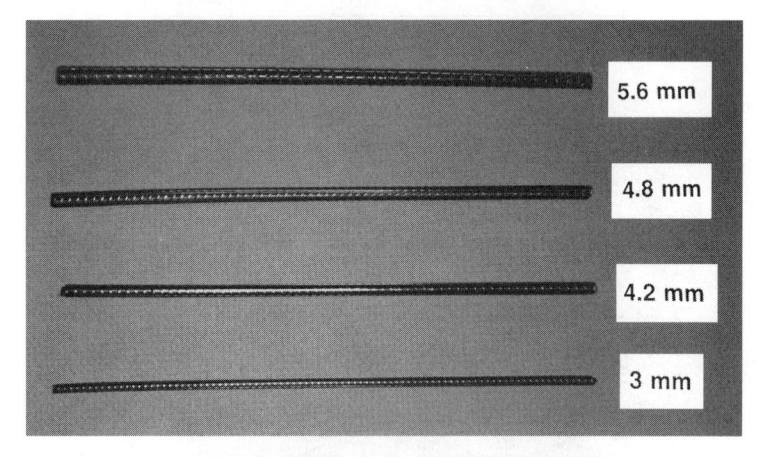

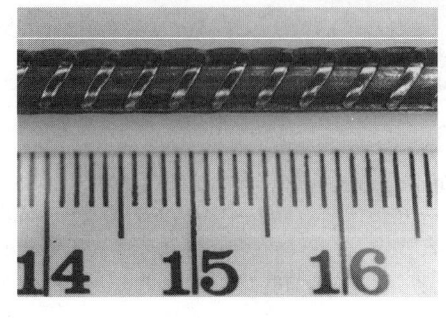

Figure 5.47 Examples of wires deformed at Drexel University.

INELASTIC MODELS: STRUCTURAL STEEL AND REINFORCING BARS

Table 5.11 Geometric Properties[a] of Model and Prototype[b] Reinforcement

Bar Designation	Size	Nominal Dimensions Diameter, mm	Cross-Sectional Area, mm²	Perimeter, mm	Maximum Average Spacing, mm	Minimum Average Height, mm	Maximum Gap (Chord 15% of Nominal Perimeter), mm
No. 3	Prototype	9.53	70.97	29.92	6.66	0.38	3.63
	⅓ scale	3.18	7.87	9.96	2.22	0.13	1.21
	Model	3.10	7.55	9.73	2.42	0.15	0.58
No. 4	Prototype	12.70	129.03	39.90	8.89	0.51	4.85
	⅓ scale	4.22	14.32	13.28	2.96	0.17	1.62
	Model	4.12	13.29	12.90	2.90	0.22	1.23
No. 5	Prototype	15.88	200.00	49.86	11.10	0.71	6.07
	⅓ scale	5.28	21.94	16.61	3.70	0.24	2.02
	Model	4.88	18.65	15.34	2.95	0.10	1.08
No. 6	Prototype	19.05	283.87	59.84	13.34	0.97	7.26
	⅓ scale	6.35	31.61	19.94	4.45	0.32	2.42
	Model	6.35	31.61	19.94	5.49	0.61	1.56

[a] Based on an average of six measurements.
[b] As specified in ASTM A-615.

Table 5.12 Mechanical Properties of ⅓-Scale Model Reinforcement Bars[a]

	Bar No.	Plain Bar (as Delivered)	Annealed and/ or Deformed	Heat Treated (or final)	COV[b] (%)	Prototype[c]
Yield strength, MPa	3	310.3	**415.8**	—	1.07	**386.8**
	4	583.5	468.9	**446.8**	3.06	**461.3**
	5	353.6	462.4	**440.6**	1.31	**441.3**
	6	**453.6**	—	—	10.01	**450.0**
Ultimate strength, MPa	3	—	**437.1**	—	0.57	**560.6**
	4	615.0	487.9	**475.1**	5.37	**734.3**
	5	369.3	478.1	**476.2**	1.25	**712.9**
	6	**624.8**	—	—	8.40	**738.7**
Ultimate strain, mm/mm	3	—	—	0.018	—	—
	4	—	—	0.031	—	—
	5	—	—	0.05	—	—
	6	—	—	0.0791	—	—

[a] Based on an average of five specimens.
[b] COV = coefficient of variation.
[c] Shear wall tests, University of Colorado at Boulder.

An illustration of modeling prototype steel rebar at ⅓ scale is given in Table 5.11. Geometric propreties of Grade 60 No. 3 (9.5 mm), No. 4 (12.5 mm), No. 5 (15.9 mm), and No. 6 (19 mm) scaled at ⅓ scale according to ASTM A615 and the actual model bar diameters are given in Table 5.11. The model No. 3 bar was a 3-mm wire, the No. 4 was a 4.1-mm wire, and the No. 5 was a 4.9-mm wire; the model No. 6 bar was a No. 2 bar purchased from the Portland Cement Association. Table 5.12 lists the mechanical properties of the deformed ⅓-scale model bars at various stages in their fabrication.

5.4.8 Heat Treatment of Model Reinforcement

5.4.8.1 General

Heat treatment is an essential process for proper simulation of reinforcing steel. Model bars will rarely have either the desired yield strength or sufficient ductility (yield plateau) without proper annealing. Also when smooth bars are cold-formed to produce the required surface deformation, their yield strength increases while their ductility decreases. This can be attributed to the state of internal strain produced by cold forming (Harris et al., 1966a; Kim et al., 1988). Heat treatment of or annealing of model bars is used to control the yield strength and to improve the yield and postyield characteristics, such as developing a sharp yield point and increasing the ductility.

The annealing process can be divided into three distinct regions: recovery or strain relieving, recrystallization, and grain growth (Figure 5.48a) (Harris et al., 1966a). At the recovery stage, the metal restores its physical properties without any significant change in its microstructure. A sharp yield point can be obtained at this stage by annealing the steel to 340°C for about 2 to 3 h. Recrystallization (which is usually defined as conventional annealing) is the replacement of the cold-worked structure by a new set of strain-free grains. From the different possible combinations of temperature and time used for conventional annealing, a temperature of 540°C for various time periods is recommended by Harris et al. (1966a) for typical steels.

5.4.8.2 Furnaces Used for Annealing

A variety of furnaces can be used for the annealing process. Comparison of two Limberg electric furnaces was carried out at Cornell University by Kim et al. (1988) with some interesting results. Furnace A (Figure 5.48b) used by these investigators was a vacuum-tube furnace and furnace B (Figure 4.48c) was an open-tube furnace. These furnaces were controlled with digital thermocouples and the furnace body was divided into three zones which could be controlled separately. The thermocouple control system was designed to give the average temperature inside the furnace.

A Lucifer economy box furnace (Figure 5.48d) is used with good results at the Structural Models Laboratory, Drexel University. This model has a chamber size 305 mm H × 305 mm W × 610 mm L and is rated for a 1260°C operation. It has a gas spring vertical spring door and heating elements mounted along both side walls. The furnace is equipped with a Honeywell digital temperature controller. Its main advantage is that it can anneal most of the model reinforcement needed for small-scale work and it is part of the laboratory equipment and thus readily available.

5.4.8.3 Annealing Processes

The main factors affecting the heat-treatment process are annealing time, annealing temperature, rate of cooling, and temperature distribution inside the furnace. While the first two factors are the control parameters in this process, the last two are furnace dependent. Prior experience at Cornell indicated that slight nonuniformity of temperature distribution inside the furnace may significantly affect the annealing results. The annealing process using either furnace A or B was repeated several times to obtain the proper temperature and annealing time for each of several model bars. The use of furnace A was as expected more time-consuming since it involved creating a vacuum in the tube and leaving the specimens to cool inside the furnace.

Several advantages were observed when using furnace B (with an open tube). In addition to the easy placement of specimens inside the furnace, the temperature distribution inside the furnace was found to be more uniform than that of furnace A. Also, furnace B allowed a precise time exposure to temperature since it was not necessary to leave the specimen inside the furnace during the heating and cooling times.

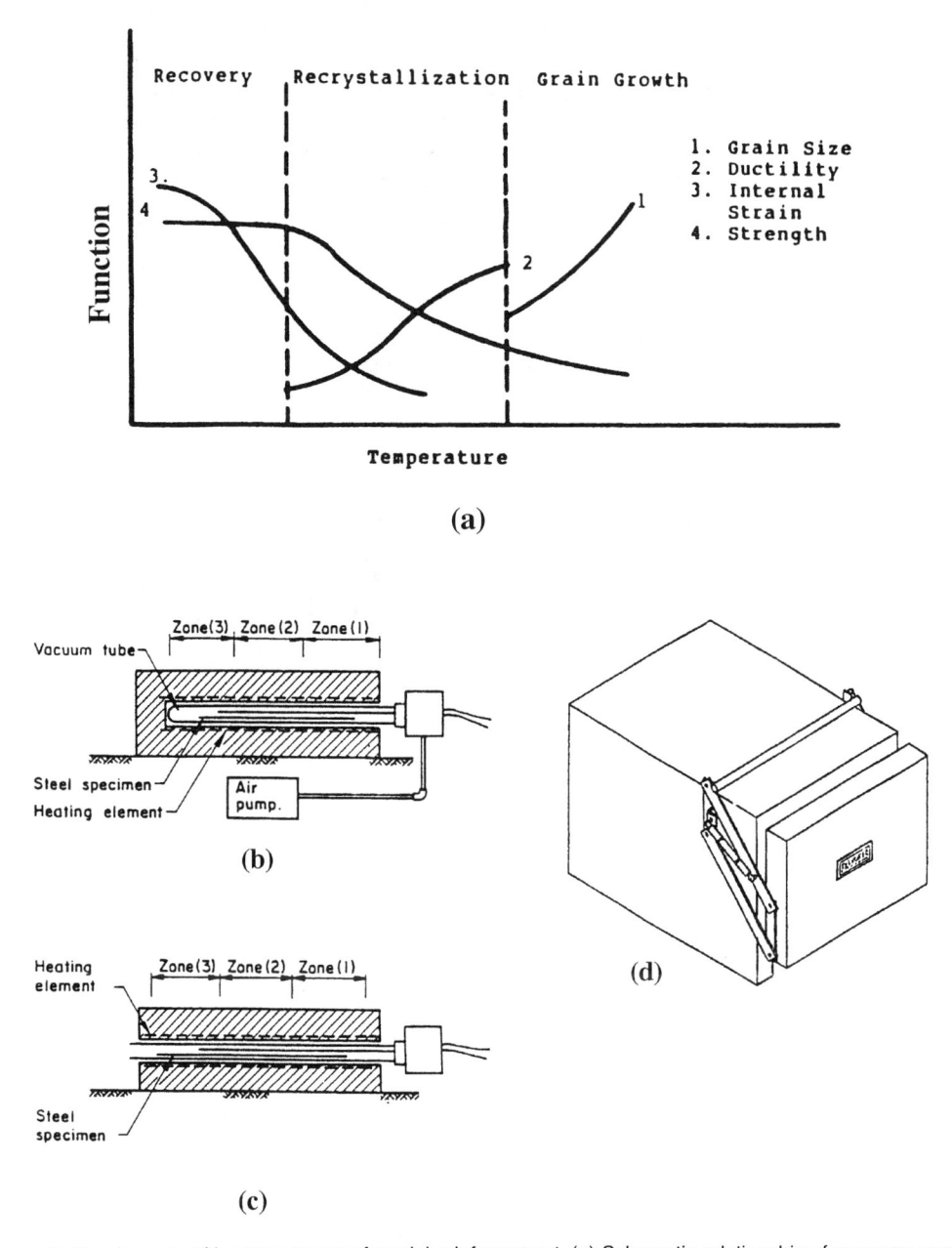

INELASTIC MODELS: STRUCTURAL STEEL AND REINFORCING BARS 251

Figure 5.48 Aspects of heat treatment of model reinforcement. (a) Schematic relationship of recovery, recrystallization, and grain growth; (b) Lindberg vacuum-tube furnace; (c) Lindberg open-tube furnace; and (d) Lucifer economy box furnace.

5.4.9 Model Reinforcement Selection

Once the major obstacles of surface deforming and annealing have been carried out, the next major concern in strength modeling of reinforced concrete structures is the correct selection of model reinforcing bars. In rare cases, the correct number and diameter of model bars may be available for the chosen model scale. More often slightly different diameter model bars must be used to model the prototype reinforcement. By taking into account the bar area and yield strength, a correctly scaled model force can be chosen. To illustrate this the following example is given.

Example 5.1

It is required to model the prototype doubly reinforced beam section shown below on the left at ⅙ scale. The section is reinforced with six No. 6 (19 mm diameter, 1703 mm² area) bars in the longitudinal direction and with No. 3 closely spaced ties. the beam main reinforcement is Grade 60 steel with an actual measured yield strength of 455 MPa. The available laboratory model deformed bars have a diameter of 2.95 mm and a cross-sectional area of 6.77 mm². The yield strength of the model bars as delivered is 758.5 MPa. It is required to select the *number* and *strength* of the model reinforcement so that similitude at ⅙ scale will be maintained.

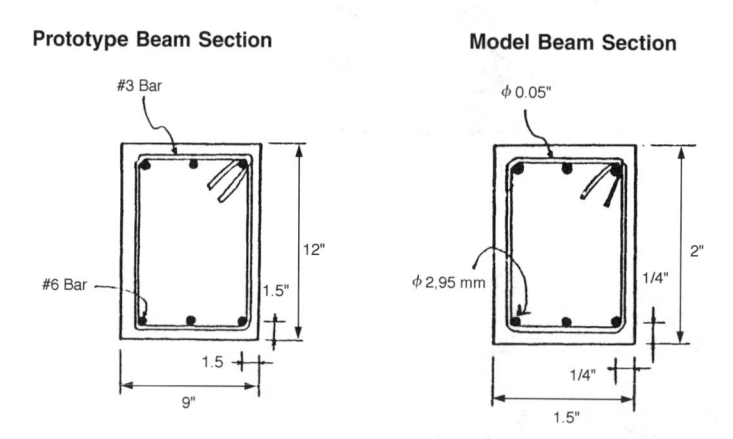

A replica model of the No. 6 bar at ⅙ scale would need to have a diameter of 3.17 mm and a yield strength equal to 455 MPa. Since the actual model bar diameter is 7.5% undersize, the model bars must be annealed to a higher yield strength than the No. 6 bar so that proper scaling of the tensile force in model and prototype beams will be maintained. The table below summarizes the values that result from these computations.

Specimen 1	Concrete 2	Longitudinal Reinforcement				
		Type 3	No. Used 4	A_s, mm² 5	f_y, MPa 6	$A_s f_y$, N 7
Prototype	Prototype concrete	No. 6 Bar	6	1703.2	415.8	774840
Model	Microconcrete	Laboratory deformed	6	40.6	491.4	21795

It should be noted that the forces in the model bars (Column 7) scale as S_l^2 by designing the model bar with a yield strength of 237.8 MPa. The model beam section at a scale of ⅙ is shown above on the right.

5.5 MODEL PRESTRESSING REINFORCEMENT AND TECHNIQUES

5.5.1 Model Prestressing Reinforcement

There are several possible substitutes for prestressing tendons, which include:

1. Individual strand prestressing wire and strands, as used in making twisted-strand cables for prototype prestressing;
2. Piano wire;

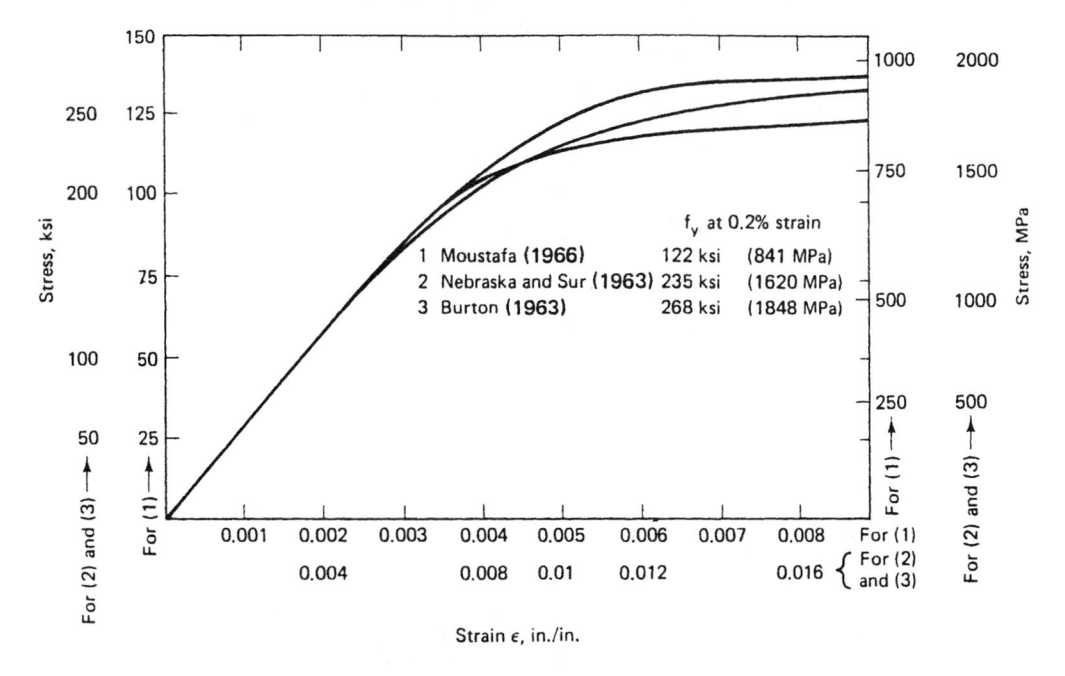

Figure 5.49 Stress–strain curves for prestressing wires.

3. Stainless steel twisted-strand cable;
4. Bicycle spoke material, complete with threaded ends (particularly useful for post-tensioning applications);
5. Stranded aircraft or marine cable.

Typical stress–strain curves for some of these model prestressing wires are shown in Figure 5.49. Prestressing wires and piano wires have been successfully used in model studies of prestressed concrete beams and slabs at McGill University (Pang, 1965). In a prestressed concrete model study at Cornell University the surfaces of individual wires were mechanically roughened with an oversize thread-cutting die to improve their bond characteristics.

Harris and Muskivitch (1977; 1979) used stainless steel twisted-strand cables in a Drexel University model study on the nature and mechanism of progressive collapse of precast prestressed large panel buildings. Some of their results are shown in Figure 5.50a. Kemp (1971) successfully used the smallest available plain wire 7-mm-diameter prestressing system in post-tensioned, non-grouted, prestressed slab models. In a Cornell university study Moustafa (1966) used 2.6-mm-diameter custom-length bicycle spokes in a prestressed concrete flat slab. Maisel (1978a) used hot-dipped galvanized 1.55-mm-diameter prestressing cables in prestressed concrete model studies at Stuttgart. A typical stress–strain curve for this cable is shown in Figure 5.50b.

5.5.2 Anchorage Systems

Labonte (1971) used seven 7-mm wire strands to model the post-tensioning system of the buttress of a prestressed concrete reactor vessel. The standard BBRV system was used to anchor the tendons in the ⅙-scale model of the buttress. Typical details of the passive and active load cells are shown in Figure 5.51.

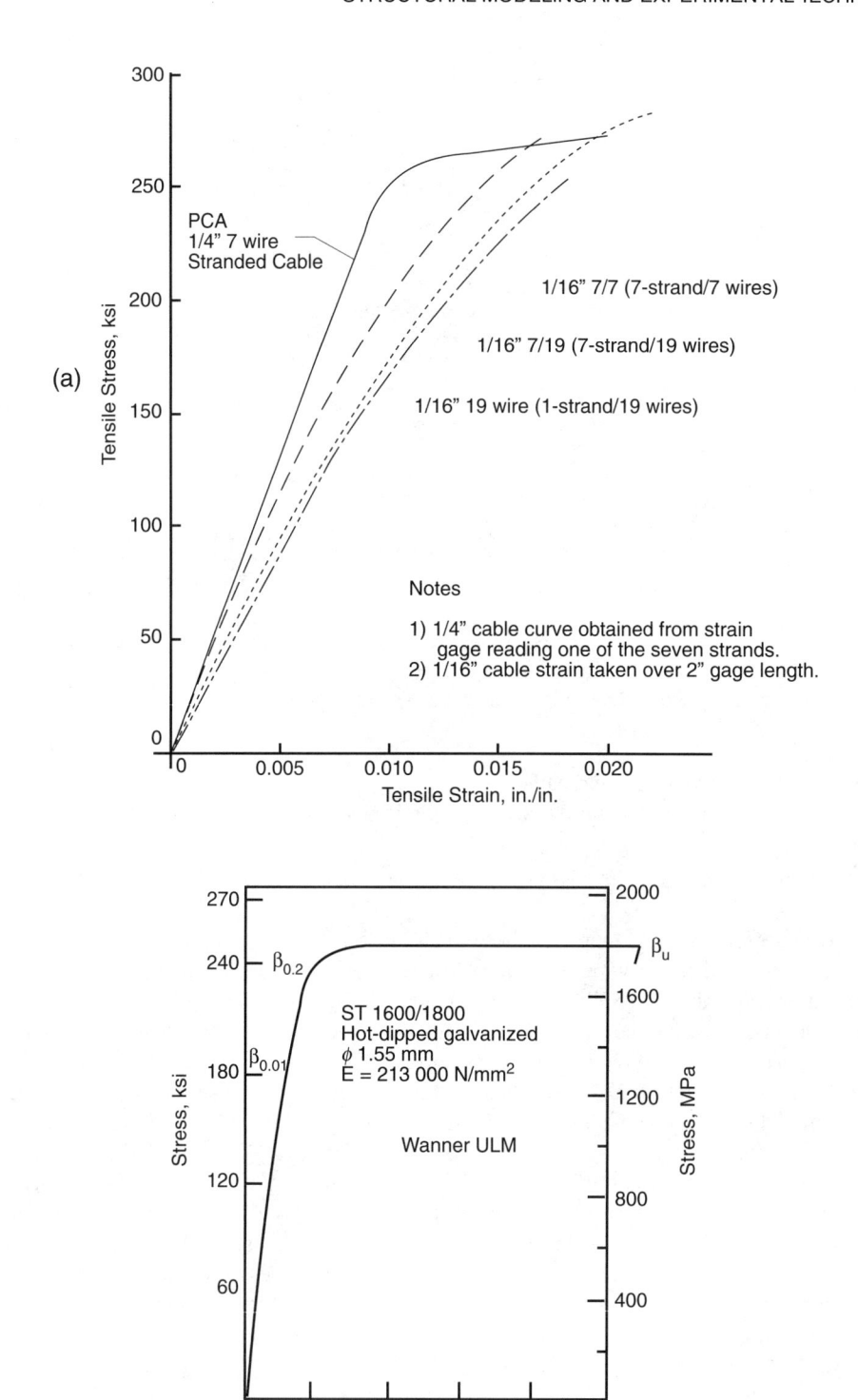

Figure 5.50 Stress–strain curves for high-strength cables. (a) Stainless steel cables for 3/32-scale model (courtesy of Drexel University, Philadelphia); (b) prestressing cables for a 1/6-scale model (from Maisel, 1978a.)

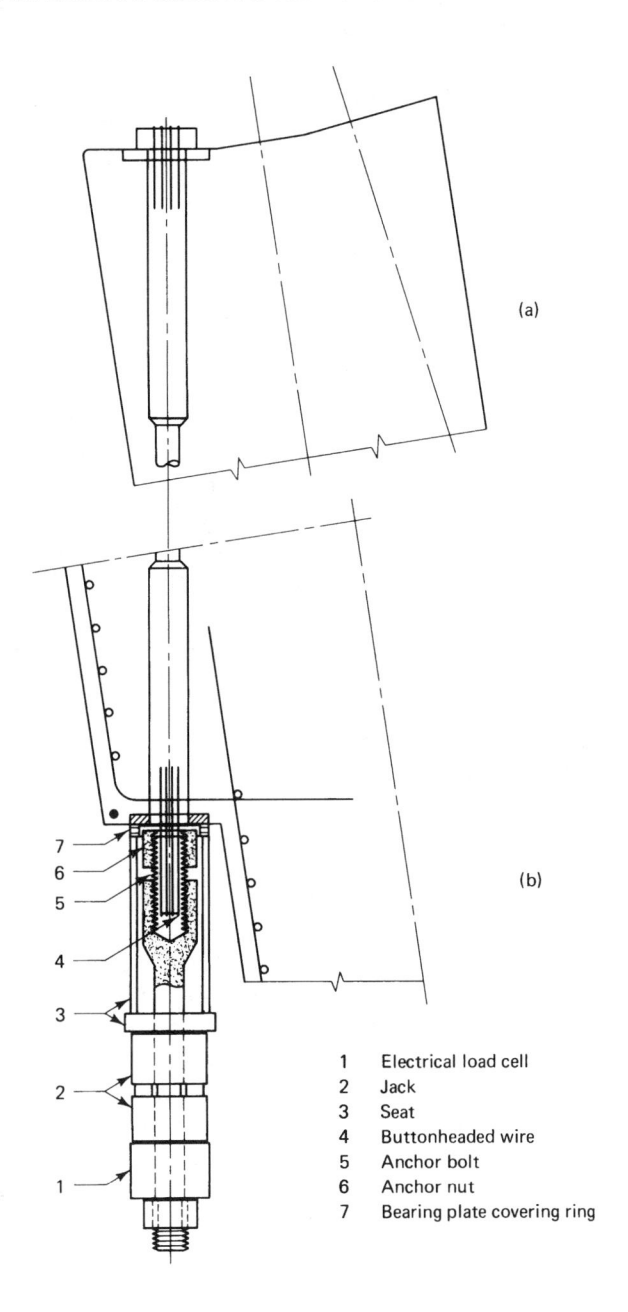

Figure 5.51 Posttensioning system. (a) Passive-end anchorage. (b) Active-end anchorage.

5.6 FRP REINFORCEMENT FOR CONCRETE MODELS

5.6.1 General

Considerable interest has been generated recently in the use of fibrous composites as reinforcement in concrete structures. Replacement of the steel reinforcement with more corrosion resistant substitutes is rapidly becoming a more economical option for constructed facilities worldwide (ACI Report 440R-96; Mufti et al., 1991; Iyer and Sen, 1991; Nanni and Dolan, 1993; Basham, 1994;

Saadatmanesh and Ehsani, (1996, 1998; El-Badry, 1996; Head, 1996). FRP of the bar type, because of its versatility, can be used in new, but especially in repaired reinforced concrete structures. In general, FRP systems (which usually consist of glass, aramid, or carbon fibers in a polymer matrix) have high strength, a wide range of moduli of elasticity, and very low ultimate tensile strains as compared with those of steel.

5.6.2 Nonductile FRP Reinforcement

The stress–strain behavior of representative GFRP (glass fiber–reinforced polymer), AFRP (aramid fiber–reinforced polymer), and CFRP (carbon fiber–reinforced polymer) systems is linear up to failure which is brittle. A comparison of the stress–strain behavior of the most common FRP fibers with that of steel reinforcement with its ductile behavior is shown in Figure 5.52. Reinforcements made of these nonductile FRP fibrous composite systems are being used at an accelerating rate in the civil infrastructure as reinforcement to flexural members where corrosion resistance, light weight, and nonmagnetic properties are of primary concern. In the development of these new systems, models of structures reinforced or prestressed with any of the common FRP systems can be tested in the laboratory using small-diameter prototype reinforcement. These usually large-size models are being used extensively to generate the design data required to characterize the structural behavior of the reinforcing systems. Material ductility is lacking in all the above fibrous composite systems, prompting the need to develop a new ductile bar reinforcement that can mimic the desired qualities of reinforcing steel. A project using the small-scale modeling technique as a product development approach was conducted at Drexel University to develop a ductile composite rebar. In the laboratory, 3- and 5-mm bars were developed with the logical extension to larger, prototype-size bars being funded by the National Science Foundation. The results of this novel modeling work are described below.

5.6.3 Ductile Hybrid Fiber-Reinforced Polymer

Ductility is the ability of a material, or structural member, to undergo large inelastic deformations without distress. In the extreme event of a structure loaded to failure, it should be able to undergo large deflections at near its maximum load-carrying capacity. This will give forewarning of failure and prevent total collapse and may save lives. In statically indeterminate concrete structures, ductility of the members at the critical sections is necessary to allow moment redistribution to take place. In order to achieve ductility in reinforced concrete structures without using conventional steel rebar, a new design methodology was introduced to identify suitable fibrous composite materials that mimic the stress–strain characteristics of steel. Taking advantage of the design flexibility, braided structures were employed as the primary fiber architecture for the construction of the reported ductile hybrid FRP (D-H-FRP) rebar system (Harris et al., 1997; 1998a,b; Ko et al., 1997; Somboonsong, 1997; Huesgen, 1998). The methodology of braiding, as detailed by Ko (1989), is a well-established technology which intertwines three or more strands of yarns to form a tubular structure with various combinations of linear or twisted core materials. By judicious selection of fiber materials and fiber architecture for the braid sleeve and the core structure (Figure 5.53), the load–deformation behavior of the braided fibrous assembly can be tailored (Somboonsong et al., 1998).

A small-scale modeling approach was used to develop the D-H-FRP reinforcement at Drexel University. The laboratory technique developed (Figure 5.54) allowed the production of 3- and 5-mm nominal diameter bars for strength evaluation. Figure 5.55 shows the 5-mm D-H-FRP bar produced as a scale model of the yet-to-be-developed prototype bar. This use of scale models, not covered in the first edition of this book, is a very powerful technique and is fairly common in industry for new product development.

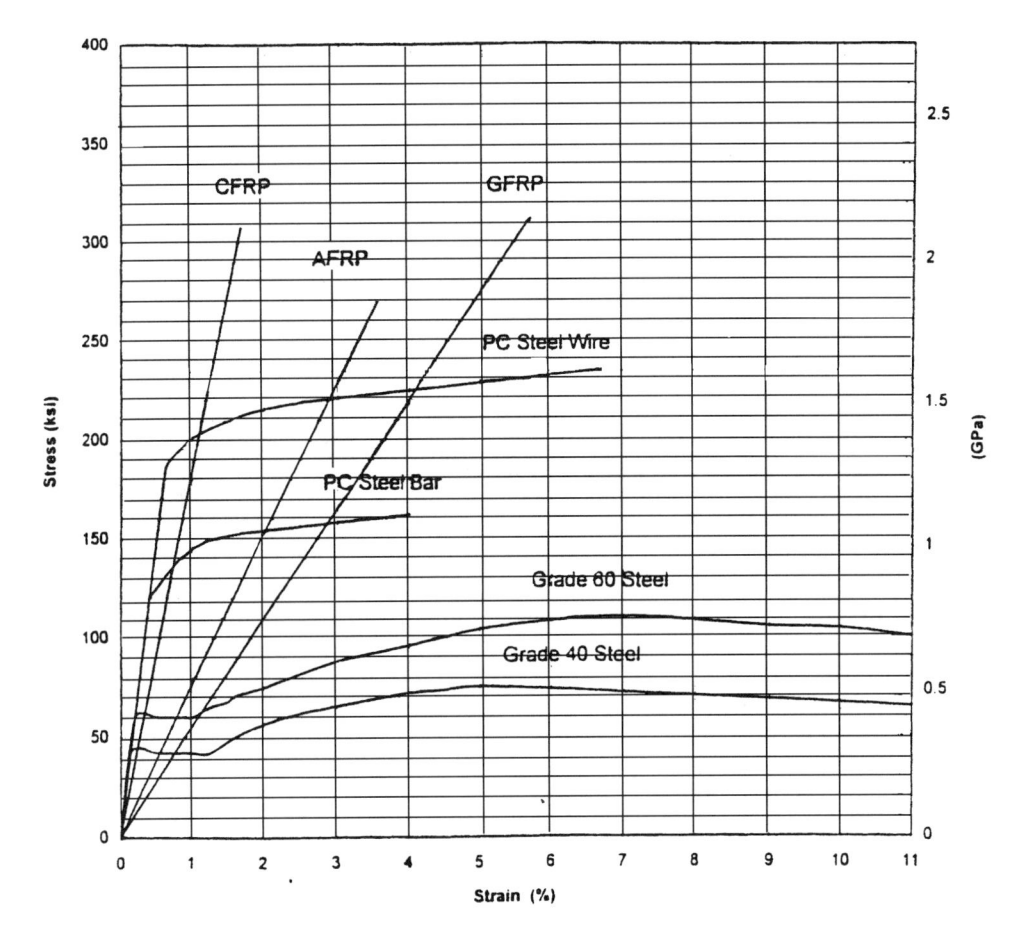

INELASTIC MODELS: STRUCTURAL STEEL AND REINFORCING BARS 257

Figure 5.52 Stress–strain characteristics of common FRPs compared to steel.

Tensile strength characterizations were carried out on the 3- and 5-mm diameter D-H-FRP bars produced in the laboratory using a bench-type 44.5 kN capacity displacement-controlled Tinius-Olsen T10000 universal testing machine. Tensile specimens of the 3-mm bars were 250 mm long and had 50-mm-long hard paper tabs glued with epoxy; the tensile specimens for the 5-mm bars were 405 mm long and had 50-mm-long tabs (Figure 5.56). For the 5-mm bars, special fiberglass-reinforced epoxy tabs were cast in tapered aluminum molds and were allowed to cure at room temperature. The tab geometry was such as to fit the grips of the testing machine. The tests were conducted at a strain rate of 0.02 mm/mm/min (ASTM D3039, 1993) using a Tinius-Olsen extensometer with a 50-mm gage length. Because of the higher yield loads of the 5-mm bar and the difficulty of preventing slip of the Tinius-Olsen extensometer when fiber fracture occurred, a specially fabricated extensometer was made consisting of a fixture clamped on to the specimen and accommodating two 100-mm LVDTs with extensions resulting in a 200-mm gage length. The stress–strain characteristics for the 5-mm bar are shown in Figure 5.57.

Stress–strain behavior in FRP bars, such as that shown in Figure 5.57, is equivalent bilinear. A lower bound to the stress–strain curves obtained from the tensile test data (Figure 5.58) can be used for flexural design purposes (Harris et al., 1998a,b). Such lower-bound stress–strain curves were used to design small-scale beams using the demonstration 3- and 5-mm D-H-FRP bars. Their behavior was directly compared with companion steel-reinforced beams. Results of these tests are given in Chapter 10.

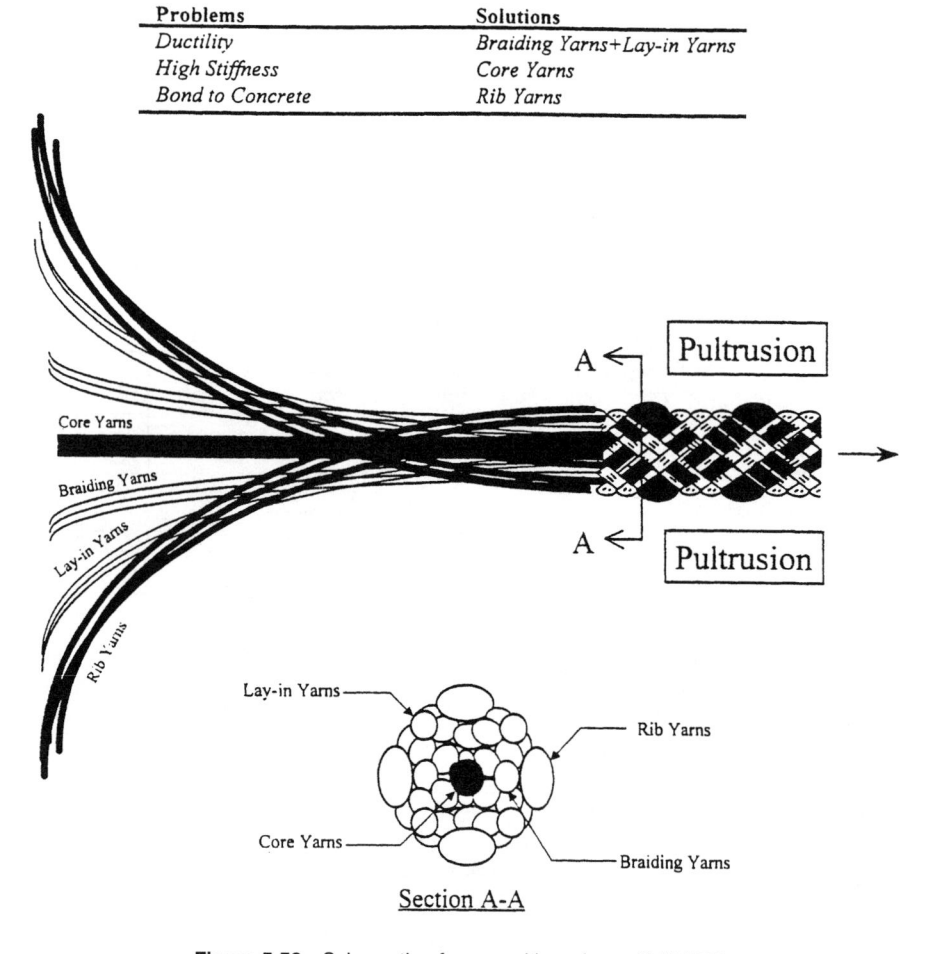

Problems	Solutions
Ductility	Braiding Yarns+Lay-in Yarns
High Stiffness	Core Yarns
Bond to Concrete	Rib Yarns

Figure 5.53 Schematic of composition of new D-H-FRP.

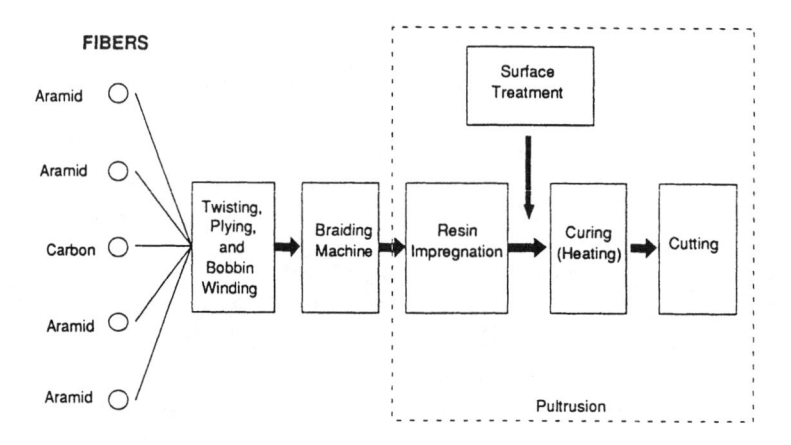

Figure 5.54 Schematic of production of D-H-FRP bars.

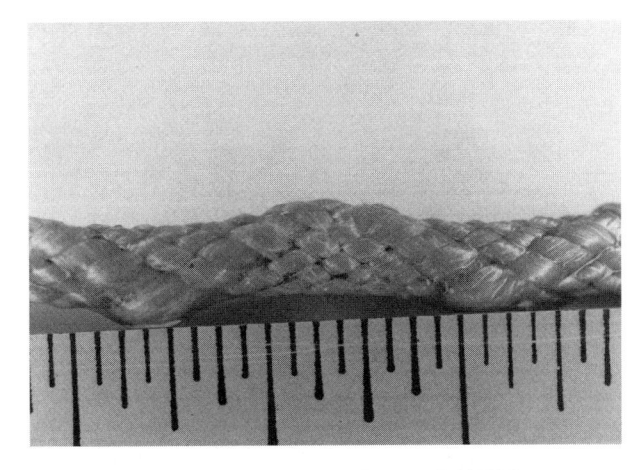

Figure 5.55 Close-up of 5-mm D-H-FRP.

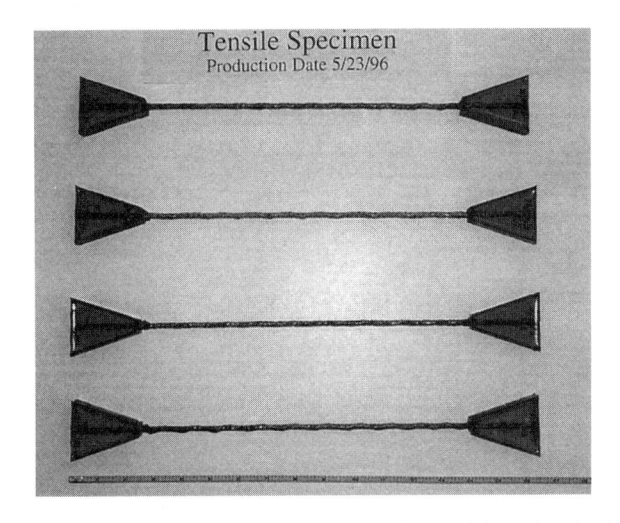

Figure 5.56 Tensile specimens with fiber glass-reinforced end tabs.

5.7 BOND CHARACTERISTICS OF MODEL STEEL

The requirements of bond similitude are discussed in Section 2.5.1. Some of the highlights of the above discussion are summarized below:

1. The use of standard deformed bars does not necessarily ensure true bond modeling because of the difference in bond strength of small- and large-size bars.
2. Modeling of bond is seriously complicated by our limited knowledge of the bond mechanism in prototype members.
3. The ultimate bond strength per unit length of the bar is proportional to $\sqrt{f'_c}$. It would be grossly misleading to model the dimensionally nonhomogeneous relationship for bond with small-size reinforcement and wires.
4. "True" bond–stress distribution bears little resemblance to the computed average unit bond stress, thereby making it difficult to model such a poorly defined quantity.
5. The effect of concrete cover upon bond strength and the increased bond strength afforded by stirrups and other steel, which tends to prevent splitting from developing adjacent to bars under high bond stress, are difficult to predict and to model.

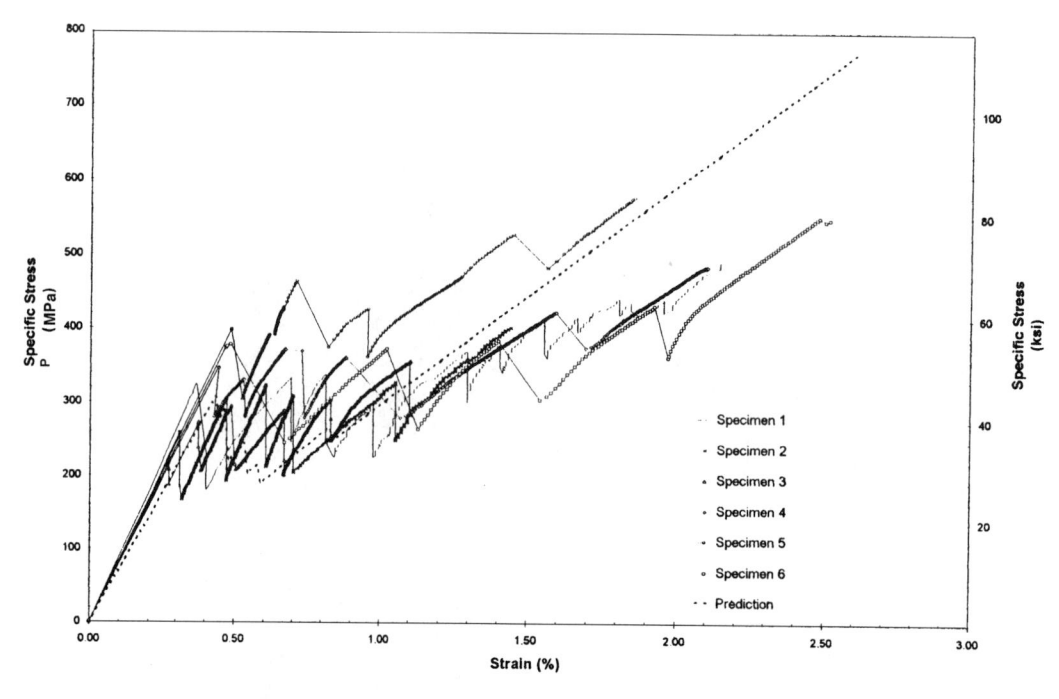

Figure 5.57 Stress–strain characteristics of demonstration 5-mm FRP bars.

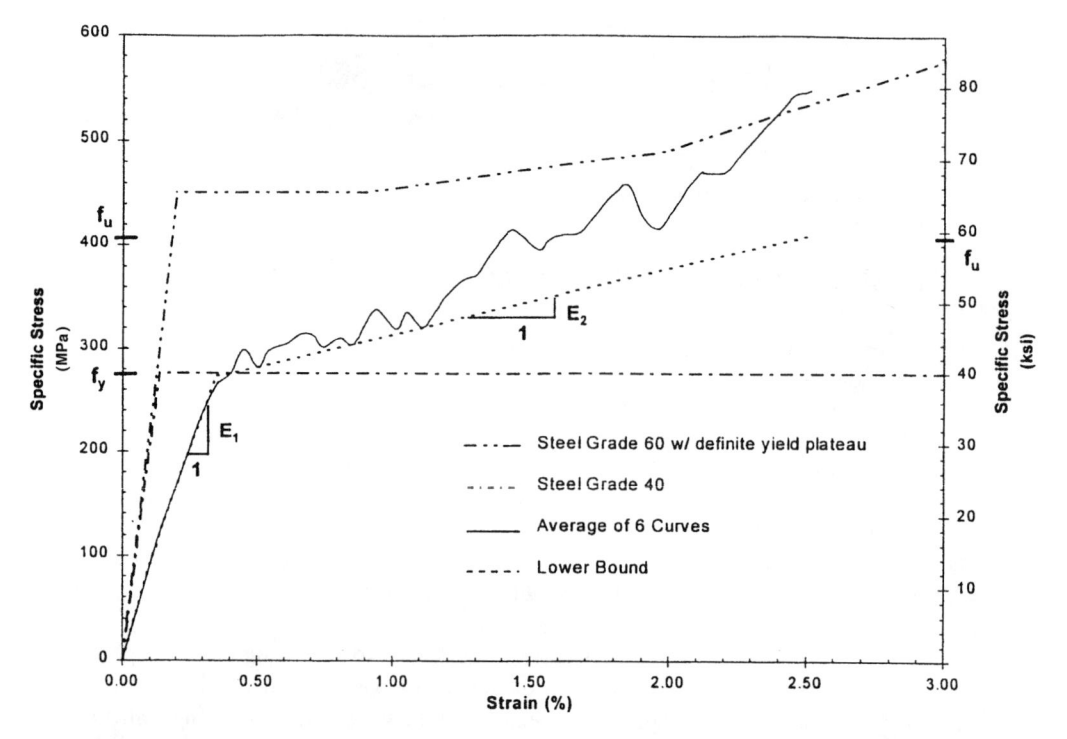

Figure 5.58 Average and lower-bound bilinear stress–strain curves of 5-mm FRP bars.

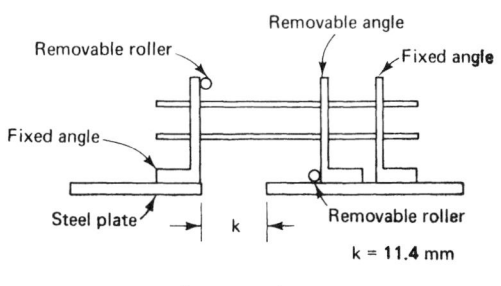

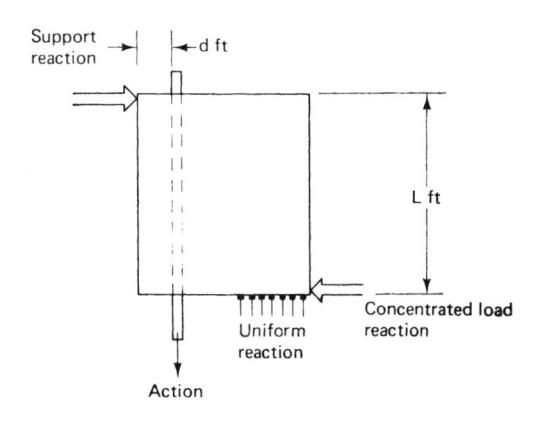

INELASTIC MODELS: STRUCTURAL STEEL AND REINFORCING BARS 261

Figure 5.59 Eccentric pullout test details.

Several different types of specimens and test procedures may be employed in studying the bond strength of reinforcement, including the concentric and the eccentric pullout tests (Figure 5.59), the embedded bar tensile test, and various types of flexural specimens, e.g., the University of Texas beam (Figure 5.60), the National Bureau of Standards beam, and the McGill doubly symmetrical bond beam (Figure 5.61). The flexural tests are preferable because flexural bond is of prime concern and these tests do simulate the practical situation more closely. The concentric pullout test is economic, simple, and less time-consuming; however, the main disadvantage of this type of test is that the concrete at the loaded end is in compression and eliminates or significantly decreases the transverse cracking. This also results in an increased relative slip between the concrete and the steel. Earlier investigations of bond between plain wire and model concrete (Harris et al., 1963; Taher, 1963; Aldridge, 1966; Lim et al., 1968) showed that for black annealed wires, an embedment length-wire diameter ratio of about 20 to 25 was adequate to cause the wire to yield. As expected, the smaller-diameter wire (16 and 18 SWG) exhibited better bond characteristics than larger-diameter wires (12 SWG and larger); this trend is similar to that observed in prototype bars. Using smooth wires in pullout and flexure specimens, Taher (1963) found that rusting increased the bond resistance, so that for 12 SWG wires and smaller, a yield point of 345 MPa could be developed with length over 20 diameters, corresponding to a bond strength of 4.3 MPa. When rusted wire is used, caution must be taken to prevent deterioration of the wire as a result of excessive rusting. Harris et al. (1963) also noted that rusting the wires for 7 days improved their bond characteristics. Unfortunately, smooth or rusted wires will not be adequate if the purpose of the model study is to investigate bond failure; number, size, and distribution of cracks; postcracking deflections; effects of reversed or repeated loads; or redistribution of internal stresses. To simulate bond characteristics accurately, the model reinforcement should have surface deformation similar to those of prototype reinforcement. In a comprehensive study, Harris et al. (1966a) investigated the bond characteristics of plain wire, commercially available deformed wire, threaded rod, and laboratory-deformed steel

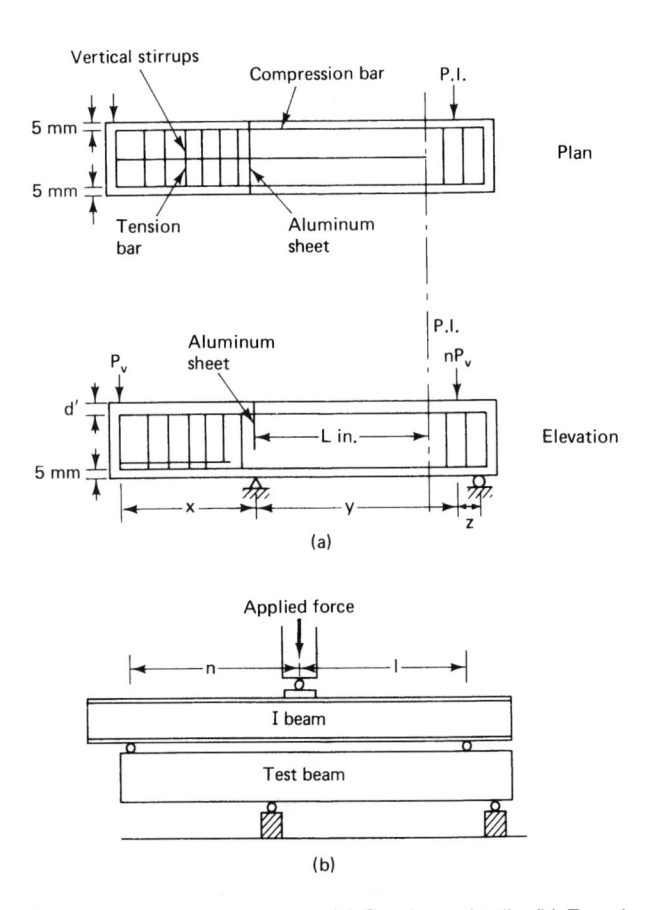

Figure 5.60 University of Texas bond beam. (a) Specimen details. (b) Experimental setup.

wires with cement and gypsum-based model concretes using concentric pullout, tension, and flexure tests, and compared the results with similar tests conducted on prototype specimens. It was possible to develop the yield strength of a 1-mm-diameter deformed steel wire with length–diameter (L/D) ratio of 15, and to develop yielding of a 1.5-mm-diameter deformed wire with an L/D ratio of 8.

The pullout test used by Harris et al. (1966b) consisted of a model bar embedded in a 25 × 50 mm cylindrical specimen, and this test was used to study the bond characteristics of the various model reinforcing wires used in model work. Results of a number of tests are plotted in Figure 5.62, along with other model and prototype data. Plain wires showed a marked decrease in average ultimate bond stress with increasing L/D ratio. Deformed model wires had an ultimate bond stress comparable with large prototype bars, with best agreement at the lower L/D ratios. The comparison of ultimate bond stresses as presented in Figure 5.62 indicates that suitably deformed wires will have pullout bond strengths reasonably close to those measured for prototype bars.

Hsu (1969) investigated the influence of concrete strength, clear cover, end anchorage, vertical stirrups, and rust on bond between plain and deformed bars and model concrete. The average ultimate bond stress values for the eccentric and doubly symmetrical bond beam tests were in good agreement with each other for both plain and deformed bars. This behavior was anticipated since the free end of the test bar was subjected to concentrated loads in both cases. Average ultimate bond stress values calculated for the University of Texas beams were generally higher than the results for the eccentric pullout and the doubly symmetrical bond beam tests. The University of Texas specimens provided for a point of contraflexure at the free end of the test bar, thus eliminating the disturbance caused by the concentrated loads in other types of bond specimens and resulted in higher value of the average ultimate bond stress.

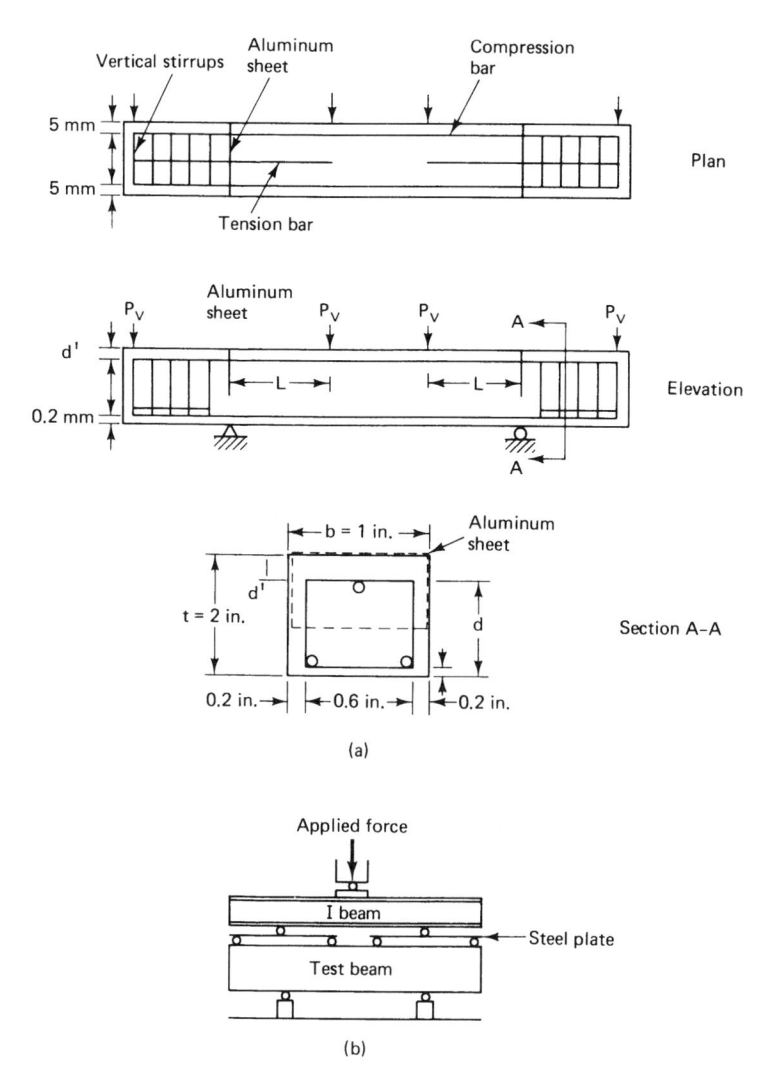

INELASTIC MODELS: STRUCTURAL STEEL AND REINFORCING BARS   263

Figure 5.61 Symmetrical bond beam test. (a) Specimen details. (b) Experimental setup.

As expected, the bond between steel and concrete improved with an increase in the concrete compressive strength, and the average ultimate bond stress was approximately proportional to $\sqrt{f'_c}$. The average ultimate bond stress increased with the embedment length to bar diameter (L/D) ratio up to 15, after which the average ultimate bond resistance decreased gradually. It was noted that for D2- and D4-size wires, and a 21-MPa concrete, an embedment length of approximately 15 diameters was adequate to cause these wires to yield at steel stresses between 439 and 528 MPa and an L/D ratio between 25 and 30 necessary to cause the steel wire to fracture. It was also observed that as the concrete compressive strength decreased from 35 to 21 MPa, the L/D ratio required to cause the steel wire to fracture increased from approximately 12 to about 25 (Figure 5.63). It was concluded from these various experiments that deformed bars exhibit bond characteristics that are comparable to those of the prototype reinforcing bars, as was observed in the Cornell University tests.

White and Clark (1978) conducted concentric pullout tests on specimens reinforced with knurled and rolled wires and also, for comparison purposes, on plain wire, plain rusted wire, and threaded rod. By instrumenting the pullout specimen with a transducer and a load cell, a direct plot of bond

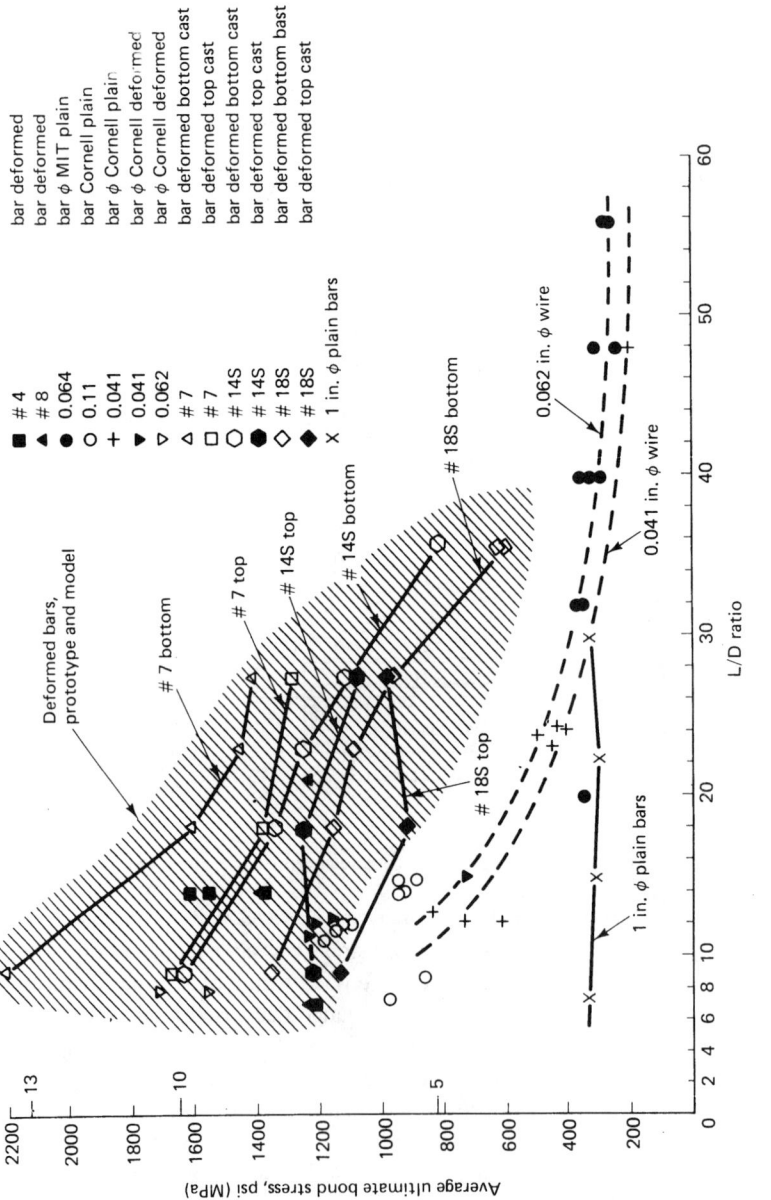

Figure 5.62 Average ultimate bond stress vs. L/D ratio from pullout tests. Comparison of prototype and model results.

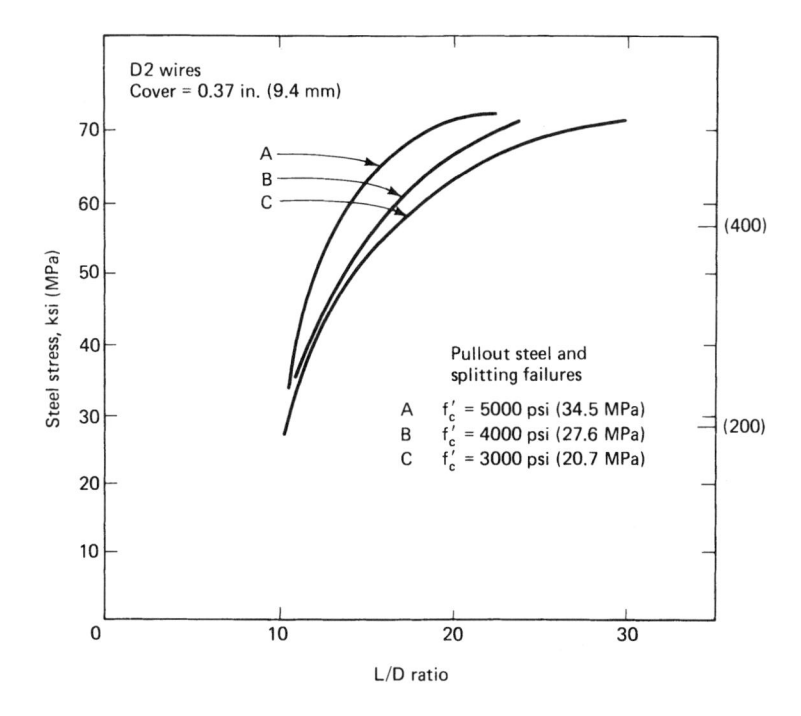

Figure 5.63 Variation of steel stress developed with L/D ratio. (After Hsu, 1969.)

force–slip relationship was obtained. The resulting bond stress–slip curves are shown in Figure 5.64. The principal conclusions can be summarized as follows:

1. Both types of deformed wires gave higher average ultimate bond strengths with the knurled-type reinforcement showing a stiffer bond strength–slip relationship. The failure mechanisms were significantly different from those obtained in prototype tests.
2. As a result of tests with mixes of different single aggregate sizes, it was observed that for all types of wires, "coarse" aggregates gave higher ultimate bond strengths and stiffer bond behavior than "fine" aggregate mixes. When compared with plain wires, the aggregate size had a greater influence on bond with deformed wires. However, experiments with two different aggregate shapes showed that the aggregate shape did not appear to have a significant influence on bond characteristics.
3. Post-test examination of the model concrete–reinforcement interface showed that in many of the specimens the concrete on the underside of the reinforcement (as cast) was badly compacted. This was due to air becoming trapped under the reinforcement during casting. In an attempt to solve this problem, the specimens were cast in a vertical position. The effect of direction of casting was investigated by Rehm (1961), who showed that bond stresses obtained for horizontally cast specimens were as low as 25% of the value obtained for the vertically cast specimens. The model tests at the Cement and Concrete Association (C & CA), U.K. have not agreed with Rehm's results. Little difference was observed in the bond strength and behavior of the horizontally and vertically cast specimens.

According to White and Clark (1978), compaction is one of the main factors influencing the bond characteristics. Results of the C & CA tests indicated an improvement in bond by using less workable mixes that require long vibration times and by using molds of greater mass.

Figure 5.64 Bond stress distributions.

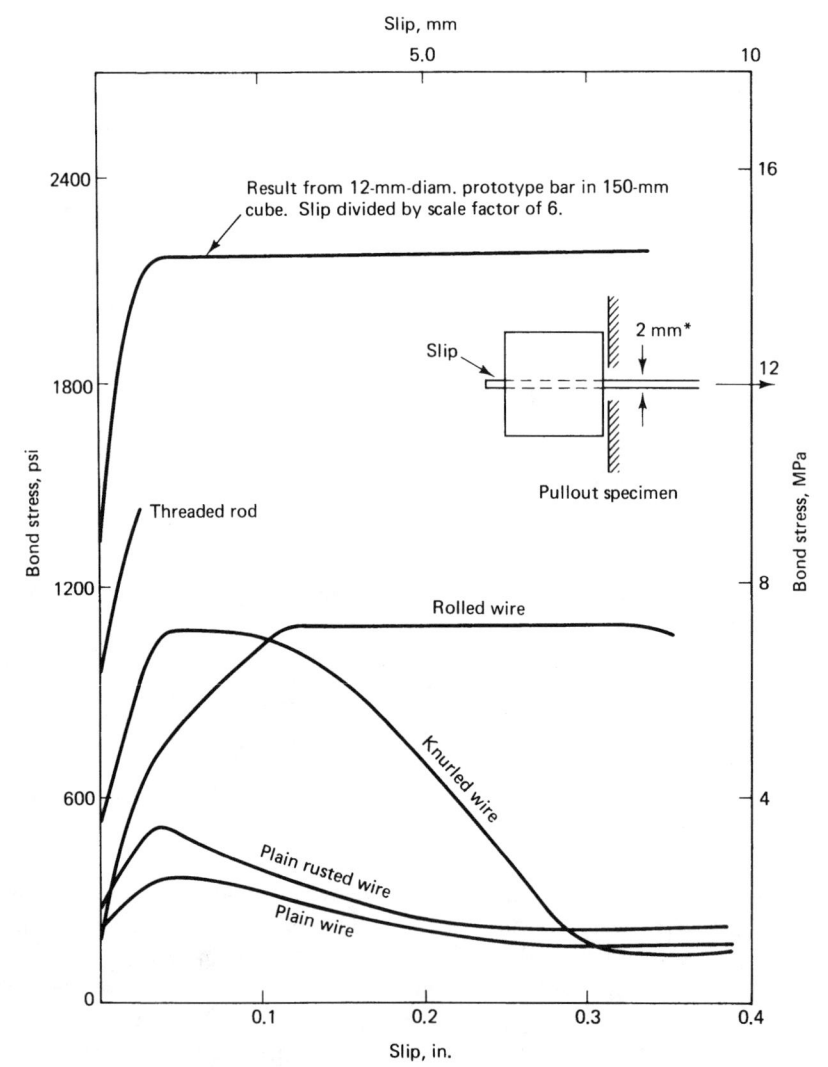

*Bond stress calculations are based on an average diameter of 2 mm.

5.8 BOND SIMILITUDE

Although some investigators have studied the bond characteristics of plain and deformed steel wires, experimental data on the similitude of bond is almost nonexistent. Harris et al. (1966a) modeled two prototype specimens (scale 1:8.33) to study the effect of wire deformation on bond and cracking in reinforced concrete models. The crack patterns in the prototype and its models reinforced with deformed wires were very similar with respect to secondary and tertiary cracks. The primary cracks in these models compared excellently with the primary cracks observed in the prototypes and predicted using Brom's theoretical work (1965). However, the number of cracks in the models reinforced with plain wires was between 25 and 50% of the number of cracks in the prototype.

INELASTIC MODELS: STRUCTURAL STEEL AND REINFORCING BARS 267

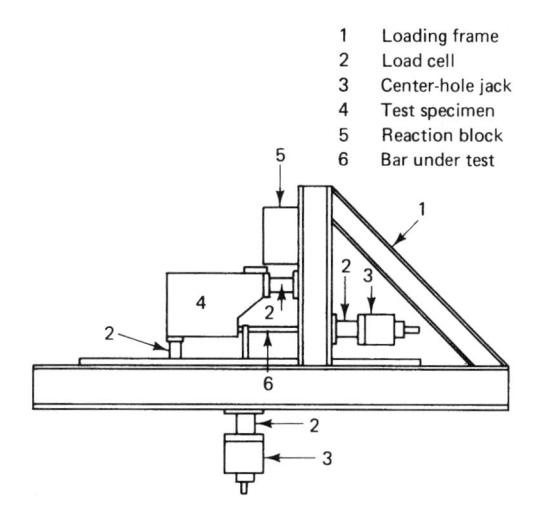

1	Loading frame
2	Load cell
3	Center-hole jack
4	Test specimen
5	Reaction block
6	Bar under test

Figure 5.65 Bond similitude arrangement.

Stafiej (1970) tested 23 direct models of the prototypes tested by Gergely (1969) to study the interaction between the applied pullout and dowel forces in causing splitting cracks along the main reinforcing bars in the beam end zones. Three scale factors ($1/5.95$, $1/4.83$, and $1/4.03$) were used to simulate the No. 6, No. 8, and No. 10 prototype deformed bars with locally available deformed steel wires that conformed to ASTM A496 specifications.

The test procedure consisted of leveling the specimen on a loading frame (Figure 5.65) and applying the dowel and pullout forces by using two center-hole jacks bearing against plates as shown. The wires were gripped with standard prestressing wire grips. The pullout force, the horizontal reaction in the compression zone, the vertical reaction, and the dowel force were monitored with a strain indicator and four load cells.

The specimens failed in one of the following five modes and showed excellent correlation with the cracking modes observed in the prototypes.

1. Steel yielding and fracture
2. Concrete splitting at the bottom only
3. Concrete splitting at the sides only
4. Concrete splitting at both the bottom and the sides
5. Complete bar pullout with no splitting

Similitude relations were developed from dimensional analysis, and the resulting prediction equations were used to compare the pullout force values in the prototypes and their models. Most of the predicted values were within a range of ±15% of the experimental results. A statistical analysis of the experimental data showed that the mean and the standard deviation of the experimental strength–predicted strength ratio was 1.04 and 0.128, respectively.

5.9 CRACKING SIMILITUDE AND GENERAL DEFORMATION SIMILITUDE IN REINFORCED CONCRETE ELEMENTS

The inelastic load–deflection response of a reinforced concrete structure is often strongly dependent upon the degree and manner of cracking (Zia et al., 1970). Cracking modes can also influence behavior under reversed or repeated loading, moment and force redistribution in indeterminate systems, and occasionally even the final mode of failure. The modeling of cracking is just as difficult as modeling of the bond; the two are intimately related phenomena. Moreover, an incomplete understanding of the cracking mechanism contributes greatly to the difficulties. It must also be noted that the cracks that first become visible in a prototype structure are not visible to the naked eye at the cracking load level scaled from the prototype because the width of these cracks gets scaled down, thus rendering these cracks "invisible." Normally, these can be detected by using a magnifying glass with an appropriate magnification factor and by the sudden change in the slope of the load–deformation curve.

Borges and Lima (1960) analyzed the similitude conditions for cracking and deformation in reinforced concrete both analytically and experimentally. They reported good success in modeling cracking in beams; however, their smallest model was a ¼-scale model of a beam 1 m deep.

Kaar (1966) presented a study on similitude of flexure cracking in T-beam flanges in which ½- and ¼-scale models of a prototype T beam 1016 mm deep with 178 × 2286 mm flanges were tested to failure. Deformed bars were used in all specimens, with No. 8 bars in the prototype and No. 2 bars in the ¼-scale model. Crack patterns at identical steel stress levels in the three specimens revealed that the total number of cracks decreased with decreasing beam size. However, overall cracking patterns were similar, and the scaled load–deflection curves for the prototype beam and its two models were practically identical. Thus, it appears that overall deformational similitude was achieved in Kaar's tests despite the differences in number of cracks.

Sabnis (1967) modeled two beams C2 and C5 from Mattock's tests (1964) at the Portland Cement Association Laboratories. The geometric scale factor used was ¹⁄₁₀, and all beams were 15 × 28 mm in cross section. Of the four beams, two had a steel ratio of about 0.0145 and two had a steel ratio of 0.025. All main reinforcement was from the same batch of 2-mm-diameter laboratory-deformed (Cornell deformeter) wires. The mix properties were 1:1:0.35 by weight of Ultracal, sand, and water, respectively, and gave a strength of approximately 21 MPa at 24 h. The crack patterns obtained in the model beams showed a striking resemblance to those of the prototype beams. A high consistency of cracking for identical models was also most encouraging.

Further work at Cornell on ¹⁄₁₀-scale gypsum plaster models of eccentrically compressed columns and a ⅙-scale gypsum plaster model (Figure 5.66) of a portal frame indicated good agreement between the experimental load–deflection and the predicted values. It was inferred from the moment–deflection relations that the overall behavior of the models was adequate and the crack patterns obtained were quite satisfactory. It must be noted that the model frame with minimum cross-sectional dimensions of 16.9 × 25.4 mm reinforced in each corner with one 1.6-mm-diameter annealed laboratory-deformed wire and 0.5-mm-diameter stirrups is a ¹⁄₁₅-scale model of a full-scale structure since the full-size frame tested was, in itself, a reduced-scale structure. Although there were fewer cracks in the models (Figure 5.67), a linear scaling of the load–deflection curves for the models gave results nearly identical to those of the prototype (Figure 5.68). Therefore, there must be many very small cracks, invisible to the naked eye, that give proper stiffness scaling in the model load–deformation relationships. A reasonable degree of confidence can, therefore, be placed in the use of very small-scale models in limit design studies (Sabnis, 1967).

Mirza (1967) tested three prototype beams and their quarter-scale models under combined bending, torsion, and shear and obtained excellent similitude between cracking patterns, failure mechanisms, and strengths at initial cracking and failure. Pang (1965) also observed satisfactory similitude of cracking and mechanism of failure between post-tensioned prestressed concrete prototype beams and their ½- and ¼-scale models subjected to combined bending and shear (see also Mirza and McCutcheon, 1971).

Syamal (1969) and Mirza (1978) reported results from tests on 56 reinforced concrete specimens that included 11 prototypes (built at a local precasting company) and their 45 direct models, at different geometric scale factors (½, ¼, and ⅙), constructed at McGill laboratories using a model

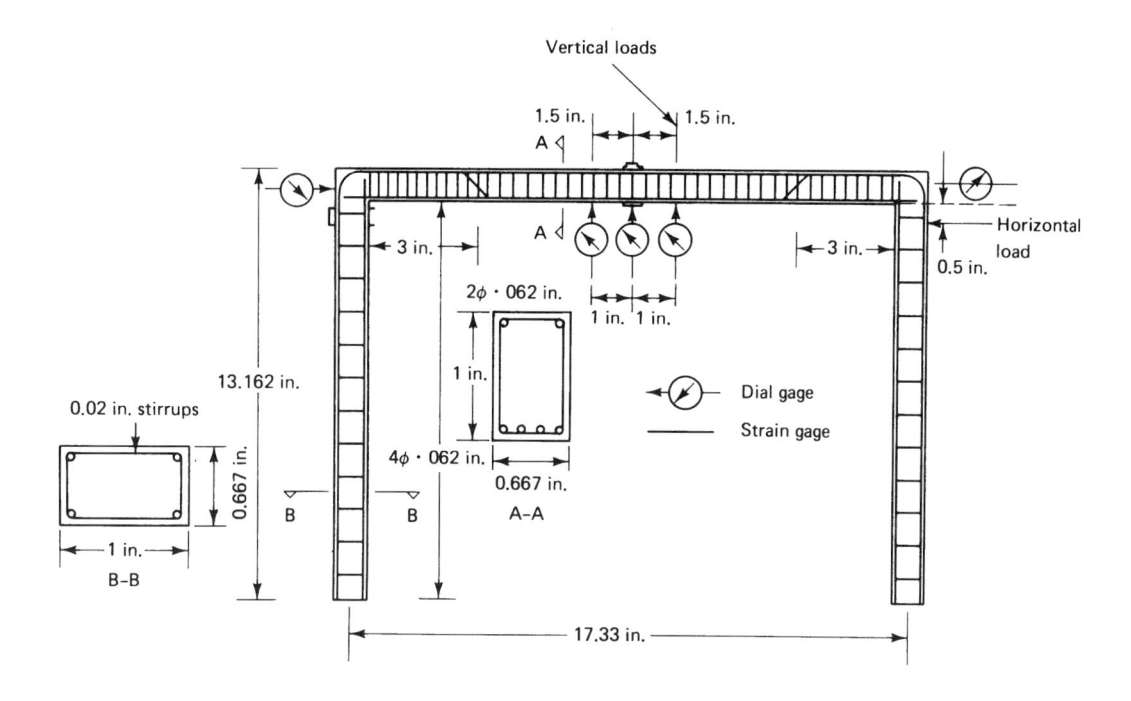

INELASTIC MODELS: STRUCTURAL STEEL AND REINFORCING BARS

Figure 5.66 Details of portal frame.

Figure 5.67 Crack pattern in the model frame.

concrete mix with a nominal strength of 21 MPa and locally available deformed and plain steel wires. The tests can be grouped in the following three categories:

1. Prototype and model specimens under concentric and eccentric loads (combined axial force and bending);
2. Prototype and model specimens under combined bending and shear;
3. Prototype and model specimens under pure torsion and combined bending and torsion.

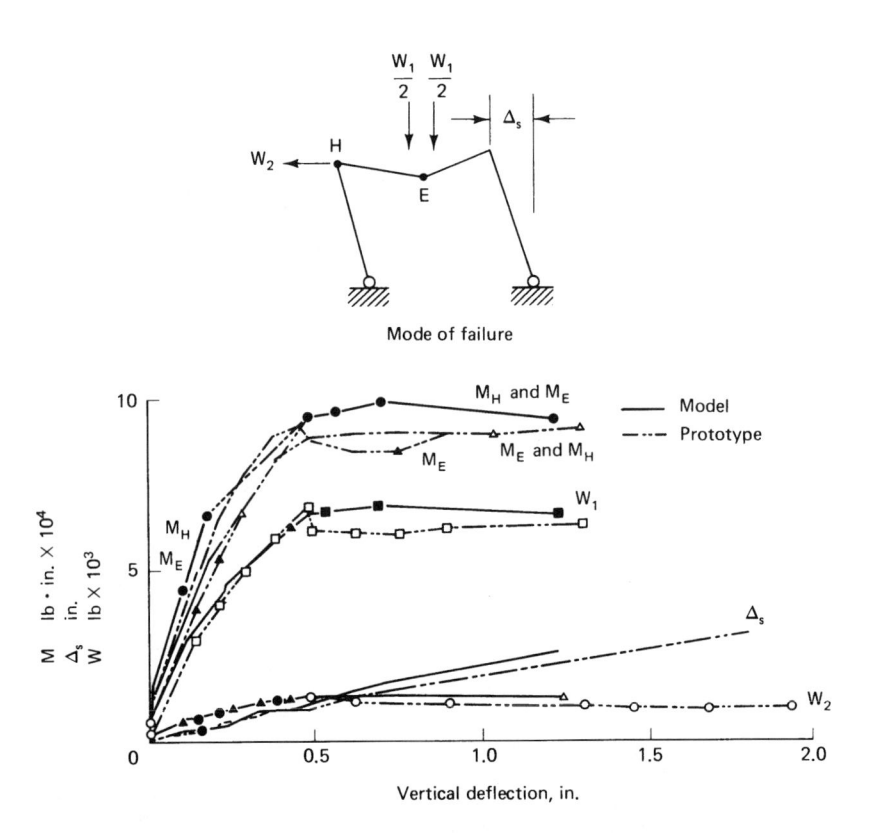

Figure 5.68 Behavior of model frame and comparison with prototype results.

For all loading conditions, it was observed that the crack patterns were reasonably similar except that more cracks appeared in the prototype and the ½-scale models than in the ¼-scale and the ⅙-scale models. Comparison of the crack patterns between two identical specimens, one reinforced with plain wires and the other with deformed wires of mild steel, showed that the beam reinforced with the plain galvanized wires was poor in reproducing the crack patterns; however, the beam reinforced with deformed wires reproduced crack patterns of the prototype reasonably well, although the number of cracks was generally less than in the prototype. Excellent similitude was obtained for deflections in short and normal beams, but there was a variation of ±30% in deflections for the deep beams at cracking. However, this scatter decreased to between 10 and 15% at 80% of ultimate load. The higher tensile strength of model concrete delayed the formation of cracks and led to discrepancies in deflections at cracking load for deep beams. For the specimens tested under pure torsion and combined bending and torsion, excellent deformation similitude was noted up to cracking load values of 0.4 to 0.45 of the ultimate load. However, postcracking torque–twist similitude was not as good, and the difference between the measured and the predicted values of angles of twist ranged between 10 and 40%. Some nondimensional moment–deflection and torque–twist curves are shown in Figures 5.69 and 5.70, respectively. It was concluded that it is possible to obtain reasonably good deformation similitude (deflections, twists, etc.) for the entire loading range between the prototype and its small-scale micro-concrete models (minimum dimension 50 mm) reinforced with deformed steel wires.

Work on models of reinforced concrete slabs showed that size effects of the model slabs are reflected not only in ultimate load but also in the load–deflection relationships. Mastrodicasa (1970) noted that discounting the fact that exact similitude was not obtained in model material properties, the relative model deflections tended to be larger as the size of the model slab decreased from 1410 × 1410 × 47 mm to 353 × 353 × 11.8 mm. However, the crack pattern and crack spacing in the models were similar to those in the prototype slabs. Also, the modes of failure in the prototype

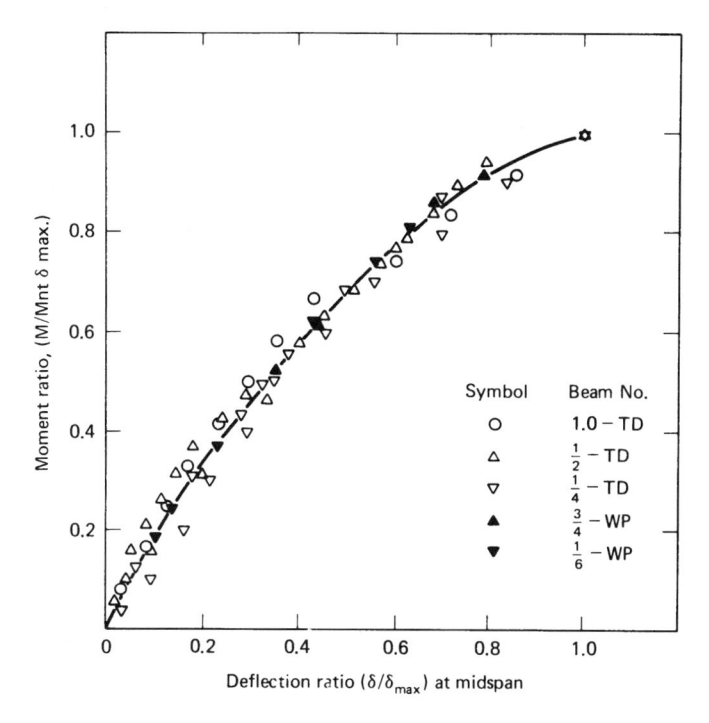

INELASTIC MODELS: STRUCTURAL STEEL AND REINFORCING BARS

Figure 5.69 Moment–deflection characteristics, Series B5. (From Mirza, 1978.)

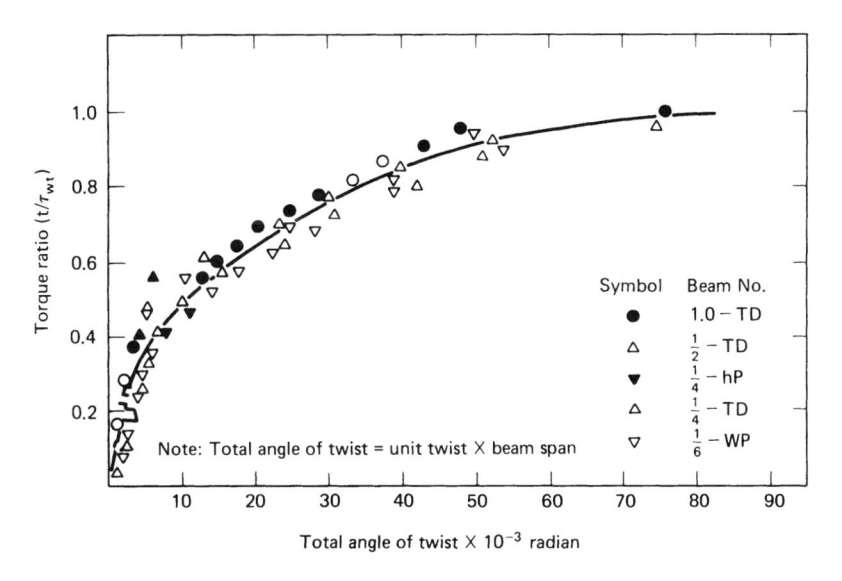

Figure 5.70 Torque–twist characteristics (M/T = 0). (From Mirza, 1978).

and the model slabs were identical. Clark (1971) observed from the results of a series of tests on eight ⅓.₇-scale model one-way slabs that cracking did not scale. The main reason for this was that the model specimens, as compared with the prototypes, exhibited greater cracking strains and more ductility in tension. This scale effect was shown to be a function of the absolute size of the model, in particular its total depth, rather than a function of the model scale. He concluded that prototype strains and crack widths can be predicted accurately from a model test only if the material properties of both the model and the prototype are known accurately.

More research is needed on the fundamental properties of materials used in modeling reinforced concrete. The property that has the greatest influence on the correlation of model and prototype behavior at the service load level is the tensile stress–strain relationship for concrete, for which very few data are available for either normal-size concrete or model concrete. It is clear that an investigation is needed to establish a tensile stress–strain relationship for concrete that takes into account variations in maximum aggregate size, aggregate grading and type, and mix proportions.

Finally, it must be pointed out that most of the investigators have observed good to excellent similitude between load–deformation characteristics of the prototype structures and their small-scale models. Each investigator observed a smaller number of cracks in the model than in the prototype. However, the overall pattern of cracking was faithfully reproduced in each case.

5.10 SUMMARY

Materials suitable for modeling structural steel and for modeling reinforcement in reinforced and prestressed concrete structures are presented in this chapter. Structural steel models are added to this second edition because of the need to test small-scale models in educational applications and in research under dynamic effects, especially earthquakes. Wide-flange sections milled from ASTM C1020 hot-rolled bar stock material and construction methods using TIG welding can be used to model at small-scale A36 structural steel frameworks. Properties of commercially available plain and deformed wires and the laboratory-deformed wires and the techniques for annealing these wires to simulate prototype steel yield and ultimate strengths, ductility, etc. are discussed. Various laboratory wire-deforming methods are presented and the types of deformations discussed. Non-metallic fiber-reinforced polymer (FRP) for reinforcing concrete structures is presented and the use of small-scale modeling techniques to develop a ductile FRP bar is discussed in this second edition.

Bond characteristics of available steel wires used to model reinforcing steel are presented. The tests used by various investigators and a comparison of model test results with prototype test data are also presented. Influence of different parameters on bond characteristics of steel wires are given. An exploratory investigation on bond similitude showed that certain phenomena, including bond, can be modeled with reasonable reliability if care is exercised in selecting model materials and in constructing and testing the model.

Experimental results show that the total number and width of cracks decrease as the model size is decreased. Existing information shows that in spite of the difficulty of simulating the number and the width of cracks, the overall load–deformation characteristics (load–deflection, moment–rotation, torque–twist, etc.) can be reproduced with reasonable accuracy in small-scale models built from microconcrete or gypsum plaster mixes.

PROBLEMS

(*Note:* Some of these problems assume familiarity with strain measurement techniques covered in Chapter 7.)

5.1 A ¼-scale model of a steel cantilever plate-girder beam fails by buckling of the compression flange when the load is 45 N. Perform a dimensional analysis and answer the questions: At what load will the prototype fail? How much will the prototype be deflected at the ultimate load if the maximum deflection of the model is 8 mm? Would this model study have been possible to perform using any other material than steel?

5.2 Determine load–deflection and load–bending stress curves for a steel bar 24 in. (600 mm) long used as a beam with central loading on a simple span of 20 in. (500 mm). Use a minimum number of strain gages. Suggested load increments: Estimate yield moment by

INELASTIC MODELS: STRUCTURAL STEEL AND REINFORCING BARS

assuming $f_y = 40$ ksi (270 MPa) for the steel. No less than four increments of load should be used from $P = 0$ to $P = P_{yield}$.

5.3 Given: Aluminum beam section, strain gages and strain indicator equipment, dial gages, loading device, load cell, and other accessories. For loading in bending to some stage substantially beyond yield, determine:
  a. Load–deflection behavior (deflection measured at center of span).
  b. Load–strain behavior (strain measured at or near center of span; at least one strain gage).
  c. E from both the measured deflection and the measured strain. How do these differ?

5.4 In order to determine the buckling behavior of a steel-tied arch ($E = 30 \times 10^6$ psi), a true-scale model is made from aluminum ($E = 10 \times 10^6$ psi). The span length of the model arch is 10 ft, and the prototype arch 200 ft. All dimensions are to the same geometric scale. A sketch of the model is given below.

The load is applied to the model in the form of a horizontal force H at the boundaries, and the response of the model arch is measured by a transverse micrometer that measures δ, the *transverse* deflection movement of the arch at midspan.

The experimental results of this test are given in the table.

  a. What can you deduce about the buckling characteristics of the arch model from this data? In particular what can you say about its behavior at small and large values of loading?
  b. What is the small displacement elastic buckling load of the model?
  c. What is the limit of small displacement response of the model, for acceptable accuracy?
  d. What is the prediction for the elastic buckling load of the prototype?
  e. What is the prediction for the limit of small dsplacement response of the prototype?

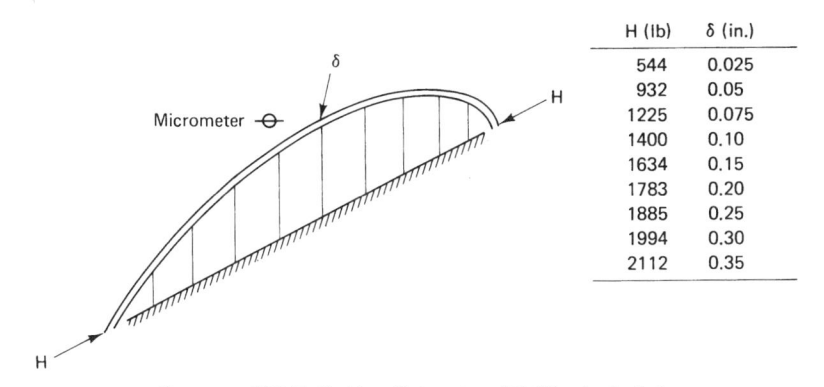

H (lb)	δ (in.)
544	0.025
932	0.05
1225	0.075
1400	0.10
1634	0.15
1783	0.20
1885	0.25
1994	0.30
2112	0.35

Courtesy of W. G. Godden, University of California, Berkeley.

5.5 Thin webbed metallic beams used in many industrial applications are frequently stiffened by transverse ridges (beads) that are pressed into the webs as shown below. The web shear stress τ that will "cripple" the beads when the member buckles depends on the shear force in the beam V, the depth of the web h, the spacing b between beads, the depth d of the bead, the thickness of the web material t, its yield stress σ_y, and its Young's modulus E.
  a. Make a dimensional analysis of the problem.
  b. Design an aluminum 1/10-scale model for a simply supported 2-m-long steel beam with a depth of 250 mm loaded by a concentrated central load of 25 kN. The thickness is 5 mm, $d = 200$ mm, and $b = 150$ mm.
  c. Determine the critical shear stress, τ, in model and prototype beams.

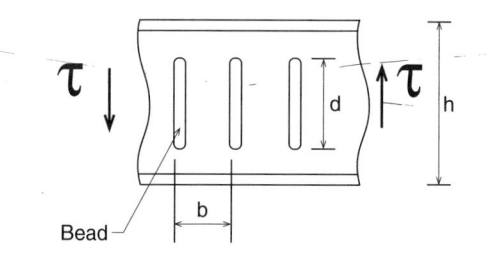

5.6 A W 24 × 100 beam, with a depth of 24.00 in., and an I of 2990 in.⁴ supports a concentrated load of 20 kips at the center of a 30-ft span.

    a. Design a distorted direct model for analysis in accordance with the following data: $S_L = 10$; aluminum alloy plates to be 0.100 in. thick; width of flange plates to be 1 in. The overall depth of the model is arbitrarily selected as 2 in. Strains are to be measured with strain gages. $E_s = 29,000$ ksi, and $E_a = 10,000$ ksi.

    b. Determine the value of the concentrated load that should be applied to the model.

    c. What is the value of the prediction factor by which the observed strains should be multiplied to correct for the fact that the requirements for similitude have not been satisfied relative to the depth of the model?

    d. Check your model design by computing and comparing midspan deflections, stresses, and strains of the model and prototype.

    e. Discuss the suitability of the model for predicting lateral buckling resistance of the prototype.

    f. Choose suitable strain gages for the model, and describe thoroughly the instrumentation you would use, including the disposition of the temperature-compensating gage(s). (Hint: See Chapter 7.)

5.7 The reinforced concrete beam/slab shown below is to be modeled at ⅓ scale using the available model bars whose properties are given in the Table 5.12. Assume that the model concrete has been selected and the prototype bars are a nominal Grade 60 steel (yield strength of 413.7 MPa). Choose the number of bars and spacing.

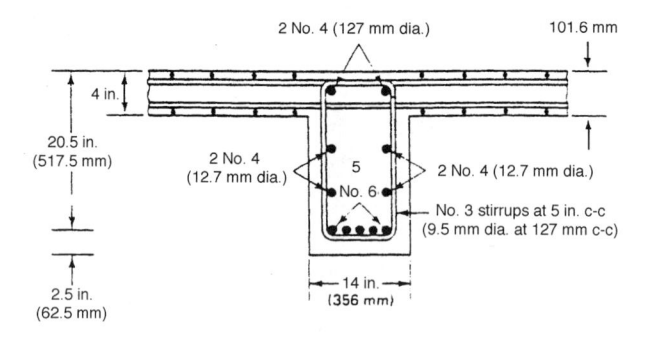

REFERENCES

ACI Report 440R-96, *State-of-the-Art Report on Fiber Reinforced Plastic Reinforcement for Concrete Structures,* Reported by ACI Committee 440, Report Chair M. R. Ehsani, American Concrete Institute, Box 19150, Redford Station, Detroit, MI.

Aldridge, W. W. (1966). Ultimate Strength Tests of Model Reinforced Concrete Folded Plate Structures, Ph.D. thesis, the University of Texas, Austin.

Antebi, J., Smith, H. D., Sharma, H. D., and Harris, H. G. (1962). Evaluation of Techniques for Constructing Model Structural Elements, Research Report R62-15, Department of Civil Engineering, Massachusetts Institute of Technology, Cambridge, May.

Baker, J. F., Horne, M. R., and Heyman, J. (1956). *Plastic Behavior and Design,* Vol. 2 of *The Steel Skeleton,* Cambridge University Press, New York.

Basham, K. D., Ed. (1994). *Infrastructure: New Materials and Methods of Repair, Proc. Third Materials Engineering Conf.,* San Diego, Nov. 13–16, American Society of Civil Engineering, New York.

Brock, G. (1959). Direct models as an aid to reinforced concrete design, *Engineering* (London), 187 (April), 468–470.

Broms, B. B. (1965). Crack width and crack spacing in reinforced concrete members, *Proc. Am. Conc. Inst.,* 62(10), October, 1237.

Borges, J. F. and Lima, J. A. E. (1960). Crack and deformation similitude in reinforced concrete, *RILEM Bull.* (Paris), New Series no. 7 (July), 79–90.

Burton, K. T. (1963). A Technique Developed to Study the Ultimate Strength of P/C Structures by the Use of Small scale Models, M.Sc. thesis, Cornell University, Ithaca, NY.

Chao, N. D. (1964). Ultimate Flexural Strength of Prestressed Concrete Beams by Small Scale Models, unpublished M.S. thesis, Cornell University, Ithaca, NY.

Cherry, S. and Mosborg, R. J. (1950). Tests of Structural Steel Columns, Progress Report No. 2, Engineering Experiment Station, Structural Research, University of Illinois.

Cherry, S., Chow, P. Y., and Austin, W. J. (1952). Experimental Studies of Columns, Progress Report No. 4, Civil Engineering Studies Structural Research Series No. 34, University of Illinois Engineering Experiment Station, Urbana, IL, 1 July.

Chowdhury, A. H. and White, R. N. (1971). Inelastic Behavior of Small Scale Reinforced Concrete Beam-Column Joints under Severe Reversing Loads, Report No. 342, Department of Structural Engineering, Cornell University, Ithaca, NY. October, 135 pp.

Clark, L. A. (1971). Crack Similitude in 1:3.7 Scale Models of Slabs Spanning One Way, Technical Report, Cement and Concrete Association, London, March.

Dukakis, C. (1980). A study of the structural behavior of a model wide flange beam under two-point load, *Drexel Tech. J.,* 43(1), Drexel University, Philadelphia, PA, Fall.

El-Attar, A. G., White, R. N., and Gergely, P. (1991). Shaking Table Test of a 1/8-Scale Three-Story Lightly Reinforced Concrete Building, Technical Report NCEER-91-0018, National Center for Earthquake Research, SUNY, Buffalo, February 28.

El-Attar, A. G., White, R. N., and Gergely, P. (1997). Behavior of gravity load designed reinforced concrete buildings subjected to earthquakes, *ACI Structural Journal,* Vol. 94, No. 2, March-April, pp. 133-145.

El-Badry, M. M., Ed. (1996). *Advanced Composite Materials in Bridges and Structures,* The Canadian Society of Civil Engineering, Montreal.

Evans, D. J. and Clark, L. A. (1978). A machine for cold-rolling deformed reinforcing bars for model tests, *Mag. Concrete Res.,* 30(102), March, 31–34.

Galambos, T. V., Lin, F. J., and Johnston, B. G. (1996). *Basic Steel Design with LRFD,* Prentice-Hall, Upper Saddle River, NJ.

Gaylord, Jr., E. H., Gaylord, C. N., and Stallmeyer, J. E. (1992). *Design of Steel Structures,* 3rd ed., McGraw-Hill, New York.

Gergely, P. (1969). Splitting Cracks Along the Main Reinforcement in Concrete Members, Cornell University Report to the Bureau of Public Roads, U. S. Department of Transportation, April.

Harris, H. G. and Muskivitch, J. C. (1977). Report 1: Study of Joints and Sub-Assemblies—Validation of the Small-Scale Direct Modeling Techniques, Nature and Mechanism of Progressive Collapse in Industrialized Buildings, Office of Policy Development and Research, Department of Housing and Urban Development, Washington, D. C., October, 165 pp. and Department of Civil Engineering, Drexel University, Philadelphia, PA.

Harris, H. G. and Muskivitch, J. C. (1980). Models of precast concrete large panel buildings, *ASCE J. Struct. Div.,* 106, No. ST2 (February), Proc. Paper 15218, 545-65.

Harris, H. G., Pahl, P. J., and Sharma, S. D. (1962). Dynamic Studies of Structures by Means of Models, Report R63-23, Department of Civil Engineering, Massachusetts Institute of Technology, Cambridge.

Harris, H. G., Schwindt, R., Taher, I., and Werner, S. (1963). Techniques and Materials in the Modeling of Reinforced Concrete Structures under Dynamic Loads, Report R63-54. Department of Civil Engineering, Massachusetts Institute of Technology, Cambridge, December, also NCEL-NBY-3228, U.S. Naval Civil Engineering Laboratory, Port Hueneme, CA.

Harris, H. G., Sabnis, G. M., and White, R. N. (1966a). Small Scale Direct Models of Reinforced and Prestressed Concrete Structures, Report No. 326, Department of Structural Engineering, Cornell University, Ithaca, NY, September, 362 pp.

Harris, H. G., Sabnis, G. M., and White, R. N. (1966b). Reinforcement for Small Scale Direct Models of Concrete Structures, paper No. SP 24-6, *Models for Concrete Structures,* ACI SP 24, American Concrete Institute, Detroit, MI, 141–158.

Harris, H. G., Somboonsong, W., and Ko, F. K. (1997). A new ductile hybrid fiber reinforced polymer (FRP) reinforcement for concrete structures, in *Proceedings of the 1997 International Conference on Engineering Materials,* 8–11 June, Ottawa, Canada, Vol. I, 593–604.

Harris, H. G., Somboonsong, W., Ko, F. K., and Huesgen, R. (1998a). A second generation ductile hybrid fiber reinforced polymer (FRP) for concrete structures, in *Proc. Second International Conf. on Composites in Infrastructure,* Tucson, Jan. 15–17, Department of Civil Engineering and Engineering Mechanics, University of Arizona.

Harris, H. G., Somboonsong, W., and Ko, F. K. (1998b). A new ductile hybrid fiber reinforced polymer (FRP) reinforcing bar for concrete structures, *ASCE J. Composites Construction,* 2(1), February, 28–37.

Head, P. R. (1996). Advanced composites in civil engineering — a critical overview at this high interest, low usage stage of development, in *Advanced Composite Materials in Bridges and Structures,* El-Badry, M., Ed., The Canadian Society of Civil Engineering, Montreal, 3–15.

Hsu, C. T. (1969). Investigation of Bond in Reinforced Concrete Models, M.Eng. thesis, Structural Concrete Series No. 14, McGill University, Montreal, Canada, April.

Huesgen, R. (1998). Flexural Behavior of Ductile Hybrid FRP Rebars in Singly Reinforced Concrete Beams, M.Sc. thesis, Department of Civil and Architectural Engineering, Drexel University, Philadelphia.

Iyer, S. L. and Sen, R., Eds. (1991). *Advanced Composite Materials in Civil Engineering Structures, Proc. Specialty Conf.,* Las Vegas, Jan. 31–Feb. 1, American Society of Civil Engineering, New York.

Kaar, P. H. (1966). High strength bars as concrete reinforcement. Part 8, Similitude in flexural cracking of T-beams flanges, *J. PCA Res. Dev. Lab.,* 8(2), May, 2–12.

Kabe A. M. and Rea, D. (1980). Earthquake-Induced Inelastic Deformations in Small Shear-Type Steel Structures, Report UCLA-ENG-8036, School of Engineering and Applied Science, University of California, Los Angeles, July.

Kemp, G. (1971). Simply Supported, Two Way Prestressed Concrete Slabs under Uniform Load, M. Eng. thesis, Structural Concrete Series No. 71-4, McGill University, Montreal, August.

Kim, W., El-Attar, A., and White, R. N. (1988). Small-scale modeling techniques for reinforced concrete structures subjected to seismic loads, Rept. No. NCEER-88-0041, National Center for Earthquake Engineering Research, State University of New York at Buffalo, Nov.

Ko, F. K. (1987). Braiding, *Engineered Materials Handbook,* Vol. 1: *Composites,* ASM International, Menlo Park, MI, 519–528.

Ko, F. K., Somboonsong, W., and Harris, H. G. (1997). Fiber architecture based design of ductile composite rebars for concrete structures, in *Proceedings of the International Conference on Composite Materials,* Scott, M. L., Ed., July 14–17, Gold Coast, Australia, Vol. VI. *Composite Structures,* VI-723–VI-730.

Labonte, L. R. S. (1971). An Investigation of Anchorage Zone Behavior in Prestressed Concrete Containments, M.Eng. thesis, Structural Concrete Series No. 71-6, McGill University, Montreal, September.

Lee, S. J. and Lu, L. W. (1984). Studies on full-scale composite beam–column components, paper presented at the ASCE Annual Convention, San Francisco, October.

Lim, S. N., Syamal, P. K., Khan, A. Q., and Nemec, J. (1968). Development Length in Pullout Tests, McGill University, Montreal, January.

Litle, W. A. (1966). Small scale models for steel frameworks, *AISC Eng. J.,* American Institute of Steel Construction, Inc., New York, July.

Litle, W. A. and Foster, D. C. (1968). Fabrication techniques for small-scale steel models, Bulletin No. 10, *M.I.T. Structural Behavior of Small-Scale Steel Models,* Steel Research for Construction, American Iron and Steel Institute, New York, April.

Litle, W. A., Falcone, P. A., and Reimer, R. B. (1968). Ultimate strength behavior of small-scale steel frameworks, Bulletin No. 10, *M.I.T. Structural Behavior of Small-Scale Steel Models,* Steel Research for Construction, American Iron and Steel Institute, New York, April.

Lord, W. D. (1965). Investigation of the Effective Width of the Slab of Reinforced Concrete T-Beams, M.Sc. thesis, Cornell University, Ithaca, NY.

Maisel, E. (1978a). Reinforced and Prestressed Microconcrete Models, in *Proceedings of the Joint Institution of Structural Engineers/Building Research Establishment Seminar on Reinforced and Prestressed Microconcrete Models,* Garston, England, May.

Maisel, E. (1978b). *Mikrobeton für Modellstische Untersuchungen II,* Institute für Modellstatik, Universitat Stuttgart, Stuttgart, Germany.

Mastrodicasa, A. G. (1970). Size Effects in Models of Reinforced Concrete Slabs, Research Report R70-57, Massachusetts Institute of Technology, Cambridge, August.

Mattock, A. H. (1964). Rotational capacity of hinging regions in reinforced concrete, ACI SP 12, in *Flexural Mechanics of Reinforced Concrete,* American Concrete Institute, Detroit, 143–182.

Mills, R. S., Krawinkler, H., and Gere, J. M. (1979). Model Tests on Earthquake Simulators Development and Implementation of Experimental Procedures, Report No. 39, The John A. Blume Earthquake Engineering Center, Department of Civil Engineering, Stanford University, Stanford, CA, June.

Mirza, M. S. (1967). An Investigation of Combined Stresses in Reinforced Concrete Beams, Ph.D. thesis, Department of Civil Engineering and Applied Mechanics, McGill University, Montreal, March.

Mirza, M. S. (1972). Heat Treatment of Deformed Steel Wires for Model Reinforcement, Structural Concrete Series No. 72-4, McGill University, Montreal, March.

Mirza, M. S. (1978). Reliability of structural models, Proceedings of the Joint Institution of Structural Engineers/Building Research Establishment Seminar on Reinforced and Prestressed Microconcrete Models, Garston, England, May.

Mirza, M. S. and McCutcheon, J. O. (1971). Bond Similitude in Reinforced Concrete Models, Paper presented to the National Structural Engineering Meeting of the ASCE, Baltimore, April.

Mirza, M. S. and McCutcheon, J. O. (1978). Direct models of prestressed concrete beams on bending and shear, *Build Int.* (London), 7(2), March, 99–125, 1974. Reprinted in the *Proceedings of the Joint Institution of Structural Engineers/Building Research Establishment Seminar on Reinforced and Prestressed Microconcrete Models,* Garston, England, May.

Moustafa, S. E. (1966). A Small Scale Model Study of a Prestressed Concrete Slab, M.Sc. thesis, Cornell University, Ithaca, NY.

Mufti, A., Erki, M. A., and Jaeger, L., Eds. (1991). *Advanced Composite Materials with Application to Bridges,* Canadian Society of Civil Engineers, Montreal.

Murayama Y. and Noda, S. (1983). Study on Small Scale Model Tests for Reinforced Concrete Structures — Small Scale Model Tests by Using 3 mm Diameter Deformed Rebars, Report No. 40, Kajima Institute for Construction Technology, Tokyo, Feb.

Nanni, A. and Dolan, C. W., Eds. (1993). *Fiber-Reinforced-Plastic Reinforcement for Concrete Structures, Proceedings of International Symposium,* Vancouver, ACI SP-138.

Nebraska, J. E. and Sur, L. M. (1963). Behavior of Miniature Prestressed Concrete Beams, Report of a project sponsored by the National Science Foundation. Undergraduate Research Program, Department of Civil Engineering, University of Illinois, Urbana, IL, June.

Noor, F. A. and Goodman, J. (1979). Discussion of A machine for cold-rolling deformed reinforcing bars for model tests, by D. J. Evans and L. A. Clark, MCR 102, *Mag. Concrete Res.,* 31(106), March.

Pang, C. L. (1965). Reliability of Models in the Analysis of Prestressed Concrete Beams in Flexure, M.Eng. thesis, McGill University, Montreal, April.

Robinson, J. M., Jr. (1970). Model Analysis of Frame Buckling, B.Sc. thesis, Department of Civil Engineering, Drexel University, Philadelphia, December.

Saadatmanesh, H. and Ehsani, M. R., Eds. (1996). *Fiber Composites in Infrastructure, Proc. First International Conf. on Composites in Infrastructure,* Tucson, Jan. 15–17, Department of Civil Engineering and Engineering Mechanics, University of Arizona.

Saadatmanesh, H. and Ehsani, M. R. (1998). Editors, Proceedings of the Second International Conference on Composites in Infrastructure, Department of Civil Engineering and Engineering Mechanics, University of Arizona, Tucson, January 15-17.

Sabnis, G. M. (1967). Investigation of Reinforced Concrete Frames Subjected to Reversed Cyclic Loading Using Small Scale Models, Ph.D. thesis, Cornell University, Ithaca, NY, June.

Salmon, C. G. and Johnson, J. E. (1996). *Steel Structures — Design and Behavior,* 4th ed., Harper Collins, New York.

Somboonsong, W. (1997). Development of Ductile Hybrid Fiber Reinforced Polymer (D-H-FRP) for Concrete Structures, Ph.D. thesis, Department of Civil and Architectural Engineering, Drexel University, Philadelphia, December.

Somboonsong, W., Ko, F. K., and Harris, H. G. (1998). Ductile hybrid fiber reinforced plastic (FRP) rebar for concrete structures: design methodology, *ACI Mater. J.,* Vol. 95, No. 6, pp. 655-666, Nov.–Dec.

Stafieji, A. P. (1970). Bond Similitude in Reinforced Concrete Models, B.Sc. (Hon.) thesis, Department of Civil Engineering and Applied Mechanics, McGill University, Montreal, April.

Subedi, N. K. and Garas, F. K. (1980). Bond characteristics of small diameter bars used in microconcrete models, in *Reinforced and Prestressed Microconcrete Models,* Garas, F. K. and Armer, G. S. T., Eds., The Construction Press, New York.

Taher, I. (1963). A Study of Bond Characteristics in Wire Reinforced Specimens, M.Sc. thesis, Massachusetts Institute of Technology, Cambridge, August.

Wallace B. J. and Krawinkler, H. (1985a). Small-Scale Model Experimentation on R/C Assemblies U.S.–Japan Research Program, Report No. 74, The John A. Blume Earthquake Engineering Center, Department of Civil Engineering, Stanford University, Stanford, CA, June.

Wallace B. J. and Krawinkler, H. (1985b). Small-Scale Model Experimentation on Steel Assemblies U.S.–Japan Research Program, Report No. 75, The John A. Blume Earthquake Engineering Center, Department of Civil Engineering, Stanford University, Stanford, CA, June.

White, I. G. and Clark, L. A. (1978). Bond similitude in reinforced microconcrete models, in *Proceedings of the Joint Institution of Structural Engineers/Building Research Establishment Seminar on Reinforced and Prestressed Microconcrete Models,* Garston, England, May.

Zia, P., White, R. N., and Van Horn, D. A. (1970). Principles of model analysis, in *Models for Concrete Structures,* ACI SP-24, American Concrete Institute, Detroit, pp. 19–39.

CHAPTER 6

Model Fabrication Techniques

CONTENTS

6.1	Introduction		280
6.2	Basic Cutting, Shaping, and Machining Operations		281
	6.2.1	Cutting of Metal, Plastic, Wood, and Paper Products	281
		6.2.1.1 Metal and Plastic Materials	281
		6.2.1.2 Wood and Paper Materials	281
	6.2.2	Shaping and Machining Operations	281
	6.2.3	Drilling and Milling	282
	6.2.4	Lathe Turning and Boring	283
6.3	Basic Fastening and Gluing Techniques		283
	6.3.1	Mechanical Fastening	283
	6.3.2	Soldering	284
	6.3.3	Spot Welding	285
	6.3.4	Glues and Adhesives	286
	6.3.5	Epoxy Resins	286
6.4	Construction of Structural Steel Models		287
	6.4.1	General	287
	6.4.2	Silver Soldering	287
	6.4.3	Tungsten Inert Gas Welding	288
6.5	Construction of Plastic Models		288
	6.5.1	Fabrication Considerations	288
	6.5.2	Capillary Welding	292
	6.5.3	Thermal Forming Processes	292
	6.5.4	Drape or Gravity Forming and Drape Molding of Shell Models	293
	6.5.5	Vacuum Forming	294
	6.5.6	Fabrication Errors in Thermal Forming	295
	6.5.7	Casting of Plastic Models	296
	6.5.8	Spin Forming of Metal Shells	296
6.6	Construction of Wood and Paper Models		296
	6.6.1	Balsa Wood Models	296
	6.6.2	Structural Wood Models	298
	6.6.3	Glue-Laminated (Glulam) Beams	298
	6.6.4	Cartridge or Manilla Envelope Paper Models	299
6.7	Fabrication of Concrete Models		299
	6.7.1	Reinforced Concrete Models	299

		6.7.1.1	General	299
		6.7.1.2	Fabrication Methods	299
			1. Bending of Reinforcement	299
			2. Use of Ties	300
			3. Welding of Reinforcement	300
			4. Use of Epoxies	300
			5. Use of Soldering	300
		6.7.1.3	Accuracy of Reinforcement Placement	300
	6.7.2	Forms for Casting Reinforced Concrete Models		302
		6.7.2.1	Plexiglas Forms	302
		6.7.2.2	Plywood Forms	302
	6.7.3	Prestressed Concrete Models		302
		6.7.3.1	General	302
		6.7.3.2	Pretensioning Technique — Single Wires in Beams	302
		6.7.3.3	Pretensioning Technique — Multiple Wires in Beams	307
		6.7.3.4	Post-tensioning Technique — Multiple Wires in Slab	307
6.8	Fabrication of Concrete Masonry Models			309
	6.8.1	General		309
	6.8.2	Drexel University/NCMA Block-Making Machine		309
	6.8.3	New Drexel Model Block-Making Machine		309
	6.8.4	Building Model Masonry Components and Assemblies		309
6.9	Summary			312
Problems				312
References				315

6.1 INTRODUCTION

Models of buildings, bridges, and other structures are miniature structures in every sense, and, as in the prototype, their construction requires careful planning and considerable skill and experience. Model construction requires good knowledge of the mechanical properties of the materials selected along with their limitations and the resulting difficulties in the construction process. The following factors must be given proper attention in the construction of any model: availability of proper working tools and trained personnel and limitations of available time and cost. Pahl and Soosaar (1964) and Cowan et al. (1968) classify the available methods for construction of elastic models into the following five basic techniques:

1. Cutting or carving of the model from a continuous piece
2. Assembly of two or more individual components
3. Thermal forming process
4. Casting process
5. Spin forming of soft-metal models

Inelastic models on the other hand are usually the result of many repetitions of steps 2 and 4 above with adaptation of additional construction techniques that are applicable to the model structural type, the scale chosen, model materials used, and type of loading to which the structure is subjected.

The successful implementation of a particular model study may very well hinge on the techniques adopted in the model fabrication. Dimensional accuracy is much more critical in small-scale model work than in prototype structures because errors tend to amplify in direct relation to the scale chosen. There are sources of systematic error and deviations inherent in any given construction process, and these can significantly influence some types of structural behavior, such as the buckling phenomenon of shells that may be very sensitive to geometric imperfections. One must therefore carefully assess these potential sources of error before constructing the model. This chapter assembles and greatly amplifies the information that was scattered throughout the first edition of this book. This new added chapter to the second edition is included as an aid to the student and the model designer who needs a ready reference of information on model construction. In addition, the information provided can be considered as part of the necessary library that must accompany a well-stocked minimachine shop forming part of any successful structural models laboratory.

6.2 BASIC CUTTING, SHAPING, AND MACHINING OPERATIONS

6.2.1 Cutting of Metal, Plastic, Wood, and Paper Products

6.2.1.1 Metal and Plastic Materials

Metals and plastics can be cut and shaped using standard tools and machines found in any well-equipped structural models laboratory. Such a facility should be available to students carrying out research involving physical models.

Cutting of metals such as aluminum and brass, which are used in making elastic models or forms for casting concrete models, requires a variety of cutting equipment. These involve (1) band saws for cutting metals and plastics, (2) reciprocal saws for cutting mostly metals but that can be easily adapted to also cut most plastics, (3) circular saws for cutting mostly wood, and (4) jigsaws for cutting small sheets or sections of plastics or wood.

Hand tools can also be used to cut all of these materials so that a well-stocked student machine shop may be all that is required if the materials are in the form of strips, tubes, or shapes.

6.2.1.2 Wood and Paper Materials

Balsa wood is probably the easiest modeling material to work with, hence its enormous popularity with model makers of all types. Its use in making elastic structural models was discussed in Chapter 3. Cutting of balsa wood is easily performed with sharp knives or blades and miniature saws similar to the ones illustrated in Figure 6.1. The cutting of balsa wood truss members at various angles in order to fit properly can be easily accomplished using these techniques.

Wooden models other than balsa wood are usually cut to size using circular saws and finished using standard woodworking techniques. Strips as small as 5 mm in thickness can be cut on a good circular saw to fabricate glue-laminated beam models.

Cartridge and good quality cardboard paper used in making elastic and architectural models can be cut using scissors or blades and sharp knives and a straightedge.

6.2.2 Shaping and Machining Operations

Shaping and machining operations involve the use of machine tools such as the lathe and the milling machine. These tools are indispensable for the fabrication of models and forms for making models. Although individual lathes and milling machines can be found in the support machine shop run by experienced machinists and servicing the typical structural models laboratory, these machines are expensive to purchase and maintain for the student machine shop. A smaller-size combination lathe and milling machine, as well as drill press similar to the one shown in Figure 6.2, can be purchased at a reasonable price for the small machine and tool shop which is necessary for a well-functioning structural models laboratory. Such a multipurpose machine can be used to drill, mill, turn, or bore metal, plastic, or wood pieces for constructing scale models.

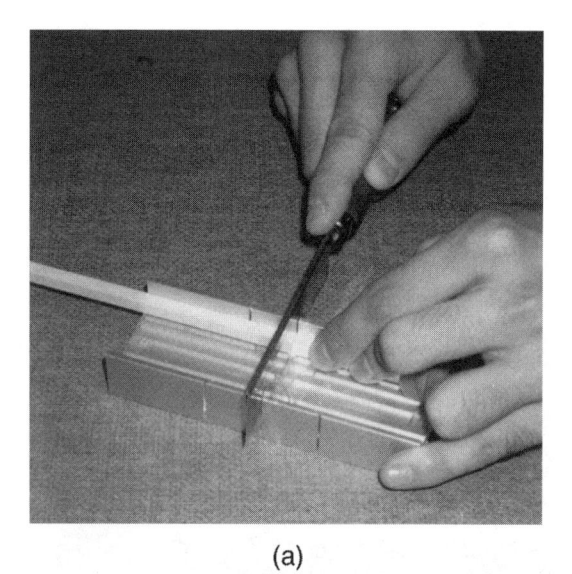

Figure 6.1 Cutting details of balsa wood models. (a) Miniature saw blade. (b) X-Acto knife blade.

It should be strongly emphasized that for reasons of safety only properly trained students should be allowed to use machining equipment.

6.2.3 Drilling and Milling

The drill press portion of the multipurpose machine can be used interchangeably with the mill without the need to reset the workpiece. The drill offers all the capabilities of a standard drill press, including a lock feature for accurate cut depths. The mill can be used to cut grooves, mill flat

MODEL FABRICATION TECHNIQUES   283

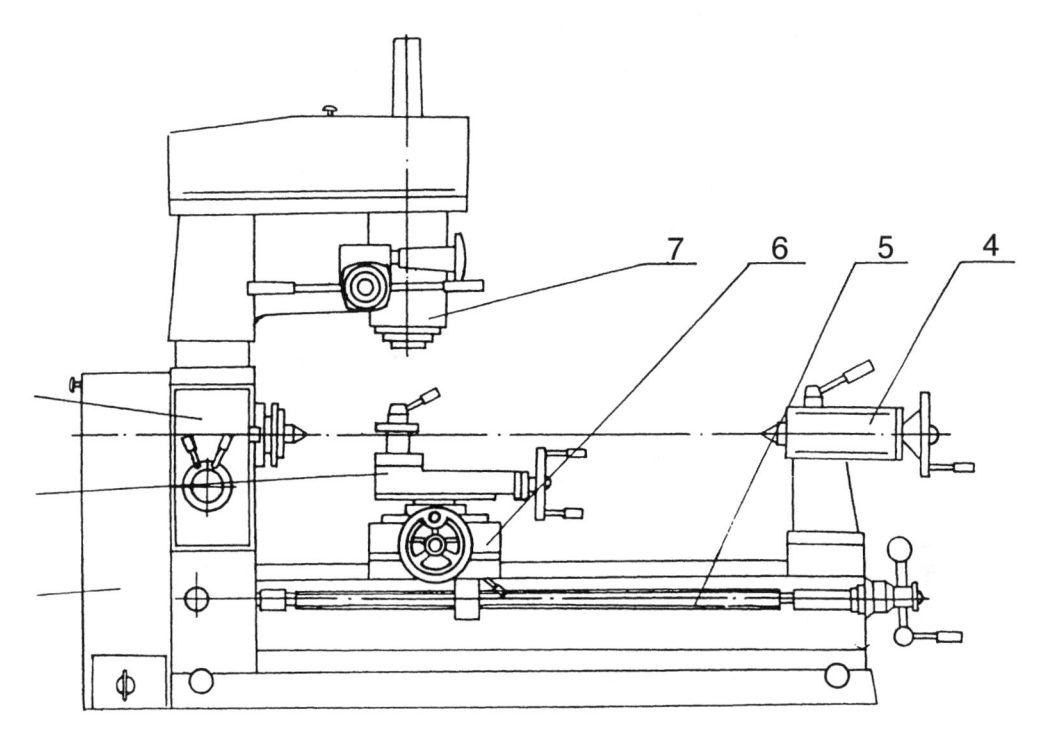

Figure 6.2 Multipurpose machine tool ideal for model machine shop. (Courtesy of Smithy Corporation, The Dalles, OR.)

surfaces, and make plunge cuts. By combining the tilting table with vises, precision cuts at nearly any angle can be made.

6.2.4 Lathe Turning and Boring

The lathe is a versatile tool with a variety of uses in model making. One can use the lathe portion of the multipurpose machine to turn materials as hard as high-carbon steel or as soft as wood. Standard size SAE or metric threads can also be cut on the lathe. One can also use the lathe or mill to bore large-diameter holes.

6.3 BASIC FASTENING AND GLUING TECHNIQUES

6.3.1 Mechanical Fastening

Mechanical fastening occurs usually in the modeling of structural steel and wooden structures. Duplication of exact jointing details of the prototype structure can present severe restrictions at small scales. However, with proper care, accurate duplication of prototype joints can be achieved. A large variety of mechanical fasteners are used in prototype construction. In modeling a small-scale replica model of a mechanically fastened structure, the range of prototype bolts, screws, or nails does not usually permit off-the-shelf purchases of the required small sizes. If large quantities of particular fasteners are required for a project, these may be specially ordered from a manufacturer. The models engineer must exercise considerable care and ingenuity in duplicating details and achieving accurate results.

Some examples of mechanical connections are shown below to illustrate some of the possibilities at small scale. A bolted moment connection of a structural steel frame is shown in Figure 6.3a.

Figure 6.3 Mechanical fasteners in steel bolted joint. (a) Prototype; (b) scaled model.

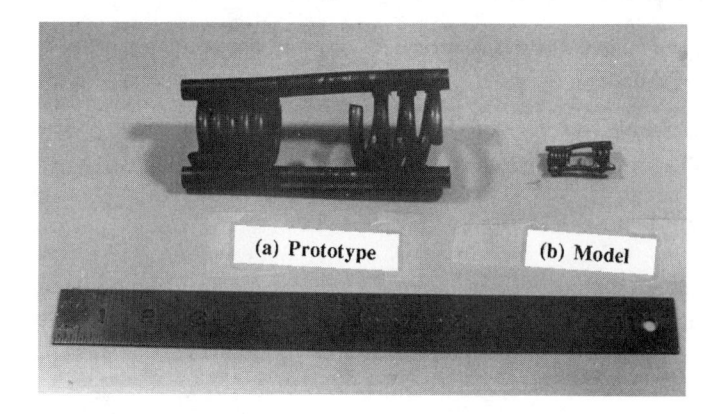

Figure 6.4 Precast concrete large panel mechanical connection. (a) Prototype; (b) model.

Mills et al. (1979) studied this frame at ⅙ scale and produced the connection shown in Figure 6.3b. As can be seen from Figure 6.3 and the test results of their investigation discussed in Chapter 11, the small-scale mechanically fastened joints performed very well. Another example (Woo, 1985) that illustrates the small-scale modeling technique is a mechanically fastened connection drawn from precast concrete large panel (LP) construction and is shown in Figure 6.4. This figure shows the prototype insert on the left and the ³⁄₃₂-scale model counterpart on the right.

6.3.2 Soldering

Soldering is the quickest and easiest method of fastening metal parts such as steel and copper base alloys such as brass and bronze. It involves using appropriate-diameter soldering wire, a heat source such as an electric wire-soldering gun and a borax flux. The edges to be soldered must be held in place firmly and clean of burs, dirt, or grease. Considerable detail can be achieved in soldering joints but the strength is a problem. For example, in steel soldered joints, the full strength of the metal cannot be achieved.

MODEL FABRICATION TECHNIQUES

(a)

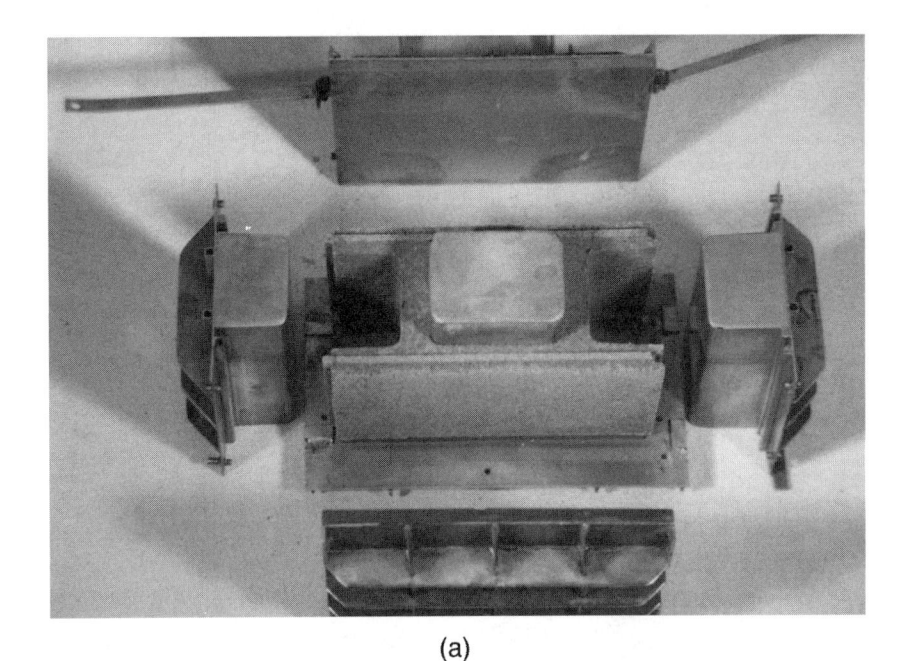

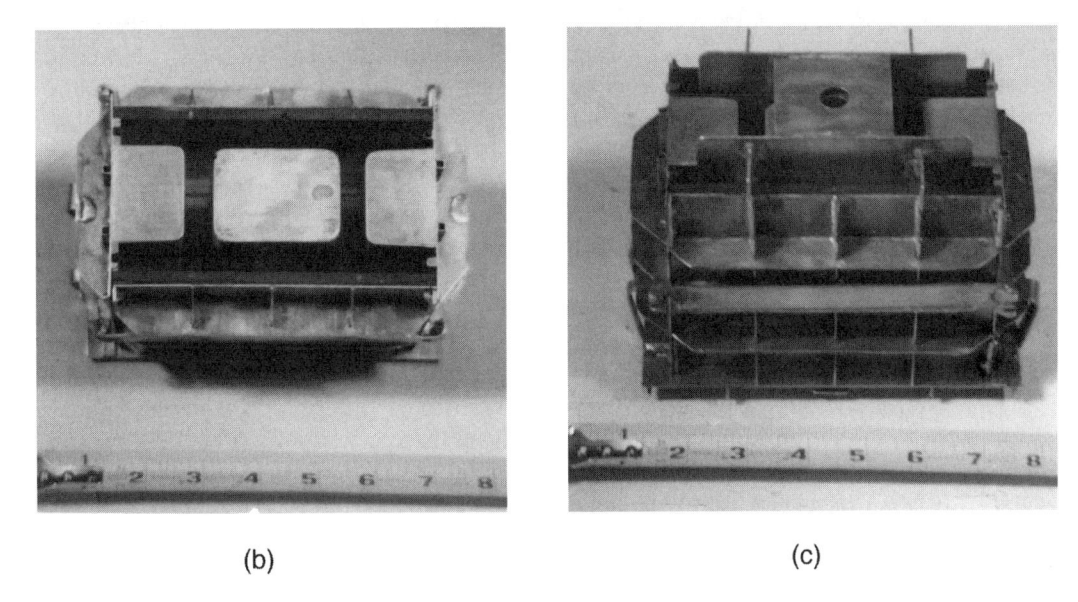

(b)   (c)

Figure 6.5 Soldered brass mold for making ⅓-scale blocks. (a) Exploded view; (b) mold ready for casting; (c) top cover on.

Soldering is a very efficient way to fabricate metal forms or molds, as illustrated in the brass mold for a ⅓-scale interlocking mortarless masonry block shown in Figure 6.5 (Harris and Oh, 1993). The disassembled soldered brass mold is shown in Figure 6.5a, the mold ready for casting without the cover in Figure 6.5b, and the fully assembled mold ready for vibrating the very dry mix in Figure 6.5c.

6.3.3 Spot Welding

Fabrication of small-scale models is greatly facilitated by the use of a spot welder. Such an electronic microwelder manufactured by Raytheon is illustrated in Figure 6.6 to weld small diameter

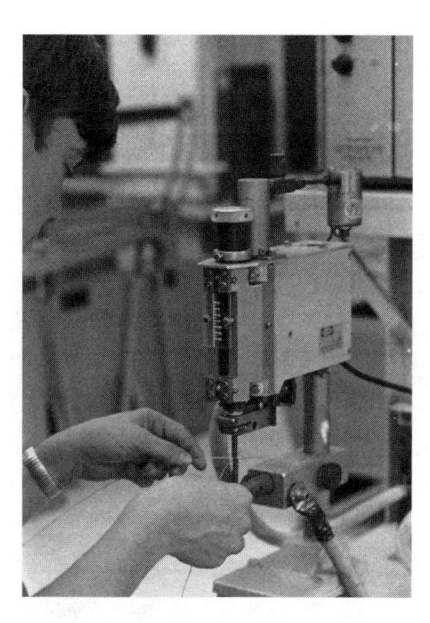

Figure 6.6 Spot welder in use.

wires. In many cases, spot welding considerably improves the accuracy of reinforcement placement and increases the speed with which small-scale models can be fabricated. To prevent reduction in the tensile strength of the model reinforcement, care must be taken to prevent damage caused by excessive weld heat.

6.3.4 Glues and Adhesives

Some models are assembled using glues or adhesives that are normally different from the material being joined. Incompatibility problems may arise from differences that may be elastic, chemical, or viscoelastic in nature. The adhesive must be selected carefully because occasionally the solvent in the adhesive, which must evaporate to set, dissolves certain chemicals in the plastics and the joint is thus weakened. Although the areas of the model stiffened by the adhesive initially accept a greater proportion of the loads, if the adhesive is strongly viscoelastic, with time all load will be transferred to the adjoining members and the joint may possibly distort out of shape. An ideal adhesive must be capable of resisting the high level of stresses and strains that the model is expected to carry. This means that the failure criteria for the adhesive must be identical to those of the members it joins. This condition is seldom fulfilled in practice. Furthermore, the evaporation of the solvent, reduction in process temperature, or polymer network formation normally causes shrinkage in the adhesive, and parasitic stresses may result as a consequence. Therefore, special attention must be paid to the choice of an acceptable adhesive.

6.3.5 Epoxy Resins

Epoxy resins are used to assemble models because of their relative ease of application, their very high strengths, and the relative ease with which their elastic properties can be adjusted. By varying the proportions of the plasticizer and/or the accelerator or hardener (the two components of an epoxy resin), the elastic modulus can be adjusted from 207 to 41,400 MPa, thus permitting the joints to be compatible with the model material stiffness. Another advantage is that the setting of the adhesive does not release any volatiles, and therefore there is a minimal effect on the adjacent member material.

Considerable care must be exercised in cleaning the adherent surfaces before using any adhesives, especially for epoxy resins. Adhesives do not penetrate oil, grime, or oxidized layers, and these must be removed to obtain a good joint.

6.4 CONSTRUCTION OF STRUCTURAL STEEL MODELS

6.4.1 General

Scaled model fabrication is concerned with the replication of prototype geometry and initial state of stress and with the ability to reproduce faithfully all details that may significantly influence the structural behavior. It must be recognized therefore that certain characteristics of steel structures, and structures in general, are easier to simulate than others, and some will be almost impossible to replicate at small scales. In carefully designed and fabricated scale models of steel structures, phenomena such as initiation and propagation of yielding, buckling of members and frames, and local and lateral torsional buckling can be properly simulated. On the other hand, localized problems at connections can hardly ever be studied using small-scale models. For example, at welded beam–column joints where in small-scale models welds usually will be oversized, it would be inappropriate to study such localized characteristics as weld fracture or column flange distortion. These structural effects are better investigated with full-size subassembly tests.

Standard tolerance limits for steel structures should be scaled geometrically in order to provide acceptable tolerances for small-scale model construction and element fabrication. The ability to produce a small-scale model within these acceptable tolerances will often place an upper limit on the usable model scale. It cannot be overemphasized that the ability to achieve close tolerances in scaled model work requires considerable technical skill and experience.

Eight possible techniques can be considered for fabricating small-scale steel beams, columns, and frames. These include soldering, silver soldering, or gluing thin strips of precut sheet metal, hot rolling, die extruding, electron beam welding, resistance welding, and milling bar stock. Of these, the first three can be excluded because of lack of strength if testing of the model beyond yielding is anticipated. Hot rolling and die extruding are prohibitively expensive and electron beam welding and resistance welding have been found lacking in accuracy. Milling bar stock of the proper chemical composition and heat treatment has been shown to be the best method of fabricating scaled steel sections.

6.4.2 Silver Soldering

Alloys having copper as a base are on the whole the most readily machinable metals and the ease with which they can be milled into various shapes may be of great advantage in model making, In addition, the ease of jointing by the relatively simple process of silver soldering adds greatly to their advantage where models limited just to yielding and for educational purposes are considered. The reason for this is that, although annealed "free cutting" phosphor bronze does have a definite yield point, it is very limited compared to that of structural steel. The technique of silver soldering has been found to have adequate strength in joints of milled phosphor bronze sections (Harris et al. 1962) but not in milled steel sections (Litle, 1966). This method of jointing is therefore only recommended for jointing copper base alloys.

The process of silver soldering is extremely simple, with a minimum of equipment necessary. All that is required is a gas torch, silver solder wire, and a borax flux, as shown in Figure 6.7. The technique has been applied to $1/25$-scale models of steel structures (Harris et al., 1962) and some of the fabricating results are illustrated in Figure 6.8. A $1/25$-scale model of a wide-flange beam ready for silver soldering the stiffeners is shown in Figure 6.8a. Typical beam–column joints ready for silver soldering are shown in Figure 6.8b. The basic elements and the silver soldered single-story frame are shown in Figure 6.8c.

Figure 6.7 Silver soldering operation.

6.4.3 Tungsten Inert Gas Welding

The TIG (tungsten inert gas) process was found by Litle and Foster (1968) to give the best welds for model steel joints. It produced smooth, clean welds and good control was achieved. The tungsten electrode is not consumed in this welding process, but rather a filler wire is fed into the arc, melted, and propelled toward the joint being formed. Shielding of the arc is obtained with inert gases which prevent oxidation of the weld. The gas atoms are ionized and carried by the arc from the electrode to the work as shown in Figure 6.9a. Argon and helium are the two inert gases most commonly used, with the argon used alone or in combination with the helium. These two gases provide a smooth arc and hence cause little spatter. Argon operates at a lower voltage than most gases and thus lessens the chance of burnthrough when welding thin material. Argon also has been found to be a good backup gas to prevent oxidation on the back sides of the material being welded (Mills et al., 1979). The equipment required to produce TIG welds include generator, gas tanks, welding torch (Figure 6.9b), electrodes, and safety masks. A typical wide-flange beam with web stiffeners is shown in Figure 6.9c to demonstrate the quality of the welds that can be achieved by the TIG process. As in all small-scale work, it requires experience to produce quality welds.

The fabrication of a beam–column connection by TIG welding is illustrated in Figure 6.9d. The model frame is positioned horizontally with the members clamped in place prior to welding as shown in Figure 6.10a. Even with the use of argon gas to minimize arc voltage, localized distortions due to heat effects are difficult to control. For this reason, clamping and stress relieving of the welded frame are usually required to satisfy scaled tolerances. The clamping mechanism should provide minimum confinement required to reduce distortions during welding (Figures 6.10a and b). The stress relieving is performed at a temperature below any phase change or annealing temperature of the material. The effect of stress-relieving heat treatment is illustrated in Figure 6.10c and d on a planar model frame.

6.5 CONSTRUCTION OF PLASTIC MODELS

6.5.1 Fabrication Considerations

The techniques of cutting, milling, gluing, welding, and drilling are normally used for relatively simple planar elastic models. Acrylic plastics possess the desirable characteristics of low elastic

MODEL FABRICATION TECHNIQUES 289

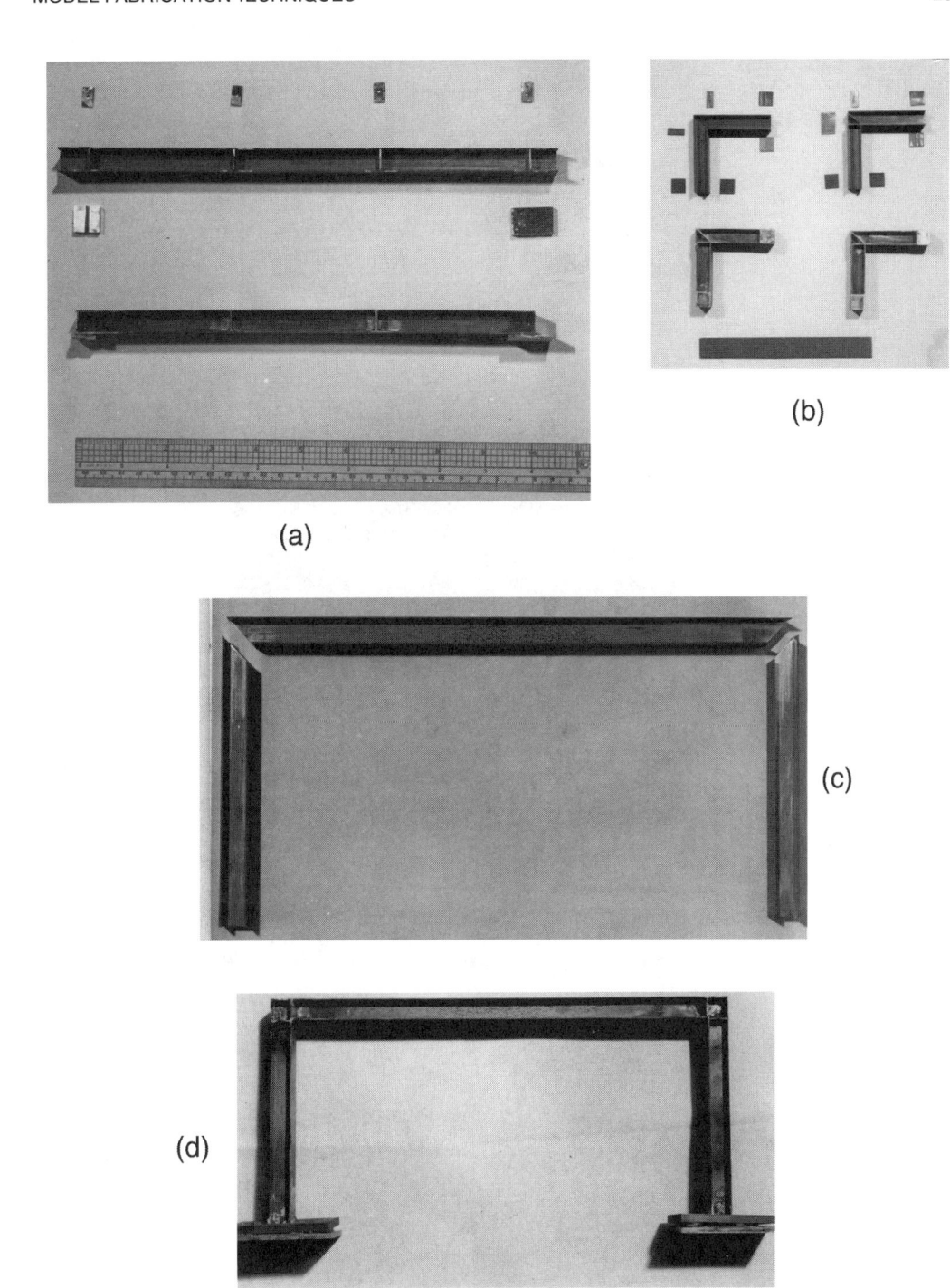

Figure 6.8 Examples of ¹/₂₅-scale model steel structures. (a) Stiffened wide-flange beam; (b) beam–column joints; (c) cut model frame; (d) silver soldered frame.

moduli, extensive linear ranges, good machinability, and relatively low cost and therefore are used very widely for the construction of elastic models. Plexiglas, Perspex, and Lucite, the more commonly used acrylics, are available in the form of rods, tubes, and sheets, with thickness varying from 0.8 to 100 mm. The sheets are available in various sizes up to 1.5×2.5 m.

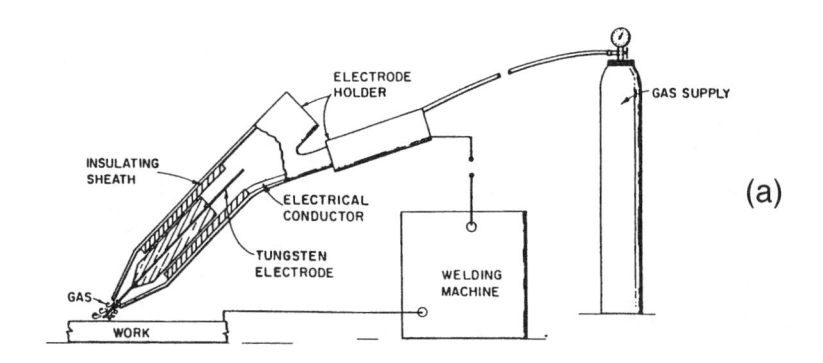

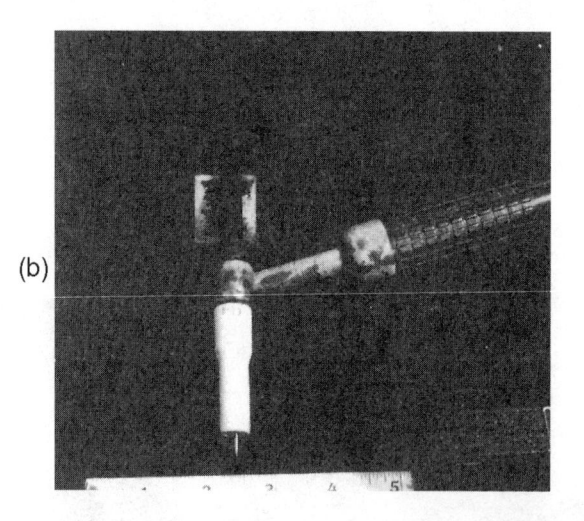

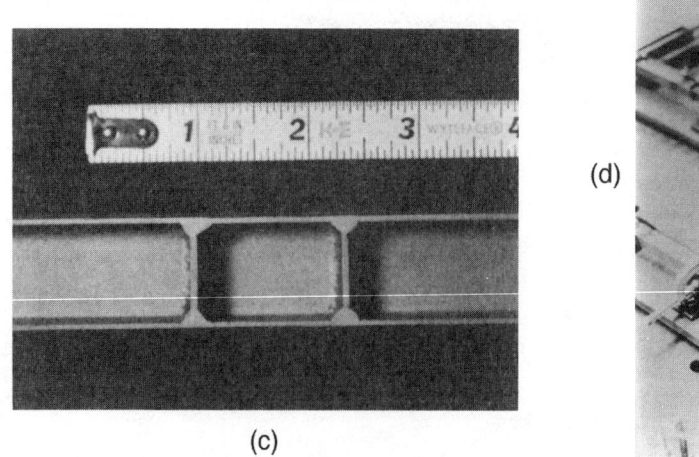

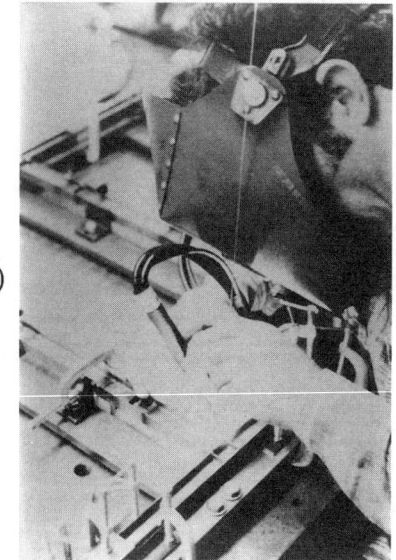

Figure 6.9 Tungsten inert gas (TIG) heliarc welding proess. (a) Schematic; (b) heliarc welder; (c) welded web stiffener; (d) heliarc welding operation.

MODEL FABRICATION TECHNIQUES 291

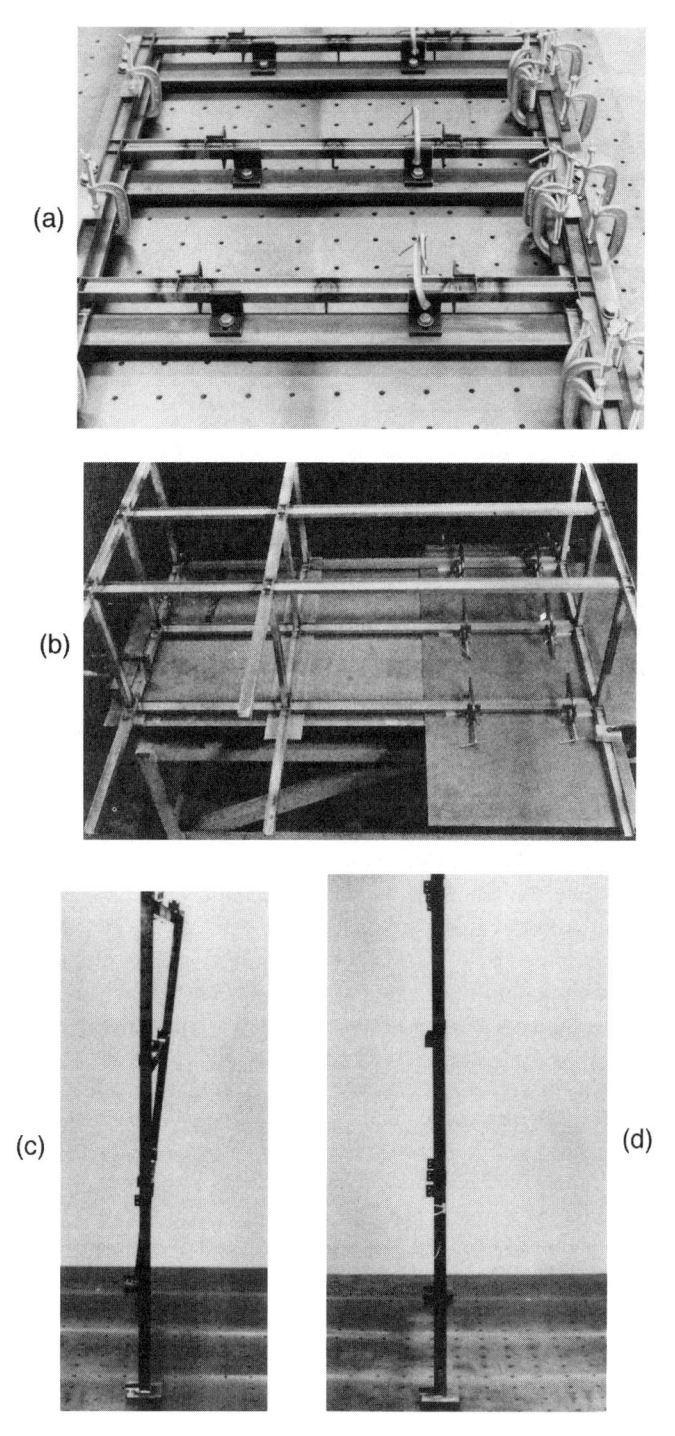

Figure 6.10 Heliarc welded model steel frames. (a) Frame positioned and clamped; (b) fabricating three-dimensional frame; (c) before heat treatment; (d) after heat treatment.

Most acrylic plastics are subject to inception and growth of crazing, which is defined as visible mechanical cracks, to submicroscopic failures that result in noticeable blushing of an otherwise transparent material, or to fine cracks that may extend in a network over or under the surface of or through a plastic (Lever and Rhys, 1978). Crazing is normally dependent on the duration of stress application,

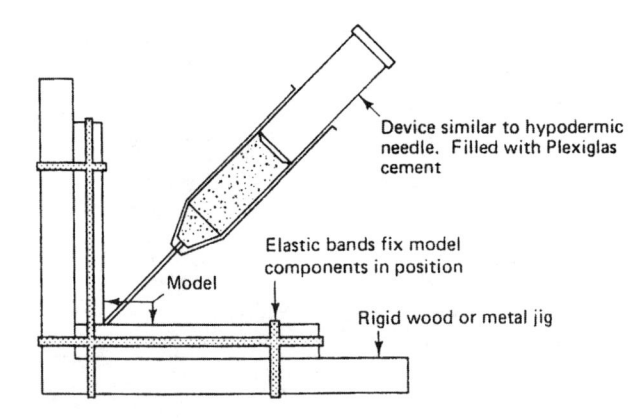

Figure 6.11 Capillary cementing of acrylic plastic models. (After Pahl, P. J. and Soosaar, K., 1964.)

but it can also occur under impact loads. Commonly used acrylic plastics are subject to crazing at stresses of the order of one tenth of their tensile strength. Thus, the stresses during testing must be limited to a maximum of 10.3 MPa at normal room temperatures. Also, proper care must be exercised in machining acrylic plastics using ordinary working tools (Rohm and Haas, 1979).

6.5.2 Capillary Welding

A majority of structural models are constructed from two or more individual components. In any structure, the joints are required to behave in a manner similar to the rest of the model. This may involve consideration of elastic, rheological, and ultimate load behavior. Capillary welding for acrylic plastics is best achieved as follows. Adjoining pieces are held together firmly and solvent is applied to the joint with a hypodermic needle (Figure 6.11). Acetone is usually used as solvent; however, any ketone or ester and some alcohols can also be used. The solvent is drawn into the joint by capillary action; it dissolves the plastic at this joint. With time, the solvent evaporates, leaving a continuous and homogeneous joint that acts monolithically with the rest of the structure. This technique has an advantage in that it requires no application of heat to aid polymerization. Also, the inspection of the joint in transparent plastics is very simple. A good continuous joint is clear and transparent, while a poor one is either cloudy or full of bubbles.

6.5.3 Thermal Forming Processes

Thermal forming techniques are used exclusively with thermoplastics which possess the basic characteristic of glass-transition temperature, i.e., the temperature above which the plastic ceases to be hard and becomes rubberlike and easily deformed. Although the material does not melt, it becomes rubbery, and significant deformations can be applied to conform to any imposed shape with relative ease. It is possible to heat a thermoplastic sheet to just beyond its glass-transition temperature and force it against a prepared mold. The temperature is then reduced below the transition level, which will cause the plastic to cool in the shape of the mold.

It is important to fulfill the fundamental requirements of strength, stiffness, and creep characteristics in any thermoplastic selected for model construction. In addition, the glass-transition temperature must be a good deal above the room temperature to avoid accidental heating by radiators, sunlight, or electrical equipment. Moreover, this temperature must not be too high to impose unnecessarily large heating requirements. There must also be a sufficient spread between the glass-transition and melting temperatures to eliminate sudden transformations from the rubbery state to the molten state as a result of slight overheating. Extreme care must be used in selecting the model material, as not all polymers satisfy the above criteria.

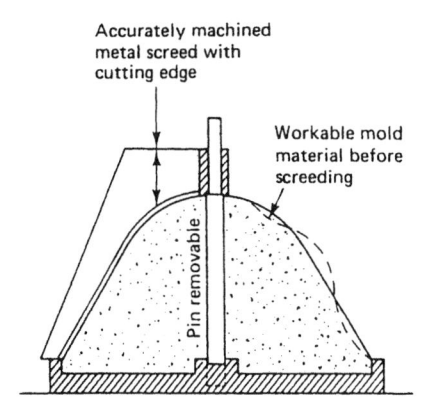

Figure 6.12 Plaster mold formation for surfaces of revolution. (After Pahl, P. J. and Soosaar, K., 1964.)

It is important that the elongation required to form a given model does not exceed the ductility available in the rubbery stage between the glass-transition and melting temperatures. Any overstraining is visibly apparent as tearing and discoloration. The ductility is a function of the composition and structure of the plastic. Polyvinylchloride (PVC) and polymethyl methacrylate are readily available commercially and have been successfully used in thermal forming of structural models. Pahl and Soosaar (1964) suggest that polymethyl methacrylate is successful in drape-forming shells of single curvature; however, in vacuum forming, where stretching of the middle surface is required, the relatively low elongation available is a potential problem, and it is normally manifested in tearing of the sheets. PVC is more commonly used because it is easily formable, has an adequate temperature range, and has sufficient ductility. It is available as calendered sheets that have reasonably isotropic elastic properties. It is also available in several combinations with plasticizers and fillers and has many industrial applications.

6.5.4 Drape or Gravity Forming and Drape Molding of Shell Models

Gravity forming or drape forming of shells of single curvature is a relatively simple model-manufacturing technique. A mold is fabricated from wood, metal, foamed plastics, or plaster to reproduce the desired geometric configuration. It must be noted that all geometric deviations and imperfections are reproduced in the formed shell, and therefore a high degree of precision is required in the fabrication of the mold. Wooden molds are dimensionally unstable on account of changes in humidity and temperature, and machined metal molds can be expensive and time-consuming. Quick-setting plasters can be used to fabricate accurate molds, especially for surfaces of revolution as shown in Figure 6.12. A skilled technician can limit the mold imperfections to approximately 0.025 mm. Molds for hyperbolic paraboloid model structures can be formed by screeding as shown in Figure 6.13.

In drape molding, the plastic sheet and the mold are set in controlled ovens, and the temperature is raised to just beyond the glass-transition temperature of the plastic. The plastic sheet becomes rubbery and unable to support itself and drapes onto the mold. The oven temperature is slowly lowered to eliminate any bowing of the model because of uneven cooling, and the plastic is allowed to cool in the form of the mold.

This technique can be used for forming shells of single curvature and very shallow shells of double curvature. For models with sudden changes in curvature, such as sharp corners or deep depressions, the rubbery plastic will normally not conform to sharp changes in geometry, and an additional mating mold must be used to force the rubbery plastic into the sharp corners (Figure 6.14). Proper care must be taken to ensure that the elongation limit is not exceeded and that the plastic does not tear. Polymethyl methacrylate is well suited for the process of forming shells of single

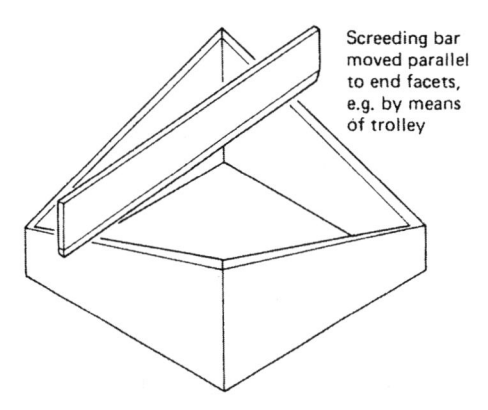

Figure 6.13 Timber mold formation for hyperbolic paraboloids. (After Pahl, P. J. and Soosaar, K., 1964.)

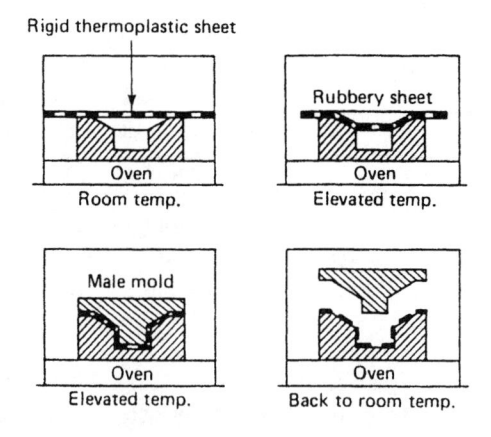

Figure 6.14 Two-mold drape forming of models with sharp corners. (After Pahl, P. J. and Soosaar, K., 1964.)

curvature and very shallow double curvature and is not suitable for forming models with sharp changes in curvature. Other plastics such as PVC are more suitable for such models.

6.5.5 Vacuum Forming

Vacuum forming is used to form doubly curved models since the plastic will not stretch spontaneously in several directions under its own weight. Various techniques can be used, including blowing the rubbery sheet or vacuuming it onto a male or female mold. One technique (Figure 6.15) consists of clamping a plastic sheet to a fixed frame and heating it to its glass-transition temperature using a movable electric coil heater. A platform with a mold is then raised and pushed against the rubbery plastic to form an airtight seal. A vacuum is applied from within the platform, forcing the plastic to conform snugly to the mold. The heating system is usually removed before applying the vacuum, and the model is allowed to cool on the mold. For models with sudden changes in curvature, a companion mold can be used to produce local changes in geometry. Also, it may be necessary to provide additional access for the vacuum by drilling holes in the mold.

Uneven heating of the plastic sheets causes varying cooling rates in different areas of the model, which results in large residual stresses. Careful distribution of heating sources can minimize these effects. On account of the low thermal conductivity of plastics, it is possible to have a significant temperature difference between the two surfaces, which can cause tearing of the plastic sheet during

MODEL FABRICATION TECHNIQUES 295

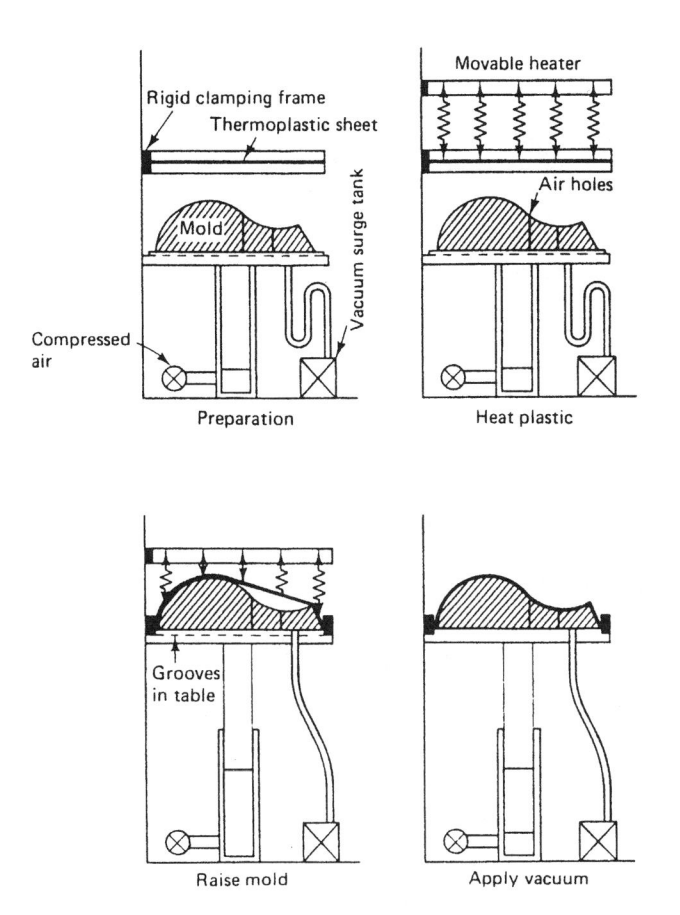

Figure 6.15 Vacuum forming sequence. (After Pahl, P. J. and Soosaar, K., 1964.)

forming. Pahl and Soosaar (1964) recommend that in forming models with thicknesses larger than 1.3 mm both sides of the sheet should be heated. Also, if some parts of a model are steeper than others, these parts require greater elongation during the forming process, which can be achieved by concentrated heating of these areas. A practical solution is to preheat the entire sheet with the regular heating elements, shielding the areas where the extra elongation is not required.

6.5.6 Fabrication Errors in Thermal Forming

The heating of the plastic sheet and its stretching during the forming process result in some manufacturing errors in the model. Thermal forming processes cause shrinkage in plastics upon cooling. This shrinkage has two basic components. Further heating after the primary heating cycle can cause approximately 1% contraction, and the high coefficient of thermal expansion results in another 2 to 3% contraction. In some cases, if the plastic is allowed to cool in contact with the restraining mold, the resulting contraction is usually reduced.

Vacuum forming of doubly curved shells causes the plastic sheet to stretch and become thinner in certain regions of high curvature. Litle (1964) noted that the extent of stretching and thinning is largely governed by the shape, the curvatures, and the sudden changes in the geometry of the structure. He observed very small behavioral differences, only of the order of a few percent, in some models formed using spherical and concave molds. However, structures with very pronounced geometric changes can be expected to exhibit larger deviations in behavior.

6.5.7 Casting of Plastic Models

Some shell structures have varying thickness and thus are not amenable to precise analytical solutions. Gravity or drape and vacuum forming techniques produce shells of almost uniform thickness and cannot be used for shells of variable thickness. Although it is possible to construct a variable-thickness shell by gluing additional layers to a thermally formed model, the results from such models must be interpreted with reservation because of the problems associated with the gluing process (Pahl and Soosaar, 1964).

Variable thicknesses can be achieved by using a pair of matching and interlocking male and female molds and hand finishing with high precision to achieve good accuracy. For some structural models (Stevens, 1959), a technique of building layer upon layer of fiberglass cloth over a single mold and binding with resin is used. The dimensional control for this technique is normally not better than ±0.2 mm. The larger difference between the elastic moduli of the glass fibers and the resin matrix results in considerable anisotropy locally, and therefore strains measured by a short gage can be significantly different from the strains measured by a long gage. Stress concentrations also result from air bubbles trapped in the resin.

6.5.8 Spin Forming of Metal Shells

Structural models of shells can be fabricated commercially by the spin-forming technique using soft metals such as aluminum alloys, copper, and some low-carbon steels. A flat disk of the selected metal is cold-formed into the desired shape, which is a surface of revolution, by spinning it at a very rapid rate while pressing a wooden or metal mold against it. This cold working does not affect the elastic properties appreciably, provided that the spinning is followed by annealing of the model. Also, curvatures in the model must be gradual relative to the material thickness to avoid local overstraining and tearing of the material.

Spin forming requires quite precise and heavy mechanical equipment, and consequently the fabrication, which can be expensive, is usually done commercially.

6.6 CONSTRUCTION OF WOOD AND PAPER MODELS

Construction of wood and paper models requires two distinct categories of tools: light hand tools for the soft balsa wood and paper and regular woodworking tools for the harder wood species. The most useful tools needed for fabricating models are described in Section 6.2 above.

6.6.1 Balsa Wood Models

The set of cutting tools is all that is needed to cut and shape components for a balsa wood model. Rectangular or square members can be cut and shaped to very accurate tolerances as illustrated in Figure 6.1. Figure 6.1a illustrates the sawing of small sections and shows cutting with a sharp blade. Built-up sections such as angles, I beams, wide-flange beams, T sections, tubes, and practically any shape consisting of a series of straight pieces can be glued together to extend the large variety of available shapes (Figure 3.15). The ease and versatility of working with balsa wood is the reason for its popularity and widespread use for elastic models.

The construction of balsa wood model structures is very straightforward and can easily be handled by the novice. Construction of some examples of truss, frame, and plate girder model structures made from balsa wood is described in what follows. First, the members are cut to size and the ends finished to the correct angles using a sharp razor knife or saw blade. Drilling of the ends of the members (if gusset plates are to be used) is done next. Small cardboard gusset plates can be cut to the appropriate size and holes punched with a steel awl. Small wires can be used for

MODEL FABRICATION TECHNIQUES 297

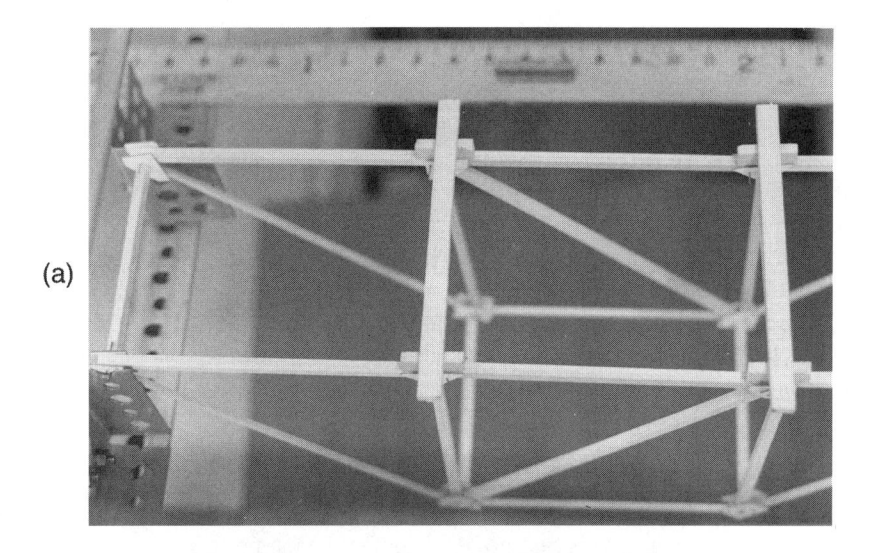

Figure 6.16 Balsa wood model truss joints. (a) Gusseted; (b) glued.

the pins connecting the gusset plates to the balsa wood members. A balsa wood truss bridge constructed in this manner with paper gusset plates is shown in Figure 6.16a.

Gluing of balsa wood joints is easily done using Elmer's glue or epoxy which ensures joints stronger than the wood. An example of a balsa wood truss bridge model with glued joints at the gusset plates is shown in Figure 3.20a simulating a welded steel construction. Welded steel tubes were used to construct the prototype three-dimensional microwave antenna modeled using glued balsa wood (Kopatz, 1982) at ⅑ scale (Figure 3.20c). A close-up of the joints of the ⅑-scale balsa wood model is shown in Figure 6.16b. In this model, a fast curing epoxy was used to speed up the construction process.

A ¹⁄₁₂-scale balsa wood buckling study model of a steel walkway was illustrated in Figure 3.21 (Simonetti, 1982). The fabrication of the walkway model can be characterized by the following steps:

1. Squaring of the balsa wood material.
2. Precutting of the web, flanges, bracing members, web stiffeners, and instrumentation target pieces.
3. Gluing the flanges to the web of the beams and installing the stiffeners and lateral bracing.
4. Installation of temporary bracing to the bottom flanges of the beams.

In step 3, Elmer's carpenter's glue was applied to the edges of the balsa wood web piece and attached to the flange pieces while assembly was in a holding jig, which prevented excessive warping due to drying of the glue. This kept the section uniform throughout the length of the member. After the initial gluing stage, the beams were examined by passing a light through the glued joints. If any cracks existed, these were eliminated by application of more glue using a toothpick. Next to be installed were the tight-fit web stiffeners at the two load and two support points. Again, a bead of glue was applied to the three edges of the stiffener that would attach to the flanges and web of the beam (Figure 3.21). The lateral bracing members were attached to the underside of the compression flanges of the two walkway beams using 5-min epoxy to speed the fabrication process. Temporary strips of balsa wood were attached to the underside of the tension (bottom) flanges to maintain section continuity and protect the walkway system during installation of the instrumentation and testing apparatus. These strips were removed prior to actual testing.

6.6.2 Structural Wood Models

Fabrication of wooden models of prototype structures made from a particular species of wood is carried out using the same species and grade of lumber as the prototype. A series of ¹⁄₁₀-scale wooden roof trusses (Figure 3.22) were fabricated from clear white pine for all members and birch veneer plywood, three-ply, 1.5-mm-thick, plied laminate, for the gusset plates (Heckler, 1980). The clear white pine was milled using a 3-mm upright table saw. The members were cut, squared, and surfaced by means of a hollow-ground planer blade from 25 × 50 mm stock.

When all the members were milled and cut to size, the layout of the truss began. The layout surface was a piece of plywood approximately 200 mm wide by 950 mm long. On the plywood surface, member-edge lines were drawn; then blocks of wood were glued on the outside edges of the truss outline to act as stops when gluing commenced. This formed a jig that could be used for constructing the trusses. The chord and web members of each truss were then cut at the approximate intersecting angles and set aside. The gusset plates were cut from the birch veneer plywood sheet according to their dimensions. The chords and webs were placed on the jig and pinned in place. Glue was applied sparingly to the plywood gusset plates and members. The members and gusset plates were then clamped or weighted in position until dry. It is important to mention that the joints on the first side of the truss should not be all glued simultaneously. Since accuracy of the joint intersection is of primary importance, only one or two joints should be set in place at a time. When the first side is complete, the form of the truss is established. The joints on the second side can be glued more than one or two at a time.

6.6.3 Glue Laminated (Glulam) Beams

Models of glulam beams are fabricated in a similar manner to the prototype beams, by gluing thin strips of good-quality wood together. The scale that can be safely used for the model is usually limited by the smallest thickness of individual strips available. A thickness of 3 mm is the thinnest that can be cut with ordinary table saws while maintaining uniformity along the length. Both a resorcinol glue and fast curing epoxy can be used for gluing the various strips. A ¹⁄₁₂-scale tapered glulam beam model made of Douglas fir is shown in Figure 3.23 (Kopatz, 1984). An example to determine the feasible model dimensions of a typical glulam prototype beam is given below.

Example 6.1

Given a double-tapered glulam beam 9.14 m long, 130.2 mm wide, and 1066.4 mm in depth at the center (304.8 mm in depth at the end) having the configuration shown in Figure 3.23c, determine the smallest possible scale which will still ensure accurate reproduction of the individual glued lamina which in the prototype are 38.1 mm thick. The static behavior of the beam under concentrated load can be expressed as a function of the following form:

$$F(Q, \sigma, \varepsilon, E, L, \delta, A, I) = 0 \tag{6.1}$$

where: Q = concentrated load, N
σ = stress, Pa
ε = strain
E = modulus of elasticity, Pa
L = length, m
A = cross-sectional area, m^2
I = moment of inertia, m^4

A dimensional analysis converts Equation 6.1 to dimensionless products of the form:

$$\delta = L\,(Q/EL^2,\ \sigma/E,\ \varepsilon,\ A/L^2,\ I/L^4) \tag{6.2}$$

The nonlinearity of the prototype material and the desire to test the model to failure dictate that the model material to be used must have the same stress–strain characteristics as the prototype; hence it must be the same species of wood. In this case, the limitation on lamina thickness that one can work with ease in the fabrication of the beam will determine the smallest size model feasible. A 3.2-mm minimum size wood strip for the model lamina determines the model scale ($S_l = 12$). The other dimensions and scale factors of the model are therefore established.

6.6.4 Cartridge or Manila Envelope Paper Models

Cartridge or Manila envelope paper is used in combination with balsa wood to construct small-scale models for demonstrating certain types of structural behavior such as elastic buckling. For example, a series of models of a plate girder can be fabricated with different ratios of flange-to-web stiffness to illustrate the range of behavior. Figure 3.24 illustrates the wood and paper model of a plate girder with different forms of buckling.

6.7 FABRICATION OF CONCRETE MODELS

6.7.1 Reinforced Concrete Models

6.7.1.1 General

Fabrication of reinforcing in small-scale models forms an important part of the modeling process. One might think of bending the wires as is done in prototype bars and tying them with very thin wires as in the case of prototype reinforcement cages. Although this can be done in the usually larger size models, it proves to be extremely difficult for very small-scale structures. Other methods used for positioning the wires during fabrication include welding, the use of epoxies, and soldering. The following sections describe the methods and techniques used at MIT, Cornell, McGill, Stanford, University of California, Berkeley, and Drexel Universities.

6.7.1.2 Fabrication Methods

1. Bending of Reinforcement — Bending of reinforcement must be done carefully so that all the wires are accurately located and placed. The main reinforcement for beams can be easily bent over a template after marking the bending positions accurately. In order to avoid very small hooks at the ends, extra lengths covering the length of hook need to be provided. In bending stirrups a template of Plexiglas or steel is made and all the stirrups are bent to give the desired uniform shape. Commercially available bar benders have also been used to bend the larger size bars and stirrups. Figure 6.17a shows the technique of fabricating model stirrups using a steel mandrel of the correct size to bend the wire to the correct shape.

2. Use of Ties — For the larger-size models, ties made from thin hobby or screen wires have been successfully used to tie the stirrups to the main reinforcement. A fine needle-nose pliers is required to make the tight connections as illustrated in Figure 6.17b. A wire cutter is used to cut the excess material at each joint. Usually not every intersection needs to be tied to the main reinforcement and a rule of thumb to follow is to place enough ties to keep the reinforcement cage free from changing shape while being handled.

3. Welding of Reinforcement — The welding of model reinforcing enables easy fabrication of complicated reinforcing cages in a short time. It also helps to keep better control on forming joints. The Raytheon spot welder (Weldpower-Model: QB) can be used for welded fabrication (Raytheon Company, 1962). Straight wires and small stirrups are welded using the main unit. For curved reinforcing, as in shells, a hand unit (Weldplier-Model: 2-205) can be conveniently used. A complete assembly is shown in Figure 6.17c.

While using a welder, however, proper precaution should be taken to avoid excess of heat input at the joint. This becomes very important for welded cages with closely spaced wires. Tension tests were conducted on wires with cross welds of different spacing to study this effect. The main wire was 1.6 mm in diameter, and the cross wire pieces of 1.4 mm diameter were welded at various spacings of 6, 12, 20, and 25 mm, respectively. These were compared with a specimen with no welds. A reduction of about 15, 8, 5, and 0% in the yield strength of wires was found for the above spacings.

Mirza (1967) tested several coupons of 5 and 8 SWG steel wires with one, two, or three 11, 13, or 16 SWG wire pieces spot-welded across the coupon using a low heat cycle. He compared their yield strength with those of the corresponding unwelded wires and noted that there was no reduction in the strengths of the 5 SWG wires. However, there was a strength decrease between 7 and 10% in the cross-welded 8 SWG wire coupons.

4. Use of Epoxies — There are various epoxies that are useful in reinforcement fabrication. One can select a good metal-to-metal epoxy, with a curing period that varies from a few minutes to a few hours. In beams, main reinforcing wires may be held by end blocks and stirrups positioned as desired. The epoxy is then placed at the stirrup–bar joints using a small steel wire or nail. The disadvantages of this technique are that the connections tend to be larger than those made with the other techniques and that they are time-consuming.

5. Use of Soldering — Soldering of reinforcement cage joints is probably the oldest technique of holding the various steel reinforcements in place before casting. This technique is time-consuming as in the case of epoxy application. Soldered of joints in the reinforcement cages of spherical shells have been carried out at MIT as shown in Figures 11.40 and 11.41.

6.7.1.3 Accuracy of Reinforcement Placement

It is necessary to ensure that the reinforcement is placed in the proper position in the model. Accurate placement is one of the more difficult problems met in fabricating small-scale models. The wires should be straight; they should be held in place with miniature bar chairs (fabricated

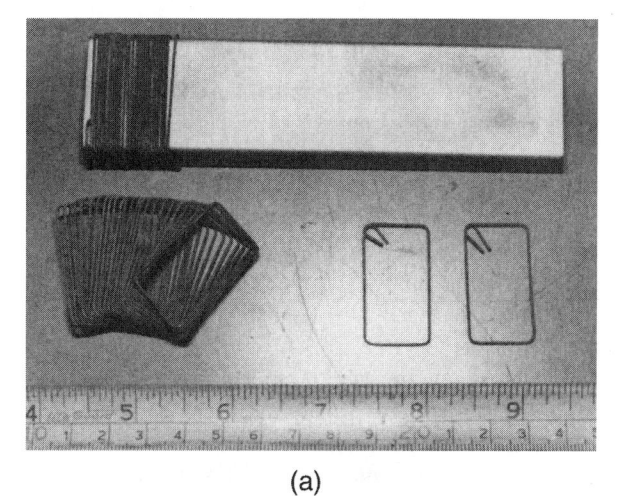

MODEL FABRICATION TECHNIQUES 301

(a)

(b)

(c)

Figure 6.17 Steel reinforcement fabrication techniques. (a) Stirrup fabrication (Courtesy Stanford University); (b) making ties (Courtesy Drexel University); (c) welding (Courtesy Cornell University).

from wire) along the length of the member. Exact positioning of main reinforcement in beams is achieved by passing it (reinforcement) through the end blocks of the forms. Accurate placement is easier to achieve in prestressed beams because the initial tension in the wire holds it in place during casting. The load in the wire also removes any kinks or curvature along its length, and accurate end positioning is sufficient to produce adequate models.

6.7.2 Forms for Casting Reinforced Concrete Models

6.7.2.1 Plexiglas Forms

The casting forms for reinforced concrete models must be nonabsorbent and leak proof so that the water content of the model concrete mix will remain constant from batch to batch. In addition, when many casting repetitions are required, a material that does not change dimensions or warp is needed. For these reasons, extensive use of easily machinable noncorrosive materials such as acrylic plastics and aluminum is made for model forms. The aluminum helps to stiffen the plastic and provide strong end attachments for the reinforcing wires. The added advantage of transparent plastics is that one can determine the quality of the model concrete during the casting operation. An example of a Plexiglas and aluminum form (Harris and Muskivitch, 1977) for precast concrete model panels is shown in Figure 6.18a. Prior to the introduction of the reinforcement in the forms, the forms are covered with a coating of light mineral oil to act as a mold release. The narrow Plexiglas strips on top of the form in Figure 6.18a are holders for short pieces of 6-mm-diameter plastic tubing that when removed from the cast model form the loading holes in the concrete slab.

The forms are clamped to a variable-frequency FMC Corp. vibrating table during the casting operation to ensure uniformity and a minimum of entrapped air in the model concrete (Figure 6.18b). Enough model concrete is mixed during each casting session to fill the prepared forms and the necessary control specimens. After casting, the molds and control cylinders are placed in a moist environment. After 1 day, the cast components are removed from the forms and returned to the moist room until used for assembling a larger model component or for testing.

6.7.2.2 Plywood Forms

When only a few repetitions from the mold are required or the added expense of Plexiglas cannot be justified, the mold can be constructed from marine plywood sheets sanded and covered with several coats of shellac or clear polyurethane. An example of a plywood form for the horizontal casting of a ⅙-scale three-story lightly reinforced model frame (Harris et al., 1993), cast integrally with its fairly massive footing, is shown in Figure 6.19a. The casting operation usually requires several hands (Figure 6.19b) depending on the model size and complexity.

6.7.3 Prestressed Concrete Models

6.7.3.1 General

In prestressed concrete models, the reinforcing technique can be divided into two categories: (1) pretensioning and (2) post-tensioning. As the nature of problems in the two cases is different, they are discussed separately in the following sections.

6.7.3.2 Pretensioning Technique — Single Wires in Beams

An easily used pretensioning technique was developed at Cornell University (Burton, 1963; Chao, 1964) for single wires in beams. A prestressing bed was designed for pretensioning the steel wire as shown in Figure 6.20. The tension is applied through a lever arm, with the other end of

MODEL FABRICATION TECHNIQUES

(a)

(b)

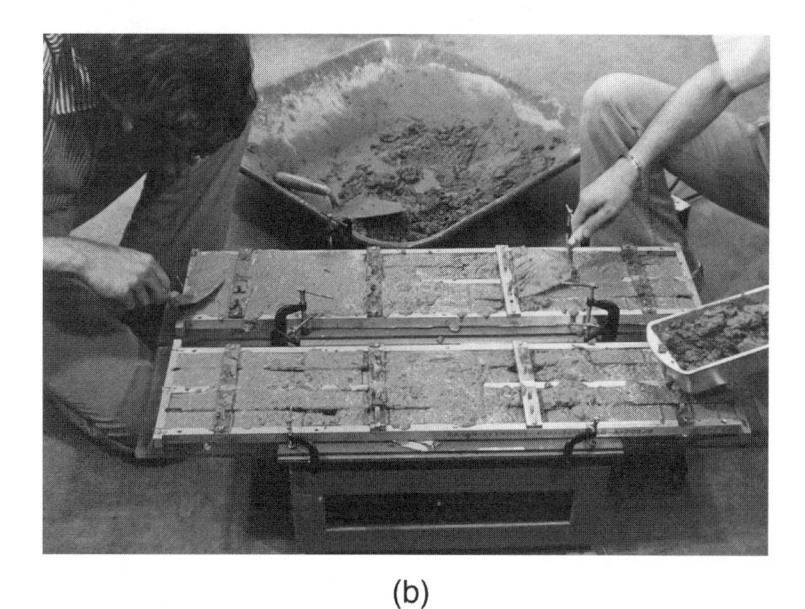

Figure 6.18 Plexiglas forms for casting reinforced concrete models. (a) Plexiglas and aluminum forms; (b) casting on vibrating table (Courtesy of Drexel University).

the wire clamped. Slipping of wire at the tensioning end is prevented by means of a clamp, details of which are shown in Figure 6.20. A detail of the end of the beam is shown in Figure 6.21. The desired load is applied to the lever (with an arm ratio of 4), and the beam is cast. The loads are

Figure 6.19 Plywood forms for casting reinforced concrete models. (a) Reinforcement in place; (b) horizontal casting. (Courtesy of Drexel University).

removed when the concrete has attained the required strength to take the prestress. The main shortcoming of this method is in measuring the prestress force accurately. This could be overcome by placing a dynamometer at the clamped end and measuring the true load in the wire.

Another simple prestressing device consisting of three parts is shown in Figure 6.22. It was developed at McGill University (Pang, 1965; Mirza and McCutcheon, 1974; 1978) to pretension model prestressing wires. The jaw shown in Figure 6.22a was designed so that the button (Figure 6.22c) would just fit into the slot provided at one end. The jaw also contained a long, smooth rod welded to a rectangular steel block designed to slide in the frame shown in Figure 6.22b. Two strain gages were installed 180° apart on this smooth rod and were calibrated to read the applied prestressing force.

The free end of the prestressing wire was flattened by passing it through the hole to develop the ultimate tensile strength of the wire. Other attempts to anchor the wire (without the flattened end) using

MODEL FABRICATION TECHNIQUES

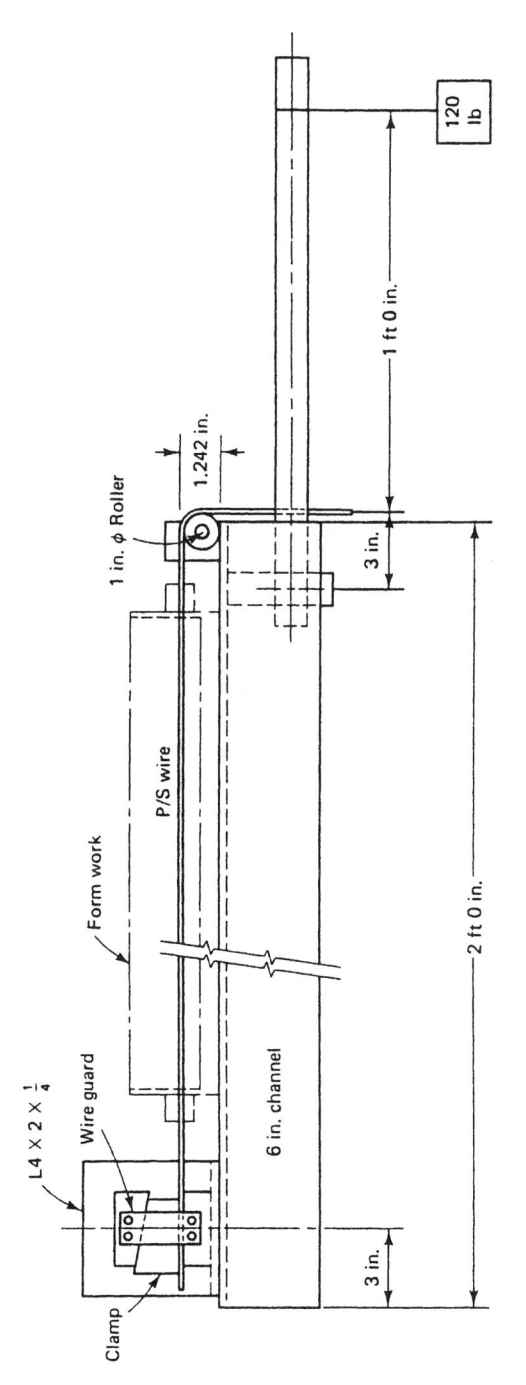

Figure 6.20 Detail of prestressing system.

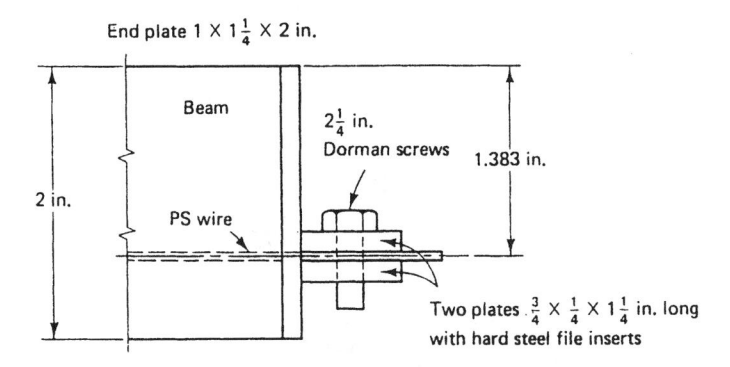

Figure 6.21 End detail of beam.

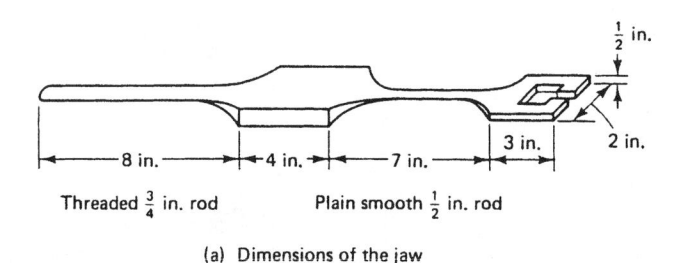

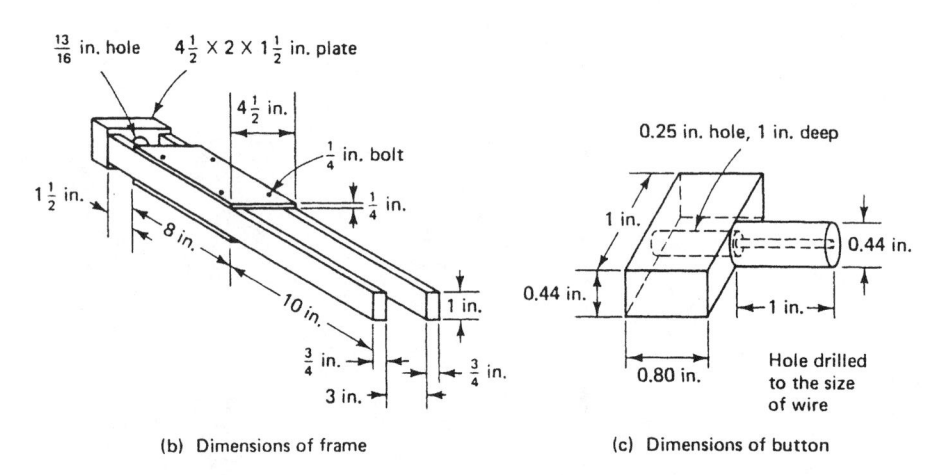

(a) Dimensions of the jaw

(b) Dimensions of frame

(c) Dimensions of button

Figure 6.22 Detail of the prestressing equipment. (From Pang, C. L., M. Eng. thesis, McGill University, Montreal, April, 1965. With permission.)

techniques that involved welding, soldering, and cementing with an epoxy glue were not successful. It is suggested that an epoxy glue or low-temperature soldering be used in the 6-mm-diameter hole of the button to augment the anchorage capacity of the flattened end and to prevent any slip.

The button (with the wire) is slipped into the appropriate end of the jaw, which is then slid into the frame, and a nut is mounted at the free end of the threaded rod. The desired prestressing force can be applied by tightening this nut and reading the load from the calibrated strain gages on the jaw.

The model prestressing bed consisted of a W10 × 21 with four 175 × 100 × 22 mm unequal leg angles as bulkheads bolted firmly to the top flanges (Figure 6.23). The position of the 175 × 100 × 22 mm angles could be varied to suit the length of the model. Threaded rods, 75 mm long and 10 mm in diameter, were used to allow the wires to pass through a 3-mm-diameter hole along the rod axis, and wires were tensioned to the desired load level, one at a time, using the prestressing

MODEL FABRICATION TECHNIQUES

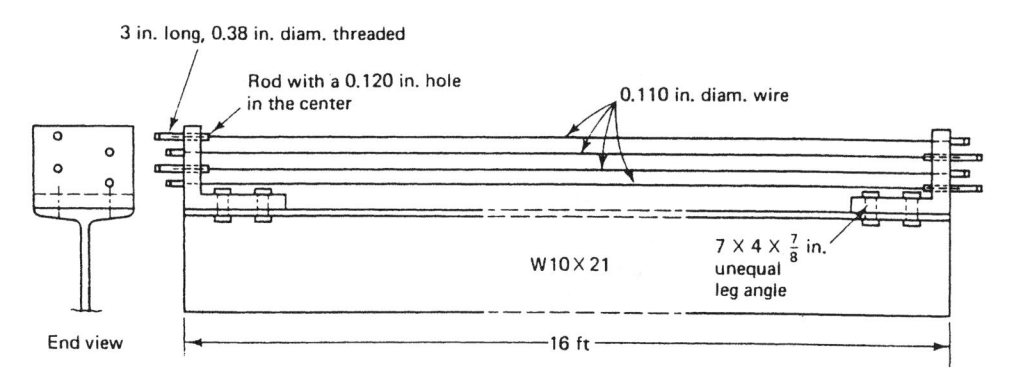

Figure 6.23 Prestressing end systems for the prestressed model. (From Pang, C. L., M. Eng. thesis, McGill University, Montreal, April, 1965. With permission.)

device (Figure 6.22). The appropriate threaded rod on the bulkhead was then adjusted to transfer the prestress to the abutment and release the prestressing device for the next tensioning operation.

6.7.3.3 Pretensioning Technique — Multiple Wires in Beams

A mechanical prestressing frame for model beams with multiple tendons has been used at the University of Illinois (Nebraska and Sur, 1963). As shown in Figure 6.24, the device utilized steel grips capable of accommodating up to ten 1.2-mm stranded cables. Prestressing was applied by moving the jacking plate and monitoring the applied force with a dynamometer consisting of a short length of tubing instrumented with electrical resistance strain gages. The frame is sufficiently long to permit simultaneous casting of three beams.

6.7.3.4 Post-tensioning Technique — Multiple Wires in Slab

As mentioned in Section 5.5.1, Moustafa (1966) used 2.6-mm custom-length bicycle spokes in a prestressed concrete flat slab model study at Cornell University. In order to prevent bond between the wire and the surrounding concrete, the wire was coated with grease just before casting of the slab. As there was free movement of the wire after the concrete was set, this condition of no bond was achieved. Prestressing was accomplished by torquing one nut against a steel bearing plate. The amount of torque required for the desired tension in the wire was determined by precalibration. The torque wrench calibration was done in the testing machine, tensioning the wire and noting the corresponding torque shown by the torque wrench. Thus, for any desired pretension the amount of torque required could be fixed prior to post-tensioning. Transfer of prestress was through steel bearing plates fixed to the slab. The prestress induced in the concrete was checked by electrical strain gages fixed on both surfaces of the slab along the length of prestressing wires.

In an investigation on simply supported, post-tensioned, nongrouted slab models tested under uniformly distributed loads, Kemp (1971) used the Freyssinet monowire system with 7-mm-diameter wires anchored by cone anchorages. Again, to prevent bond between the prestressing wire and the concrete, the wires were coated with grease before casting of the slab concrete. The monowire jack allowed the wire to pass through the center of the piston; each wire was stressed individually by the jack and then anchored by hammering in the radial friction-type wedges. Mild steel bearing plates of size $100 \times 100 \times 6$ mm were used at each anchorage. To reduce the tensile strains resulting from the jacking operation, two 3-mm-diameter mild steel deformed bars were placed behind the plates, along the entire slab edge, and spaced 6 mm from each face of the slab. These prevented spalling of the concrete edges during handling and also, because of their small sizes and locations, had no effect on the load–deformation characteristics of the simply supported slab.

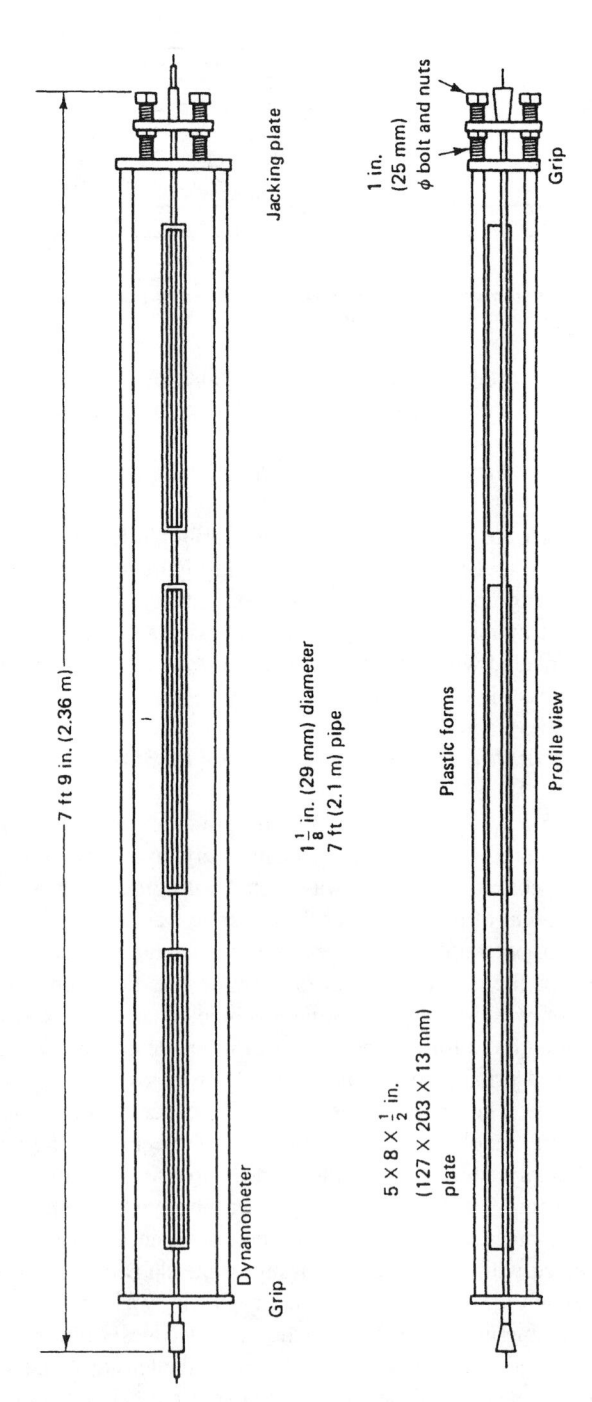

Figure 6.24 Prestressing frame. (Nebraska and Sur, 1963).

MODEL FABRICATION TECHNIQUES

6.8 FABRICATION OF CONCRETE MASONRY MODELS

6.8.1 General

Concrete masonry models can be fabricated using procedures closely resembling prototype masonry construction. In this section, the fabrication of ⅓-scale and ¼-scale concrete masonry blocks is described as well as the construction of small-scale models of block masonry structures. The results of selected reinforced block masonry model studies under static and dynamic loads are discussed in Chapters 10 and 11, respectively.

6.8.2 Drexel University/NCMA Block-Making Machine

A single-unit block-making machine was adapted to produce the ¼-scale blocks described in Section 4.11.2.2 at the National Concrete Masonry Association (NCMA) in the 1960s. The way this was accomplished was to replace the original molding box of the machine that produced a single 200 × 200 × 400 mm block with a molding box that produced four ¼-scale blocks 50 × 50 × 100 mm in size. In the late 1970s this machine was donated to Drexel University where additional molding boxes were made producing three new ¼-scale model blocks and a ⅓-scale block. The procedure used in manufacturing model blocks by the Drexel block-making machine in shown schematically in Figure 6.25. A close-up of the mold box producing the ⅓-scale model blocks is shown in Figure 6.26a. The five ⅓-scale model units extracted from the mold box are seen in Figure 6.26b.

6.8.3 New Drexel Model Block-Making Machine

In order to manufacture the ⅓-scale interlocking mortarless model blocks described in Section 4.11.2.2, a new block-making machine was developed at Drexel University. Unlike the NCMA model block making machine, which uses a plunger system to push the core of the mold (Figure 6.27), the new machine is modeled after the Besser block-making process also shown schematically in the same figure. The laboratory prototype model block-making machine is shown in operation in Figure 6.27a and the extracted single ⅓-scale WHD block fabricated in Figure 6.27b.

6.8.4 Building Model Masonry Components and Assemblies

A vertical casting technique (Becica and Harris, 1977) has been developed for ¼-scale concrete masonry work. Unfortunately, the accuracy of this method is greatly dependent on workmanship. To eliminate as much as possible the workmanship-related errors from the model, it was decided to have an experienced and qualified mason construct all model specimens at Drexel University. The techniques used in assembling all model masonry specimens were similar to the ones used in the field. A special steel frame (Figure 6.28a) is used to aid the mason in the construction of the model specimens. The frame provides vertical alignment and stability of the specimens during the construction process. The construction procedures are simple and straightforward. The model units are placed flush against the upright side of the frame as shown in Figure 6.28a. Mortar is then troweled all around the top of the outer shell of the unit. A unit is then placed with firm pressure on top of the mortar, causing excess mortar to squeeze out of the joints. Gentle tapping with the wooden end of the trowel is used to consolidate the joints further to the required 2.4 mm thickness and to ensure horizontal laying of the units. The final horizontal and vertical alignment is established by use of a spirit level. Excess mortar around the joints is cleaned from the sides and cores during construction. Once the specimen is cast, it is removed from the casting frame and the joints are flushed from all sides. A wallette ready to be removed from the frame is shown in Figure 6.28b. Typically, block masonry models are air-cured in the laboratory at 20°C and 50% humidity (Figure 6.28c). Control specimens are constructed along with the model masonry and are air-cured in the laboratory under the same environmental conditions.

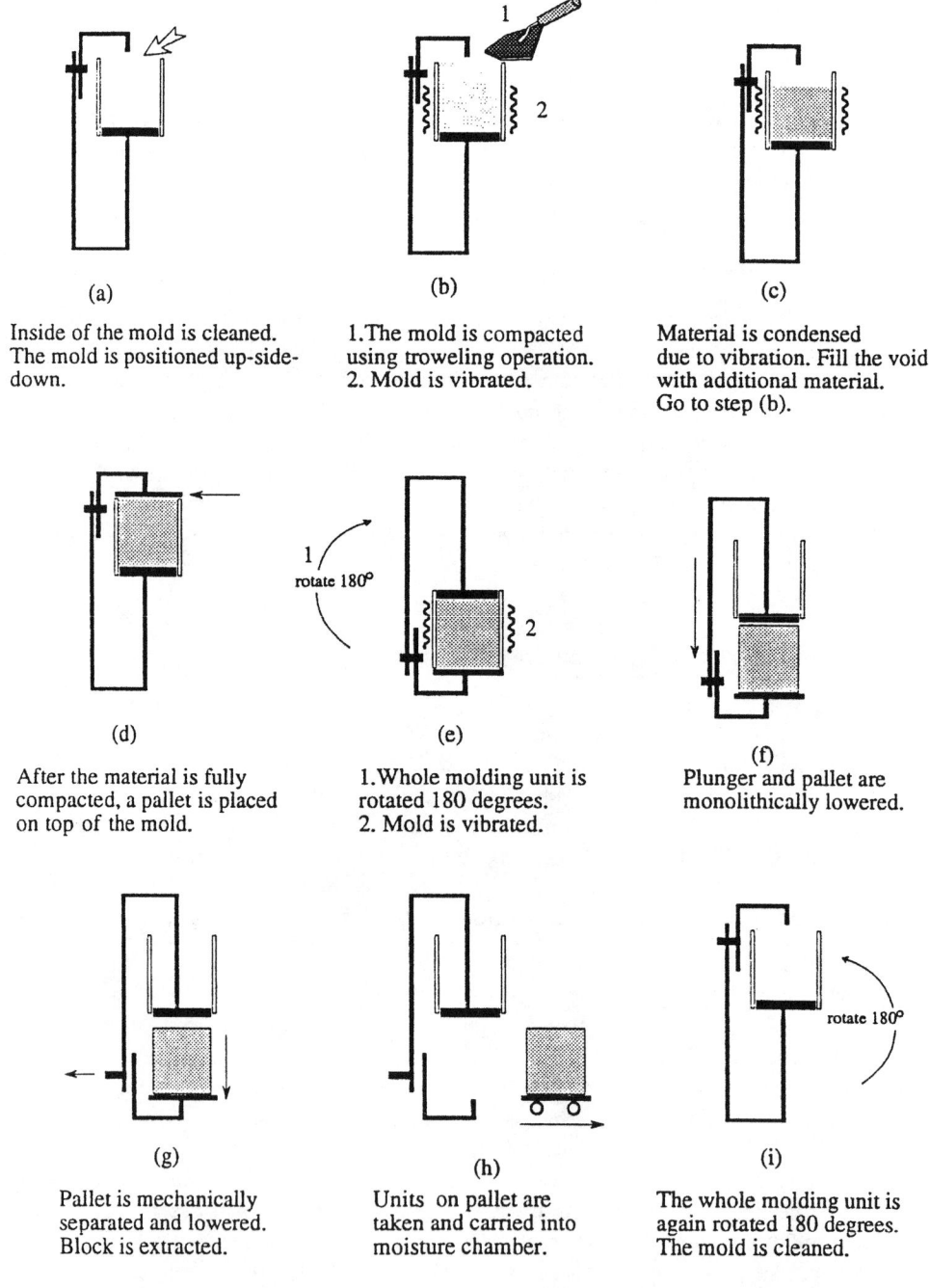

(a) Inside of the mold is cleaned. The mold is positioned up-side-down.

(b) 1. The mold is compacted using troweling operation. 2. Mold is vibrated.

(c) Material is condensed due to vibration. Fill the void with additional material. Go to step (b).

(d) After the material is fully compacted, a pallet is placed on top of the mold.

(e) 1. Whole molding unit is rotated 180 degrees. 2. Mold is vibrated.

(f) Plunger and pallet are monolithically lowered.

(g) Pallet is mechanically separated and lowered. Block is extracted.

(h) Units on pallet are taken and carried into moisture chamber.

(i) The whole molding unit is again rotated 180 degrees. The mold is cleaned.

Figure 6.25 Schematic of Drexel model block-making machine operation. (From Oh, 1994.)

If grouting of the masonry models is required, this is carried out 24 h after construction. The grout is poured into the cavities and is well puddled with a long rod (Figure 6.28d). It is important to reconsolidate the grout to provide good compaction and good bond to the masonry. Also, to increase workability and bonding to the model block, an appropriate admixture can be added to the grout.

The construction of infilled frames requires multiple steps (Ballouz, 1993; Harris et al., 1993). After casting of the ⅙-scale model planar frame (Figure 6.29), the model masonry is constructed

MODEL FABRICATION TECHNIQUES 311

(a)

(b)

Figure 6.26 Drexel University model block-making machine. (a) Tilting of the mold box; (b) mold box with five $1/3$-scale units.

by the same experienced and qualified mason who has constructed all masonry structures at Drexel. The bare frame before application of the infill is shown in Figure 6.29a. The infill is constructed from the bottom up starting with the first story (Figure 6.29b). Construction of the second-story level is shown in Figure 6.29c, and the finished infilled frame prepared for quasi-static seismic load testing is shown in Figure 6.29d.

Consistency and quality of the model masonry construction is an important factor in the success of a masonry model study. In most areas, the local masonry associations working with the NCMA can help in the location of experienced and qualified masons who can perform small-scale masonry model work and whose labor may even be subsidized by the local masonry contractors as part of the continuing interest of the masonry industry in education and research.

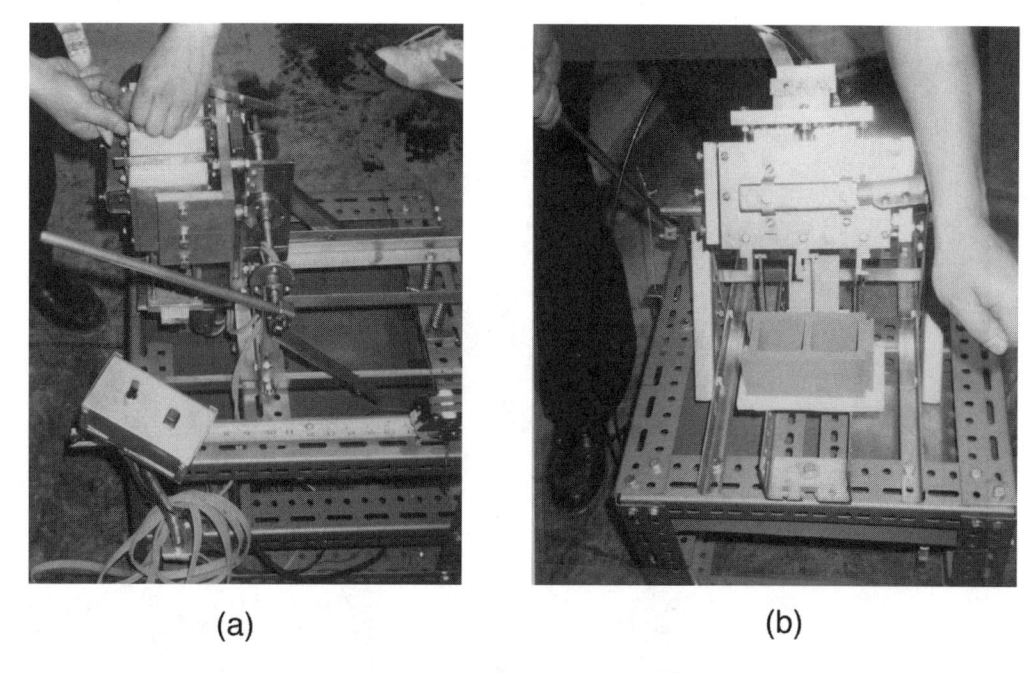

Figure 6.27 Prototype of new Drexel model block-molding machine. (a) Mold with a block-carrying plate clamped. (b) Mold extracting a WHD block.

6.9 SUMMARY

The basic processes and techniques for fabricating small-scale structural models are presented along with basic cutting, shaping, and machining operations of small-scale models. Techniques of jointing the various structural materials used for models are discussed, and the fabrication of small-scale models of structural steel structures are also presented. Elastic models of structures made of plastic, wooden, and paper materials are illustrated, as well as construction techniques for ultimate-strength models of wooden structures. Construction of small-scale models of reinforced and prestressed concrete structures are discussed. Fabrication techniques for small-scale model concrete masonry blocks and the construction of model concrete masonry structures are presented.

PROBLEMS

6.1 A model made of balsa wood rectangular members and thin cardboard or thin balsa wood sheets is to be constructed to demonstrate the principle of composite action of a slab bridge. Design such a model stating your choice of materials and the method you would use to transfer the shear stresses between the slab and the beams. Now go to the laboratory and demonstrate this principle by testing a simple model that you have designed.

6.2 A model of an aluminum riveted joint in single shear is shown below. The model is to be tested to demonstrate the two most important modes of failure: shear failure of the rivet and bearing (crushing) failure of the sheet. You are asked to design this model and test it in a well-equipped structural models laboratory. Perform a dimensional analysis as a guide to enable you to determine the most important variables influencing the strength of the joint. Use the dimensions on the drawing and assume the ultimate tensile strength of the aluminum plates is σ_u, the plate thickness is t, and the joint failure stress is σ_f.

Answer the following:

MODEL FABRICATION TECHNIQUES 313

(a)

(b)

(c)

(d)

Figure 6.28 Fabrication technique for model masonry structures. (a) Laying blocks against steel frame; (b) finishing model wallette; (c) ⅓-scale model masonry conponents; (d) grouting the model block cells (Courtesy of Drexel University).

    a. What materials would you use for the plates and the rivets?
    b. What are the limitations on constructing such a model?
    c. If the plates to be riveted are only 5 mm thick, how big would the scale be?

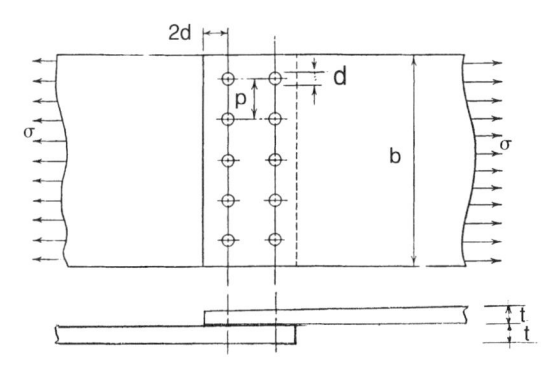

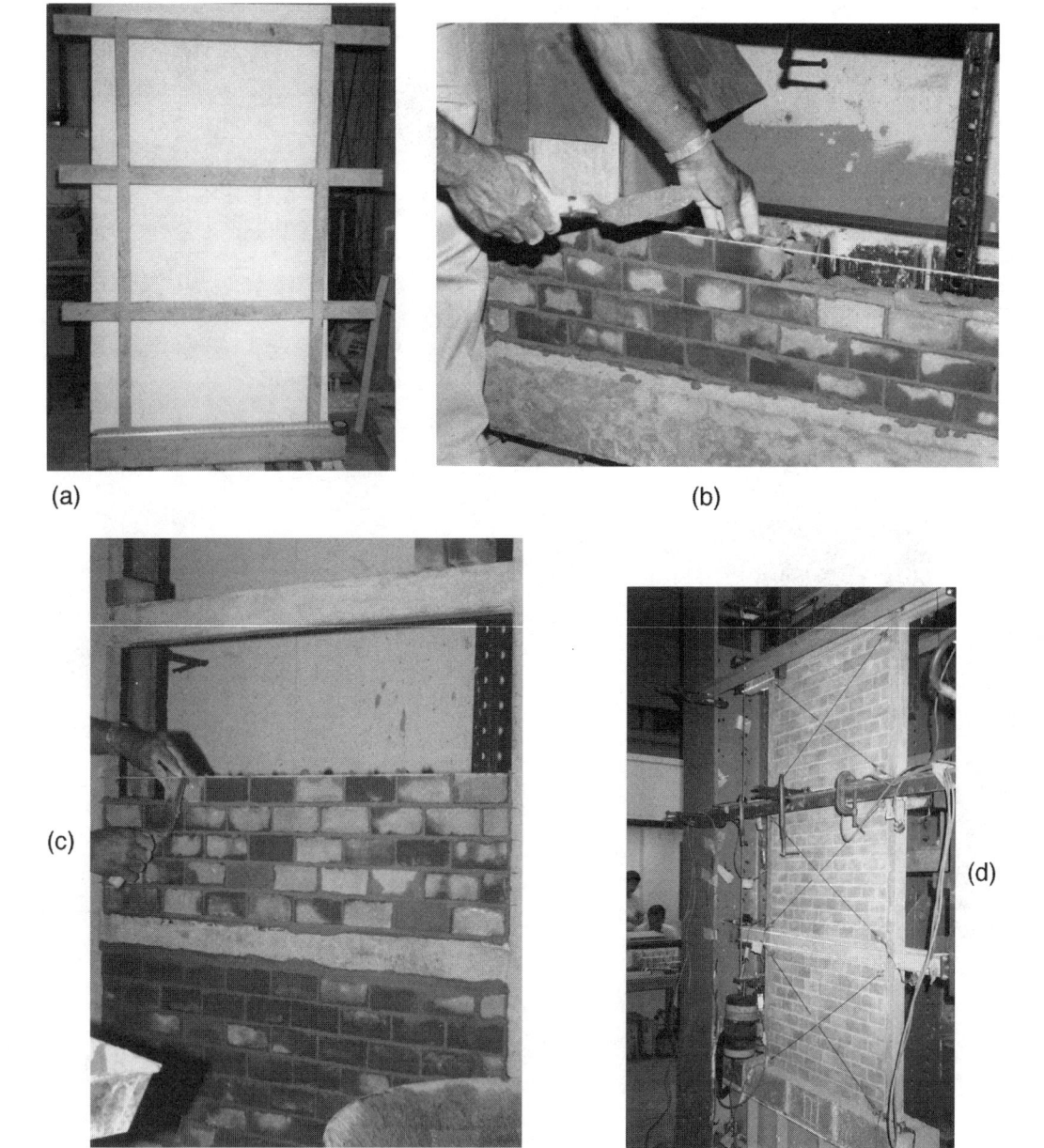

Figure 6.29 Building ⅙-scale R/C (reinforced concrete) block masonry infilled frames. (a) Bare three-story frame; (b) infilling first story; (c) infilling second story; (d) completed infilled frame. (Courtesy of Drexel University).

6.3 In casting model concrete it is imperative to use nonabsobent materials for the forms such as the various plastics and noncorrosive metals. However, oftentimes we must settle for the use of wooden forms. Investigate the different methods of waterproofing wooden forms and determine the number of unrestricted uses that each method can produce in different types of wood.

6.4 You are asked to do a stress and deflection analysis of a four-section hyperbolic paraboloid roof structure by means of model analysis. The prototype structure spans 80 ft and is built of 3000-psi concrete. It is 5 in. thick with edge beams 18 in. deep and 9 in. wide.

The model is to be no more than 3 ft square and is to be built of Plexiglas ($E = 450{,}000$ psi). The critical prototype loading of snow plus dead load is to be simulated by a series of uniformly spaced concentrated loads.

Design the model and determine the required model loads for a 100% overload on the prototype (basic snow load may be taken at 40 lb/ft^2).

Pay particular attention to proper simulation of the deadweight effects, including the edge members. That is, use similitude to establish the required model loads to simulate dead load properly, and specify clearly what the intensity of model loads must be.

Plexiglas sheets are available in thicknesses of 0.06, 0.08, 0.10, 0.125, 0.15, or 3/16 in. for this particular project.

Specify all scale factors, including prediction equations for deflections, stresses, and strains. You may assume $E = 3{,}000{,}000$ psi for the concrete and specific weight = 150 lb/ft^3.

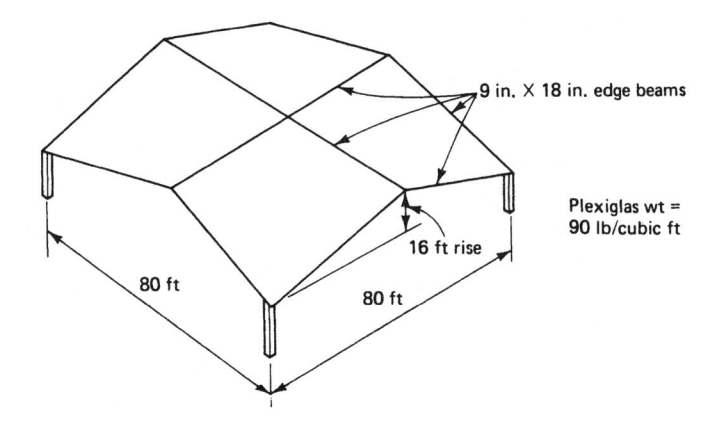

[Figure: Roof structure with 9 in. × 18 in. edge beams, Plexiglas wt = 90 lb/cubic ft, 16 ft rise, 80 ft × 80 ft plan]

6.5 The model of Problem 6.3 is made of plastic. Why do we use plastic so often as a model material? What are the main advantages and disadvantages of using plastic for elastic models? How is the creep problem handled with respect to application of loads? (Your discussion should be complete but concise.)

REFERENCES

Ballouz, G. R. (1993). Testing of lightly reinforced concrete frames infilled with block masonry under lateral loads, M.Sc. thesis, Department of Civil and Architectural Engineering, Drexel University, Philadelphia, PA, June.

Burton, K. T. (1963). A Technique Developed to Study the Ultimate Strength of P/C Structures by the Use of Small scale Models, M.Sc. thesis, Cornell University, Ithaca, NY.

Chao, N. D. (1964). Ultimate Flexural Strength of Prestressed Concrete Beams by Small Scale Models, unpublished M.Sc. thesis, Cornell University, Ithaca, NY.

Cherry, S., Chow, P. Y., and Austin, W. J. (1952). Experimental Studies of Columns, Progress Report No. 4, Civil Engineering Studies Structural Research Series No. 34, University of Illinois Engineering Experiment Station, Urbana, 1 July.

Cowan, H. J. et al. (1968). *Models in Architecture,* Elsevier, New York, 228 pp.

Dukakis, C. (1980). A study of the structural behavior of a model wide flange beam under two-point load, *Drexel Tech. J.,* 43(1), Drexel University, Philadelphia, PA, Fall.

Harris, H. G. and Muskivitch, J. C. (1977). Study of joints and sub-assemblies — validation of the small-scale direct modeling techniques, in *Nature and Mechanism of Progressive Collapse in Industrialized Buildings,* Report 1 to U.S. Department of Housing and Urban Development, Department of Civil Engineering, Drexel University, Philadephia, October.

Harris, H. G., Pahl, P. J., and Sharma, S. D. (1962). Dynamic Studies of Structures by Means of Models, Technical Report to the Defense Atomic Support Agency, Massachusetts Institute of Technology, Department of Civil Engineering, Cambridge, September.

Harris, H. G., Ballouz, G. R., and Kopatz, K. W. (1993). Preliminary studies in seismic retrofitting of lightly reinforced concrete frames using masonry infills, in *Proceedings of the Sixth North American Masonry Conference,* June 6–9, Vol.1, pp. 383–395.

Heckler, G. F. (1980). The Analysis, Fabrication, and Testing of a Single Fink Truss, a Modified Single Fink Truss, and a Raised Collar Truss and Similitude Study, Unpublished report for graduate course: Model Analysis of Structures, Department of Civil and Architectural Engineering, Drexel University, Philadelphia, January.

Kabe, A. M. and Rea, D. (1980). Earthquake-Induced Inelastic Deformations in Small Shear-Type Steel Structures, Report UCLA-ENG-8036, School of Engineering and Applied Science, University of California, Los Angeles, July.

Kemp, G. (1971). Simply Supported, Two Way Prestressed Concrete Slabs under Uniform Load, M.Eng. thesis, Structural Concrete Series No. 71-4, McGill University, Montreal, August.

Kopatz, K. W. (1982). Deflection investigation of the 30/20 GHz antenna using a wooden model, Term project, Course G218: Model Analysis of Structures, Drexel University, Philadelphia, May 5.

Kopatz, K. W. (1984). Model analysis of double-tapered glulam wooden beams, Term project, Course G280: Experimental Analysis of Nonlinear Structures, Drexel University, Philadelphia, September 26.

Lever, A. E. and Rhys, J. A. (1968). *The Properties and Testing of Plastic Materials,* International Scientific Series, CRC Press, Cleveland, OH, 445.

Litle, W. A. (1964). *Reliability of Shell Buckling Predictions,* Research Monograph No. 25, Massachusetts Institute of Technology, Cambridge, 149.

Litle, W. A. (1966). Small scale models for steel frameworks, *AISC Eng. J.,* American Institute of Steel Construction, Inc., New York, July.

Litle, W. A. and Foster, D. C. (1966). Fabrication Techniques for Small Scale Steel Models, Department of Civil Engineering, Massachusetts Institute of Technology, Cambridge, October.

Litle, W. A., Falcone, P. A., and Reimer, R. B. (1968). Ultimate strength behavior of small-scale steel frameworks, *Steel Research for Construction,* American Iron and Steel Institute, New York, April.

Mills, R. S., Krawinkler, H., and Gere, J. M. (1979). Model Tests on Earthquake Simulators Development and Implementation of Experimental Procedures, Report No. 39, The John A. Blume Earthquake Engineering Center, Department of Civil Engineering, Stanford University, Stanford, CA, June, 272 pp.

Mirza, M. S. (1967). An Investigation of Combined Stresses in Reinforced Concrete Beams, Ph.D. thesis, Department of Civil Engineering and Applied Mechanics, McGill University, Montreal, March.

Mirza, M. S. and McCutcheon, J. O. (1971). Direct models of prestressed concrete beams on bending and shear, *Build Int.,* (London) 7(2), March, 99–125, 1974. Reprinted in the *Proceedings of the Joint Institution of Structural Engineers/Building Research Establishment Seminar on Reinforced and Prestressed Microconcrete Models,* Garston, England, May.

Moustafa, S. E. (1966). A Small Scale Model Study of a Prestressed Concrete Slab, unpublished M.Sc. thesis, Cornell University, Ithaca, NY.

Oh, K. H. (1994). Development and Investigation of Failure Mechanism of Interlocking Mortarless Block Masonry Systems, Ph. D. thesis, Department of Civil and Architectural Engineering, Drexel University, Philadelphia, December.

Pahl, P. J. and Soosaar, K. (1964). Structural Models for Architectural and Engineering Education, Report No. R64-3, Department of Civil Engineering, Massachusetts Institute of Technology, Cambridge, 269 pp.

Pang, C. L. (1965). Reliability of Models in the Analysis of Prestressed Concrete Beams in Flexure, M.Eng. thesis, McGill University, Montreal, April.

Nebraska, J. E. and Sur, L. M. (1963). Behavior of Miniature Prestressed Concrete Beams, Report of a project sponsored by the National Science Foundation Undergraduate Research Program, Department of Civil Engineering, University of Illinois, Urbana, IL, June.

Raytheon Company (1962). A Discussion of Module Welding Techniques, presented for General Dynamics Corporation, Raytheon Company, Waltham, MA, August.

Robinson, J. M., Jr. (1970). Model Analysis of Frame Buckling, B.Sc. thesis, Department of Civil Engineering, Drexel University, Philadelphia, December.

Rohm and Haas (1979). Fabrication of Plexiglas, Bulletin No. PL-383-C, Philadelphia.

Simonetti, R. K. (1982). Model Analysis of Steel Walkway System Using Balsa Wood Models, Unpublished report for graduate course: Model Analysis of Structures, Department of Civil and Architectural Engineering, Drexel University, Philadelphia, April 25.

Stevens, L. K. (1959). Investigations on a Model Dome with Arched Cut-Outs, *Mag. Concr. Res.* (London), 11(31), March, 3–14.

Woo, J. (1985). Earthquake Resistance Behavior of Mechanical Connections in Precast Large Panel Buildings, Ph.D. thesis, Department of Civil and Architectural Engineering, Drexel University, Philadelphia, June.

CHAPTER 7

Instrumentation — Principles and Applications

CONTENTS

7.1 General 320
7.2 Quantities to be Measured 320
7.3 Strain Measurements 322
    7.3.1 Mechanical Strain Gages 322
    7.3.2 Electrical Strain Gages 323
        7.3.2.1 Electrical Resistance Strain Gages 323
        7.3.2.2 Other Electrical Strain Gages 326
        7.3.2.3 Choosing Electrical Resistance Strain Gages and Cements 328
    7.3.3 Resistance Strain Gage Circuitry and Applications 332
        7.3.3.1 Circuit Analysis 333
        7.3.3.2 Gage Factor 334
        7.3.3.3 Temperature Compensation in Strain Gage Circuits 335
        7.3.3.4 Self-Temperature Compensating Gages 335
        7.3.3.5 Practical Circuits and Their Applications 337
        7.3.3.6 Gage Configurations 338
        7.3.3.7 Calculation of Stresses from Measured Strains 340
7.4 Displacement Measurements 348
    7.4.1 Mechanical Dial Gages 349
    7.4.2 Linear Variable Differential Transformer (LVDT) 349
    7.4.3 Linear Resistance Potentiometers 350
7.5 Full-Field Strain Measurements and Crack Detection Methods 351
    7.5.1 Brittle Coatings 352
    7.5.2 Photoelastic Coatings 352
    7.5.3 Other Crack Detection Methods 353
7.6 Stress and Force Measurement 353
    7.6.1 Load Cells — Types and Sizes 354
    7.6.2 Embedded Stress Meters and Plugs 357
    7.6.3 Other Measuring Devices 360
7.7 Temperature Measurements 361
7.8 Creep and Shrinkage Characteristics and Moisture Measurements 362
7.9 Data Acquisition and Reduction 364
    7.9.1 Types of Data Recording 364
    7.9.2 Various Data Acquisition Systems 365

7.10　Fiber Optics and Smart Structures 367
　　　7.10.1　Types of Fiber-Optic Sensors for Smart Structures 368
　　　　　7.10.1.1　The Fiber Fabry–Perot Interferometer................................368
　　　　　7.10.1.2　The Fiber Bragg Grating Sensor 368
　　　　　7.10.1.3　The Elliptic-Core Two-Mode Fiber-Optic Sensor..............371
　　　　　7.10.1.4　The Polarimetric Fiber-Optic Sensor..................................371
　　　　　7.10.1.5　The Mach–Zehnder Interferometer.....................................372
　　　7.10.2　Criteria and Selection of Fiber-Optic Strain Sensors.........................372
　　　7.10.3　Integration of Fiber-Optic Sensors into Concrete Structures: Case Studies........373
7.11　Summary 377
Problems...377
References 381

7.1 GENERAL

Earlier chapters dealt with similitude analysis, material properties, and model fabrication techniques, which essentially enable one to conceive and construct a model. The techniques of loading and testing model structures are discussed in Chapters 8 and 10 for static loadings and Chapter 11 for dynamic loadings. Meaningful interpretation of such model tests is not possible unless proper instrumentation is used for measuring the many important quantities related to the behavior of the structure.

The instrumentation process includes careful identification of the quantities to be measured; selection of the appropriate sensors and the necessary auxiliary equipment; installation of the sensors on the completed model; calibration of sensors and checkout of equipment prior to the model test; acquisition of data; and reduction of data into meaningful stresses, forces, and force–deformation relationships. This process can be quite demanding, particularly for the more-sophisticated models involving dynamic response, thermal loading, high internal pressure, and other complex loadings. Instrumentation can be the most time-consuming and expensive part of some modeling studies, and the serious models engineer must know not only how to measure everything in a given experiment, but also how to strike the rather delicate balance of "getting enough data" vs. overinstrumenting and running up the costs unnecessarily.

This chapter deals with the major aspects of instrumentation, including quantities to be measured, means and techniques of measurements, and the underlying theory. Applications to actual case studies are only referenced here and are treated in more detail in Chapters 10 through 12.

7.2 QUANTITIES TO BE MEASURED

As mentioned earlier, the behavior of a structure is reflected in the forces and deformations that result from subjecting it to the different loading conditions. These are measured through the instrumentation on the structure, on the surface, at the boundaries and loading points, and sometimes inside the model. In general, in a reinforced or prestressed concrete structure or its model, the following quantities need measurement:

1. *Strain:* Its distribution across the section under consideration. Strain may be measured in concrete, either by instrumenting the surface *or* by suitably embedding gages inside, or on the steel reinforcement and the prestressing strands. Knowing the stress–strain

characteristics, the stresses associated with these strains in a structure can then be determined.
2. *Deflection:* Its distribution along the structure and its variation with the applied load and magnitude in a structure or a constituent element. Deflection measurements are needed to define the load–deformation characteristics and can be helpful in determining the limits of elastic behavior, curvature, and changes in curvature.
3. *Cracks:* Their locations, patterns, and widths related to the loading. This information is used to determine satisfactory service load conditions and also to obtain the ultimate or limit load stress conditions.
4. *Forces:* Their magnitudes and nature in the concrete or the steel reinforcement, at the boundary supports, and sometimes at loading points. Knowledge of these internal forces, which are in equilibrium with the applied forces, is especially useful in the study of indeterminate structures.
5. *Temperature:* Its distribution within the mass of concrete, where the structure is subjected to differential temperature conditions.
6. *Creep and shrinkage:* Their measurements in a structure subjected to sustained loading. These are similar to item 1 above, but care must be exercised to ensure that the instrumentation is stable over the entire period of measurement.
7. *Properties of materials:* They must be determined in order to translate other measurements (such as strains) into overall structural behavior, and to correlate test results with theory. Measurement of properties of concrete are particularly important since they are subject to variations from environmental conditions, such as relative humidity and temperature.
8. *Dynamic response:* Various types of responses of a structure when subjected to dynamic loads, e.g., impact, blast, seismic, fatigue, and repeated loadings. Accelerations, velocities, and displacements are measured.

In most other types of models, items 1, 2, 4, 7, and 8 above are the important quantities to be measured. The equipment to measure the above quantities varies from simple hand instruments to the more-sophisticated electronic devices. The former are used manually from point to point, and the latter, although they work much faster, require elaborate setups for monitoring. The readout instruments accompanying these measuring devices also vary from hand-operated to continuous scanning, recording, and monitoring systems. The choice of using one or the other will depend on the type of quantities to be measured, loading, reliability of measurements, and economics.

The outcome of any experimental program depends significantly on the accuracy and reliability of measurements. In the case of small-scale models, the quantities to be measured are much smaller in magnitude, based on the principles of similitude, thus magnifying the error possibility and the associated need for accuracy. To achieve the same level of accuracy as in the prototype tests, accuracy in measurements of model quantities must be achieved, theoretically, to the order of at least the scale factor. For example, if in the case of a prototype beam, the deflection is read to within ±0.25 mm, then the deflection in its $1/10$-scale model should be read within ±0.025 mm to achieve the same order of accuracy and the corresponding reliability of the model test results. More about the accuracy and reliability of modeling will be discussed in Chapter 9.

The foremost measurement in all model testing is that of the strain; the basic reason behind this is that stress and strain are related to each other by a fundamental relation, the modulus of elasticity, for linearly elastic materials. Although strain is a fundamental physical quantity and stress is a derived quantity, more use is made of the word *stress* to express the ability of a material to support applied loads or forces because it is easier for the engineer to visualize in terms of stress rather than in terms of strain. Because of its basic importance, strain measurement should be carried out with the utmost care and accuracy.

7.3 STRAIN MEASUREMENTS

Strain is measured with a strain gage, which in essence is a means of magnifying the change in length over a given length; strain gages are therefore classified according to the type of magnification system they use. To achieve accurate strain measurements, all strain gages must fulfill the following basic requirements, which define the "perfect universal strain gage" according to Perry and Lissner (1962).

1. It must be extremely small and of insignificant mass (inherent need for this exists because strain should be measured at a "point," and dynamic response of the structure may be affected by any additional mass attached to it).
2. It must have a high sensitivity to strain and attach easily to the structure.
3. It must not be susceptible to variation in ambient conditions such as temperature, vibration, or humidity that will be encountered in the test.
4. It must be capable of indicating both static as well as dynamic strains by remote indication and recording.
5. It must be inexpensive and convenient to use.

Unfortunately, if one were to develop the "perfect" strain gage with all of the above characteristics, the cost of such a gage would be astronomical. As a compromise, over the last few decades efforts have been made to achieve many of these properties with emphasis on economy and convenience in use.

The major types of strain gages are

1. Mechanical
2. Optical
3. Electrical

The optical strain gage has become obsolete because of advancements in electrical gages and will not be discussed here. Some techniques using optics have been developed recently; the details may be obtained from Photoelastic, Inc.

7.3.1 Mechanical Strain Gages

The mechanical strain gage in its basic form uses mechanical systems such as levers, gears, or similar means for magnification of strain. Although this appears simple, the magnification from one gear and/or lever to another causes mechanical interaction, such as friction, lost motion, inertia, and flexibility of the parts, and, if not overcome, some of these shortcomings cause reduction in accuracy. In most instances, mechanical strain gages are limited to static measurements of strain, since their size and inertia rule out any reasonable frequency response, which is required in the dynamic applications. In spite of some of the disadvantages of mechanical gages, they are used often, primarily because they are self-contained. The strains to be measured are shown on scales or dials, and no additional equipment is required for readouts. They are reusable, which makes them economical.

The Whittemore gage, manufactured by Baldwin Locomotive Works, has been in use for many years. As shown in Figure 7.1, it is self-contained and consists essentially of two frame members connected together by two elastic hinges, which provides a parallel frictionless motion. Conical points are attached to the frame legs, which are inserted into the attachment holes on the structure; these points define the gage length. The strains are measured with an integral dial indicator.

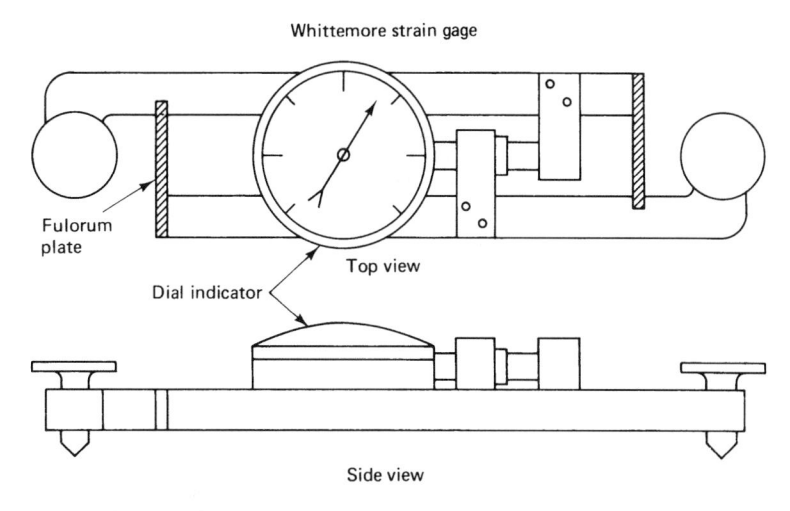

Figure 7.1 Whittemore gage showing its various components.

The main disadvantage of this gage is the potential error induced when the gage is repositioned on the structure for each strain reading. This error is minimized by having the operator that reads the gage develop a consistent technique and also by having the same operator read the same gage throughout a given test. In spite of this disadvantage, the gage is extremely useful for long-term (creep) measurements on concrete members, for measuring distortion in shear panels, and in other similar applications where measurement over a relatively long gage length is permissible.

7.3.2 Electrical Strain Gages

Electrical strain gages use the principle of change in some electrical characteristic of the gage material caused by strain in the structure. The electrical variables commonly used are resistance, capacitance, and inductance. The important advantage of electrical strain gages is the relative ease with which the output can be amplified, recorded, and displayed. Of the various kinds of electrical strain gages, the resistance type is the most commonly used because of its many advantages, and this type will be discussed separately.

7.3.2.1 Electrical Resistance Strain Gages

The resistance type of strain gage functions as a resistance element in an electric circuit. Strain produces a change in the magnitude of the resistance associated with the gage. This change is recorded and related to strain during calibration, after the gage is bonded to the structure.

Although the principle of metallic resistance was observed by Kelvin in 1856, the first use of this principle related to strain measurement was by Carlson and Eaton in 1930. This first gage developed was unbonded, of the metallic type, with a single wire wound over two pins that were embedded in the structure. This gage had appreciable mass, size, and gage length, and consequently it was not used very much as a practical strain gage. Simmons and Ruge in 1938 independently conceived the idea of bonding the wire either directly to the test specimen or to a thin paper backing which was in turn bonded to the specimen.

Since that time, a considerable amount of development has taken place in bonded electrical resistance strain gages, using etched foil circuits as well as wire circuits (Figure 7.2). Today, there are strain gages in a large variety of shapes and sizes, and gage lengths as small as 0.4 mm. These resistance-type gages, such as the SR-4 strain gage (so designated to embody the initials of both

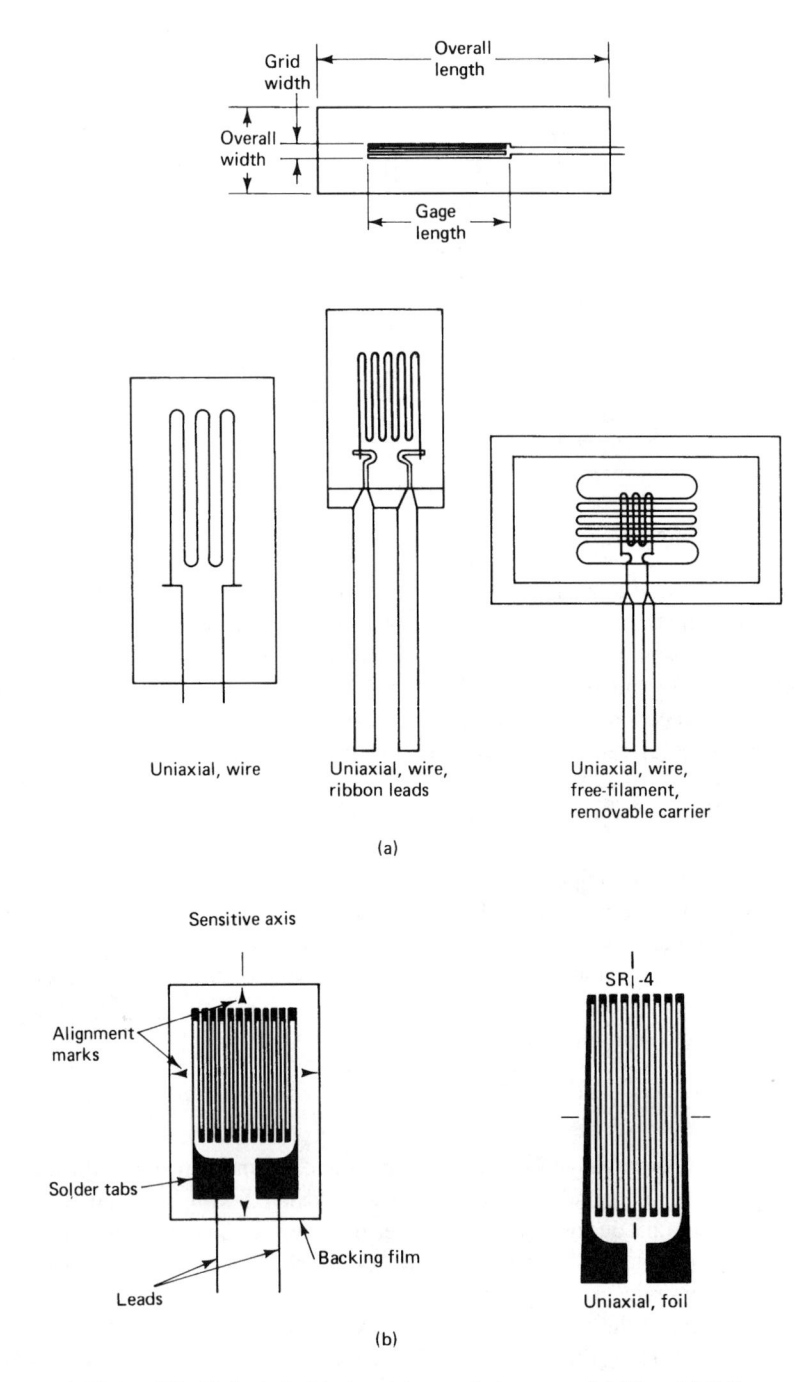

Figure 7.2 Typical electrical resistance strain gages. (a) Wire. (b) Foil.

inventors, Simmons and Ruge), are the best tools for strain measurements in all types of structures and structural elements.

Because of their relatively low cost and lightness, a large number of these gages can be used to investigate the behavior of a structure. Electric resistance strain gages have achieved the widest applications for strain measurements, including use in airplane parts, boat hulls, structures, buildings, bridges, and machine parts. They have a number of advantages that make them extremely versatile, as shown in the following table.

Property	Comment
Size	Very small
Weight	Insignificant
Ease of attachment	Relatively simple
Sensitivity to strain	Good; higher output very advantageous
Static and dynamic strain measurement	Equal ease
Remote indication and recording	Easily accomplished
Expense	Relatively inexpensive
Gage length	As small as 0.4 mm
Sensitivity to ambient variables	Slightly affected, but gages can be usually protected or variables involved can be properly compensated for
Linearity of output	Good

The main disadvantages of electrical resistance gages are the drift that can occur over a long period of time because of instability of the mounting cement, and some sensitivity to ambient environmental conditions. This gage also needs appropriate signal conditioning and readout equipment (either a manual or automatic strain indicator system). These disadvantages are rather minor as compared with their many advantages, however. The sensitivity of this type of gage can be as high as 0.000001 mm/mm (commonly called one microstrain, 1 $\mu\varepsilon$).

Weldable resistance strain gages have been devised for use in testing jet aircraft, nuclear containment vessels, rockets and missile surfaces, and heavy construction projects in severe environments, such as shock, vibration, steam, and saltwater. These testing applications call for rugged and stable gages capable of withstanding temperature variations up to as high as 649°C over a period of time. The weldable wire resistance strain gage can be installed in a matter of minutes. This unique technique, utilizing capacitive discharge spot-welding equipment, eliminated the need for all bonding materials and overcame the weaknesses of the bonding processes used for conventional electrical resistance strain gages. The gage consists of a filament configuration in which the strain-sensing filament and lead-out wire are of unitized construction. The lead-out section is electroformed (using gold) to a diameter of 0.18 mm. A controlled taper between this diameter and that of the strain wire is provided. Such a construction technique eliminates joint fatigue problems or instabilities due to erratic electrical connections and contact resistance problems. Strain-sensitive alloys generally used are nickel–chrome and platinum–tungsten, both of which are excellent for strain measurements. The former is used for temperatures up to 326°C and the latter at higher temperatures up to 613°C.

Two geometric configurations are used for the active and dummy filaments. The active element is a V-shaped simple element, while the dummy is an identical filament wound in a helix. The gage factor is adjusted to zero by selecting a proper pitch angle of the helix. The Poisson's ratio is matched so that no net dimensional change results when the gage is strained. Thus, a temperature compensation is provided by the dummy element, which is in the same environment, but without the applied strain. The strain filament is encased in a strain tube that is made by welding a tubular metallic shell to a flat flange stock. The filament is mechanically coupled to the strain tube but electrically isolated from it by highly compacted (metallic) high-purity magnesium oxide powder. This compaction is obtained by using a programmed high-speed centrifuge and a swaging operation.

When the gage is welded to a specimen and the test specimen put into a stress field, the stress is transmitted through the welds to the mounting flange, into the strain tube, and through the magnesium oxide powder. Because of the high compaction of the powder, the strain is transmitted to the sensing element all along its length, with no slippage. The property of stable, reversible, strain transfer is due to the large value of the ratio of surface area to the cross-sectional area. Weldable gages (Figure 7.3) are equipped with a thin flange welded to the strain tube; the flange is spot-welded to the test structure, thus providing the bond required for strain transfer. A guide for the selection of weldable gages for different applications is given in Table 7.1.

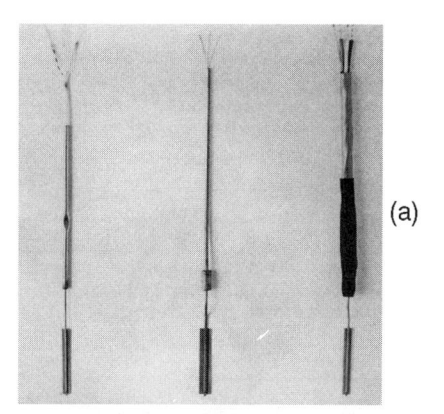

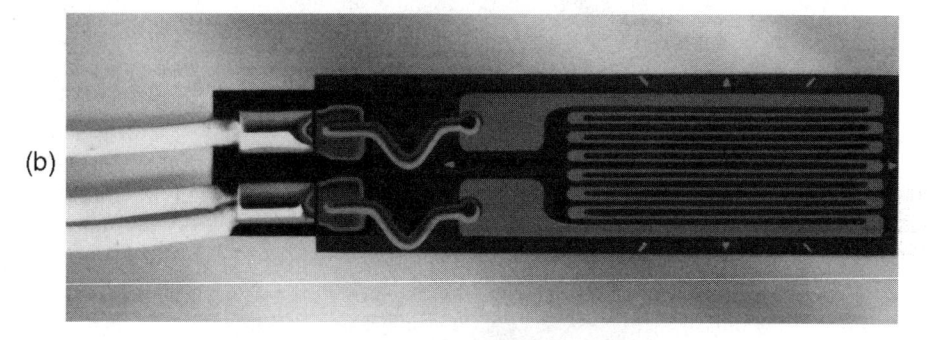

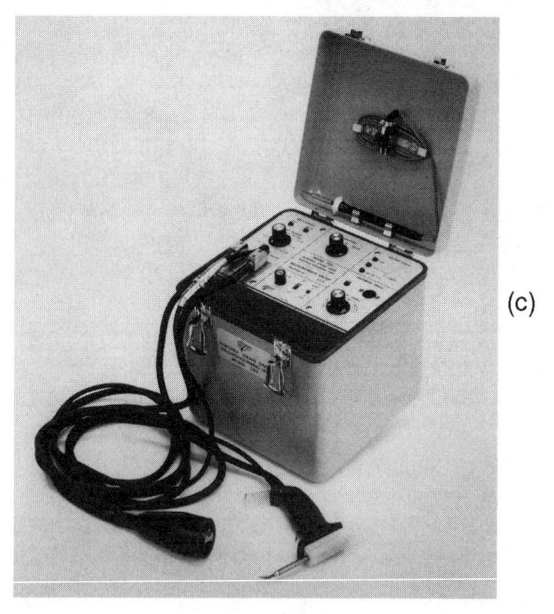

Figure 7.3 Weldable gages. (a) Gages. (Courtesy of Ailtech.) (b) Weldable gage type LWK. (Courtesy of Measurements Group.) (c) Strain gage welding unit. (Courtesy of Measurements Group.)

7.3.2.2 Other Electrical Strain Gages

Other types of electrical gages make use of properties affected by straining, namely, capacitance, inductance, and the piezoelectric effect. In the capacitance gage, change in capacitance of the condenser is observed as strain varies. The condenser consists basically of two plates separated by

Table 7.1 Strain Gage Selection Chart[a]

Applications	Preferred Gage Type[b]	Filament Material	Description
Strain Measurement in the Field — bridges, buildings, pilings — whenever humidity, rain, direct immersion, or a simple, rugged installation procedure is required	SG 189	Nickel chrome alloy	Flexlead Strain Gage[c] — $\frac{1}{4}$ bridge, 120 Ω, self-compensated gages with prewaterproofed, vinyl insulated, three-conductor, shielded cable
Extended Temperature Ranges — to 650°F static or 1500°F dynamic even when water, steam or corrosive media surrounds gages	SG 125 SG 128	Nickel chrome alloy	Integral Lead Strain Gage[c,d] — $\frac{1}{4}$ bridge, 120 Ω self-compensated gages with hermetically sealed, stainless steel jacketed cable; leads insulated with fiberglass (SG 128) or magnesium oxide (SG 125) for the most severe applications
Cryogenic Temperatures — up to 650°F — coupled with water, steam, corrosive media	SG 325 SG 328	Nickel chrome alloy	Integral Lead Strain Gages[c,d] — $\frac{1}{2}$ bridge versions of SG 125 or SG 128 gage provide optimum thermal compensation in the cryogenic or subzero ranges; also available for minimizing drift at 650°F
Ultra-temperature Environments to 950°F static (1200°F for short times) with protection from corrosive media	SG 425	Platinum tungsten alloy	Integral Lead Strain Gages — $\frac{1}{2}$ bridge Pt-W gage for static strain measurements to 1200°F, equipped with stainless steel jacketed, magnesium oxide insulated integral leads
Embedment in Concrete — for models or full size structures	CG 129	Nickel chrome alloy	Flexlead Strain Gage — designed for direct embedment in concrete, 0-180°F; other temperature ranges available

[a] Courtesy of Ailtech, Inc. — Hydrostatic testing, special temperature ranges, etc. are available.
[b] Basic gage types (without integral leads) are available. They provide features of ruggedness and simplicity of installation when integral lead, hermetically sealed cables are not required. Some basic gages are useful to 1800°F for dynamic measurements.
[c] 350 Ω gages are available.
[d] Short length gages are available.

a short distance and insulated from each other. The strain gage is attached to the structure at the point where strain will alter either the spacing or the arc of the above two plates. This change in turn causes the change in the impedance of the AC circuit and is a function of the capacitance of the condenser. Because of the separation of the two plates, it has inherent disadvantages, which include oversensitivity to vibration and difficulty in attaching the gages. This has considerably restricted the commercial development of these gages.

The inductance strain gage has an iron-core coil, whose inductance can vary with the applied strain. The variation in the inductance is achieved by changing the position of the armature with respect to the coil or by altering the space of the air gap between the two. This type of strain gage has disadvantages similar to the capacitance gage and it is heavier. However, it has a very high electrical output compared with a resistance strain gage, which helps simplify the readout circuitry. Because of its bulk and relatively large gage length, it is used more as a displacement measuring device in the form of a variable inductance transducer rather than as a strain gage.

The piezoelectric principle, related to the ability of a material to generate an electric charge when strained, was observed first by Curie. Some crystalline materials are naturally piezoelectric; other polycrystalline ceramics can be given this property artificially. In the last 20 years, this principle has been used in various kinds of transducers for pressure and acceleration measurements. Mark and Goldsmith (1973) have reported a successful use of barium titanate as a piezoelectric material in strain gages. The main advantage of this type of gage lies in the very high signal produced per unit strain and also in the fact that it is self-generating, i.e., no external power supply is needed. Their main disadvantages are (1) they are not suitable for static strains, hence they are difficult to precalibrate; (2) they cannot be precalibrated because of oversensitivity while mounting; and (3) they produce considerable reinforcement at the point of attachment.

The development of these various gages has remained somewhat specialized and limited because of the success and versatility achieved by the resistance-type strain gages.

In view of the short space, only the basic principles have been discussed here. A number of leading manufacturers of the various types of strain gages maintain up-to-date literature and the specialized circuitry for specific types of gages; this information is readily available to the user. A list of these manufacturers is provided in Table 7.2.

7.3.2.3 Choosing Electrical Resistance Strain Gages and Cements

Various types of resistance strain gages with different properties and configurations have been developed for a variety of environmental conditions. Gage length is determined by the particular application; in general, the longest feasible gage is used to make installation easier, but in cases where the strain gradient is high, very short gages may be necessary. Gages on the surface of concrete should have a gage length of at least several times the maximum aggregate dimension to avoid stress concentration effects. The range of strain to be measured is an important factor in selecting the type of gage, as is the type of test (static vs. dynamic). The user is advised to keep up-to-date copies of manufacturers' catalogs on hand and to consult them carefully before making a choice of gage type and length. Additional discussion on gage configuration is given in Section 7.3.3.6.

Strain Sensing Filaments — Copper–nickel alloy (constantan alloy 400) is primarily used in static strain measurements because of its low and controllable temperature coefficient. This alloy may be operated safely between –22 and 205°C. Nickel–chrome alloy (Nichrome alloy 200) can be used for high-temperature static and dynamic strain measurement of up to 929°C with the use of a proper ceramic cement. Nickel–iron alloy (Danaloy alloy 600) is recommended for dynamic tests where its larger temperature coefficient minimizes the temperature compensation problem. The higher gage factor (see Section 7.3.3.2), which leads to higher electrical output, is an advantage for measuring dynamic strains of small magnitude. Platinum alloy (alloy 1200) is used for its excellent stability and fatigue resistance at elevated temperatures. Its only disadvantage is that this alloy cannot be adjusted for self-temperature compensation and, therefore, as will be shown in Section 7.3.3.3, an additional (dummy) gage must always be used along with each active (measuring) gage. A typical strain gage selector chart from Micro-Measurements Division, Measurements Group, Inc., is given in Table 7.3.

Cements (Adhesives) — A large number of cements are available for use in various types of measurements. More commonly used cements are: Duco,* M-Bond,** and various epoxies. Duco (single-component, nitrocellulose, room-temperature curing cement) is used particularly with the paperbacked strain gage and has a maximum operating temperature of 66°C. M-Bond or its equivalent, single-component contact cement, fast-curing at room temperature, is especially used with polyimide-backed strain gages and is excellent for short-term testing. Fast-curing, two-part epoxy resins, available at local hardware stores, are also adequate for bonding strain gages and work particularly well on concrete surfaces where the M-Bond type cements are less satisfactory because of the porous nature of concrete. Other special types of cements include epoxy and ceramic base cements that have better long-term stability characteristics; their use is recommended on some types of transducers, for long-term strain measurement, and under thermal environments. It cannot be overemphasized that the cement is the only medium available to transfer the strain in the structure into the gage sensing element. The cement layer should be as *thin* as possible. A poor cementing operation means inaccurate strain measurement. As in selecting gages, up-to-date literature from the manufacturer should be studied carefully before choosing a cement. A typical cement-selection chart from Micro-Measurements Division, Measurements Group, Inc., is shown in Table 7.4.

---

* Registered trademark of E. I. duPont de Nemours and Co., Wilmington, DE.
** Registered trademark of the Measurements Group, Inc., Raleigh, NC.

INSTRUMENTATION — PRINCIPLES AND APPLICATIONS

Table 7.2 Partial List of Manufacturers of Different Types of Instruments

Manufacturer	Address	Types of Accessories
A.L. Design, Inc.	1411 Military Rd. Buffalo, NY 14217-1395 tel: (716) 875-6240 fax: (716) 895-2404	Sensors, transducers, and instrumentation.
BLH Electronics, Inc.	75 Shawmut Rd. Canton, MA 02021 tel: (617) 821-2000 fax: (617) 828-1451	Strain gages, accessories, transducers
Eaton Corp.	1728 Maplelawn Rd. P.O. Box 1089 Troy, MI 48099 tel: (810) 643-0220 fax: (810) 643-0259	Load cells
Instron Corp.	100 Royall St. Canton, MA 02021 tel: (617) 828-2500	Testing machines and load accessories
JP Technologies, Inc.	42 north Benson Ave. P.O. Box 1168 Upland, CA 91786 tel: (714) 946-1000 fax: (714) 946-6267	Strain gages and accessories
Lucifer	2048 Bunnell Rd. Warrington, PA 18976 tel: (215) 343-0411 fax: (215) 343-7388	Annealing furnaces
Magnaflux Corp.	Div. of ITK Fluid Prod. Group 3624 W. Lake Ave. Glenview, IL 60025 tel: (847) 657-5300	Stress coats of different types
Micro-Measurements D.V. Measurements Group, Inc.	P.O. Box 27777 Raleigh, NC 27611 tel: (919) 365-3800 tel: (919) 365-3945 www.mearurementsgroup.com	All types of strain gages and related accessories for strain reading; data acquisition systems
MTS Systems Corp.	Box 24012 Minneapolis, MN 55424 tel: (612) 937-4000 fax: (612) 937-4515	Servo-controlled loading systems
National Instruments	6504 Bridge Point Rd. Austin, TX 78730-5039 tel: (512) 794-0100 fax: (512)794-8411 www.natins.com	Data acquisition systems
OMEGA Engineering, Inc.	An OMEGA Technologies Co. P.O. Box 2721 Stamford, CT 06906 tel: (203) 359-1660 fax: (203) 359-7807	Strain gages, load cells, and all related accessories for reading, strain, pressure, and force
PCB Piezotronics, Inc.	3425 Walden Ave. Depew, NY 14043 tel: (716) 684-0001	Quartz transducers for quality measurement
SATEC Systems, Inc.	900 Liberty St. Grove City, PA 16127 tel: (412) 458-9610	Electro mechanical testing systems
TRANS-TEK	P.O. Box 338 Rt. 83 Ellington, CT 06029 tel: (860) 872-8371 tel: (800) 828-3964 fax: (860) 872-4211	Linear displacement transducers
Tinius-Olsen Testing Machine Co., Inc.	Easton Rd. P.O. Box 429 Willow Grove, PA 19090-0429 tel: (215) 675-7100 fax: (215) 441-0899	Testing machines and accessories

Table 7.3 Standard Strain Gage Series Selection Chart (Courtesy Measurements Group, Inc.)

Gage Series	Description and Primary Application	Temperature Range	Strain Range	Fatigue Life Strain Level in µε	No. of cycles
EA	Constantan foil in combination with a tough, flexible, polyimide backing; wide range of options available, primarily intended for general-purpose static and dynamic stress analysis; not recommended for highest accuracy transducers	Normal: −75 to +175°C Special or short-term: −195 to +205°C	±3% for gage lengths under 3.2 mm ±5% for 3.2 mm and over	±1800 ±1500 ±1200	10^5 10^6 10^8
CEA	Universal general-purpose strain gages; constantan grid completely encapsulated in polyimide, with large, rugged copper-coated tabs; primarily used for general-purpose static and dynamic stress analysis; 'C'-feature gages are specially highlighted throughout the gage listings	Normal: −75 to +175°C Stacked rosettes limited to +65°C	±3% for gage lengths under 3.2 mm ±5% for 3.2 mm and over	±1500 ±1500	10^5 10^{6a}
N2A	Open-faced constantan foil gages with a thin, laminated, polyimide-film backing; primarily recommended for use in precision transducers, the N2A Series is characterized by low and repeatable creep performance; also recommended for stress analysis applications employing large gage patterns, where the especially flat matrix eases gage installation	Normal static transducer service: −75 to +95°C	±3%	±1700 ±1500	10^6 10^7
WA	Fully encapsulated constantan gages with high-endurance lead wires; useful over wider temperature ranges and in more extreme environments than EA Series; Option W available on some patterns, but restricts fatigue life to some extent	Normal: −75 to 205°C Special or short-term: −195 to +260°C	±2%	±2000 ±1800 ±1500	10^5 10^6 10^7
SA	Fully encapsulated constantan gages with solder dots; Same matrix as WA Series; same uses as WA Series but derated somewhat in maximum temperature and operating environment because of solder dots	Normal: −75 to +205°C Special or short-term: −195 to +230°C	±2%	±1800 ±1500	10^6 10^7

EP	Specially annealed constantan foil with tough, high-elongation polyimide backing; used primarily for measurements of large post-yield strains; available with Options E, L, and LE (may restrict elongation capability)	−75° to +205°C	±10% for gage lengths under 3.2 mm ±20% for 3.2 mm and over	±1000	10^{4b}
ED	Isoelastic foil in combination with tough, flexible polyimide film; high gage factor and extended fatigue life excellent for dynamic measurements; not normally used in static measurements due to very high thermal output characteristics	Dynamic: −195 to +205°C	±2% — nonlinear at strain levels over ±0.5%	±2500 ±2200	10^6 10^7
WD	Fully encapsulated isoelastic gages with high-endurance lead wires; used in wide-range dynamic strain measurement applications in severe environments	Dynamic: −195 to +260°C	±1.5% — nonlinear at strain levels over ±0.5%	±3000 ±2500 ±2200	10^5 10^7 10^8
SD	Equivalent to WD Series, but with solder dots instead of lead wires	Dynamic: −195 to +205°C	±1.5%	±2500 ±2200	10^6 10^7
EK	K-alloy foil in combination with a tough, flexible polyimide backing; primarily used where a combination of higher grid resistances, stability at elevated temperature, and greatest backing flexibility is required	Normal: −195 to +175°C Special or short-term: −269 to +205°C	(see above note) ±1.5%	±1800	
WK	Fully encapsulated K-alloy gages with high-endurance lead wires; widest temperature range and most extreme environmental capability of any general-purpose gage when self-temperature compensation is required; Option W available on some patterns, but restricts both fatigue life and maximum operating temperature	Normal: −269 to +290°C Special or short-term: −269 to +400°C	±1.5%	±2200 ±2000	10^6 10^7
SK	Fully encapsulated K-alloy gages with solder dots; same uses as WK Series, but derated in maximum temperature and operating environment because of solder dots −269° to +260°C	Normal: −269 to +230°C Special or short-term:	±1.5%	±2200 ±2000	10^6 10^7
S2K	K-alloy foil laminated to 0.025-mm-thick, high-performance polyimide backing, with a laminated polyimide overlay fully encapsulating the grid and solder tabs; provided with large solder pads for ease of lead wire attachment	Normal: −75 to +120°C Special or short-term: −185 to +150°C	±1.5%	±1800 ±1500	10^6 10^7

[a] Fatigue life improved using low-modulus solder.
[b] EP gages show zero shift under high-cyclic strains.

Table 7.4 Recommended Adhesives for Different Strain Gage Series (Courtesy Measurements Group, Inc.)

Type of Test or Application	Operating Temperature Range, °C	M-M Gage Series[a]	M-Bond Adhesive[b]
General static or static-dynamic stress analysis	−45 to +65	CEA, EA	200 or AE-10 or AE-15
		WA, SA, WK, SK	AE-15 or 610
	−45 to +205	WA, SA, WK, SK	600 or 610
	−269 to +230	WK, SK	610
	<315	WK	610
High elongation (postyield)	−45 to +65	CEA, EA	200 or AE-10
		EP	AE-15 or AE-12
Dynamic (cyclic) stress analysis	−75 to +65	ED	200 or AE-10
		WD	AE-10 or AE-15
	−195 to +260	WD	600 or 610
Transducer gaging	−45 to +65	CEA, EA	AE-10 or AE-15
	−45 to +95	N2A, J2A	600 or 610 or 43-B
	−45 to +150	WA, SA, TA, TK, J5K	610, 450
	−195 to +175	WK, SK, TK, J5K	610, 450

[a] See Table 7.3.
[b] See Table 7.5.

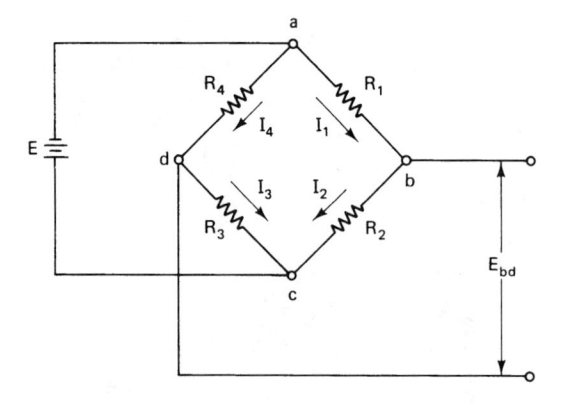

Figure 7.4 Wheatstone bridge circuit.

7.3.3 Resistance Strain Gage Circuitry and Applications

The theory of resistance strain gages is based on the Wheatstone bridge principle with four resistances, R_1, R_2, R_3, and R_4, as shown in Figure 7.4. A summary of the following is presented in this section.

1. Circuit analysis
2. Gage factor and transverse sensitivity
3. Temperature compensation (external and internal)
4. Typical applications
5. Types of bridges (full, half, and quarter)
6. Various configurations
7. Reduction of stresses from strain measurements

Details of some of the above items may be found in books by Perry and Lissner (1962), Dally and Riley (1965), Dove and Adams (1964), and Omega Engineering, Inc. (1992).

7.3.3.1 Circuit Analysis

The circuit most commonly used with resistance strain gages is a four-arm bridge with a constant voltage excitation E (Figure 7.4). In this circuit, the condition of balance in the bridge is first established; this is followed by determining the imbalance due to change in resistance (i.e., change in strain) in one or more gages mounted at positions R_1, R_2, R_3, and R_4.

For a balanced condition, the potential E_{bd} across the diagonal bd must be equal to zero, or the voltage at d equals the voltage at b. Therefore, the voltage drop from a to d ($I_4 R_4$) must equal the drop from a to b ($I_1 R_1$). Similarly, the voltage drops in the two legs in the bottom half of the bridge must be equal. These conditions can be summarized in

$$I_3 R_3 = I_2 R_2 \quad \text{and} \quad I_4 R_4 = I_1 R_1 \tag{7.1a}$$

In addition, with the bridge balanced, $E_{bd} = 0$, and thus the current I_1 must be the same as I_2, and I_3 must equal I_4, or

$$I_1 = I_2 \quad \text{and} \quad I_3 = I_4 \tag{7.1b}$$

By substituting Equation 7.1b into 7.1a and forming the ratio of the two new equations to eliminate the values of current I,

$$\frac{R_4}{R_3} = \frac{R_1}{R_2} \tag{7.2}$$

which is the condition for a balanced bridge.

Let us establish the expression for a change in E_{bd} due to change in resistance, say, R_1, with all other parameters held constant. From Figure 7.4:

$$E = I_1(R_1 + R_2) \quad \text{or} \quad I_1 = \frac{E}{R_1 + R_2} \tag{7.3}$$

which gives

$$E_{ab} = R_1 I_1 = \frac{E R_1}{R_1 + R_2}$$

$$\frac{dE_{ab}}{dR_1} = \frac{(R_1 + R_2)E - ER_1(1)}{(R_1 + R_2)^2} = \frac{R_2 E}{(R_1 + R_2)^2}$$

or

$$dE_{ab} = \frac{R_2 dR_1}{(R_1 + R_2)^2} E$$

But

$$E_{bd} = E_{ad} - (E_{ad} + dE_{ab}) = -dE_{ab}$$

which results in

$$E_{bd} = -\frac{R_2 dR_1}{(R_1 + R_2)^2} E \tag{7.4}$$

Similar expressions can be obtained for changing each of the other resistances R_2, R_3, and R_4. If all these resistances are changed simultaneously, then

$$E_{bd} = \sum_{i=1}^{4} E_{ibd}$$

By using $F = dR/R/\varepsilon$ (where F is defined as a gage factor* for each resistance R and the strain ε),

$$E_{bd} = E\left[\frac{-R_2 R_1 F_1 \varepsilon_1}{(R_1 + R_2)^2} + \frac{R_1 R_2 F_2 \varepsilon_2}{(R_1 + R_2)^2} - \frac{R_4 R_3 F_3 \varepsilon_3}{(R_3 + R_4)^2} + \frac{R_3 R_4 F_4 \varepsilon_4}{(R_3 + R_4)^2}\right] \quad (7.5)$$

In the case when all resistances and gage factors are equal, we obtain

$$E_{bd} = \frac{FE}{4}(-\varepsilon_1 + \varepsilon_2 - \varepsilon_3 + \varepsilon_4) \quad (7.6)$$

This means that the imbalance of the bridge is proportional to the sum of the strains in opposite arms and to the difference of strains in adjacent arms of the bridge.

7.3.3.2 Gage Factor

The term *gage factor* is used to describe the sensitivity of output characteristic of the bonded resistance strain gage. It is defined as

$$F = \frac{\frac{\Delta R}{R}}{\frac{\Delta L}{L}} = \frac{\frac{\Delta R}{R}}{\varepsilon} \quad (7.7)$$

where: F = gage factor
R = electrical resistance
L = length of strain sensitive element
ε = normal or axial strain

Thus, gage factor is the ratio of the change in resistance per unit of original resistance to the applied strain. The gage factor, which has a value close to 2 for gages commonly used in static strain analysis, is quite important since the higher the gage factor, the higher will be the electrical output and the higher the sensitivity. Successful attempts have been made to achieve strain gages with relatively high gage factors. The limiting factor in determining the gage factor is the conductor material. Materials that provide for high gage factors have other undesirable characteristics to render them less suitable as strain gage material.

In wire gages, when the filament (or wire) of the strain gage is wound back and forth to form a grid, there are a large number of "turnarounds" or bonded ends, which results in a significant decrease in ΔR, thereby causing variation in the value of the gage factor. To account for this, proper calibration of the gage factor is performed in the commercially available gages by installing several identical gages on a specimen bar and applying a known mechanical strain. Statistical averages of

---

* Defined in next section.

all the gages are then proportioned to the value of the specimen strain to obtain the gage factor value for that type of gage.

7.3.3.3 Temperature Compensation in Strain Gage Circuits

Throughout the above discussion, we have considered that the temperature is constant during the test in which strains are measured. However, temperature influences the electrical properties of the metals used in constructing the gages, and we must account for this effect as well as for the thermal strains that will result in the structure itself if the temperature of the testing area changes during the test. The separation of mechanical strains (induced by the physical loading of the structure) from unwanted thermally induced strains is an important part of the experimental stress analysis process.

Temperature compensation can be achieved by forcing adjacent arms of the Wheatstone bridge to have identical thermal strain components. Since the bridge output is proportional to the algebraic difference of strains in adjacent arms, the thermal strain increments will cancel each other out. The objective is accomplished by mounting a gage identical to the active strain gage on an unstressed piece of the same type of material as in the structure. If the active gage is R_1 in the bridge, then the so-called "dummy" gage is placed in either position R_2 or R_4. Writing Equation 7.2 for the case with R_2 being the dummy gage and R_1 the active gage,

$$\frac{R_4}{R_3} = \frac{R_1 + \Delta R_1}{R_2 + \Delta R_2} \tag{7.8}$$

where the changes in R_1 and R_2 are produced by thermal effects. It is obvious that the bridge is still balanced and that the output is zero. The reader should verify that making R_4 the dummy gage will also lead to a fully balanced bridge and hence inherent temperature compensation for the active gage R_1.

Thermal problems produced in plastic models by local heating from the bridge current are discussed in Section 3.5.4.

7.3.3.4 Self-Temperature Compensating Gages

As described in the last section, it may not always be possible to find a suitable location for installing a compensating gage *or* the temperature variation in both active and compensating gages may not be identical. In such situations, self-temperature-compensated (STC) gages should be used. The term *STC* is applied to strain gages in which the resistance change of the gage due to temperature effects is held to a very low value. The available STC gages can be classified into three basic types: (1) the two-element, (2) the single-element, and (3) the universally compensated gage.

The two-element gage is made with two wire element grids in series. One element has a negative temperature coefficient of resistance, while the other has a positive coefficient of resistance. The lengths of two grids are adjusted such that the net temperature coefficient of resistance will compensate for this change in resistance as the result of a differential coefficient of expansion when the gage is mounted on a particular material. The limitation of this gage is that a particular design is restricted to the material with the corresponding coefficient of expansion, and a variety of gages must be kept to accommodate the common construction materials.

The single-element gage is made by producing a grid where the change in resistance due to a change in specific resistance will be just equal in magnitude but opposite in sign to the change due to the difference in the coefficients of linear expansion of the gage and the material on which it is mounted.

The universal-type STC gage is also a two-element gage, but the relative effect of the two elements can be adjusted by making changes in the external circuitry. Figure 7.5 shows a gage of

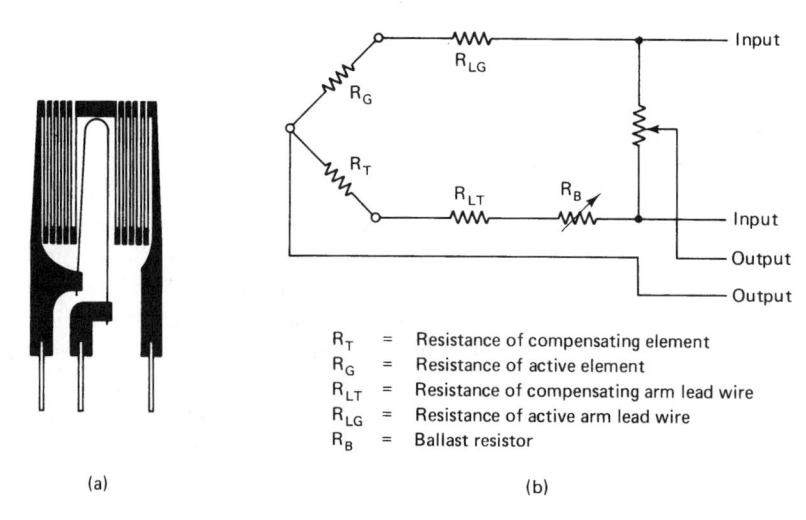

Figure 7.5 Universal temperature compensating gage. (a) Two-element STC gage. (b) Circuitry for installation. (Courtesy of BLH Electronics, Inc.)

R_T = Resistance of compensating element
R_G = Resistance of active element
R_{LT} = Resistance of compensating arm lead wire
R_{LG} = Resistance of active arm lead wire
R_B = Ballast resistor

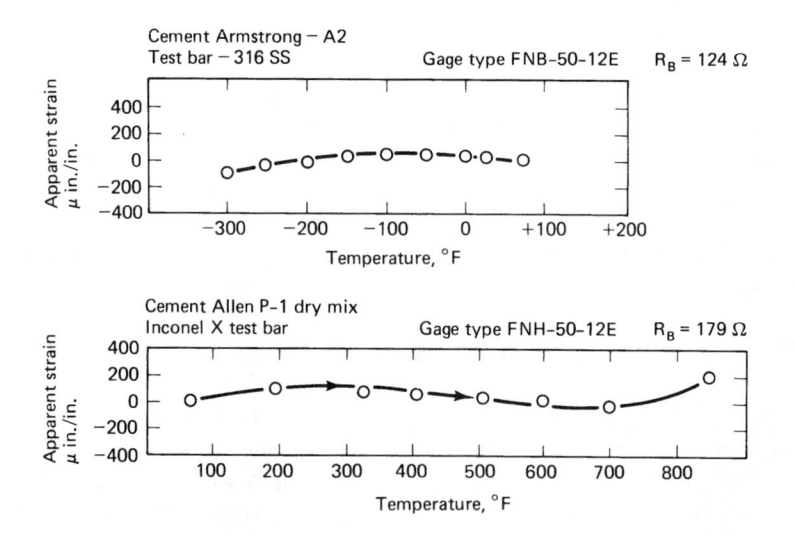

Figure 7.6 Apparent strain vs. temperature characteristic. (Courtesy of BLH Electronics, Inc.)

this type and the circuitry that goes with it. In this circuit, R_G and R_T are the active compensating elements, while R_{LG} and R_{LT} are the respective lead wire resistances. R_B is the variable resistor, which is adjusted to give least apparent strain over the range of interest. Figure 7.6 shows a typical plot between the apparent strain and temperature for a gage circuit of this type.

For further information on commercially available STC gages, the reader is referred to the manufacturer's literature in Table 7.2.

Another problem that arises is due to temperature variation in lead wires of different lengths. An example is shown in Figure 7.7, where the single active gage R_1 is a STC gage with No. 18 copper lead wires totaling 6.1 m in length, and the other three gages are internal resistances in the strain indicator system. The lead wire resistance R_w is about 0.12 Ω. If the wires are subjected to a temperature increase of 16.67°C, the change in wire resistance is

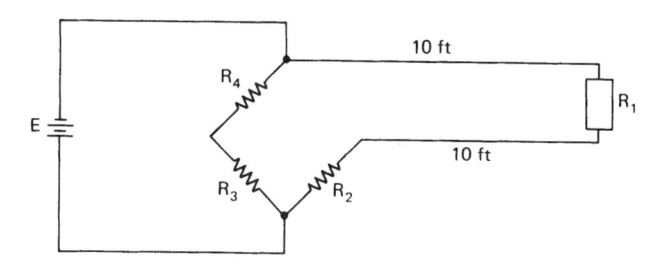

INSTRUMENTATION — PRINCIPLES AND APPLICATIONS

Figure 7.7 Temperature effect on lead wire resistance.

$$\Delta R = \text{(Temperature coefficient of resistance)}(\Delta T)(R)$$
$$= (0.00396/°C)(16.67)(0.12) = 0.00792 \ \Omega$$

With a gage factor $F = 2.1$, and the gage resistance $R_1 = 120 \ \Omega$, the apparent strain is

$$\varepsilon = \frac{\Delta R}{(R/F)} = \frac{0.00792}{(120.79/2.1)} = 0.000031$$

If the gage is mounted on a steel structure, with $E = 207$ GPa, then the apparent strain produced by the temperature change is $E\varepsilon = 6.5$ MPa.

It should be apparent to the reader that the same problem exists in circuits with active gages having long lead wires and dummy gages having short lead wires.

7.3.3.5 Practical Circuits and Their Applications

Although the basic Wheatstone bridge has four arms, as discussed earlier, it is often necessary to use only one or two arms with active gages to measure individual strains. Furthermore, Equation 7.6, which established the relation between the strains ε_1, ε_2, ε_3, and ε_4 and the change of potential E_{bd} enables one to make use of three practical circuitries as follows:

1. Full bridge (four active gages)
2. Half bridge (two active gages)
3. Quarter bridge (one active gage)

Full bridge is one in which all four arms of the bridge are active, as shown in Figure 7.8a, and the bridge output E_{bd} is given by $E_{bd} = (FE/4)(-\varepsilon_1 + \varepsilon_2 - \varepsilon_3 + \varepsilon_4)$. The inherent differences in sign for the strain terms in this expression may be used to good advantage in a load cell. The compressive cylindrical load cell in Figure 7.9a is instrumented with gages R_1 and R_3 in the longitudinal direction and with gages R_2 and R_4 around the circumference. The latter gages will go into tension because of the Poisson effect. Hence, E_{bd} becomes $(FE/4)[-\varepsilon_1 + \nu(-\varepsilon_1) - \varepsilon_3 + \nu(-\varepsilon_3)]$(strain in either gage 1 or 3), and the output of the bridge circuit is magnified by a factor of $2(1 + \nu)$ over that of a single gage. Temperature compensation is automatically achieved because of the relation

$$\frac{R_4 + \Delta R}{R_3 + \Delta R} = \frac{R_1 + \Delta R}{R_2 + \Delta R} \tag{7.9}$$

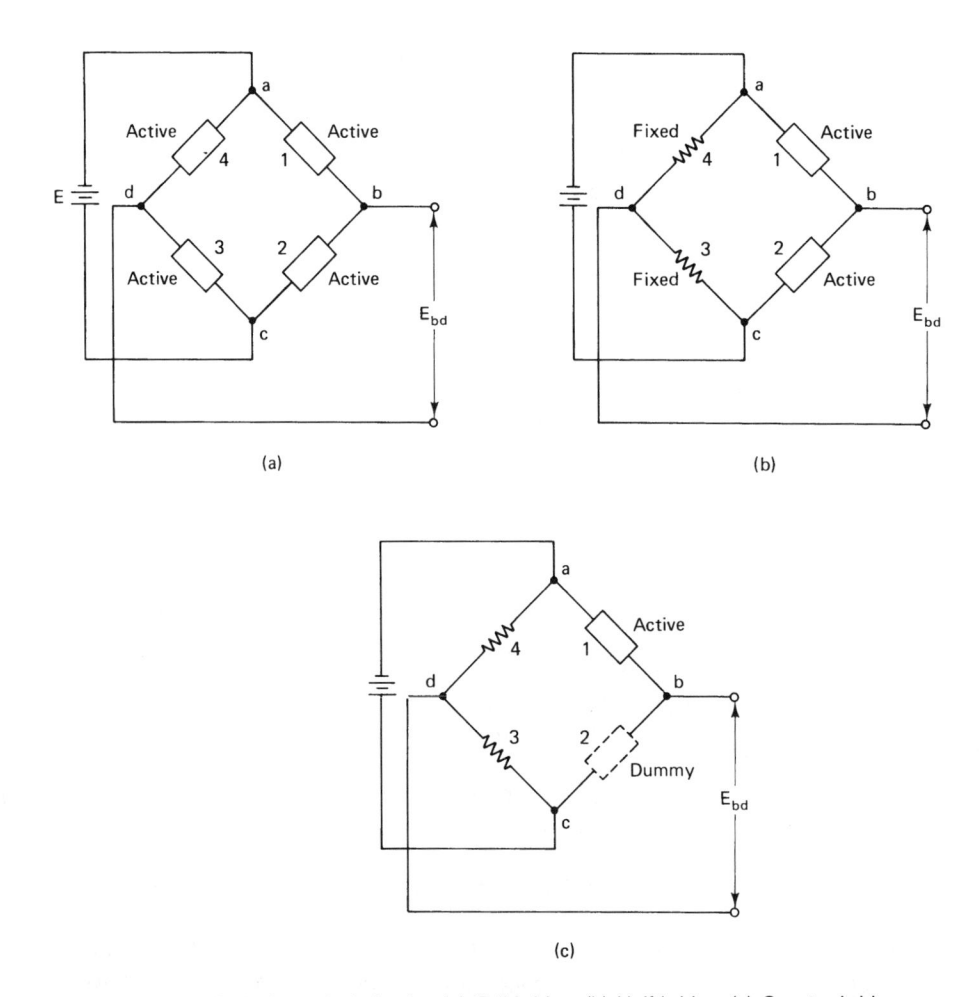

Figure 7.8 Typical practical circuits. (a) Full bridge. (b) Half bridge. (c) Quarter bridge.

Half bridge is one with only two active gages R_1 and R_2 (or any two adjacent arms of the bridge) and two other fixed resistances to complete the bridge. The bridge output is $E_{bd} = (FE/4)(-\varepsilon_1 + \varepsilon_2)$. For example, a test on a cantilever beam to determine bending strains as shown in Figure 7.9b may be instrumented with R_1 and R_2 so that E_{bd} is proportional to 2ε. The sensitivity of the circuitry is thus improved, and temperature compensation is simultaneously provided for.

Quarter bridge has only one active gage and is used when single strains in a stress field are to be measured. It can immediately be realized that proper precaution for temperature compensation must be exercised by using a dummy gage and lead wires, of proper length, as discussed earlier.

7.3.3.6 Gage Configurations

Various configurations of strain gages are shown in Figures 7.10 and 7.11. Gage geometries can vary from general-purpose, single grids to two-element and three-element rosettes that measure strains in different directions. Single grids (Figures 7.11a and b) are used for measuring strain in one direction only; the length and width needed for a particular application depends on the strain gradient, the specimen size, the material used, etc. The usual range of lengths is from 0.4 to 100 mm, and the width varies from 0.76 to 6.4 mm.

The two-element rosettes in Figures 7.10b and d and 7.11c and d are used to measure strains in a general state of strain where the direction of principal strains is known. Two different types

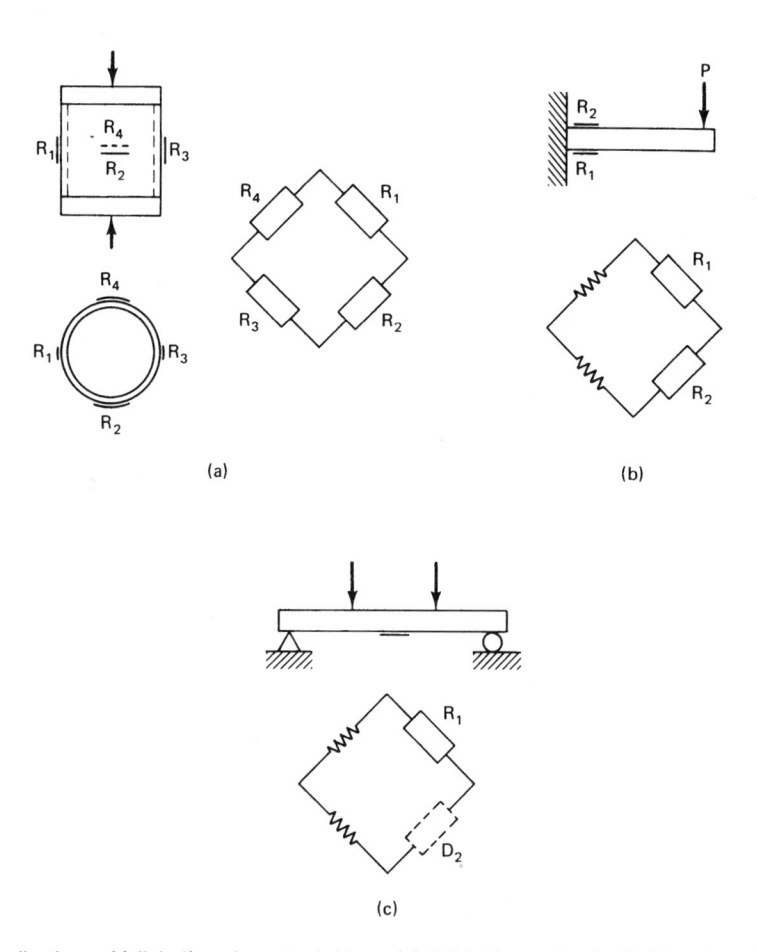

INSTRUMENTATION — PRINCIPLES AND APPLICATIONS

Figure 7.9 Applications of full, half, and quarter bridges. (a) Full bridge on load cell. (b) Half bridge. (c) Quarter bridge.

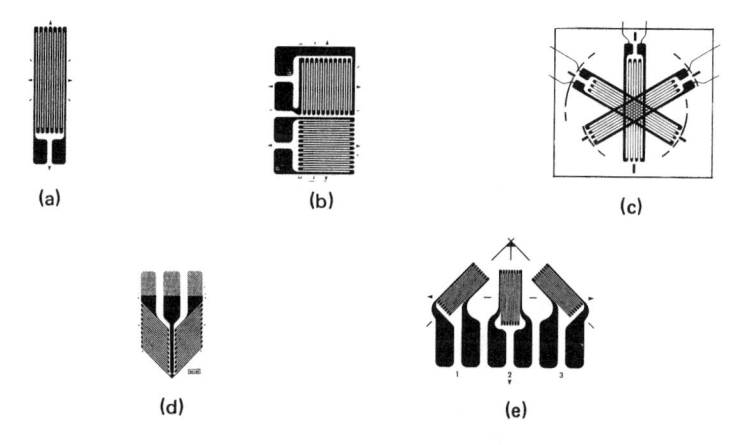

Figure 7.10 Various strain gage configurations.

of three-element rosettes are shown in Figures 7.10c and e and 7.11e and f; these gages are used when nothing is known about the direction or magnitudes of strain in a general strain field, and three independent values of strain are needed to define the two principal strains and their direction

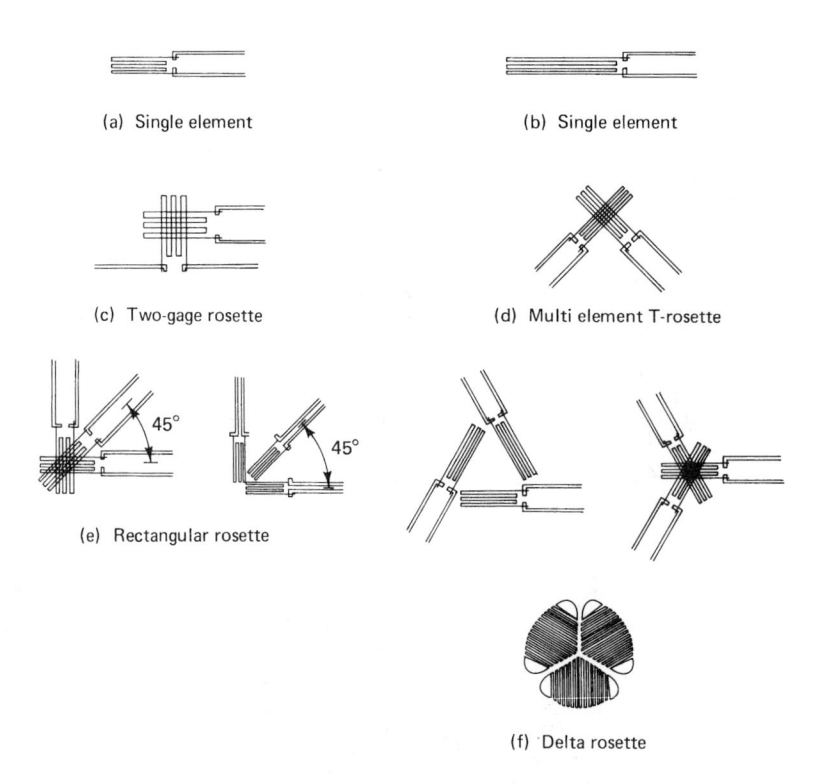

Figure 7.11 Various strain gage configurations.

(as will be discussed in the Section 7.3.3.7). *Delta* rosettes with angles between gages of 120° (or 60°) are shown in Figures 7.10c and e and 7.11f. *Rectangular* rosettes with two gages at 90° to each other and the third bisecting this angle are illustrated in Figure 7.11e. In the second figure the gages are located adjacent to one another while in the first figure they are placed on top of each other to facilitate measurement of all three strain components at a single point.

7.3.3.7 Calculation of Stresses from Measured Strains

Frequently, the objective of the model study is to determine the stress distribution in a structure from strain measurements, using the techniques described earlier. The development of the basic relations between stresses and strains may be found in any textbook on the theory of elasticity (e.g., Timoshenko and Goodier (1965). In this section, these relations will be presented directly without proof and with the various procedures for their rapid solution. These can be classified as (1) direct reduction (algebraic), (2) graphic (geometric or nomographic) reduction, and (3) computer reduction. The first two are suitable only for a few point readings; the last one can be used for large numbers of reductions.

Before discussing these methods, it must be pointed out that the strain readings are made on a finite area whose dimensions are governed by the size of the gage. However, the measurements are assumed to be at the center of the gage, and it is normally assumed that variations in the strains are linear in the area. Special care is therefore necessary to plan the number and size of gages in the areas of stress concentration. Furthermore, the number of strain measurements necessary for the stress determinations at the point of strain measurement depends on the knowledge (or lack of knowledge) of the directions of principal stresses (or strains). If the directions are known, only *two* measurements are necessary. However, if they are unknown, which may generally be the case, then *three* strain measurements are necessary to determine both the magnitudes and direction of the principal strains and stresses. The discussion thus relates to the various rosette arrangements of strain measurements as noted in the previous section.

According to the theory of elasticity, the direct stress on a surface in any direction x is given by

$$\sigma_x = \frac{E}{(1-v^2)}(\varepsilon_x + v\varepsilon_y) \qquad (7.10a)$$

where: E = modulus of elasticity
v = Poisson's ratio
ε_x = unit strain in x direction
ε_y = unit strain in y direction perpendicular to x direction

and the shearing stress is given by

$$\tau_{xy} = \gamma_{xy} G \qquad (7.10b)$$

where

$$G = \text{shearing modulus} = \frac{E}{2(1+v)}$$

In these expressions it is assumed that effects from pressure applied normal to the surface may be neglected.

If the x and y directions are known principal stress axes (known from symmetry considerations, for example), then two measurements of ε_x and ε_y are sufficient to determine the principal stresses σ_1 and σ_2 from the following expressions given in Equation 7.11:

$$\sigma_1 = \frac{E}{(1-v^2)}(\varepsilon_1 + v\varepsilon_2)$$

$$\sigma_2 = \frac{E}{(1-v^2)}(\varepsilon_2 + v\varepsilon_1) \qquad (7.11)$$

and

$$\tau_{max} = \frac{E}{2(1+v)}(\varepsilon_1 - \varepsilon_2) = \frac{\sigma_1 - \sigma_2}{2} \qquad (7.12)$$

On the other hand, without a prior knowledge of direction of the principal stresses, it is necessary to measure strains in three arbitrary directions and then solve for ε_x, ε_y, and γ_{xy} for use in the stress Equations 7.10. The strain ε_θ at any direction θ from the x axis (Figure 7.12) is related to the orthogonal normal strains ε_x and ε_y and the shearing strain γ_{xy} by the expression

$$\varepsilon_\theta = \varepsilon_x \cos^2\theta + \varepsilon_y \sin^2\theta + \gamma_{xy} \sin\theta\cos\theta$$

i.e.,

$$\varepsilon_\theta = \frac{\varepsilon_x + \varepsilon_y}{2} + \frac{\varepsilon_x - \varepsilon_y}{2}\cos 2\theta + \frac{\gamma_{xy}}{2}\sin 2\theta \qquad (7.13a)$$

From Equation 7.13a it is obvious that three unique values of ε_θ (three separate strain gage readings) are needed to solve for ε_x, ε_y, and γ_{xy}.

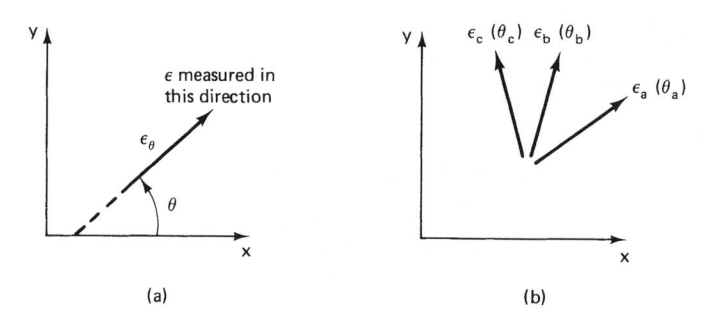

Figure 7.12 Measurements of strains and reference axes. (a) Location of strain. (b) Measured from x axis.

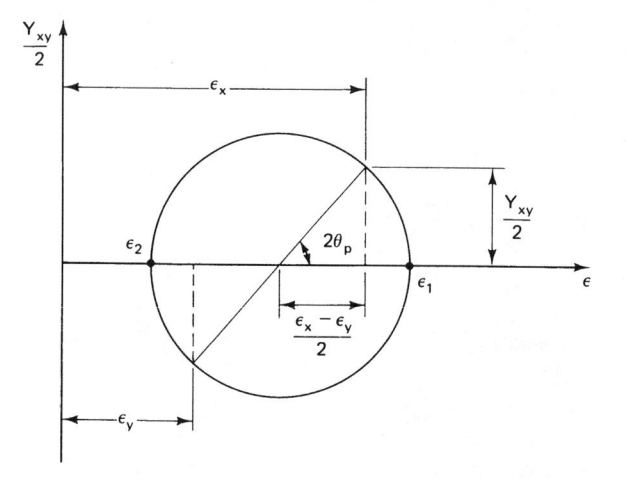

Figure 7.13 Mohr circle for strains.

If three strain gages a, b, and c were applied at a particular point with directions θ_a, θ_b and θ_c from an arbitrarily established x axis, forming a rosette, as shown in Figure 7.12b, Equation 7.13a can be written for each gage as follows:

$$\varepsilon_{a,b,c} = \frac{\varepsilon_x + \varepsilon_y}{2} + \frac{\varepsilon_x - \varepsilon_y}{2}\cos 2\theta_{a,b,c} + \frac{\gamma_{xy}}{2}\sin 2\theta_{a,b,c} \tag{7.13b}$$

These equations can be solved for $\varepsilon_x, \varepsilon_y$, and γ_{xy}. If $\varepsilon_x, \varepsilon_y$, and γ_{xy} are known, the principal strains ε_1 and ε_2 can be obtained from the following equations:

$$\varepsilon_{1,2} = \frac{\varepsilon_x + \varepsilon_y}{2} \pm \left[\left(\frac{\varepsilon_x - \varepsilon_y}{2}\right)^2 + \left(\frac{\gamma_{xy}}{2}\right)^2\right]^{1/2} \tag{7.14a}$$

$$\tan 2\theta_p = \frac{\gamma_{xy}}{\varepsilon_x - \varepsilon_y} \tag{7.14b}$$

Equation 7.14 can be derived either by maximizing ε in Equation 7.13a or using Mohr's circle for strain as shown in Figure 7.13. These strains can be converted into stresses using the stress–strain relations of Equation 7.11.

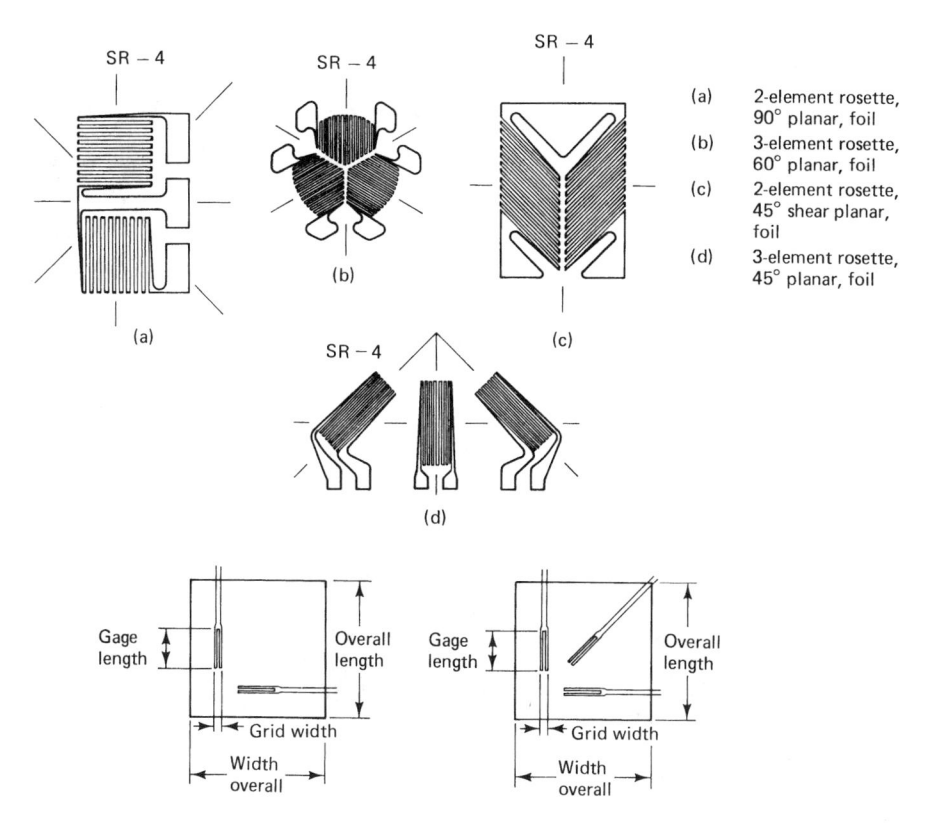

INSTRUMENTATION — PRINCIPLES AND APPLICATIONS 343

Figure 7.14 Various types of rosettes. (a) Two-element rosette, 90° planar, foil. (b) Three-element rosette, 60° planar, foil. (c) Two-element rosette, 45° shear planar, foil. (d) Three-element rosette, 45° planar, foil.

The general procedure is now established. One places three strain gages in some arbitrary (but known, measurable, and different) directions. Corresponding strain measurements ($\varepsilon_a, \varepsilon_b, \varepsilon_c$) in these three directions are made and used in Equation 7.13, which is then solved for any orthogonal strains $\varepsilon_x, \varepsilon_y$, and γ_{xy}. The values of $\varepsilon_x, \varepsilon_y$, and γ_{xy} are substituted into Equations 7.14 to determine ε_1, and ε_2 and the direction of principal planes given by angle θ. These are then used in Equation 7.11 to obtain the magnitude of the two principal stresses along the principal planes given by the angles θ_p and ($\theta_p + 90°$) from the x and y axes.

For practical purposes, the preferred orientations for gages a, b, and c are the rectangular or 45° rosette and the delta or 60° rosette discussed earlier. Figure 7.14 illustrates typical commercial rosette gages.

Figure 7.15 shows the rectangular rosette geometry and the corresponding sine and cosine functions. Thus, from Equation 7.13, one obtains

$$\varepsilon_a = \varepsilon_x$$
$$\varepsilon_b = \frac{\varepsilon_x + \varepsilon_y}{2} + \frac{\gamma_{xy}}{2} \qquad (7.15)$$
$$\varepsilon_c = \varepsilon_y$$

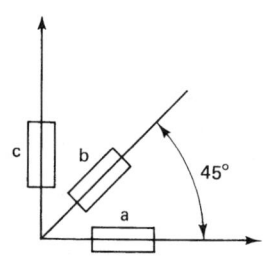

Double angle functions:
$2\theta_{ab} = 90° \rightarrow \cos = 0$ and $\sin = 1$
$2\theta_{ac} = 180° \rightarrow \cos = -1$ and $\sin = 0$

Figure 7.15 Geometry of rectangular rosette.

Solving,

$$\varepsilon_x = \varepsilon_a$$

$$\varepsilon_y = \varepsilon_c$$

$$\gamma_{xy} = -\varepsilon_a + 2\varepsilon_b - \varepsilon_c = 2\varepsilon_b - (\varepsilon_a + \varepsilon_c) \tag{7.16}$$

Substituting into Equations 7.14

$$\varepsilon_{1,2} = \frac{\varepsilon_a + \varepsilon_c}{2} \pm \frac{1}{2}\sqrt{(\varepsilon_a - \varepsilon_c)^2 + [2\varepsilon_b - (\varepsilon_a + \varepsilon_c)]^2}$$

$$\tan 2\theta_p = \frac{2\varepsilon_b - (\varepsilon_a + \varepsilon_c)}{\varepsilon_a - \varepsilon_c} \tag{7.17}$$

These results can in turn be substituted into Equations 7.11 and 7.12 to obtain:

$$\sigma_{1,2} = E\left\{\frac{\varepsilon_a + \varepsilon_c}{2(1-\nu)} \pm \frac{\sqrt{(\varepsilon_a - \varepsilon_c)^2 + [2\varepsilon_b - (\varepsilon_a + \varepsilon_c)]^2}}{2(1+\nu)}\right\}$$

$$\theta_p = \frac{1}{2}\tan^{-1}\frac{2\varepsilon_b - (\varepsilon_a + \varepsilon_c)}{\varepsilon_a - \varepsilon_c}$$

$$\tau_{max} = \frac{E}{2(1+\nu)}\sqrt{(\varepsilon_a - \varepsilon_c)^2 + [2\varepsilon_b - (\varepsilon_a + \varepsilon_c)]^2} \tag{7.18}$$

Relations between the principal stresses and strains for delta and T rosettes can be derived in a similar fashion; results for these three major types of rosettes are summarized in Table 7.5. In addition to these algebraic solutions, which are easily programmed, there are a number of graphical and nomograph-type solutions as discussed by several authors, such as Murphy (1945).

INSTRUMENTATION — PRINCIPLES AND APPLICATIONS 345

Table 7.5 Basic and Special Purpose Strain Gage Adhesives (Courtesy Measurements Group, Inc.)

Type	Principle Features
200	Most widely used general-purpose adhesive; easiest to handle; fast room-temperature curing
AE-10	General-purpose adhesive that is highly resistant to moisture and most chemicals; room-temperature curing
AE-15	Similar to AE-10; recommended for more critical applications, including transducer gaging; elevated-temperature curing
610	Used primarily in applications over wide temperature range; widely used in tranducer gaging; elevated-temperature curing
600	Similar to 610, but faster reacting; can be cured at lower temperatures than 610
43-B	Normally used in transducer gaging; highly resistant to moisture and chemical attack; elevated-temperature curing
GA-2	General-purpose adhesive primarily used on very rough or irregular surfaces; room-temperature curing
GA-61	Similar to GA-2, but more viscous; also used to fill irregular surfaces and to anchor leadwires; elevated-temperature curing
GA-100	Ceramic cement for use with special-design strain gages operating at high temperatures; elevated-temperature curing
A-12	Very high-elongation adhesive; used only when other adhesives cannot meet elongation requirements; elevated-temperature curing
300	Polyester adhesive used primarily when low-temperature curing is required; sensitive to solvents; not recommended as a general-purpose adhesive
450	High-performance epoxy for high-temperature transducer applications

Table 7.6 Strain Rosette and Principle Stresses

	Rectangular	Delta	T Delta
Pattern	(figure)	(figure)	(figure)
Strains	Three strains measured 45° apart	Three strains measured 60° apart	Three strains measured 60° apart and a fourth one 90° to one of them
Relations	$\sigma_{max} = K_1 A + K_2 \sqrt{B^2 + C^2}$ $\sigma_{min} = K_1 A - K_2 \sqrt{B^2 + C^2}$ in which $\theta_p = \tfrac{1}{2}\tan^{-1}\left(\dfrac{C}{B}\right)$ $K_1 = \dfrac{E}{1-\nu}$ $\tau_{max} = K_2 \sqrt{B^2 + C^2}$ $K_2 = \dfrac{E}{1+\nu}$		
A	$\dfrac{\varepsilon_a + \varepsilon_c}{2}$	$\dfrac{\varepsilon_a + \varepsilon_b + \varepsilon_c}{3}$	$\dfrac{\varepsilon_a + \varepsilon_d}{2}$
B	$\dfrac{\varepsilon_a - \varepsilon_c}{2}$	$\varepsilon_a - \dfrac{\varepsilon_a + \varepsilon_b + \varepsilon_c}{3}$	$\dfrac{\varepsilon_a - \varepsilon_d}{2}$
C	$\dfrac{2\varepsilon_b - (\varepsilon_a + \varepsilon_c)}{2}$	$\dfrac{\varepsilon_c - \varepsilon_b}{\sqrt{3}}$	$\dfrac{\varepsilon_c - \varepsilon_b}{\sqrt{3}}$

Example 7.1

A delta rosette on a model of a steel bin structure has the following strain values: $\varepsilon_{a,b,c}$ −78, +135, and +173 microstrain, respectively. Determine the principal stresses and the direction of the maximum principal strain from an axis coincident with gage *a*. $E = 207$ GPa and $\nu = 0.3$.

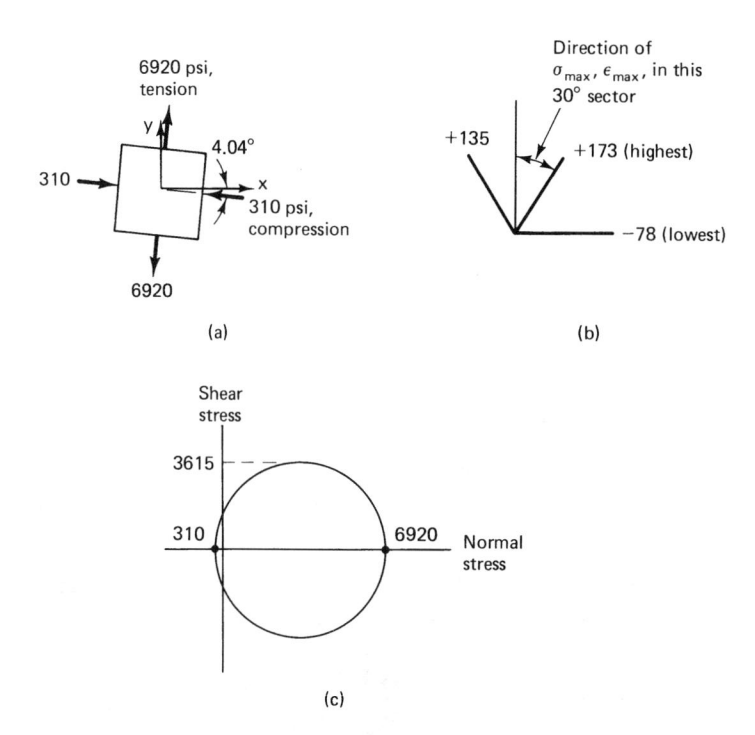

Figure 7.16 Delta rosette stress calculations. (a) Principal stresses. (b) Rosette with highest strain on intermediate gage position. (c) Mohr's circle for stress.

From Table 7.6,
$$K_1 = \frac{207}{1-0.3} = 296 \text{ GPa}$$

$$K_2 = \frac{207}{1+0.3} = 159 \text{ GPa}$$

$$A = \frac{-78+135+173}{3} = +77 \text{ microstrain}$$

$$B = -78 - 77 = -155 \text{ microstrain}$$

$$C = \frac{+173-135}{\sqrt{3}} = +22 \text{ microstrain}$$

and

$$\sigma_{max,min} = K_1 A \pm K_2 \sqrt{B^2 + C^2} = +47.7, \; -2.1 \text{ MPa}$$

$$\theta_p = \tfrac{1}{2} \tan^{-1}(22/-155) = -4.04° \text{ (clockwise)}$$

A key question is whether the angle θ_p is from the x axis to the minor principal strain or to the major principal strain. In this case it is to the minor principal strain (or stress), as shown in Figure 7.16a. It can be shown rather easily that the direction of the maximum principal strain will always be within ±30° from the direction of the gage with largest (in a positive sense) strain value; this handy rule applies only to delta rosettes. If the delta gage configuration is redrawn with the largest positive strain value being on the intermediate gage, with the other gages each 60° from this direction, then the maximum principal strain (or stress) direction will always be within the 30° sector that lies on the side closest to the second highest gage (the ±135 gage in this case), as shown in Figure 7.16b. Users of rosette gages are encouraged to use these simple relationships to check calculated results.

INSTRUMENTATION — PRINCIPLES AND APPLICATIONS

Mohr's circle for stress is plotted in Figure 7.16c.

Murphy (1945) demonstrated a graphical solution by constructing a strain circle to convert measured strains into principal strains. Equation 7.13 can be put into the following alternate form to indicate that a Mohr's circle for strains is possible.

Since all the points $\varepsilon_a, \varepsilon_b, \varepsilon_c$ must lie on a circle, they must be separated by angle 2θ (i.e., twice the angle in the rosette) and have the same orientation. Thus,

$$\varepsilon_{a,b,c} = \frac{\varepsilon_1 + \varepsilon_2}{2} + \frac{\varepsilon_1 - \varepsilon_2}{2} \cos 2\theta_{a,b,c} \tag{7.19}$$

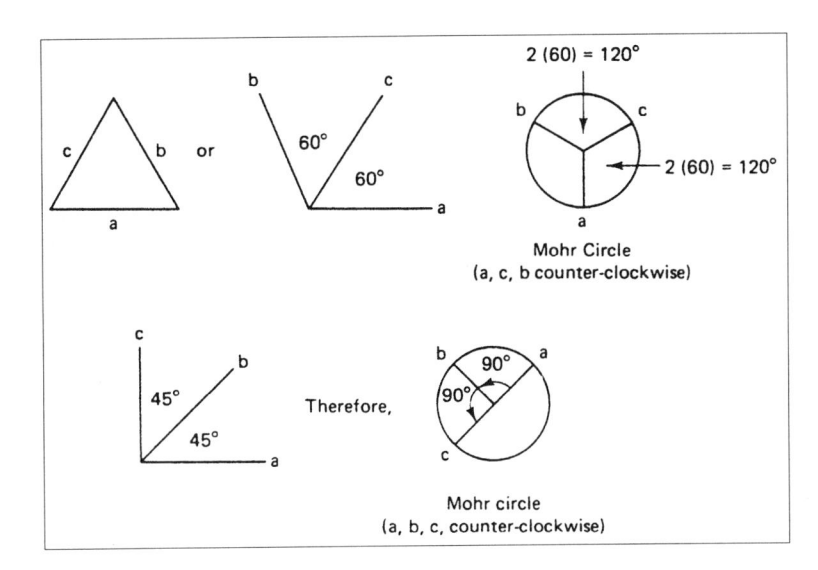

Example 7.2

Construct Mohr's circle for strain for the rosette with strain values of $\varepsilon_{a,b,c} = +100, +150$, and $+300$ microstrain, respectively. The solution proceeds as follows:

1. Plot the values of strain $\varepsilon_a, \varepsilon_b, \varepsilon_c$ on a strain $(\varepsilon-\gamma/2)$ axis system as shown in Figure 7.17a. We know that point A on Mohr's circle for strain must lie on line ε_a, point B on line ε_b, and point C on line ε_c. The problem of finding the center of Mohr's circle is addressed in the next four steps.
2. Redraw the rosette configuration to have the gage with intermediate strain value (gage b in this case) between the other two gages (a and c), and with the total included angle of the new configuration not greater than 180°. This configuration is shown in Figure 7.17b.
3. Choose an arbitrary point P on the intermediate line (ε_b for this combination of strain results) and draw a replica of the gage configuration from Step 2 on the intermediate line. The gage direction must always be vertical. Extend the gage configuration lines such that the gage a extension intersects line ε_a at point A and gage c intersects line ε_c at point C. (Note that this step often involves plotting the rosette configuration pointing downward in order to have the extensions of the gage configuration lines intersect their corresponding strain lines.)
4. Construct perpendicular bisectors to the lines PA and PC as shown in Figure 7.17d; their intersection defines the center of Mohr's circle for strain, and the construction is essentially completed.

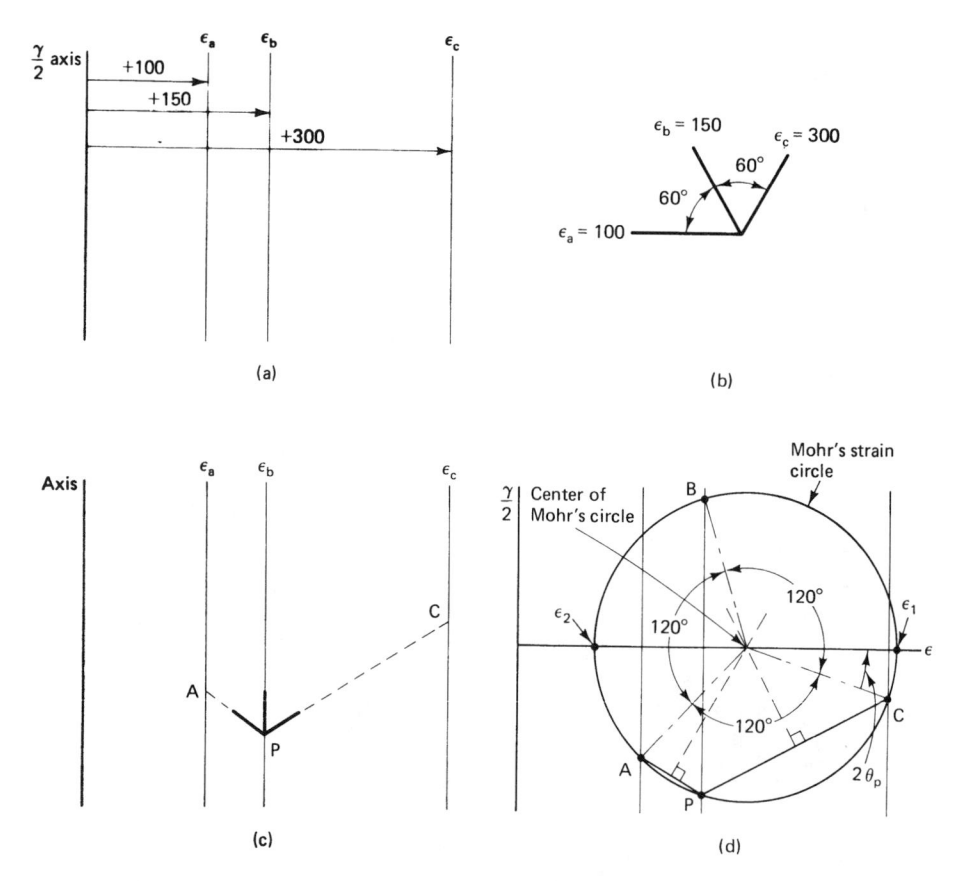

Figure 7.17 Murphy's method for plotting strain rosettes. (a) Plot strain lines. (b) Rosette drawn with intermediate strain value in intermediate position. (c) Plot rosette configuration on intermediate strain line and define points A and C. (d) Complete Mohr's circle.

Several points may be made about the completed Mohr's circle in Figure 7.17d. The angular orientation of points A, B, and C is clockwise, just as in the gage configuration in Figure 7.17b. The angles between A and B, B and C, and C and A can be shown to be 120°, thus maintaining the double angle relationship that always exists between Mohr's circle plots and real directions of stress and strain.

The angle θ_p is defined by the double angle $2\theta_p$ in Figure 7.17d. Maximum principal stress direction is θp (about 5°) counterclockwise from the direction of gage c.

Stresses are calculated by measuring the maximum and minimum values of strain, ε_1 and ε_2, and then using Equations 7.11 and 7.12.

This plotting procedure is equally useful for any type of rosette and is particularly convenient to use when a "homemade" rosette is used that does not have the usual angular orientations (the standard algebraic equations in Table 7.6 will not work in such cases). The other advantage of the approach is that it forces the user to think about actual strain values and directions. Hence, it is particularly useful for the new user of rosette gages who wants to get a better feel for what is going on. Quick freehand sketching of Mohr's circle with Murphy's method will give surprisingly accurate results and will also fully clarify any questions about directions of principal strain (stress).

7.4 DISPLACEMENT MEASUREMENTS

Deflection measurements on a scaled model need more care than on a full-scale structure because the model displacements are reduced by the geometric scale factor in comparison to the

Figure 7.18 A typical dial gage.

full-scale displacements. The problem becomes particularly acute with very small scale models where displacements on the order of a few hundredths of a millimeter must be measured accurately.

Deflections may be measured using mechanical, electrical, or optical techniques, and recording methods vary from an individual reading by eye to a sophisticated continuous recording system. Mechanical dial gages are popular because of their low initial cost, ease of attachment, and manual recording, but they tend to be less accurate than some of the electrical methods and cannot be recorded continuously. Optical methods are relatively specialized and will not be discussed here.

7.4.1 Mechanical Dial Gages

Deflections of external surface points of a structure can be measured economically using mechanical dial gages. These are quite compact, easy to apply, self-sufficient (for readouts), and accurate. Because of these advantages they can also be used for strain measurement over a fairly large gage length, as discussed in Section 7.3.1.

In general, a dial gage consists of an encased gear train, which is actuated by the rack cut out in a spindle and follows the motion to be measured. A spring attached to this spindle maintains a small force to maintain a positive contact with the structure (Figure 7.18). The gear train is connected to a pointer that indicates spindle travel on a graduated dial, generally with 0.025 or 0.0025 mm divisions, and with ranges of travel from 6.25 to 150 mm.

Sometimes more than one gage may be selected to achieve the acceptable accuracy; e.g., if the range is large, say, 25 mm, this accuracy might only be 0.025 mm. In this case, if 25 mm travel includes postyield deflections, it would be preferable to use two gages, first one with a range of 5 mm and an accuracy of 0.0025 mm, then another one of 25 mm range to measure larger deflections. Precaution should be taken to ensure that at no stage of the test does the dial gage itself exert a significant force on the structure, particularly the very small-scale models made of light-gage material. By the nature of this measurement technique, it is not possible to measure dynamic deflections with mechanical dial gages.

7.4.2 Linear Variable Differential Transformer (LVDT)

The LVDT is a reasonably compact electrical device that can be used for precision measurement of displacements (Herceg, 1976). It is a transformer with outer coils (a central primary and two

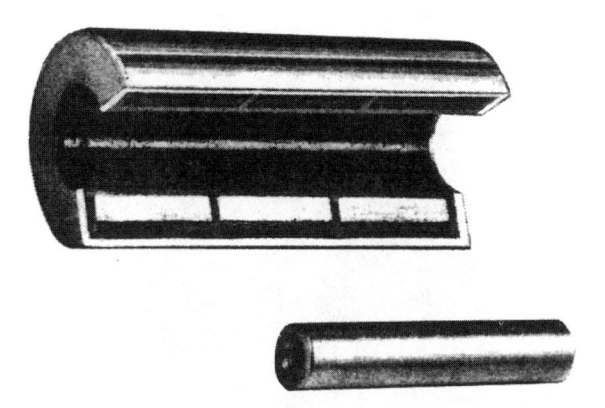

Figure 7.19 Cutaway photograph of an LVDT. (From Herceg, D. E., *Handbook of Measurement and Control — An Authoritative Treatise on the Theory and Application of LVDT*, rev. ed., Schaevitz Engineering, Pennsauken, NJ, 1976. With permission.)

symmertical secondaries) and a movable central core (Figure 7.19). As the core is moved from a structural displacement, it produces an electrical signal that can be recorded remotely and continuously, for both static and dynamic response. LVDTs being used to determine displacements of a model frame, with the outer casing of each LVDT attached to an auxiliary supporting frame and the movable cores attached in a spring-loaded fashion to the model, are shown in Figure 7.20a. The LVDT can also be adopted to act as a strain gage to monitor strains over a moderate or extended gage length as shown in Figure 7.20b.

Hanson and Curvitts (1965) discuss the details of LVDT instrumentation and Herceg (1976) gives considerable information on the theory and application of the LVDT. This instrument does require proper signal conditioning plus calibration before use. The LVDT has many commendable features that make it useful for a wide variety of applications. Some of these features, such as its frictionless measurement and infinite resolution, are unique to the LVDT and are not available in any other transducer. These features arise from the basic fact that the LVDT is an electrical transformer with a separable noncontacting core. A similar device, the DCDT (direct current differential transformer) or DC-LVDT also has wide usage in modern experimental work. Special rotation meters (RVDTs) for measuring the rotation on a beam or column are also available. These inclinometers operate on the same principles as the linear displacement LVDTs and DCDTs.

7.4.3 Linear Resistance Potentiometers

This typical transducer element is principally useful for measuring relatively large linear displacements because most potentiometers use wire-wound resistors that limit resolution; this can be increased if a film resistor is used. The potentiometer circuit, Figure 7.21, has a constant voltage input E_i applied to the ends of a fixed resistance that has a movable slider contacting the resistance element. The output voltage E_o is taken from the slider and from one of the other terminals into a high input resistance indicator.

A drawback of the potentiometer is the variation of output impedance with slider position. The signal conditioning equipment must have high input impedance. A change in the applied voltage or the shunting effects of a low resistance in the output circuit changes the current through the resistance element, becoming nonlinear. Consequently, linearity may be difficult to attain when a simple meter is used to indicate slider position.

The inertia and friction present in the mechanical portion of the linear resistance potentiometer elements represent significant restrictions to the use of these devices for dynamic measurements. Furthermore, the sliding contact wears rapidly in continuous service, severely reducing reliability.

INSTRUMENTATION — PRINCIPLES AND APPLICATIONS

Figure 7.20 Use of LVDTs in measuring displacements and strains. (a) LVDTs for deflection measurements. (Courtesy of Cornell University.) (b) LVDTs for strain measurements. (Courtesy of Drexel University.)

7.5 FULL-FIELD STRAIN MEASUREMENTS AND CRACK DETECTION METHODS

The electrical resistance strain gage is unequaled for measuring strain at a designated point, but in many situations the location of peak strain is not known. This situation calls for techniques that give a picture of strain distribution over a complete region of a structure or model. Such techniques are called *full-field strain measurements methods,* and include the following:

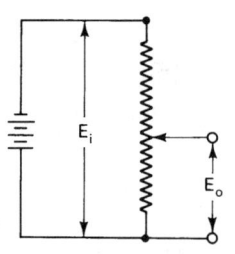

Figure 7.21 Circuit diagram for a q resistance potentiometer. (After Herceg, D. E., 1976.)

1. Whitewash (lime and water) brushed on the specimen before testing.
2. Brittle lacquers that crack at certain tensile strain levels.
3. Photoelastic coatings cemented on the surface of the structure or model.
4. Optical methods (moiré fringe and speckle holography) in which strain fields may be interpreted from optical patterns produced by the distorted structure. These methods will not be discussed here.

Most of these techniques (especially the first) are also useful in detecting cracking in concrete models and structures. Two levels of accuracy are needed for crack studies in concrete: detection and location of cracks, and definition of actual crack widths.

7.5.1 Brittle Coatings

A simple solution of lime and water brushed on the surface of any structure prior to testing will help interpret the behavior. Cracks in concrete will be accentuated and yield lines in steel structures will become visible as the whitewash coating cracks and peels under high strains. A particularly useful technique for studying cracking in wall-type specimens is to apply the whitewash and then to superimpose a gridwork of black lines to help establish a background reference system for studying motion of the cracked segments of the wall under high strain fields.

A more-sophisticated brittle coating is a spray-on commercially available lacquer that will crack at a specified level of tensile strain. The coating cracks normal to the direction of principal strain, thereby defining the directions of the principal stresses in the specimen or model. This makes the brittle lacquer a useful preliminary testing method prior to application of electrical resistance strain gages; with known directions it is possible to use two-element rosettes rather than three-element rosettes.

The brittle lacquer method has a number of advantages. Its effective gage length is essentially zero. It can be used on any surface of the structure, regardless of material, shape, or type of loading. When properly used in an environment with controlled humidity and temperature, cracking can be quite closely correlated with a certain tensile strain level, thus providing quantitative as well as qualitative information on the development of strain (and stress) with loading. Since the coating material relaxes under load, it is possible to detect high compressive strains by loading the coated structure, leaving the load in place for a number of hours, and then unloading very suddenly. The relaxed coating sees the unloading from heavy compression as a tensile load and cracks.

Brittle coatings are useful for crack detection in concrete only if the concrete cracks at a strain level below the natural cracking strain threshold of the coating.

7.5.2 Photoelastic Coatings

Birefringent materials may be bonded to the surface of structures and examined with a polariscope after the structure is loaded. The resulting optical patterns show directions of principal stresses and magnitude of shear stress (difference in principal stresses). With proper use, accurate stress analysis may be done with this method.

Several manufacturers can provide information on this technique (see list in Table 7.2). The coatings used are in the form of a very thin, birefringent plastic sheet, which is attached to the surface with an epoxy adhesive. in the case of PhotoSress® (Measurements Group, Inc.), a liquid form of the coating can be cast on a flat-plate mold, allowed to partially polymerize, and then formed by hand to the shape of the test surface. It is recommended that preliminary tests be conducted on a simple flexure specimen to get some experience with the technique before actually using it on a complicated structure.

Photoelastic coatings have been used quite successfully for detecting cracking in concrete structures before the cracks are visible to the naked eye (Abeles, 1966; Corum and Smith, 1970). The basic mechanism of failure of concrete (or model concrete) under load is precipitated by the initiation and propagation of small cracks that extend and interconnect to form major structural cracks. As these microcracks form, the resulting strain concentrations over a very short gage length produce extremely high local stresses in the photoelastic coatings bonded to the concrete surface, and a literal explosion of optical effects (fringes) results when a crack forms.

7.5.3 Other Crack Detection Methods

Recent research has led to the development of a crack detection coating. The coating will crack whenever the underlying structure cracks; this is evidenced both visually and electrically. The crack detection coating is so designed that, when a crack occurs, an electrically conductive liquid is released and the crack is immediately visible. This conductive medium presents a low resistance path, so that with an electrical connection at the surface of the coating, a change in resistance or current flow between the metal base and the electrical connection can be detected by a simple ohmmeter. Its advantage is in its capability for remote observation, which can be followed (after electrical response indicates failure) by visual observation of the precise location of the crack. The technique is particularly useful for coated areas that are not readily accessible or that consist of a large area, such as a long, welded joint.

The simplest method of crack detection in concrete structures is by visual inspection, either by the naked eye or with a magnifying glass or a handheld microscope. A graduated handheld microscope is commercially available that enables one not only to detect cracks but also to measure their widths.

The individual strain gages of single-element type can be used for crack detection by mounting on a crack detection strip. They are connected to a continuous strip chart recorder and monitored remotely. Such a setup is suitable for crack detection of the internal surface of a vessel when it is subjected to pressure test and is not accessible during testing. It may be noted that these gages are also convenient for cracking strain measurement in concrete because the nonlinearity due to the cracking will be indicated on the recorder.

Similar to the electrical resistance gages, short-length LVDTs, described earlier, can be used for crack detection. Because of its more elaborate setup and initial adjustments, the LVDT is more suitable for mounting on the external surface of the structure and where automatic recording is desired.

The last crack detection method to be mentioned here relies on sensing and measurement of the noise generated by the cracking process. The *acoustic emission* technique uses rather sophisticated electronic equipment. It has been used quite extensively in fracture mechanics testing and in mechanical engineering applications. It has great potential for further applications in structural testing and structural model analysis.

7.6 STRESS AND FORCE MEASUREMENT

For complete analysis in model testing, measurements of both forces and stresses are needed. Forces are either measured directly or obtained using equilibrium principles, whereas the stresses in a structure are derived from the measured strains at a number of locations.

Various types of instrumentation are available for directly measuring different forces, such as compression or tension. They include load cells (for measuring reactions and external forces), embedded stress plugs or meters (for measuring stresses and strains inside a concrete structure), and stress sensitive paints (between washers to measure forces by the electrical resistance of these paints). Although most of these are available commercially, often their use is precluded because of economic factors and the nature of the experiment; e.g., a load cell required for measuring reactions in a small-scale beam test may not be available in that small size or else may not fit in the available space for the measurement. In such a case, available laboratory equipment can be easily used to fabricate the required load cell. The following discussion will be focused on both commercially available and laboratory-made load cells to show that a proper combination may be made to achieve the best and most economical results.

7.6.1 Load Cells — Types and Sizes

Load cells are used for measuring loads and reactions and other forces and can be classified into three major categories, depending on the type of loading. Accordingly, they are called *compression, tension,* or *universal* (to measure either tension or compression) load cells. Tension cells are the simplest because of their innate stability in a state of tension and can vary from simple tension bolts, for rough estimation of forces, to more accurate and expensive load cells.

A basic load cell of any of these kinds consists of a complete strain gage bridge shown in Figure 7.9a. The strain gages 1, 2, 3, and 4 are arranged so as to eliminate the effect of the undesired stress components. From bridge theory, the output of the bridge may be expressed as

$$\varepsilon_o = -\varepsilon_1 + \varepsilon_2 - \varepsilon_3 + \varepsilon_4 \tag{7.20}$$

where ε_1 through ε_4 are strains of the different arms shown in Figure 7.9a.

If the strain gages reading strains ε_2 and ε_4 are placed in the load cell so as to read strains opposite in sign to strains ε_1 and ε_3, the sensitivity and accuracy of the load cell improves. This is accomplished by placing R_1 and R_3 in the axial direction of the applied force, and R_2 and R_4 in the transverse direction, as shown earlier in Figure 7.9a. Thus

$$\varepsilon_0 = K\varepsilon_1 \tag{7.21}$$

where K = bridge multiplication factor. $K = 2.6$ if Poisson's ratio is 0.3. It may further be noticed that any torsion in this cell and any bending during the applied loading is automatically eliminated and thus will not influence the output.

An ideal load cell should be compact in shape and size for a given load, be inexpensive and easy to use, have high sensitivity, have elastic behavior (for repeated use) at the upper end of the loading range, have inherent stability for long-term loading, and be relatively rigid. Several of these qualifications are generally found in the available load cells, and force measurement does not pose a problem, at least for static model testing. However, the above characteristics are discussed to help readers plan their own load cell fabrication when required.

Sensitivity of a load cell may be expressed in units of strain per unit load. Thus, it is directly proportional to the maximum stress used in the design of the cell and inversely proportional to its maximum load capacity. By rewriting Equation 7.21,

$$\varepsilon_0 = K\varepsilon_1 = K\frac{\text{design stress}}{E} \tag{7.22}$$

GENERAL PURPOSE
FEATURES
- 5 lbs. to 20,000 lbs.
- Low profile
- Low sensitivity to extraneous loads

MODEL 3397
Capacities available
25 to 300 lbs.

MODEL 3169
3169 (English Thd.)
MODEL 3169-106
(Metric Thd.)

MODEL 3171
(English Thd.)
MODEL 3171-106
(Metric Thd.)

FATIGUE RESISTANT
FEATURES
- 200 lbs. to 2,000,000 lbs.
- Resists fatigue failure
- High resistance to bending movements and side loads

MODEL 3116
Capacities available
5K, 10K and 20K lbs.

MODEL 3156
Capacities available
25K to 150K lbs.

MODEL 3161
Capacities available
2K, 5K, 10K and 25K lbs.

Figure 7.22 Load cells of various capacities. (Courtesy of Eaton, Lebow® Products.)

Then

$$\text{Sensitivity} = \frac{\varepsilon_0}{\text{design load}}$$

$$= \frac{\text{design stress}}{\text{design load}} \cdot \frac{K}{E} \qquad (7.23)$$

This means that for a given design stress and design load the optimum sensitivity will result from a maximum value of K and a minimum value of E. Although the value of K can be as high as 4, in a load cell of usual design and with the advantages mentioned earlier it has a value of 2.6 with a Poisson's ratio value of 0.3. Various metals, such as aluminum and steel, are used for fabricating load cells. An aluminum load cell is three times as sensitive as a steel load cell because of its lower modulus of elasticity. But, it has a disadvantage in that aluminum has a nonlinear stress–strain relationship at a smaller design stress level than does steel.

Typical load cells, commercially available, are shown in Figure 7.22. They consist of an enclosed tube that is instrumented with strain gages as explained earlier. The cross section of the tube depends on the highest permissible level of stress in the cell material and on the yield properties of the material. The outside cover essentially protects the core, which may be damaged during testing or as a result of the environment. At both ends of the universal gage, there are threaded ends in the core tube to provide suitable attachments for either tension or compression force measurements. A wide range of load capacities is available, ranging from loads as small as 10 N to as high as 7000 kN.

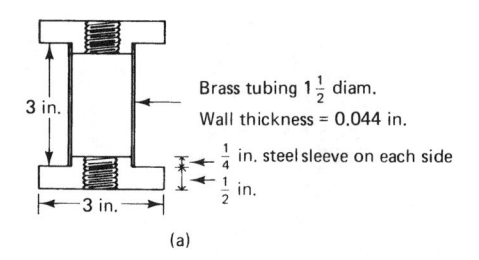

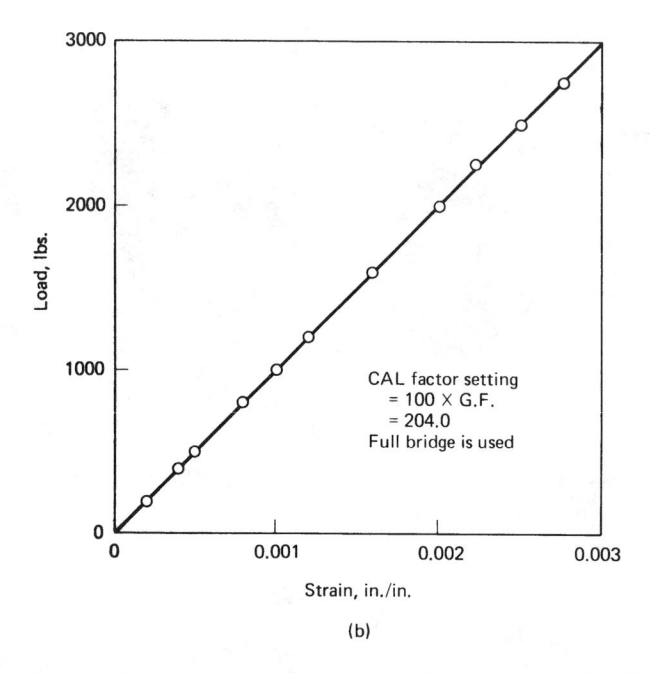

Figure 7.23 Load cell and its characteristics made in Cornell Laboratory. (a) Details of load cell. (b) Calibration chart for 3000 lb load cell.

For very small-scale models, the commercially available load cell may pose problems with regard to size or perhaps cost, and one may construct his or her own cell. The load cell should be cycled through its design range after construction and prior to use in an experiment in order to eliminate any hysteresis in its behavior. Careful consideration should be given to the material used and to the stability of the strain gage cement as well as to the instrumentation needed over the designated loading period. Heat-cured epoxy resin cements are often used in load cell strain gaging because of their superior stability. All load cells should be recalibrated periodically to check linearity and the calibration constant. A typical load cell made at the Cornell Models Laboratory to measure reaction is shown in Figure 7.23. It was made using a machined brass tube 38 mm in diameter and instrumented with four SR-4 strain gages to form a complete bridge. A calibration chart that gives the different ranges of accuracy for a particular set of measurements is also shown. Commercially made load cells typically specify a variety of basic characteristics. For load cell Model 3116-161, 100 kN from Eaton Corporation, Lebow® Products, Troy, MI, for example, these data are

Full-scale range	20 kN to 100 kN
Nonlinearity	±0.2% of rated output
Repeatability	±0.05% of rated output
Hysteresis	±0.20% of rated output
Safe operating temperature	−54 to +93°C

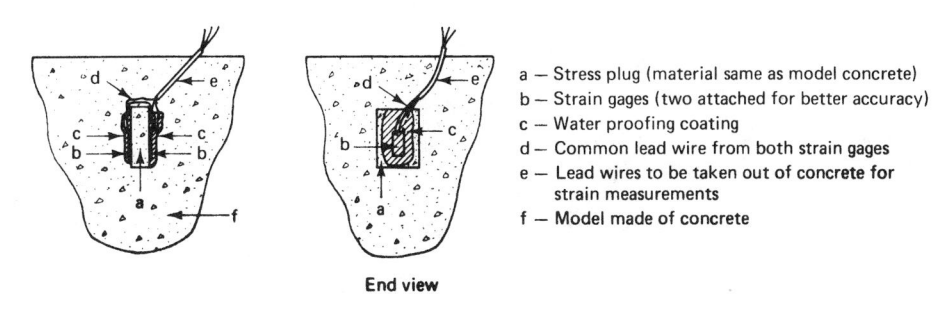

INSTRUMENTATION — PRINCIPLES AND APPLICATIONS

a — Stress plug (material same as model concrete)
b — Strain gages (two attached for better accuracy)
c — Water proofing coating
d — Common lead wire from both strain gages
e — Lead wires to be taken out of concrete for strain measurements
f — Model made of concrete

End view

Figure 7.24 Stress plug details (typical).

7.6.2 Embedded Stress Meters and Plugs

In order to obtain information in the core of a concrete structure (such as in dams or thick pressure vessels), one has to resort to embedded instrumentation. This is particularly important because the triaxial stress condition that exists is hard to investigate from surface strains, especially if material-related effects such as creep and shrinkage must also be studied. The embedded gage gives strain and/or stress readings at interior points in the structure to enable one to correlate the experimental work with theoretical predictions.

The problem with this kind of instrumentation is basically related to the heterogeneity that is introduced into the system because of these embedments. In addition to changing the properties of the structure in the immediate area, the instrumentation is susceptible to damage during the construction stage and requires a good deal of protection. Because of the time needed for curing of the structure, time-dependent stability of this instrumentation must be carefully considered. Reports by Corum and Smith (1970) and Geymeyer (1967) treat embedded gages.

In the *stress plug method* (Brownie and McCurich, 1967), a small volume of concrete is replaced by a plug, also made of concrete (or mortar), which is instrumented with strain gages. The plug shape and characteristics of its material are determined so that the strains can be converted into stresses. It is interesting to determine the error that may be introduced by the different properties of plug material and the surrounding concrete. Figure 7.24 shows a plug embedded in concrete. Meter and concrete properties are designated by subscripts m and c, respectively. By means of Hooke's law:

$$\sigma_c = E_c \varepsilon_c, \quad \sigma_m = E_m \varepsilon_m \quad (7.24)$$

from which it can be shown that

$$\sigma_m = \sigma_c (1 + C_s) \quad (7.25)$$

and

$$\varepsilon_m = \varepsilon_c (1 + C_e) \quad (7.26)$$

where: C_s = stress concentration factor
C_e = strain increment factor

Constants C_s and C_e are essentially the errors introduced in these measurements as a result of imperfect matching of the physical properties of the concrete and the meter. Thus, the error will be magnified because of the difference in moduli, Poisson's ratio, and thermal expansion. Although the error can be significant, by proper calibration and by keeping the mismatch to a low value, it can be reduced to a minimum. Tests have indicated reasonable accuracy, less than 0.5% variation for stresses up to 48 MPa, with a negligible drift over a 1-year period.

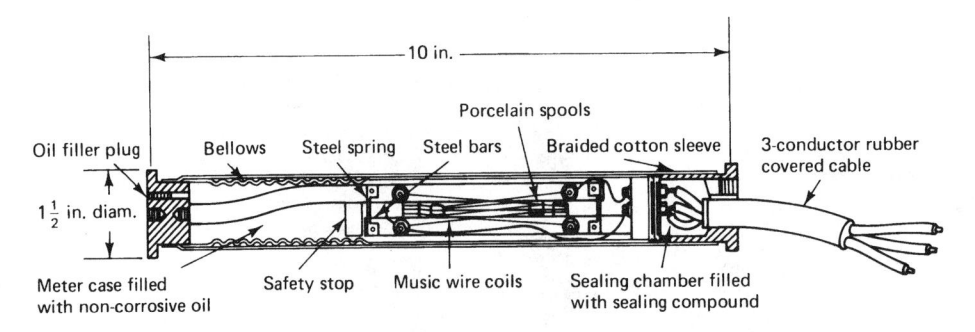

Figure 7.25 Carlson meter.

The *Carlson meter*, which is commercially available, has been used in the U.S. over a long period of time for strain and stress measurements in mass concrete. Although the original meters are somewhat too large for use in small-scale model structures, recent advances have reduced their size for successful use in large-scale models. The description as presented by Geymeyer (1967) applies to both small and large meters. The strain meter, in general, consists of a long cylinder with anchors that engage in the surrounding concrete. Inside, there are two equal coils of very fine steel music wire, 0.064 mm in diameter, wound under a tensile stress of 6895 MPa. When the displacement at the ends of the tube (or meter) takes place, these coils undergo equal but opposite changes in resistance. When they are connected in a suitable position to a Wheatstone bridge, strain can be measured within 5μ strain. It also measures temperature variation up to $0.06°C$. A cross section and some of the details of this meter are shown in Figure 7.25.

The Carlson stress meter is very similar to the strain meter, except that the pressure acting on a circular plate (Figure 7.26) is measured. This plate is a mercury-filled diaphragm, designed so that the pressure in the mercury is always substantially equal to the pressure in the concrete normal to the plate. The center portion of the plate is made slightly flexible by cutting away part of its thickness. Since the mercury is in contact with the more flexible central part, it deflects in direct proportion to the intensity of the applied stress. The sensitivity and stability of the meter are excellent over a period of time.

Embedded strain gages as used by Corum and Smith (1970) are generally encapsulated in plastic, resin, or an epoxy protection cover and may be used as single gages or rosettes depending upon the application. A typical rosette assembly is shown in Figure 7.27. Thus, foil gages are used as sensing elements and attached on each side of the epoxy strip so as to average the strains. Additional epoxy coating is given to the gage for further protection. The sensing and performance of these gages is similar to the electrical resistance gages described earlier. The embedded gage shown in Figure 7.27 is laboratory made since commercially available gages for small-scale models are lacking.

Another economical solution to the use of embedded gages is the application of a waterproof plastic cover to provide adequate protection to the regular strain gage. Many researchers have used these gages, however, with contradicting opinions and results. The problems inherent in using this type of gage are threefold: (1) the plastic material is not fully waterproof and might cause some drift in strain readings; (2) the mismatch of thermal expansion between the plastic and the concrete and also the self-heating effect from the gage itself produces problems; and (3) creep and relaxation of the plastic relative to the surrounding concrete could result in an apparent strain indication. Plastic-encapsulated gages, however, have a fairly good short-term stability.

Two experimentally produced embedded gages are shown in Figure 7.28. In each gage the sensing element is a weldable resistance gage, welded to a metal strip. The roughness of the strip permits a strong bond with the surrounding concrete and a transmission of the concrete strain field into the gage.

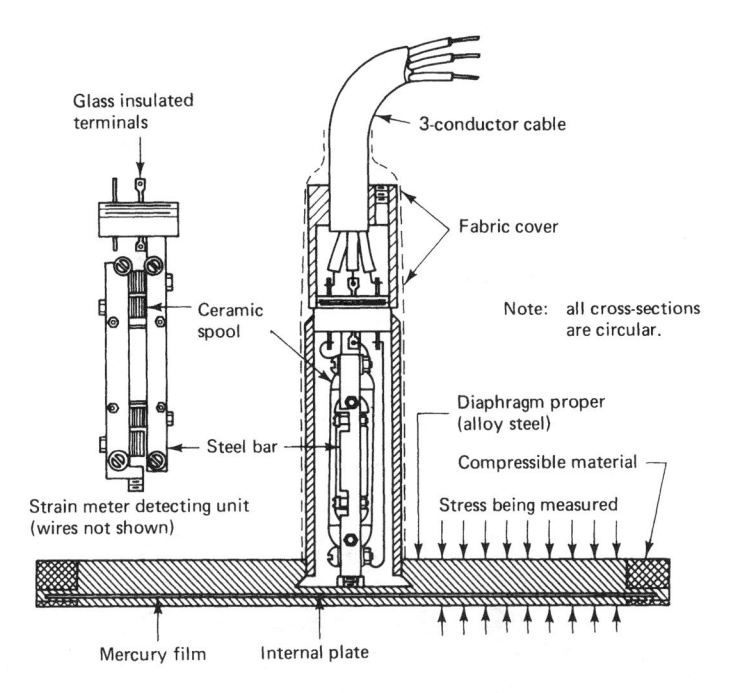

INSTRUMENTATION — PRINCIPLES AND APPLICATIONS

Figure 7.26 Carlson stress meter.

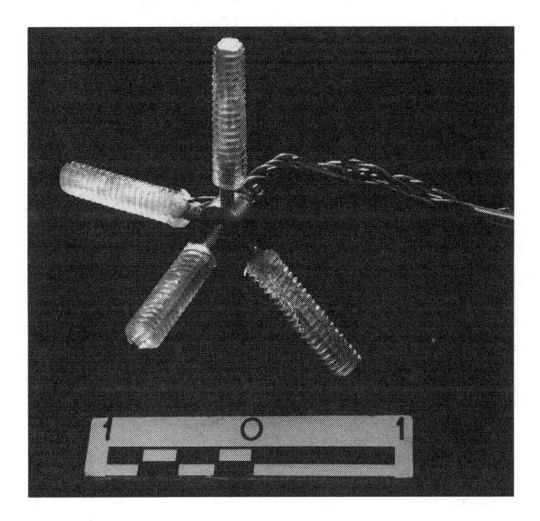

Figure 7.27 Embedded strain gage. (From Corum, J. M. and Smith, J. E., 1970.)

The *vibrating wire gage* shown in Figure 7.29 operates on the principle of the natural frequency of a tensioned wire. The steel wire enclosed in the middle of the metal tubular body of the gage can be "plucked" with the actuation device at its midlength, and the resulting vibrations are sensed by the same device and transmitted to a recorder. With the circular end of the gage plates cast into concrete, any strain in the concrete will result in a relative motion of the end plates and hence a change in the initial tension in the wire. The change in vibration frequency of the wire is then translated into an equivalent change in length of wire and then into effective strain.

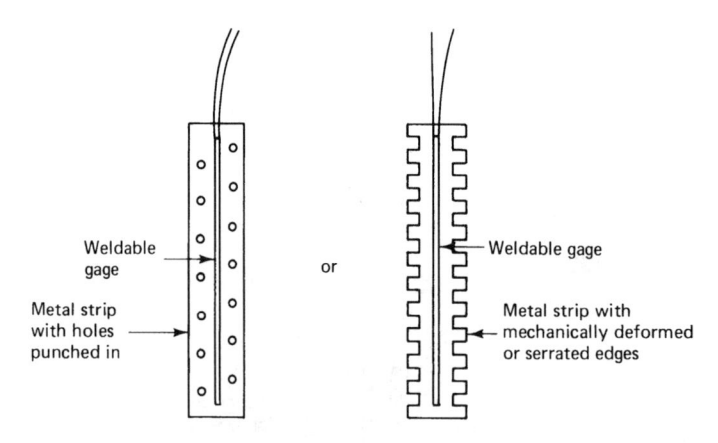

Figure 7.28 Weldable gage for embedment in concrete.

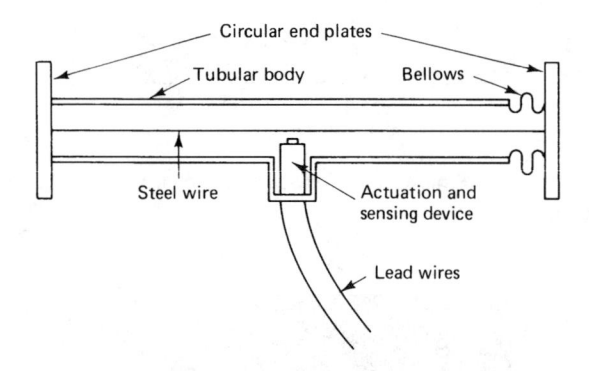

Figure 7.29 Vibrating wire strain gage for embedment in concrete.

This type of gage has seen extensive use in thick-walled prestressed concrete reactor vessels, particularly in Europe. A welded stainless steel vibrating wire gage has been fabricated and tested to have the following characteristics: resolution of 0.1 microstrain with the best available recording equipment, drift of one microstrain per year at room temperature, and drift of about 10 microstrain per year at 93°C. These are rather remarkable specifications, particularly at low temperatures, and the chief disadvantage of the gage is its relatively high cost.

7.6.3 Other Measuring Devices

The proving ring, a device resembling a short length of metal pipe, is used to measure loads by sensing its change in diameter when compressed or extended. The diameter change is measured with a dial gage. Provided that applied loads are kept within the elastic limit, this type of device is very accurate and repeatable.

A newer stress-measuring device is based on stress-sensitive paints that generate an electrical output when loaded in compression. The paint is applied to either washers or wafers made of steel or aluminum. A typical wafer-type device (called *Microducer*) is shown in Figure 7.30; also shown is its calibration chart between the current measured in milliamperes and the applied load. The circuit is similar to a balanced Wheatstone bridge. This is basically a miniature load cell and can be used successfully for measuring very small loads. Recently, washer-type load cells have been developed to measure loads up to 1 kN. The washers (or miniature load cells) have a diameter of 25 mm and a thickness of only 3.8 mm.

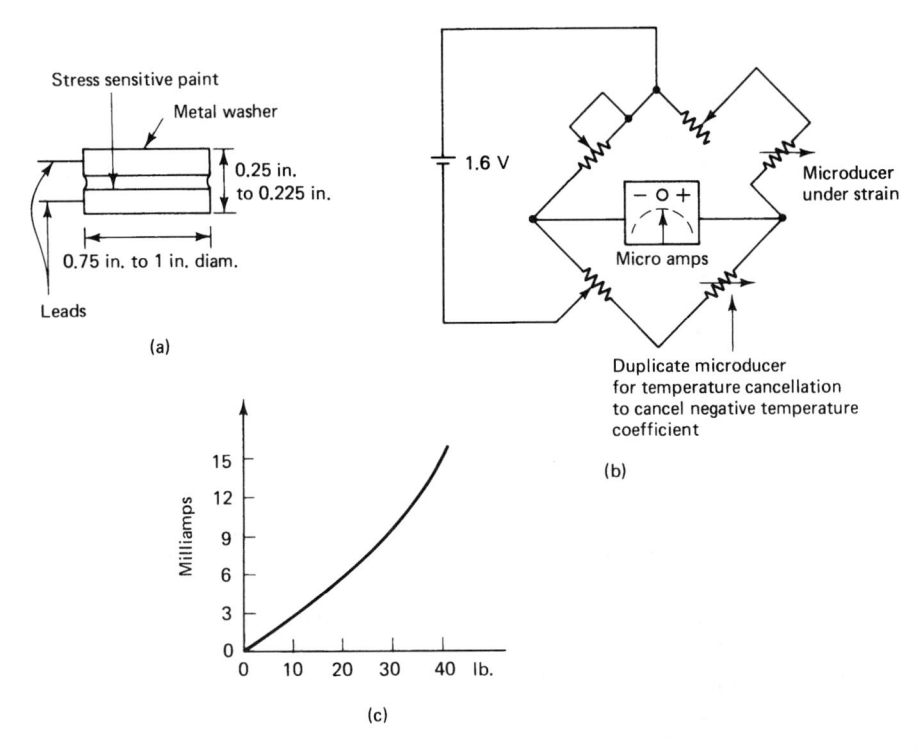

INSTRUMENTATION — PRINCIPLES AND APPLICATIONS

Figure 7.30 Microducer used as a load washer. (a) Microducer. (b) Circuitry. (c) Relation between load and current.

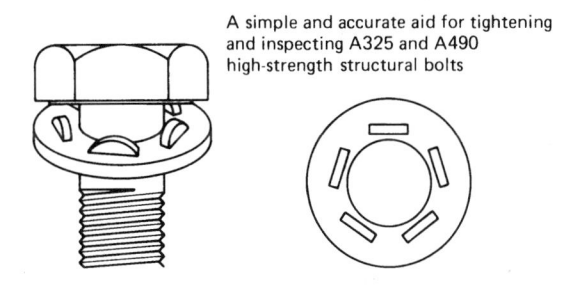

Figure 7.31 Load indicator washer. (Courtesy of Bethlehem Steel Co.)

The load-indicating washer (shown in Figure 7.31) is very similar to the regular washer. They have three small protrusions that are compressed elastically when the load is applied; hence they can be calibrated and used repeatedly. They are commercially available for full-scale load measurements; however, this principle can be extended to smaller washers for measuring loads in small-scale models.

7.7 TEMPERATURE MEASUREMENTS

Temperature is an important quantity, particularly in the case of prestressed concrete reactor vessels or containments. In these structures, the temperature variations in the concrete is of interest to the researcher. Since it is difficult to calculate temperature inside the mass, it is determined by actual measurement.

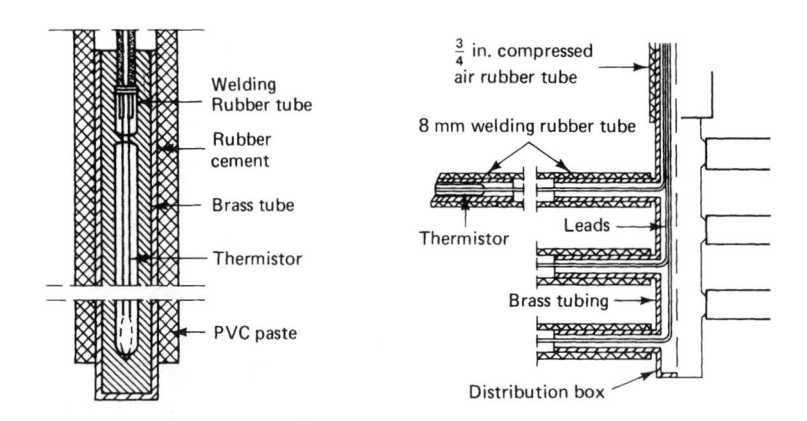

Figure 7.32 Installation of thermistors in concrete.

Thermocouples and thermistors are generally used for measuring internal temperatures in concrete. The basic principle of the thermocouple is that an electric current is maintained in a circuit of two dissimilar metals when their junctions are held at different temperatures. The substantial variation in the electromotive force with temperature gives the thermocouple practical importance in that it provides a device with an accurately measurable electrical potential over a wide range of temperatures.

A thermistor, in principle, is a thermally sensitive electric resistor based on the semiconductor effect. The resistance of a thermistor decreases with an increase in the temperature. Advantages of the thermistor are the reduced degree of amplification required, reduced size, and the thermal inertia of the sensing unit compared with conventional thermal resistors.

A more recent development in this area is the thermistor-type thermometer and the method developed for embedding it in concrete. The thermistor is enclosed in a brass tube surrounded by a length of rubber welding hose, which protects the leads against mechanical damage and moisture movement. The whole assembly is moisture proofed and has moisture-proof leads. Moisture proofing is accomplished by sealing the thermistor and the polyvinylchloride (PVC) insulated leads in a thermoplastic material and applying a layer of rubber cement between the brass tube and the moisture-tight rubber welding hose (Figure 7.32). The whole assembly is prepared, calibrated, and ready-mounted in the required sections. All thermometer cables belonging to each section are enclosed in a 19-mm compressed-air rubber tube, connected from the section to the readout unit. A separate thermometer is connected to an individual cable via the distribution box at the other end of the compressed-air tube. These are mounted on each section during concreting. It is obvious that the thermistors are not recoverable after the test.

Sometimes it is necessary to create differential temperature conditions across the section of concrete under test to reproduce operating or extreme accident conditions. Since concrete has a small thermal conductivity, the desired gradient cannot be obtained by merely artificially heating one surface, and cooling has to be done internally by artificial means. In several tests by Labonte (1971) on models of the anchorage zone of a prestressed concrete containment, the desired gradient was successfully created by embedding thin copper tubing across the wall and running boiling water through the tubes until the operating temperature was reached.

7.8 CREEP AND SHRINKAGE CHARACTERISTICS AND MOISTURE MEASUREMENTS

Under long-time loading on a concrete structure, in addition to strains from the applied load, there are additional strains due to creep and shrinkage. Simultaneous observations should, therefore,

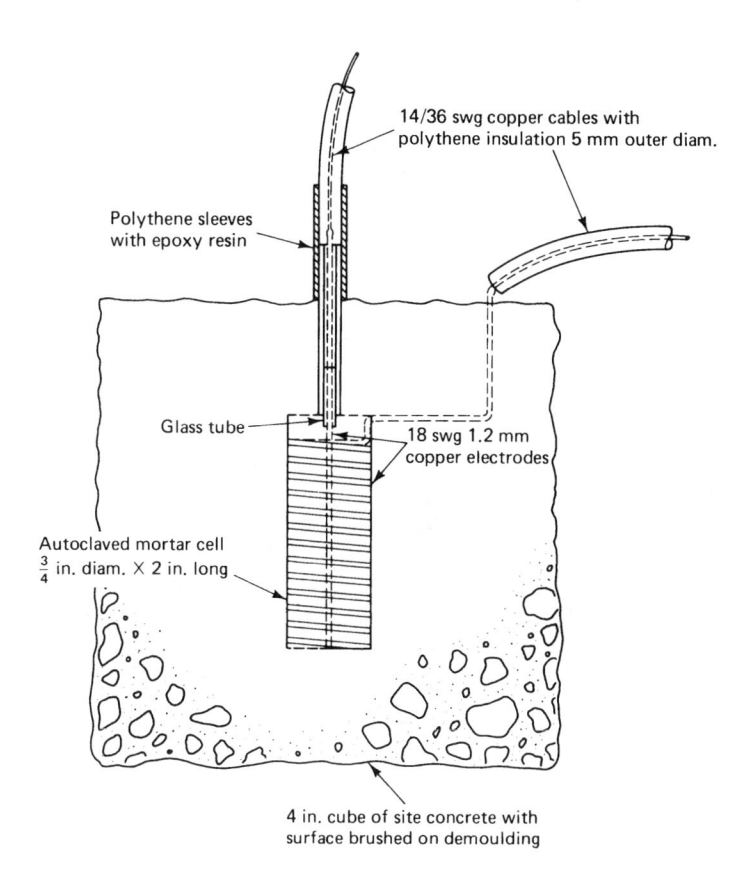

Figure 7.33 Details of moisture gage embedded in concrete.

be taken to isolate the effect of these quantities. As described in the earlier sections, the necessary strains can be measured without too much difficulty. Very often, only mechanical gages are used since the electrical ones are usually not reliable for measuring strains over long periods.

Information on the progress of drying in massive concrete structures is of considerable value, not only in assessing the consequent changes in concrete properties (e.g., effectiveness as a neutron shield, thermal expansion, etc.) but also in determining its shrinkage characteristics. The latter may lead to stress development, cracking or dimensional changes, and loss of prestress.

Electrical resistance moisture gages have been used to measure the changing moisture distribution within concrete structures. The basic principle in this method is the relation between the resistance of the concrete and the concrete moisture content. In general, the contact resistance between the electrodes and the concrete may exceed the actual resistance of the moist concrete by a considerable magnitude. To overcome this difficulty, the absorbent type of gage shown in Figure 7.33 has been used. The electrodes, instead of being placed directly in the concrete, are enclosed in an absorbent material, which in turn is buried in the concrete. To eliminate the contact resistance between the absorbent material and the concrete, a concentric electrode system is used. With such an arrangement, the relationship between the gage resistance and the concrete moisture content is no longer a direct one. Any change in the moisture content of the concrete is normally accomplished by a change in partial pressure. A pressure gradient between the gage and the concrete is thereby created, which in turn causes a change of moisture content within the absorbent material and a corresponding change in the electrical resistance.

7.9 DATA ACQUISITION AND REDUCTION

Recording of data is one of the crucial parts of any experiment, and it must be carefully planned and tested well in advance of the experiment. Modern data acquisition equipment ranges from a simple, manually operated strain indicator box to sophisticated automatic systems that digitally record data continuously. The range of satisfactory equipment for a static test is very broad because recording time is usually not crucial except when failure is approached and strains and displacements are changing rapidly. On the other hand, dynamic tests require a data acquisition system that is capable of monitoring and recording many channels simultaneously in a fraction of a second. The rather specialized topic of large-scale dynamic instrumentation systems is beyond the scope of this book, not only because of the complexities of the topic, but also because state-of-the-art systems change so rapidly that standard book treatments quickly become obsolete. Information from equipment manufacturers is of crucial importance in selecting a new data acquisition system.

Reduction of data must be considered in selecting data acquisition systems. Methods of data acquisition in which numbers (strains, displacement, etc.) that are written down by hand and later reduced by hand or after putting the data manually into computer storage are now obsolete. Both processes are prone to human error and should be avoided whenever possible. Fortunately, the electronics revolution has made it possible for even a modest-budget laboratory to have a data acquisition system tied to a computer. Powerful computers of small size and modest cost can perform both the function of data acquisition and reduction as well as controlling the loading machines. These systems not only make data acquisition itself much easier, but perhaps even more importantly, the data are stored permanently on the hard drive or auxiliary storage, ready for reduction and conversion into stresses, stress resultants, and plots and tables that can be used directly in reports on the experiments. Key data can be monitored and plotted in real time as the test progresses.

7.9.1 Types of Data Recording

Data acquisition systems may be classified as follows:

1. Intermittent
2. Semicontinuous
3. Continuous

Intermittent recording indicators are usually manual; i.e., each strain is read manually and recorded separately, one at a time. The strain indicator is usually connected to switch and balance units to permit multiple channels to be read by a single indicator. The indicator contains Wheatstone bridge circuitry, internal resistances for completing the bridge when half or quarter bridges are used without external dummy gages, and other electronics and mechanical equipment needed to convert the electrical signals into numerical values of strain. Each switch and balance unit accommodates a number of strain gages (usually 10, 20, or 30). It also has bridge completion resistances and terminals that permit the use of external dummy gages when needed. The balancing capability permits zero values to be set for each channel to facilitate later reduction of data. Commercially available indicator units are very compact and easy to use, and are employed extensively when a small number of readings are to be taken. A typical indicator is shown in Figure 7.34. The reader is referred to commercial literature for more details.

Semicontinuous recording systems permit a large number of measurements from strain gages or other sensing devices to be made rapidly. Commercially available units have up to hundreds of channels and consist of a power supply, signal conditioning circuitry such as bridge completion units for strain gages, a multiplexer or scanner, a digital voltmeter with a display unit, and a data

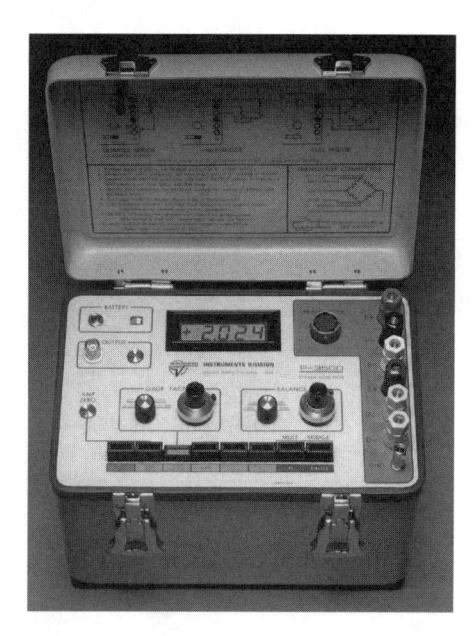

Figure 7.34 Portable digital strain indicator. (Courtesy of Measurements Group, Inc.)

storage capability. The scanner sequentially scans many channels per second (typically 10 or 20) and controls the power supply and voltmeter for each channel. Sensing devices that may be used include strain gages, LVDTs, DCDTs, potentiometers, load cells, thermocouples, and any other device that generates a voltage. A printer and at least one plotter are the usual accessory devices for this type of system. While it may be tempting to let this automatic system acquire all data during a test without spending any time looking at the data, this is the wrong approach. Critical quantities (pressure during a pressure test; maximum displacement during a beam, frame, or slab test; load during a prestressing operation; etc.) should be output in real time so that the progress of the test can be continuously monitored while it is running.

A continuous recording system records measurements continuously in a number of digital storage devices for later reduction and analysis. This type of system also accepts essentially any electrical device that generates a voltage. Important applications in structural testing include dynamic testing (where the analog-to-digital conversions are then done later), recording of complete moment–rotation relationships for beams and other structural components subjected to reversing loads, and monitoring of instruments as failure occurs. Before considering a specific application in data acquisition systems, it is worth pointing out that tape recorders, digital and video cameras are often useful in helping document experiments. The audio tape recorder may be used for storing visually obtained data and for recording remarks by the test engineer on the progress of the test. A video recorder is useful in capturing the behavior of a test specimen, particularly at the moment of failure. Stop-frame analysis of a video recording after the experiment is over may well lead to a greatly improved understanding of the precise behavior modes during the failure process.

7.9.2 Various Data Acquisition Systems

Many different commercially available data acquisition systems are in use in structural testing laboratories. Some are used in the as-furnished condition, but most have at least some custom electronic components and interfaces to suit the needs of the individual users. The services of a good electronics technician are usually indispensable in getting a modern data system operating properly.

Figure 7.35 The System 5000 data acquisition system. (Courtesy of Measurements Group, Inc.)

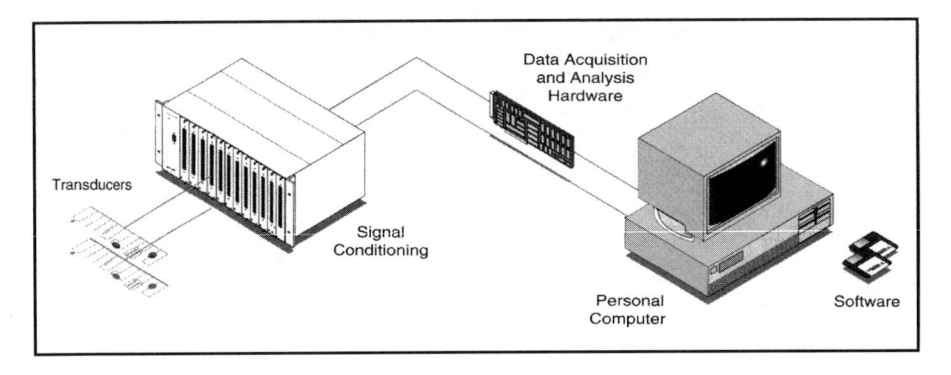

Figure 7.36 Data acquisition system. (Courtesy of National Instruments.)

A system that can be easily adapted to structural model testing made by Measurements Group, Inc. is the System 5000 shown in Figure 7.35. This is a Microsoft Windows-based software system addressing virtually every variable that must be considered in model testing: from initial data entry, to acquisition and conditioning, to online and off-line correction, reduction, and presentation of results. The System 5000 will accept inputs from strain gages, thermocouples, LVDTs, load cells, and other transducers.

The scanner accepts up to four cards (five channels per card and up to 20 channels per scanner). A maximum of 60 scanners (1200 channels) can be accommodated. A 16-bit (15-bit plus sign) successive approximation converter with a 15-bit usable resolution and 40 μs conversion time per reading is used. The scan rate is 1 ms/scan with ten complete scans per second typical usage and concurrent scanning for all scanners. Input channels in each scanner are scanned sequentially at a rate of 25,000 samples per second and stored in random access memory within a 1-ms window.

Today, most engineers involved in structural testing use personal computers with expansion buses that can be adapted for laboratory research and all forms of measurement. A versatile system for data acquisition, measurement, and virtual instrument programming has been pioneered by National Instruments (1998). Obtaining proper results from a PC-based data acquisition (DAQ) system depends on each of the system elements shown in Figure 7.36. These include (1) personal computer, (2) transducers, (3) signal conditioning, (4) DAQ hardware, and (5) software. Most sensors require some form of signal conditioning for use with plug-in DAQ boards. These requirements may include amplification, excitation, filtering, high-voltage isolation, or multiplexing. Because most data acquisition applications involve a unique combination of sensors and signals, National Instruments offers a variety of signal conditioning products and data acquisition hardware.

National Instruments introduced in 1986 LabVIEW,® a graphical programming package, with the goal of providing a software tool to enable engineers to develop customized systems for analyzing a diversity of engineering problems. The success of this program, initially for the Macintosh, was followed with versions for PCs, workstations, Windows versions, and translations into several foreign languages. This innovation started the virtual instrumentation revolution, which is changing instrumentation in the test and measurement world as well as the industrial markets. LabVIEW® can transform a personal computer to any number of virtual instruments. The computer then becomes a powerful, multipurpose laboratory tool that can replace older, outmoded equipment. Innovative engineering curricula developed at several U.S. Universities, including the one developed at Drexel University, use LabVIEW as the main laboratory software (Tanyel and Quinn, 1993, 1994; Tanyel, 1994).

Similar data acquisition and data reduction systems are available from a number of electronic hardware/software manufacturers including Daytronic, Hewlett Packard, and others. Capabilities of these systems vary with each supplier, and care must be exercised in designing and ordering such systems.

Such systems have several key characteristics that are indispensable in performing experiments with many data sensors: (1) high scan rate to capture a complete set of data in a minimum elapsed time; (2) data storage and display capabilities; (3) completely flexible control by an easily programmed computer; (4) sufficient calculation capabilities to permit subsequent reduction, tabulation, and plotting of results; and (5) portability and no need for air conditioning (although the environment should be as clean as possible).

The computers used in the systems described above are also capable of being interfaced with the loading equipment to control the application of load completely. This is particularly important in experiments that involve repeated or reversed loading cycles. This capability affords an important new dimension in testing capability, freeing the test engineer from the many routine (and energy-sapping) tasks involved in loading and measuring the response of a structure or model.

7.10 FIBER OPTICS AND SMART STRUCTURES

A "smart" material or structure is one that has the ability to sense environmental changes within or around the structure and has the ability to interpret and react to these changes. In order to accomplish this, a "nervous system" capable of sensing change in the material or structure is required. Fiber-optic sensor technology has enabled the implementation of this nervous system by (1) providing sensors that are small and rugged enough that they can be incorporated into materials which are part of the structure, (2) enabling sensors to be multiplexed in substantial numbers along a single line allowing reduction of weight, (3) providing electric and magnetic isolation, and (4) supporting the high bandwidth necessary for large numbers of high-performance sensors (Udd, 1995).

The revolution in the fiber-optic telecommunication and optoelectronic industries has enabled the development of fiber-optic sensors that have unique advantages over conventional electrical sensors. These sensors offer a series of advantages, including small size and weight; immunity to electromagnetic interference thus reducing dramatically the cost of shielding; environmental ruggedness; high multiplexing potential; and the prospect of reduced cost. The advent of fibrous composite structures is another reason for the recent impetus in the rapid development of fiber-optic-based sensors and instrumentation. Advances in the application of composite materials in many industries, including the civil infrastructure, have stimulated interest in innovative characterization methods for their structural evaluation. Surface-mounted gages, customarily used to determine stress and strain in metals, concrete, and masonry structures, do not provide adequate information about the state of composite materials, which have more complex material configurations. However, optic fiber–based embedded sensors for *in situ* measurements can be used for the real-time monitoring of the state of composites during their processing, fabrication, and installation, as well as in service (Technological Advances, 1994).

7.10.1 Types of Fiber-Optic Sensors for Smart Structures

In this section, brief descriptions are given of only a few fiber-optic sensors that have been used in experiments related to smart materials and structures and which show great potential. These are all single fiber sensors of the interferometric type. They include:

1. The fiber Fabry–Perot interferometer (FFPI)
2. The fiber Bragg grating sensor (FBGS)
3. The two-mode elliptic-core fiber-optic sensor (TM)
4. The back-reflective polarimetric fiber-optic sensor (P)
5. The Mach–Zehnder intenferometer (MZI)

Fiber-optic sensors are often categorized as being either *extrinsic* or *intrinsic*. *Extrinsic* fiber-optic sensors have an optical fiber that carries a light beam to and from a "black box" that modulates the light beam in response to strain or other environmental effects. Intrinsic fiber-optic sensors, on the other hand, are made totally of fiber-optic construction and measure the modulation of light by an environmental effect within the fiber.

It should be pointed out that there is currently a considerable amount of ongoing research and development in fiber-optic sensors and it is not clear which of the many systems proposed will become commercially available in the future.

7.10.1.1 The Fiber Farby–Perot Interferometer

The principle of the Fabry–Perot interferometer goes back to 1899 (Fabry and Perot). The first fiber-optic versions appeared in the literature in the early 1980s and the first uses of the FFPI in sensing temperature, strain, and ultrasonic pressure in composite materials were reported in the late 1980s (Lee and Taylor, 1995). The FFPI sensor can be configured as either intrinsic (Figure 7.37a) (Holst, A. et al., 1992) or extrinsic (Figure 7.37b) (LeBlanc, M. and Measures, R. M., 1992.) The extrinsic configuration has the advantage of being more protected and thus less subject to disturbances. Besides being more robust, the latter can be also used as a strain rosette as shown in Figure 7.38. The extrinsic sensor has less transverse coupling and can therefore evaluate more directly axial strains. This latter sensor can also be built with long gage lengths because of the separation between gage length and cavity length (Figure 7.37b).

Externally mounted FFPIs have been used to sense strain, temperature, ultrasonic pressure, and gas pressure in an internal combustion engine. FFPIs have been embedded in composites and in aluminum, where they have been used to sense temperature, strain, and ultrasonic pressure. The FFPI is a strong candidate for smart material and structure applications because it is extremely sensitive, provides point-sensing capability, has excellent mechanical properties, and its output is easy to process. However, the FFPI is difficult to make rugged enough for the harsh construction environment (especially embedding in concrete).

7.10.1.2 The Fiber Bragg Grating Sensor

The fiber Bragg grating sensor (FBGS) is illustrated in Figure 7.39. United Technologies Research Center has developed a method (Dunphy et al., 1995) to write permanent optical Bragg grating filters into the core of conventional optical fibers (Figure 7.40). The outstanding feature of the above device is that the index pertubations are exposed into the core of the fiber from the side, extending over a limited length of fiber. Through this approach, a length of fiber is created in which a modulated disturbance in the index of refraction of the core is made. Input and output leads are easy to make and the sensor is easily interfaced with an appropriate optical system with minimum power losses.

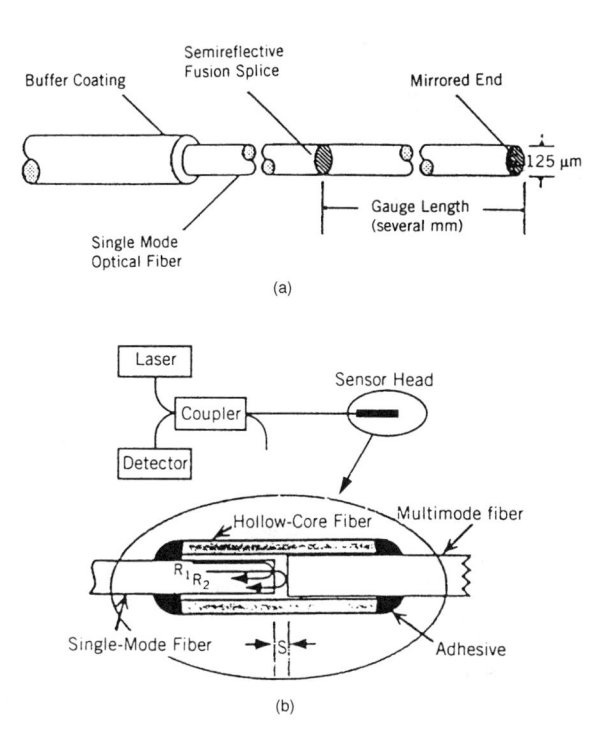

INSTRUMENTATION — PRINCIPLES AND APPLICATIONS

Figure 7.37 Fiber Fabry–Perot interferometer (FFPI). (a) Intrinsic; (b) extrinsic. (From Measures, R. M., in *Fiber Optic Smart Structures*, Udd, E., Ed., John Wiley & Sons, New York, 1995, 171–247. With permission.)

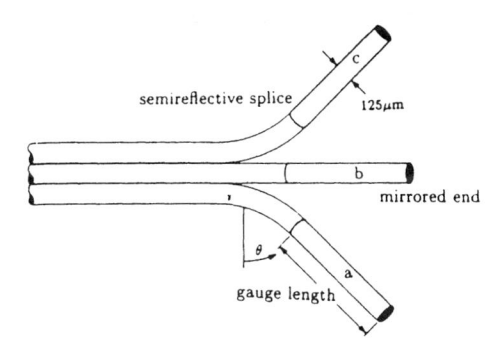

Figure 7.38 Three-axis strain rosette with FFPI sensors. (From Lee, C. E. and Taylor, H. F. in *Fiber Optic Smart Structures*, Udd, E., Ed., John Wiley & Sons, New York, 1995, 249–269. With permission.)

The FBGS described by Dunphy et al. (1995) is of interest to sensor systems because, according to them:

1. Optical fiber gratings respond to strain and can be applied as all-dielectric replacements for conventional strain gages.
2. A large number of FBGS's can be placed at predetermined locations on a single fiber string to create an array of quasi-distributed, quasi-point sensors.
3. The FBGS's on the fiber string can be interrogated in transmission (a two-connector system) or in reflectance (a one-connector system).
4. Wavelength-division and time-domain multiplexing instrumentation can be configured to address a number of fiber grating sensors on a fiber string simultaneously.
5. Optical fiber grating filters can be used as components in instrumentation for demultiplexing and signal processing of signals from remotely located fiber grating sensors.

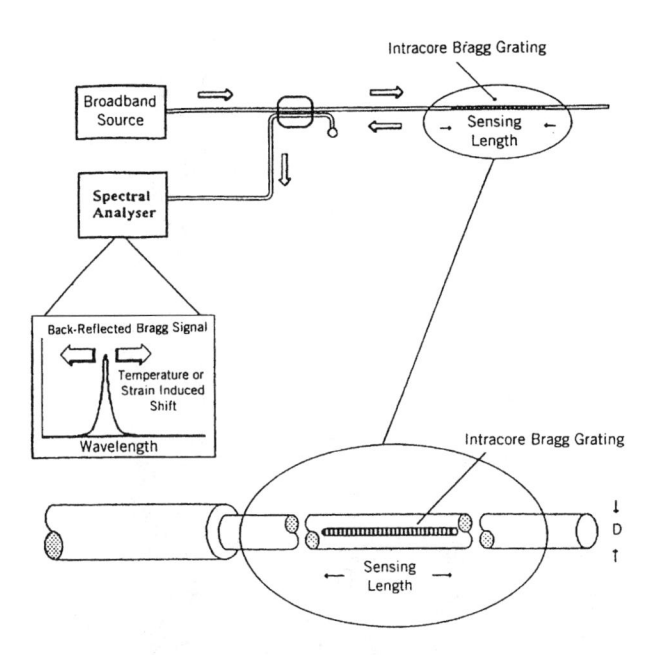

Figure 7.39 Intracore Bragg grating fiber-optic sensor. Also indicated is the backreflected narrow peak Bragg spectrum. (From Measures, R. M., in *Fiber Optic Smart Structures*, Udd, E., Ed., John Wiley & Sons, New York, 1995, 171–247. With permission.)

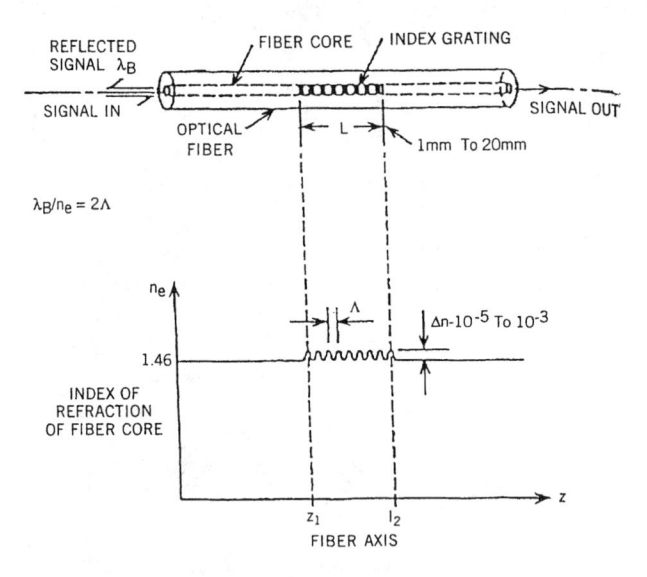

Figure 7.40 Detail of fiber-optic Bragg grating. (From Dunphy, J. R. et al., in *Fiber Optic Smart Structures*, Udd, E., Ed., John Wiley & Sons, New York, 1995, 271–285. With permission.)

6. The FBGS fabrication method can be extended to provide relatively low-cost, highly available devices.
7. FBGS devices have material and geometry characteristics that make them compatible with many diagnostic applications, such as embedded in composite smart structures.

Like in all of the current optic fiber sensor development, an intense effort is under way to increase the availability of the FBGS, lower its unit cost, further enhance its durability, and develop the necessary associated instrumentation concepts.

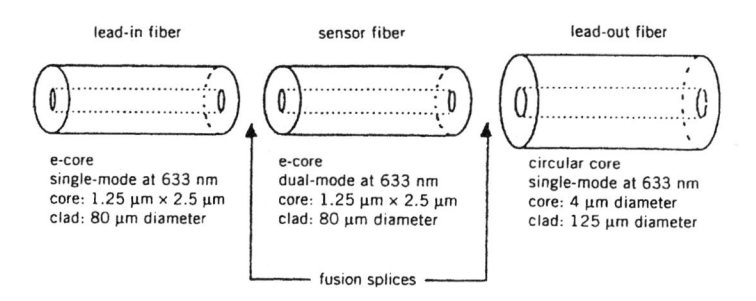

INSTRUMENTATION — PRINCIPLES AND APPLICATIONS

Figure 7.41 Two-mode elliptic-core fiber-optic sensor configuration. (From Murphy, K. A. et al., *Proc. SPIE Int. Soc. Opt. Eng.*, 1170, 566-573, 1990. With permission.)

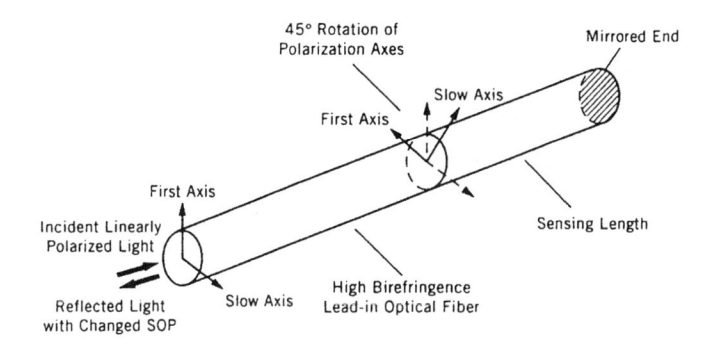

Figure 7.42 Backreflective polarimetric fiber-optic sensor using a 45° rotation of the polarization eigenaxes for localization of sensing segment. (From Measures, R. M., in *Fiber Optic Smart Structures*, Udd, E., Ed., John Wiley & Sons, New York, 1995, 171–247. With permission.)

7.10.1.3 The Elliptic-Core Two-Mode Fiber-Optic Sensor

The concept (Figure 7.41) and theoretical background for the elliptic-core two-mode (TM) sensor is given by Murphy et al. (1990). This elleptic-core TM sensor, also known as a form of modalmetric sensor, depends on the transverse spatial mode distribution of light within the optic fiber. The behavior of an elliptic-core TM fiber grating has been analyzed experimentally, and a TM elliptic-core fiber with a chirped grating, with the sensitivity varying as a function of length, has been demonstrated as a selective vibration-mode sensor (Murphy et al., 1995).

7.10.1.4 The Polarimetric Fiber-Optic Sensor

The back-reflective polarimetric (P) fiber-optic sensor using a 45° rotation is shown in Figure 7.42 (Measures, 1995). According to Measures,

> In a polarimetric sensor, changes in the state of polarization of light traveling in a single-mode optical fiber are used to determine the strain imposed on the sensor. To make the sensor reliable, special polarization maintaining optical fiber is used, and it is the change in the phase of the two orthogonal polarization eigenmodes of this high-bifrigence fiber that is used to evaluate the strain. Localization of the sensor is achieved by rotating the polarization eigenaxes through 45° in the sensing section relative to their orientation in the lead in/out sections of optical fiber (Figure 7.42) and linearly polarized light is launched into one of the polarization axes of the lead-in optical fiber.

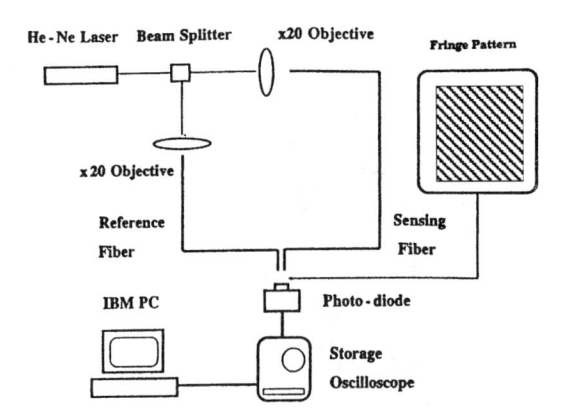

Figure 7.43 Fiber-optic sensor using MZI technique. (From Narendran, N. et al., *Eng. Fracture Mech.*, 38(6), 491–498, 1991. With permission.)

7.10.1.5 The Mach–Zehnder Interferometer

The Mach-Zehnder interferometer (MZI) technique as used by Narendran et al. (1991) is shown in Figure 7.43. The beam from an He-Ne laser is split into two parts by a beam splitter and each is focused and then injected into an optical fiber. One fiber is the sensor exposed to the applied strain to be measured, while the other serves as a reference path. The two beams are recombined and made to interfere, as shown in Figure 7.43. A strain applied to the sensing fiber causes a phase shift and a corresponding displacement of the fringe pattern. Fringes are detected by photodiode and the data can be stored on an oscilloscope and later processed. By counting the number of fringes passing a given point, the strain can be computed from precalibration curves.

7.10.2 Criteria and Selection of Fiber-Optic Strain Sensors

Criteria for assessing the potential suitability of any fiber-optic sensor for use in making strain measurements in smart materials and structures have been discussed by Turner et al. (1990). According to Measures (1995), the ideal fiber-optic strain sensor for smart materials and structures should be:

1. Intrinsic in nature for minimum perturbation and stability.
2. Localized, so that it can operate remotely with insensitive leads.
3. Able to respond only to the strain field and discern any change in direction.
4. Well behaved, with reproducible response.
5. All-fiber for operational stability.
6. Able to provide a linear response.
7. A single optical fiber for minimum perturbation and common-mode rejection.
8. Single ended for ease of installation and connection.
9. Sufficiently sensitive with adequate measurement range.
10. Insensitive to phase interruption at the structural interface.
11. Nonperturbative to the structure and robust for installation.
12. Interrupt immune and capable of absolute measurement.
13. Amenable to multiplexing to form sensing networks within structure.
14. Easily manufactured and adaptable to mass production.

To this list should also be added that it must have low cost. Although many fiber-optic sensors have been reported for strain measurements, only the FFPI, the FBGS, the TM, and the P meet the criteria 1, 2, 4, 5, 7, and 10. Their adherence to the rest of the criteria listed above are summarized in Table 7.7. At the present time, the first two sensors appear to have more advantages than the

INSTRUMENTATION — PRINCIPLES AND APPLICATIONS

Table 7.7 Comparison of Fiber-Optic Sensors

	P	FFPI	FBGS	TM
Localized	Y	Y	(1)	(2)
Responds only to strain	(3)	(3)	(3)	(3)
Direction change response	(4)	Y	(4)	(4)
Linear response	(4)	Y	(4)	(4)
Single-ended	Y	Y	(5)	(5)
Adequate sensitivity and range	Y	(6)	(7)	(7)
Absolute measurement	(8)	Y	(8)	(8)
Multiplexing within structure	(9)	Y	(9)	(9)
Potential for mass production	(11)	Y	(10)	(10)

Note: Y stands for "yes."

Key:
(1) Sensing length is two-mode while lead-in/lead-out optical fibers are single mode.
(2) Polarization eigenaxes are rotated by 45° at start and sensing length ends in mirror.
(3) If temperature changes, sensor must be able to compensate for thermal apparent strain.
(4) Requires *quadrature* detection and suitable signal demodulation.
(5) Requires a mirror at the end of the sensing length.
(6) Limited by demodulation system, but 1 microstrain should be achievable with laser sensor.
(7) Sensitivity is about 1% that of the FFPI. This would restrict spatial resolution to many centimeters.
(8) Requires special demodulation system, so absolute measurements is not available at all times.
(9) Difficult, except for large structures, where time-division multiplexing can be used.
(10) Requires at least one fusion splice and therefore some degree of handling.
(11) Recently, FFPI sensors have been made from two Bragg gratings. Under these circumstances no fusion splice is required. Also extrinsic Fabry-Perot sensors do not require a fusion splice.

From Measures, R. M., in *Fiber Optic Smart Structures*, Udd, E., Ed., John Wiley & Sons, New York, 1995, 171–247. With permission.

latter two. Intensive development will determine which of these four will have the greatest advantage in practice. The MZI sensor, although not compared in Table 7.7, is also under intense development.

7.10.3 Integration of Fiber-Optic Sensors into Concrete Structures: Case Studies

Civil engineering structures, mostly of reinforced and prestressed concrete, present potential applications of fiber-optic sensors. A review of current trends in this area (Huston and Fuhr, 1995) reveals only very few actual cases. Embedment of fiber-optic strain sensors in concrete structures presents one of the most challenging tasks to the instrumentation engineer because of the harshness of the construction environment. Very few studies have been made of the nature of the compatibility of the optical fiber and the concrete and the nature of the bonding and stress transfer between the fiber and the surrounding concrete material (Nanni et al., 1991). A few examples of such applications are given below.

1. High-strength concrete beams reinforced with prestressed prisms and instrumented with conventional strain gages and LVDTs as well as FBGS were studied by Chen and Nawy (1994) in a series of simple spans and by Nawy and Chen (1998) in a series of two-span continuous members. The longitudinal section and cross sections of 4 of the 13 beams tested in the first series (Chen and Nawy, 1994) is shown in Figure 7.44. The schematic of the FBGS instrumentation is shown in Figure 7.45a and a typical wavelength shift between two closely spaced points resulting from strain on the fiber-optic sensor is shown in Figure 7.45b. Typical load–strain correlation between the FBGS and foil strain gages was very good, as shown in Figure 7.46a. Average crack width also showed very good comparison between the fiber-optic sensors and the micrometer (Figure 7.46b).

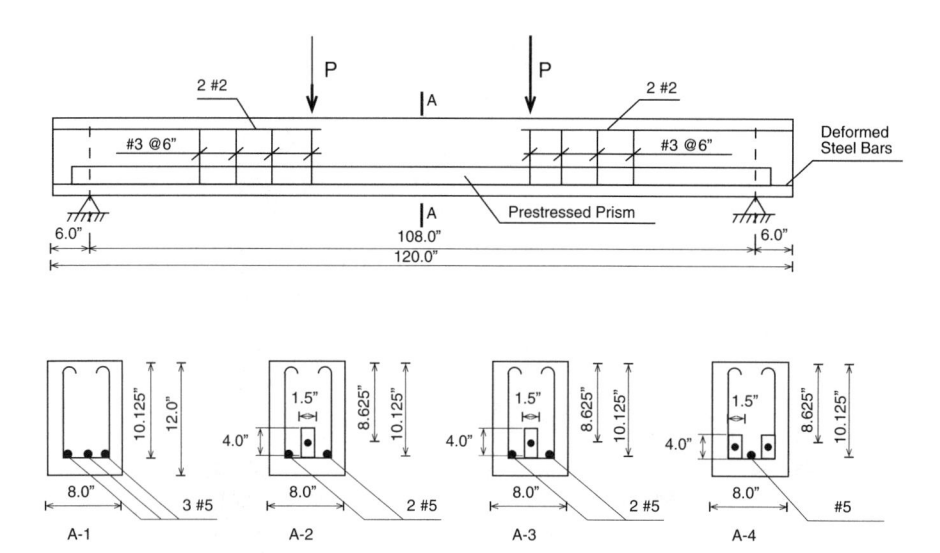

Figure 7.44 Elevation and sections of simply supported R/C beams. (From Chen, B. and Nawy, E. G., *ACI Struct. J.*, 91(6), Nov.–Dec., 708–718, 1994. With permission.)

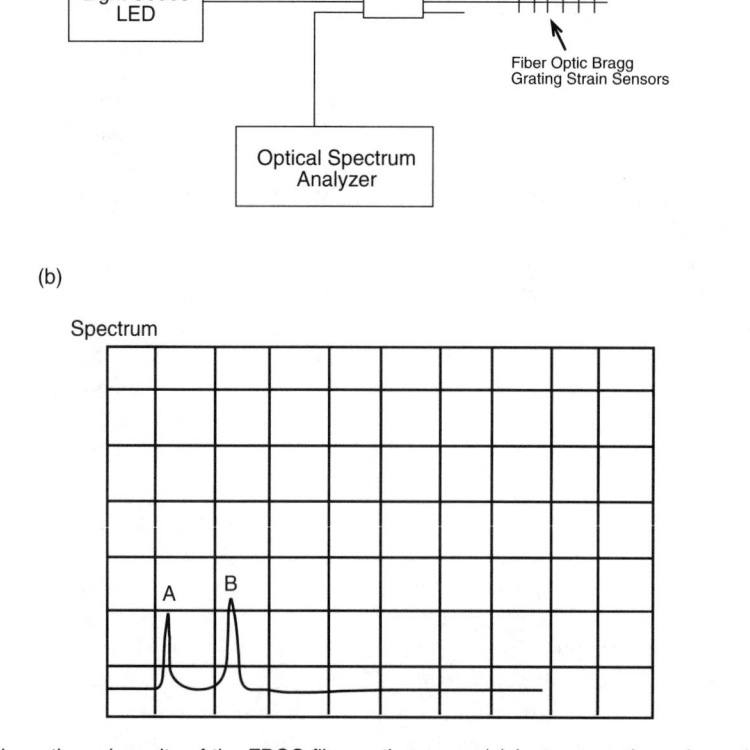

Figure 7.45 Schematic and results of the FBGS fiber-optic sensor. (a) Instrumentation schematic. (b) Typical wavelength shift from point *A* to point *B* resulting from strain on FBGS. (From Nawy, E. G. and Chen, B., *ACI Struct. J.*, Jan.–Feb., 51–60, 1998. With permission.)

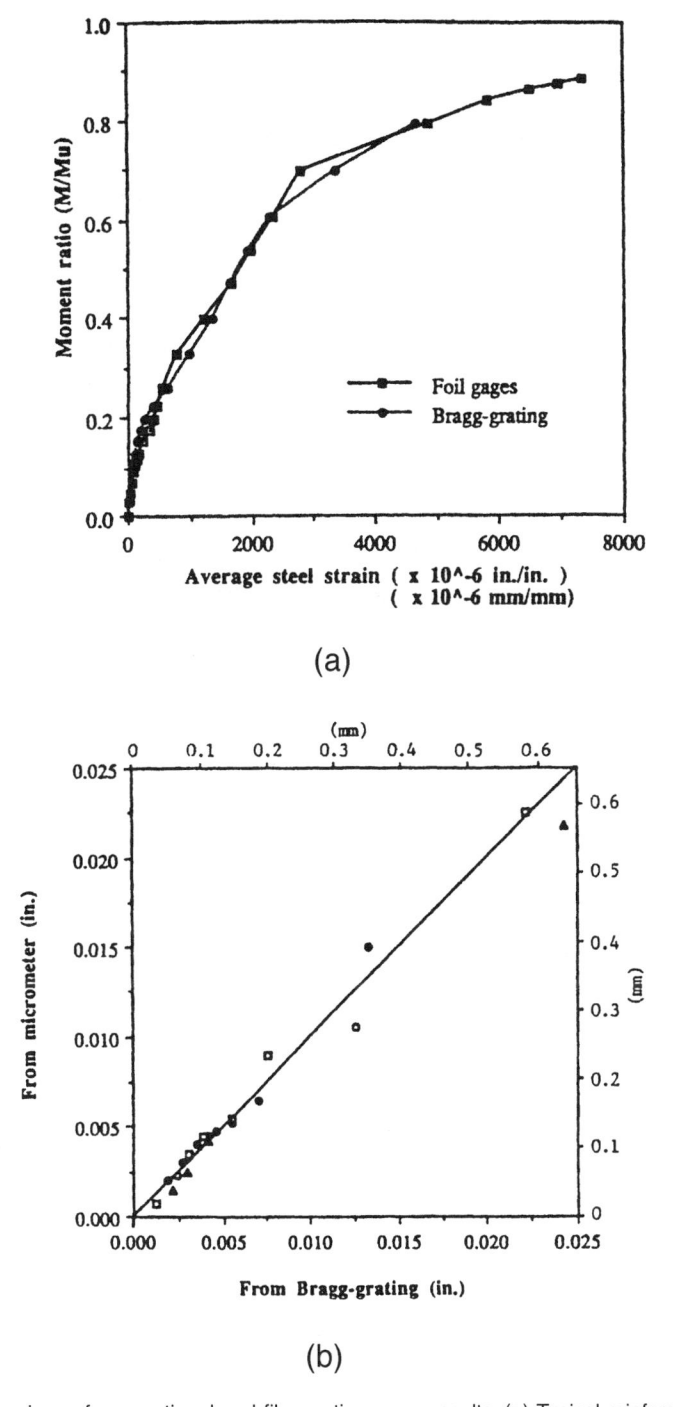

INSTRUMENTATION — PRINCIPLES AND APPLICATIONS 375

Figure 7.46 Comparison of conventional and fiber-optic sensor results. (a) Typical reinforcement strain measurements by FBGS and foil gages, Beam A-3. (b) Average crack width measurement obtained from FBGS and micrometer. (From Chen, B. and Nawy, E. G., *ACI Struct. J.*, 91(6), Nov.–Dec., 708–718, 1994. With permission.)

The configuration of the four two-span beams tested (Nawy and Chen, 1998) is shown in Figure 7.47a. The same type of FBGS used in the single-span beams were also used in this series with comparable good correlation of strains (Figure 7.47b).

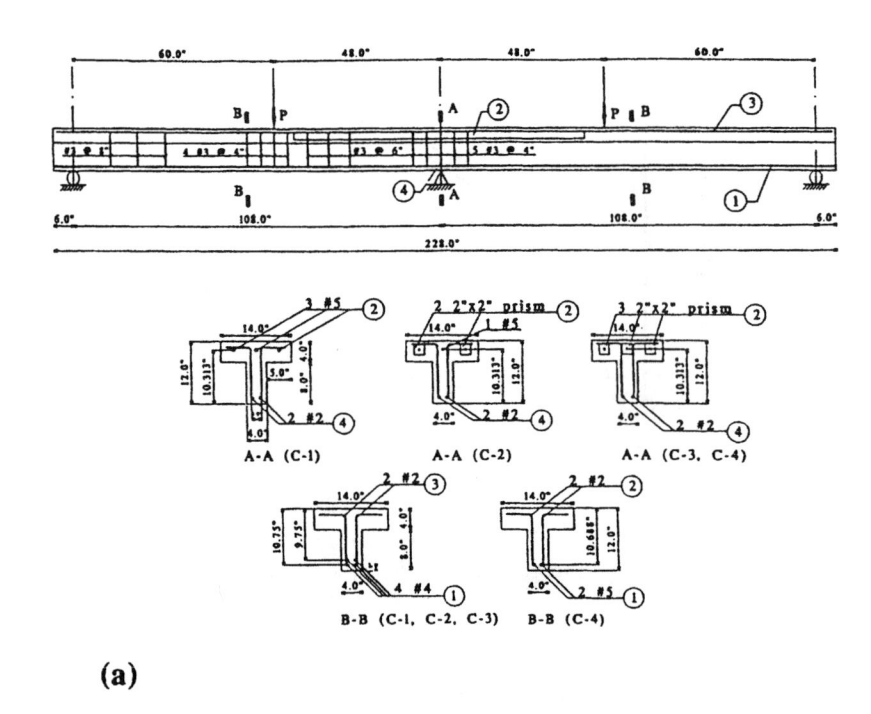

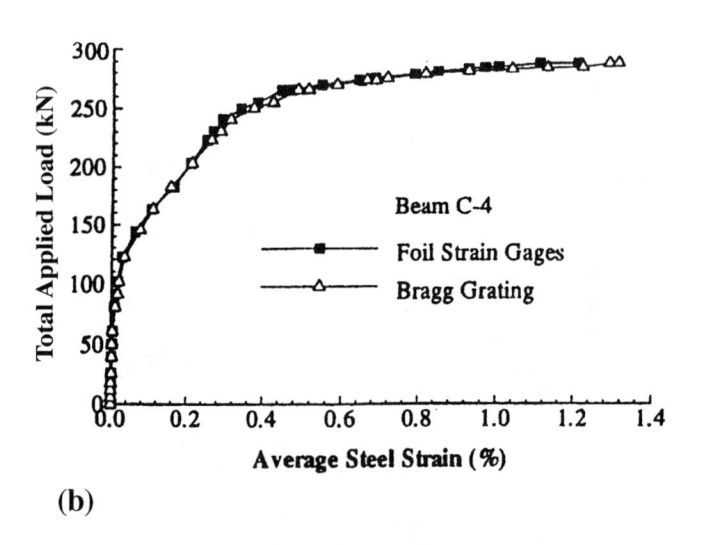

Figure 7.47 Two-span R/C beams with FBGS fiber-optic sensors. (a) Sectional details of the test specimens. (b) Load vs. average strain measurement in the reinforcement over the center support, Beam C-4. (From Nawy, E. G. and Chen, B., *ACI Struct. J.*, Jan.–Feb., 51–60, 1998. With permission.)

2. One of the earliest applications of fiber-optic sensors in concrete structures was the Schiessbergstrasse prestressed concrete three-span (53 m long, 10 m wide) bridge in Leverkusen, Germany (Wolf and Miesseler, 1992). Built-in fiber-optic sensors (combination of microbend and elongation) were designed to evaluate the load effects on the tendons and monitor the strain in the concrete deck (Figure 7.48).

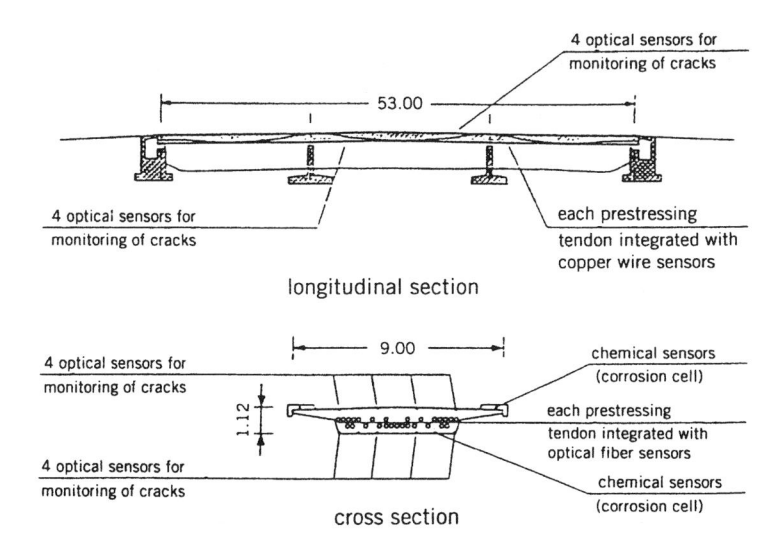

Figure 7.48 Arrangement of optical fiber sensors and chemical sensors in the triple-span Schiessbergstrasse Road bridge. (From Wolff, R. and Miesseler, A. J., *Proc. SPIE Int. Soc. Opt. Eng.*, 1777, 23–29, 1992. With permission.)

7.11 SUMMARY

An introduction to the instrumentation needed for a successful structural model study has been given. Techniques for the measurement of strains, deflections, and temperatures and some examples of commercially available equipment have been presented. The measurement of strains and their interpretation to obtain stresses in the model have been discussed. It has been emphasized that the correct measurement and interpretation of relevant physical quantities is a crucial step in the modeling process. Several large-scale automated instrumentation systems have been described, which help interpretation and the data reduction process. A brief introduction to fiber-optic sensors and their use in smart materials and structures has been presented.

PROBLEMS

7.1 A cantilever beam is loaded at its end by a load of unknown magnitude and angle of inclination.

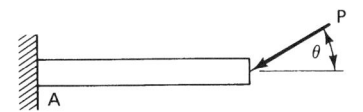

Describe strain gage location and type of circuitry to measure:
a. Axial load component
b. Shear force
c. Bending moment at A
All circuits should have inherent temperature compensation.

7.2 A tubular steel section (round pipe) is to be used for a transducer to measure applied torque. Design the layout of strain gages (number and location) and the circuitry, and express the voltage output of the bridge as a function of the strain change seen by the gage. Is the system you have selected sensitive to axial force? Is it sensitive to bending moment? The circuit should have inherent temperature compensation.

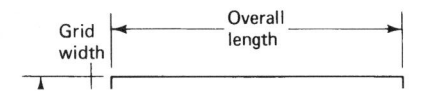

7.3 Establish a circuit using a single Wheatstone bridge to measure the value of P on the cantilever beam below, where P is a load applied somewhere in the outer half of the beam length. *Hints:* Shear may be expressed as differences in moment, that is, $V = dM/dx$. Provide temperature compensation.

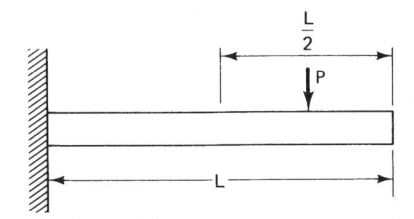

7.4 Derive equations for principal stresses σ_1 and σ_2, maximum shear stress τ_{max}, and the direction of principal stress θ_p for a rectangular strain gage rosette with gages a, b, and c as shown:

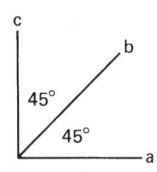

7.5 A delta rosette has the strains shown below. Sketch freehand an approximate solution to the principal strains (using Murphy's graphical method), and indicate the approximate principal strains and their directions on a sketch of a free body element properly oriented in direction. (You need not take the time to scale values and make an accurate plot of Mohr's circle for strain — a quick sketch is fully adequate.)

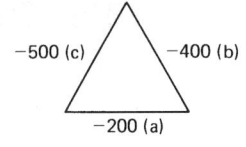

strains shown in microstrain

7.6 A delta rosette has the strains as shown below. Compute principal stresses and θ_p from the expressions derived in text. Then apply Murphy's method and compare results. Use $E = 69$ GPa and Poisson's ratio, $\nu = 0.32$. What are the stresses at an angle of $+45°$ from the x axis? Give answers on an element sketch.

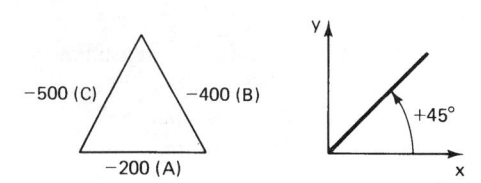

INSTRUMENTATION — PRINCIPLES AND APPLICATIONS 379

7.7 A rosette made of three separate gages has the angular orientation and strain values as shown below. For $E = 200$ GPa and $\nu = 0.3$, determine the values and directions of principal stress and the maximum shearing stress. Give answers on elements at the proper orientation.

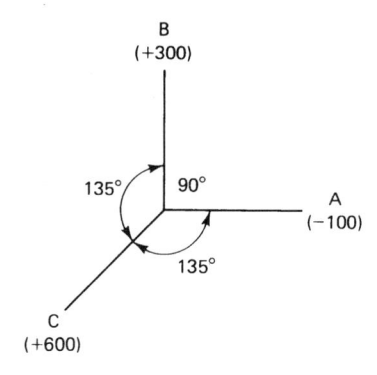

```
           B
        (+300)

     135°  │  90°
        ╲  │         A
         ╲ │      (−100)
          ╲│
           ╲
            ╲  135°
             ╲
              C
           (+600)
```

7.8 A deep, thin-webbed plate girder is instrumented with strain gages aligned in the x direction to measure the strain distribution at high load levels when it may become nonlinear. Gage 1 is located on the top flange. The beam is subjected to combined bending and shear.

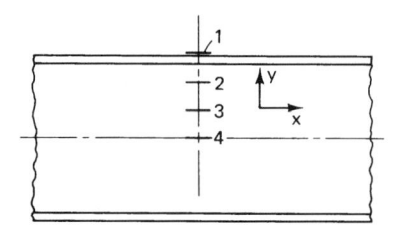

Discuss the effects of combined bending and shear on the strains measured by each gage. *If bending strains only are needed*, is the gage layout satisfactory, or does shearing strain have an effect on the strain output of the gages? Assume a vertical loading was applied to the beam in the vicinity of the gages. How does this affect the bending strains? Be specific.

Your assistant suggests that gages be installed at ±45° to measure shear effects in the beam. Would this arrangement be satisfactory, or would the readings be influenced by the bending strains present in the beam?

7.9 For the rosettes shown below:
a. Apply the equations for principal stresses and θ_p.
b. Check your results by constructing Mohr's circle for strain.

$$E = 200 \text{ GPa, Poisson's ratio} = 0.30$$

Show your results on sketches properly oriented with respect to stress direction. Also show the stress state that produces maximum shearing stress.

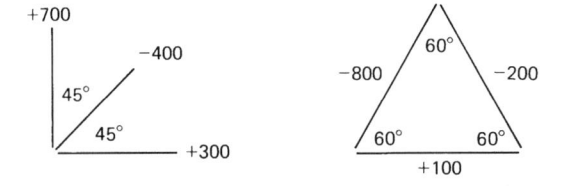

```
   +700
    │        −400
    │       ╱                      60°
    │      ╱                  −800 ╱╲ −200
    │ 45° ╱                       ╱  ╲
    │    ╱                       ╱    ╲
    │45°╱                        ╱ 60° 60°╲
    │  ╱                        ───────────
    └─────── +300                   +100
```

7.10 The rosettes below were used to measure strains on a Plexiglas model with $E = 3.1$ GPa and $\nu = 0.35$.
   a. Delta rosette

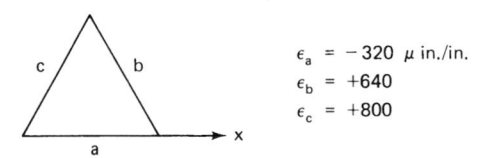

$\epsilon_a = -320$ μ in./in.
$\epsilon_b = +640$
$\epsilon_c = +800$

Solve for $\varepsilon_1, \varepsilon_2, \sigma_1, \sigma_2, \tau_{max}$, and θ_{1-x} by (1) derived expressions and (2) graphical method.
   b. Rectangular rosette

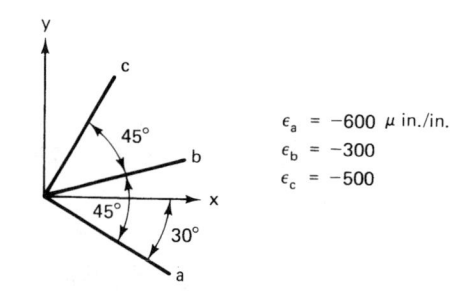

$\epsilon_a = -600$ μ in./in.
$\epsilon_b = -300$
$\epsilon_c = -500$

   i. Using the graphical method, solve for ε_1, ε_2, ε_x, σ_x, τ_{max}, and θ_{1-x}.
   ii. What are the stresses at an angle of 45° counterclockwise from the x-axis? Give your answers on a sketch of an element.

7.11 Using graphical construction, *sketch* a solution to the strain field as measured by the rosette shown below. You can do this *freehand*, without accurate scaling or use of a compass. Estimate the principal strains and show them in their proper orientation and directions on a sketch of an element.

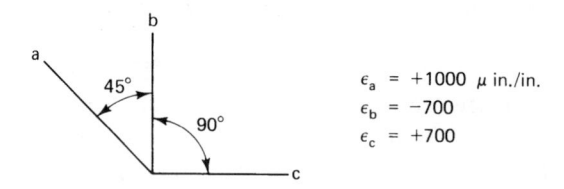

$\epsilon_a = +1000$ μ in./in.
$\epsilon_b = -700$
$\epsilon_c = +700$

7.12 One of the difficult problems faced in strain gage instrumentation is the extremely small change in resistance produced by strain in the gage. Using the basic definition of gage factor F in Section 7.3.3.2, determine the change in resistance produced in a 120 Ω resistance gage with a gage factor $F = 2.1$ when the gage is applied to a steel member and is stressed by 0.7 MPa.

7.13 A load cell is made from a 50-mm-outside-diameter steel pipe with 2-mm-thick walls and is instrumented with a full bridge (two gages in the axial direction and two gages in the Poisson configuration). If the bridge is powered with 2 V, gages are 120 Ω, and gage factor is 2.05, what is the output of the bridge in volts? Use modulus = 200 GPa.

7.14 The blade of a metal screwdriver is to be instrumented with electrical resistance strain gages to measure the ratio of torque to axial force needed to drive screws into various species of wood. Using no more than two separate circuits with no more than four active strain gages, show the location of gages and the circuits needed to measure the axial force and the torque, and prescribe the Wheatstone bridge output in terms of the strain values ε_1, ε_2, ε_3, and ε_4. Each circuit should have inherent temperature compensation. You may assume that any accidental bending strains may be neglected.

7.15 The modulus of elasticity of thin rigid planished vinyl plastic sheeting was measured two ways — by using foil electrical resistance strain gages, and with an extensometer mounted on the tensile specimen to measure the increase in length with increasing tensile load. The results were as follows:

Thickness	*Modulus E*
0.010 in.	728,000 psi (with strain gages)
	425,000 psi (with extensometer)
0.015 in.	544,000 psi (with strain gages)
0.020 in.	493,000 psi (with strain gages)
	425,000 psi (with extensometer)

The differences are produced by the stiffening effect of the gage on the plastic. Both the metallic strain-sensing material (foil) and the epoxy backing of the gage have elastic modulus values higher than that of the vinyl model material. The problem is accentuated with extremely thin material, such as that listed above (which was used in building a model of a steel bin structure).

What value of E would you use in reducing strain gage readings to stresses?

REFERENCES

Abeles, P. W. (1966). Cracking and bond resistance in high strength reinforced concrete beams, illustrated by photoelastic coating, *J. Am. Concr. Inst.,* November, 1265–1278.

Brownie, R. D. and McCurich, L. H. (1967). Measurements of strain in concrete pressure vessels, Paper 52, Group I, *Conference on PCRV,* London, March.

Chen, B. and Nawy, E. G. (1994). Structural behavior evaluation of high-strength concrete beams reinforced with prestressed prisms using fiber optic sensors, *ACI Struct. J.,* 91(6), Nov.–Dec., 708–718.

Corum, J. M. and Smith, J. E. (1970). Use of Small Models in Design and Analysis of PCRV's, ORNL-4346, Oak Ridge National Laboratory, Oak Ridge, TN.

Dally, J. W. and Riley, W. F. (1965). *Experimental Stress Analysis,* McGraw-Hill, New York.

Dove, R. C. and Adams, P. J. (1964). *Experimental Stress Analysis and Motion Measurements,* Prentice-Hall, Englewood Cliffs, NJ, 515 pp.

Dunphy, J. R., Meltz, G., and Morey, W. W. (1995). Optical fiber Bragg grating sensors: a candidate for smart structure applications, in *Fiber Optic Smart Structures,* Udd, E., Ed., John Wiley & Sons, New York, 271–285.

Eaton Corp. (1994). *Lebow Load Cell and Torque Sensor Handbook No. 600A,* Lebow Products,1728 Maplelawn Road, Troy, MI 48099.

Fabry, C. and Perot, A. (1899). *Ann. Chim. Phys.,* 16, 115.

Geymeyer, H. G. (1967). Strain Meters and Stress Meters for Embedment in Models of Mass Concrete Structures, Technical Report No. 6-811, U.S. Corps of Engineers, Vicksburg, MS, March, 60 pp.

Hanson, N. W. and Curvitts, O. A. (1965). Instrumentation for Structural Testing, *J. PCA Res. Dev. Lab.,* May, 24–39.

Herceg, D. E. (1976). *Handbook of Measurement and Control — an Authoritative Treatise on the Theory and Application of the LVDT,* rev. ed., Schaevitz Engineering, Pennsauken, NJ.

Holst, A. Habel, W., and Lessing, R. (1992). Fiber optic intensity modulated sensors for continuous observations of concrete and rock-filled dams, *Proc. SPIE Int. Soc. Opt. Eng.,*1777, 223-226.

Huston, D. R. and Fuhr, P. L. (1995). Fiber optic smart civil structures, in *Fiber Optic Smart Structures*, Udd, E., Ed., John Wiley & Sons, New York, 647–665.

Labonte, L. R. S. (1971). An Investigation of Anchorage Zone Behavior in Prestressed Concrete Containments, M.Eng. thesis, Structural Concrete Series No. 71-6, McGill University, Montreal, September.

LeBlanc, M. and Measures, R. M., Impact damage assessment in composite materials with embedded fiber optic sensors, *J. Composite Eng.*, 2, 729-731.

Lee, C. E. and Taylor, H. F. (1995). Sensors for smart structures based on the Fabry-Perot interferometer, in *Fiber Optic Smart Structures*, Udd, E., Ed., John Wiley & Sons, New York, 249–269.

Mark, J. W. and Goldsmith, W. (1973). Barium titanate steam gages, *Proc. Exp. Stress Anal.*, 13(1), 139–150.

Measurements Group, Inc. (1998). *Product Information Manual*, P.O. Box 27777, Raleigh, NC 27611.

Measures, R. M. (1995). Fiber optic strain sensing, in *Fiber Optic Smart Structures*, Udd, E., Ed., John Wiley & Sons, New York, 171–247.

Murphy, K. A., Miller, M., Vengsarkar, A. M., Claus, R. O., and Lewis, N. E. (1990). Embedded modal domain sensors using elliptical core optical fibers, *Proc. SPIE Int. Soc. Opt. Eng.*, 1170, 566-573.

Murphy, K. A., Vengsarkar, A. M., and Claus, R. O. (1995). Elliptical-core two-mode optical fiber sensors, in *Fiber Optic Smart Structures*, Udd, E., Ed., John Wiley & Sons, New York, 287–317.

Nanni, A., Yang, C. C., Pan, K., Wang, J., and Michael, R. R., Jr., (1991). Fiber-optic sensors for concrete strain/stress measurement, *ACI Mater. J.*, 88(3), May–June, 257–264.

Narendran, N., Shukla, A., and Letcher S. (1991). Application of fiber optic sensors to fracture mechanics problems, *Eng. Fracture Mech.*, 38(6), 491–498.

National Instruments (1998). *Instrumentation Reference and Catalogue*, U.S. Corporate Headquarters, 6504 Bridge Point Parkway, Austin, TX 78730-5039.

Nawy, E. G. and Chen, B. (1998). Deformational behavior of high performance concrete continuous composite beams reinforced with prestressed prisms and instrumented with Bragg grating fiber optic sensors, *ACI Struct. J.*, Jan.–Feb., 51–60.

Omega Engineering, Inc. (1992). *The Pressure Strain and Force Handbook*, Vol. 28, P.O. Box 4047, Stamford, CT 06907-0047.

Perry, D. C. and Lissner, H. R. (1962). *The Strain Gage Primer*, McGraw-Hill, New York, 332 pp.

Stephen, R. M. and Bouwkamp, J. G. (1975). New general purpose component test system, paper presented at the ASCE National Structural Engineering Convention, New Orleans, LA (Preprint No. 2475), 27.

Tanyel, M. (1994). *Engineering Explorations with LabVIEW,*® Harcourt Brace, Orlando, FL.

Tanyel, M. and Quinn, R. (1993). *Introduction to Experimentation*, McGraw-Hill, New York.

Tanyel, M. and Quinn, R. (1994). *Introduction to Engineering Experimentation*, Harcourt Brace, Orlando, FL.

Technological Advances (1994). *Mater. Technol.*, 9(7/8), July/August, 141–153.

Timoshenko, S. and Goodier, J. N. (1951). *Theory of Elasticity* 2nd ed., McGraw-Hill, New York.

Turner, R. D., Valis, T., Hogg, W. D., and Measures, R. M. (1990). Fiber optic strain sensors for "smart" structures, *J. Intell. Mater. Syst. Struct.*, 1, 26–49.

Udd, E., Ed. (1995). *Fiber Optic Smart Structures*, John Wiley & Sons, New York.

Wolf, R. and Miesseler, H. J. (1992). Monitoring of prestressed concrete structures with optical fiber sensors, *Proc. SPIE Int. Soc. Opt. Eng.*, 1777, 23–29.

CHAPTER **8**

Loading Systems and Laboratory Techniques

CONTENTS

8.1	Introduction	383
8.2	Types of Loads and Loading Systems	384
	8.2.1 Load Reaction Systems	384
	8.2.2 Loading Devices — Discrete Loads	384
	8.2.3 Loading Systems — Pressure and Vacuum	386
8.3	Discrete vs. Distributed Loads	389
8.4	Loadings for Shell and Other Models	390
	8.4.1 Vacuum and Pressure Loadings	392
	8.4.2 Discrete Load Systems	395
	8.4.3 Effects of Load Spacing	400
8.5	Loading Techniques for Buckling Studies and For Structures Subject to Sway	400
	8.5.1 Shell Instability	401
	8.5.2 Structures Undergoing Sway	403
8.6	Miscellaneous Loading Devices	404
	8.6.1 Thermal Loads	406
	8.6.2 Self-Weight Effects	406
8.7	Summary	407
Problems		407
References		409

8.1 INTRODUCTION

Prototype structures are normally designed for either concentrated forces or uniformly distributed loads. If using direct models, the concentrated loads must be scaled down to small concentrated loads on the model, and the uniform loading may be represented on the model by either a series of discrete loads or by suitably scaled pressure. Various loading methods will be covered in this chapter. Errors introduced by discretization of distributed loads will be examined. Particular attention will be given to modeling of loading for shell structures and for buckling studies. Brief comments will be made on thermal models and on special techniques for simulation of self-weight effects. The careful reader will recognize that development of adequate yet simple loading techniques requires a great deal of common sense as well as a certain degree of ingenuity. Ideas developed for one type of loading usually can be refined and adapted to situations that appear to be totally different.

This chapter will concentrate on static loadings, with some discussion of quasi-static representation of earthquake effects. Dynamic loads will be treated in more depth in Chapter 11.

Any model loading system should:

1. Accurately represent the prototype loads (both magnitude and direction);
2. Be easy to apply to the structure, and to remove and reapply;
3. Present no undue safety hazards;
4. Offer no restraint to the model, particularly for buckling studies;
5. Be capable of being "caught" if the model should fail suddenly and catastrophically.

8.2 TYPES OF LOADS AND LOADING SYSTEMS

Loading systems for models must be very carefully designed and constructed if they are to function properly throughout the model test. Prior to considering detailed provisions for several categories of loading systems, some fundamental ideas and concepts of laboratory-applied loadings will be reviewed.

8.2.1 Load Reaction Systems

Discrete or point loads may be applied mechanically or hydraulically, or with deadweights. Mechanical and hydraulic loads must react against another structure that is substantially stronger and stiffer than the model structure, and the provision of such reaction structures is often the most expensive portion of a new loading system. Many modern structural testing laboratories are built on a strong floor that serves as the reaction system for loading. The system in the Portland Cement Association Laboratory in Skokie, IL, is illustrated in Figure 8.1, which shows a ½-scale prestressed concrete bridge girder being loaded with a series of hydraulically applied loads through the transverse beams that are spaced along the girder. The beams are loaded by tensioning the steel tie bars that extend through the test floor to another cross beam, which in turn is jacked away from the underside of the floor with a hydraulic ram. This system and similar systems in other laboratories have been used to load many large models.

The concept is used in the Cornell University Structural Models Laboratory for loading small models (Figure 8.2). Each testing table (White, 1972) is made from steel bridge deck and is supported 0.9 m off the floor by tubular steel legs. Aluminum reaction frames are bolted to the table, along with support devices for the model, and the applied loads react against the frame and transmit all forces back into the table. The distinguishing feature of both the strong floor laboratory and the stiff models testing table is that the force system is self-equilibrated by reacting against the structure and integral parts of the testing facility. New structures are tested simply by changing the locations of the loading devices.

8.2.2 Loading Devices — Discrete Loads

Hydraulic and electrically driven universal testing machines are perhaps the most common method of applying a single, discrete load to a structure or test specimen. With suitable load–distribution devices, these machines can also produce a series of concentrated loads on a beam span.

Individual hydraulically actuated rams (or jacks) are familiar pieces of loading equipment in any laboratory, with load capacities ranging from less than 4.5 kN to hundreds of kilonewtons. They can be attached to loading frames and used to apply loads in any direction. Quick-release hose connectors are used with manual or electric pressure systems operating at pressures from 21 to 69 MPa. Approximate load intensities may be obtained from a pressure gage and the known area

LOADING SYSTEMS AND LABORATORY TECHNIQUES

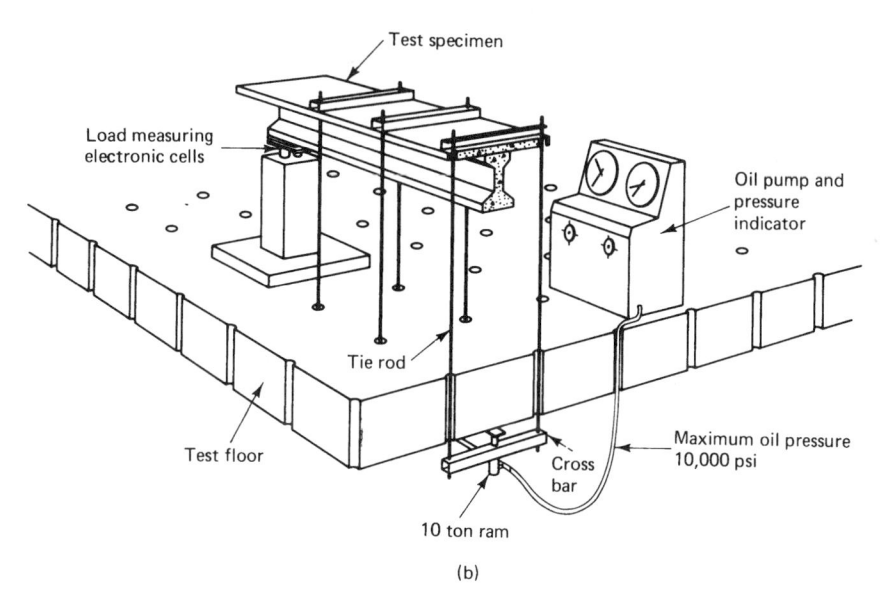

Figure 8.1 Strong floor loading system. (From Hognestad et al., 1959). (a) (Courtesy of Portland Cement Association.) (b) Application of vertical load.

of the piston in the hydraulic ram; load cells placed between the ram and either the structure or the reaction system must be used to obtain more accurate values of load.

Mechanical loading devices such as those being used in Figure 8.2 are very convenient for many model tests. They are operated by turning the crank, and feature zero backlash (no lost motion in either loading or unloading). The applied load must be measured with a load cell such as the small compression load cell attached to the mechanical loading devices in Figure 8.2. This type of load cell uses electrical resistance strain gages as the sensing elements (see Chapter 7).

Suspended deadweights are often used for both augmenting the scaled dead load and for live loads. For small suspended loads (say, up to 1 N/load point), one can use either lengths of steel bars with hooks at both ends or tin cans filled with lead shot or steel punchings. Both types of load are shown in Figure 8.3a. When heavier loads are needed, such as in the modeling of deadweight

Figure 8.2 Stiff testing table system for small models. (Courtesy of Cornell University, Ithaca, NY.)

during construction of a prestressed concrete bridge model, suspended weights made from household bricks or blocks of concrete are normally the least expensive to use.

Often a single concentrated load is needed to load bridge models or similar structures at many points. A convenient dead load for these applications is shown in Figure 8.3b. Mechanical loading systems (see Figures 1.10 and 1.11) can also be built for these models, but at considerably more cost. The convenience and cost of the mechanical system must be balanced against the large amount of physical effort involved in using the suspended deadweights. A mechanical system for moving stacks of lead sheets to different locations on another bridge model is shown in Figure 8.3c.

Dead weights may also be applied to the surface of the model. Figure 8.3d illustrates the application of dead loads to a slab through a layer of sand, which helps produce a uniform distribution of load. However, care must be taken to prevent any arching action. Sand, prepacked into plastic bags, is often convenient. But one must beware of changes in moisture content, which affect the weight of any common building material such as sand, concrete, or brick.

Substantial difficulties are met with suspended deadweights when a large number of load points must be loaded simultaneously, particularly in controlling the simultaneous application of loads, in unloading the model quickly, and even in reaching into the system to apply incremental loadings.

The whiffle-tree load system of Figure 8.3e is ideal for applying a large number of identical discrete loads. This system, which involves an articulated set of load distribution bars and either one large dead load or a jacking system to apply a single concentrated load, is further described in Section 8.4.2.

8.2.3 Loading Systems — Pressure and Vacuum

Either air pressure or vacuum loading systems are usually used to apply uniformly distributed loads normal to the surfaces of a model. Again the provision of a strong reaction structure is the most difficult part of design and construction of loading equipment.

The use of a pressurized gas directly against the model (Figure 8.4a) is usually not recommended in static tests because of construction difficulties in making a container that does not restrain the model at the connection between the model and the container. The use of an air bag between a reaction device and the model (Figure 8.4b) is a better adaptation of using air pressure to distribute a single applied force over a structure.

LOADING SYSTEMS AND LABORATORY TECHNIQUES 387

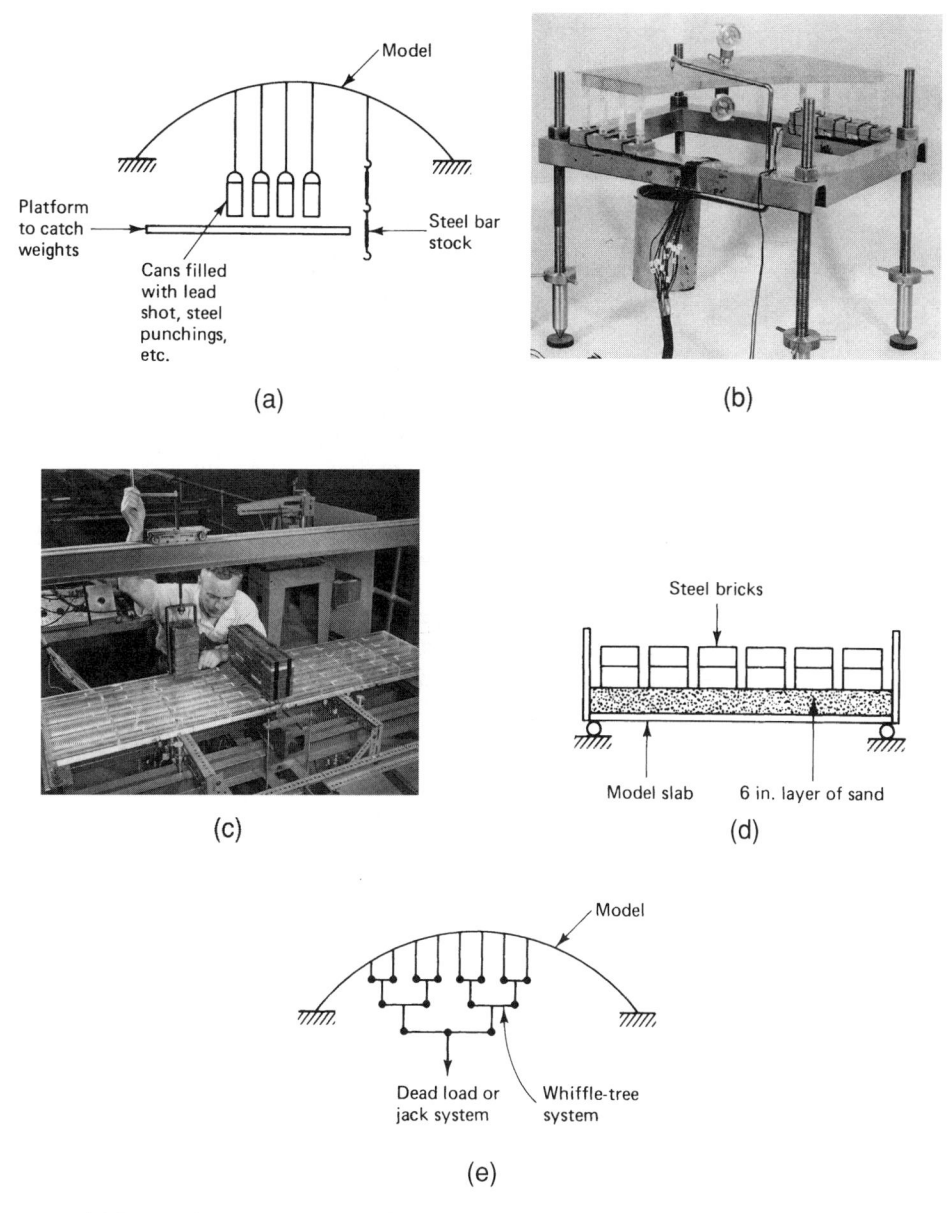

Figure 8.3 (a) Suspended deadweights. (b) Single point load. (Courtesy of LNEC.) (c), (d) Dead loads applied on top of model. (Courtesy of Portland Cement Association.) (e) Whiffle-tree system.

Vacuum loadings are more widely used because they are safer, the edge-sealing problem is easier to deal with, and it is easy to generate the small vacuum load needed using only a commercial-size shop vacuum. The following procedure is recommended for vacuum loading of a model structure:

1. Any commercial-size vacuum cleaner will suffice for small-scale work in which load requirements are less than 14 to 19 kN/m^2.
2. The space between the model and vacuum chamber is left at about 3 to 6 mm and is closed off with polyethylene film and petroleum jelly as shown in Figure 8.5.

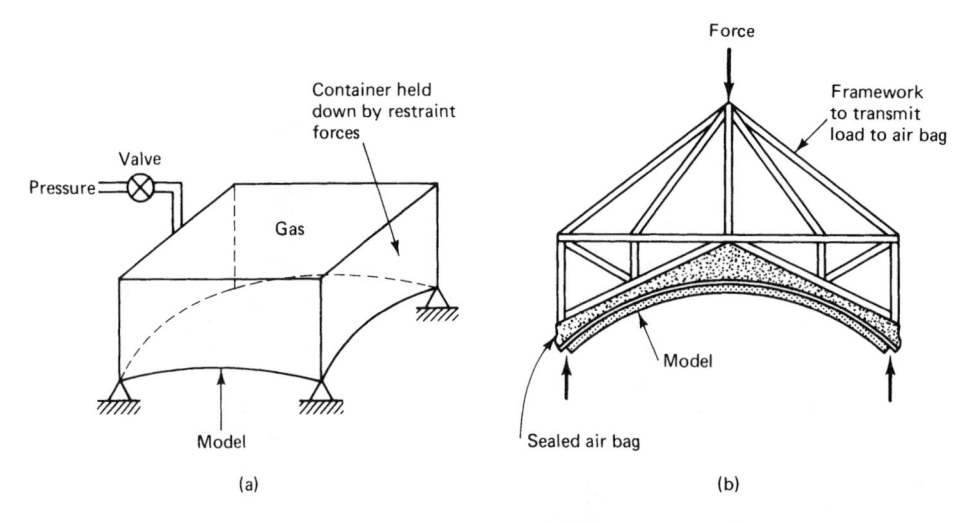

Figure 8.4 Pressure loading system for uniform load application. (a) Use of gas pressure as loading. (b) Air bag loading.

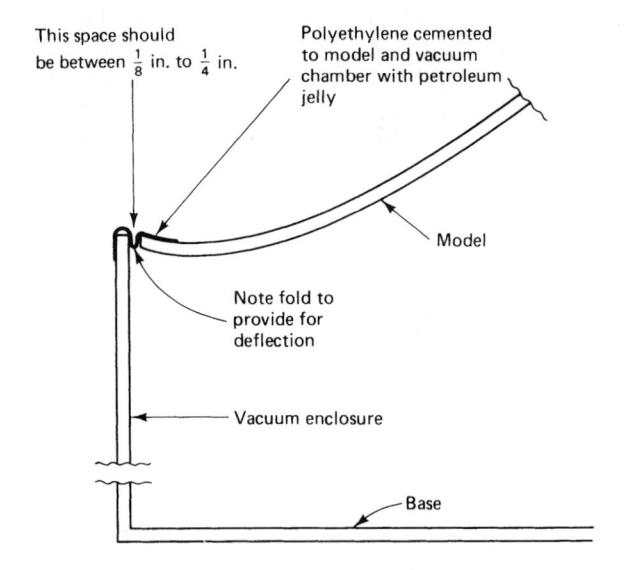

Figure 8.5 Vacuum loading system.

3. A simple open water manometer is the best method for measuring the negative pressure. The addition of colored water makes it easier to read the instrument. An electric pressure gage should be used if automatic recording is needed.
4. An adjustable opening in the vacuum chamber wall is necessary to control the load.

Applications of the vacuum loading method will be seen later in this chapter and in the case studies of model structures presented later in the text.

Air pressure may also be utilized for modeling blast effects on structures. An ingenious design for blast effects on model slabs, perfected at MIT, is shown in Figure 8.6. This device permits control of both rise time and pressure dropoff on the slab. Both chambers are initially pressurized

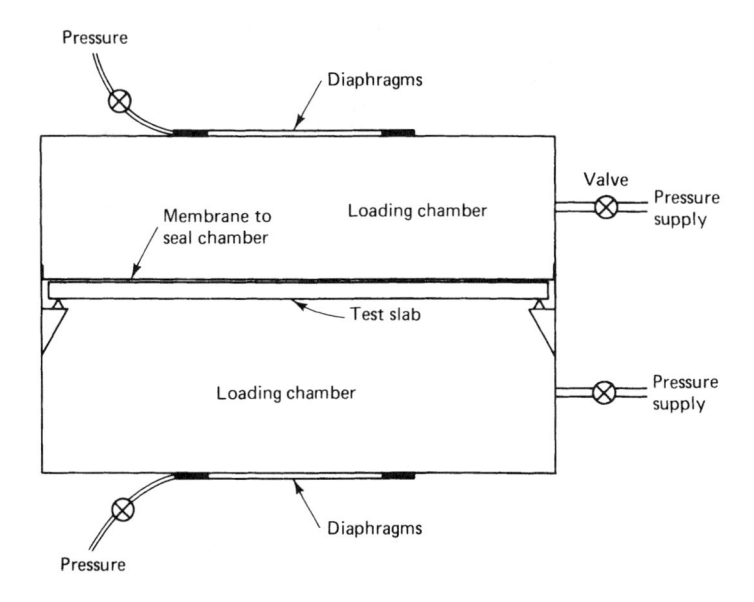

Figure 8.6 Pressure system for dynamic loading of slabs.

to the same level. Then one of the two-element diaphragms covering the port of one chamber is ruptured by pressurizing the diaphragm with a separate pressure line. This permits the gas to escape from one chamber, at a rate controlled by the initial pressure and port size, and the slab "feels" the pressure from the other loading chamber. Unloading is controlled by rupturing the diaphragm in the other loading chamber.

8.3 DISCRETE VS. DISTRIBUTED LOADS

A recurring question facing the models engineer is: How many discrete loads are needed to adequately represent a uniformly distributed load? Two points are important here: (1) accuracy of bending moments and other stress resultants induced by the loading and (2) local effects, such as local bending in thin shells. The latter point is particularly important in shell stability studies, where the local deformations induced by discrete loads may influence buckling behavior very strongly.

One can often profit by analytical study of effects of load spacing on internal forces in the model. By choosing simple structures that are representative of the actual model, it is possible to perform relatively quick analyses to give insight into the degree of inaccuracies produced with discrete loads. Litle et al. (1970) studied several cases of discrete load spacing effects on beams and arches. The first, shown in Figure 8.7, gives the bending moment at the center of a simple span beam as a function of the number of equal-spaced concentrated loads used to represent the uniformly distributed load w. It is seen that any even number of loads produces $M = wL^2/8$ at midspan, while the use of only three loads leads to an error of $+11\%$ in bending moment.

A more revealing study is summarized in Figure 8.8, where a parabolic two-hinged arch with a rise of $0.3L$ is analyzed for extreme fiber stresses produced by three different load cases: uniform load on the horizontal, 15 concentrated loads, and 5 concentrated loads. The highly variable stresses produced by 5 loads bear no resemblance to the uniform state of compression from a uniform loading. Even with 15 loads with stress, variation remains quite severe. The question of how severe such variations are depends upon what one is expecting from the model. Clearly an experimental

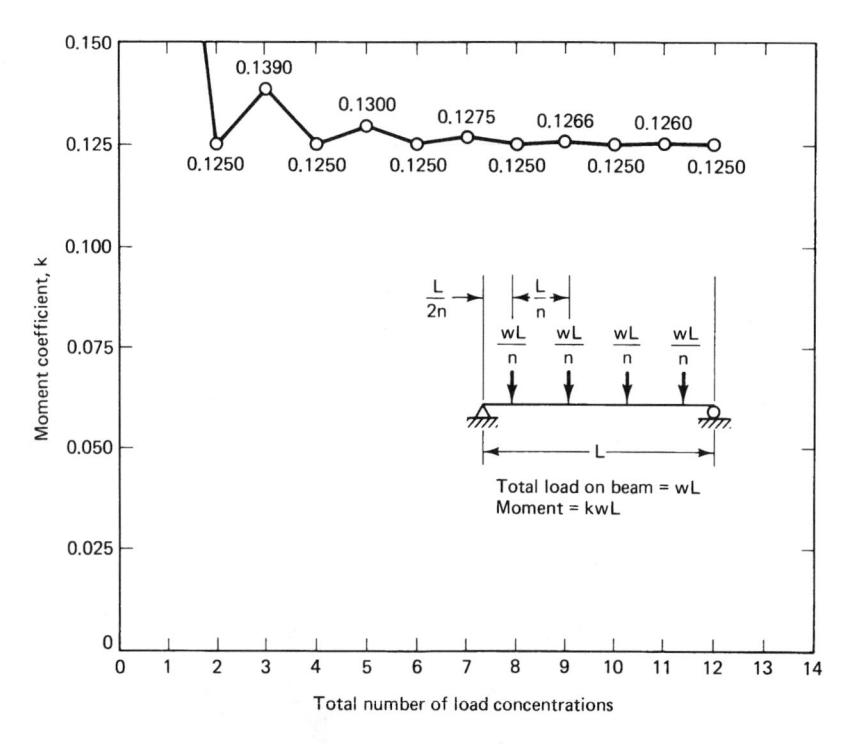

Figure 8.7 Effect of load discretization on simple beam bending.

stress analysis done on such a structure with discrete loading would be quite misleading if the prototype did have a truly uniform loading. Some ideas of potential problems to be met in interpreting strains measured on the surface of a discretely loaded shell structure model should be gained from this simple example.

The third study, which was done to determine the influence of discrete loading on buckling of a flat, parabolic fixed-base arch with rise = $0.1L$, is given in Figure 8.9. The buckling load reduces as fewer loads are used because of the more severe departure from the desired state of uniform compression. Using 10 loads gives a predicted buckling capacity of 90% of the true value. Even with 20 loads the predicted capacity is about 5% too low.

These studies on the arch structures are not meant to be extrapolated to other cases, but instead are intended to illustrate the desirability of performing analytical studies to help assess the inaccuracies met in using discrete loads for modeling uniformly distributed loads. Additional discussion of local load effects in shells is given in Section 8.4.2.

The other main problems encountered in using distributed loads is that of modeling gravity loads (such as self-weight and snow) on inclined roof surfaces with vacuum or pressure loads that act normal to the surface rather than in a desired gravity mode. This problem is discussed further in the next section.

8.4 LOADINGS FOR SHELL AND OTHER MODELS

Prototype shells transmit distributed surface (deadweight, snow, and wind) to discrete support locations. Proper modeling involves not only the correct magnitude of loads but also the correct distribution. Loading can be interpreted as also including reactions to applied loads, or the particular support and boundary conditions for a given shell. It is easy (but dangerous) to overlook proper modeling of the boundary conditions as well as the load system. Litle (1964), Wang et al. (1966),

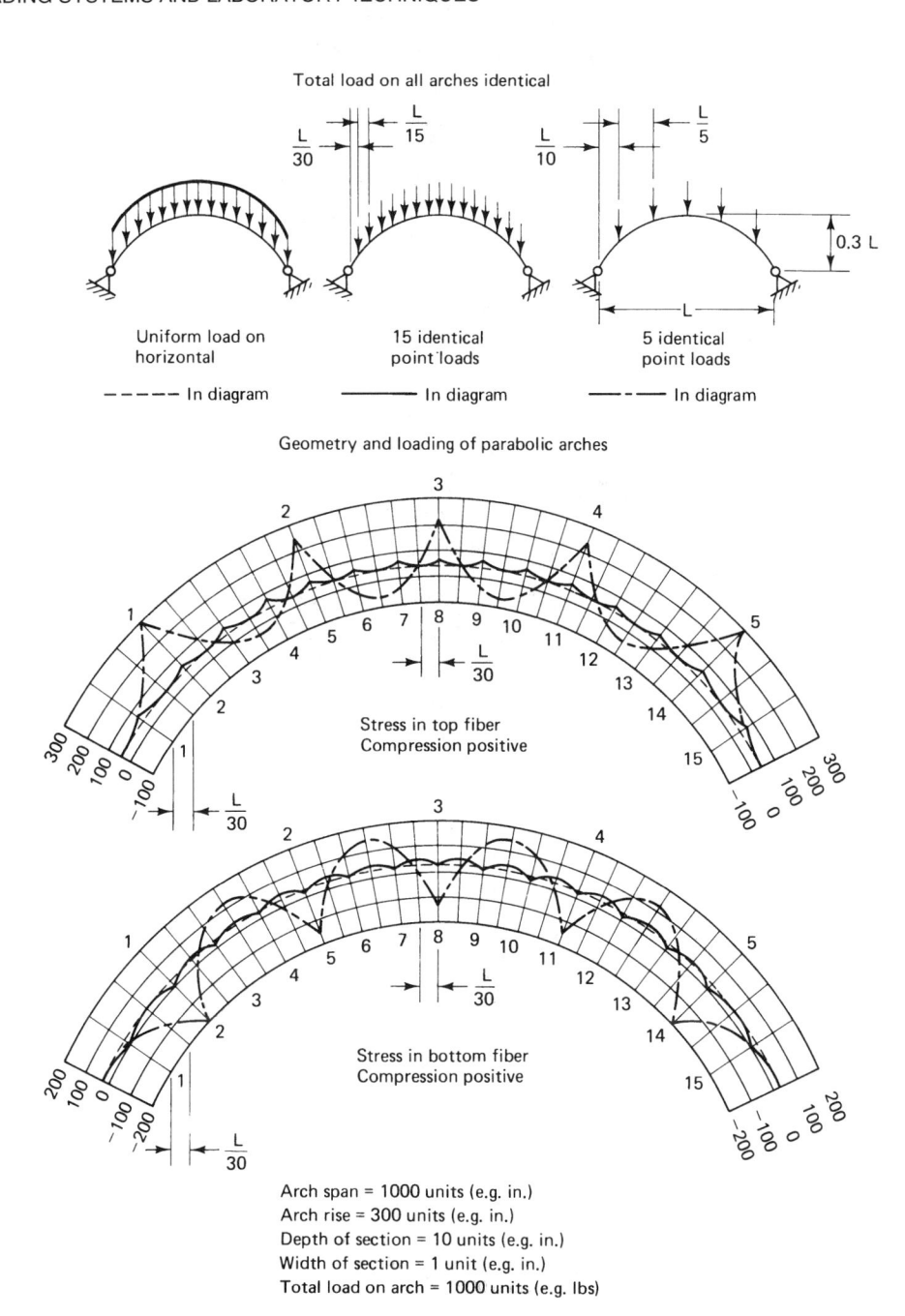

Figure 8.8 Effect of load discretization on stress distribution in a parabolic arch.

Wang (1967), and others have investigated the effect of varying boundary conditions on shell behavior; some conclusions are presented in Section 8.5. Loading methods for investigating conrete shell buckling has been reported by Billington and Harris (1981).

As described earlier, loading systems for surface loads on shells are usually one of three types: discrete, pressure, or vacuum. The choice of one system over the other is often based on the personal preference of the investigator and on the type of loading equipment available in the particular structural models laboratory. The vacuum system is certainly one of the best ways to load shells. Load can be applied and released quickly. The sole disadvantage of the vacuum loading technique

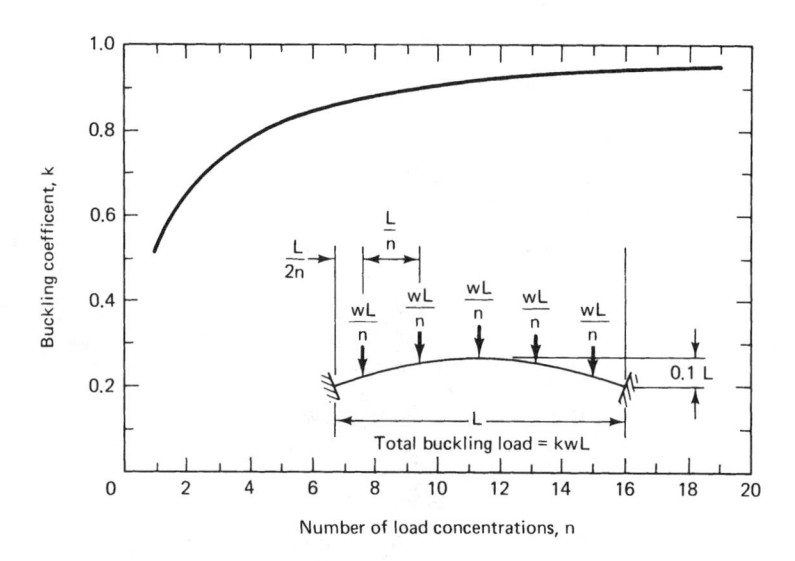

Figure 8.9 Effect of load discretization on buckling of a parabolic arch.

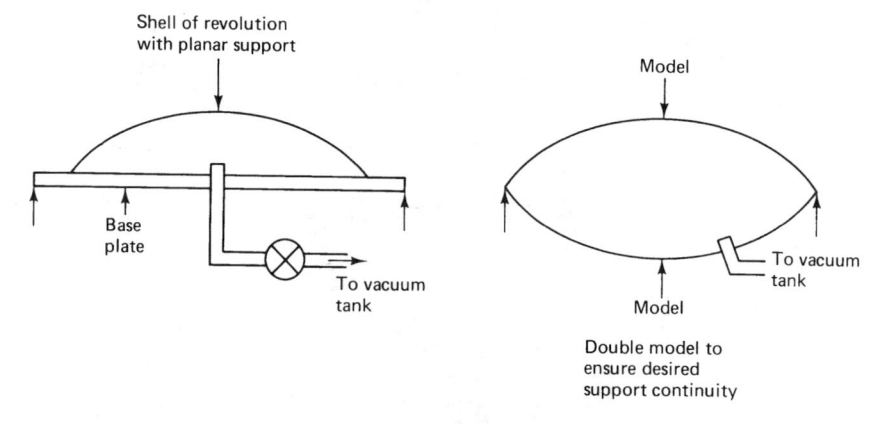

Figure 8.10 Vacuum loading of shell models.

for shell models is that it exerts a uniform pressure normal to the surface instead of a gravity-type loading. Thus, its use on steep shells may be questioned, and it cannot be used where a partial snow load or variable wind pressure is to be considered. The errors associated with not having the loading vertical are not appreciable for most practical shells where the "idealization" of the loading certainly seems reasonable when compared with the accuracy with which we know the assumed loadings or with the types of assumptions we make whenever an analytical approach is used. This is not, however, the case when buckling is a crucial factor.

8.4.1 Vacuum and Pressure Loadings

Two vacuum-loaded shell modeling studies are illustrated in Figure 8.10. In the first model, the shell was supported directly on a metal base plate and the load was applied by applying a vacuum to the enclosed space defined by the shell and the base plate. Lubricating oil around the base of the shell formed the seal. In the second model, where nonrotational edge member conditions were desired, two identical models were cemented together and the vacuum was applied in the defined cavity of the two models. This loading method also illustrated another important concept — that

Figure 8.11 Air bag loading on hyperbolic paraboloid shell models. (Courtesy of Cornell University, Ithaca, NY.)

of using symmetry to dictate what is happening at a support condition. Since each half of the double model tends to have the same rotational tendencies at the support, the effects cancel out and a clamped edge condition is achieved.

An air bag pressure loading system is illustrated in Figure 8.11 for a hyperbolic paraboloid umbrella shell (White, 1975). The air bag was built to conform to the shape of the hypar shell and then pressurized slightly to maintain its shape. The upper surface of the bag was formed from a sheet of plywood to serve as the loading surface. Deadweights placed on it transmitted a uniform pressure through the air bag to the shell surface. No instrumentation is needed to measure this type of loading since it is precisely controlled by the amount of deadweight applied to the top of the air bag. Precise balancing of load is essential, however, to prevent "drooping" of one side of the bag, and membrane effects at the edge of the model must be minimized.

A unique ring loading device for loading shell models was developed at McGill University. Harris (1964) used two simple techniques to load spun aluminum spherical shells 711 mm in diameter and fixed at the ends (shell thickness 2.03 mm). To apply a load of 179 N on a circular area 37.24 mm in diameter and centered on the shell crown, he used 22-N weights on a loading platform (Figure 8.12a); this was designed to eliminate any horizontal force between the platform and the loading device and to ensure that the load applied to the shell was truly vertical. Spring balances were used to keep the platform stable under full load.

The two ring loads (2.32 kN on a ring 12.22 mm wide and with a mean diameter of 247 mm; 4.11 kN on a ring 11.7 mm wide and with a mean diameter of 486.5 mm) were applied using loading rings (Figure 8.12b) and a 682-l domestic fuel oil tank that could be filled with water. The tank was calibrated on a scale to an accuracy of 224 N in the lower load range and 448 N in the upper range, and the loads were read using a glass manometer. The conical points on the loading rings were heat-treated to make them very hard and to prevent blunting by the loads. This also eliminated any appreciable bending moments from being transmitted from the loading platform or tank to the loading rings. The loading rings were fitted with a 3-mm-thick neoprene gasket to distribute the load evenly. The oil tank was carefully balanced with four horizontal strings and spring balances at the top. The entire assembly was supported by a framework made of perforated Dexion 225 standard angles and Dexion punched straps (Figure 8.12c).

Mufti (1969) successfully used a similar ring loading device to apply an intermittent ring load to a similar spherical cap with a hole at the crown.

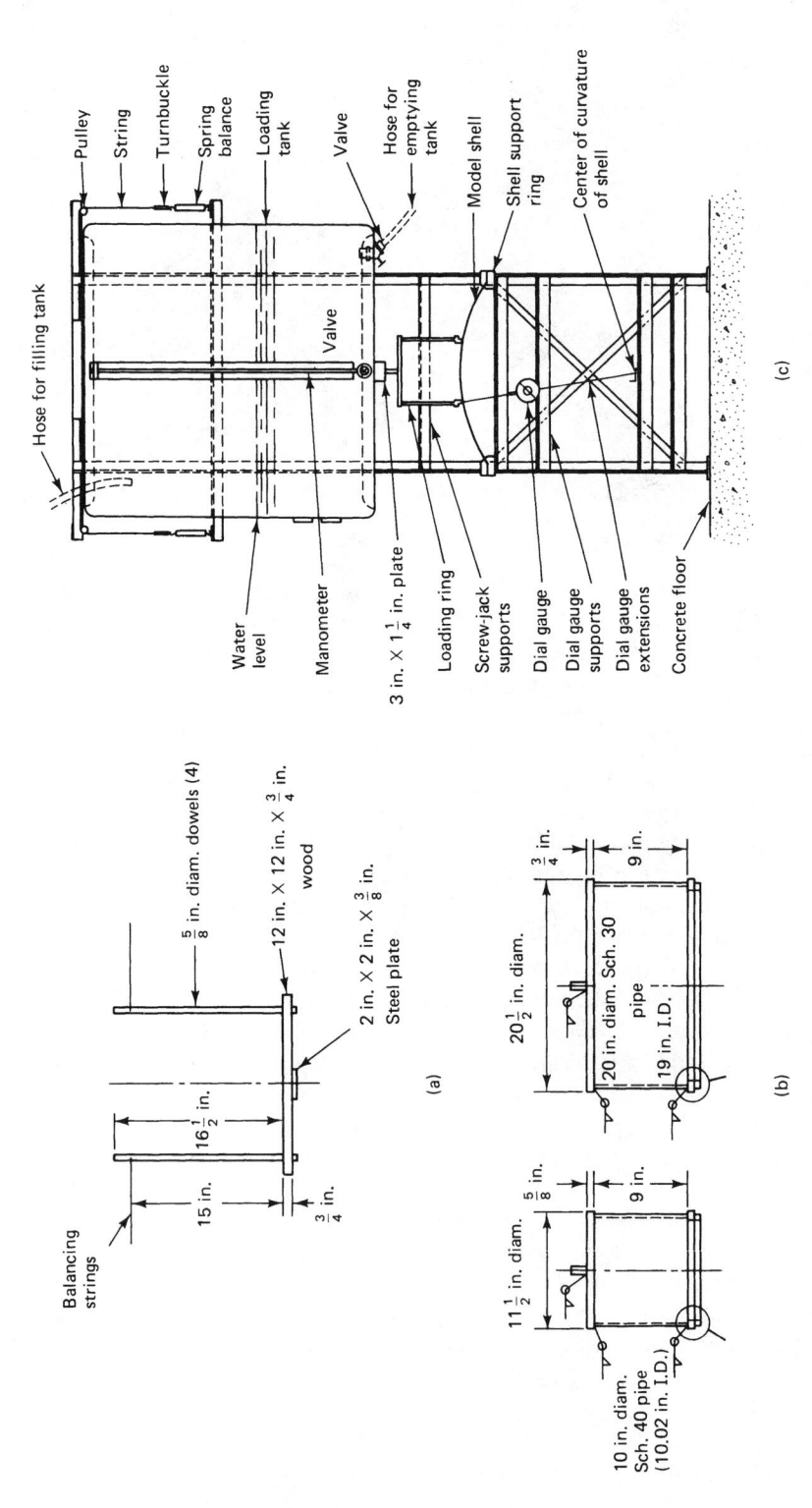

Figure 8.12 General arrangement of shell test apparatus. (a) Loading platform. (b) Loading rings. (c) Test setup. (After Harris, P. J., 1964.)

LOADING SYSTEMS AND LABORATORY TECHNIQUES 395

Figure 8.13 Suspended dead loads on shell model, with a catching mold for shell and weights. (Courtesy of MIT, Cambridge.)

8.4.2 Discrete Load Systems

Suspended deadweights are often utilized in discrete loading. Any deadweight system presents the problems of how to unload the model quickly and how to catch the loading system if the model fails suddenly. A typical deadweight system for shell loading used at MIT is illustrated in Figure 8.13, where a supporting mold fits the underside of the shell, with the weight strings passing through the mold. Before loading, or to unload, the mold is jacked up against the shell, transmitting the load to the mold and its supports. Loading occurs when the supporting mold is lowered away from the model. The mold is always kept close to the shell lower surface (but not in contact with it) so as to catch the model as buckling or other large displacements occur.

Figure 8.14a and b illustrate other shell modeling studies that utilized suspended discrete loads. In the second case the loads are not meant to be removed quickly, nor can the structure be caught if it begins to fail. In another variation of suspended loading, cans are suspended from the model and are temporarily supported by a movable table; the table is lowered to apply the loads to the shell. This system was employed for a buckling study described in Section 8.5.

Three different types of load–distribution pads are shown in Figure 8.14c. In each type the underlying principle is to minimize the effect of the concentrated load by spreading it out to several loading feet or legs, or to make it even more uniform by transmitting the load through a foam rubber pad.

A 64-point whiffle-tree load system for ultimate-strength modeling of hyperbolic paraboloid shells (White, 1975) is shown in Figure 8.15a. A similar system for cylindrical shells is shown in Figures 8.15b and c (Harris and White, 1972). One potential disadvantage of the whiffle-tree system is its relatively high deadweight, which cannot be removed from the shell. Since deadweight augmentation is needed to simulate gravity stresses in ultimate-strength models made from the same materials as the prototype, the weight of the whiffle tree can be used as part of the dead load. In addition, a well-constructed whiffle tree converts a single load into a large number of very accurate equal loads that will remain almost equal even under severe deformations of the shell (Figure 8.15c).

An interesting whiffle-tree loading system was developed at Drexel University to load a three-dimensional model of a precast, large-panel concrete building (Muskivitch and Harris, 1979, 1980a,b; Harris and Muskivitch, 1980). It consists of an interior whiffle tree in compression spreading a point load supplied by a pivoted I beam to 16 bearing pads (Figure 8.15d). There are four of these compression arrangements per bay of the three-bay six-story model (Figure 8.15e). Each of the four compression whiffle trees are collected by a tension whiffle-tree arrangement to

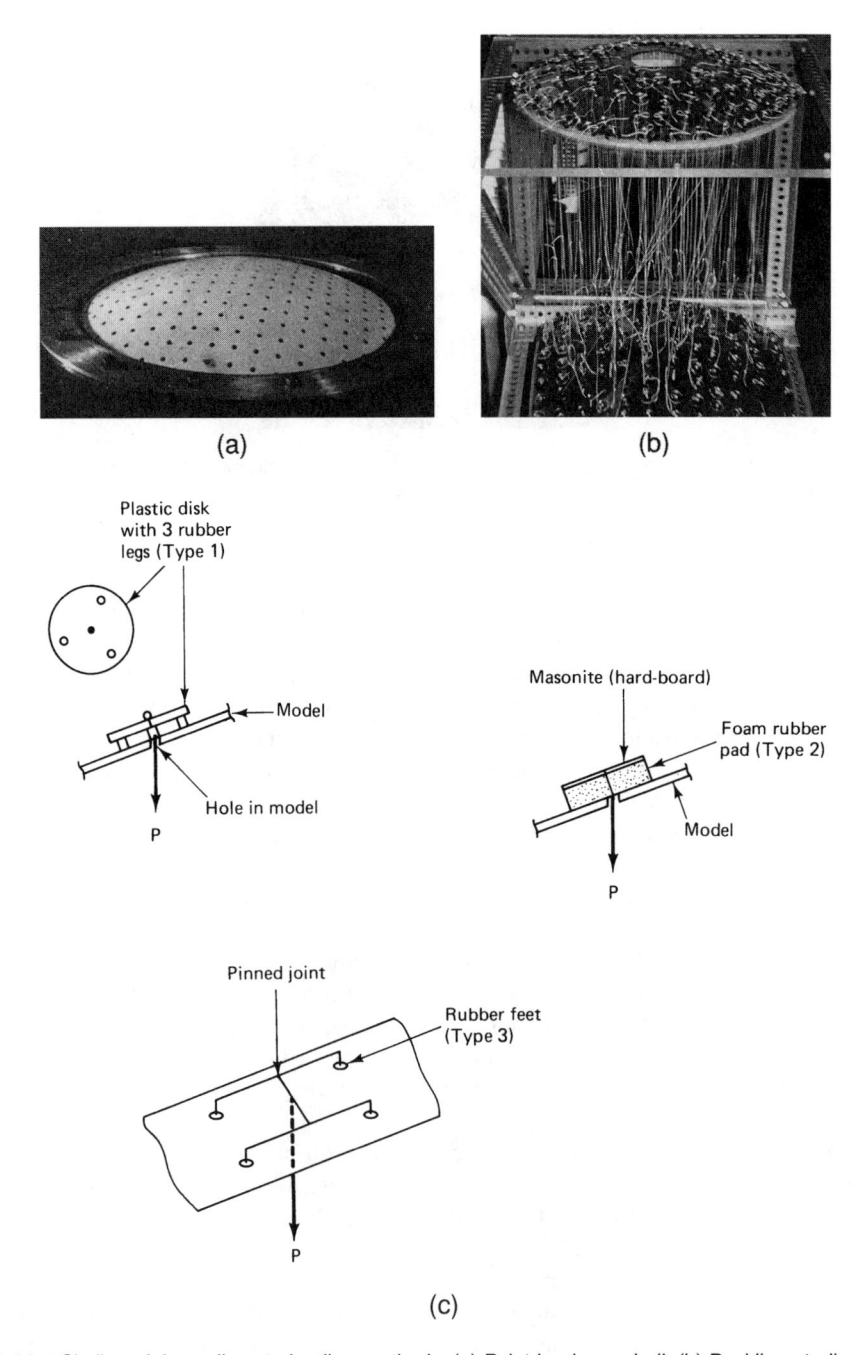

Figure 8.14 Shell models — discrete loading methods. (a) Point loads on shell. (b) Buckling studies on shells (a and b courtesy of MIT, Cambridge.) (c) Load pads.

a single tension jack. The resulting 18 jacks attached to a common hydraulic system are shown in Figure 8.15f. An overall view of the discrete point mechanical-hydraulic loading system for applying gravity loads to the $3/32$-scale model building is shown in Figure 8.15g. Thus, whiffle-tree load systems are most useful for ultimate-strength models where applied loads are high and the dead-weight of the whiffle tree is not an important factor but can be used to advantage. Whiffle trees of the size shown in Figure 8.15 would be useless for thin elastic models made of plastic; the shell would most likely be overloaded from the weight of the whiffle tree alone.

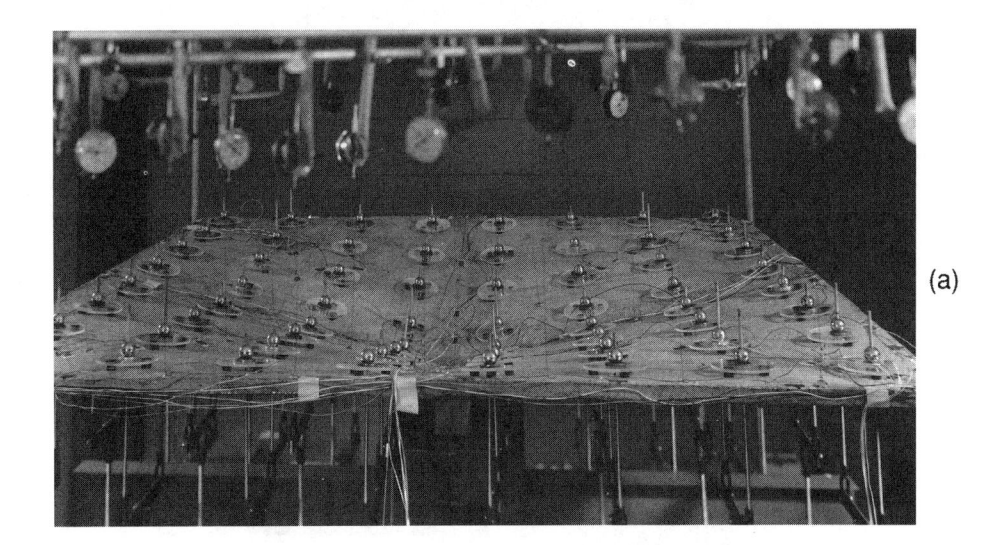

LOADING SYSTEMS AND LABORATORY TECHNIQUES 397

Figure 8.15 Whiffle-tree systems for plates and shells. (a) 64-point whiffle-tree system on hypar shell. (b) Whiffle-tree system on medium-length cylindrical shell. (c) Whiffle-tree system on short cylindrical shell (a to c courtesy of Cornell University, Ithaca, NY.) (d) Details of inner (compression) whiffle tree. (e) Close-up of inner whiffle tree. (f) Front view of loading system (e and f courtesy of Drexel University, Philadelphia.) (g) Loading system for three-dimensional model.

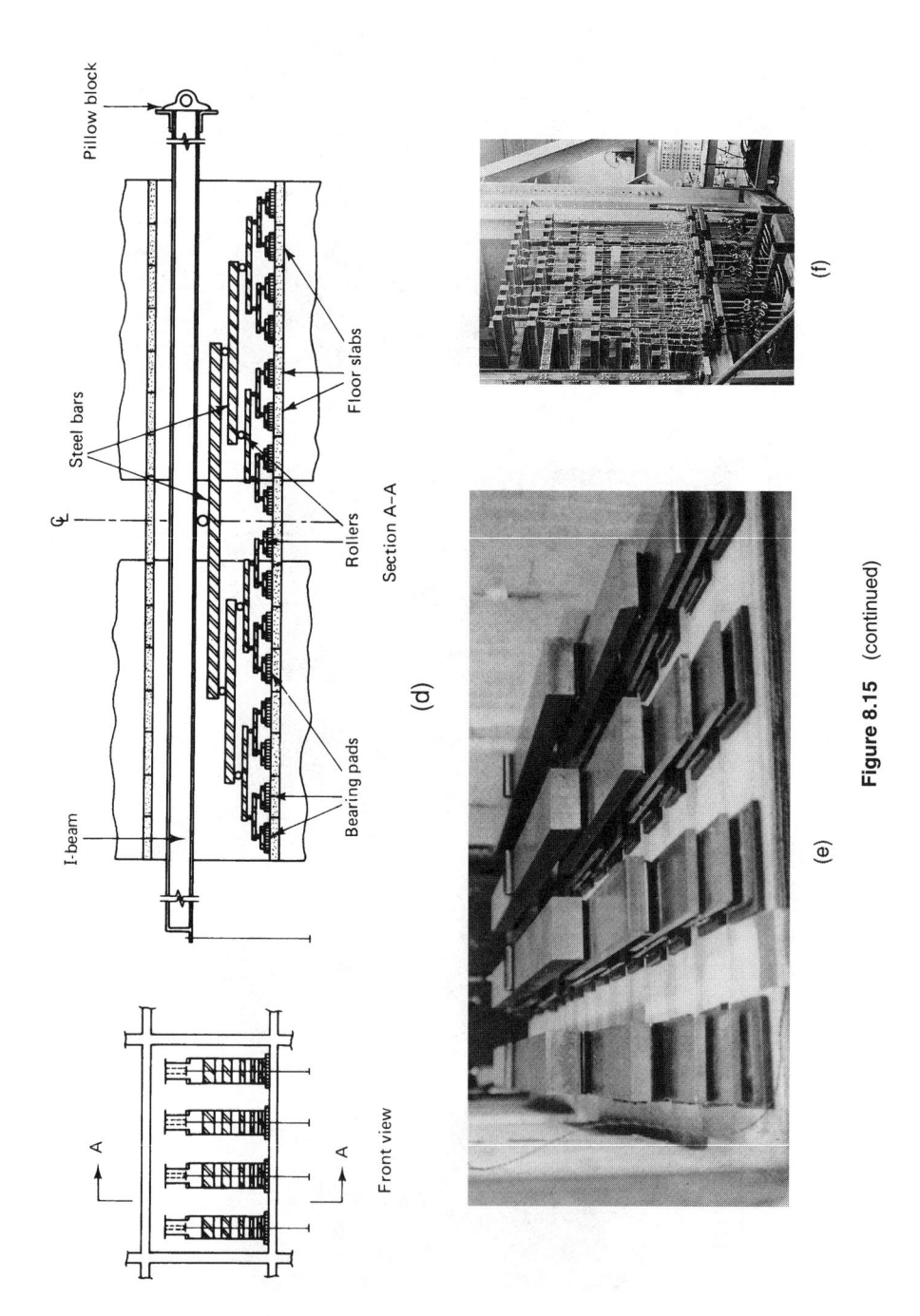

Figure 8.15 (continued)

LOADING SYSTEMS AND LABORATORY TECHNIQUES

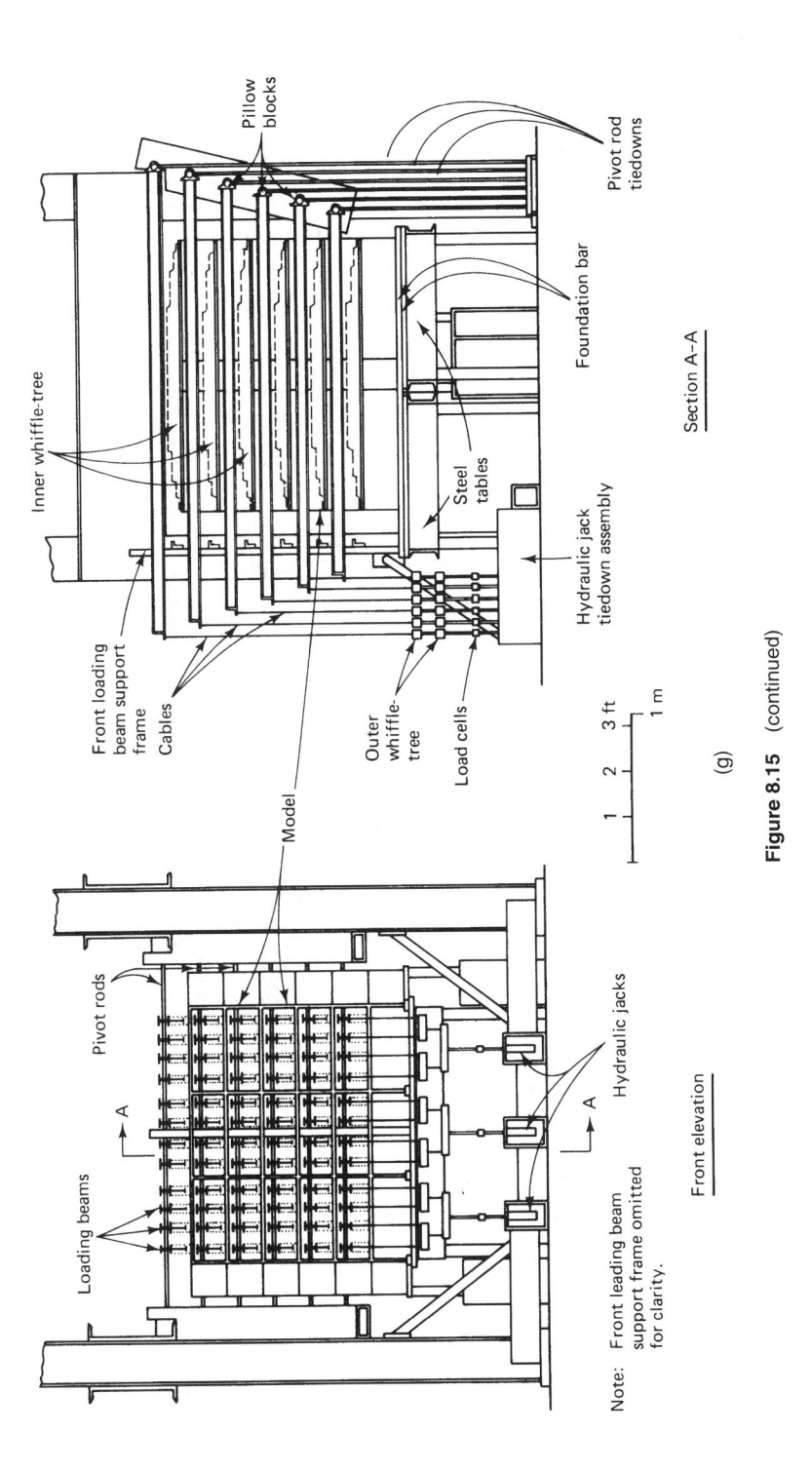

Figure 8.15 (continued)

Some examples on loading calculations of models are given below.

Example 8.1

Given a uniform line load q (kN/m) to be appied to a portion of a model structure over a length L. The load is to be applied as three equal discrete loads. Design the whiffle-tree loading system to bring the total load to one jacking point.

The solution to this problem is easily carried out if one realizes that the articulated system of the whiffle tree is statically determinate. The three individual point loads act at the center of the three segments of the length $L/3$ shown below. Loads 1 and 2 are combined with a loading beam (usually a steel bar of appropriate strength) at one level below the test structure and having a length of $L/3$ plus small distances on either end for the attachment cables or eye bolts. At a level below this, another loading beam (usually of steel and larger than the beam above) of length $3L/2$, plus end distances, carries a load of $2P$ on one end and P on the other with a resultant of $3P$ acting at $L/3$ from the left as shown in the figure. The resultant passes through the midpoint of the loaded length L as expected. For given values q and L, the sizes of the loading beams are easily determined using an appropriately large factor of safety of two to three. In case the test structure is overstrength, one wants the test structure to fail and not the loading system!

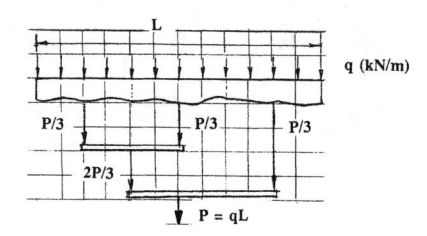

8.4.3 Effects of Load Spacing

The effects of load spacing were examined at MIT by Soosar (1963) and later summarized by Litle (1964). They showed that replacement of a continuous loading with discrete loads may lead to a very large systematic error in shell stress distribution. However, discrete loading does not affect buckling pressure if the grid spacing is sufficiently small. It should not be necessary to go to less than a 25-mm grid spacing for shell stability studies, and the spacing can usually be higher.

The ultimate-strength hypar models of Figure 8.15a had discrete loads at a 150-mm grid spacing, which is one-eighth of the shell-span dimension. Each load in turn was distributed to three points on the shell surface (Figure 8.14c, Type 1). While this spacing certainly influenced the local shell stresses, it did not seem to affect the development of major inelastic action and the failure mode. The precise definition of satisfactory discrete load spacings needs additional exploration. It is felt that the examples given here represent acceptable practice for the particular model being studied. One cannot overemphasize the danger of looking at someone else's loading system, which might have been designed for heavy loads on ultimate-strength models, and then using a similar setup for light loads on elastic models. A final point on load systems is that they must not exert significant restraint on the model. At the same time, the supports for a model must not move if they are intended to remain fixed in position. Proper planning of supporting fixtures for shell models requires considerable thought and common sense.

8.5 LOADING TECHNIQUES FOR BUCKLING STUDIES AND FOR STRUCTURES SUBJECT TO SWAY

The major requirement for buckling- and sway-prone structures is that the loading must not restrain the movement of the structure because such restraint will tend to increase the load capacity.

Figure 8.16 Frame fastened to models testing table. (Courtesy of Cornell University, Ithaca, NY.)

Plane trusses, frames, space trusses, and similar structures that are being modeled for instability are usually loaded with gravity loads if the applied load is not excessively high. If it appears unfeasible to use dead load, one may have to resort to a gravity load simulator of the type shown in Figure 8.16. This simulator is merely a mechanism that permits transverse motion without any vertical motion; thus it is capable of maintaining a constant vertical load while moving laterally. Six such devices were used in a frame test to be described subsequently.

8.5.1 Shell Instability

Shell structures have been known to suffer various forms of instability. A large dome covering a market in Bucharest reportedly collapsed from creep buckling action. The failure mode of the Ferrybridge hyperbolic cooling towers in England included a snap-through type of behavior that had been demonstrated in earlier model studies by Der and Fidler (1968). A similar type of buckling action occurred after extensive inelastic deformation in some of the intermediate-length cylindrical shell tests reported by Harris and White (1972).

The very thin shell surfaces utilized in metal skin construction are almost always subject to buckling failure modes, and light-gage steel hypars and folded plates often are limited in strength by the buckling capacity of the corrugated decking. Instability can also be a serious problem for certain folded-plate geometries.

Loading for a buckling model must be very carefully planned in order to prevent any restraint against motion that might occur when buckling takes place. Swartz et al. (1969) utilized a whiffle-tree loading system fastened to pull-type hydraulic rams that produced vertical concentrated loads on a 84 × 84 mm square grid (Figure 8.17) over the surface of a folded-plate model. Horizontal movement of the hydraulic rams during buckling was accomplished by having the rams anchored to thrust bearings that permitted the necessary movement.

The shell roof of the Providence, RI, post office building (described by Litle and Hansen, 1963) was designed for suitable stiffness against instability failures by using plastic models loaded with dead loads (Figure 8.18). The concept of using a "catching device" under the model was also employed here. This not only prevents the destruction of the model but also permits its retesting if the instability mode does not produce stresses beyond the linear range of behavior for the plastic. The plastic models shown in Figures 8.13 and 8.14a were also intended primarily as instability models.

In his definitive work on the reliability of models for predicting shell stability, Litle (1964) presents considerable practical information on loading techniques and on other laboratory techniques needed

Figure 8.17 Stability model for folded-plate structure. (Courtesy of IIT (Illinois Institute of Technology), Chicago.)

Figure 8.18 Shell roof model. (Courtesy of MIT Cambridge.)

in shell modeling. The reader is urged to study this important reference before embarking on any shell modeling problem. Litle emphasizes that buckling behavior is extremely sensitive to changes in boundary restraint. If the model loading rig induces initial edge bending, the buckling behavior will almost certainly be affected. Any buckling model must reproduce prototype boundary conditions as closely as possible. When the prototype boundary conditions are not well defined, it is recommended that alternate model tests be conducted in which the upper and lower bounds on edge restraints are both modeled to give bounds on the actual prototype conditions.

Litle also recommends that closely spaced hanging loads be used instead of pressure (or vacuum) loads for shells that have any appreciable slope, since buckling can be quite dependent upon load direction as well as magnitude.

8.5.2 Structures Undergoing Sway

A model study for the Australia Square Tower in Sydney (Gero and Cowan, 1970) is shown in Figure 8.19 for a design wind condition that produces substantial sway. A whiffle tree is used in an ingenious fashion for this uniform loading, and a hydraulic jack is used to produce the single horizontal force. Note that a mechanical lever is used to amplify the jack force and that the entire weight of the whiffle tree is suspended from a support. Each line of the tree is adjustable to align the applied forces in the horizontal direction. Although the gravity loads are not applied simultaneously with wind effects in this model, a gravity load system could be superimposed and the two systems could act independently without restraining each other.

The three-story, two-bay frame model (Chowdhury and White, 1977) of Figure 8.20b illustrates several important features of loading, including:

1. Testing a vertical structure in a horizontal plane;
2. Combined gravity and simulated static lateral seismic loads;
3. Combined axial force and reversing bending loads on a component of the frame (Figure 8.20a).

The frame has fixed bases that are modeled by bolting the heavy base beam to the stiff steel testing table. The frame was tested in the horizontal plane because it was relatively easy to provide the needed out-of-plane restraint with the testing table. Also, the table was a convenient attachment surface for the many gages and loading devices employed in the experiments.

The gravity loads are applied at the quarter points of each of the six beams with six gravity load simulators. These mechanisms are fixed to the table and fastened to the beams as shown. They are activated by small hydraulic rams connected on each level to a common hydraulic pressure source to give a fixed gravity load on each floor. This gravity load, which is measured with tensile load cells, is maintained constant during the test, and permits side sway of 75 mm.

The reversing simulated seismic loads are applied at each floor level with hand-cranked mechanical loading devices. The major difficulty met with in this type of load system is in achieving the proper lateral loads at each floor level. The load induced at each level is dependent on loads at other levels, and considerable adjustments are needed, particularly at high loads when the frame is approaching failure.

Prior to the frame test, component testing was done on the beam–column joint specimen of Figure 8.20a. This specimen represents an exterior connection with its half columns and half beam that would extend to the normally assumed inflection points when the frame is under lateral load. The loading on this specimen was constant axial force on the column and fully reversing bending moment on the cantilever beam section. This rather complex load condition was achieved quite simply by some minor adaptations to an existing 133 kN capacity universal testing machine. Column axial load was applied vertically with the testing machine. The force to create moment was applied vertically through a double-acting hydraulic ram that was anchored to a steel pipe column spanning between the floor and ceiling, immediately in front of the machine. The horizontal reactions at the column ends induced by the bending force were taken back to a building column immediately behind the machine, through steel angles fastened to the end reaction devices on the model column. This loading system functioned extremely well and was inexpensive. It is another example of how ingenuity must be used to solve rather complex loading designs.

Gravity loads can also be applied to frames that are subject to sway by using long, flexible tensile elements that span between the load point and the floor (base of the frame). These elements offer very little resistance to lateral load as long as the sway is small, but one set of tensile elements may interfere with others when the frame is more than a single story high.

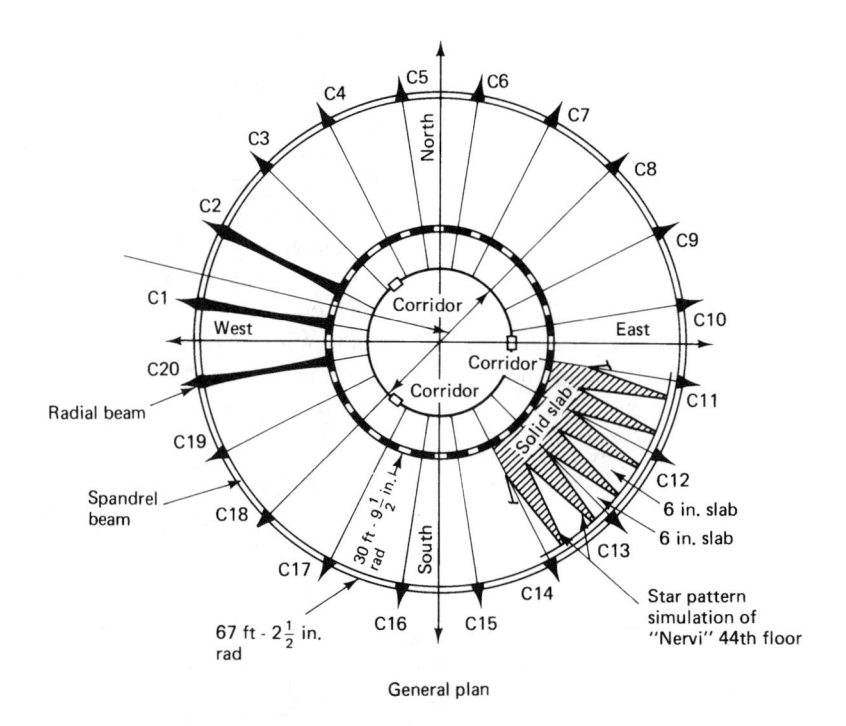

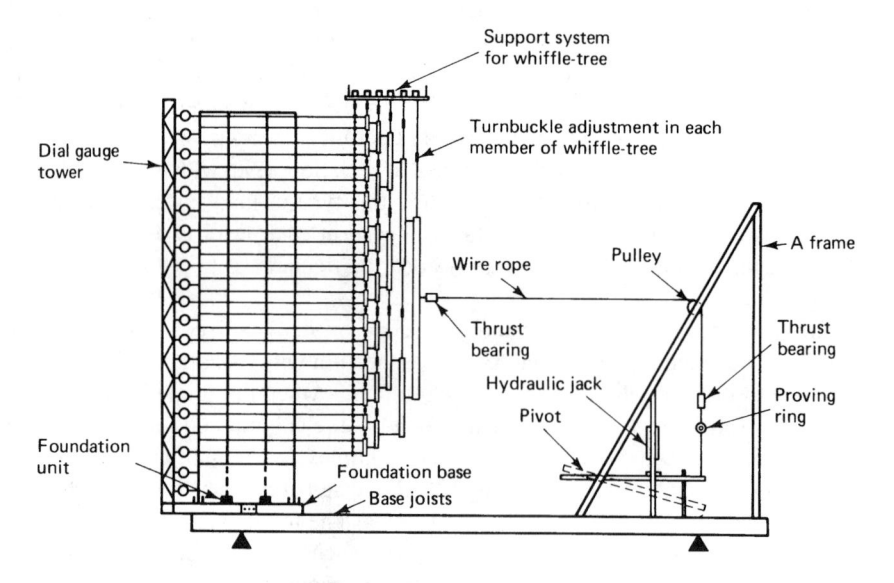

Figure 8.19 Lateral load system for high-rise building frame.

8.6 MISCELLANEOUS LOADING DEVICES

There are many special situations met in devising loading schemes for structural models, including thermal effects in dams, nuclear reactors, and other structures; self-weight effects where additional surface loadings do not produce the necessary state of deadweight stresses; internal pressure loadings for nuclear reactor vessels and other pressure vessels; and dynamic loads for earthquake, wind, and oscillating machinery effects. Several of these topics will be discussed briefly

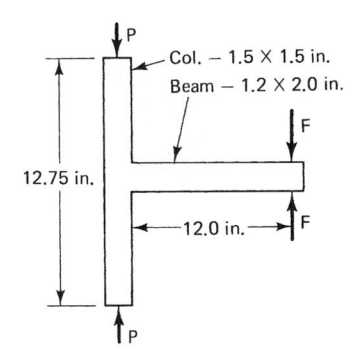

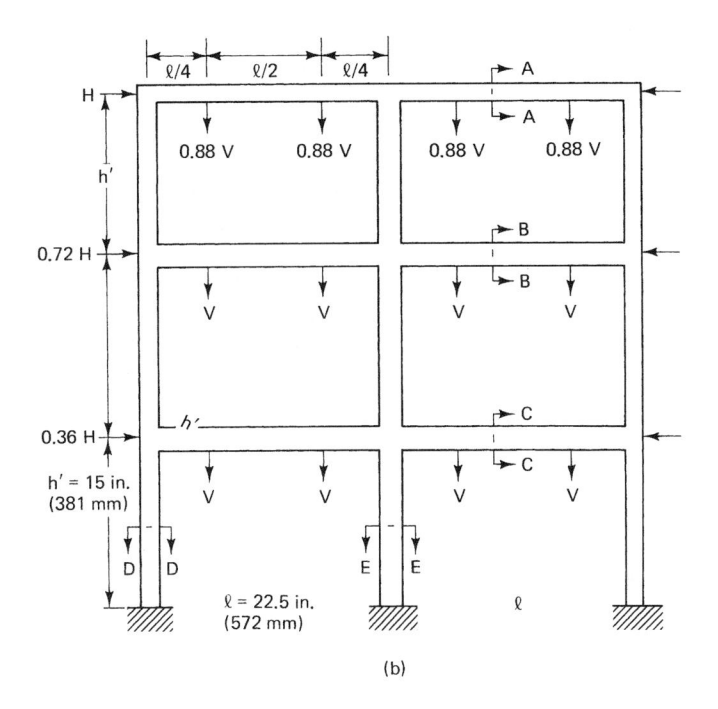

LOADING SYSTEMS AND LABORATORY TECHNIQUES 405

Figure 8.20 Simulated seismic and gravity loads on 1/10-scale model frame. (a) Beam–column joint. (b) Model frame. (c) Test setup for frame. (Courtesy of Cornell University, Ithaca, NY.)

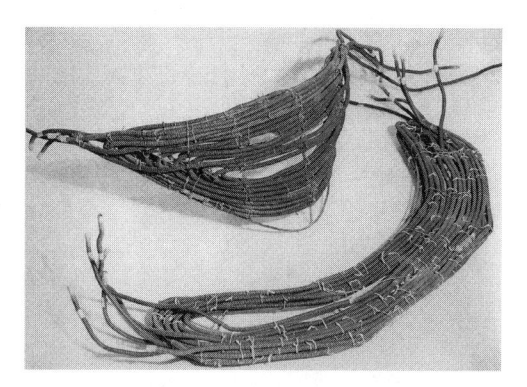

Figure 8.21 Upstream and downstream heating coils. (Courtesy of LNEC.)

here and others will be returned to in Chapter 11 and in case studies and special modeling applications later in the book.

8.6.1 Thermal Loads

Thermal loads require a heat source and a medium to distribute the heat properly to the model. Electric heating elements are the most common heat source because they are compact and can be controlled with a high degree of accuracy. Typical heating coils for modeling thermal effects in dams are shown in Figure 8.21. As described by Rocha (1961), the dam is first covered with a waterproof membrane, and the shaped coil assemblies are then inserted into upstream and downstream cavities that contain water. The coils are used to heat the water to the proper temperatures.

It should be noted that the normal critical factor in thermal modeling is the presence of a thermal gradient across the model thickness. The absolute temperatures are normally not important, which means that cooling can also be used to induce the proper gradients. Thus, room temperature on one face of a model and a refrigerated condition on the other face could produce the desired gradient.

Thermal effects in reactor structures are also usually generated by electric heating elements and either an oil or water bath to properly disperse the heat. Local hot spots might be modeled with heating elements alone, with the thermal load being distributed by the material itself.

8.6.2 Self-Weight Effects

Models have been tested in a centrifuge to induce a higher acceleration of the mass of the model, thus satisfying the deadweight similitude condition. A large number of centrifuges have been built and operated in the last two decades. These powerful testing devices are described in Chapter 11. Many structures that interact with the soil medium in which they are founded have been modeled by this method. This approach may be the only feasible method of getting proper deadweight effects in soils models.

In using small-scale models for structural engineering studies, the deadweight stresses are not properly modeled in most situations. The only exception is where the stress scale can be chosen in such a way as to satisfy the dead-load similitude conditions, but this is uncommon in most models for reinforced concrete structures, and even other types of structures. Several methods have been used to satisfy, at least approximately, deadweight stresses in models. In dams, for instance, where the final stress state is a strong function of deadweight, models are sometimes built in many lifts (layers), and prestressing is applied to each lift to approximate the state of stress that is desired during the construction phase. This is not entirely satisfactory because of the stress concentrations and disturbances created by the prestressing elements and anchorages.

Durelli et al. (1970) studied gravity and surface load stresses in hollow spheres by using the immersion technique. In this method the model of the structure being studied for gravity deadweight stresses is placed upside down in a fluid that has a specific gravity that is substantially higher than the model itself. The model, which tends to float, is held down at those points where it would normally rest in its upright position. The resultant load on the model becomes the weight of the model times the factor $(k-1)$, where k is the ratio of the density of the fluid to the density of the model material. This rather ingenious technique is discussed in more detail in the reference cited above.

8.7 SUMMARY

Loading systems for structural model tests take on many forms and require great care and substantial ingenuity in their design and construction. They range from the very simple, single hanging concentrated deadweight system to automated, computer-controlled systems. The reader is urged to study carefully the many experimental studies reported in the journals of *ASCE, ACI, SESA (SEM)*, and other technical societies, and in special models publications and conference proceedings. Most of the references on models at the end of each chapter include descriptions of loading systems. The development of an ability to design efficient and inexpensive loading systems for complex models can be extended directly to tests for large structures and thus represents a highly valuable skill.

PROBLEMS

8.1 A whiffle-tree loading arrangement is to be devised for an equivalent earthquake loading system for the two-story frame shown below. Design a horizontal acting whiffle-tree arrangement that will be equivalent to a triangular loading representing the first mode inertial loading of the structure. Develop a technique of supporting the loading elements.

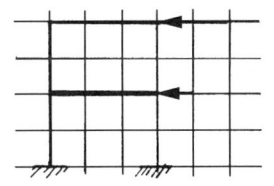

8.2 Discuss the uniform hydrostatic loading arrangement of the $\frac{1}{30}$-scale strength model made of reinforced microconcrete model of a submerged highway tunnel that is planned under the Delaware River crossing between Philadelphia, PA and Camden, NJ. Design an articulated loading system and instrumentation for critical strains and deformations of the circular configuration.

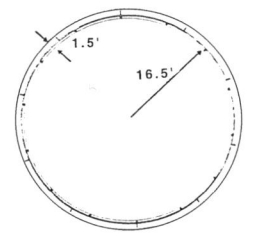

8.3 Find the number of concentrated loads required to approximate the peak bending moment of a simply supported and uniformly loaded model beam of length L to an accuracy of 10%.

8.4 A section of the loading system that has been developed for applying a uniform load on a simply supported reinforced concrete slab in order to study its yield line behavior is shown below. In the transverse direction only two of these whiffle trees shown are combined and brought to one jacking point. Comment on the accuracy of the setup for slab aspect ratios $b/a = 1$ (square slab) and $b/a = 4$ and the possible improvements that could be made in order to approximate the bending moments more accurately.

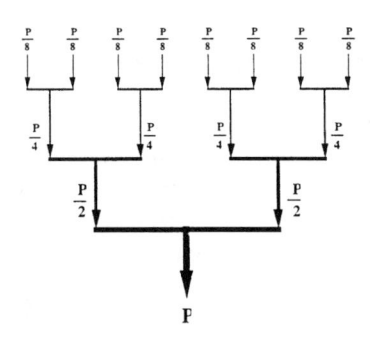

8.5 A model beam of length L is to be loaded with one hydraulic jack by pulling down on it from below. The design load on the beam is linearly varying from zero on the left support to a value q (N/m) on the right support. Design a whiffle-tree arrangement to load the beam with an equivalent total load with:
  a. Two equally spaced pulling points.
  b. Three equally spaced pulling points.

8.6 Given a three-story concrete infilled model frame to be tested under lateral uniform load simulating the wind condition; see figure. Only one horizontal loading jack is available to provide the loading to the frame. Design a whiffle-tree arrangement to distribute the load equally to the second, third, and roof levels. Determine the length of the loading members and method of keeping them in place during loading.

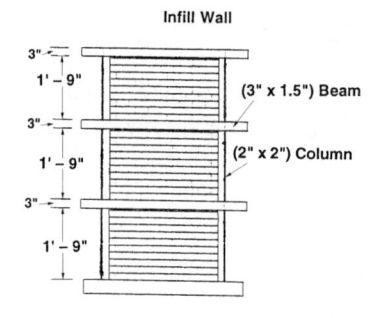

REFERENCES

Billington, D. P. and Harris, H. G. (1981). Test methods for concrete shell buckling, in *Concrete Shell Buckling,* E. P. Popov and S. J. Medwadowski, Eds., Publication SP-67, American Concrete Institute, Detroit, 187–231.

Chowdhury, A. H. and White, R. N. (1977). Materials and modeling techniques for reinforced concrete frames, *Proc. Am. Concr. Inst.,* 74(11), Nov., 546–551.

Der, T. J. and Fidler, R. (1968). Model study of the buckling behavior of hyperbolic shells, *Proc. Inst. Civ. Eng.* (London), 41, 105–118.

Gero, J. E. and Cowan, H. J. (1970). Structural concrete models in Australia, in *Models of Concrete Structures,* ACI, SP-24, American Concrete Institute, Detroit, MI, 353–386.

Harris, H. G. and Muskivitch, J. C. (1980). Models of precast concrete large panel buildings, *ASCE J. Struct. Div.,* 106, no. ST2 (February), Proc. Paper 15218, 545–565.

Harris, H. G. and White, R. N. (1972). Inelastic behavior of reinforced concrete cylindrical shells, ASCE *J. Struct. Div.,* 98, no. ST7 (July), Proc. Paper 9074, 1633–1653.

Harris, P. J. (1964). The Analysis of Axially Symmetric Spherical Shells by Means of Finite Differences, Ph.D. thesis, McGill University, Montreal, July, 168 pp.

Litle, W. A. (1964). *Reliability of Shell Buckling Predictions,* Research Monograph No. 25, Massachusetts Institute of Technology, Cambridge, 149.

Litle, W. A., Cohen, E., and Sommerville, G. (1970). Accuracy of structural models, in Models for Concrete Structures, ACI SP-24, American Concrete Institute, Detroit, MI, pp. 65–124

Litle, W. A. and Hansen, R. J. (1963). The use of models in structural design, *J. Boston Soc. Civ. Eng.,* 50(2), 59–94.

Mufti, A. A. (1969). Matrix Analysis of Thin Shells Using Finite Elements, Ph.D. thesis, McGill University, Montreal, July, 141 pp.

Muskivitch, J. C. and Harris, H. G. (1979). Behavior of large panel precast concrete buildings under simulated progressive collapse conditions, Nature and Mechanisms of Progressive Collapse in Industrialized Buildings, Report No. 2, Office of Policy Development and Research, Department of Housing and Urban Developent, Washington, D. C., January, 207 pp.

Muskivitch, J. C. and Harris, H. G. (1980a). Behavior of large panel precast concrete buildings under simulated progressive collapse conditions, in *Proceedings, International Symposium, Behavior of Building Systems and Building Components,* Vanderbilt University, March 8–9, Nashville, TN.

Muskivitch, J. C. and Harris, H. G. (1980b). Report 2: Behavior of precast concrete large panel buildings under simulated progressive collapse conditions, in *Nature and Mechanism of Progressive Collapse in Industrialized Buildings,* Office of Policy Development and Research, Department of Housing and Urban Development, Washington, D.C., January, 207 pp., also Department of Civil Engineering, Drexel University, Philadelphia.

Rocha, M. (1961). Determination of thermal stresses in arch dams by means of models, *RILEM Bull.,* (Paris), (10), 65.

Soosaar, K. (1963). Systematic Errors in the Loading of Stress Models of Shells, Report T63-6, Department of Civil Engineering, Massachusetts Institute of Technology, Cambridge.

Swartz, S. E., Mikhail, M. L., and Guralnick, S. A. (1969). Buckling of folded plate structures, *Exp. Mech.,* 9(6), 269–274.

Wang, L.R.-L., Rodriquez-Agrait, L., and Litle, W. A. (1966). Effect of boundary conditions on shell buckling, *J. Eng. Mech. Div., ASCE,* EM6, December.

Wang, L. R.-L. (1966). Discrepency of experimental buckling pressures of spherical shells, *J. Am. Inst. Aeronaut. Astronaut.,* 5(2), February, 357–359.

White, R. N. (1972). Structural Behavior Laboratory—Equipment and Experiments, Report No. 346, Department of Structural Engineering, Cornell University, Ithaca, NY.

White, R. N. (1975). Reinforced concrete hyperbolic paraboloid shells, *ASCE J. Struct. Div.,* September, 1961–1982.

CHAPTER 9

Size Effects, Accuracy, and Reliability in Materials Systems and Models

CONTENTS

9.1	General	412
9.2	What Is a Size Effect?	414
9.3	Factors Influencing Size Effects	414
9.4	Theoretical Studies of Size Effects	415
	9.4.1 Classical Theory of "Bundled Strength"	415
	9.4.2 The Weakest Link Theory	416
	9.4.3 Other Theoretical Studies	418
	9.4.4 Fracture Mechanics Approach	418
	9.4.5 Evaluation of Theoretical Studies	420
9.5	Size Effects in Plain Concrete — Experimental Work	420
	9.5.1 Experimental Factors Influencing Size Effects	420
	9.5.1.1 Random Strength	420
	9.5.1.2 Diffusion Effect Due to Drying	420
	9.5.1.3 Wall Effect	421
	9.5.1.4 Aggregate Size	421
	9.5.1.5 Matrix Strength and High-Strength Concretes	421
	9.5.1.6 Compaction Density and Loss of Water	421
	9.5.1.7 Curing and Drying	422
	9.5.1.8 Strain Rate	422
	9.5.1.9 The State of Stress	422
	9.5.1.10 Testing Machine and Loading Platens	422
	9.5.2 Experimental Research on Size Effects	423
	9.5.3 Evaluation of Experimental Research	423
	9.5.4 Tensile and Flexural Strength	426
	9.5.5 Evaluation of Experimental Research on Tensile Strength	429
	9.5.6 Size Effects in Long-Term Properties of Concrete	429
	9.5.6.1 Shrinkage	429
	9.5.6.2 Creep	430
	9.5.7 Size Effects in Gypsum Mortar	430
9.6	Size Effects in Reinforced and Prestressed Concrete	431
	9.6.1 Bond Characteristics	431
	9.6.2 Cracking Similitude (Service Conditions)	432
	9.6.3 Ultimate Strength (Load–Deflection Behavior)	432

9.7	Size Effects in Metals and Reinforcements	433
9.8	Size Effects in Masonry Mortars	434
9.9	Size Effects and Design Codes	435
9.10	Errors in Structural Model Studies	437
9.11	Types of Errors	439
	9.11.1 Blunders	439
	9.11.2 Random Errors	440
	9.11.3 Systematic Errors	440
9.12	Statistics of Measurements	441
	9.12.1 Normal Probability Density Function	441
9.13	Propagation of Random Errors	444
9.14	Accuracy in (Concrete) Models	450
	9.14.1 Dimensional and Fabrication Accuracy	451
	9.14.2 Material Properties	454
	9.14.3 Accuracy in Testing and Measurements	455
	9.14.4 Accuracy in Interpretation of Test Results	456
9.15	Overall Reliability of Model Results	457
9.16	Influence of Cost and Time on Accuracy of Models	458
9.17	Summary	458
Problems		459
References		460

9.1 GENERAL

This book has so far presented basic information on the laws of similitude, various materials for model fabrication, instrumentation, etc., needed by the models engineer to understand and to undertake a research or design project that will enable one to predict the behavior of a prototype structure. There are a number of factors that influence the strength of materials and the structures built from these materials. In prototype structures the basic material strength properties are usually regarded as measured from control specimens, and as they occur in the members, to be independent of the absolute size of the structural members. For scaled models, in particular, the size of the structure becomes more important since the eventual result of a model testing is to enable the experimenter to predict the strength of the prototype. In the case of concrete, which is highly heterogeneous, the reduction in the size of a specimen may change its properties significantly. The presence of reinforcing or prestressing bars acting in conjunction with concrete poses additional problems in interpreting the results because of the difference in size of the model and prototype.

However, users of structural models may raise questions about the reliability of such models. Such questions relating to the reliability and accuracy include

How accurate must the results of a model study be to satisfy the objectives of the program?
What can one do to optimize the chances that the results from any particular model study will be accurate?
What are the various factors that affect an acceptable accuracy?

Furthermore, the value of the results, either from a research or a design model study, is related to the confidence that the engineer can place in the experimental results. This confidence level will in turn depend on the errors that are introduced at various stages of the modeling process. These errors are summarized in Figure 9.1. This chapter deals with this aspect of the model study, which relates to size and scale effects of the model results and the various types of errors and their causes.

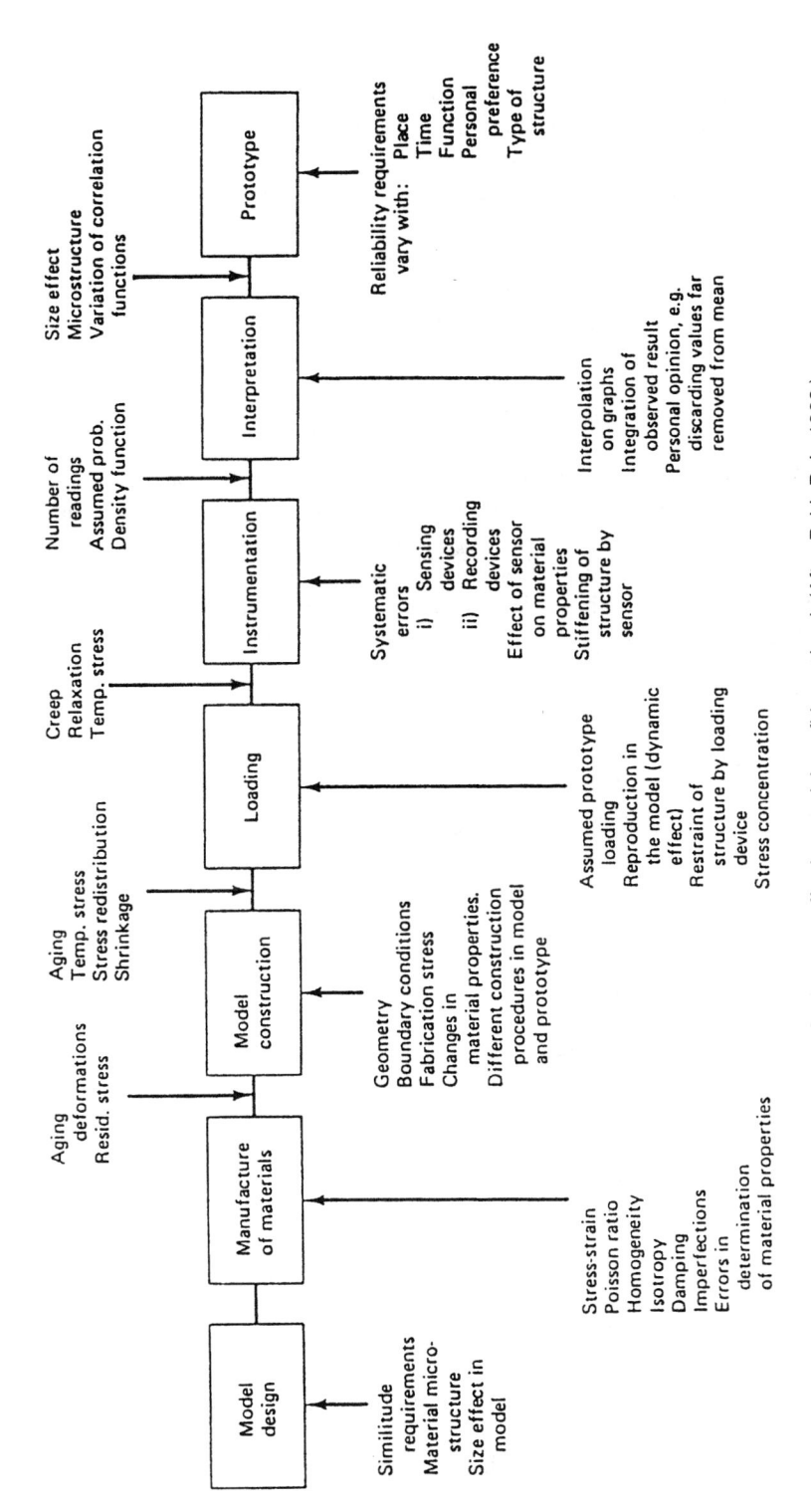

SIZE EFFECTS, ACCURACY, AND RELIABILITY IN MATERIALS SYSTEMS AND MODELS

Figure 9.1 Parameters affecting model confidence level. (After Pahl, P.J., 1963.)

9.2 WHAT IS A SIZE EFFECT?

Size effect, a phenomenon observed by many researchers, is related to the change, usually an increase, in strength that occurs when the specimen size is decreased. Sabnis (1980) reviewed a large number of theoretical studies available in the literature during the last several decades; these include Weibull (1939), Tucker (1941), and others. Various experimental investigations into size effects have been summarized by Sabnis and Aroni (1971) and later Sabnis and Mirza (1979). Recently, Bazant and Kazemi (1988) and Bazant (1992) have demonstrated the use of the energy release and *fracture mechanics* approach.

Why are size effects important? One reason is the use of various-size specimens, in addition to different shapes, in various countries and at various times for the characterization of material properties. However, a much more important reason is in connection with model studies. Models tested to failure to determine ultimate load capacity and their use in standards and codes may be subject to size effects which would give nonconservative predictions of prototype (and design) strengths. Thus, in addition to efforts in minimizing size effects, there is a need to be able to predict them.

How should the size effect be considered in interpreting the experimental data from the perspective of "Standards (Code)" for concrete such as ACI 318 (1995)? These consider three main aspects of experimental work, namely, planning, testing, and interpreting the results affecting the final outcome. Both the material as well as structural actions are considered systematically during the above three stages. This approach is necessary to guide design engineers in proper use of concrete in different size and shape concrete structures. As an initial step, ACI 444, *Experimental Analysis for Concrete Structures* made recommendations for the use of size of specimen in modeling of structures. A somewhat empirical (but very sound) practice is to use a suitably small specimen to measure model material properties. This tends to compensate for the size effect of the model itself.

9.3 FACTORS INFLUENCING SIZE EFFECTS

A number of factors influence the strength properties and hence the behavior of material systems. The strength properties include compressive and tensile strengths, bond and fatigue strengths, and creep and various dimensional changes. Along with these properties, the nature of the material and the geometric configuration of specimens are also important. The materials range from naturally occurring timber and rocks to manufactured materials, such as concrete, steel, plastics, etc.

Some of the properties of the above materials are affected more by changes in size than by other variables. Some properties may not influence the final interpretation of a model investigation to the same degree as the size effect because of their minor influence on the behavior of a structure; e.g., in the case of reinforced concrete, the change in compressive strength is not as important as the yield strength of the reinforcement in an underreinforced or lightly reinforced beam. On the other hand, in the investigation of shear strength of a slab or heavily reinforced beam, the compressive strength (as related to its tensile strength) plays a direct and important role. If not considered properly, the observed difference in the strength of specimens of two scales might be attributed incorrectly to the size effect.

Generally, theoretical studies treat the behavior of material systems and their physical results on a statistical basis, the basic philosophy being that the failure in heterogeneous materials is a statistical phenomenon. Thus, the larger the volume, the greater the chances for failure which will result in lower strength.

In general, variations in strength of similar shape but different size specimens of concrete are caused by the following factors:

1. Differential curing rates of the various size specimens;
2. Differences in the quality (density) of the material cast into the various size molds;
3. Change of quality of the cast material as a result of the water gain of the top layers and water leakage through the forms;
4. Differential drying of the various size specimens during testing;
5. Difference in induced stress conditions because of variation of quality of end capping of different size compressive specimens;
6. Statistical variations in strength as a result of volume effects;
7. Loading rate and method;
8. Strain gradient effects in flexural specimens.

As can be seen from this list, the purely material volume effect is only one of the causes.

9.4 THEORETICAL STUDIES OF SIZE EFFECTS

Generally, specimens of smaller size are observed to have higher strength. Also, the scatter in strength is generally greater in the smaller specimens. This phenomenon of size effect and scatter led to theories for explaining this behavior. The basic approach in statistical theories of strength is to evolve a statistical distribution function that adequately characterizes the random heterogeneity of materials and the variation (scatter) in their strength. In effect, it identifies admissible forms of distribution functions with suitable parameters that reflect accurately the true material behavior and a realistic mechanism of failure for which the distribution function is applied.

Among the theoretical studies of size effects are those of Weibull (1939), Tucker (1941), Wright and Garwood (1952), Nielsen (1954), and Glucklich and Cohen (1968). Basically, two distinct approaches were used to study the statistical aspects of strength of materials and size effect. These approaches are based on the classical bundle concept, as presented by Freudenthal (1968), and the weakest link concept. In the classical bundle concept, the strength is not only determined by that of the weakest element but is dependent on the strength of the elements in the neighborhood. In this model the specimen is assumed to be made up of parallel fibers (elements), and in such a situation the gross strength at failure is influenced by the strength of all constituent fibers.

In the weakest link concept, the presence of a single severe defect in any of the constituent elements is adequate to cause failure of the total material. Consequently, overall strength of a specimen subjected to uniform stress is determined by the strength of the weakest element present. These two approaches are, in a way, idealizations used to make the problem tractable. However, in reality the actual characteristics of materials fall in between the two theories. A summary of these theories is presented to give the reader a basic idea of the size effect phenomenon from the theoretical point of view.

ACI446 (1989) demonstrated that these theories ignore the size effect caused by the stress redistribution prior to failure. He further explained this aspect using fracture mechanics and its relation to the energy release during the crack propagation and emphasized that it would exist even if the material behavior were deterministic. He concluded that for concrete (and concrete structures) size effect data are better fitted to fracture mechanics theory first and the remaining size effect may be attributed to statistical phenomena.

9.4.1 Classical Theory of "Bundled Strength"

The concept of this theory lies in the assumption that the bulk specimen consists of parallel elements and that the instability or failure of one element will not lead to total fracture; rather it will be stopped before it propagates from local to bulk scale. This means that the weakest volume element containing the failure is surrounded by elements of such high local strength that the stress carried by the weakest element prior to its failure can indeed be transferred to the elements. The statistical model according to Freudenthal (1968) replaces the bulk specimen by a classical bundle and this discussion is based on his work. This bundle consists of a large number of parallel filaments of identical length L and cross section A, all of which come from a common source, so that the statistical distribution of local filament strength is homogeneous and constant; a random filament is the weakest link of the specimen. The strength of this classical bundle model, as well as of the bulk specimen that it is designed to represent, is represented by the forces under which a chain t-reaction process of consecutive filament failures resulting from the successive overload carried by the surviving filaments leads to final failure of all filaments. The fracture process starts at the weakest point in the bundle, but, contrary to the weakest link model, it does not necessarily propagate unless, in a bundle of total filaments (n) of (weakest link) strength $\sigma_n, \sigma_{n-1},...\sigma_2, \sigma_1$, arranged in the order of their consecutive failure, the following conditions are satisfied:

$$0 \leq \sigma_n \leq \frac{S}{nA} = s_n$$

$$\sigma_n \leq \sigma_{n-1} \leq \frac{S}{(n-1)A} = s_{n-1}$$

$$\sigma_3 \leq \sigma_2 \leq \frac{S}{2A} = s_2$$

$$\sigma_2 \leq \sigma_1 \leq \frac{S}{A} = s_1$$

(9.1)

where: S = total applied force
s_i = stress in the individual element

This theory of bundle strength results in a bulk specimen with a fracture process represented by a bundle model, no pronounced effect of volume on either mean strength or variance is anticipated, except that the variance might show a tendency to decrease with increasing specimen size. In particular, it is this last trend (not usually found in tests leading to brittle fracture) which suggests that the bundle model is applicable only to the description of fracture processes in materials (1) in which bundles such as groups of long-chain molecular filaments physically exist or (2) in which fracture takes place following large strain, and lets one follow the weakest link theory for additional explanation.

9.4.2 The Weakest Link Theory

The weakest link concept has been used widely in developing various statistical strength theories, which differ from each other only in the way in which the use of the concept is justified or in which the form of distribution function of local strength is assumed. Early attempts were made by many to formulate a theory for the strength of cotton yarn and later for strengths of solid volumes. The development put forth by Weibull (1939) is a culmination of these earlier efforts.

Weibull pointed out the inadequacy of specifying material strength by a single quantity as is usually done in deterministic approaches. In developing his theory Weibull used additional parameters to characterize the strength of a material. He regarded a specimen as an ensemble of a very large number of primary elements. He considered the failure of the total material the same as that of any one of the primary elements or of the weakest link.

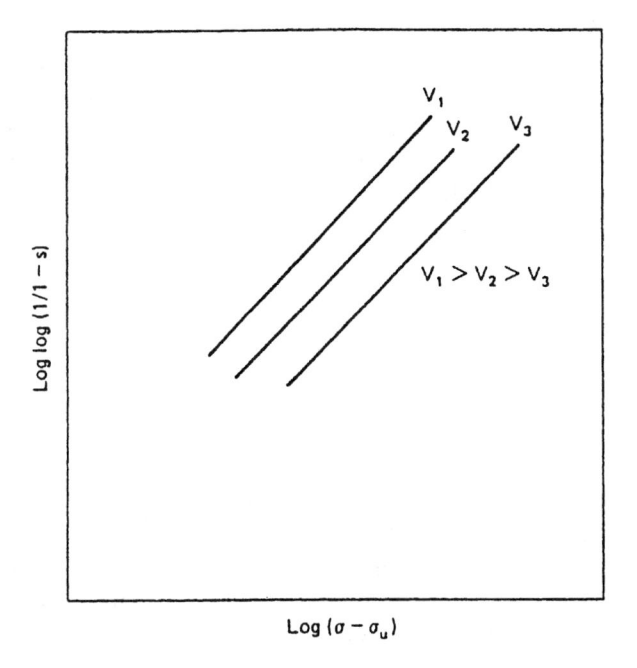

Figure 9.2 Weibull's plots for different specimen sizes (schematic).

If the probability of failure of the primary element for a stress between 0 and σ is S_0, then the probability of survival of the element is given by $(1 - S_0)$. Also, if S denotes the cumulative probability of failure of a specimen of total volume V, then the probability of survival of the total specimen, assuming statistical independence between elements, is given by

$$(1-S) = (1-S_0)^V \tag{9.2}$$

This basic expression can be converted into a form given by the following two equivalent equations:

$$\frac{1}{1-S} = \exp\left[V\left(\frac{\sigma - \sigma_u}{\sigma_0}\right)^m\right] \tag{9.3}$$

or

$$\underbrace{\log\log\frac{1}{1-S}}_{y} = m\underbrace{\log(\sigma - \sigma_u)}_{x} - m\log\sigma_0 + \log V \tag{9.4}$$

These expressions indicate that as specimen size increases, the mean strength and variance decrease, which is consistent with the experimental observations of size effect on some materials. Thus, with Weibull's theory a linear relationship can be developed between $\log 1/(1 - S)$ and $\log (\sigma - \sigma_u)$; this is shown in Figure 9.2. Weibull applied his theory to a number of cases, including the strength of glass rods in tension, the bending strength of porcelain, and the tensile strengths of Portland cement, wood, plaster of Paris, malleable iron castings, and other materials. In all cases, he verified the applicability of the linear fit of the distribution function with test data. The *weakest link concept* has since been used by several researchers.

9.4.3 Other Theoretical Studies

Tucker's (1941) *strength summation theory* is based on the assumption that the strength of a specimen is equal to the sum of the strengths contributed by the component parts, or elements. In this theory the specimen is assumed to have a parallel link assembly. Thus, it is implied, for example, that the flexural strength of a beam specimen decreases with increasing length and increasing depth, as governed by the weakest link theory, but is independent of the width. Nielsen (1954) assumed in his *surface theory* that failure of a beam is determined solely by the thin surface layer of the tension zone, and if this surface layer is assumed to be brittle, Weibull's theory may be applied to the tensioned surface. This implies that the flexural strength would decrease with increasing length and width but should be independent of depth. Since the strain distribution in a beam is triangular and since only the extreme fibers govern the flexural strength, Johnson (1962) showed the influence of depth to be very small. Nielsen's tests on concrete beams cured in water until testing show no influence of beam depth on the tensile flexural strength. This process essentially prevented any drying and thus caused the modulus of rupture to be independent of depth. Pahl and Soosaar (1964) indicate in their work that concrete and mortar are fairly brittle materials, so that failure at a few points in the mass is soon followed by overall collapse. A small specimen has fewer points than a large specimen at which failure can be initiated by local weaknesses, so that the strength of the small specimens on the average might be larger than that of the larger specimens. They suggest that this size effect can be represented by a best-fitting equation:

$$f = a + bV^{-c} \tag{9.5}$$

where: f = strength of concrete
V = volume of specimen
a, b, c = positive constants depending on the concrete mix

Glucklich and Cohen (1968) drew attention to the influence of size, in certain materials, on the brittle–ductile transition and strength, with both ductility and strength decreasing with increasing size. In most metals, ductility is mainly due to plastic deformation prior to nucleation of cracks, which is coincident with fracture. These are Griffith-type materials, and Weibull's statistical theory describes the size effect of crack nucleation. However, materials such as concrete exhibit no permanent deformation prior to crack nucleation but appreciable permanent deformation during slow crack growth. Thus, crack nucleation and fracture are not coincident, and Weibull's weakest link theory is not sufficient to explain the behavior during crack growth. Crack propagation depends on equilibrium between the respective rates of strain energy release due to cracking, and the energy demand for continued propagation of cracking. Any sudden drop in the energy demand, for example, as a result of the propagating crack encountering preexisting cracks in its plane of advance, will create an excess of released energy. This energy, converted to kinetic energy, can work against the remaining uncracked material and bring about reduced ductility and premature fracture. If there is a large amount of stored energy, the rate of energy adsorption is low and a premature fracture will occur, as described above. Specimen size thus enters into the picture not as a size effect per se, but insofar as it governs the amount of stored energy.

9.4.4 Fracture Mechanics Approach

The size effect in two dimensions is defined in terms of nominal stress at failure

$$\sigma_N = c_n \, (P_u / bd) \tag{9.6}$$

SIZE EFFECTS, ACCURACY, AND RELIABILITY IN MATERIALS SYSTEMS AND MODELS

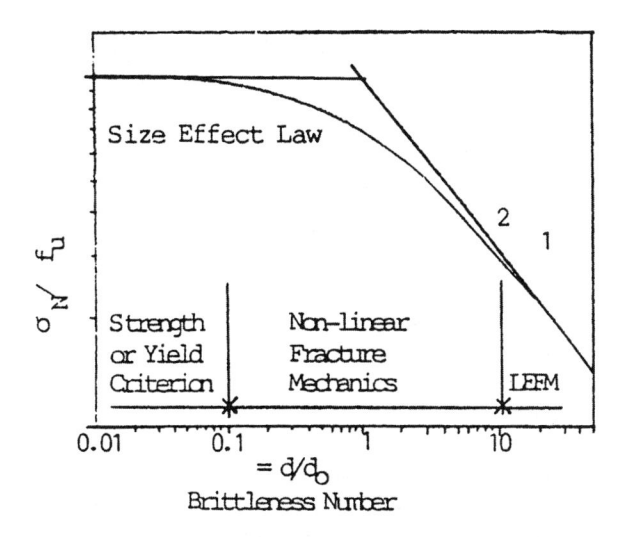

Figure 9.3 Size effect law presented in its general form.

in which P_u = maximum (ultimate) load, b = thickness of specimen or structure, d = characteristic dimension, and c_n = coefficient. This is rewritten in terms of size effect of fracture mechanics type in the form of

$$\sigma_N = B f'_c \big/ \sqrt{(1+\beta)}; \quad \beta = d/d_0 \tag{9.7}$$

in which f'_c = a measure of material strength and B and d_0 are empirical constants; coefficients B and d_0 represent specimen shape and size. This assumes thickness b = constant for different d and also the specimen proportions are constant for all sizes, e.g., d/s = constant. This equation is shown in Figure 9.3, conveniently on a log–log basis for different ranges of applications. When the specimen is small, plasticity is also small ($\beta < 0.1$), there is no size effect at these smaller values of β, and essentially results in a horizontal curve; at intermediate values say, ($0.1 < \beta < 10$), there is a smooth transition and in the case of *linear elastic fracture mechanics* (LEFM) at large sizes ($\beta \to \infty$) at which size effect is very pronounced it approaches the asymptote with slope of 1:2 (Shah et al., 1995). Bazant and Cao (1987) has shown that this expression can be used in the limit-analysis-based code formulas, taking into account nonlinear fracture mechanics, by replacing the nominal stress say, f'_c in Equation 9.7. The material strength f'_c can be replaced by other forms in shear and flexure equations. Equation 9.7 also has an advantage that it can be transformed to a linear regression plot in the form of $y = mx + c$, where $x = d$, $y = (f'_c/\sigma_N)^2$, $B = c^{-1/2}$ and $d_0 = c/m$.

Bazant and Kazemi (1988) have shown that this equation can be modified to involve material parameters. Such a form can be written as

$$\rho_N = \sqrt{\{EG_f/(c_f + D)\}} \tag{9.8}$$

where G_f and c_f are material fracture parameters and ρ_N is called as shape-independent nominal strength, defined by

$$\rho_N = (P_u/bd) \sqrt{\{g'(\alpha_0)\}} \tag{9.9}$$

where $g'(a_0)$ is the value of the first derivative of $g(a/b)$ with respect to a/b and D represents the shape-independent characteristic dimension of the structure, defined as

$$D = \{(g(\alpha_0)/(g'(\alpha_0))\} d \qquad (9.10)$$

and that $g'(\alpha_0)$ is evaluated at $\alpha = \alpha_0$. For a size range of 1:20, in which the approximate size effect law (Equation 9.2) is applicable, the values of g and g' sufficiently take into account the shape of the structure. The term β in Equation 9.7 characterizes *brittleness* and allows three different ranges of b to distinguish the behavior taking into account size effect as was indicated in Figure 9.3. It must be noted that most experimental work and its application in design codes also generally falls in the geometric scale range of 1:20.

9.4.5 Evaluation of Theoretical Studies

It becomes apparent that the theoretical investigations discussed above are based solely on the statistical treatment of heterogeneity of the material. The weakest link theory focuses attention only on the most critical or potential flaw, but disregards the interaction between the flaws that will exist in continuous systems. The fracture mechanics approach is also theoretically sound and has been supported by experimental evidence. On the other hand, experimental results have been used extensively to explain size effects, and therefore should also be examined.

Thus, the theoretical investigations presented in this chapter are all valid from the theoretical point of view, but should be considered with caution. For concretes, in particular, the weakest link theory applies extremely well. However, such a conclusion should be confirmed by very carefully executed experiments, with controlled curing, and considering the factors mentioned earlier.

9.5 SIZE EFFECTS IN PLAIN CONCRETE — EXPERIMENTAL WORK

The heterogeneity of any material leads to the theoretical arguments presented in the last section that predict the existence of size effects. When investigating the behavior of concrete, it is important to recognize factors that contribute directly to the observed changes in properties with size. An understanding of these factors can help minimize the effect of size, before the intrinsic variability is ineptly used to explain the observed data.

9.5.1 Experimental Factors Influencing Size Effects

9.5.1.1 Random Strength

Statistical heterogeneity of the material plays an important role in the micromechanisms determining its strength. It describes the effect of the size distribution of the flaws in the microstructure on the material strength. However, the randomness of strength due to the heterogeneity of the material based on Weibull-type probabilistic models alone, does not seem to explain fully the size effect observed in brittle failures of most concrete structures, except those where the maximum stress is uniform over a large part of the structure (e.g., a long uniformly stressed specimen in tension), or where the structure fails at the first crack initiation.

9.5.1.2 Diffusion Effect Due to Drying

Significant structural size effects can be obtained due to the diffusion process of drying of concrete in structures, the conduction of heat produced by hydration, and the non-uniformities of creep produced by differences in temperature and moisture content throughout the structure. More work has to be done to isolate these effects from the rest of the size effect.

9.5.1.3 Wall Effect

Another type of size effect is caused by the fact that a boundary layer near the surface of concrete inevitably has a different composition and strength than the interior of the concrete structure. This layer, whose thickness is about one aggregate size, contains a lower percentage of large aggregates and a higher percentage of mortar. This phenomenon is known as "wall effect." In a small structure, the effect of this layer is larger than in a large structure because the boundary layer thickness is independent of structure size. For very thick cross sections, this effect becomes negligible. One must separate this effect for a complete understanding of size effect.

9.5.1.4 Aggregate Size

The fracture parameters as well as the size effect law are valid only for structures made for one and the same concrete, which implies the same aggregate size. If the aggregate size is changed, the fracture parameters and the size effect law parameters change. Bazant and Kim (1984) recommend that the size effect Equation 9.6 can be modified to include this effect as follows:

$$\sigma_N = B f_t' / \sqrt{1+\beta}; \quad f_t' = f_t^0 \left(1 + \sqrt{(c_o/d_a)}\right); \quad \beta = d/d_o \tag{9.11}$$

in which f_t^0 is the direct tensile strength for a chosen reference concrete, f_t' is the direct tensile strength for maximum aggregate size d_a, and c_o is an empirical constant. This equation is analogous to the Petch formula for the effect of grain size on the yield strength of polycrystalline metals, which is derived by the dislocation theory. Equation 9.11 has been shown to agree reasonably well with the fracture test data. For a sufficiently small specimen size, a higher nominal strength is obtained with smaller aggregate sizes d_a, while for a sufficiently large specimen size, a higher nominal strength is obtained with larger aggregate sizes. For intermediate structure sizes, the aggregate size makes little difference. Other, more general fracture mechanics techniques are discussed in Shah et al. (1995).

9.5.1.5 Matrix Strength and High-Strength Concretes

The higher strength of concrete has been achieved mainly by increasing the matrix strength and the aggregate–matrix bond. In high-strength concretes, the differences between *the strength and elastic modulus of the aggregate and the matrix* (*mortar*) are much smaller than they are for normal-strength concrete. Consequently, high-strength concrete behaves as a more homogeneous material, and the result is that the fracture process zone becomes smaller. In view of the discussion of the size effect (especially the fact that coefficient d_0 in the size effect equation is related to the size of the fracture process zone), it is clear that, for the same structural size, the behavior of high-strength concrete is closer to linear elastic fracture mechanics, i.e., more brittle, than the behavior of the same structure made of normal concrete. Therefore, fracture mechanics analysis and size effects are much more important for high-strength concretes than for normal-strength concretes.

9.5.1.6 Compaction Density and Loss of Water

Compaction is an important variable influencing concrete strength, yet cannot be scaled, and, therefore, smaller specimens will tend to achieve better compaction, higher density, and thus higher strength. This is particularly true when standard compaction procedures are followed, involving a given time of vibration or specific number of tampings. Larger specimens will undoubtedly have more internal voids and entrapped air. When uniform compaction is achieved, the size effect due to this factor will be minimized. This has been confirmed for gypsum mortars by Loh (1969).

Water loss from specimens during casting can vary with size and cause different quality in the cast material. To minimize this source of variability and size effects, a controlled humidity room during casting and watertight molds such as Plexiglas or polyvinylchloride (PVC) are needed.

9.5.1.7 Curing and Drying

Curing is another important factor. Curing of two specimens of different size will take place at different rates, because the surface–volume ratio increases with decrease in specimen size, and the length of moisture migration paths will differ. The strength of the material will vary from the surface of the specimen to its center, depending on its size, since hydration may not be uniform throughout the specimen at the time of testing. Studies summarized by Sabnis and Aroni (1971) indicate that if curing is controlled, for example, by sealing (discussed later in detail) the surfaces of the test specimens, the increase in strength due to the reduction in size can be minimized. Tests on cores drilled from a massive concrete dam after a period of 5 years showed an insignificant difference in strengths between 250 and 560 mm diameter cylinders. This is probably due to the more uniform curing conditions inside the dam as well as the advanced age at which hydration was almost complete. The density of the cores would be identical.

Drying of the specimen also results in higher strength and will depend on the surface-to-volume ratio, which varies inversely with the specimen size. Slower surface drying of larger specimens will result in smaller flow gradients and greater resistance to drying due to the longer distance to the surface. This also influences size effects in sustained loading (creep).

9.5.1.8 Strain Rate

High rates of loading lead to higher strengths. In a given testing machine, with the rate of crosshead movement kept constant, smaller specimens will experience higher strain rates. As the specimen size is decreased, the crosshead movement rate should be reduced accordingly. However, with very small specimens it is not always possible to achieve the required low rates of crosshead movement with existing testing equipment, which contributes to their apparent increased strength. Of course, in normal testing the influence of increased strain rate is not critical, except in cases of dynamic loading (see Chapter 11).

9.5.1.9 The State of Stress

The stress state, such as compression, tension, and flexure, influences the strength of the specimen. The strength of compressive specimens depend on the accuracy of the loaded ends, and on parallelism, if rotating heads are not used. It is possible to achieve a higher level of capping accuracy in smaller cylinders, which result in higher strength. Tests by Wright and Garwood (1952) showed that flexural stress increased with a decrease in the specimen size. The effect of strain gradient was discussed in detail in Chapter 4.

9.5.1.10 Testing Machine and Loading Platens

In addition, the properties of the testing machine and, in particular, the stiffness of the loading platens at the ends of the test cylinders have a significant effect on test results. Stiff end platens tend to apply uniform strain conditions to the specimen under test and result in higher strength than thinner platens, which tend to lead to a state of uniform stress. Also, end platens restrain lateral movements of the specimen and induce lateral stresses at both ends. The higher the lateral restraint, the higher the compressive strength will be. The mode of failure may also change with end conditions.

9.5.2 Experimental Research on Size Effects

A review of experimental work was done by Sabnis and Mirza (1979). Gonnerman (1925) conducted the earliest study on size effects in concrete with an extensive investigation into the compressive strength of cylinders with height/diameter ratio of 2. He varied the cylinder diameter from 100 to 250 mm and examined the influence of age, cement–aggregate ratio, relative consistency, and aggregate fineness as shown in Figure 9.4. Note that the aggregate size is less than 40% of cylinder diameter. Each point in Figure 9.4 is an average of 5 to 30 tests. Later research was done by Johnson (1962, influence of scaling of aggregates on cylindrical and cube specimens); Harris et al. (1963, cylinders to investigate the effect of size on the compressive strength at various ages); Neville (1966, types of concretes, curing and age); Sabnis and White (1967) and Fuss (1968) (solid and hollow cylinders of different sizes with constant moisture by surface sealing); Meininger (1968, cores of various diameter specimens from a slab, and a thick concrete wall and moist-cured for 3 months); and Mirza et al. (1972, size effects with a major emphasis on curing and compaction). Neville (1966) developed a relationship using regression analysis:

$$\frac{P}{P_6} = 0.56 + 0.697 \frac{d}{(V/6h) + h} \tag{9.12}$$

as shown in Figure 9.5. Note that d and h are in inches and V is in cubic inches and the subscript 6 refers to a 6 in. cube of the same concrete chosen as a standard specimen for comparison. Other results indicated no significant effect of size of cylinder or core on strength if there was controlled curing.

Mirza et al. (1972) tested more than 500 cylinders ranging from 25×50 mm to 150×300 mm at ages of 3, 7, and 14 days. Curing procedures in which cylinders were kept in a continuously moist environment were equivalent for all sizes and were more effective than coating procedures and air drying. The effectiveness of moist curing compared with air drying is indicated by a gradual decrease in the compressive strength with a decrease in cylinder size. This is due to the comparatively more rapid moisture migration from smaller cylinders. Curing and sealing with chemical coatings allowed some exchange of moisture with the environment and resulted in strengths intermediate between fully cured and air-dried conditions. Spray paint (lacquer or polyurethane) was ineffective in maintaining internal moisture and also caused deterioration of the concrete. It was shown for their mixes that 75- and 100-mm-diameter cylinders exhibited strength increase of approximately 5 to 15% and for 50-mm-diameter cylinders strength increased up to 40% over the strength of 150-mm-diameter cylinders.

9.5.3 Evaluation of Experimental Research

Size effect is usually not an important factor in the selection of prototype test cylinder size, since all cylinders above 50×100 mm fall on the flat section of the strength curve. However, for small-scale models, the selection of smaller-size control cylinders can have a considerable variation in the compressive strength. Based on the available data to date, the relative strengths of various cylinder sizes that have typical curing and drying histories with no surface sealing are as shown in Figure 9.6.

In general, the scale of cylinder to measure the compressive strength of concrete in the structure should be consistent with the scale factor of the model structure itself. ACI Committee 444 (1979)

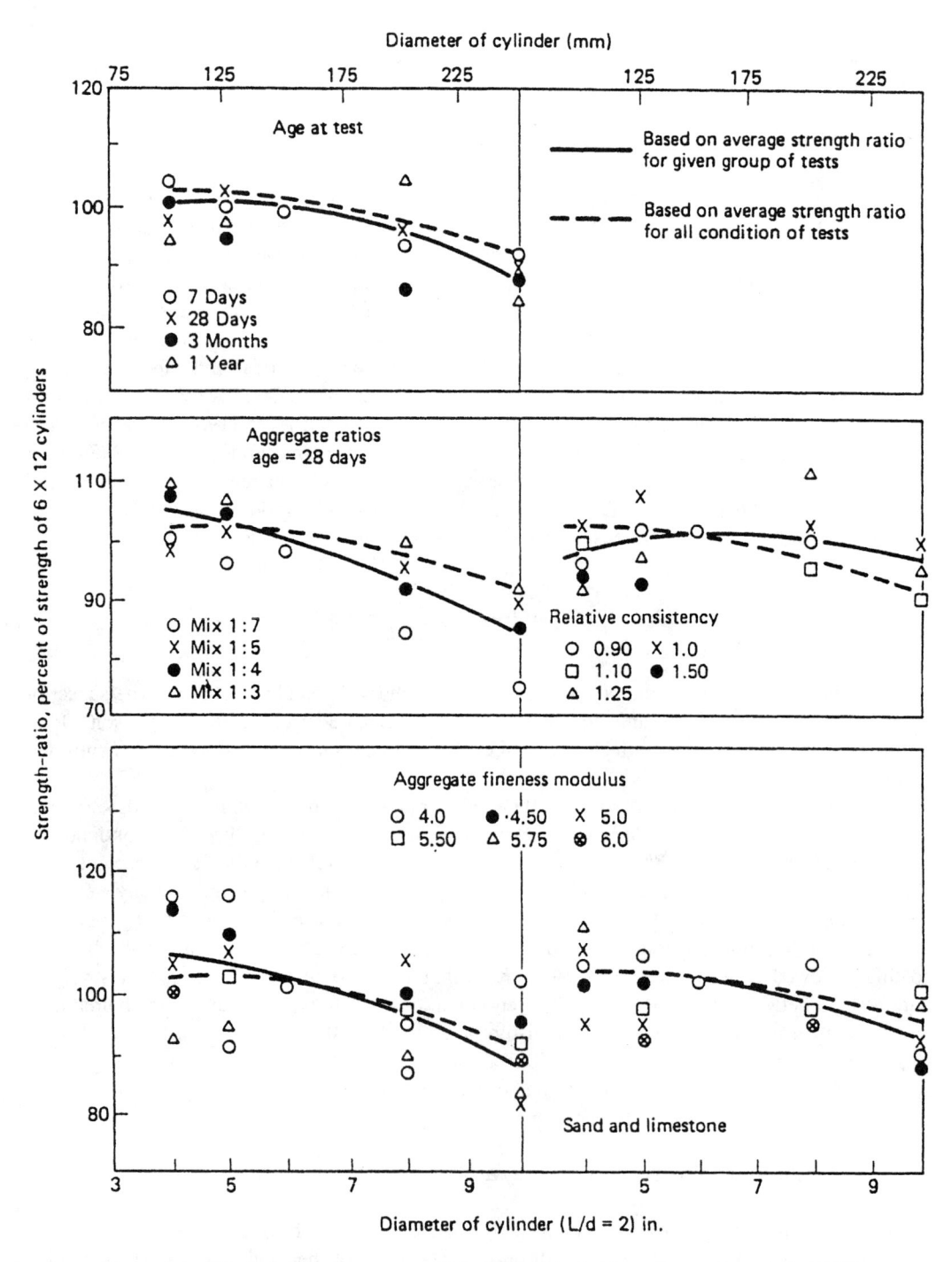

Figure 9.4 Effect of size on compressive strength. (Adapted from Gonnerman, J. F., 1925.)

recommends that 50 × 100 mm cylinders be accepted as a standard for comparing model concrete mixes. In addition, ACI Committee 444 (1979) suggests that for model elements of very small minimum characteristic dimension (less than 12.5 mm), the apparent strength of the control cylinders may not be representative of the strength of the model material. In such situations, additional tests should be conducted on model cylinders with a length diameter ratio of 2 to 1 and diameter

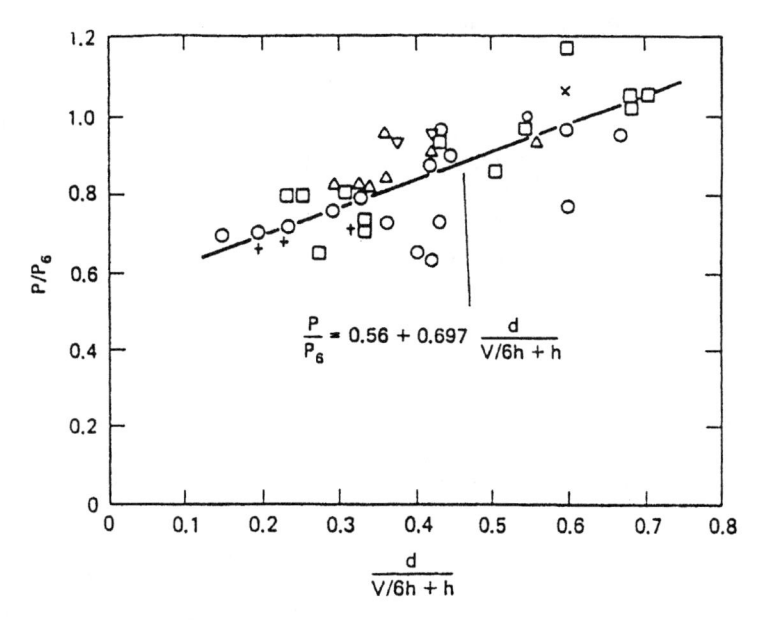

SIZE EFFECTS, ACCURACY, AND RELIABILITY IN MATERIALS SYSTEMS AND MODELS 425

Figure 9.5 Relation between P/P_6 and $d/[(V/6h) + h]$. (From Neville, A. M., *Proc. Am. Concr. Inst.*, 63, 1095–1110, 1966. With permission.)

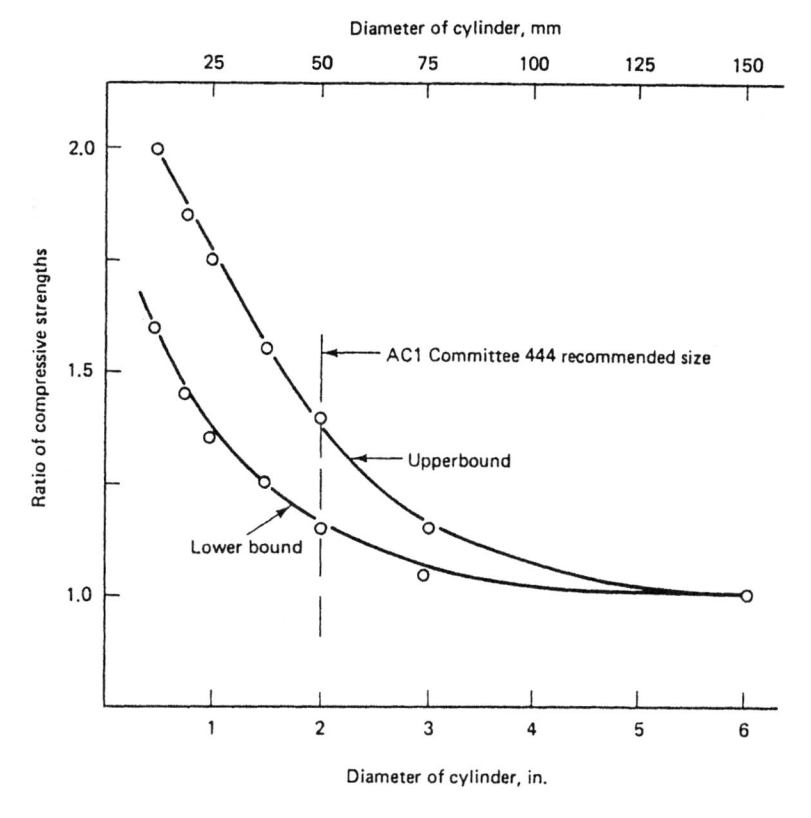

$$\frac{P}{P_6} = 0.56 + 0.697 \frac{d}{V/6h + h}$$

Figure 9.6 Relative strengths of different size cylinders with 6-in. cylinder as a unit. *Note*: All cylinders have height equal to twice the diameter.

equal to the characteristic dimension of the model. Pahl and Soosaar (1964) suggest the following empirical rules:

1. The strength of the model material should be equal to that of a test cylinder with diameter equal to the minimum dimension of the structure in the region of failure, e.g., the shell thickness or the width of the beam.
2. The size of the largest sand particle used in the mix may not be larger than one fifth of the cylinder diameter, nor larger than 80% of the clear distance between reinforcing bars in the model.

9.5.4 Tensile and Flexural Strength

The tensile strength of concrete is a fundamental property, and it has a significant influence on several important phenomena in reinforced concrete elements, such as shear strength, bond strength of deformed bars, cracking load and crack patterns, effective moment of inertia, and nonlinear response. Several tension tests are available, and the best choice depends on the strain distribution existing in the member; for example, uniform tension requires direct tension tests. These are difficult to perform and are seldom used. The indirect tension test (split cylinder test) and the torsion test are used for sections in which the principal compressive and tensile stresses are of the same order. The flexure test (modulus of rupture) would be used in connection with reinforced and prestressed concrete flexural members, pavements, etc.

The strain distribution was studied by Blackman et al. (1958) on three identical specimens: one loaded axially, one in pure flexure, and one in between these two conditions (Figure 9.7). Although each of the stress distributions caused a failure at the tensile strength of the material, the influence of a strain gradient became obvious in that the ultimate tensile strain increased with the applied gradient. The situation is exaggerated in smaller beams, which have higher strain gradients, when subjected to a similar flexural loading. Beam tests by Wright and Garwood (1952) indicated increase in the flexural tensile strength with an increased strain gradient, i.e., with a reduction in depth of the beam.

With regard to the method of casting and testing, experimental results in flexural strength and size effects have been shown for both gypsum and cement mortars, for beams cast on their sides (horizontal position), and in the usual manner, on their bottom faces (vertical position). Results from tests by White and Sabnis (1968) shown in Figure 9.8 clearly indicate that the strength of the horizontally cast beams is consistently higher than that of those cast in the vertical position. In horizontally cast beams, the tensile face of the beam, as tested, contains material cast at various depths and is thus more heterogeneous, causing the observed larger scale effect.

Abrams (1922) was probably the first to report extensive flexural tests on concrete; his tests considered a number of variables associated with strength of concrete. Although his objective was not to investigate the effect of size on flexural strength, his results on beams with depth varying from 100 to 250 mm indicated that deeper beams had lower strength, the variation being approximately 10% from shallower to deeper beams. Some notable research was done by Blackman et al. (1958, axial tensile and flexural specimens to investigate the effect of strain distribution); Harris et al. (1963, influence of size/depth on the flexural tensile strength); Harris et al. (1966, effect of strain gradient on strength); Mirza (1967, tensile splitting and square beams); Kadlecek and Spetla (1977, size effect in cylinder and prism tests in direct tension); Malhotra (1969, size effect on tensile strength using direct tension, ring tension, and splitting tension tests); and Mirza et al. (1972). Kadlecek and Spelta (1967) developed the following relation:

$$f_t = AV^{-B} \qquad (9.13)$$

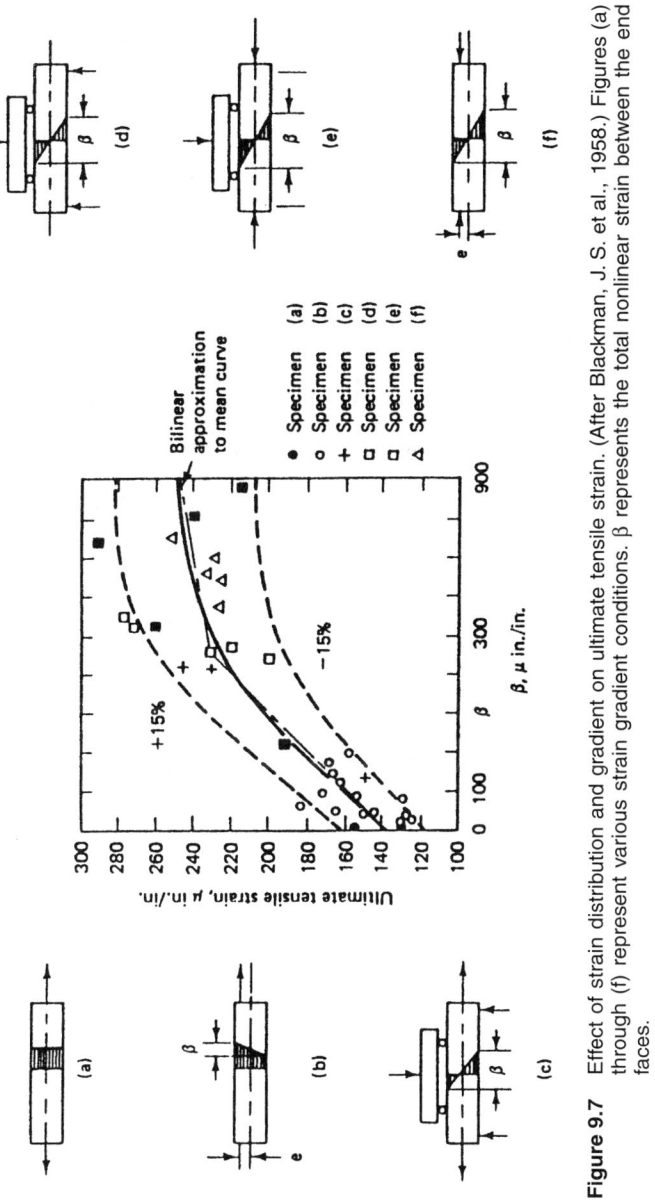

Figure 9.7 Effect of strain distribution and gradient on ultimate tensile strain. (After Blackman, J. S. et al., 1958.) Figures (a) through (f) represent various strain gradient conditions. β represents the total nonlinear strain between the end faces.

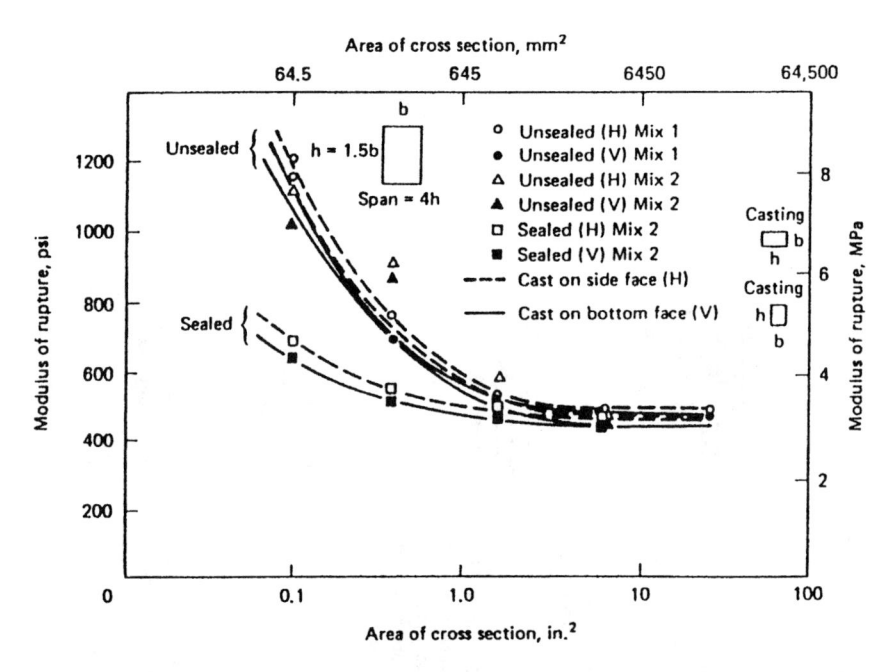

Figure 9.8 Variation of modulus of rupture with size for gypsum mortar. (After White, R. N. and Sabnis, G. M., 1968.)

where: f_t = tensile strength in kilograms per square centimeter
V = test volume of specimen in cubic centimeter $\times\ 10^{-3}$
A, B = constants for best fit data

The values of A and B ranged between 23.32 to 29.56 and 0.021 to 0.041, respectively. This relation implies that tensile strength decreases indefinitely as the size of the specimen increases. A correction made by Rao (1972) improved the relation with a finite limitation to the form:

$$f_t = f_{\text{limit}}\left(1 + CV^{-D}\right) \tag{9.14}$$

where: f_t = minimum value of the tensile strength
f_{limit} = the strength of standard specimen in the investigation
C, D = experimental constants

Mirza et al. (1972) observed that the indirect tensile strength f_{sp} was observed to be 10% of the compressive strength f'_c; and f_{sp}, the split tensile strength, was related to f'_c by the following empirical equation:

$$f_{sp} = 6.45\sqrt{f'_c} \tag{9.15}$$

where, f_{sp} and f'_c are expressed in psi.

Table 9.1 Relative Tensile Strength of Prototype and Model Concrete

Concrete Compressive Strength, MPa (1)	Type of Test (2)	150 × 300 mm cylinder (3)	50 × 100 mm cylinder (4)	25 × 50 mm cylinder (5)
14	Direct tension	9.2	11	12.2
	Split cylinder	11.8	12	12
	Modulus of rupture	19.7	20	—
28	Direct tension	8.5	10	12
	Split cylinder	9.8	11	13
	Modulus of rupture	15.7	17	20[a]
42	Direct tension	8.3	9	11
	Split cylinder	9	10	11
	Modulus of rupture	13	17	30[a]

[a] Gypsum mortar, the others being cement mortar.

9.5.5 Evaluation of Experimental Research on Tensile Strength

Results of various tensile strength tests indicate that the three different types of tensile specimens give different values which also vary with the compressive strength (Table 9.1). For model concretes, the variation with compressive strength is not as pronounced. In addition, the relative tensile strength is found to be a higher percentage of the compressive strength.

Direct modeling is hard to achieve in modeling the prototype tensile strength. This affects both the strength and deflections in the model structure because the tensile strength controls cracking; further research is needed in this area of direct modeling.

9.5.6 Size Effects in Long-Term Properties of Concrete

The long-term properties of concrete involve creep and shrinkage. They affect structural behavior in many ways; creep is especially important in long-term deflections. An understanding of these properties in model materials will undoubtedly enhance the use of the direct model approach for studying the sustained load behavior of structures. At the present time, however, models are not used to investigate these properties. The influence of size effects, which was discussed earlier for short-term loading, should also be investigated for creep and shrinkage. There has been only exploratory research done in this field, and the discussion presented here merely indicates some potential uses, as well as problems, in the application of direct models to study long-term effects.

9.5.6.1 Shrinkage

The size and shape of a concrete member will influence the rate at which moisture moves to or from the concrete, and will therefore affect the rate of shrinkage. Carlson (1937) showed that at a relative humidity of 50%, drying would only be felt to about 75 mm from the surface of a large concrete member during the first month of exposure. After 10 years, the drying will be felt about 600 mm from the surface. This means that a small or slenderly proportioned test specimen will shrink much more rapidly and uniformly than a large and bulky member. Ross (1944) suggested that the most suitable parameter in comparing the shrinkage of members of different sizes and shapes is the ratio of exposed surface area S to the volume V of the member. He showed that the ratio of exposed surface area S to the volume V can be used to correlate the shrinkage at three different ages of a series of small mortar specimens having rectangular, circular, triangular, and annular cross sections. Hansen and Mattock (1966) tested three different-size specimens with two different aggregates to investigate the influence of specimen size on shrinkage and creep. They observed that both the rate and the amount of shrinkage at a given age decreased as the size of the specimen increased. The sealed specimens stored at 100% relative humidity had a negligible amount of shrinkage (about 5% or less than that of the exposed unsealed specimens of corresponding size). This is an important conclusion to consider if one were to undertake tests on small-scale models under sustained load.

9.5.6.2 Creep

While shrinkage is caused by the loss of water from cement gel and takes place under no load, creep of concrete under a sustained load can occur in fully saturated concrete and in concrete sealed to prevent loss of moisture. The rate of creep is observed to increase if there is simultaneous moisture movement, either into or out of concrete. Creep that occurs without the exchange of moisture between the concrete and its surrounding environment has been called basic creep. Thus, the basic creep rate independent of moisture movement will not be subject to the size of specimen. It may, however, be a function of applied stress and a function of the strength of concrete, which is influenced by the size of the specimen.

No tests are available on creep effects in specimens smaller than 75 mm thick, which approximates the upper limit in terms of practical small-scale model dimensions. A general conclusion from the available tests indicates that even in larger sizes, creep decreases with the increase in specimen size; the largest difference occurs during early days of curing. Surface drying is also important because it involves moisture transfer. Tests on sealed specimens or cores from large dams show much smaller variation in creep as a result of change in the size of specimen. Tests by Hansen and Mattock (1966), who were investigating the size effect in creep, showed that for large specimens, creep reduces to the value of basic creep corresponding to a sealed specimen.

9.5.7 Size Effects in Gypsum Mortar

Gypsum mortar has been used successfully as a substitute for cement mortar to model prototype concrete. Investigators have considered mortars made of gypsum because of two main disadvantages of Portland cement mortars from the modeling point of view. First, cement-based model concrete tends to have excessive tensile strength compared with prototype concretes, and this becomes even higher as the size of specimen is reduced; second, cement-based model concretes require a longer curing period to develop the required stable strength level. The very short curing time (less than 1 day) and better control of tensile strength of gypsum mortars make them an attractive substitute for model concretes.

The major variables on which the size effect of gypsum mortar should be evaluated are the same as those for cement mortar. The earliest reported work on this type of model material was done at Cornell University, where gypsum mortars were used extensively for the first time for direct modeling purposes (White and Sabnis, 1968; Loh, 1969). They presented experimental evidence and the related techniques that could be used to minimize the so-called size effect phenomena.

Although the major factors influencing size effects were discussed earlier for cement mortars, some of these are repeated with particular reference to gypsum mortar.

1. *Differential drying of specimens, both before and during testing* (if the latter is extended over a period of many hours). The time required to gain a certain strength level will increase with increasing specimen size because the lower ratio of surface area to volume retards the rate of drying and strength. Different-size specimens will have a different drying history if they are to have the same strengths at a given age.

2. *Difference in quality of material as cast in various-size molds.* It is not possible to scale precisely the compaction process; also water gain in the upper portion of the specimens as well as its loss from imperfect molds will be different for different sizes. This is crucial because gypsum gains strength only as the specimen water content decreases.
3. *Statistical variations in strength because of differences in volume.* It should be pointed out that one must be very careful in attempting to explain size effects with statistical strength theories in the case of gypsum mortar because of the strong possibility of including other effects (as in 1 and 2 above) that have all the appearances of a statistical strength variation.

An attempt was made to separate the variables affecting size effects on the behavior in compression and flexure (Loh, 1969). Based on the results presented in Chapter 4, the following conclusions were drawn:

1. *Compression and split cylinder tests.* It was concluded that when the density, moisture content, and loading conditions are identical for various-size specimens, size effects both in uniaxial compressive and split cylinder tensile strength are negligible.
2. *Modulus of rupture tests.* It was found that there exists some size effect in flexural strength; this was mainly attributed to the strain gradient (see Figure 9.7).

9.6 SIZE EFFECTS IN REINFORCED AND PRESTRESSED CONCRETE

Overall size effects in reinforced and prestressed concrete models are important since the behavior of the model is to be extrapolated for predicting prototype behavior. With this in mind, three types of behavior are important in comparing model and prototype structures made of reinforced and prestressed concrete.

1. Bond characteristics
2. Cracking similitude (service conditions)
3. Ultimate strength and deformation

Relatively few tests have been reported with the specific objectives of studying size effects per se in reinforced concrete. Details of tests on reinforced concrete elements are presented elsewhere in the text. The general discussion given here is to make the reader aware of the importance of size effects in the behavior of model structures.

9.6.1 Bond Characteristics

Investigation into the bond characteristics is complicated by limited knowledge of the bond phenomenon in the prototype concrete. The bond strength of prototype deformed bars is mainly due to the mechanical wedge action and eventual cracking of concrete stressed by the deformations. This action reduces the size effect on bond to a certain degree, if bar deformations are reproduced in the model reinforcement.

A limited number of pullout tests by Aroni (1959) on smooth and square-twisted bars, indicated the existence of scale effects in bond strength of different-size bars tested with the bars in their usual condition; however, when the surface was polished, this scale effect disappeared, suggesting that it was related to the surface condition associated with a given size, rather than the size itself. Alami and Ferguson (1963) concluded from beam tests that models fail to predict the behavior of reinforced concrete prototypes as the result of inadequate bond when it is the primary reason of failure, thus casting doubt on the use of models in the cases where the bond may be the expected cause of failure. For investigations concerned with precracking behavior, or if the flexural or shear resistance is required, it is not necessary to satisfy all the requirements of bond similitude. It is sufficient to ensure that there is sufficient bond resistance so that premature cracking or bond failure does not occur. This can be achieved by providing sufficient embedment length to develop the yield strength of the bar.

The results of tests by Harris et al. (1966), Mirza (1967), and many others indicate that certain phenomena involving the bond as the primary cause for failure can be modeled with reasonable reliability if these models are constructed carefully to eliminate any variation and also if the variables such as the concrete strength, steel yield strength, and mechanical deformations are controlled accurately. Clark (1971) showed that the bond between concrete and reinforcement has a significant effect on service load behavior; this is so particularly in small-scale models where the reinforcement may take a variety of forms, such as threaded rods or deformed wires, which would exhibit entirely different bond characteristics.

From these various experiments, it may be concluded that certain deformed model bars exhibit bond characteristics that are comparable to those of the prototype reinforcing bars (see Chapter 5).

9.6.2 Cracking Similitude (Service Conditions)

The inelastic load–deflection response of a reinforced or (partially) prestressed concrete structure is often strongly dependent upon the degree and manner of cracking. Cracking modes can also influence behavior under reversed or repeated loading, moment and force redistribution in indeterminate systems, and service load conditions. The existence of size effects in cracking is defined as follows: crack width should vary with the size of model, and the number of cracks will be reduced with decreased model size. Initiation of cracking is a function of the tensile strength of concrete. As seen earlier, tensile strength increases as the size of specimen is reduced. Therefore, it may be stated that on reducing the size of a structure the load level at which the first crack forms will be somewhat higher. Crack spacing and width will both be dependent on the bond between the two materials. In the case of deformed bars or wires in models, the crack spacing will be evenly distributed depending on the distribution of mechanical lugs or deformations. If the bond properties are inadequate, there will be a reduced number of cracks with relatively fewer and wider cracks. Variation of strain-gradient can also affect cracking, but very little research has been done on this parameter.

Many tests conducted on small-scale reinforced concrete beam specimens reveal that the total number of major visible cracks decreases with decreasing beam size; however, the overall cracking patterns are found to be similar, and load–deflection behavior is properly modeled. These tests also indicate that only a small size effect is associated with cracking in scaled models, provided that the other conditions of similitude (mainly the properties of materials and bond strength) are satisfied.

9.6.3 Ultimate Strength (Load–Deflection Behavior)

In many model studies, the objective is to obtain the ultimate strength as well as the load–deflection behavior of a scaled model.

The different types of behavior in reinforced concrete include:

1. Underreinforced beams (both simple spans and two spans) to study the entire behavior due to redistribution of stresses, and failure;
2. Overreinforced beams;
3. Underreinforced slabs;
4. Punching shear behavior of slabs;
5. Behavior under combined axial force and bending, and bending and torsion;
6. Seismic (or reversed) loading tests on ductile frames.

Although a large number of tests are given in the literature, only items 2, 4, 5, and 6 are discussed here because of their greater dependence on the concrete properties.

Tests of overreinforced beams have been reported by Sabnis (1969) with models at scale factors of 1/10 and 1/6 to compare the prediction of ultimate loads, moment–rotation behavior, and effectiveness of helical binders (for confinement) in the compression zone. The cross section of specimens were 15 × 28 mm and 25 × 47 mm to match the scale of the available reinforcement. Comparable-size cylinders were tested to determine the various properties of the model concrete. It was found from moment–rotation curves that for prototype and two different sizes of models there was no size effect in these beams and that the predictions of other related behavior was within ±10% for models of both scales.

In punching shear investigations of slabs, Sabnis and Roll (1971) used a scale factor of 2.5 for model slabs as well as for the control cylinders. Since this type of failure is directly related to the tensile strength of concrete, great care in determining tensile strength is required in such tests. Predictions of the parameter $P_u/bd\sqrt{f_c'}$ and the punching shear strength of slabs were excellent, with no scale effect observed.

From tests on specimens tested under pure torsion and combined bending and torsion, Syamal (1969) reported excellent deformation similitude to cracking load values of about 40 to 45% of the ultimate load. However, postcracking torque–twist and the predicted values of angles of twist showed considerable variation. He concluded that it is possible to obtain reasonably good deformation similitude (deflections, twists, etc.), for the entire loading range, between the prototype and its small-scale models (minimum dimensions = 50 mm) reinforced with deformed steel wires without excessive size effects.

Chowdhury (1974) tested model beam–column joints at the scale factor of 10. His tests were successful in predicting the complete behavior of reinforced concrete joints, subjected to fully reversed loads; this included flexural and shear behavior and load–deflection response. Cracking patterns obtained were also very similar to those in the prototype.

9.7 SIZE EFFECTS IN METALS AND REINFORCEMENTS

Compared with the detailed discussion given above on size effects in brittle or semibrittle materials, such as concrete, relatively little experimental evidence exists for the investigation of size effect in metals. One reason for this is the homogeneity of ductile materials (metals). This section considers the basic size effects in metal specimens.

Morrison (1940) carried out tests on small steel beams to investigate the influence of specimen size on load-carrying capacity. He concluded that the magnitude of the upper yield point stress at which a beam yielded was increased with a decrease in the beam size. Davidenkov et al. (1947) investigated the effect of size on the embrittlement of steel in a liquid–air environment. They observed that the strength and the standard deviation increased with a decrease in the specimen size. The variation of strength and dispersion in the results was explained by Weibull's theory (1939). Sidebottom and Clark (1954) reported test data using steel beams with rectangular cross section. In this study a total of 18 beams with depths of 75, 25, 12, and 6 mm, respectively, were used. The experimental moment was 11, 8, and 10% below the theoretical plastic moment for the first three depths, respectively, and for the last one it was 1.5% above the theoretical. They concluded that there was a definite increase in the load-carrying capacity as a result of the decrease in depth. The increase was attributed to the higher stress gradient at ultimate moment stages in the beam as the depth was decreased. Richards (1954) investigated size effect on the tensile strength of mild steel; he concentrated on both upper and lower yield points. He differentiated between the two by considering the mild steel model to consist of two components, one brittle and the other ductile. Under the increasing rate of loading, the system behaves elastically up to the upper yield point, at which time the brittle component fails suddenly and the load drops to the lower yield point, which

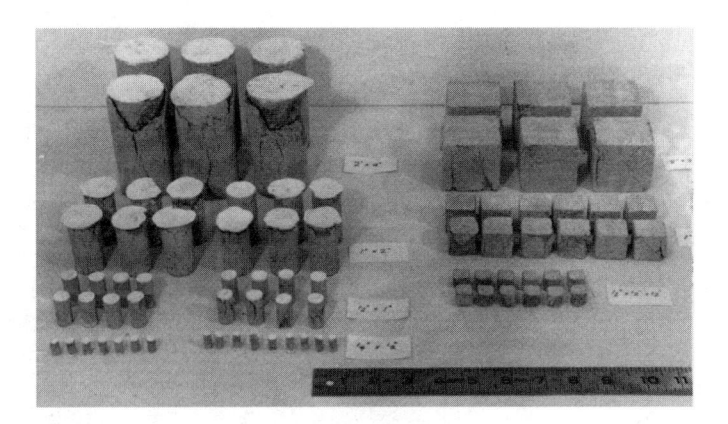

Figure 9.9 Model companion specimens to study scale effects of model masonry mortars. (After Becica, I. J. and Harris, H. G., 1977.)

is then taken by the ductile portion. With this proposition, he concentrated on upper yield (the brittle fracture), which would depend on nonhomogeneity, microcracks, and so on. He tested bars 3, 12, and 37 mm in diameter with a corresponding volume ratio of 1:64:1000 and demonstrated that the upper yield strength was an inverse function of the stressed volume. In his later work, Richards (1958) tested beams of mild steel to investigate size effect on yielding in flexure. These beams were dimensionally similar, in five different sizes and the ratio between the largest and smallest dimensions was 6.3. The results indicated that the upper yield point of mild steel in flexure was influenced by size effect.

9.8 SIZE EFFECTS IN MASONRY MORTARS

Harris and his associates at Drexel University observed in modeling of masonry structures that 50 mm cubes and 25 × 50 mm cylinders of model masonry mortar were not representative of model joint strength in compression. Tests on the prototype compression specimens 50 × 100 mm cylinders or 50 mm cubes are also poor representatives of actual joint strength. However, the goal is usually to verify modeling techniques by correlating model data with prototype data. To examine the effect of specimen size on mortar strength for both companion specimens and mortar joints, a study (Becica and Harris, 1977) was designed incorporating a range of cube and cylinder sizes. Cylinders having the dimensions of 50 × 100, 25 × 50, 12 × 24, and 6 × 12 mm and cubes of 50, 25, and 12 mm were cast of mortar mix 1:1:4. These were tested at 7 days after 4 days of continuous moist curing. Figure 9.9 shows the population size and relative dimensions of the specimens. Strength vs. size effects from this test are presented in Figure 9.10. Note that the volume of the 6 × 12 mm cylinder most closely approaches the unit volume of an ideal model mortar joint. The results of these tests tend to support the previous finding (Harris et al., 1963); that is, with decreasing volume, the strength is increased. From the modeling standpoint, this effect greatly influences the behavior of masonry assemblages (Khoo and Hendry, 1973).

In reducing the data of Figure 9.10 it was noted that the quality of specimen (workmanship) plays an important role in strength of smaller specimens. For the 6-mm-diameter cylinders, an increase in strength of 14% was observed between two experienced workers. For the 25 mm diameter cylinders, however, only a 3% increase was observed. The difficulty of accurately casting the 6 × 12 mm cylinders is the main cause of scatter in the strength values obtained.

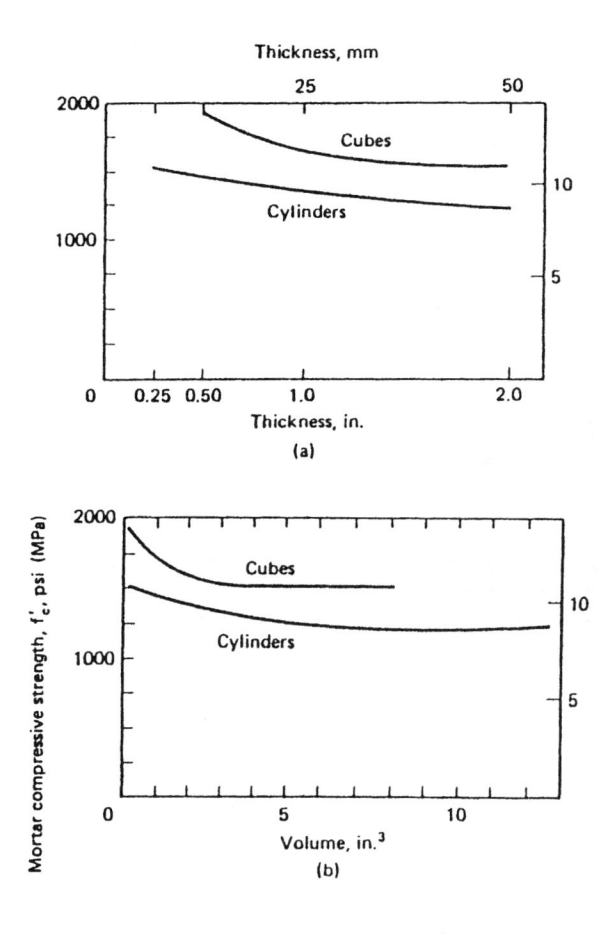

SIZE EFFECTS, ACCURACY, AND RELIABILITY IN MATERIALS SYSTEMS AND MODELS 435

Figure 9.10 Strength–volume relations for model mortar.

9.9 SIZE EFFECTS AND DESIGN CODES

Many formulas in design codes have been developed empirically and are based on experimental evidence. Even with analytical verification of experimental results, structural safety often needs to be verified for structures, which are often of unusual size. With recent advances in the development of large infrastructures, one must pay special attention to the phenomenon of size effect. Until recently, only a few codes considered size effect. Since the flexural capacity of large-scale reinforced concrete is often satisfied with lightly placed reinforcement with large distance between the main steel and extreme compression fiber of concrete, both the size-dependent nominal shear strength of members and the postcracking flexural ductility become important. The importance of such interaction is shown in Figure 9.11.

Size dependence is an important aspect to be considered, when rational and safe design is to be accomplished. Kani (1967) demonstrated that the nominal shear strength of slender beams without shear reinforcement, $v_u = V_u/bd$, decreased with increasing cross section depth, d. This effect is quite significant. Kani showed a reduction of 50% in v_u, when the depth was increased 400% from 200 mm. Later, Shioya et al. (1989) expanded Kani's work further by increasing the depth to 3,000 mm and a further reduction of 50%.

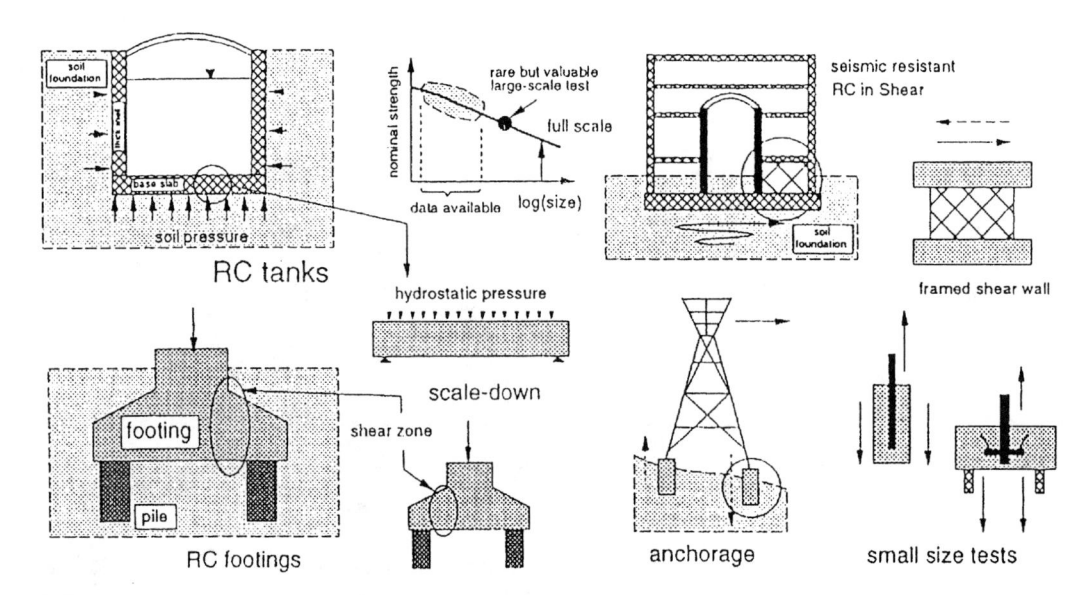

Figure 9.11 Large-size structures where size effect is problematic in design on ultimate limit state and its dependency on the size. (From Okamura, H. and Maekawa, K., in *Size Effect in Concrete Structures*, Mihashi, H., Ed., 1994, 1–24, E & FN SPON, an imprint of Chapman & Hall, London, U.K. With permission.)

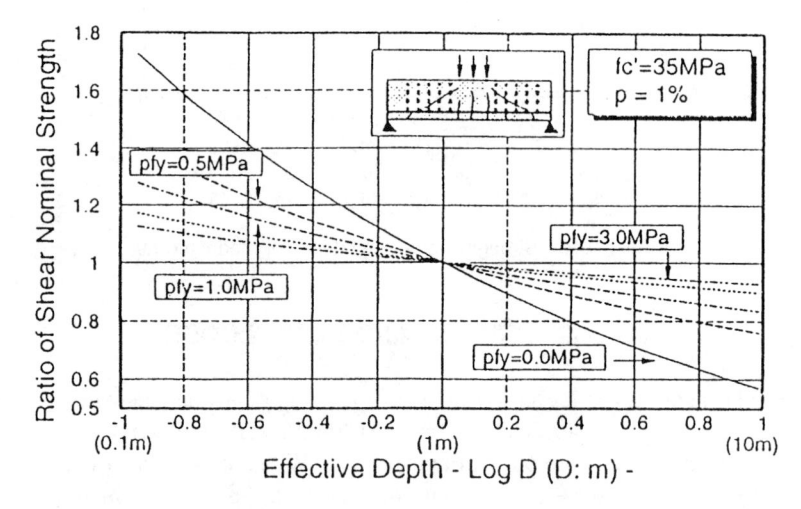

Figure 9.12 Size sensitivity of web-reinforced concrete beam in shear (1 m effective depth case is referential). (From Okamura, H. and Mackawa, K., in *Size Effect in Concrete Structures*, Mihashi, H., Ed., 1994, 1–24. With permission.)

It is also important to note that the researchers performed testing on the specimens which are relatively small size. Results of such tests combined with statistical analyses often result in code equations. Such code equations, when applied to large structures, may result into a reduced strength and might prove to be disastrous. This again brings out the point that size effect should be considered not only for research purposes but also from the viewpoint of safety and code requirements.

As was pointed out earlier, the size effect is pronounced when the failure is related to concrete rather than steel (i.e., more in shear or tension) and therefore attention should be focused on typical sections prone to failure in shear. These were demonstrated in Figure 9.12. Therefore, the most appropriate place to include size effect in the code is shear provisions.

Table 9.2 Size Factors in Various Codes

Code	Size Factor for Shear	Comment
CEB/FIP MC 90	$k = 1 + \sqrt{200/d}$	d in mm
Eurocode	$k = 1.6 - d > 1$	d in m
NEN (6720) (Netherlands)	$k = 1.6 - d > 1$	d in m
NS 3473 (Norway)	$k = 1.5 - d > 1$	d in m
BS 8110 (England)	$k = (400/d)^{0.25} > 1$	d in mm
DIN 1045 (Germany)	$k = \dfrac{0.2}{d} + 0.33 > 0.5 < 1.0$	d in m
BBK 79 (Sweden)	$k = 1$	For $d < 200$ mm
	$k = 1.6 - d$	For $200 < d < 500$ mm
	$k = 1.3 - 0.4d$	For $d > 500$ mm
Can. Std. '84	—	
ACI	—	
SIA (Switzerland)	—	
DSA 411 (Denmark)	—	

Present shear provisions in ACI 318 (1995) code have been based on experimental and empirical work. The provisions with minor changes have been in force for many years and do not include any size effect–related provisions to correlate the member performance. ACI-ASCE committee on shear noted for some time that ACI 318 provisions did not predict the actual behavior. Other codes, such as Comité European du Béton, CEB, (1993) and Japan Society of Civil Engineers, JSCE, (1986) have included provisions, which contain parameters that adjust for member size and its effect on shear performance. Although empirical, they signify the essential fact that deeper members are less ductile and have less shear capacity than that predicted by simply scaling up of the shear force based on the size alone. Recent research in the U.S., Europe, and Japan indicate that the fracture mechanics approach can be used with success to demonstrate the behavior and failure of members under shearing forces and provide a simplified but accurate approach that can be readily used in the design office.

Comparison of current provisions in many building codes indicates that a parameter "k" can be introduced in the shear equation and can be evaluated as shown by Walraven (1995). This is presented in Table 9.2. As discussed earlier, ACI does not have any k-factor, which CEB, Eurocode, and others have as some finite value. On a quantitative basis, they can be evaluated for a common depth d, as is shown in Figure 9.13 for $d = 200$ mm. This comparison clearly shows that all codes underestimate size effects.

This brief discussion on the codes and size effect is presented to indeed show that in the future there will be some treatment of size effect in the ACI code as well as in the others.

9.10 ERRORS IN STRUCTURAL MODEL STUDIES

The value of results, either from a research or a design model study, is related to the confidence that the engineer can place in the model study. This confidence level will in turn depend on the errors which are introduced at various stages of the modeling process. The parameters affecting model confidence level were summarized in Figure 9.1, taken from Pahl (1963). It is seen that certain types of errors and their causes occur in the total study. Various types of these errors are discussed in detail in this and later sections.

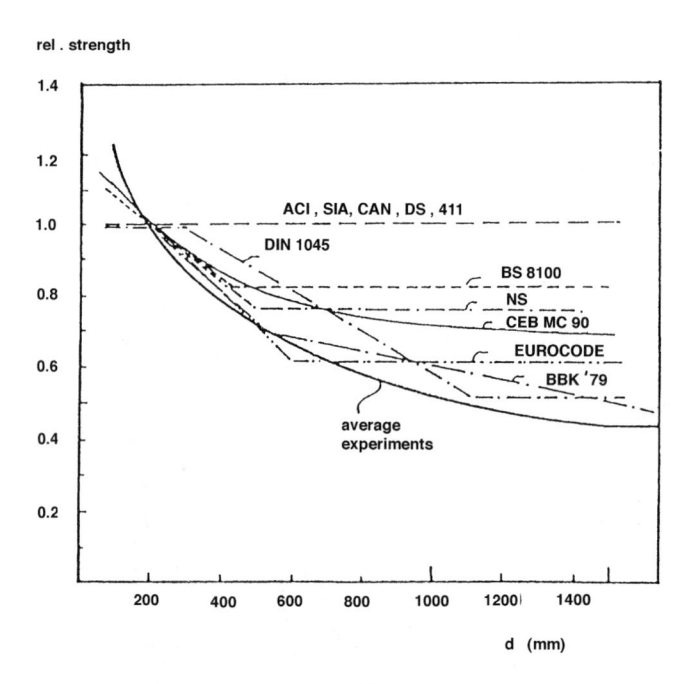

Figure 9.13 Size factors as used in a number of building codes.

It has been shown that structural model analysis can be considered as a process incorporating a sequence of five steps, namely, (1) planning, (2) fabrication, (3) loading, (4) data recording, and (5) interpretation and/or extrapolation to the prototype. Errors may enter in each of these five steps, and the following is a list of possible error sources. It is not exhaustive but merely illustrative.

1. Planning
   a. Mistake in dimensional analysis
   b. Failure to recognize a relevant variable
   c. Mistake in proportioning model
   d. Choice of inadequate material

   Planning is very important, since all variables involved in the model project should be accounted for as adequately as possible at this stage. Some variables may be eliminated, however, after preliminary testing, dimensional analysis, or other relevant considerations are made. In the planning phase, the models engineer must define the scope and, thus, the acceptable accuracy of the rest of the project. *Errors* in this phase are of major concern and should be avoided.

2. Fabrication
   a. Geometry: thickness, length, etc.
   b. Material properties
      i. Poisson's ratio, for example, $\nu_{plastic} = 0.3$ to 0.5 as compared with $\nu_{concrete} = 0.15$ to 0.2
      ii. Modulus of elasticity
      iii. Complete stress–strain–time characteristics
      iv. Coefficient of thermal expansion
      v. Microscopic and macroscopic structure
      vi. Creep characteristics
      vii. Initial stresses

   In the *fabrication process*, material properties should be very carefully modeled. If the experiment is short term and loads do not exceed service load conditions, the behavior of the structure is approximately elastic; therefore, material properties become somewhat secondary. The geometry should be reproduced with minimal error to give maximum accuracy. In ultimate-strength models, the correctness of geometry or dimensions become paramount, especially with respect to the placement of steel reinforcement in concrete structures.

3. Loading
   a. Boundary conditions
   b. Magnitude of load
   c. Direction of load
   d. Distribution of load
   e. Time history of load
   f. Effect of gravity loading

   The *loading phase* is equally important, because all loading conditions of interest for the prototype structure must be reproduced as faithfully as possible. Any errors in applying the loading will be reflected in the results, and therefore in the prediction of prototype behavior.

4. Instrumentation and Data Recording
   a. Error in writing down data and readings
   b. Electrical resistance gages
      i. Incomplete bonding of adhesives
      ii. Chemical attack on plastics by adhesives
      iii. Temperature compensation
      iv. Calibration errors
      v. Inherent recording instrument error
      vi. Gage factor error
      vii. Transverse sensitivity
      viii. Current heating effect on plastic materials
      ix. Gage stiffening of plastic materials
   c. Displacements
      i. Judgment errors in smallest division of instrument
      ii. Support system of recording device not compatible with magnitude of displacements
      iii. Calibration errors
      iv. Inherent recording instrument error
   d. Pressure
      i. Meniscus corrections in a liquid manometer
      ii. Improper calibration and/or reading of pressure gages

5. Interpretation
   a. Incorrect assumption for transformation of measured surface strains into stresses, e.g., assumption of plane strain instead of plane stress
   b. Error in reduction of data

The above listing includes a wide variety of errors, some integration of which determines what is commonly referred to as *experimental error*. In a more specific sense, however, each of the errors listed above may be considered to fall into one of three general error categories: (1) blunders, (2) random errors, and (3) systematic errors. These are discussed in the next section.

9.11 TYPES OF ERRORS

9.11.1 Blunders

These are errors that have no place in scientific experiments. They are outright mistakes and should be eliminated by care and repetition of measurements. Examples of blunders would be:

1. Using incorrect logic in dimensional analysis
2. Misreading an instrument
3. Making a mistake in dimensional units
4. Mounting a strain gage in incorrect position
5. Making a mistake in loading

9.11.2 Random Errors

It is impossible to give a rigorous operational definition of randomness; however, the nature of the concept is associated with the fact that a random phenomenon is characterized by the property that its empirical observation under a given set of circumstances does not always lead to the same observed outcome but rather to different outcomes in such a way that there is statistical regularity among these different outcomes. In view of this vagueness, it is not surprising that several meanings have been advanced for random errors. The differences in such meanings are rather subtle, however, and one can think of a random error as the difference between a single measured value and the "best" value of a set of measurements whose variation is random. What constitutes the best value depends on one's purpose, but the best value here will always be taken as the *arithmetic mean* of all the actual trial measurements. It should be noted that the algebraic sign of a random error can be either positive or negative.

Random errors may arise in two rather different contexts. First, there are random phenomena associated with the statistical nature of the physical model or the property being measured. For example, the depth of 1000 W6 × 8.5 (152.4 mm) steel beams would not each be expected to equal the nominal value of 152.4 mm. In fact, the steel companies specify a tolerance of ±3.18 mm so that one might expect to find a range of depths, perhaps the great majority lying between 149.86 and 154.94 mm but with some limited, exceptional ones falling outside these limits. Similarly, the yield stress in a certain portion of each of the 1000 beams would vary over a range of values, perhaps between 193 and 331 MPa. Second, random errors may be introduced directly as a part of the measuring process. Examples of these errors would be (1) variation inherent in estimating the smallest division on some measuring instrument and (2) the fluctuation in apparent strain due to random supply voltage changes in an electrical resistance gage circuit.

9.11.3 Systematic Errors

Suppose now that the best value of the depth of the W6 × 8.5 (152.4 mm) beams is 152.15 mm. Now someone comes along with an old ruler graduated in hundredths of an inch, but the ruler has been used so much that the ends have been worn very considerably. He measures the 1000 beams and finds a range of depths between 146.30 and 152.65 mm. It is seen that, in addition to the inherent random error, an error that always has the same algebraic sign (in this case about 2.79 mm) has been inserted. Such an error is called a systematic error.

If the systematic error is always of constant magnitude, it merely shifts the entire range of values either up or down the scale. If it changes in magnitude during the course of the experiment, the relation of the measurements, one to another, is altered. In the limit, as the changes become more and more random, the systematic error may also be considered random.

Other examples of systematic error would be

1. Improper bonding of electrical resistance strain gage;
2. Support which offers moment restraint when a hinge is desired;
3. Incorrect calibration of a measuring instrument;
4. Use of radial pressure in place of vertical pressure;

5. Effect of unknown residual stresses on the buckling of a compression element;
6. Use of wrong E or ν in converting strain ε to stress or σ;
7. Gage stiffening effects on plastics.

9.12 STATISTICS OF MEASUREMENTS

A rather extensive mathematical theory has been formulated that enables the engineer to make logical quantitative statements concerning the behavior of a structural system that is influenced by random fluctuations. It has already been stated that many of the errors involved in an experimental small-scale model study may be of a systematic nature, and hence the model results may not be amenable to statistical argument. Nevertheless, there are many experimental phenomena that are random, and the models engineer should certainly be aware of the basic techniques for the statistical treatment of random phenomena. A brief introduction is given below, and more complete treatments of this subject will be found in the works by Ang and Tang (1975; 1984), Beers (1957), Parratt (1961), Parzen (1960), and Wilson (1952).

9.12.1 Normal Probability Density Function

The most common continuous probability density function arising in the field of direct measurements is the normal (Gaussian) density function. Books on probability and statistics are full of discussions on the origin, derivation, applicability, and use of the normal density function; however, only certain results will be presented here.

The mathematical equation for the normal density function is

$$p(x) = \frac{1}{\sigma\sqrt{2\pi}} e^{-\left[(x-\mu)^2/2\sigma^2\right]} \tag{9.16}$$

where: x = random variable (observed deflection or strain for example).
$p(x)$ = probability density function.
m = mean (or average) value for the distribution of the whole universe of measurements. Denoted by $\overline{X}$ when it refers to the mean value of any finite sample.
σ = a measure of the dispersion about the mean value. Known as the standard deviation for the distribution of the whole universe of measurements. Denoted by S when it refers to the standard deviation of any finite sample.

The mean value of any series of measurements is determined by

$$\overline{X} = \frac{\sum_{i=1}^{n} x_i}{n} \tag{9.17}$$

whereas the value of the standard deviation of a series of measurements is

$$S = \sqrt{\frac{\sum_{i=1}^{n}(x_i - \overline{X})^2}{n}} \tag{9.18}$$

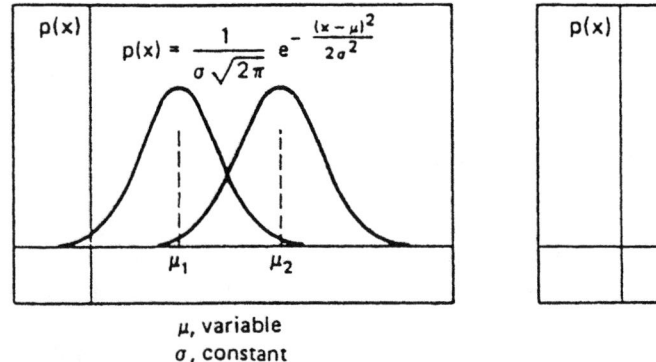

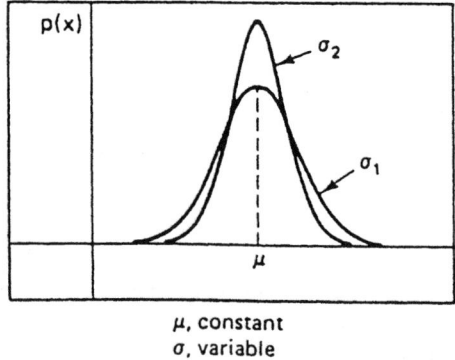

Figure 9.14 Parameter variation with normal density function.

In reality, S comes from the variance of a series of measurements. The variance is defined to be the average of the sum of the squares of the individual deviations from the mean value $\overline{X}$,

$$\text{Variance} = S^2 = \frac{\sum_{i=1}^{n}(x_i - \overline{X})^2}{n} \tag{9.19}$$

In passing from the sample mean and standard deviation to the mean and standard deviation of the entire population or universe of measurements, it is usually stated that the best values of universe mean and universe standard deviation are given by

$$\mu \doteq \overline{X} \tag{9.20}$$

$$\sigma \doteq S\sqrt{\frac{n}{n-1}} \tag{9.21}$$

and the square root in Equation 9.21 has been included, so that the bias of the standard deviation of the measurements is reduced.

The determination of these best values requires the introduction of some measure of what is best; the method of maximum likelihood provides such a measure and is generally accepted for all practical situations occurring in the measurements of engineering phenomena.

Equation 9.16 involves the two parameters μ and σ. It is instructive to note the manner in which the probability density changes for variations in the two parameters and these variations are shown in Figure 9.14. We also note that

$$0 \leq p(x)dx \leq 1 \tag{9.22}$$

$$\int_{-\infty}^{\infty} p(x)dx = 1 \tag{9.23}$$

i.e., that the probability of a certain event is equal to 1; thus, because it is certain that the random variable will assume values between $(+\infty, -\infty)$, the associated probability is equal to 1. If one had a random variable x that obeyed the normal law given by Equation 9.16, one could determine the

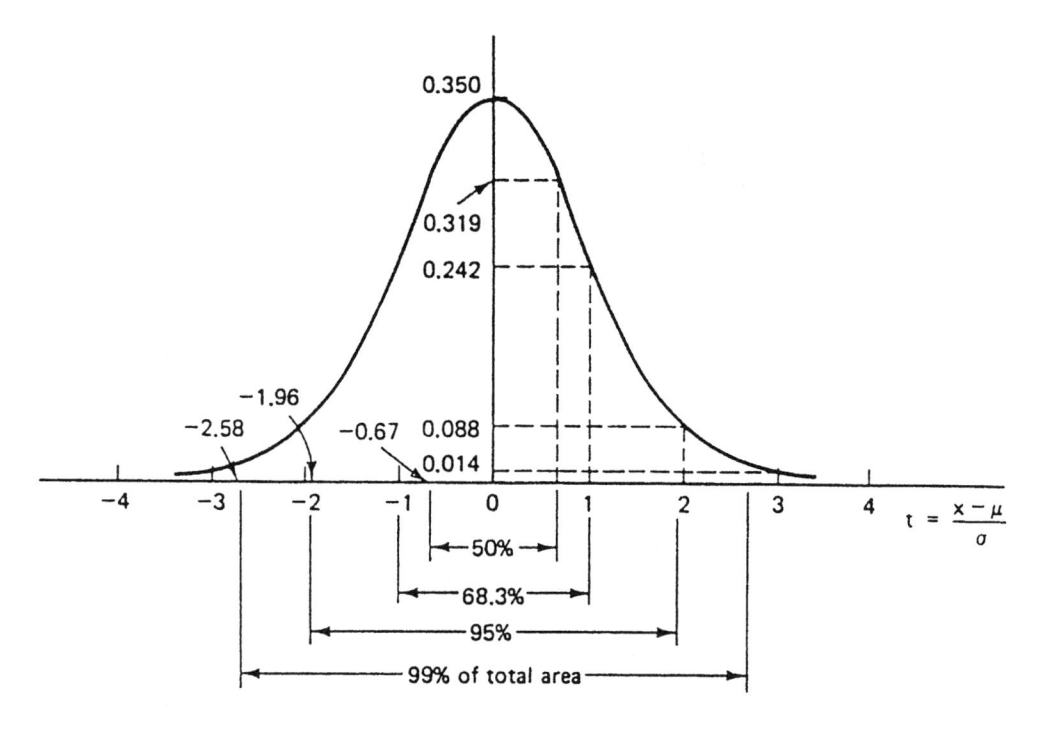

SIZE EFFECTS, ACCURACY, AND RELIABILITY IN MATERIALS SYSTEMS AND MODELS

Figure 9.15 The normal density function.

probability that x was less than some value, say t, merely by integrating the probability density function as follows:

$$p(x < t) = \Phi(t) = \int_{-\infty}^{t} \frac{1}{\sigma\sqrt{2\pi}} e^{-(1/2)[(x-\mu)/\sigma]^2} dx \quad (9.24)$$

If the change of variable $\tau = (x - \mu)/\sigma$ is made, Equation 9.24 reduces to

$$\Phi(t) = \frac{1}{\sqrt{2\pi}} \int_{-\infty}^{t} e^{-(\tau^2/2)} d\tau \quad (9.25)$$

Equation 9.25, which is denoted as a cumulative probability distribution function in that it accumulates all the probability density from $-\infty$ up to t, can alternatively be set in several forms. One common form is by means of the error function, $erf(t)$

$$erf(t) = \frac{1}{\sqrt{2\pi}} \int_{-\infty}^{t} e^{-(\tau^2/2)} d\tau \quad (9.26)$$

The probability density function, which is the parent of the cumulative probability distribution function, is shown in Figure 9.15, and a very short table of numerical values for Equation 9.25 is given in Table 9.3.

The measurements of engineering phenomena are fitted fairly well by a normal density function. In addition, the law is well known and tables that provide for quantitative predictions of the probability of any event are readily available; the law allows convenient equations for curve fitting,

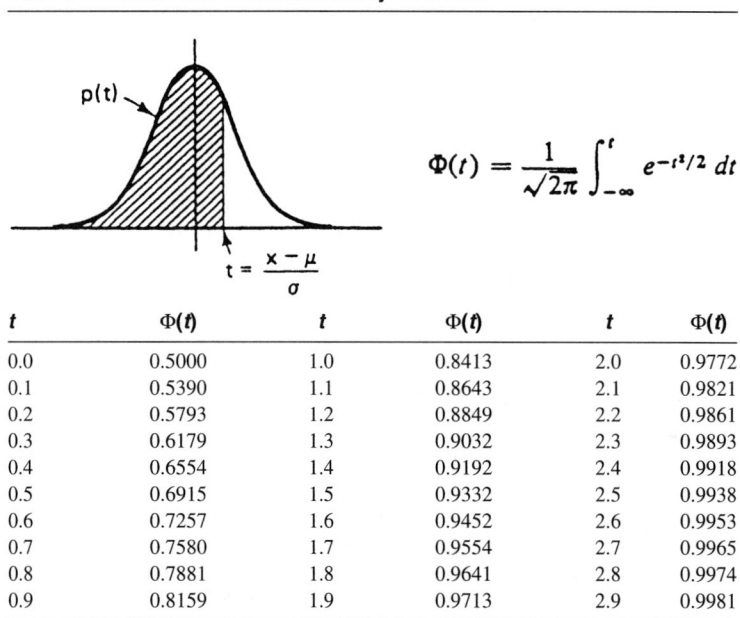

Table 9.3 Area Under Normal Density Function

$$\Phi(t) = \frac{1}{\sqrt{2\pi}} \int_{-\infty}^{t} e^{-t^2/2}\, dt$$

$$t = \frac{x - \mu}{\sigma}$$

t	$\Phi(t)$	t	$\Phi(t)$	t	$\Phi(t)$
0.0	0.5000	1.0	0.8413	2.0	0.9772
0.1	0.5390	1.1	0.8643	2.1	0.9821
0.2	0.5793	1.2	0.8849	2.2	0.9861
0.3	0.6179	1.3	0.9032	2.3	0.9893
0.4	0.6554	1.4	0.9192	2.4	0.9918
0.5	0.6915	1.5	0.9332	2.5	0.9938
0.6	0.7257	1.6	0.9452	2.6	0.9953
0.7	0.7580	1.7	0.9554	2.7	0.9965
0.8	0.7881	1.8	0.9641	2.8	0.9974
0.9	0.8159	1.9	0.9713	2.9	0.9981

although such equations are not covered here, and it has the property that certain functions of normally distributed quantities are themselves normally distributed.

9.13 PROPAGATION OF RANDOM ERRORS

In Section 9.12 the concern was with the statistics of measurements, i.e., with determining the nature of the probability density function from which the sample of measurements was drawn and further with determining in a probabilistic way the likely outcome of an additional, as yet unmeasured, outcome. There are many cases, however, in which it is not enough merely to know something about the measured phenomenon. For example, if one wanted to deduce experimentally information regarding the physical quantity known as stress, the customary procedure is to measure a strain and certain mechanical properties of the loaded material. Thus, if the material were linearly elastic, a surface stress would be determined from the well-known formula

$$f_1 = \frac{E}{1 - v^2}(\varepsilon_1 + v\varepsilon_2)$$

which reduces in the case of uniaxial stress conditions to

$$f_1 = E\varepsilon_1 \tag{9.27}$$

The question now arises whether it is possible to determine statistical relationships regarding f_1, where in fact we have no measurements of the quantity itself. Of course, the intent is to answer this question for a much more general class of situations than the simple product relationship in Equation 9.27.

Suppose that one has a derived quantity that is related to the directly measured values of two random variables x and y. The functional relationship might have the general form

SIZE EFFECTS, ACCURACY, AND RELIABILITY IN MATERIALS SYSTEMS AND MODELS

$$v = f(x, y) \tag{9.28}$$

If the deviations about the mean $\delta x = x - \overline{X}$ and $\delta y = y - \overline{Y}$ are small, then they might be considered as differentials, and the chain rule of partial differentiation may be utilized. With this assumption one obtains

$$v - \overline{v} = \delta v = \frac{\partial f}{\partial x}\delta x + \frac{\partial f}{\partial y}\delta y = \frac{\partial f}{\partial x}(x - \overline{X}) + \frac{\partial f}{\partial y}(y - \overline{Y}) \tag{9.29}$$

remembering that the equality is justified only because it is assumed that the higher-order terms can be neglected in view of the smallness of $\delta x = x - \overline{X}$ and $\delta y = y - \overline{Y}$. It should be noted that Equation 9.30 implicitly sets $\overline{v} = f(\overline{X}, \overline{Y})$ in place of $\overline{v} = \sum_{i=1}^{n} v_i / n$. When the deviations are small, the two expressions for $\overline{v}$ are equivalent. For example, suppose that $v = xy$, then

$$v_i = x_i y_i = (\overline{X} + \delta x_i)(\overline{Y} + \delta y_i) = \overline{X}\overline{Y} + \overline{X}\delta y_i + \overline{Y}\delta x_i + \delta x_i \delta y_i$$

If $\delta x_i, \delta y_i$ can be neglected in this expression, then

$$v_i \doteq \overline{X}\overline{Y} + \overline{X}\delta y_i + \overline{Y}\delta x_i$$

$$\overline{v} = \frac{\sum_{i=1}^{n} v_i}{n} \doteq \frac{\sum_{i=1}^{n}(\overline{X}\overline{Y} + \overline{X}\delta y_i + \overline{Y}\delta x_i)}{n} = \frac{\sum_{i=1}^{n} \overline{X}\overline{Y}}{n} + \frac{\sum_{i=1}^{n} \overline{X}\delta y_i}{n} + \frac{\sum_{i=1}^{n} \overline{Y}\delta x_i}{n}$$

Now by definition

$$\sum_{i=1}^{n} \delta y_i = \sum_{i=1}^{n} \delta x_i = 0$$

so that $\overline{v} = \overline{X}\overline{Y}$.

It was shown in Section 9.12.1 that the standard deviation of a set of measured quantities is defined as

$$S = \sqrt{\frac{\sum_{i=1}^{n}(x_i - \overline{X})^2}{n}} = \sqrt{\frac{\sum_{i=1}^{n} \delta x_i^2}{n}}$$

According to Equation 9.19, the standard deviation of v is

$$S_v = \sqrt{\frac{(\partial f/\partial x)^2 \sum_{i=1}^{n} \delta x_i^2 + 2(\partial f/\partial x)(\partial f/\partial y)\sum_{i=1}^{n} \delta x_i \delta y_i + (\partial f/\partial y)^2 \sum_{i=1}^{n} \delta y_i^2}{n}} \tag{9.30}$$

It should be noted that δx_i^2 and δy_i^2 are always positive, whereas δx_i and δy_i may be either positive or negative. Consequently, no matter how large n is, $\Sigma \delta x_i^2$ and $\Sigma \delta y_i^2$ are positive, but $\Sigma \delta x_i \delta y_i$ may take on a variety of values. In particular, if x_i and y_i are completely *independent*, then in the limit, as n becomes large, one should expect that $\Sigma \delta x_i \delta y_i \to 0$, since $\delta x_i \delta y_i$ is just likely to be positive or negative. If x_i and y_i are not independent, it becomes necessary to introduce the notion of correlation between x_i and y_i. Such correlation is measured by a quantity known as the correlation coefficient, which can vary between ±1. When x and y are independent, the correlation coefficient vanishes.

The previous results can easily be generalized so that, for *j independent random* variables with mean values $\overline{X}_1, \overline{X}_2, ..., \overline{X}_j$, and standard deviations $S_1, S_2, ..., S_j$, it can be said that

$$\overline{v} = f(\overline{X}_1, \overline{X}_2, ..., \overline{X}_j) \tag{9.31a}$$

$$S_v = \sqrt{\sum_{i=1}^{j} \left(\frac{\partial v}{\partial x_i}\right)^2 \delta x_i^2} \tag{9.31b}$$

In illustration of Equation 9.31b, if

$$v = x_1 \pm x_2 \pm ... \quad \text{then} \quad S_v = \sqrt{\delta x_1^2 + \delta x_2^2 + ...}$$

$$v = x_1 x_2 \quad \text{then} \quad S_v = \sqrt{\overline{X}_2^2 \delta x_1^2 + \overline{X}_1^2 \delta x_2^2}$$

$$v = x_1^a x_2^b \quad \text{then} \quad S_v = \sqrt{a^2 \overline{X}_1^{2(a-1)} \overline{X}_2^{2b} \delta x_1^2 + b^2 \overline{X}_1^{2a} \overline{X}_2^{2(b-1)} \delta x_2^2}$$

where it is noted that the partial derivatives are to be evaluated at the mean values $\overline{X}_1$, and $\overline{X}_2$, and consequently are constants. It should be noted that the derivation that led to Equation 9.31 did not require a specification of the probability density functions of the independent random variables $x_1, x_2, ..., x_j$. However, having the knowledge of the mean and standard deviation of the derived variable does not imply knowledge of the probability density function of the derived variable even when the density functions of $x_1, x_2, ..., x_j$ are known. If one wants to know this additional information, then one must resort to the use of convolutions, or generating functions, or other less elementary techniques of the theory of probability. It may be stated here that if each of the independent random variables $x_1, x_2, ..., x_j$ are normally distributed, then it is true that the derived random variable of a sum or difference of normal variates is also normally distributed. A similar statement cannot be made when the derived variable is a product, logarithm, square root, etc. of normally distributed variables; there are, however, techniques that estimate the probability density functions of functions of random variables in certain cases (e.g., Ang and Tang, 1984).

Example 9.1

Suppose that one were considering a prototype structure that was a simple-span prismatic beam (Figure 9.16), the material of which is linearly elastic. The beam is subjected to a midspan load P, and it is of interest to determine the tensile stress in the bottom fibers of the midspan cross section.

Analytical Solution — It is well known that the mathematical formulation of this problem is expressible as $f = Mc/I$, or

$$f = \frac{Mc}{I} = \frac{3}{2} \frac{PL}{bd^2} \tag{9.32}$$

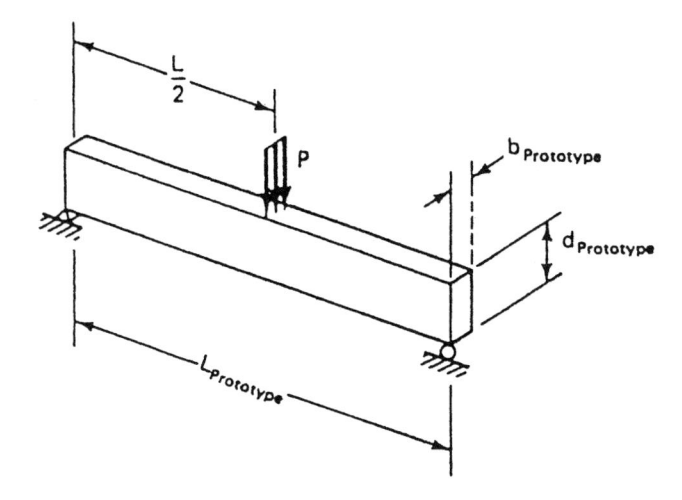

SIZE EFFECTS, ACCURACY, AND RELIABILITY IN MATERIALS SYSTEMS AND MODELS

Figure 9.16 Prismatic beam.

It is seen that the stress is a function of four variables, three geometric and one loading. If these four variables are considered to be independent random variables with known means and known standard deviations, it would be possible to determine the mean and standard deviation of the derived quantity f by utilizing Equation 9.31. Of course, it would be highly desirable to be able to readily obtain the probability density function of f but such information cannot be obtained from the theory of error propagation embodied in Equation 9.31. Thus, Equation 9.31 can be used in this example problem even if the random variables P, L, b, and d each have different types of probability density functions (e.g., uniform, normal, log normal, triangular). The equations yield no information regarding the type of probability density function of f even in the case where P, L, b, and d all have the same type of probability density function (e.g., all normal). With these limitations in mind, one can write

$$\bar{f} = \frac{3}{2} \frac{\overline{PL}}{\overline{bd}^2} \qquad (9.33)$$

$$\sigma_f = \sqrt{\left(\frac{3}{2}\frac{\overline{L}}{\overline{bd}^2}\right)^2 \sigma_P^2 + \left(\frac{3}{2}\frac{\overline{P}}{\overline{bd}^2}\right)^2 \sigma_L^2 + \left(\frac{3}{2}\frac{\overline{PL}}{\overline{b^2d}^2}\right)^2 \sigma_b^2 + \left(\frac{3\overline{PL}}{\overline{bd}^3}\right)^2 \sigma_d^2} \qquad (9.34)$$

A quantitative probabilistic prediction regarding the stress at the midspan cross section could now be made.

Experimental Solution — If one were to adopt an experimental investigation on a small-scale structural model as a means of determining the stress in the prototype beam, a dimensional analysis of the problem should be performed first. The physical quantity of interest is the stress, but to measure stress directly is not possible, and measurements of strain are usually taken. Then with knowledge of the elastic properties of the model material the strain measurements are transformed into stress. Equation 9.27 can be used to make the transformation in this plane stress problem. Thus, it is necessary only that the model material be elastic and that we have some knowledge of the magnitude of the modulus of elasticity. Then

$$F(f, L, b, d, P) = 0$$

which according to Buckingham's theorem can be reduced to

$$\Phi\left(\frac{fL^2}{P}, \frac{L}{b}, \frac{L}{d}\right) = 0$$

or, in the solved form,

$$f = \frac{P}{L^2} \phi\left(\frac{L}{b}, \frac{L}{d}\right)$$

Of course, in this simple problem it is known that

$$\phi\left(\frac{L}{b}, \frac{L}{d}\right) = \frac{3}{2}\frac{L}{b}\left(\frac{L}{d}\right)^2 = \frac{3L^3}{2bd^2}$$

However, such information cannot be obtained from the dimensional analysis alone. Thus, the model restrictions and model-to-prototype extrapolation are given by

$$\left(\frac{L}{b}\right)_m = \left(\frac{L}{b}\right)_p \quad \text{and} \quad \left(\frac{L}{d}\right)_m = \left(\frac{L}{d}\right)_p \tag{9.35}$$

$$f_p = f_m \frac{P_p L_m^2}{P_m L_p^2} \tag{9.36}$$

Now actually the model strain is measured, perhaps by means of an electrical resistance strain gage. If the model is carefully constructed according to Equation 9.35, and several measurements (say five) are made of the strain, there will be some variation among the individual measurements. This variation is certainly not due to variations in L, b, or d, and most likely not due to variations in P either. Thus, the major part of the variation observed in the five measurements may be due to errors in the measuring system itself a factor not present in the prototype structure. If a second model were constructed to duplicate the first model, five new strain measurements would again show some dispersion about a mean value; but in all likelihood the mean value would not be the same as the mean value obtained from the first model beam. If this procedure were repeated ten times, a possible set of results might be as indicated in Figure 9.17.

The question arises of how such a set of data can and should be interpreted, keeping in mind that the real quantity of interest is the stress in the prototype beam. Further, is it possible to deduce quantitative probabilistic results with regard to the prototype stress in the same way as Equations 9.33 and 9.34. Although a closed-form solution cannot be given to the first question, the following points should be mentioned:

1. The results indicated in Figure 9.17 may be clouded with all three error types: blunders, systematic errors, and random errors. Equations 9.31 can be applied only to random phenomenon so every effort should be made to eliminate blunders and systematic errors. In this respect the results obtained on model 5 are certainly suspect and perhaps should be eliminated.
2. Since it is felt that the variation within each model may be due largely to errors in the measuring system, a more realistic determination of the model strain could be obtained

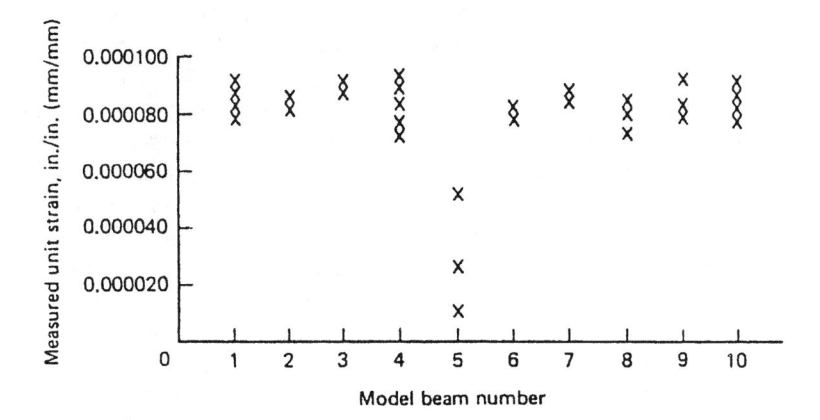

SIZE EFFECTS, ACCURACY, AND RELIABILITY IN MATERIALS SYSTEMS AND MODELS

Figure 9.17 Possible results of model tests.

by taking the mean value and the standard deviation of the nine remaining means. Of course, there may be a systematic error present in all models (e.g., a consistent error in the setting of the gage factor dial on the strain indicator), which would lead to an equal error in each of the nine individual means. Such an error cannot be suspected merely by inspection of the experimental data.

3. In this particular problem the strain was measured only in order to be able to compute stresses. Therefore, to get the best possible value for the mean and standard deviation of the stress, test coupons of each model should be tested to determine the modulus of elasticity of the model material. After computing $f_i = E_i \varepsilon_i$ for each of the nine valid model beams, one could compute the mean of the means and the standard deviation of the means. Thus, from Equations 9.20 and 9.21

$$\bar{f}_{model} \doteq \frac{\sum_{i=1}^{9} \bar{f}_i}{9} \quad \text{and} \quad \sigma_{f\,model} \doteq \sqrt{\frac{\sum_{i=1}^{9} (\bar{f}_i - \bar{f}_{model})^2}{8}} \tag{9.37}$$

As to how one can use the information obtained in Equation 9.37 to obtain quantitative probabilistic results with regard to the stress in the prototype, attention should be focused on the extrapolation equation deduced from the dimensional analysis. Equation 9.36 was deduced by setting

$$\left(\frac{fL^2}{P}\right)_{model} = \left(\frac{fL^2}{P}\right)_{prototype}$$

But Equation 9.35 states that

$$\left(\frac{L}{b}\right)_{model} = \left(\frac{L}{b}\right)_{prototype} \quad \text{and} \quad \left(\frac{L}{d}\right)_{model} = \left(\frac{L}{d}\right)_{prototype}$$

so that Equation 9.36 is certainly not a unique extrapolation equation. It could be written in several ways, for example,

$$f_p = f_m \frac{P_p L_m^2}{P_m L_p^2} = f_m \frac{P_p b_m^2}{P_m b_p^2} = f_m \frac{P_p L_m^5 d_p^3}{P_m L_p^5 d_m^3} = f_m \frac{P_p L_p b_m d_m^2}{P_m L_m b_p d_p^2} = \cdots \tag{9.38}$$

Clearly all of the above expressions will yield the same result if they are used to extrapolate the observed mean values to obtain an approximation of the mean value of the prototype stress. On the other hand, each expression will lead to a different value of the standard deviation of the prototype stress. In fact, none of these expressions could lead to the result given by Equation 9.33, which is known to be correct for this simple problem. Thus, we have seen that *while experimental results obtained from small-scale structural models can be used to predict the average or mean value of a particular physical quantity in a prototype structure, it is in general not possible to determine anything regarding the possible dispersion about this derived mean value.* In other words, even if the mean and standard deviation of f, P, L, b, and d were obtained by measurements on a series of model experiments and if the mean and standard deviation of P, L, b, and d in the prototype were known, it would be impossible to determine the standard deviation of the prototype stress unless the correct analytical formulation given by Equation 9.32 were known — and in that case there is often no advantage to be gained in conducting a model study. Two further points are worthy of mention. First, in the simple beam problem that has just been discussed there is no ambiguity regarding what is meant by the depth d. As it related to this problem it was clearly the depth at the cross section under investigation. Similarly, there was no indefiniteness involved in the definition of P, L, or b. It should be realized, however, that as soon as one enters into the field of statically indeterminate structures, real problems of definition arise. For example, where should the depth d be measured in a two-span beam, since the internal moment acting at the cross section of interest is some function of the depth existing along the entire beam? Or, going one step further into the woods, what is meant by the thickness of a small-scale, thin-shell model when that thickness varies in an irregular manner over the entire shell surface?

Finally, it would only be the rare occasion when ten completely separate models would be fabricated and tested. In fact, perhaps the most serious immediate problem facing the experimental designer is that the time and cost involved in even a single model may be prohibitive. If only one model is constructed and tested, how are the results from that single test to be interpreted? What if in the simple beam problem only one model had been tested, and it had been beam number 5? It has been shown that the only advantage of fabricating and testing more than one model is to obtain a better approximation of the true model mean. If the mean is taken to be the result obtained from a single model, then it is particularly important that the investigator be convinced of the absence of blunders and major systematic errors in that single model study. Such systematic errors can enter into the model results through a variety of means, e.g., through the physical means of providing for the boundary supports, through incomplete bonding of a strain gage, through cementing a strain gage to a plastic material that is not resistant to the solvents in the cement, through switching circuits incorporating large switching resistances, through battery decay in a recording device, and through the use of radially applied loads in place of actual gravity loads. A partial check against many systematic errors can be obtained by building into the model certain internal checks. For example, it may be possible to provide for several static checks within the model. Such considerations are considered in more detail in the following section. Other indicators of blunder or systematic error are trends in the data, jumps in the data, a periodicity in the data, or a change in precision of the data; however, when only one model is to be studied, such indicators can seldom be used. Finally, it would be extremely useful to know that an extensive set of tests had been successfully carried out on a similar problem using techniques of fabrication, loading, and instrumentation similar to those proposed. Thus, the great background of experience underlying the use of electrical resistance strain gages on metallic materials leads one to have confidence in such results, whereas results obtained on foamed plastic materials might be suspect.

9.14 ACCURACY IN (CONCRETE) MODELS

Following the above general discussion of errors in structural models, the remaining part of the chapter deals with those problems specifically related to concrete models. The topics discussed are:

1. Dimensional and fabrication accuracy
2. Accuracy-related material properties
3. Accuracy in testing and taking measurements
4. Accuracy in interpretation of test results

9.14.1 Dimensional and Fabrication Accuracy

Dimensional accuracy is an important aspect of modeling work and affects the fabrication of concrete models in two ways: first, in the making of forms and, second, in the actual concreting process. There is always a question of how accurately the fabrication process should be controlled. As geometric scale becomes smaller, so do the absolute fabrication tolerances, if the model is to represent a physically larger specimen appropriately. In the case of shells, for example, beyond a certain scale it may just become impractical to construct an accurate mortar model. In general, a philosophy to maintain approximately the same degree of accuracy as in the prototype is desirable; a tolerance of ±5% is recommended to achieve sufficient accuracy. To achieve this,

1. The forms should be machined accurately from a material that maintains its dimensions with time and wet environments. Plexiglas is considered to be a very effective and relatively inexpensive material for making molds. Plastic-coated wooden forms may also be used to reduce costs. Aluminum sections can also be used for model forms, especially where stiffness is a problem. Aluminum works well in combination with Plexiglas but tends to be expensive.
2. Even after the forms are machined accurately, there may be some errors in dimensions during casting and screeding the material in the forms to obtain the required thickness. Although this does not pose a problem for linear models (frames, beams, and columns), it is very important in spatial structures (slabs, shells).

The accurate measurement of cross sections and thicknesses, therefore, must be recorded and considered for the correlation between the model and the prototype behavior. Experience indicates that model dimensions tend to be slightly greater than design values, even when using accurately machined forms.

The dimensional accuracy in the fabrication of concrete models appears not only in the actual concreting process but also in the placement of the reinforcement. Compared with the large-scale models or prototype structures, extra care is required in this respect because of the flexibility of reinforcement and its response to the casting and vibration process during casting. In the case of underreinforced sections, wherein steel governs the failure of the structure, the accuracy will depend on the positioning of the reinforcement. Any change in position will directly influence the load capacity of the element.

Consider, for example, the concrete dimensions of two types of structures: linear and two-dimensional. In the case of a beam or column, screeding of the model in one direction during casting is easier than in the case of a slab or a shell, and will be reflected in the thickness measurements of these elements.

Typical tolerances obtained in cement mortar and gypsum mortar beams are shown in Table 9.4 (Harris et al., 1966). As shown in Table 9.4, the ratio of measured to desired values of depth for these modulus of rupture specimens ranged from 1.00 to 1.05, with an average of 1.02. The smallest beams, 10 mm deep, were oversize by an average of 3%. Plexiglas forms were used for all model beams.

Slab thickness is relatively easy to control provided accurate forms are used and reasonable care is taken in the screeding the slab surface. Dimensional variations measured in several slab models are given in Table 9.5. It should be noted that all slabs were cast by students who were engaged in model studies for the first time. It is seen that slab thicknesses of the order of 12 mm

Table 9.4 Details of Mortar Beam Dimensions

Beam	Depth, Horizontal Cast, mm			Depth, Vertical Cast, mm		
	Designed	Actual	Ratio	Designed	Actual	Ratio
1	9.525	9.601	1.008	9.525	9.627	1.01
2	9.525	9.703	1.018	9.525	9.652	1.029
3	9.525	9.677	1.016	9.525	9.804	1.029
4	9.525	9.652	1.013	9.525	9.982	1.048
5	9.525	9.677	1.016	9.525	10.008	1.05
6	9.525	9.703	1.018	9.525	9.906	1.04
7	19.050	19.228	1.009	19.050	19.355	1.01
8	19.050	19.101	1.003	19.050	19.456	1.013
9	19.050	19.126	1.004	19.050	19.456	1.013
10	19.050	19.101	1.003	19.050	19.533	1.025
11	19.050	19.228	1.009	19.050	19.558	1.027
12	19.050	18.999	0.997	19.050	19.355	1.016
13	38.100	38.100	1.000	38.100	38.354	1.006
14	38.100	38.354	1.007	38.100	38.354	1.006
15	38.100	39.116	1.026	38.100	38.354	1.006
16	38.100	38.100	1.000	38.100	38.100	1.000

Table 9.5 Model Slab Thickness Measurements

Nominal Slab Dimensions, mm	Number of Points	Measured Thickness, t, mm			Measured t / Design t
		Maximum t	Minimum t	Average t	
$12.7 \times 300 \times 300$	25	13.462	12.090	12.802	1.01
$19.1 \times 450 \times 450$	25	20.422	18.999	19.812	1.04
$19.6 \times 700 \times 700$	25	22.885	18.821	21.590	1.10[a]
$19.6 \times 700 \times 700$	25	21.895	19.355	20.879	1.07

[a] It was realized during the casting process that this slab was excessively thick.

may be achieved to an accuracy of approximately 5% even by inexperienced personnel using proper care. Thinner slabs, of the order of 6 mm thick, are substantially more difficult to produce, as are curved shell sections. Harris and White (1967) investigated a series of reinforced mortar cylindrical shells having a design thickness of 3.5 mm which were fabricated and tested to failure. Figure 9.18 shows a thickness contour map of one of these shells with two layers of reinforcement. The average measured thickness was 3.12 mm.

These studies indicate that thickness tolerances no greater than those found in Plexiglas sheets (±15%) can be achieved. It is of interest to note that the "errors" in even the smallest models were of the same relative magnitude as those in the prototype beams.

As was pointed out earlier, placement of model reinforcement is also important. Placement of reinforcement in a model can be done in a similar way to that in the prototype by the use of chairs (or using short pieces of wire) underneath. Because of the small size of models, very often holes are accurately drilled in the end blocks of the form and the reinforcement passed through to obtain exact placement. A slight tension placed on the model wires helps to maintain their correct alignment. An error of ±5% in positioning of reinforcement is considered acceptable. The accuracy of the process can be evaluated using actual test results as presented in Table 9.6. A series of three $25 \times 38 \times 265$ mm beams having three types of model wires and reinforced identically were cast to measure the final position of the reinforcement after curing. These beams were cut along the centerline section using a diamond blade saw, and measurements for each wire were made from the bottom of the beam on both sides of the cut. The designed concrete cover from the bottom of the beam to the centerline of the reinforcing wires was 4.57 mm; the average measured value was 4.65 mm, within less than +2%. These dimensions (Table 9.6) are in remarkably close agreement,

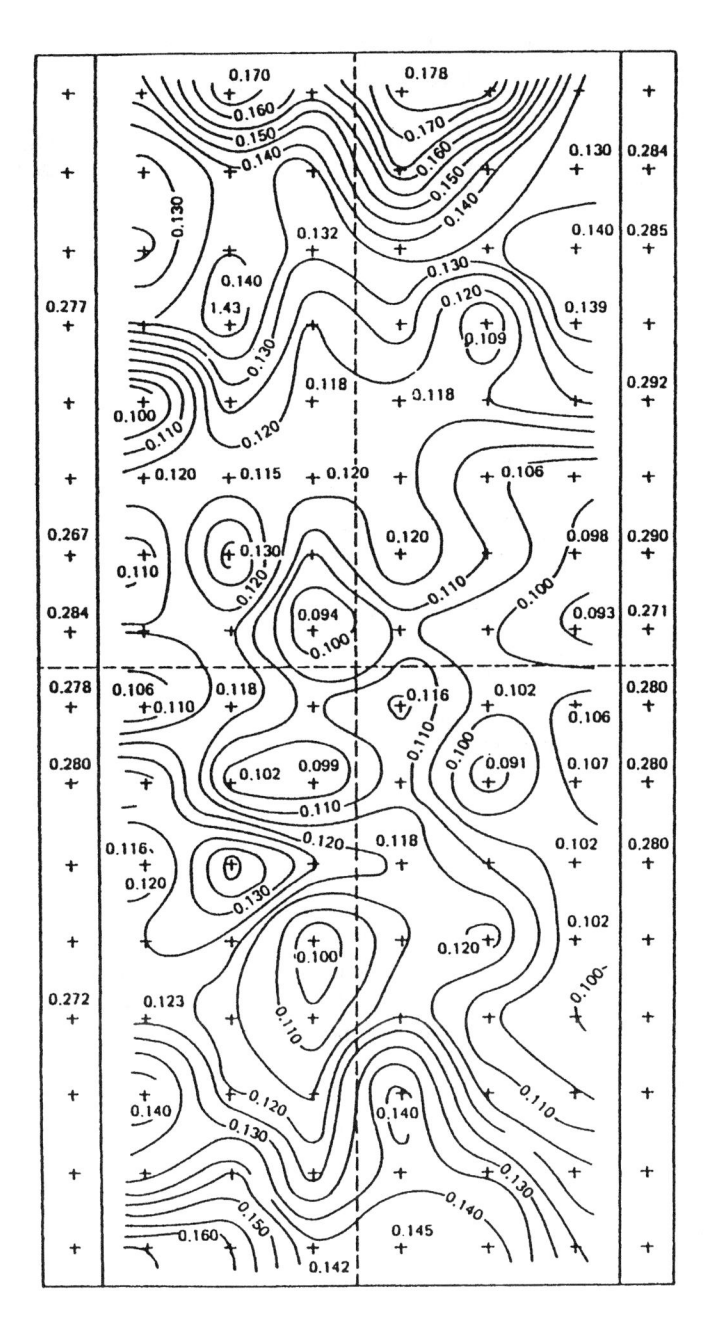

Figure 9.18 Thickness control (in.) in a model reinforced concrete shell. (From Harris and White, 1967.)

showing that with good quality of work model specimens can be cast with accuracies comparable to those achieved in prototype beams. In the final analysis of these beams, the above-measured discrepancies will cause approximately 2 to 3% error in the depth of steel or, in turn, on the lever arm, and will introduce error of similar magnitude in the calculation of loads. A correction is highly recommended where possible, and should be made by checking the position of reinforcement at the end of the test, before analyzing the data.

Some insight into the anticipated stress conditions will also help in determining the degree of accuracy; e.g., in the case of a column or tie where the stress condition is fairly axial or uniform, the location of longitudinal reinforcement in a corresponding reinforced concrete model column

Table 9.6 Measured Positions of Reinforcement in Model Beams

Type of Wire	Design Depth	Face 1 of Cut Wire 1	Face 1 of Cut Wire 2	Face 2 of Cut Wire 1	Face 2 of Cut Wire 2	Average No. 1	Average No. 2	C.G. Location	Error, %
Plain	4.57	4.597	4.140	4.699	4.420	4.648	4.267	4.470	−2.2
Deformed	4.57	4.470	4.902	4.242	4.801	4.343	4.851	4.597	+0.5
Fabribond	4.57	4.597	5.131	4.597	5.182	4.597	5.156	4.877	+6.6

has a much smaller effect on the accuracy of the results compared with a similar situation of misplaced steel in a flexural specimen.

In small-scale models there is always the question about how accurately the fabrication process can be controlled. As geometric size becomes smaller and smaller, the absolute fabrication tolerances become less and less, if the model is to represent a physically larger specimen appropriately. While it is obvious that there becomes a size below which it just becomes impractical to construct accurate models, reinforced mortar models of plate- and shell-type structures have been adequately fabricated, and at reasonable cost, down to 3 mm thicknesses, and beam-type models ranging down to about 6 mm are possible.

9.14.2 Material Properties

Material properties are very important in the case of ultimate-strength or inelastic models, in which the properties generally have to be identical to the prototype for investigating the true behavior of a structure. In case of elastic models (e.g., materials, PVC, Plexiglas, etc.), a number of similitude requirements can be relaxed without affecting the accuracy.

In an extensive study of elastic buckling of spherical shells (Litle, 1964), the models were vacuum-formed from flat sheets of PVC. The variability of the modulus of elasticity and its sensitivity due to the manufacturing process were examined. In Figure 3.12, test values of the modulus of elasticity from four specimens taken from each of 20 shell models are shown. Since the fabrication process in such materials involved heat forming, it was of interest to know the effect of annealing. By necessity, the vacuum-forming process causes the material to stretch in going from the initial flat sheet to the contour of the model mold. Figure 3.13 presents Litle's (1964) results to illustrate the influence of the annealing temperature on the bending modulus of the material.

On the other hand, when considering the behavior of structures at or near collapse, other material properties such as the tensile and compressive strengths, the ductility, and possible changes in the elastic constants become extremely significant. The ultimate compressive stress–strain curve can be modeled with a sufficient accuracy for any practical strength of prototype concrete (discussed in detail in Chapter 4) using either cement or gypsum mortar, as shown in Figure 4.5. As was discussed in Chapter 2, the similitude requirement for modulus and for ultimate compressive strain are less satisfactory for concrete. A slight distortion of the strain scale will be a general condition of the modeling process, and corrections and allowances for this will have to be made for certain problems in which the total amount of strain is of prime importance in the overall behavior of the structure. Tensile strength (Figure 4.21) and ultimate tensile strain tend to be higher than desired in the model material at a given compressive strength level, but the effects of this distortion can usually be accounted for in most model studies, without sacrificing accuracy to any significant degree.

Normally, it is not possible to simulate accurately the effects of shrinkage or creep in a microconcrete model. Although both model and prototype materials have similar shrinkage characteristics and identical construction techniques are used where possible, the effect of member size on shrinkage cannot be easily simulated. Similarly, the creep characteristics of microconcrete will qualitatively be similar to that of prototype concrete; only a limited amount of evidence exists to indicate the reliability of any quantitative similarities.

The mechanical strength properties are extremely sensitive to test procedures and imperfections in specimens, with the sensitivity increasing as the specimen size decreases. The errors associated with these properties can be classified as fabrication or testing errors.

In case of reinforcement, it was shown in Chapter 5 that careful choice of model reinforcing wires and suitable annealing processes may be used to produce deformed wire with surface properties similar to the prototype reinforcement. The desired accuracy of model tests can be achieved by modeling the stress–strain characteristics and the surface properties of reinforcement.

The properties of steel reinforcing can be controlled well in the factory. However, since the strength of steel is generally specified as "the minimum" only, actual strengths may vary considerably from batch to batch and from source to source. This will affect the accuracy of the final results of model tests, especially in the usual case of underreinforced members. Some of the remedies used to increase the accuracy of steel properties are:

1. Use reinforcement from the same batch and source. For model tests, it is a relatively minor problem since the quantities required are small and the batch is large enough to outlast the project needs.
2. Make frequent tests on samples of the specimens used to reveal all possible variations of strength properties.

The effect of steel strength variability can be considered in the strength calculations without any problem.

9.14.3 Accuracy in Testing and Measurements

Accuracy in testing is reflected by the accuracy of both the testing setup and the loading techniques. Scaling down of some of the items associated with testing can be quite involved, and proper attention must be given to such aspects. Even in a simple test setup, the reproduction of the support system for a beam by mere scaling might pose a problem, and sometimes an alternate scheme will work out much more accurately. The judgment on such variations will have to be made only with experience in the testing laboratory.

The success of a model study will also depend on the accuracy of the loading techniques. As discussed in Chapter 8, the prototype load should be represented as truly as possible. For example, a uniform loading in the prototype is simulated by discrete loads on the model for two reasons: first, the simplicity involved in applying them (such as whiffle-tree-type loading) and, second, the feasibility. To show the accuracy of the discretization, a number of studies have been made. In a simple beam (see Litle et al., 1970) and two-hinged arch (see Pahl et al., 1964), the effect of a uniformly distributed load replaced by a number of concentrated loads was studied. Figure 8.7 indicates the effect of discretizing the uniform load on the beam. Although the convergence of the moment coefficient k is obvious as the number of loading points is increased, only the even number of loads will give better accuracy. While considering the accuracy, other types of behavior (e.g., deflection coefficient) should also be considered for better representation of the prototype loading. Figure 8.8 presents a similar analysis of load discretization with reference to arch bending [Pahl et al. (1964)]. In this case, it becomes apparent that such a system may be entirely unsatisfactory unless a large number of points are used, and an alternate scheme should be worked out. In case of buckling experiments, the loading may produce totally acceptable results. Figure 8.9 from Kausel (1967) shows that the buckling load of an arch is not as badly degraded by load discretization as is the stress distribution. Whereas the 5- and 15-point load representations caused errors of 600 and 100%, respectively, in the maximum stress level in the shallow arch; these same loading patterns lead to errors of only 17 and 6%, respectively, in the magnitude of the buckling load.

In addition to the loading, instrumentation should also be considered in the accuracy of the results. The improper location will not only affect the measurements but will also influence their interpretation. Increased care is required in locating the measuring devices on the model structure as its size is reduced, particularly in the case where a strain gradient exists. Electrical resistance strain gages, which are so popularly used, should not only be located (both placing and direction) with extreme care but also should be mounted (i.e., bonded) properly. Improperly mounted gages may cause false strains and lead to erroneous conclusions.

Measurements in an experiment, whether on a model or a prototype, form an important aspect of accuracy. The success of an entire testing program will depend on how accurately measurements are made. Measurements range from loads, reactions, and deflections in simple tests to strains, curvatures, and cracking (both location and width) in ultimate-strength models. The basic philosophy should be to achieve as much accuracy as possible but in no case less than that obtained or expected in a full-scale test. This means that a proper choice of instruments has to be made to obtain the best results. Most manufacturers specify the accuracy and range of applicability for their equipment, which should help the models engineer considerably in successfully achieving the desired degree of accuracy.

Loads are measured using deadweights or load cells. Good accuracy can be achieved using either method of measurements. Deadweights are particularly suitable for measurements in the case of slab- or shell-type structures, since a number of load cells will be required to achieve loading and may become extremely expensive. In case of a deadweight system, a proper count should be kept while loading and unloading to minimize any error. Load cells are generally accurate up to 0.5% of the range, but accuracy in actual measurements will depend on the useful range; this is done by selecting the proper-size load cells.

Accuracy of deflection measurements will depend on the type of instruments, such as dial gages or transducers (LVDTs). Accuracy of dial gages varies between 0.025 and 0.0025 mm, and some of the transducers can be read continuously up to an accuracy of 0.00025 mm.

In the measurement of strains and curvatures, electrical resistance strain gages are probably the most widely used. The limitations on the accuracy should therefore be recognized fully, since it will depend on the material used in the structure as well as the gage itself. In plastic models, for example, the problems associated with gage stiffening and gage heating should be carefully examined. A certain time duration of between 5 and 15 min should be allowed to elapse before the readings are taken in the steady state; otherwise errors of the order of 15 to 50% can occur. Such errors, however, take place as a result of the material property rather than the strain gage itself. Electrical gages have been successfully used on models made of cement and gypsum mortar.

9.14.4 Accuracy in Interpretation of Test Results

Accurate and reliable test results and their interpretation are absolutely essential to a successful model technique. The experience so far on small-scale models indicates that direct ultimate-strength models will predict the failure mode and ultimate load of a prototype structure within an acceptable tolerance of the order of 10%. This degree of accuracy is completely dependent upon an intimate knowledge of the material properties of the model as was described earlier in this chapter. It is also important to realize that the results of several model tests compared with a single prototype test will normally show some difference because the prototype itself is subject to variations in strength; the single prototype might well be 10% over- or understrength as compared with the entire population.

When interpreting an experimental result, it is desirable to compare it with some available theory. In case of difference between the two, the question arises whether an experimental or analytical model is a better predictor of the structural behavior. First, an attempt should be made to determine why the theory was not able to predict the test result. Perhaps there was a measurement error in the test, but in the absence of such improprieties, and in the absence of other experimental evidence to the contrary, one should depend on the experimental work and not on the theory if this is not very well established (especially in the case of concrete structures).

Serious difficulties are encountered in test results as well as in their interpretation only when the model material properties deviate from the prototype properties to such an extent that failure modes are changed; for example, this could happen if the tensile strength of model concrete were sufficiently high to prevent an expected diagonal tension failure. Bond-critical structures remain difficult to model with any degree of confidence, but it is believed that further development of deformed wires will markedly improve this situation.

9.15 OVERALL RELIABILITY OF MODEL RESULTS

One unfortunate aspect of model study programs is that there is sufficient time and money for only one or a few tests. In order to guard against loss of accuracy as a result of systematic errors in one or more of the techniques employed, procedures can usually be followed to detect any potential error sources before the results are properly interpreted. These procedures might include:

1. Calibration procedures prior to testing
2. Statics checks during and after testing
3. Symmetry checks
4. Repeatability checks
5. Comparisons with analytical predictions
6. Observation of trends in the data, e.g., linearity
7. Gross behavior observations

Calibration procedures prior to testing must be emphasized; they are an absolute necessity. Similar to full-scale testing, model test setups need debugging. Such calibration, similar to debugging of computer programs, can save both time and money and is helpful in obtaining good results and interpreting them properly.

Statics checks performed during and after testing can give the engineer confidence in the results, or in those cases where the check turns out to be negative, it is at least helpful to learn this fact as early as possible. It may even be possible to stop the test and correct the situation before the model structure and associated equipment are damaged. The provision for these checks must be built into the model study, as instrumentation must be positioned properly and must be adequate to provide the required data. Symmetry checks, like statics checks, can indicate the reliability of the test data.

Repeatability in any one test or (better yet) from one test specimen to another or (best of all) from one test to a completely different test using a different procedure can help the engineer's confidence.

Earlier, examples were cited to demonstrate the importance of various basic quantities and the degree of accuracy that can be achieved. It always is of interest for the engineer to have an "overall" reliability of the results obtained from the model tests. Using the same examples presented in earlier sections, the overall accuracy will be illustrated. In Table 9.7 the accuracy of measurements from beam model studies are compared. The correlation for these $1/10$-scale models is shown to be very good. Results of these basic tests are presented in the following manner:

1. Comparison of ultimate load predictions
2. Comparison of load–deflection or moment–curvature (rotation) behavior

Table 9.7 compares the prediction values for beams from two model beams for each of the two prototypes.

Table 9.7 Comparison of Behavior of Beams

Beam (1)	Quantity (2)	Model Theoretical (3)	Model Experiment (4)	Col. (4)/Col. (3) (5)	Prototype Theoretical (6)	Prototype Model (7)	Prototype Test (8)	Col. (7)/ Col. (6) (9)	Col. (8)/ Col. (6) (10)
B_{21}	P^a	783	852	1.09	58.1	62.7	60.5	1.08	1.04
	M_y^b	54.7	59.5	1.09	40.6	43.7	42.1	1.08	1.04
B_{22}	P	783	903	1.15	58.1	66.9	60.5	1.15	1.04
	M_y	54.7	62.9	1.15	40.6	46.8	42.1	1.15	1.04
B_{31}	P	1197	1272	1.06	110.8	117.9	121.9	1.07	1.10
	M_y	83.6	88.7	1.06	77.3	82.2	85.1	1.07	1.10
B_{32}	P	1197	1388	1.16	110.8	128.5	121.9	1.16	1.10
	M_y	83.6	96.6	1.16	77.3	89.7	85.1	1.16	1.10

[a] P = N for model and kN for prototype.
[b] M_y = N•m for model and kN•m for prototype.

It may be concluded, based on this discussion, that the overall accuracy of the models is less (approximately 10%) compared with errors usually encountered in the basic dimensions, which were within ±2%. The other error-causing factors, discussed previously, must interact and thus influence the overall accuracy.

9.16 INFLUENCE OF COST AND TIME ON ACCURACY OF MODELS

Cost and time are important to the extent that a project may not be carried out because of their limitations. Very often satisfactory solutions using model studies can be obtained if proper time and cost are allotted, in comparison to the other types of solutions. From the design engineer's point of view, even a very accurate solution is no good if the information cannot be obtained in a reasonable amount of time; ideally, the design engineer will always want the cheapest, most accurate solution in the shortest possible time. The research engineer is, on the other hand, generally less concerned about the time factor and more concerned with accuracy and cost.

Similar to any other project, cost and time will undoubtedly affect accuracy of model results. Generally, the cost of an experimental structure or model will be more than the cost of a theoretical solution; but as was pointed out in Chapter 1 of this book, a physical model is more useful and desirable, especially in the case of structures whose behavior is to be studied up to and including the ultimate load stage and where questions of design criteria and loading are not resolved. Provided that the previous conditions of accuracy of modeling are satisfied, experimental results will be within an acceptable engineering range.

In the case of important or special structures, a test on a physical model is particularly helpful. In such projects, time will be an important aspect; but costs will play a secondary role to achieve the desired accuracy. The cost, among other things, will determine the scale factor. It is a general belief that the cost of a model will be reduced as the scale is reduced; this is true only to a certain extent.

9.17 SUMMARY

Theoretical studies on scale effect have been described, based primarily on the heterogeneity of materials and their behavior. Experimental evidence has been presented that indicates the higher strength of smaller specimens of various types of materials. With particular reference to concrete, various physical factors have been identified which produce differences in quality of cast material and of testing, and hence contribute to the size effect. Procedures have been indicated that can lead to a reduction of this effect.

A discussion on errors, accuracy, and reliability in model studies with several examples has been presented. An introduction to the theory and propagation of random errors has been presented. The implications of different types of errors on the experimental results has been discussed. Although a lot is yet to be learned about structural reliability, techniques and materials are available today to apply physical modeling with associated random variability of the parameters with confidence to many structural problems. In the case of ultimate-strength models, limitations on materials, fabrication, and loading will probably govern the optimum scale size.

It must be emphasized that confidence levels in model testing can best be established by comparing similar members or structures of different size. As is indicated by several examples, concrete models can effectively predict deflections, strains, modes of failure, and failure loads for beams, columns, slabs, and shell structures.

PROBLEMS

9.1 List the various errors that you have encountered in your experimental work. Give one example of each type from your own data. Why is it difficult to determine if a systematic error is occurring in your experimental work?

9.2 The compressive strength of three sets of microconcrete cylinders is given below. Find the mean stress, the standard deviation, and the 90% bound on the mean assuming a normal distribution of strength. What is the ratio of the bound to the mean for each size cylinder?

Cylinder Test, in.	Size No.	Fail Load, lb
6 × 3	1	24 100
	2	24 900
	3	27 000
	4	24 100
4 × 2	1	12 650
	2	11 300
	3	12 300
	4	12 200
	5	10 000
	6	12 600
2 × 1	1	3 200
	2	3 250
	3	3 750
	4	3 200
	5	3 300
	6	3 520

9.3 A model shell is loaded with 20 weights which are intended to be of equal magnitude. Assume that the weight differed varying amounts from the intended value thought to be 453.6 g.
    The weights are

404	420	390	422
386	398	400	408
414	402	408	396
410	416	396	398
394	392	380	418

Predict the population mean of the weights with 90% confidence. What is the systematic error if all 20 weights are needed to cause failure in the model?

9.4 A PVC model in a state of plane strain is used to predict the elastic behavior of concrete prototype structure. At a point in the structure the strains are measured in two orthogonal directions:

$$\varepsilon_x = 100 \text{ microstrain}, \quad \varepsilon_y = 50 \text{ microstrain}$$

The material properties are

$$E_p = 13.79 \text{ GPa} \quad \nu_p = 0.15$$

$$E_m = 2.76 \text{ GPa} \quad \nu_m = 0.40$$

Find the systematic error in the predicted prototype stresses. For this situation assume that the stresses are related to the strain by the following equation for a structure in plane strain.

$$f_1 = \frac{E}{1-\nu^2}(\varepsilon_1 + \nu\varepsilon_2)$$

Also, assume the stresses in the prototype from similitude considerations are

$$\sigma_p = E_p/E_m \, \sigma_m$$

[*Hint:* compare the prototype strains obtained by the formula and also from the extrapolated prototype stresses using the relationship

REFERENCES

Abrams, D. A. (1922). Flexural strength of plain concrete, *Proc. Am. Concr. Inst.*, 18, pp. 20–50.
ACI 444 (1979). Models of concrete structures — state-of-the-art, *Concr. Int. Des. Conctr.*, 1(1), January.
ACI 446, (1989). Fracture mechanics of concrete: concepts, models and determination of material properties, in *Fracture Mechanics,* Bazant, Z. P., Chairman. American Concrete Institute, Detroit, MI.
ACI 318 (1995). *Building Code Requirements,* American Concrete Institute, Detroit, MI.
Alami, Z. X. and Ferguson, P. M. (1963). Accuracy of models used in research on reinforced concrete, *Proc. Am. Concr. Inst.*, 60(11), November, 1643–1663.
Ang, A. H.-S. and Tang, W. H. (1975). *Probability Concepts in Engineering Planning and Design, Vol. I — Basic Principles,* John Wiley & Sons, New York.
Ang, A. H.-S. and Tang, W. H. (1984). *Probability Concepts in Engineering Planning and Design,* Vol. II — *Decision, Risk and Reliability,* John Wiley & Sons, New York.
Aroni, S. (1959). Pullout resistance of square twisted and plain round bars, Department of Civil Engineering Rept. (unpublished), University of Melbourne, Australia.
Bazant, Z. P. (1992). Private correspondence on recommendations of use of fracture mechanics and size effects in the building code.
Bazant, Z. P. and Cao, Z. (1987). Size effect in punching shear failure of slabs, *ACI Struct. J.*, 84, 44–53.
Bazant, Z. P. and Kazemi, M. T. (1988). Size dependence of concrete fracture energy determined by RILEM work-of-fracture method, *Int. J. Fracture*, 121, 121–138.
Bazant, Z. P. and Kim, J. K. (1984). Size effect in shear failure of longitudinally reinforced beams, *ACI J.*, 81, 456–468.

Becica, I. J. and Harris, H. G. (1977). Evaluation of Techniques in the Direct Modeling of Concrete Masony Structures, Structural Models Laboratory Report no. M77-1, Department of Civil Engineering, Drexel University, Philadelphia, June.

Beers, Y. (1957). *Introduction to the Theory of Error,* Addison-Wesley, Reading, MA.

Blackman, J. S., Smith, D. M., and Young, L. E. (1958). Stress distribution affects ultimate tensile strength, *J. Am. Concr. Inst.,* 55, 675–684.

Carlson, R. W. (1937). Drying shrinkage of large concrete members, *J. Am. Concr. Inst.,* 33, January–February, 327–336.

Chowdhury, A. H. (1974). An Experimental and Theoretical Investigation of the Inelastic Behavior of Reinforced Concrete Multistory Frame Models Subjected to Simulated Seismic Loads, Ph.D. dissertation, Cornell University, Ithaca, NY.

Clark, L. A. (1971). Crack Similitude in 1:3.7 Scale Model of Slabs Spanning One Way, Technical Report, Cement and Concrete Association, London, March.

Davidenkov, N. et al. (1947). The influence of size on the brittle strength of steel, *J. Appl. Mech.,* 14, March, 63–67.

Fruedenthal, A. M. (1968). Statistical approach to brittle fracture, in *Fracture,* vol. 2, H. Liebowitz, Ed., Academic Press, New York, chap. 6, 591–619.

Fuss, D. S. (1968). Mix Design for Small-Scale Models of Concrete Structures, Report No. R.-564, Naval Civil Engineering Laboratory, Port Hueneme, CA.

Glucklich, J. and Cohen, L. J. (1968). Strain energy and size effects in a brittle material, *Mater. Res. Stand.,* 8, 17–22.

Gonnerman, J. F. (1925). Effects of size and shape of test specimen on compressive strength of concrete, *Am. Soc. Test. Mater. Proc.,* 25, 237–250.

Hansen, T. C. and Mattock, A. H. (1966). Influence of size and shape of member on the shrinkage and creep of concrete, *ACI J. Am. Concr. Inst.,* 63, February, 267–290.

Harris, H. G., Schwindt, R., Taher, I., and Werner, S. (1963). Techniques and Materials in the Modeling of Reinforced Concrete Structures under Dynamic Loads, Report R63-54, Department of Civil Engineering, Massachusetts Institute of Technology, Cambridge, December; also NCEL-NBY-3228, U.S. Naval Civil Engineering Laboratory, Port Hueneme, CA.

Harris, H. G., Sabnis, G. M., and White, R. N. (1966). Small Scale Direct Models of Reinforced and Prestressed Concrete Structures, Report No. 326, Department of Structural Engineering, Cornell University, Ithaca, NY, September, 362 pp.

Harris, H. G., Sabnis, G. M., and White, R. N. (1970). Reinforcement for small-scale direct models of concrete structures, paper no. SP-24-6, *Models for Concrete Structures,* ACI SP-24, American Concrete Institute, Detroit, MI, 141–158.

Harris, H. G. and White, R. N. (1967). The Inelastic Analysis of Concrete Cylindrical Shells and Its Verification Using Small Scale Models, Report No. 330, Department of Structural Engineering, Cornell University, Ithaca, NY, May.

Johnson, R. P. (1962). Strength tests on scaled-sown concretes suitable for models, with a note onmix design, *Mag. Concr. Res.,* 14(40), March.

Kadlecek, V. and Spetla, Z. (1977). How size and shape of specimens affect the direct tensile strength of concrete, *Tech. Dig.* (Prague), 9, 865–872.

Kani, G. N. J. (1967). How safe are our large reinforced concrete beams? *ACI J.,* 64, 3, 128–141, Mar.

Kausel, E. (1967). Teoria general de la estabilidad elastica de arcos planos, M. Eng. thesis, Universidad de Chile, Santiago, September.

Khoo, C. L. and Hendry, A. W. (1973). Strength tests on brick and mortar under complex stresses for the development of a failure criterion for brickwork in compression, *Proc. Br. Ceram. Soc.,* No. 21 (April), Load Bearing Brickwork (4), 51–66.

Loh, G. (1969). Factors Influencing the Size Effects in Gypsum Mortar, M.S. thesis, Cornell University, Ithaca, NY., September.

Litle, W. A. (1964). Reliability of Shell Bucking Predictions, Research Monograph No. 25, Massachusetts Institute of Technology, Cambridge, p. 149.

Litle, W. A., Cohen, E., and Sommerville, G. (1970). Accuracy of structural models, in Models for Concrete Structures, ACI-SP24, American Concrete Institute, Detroit, MI, 65–124.

Malhotra, V. M. (1969). Effect of Specimen Size on Tensile Strength of Concrete, Report of Department of Energy, Mines and Resources, Ottawa, Canada, June, 9 pp.

Meininger, R. C. (1968). Effect of core diameter on measured concrete strength, *J. Mater.,* 3(1), March, 320–336.

Mirza, M. S.(1967). An Investigation of Combined Stresses in Reinforced Concrete Beams, Ph.D. thesis, Department of Civil Engineering and Applied Mechanics, McGill University, Montreal, March.

Mirza, M. S., Labonte, L. R. S., and McCutcheon, J. O. (1972). Size effects in model concrete mixes, paper presented at the ASCE National Convention, Cleveland, April.

Morrison, J. L. M. (1940). The yield of mild steel with particular reference to the effect of size specimen, *J. Proc. Inst. Mech. Eng.,* 142(3), January, 193–223.

Neilsen, K. E. C. (1954). Effect of various factors on the flexural strength of concrete test beams, *Mag. Concr. Res.* (London), 15, 105–114.

Neville, A. M. (1966). A general relation for strengths of concrete specimens on different shapes and sizes, *Proc. Am. Concr. Inst.,* 63, 1095–1110.

Okamura, H. and Maekawa, K. (1994). Reinforced concrete design and size effect in structural non-linearity, in *Size Effect in Concrete Structures,* Mihashi, H., Ed., 1–24, E & FN SPON, an imprint of Chapman & Hall, London, U.K.

Pahl, P. J. (1963). Confidence Levels for Structural Models, Report No. 63-05, Department of Civil Engineering, Massachusetts Institute of Technology, Cambridge, February.

Pahl, P. J. and Soosaar, K. (1964). Structural Models for Architectural and Engineering Education, Report No. R64-3, Department of Civil Engineering, Massachusetts Institute of Technology, Cambridge, 269 pp.

Parratt, L. G. (1961). *Probability and Experimental Errors in Science,* John Wiley & Sons, New York.

Parzen, E. (1960). *Modern Probability Theory and Its Applications,* John Wiley & Sons, New York.

Rao, C. V. S. K. (1972). Some Studies on Statistical Aspects of Size Effects on Strength and Fracture Behavior of Materials and Fracture Resistant Design, Ph.D. thesis, Indian Institute of Technology, Kanpur, India, June, 225 pp.

Richards, C. W. (1954). Size effect in the tension test of mild steel beams, *Am. Soc. Test. Mater. Proc.,* 54, 995.

Richards, C. W. (1958). Effect of Size on the Yielding of Mild Steel Beams, Preprint 75, ASTM, 61st Annual Meeting, June.

Ross, A. D. (1944). Shape, size and shrinkage, *Concr. Constr. Eng.,* 38, August, 193–199.

Sabnis, G. M. (1969). Behavior of over-reinforced concrete beams: a small-scale model approach, *Indian Concr. J.,* 43(1), January, 13–24.

Sabnis, G. M. (1980). Size effects in material systems and their impact on model studies: a theoretical approach, *Proc. of SECTAM X Conference,* Knoxville, TN, 649–668.

Sabnis, G. M. and Aroni, S. (1971). Size effects in material systems, the state-of-the-art, paper no. 12, in *Structure, Solid Mechanics and Engineering Design, The Proceedings of the Southhampton 1969 Civil Engineering Materials Conference,* M. Te'eni, Ed., Wiley Interscience, New York, 131–142.

Sabnis, G. M. and Mirza, M. S. (1979). Size effects in model concretes? ASCE *J. Struct. Div.,* 105(ST6), June, 1007–1020.

Sabnis, G. M. and Roll, F. (1971). Importance of scaled compressive strength cylinders in shear resistance of reinforced concrete slabs, *Proc. Am. Concr. Inst.,* 68(3), March, 218–221.

Sabnis, G. M. and White, R. N. (1967). A gypsum mortar for small-scale models, *Proc. Am. Concr. Inst.,* 64(11), November, 767–774.

Shah, S. P., Swartz, S. E., and Ouyang, C. (1995). *Fracture Mechanics of Concrete: Applications of Fracture mechanics to Concrete, Rock, and Other Quasi-Brittle Materials,* John Wiley & Sons, New York.

Shioya, T., Iguo, M., Nojiri, Y., Akiayama, H., and Okada, T. (1989). Shear strength of reinforced concrete beams, *ACI J.,* 118, 259–279.

Sidebottom, O. M. and Clark. M. E. (1954). The effect of size on the load-carrying capacity of steel beams subjected to dead loads, ASME Meeting, Milwaukee, WI.

Syamal, P. K. (1969). Direct models in combined stress investigations, M.Eng. thesis, Structural Concrete Series No. 17, McGill University, Montreal, July.

Tucker, J., Jr. (1941a). Effect of dimensions of specimens upon precision of strength data, *ASTM Proc.,* 45, 592–659.

Tucker, J., Jr. (1941b). Statical theory of the effect of dimensions and method of loading on the modulus of rupture of beams, *Am. Soc. Test. Mater. Proc.,* 41, 1072–1088.

Walraven, J. C. (1994). Size effects: their nature and their recognition in building codes, in *Size Effect in Concrete Structures,* Mihashi, H., Ed., 375–394, E & FN SPON, an imprint of Chapman & Hall, London, U.K.

Weibull, W. (1939). A statistical theory of the strength of materials, *Royal Swedish Proc.*, 151, 152.

White, R. N. and Sabnis, G. M. (1968). Size effects in gypsum mortar, *J. Mater.,* 3(1), March, 163–177, ASTM, Philadelphia.

Wilson, E. B. (1952). *An Introduction to Scientific Research,* McGraw-Hill, New York.

Wright, P. J. F. and Garwood, F. (1952). The effect of the method of test on the flexural strength of concrete, *Mag. Concr. Res.* (London), 11, 67–76.

CHAPTER 10

Model Applications and Case Studies

CONTENTS

10.1 Introduction 466
10.2 Modeling Applications 466
    10.2.1 Building Structures 466
        10.2.1.1 Structural Steel Frames 466
            1. General 466
            2. Quasi-Static Earthquake Loading 467
                a. Model Materials and Fabrication..469
                b. Test Results 470
        10.2.1.2 Earthquake-Resistant Masonry Buildings.....................................473
            1. Walls in Out-of-Plane Bending..473
            2. Masonry Shear Walls 475
            3. Horizontal Joints 476
            4. Masonry Buildings 482
        10.2.1.3 Precast Concrete Large-Panel Buildings 486
            1. General 486
            2. Horizontal Joints 488
            3. Cantilever Wall Component Assembly 488
            4. Three-Dimensional Model 490
    10.2.2 Bridge Structures...497
        10.2.2.1 Zarate-Brazo Largo Highway-Railway Bridge.............................497
        10.2.2.2 Earthquake-Resistant Bridge Piers 502
    10.2.3 Special Structures..507
        10.2.3.1 Prestressed Concrete Reactor Vessels...507
        10.2.3.2 Shell Structures 512
        10.2.3.3 Concrete Slabs with Penetrations 517
        10.2.3.4 Beams Reinforced with Ductile Hybrid FRP (D-H-FRP) 522
            1. Background 522
            2. Results of Beam Tests..524
10.3 Case Studies 529
    10.3.1 Case Study A, TWA Hangar Structures 529
        10.3.1.1 Introduction 529
        10.3.1.2 Wind Effects Model A1 (Phase I) 531
        10.3.1.3 Wind Effects Model A2 (Phase II) 532
        10.3.1.4 Elastic Model A3..534

   10.3.2 Case Study C, Reinforced Concrete Bridge Decks 537
     10.3.2.1 Introduction 537
     10.3.2.2 Static Deck Behavior 538
     10.3.2.3 Fatigue Deck Behavior 540
   10.3.3 Case Study E, Prestressed Wooden Bridges 542
     10.3.3.1 Introduction 542
     10.3.3.2 Results of Long-Term Effects Model E1 545
     10.3.3.3 Results of Ultimate-Strength Model E2 548
       1. Linearly Elastic Response 548
       2. Ultimate-Strength Test 550
   10.3.4 Case Study F, Interlocking Mortarless Block Masonry 556
     10.3.4.1 Introduction 556
     10.3.4.2 Modified H-Block Unit 556
     10.3.4.3 The WHD Block Unit 557
     10.3.4.4 Tests on Assemblages 557
   10.3.5 Case Study G, Pile Caps, a Link between the Foundation and
     the Superstructure 560
     10.3.5.1 Introduction 560
     10.3.5.2 Details of Test Program 563
     10.3.5.3 Test Results and Discussion 564
   10.3.6 Case Study H, Externally/Internally Prestressed Concrete Composite
     Bridge System 568
10.4 Summary 572
Problems 573
References 579

10.1 INTRODUCTION

A large number of successful modeling studies has been reported in the literature in the past few decades. These fall into the categories of educational, research, and design models and are far too numerous to even list here. As an alternative, only a small sample of model applications is examined for the normal structural classifications encountered in civil and architectural engineering construction. A new category of structural models, those used for product development, has also been used successfully and is illustrated in this second edition. Examples of typical design and research studies are illustrated with varying degrees of completeness. In all of these applications, the predominant need is to study behavior nonlinearities, both material and structural. This is an advantage well suited to the physical modeling technique.

Six case studies described briefly in Chapter 1 are discussed in greater detail in this chapter. These consist of two design applications, two research applications, and two product and concept development applications. Educational models including case studies are discussed in detail in Chapter 12.

10.2 MODELING APPLICATIONS

10.2.1 Building Structures

10.2.1.1 Structural Steel Frames

1. General — Studies of steel frame buildings under gravity and lateral loading have been carried out using $1/15$-scale models (Litle and Foster, 1966). These studies established the feasibility of

MODEL APPLICATIONS AND CASE STUDIES 467

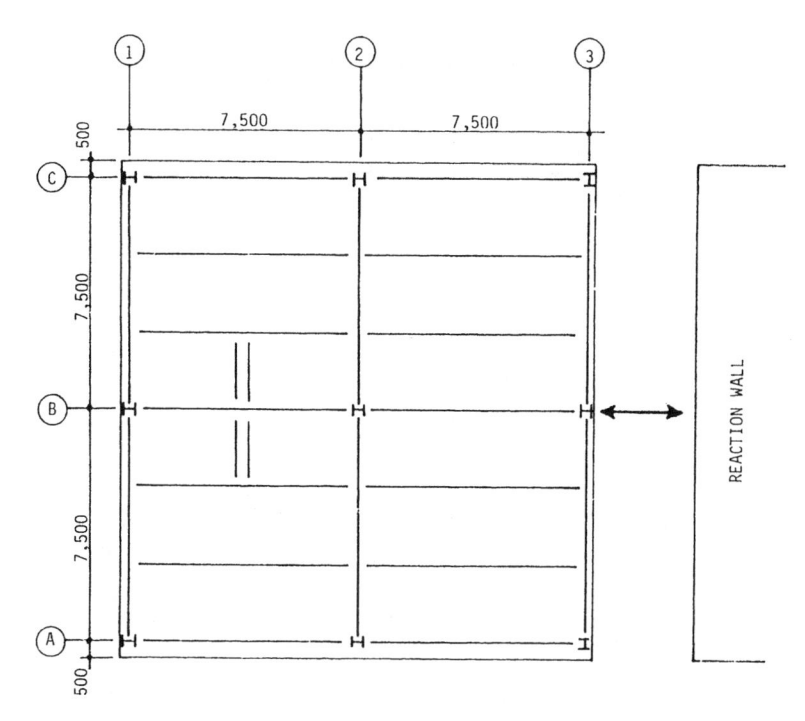

Figure 10.1 Plan view of prototype steel structure. (From Wallace, J. B. and Krawinkler, H., Rept. No. 75, Small-Scale Model Experimentation of Steel Assemblies, U.S.–Japan Research Program, The John A. Blume Earthquake Engineering Center, Department of Civil Engineering, Stanford University.)

using small-scale steel frameworks for design and research purposes. The advent of computer advances in the last decades has been astounding perhaps making the use of such models obsolete for predominantly static loadings. In the case of dynamic loading, especially earthquake loading studies using static or shaking table simulation, as well as for educational purposes, model studies of steel frameworks are still very feasible. Two examples of small-scale model studies of steel frameworks are summarized in this second edition. These both deal with earthquake loading; one in which the loads are applied quasi-statically (Wallace and Krawinkler, 1985) is described below, and the other model study of a steel structure (Mills et al., 1979), tested on a shaking table, is described in Chapter 11.

2. Quasi-Static Earthquake Loading — The prototype test specimen for the steel phase of the U.S.–Japan cooperative research program consisted of a six-story braced and two unbraced steel frames as shown in plan view in Figure 10.1. A major portion of the lateral loads was resisted by the K-braces in the frame on column line B which is shown in Figure 10.2. The elevation of the unbraced frames along column lines A and C is shown in Figure 10.3. Floors were 165 mm thick and consisted of lightweight concrete on 16-gage composite steel deck with corrugations 75 mm deep. Floor beams were of a composite design, using 22-mm-diameter welded studs as shear connectors to the slab. All beams and columns were made of standard wide-flange sections using a material comparable to A36 structural steel. The bracing members consisted of square tubing made of A500 grade B steel.

A coordinated program of testing consisting of full-scale and large-scale models ($S_1 = 3.3$) of components, assemblies, and the complete structure was carried out by various institutions in Japan and the U.S. A 0.3-scale model of the six-story concentrically braced steel structure (Uang and Bertero, 1986) and a 0.3-scale model of the six-story eccentrically braced steel structure (Whittaker et al., 1987) were both tested on the University of California, Berkeley shaking table. The focus

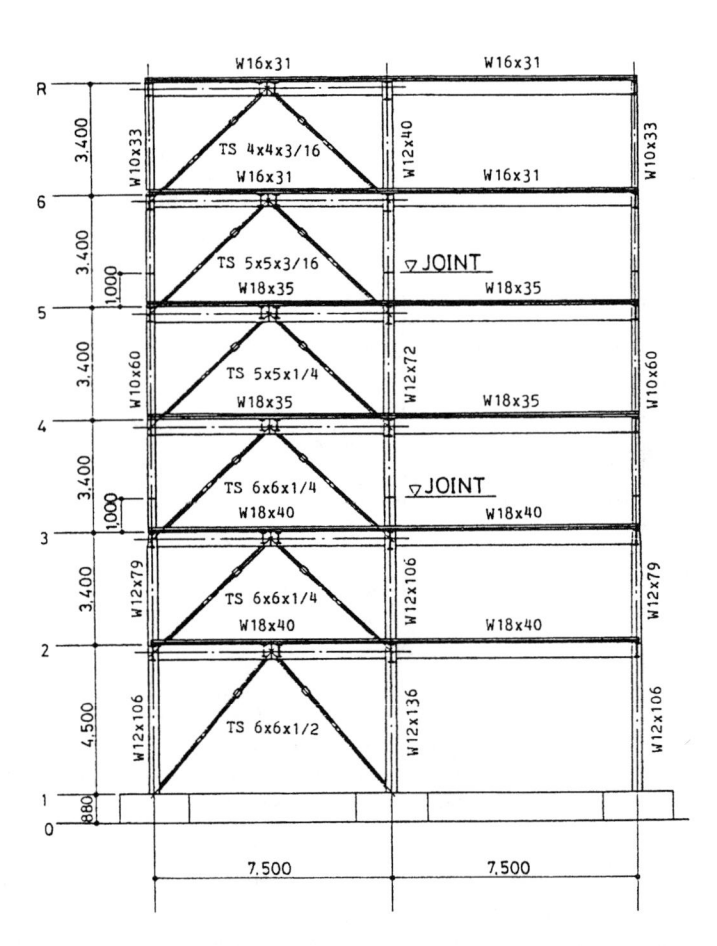

Figure 10.2 Elevation of braced frame unit at line B. (From Wallace, J. B. and Krawinkler, H., Rept. No. 75, Small-Scale Model Experimentation of Steel Assemblies, U.S.–Japan Research Program, The John A. Blume Earthquake Engineering Center, Department of Civil Engineering, Stanford University.)

of this section is on the results of a ¹⁄12.5-scale model study conducted at Stanford University of beam–column assemblies and the central braced frame of the prototype structure (Wallace and Krawinkler, 1985). The objectives of this study were (1) to assess the feasibility and limitations of scale model testing in earthquake engineering research, (2) to study the simulation accuracy of specific failure modes in small-scale models, (3) to correlate results of tests at different scales to assess prototype response prediction from experimental studies, and (4) to study the seismic behavior of components and assemblies of braced frame structures.

In order to fulfill these objectives, all model test specimens were made to be "exact" replicas of the prototypes tested in Japan or of full-scale components tested by others. In the small-scale experiments, the geometric and material properties as well as the loading histories of the reference tests are simulated as closely as practical. Tests were conducted on small-scale specimens of beam–column assemblies, a braced frame, and an unbraced frame. All of these specimens included floor slabs which consisted of model concrete reinforced with wire mesh and were cast onto metal deck. Shear studs welded to the beams created composite structural action. The beam–column assemblies (Figure 10.4) represent portions of the frame on the center column line (see Figure 10.2) and on the exterior frames (see Figure 10.3).

The braced frame specimen (Figure 10.5) represents a model of the center unit of the concentrically braced prototype test structure including the floor slabs. The unbraced frame specimen is

MODEL APPLICATIONS AND CASE STUDIES 469

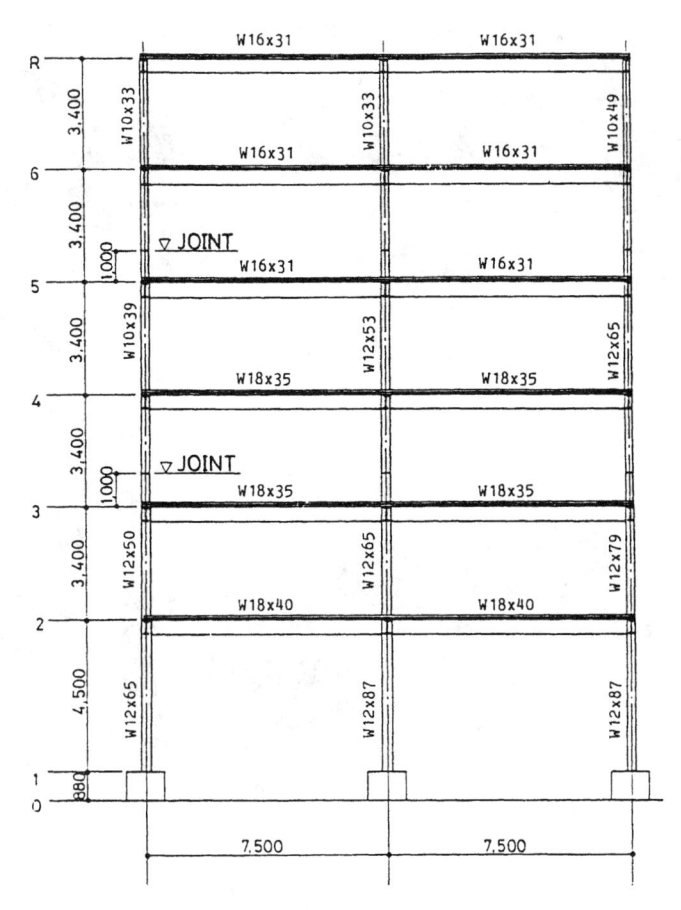

Figure 10.3 Elevation of unbraced fames at lines A and C. (From Wallace, J. B. and Krawinkler, H., Rept. No. 75, Small-Scale Model Experimentation of Steel Assemblies, U.S.–Japan Research Program, The John A. Blume Earthquake Engineering Center, Department of Civil Engineering, Stanford University.)

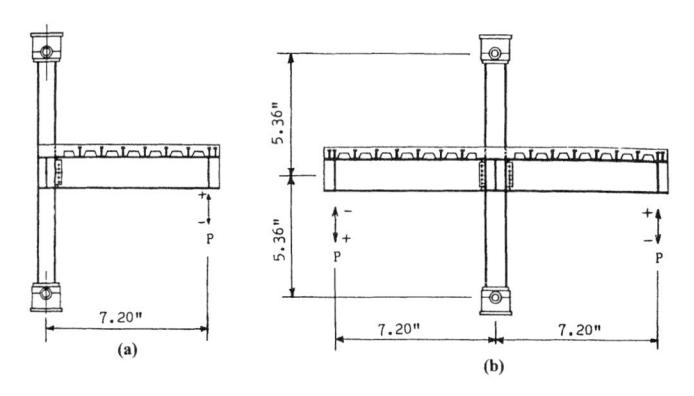

Figure 10.4 Exterior and interior model connections.

the braced frame specimen with the braces removed. This specimen was tested to give information about the resistance of the moment frame in the braced frame specimen.

a. **Model Materials and Fabrication** — Wide-flange steel shapes used in the beam–column assemblies, and braced frame specimen were milled from a single piece of A36 steel plate.

Figure 10.5 Braced frame specimen.

Measurements of the model beams converted to the prototype domain are compared to the theoretical and shown in Table 10.1. Tensile coupons cut from flanges and webs of beams had an average yield strength of 306.0 MPa and an ultimate strength 477.0 MPa. Typical stress–strain characteristics of the model beam and column material are shown in Figure 5.23. The $1/12.5$-scale model studs were custom made by a manufacturer of welded studs and welded to the beams with a capacitive discharge stud-welding machine. Square steel tubing used for the bracing members was fabricated from round ANSI-4130 steel tubing. Composite metal deck with a depth of 75 mm and a thickness of 16 gage was used in all floor slabs of the prototype structure. Models of this deck were manufactured of mild steel shim stock. Corrugations were formed in the shim stock by a custom-made die in a press brake. The steel-braced model with the steel deck prior to casting of the concrete slab is shown in Figure 5.22.

b. Test Results — Typical results of the beam–column assemblies tested at full scale (Lee and Lu, 1984) and the $1/12.5$-scale models are presented. The loading (applied deflection) history used in model and prototype tests consisted of groups of three equal cycles, with each group of a larger amplitude than the previous groups (Figure 10.6). The loading history shown in Figure 10.6 was applied simultaneously to both beams of the interior joint assembly IJ-FC specimen (Figure 10.4b), with one beam forced upward while the other beam was forced downward. Almost all of the inelastic deflection of the specimen IJ-FC was due to distortion of the column web panel zone. The test results, along with the results from the prototype tests are shown in Figure 10.7. The close similarity of the model and prototype results is clearly apparent in this figure. In both the model and prototype, failure occurred at very large inelastic deformations in the joint.

The $1/12.5$-scale model of the braced frame along column line B (see Figure 10.1) consists of the steel framing and portion of the floor slab extending to the nearest floor beams which are located

MODEL APPLICATIONS AND CASE STUDIES 471

Table 10.1 Section and Material Properties of Assembly Specimens

Assembly	Member	d, mm	t_w, mm	b_f, mm	t_f, mm	F_{yweb}, MPa	$F_{yflange}$, MPa
Model EJ-FC	W10×60	260.6	12.70	256.8	17.78	306.1	306.1
Proto. EJ-FC	W10×60	259.8	10.67	256.0	16.74	236.5	244.1
Model EJ-FC	W18×35	450.9	10.47	152.4	11.13	306.1	306.1
Proto. EJ-FC	W18×35	450.3	7.85	150.9	10.72	253.0	260.6
Model EJ-WC	W12×65	308.1	8.89	304.3	15.88	306.1	306.1
Proto. EJ-WC	W12×65	308.1	10.67	306.6	14.83	251.0	270.3
Model EJ-WC	W18×35	450.9	9.20	152.4	11.13	306.1	306.1
Proto. EJ-WC	W18×35	450.3	7.85	150.9	10.72	253.0	260.6
Model IJ-FC	W12×72	308.9	9.83	305.6	17.78	306.1	306.1
Proto. IJ-FC	W12×72	310.9	10.97	305.3	16.82	242.0	262.0
Model IJ-FC	W18×35	450.9	10.16	152.4	10.47	306.1	306.1
Proto. IJ-FC	W18×35	450.3	7.85	150.9	10.72	253.0	260.6

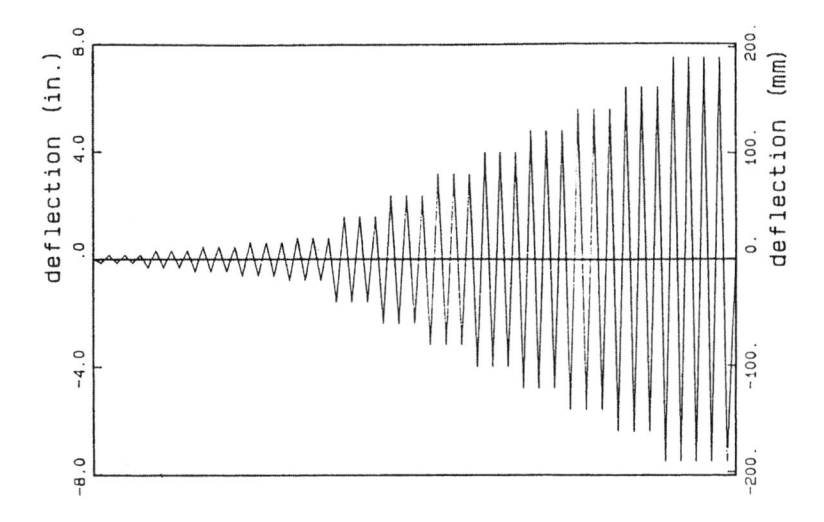

Figure 10.6 Loading history for IJ-FC specimen.

2.5 m from the frame centerline. Basic dimensions of the model specimen are shown in Figure 5.21 and the test setup is shown in Figure 10.5. The main purpose of the model study was to correlate test results at different scales with the prototype as the primary reference. In order to subject the model structure to similar damage as the prototype, the roof displacements of the elastic, moderate, and final pseudodynamic tests (Chapter 8), to which the prototype test structure was subjected, were simulated in the loading program for the model tests. The final loading roof displacement history is shown in Figure 10.8. The results of the final loading test are illustrated in Figure 10.9 which shows very good correlation between the model and prototype hysteretic behavior.

The major conclusion from this study (Wallace and Krawinkler, 1985), is that small-scale models can give good predictions of global response and member deformation modes. Predictions of localized failure modes at connections are harder to simulate because of fabrication and material limitations. Based on their results, the following major conclusions on small-scale modeling and structural behavior of steel frame structures can be drawn (paraphrasing the authors):

1. Overall cyclic inelastic load–deformation response of components and structures can be reproduced adequately in tests of carefully designed and detailed small-scale models.
2. Inelastic panel zone distortion and beam flexure can be properly simulated at small scales. In addition, composite action between steel beams and concrete floor slabs can be simulated with good accuracy.

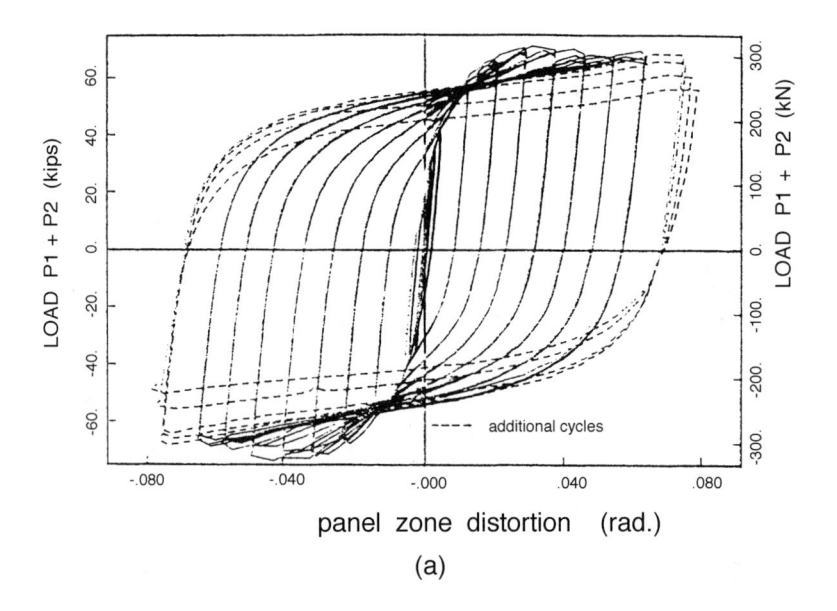

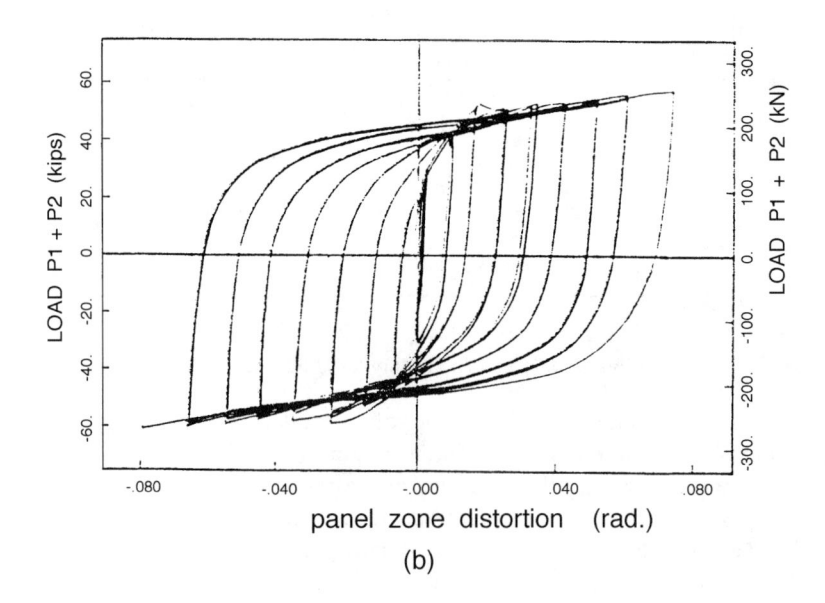

Figure 10.7 Shear distortion of panel zones, IJ-FC specimens. (a) Model specimen. (b) Prototype specimen.

3. Postbuckling response of complex bracing systems can be reproduced in small-scale model tests. However, inelastic buckling strength is sensitive to material properties and connection details which become difficult to simulate at small model scales.
4. Size effects in small-scale model tests under cyclic inelastic loading will usually result in high predictions of strength (because of strain gradient effects) and ductility (because of improved fracture properties). These factors and other fabrication difficulties limit the use of small-scale models for predicting failure modes in structural connections.

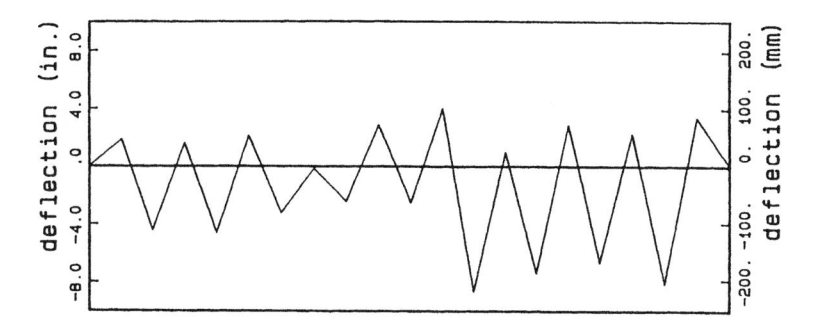

Figure 10.8 Roof displacement history, final test.

The small-scale model tests led to the same major conclusions on structural behavior as the prototype tests. The following must be emphasized:
  a. Panel zone shear strength may limit the capacity of steel beam–column joints. Inelastic shear deformations of the panel zone are usually very ductile, absorbing considerable energy before failure occurs. Excessive shear distortions in joint panel zones may cause fracture of the beam flange welds due to the high curvatures at the joint corners.
  b. Composite floor slabs increase the strength of beams and joints in steel frames. At beam–column joints this increase in strength is limited by the bearing capacity of the concrete that transfers the slab stresses to the column. Under severe deformations, the composite floor action is seen to deteriorate as the concrete surrounding the column is crushed.
  c. The strength of "K" braces in a braced frame deteriorates rapidly after first buckling. This will cause the formation of a "soft" story which may then be subjected to excessive drifts.
  d. Tubular braces are susceptible to severe local buckling after brace buckling. Under load reversals, this local buckling may lead to crack initiation and fracture of braces.

10.2.1.2 Earthquake-Resistant Masonry Buildings

An explosion of masonry research has taken place since the first edition of this book. The modeling technique, in its infancy at the writing of the first edition, has matured to a level of high interest and application in all phases of masonry construction but especially in earthquake-resistant design. Studies of masonry structures (reinforced, unreinforced, concrete block, stone, and clay units) too numerous to list here have been conducted all over the world in the last 15 years. A mere sampling is discussed in this and Chapter 11, much from the authors' own experience.

1. Walls in Out-of-Plane Bending — One of the earliest in-depth studies of masonry wall structures using direct scale models was reported by Abboud (1987) using the ¼-scale blocks shown in Figure 4.39. A total of thirteen ¼-scale models of reinforced block masonry walls was tested under out-of-plane monotonic and cyclic reversed lateral loads with the variables listed in Table 10.2. In order to achieve the desired slenderness ratio (H/t) of 32 and 40 and maintain a wall thickness of 49 mm, the model masonry walls were built three units long (302 mm) and either 30 or 37 units high (1524 mm and 1880 mm, respectively). Type S model mortar and normal strength model grout (Chapter 4) were used in the construction of the model walls (Figure 10.10a). The model walls were fully grouted and reinforced with laboratory deformed No. 11 gage reinforcing bars whose deformation and stress–strain characteristics are given in Chapter 5.

The main objectives of the study were the investigation of the postyield characteristics and the hysteretic behavior of the slender walls. Comparisons to companion reinforced concrete walls were

(a) Model Specimen

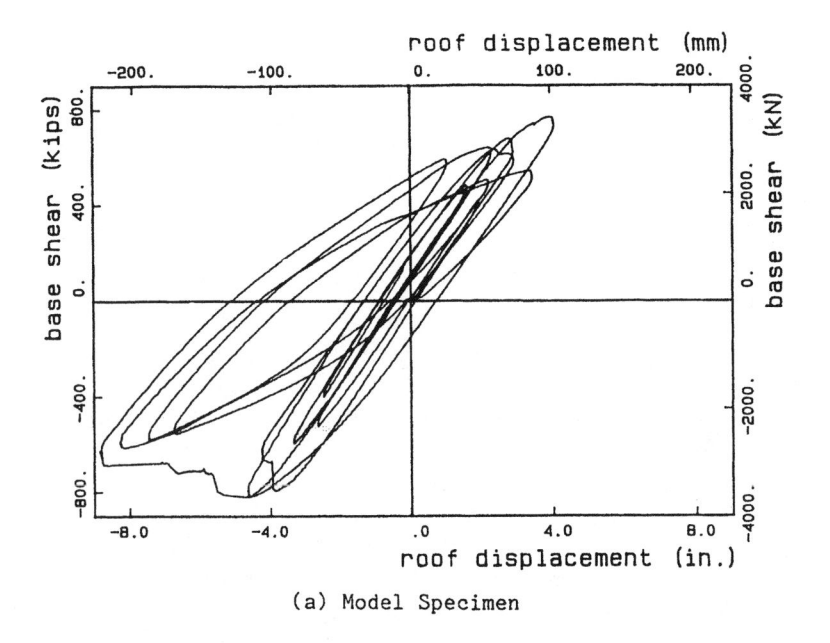

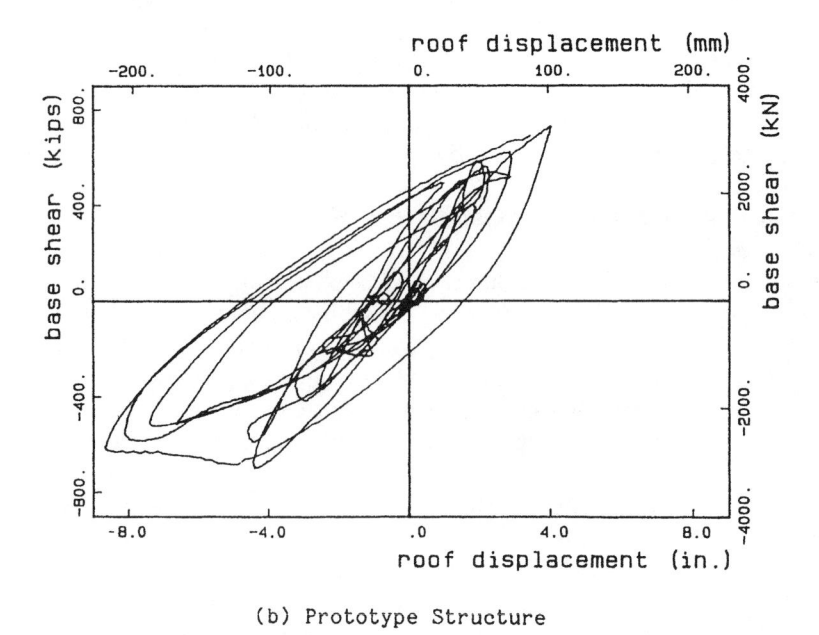

(b) Prototype Structure

Figure 10.9 Base shear vs. roof displacement, final test.

also made. Of primary concern was the use of the test results to develop analytical models for determining deflection, strength, and ductility and the evaluation of design code recommendations. The test setup for the ¼-scale slender walls is shown in Figure 10.10b. Two symmetrically placed line loads were applied by the displacement-controlled actuator giving a middle region of constant moment. For the cyclically loaded specimens the loading history consisting of three cycles of steadily increasing displacement (Figure 10.10c) was applied.

MODEL APPLICATIONS AND CASE STUDIES

Table 10.2 Summary of Model Wall Tests

Type of Specimen	Wall Group Designation	Number of Specimens	Type of Loading	H/t[a] Ratio	Reinforcement Ratio, ρ_g,[b] %	Level of Axial Load, N
Reinforced	W1	3	Monotonically	32	0.27	—
Concrete	W2	2	increasing	40	0.27	—
Block	W3	2	out-of-plane	32	0.27	356
Masonry	W4	2	loads	32	0.27	956
Wall	W5	1		40	0.27	956
	W6	2	Cyclically	32	0.27	—
	W7	1	reversed out-of-plane loads	32	0.27	956
Reinforced Concrete Wall	WCON	1	Monotonically increasing out-of-plane loads	32	0.27	—

[a] Height-to-thickness ratio, based on actual wall dimensions.
[b] Based on cross-sectional area, where $\rho_g = A_s/bt$.

Typical results from this study indicated for the first time (prior to full-scale tests under the U.S.–Japan Coordinated Program for Masonry Building Research) the unique pinched hysteretic behavior of centrally reinforced block masonry walls (Figure 10.11). From these tests, analytical models were developed for predicting the inelastic behavior of block masonry walls. For the range of parameters investigated, the test results (Figure 10.12) showed no basic deterioration in the load–deflection characteristics caused by cyclic loading (Abboud et al. 1990).

2. Masonry Shear Walls — In the design of masonry structures, one of the most important considerations is their ability to withstand lateral loads. This depends mainly on the behavior of their shear walls. For relatively high-aspect-ratio shear walls, it has been shown that the behavior is governed by flexure and their flexural strength can be accurately predicted by ordinary beam theory. In the case of low-aspect-ratio shear walls, shear deformations govern the behavior. In a study (Larbi, 1989; Larbi and Harris, 1990, 1990a), tests were conducted on a series of nine low-rise ⅓-scale shear walls to determine the effect of the amount of vertical and horizontal reinforcement on the shear and flexural strength.

The model shear walls (Table 10.3) were 0.61 × 0.61 m ⅓-scale replicas of full-scale walls tested by Shing et al. (1988; 1989) and are shown in Figure 10.13. The model blocks used were the ⅓-scale units shown in Figure 4.44. The walls were of single-wythe fully grouted construction with laboratory-deformed reinforcing bars whose properties are given in Table 5.12.

The test setup (Figure 7.20b) with a failed specimen in place is shown in Figure 10.14. The loading history, duplicating the loading to which the full-scale walls were subjected, is shown in Figure 10.15. This loading conforms with the sequential phased displacement (SPD) recommended by the Technical Coordinating Committee for Masonry Research (TCCMAR) committee (Porter, 1987). Typical results of the final cracking patterns are shown in Figure 10.16. The results of the model wall tests are in good agreement with those of the prototype walls. All conclusions on the behavior of low-aspect-ratio block masonry shear walls could be predicted from the tests on ⅓-scale models. This is borne out in the correlation of model and prototype results shown in Table 10.4. The following was concluded from this study:

1. Flexural and shear strengths increased with the amount of vertical reinforcement.
2. Increasing the vertical reinforcement ratio favored the shear failure, and therefore, decreased the wall ductility.

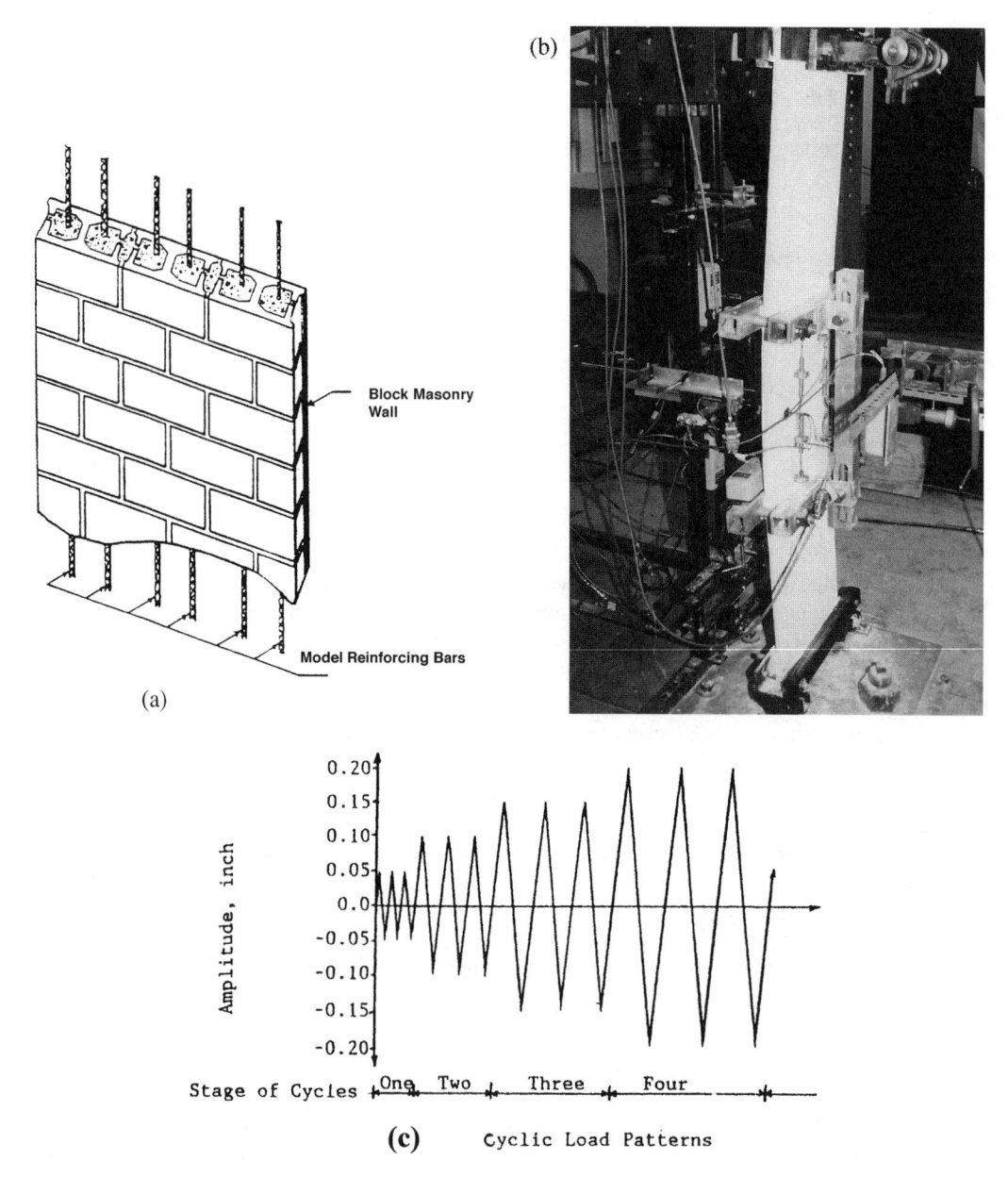

Figure 10.10 One-fourth-scale masonry wall tests. (a) Model walls; (b) test setup; (c) cyclic load patterns.

3. Increasing the horizontal reinforcement changed the wall behavior from a shear (brittle) failure to a flexural (more ductile) failure. Hence, with proper amount of horizontal reinforcement, the diagonal tensile failure (brittle) can be inhibited and consequently an improved behavior would result.
4. Flexural strength can be predicted with a reasonable accuracy if the masonry and steel mechanical properties are accurately modeled.

3. Horizontal Joints — Masonry buildings which are basically bearing wall structures have rigid precast concrete floor and roof systems usually with a cast-in-place concrete topping to enable the whole to work as a diaphragm. A critical connection in such structures that must preserve both the

MODEL APPLICATIONS AND CASE STUDIES

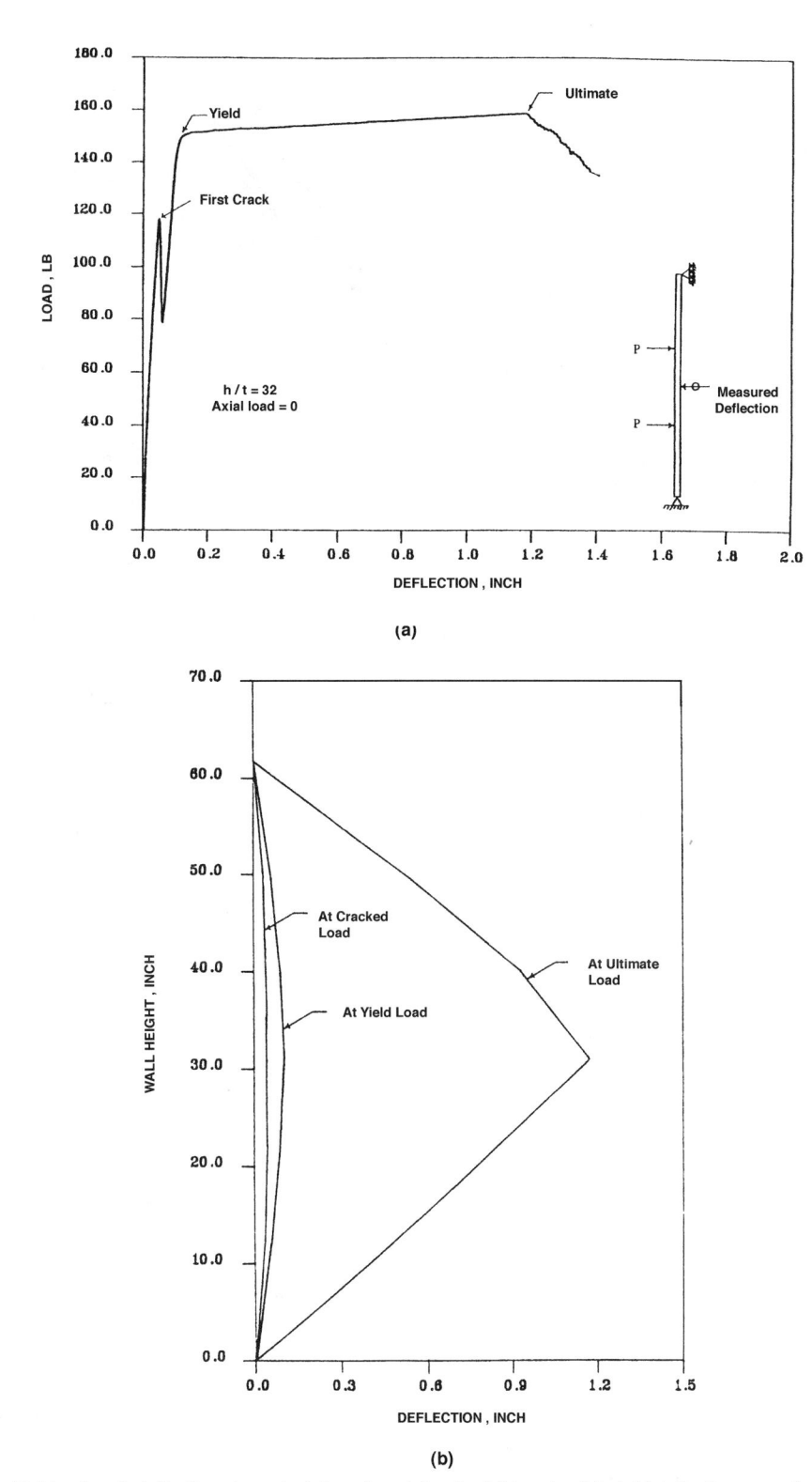

Figure 10.11 Load–deflection characteristics of model walls. (a) Load–midheight deflection curve for wall W1B. (b) Deflected shapes for wall W1B. (c) Hysteresis load–midheight deflection curves for wall W7A, Cycle 1-24.

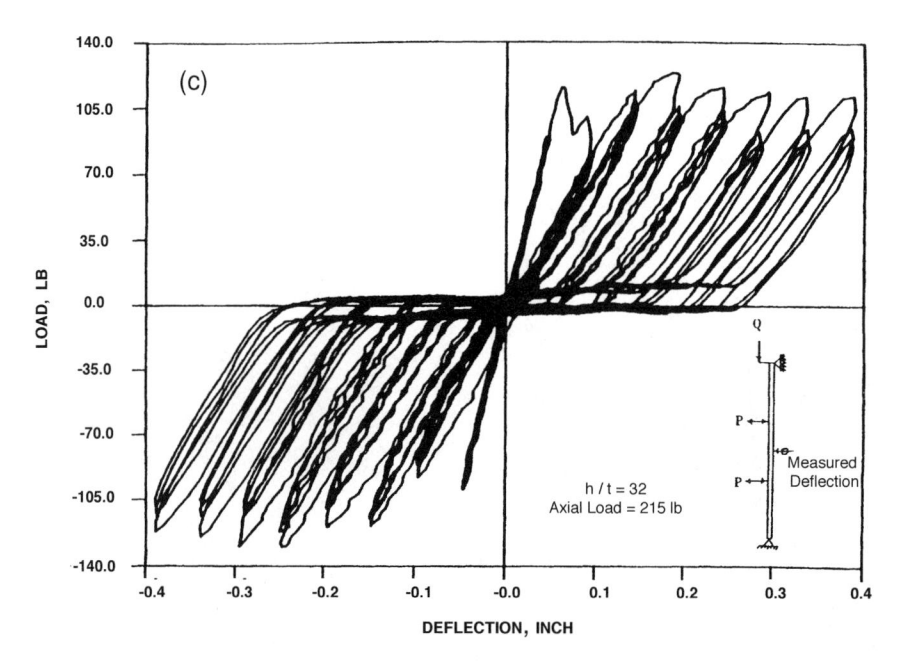

Figure 10.11 (continued)

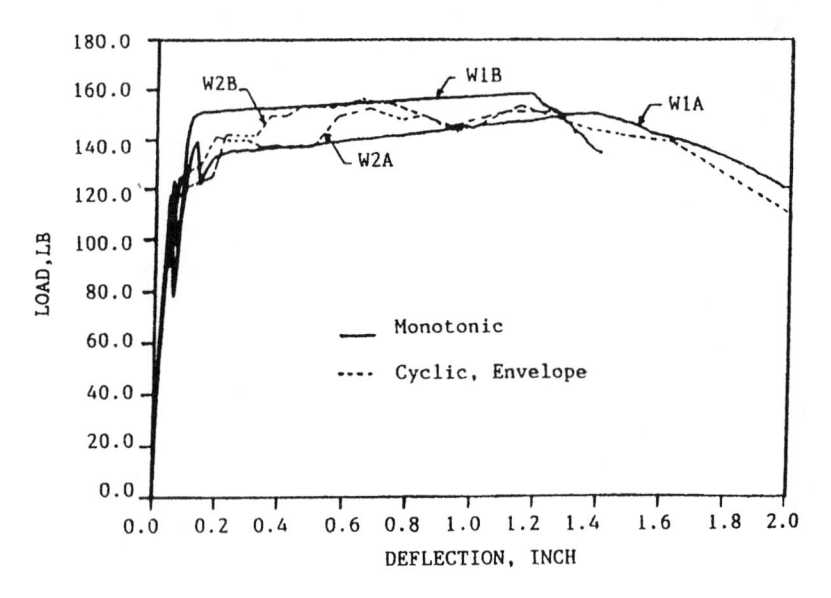

Figure 10.12 Comparison of the load–deflection envelope relationships under monotonic and cyclic loads.

vertical and horizontal continuity is the so-called horizontal joint between the wall and floor elements. Such exterior and interior horizontal joints are illustrated in heavy shading in Figure 10.17. A test program to study the cyclic behavior of horizontal joints of block masonry wall and precast concrete floor construction was carried out (Oh, 1989, Oh and Harris, 1990; Harris and Oh, 1990) using the two types of specimens shown in Figure 10.18. The test matrix for this study is summarized in Table 10.5.

The model blocks used were the ⅓-scale units shown in Figure 4.44. Type S mortar and a coarse ⅓-scale grout were used as given in Chapter 4. Hollow-core floor plank units were ⅓-scale replicas of those used in the prototype joints (Anvar et al., 1983). Grade 40 prototype reinforcing

MODEL APPLICATIONS AND CASE STUDIES

Table 10.3 Test Matrix

Wall Designation	Vertical Reinforcement	Horizontal Reinforcement	Axial Stress, MPa	Loading
Model SW1 A, B, D	5 #5 (0.33%)	5 #3 (0.13%)	1.86	Monotonic
Model SW1 C	5 #5 (0.33%)	5 #3 (0.13%)		Cyclic
Prototype W9	0.38%	0.14%	1.86	Cyclic
Model SW2 A, B	5 #5 (0.33%)	5 #4 (0.23%)	1.86	Monotonic
Model SW2 C	5 #5 (0.33%)	5 #4 (0.23%)		Unsuccessful
Model W1	0.38%	0.24%	1.37	Cyclic
Model SW3 A, B	5 #6 (0.55%)	5 #3 (0.13%)	1.86	Monotonic
Model SW3 C	5 #6 (0.55%)	5 #3 (0.13%)		Cyclic
Prototype W14	0.54%	0.14%	1.86	Cyclic

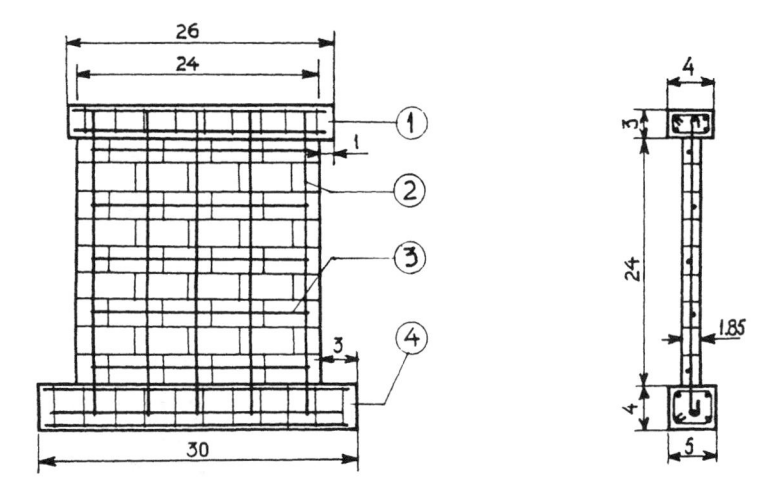

Figure 10.13 Typical masonry wall panel. (1) Top reinforced concrete beam; (2) vertical reinforcement; (3) horizontal reinforcement; (4) reinforced concrete footing. All dimensions in inches.

Figure 10.14 Model shear wall.

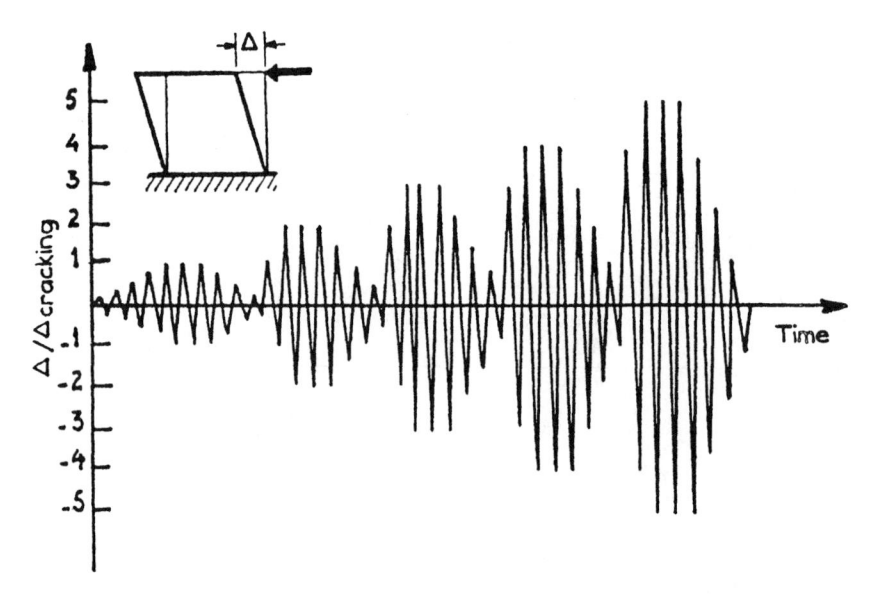

Figure 10.15 Displacement history for the cyclic tests.

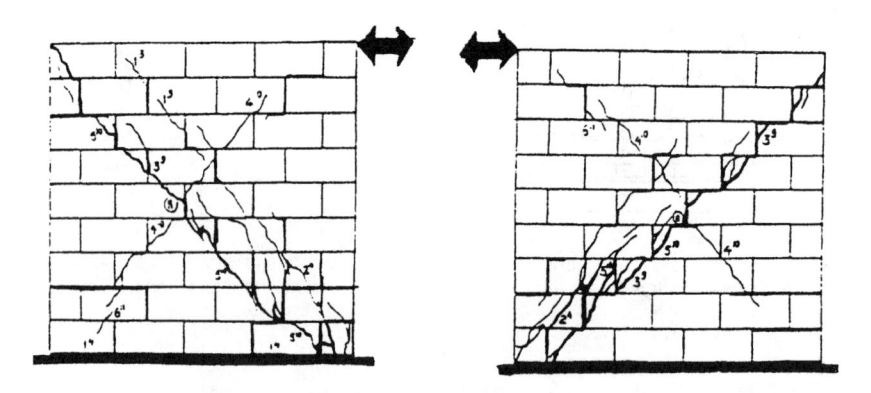

Figure 10.16 Crack pattern of failed specimen SW3C.

Table 10.4 Model and Prototype Wall Strength Properties

Specimen	ρ_v, %	ρ_h, %	Axial Stress, MPa	Yield Stress, MPa	Diagonal Cracking Strength, MPa	Ratio Model/ Prototype	Ultimate Strength, MPa	Ratio Model/ Prototype	Mode of Failure
Model SW1	0.33	0.13	1.86	1.06	1.45	0.93	1.73	1.06	Shear
Prototype W9	0.38	0.14	1.86	1.29	1.56		1.63		Shear
Model SW2	0.33	0.23	1.86	1.06	1.51	1.02	1.55	1.11	Flexure
Prototype W1	0.38	0.24	1.37	1.02	1.47		1.40		Flexure
Model SW3	0.55	0.13	1.86	1.33	1.61	0.93	1.95	1.09	Shear
Prototype W14	0.54	0.14	1.86	1.44	1.73		1.78		Shear

MODEL APPLICATIONS AND CASE STUDIES 481

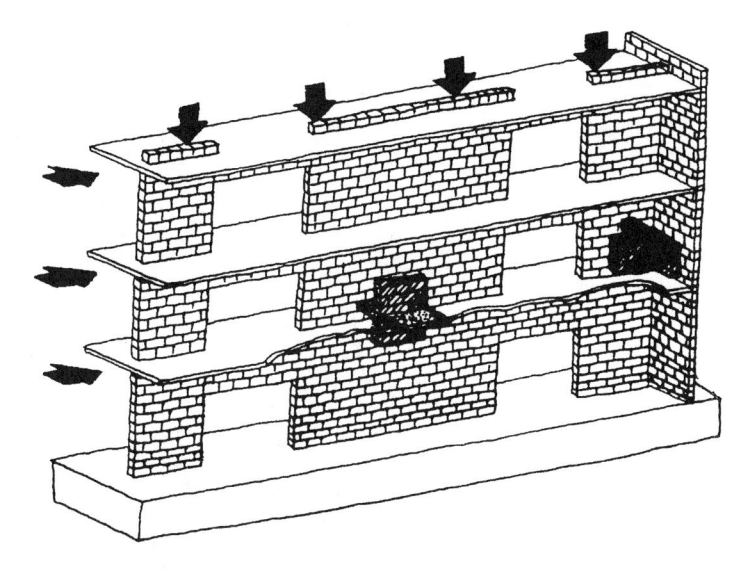

Figure 10.17 Floor-to-wall and wall-to-wall components of typical masonry shear wall.

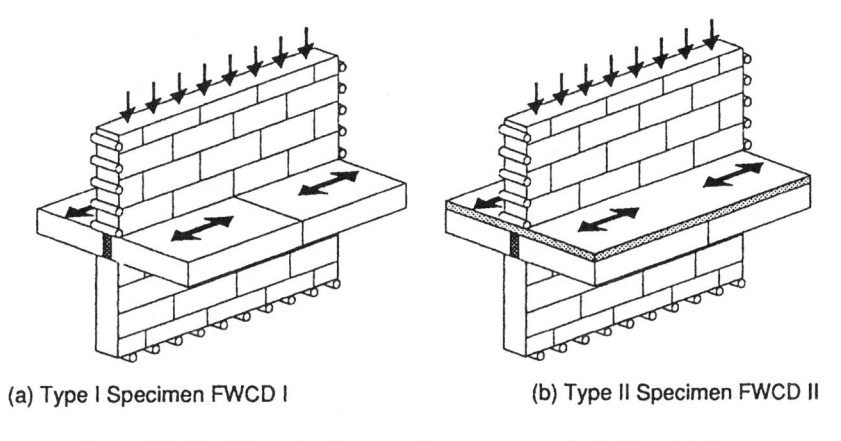

(a) Type I Specimen FWCD I (b) Type II Specimen FWCD II

Figure 10.18 Schematic diagram of specimens under inplane loading.

Table 10.5 Test Matrix

Type	Loading	Specimen	Interface	Remark
Type I	Cyclic	FWCD I-S1	Upper	Pilot test
	Cyclic	FWCD I-S2	Upper	Specimen damaged
	Cyclic	FWCD I-S3	Upper	
	Cyclic	FWCD I-S4	Lower	
	Cyclic	FWCD I-S5	Upper	
	Monotonic	FWCD I-S6	Upper	
Type II	Cyclic	FWCD II-S1	Upper	
	Cyclic	FWCD II-S2	Upper	
	Monotonic	FWCD II-S3	Upper	
	Monotonic	FWCD II-S4	Upper	

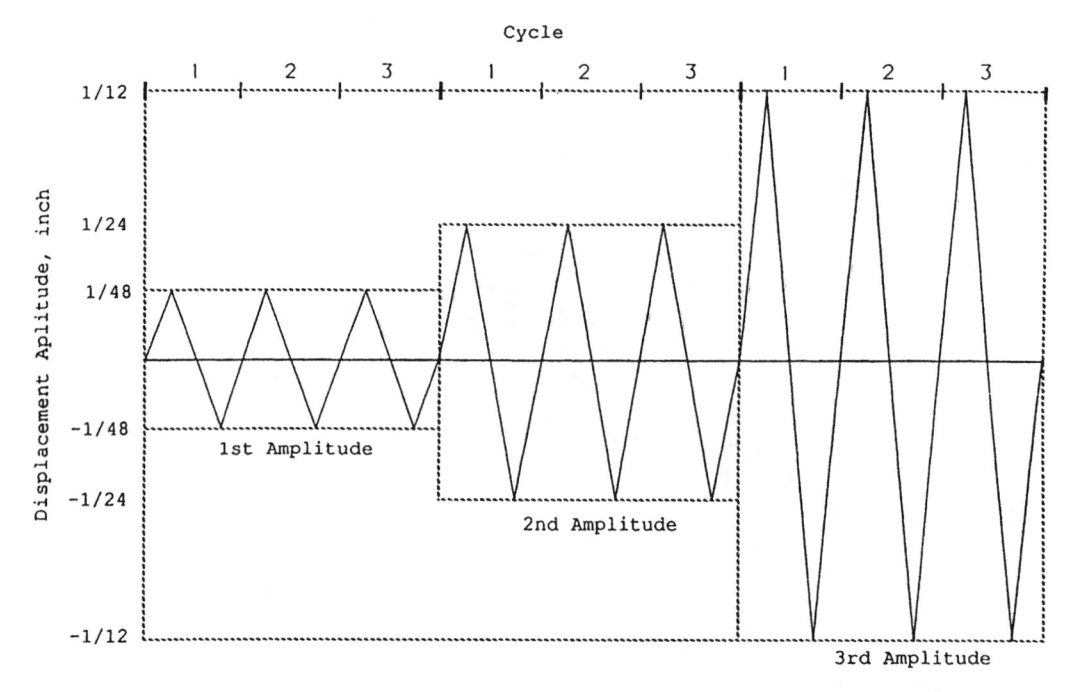

Figure 10.19 Displacement control loading history.

bars of size No. 3, No. 4, and No. 5 were modeled at ⅓ scale using the Drexel University wire-deforming machine (Figure 5.45).

The model joints were subjected to displacement-controlled shear forces applied according to the loading history shown in Figure 10.19. Each individual specimen was subjected to six complete tests with six different levels of precompression: 345 to 1379 kPa. After the first complete test with 345 kPa precompression, the specimen was already cracked and the concrete surrounding the reinforcement was already fully deteriorated. However, the five additional tests were carried out on the same specimen to investigate the effect of precompression on the connection interface. Typical results from tests of the Type I joint are shown in Figure 10.20. The effect of precompression on the shear force from model and prototype horizontal connection tests is compared in Figure 10.21. As can be seen from this figure, the ⅓-scale model connections can predict the behavior of the prototype very adequately. Specific observations made from the model tests can be listed as follows:

1. The microslip behavior observed in prototype tests could be observed in all the model specimens tested and was identified by a very sudden drop in stiffness.
2. Under the initial precompression of 345 kPa, the hysteretic behavior of the joint showed appreciable force and stiffness degradation which was identical in model and prototype; at all subsequent precompression values, the hysteresis loops stabilized (see Figure 10.20).
3. Comparison of cyclic and monotonic failure envelopes showed a decrease in strength due to cyclic loading.

4. Masonry Buildings — Reinforced concrete masonry models at ¼ scale have been tested at the University of Illinois (Paulson and Abrams, 1990) under static earthquake simulation and on a shaking table. The dynamic tests (Abrams and Paulson, 1990; 1991) will be described in Chapter 11. The three-story test model (Figures 10.22 and 10.23) consisted of two perforated flanged shear walls tied to rigid reinforced concrete floor slabs in a doubly symmetric arrangement. The model

MODEL APPLICATIONS AND CASE STUDIES 483

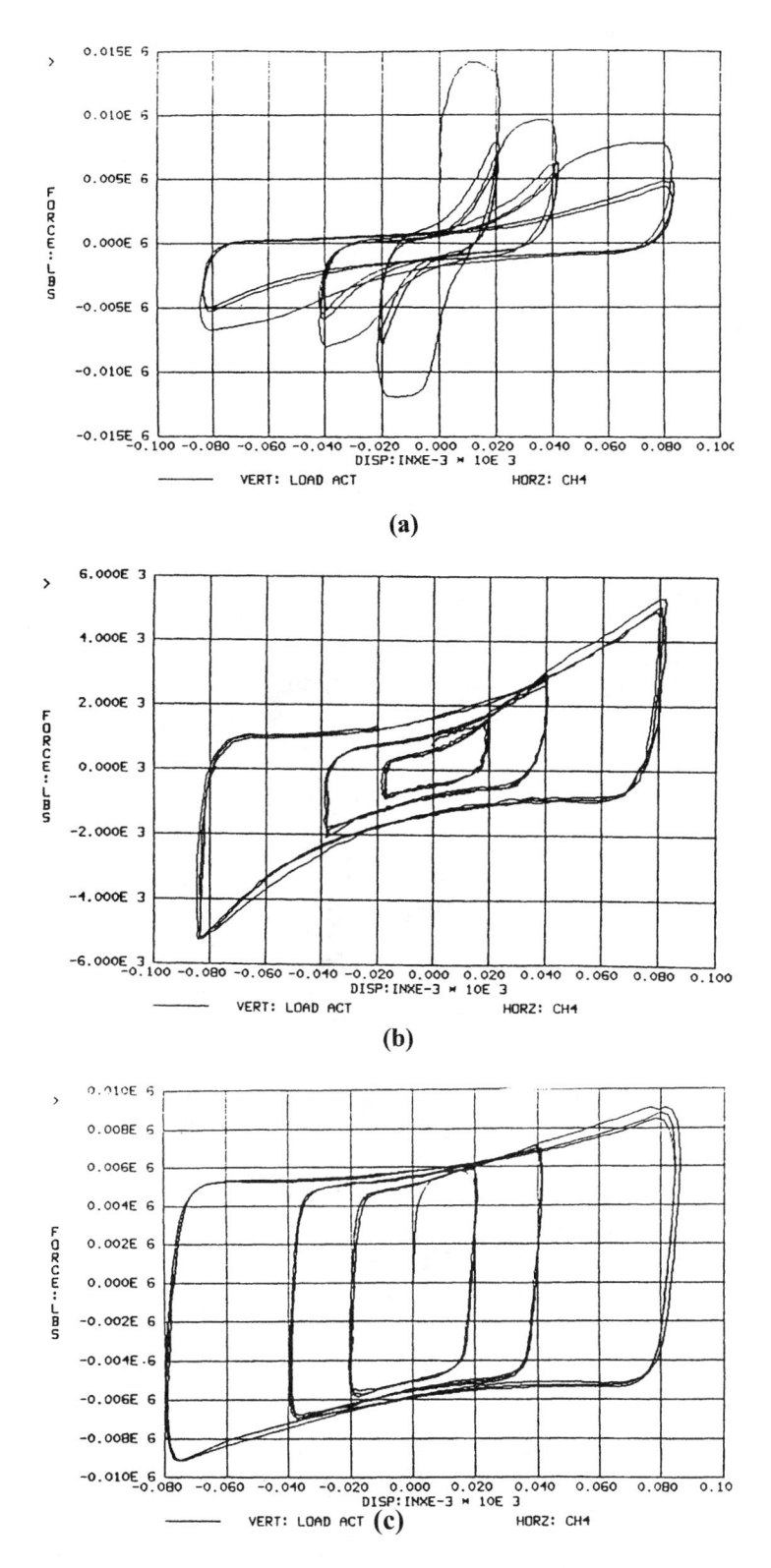

Figure 10.20 Shear force vs. relative displacement curves for specimen FWCD I-S5. (a) RUN 5A, 50 psi precompression; (b) RUN 5C, 75 psi precompression; (c) RUN 5F, 200 psi precompression.

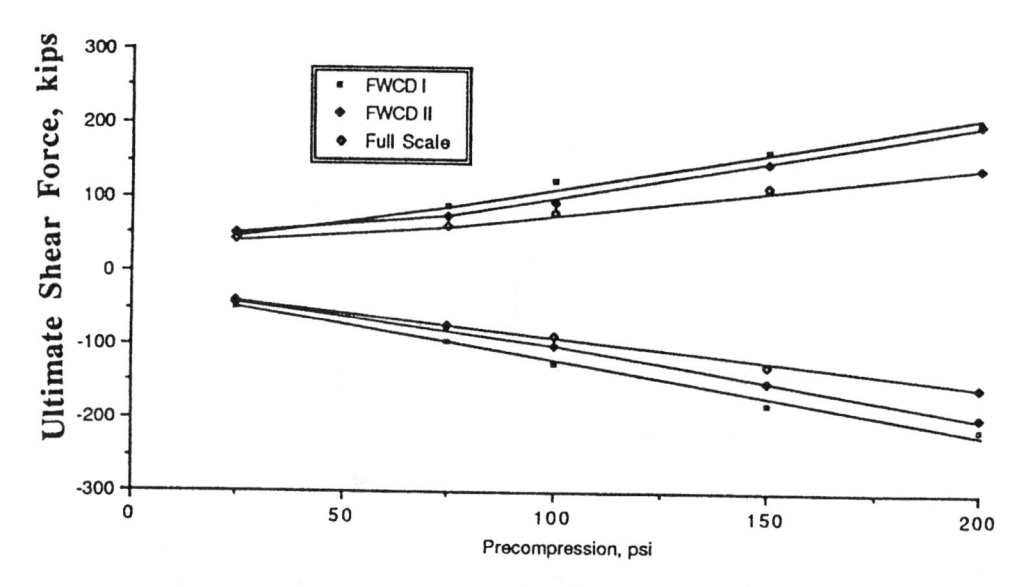

Figure 10.21 Effect of precompression on connection response for the second displacement amplitude.

Figure 10.22 One-quarter-scale reinforced concrete masonry test structure. (Courtesy of Prof. D. Abrams, University of Illinois, Urbana-Champaign.)

specimen was designed so that yielding of the vertical reinforcement would be followed by crushing of the masonry (flexural failure modes). This required sufficient horizontal reinforcement (Figure 10.24) to preclude shear failure of the piers between openings. In addition, the model vertical reinforcement chosen, which was a smooth No. 11 gage wire (Figure 10.25), ran continuously from the base anchorage to the top. The ¼-scale model masonry grouted masonry construction used (Figure 10.26) followed basic modeling techniques described in Chapters 4 and 6.

MODEL APPLICATIONS AND CASE STUDIES

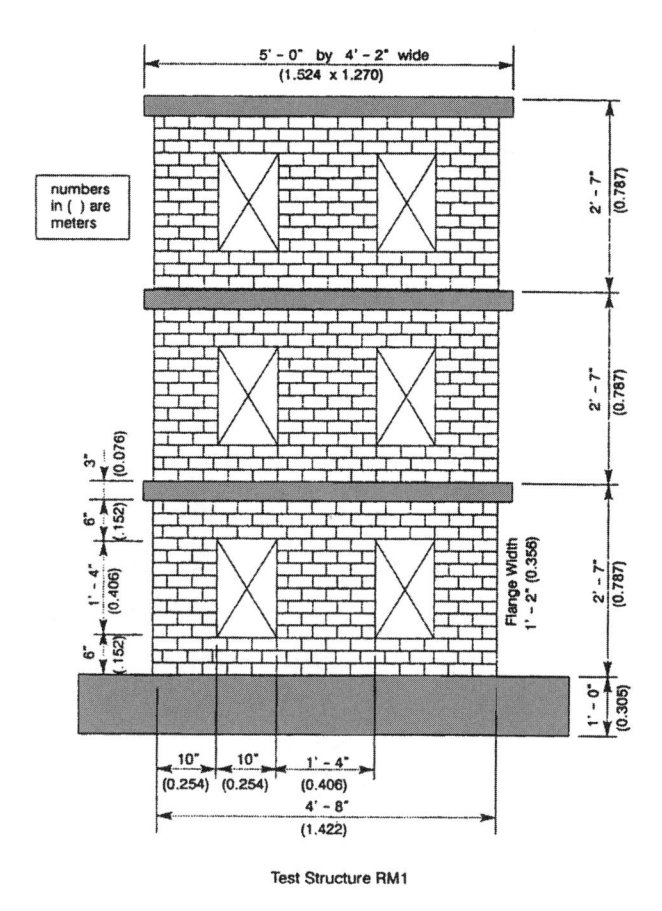

Figure 10.23 Structural configuration of test structures.

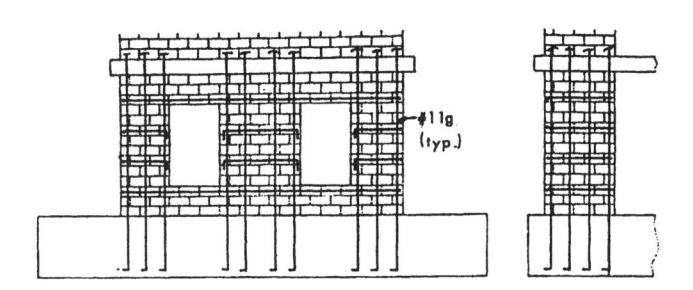

Figure 10.24 Reinforcement layout (typical for each story).

The static test setup is shown in Figure 10.27. A separate controller was used with each of the 111 kN actuators attached to each floor. The top actuator was operated in displacement control using the displacement records of a companion specimen tested on the shaking table. Actuators at the first and second levels were operated in force control such that an inverted triangular force distribution would result. A computerized loading system was developed for the static tests which gave essentially the same top floor displacement, base shear, and base moment as in the dynamic tests.

Base moment vs. top-level deflections were plotted for three of the most severe test runs. According to the authors: "Marked reductions in both strength and stiffness were observed when

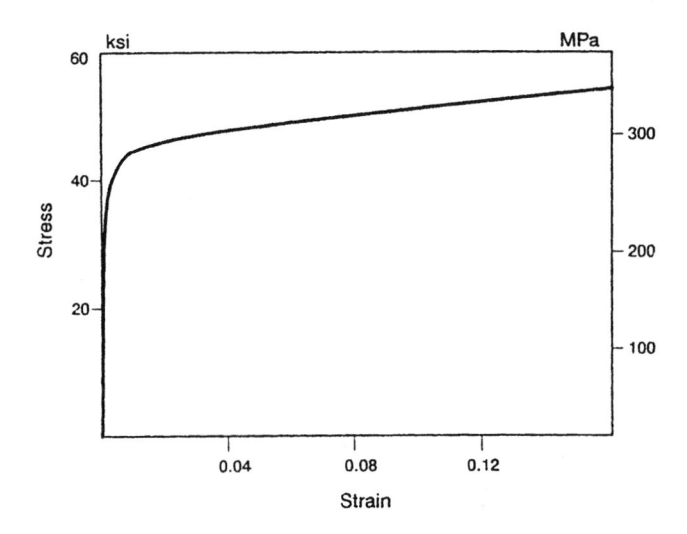

Figure 10.25 Typical measured stress–strain curve for model reinforcement.

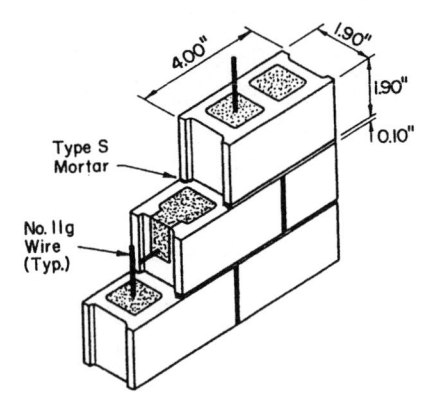

Figure 10.26 Typical construction of reduced-scale masonry (1 in. = 25.4 mm).

the specimen was loaded statically rather than dynamically. Significant differences in stiffness were observed for the test runs that resulted in flexural cracking (runs one and two), and little difference was observed once the structure was cracked. The hysteresis curves for run three were essentially same for the each specimen. However, large differences in stiffness characteristics returned during run four when substantial diagonal tension cracking was observed." Typical results of base moment vs. top-level deflection from run four are shown in Figure 10.28.

The focus of the model study described above was the correlation in behavior of the same structural system subjected either to inertial loading as a result of shaking or to equivalent static forces. The ¼-scale model structures tested did not represent any prototype structure. As stated by the authors, "absolute direct scaling of masonry (and reinforcement) mechanical properties was not the primary intent of the research."

10.2.1.3 *Precast Concrete Large-Panel Buildings*

1. General — This research investigation, dealing with the behavior of precast concrete large-panel (LP) buildings under simulated progressive collapse conditions, utilized three types of ultimate-strength models: joint details, planar components, and three-dimensional models. Progressive

MODEL APPLICATIONS AND CASE STUDIES

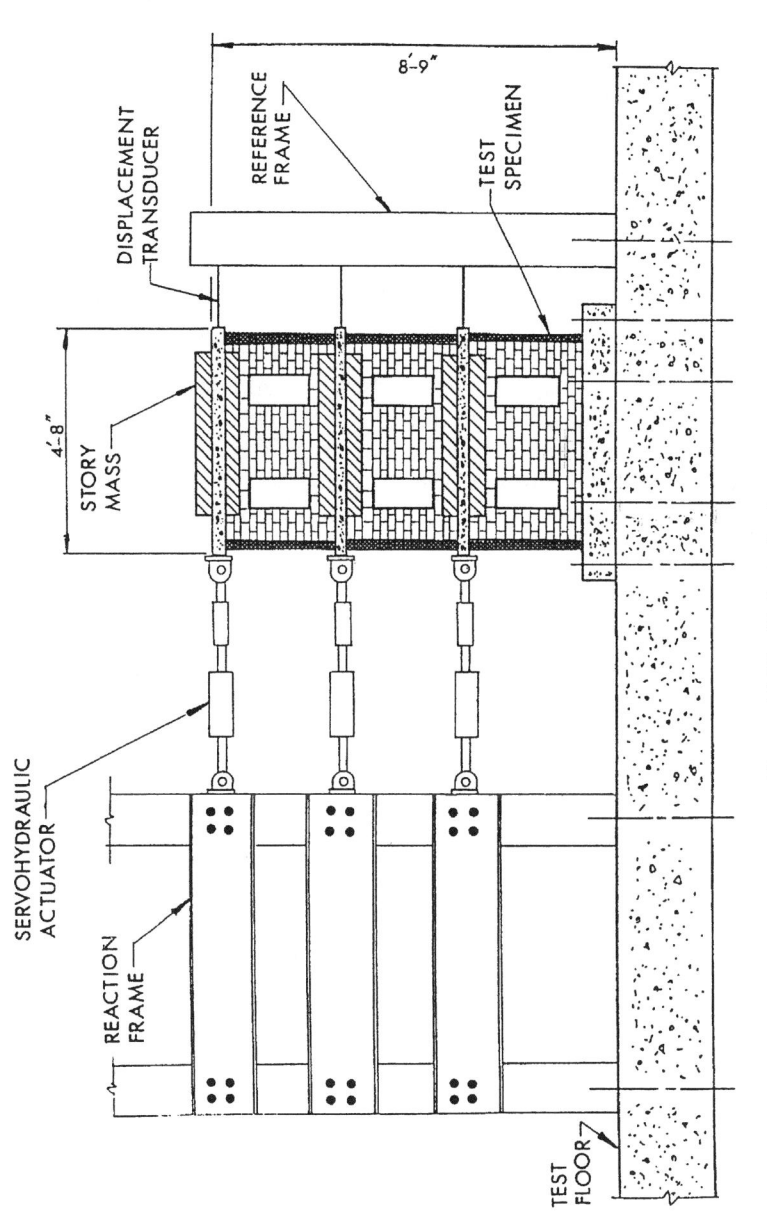

Figure 10.27 Static test setup.

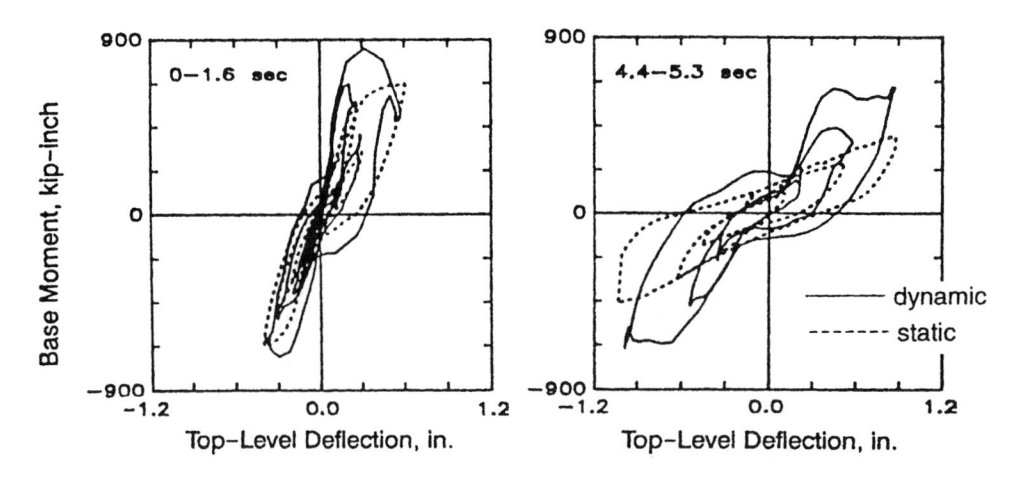

Figure 10.28 Static and dynamic response for ¼-scale masonry test structures.

collapse, usually initiated by an abnormal load such as a blast or impact, may be defined as a chain reaction of failures following damage to only a relatively small part of the structure (Leyendecker and Burnett, 1976). LP buildings, which consist of large numbers of precast floor and wall panels, are basically bearing wall structures arranged in box-type layouts (Figure 10.29). They have inherent lines of weakness in their many cast-in-place joints and are susceptible to progressive collapse (Breen, 1975) if not properly tied together. A system of ties in the three orthogonal directions has been recommended by the Portland Cement Association to enable the LP structure to bridge over local damage (Fintel et al., 1976). Direct models were used in the study to check the adequacy of the proposed tie system (Harris and Muskivitch, 1977; 1980). This research, although carried out 20 years ago, is still relevant in light of the threats of terrorist blast attacks on public and government buildings. The dynamic aspects of blast loading on structures are covered in Chapter 11.

2. Horizontal Joints — The horizontal joint, which connects the load-bearing wall panels to the horizontal floor slabs, is the most critical joint in LP structures. It must allow the wall and floor loading to be channeled to the foundations and in addition enable the structure to redistribute the gravity loads by cantilever action in case of loss of a bearing wall member.

A series of tests were conducted (Harris and Muskivitch, 1977; Dow and Harris, 1978) on ¼-scale models of an interior horizontal joint loaded in axial compression (Figure 10.30). The 30 models tested consisted of two wall panels, two hollow-core floor panels, joint mortar, and dry pack material. The specimens were assembled from precast elements. Each specimen was instrumented with dial gages, which were used to measure wall and dry pack expansion and overall joint shortening, and strain gages, which were used to measure wall compressive strains.

The main variables in the model and prototype tests were the amount of wall reinforcement to prevent wall splitting and the strength of the joint grout. The ultimate strength of all model specimens, consisting of average values from three identical specimens, is compared with the full-scale prototype tests conducted at the PCA (Johal and Hanson, 1978) in Figure 10.31. As can be seen from Figure 10.31, the complete range of prototype joint behavior could be predicted from the model tests. Typical crack patterns at failure of specimens with weak, intermediate strength, and strong joint grout are shown in Figure 10.32.

3. Cantilever Wall Component Assembly — A second group of three models at ³⁄₃₂-scale consisted of planar wall and floor slab assemblies with one first-story wall missing. This test simulates the condition of a loss of a critical load-bearing member (Figure 10.33). Its main objective was to verify that the damaged structure can absorb the dead load (DL) plus one half the live load (0.5 LL)

MODEL APPLICATIONS AND CASE STUDIES 489

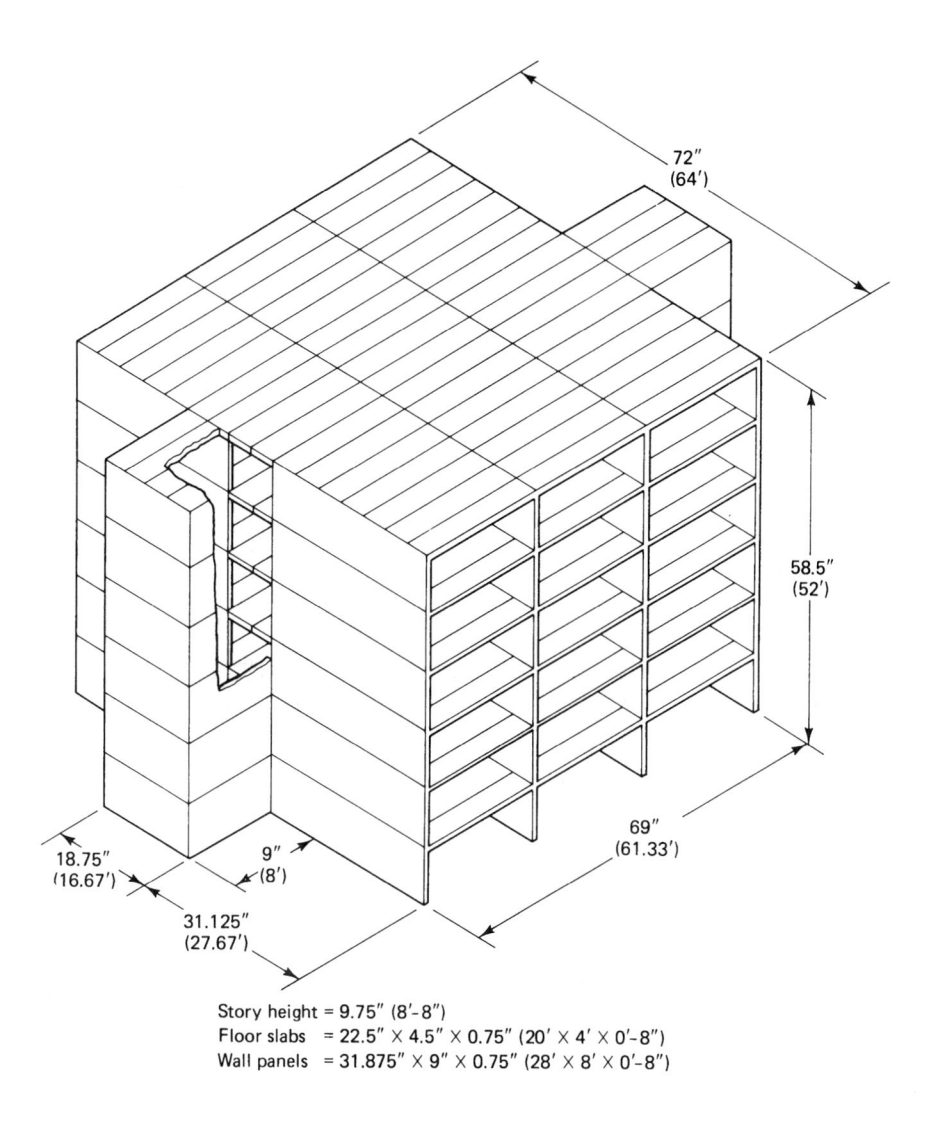

Figure 10.29 Isometric view of six-story, three-bay 3/32-scale LP model showing model (prototype) dimensions.

used in design by means of a cantilever action of the wall panels above the damaged region. Two models were constructed with the same amount of transverse reinforcing in the horizontal joints, to demonstrate the repeatability of two identical models. A third model contained one half the amount of the transverse reinforcing to determine its effect on the behavior.

Deflections, joint openings, and horizontal joint slip were measured and compared with test results obtained by Hanson (1975) at the Portland Cement Association (PCA) on larger, 3/8-scale models used as the prototype. Overall, the small-scale model cantilever walls exhibited a nonlinear load–deflection behavior similar to that found in the prototype structure (Figure 10.34). Vertical joint openings, measured at top and bottom of each story level, are compared for models and prototype in Figure 10.35. The range of values obtained from several model tests to service load levels follow the prototype measured values very closely, as can be seen from Figure 10.35. All models, irrespective of their failure mode, exhibited the same ability to undergo large deflections. The ultimate load was also found to be directly related to the amount of transverse joint reinforcement (Harris and Muskivitch, 1977).

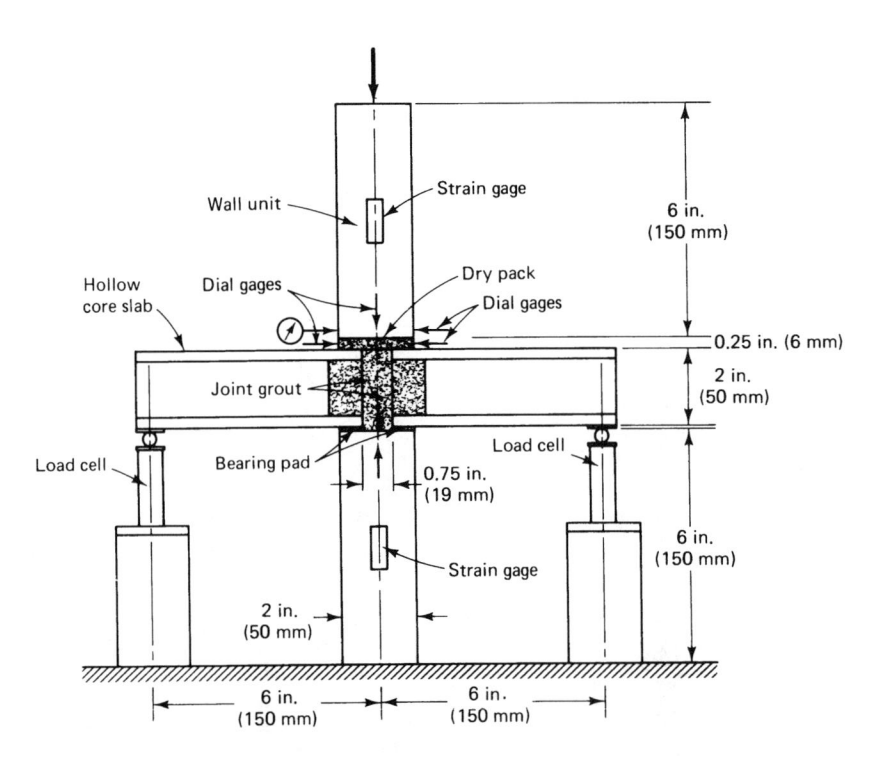

Figure 10.30 Dimensions and load setup for ¼-scale model horizontal joints (E1).

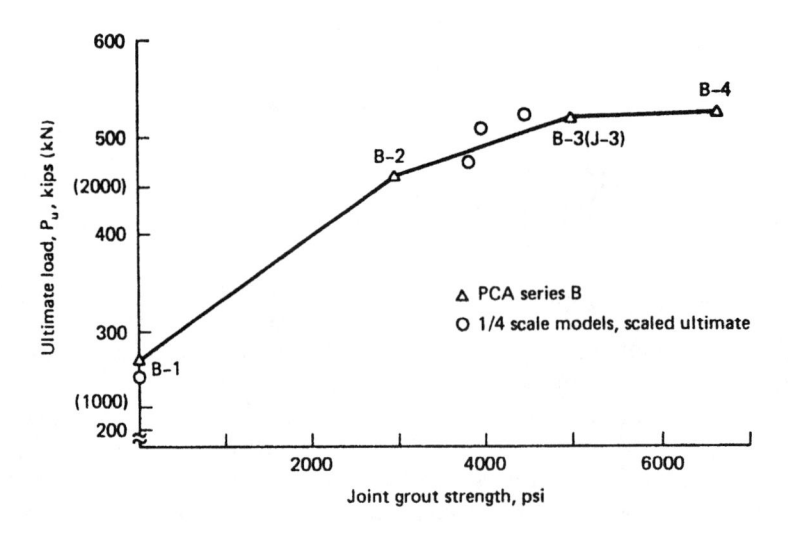

Figure 10.31 Comparison of medel and prototype horizontal joint results. (After Harris, H. G. and Muskivitch, J. C., 1977.)

4. Three-Dimensional Model — A final and three-dimensional model of a six-story, three-bay cross-wall structure was constructed (Harris and Muskivitch, 1980) in the same way as a typical precast concrete LP building with American-type details (see Figure 10.29). The properties of the model material were chosen to be related as closely as possible to those used in typical prototype construction. Mild steel was used for the reinforcement of the wall panels and floor slabs and continuous-thread steel–threaded rod and stainless steel cables for the system of vertical and

MODEL APPLICATIONS AND CASE STUDIES

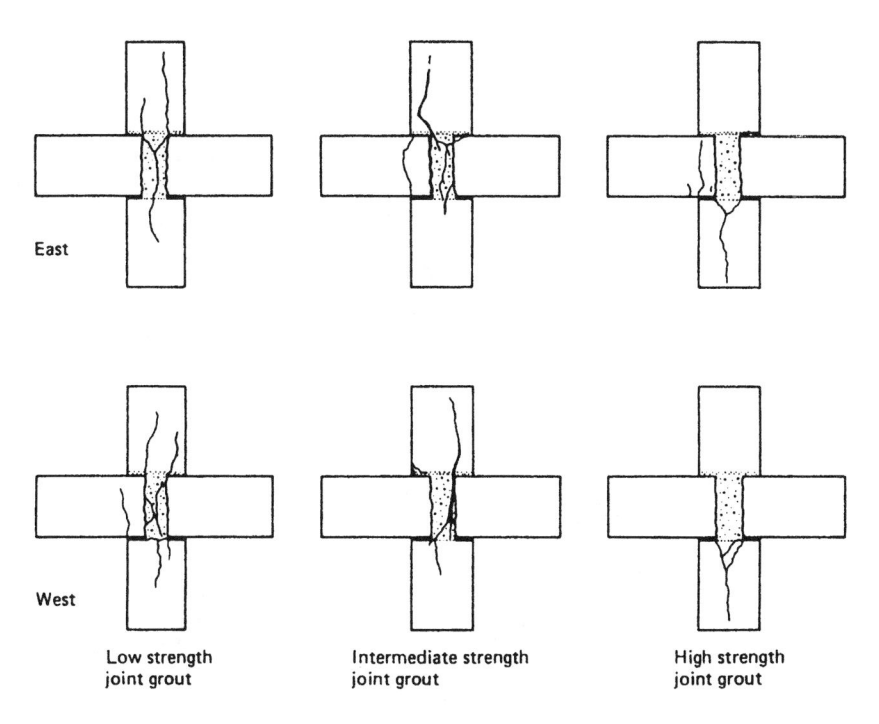

Figure 10.32 Crack patterns at failure, ¼-scale models.

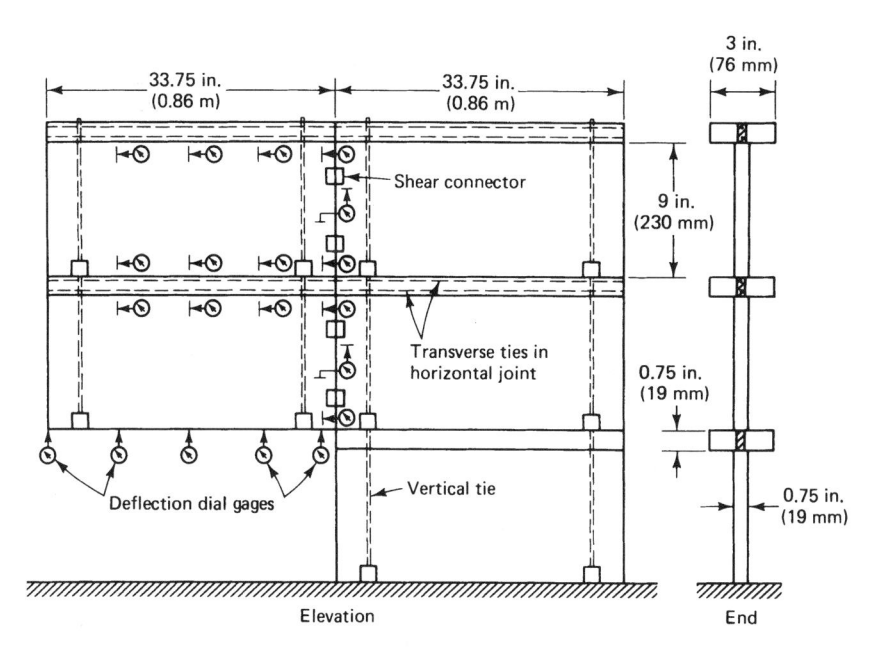

Figure 10.33 Two-story cantilever wall model (E2a) showing tie arrangement and instrumentation at 3/32 scale.

horizontal ties, respectively. The model was built with four (three exterior and one interior) removable wall panels at different critical locations (Figure 3.36). Thus, a damaged configuration could be studied by removal of a panel without hampering the loading operation. Assembly of the model from the various precast components followed the procedures used for full-scale structures of this type. Load was applied through a system of "whiffle trees" and loading beams connected to a system of hydraulic tension jacks described in Chapter 8.

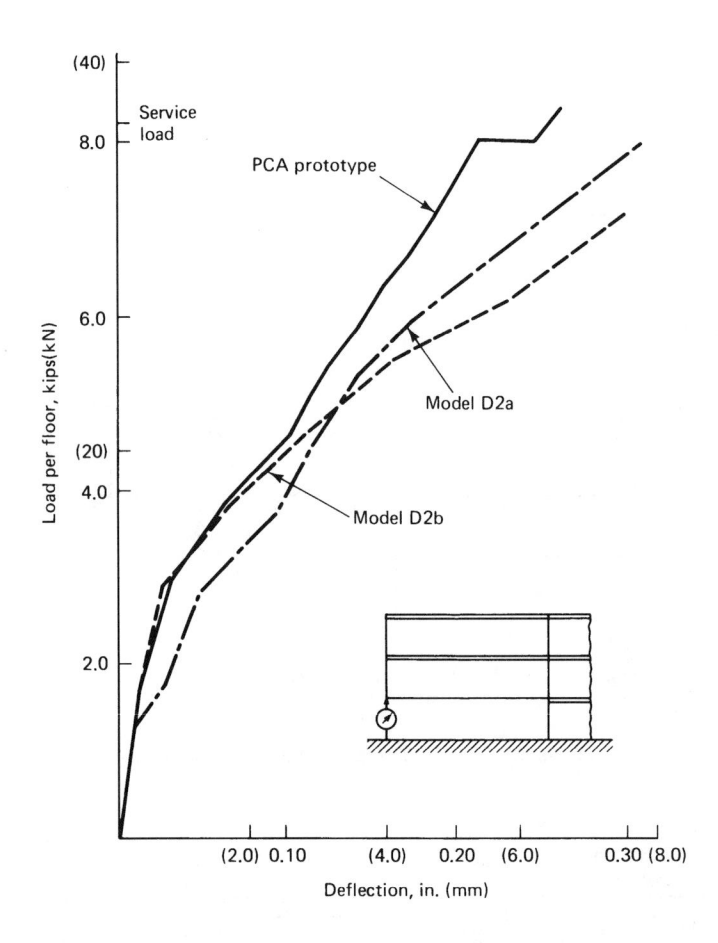

Figure 10.34 Load vs. deflection comparison of prototype and two extrapolated $1/10.67$-scale models. (After Harris, H. G. and Muskivitch, J. C., 1977.)

The results of the several tests of the three-dimensional model support the recommendations of the PCA for the use of a three-dimensional tensile tie system (Figure 10.37) to enable an LP structure to bridge over local damage. The function of a tie system is to enable the structure to "bridge" over a wall failure and limit the extent of the damage to a localized area. Major factors involved in the design of the tie system are the type of tie, the location of the tie within the structure (i.e., story level), the average story weight and height, and the length of wall panel damage. In most cases, ties (T) (Figure 10.37) consist of high-strength stranded wire cable because of its ability to develop large forces, and also to facilitate placement in joints. The criteria used in the design of the tie system were a combination of many of the recommendations proposed by the PCA with current local accepted practice. The vertical tie (V) system (Figure 10.37) provides continuous vertical continuity between successive wall lifts. In the model, two continuous-thread threaded steel rods were used in each wall panel to provide the story-to-story connection.

The longitudinal (L) tie system ties the floor and roof elements together to prevent excessive debris loading. Standard cable, grouted into the joints (T) between adjacent floor slabs, provides the necessary tensile continuity and ductility to permit a suspension system to form above a damaged wall. The peripheral tie (P) is provided to establish a continuous tensile ring around the floor and roof systems of an LP structure. This tensile ring ensures the integral action of the system aiding the floor and roof to act as diaphragms. Requirements for the transverse and longitudinal ties will, in most cases, control peripheral tie design.

MODEL APPLICATIONS AND CASE STUDIES 493

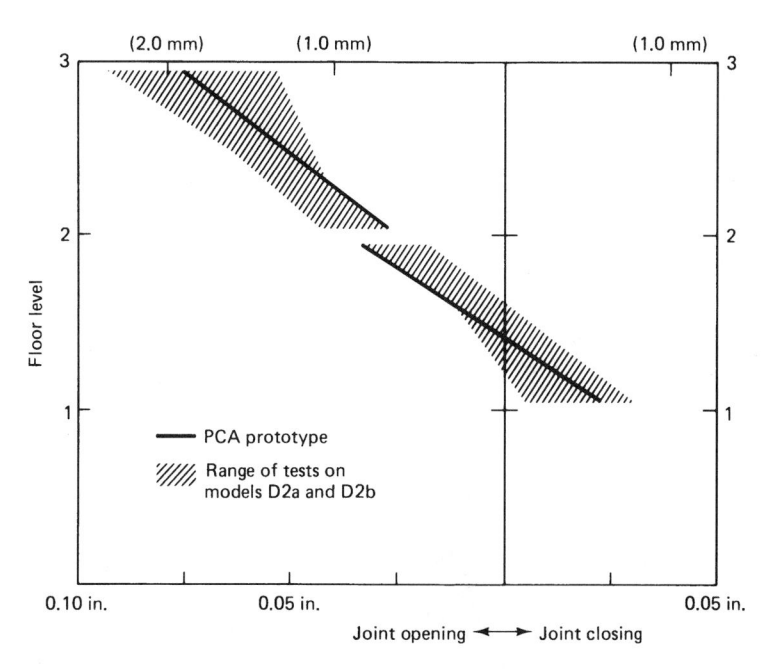

Figure 10.35 Comparison of extrapolated model and prototype joint openings. (After Harris, H. G. and Muskivitch, J. C., 1977.)

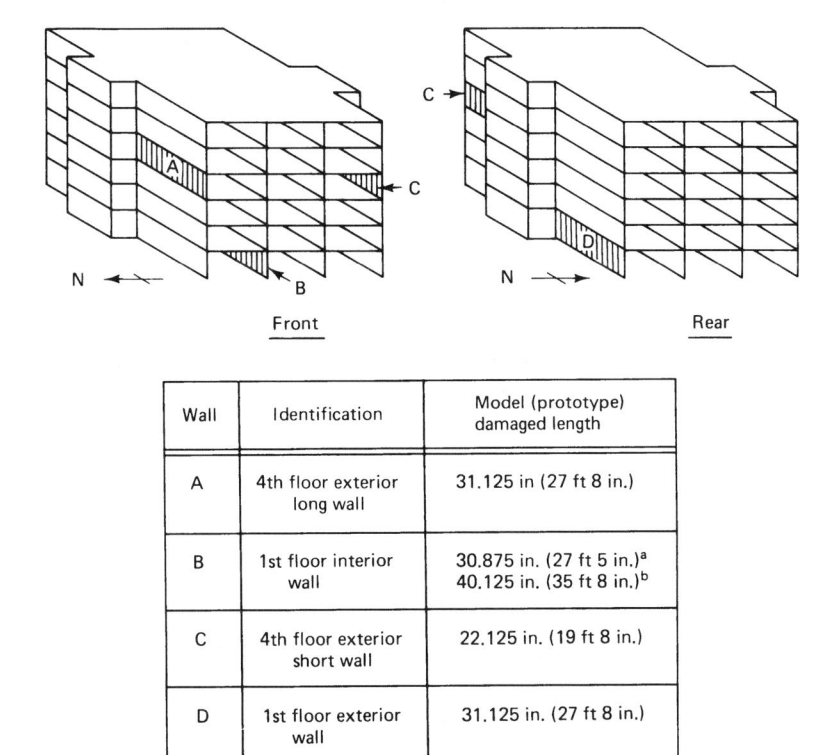

Wall	Identification	Model (prototype) damaged length
A	4th floor exterior long wall	31.125 in (27 ft 8 in.)
B	1st floor interior wall	30.875 in. (27 ft 5 in.)[a] 40.125 in. (35 ft 8 in.)[b]
C	4th floor exterior short wall	22.125 in. (19 ft 8 in.)
D	1st floor exterior wall	31.125 in. (27 ft 8 in.)

[a] with secondary support element

[b] with no secondary support element – includes width of corridor — 8.25 in. (7 ft 4 in.)

Figure 10.36 Location of removable wall panels, model D3. (After Harris, H. G. and Muskivitch, J. C., 1980.)

Figure 10.37 Three-dimensional model D3 configuration showing tie system and model (prototype) dimensions (1 in. = 2.54 cm; 1 ft = 0.305 m). (After Harris, H. G. and Muskivitch, J. C., 1980.)

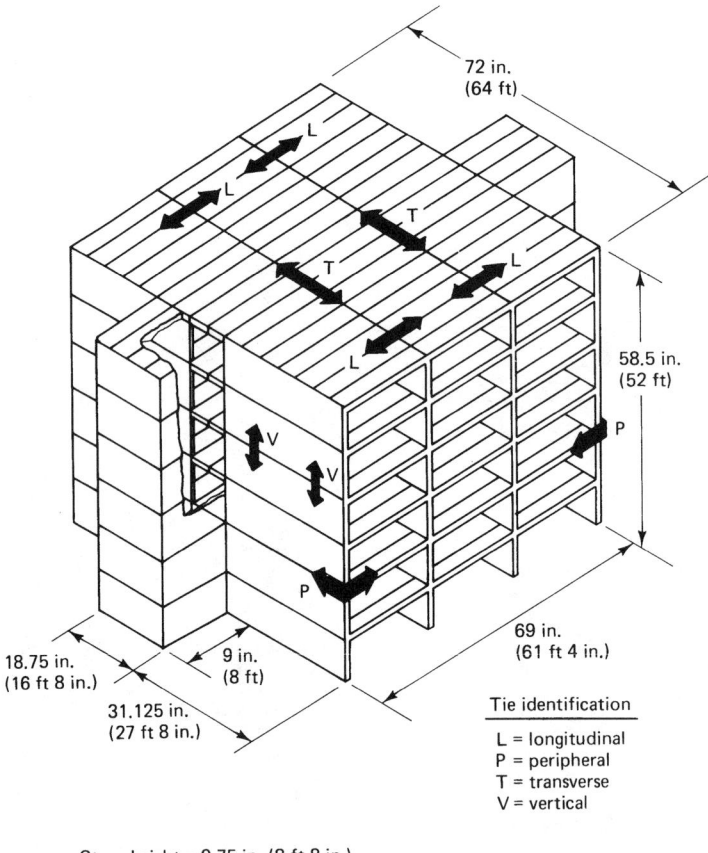

A number of displacement dial gages were placed at various locations to monitor floor slab deflections; overall axial, lateral, and support movements; and cantilevered wall deflections in the damaged configurations. The dial gage locations were chosen in an attempt to characterize the three-dimensional aspects of the behavior during each test. All major bearing wall panels and several floor slabs near the simulated damaged regions of the model were instrumented with electrical resistance strain gages.

The load was applied to the three-dimensional model through a system of whiffle trees and loading beams (Chapter 8) connected to a system of hydraulic tension jacks. A series of tests was conducted on the intact structure and on the structure in each of its damaged configurations to determine its behavior when subjected to design load and overloads. The various damaged configurations were simulated through the removal of an entire wall panel. The basic test procedure involved the loading of the model, through the hydraulic whiffle tree system, in a number of increments up to a chosen load level. At each load increment, instrumentation readings were recorded. Because of similitude requirements, it was necessary to compensate for the dead-load effects of the prototype wall and floor panels through the application of additional load to the model floor system.

The extent of damage within a damaged structure can be determined through the examination of the deflection pattern of the floor system in the vicinity of the damaged wall. Figure 10.38 shows

MODEL APPLICATIONS AND CASE STUDIES 495

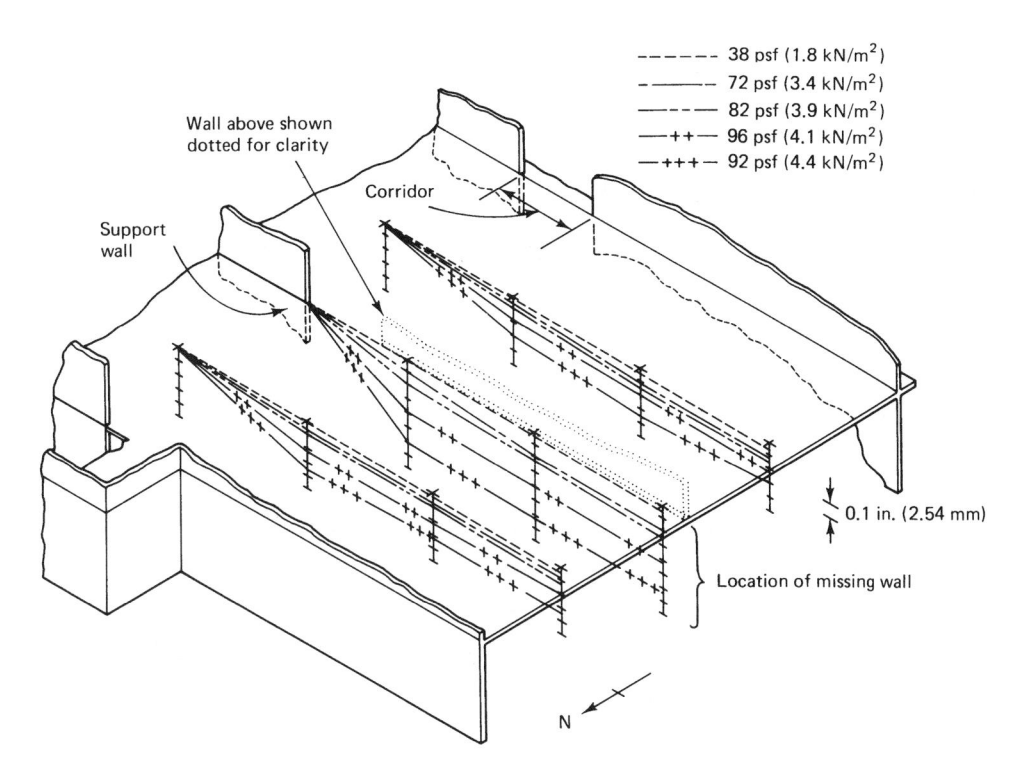

Figure 10.38 Floor system deflection pattern above missing interior wall panel with no secondary support. (After Harris, H. G. and Muskivitch, J. C., 1980.)

the deflection pattern of the floor system in the vicinity of the missing first-floor interior wall. Large deflections occurred in the immediate area of the missing wall and in the adjacent floor slabs. The large floor slab deflections were reduced to normal bending deflections in the region where the floor was supported by the rear wall. This demonstrates the apparent nature of large deflections being limited to the localized areas of damage. One recommendation of the PCA design criteria that was not included in the design of the model was the design criterion that requires the use of secondary support elements (integral columns or return walls). These elements, usually located at the innermost portion of the wall, provide the necessary vertical support to permit the cantilever mechanism to develop, but are not usually incorporated into current practice.

The need for these elements is demonstrated by the results of a second test on the first floor interior damaged configuration. During this test, the innermost vertical member of the removable wall panel was left in position to simulate a secondary support element. A comparison of the tests on the first-floor interior damaged configuration with and without secondary support is shown in Figure 10.39. This figure shows the dramatic effect that the provision of secondary vertical support has on both minimizing end deflections and enabling the structure to mobilize the cantilever mechanism and to sustain higher loads. Specifically, crack patterns observed in the joints indicated that the tie system was successfully utilized to provide the necessary support and continuity for both the cantilever and floor suspension mechanisms to develop (Muskivitch and Harris, 1979a). An analytical model was developed for studying the mechanism of progressive collapse in precast concrete LP buildings (Muskivitch and Harris, 1981).

A major concern in any experimental program that involves structural models is how the results of the model tests can be used to predict the behavior of the prototype. Table 10.6 shows the values of the various model cantilever end deflections extrapolated to the prototype. Although there are no prototype test data available at present for comparisons, the ability of the structure to undergo

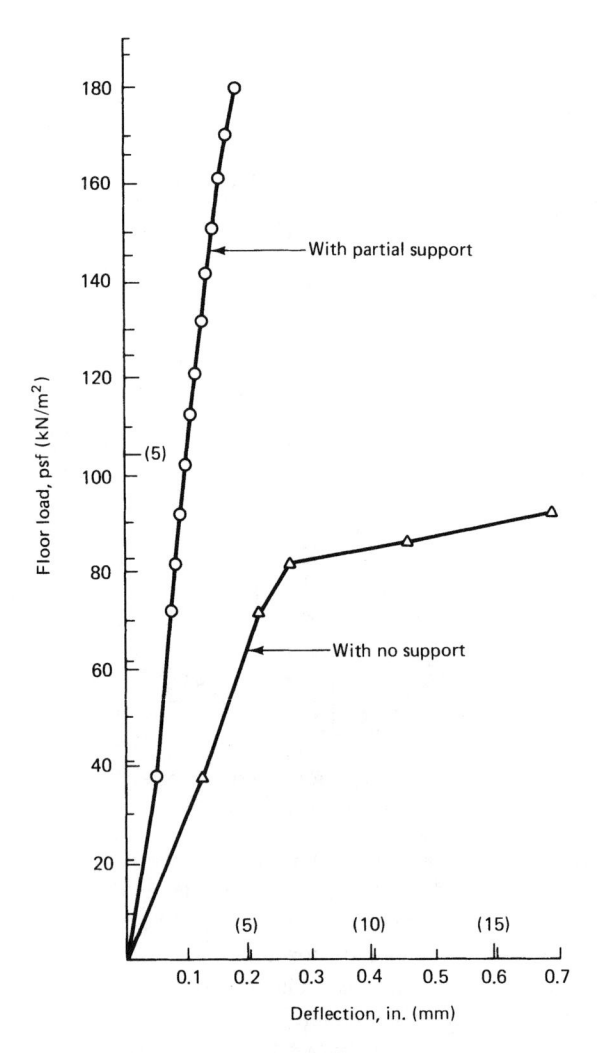

Figure 10.39 Comparison of load vs. deflection curves for first-floor interior wall assembly with and without secondary support. (After Harris, H. G. and Muskivitch, J. C., 1980.)

Table 10.6 Extrapolated Wall Assembly End Deflections

	1st Floor Exterior Wall	Damaged Configuration 1st Floor Interior Wall With Secondary Support	Damaged Configuration 1st Floor Interior Wall Without Secondary Support	4th Floor Exterior Long Wall	4th Floor Exterior Short Wall
Tie design, kN/m²	6.66	5.2	4.9	5.3	5.3
Test load/tie design load	1.21	1.66	0.91	1.53	2.07
Model deflection at test load, mm	6.8	4.4	17.5	0.36	3
Extrapolated prototype deflection, mm	8.9	46.5	187	3.8	32

MODEL APPLICATIONS AND CASE STUDIES 497

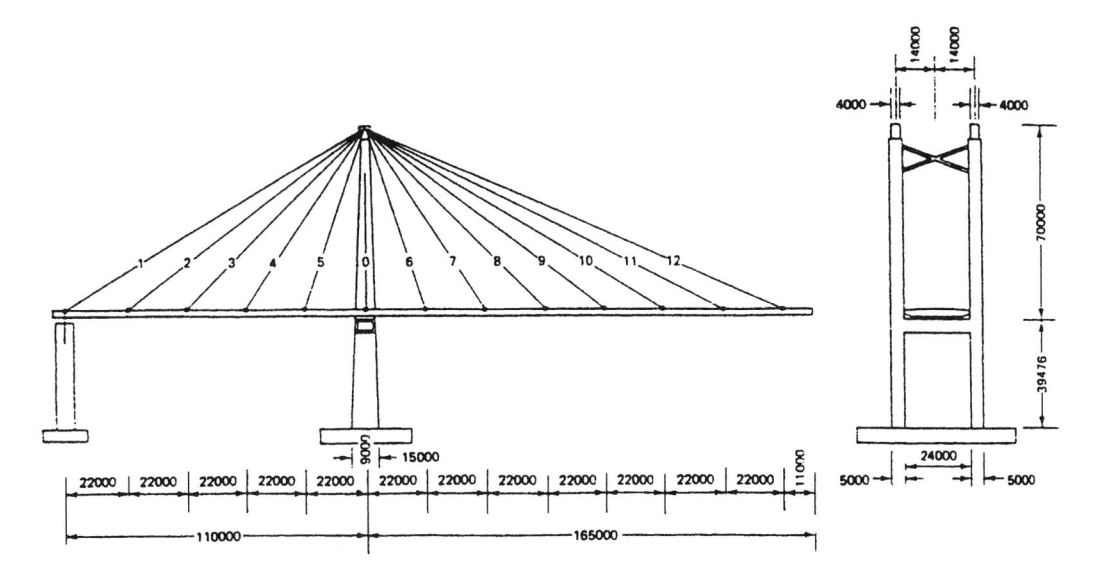

Figure 10.40 General layout of Zarate-Brazo Largo Bridges. (After Baglietto, E. et al., 1976.)

Harris and Muskivitch (1980) conclude from their tests on the three-dimensional model that:

1. Various typical prototype construction details (i.e., materials, reinforcement, connections) of a three-dimensional, precast concrete, LP structure can be accurately modeled at 3/32 scale.
2. The proper choice of the location of instrumentation will give data that will aid in the characterization of the behavior of a damaged structure.
3. The use of steel frame removable wall panels facilitated their removal from and replacement in the model.
4. The tie system, designed according to PCA recommendations (Schultz et al., 1977) allowed the three exterior-damaged configurations to satisfactorily sustain the tie design load enabling the structure to bridge over local damage.
5. In some cases, where a whole interior panel is assumed to be removed, tests on the model indicate that serious problems may arise if there is an absence of secondary support elements for interior walls. The PCA recommendations call for the use of this type of element.

10.2.2 Bridge Structures

10.2.2.1 Zarate-Brazo Largo Highway-Railway Bridge

The response of the elastic model of the Zarate-Brazo Largo Highway-Railway Bridge near Buenos Aires, Argentina, to static loads is presented in this section. Natural mode shapes and frequency measurements are described in Chapter 11. The prototype structures consist of two cable-stayed bridges over Parana de las Palmas and Parana Ciuazo Rivers approximately 80 km from Buenos Aires. The 550-m-long bridges have identical superstructures (central span of 330 m and two side spans of 110 m each) but different pile foundations.* The bridge has long prestressed

---
* Each bridge accommodates four highway lanes, two sidewalks, and an asymmetric rail track. The asymmetrical position of the railway loads over a large span of 330 m is an unusual characteristic, which had to be considered in analysis.

concrete approach viaducts (Baglietto et al., 1976). The general layout of the bridge consists of the following elements (Figure 10.40):

1. Four reinforced concrete piers. Both central piers have a rectangular, hollow cross section and are 105 m high above the rigid pile cap. Towers are connected by a concrete beam at the bridge deck level and are connected by a steel cross near the top, thus forming a statically indeterminate frame laterally. The bridge cables are anchored to steel caps at the tower tops. The remaining two piers are also statically indeterminate reinforced concrete frames laterally.
2. A fan-shaped cable system, emanating from the steel cap at the tower top and anchored to the steel deck at every 22 m. The cables consist of parallel 7-mm high-strength wires.
3. A steel deck (Figure 10.41) consisting of
   a. Two longitudinal trapezoidal box beams that provide anchorage for the cables at its external face (22-m centers).
   b. A 10-mm-thick plate deck stiffened by closed trapezoidal section ribs is connected to the two box beams by site welding. Transverse plate girders attached to the box beams at 22-m centers support the deck and resist bending moments due to the eccentric cable anchorages in the box beams. Transverse truss beams at 3.15-m spacings provide additional torsional stiffness.
   c. A truss bracing at the lower level of the plate girders and transverse truss beams to close the transverse section and to provide adequate torsional stiffness.

The cross section provides an adequate aerodynamic shape, and its torsional stiffness, along with the contribution of the suspension cables, contributes to the aerodynamic stability of the bridge. This was verified by wind tunnel tests.

A 100-mm-thick concrete pavement is provided on top of the steel deck. This concrete pavement does not interact with the steel deck; however, it assists in stiffening the deck locally.

The asymmetrical disposition of the railway load results in an asymmetrical structural design with the cables and the box beams on the railway side having twice the cross-sectional area of otherwise identical elements on the highway side.

The consulting engineers conducted a very careful examination of the computer-based design calculations for this complex structure. However, despite using the best available analysis and design tools, they decided to establish reliability of design assumptions and calculations by checking the assumed behavior by tests on a structural model. Although the design criteria were technically and economically valid, a number of problems required information that could be obtained only from suitable experimental work. The areas needing attention were torsional stiffness of the highly indeterminate bridge deck, its interaction with the cables, their fatigue behavior and the influence of secondary cables under the action of highly eccentric high-speed railway loads, and evaluation of impact coefficient.

Similitude Criteria — The following considerations were made in designing the model:

1. Correct reproduction of the characteristics of the structure with respect to the established goals of the model;
2. Limitations imposed by the availability of suitable materials, skilled personnel, laboratory space, and capacity of the testing equipment, etc.;
3. Limited funds available for the project.

As a compromise between all above considerations, a length scale factor of 33.3 was selected. The acceleration scale factor S_a was set equal to 1, because a correct simulation of the static and dynamic

MODEL APPLICATIONS AND CASE STUDIES 499

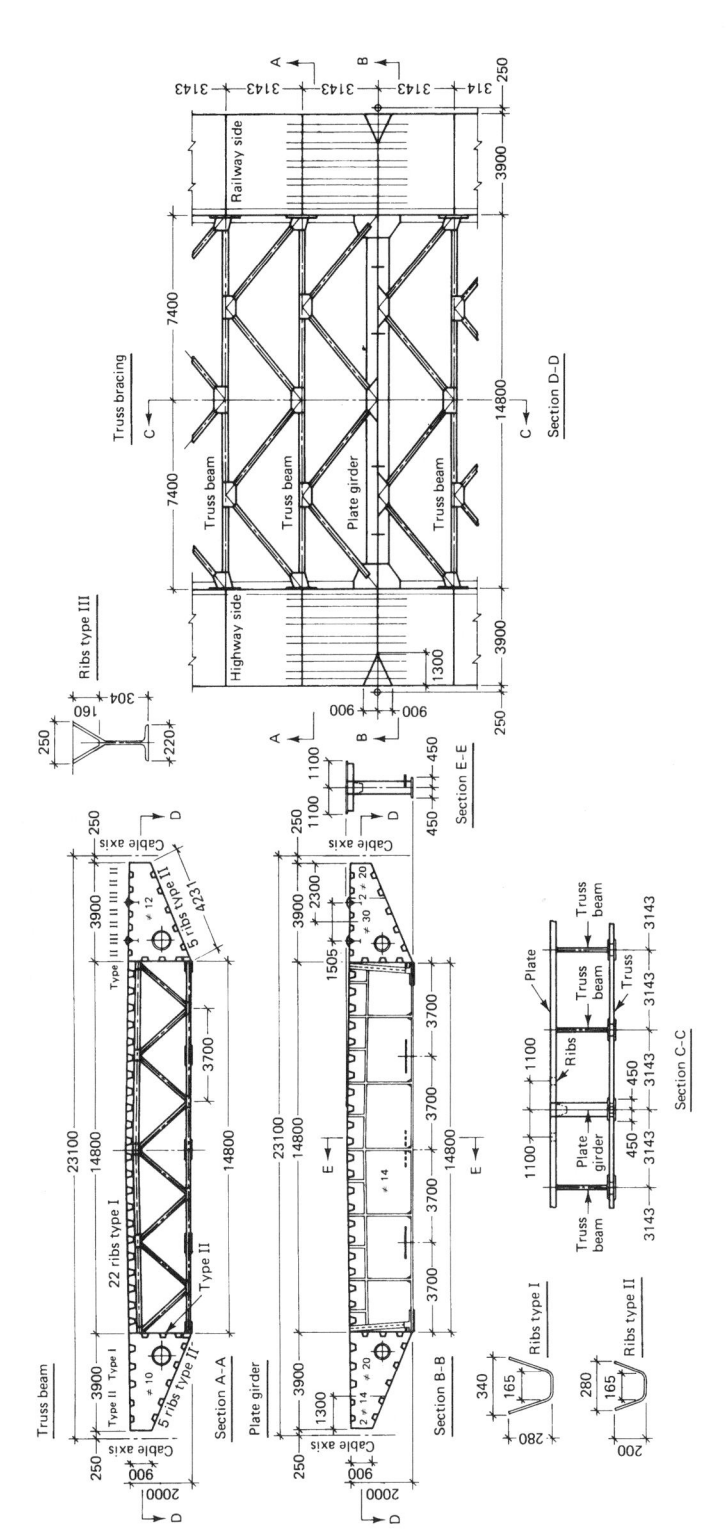

Figure 10.41 Typical cross section and truss bracing. (After Baglietto, E. et al., 1976.)

behavior of the bridge requires that the cable tension and therefore the geometric coefficient be correctly reproduced. This can only be achieved by reproducing the dead-load effects on the model. The choice of a material immediately establishes the scale factors S_E and S_ρ, and thus a material cannot be chosen arbitrarily. Moreover, since $S_a = 1$,

$$S_E = S_L S_\rho \qquad (10.1)$$

and any chosen material must satisfy Equation 10.1. Commercially available steel sheets were used for the bridge deck model, thus making $S_E = 1$. The model was suitably ballasted to satisfy the condition $S_\rho = 1/33.3 = 0.03$.

The Model — For reasons of practicality, the following modifications were necessary in the model:

1. The hollow cross section of the prototype piers was not reproduced. However, the flexural rigidity *EI* was correctly scaled.
2. The number and shape of the ribs were not reproduced. However, the axial rigidity *EA* was correctly scaled.
3. In some cases the deck was simulated by an aluminum sheet glued on a steel sheet, which distorted the model thickness. However, the axial rigidity *EA* was correctly simulated.
4. It was not possible to find exact scaled diameters of high-strength steel wires. However, the total elongation of the prototype cables was simulated by using two pieces of suitable lengths and diameters joined together.
5. The dead load was simulated by a series of lead weights distributed on the exterior and the interior of the bridge, arranged so as to allow passing of the train and to reproduce as accurately as possible the distribution of the masses, the position of the center of gravity, and the polar moment of inertia without modifying the overall stiffness of the structure.
6. Small steel cylinders were coaxially assembled on each cable to simulate its dead load without changing its axial stiffness.
7. The dimensions of cable anchorages, welding, and connections in the various parts of the model were significantly different from those of the prototype.

The above simplifications impose the following limitations on the model:

1. The model can satisfactorily simulate the overall behavior of the prototype. However, the local behavior may differ considerably, and this fact must be considered in any analysis.
2. As the limit of proportionality and the strength characteristics of the materials used and of the welds and the connections in the model are different from those of the prototype, the model results are valid only in the linear elastic range.
3. It was not possible to assess the similitude of damping achieved in the model. There were uncertainties about the damping characteristics of the prototype itself.

Details of the bridge construction are shown in Figure 10.42 and the competed model ready for testing is shown in Figure 10.43.

Static Tests: Loading and Instrumentation — The following series of static tests were conducted on the model (using prototype dimensions):

MODEL APPLICATIONS AND CASE STUDIES   501

Figure 10.42 Welding of deck elements. (After Baglietto, E. et al., 1976.) (Courtesy of ISMES, Bergamo, Italy.)

1. Influence lines for the railway side. Concentrated unit load was placed in turn at every 22 m along the railroad axis.
2. Influence lines for the outside highway lane repetition of item (1) above.
3. Railroad and highway load applied to the 110-m side span.
4. Railroad and highway load applied to the 330-m central span.
5. Railroad and highway loads 66 m long were placed along the bridge length to produce the maximum stress in the deck section.

The loads for static tests consisted of a series of steel cylinders of suitable weight that could be easily manipulated into the required positions using the available lifting equipment. The measuring and data acquisition equipment were quite complex, with numerous electrical resistance strain gages and rosettes and 41 displacement transducers installed at a total of 258 measuring points on the bridge deck and the cables. All measurements were recorded by an automatic data acquisition system. The signals from the various measurement transducers, connected to a series of balancing circuits, were sent to an amplifier by means of an automatic switch. The analog signal from the amplifier was digitized by an analog–digital converter and stored in the memory of the computer. The processed results were available as a printout, a punched tape, or a plot.

Test Results — All components of deck displacement and the cable tensions on both sides of the bridge were measured for the various loading conditions. Because of the asymmetry of the bridge cross section and the loading asymmetry, it was necessary to make measurements on both sides. A total of 114 influence lines for vertical displacements and cable tension on both sides were determined. Strain distribution for 13 transverse sections were obtained from 161 strain readings for each loading condition. Some of this data is reported by Baglietto et al. (1976).

Figure 10.43 Completed bridge model ready for tests. (After Baglietto, E. et al., 1976.) (Courtesy of ISMES, Bergamo, Italy.)

The model test results for vertical deflections showed a discrepancy of about 5 to 15% from the analytical calculations, and the variation for cable stresses was approximately 5 to 20%. These differences can be attributed to the fact that analytically determined influence lines were based on consideration of planar systems corresponding to the railway and highway sides, ignoring the deck torsional stiffness. However, the differences are small for all engineering purposes, and the slightly conservative analytical influence lines were considered adequate. The model test showed the importance of two planes of cables related to the deck torsional stiffness in redistributing the stresses by torsion. A comparison of analytical and experimental influence lines for stresses in a typical cable and for vertical displacements are shown in Figure 10.44.

In summary, satisfactory agreement was noted between analytical and experimental results, with most of the data showing differences of less than 10%.

10.2.2.2 Earthquake-Resistant Bridge Piers

An interesting modeling study of moment-reducing hinge details was conducted by Lim and McLean (1991) at Washington State University to study moment-reducing details in bridge columns. Two different scales were used (nine $1/5$-scale and over fifty $1/20$-scale models) to study the hinge details shown in Figure 10.45. From left to right these details represent a joint with horizontal and vertical discontinuity (WA detail), a joint with horizontal discontinuity only (CA detail), and the core column with the moment-reducing detail (CON) incorporated in all hinge details studied. Parameters studied in the small-scale tests included column aspect ratio, magnitude of axial load, amount of both longitudinal and spiral reinforcement, vertical discontinuity length, thickness of horizontal discontinuity material, longitudinal bar size, number of load cycles at each displacement level, column shape, and low-cycle fatigue characteristics.

MODEL APPLICATIONS AND CASE STUDIES 503

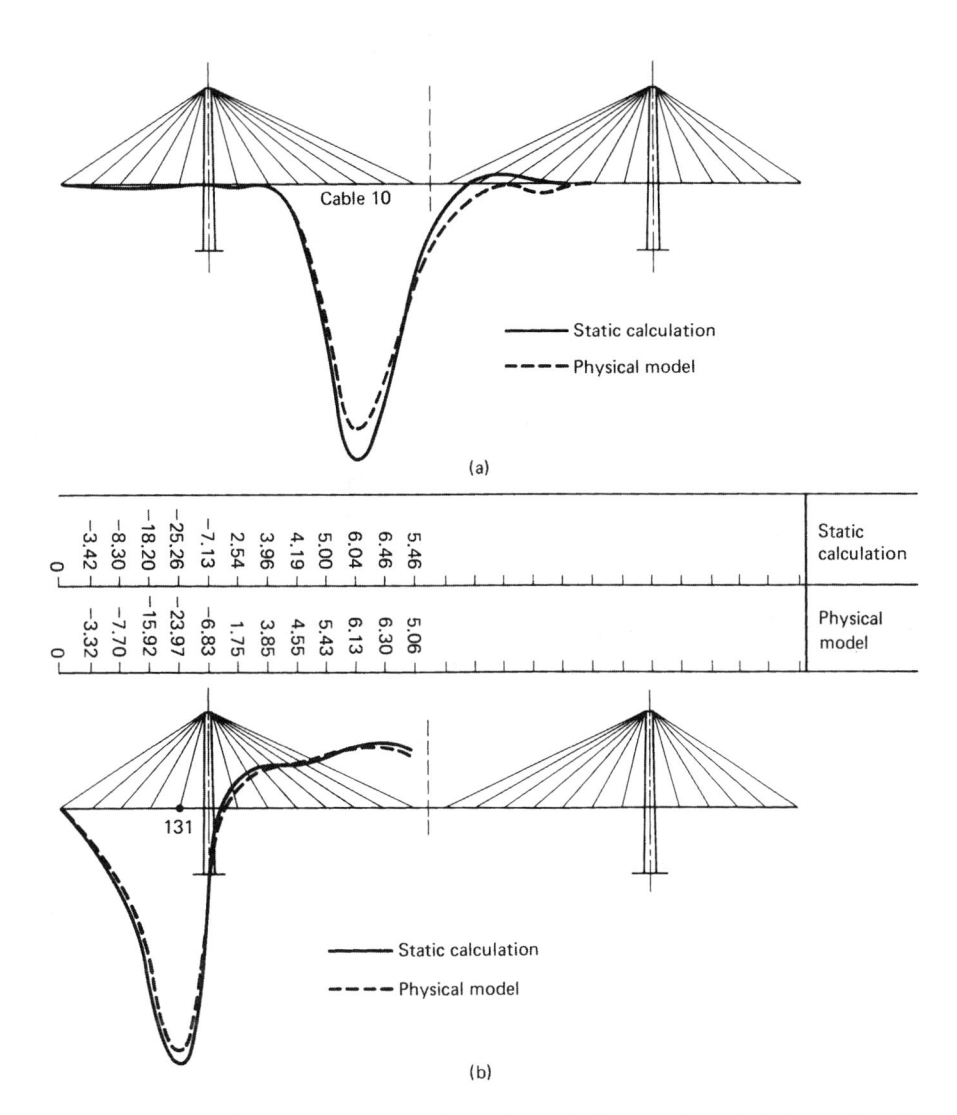

Figure 10.44 Comparison between calculated and experimental influences lines. (a) Influence lines for tension in cable 10. (b) Influence lines for displacement at deck point 131. (After Baglietto, E. et al., 1976.)

Prototype materials were used in the larger size ⅕-scale models. The ¹⁄₂₀-scale models (Figure 10.46) consisted of gypsum cement with retarder, silica sand, and water mixed in the ratio of 1:1:0.3, respectively. The properties of gypsum model concretes are discussed in Chapter 4. Longitudinal reinforcement used in the ¹⁄₂₀-scale specimens consisted of cold-drawn deformed steel bars with diameters of 3.73 and 3.10 mm which after proper heat treatment (Chapter 5) had a yield stress of 434 MPa.

Tests of the models and control specimens were conducted at 24 h using the test setup shown in Figure 10.47. The test column was bolted down to a structural models testing table (White, 1972) where lateral load was applied through a hand-operated actuator to a steel loading cap that was mounted on top of the column. The hand-operated actuator had a load capacity of 8.9 kN and a maximum stroke of 180 mm. A similar hand-operated actuator was used to provide constant axial compression to the specimen.

The typical lateral loading sequence for the ¹⁄₂₀-scale tests was two cycles at displacement ductility factors, μ (i.e., multiples of the yield displacement, Δ_y) as shown in Figure 10.48. Typical base moment–displacement factor hysteresis curves of the two types of moment reducing hinge

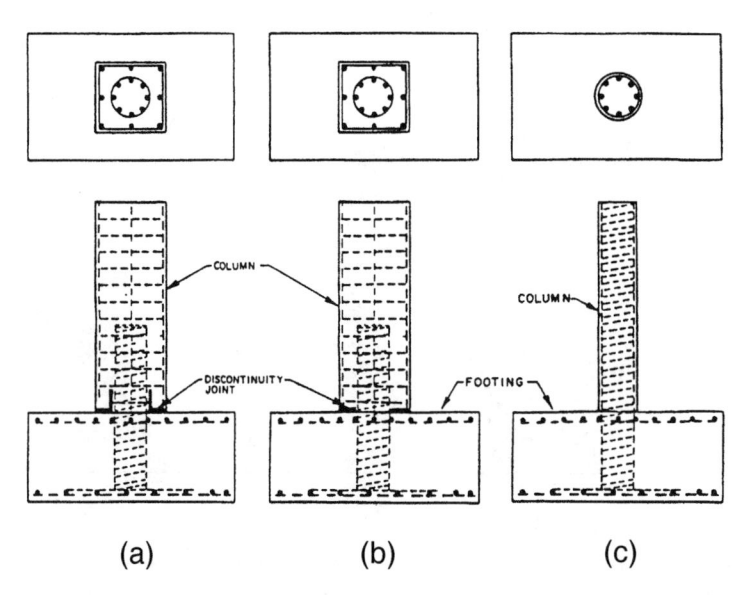

Figure 10.45 Hinge details studied. (a) WA detail; (b) CA detail; (c) CON detail. (After Lim, J. Y. and McLean, D. I., 1991.)

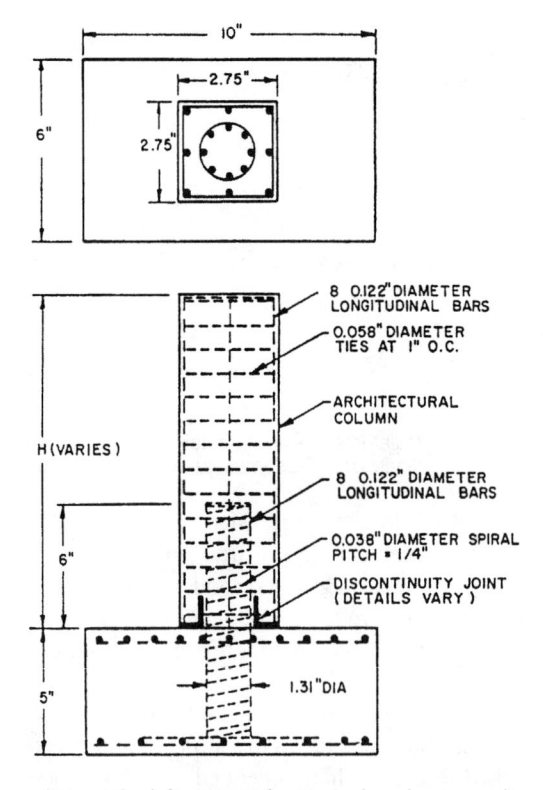

Figure 10.46 Typical dimensions and reinforcement for 1/20-scale column specimens (1 in. = 25.4 mm). (After Lim, J. Y. and McLean, D. I., 1991.)

details are shown in Figure 10.49. While the small-scale models are capable of capturing the general hysteretic behavior observed in the larger-scale tests, the behavior within the force-reversal regions of the curves varied between the 1/20- and 1/5-scale tests. Stiffness in this range reduced much more

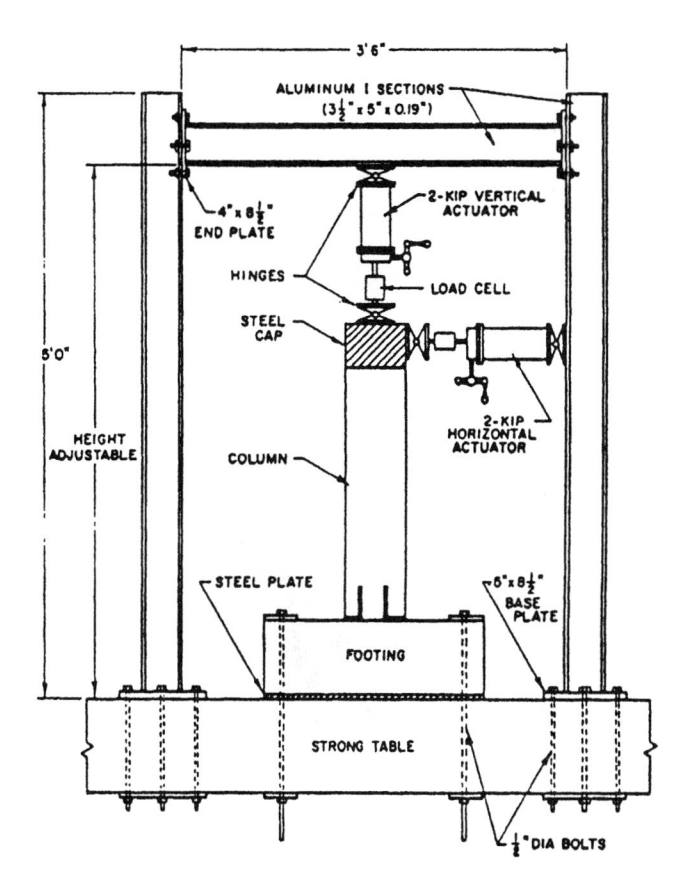

Figure 10.47 1/20-scale test setup (1 in. = 25.4 mm). (After Lim, J. Y. and McLean, D. I., 1991.)

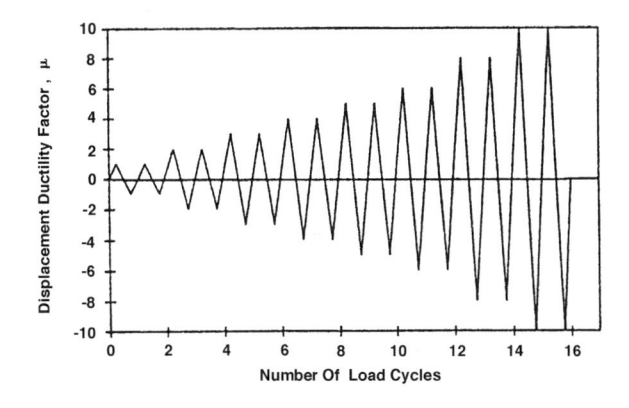

Figure 10.48 Loading history. (After Lim, J. Y. and McLean, D. I., 1991.)

rapidly for the 1/20-scale tests, a fact attributed to the differences in the bond characteristics of the No. 4 bar used in the 1/5-scale model and the underdeformed model bars used in the 1/20-scale models. The differences in bond and differences in ultimate strain (as evidenced by differences in low-cycle fatigue behavior between the 1/20- and 1/5-scale models) again point to the importance of properly modeling the reinforcement in a well-executed small-scale reinforced concrete model (Chapter 5).

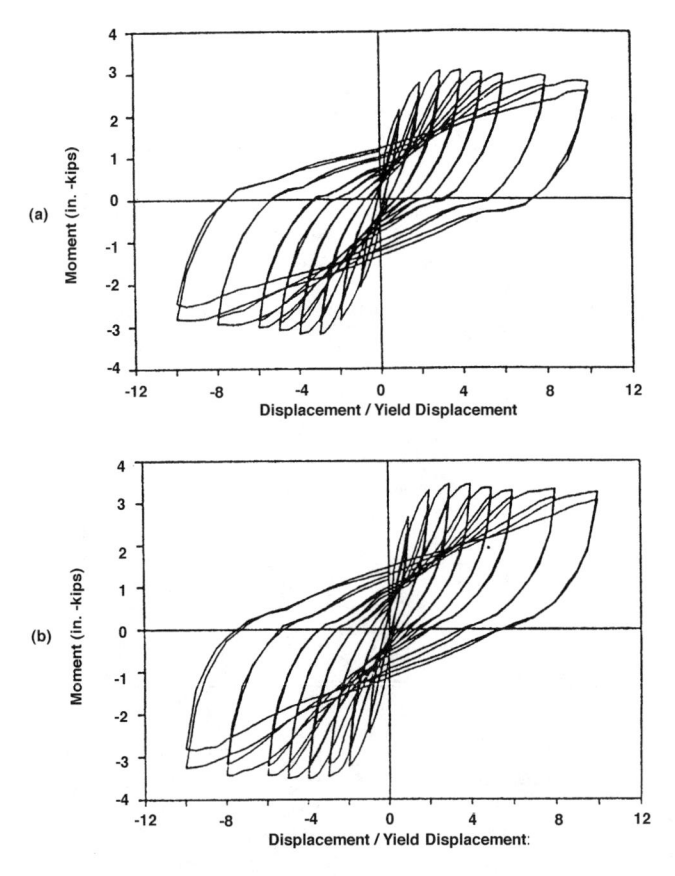

Figure 10.49 Moment–displacement factor hysteresis curves for 1/20-scale specimens: (a) 3Wc and (b) 3Cc (1 in.-kip = 0.113 kN − m). (After Lim, J. Y. and McLean, D. I., 1991.)

Table 10.7 Comparison of 1/5-Scale Moments and Scaled 1/20-Scale Moments

Hinge Detail	Average Yield Moment, kNm		Ratio, 1/20:1/5	Average Peak Moment, kNm		Ratio, 1/20:1/5
	Scaled			Scaled		
	1/5	1/20		1/5	1/20	
WA	15.26	16.39	1.07	26.56	23.28	0.88
CA	18.31	20.57	1.12	30.96	28.25	0.91
Control	10.51	16.16	1.53	23.62	20.45	0.87

The observed energy dissipation behavior of the scales of specimens tested were very similar. At low displacement ductility factors, the response of the columns with the two different details is approximately the same. However, at large displacement ductility factors, the energy dissipation effectiveness is somewhat greater for columns with the WA detail. Average scaled yield moments from the 1/20-scale tests overpredict the 1/5-scale moments by a factor ranging from 1.07 to 1.53 (Table 10.7). However, some of this variation can be attributed to inexactness in defining the yield displacement for the columns, particularly for the circular control columns. The average scaled peak moments from the 1/20-scale tests provided more consistent predictions of the 1/5-scale moments; the variaton ranged from 0.87 to 0.91.

It was observed that the columns with moment-reducing details of this study, when subjected to cycled inelastic displacements, exhibited stable moment–deflection hysteresis curves and continued to absorb energy even at displacement levels of $\mu = 12$. Flexure dominated the behavior of columns of this study, including those with an aspect ratio of 1.25. However, greater strength degradation was observed in the columns with higher aspect ratios. Columns with the moment-reducing details were only slightly affected by higher levels of axial load.

10.2.3 Special Structures

10.2.3.1 Prestressed Concrete Reactor Vessels

Prestressed concrete reactor vessels (PCRV) are complex structural forms that do not lend themselves easily to a completely analytical solution. For this reason, their development has been made possible only through extensive experimental programs using small-scale models. Table 10.8 summarizes some of the model-testing activities on PCRV structures. As can be seen from Table 10.8, a wide variety of model studies have been conducted on this type of pressure vessel all over the world. Experimental studies in the U.S. have been reported by Corum et al. (1969) on an extensive program directed by the Oak Ridge National Laboratory. Their test program attempted to investigate the suitability and accuracy of small-scale models for determining certain behavioral aspects of the PCRV.

The study encompassed tests on three different-size structures: a small concrete prototype vessel, two mortar models of the prototype, and an epoxy model of the prototype.

The prototype structure is a relatively small and simple prestressed concrete vessel. Shown in Figure 10.50, the concrete prototype represents the top half of a cylindrical vessel closed with flat heads; its dimensions are on the order of one tenth those of a full-size PCRV. The prototype has been tested to failure under internal hydraulic pressure, and the data obtained provided extensive information on the elastic and inelastic behavior of this type of prestressed pressure vessel as well as being used for direct comparison with the test results of the two smaller mortar models shown in Figure 10.50. The mortar models were exact replicas of the concrete model at a geometric scale of 1 to 2.75. The scale of the mortar models was chosen to produce a model structure that was at or near the minimum feasible size.

One of the mortar models was subjected to pneumatic pressure (nitrogen) and the other to hydraulic pressure (oil) in an attempt to distinguish any possible difference in failure behavior for the two pressurizing media.

The smallest model shown in Figure 10.50 is a $\frac{1}{50}$-scale prestressed epoxy model that has been analyzed using strain gages. Data were obtained for both prestressing and pressure loading.

The design of the prototype had many features similar to those found in an actual PCRV, including thick head and walls, unbonded post-tensioned prestressing in three directions to counteract the effects of internal pressure, and a pattern of penetrations through the head. The only conventional reinforcing elements in the structure are the two concentric rings of steel mesh at the base and small pieces of mesh under the prestressing anchor plates in the head. The prototype was designed for 3.5 MPa internal pressure, which is representative of pressures met in actual gas-cooled reactors. A conventional concrete with nominal uniaxial compressive strength of 41 MPa was used, and all prestressing elements were conventional Stressteel bars. A fiberglass-reinforced epoxy liner was applied to the interior cavity of the vessel to prevent permeation of the pressurization fluid into the concrete.

All geometric features of the prototype were reduced by a scale factor of 2.75. Model materials were chosen to duplicate prototype material properties as closely as possible. The resulting mortar models can thus be considered as true models of the prototype, with strains and stresses theoretically identical in model and prototype at any given internal pressure level. The mortar mix proportions were 1:3:1 (cement:sand:7.9 mm aggregate) with a water-to-cement ratio of 0.517.

The models were prestressed prior to pressurization using conventional Stressteel jacking equipment. Six rams (Figure 10.51) connected to a common hydraulic pressure source were used

Table 10.8 PCRV Scale Models

Organization	Test Item	Scale	Project	Number of Models	Test for[a]
French AEC	Head, PCRV	Not known	G-2, G-3	2	A, B, C
	Cylindrical PCRV	$\frac{1}{10}$	G-2, G-3	3	A, B, C
	Cylindrical vessels	I. D. 0.8 m I. H. 2.3 m	Safety studies	25	C, D
	Cylindrical vessels	Unavailable	G-2, G-3	2	A, B, C
Societe d'Etudes et	Cylindrical PCRV	$\frac{1}{6}$	EDF-3	3	A, B, C, D
d' Equipments	Cylindrical PCRV	$\frac{1}{10}$	EDF-3	1	T
d' Entreprises (SEEE),	Cylindrical PCRV	$\frac{1}{5}$	EDF-4	2	A, B, C, T
France	"Hot liner" vessel	Not known	General	1	A, B, C, T
Electricite de France	Cylindrical PCRV	$\frac{1}{5}$	Bugey I	2	A, B, C, T
(EDF), France	Two-layer cylinder	$\frac{1}{3}$	General	1	
Central Electric	Cylindrical PCRV	$\frac{1}{8}$	Oldbury	1	A, B, C, T
Research Laboratory, England	Cylindrical PCRV	$\frac{1}{8}$	Pre-Oldbury	1	B, C
Sir Robert McAlpine & Sons, England	Cylindrical PCRV	$\frac{1}{7}$	Oldbury	1	A, B, C, T, D
Taylor Woodrow	Spherical PCRV	$\frac{1}{12}$, $\frac{1}{40}$	Wylfa	2	A, B, C
Construction Ltd.	Cylindrical PCRV	Not known	Wylfa	3	A, B, C
(TWC), England	Cylindrical PCRV	$\frac{1}{10}$	Hunterston B	1	A, B
	Heads, PCRV	$\frac{1}{24}$	Several	12	A, B, C
	Multicavity PCRV	$\frac{1}{10}$	Hartlepool	1	A, B, C
	Head PCRV	$\frac{1}{13}$	Ft. St. Vrain	2	A, B, C, D
Kier Ltd., England	Spherical PCRV	$\frac{1}{12}$	Wylfa	1	A, B, C, T
Atomic Power Constr.,	Cylindrical PCRV	$\frac{1}{10}$	Dungeness B	1	A, B, C
England	Cylindrical PCRV	$\frac{1}{26}$	Dungeness B	1	B, C
	Heads, PCRV	$\frac{1}{72}$	Dungeness B	1	B, C
	Heads, PCRV	$\frac{1}{24}$	Dungeness B	3	B, C
	Heads, PCRV	$\frac{1}{26}$	Dungeness B	2	B, C
Building Research	Cylindrical PCRV	$\frac{1}{10}$	Hinkley Pt B	1	T
Station, England	Cylindrical PCRV	$\frac{1}{20}$	Hinkley Pt B	4	T
Foulness, England	Cylindrical PCRV	$\frac{1}{20}$	Study by UKAEA study group	10 models to date (30 total) 40% pneumatic	C, D
General Atomic	Cylindrical PCRV	$\frac{1}{4}$	General	1	A, B, C
	Cylindrical PCRV	$\frac{1}{4}$	Ft. St. Vrain	1	A, B, C, D, T
	Multicavity PCRV	$\frac{1}{20}$	HTGR	1	A, B, C
Oak Ridge National Laboratory	Cylindrical PCRV	$\leq \frac{1}{5}$	General	4	A, B, C
University of Illinois	Cylindrical vessels		General		C, D
University of Syndey, Australia	Head, PCRV	$\frac{1}{20}$	General	4	C, D
Siemens, Germany	Cylindrical PCRV (prefabricated blocks)	$\frac{1}{3}$		1	A, B, C
Krupp, Germany	Cylindrical PCRV	$\frac{1}{5}$	Gas-cooled reactor	1	
ENEL/ISMES, Italy	Cylindrical PCRV	$\frac{1}{20}$	HTGR	1	
Ohbayashi-	Cylindrical PCRV	$\frac{1}{20}$	HTGR	1	A, B, C
Gumi, Japan	Multicavity PCRV			1	A, B, C

[a] A, Elastic response; B, design overpressure; C, failure; D, abnormal conditions; T, long-term creep and temperature.

MODEL APPLICATIONS AND CASE STUDIES 509

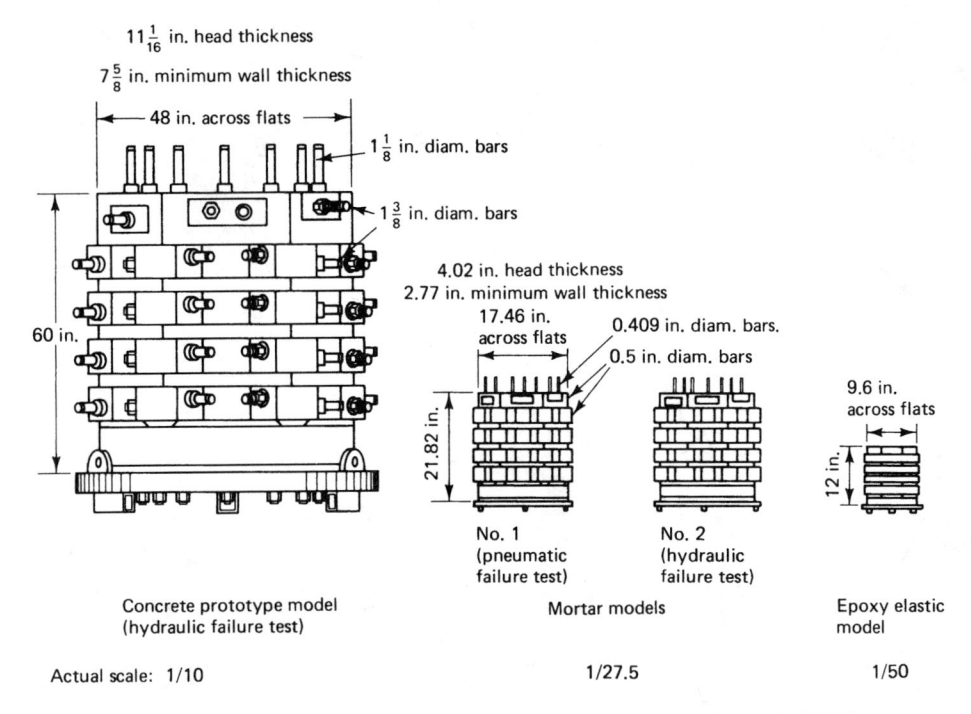

Figure 10.50 Prototype and model vessels. (After Corum, J. M. et al., 1969.)

to prestress six tendons simultaneously. The prestressing was accomplished in two phases, with all tendons stressed to half the final load level in the first phase. In each phase, the axial tendons were stressed first, then the head tendons, the first band of circumferential tendons, the second band, etc. There was a total of 12 steps plus a final adjustment and checking of stress levels. No difficulties were met in achieving an accurate level of prestress in these very short tendons.

Both the concrete prototype and the second mortar model were tested hydraulically, as opposed to the pneumatic test of the first mortar model. Both were cycled twice to a pressure of 2 MPa, once to 4 MPa, once to 6 MPa, and then to the maximum pressure dictated by liner leakage and the capacity of the hydraulic pumps used. The maximum pressure reached in the prototype by using a 60-gpm (gallons per minute) pump was 7.8 MPa. For the second mortar model, the maximum pressure reached with a 3-gpm pump was 9.6 MPa. For comparison, the maximum pneumatic pressure reached in the first mortar model was 6.7 MPa. The only apparent evidence of failure was a sudden drop in pressure and the loud hissing of escaping gas. The vessel was then filled with oil and pressurized to 7.2 MPa, which was more than 2.5 times the working pressure of 2.9 MPa. In the case of the concrete prototype, visible leakage from the model was first observed at 6.9 MPa, and by the time 7.6 MPa was reached, oil was spraying from the vessel. The corresponding pressures in the second mortar model were 7.6 and 8.3 MPa, respectively.

A comparison has been made of the measured stresses in the three models at the 3.5 MPa design pressure. The predicted and measured meridional and circumferential stresses for the concrete prototype and for the two mortar models are shown in Figure 10.52. The solid lines show the finite-element predictions for each surface, and the points show values measured with strain gage rosettes for each model. For the meridional stresses the agreement between theory and experiment for all three models is reasonably good. For the circumferential stresses, agreement is reasonably good except on the inside surface of the cylinder. The final cracking and deformation patterns of the mortar models are shown schematically in Figure 10.53.

In addition to the model tests summarized in Table 10.8, an extensive experimental program was carried out at the University of Illinois, Urbana (Paul et al., 1969), to study the structural

Figure 10.51 Prestressing circumferential tendons in model PCRV. (After Corum, J. M. et al., 1969.) (Reprinted by permission of the Oak Ridge National Laboratory, operated by Union Carbide Corporation under contract with the U.S. Department of Energy.)

response and modes of failure of PCRVs and to develop analytical procedures. In their program, 16 small-scale models were tested. The test vessels were cylindrical, 1.1 m in diameter and 1.1 m long. One end was closed by a concrete slab (the test slab), and the other end was closed by a steel plate. Circumferential prestress was provided by wrapping wire continuously. Longitudinal prestress was provided by straight Stresssteel rods or seven-wire strand. The thickness of the end slab, which had no reinforcement in it, varied from 150 to 381 mm. The wall thickness was either 5 or 7.5 in. (125 or 190 mm).

The majority of the vessels were lined with 1.6 mm neoprene sheets and pressurized internally with nitrogen. Two were tested hydraulically. The measured internal pressures at failure ranged from 1.6 to 25.4 MPa.

Appa Rao (1975) has carried out a very comprehensive test program on a 1/12-scale model of a secondary nuclear containment structure up to ultimate failure as a means of checking the linear and nonlinear analytical predictions for such structures.

Testing of physical models of PCRVs for boiling water reactors has been carried out at ISMES, Bergamo, Italy. Fumagalli and Verdelli (1976) report on one such model at 1/10-scale of a single-cavity cylindrical vessel prestressed longitudinally and circumferentially.

A 1/20-scale model of a more complex multicavity PCRV for a 1000-MW (megawatt) high-temperature, gas-cooled reactor system was constructed and tested by General Atomic, San Diego, CA (Cheung, 1969). The response of the model to increasing internal pressure was examined in four tests constituting two primary phases of testing. In the first phase, three tests were performed to demonstrate the behavior of the complete vessel under pressure. The maximum pressures

MODEL APPLICATIONS AND CASE STUDIES 511

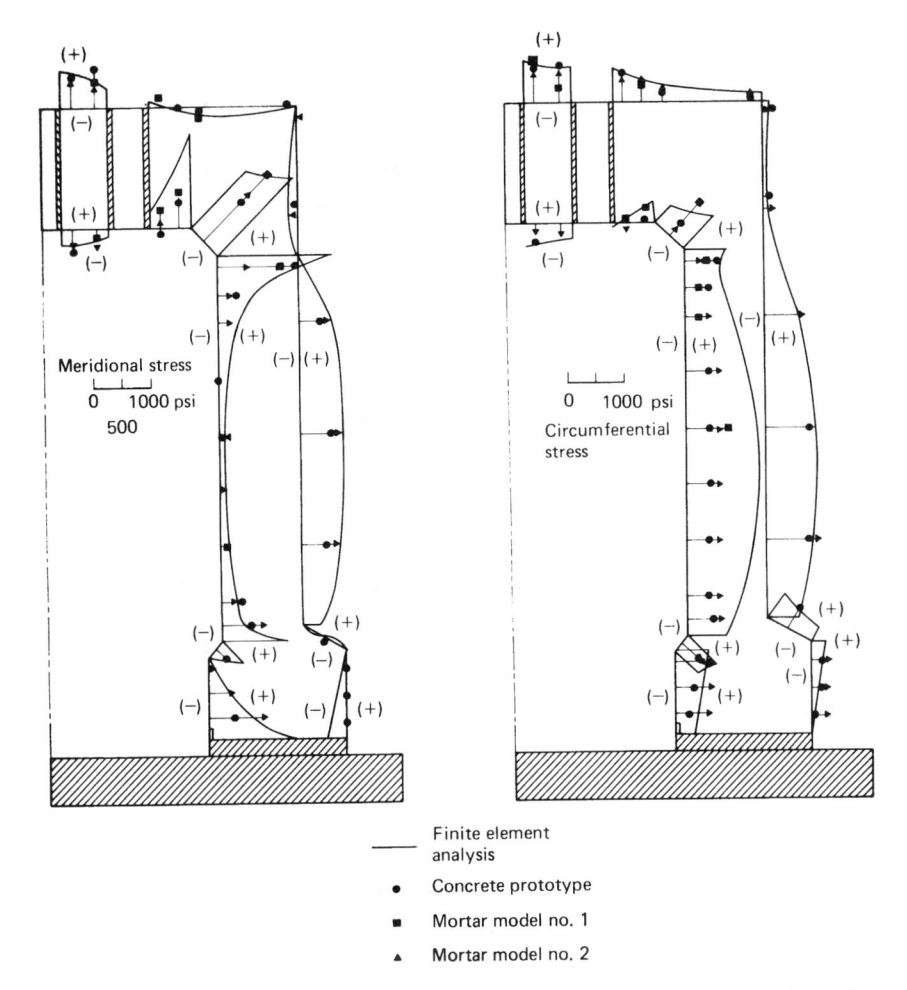

Figure 10.52 Measured and predicted meridional and circumferential stress distributions for an internal pressure of 500 psi. (After Corum, J. M. et al., 1969.)

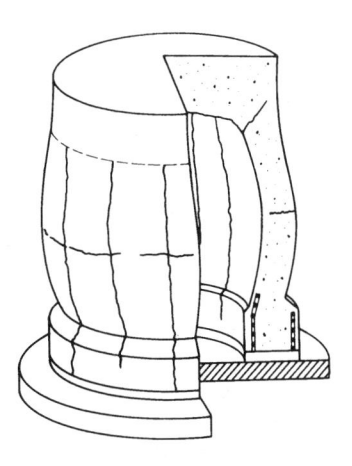

Figure 10.53 Schematic of final cracking pattern and deformatin. (Courtesy of Gulf General Atomic, Inc.)

achieved in the three tests were 11.6, 9.6, and 9.3 MPa, respectively. In the first test, separation of the head from the barrel section at the slip-plane occurred at a pressure of about 5.9 MPa (1.3 MCP, i.e., 1.3 times the main cavity pressure). At the maximum pressure of 11.6 MPa (2.6 MCP), vertical radial cracks had formed through the steam generator cavities, and the liner and rebar steel spanning the cracks had reached yield. No sign of distress was observed at the top head. In the subsequent two pressurizations, the model with cracked barrel section was tested to a maximum pressure equivalent to 2.1 MCP. The final crack pattern was a predominant flexural mode of failure.

Takeda et al. (1973) reports on tests of ¹⁄₂₀-scale single and multicavity PCRVs conducted by Ohbayashi-Gumi Ltd., Japan. The model was constructed from model concrete with stress–strain characteristics similar to those of the concrete proposed for use in actual vessel construction. Model bonded reinforcement was provided in the form of high tensile deformed wire. Top head penetrations were simulated by seventeen 60-mm-diameter standpipes. Representative steel liners for the core and steam generator cavities and cross ducts were included in the model.

Vertical prestress in the model was applied by means of high tensile bars of 12 and 16 mm diameter. Circumferential prestressing was accomplished by 2.9-mm-diameter, high-strength wire wound under tension around the surface of the model.

The model was adequately instrumented to record deflection and strain data during test. Over 300 transducers were included in the model for measurements of strain in the concrete, steel liners, representative standpipes, and bonded reinforcement. Deformation profiles of the model were measured by dial gages and linear potentiometers mounted on an independent reference frame. The main cavity and steam generator cavities together with cross ducts were pressurized hydraulically.

The deflection response clearly illustrated the gradual mode of behavior as the model approached the failure pressure. The model was shown to have essentially elastic response to a pressure equivalent to 1.8 times the design main cavity pressure. The model demonstrated an ultimate load factor exceeding 4.

10.2.3.2 Shell Structures

Plate and shell structures are one form of construction where the modeling technique has played a primary role in its widespread use and general public acceptance. The complex nature of the structural action in such systems has been made simpler for the student, the researcher, and the designer by means of physical modeling. The modeling technique has been applied successfully for a wide range of problems from elastic behavior studies (to verify and expand the existing analytical techniques) to inelastic and nonlinear studies for research purposes and finally to the ultimate load behavior studies needed by the designer to investigate the structural integrity and safety of the whole system. In this section a brief summary of model studies from each of the above categories, educational, research, and design, is presented.

In order to be able to study the inelastic behavior and the ultimate strength of plate and shell structures made from reinforced and prestressed concrete, one must resort to the use of models made from materials as close to the prototype materials as possible (Chapters 4 and 5) and having the same internal construction as that used in the prototype structure. Some examples of ultimate-strength model studies of shell structures are described below.

One of the most extensive model studies dealing with the behavior of cylindrical shell roofs under gravity loading was undertaken at the Institute for Building Materials and Building Constructions, Center for Applied Scientific Research, T.N.O., in the Netherlands (Bouma et al., 1961). A series of 11 circular cylindrical shells made from reinforced mortar and having the geometry shown in Figure 10.54 at ⅛-scale were loaded to destruction by means of increasing uniform gravity loading. The objective was to study the effects of cracking and plasticity and amount and position of the reinforcement on the stress distribution and how this deviates from the theory of elasticity at higher loads. The amount of reinforcement and its distribution in both the shell and edge beams is shown for each of the shells tested in Figure 10.55. Also shown by means of bar charts are the

MODEL APPLICATIONS AND CASE STUDIES

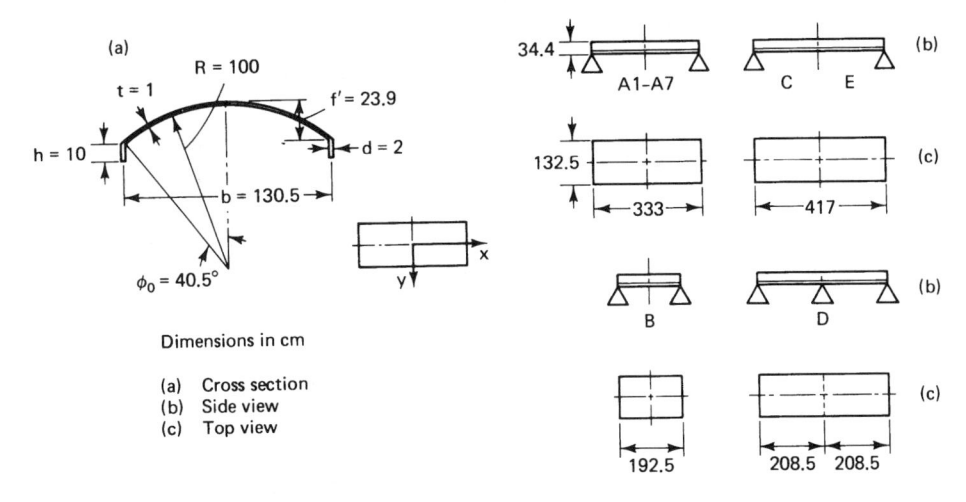

Figure 10.54 Dimensions of shell models. (After Bouma, A. L. et al., 1962.)

loading histories from initial cracking to ultimate failure of each of the models tested. Note from Figure 10.55 that the ultimate carrying capacity of all models was greater than the design load by factors ranging from 2.7 to 4.5. The load vs. deflection characteristics of the models in series A (Figure 10.56) indicate practically elastic behavior up to the design load. Beyond this load, level the deviation from elastic behavior is dependent on the amount and the distribution of the reinforcement. Typical failed specimens with longitudinal yield lines in both the positive and negative transverse moment regions are shown in Figure 10.57.

A series of smaller ($1/24$-scale) mortar shells having the same geometry as the shells in series A in Figure 10.54 were studied by Harris (1967) and Harris and White (1972). The main objective of this study was to determine the feasibility of using very small-scale "tabletop" models that required relatively little material and modest loading facilities to study the inelastic behavior of circular cylindrical shells. The $1/24$-scale models having a thickness of only 3.3 mm were made of reinforced mortar with similar properties to the larger $1/8$-scale models used as "prototype." Loading was accomplished in a scaled version of that used in the prototype shells. This consisted of a whiffle-tree arrangement with the load applied by means of pads to the outer surface of the shell (see Figure 8.15b). The inner and outer surfaces of a failed shell model are shown in Figure 10.58. The close correlation of the yield lines at failure of two small-scale models and their four times larger counterparts (prototype) are shown in Figure 10.59.

Other shell forms have also been studied extensively by means of small-scale models. Representative of this work is the testing of a variety of hyperbolic paraboloid (hypars) shell models summarized by White (1975). Both plastic and reinforced mortar models were used to study such diverse questions as:

1. General behavior and failures modes;
2. Influence of size and location of edge members on shell stiffness, strength, and bending;
3. Interaction of shell and edge members.

A much larger inverted-umbrella hypar shell, tested at the PCA laboratories, was used to establish reliability between models and prototype. Table 10.9 gives a comparison of model and prototype hypar shell properties. Figure 10.60 gives the geometry and reinforcement of the prototype and the mortar and plastic small-scale models tested at Cornell University. A typical inverted-umbrella hypar mortar model shell during testing is shown in Figure 8.15a. From this investigation it can be concluded that the tensile edge member, in the periphery of the shell, had high tensile

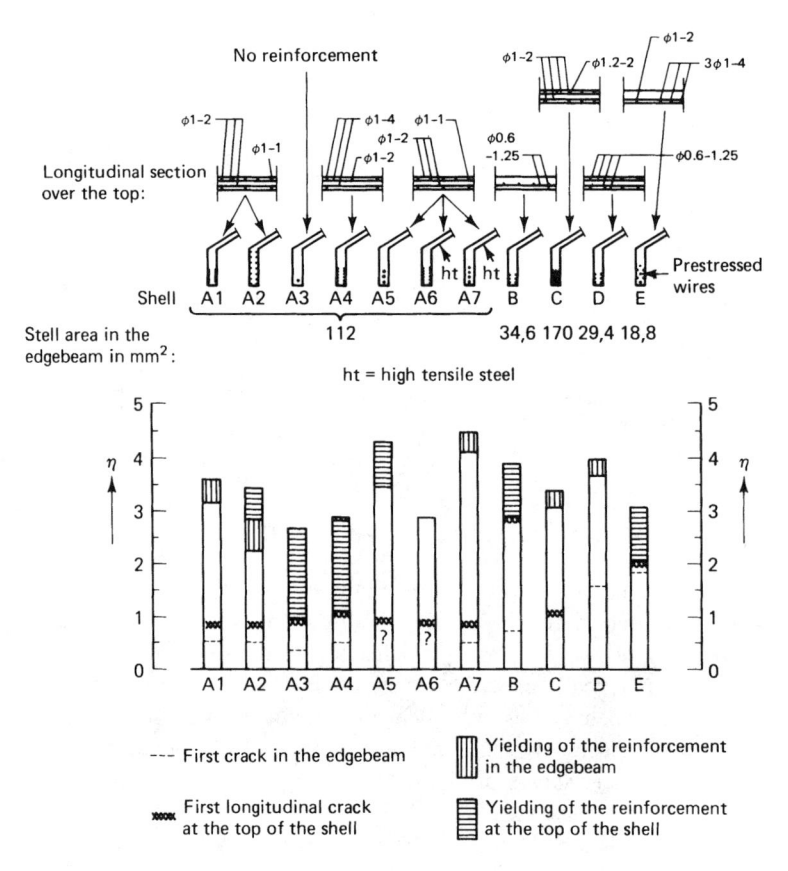

Figure 10.55 The load factors η (referred to the the design load) at which various phenomena occurred. (After Bouma, A. L. et al., 1962.)

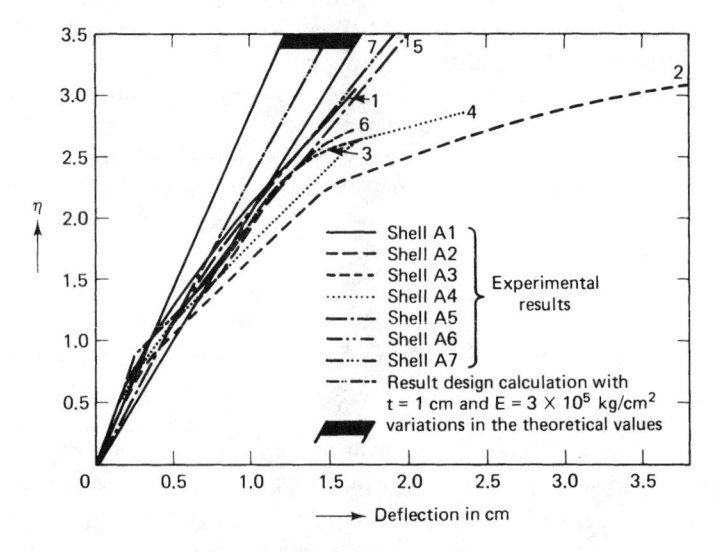

Figure 10.56 Shells of series A. Average deflection of edge beams at midspan. (After Bouma, A. L. et al., 1962.)

MODEL APPLICATIONS AND CASE STUDIES 515

Figure 10.57 Failure patterns of large model shells. (After Bouma, A. L. et al., 1962.) (Courtesy of T. N. O., Delft, Holland.)

stresses that led to severe cracking at high overloads, thus establishing this as the "weak link" of the system. Also, it was found that a portion of the shell near the edge members picks up some of the edge member load and is effective in resisting these in conjunction with the edge members. Compressive edge members were found to be very essential for carrying unsymmetric live loads. It is recommended, from this study, that the inclined compressive ribs be placed under the shell while the horizontal tensile edge members be placed above the shell surface. The combination of reinforced mortar models and plastic models is particularly useful in studying behavior of reinforced concrete shell structures. The mortar model is essential for determining postelastic response (Figure 10.61), and the plastic model is much better suited for detailed studies of strain distributions and the interrelationships between shell and edge members at working load levels. Either type is adequate for predicting working load displacements.

In addition to stress distribution and strength determination studies of plate and shell structures, small-scale models have played an important role in studies of the instability of these complex three-dimensional structures. A survey of experimental studies of concrete shell buckling has been reported by Billington and Harris (1979). A similar review for folded-plate shells has been reported by Swartz et al. (1969). Only a brief summary of studies dealing with the buckling of spherical domes will be covered here.

The most extensive experimental buckling program on concrete spherical shells was conducted at the University of Ghent, Belgium, under the direction of Dr. Vandepitte from 1967 to 1976. Results from the testing of the first seven shells has been reported by Vandepitte and Rathe (1971), and a final report covering the testing of an additional 83 shells has been given by Weymeis (1977) in an internal report. The shells used for this program were made of microconcrete. The concrete quality and the dimensions of the dome were chosen in such a manner that instability was to occur

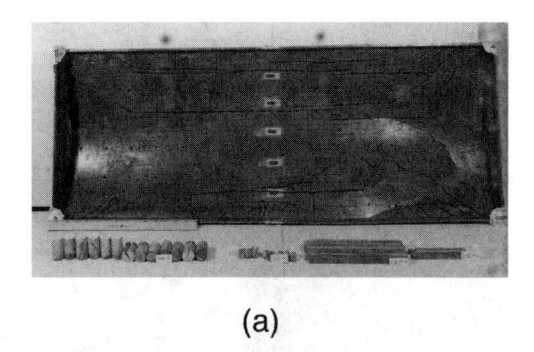

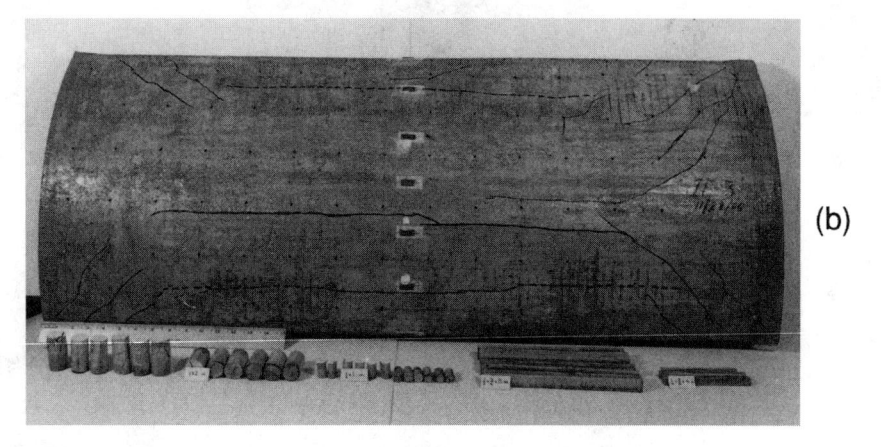

Figure 10.58 Inside and outside failure surfaces of small model. (After Harris, H. G. and White, R. N., 1972.)

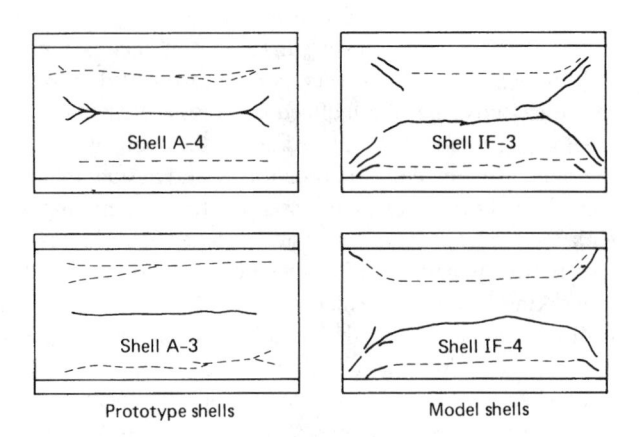

Figure 10.59 Comparison of large and small model failures. (After Harris, H. G. and White, R. N., 1972.)

before the stresses in the concrete could attain the value of the ultimate compressive strength of the material, which was about 45 MPa. In all cases, the rise of the domes and the diameter of their perimeter were 193 and 1900 mm, respectively (Figure 10.62). They were 7 mm thick, had a radius of curvature of 2431 mm, and were loaded in an upside-down manner with uniform radial pressure, using a hydraulic system. A steel ring beam support arrangement into which the domes were cast could also be prestressed during testing to minimize the edge stresses.

MODEL APPLICATIONS AND CASE STUDIES

Table 10.9 Model Behavior Compared with PCA Prototype

Structure (1)	E, MN/m^2 (2)	Poisson's Ratio, ν (3)	Stress Scale, $S_\sigma = S_E$ (4)	Length Scale, S_L (5)
Prototype	24,910	0.18	—	—
Model 2-MA	22,150	0.18	1.11	6
Model 3-MA	19,800	0.18	1.26	6
Model 5-MB	22,910	0.18	1.09	6
Models 1-PA and 2-P	2,995	0.34	8.31	8

From this very comprehensive experimental program the following conclusions were drawn by the investigators:

1. All domes failed in the same manner. A nearly circular disk is punched out of the dome, generally closer to the edge than to the center of the dome. This shows a clearly asymmetric failure pattern (Figure 10.63).
2. The average value of the coefficient c in the buckling formula for a radially loaded spherical shell, given by

$$p = cE(t/R)^2 \qquad (10.2)$$

where: E = the modulus of elasticity
 t = the thickness of the shell
 R = the radius of curvature

was found to be 0.542, with standard deviation of 9.5%. This is somewhat higher than the theoretically determined values of c given in Table 10.10.

10.2.3.3 Concrete Slabs with Penetrations

An extensive ⅒-scale model study of concrete slabs with single and multiple penetrations designed using the Hillerborg (1975) "strip method" was conducted by Sagur (1994) at Drexel University. He used a uniformly loaded simply supported rectangular slab with rectangular penetrations. A number of parameters were investigated (Figure 10.64), including size and location of penetration, number of openings, effect of corner reinforcement, and the effect of redistribution of reinforcement around the hole. Outside dimensions and reinforcement layout of the slabs tested were as shown in Figure 10.65. In slabs with penetrations, steel that was displaced was rearranged around the opening as per ACI 318-89 recommendations.

The modeling materials used in this study consisted of model concrete and deformed steel wire reinforcement as described in Chapters 4 and 5, respectively. Testing of the ⅒-scale slabs consisted of pulling an articulated statically determinate loading system from below the simply supported slab enabling a visual examination of the top and bottom slab surface during testing. Loading was carried out incrementally using a hydraulic jacking system. Deflections and cracking progress were monitored continuously. Typical deflection distributions at various load levels for a slab with a central penetration is shown in Figure 10.66. The load–deflection behavior of three identical specimens with corner penetrations (Figure 10.67) and the crack patterns at peak load (Figure 10.68) show very uniform characteristics within each group of specimens.

The typical cracking pattern of a slab with a central penetration designed with strong bands of reinforcement around the opening consisting of the displaced reinforcement caused by the hole presence is shown in Figure 10.69a. The cracks do not start at the corners but at the longitudinal

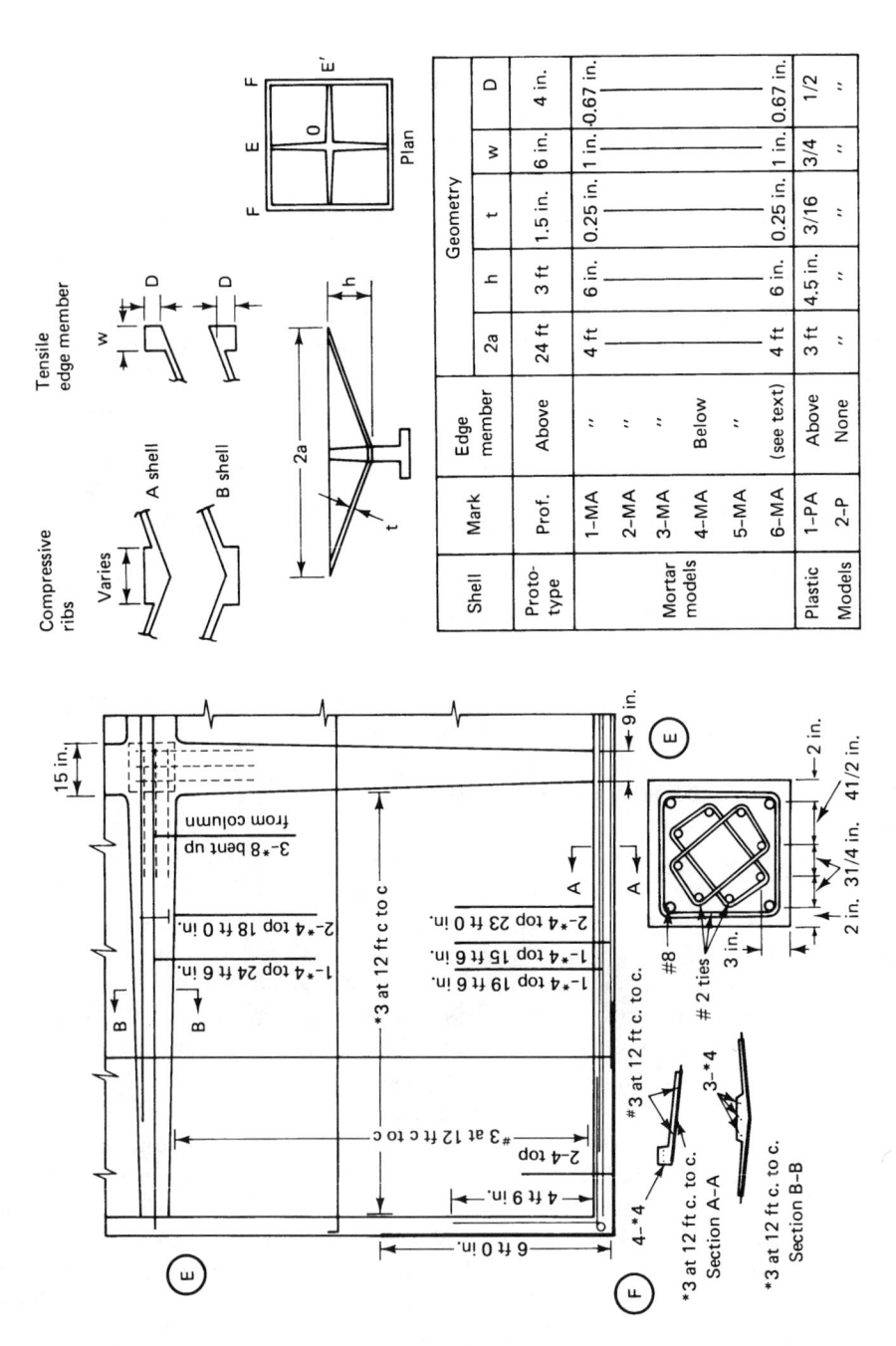

Figure 10.60 Shell geometry and reinforcing (1 in. = 25.4 mm; 1 ft = 0.305 m). (After White, 1975.)

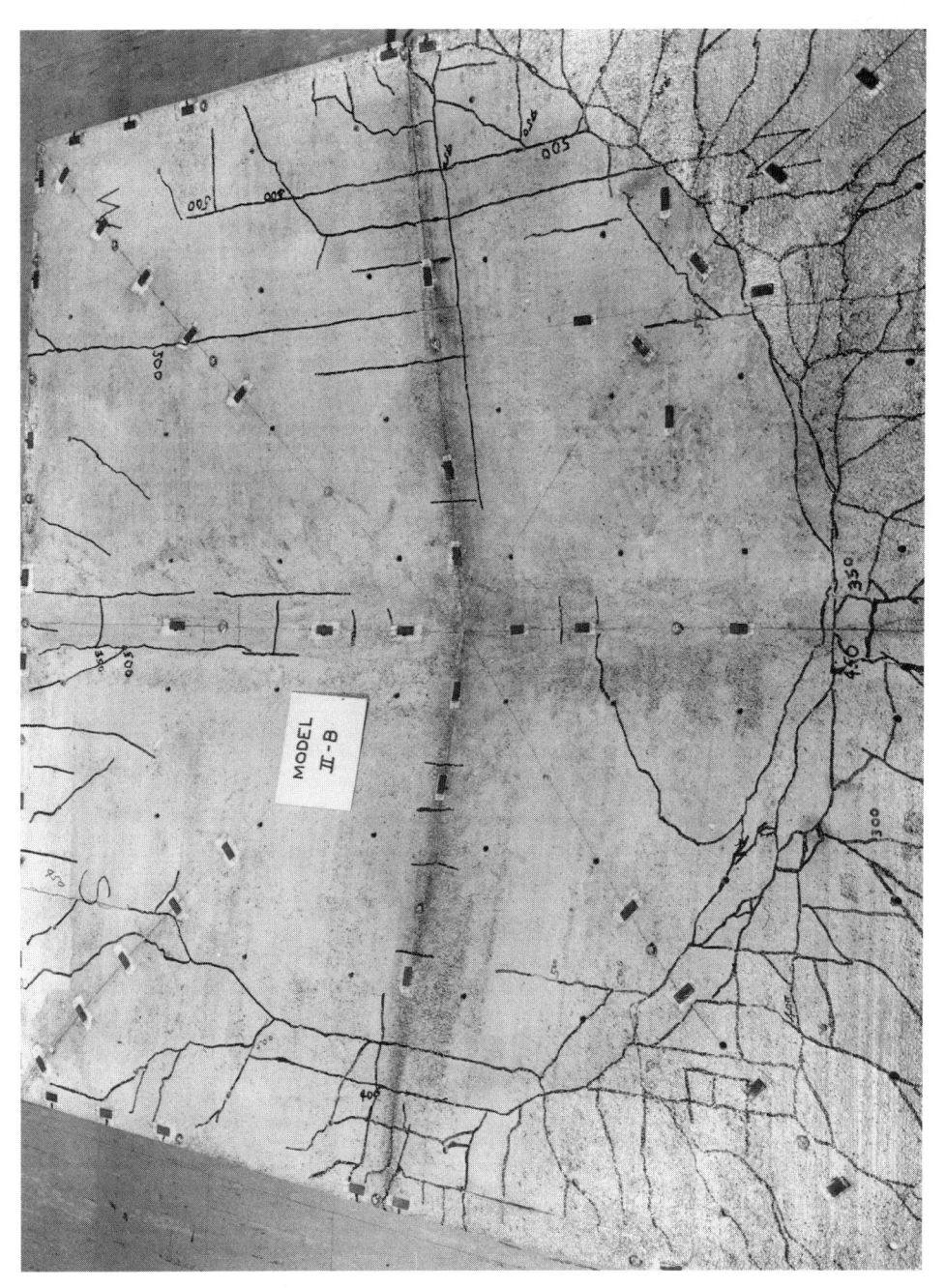

Figure 10.61 Model 4-MB after loading to failure. (After White, 1975.)

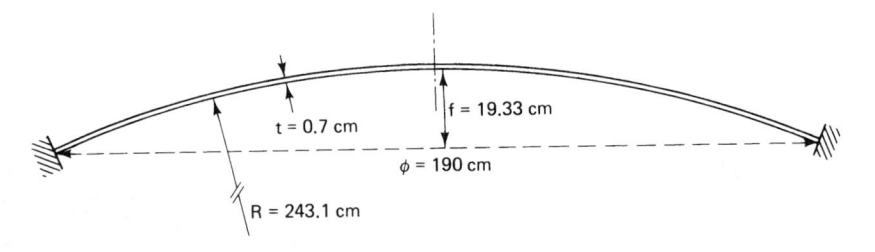

Figure 10.62 Shell geometry. (After Vandepitte and Rathe, 1971.)

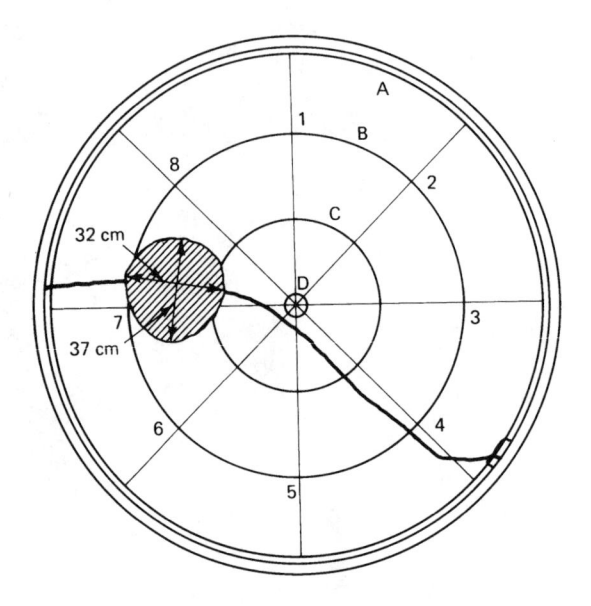

Figure 10.63 Failure mode of domes. (After Vandepitte and Rathe, 1971.)

Table 10.10 Analytically Determined Values for c in Equation 10.2

c	Ref.
0.365	Von Karman and Tsien (1939)
0.34	Tsien (1942)
0.32	Mushtari and Surkin
0.312	Feodosiev
0.31	Vreedenburgh (1966)
0.178	Classical value (n = 0.2) (not taking finite changes of the geometry into account)

edges skirting around the additional steel bars reinforcing the transverse edges of the hole (Figure 10.69b). This same cracking behavior was noted in specimens with two equal penetrations (Figure 10.70a). The idealized crack pattern shown in Figure 10.70b with the solid lines was used to find the ultimate load capacity using the strip method (Figure 10.71).

Based on the experimental study, the use of the strip method produced slabs that meet the serviceability and deflection requirements at working loads, as set by the ACI code, by providing safe, economical designs. The effect of the size of the central opening on the ultimate uniform load capacity ratio, W_u/W_o of the slab is shown in nondimensional form as a function of the opening size ratio, K in Figure 10.72. Both experimental and theoretical data are plotted in Figure 10.72. The experimental data fit the following expression:

MODEL APPLICATIONS AND CASE STUDIES

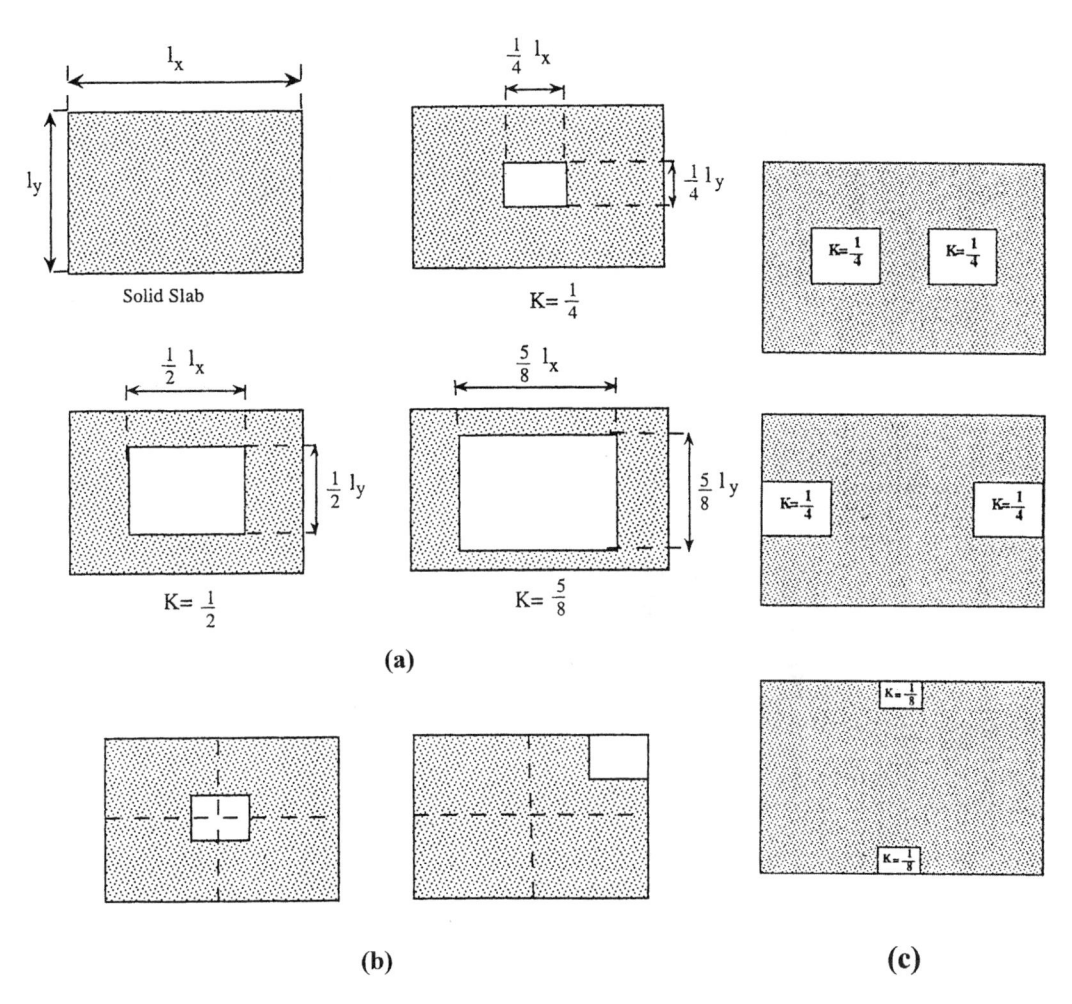

Figure 10.64 Test matrix of model slabs studied. (a) Slabs with central openings. (b) Locations of a single opening. (c) Multiple openings in a slab. (After Sagur, S., 1994.)

$$W_u/W_o = 1 + 2K - 4K^2 \tag{10.3}$$

Equation 10.3 is applicable to slabs with a central opening, where the steel displaced due to the opening is redistributed around the sides of the opening and is recommended for preliminary design. It is corroborated by the theoretical values shown in Figure 10.72 and is on the safe side.

Based on the test results, the following conclusions can be made:

1. The use of the strip method to design slabs with openings is an excellent alternative to other methods due to the fact that the steel can be placed where it is mostly needed. The test results indicate that the introduction of a central opening of size ratio $K = \frac{1}{4}$ in an isotropically reinforced solid slab, without redistributing the steel displaced due to the opening, caused up to 30% reduction in the uniform load-carrying capacity. Under similar conditions, slabs designed using the strip method did not suffer any reduction in the uniform load-carrying capacity.
2. Generally, slabs with an opening not exceeding a size ratio of 1:2 retain similar uniform load capacities as solid slabs, provided that the displaced steel due to the opening is redistributed around the opening.

(a)

(b)

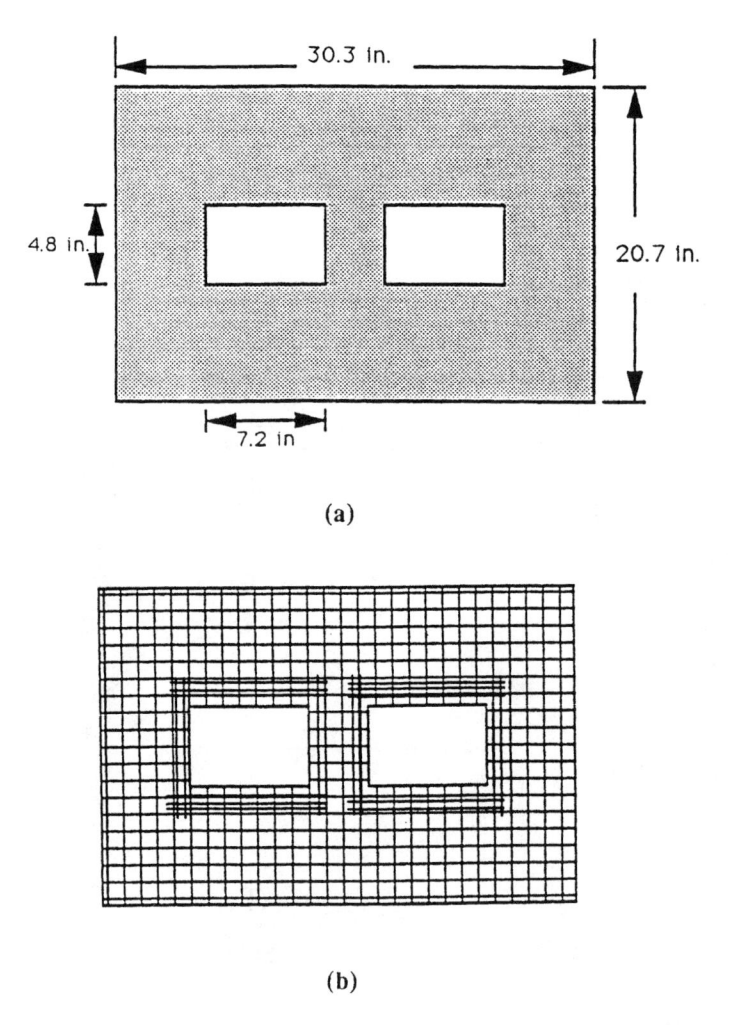

Figure 10.65 Model slab with two penetrations. (a) Plan view of the slab. (b) View of bottom reinforcement. (After Sagur, S., 1994.)

3. The stiffness of the slab, measured as the maximum deflection to slab thickness ratio, remains practically unchanged when a central opening of size ratio up to 1:3 is introduced and the displaced steel is redistributed around the opening.

10.2.3.4 Beams Reinforced with Ductile Hybrid FRP (D-H-FRP)

An experimental and analytical investigation to develop a ductile fiber-reinforced polymer (FRP) that can simulate the behavior of reinforcing steel bars has been conducted at Drexel University (Harris et al., 1997; 1998a; 1998b; Ko et al., 1997; Somboonsong et al., 1998) using small-scale modeling techniques. This application of physical modeling to develop new products (materials and structures) illustrates the great potential of this technique. The new FRP research and the development of interlocking mortarless block masonry (Case Study F, Chapter 1) are only but two examples of the many possibilities of using small-scale physical models for product development, an area not covered in the first edition of this book.

1. Background — Concrete structures reinforced with ductile hybrid FRP (D-H-FRP) bars can be designed on the basis of ultimate-strength theory similarly to concrete reinforced with steel bars. Stress–strain behavior such as that shown in Figure 5.57 is essentially bilinear. The average

MODEL APPLICATIONS AND CASE STUDIES 523

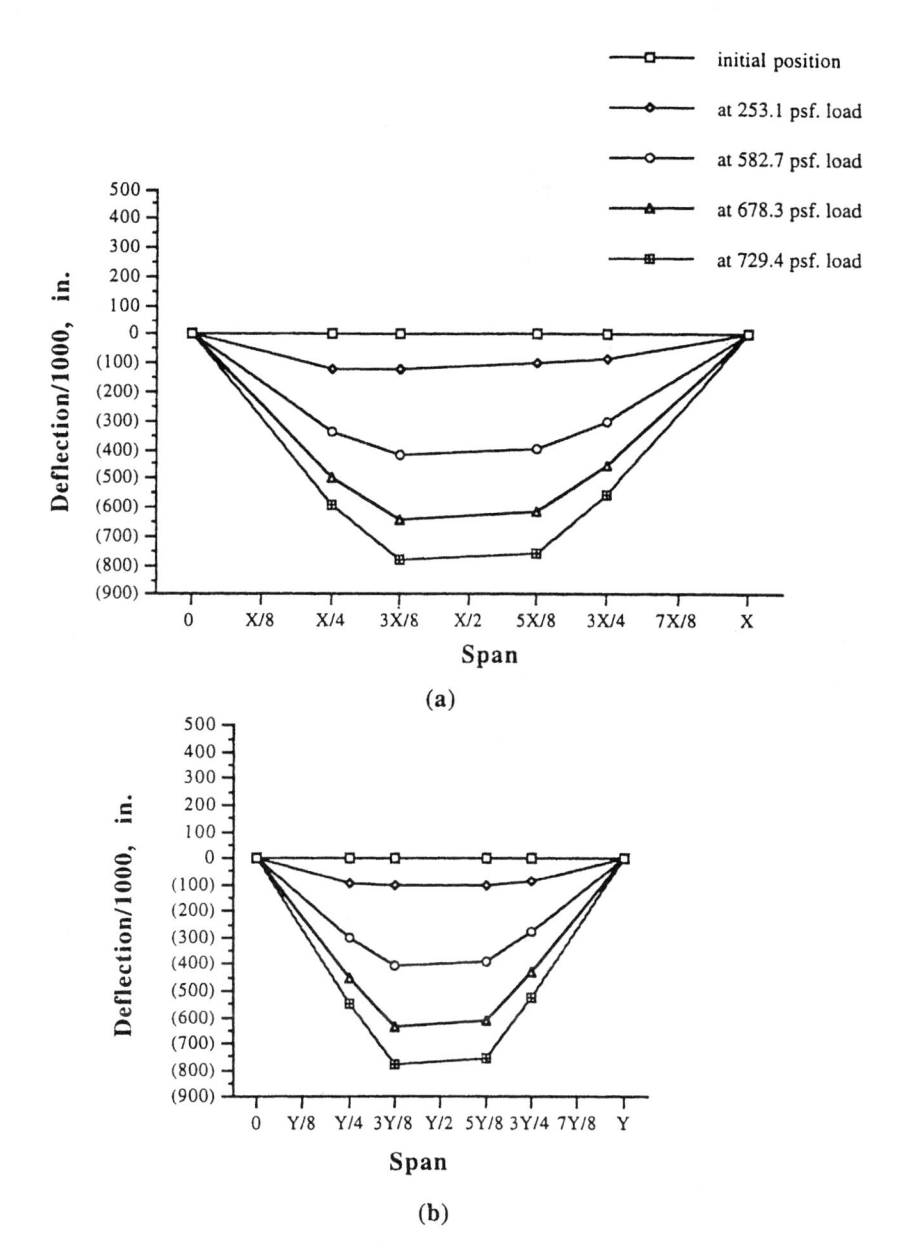

Figure 10.66 Deflection of centerline at different stages of test of slab NRCH-1. (a) X-direction deflection. (b) Y-direction deflection. (After Sagur, S., 1994.)

stress–strain curve for the six tensile samples shown in Figure 5.57 is presented in Figure 5.58. The bilinear lower-bound stress–strain curve of the mean of all the available tensile test data, also shown in Figure 5.58, can be used for flexural design purposes. The yield and ultimate strengths and the two moduli slopes of the bilinear design curve are also shown in Figure 5.58 as well as the stress–strain curves for Grade 40 and Grade 60 steel reinforcement for comparison.

Theoretical moment–curvature relationships were generated using the lower-bound bilinear design stress–strain curve (Figure 5.58) for simply supported 50 × 100 mm beams, 1.2 m long, under two-point loads. The dimensions of these beams are shown in Figure 10.73. Reinforcement consisted of 5-mm D-H-FRP and a specially deformed steel wire meeting the Grade 60 specifications (Chapter 5). The beams were designed to be underreinforced ($\rho = 0.36 \, \rho_b$ for steel and $\rho = 0.41 \rho_b$

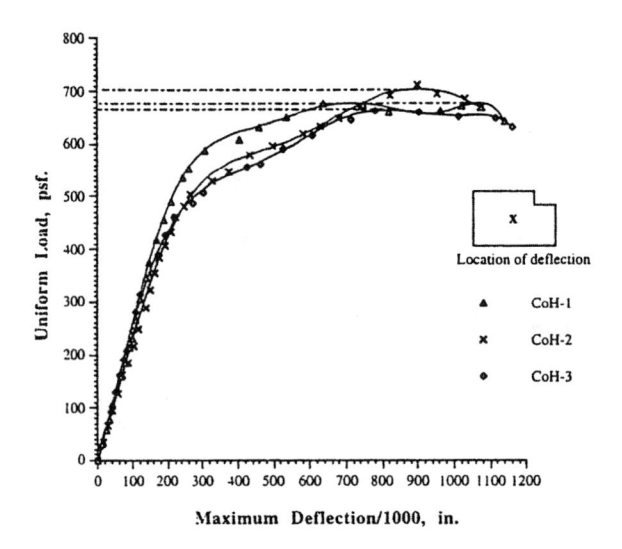

Figure 10.67 Uniform load–maximum deflection curves for CoH slabs. (After Sagur, S., 1994.)

for FRP) and to have the same ultimate moment capacity. Properties of all four beams are given in Table 10.11. As no FRP stirrups of this type were available and in order to facilitate the testing program, steel stirrups were used in both steel and FRP-reinforced beams. The predicted moment–curvature relationships for these beams are shown in Figure 10.74. Note from Figure 10.74 the close correspondence of the two beam behaviors. Because of the ductility of the steel and the new FRP, the beams use a minimum of reinforcement material and prediction of failure by secondary crushing of the concrete is expected in both.

2. Results of Beam Tests — The beams were tested using a 44.5 kN capacity displacement controlled bench-type Tinius-Olsen T10000 universal testing machine. Load–deflection and moment–curvature relationships were experimentally determined using the setup shown schematically in Figure 10.75. Typical ductile load–deflection behavior was observed in all FRP-reinforced beams (Figure 10.76). Both linear variable differential transformers (LVDTs) and regular dial gages were used to monitor continuously the deformations of the beams up to failure.

A comparison of the load–deflection behavior of the steel-reinforced and the D-H-FRP-reinforced beams is shown in Figure 10.76. Note from Figure 10.76 that the FRP-reinforced beams had very repeatable behavior with a high initial stiffness (identical to the companion steel-reinforced beam) up to cracking. The precracking behavior of all three FRP beams was identical to that of the steel-reinforced beam. The postcracking behavior of all three FRP-reinforced beams was very similar and all had a bilinear load–deflection curve up to the yield point. A maximum of five load/unload cycles was performed on each of the tested FRP-reinforced beams in the cracked and postyield ranges to study the nature of their inelastic behavior. As can be seen from Figure 10.76, the new D-H-FRP has significant energy-absorbing capabilities.

Moment–curvature relationships of all three FRP beams were computed numerically from the equally spaced deflection measurements at the midspan and are shown in Figure 10.77. As can be seen in Figure 10.77, a ductile behavior was obtained for all three beams with good reproducibility. The predicted moment–curvature relation, based on the bilinear lower-bound stress–strain curve (Figure 5.58) without tension stiffening, is plotted in Figure 10.77 and shows very close agreement with the experimental results. It should be noted that the ductile moment–curvature behavior of the new FRP is made possible by the fact that, through its special design using small-scale modeling techniques, it possesses a definite yield point, an equivalent bilinear stress–strain curve, and an ultimate strength higher than the yield.

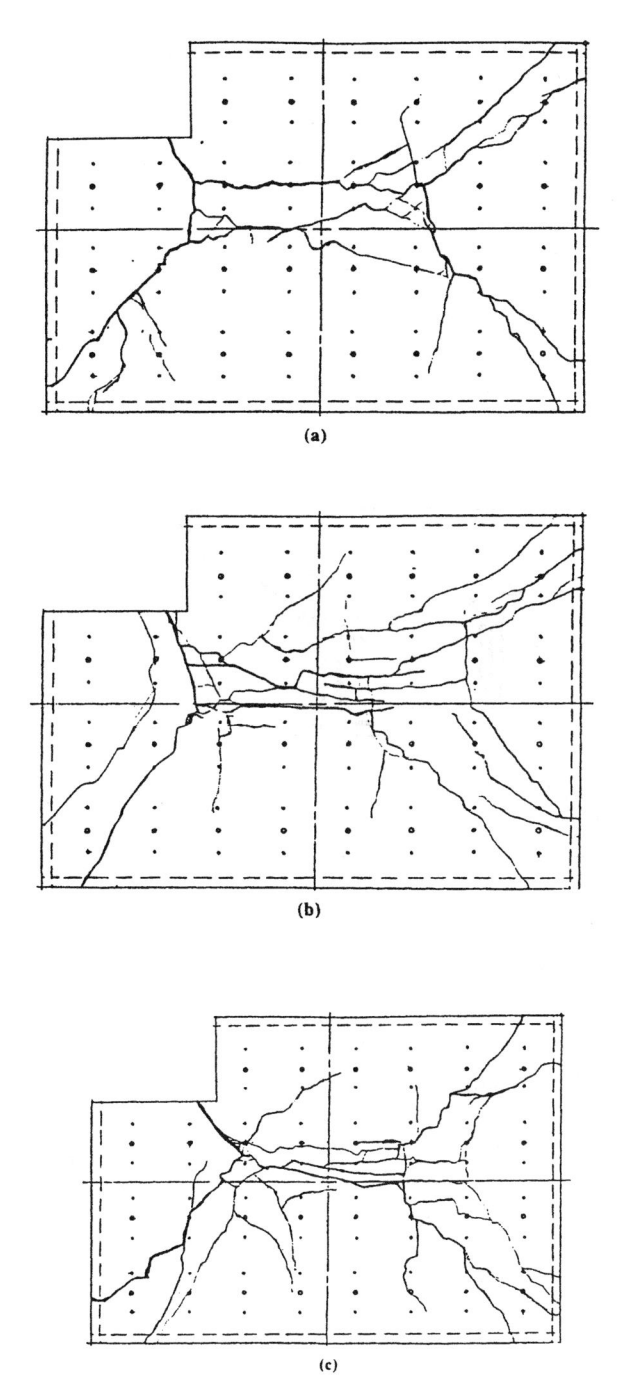

Figure 10.68 The crack pattern on the bottom surface of CoH slabs. (a) Slab CoH-1. (b) Slab CoH-2. (c) Slab CoH-3. (After Sagur, S., 1994.)

The test results of the 50 × 100 mm beams are summarized in Table 10.12. Table 10.12 shows the predicted and experimental ultimate load and bending moment values for each beam and the mode of failure. Ratios of experimental to theoretical ultimate values (columns 4 and 7) indicate that the chosen design bilinear stress–strain curve for the FRP is conservative. The ultimate loads of the three FRP-reinforced beams were remarkably close. The failure of all three FRP-reinforced

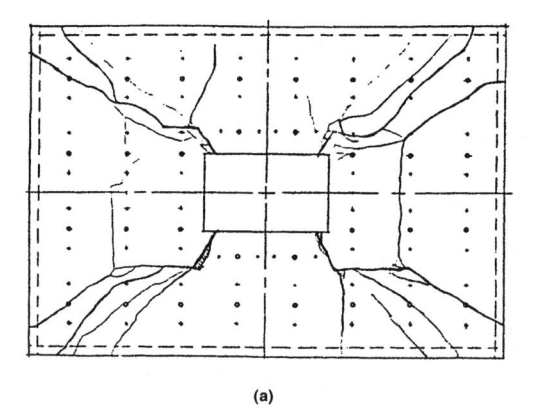

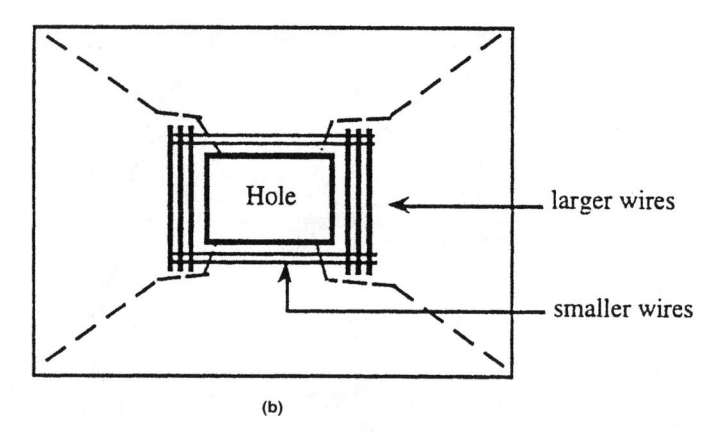

Figure 10.69 Path of the crack pattern for RCH slabs. (a) The crack pattern on the bottom surface of RCH slabs. (b) Idealized. (After Sagur, S., 1994.)

beams occurred after considerable inelastic deformation. Beam D-H-FRP No. 1 failed after fracture of two of the four FRP bars. Prior to this, failure of one of the FRP bars occurred with the beam continuing to carry the load and to show a small increase. No concrete crushing in the compression zone was observed for this beam. Beam D-H-FRP No. 2 showed concrete crushing prior to failure as shown in Figure 10.78. Beam D-H-FRP No. 3 also experienced concrete crushing prior to fracture of the FRP bars. The final cracking patterns of all four beams tested are shown in Figure 10.79.

Ductility indexes for beams are defined on the basis of deflections or curvatures as follows:

$$\mu_\Delta = \frac{\Delta_u}{\Delta_y} \quad \text{or} \quad \mu_\phi = \frac{\phi_u}{\phi_y} \tag{10.4}$$

where μ_Δ and μ_ϕ are the ductility indexes based on deflection and curvature, respectively, and Δ_y and ϕ_y are the deflection and curvature at yield, respectively, and Δ_u and ϕ_u are the corresponding values at ultimate load.

Naaman and Jeong (1995) suggested a definition of the ductility index μ_W based on energy considerations which is applicable to FRP and steel-reinforced or prestressed structures:

$$\mu_w = \frac{1}{2}\left(\frac{W_{\text{tot}}}{W_{\text{el}}} + 1\right) \tag{10.5}$$

MODEL APPLICATIONS AND CASE STUDIES 527

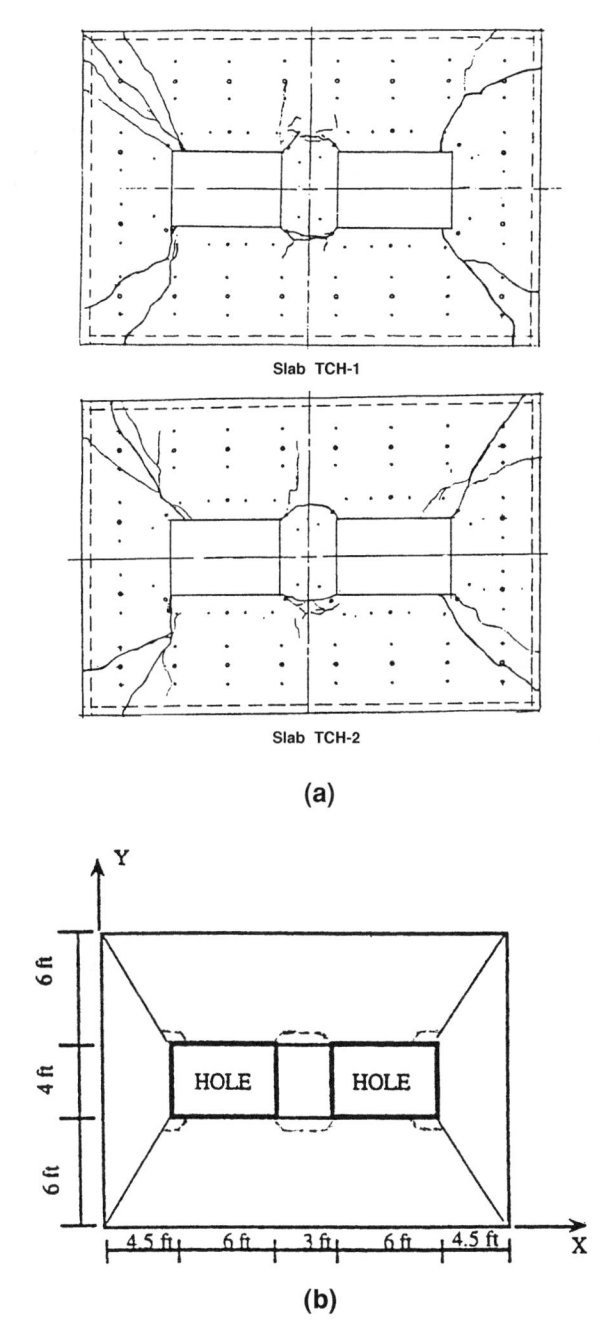

Figure 10.70 Failure mode of slab with two penetrations. (a) Actual; the crack pattern on the bottom surface of TCH slabs. (b) Idealized crack pattern of TCH slab set. (After Sagur, S., 1994.)

where W_{tot} is the total energy, computed as the area under the load defection curve up to the failure load, and W_{el} is the elastic energy which is part of W_{tot}.

Ductility indexes for the four test beams are shown in Table 10.13. Values for the FRP-reinforced beams were in the range from 4.6 to 5.4 based on deflection measurements (Figure 10.76, Equation 10.4) and compare well with the 6.1 of the companion steel-reinforced beam (Table 10.13, column 4). Ductility indexes based on measured curvature (Figure 10.77, Equation 10.4) ranged from 5.7 to 6.3 for the FRP beams and compare to 12 for the companion steel-reinforced beam

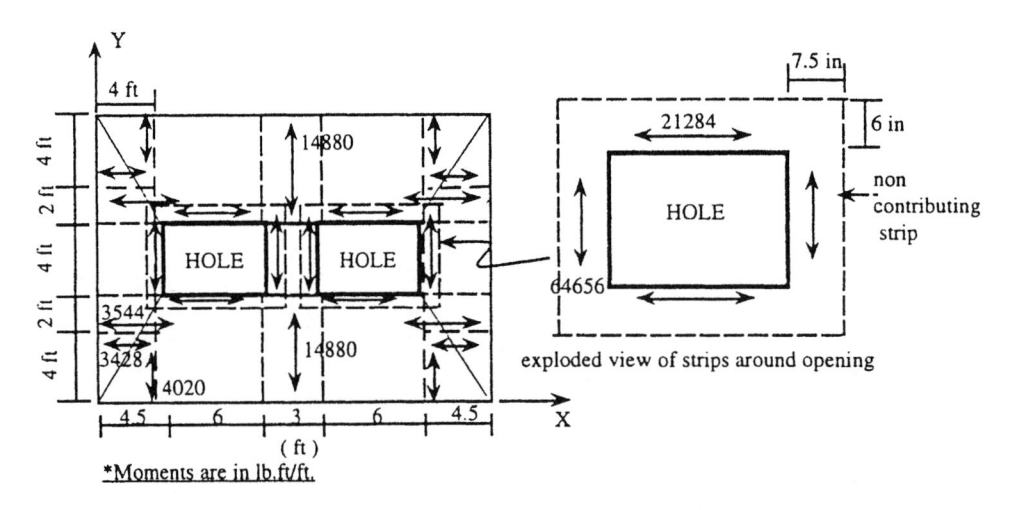

Figure 10.71 Moment capacity in strips of TCH slab set. (After Sagur, S., 1994.)

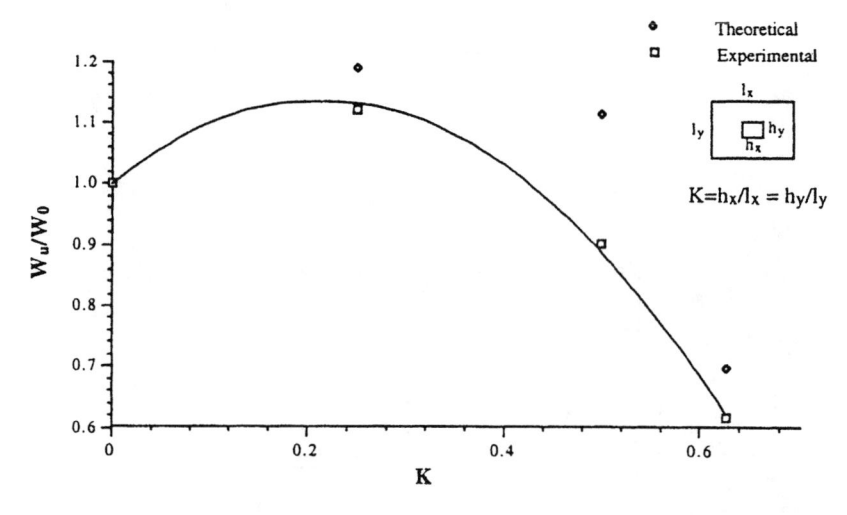

Figure 10.72 Effect of the opening size on the strength of the slab. (After Sagur, S., 1994.)

(Table 10.13, column 7). Ductility indexes computed on energy considerations as given by Naaman and Jeong (1995), (Equation 10.5) ranged from 3.4 to 3.8 for the FRP beams and compare to 4.3 for the companion steel-reinforced beam (Table 10.13, column 10). The measured ductility indexes of the FRP-reinforced and steel-reinforced beams are very similar. Such high ductility indexes for FRP-reinforced beams are not possible with any of the state-of-the-art linearly elastic FRP systems.

A new D-H-FRP bar developed at Drexel University has been presented. This new reinforcement has unique bilinear stress–strain characteristics which facilitate its use in new or repaired concrete structures. It has high strength, light weight, and is noncorrosive as concrete reinforcement in aggressive environments. Feasibility of producing the new reinforcement has been demonstrated with laboratory production of 5 mm nominal diameter bars. Tensile and flexural tests on small-scale beams show consistent stress–strain properties.

MODEL APPLICATIONS AND CASE STUDIES 529

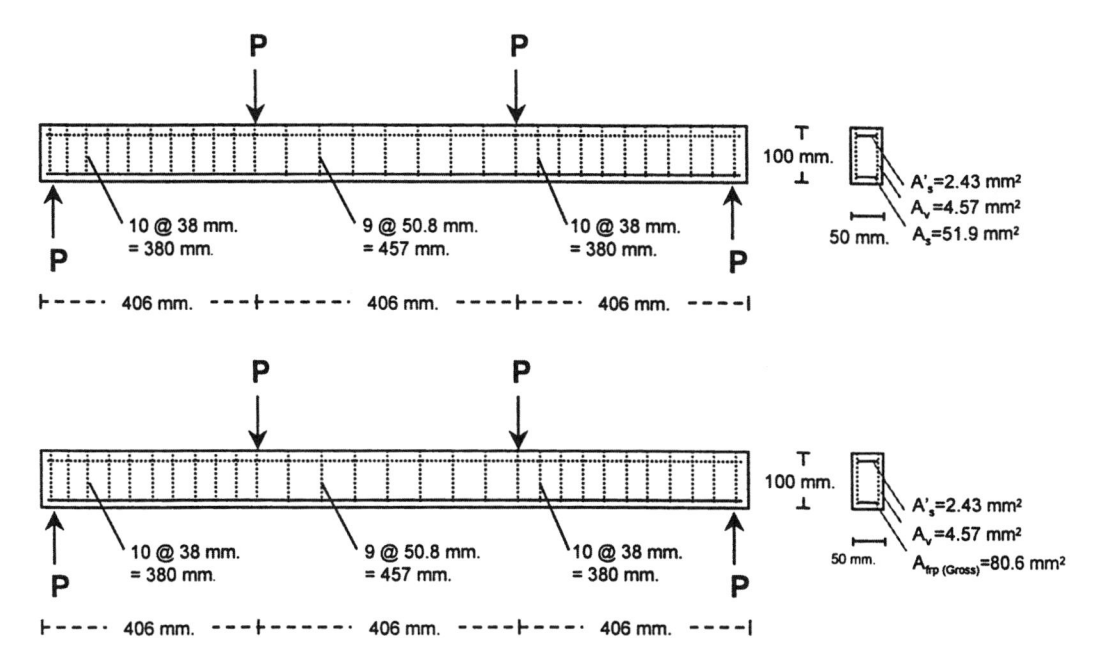

Figure 10.73 Reinforcement details of steel reinforced concrete (R/C) and FRP R/C beams.

Table 10.11 Mechanical Properties of Test Beam Materials

Beam Identification (1)	f_y, MPa (2)	f_u, MPa (3)	E_1, GPa (4)	E_2, GPa (5)	f'_c, MPa (6)	f'_t, MPa (7)	ρ,[a] % (8)
Steel	450	570	204.5	1.2	34.9	5.18	1.1
D-H-FRP-1	275	406	78.6	6.5	34.9	5.18	1.6
D-H-FRP-2	275	406	78.6	6.5	34.9	5.18	1.6
D-H-FRP-3	275	406	78.6	6.5	34.9	5.18	1.6

[a] Steel reinforcement ratio.

10.3 CASE STUDIES

10.3.1 Case Study A, TWA Hangar Structures

10.3.1.1 Introduction

This section describes the elastic and strength model studies used to assist in the design of the roof structure of the Trans World Airlines hangars at Kansas City (Guedelhoefer et al., 1972); (see also Chapter 1.)

After some discussion among the parties involved, the basic objectives of the study were defined as follows:

1. To obtain data for determination of wind forces on the prototype structure;
2. To determine stresses in the shell, edge beams, and center arch under various combinations of loadings in the design or service load range;
3. To obtain information to serve as a basis for accurately predicting short-term and long-term deflections;
4. To assess the ultimate strength and the factor of safety of the roof structure.

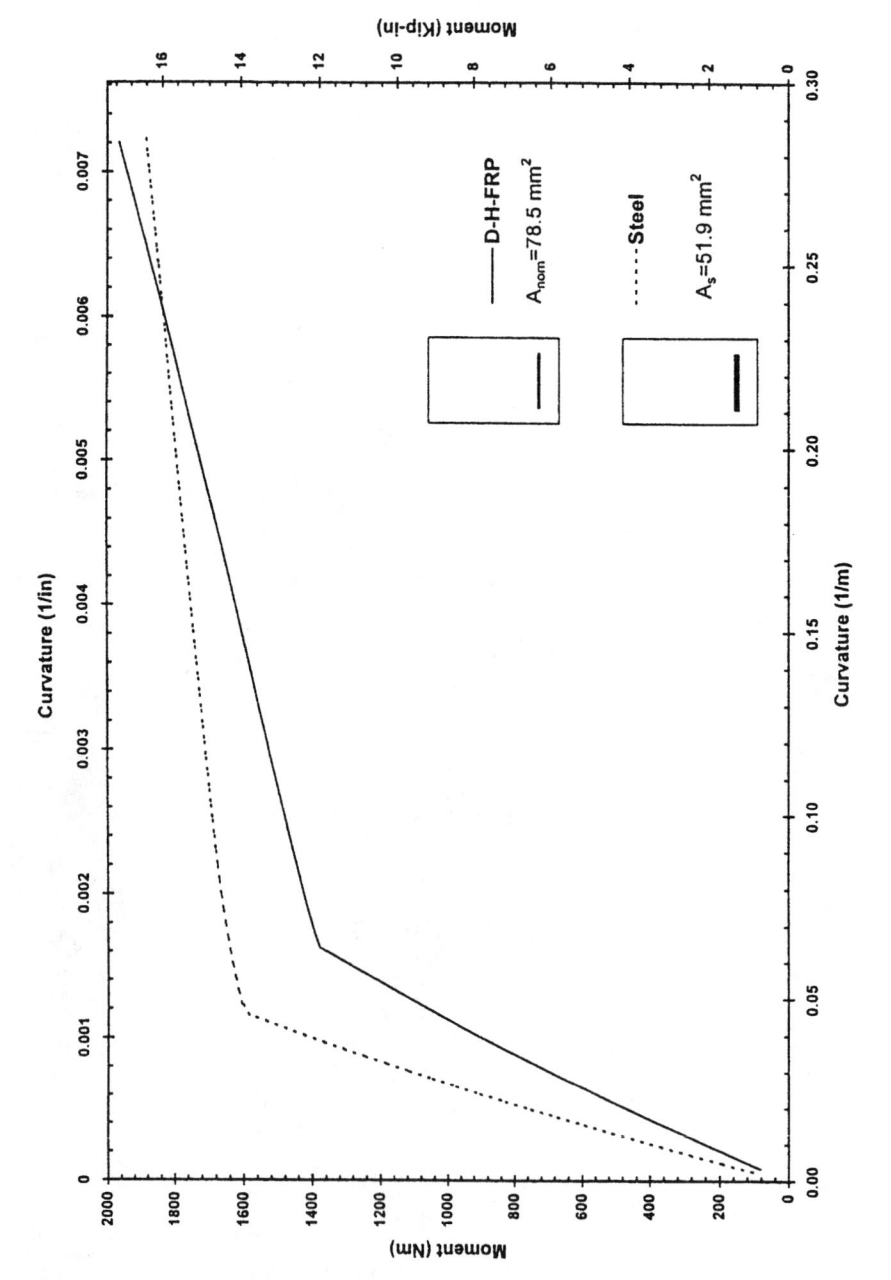

Figure 10.74 Moment–curvature predictions of beams reinforced with steel and D-H-FRP.

MODEL APPLICATIONS AND CASE STUDIES 531

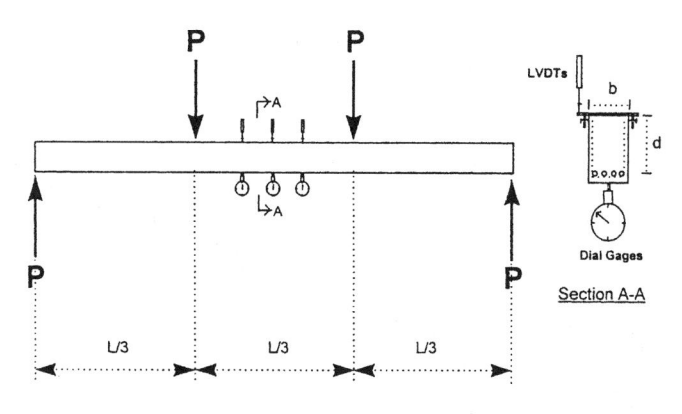

Figure 10.75 Beam test setup.

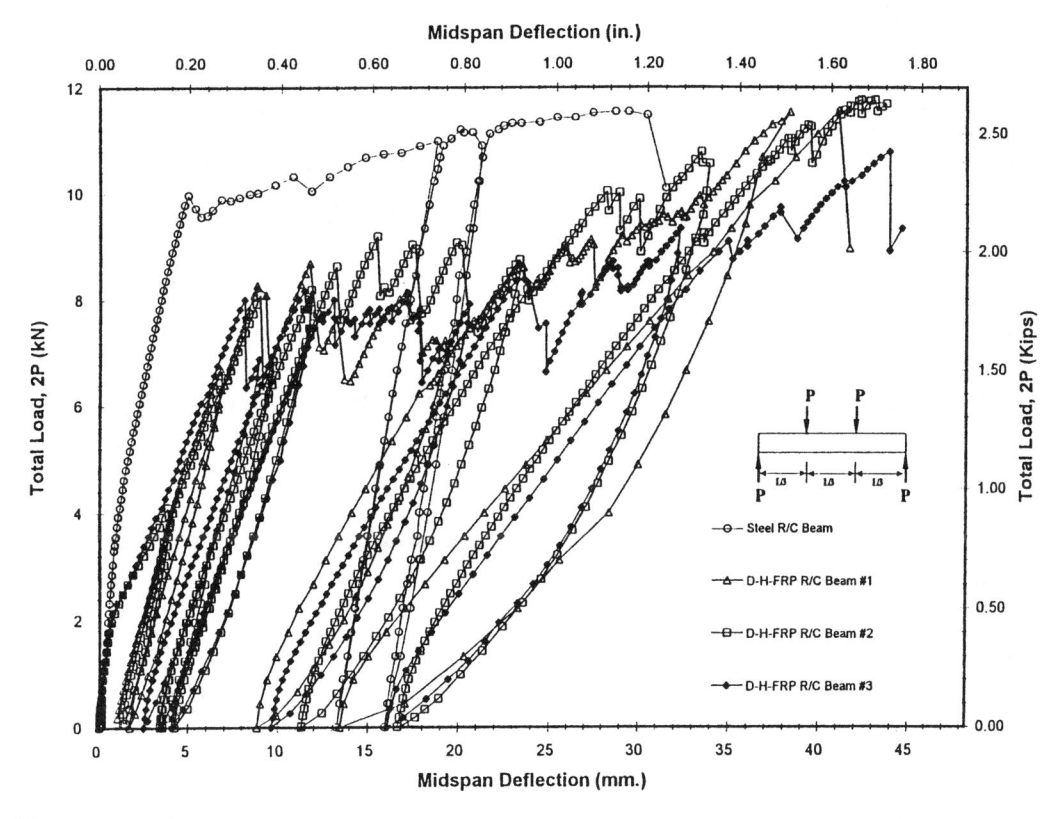

Figure 10.76 Comparison of load–deflection behavior of steel and new FRP-reinforced beams. (After Harris, H. G. et al., 1998b.)

Basic characteristics and applications of three elastic models A1, A2, and A3 and one ultimate-strength model A4 will be briefly reviewed in this section.

10.3.1.2 Wind Effects Model A1 (Phase I)

A $1/300$-scale wooden site configuration model was built reproducing accurately the shape of the entire TWA hangar complex along with certain features of the adjoining open country terrain that were considered to be influential on the distribution of external wind pressures or suction on the hangar complex (Guedelhoefer et al., 1972).

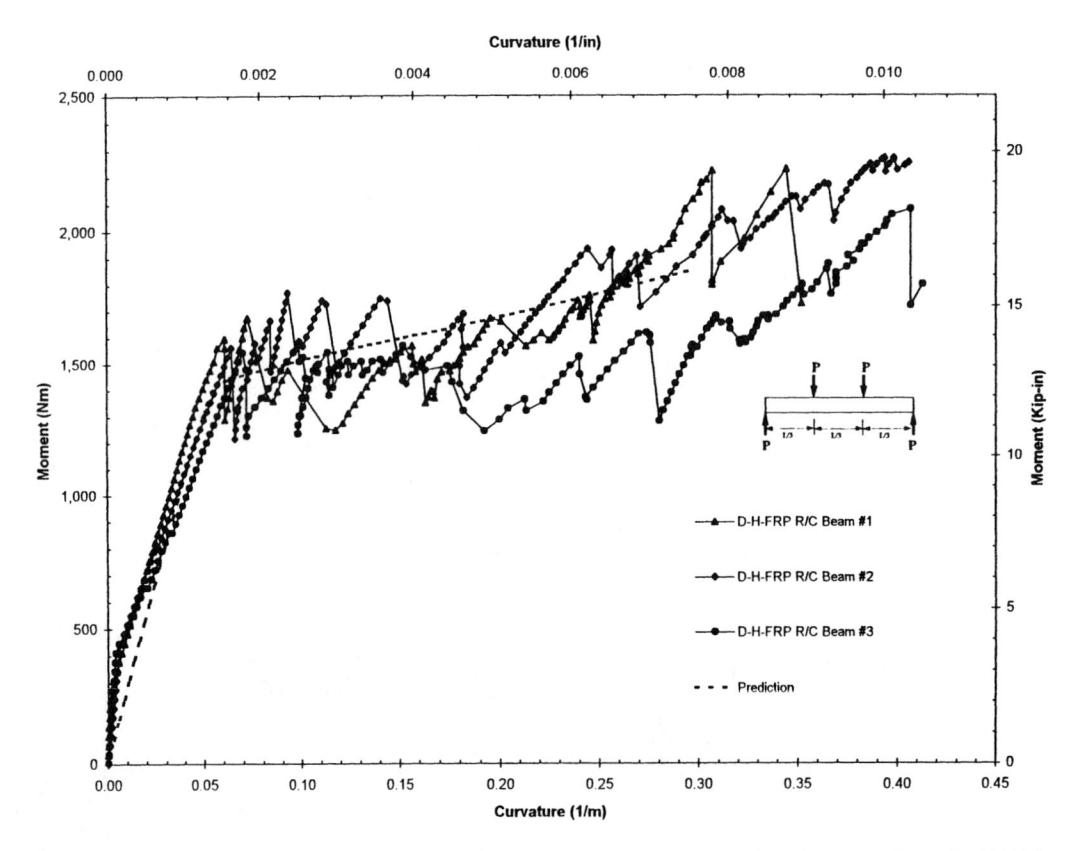

Figure 10.77 Moment–curvature behavior of D-H-FRP-reinforced beams. (After Harris, H. G. et al., 1998b.)

Table 10.12 Beam Test Results

Beam Identification (1)	Ultimate Load, N Theory (2)	Test (3)	Test/Theory (4)	Ultimate Moment, kNm Theory (5)	Test (6)	Test/Theory (7)	Failure Mode (8)
Steel	5001.0	5841	1.17	1905.0	2225.0	1.17	Yielding/concrete crushing
D-H-FRP-1	5140.1	5850	1.14	1958.0	2228.9	1.14	Yielding/FRP rupture
D-H-FRP-2	5140.1	5953.5	1.16	1958.0	2267.4	1.16	Yielding/concrete crushing
D-H-FRP-3	5140.1	5457.2	1.06	1958.0	2078.8	1.06	Yielding/concrete crushing
Average D-H-FRPs	5140.1	5753.6	1.12	1958.0	2191.7	1.12	

Four configurations were used, consisting first of the general shopping area and then successively adding hangars until the final configuration was obtained. Tufts of string mounted on the model established the airflow patterns from photographs in a wind tunnel test. Pitot tubes were installed at various elevations in the vicinity of the model hangars to establish the wind profile. The surrounding structures did not create unusual wind patterns influencing the pressures on the model. The model was mounted on a rotating table and subjected to all wind directions in 30° increments.

10.3.1.3 Wind Effects Model A2 (Phase II)

Subsequently, a 1/100-scale wooden model accurately simulating the hangar geometry details was constructed to develop more data regarding distribution of wind forces (Figure 1.3). The exterior

MODEL APPLICATIONS AND CASE STUDIES 533

Figure 10.78 Concrete compression failure in D-H-FRP-reinforced beam.

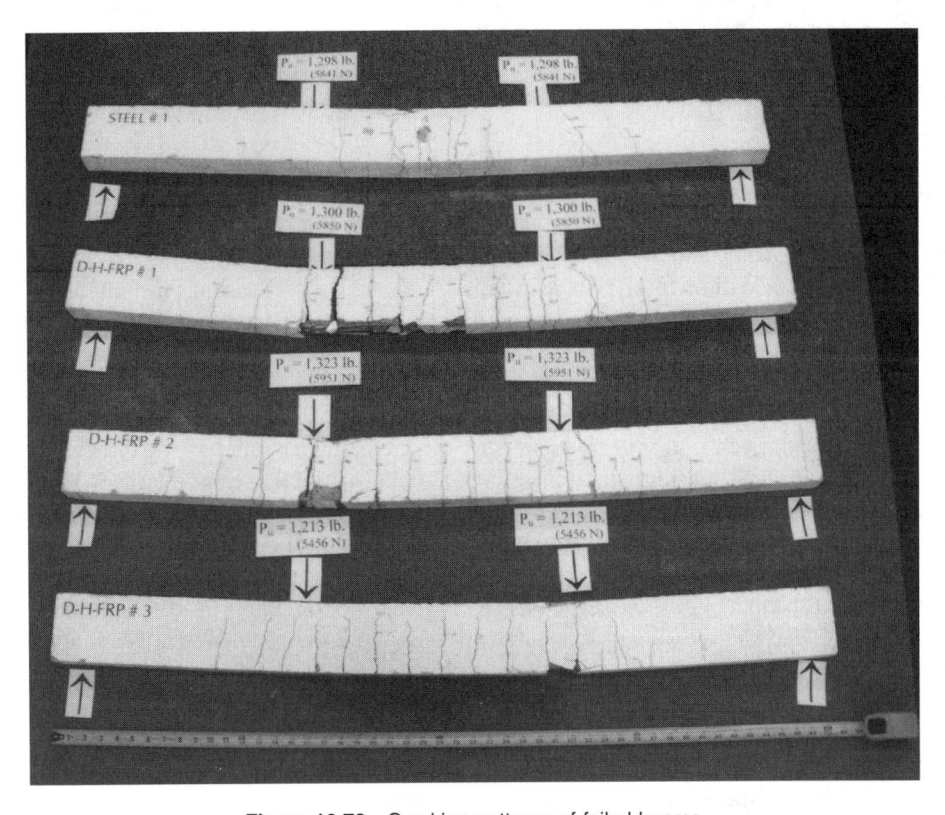

Figure 10.79 Cracking patterns of failed beams.

surface of the model was roughened to create turbulent effects expected in the prototype. The instrumentation consisted of Pitot tubes at 90 selected locations on the outside surface and five tubes in the interior of the structure. The phase II program consisted of the following three testing arrangements:

Table 10.13 Ductility Indexes of Beams Tested

Beam Identification (1)	Δ_y, mm (2)	Δ_u, mm (3)	μ_Δ, (4)	ϕ_y, rad/m (5)	ϕ_u, rad/m (6)	μ_ϕ, (7)	W_{EL}, kNm (8)	W_{TOT}, kNm (9)	μ_W, (10)
Steel	5	30.5	6.1	0.020	0.240	12.0	0.0431	0.3251	4.3
D-H-FRP-1	9	41	4.6	0.060	0.343	5.7	0.0546	0.3147	3.4
D-H-FRP-2	9	43.8	4.9	0.065	0.408	6.3	0.0609	0.3794	3.6
D-H-FRP-3	8.2	44	5.4	0.070	0.409	5.9	0.0545	0.3569	3.8
Average D-H-FRPs	8.7	42.9	5.0	0.065	0.387	6.0	0.0567	0.35033	3.6

1. The shell construction phase completed and all doors and walls absent;
2. The hangar building completely enclosed by side walls and all movable hangar doors closed;
3. The complete structure with the enclosing walls but with all hangar doors fully open.

The artificially generated wind profile represented the average of worst conditions as established from model A1. The wind tunnel could generate 112 to 144 km/h winds: all recorded data were corrected to the expected 100-year return period wind velocity of 139 km/h. Again, this model was subjected to all wind directions in increments of 30°. Similar to the observation in the smaller model, the surrounding structures were found to have no increasing effect on the pressures on the structure; in fact, the observed pressures showed a slight decrease.

10.3.1.4 Elastic Model A3

The data from the wind tunnel tests on models A1 and A2 were used to finalize the preliminary shell design and details of a 1/50-scale elastic model of the hangar structure. The elastic model was built of fiberglass and tested to study the elastic behavior of the structure under dead loads, live loads, concentrated loads, and their combinations (Figure 1.4).

Electrical resistance strain rosettes were installed at 50 stations to determine the principal stresses and the principal directions, the resulting maximum and minimum flexural and membrane stresses, and the orientation of the planes on which these stresses acted. Additional instrumentation was placed on the tie beam and on each of the two rear columns to monitor the loads in these elements.

The loading system initially consisted of 256 small containers of lead shot to represent the prototype dead load. Each lead shot container was suspended by a string passing through the shell and tied to a button that transmitted the load to the shell. This system caused large local bending stresses because the shell was extremely thin. Various other types of button and pad systems were tried in conjunction with the lead shot system, and all of these were considered unsatisfactory for representation of dead or live loads. Therefore, vacuum loading was adopted as an alternate loading system; the local bending stresses were noted to decrease considerably in the shell areas. Thus, dead load using lead shot containers was used for simulating the heavier edge beam and hangar door loads, and vacuum loading was used to reproduce the uniform portion of the dead load (Figure 1.4).

The design provided for large crane, scaffolding, and other mechanical support equipment to be suspended from the shell.

After the results of the preceding models were evaluated by the design engineers, and subsequent changes incorporated into the design, a microconcrete model was constructed to a scale of 1 to 10 (see Figure 1.5). The primary purpose of this phase of the investigation was to duplicate to scale, as accurately as possible, the prototype structure, to perform load tests to determine the load–deflection response of the structure at various points, and to determine the ultimate strength, with special attention given to mode of failure and factor of safety. Pertinent dimensions of the model are shown in Figures 10.80 and 10.81.

MODEL APPLICATIONS AND CASE STUDIES

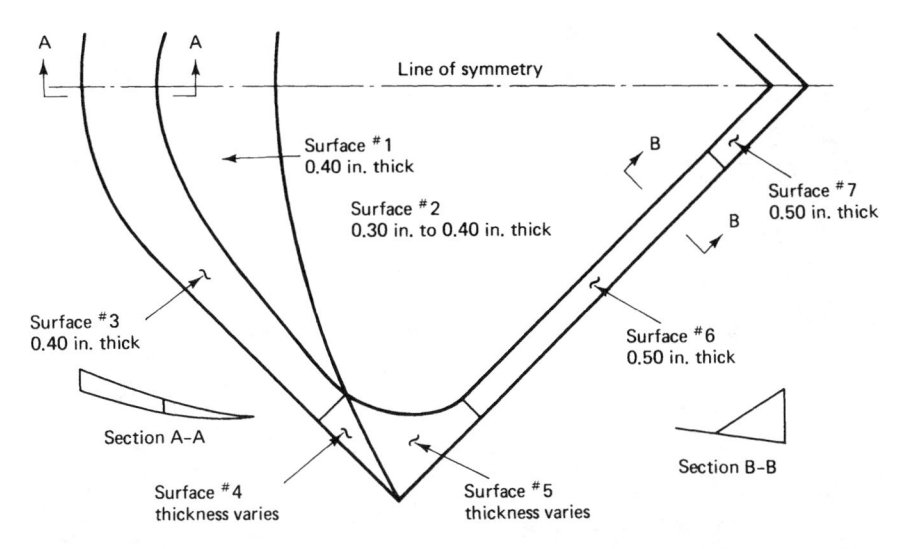

Figure 10.80 Shell surface numbering system. (After Guedelhoefer, O. C. et al., 1972.)

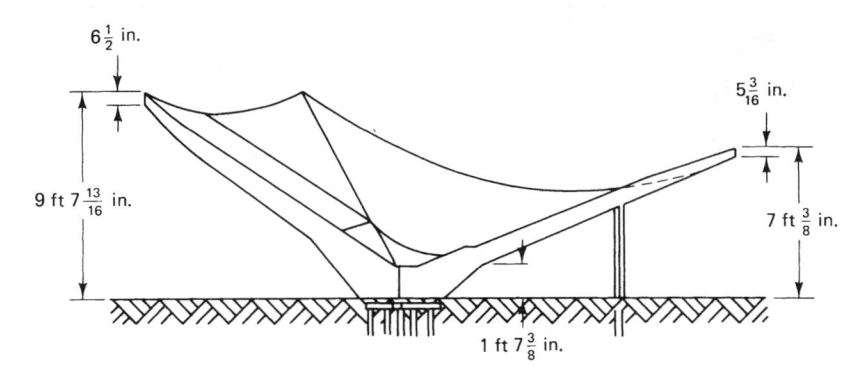

Figure 10.81 Elevation view of microconcrete model. (After Guedelhoefer, O. C. et al., 1972.)

Several trial mixes were produced in the laboratory, and 50×100 mm cylinders were tested for agreement with prototype design specifications. Reinforcing bars were simulated by using individual wires of commercially available sizes and commercially available welded wire fabrics. Precise distortion is introduced into the model by adjusting the bar locations or spacing to provide for slight differences in yield strength. With a scale factor of 1 to 10, wire sizes were available so that reasonably accurate scaling of significant reinforcing bars was possible in most cases.

Some distortion was necessary in order to simulate the shell surface designated as surface number 2 (Figure 10.80). In this region of the structure, a welded wire fabric of No. 15 gage wire was selected with spacing between wires chosen to provide the proper steel area in each of these directions. The design required reinforcing steel conforming to ASTM designation A432, with a minimum yield stress of 414 MPa throughout. The wire reinforcement (used in the abutment regions, primarily) had a yield strength much higher than 414 MPa; it was annealed moderately to reduce the yield strength. The most complex part of the model reinforcement was the front abutment wall, shown in the foreground of Figure 10.82.

Because the shells of the model were to be post-tensioned, supplementary investigations were undertaken to develop the necessary hardware and technique. Special auxiliary tests were conducted with the following objectives: to develop and test an anchorage system for the post-tensioning tendons; to develop and test a stressing system; to test a form-coating material; and to test an air-springing system to assist in decentering the formwork.

Figure 10.82 Microconcrete model during construction. (After Guedelhoefer, O. C. et al., 1972.) (Courtesy of Wiss, Janney, Elstner & Associates, Inc., Northbrook, IL.)

The actual sequence of casting the concrete was designed to be parallel to that anticipated for the prototype: footings, abutments, shell surfaces 1 and 2, vertical edge stiffeners, and then all other shell surfaces.

Screeding strips were used to control thickness in the 7.6- and 10.2-mm-thick areas of the shells. The actual casting process was continuous and concluded in 6 h. Vibration was accomplished by handheld vibrating sanders.

When the forms were removed, deflections and strains due to model weight were measured. The deflections of the rear edge were imperceptible, while the front edge deflected only 1.3 mm (i.e., 23.1 mm in the prototype).

A series of front truss lateral load tests was conducted to determine the effect of wind loading against the hanging front enclosure walls. The forces acting on the walls from either wind pressure or suctions were distributed to a series of 14 trusses, which in turn delivered the loads to the shell structure. Various combinations of pressures and suctions statically simulating conditions observed with the wind models were studied. It was found that the edge members were carrying most of these forces directly to the abutments, with only nominal stresses observed in the shell areas.

Since the uniform loads were produced by vacuum techniques, the thickened edge beams required the application of additional superimposed loads to correct for the dead loads of these members. Accurately weighed sandbags were used to apply these loads.

Dead-load data were taken from the model in two stages: first, when superimposed dead load was applied to the edge beams and, second, when the uniform dead load was applied. As testing progressed, it was decided that useful information could be obtained if the superimposed dead load was added in two controlled stages. The first stage was to place the full superimposed load on the rear stiffener, during which time complete deflection and strain data were recorded. The second stage was to place the full superimposed load on the front stiffener and to monitor instrumentation.

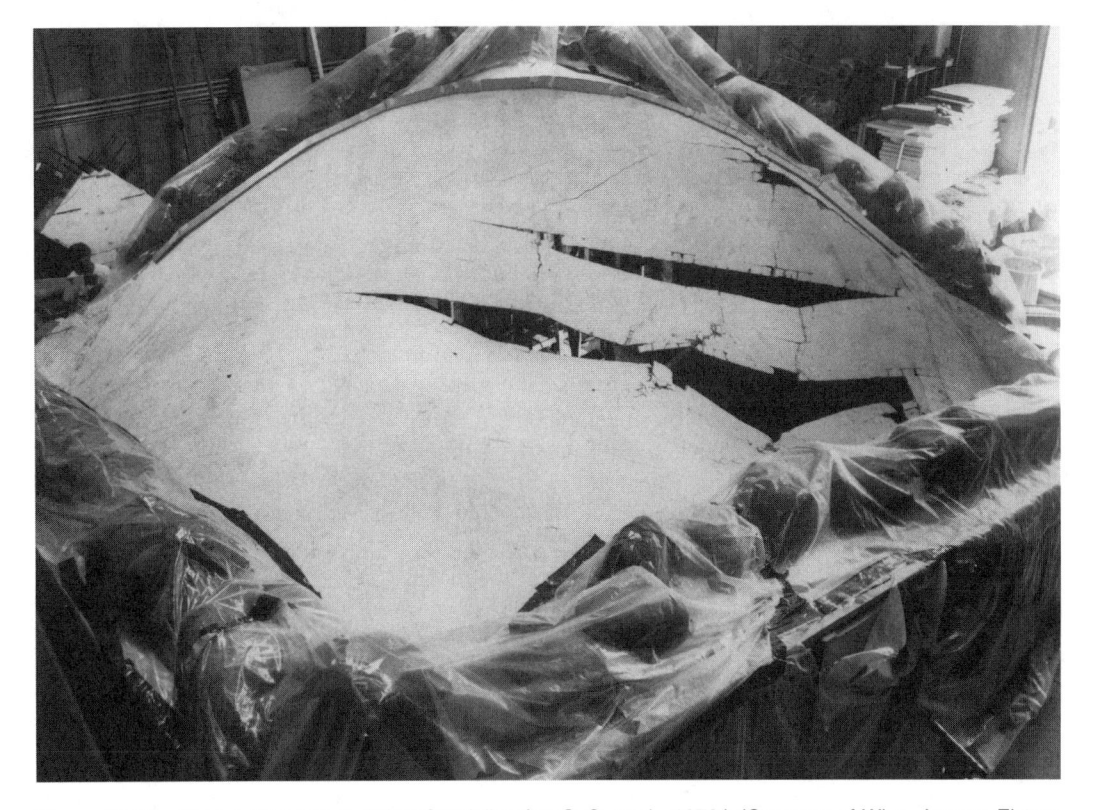

Figure 10.83 Model after failure. (After Guedelhoefer, O. C. et al., 1972.) (Courtesy of Wiss, Janney, Elstner & Associates, Inc., Northbrook, IL.)

The final test on this model consisted of loading to failure by the vacuum-loading method. As the load was applied in increments, data were recorded at the following load levels: 2.1, 4.5, and 5.4 kN/m^2. At a vacuum of 648 mm of water, or 6.3 kN/m^2, the model failed. The actual total load at this point on the structure was 6.9 kN/m^2 uniform load, plus the superimposed load of the edge beam. The ultimate failure mechanism was a pure tension (membrane) failure of the shell reinforcement, allowing the main area of surface 2 to separate and "tear" perpendicular to the line of symmetry (Figure 10.83).

It is interesting to note that the structure as a whole did not collapse, even though the center portion of the rear shell surface had failed. In fact, later when it became necessary to remove the model from the laboratory, the curved edge members and the center arch were extremely difficult to break up, even with sledgehammers, which attests to the inherent strength of the system. Model displacement during loading and the load at collapse showed excellent agreement with predicted values. Structural performance characteristics of the prototype to date have been excellent, and checked well with the model.

10.3.2 Case Study C, Reinforced Concrete Bridge Decks

10.3.2.1 Introduction

The research described in this section was conducted in the Department of Civil Engineering, Case Western Reserve University, under sponsorship from the Ohio Department of Transportation and the Federal Highway Administration (Perdikaris and Beim, 1988; Perdikaris et al., 1989; Perdikaris and Petrou, 1991; Petrou et al., 1994; Petrou et al., 1996). Tests were conducted under a concentrated static, stationary pulsating, and moving constant wheel-load on ⅓-scale (P-Series,

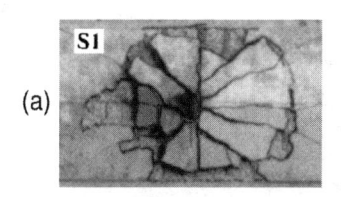

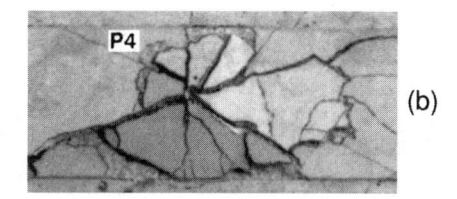

Figure 10.84 Cracking pattern of two deck regions of specimen BI3-7SP(1) subjected to concentrated static (S) or stationary pulsating wheel-load (P). (a) Concentrated static load (CC of BI3-7SP(1)). (b) Stationary pulsating wheel-load (NW of BI3-7SP(1)).

Figure 1.9b) and $1/6.6$-scale (B-Series, Figure 1.9a) physical models of a 216-mm-thick full-scale noncomposite concrete deck supported on four steel girders spaced at 2.13 and 3.05 m. The bridge was simply supported and nominally 15.24 m long. Three deck designs were studied: (1) the AASHTO design ("orthotropic" steel reinforcement), (2) the Ontario Highway Bridge Design Code (OHBDC) deck design ("isotropic" steel ratio of 0.3%), and (3) deck design with an "isotropic" steel ratio of 0.2%. The aggregates of the model concrete and the nominal diameter, mechanical properties, and size and shape of the surface ribs for the model steel reinforcement were scaled accordingly. The static test results of the deck "panel" models for the two scales of $1/3$ and $1/6.6$ showed that the size effect is minor.

10.3.2.2 Static Deck Behavior

Results from tests on seventeen $1/6.6$-scale bridge deck models (B-Series), six $1/6.6$-scale bridge deck panel models (BP-Series), and five $1/3$-scale bridge deck panel models (P-Series) were carried out in the study. Nondimensionalized load–deflection curves of the static load to static ultimate-strength ratio were plotted as a function of the net load-point deflection to thickness ratio for three reinforcing patterns (AASHTO, OHBDC, and isotropic 0.2% deck design), two prototype girder spacings, and three boundary conditions. These were defined as full continuity (C) for the central region of the bridge deck, no deck continuity (S) for the "simply supported" deck panels, and partial continuity (SC) for the corner regions of the deck. In was observed that the ultimate net deck deflections were always less than half the deck thickness. A narrower range of peak net deck deflection ($0.06h$ to $0.30h$, h = thickness) was recorded for the AASHTO full decks compared to the Ontario decks ($0.11h$ to $0.46h$) and the isotropic-0.2% decks ($0.17h$ to $0.48h$).

The concrete decks reinforced according to the AASHTO Code showed considerably higher influence of the girder spacing and restraint level on the type of failure mode than the Ontario bridge decks. It is apparent that the response of the "lighter" reinforced Ontario decks is practically independent of the girder spacing and boundary conditions investigated.

On the top deck surface, minor damage was noted around the loaded area, especially for the AASHTO decks. The damage was more for the ductile isotropic-0.2% decks. At the bottom deck surface under the loaded area, a fan-shaped pattern of radial positive cracks was observed (Figure 10.84a). No major longitudinal positive flexural cracks formed in the AASHTO decks as in the Ontario and isotropic-0.2% decks, which indicates the brittle (punching shear) nature of the failure mode in the former. In the case of the isotropic-0.2%, some Ontario decks, and the deck panels, steel yielding (especially in the transverse direction) controls causing major longitudinal bottom flexural cracks and resulted in a primary flexural failure.

The size of the punched-out cone at failure was similar for all specimens. The shape of the bottom perimeter of the cone was nearly circular for the isotropically reinforced decks and rather ellipsoidal for the orthotropically reinforced decks with the major axis in the transverse direction. The average angles, φ, of the inclined conical failure surface to the horizontal (Figure 10.85) was measured after sawing the decks transversely and longitudinally (Petrou, 1991).

MODEL APPLICATIONS AND CASE STUDIES

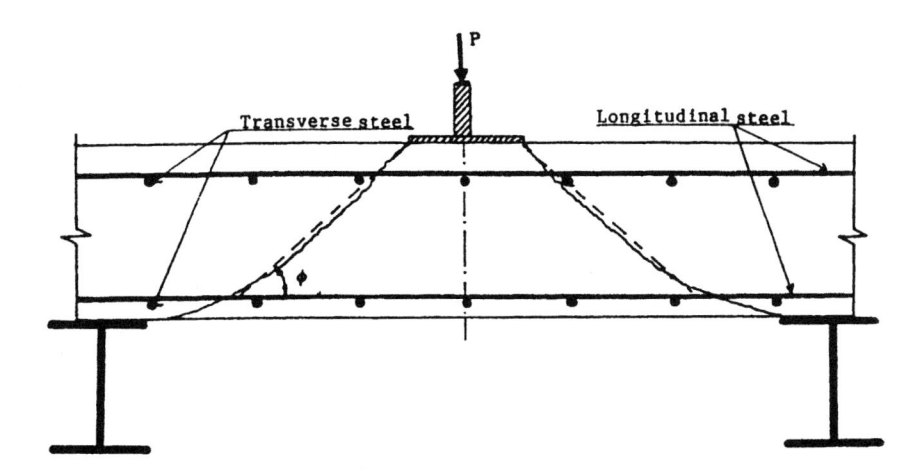

Figure 10.85 Crack angle ϕ and punched-out cone. (After Petrou, M. F., 1991.)

A major concern in any small-scale model study, from which prototype behavior is to be inferred, is the validity of the model results. Comparisons of load–deflection and load–strain measurement in the transverse and longitudinal bottom steel in the 1/3- and 1/6.6-scale models used in this case study indicate that the behavior of the two model scales is practically identical. Typically, the applied load to the static ultimate strength ratio is plotted as a function of the deflection to thickness ratio for the 1/6.6-scale (BPO-10S) and 1/3-scale (PO-10S) specimens designed under the AASHTO Code (Figure 10.86). The cracking load level for the central deck region is similar in both small-scale models. This load level corresponds to a full-scale load approximately two times the AASHTO design load of 92.5 kN. Yielding of the flexural steel under the loaded area occurs at practically the same load level in both scale models showing that the steel behavior has been properly modeled. The yielding load level, on the other hand, corresponds to approximately five times the AASHTO design load. Prototype extrapolations from the model test results are therefore valid.

Based on the investigation of the static deck behavior, the following conclusions were reported:

1. The decks designed according to the AASHTO Code and supported on steel girders spaced at 2.13 m exhibited a rather brittle behavior with a primary failure mode that of punching shear. Decks with a spacing of 3.05 m, however, showed a semibrittle behavior but appear to have failed in a punching shear mode.
2. Based on the strain measurements on the top concrete surface in the transverse and longitudinal directions, the applied load for the AASHTO decks is distributed mainly in the transverse direction (one-way slab action), while for the Ontario decks is distributed more uniformly in the two reinforcing directions (two-way slab action).
3. Punching shear failure appears to be interrelated with the snap-through mechanism activated in the deck (Figure 10.87). The predicted critical deflections at instability of a two-dimensional three-hinge compressive rigid strut mechanism restrained at the hinge supports correlate well with the measured ultimate net deck deflections.
4. For all the noncontinuous deck "panel" models tested, the ACI Code gives predictions very close to the experimental findings independently of the girder spacing and steel reinforcing pattern.
5. The ACI Code prediction for the shear capacity of the reinforced concrete bridge deck models (entire deck continuous over the steel girders) is very conservative. The predictions are more conservative in the case of decks with orthotropic steel reinforcement (AASHTO decks).

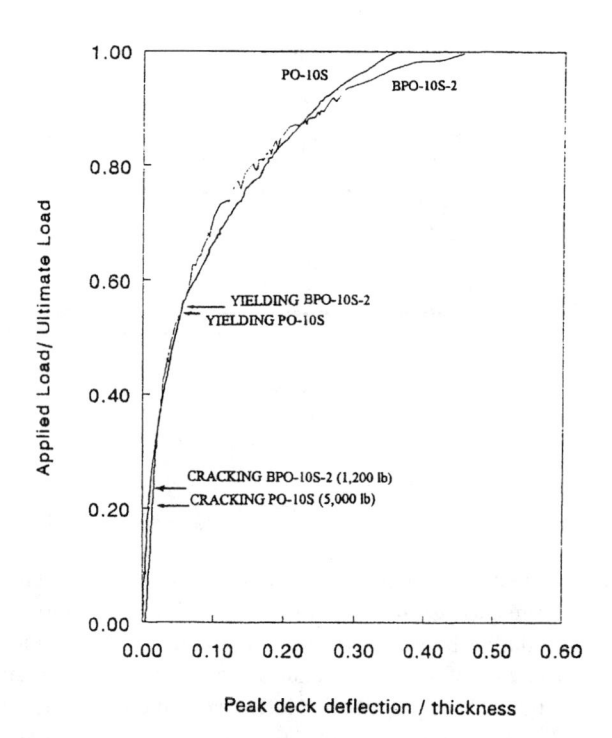

Figure 10.86 Response of BPO-10S-2 and PO-10S under static load (AASHTO design).

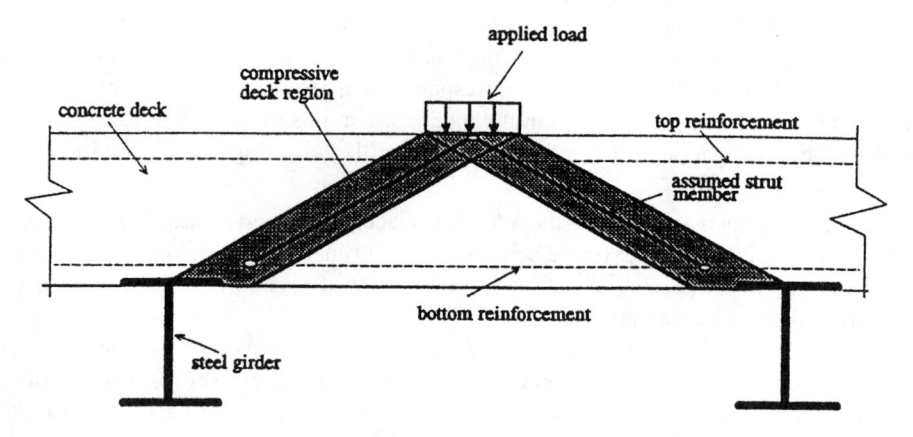

Figure 10.87 Schematic presentation of the three-hinge compressive strut mechanism in a reinforced concrete bridge deck.

6. The presence of compressive membrane forces enhances the ultimate capacity of the decks up to five times the Johansen yield-line load. The ultimate load-carrying capacity of the deck "panel" models, expected to exhibit the lowest level of membrane compressive action, is similar to the yield-line prediction.

10.3.2.3 Fatigue Deck Behavior

The cracking patterns under a stationary pulsating load are very similar to those produced by a static load (see Figure 10.84b). The bottom flexural radial cracks are subjected to an opening–closing type motion. On the other hand, fatigue under moving loads result in gridlike bottom flexural

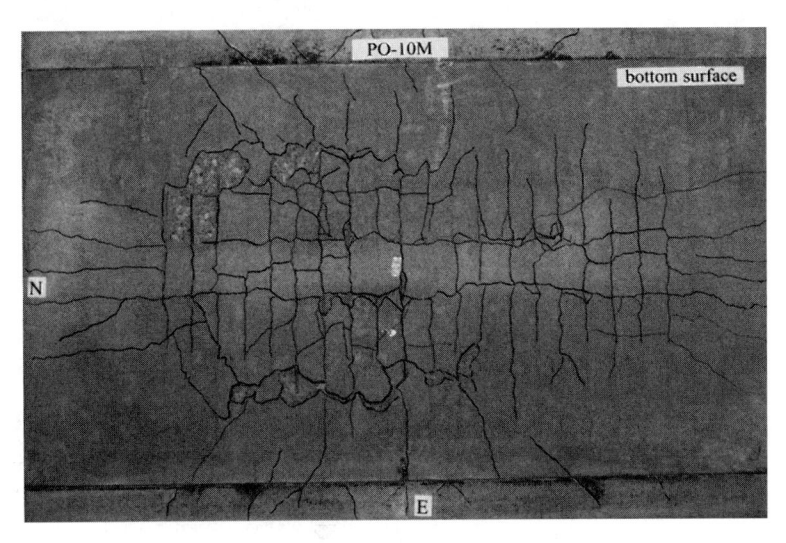

Figure 10.88 Cracking pattern of ⅓-scale model bridge deck panel, PO-10M, subjected to moving constant wheel-load.

cracks (transverse and longitudinal) following the reinforcing grid pattern (Figure 10.88). Initially, a major longitudinal flexural crack usually forms at the bottom surface of the deck along the wheel path at midspan (Figure 10. 89). The moving wheel-load causes an opening and closing of this major longitudinal crack and forces it to propagate upward, while at the same time some minor longitudinal cracks appear parallel to the steel girders. The longitudinal cracks open wider with increasing number of wheel-load passages. Then, transverse flexural cracks (perpendicular to the steel girders) form practically at the same spacing as that of the bottom transverse flexural steel reinforcement.

While the bottom longitudinal cracks open and close (flexural mode), as shown in Figure 10.89, as the wheel-load moves back and forth on the bridge deck, the bottom transverse cracks not only open and close (flexural mode) but also slide up and down (shearing mode) causing continuous rubbing of the crack interfaces (Figure 10.90). This "reversing" shear movement of the crack surfaces in the transverse cracks causes degradation of the interface shear transfer mechanism. It results also in debonding along the steel reinforcement since the cracks in both directions form usually in close proximity to the bottom steel reinforcement in a gridlike pattern (see Figure 10.88).

A total of 24 fatigue tests on concrete decks with different steel reinforcement patterns (AASHTO and OHBDC) and full-scale girder spacings (2.13 and 3.05 m) were performed at various maximum levels of stationary pulsating load. In addition, a total of 35 fatigue tests were conducted on concrete bridge deck models under constant moving wheel-load. The simulation of the traffic loads as stationary pulsating loads is not realistic because this method overestimates the fatigue life of concrete bridge decks (Figure 10.91). Exponential curve fitting of the test data from this study is also shown in Figure 10.91. The main findings of the fatigue testing are as follows:

1. For a given load level, decks subjected to a stationary pulsating load exhibited much higher fatigue strength than those fatigued under a moving constant wheel-load.
2. The predicted fatigue strength of the bridge decks under a moving constant wheel-load for 2.5 million wheel load passages is in the same range as the average flexural cracking load level of the decks, $P_{cr}/P_u = 0.21 \div 0.37$. The predicted fatigue strength of the decks under a moving constant wheel-load for 100 million passages ranges between $0.14P_u$ and $0.21P_u$ which is lower than the average flexural cracking load level.
3. The failure mode of the bridge deck models fatigued under either stationary pulsating or moving constant wheel-loads depends on the applied load level. For load levels higher

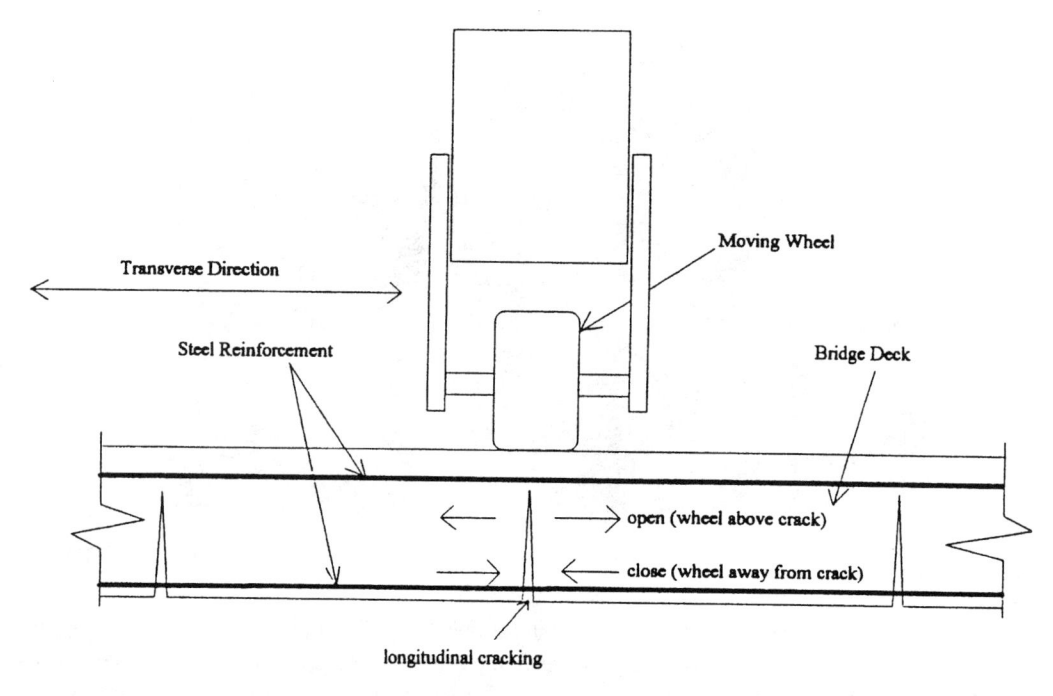

Figure 10.89 Schematic presentation of the longitudinal crack motion (flexural) as the wheel-load moves on the bridge deck in the longitudinal (traffic) direction. (After Perdikaris, P. C. et al., 1993.)

than $0.60P_w$, fatigue failure is due to punching, while for lower load levels the deck fails primarily in flexure. Independently of the primary failure mode under fatigue, almost all the decks eventually punched through.

4. Under a stationary pulsating load, the evolution of the total peak deck deflection including permanent deformations appears to have an S-shape behavior, increasing rapidly in the initial and the final (approaching failure) fatigue stages with a more gradual increase at a constant rate in the intermediate stage of fatigue. A similar general S-shape behavior was observed for the total deck deflection for the decks subjected to a moving constant wheel-load but with a much higher rate of increase for the deflection.
5. Based on the experimental fatigue results in this study, it was concluded that initiation of flexural cracking in the concrete deck is a necessary condition contributing to failure of the deck under fatigue loading conditions. Thus, cracking should be limited or precluded.

10.3.3 Case Study E, Prestressed Wooden Bridges

10.3.3.1 Introduction

The case study reported here was part of an effort to investigate the feasibility of using Pennsylvania hardwoods such as red and white oak (instead of the more commonly used softwoods) in transversely prestressed short-span rural-type bridges. Two ⅙ models of an idealized two-lane 6.71-m-long bridge design were tested at Drexel University to study the long term prestressing effects and the simulated live load and ultimate-strength behavior of this type of construction (Hoffman, 1990). Each deck model consisted of 88 12.5 × 50 × 1.22 m creosote treated red oak laminae (Hoffman, 1988). The general layout is shown in Figure 1.16. The models are based on a prototype of the dimensions shown in Table 10.14.

MODEL APPLICATIONS AND CASE STUDIES

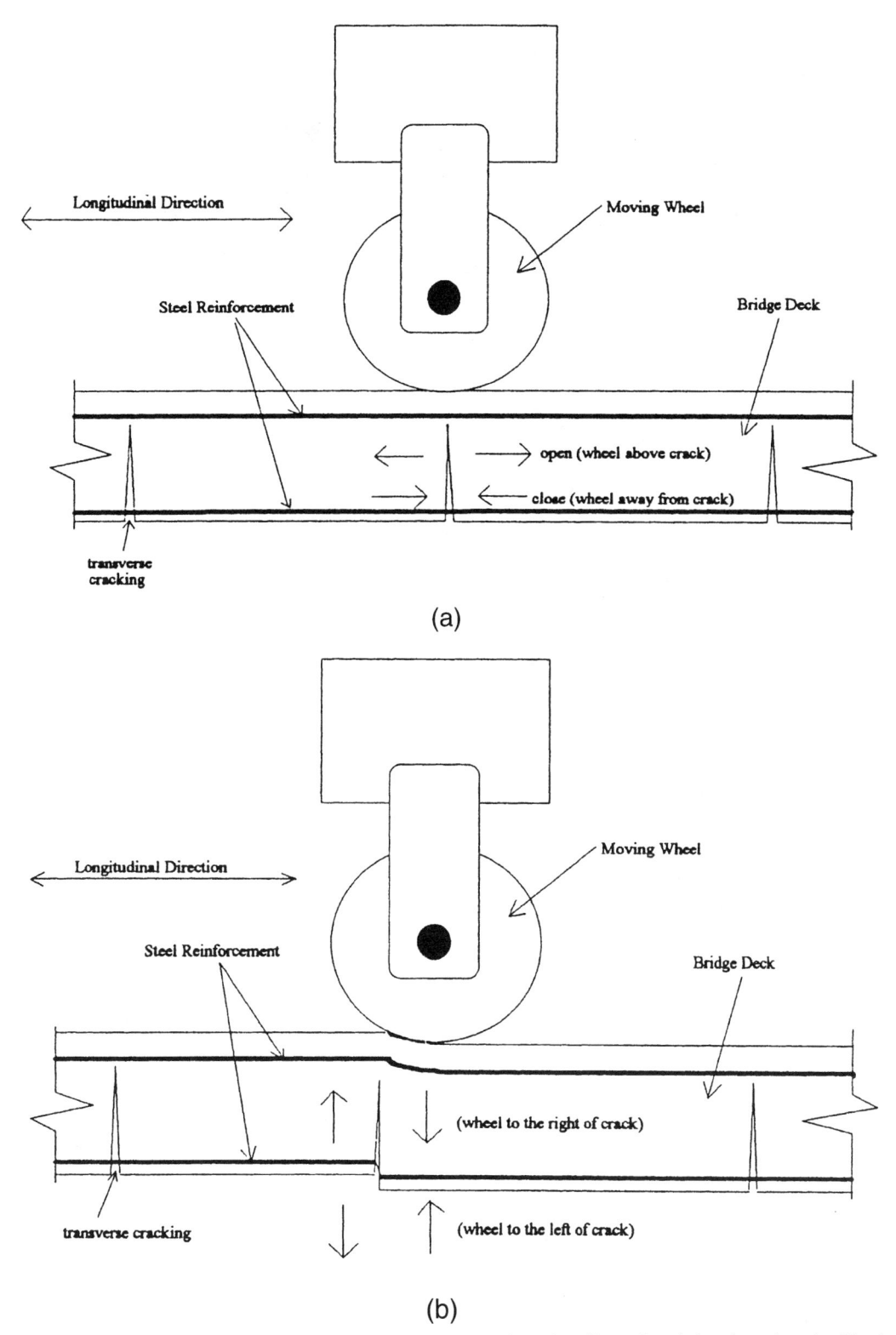

Figure 10.90 Schematic presentation of the transverse crack motion (flexural and shear) as the wheel-load on the bridge deck in the longitudinal (traffic) direction. (a) Flexural motion. (b) Shear motion. (After Perdikaris, P. C. et al., 1993.)

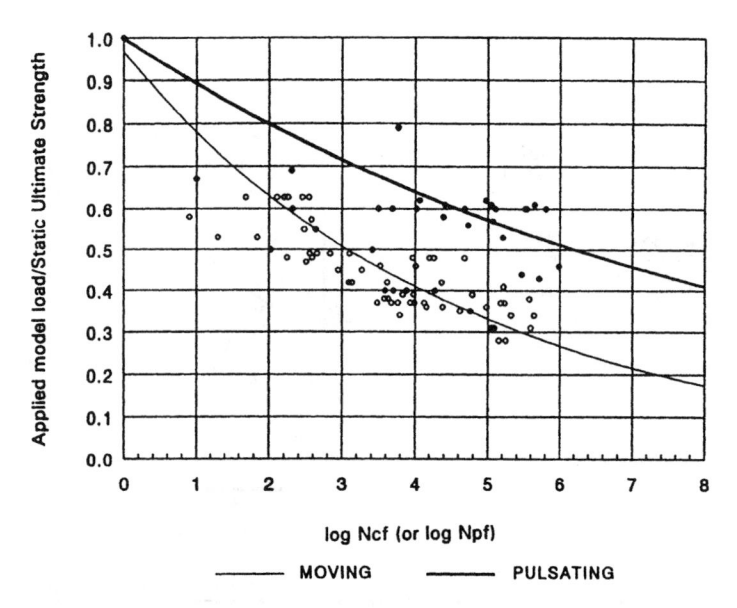

Figure 10.91 Comparison of the fatigue life of bridge decks tested under stationary pulsating load and moving constant wheel-load in terms of the applied load level to static ultimate strength vs. log N_{cf} and log N_{pf}.

Table 10.14   Model and Prototype Dimensions

Description	Prototype	Model
Width	6.71 m	1.18 m
Clear span	6.71 m	1.18 m
Overall length	7.32 m	1.22 m
Slab thickness	304.8 mm	50.8 mm
Laminae width	76.2 mm	12.7 mm
Stressing tendons		
Spacing	0.61 m	101.6 mm
Diameter	19.1 mm	3.2 mm
Strength	1034 MPa	1034 MPa
Anchor plates	228.6 mm × 228.6 mm × 31.8 mm	38.1 mm × 38.1 mm × 4.8 mm

All model materials were identical to the prototype materials. The 2.44 m lengths, as received from the treatment plant, were cut to lengths specified for the splice pattern chosen Figure 10.92. The splice pattern repeats itself every four laminae with the lengths shown in Figure 10.92.

Stressing rods consisted of 4.76-mm-diameter threaded rod with 10-32 threads. The rods have an effective solid diameter (root diameter) of 3.175 mm. In all, 11 rods spaced at 100 mm intervals as per the OHBDC (1983) in 6.35-mm holes were used in the models. Individual square steel anchor plates, similar to those used by Oliva and Dimakis (1987), were used (Figure 10.93a) in lieu of the bulkhead channels assumed in the OHBDC.

The ratio of steel area to wood area is very important because it affects the relaxation characteristics of the deck. Section 13-22.3.1 of the OHBDC requires that the steel-to-wood ratio be no greater than 0.0016 in order to control the amount of residual stress. The steel-to-wood ratio for the ⅙-scale model decks was 0.0015. The prestress is transmitted to the wood via 22 38.1 × 38.1 × 4.76 mm steel anchor plates.

The stressing force was applied to the deck via 11 small coil springs (Figure 10.93a). The entire stressing assembly is shown in Figure 10.93b. It consists of a spring centered between two large

SPLICE TYPE	MODEL	PROTOTYPE
A	8"-24"-16"	4'-12'-8'
B	10 5/8"-24"-13 3/8"	5' 4"-12'-6' 8"
C	13 3/8"-24"-10 5/8"	6' 8"-12'-5' 4"
D	16"-24"-8"	8'-12'-4'

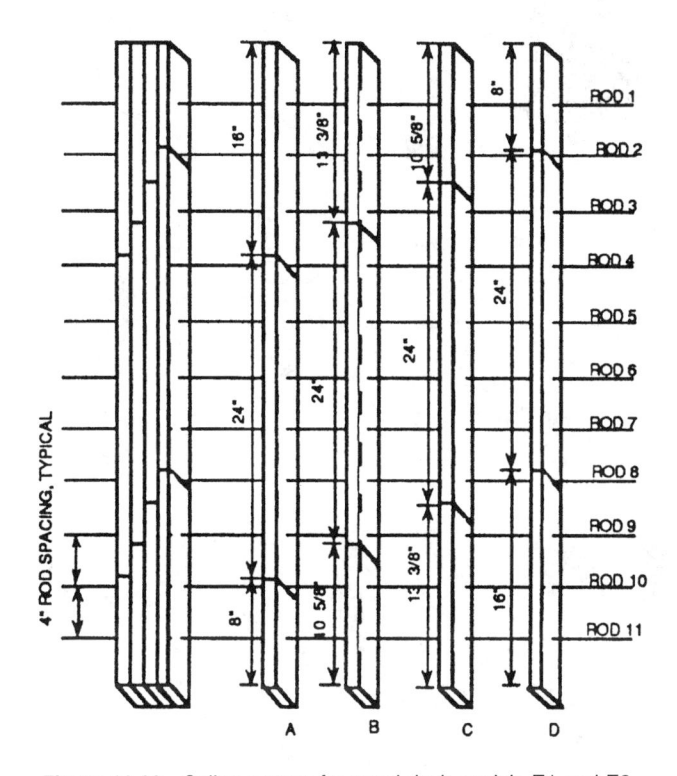

Figure 10.92 Splice pattern for wood deck models E1 and E2.

steel plates with the stressing rod passing through the middle. The one large plate bears directly against the 38.1 × 38.1 × 4.76 mm steel anchor plate. The spring is compressed between the plates by tightening the nut. There is also a nut on the opposite end of the stressing rod which bears against its anchor plate. Tightening the nuts squeezes both the wood deck and the springs and produces elongation, or strain, in the rod. The force in the rod can be determined by simply measuring the spring deformation between the two large plates and multiplying by the spring constant. The spring constant was determined to be 194.6 N/mm (Figure 10.94).

10.3.3.2 *Results of Long-Term Effects Model E1*

Model E1 (Figure 1.17) was used to study the long-term effects of stress relaxation and dead-load deformation. The deck was stressed to 0.69 MPa and the amount of relaxation with respect to time was recorded for a period of approximately 7 months. This level of prestress was chosen to preclude damage to the threaded stressing rods and the wood and is similar to the recommendations of the OHBDC (1983).

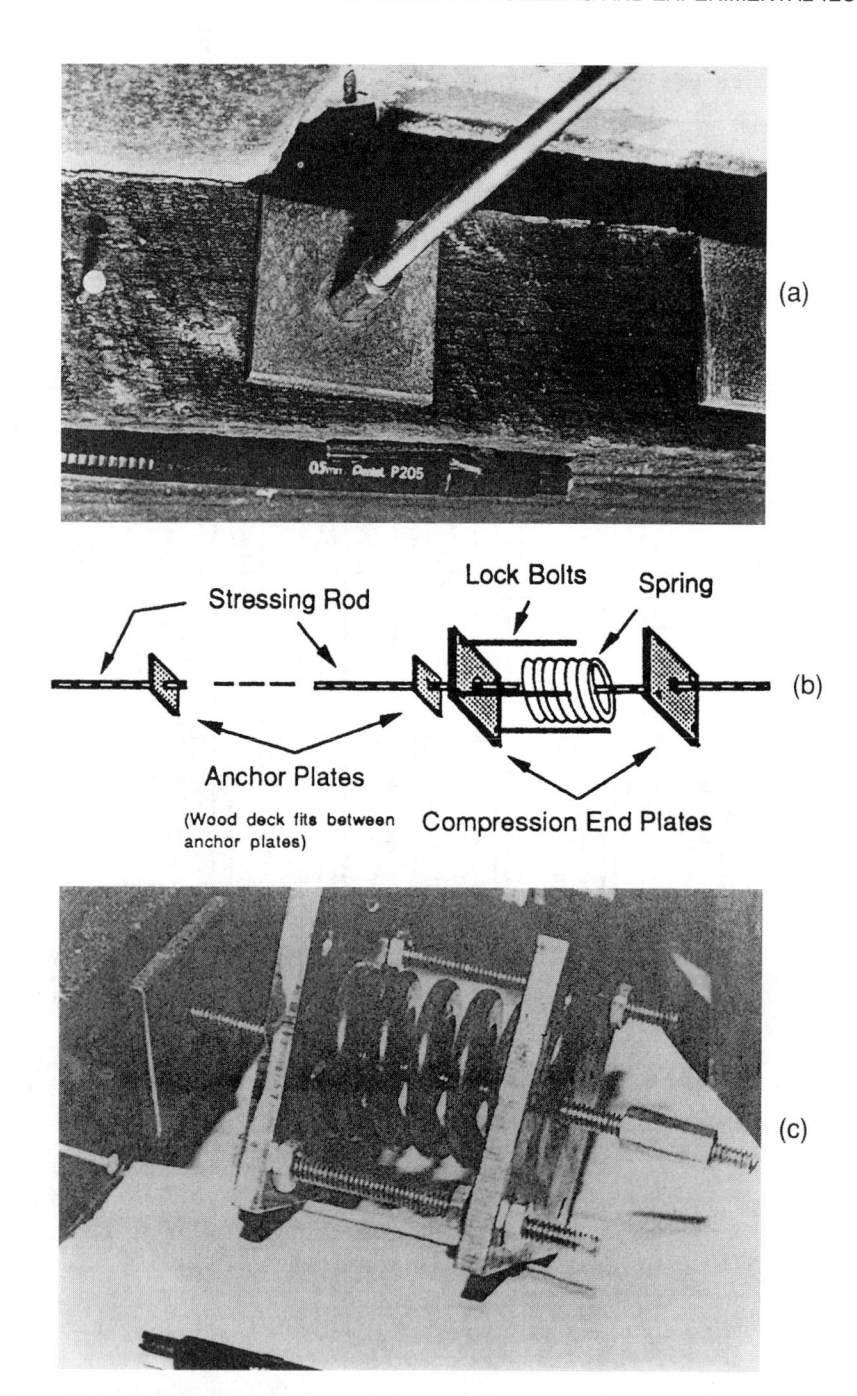

Figure 10.93 Method of applying transverse prestress to deck. (a) Anchor plate. (b) Stressing spring assembly, exploded view. (c) Stressing spring assembly.

Relaxation, or the loss of prestress due to the viscoelastic properties of the wood, must be closely monitored in this type of construction. The loss of prestress is known to be rapid during the first month after prestress application. It then reaches a stable plateau after several months. At this point as much as one half the initial prestress may be lost. The remainder stress is termed residual stress. Factors affecting the amount of residual stress include: the steel-to-wood ratio, relative humidity, and the frequency of stressing after construction. In the case of model E1, the

MODEL APPLICATIONS AND CASE STUDIES

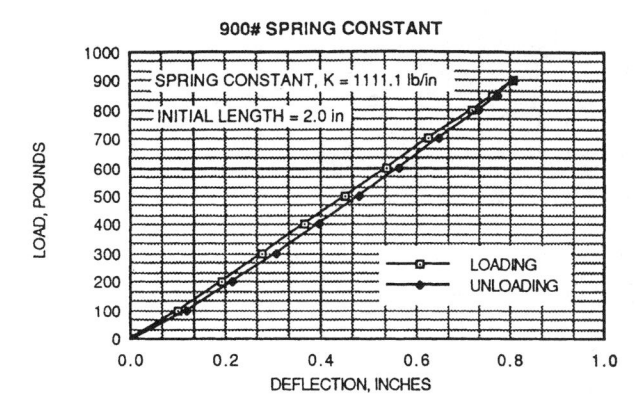

Figure 10.94 Experimental spring constant determination.

Figure 10.95 Mechanical strain gages on angle rods.

testing was conducted in an air-conditioned laboratory with steady relative humidity. However, the deck was stressed only once in order to allow collection of relaxation data in a reasonable amount of time.

Specially fabricated rod gages, functioning as mechanical strain gages, were made for measuring the elongation changes in the prestressing rods with time. Four of these were applied to rods 2, 4, 6, and 8 (Figure 10.92) and can be seen in the aerial view of model E1 shown in Figure 10.95. The apparatus consists of two 635-mm-long angle segments. One end of the first segment is bolted securely to one end of the stressing rod and the other end is attached to a dial gage. One end of the second angle segment is bolted securely to the other end of the stressing rod. The second angle segment extends to the center of the deck where it meets the dial gage. Deflection is measured between the ends of the angle segments with an accuracy of 0.0254 mm. Over the total length of the stressing rod (1.194 m) the gage can sense as little as 21.1 microstrain, which is equivalent to 34.5 N. Since the design rod force is 3.558 kN, the gage apparatus is accurate to 1%.

The deadweight stresses in the ⅙-scale wood deck were properly scaled by adding hanging weights after the application of the prestress was completed. The hanging dead load can be seen

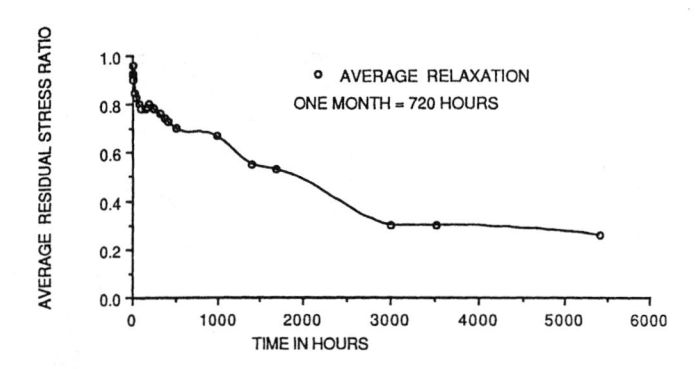

Figure 10.96 Seven month relaxation curve of model E1.

in Figure 1.17. The 7-month relaxation curve of model E1 is shown in Figure 10.96. The results clearly show that the stressing rods relaxed very rapidly in the first week. Roughly 25% of the initial 0.69 MPa stress was lost after 1 week. Loss of prestress slowed after this and by the end of 5 months, approximately 70% of the initial stress had been lost. The rate of relaxation had slowed greatly by this time but had not completely ceased (Figure 10.96).

10.3.3.3 Results of Ultimate-Strength Model E2

1. Linearly Elastic Response — Live load and ultimate-strength tests were conducted on the second ⅙-scale deck (model E2). Vertical deflections were measured along three transverse cross sections for various load configurations and prestress levels. This resulted in transverse displacement profiles from which effective distribution width can be determined.

The loading conditions used in testing model E2 attempted to simulate the AASHTO HS-20 truck loading (Figure 10.97). Four separate loading cases were investigated: maximum moment symmetric (MS), maximum moment unsymmetric (MU), maximum shear symmetric (SS), and maximum shear unsymmetric (SU). The worst moment case occurs when two trucks (passing each other in opposite directions) are on the deck and their 142.34 kN axles are located at midspan (Figure 10.98a). This loading is designated the MS loading because the deck is loaded symmetrically about its longitudinal axis. Load to the ⅙-scale deck was applied by a hydraulic jack through a whiffle tree arrangement directly to the deck via simulated truck loads (Figure 10.98b).

The next configuration also simulates a maximum moment condition with one truck crossing the bridge. The solitary truck is located in its lane with the 142.34 kN axle at midspan. This creates an unsymmetric condition about the longitudinal axis of the bridge. It is designated as the MU condition. Typical longitudinal displacement profiles for the case of 0.17 MPa prestress are shown for the MS and MU loadings, respectively, in Figure 10.99. These plots are made at five separate transverse locations across the bridge (center, quarter points, and edges). Additional loading runs for both the MS and MU conditions were also carried out for values of 0.345 and 0.518 MPa prestress.

The third loading condition simulates the maximum shear condition. Again, it occurs when two trucks are side by side on the bridge, passing each other in opposite directions as prescibed in the AASHTO Code (Figure 10.100a). The loading is designated the SS condition because it is symmetric about the longitudinal axis. The fourth loading condition also simulates a maximum shear condition. Like the MU loading, only one truck is on the deck. The truck is located in its lane and the axles are placed as in the SS case. The configuration of load is unsymmetric about both the longitudinal and transverse axes and is designated the SU case. The loading arrangements for applying the SS and SU loading to the model deck are shown in Figure 10.100b and c, respectively. The longitudinal displacement profiles at various transverse locations are plotted for the SS loading

MODEL APPLICATIONS AND CASE STUDIES 549

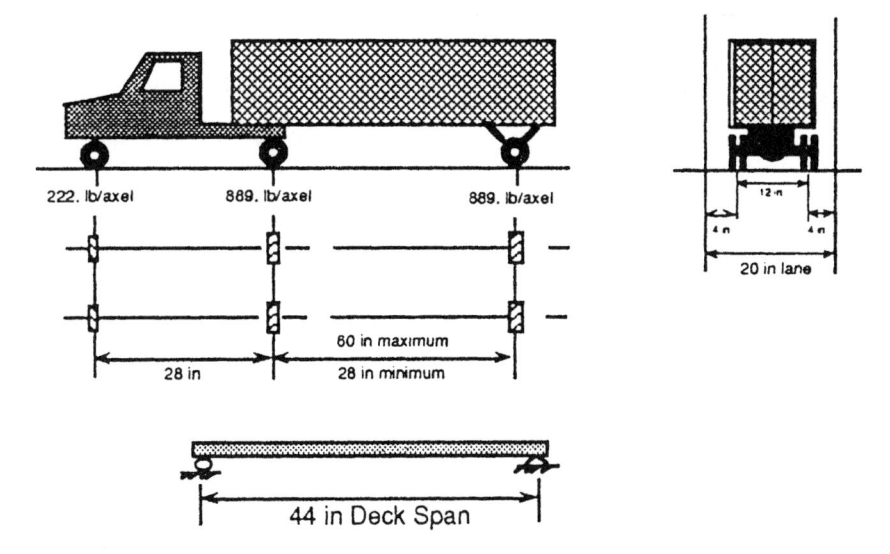

Figure 10.97 One-sixth-scale AASHTO HS-20 truck loading.

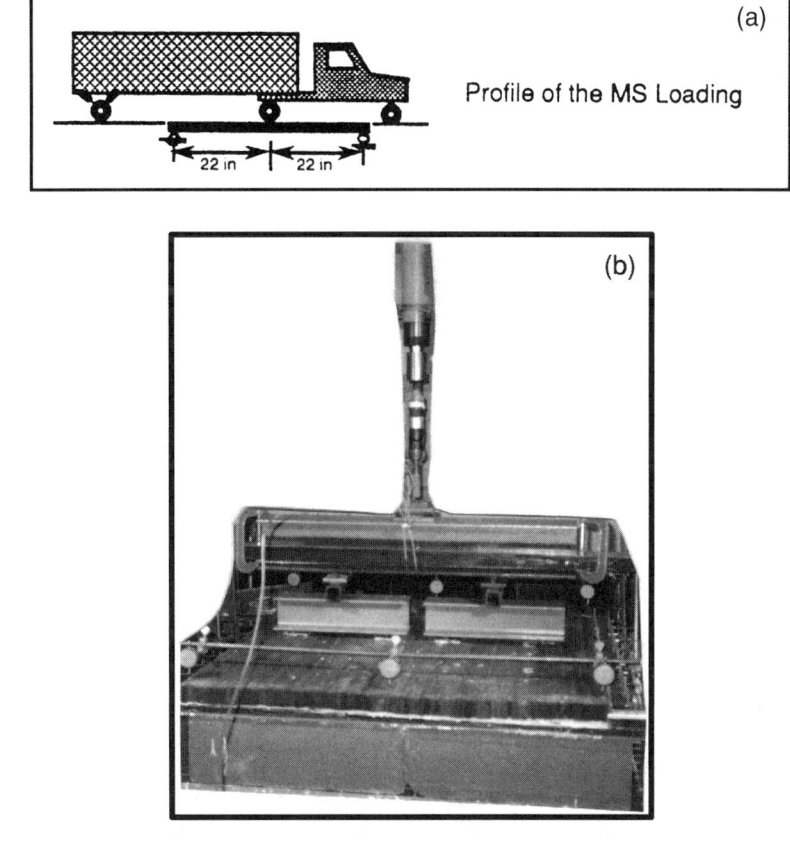

Figure 10.98 Maximum moment loading pattern. (a) Profile of the MS loading. (b) Test simulation.

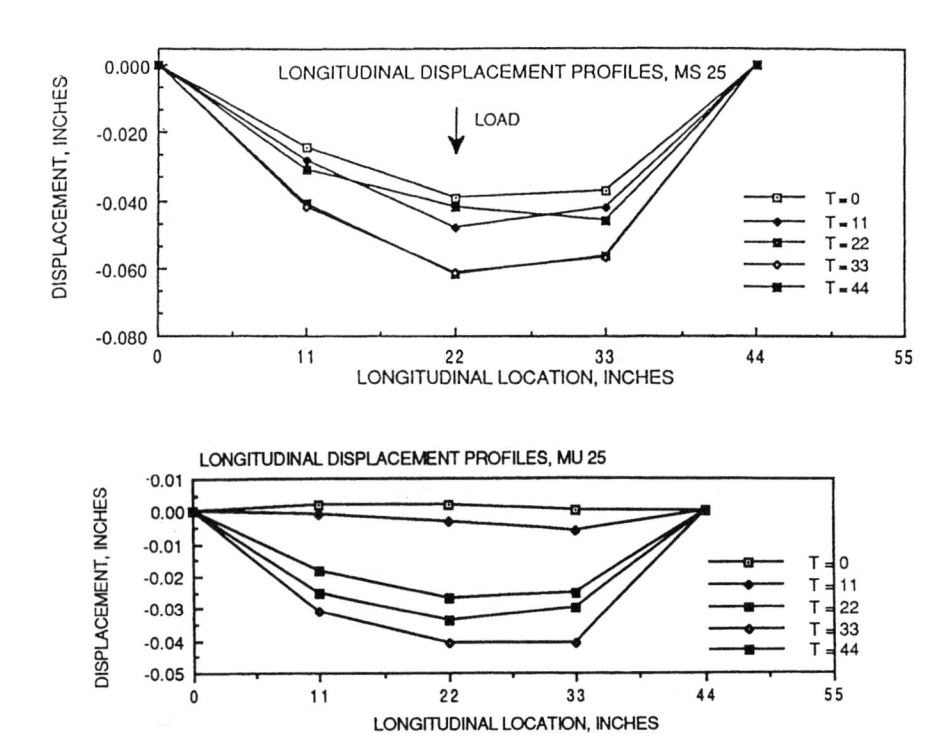

Figure 10.99 Longitudinal deflection profiles, maximum moment loading.

condition and the lowest prestress level in Figure 10.101a. Similarly, the transverse displacement profile at the center and quarter points is plotted for the SU loading condition in Figure 10.101b.

Maximum deflection (at the maximum load step) was plotted against the prestress level in the deck for the four load configurations. Results of the two maximum loading conditions MS and MU are shown in Figure 10.102. As expected, the deflection decreases with increasing stress level. Stress level had a greater impact on the deflection for the symmetric case than for the unsymmetric.

2. Ultimate-Strength Test — The inelastic response of the prestressed deck was determined by loading model E2 to destruction after all the elastic live load testing was completed. Several probable failure modes had been identified prior to testing. Important factors affecting the actual failure mode are the amount of residual post-tensioning stress, the coefficient of friction between adjacent laminae, strength of the wood, and the manner in which the load is applied to the deck. Of all the possible modes of failure, those associated with formation of mechanisms, where slip (rotation) occurs at the splices before the wood can develop its ultimate bending strength, appear to be the most plausible.

The load configuration for the ultimate-strength test was the maximum MS condition (see Figure 10.98). This configuration was chosen because it is symmetric and it loads the deck fairly uniformly in the transverse direction. It is thus assumed, and desirable for this experiment, that this loading will cause all of the spliced laminae to fail at the same time in the transverse direction.

The deck was stressed to a level of 0.345 MPa chosen to simulate the assumed steady-state condition of the prototype deck while in service. Although the results from the long-term test of model E1 indicate a final stress level below 0.207 MPa, it is assumed that this level can be increased by applying different stressing techniques such as periodic restressing.

Loading to failure proceeded in increments of approximately 2.4 kN after an initial shakedown test to 7.9 kN to stabilize the seating of the deck. Load–deflection behavior was linear up to 39 kN beyond which the deflections became steadily nonlinear (Figure 10.103). The ultimate load was

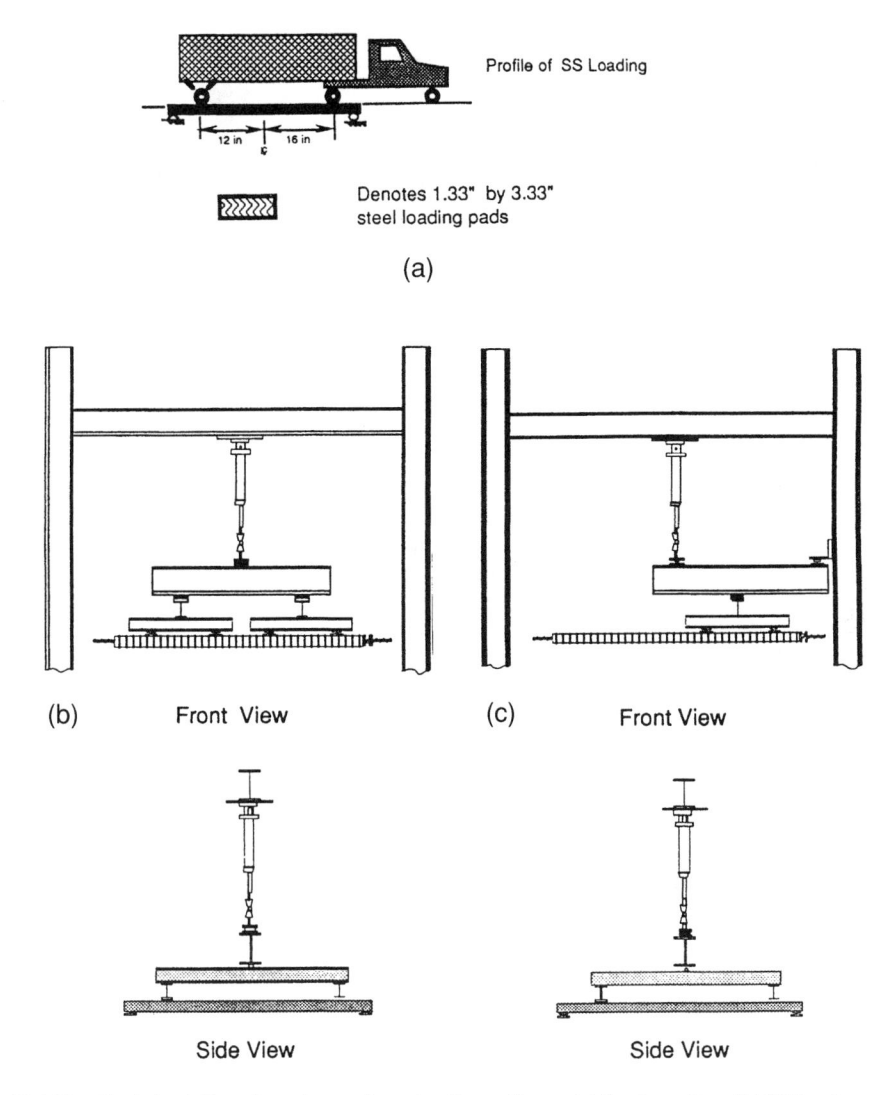

Figure 10.100 Test simulation of maximum shear loading patterns. (a) Truck position. (b) SS loading apparatus, schematic. (c) SU loading apparatus, schematic.

58.4 kN (13.13 k) after which a steady drop in load was observed. This compares very well with a theoretical prediction of 63.2 kN based on the failure mechanism. Longitudinal and transverse displacement profiles, shown at three different load levels in Figure 10.104, indicate symmetric behavior.

Failure occurred by rotation of (i.e., formation of a mechanism along) the rear splice (Figure 10.105) at the approximate quarter span. Visible failure initiated at the center of the rear splice and rapidly progressed toward the edges of the deck. The front line of splices remained intact and there was no observable evidence of rotation. None of the steel stressing rods was broken during the test. The splice pattern appears to have rotated about rod number 3 (Figures 10.92 and 10.106) which is the geometric center of the rear splice pattern. The tops of the butt ends of the laminae appear to have been crushed due to interference when they rotated while the bottoms spread apart and formed gaps (Figure 10.107). This was possible because the steel rod held the splice together longitudinally. The rod also prevented slip in the vertical direction.

The live-load and dead-load deflections measured in the model tests are extrapolated to the full-scale bridge deck in Table 10.15. The full-scale predictions for the elastic loadings and the

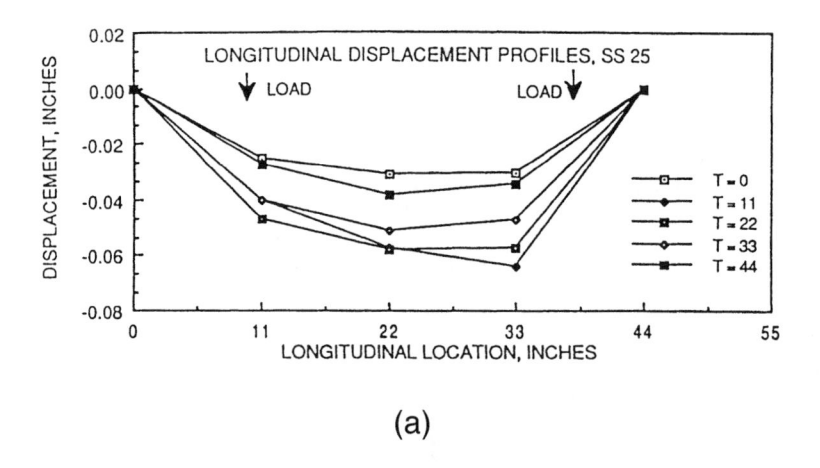

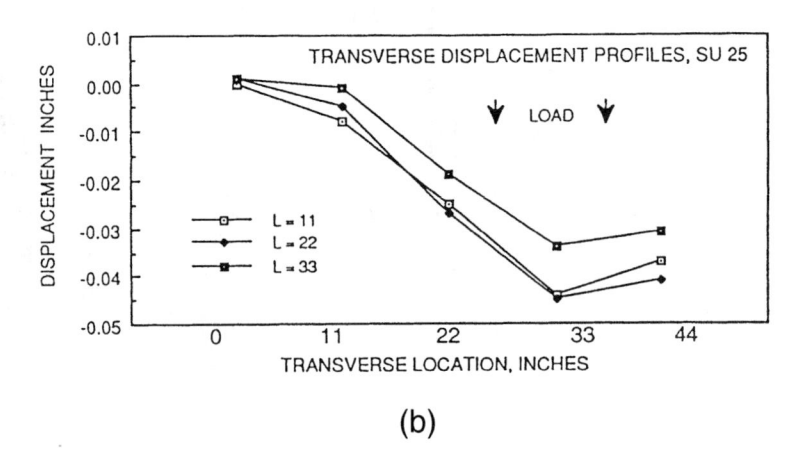

Figure 10.101 Deflections under maximum shear loading. (a) Longitudinal profile, SS loading. (b) Transverse profile, SU loading.

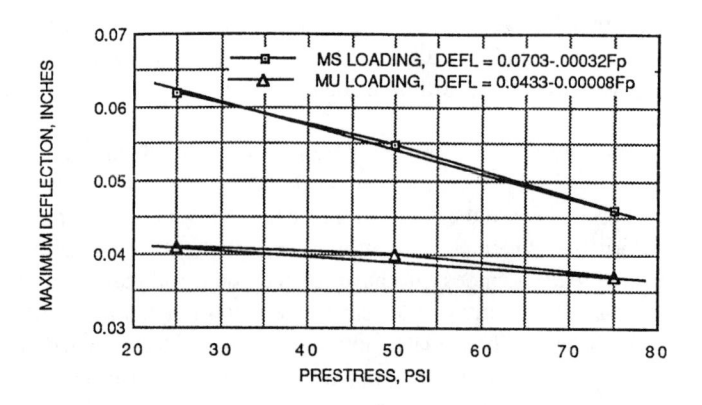

Figure 10.102 Prestress vs. deflection, moment loading.

MODEL APPLICATIONS AND CASE STUDIES 553

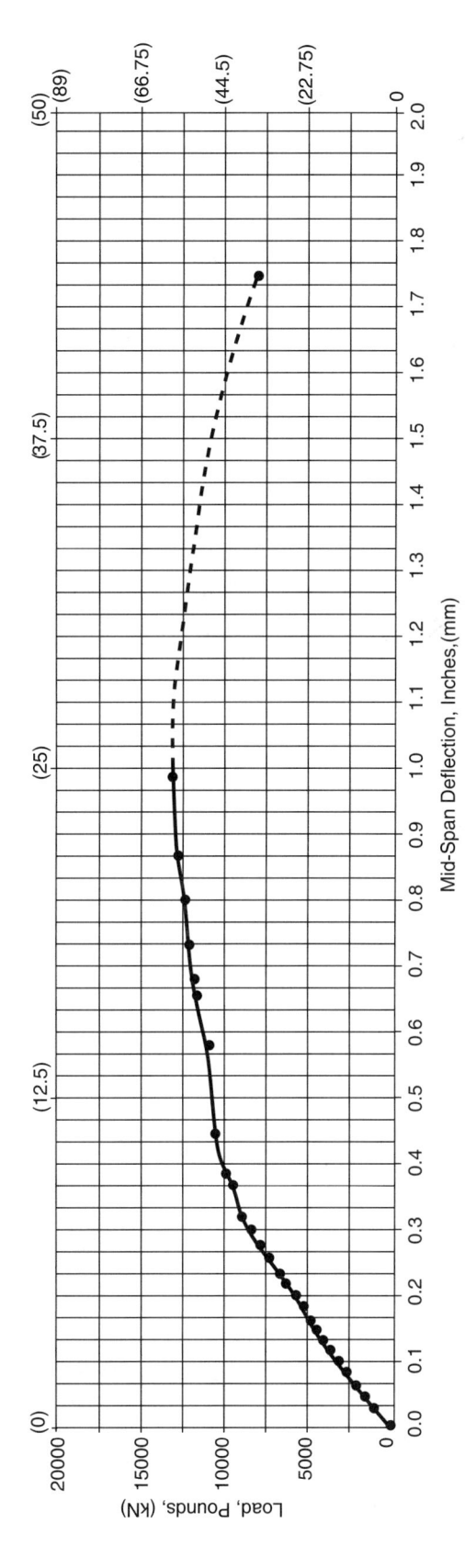

Figure 10.103 Load–deflection curve to ultimate failure.

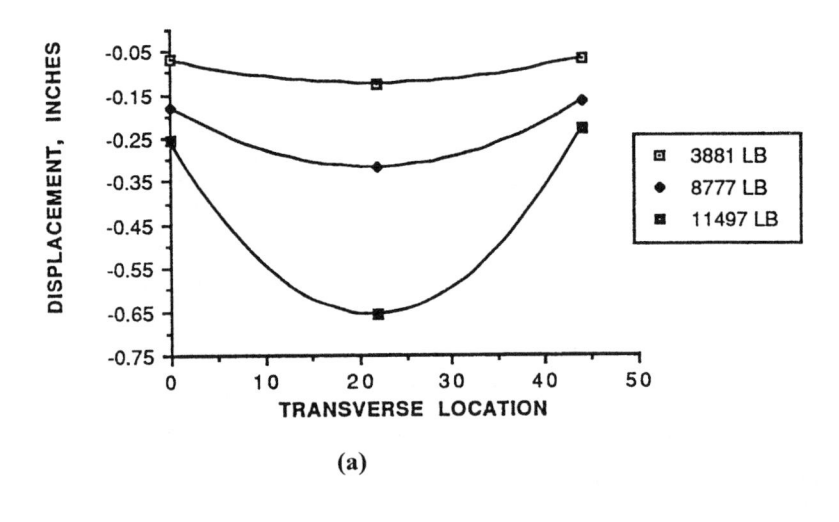

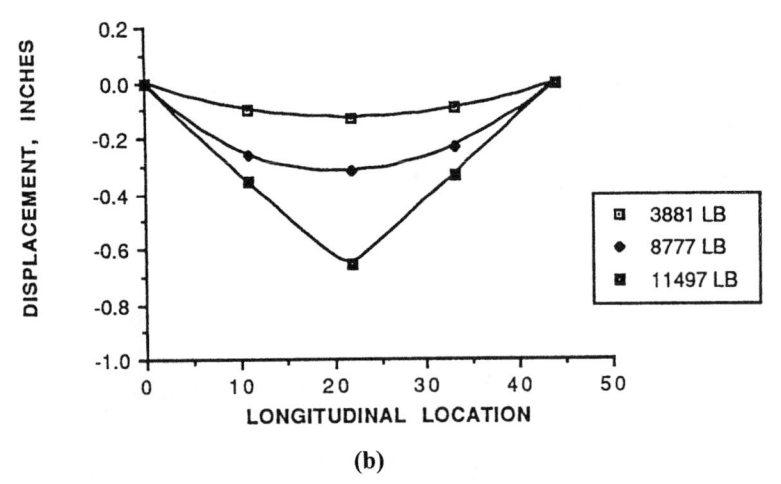

Figure 10.104 Transverse and longitudinal deflection profiles at various loads.

Figure 10.105 Rotation of splice at approximate quarter span.

MODEL APPLICATIONS AND CASE STUDIES 555

Figure 10.106 Rotation of splice, close-up.

Figure 10.107 Failure of butt joints.

failure and ultimate load are also shown in Table 10.15. The table shows that under normal AASHTO HS-20 truck loading, the maximum combined deflection of the deck (LL and DL) is 13 mm due to shear loading (condition SS). A deflection of approximately 50 mm is required to initiate failure of the deck. This would require a load of 1406 kN. There is additional deflection and load capacity after failure starts. At the full-scale ultimate load of 2104 kN the expected deflection is 150 mm.

An overload factor can be determined for the maximum MS loading. The design deck loading is 284.7 kN consisting of two trucks passing at the center of the bridge at the same time each with an axle load of 142.35 kN (AASHTO HS-20 truck loading). The overload to initiate failure is therefore 1406 ÷ 284.7 = 4.9. Similarly, the overload factor at the ultimate load is 7.4.

Table 10.15 Full-Scale Deflection Predictions

Case	Load at Full Scale	⅙-Scale Deflection, mm	Full-Scale Deflection, mm	Full-Scale Combined Deflection, mm
Dead load, gage M	0.004 MPa	0.483	3.658	—
MS-50, gage M	284.7 kN	1.397	8.382	12.040
MU-50, gage N	142.3 kN	1.016	6.096	9.754
SS-50, gage M	569.3 kN	1.549	9.296	12.954
SU-50, gage N	284.7 kN	1.194	7.163	10.820
MS at PL, gage M	1405.6 kN	8.077	48.463	52.121
MS at ULT, gage M	2103.9 kN	24.994	149.962	153.619

10.3.4 Case Study F, Interlocking Mortarless Block Masonry

10.3.4.1 Introduction

The case study described here illustrates another type of use of physical scale models, that of "product development" where the "product" is a new structural component. The new structural component desired to be developed in this case is new interlocking mortarless concrete masonry units. The need to develop efficient interlocking mortarless blocks will reduce labor costs; eliminate the most expensive component, the mortar; and allow for more automation in building construction — three important advantages. A worldwide interest in the development of masonry of this type has taken place in recent years.

Two promising interlocking block types were developed at Drexel University (Harris et al., 1992; 1993; Oh et al., 1993; Oh, 1994) for application to reinforced masonry construction including earthquake-resistant structures using ⅓-scale direct models. The main aim in this development was, first, to equal or exceed the structural performance of conventional masonry systems and, second, to provide a more economical and rational solution for the masonry system, thus leading to more competitive designs. The two interlocking block units developed are designated the modified H-Block and the WHD Block. Production of the units (facilitated by ⅓-scale reduction) was followed by strength and stiffness evaluation under compressive, bending, and, shearing loads. All comparisons were baselined to conventional mortared masonry construction. Analytical models were developed to predict behavior under load.

10.3.4.2 Modified H-Block Unit

An extensive review of worldwide available interlocking mortarless blocks has been presented by Harris et al. (1993) and Oh (1994). The large majority of the systems proposed are applicable to ungrouted or partially grouted concrete masonry construction. Very few of these systems have been developed intentionally for reinforced block masonry that could be adapted to earthquake-resistant applications. Interlocking mortarless or "drystack" masonry systems that are designed for application in seismically active or hurricane-prone areas must be reinforced. These systems must therefore meet the following criteria:

1. Provide ease of placement of vertical and horizontal reinforcement;
2. Have minimum solid area for greater continuity of grout cores;
3. Provide alignment of webs of the stacked blocks for clear core spaces.

The modified H-Block unit (Figure 1.19), with its tongue-and-groove interlock on both the bed and head joints, has been designed to meet the above criteria. A special brass mold (Figure 6.5) was made to fabricate the 120 ⅓-scale units required to study the geometric, physical, unit strength, and assembly strength of the modified H-Block system. A fine river deposit sand passing the U.S. No. 16 sieve was mixed in the ratio of 1:2.04:9.9 by weight (water:cement:sand). The fineness in particle size was adequate to achieve the tolerances needed in the interlocking tongue-and-groove details.

Geometric accuracy of the interlocking blocks, especially the levelness of the bearing surface and the uniformity of the height of the units, is vital for the optimum structural performance and the correctly aligned construction of the structure. This is because unleveled bearing surfaces of the dry joints will cause stress concentrations and hence premature failure. Unlike conventional masonry construction, where masons adjust the alignment of the masonry units as they are laying them up utilizing fresh mortar joints, for the interlocking masonry construction, one must rely on the geometric accuracy of the masonry units for alignment. Even small geometric errors of the individual unit will "grow" after many courses of stacking and will result in undesirable construction problems.

Six random samples were measured using a mechanical dial caliper with an accuracy of 0.025 mm to determine the dimensional accuracy of the ⅓-scale blocks. ASTM C90 specifies the minimum thickness of face shell and web shell shall be 31.75 and 25.4 mm, respectively. These values correspond to 10.67 and 8.38 mm for the ⅓-scale model blocks. The results shown in Figure 10.108 indicate a highly accurate geometric stability.

Unit density and absorption of the modified H-Block were also measured (Table 10.16) and found to be within the recommendations of ASTM C90 for conventional block masonry.

Unit compressive and unit tensile splitting strength for the modified H-Block are given in Tables 4.16 and 4.17, respectively.

10.3.4.3 The WHD Block Unit

The WHD Block (Figure 1.20) is the second interlocking mortarless unit developed specifically for applications where full grouting and reinforcing is required. The interlocking features of this system affect only the head joints. Using small-scale modeling techniques, 200 ⅓-scale model units were produced using the procedures and prototype machine shown in Figure 6.25 and 6.27, respectively.

Results of measurements on six randomly selected units are shown in Figure 10.109. These show high reproducibility of the desired dimensions.

Unit density and absorption of the WHD Block were also measured and found to be within the recommendations of ASTM C90 for conventional block masonry.

Unit compressive and unit tensile splitting strength for the WHD Block are given in Tables 4.16 and 4.17, respectively.

10.3.4.4 Tests on Assemblages

Tests on assemblages of the interlocking units in the ungrouted and fully grouted form were carried out and the results compared directly to specimens made in the conventional way, with mortared joints. Three strength attributes of masonry behavior were studied: compressive strength, diagonal tension (shear), and flexural bond strength as described in Chapter 4.

Compressive Strength: Grouted and ungrouted prisms (Figure 10.110) were constructed by an experienced mason according to procedures outlined in Chapters 4 and 6. A summary of the results given in Table 10.17 are the average of four specimens except for the modified H-Block which gives the average of three. In all cases the coefficient of variation for the compressive strength was

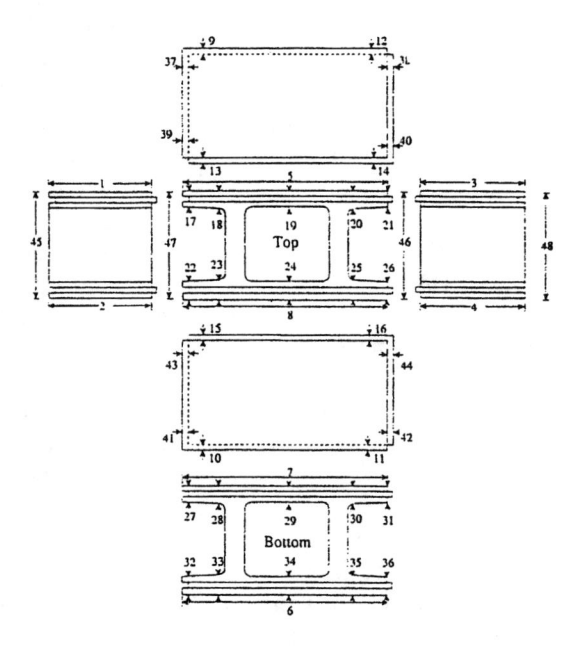

Figure 10.108 Dimensional measurements of modified H-Block.

Table 10.16 Density, Volume, and Net Area of Blocks Fabricated

Block Type	Density, kg/m³	Net Volume, m³	Gross Volume, m³	Net Area, m²	Net Area, %
Modified H	1971	2.73×10^{-5}	5.67×10^{-5}	4.31×10^{-3}	48.2
WHD	1779	3.07×10^{-5}	5.90×10^{-5}	4.50×10^{-3}	52.0
Conventional	1731	2.69×10^{-5}	5.42×10^{-5}	4.13×10^{-3}	49.7

less than 18%. Typical stress–strain results are shown in Figure 10.111. A measure of the efficiency (defined as the prism strength, f'_m divided by the block unit compressive strength, f'_b) of the interlocking units as compared to conventional masonry, some commercially available interlocking masonry, and the ACI530/ASCE/TMS Code is shown in Figure 10.112. As can be seen from this comparison the WHD unit has superior efficiency.

Diagonal Tension (Shear): The diagonal compression prism, which gives a measure of the shear strength of masonry assemblages as explained in Chapter 4, was used to compare the grouted interlocking masonry to grouted conventional masonry specimens. The test setup shown in Figure 4.67 was used. The summary of the test results are given in Table 10.18. The higher shear strength of the modified H-Block units is attributed to the higher unit tensile splitting strength (Table 4.17).

MODEL APPLICATIONS AND CASE STUDIES 559

Figure 10.109 Dimensional measurements of WHD Block.

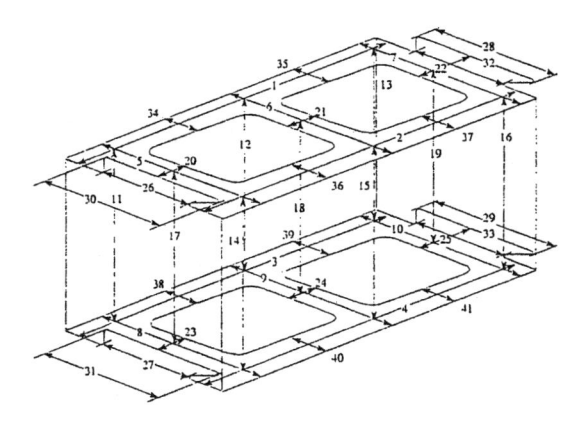

Location	Length (in.)	COV (%)	Location	Length (in.)	COV (%)	Location	Length (in.)	COV (%)
1	5.319	0.03	17	1.370	0.70	33	1.455	0.11
2	5.329	0.02	18	1.370	0.48	34	0.451	0.24
3	5.321	0.06	19	1.369	0.54	35	0.468	0.29
4	5.330	0.04	20	0.331	0.53	36	0.476	0.17
5	2.551	0.05	21	0.389	0.30	37	0.457	0.50
6	2.556	0.05	22	0.715	0.09	38	0.461	0.40
7	2.559	0.09	23	0.330	0.85	39	0.478	0.36
8	2.552	0.12	24	0.406	0.36	40	0.488	0.24
9	2.556	0.07	25	0.726	0.22	41	0.470	0.84
10	2.560	0.07	26	1.490	0.19	11'	1.363	0.54
11	2.669	0.49	27	1.492	0.18	12'	1.363	0.57
12	2.683	0.33	28	1.763	0.06	13'	1.368	0.62
13	2.693	0.36	29	1.762	0.09	14'	1.364	0.74
14	2.685	0.43	30	1.812	0.05	15'	1.366	0.49
15	2.699	0.28	31	1.808	0.26	16'	1.362	0.45
16	2.696	0.32	32	1.455	0.12			

(Note: 17-19 and 11'-16' are heights of half-WHD Blocks)

Flexural Bond: Average flexural bond strength of grouted beams representative of the horizontal and vertical directions of the wall construction were obtained using the bond wrench apparatus shown in Figure 4.59. The test results are summarized in Table 10.19. As can be seen from the table, the grouted modified H-Block construction has good bond strength in both vertical and horizontal directions. However, in the case of the WHD Block the horizontal bending is resisted only by the head joint lugs (Figure 1.20d) and thus the equivalent bond strength is very low.

In ungrouted drystacked interlocking prisms a noticeable reduction in compressive strength was observed. This reduction is directly related to seating conditions at the bed joints. To investigate this effect, measured discrepancies of the flat surface were made (Figure 10.113) and the results incorporated into a finite-element (FE) model of the prism. The FE model incorporates all constitutive relations of the various components as well as the nonlinear contact stiffness characteristics. Comparison of the test results with the FE analysis shows good correlation of the seating phenomenon (Figure 10.114).

Grouted drystacked interlocking masonry behaves in a very comparable manner to conventional mortared masonry under compressive load despite the differences noted above. In contrast to the conventional grouted masonry, strength of the block appears to have little effect on the prism strength. This is because there is little stress transfer at the dry bed joints. Prism strength is more controlled by the grout strength. In grouted drystacked masonry, the face shell serves very much like a "form" for the grout.

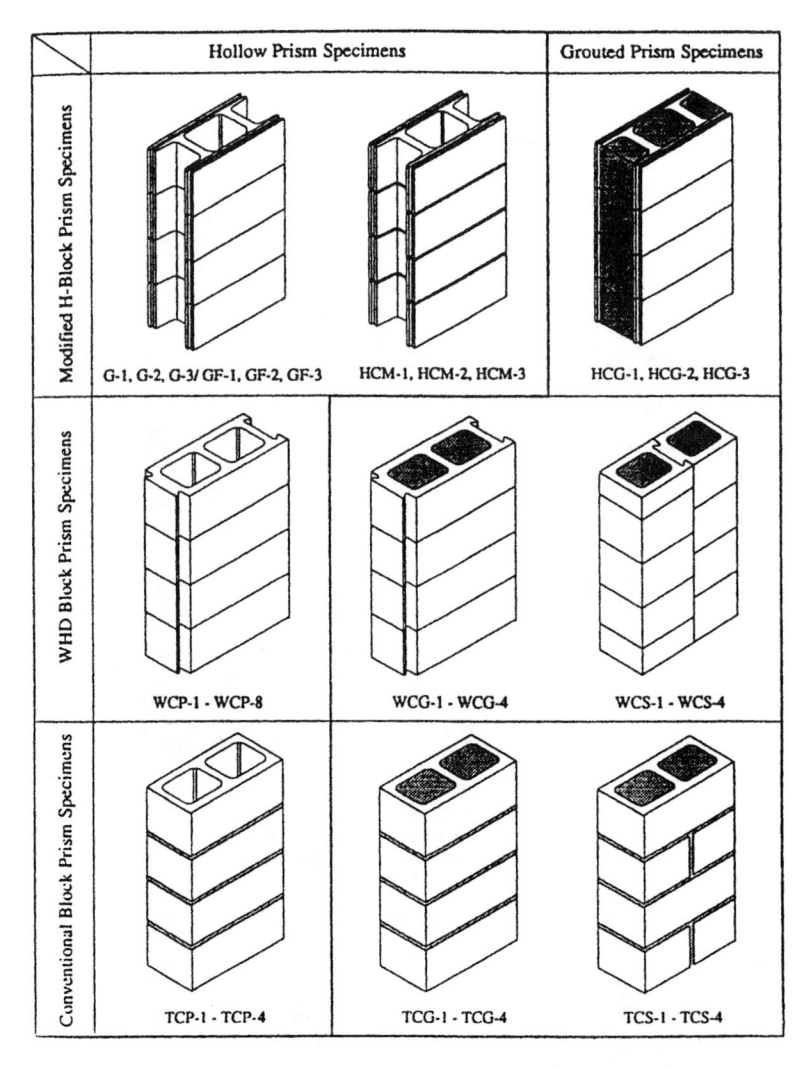

Figure 10.110 Configurations of prisms fabricated.

10.3.5 Case Study G, Pile Caps, a Link between the Foundation and the Superstructure

10.3.5.1 Introduction

The pile cap is an important element in the structural foundation, which transmits the load from columns to piles. Although an important element, it has drawn little attention. Design provisions of ACI 318 (1995) may be used only for thin pile caps, but not for thick pile caps unless suitable modifications are made. ACI Committee 445, Shear and Torsion, has looked into the behavior of deep members, such as deep beams, especially with continuity and pile caps. Very little research has been available in this area, because one needs a large facility for full-scale testing. Research at Howard University by Dagher (1988), Idowu (1979), and Ndukwe (1982) and at the University of Puerto Rico by Jimenez-Perez et al. (1986) has attempted to shed some light on this problem which was tackled using scaled models. A considerable amount of data was generated for verification of design methodology and code provisions. A wide range of parameters were considered

MODEL APPLICATIONS AND CASE STUDIES

Table 10.17 Parameters Derived From Grouted Prism Tests

Prism Type	Specimen I.D.	f'_{mt}, MPa	Peak Strain, mm/mm	E_s, MPa	E_s/f'_{mt}	f'_{mt}/f'_b, %
Modified H block prism	Average	10.894	0.0021	10445.2	970	35
(grout str. = 21.926 MPa)	COV(%)	17.7	36.4	6.0	10.8	18.0
WHD block prism	Average	14.093	0.0024	9522.0	680	108
(grout str. = 23.236 MPa)	COV(%)	13.3	12.9	7.5	10.8	13.1
	Average	13.721	0.0028	7174.9	530	105
	COV(%)	7.7	19.9	20.1	24.0	7.9
Conventional block prism	Average	12.135	0.0019	9003.5	740	94
(grout str. = 23.236 MPa;	COV(%)	10.1	8.6	5.4	6.9	10.8
mortar str. = 12.687 MPa)	Ave.	10.963	0.0016	11173.3	1020	78
	COV(%)	8.8	10.7	8.1	10.0	21.0

Note: 1. f'_{mt} was calculated by dividing maximum load by net loaded area.
2. Peak strain is strain at corresponding f'_{mt}.
3. Secant modulus of elasticity, E_s, is slope of a line connecting points of origin and 50% of f'_{mt} on stress–strain curve.
4. f'_b is block unit compressive strength.

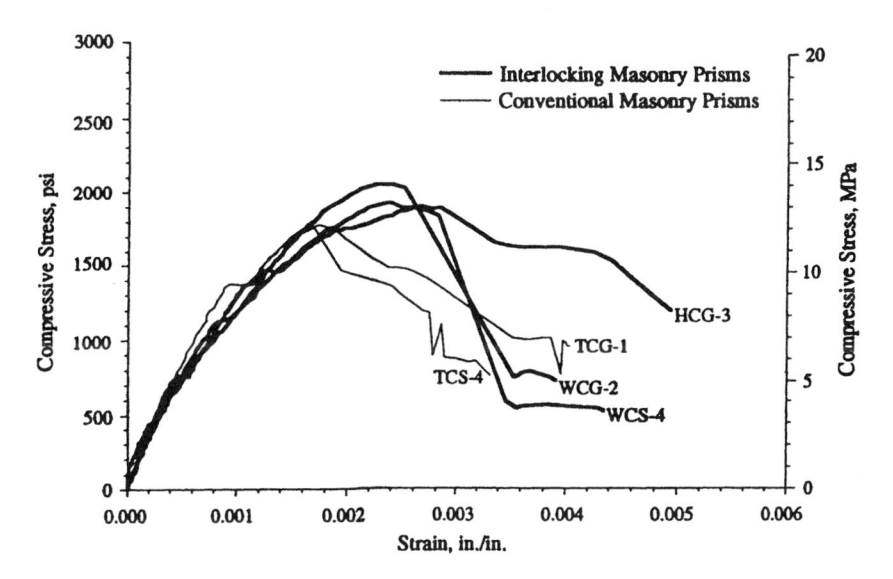

Figure 10.111 Typical compressive stress–strain curves of prisms tests.

with a particular emphasis on the variation of reinforcing steel, since it appears to be a major concern. The study used two scales, namely, ½.5 and ⅙ for modeling.

There have been significant differences in designing reinforced concrete pile caps. The design approaches range from empirical to a complex analysis approach with others falling somewhere in between. Currently, the CRSI (1996) handbook, in conjunction with the ACI Code (1995), is widely used for pile cap design in the U.S.

Research on the other hand dates back to the 1940s (OSDoT, 1947), which was conducted by the Ohio Department of Highways. In the 1960s and 1970s, only European work can be cited as reported by Clark (1973), Yan, (1954), Banerjee, (1973), Hobbs and Stein, (1957), Blevot and Fremy, (1967), and Whittle and Beattie, (1972). Much of this work is based on the truss theory, combined with the classical beam approach to determine the stresses in the steel and concrete.

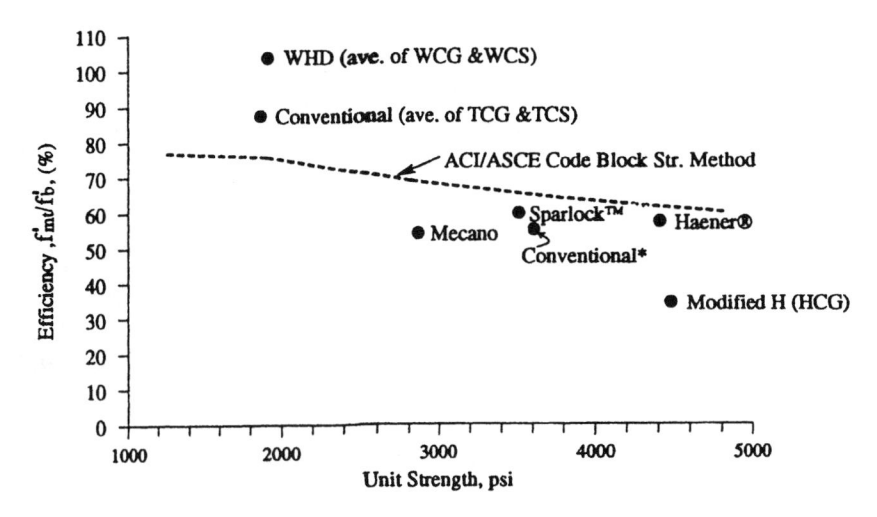

Figure 10.112 Comparison of load-bearing efficiencies of grouted prisms.

Table 10.18 Parameters Derived From Diagonal Tension Tests

Specimen Type	Specimen I.D.	Peak Shear Stress, MPa	Peak Shear Strain, mm/mm	Secant Shear Modulus, MPa
Modified H block specimen	HDB-1	1.862	0.0008	3553.7
(grout str. = 21.926 MPa)	HDB-2	1.586	0.0017	1483.1
	HDB-3	1.517	0.0009	2651.1
	Average	1.655	0.0011	2562.6
	COV(%)	10.3	44.7	40.5
WHD block specimen	WDB-1	0.827	0.0006	3235.1
(grout str. = 22.236)	WDB-2	0.690	0.0005	1723.8
	WDB-3	0.758	0.0007	1168.0
	WDB-4	0.690	0.0008	1406.6
	Average	0.758	0.0007	1883.0
	COV(%)	10.8	19.0	49.4
Conventional block specimen	TDB-1	0.758	0.0004	3045.5
(grout str. = 22.236;	TDB-2	0.896	0.0007	2309.8
mortar str. = 12.687)	TDB-3	0.896	0.0006	2865.6
	TDB-4	1.172	0.0008	2132.6
	Average	0.965	0.0006	2588.4
	COV(%)	18.5	29.5	16.9

Note: Ultimate shear stress, peak shear strain, and secant shear modulus were calculated using formulas given in Section 7 of ASTM E519-81.

Table 10.19 Average Bonding Strength of Grouted Bond Beams

Block Type	No. of Tests	Horizontal Direction Bending Strength, MPa	No. of Tests	Vertical Direction Bending Strength, MPa
Modified H	3	3.23	5	2.30
WHD	7	0.43	8	1.88
Conventional	6	1.31	10	2.29

MODEL APPLICATIONS AND CASE STUDIES 563

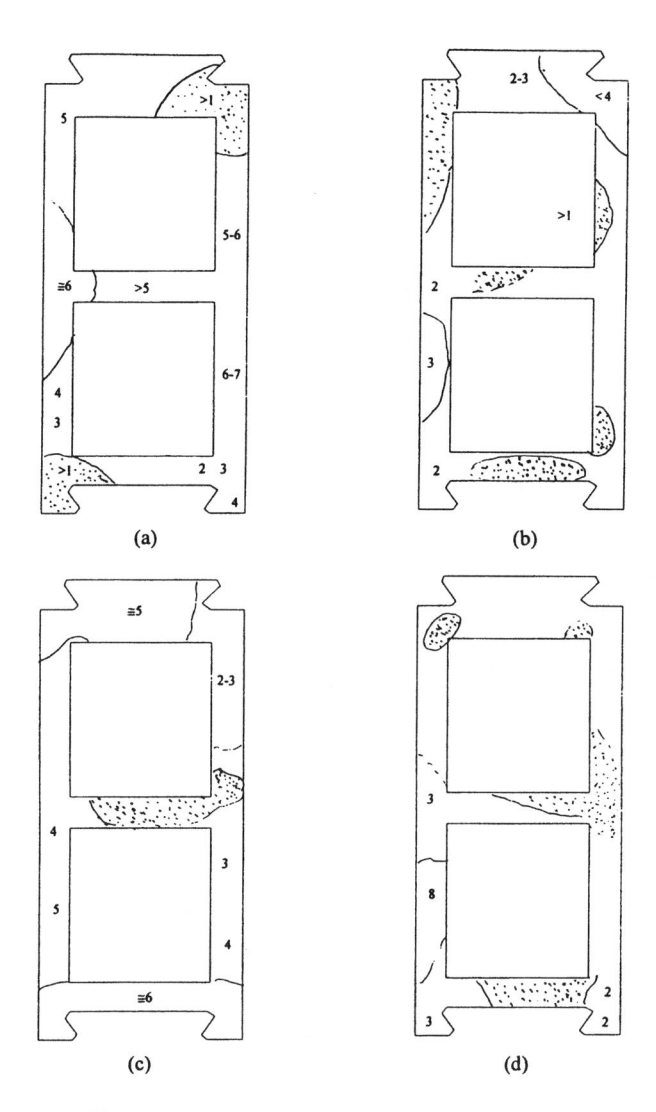

Figure 10.113 Dry joint gap estimated (unit: mils).

10.3.5.2 *Details of Test Program*

The main purpose of this case study was to investigate the behavior with varying amount of reinforcement, while other parameters, such as concrete strength and steel depth, were considered secondary based on earlier work. A scale of ⅙ was used to suit the facility at Howard University and the available reinforcement. The scale was large enough for one person to handle.

A total of 30 specimens were modeled from a four-pile configuration (Figure 1.21) given as a practical example in CRSI, with concrete dimensions as $330 \times 330 \times 152$ mm and steel placed at a depth between 102 and 114 mm from the top surface. The column and piles were simulated with 76.2 mm steel cylinders. The concrete strength was measured using 50×100 mm cylinders and ranged from 28 to 37 MPa. The reinforcement consisted of deformed steel wires and their yield strength was 517 and 655 MPa. The steel ratio varied from 0.00171 (minimum steel amount) to 0.011. The reinforcement was arranged in a grid fashion as shown in Figure 10.115. The bars were anchored to prevent bond failure and were instrumented to verify yielding.

Testing was conducted monotonically in a 890 kN Tinius-Olsen testing machine. The reinforcement was instrumented with strain gages to monitor steel behavior up to failure.

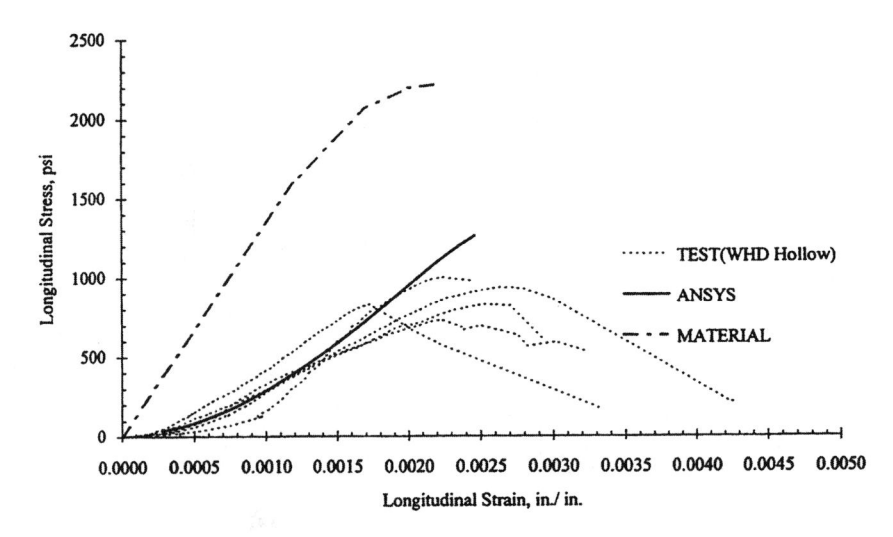

Figure 10.114 Comparison of stress–strain relations of FEM model and WHD hollow prism.

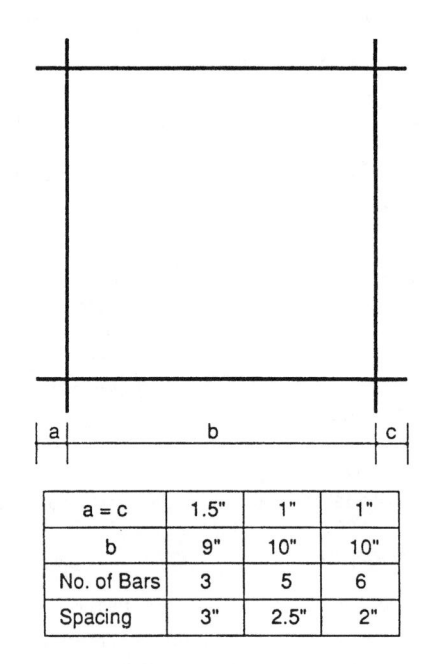

a = c	1.5"	1"	1"
b	9"	10"	10"
No. of Bars	3	5	6
Spacing	3"	2.5"	2"

Figure 10.115 Reinforement pattern.

10.3.5.3 Test Results and Discussion

Cracks formed at the bottom and spread to the four sides with the increasing load. These cracks extended vertically and on all faces of the specimen. Strain measurements indicated yielding of the reinforcement at failure, which may be classified as typical punching shear failure. The distinct feature among all specimens was a punching shear failure, with no bond or anchorage failure observed. Typical crack patterns for the two scales are shown in Figure 10.116; this figure clearly indicates the similarity and a successful use of scaled models.

MODEL APPLICATIONS AND CASE STUDIES

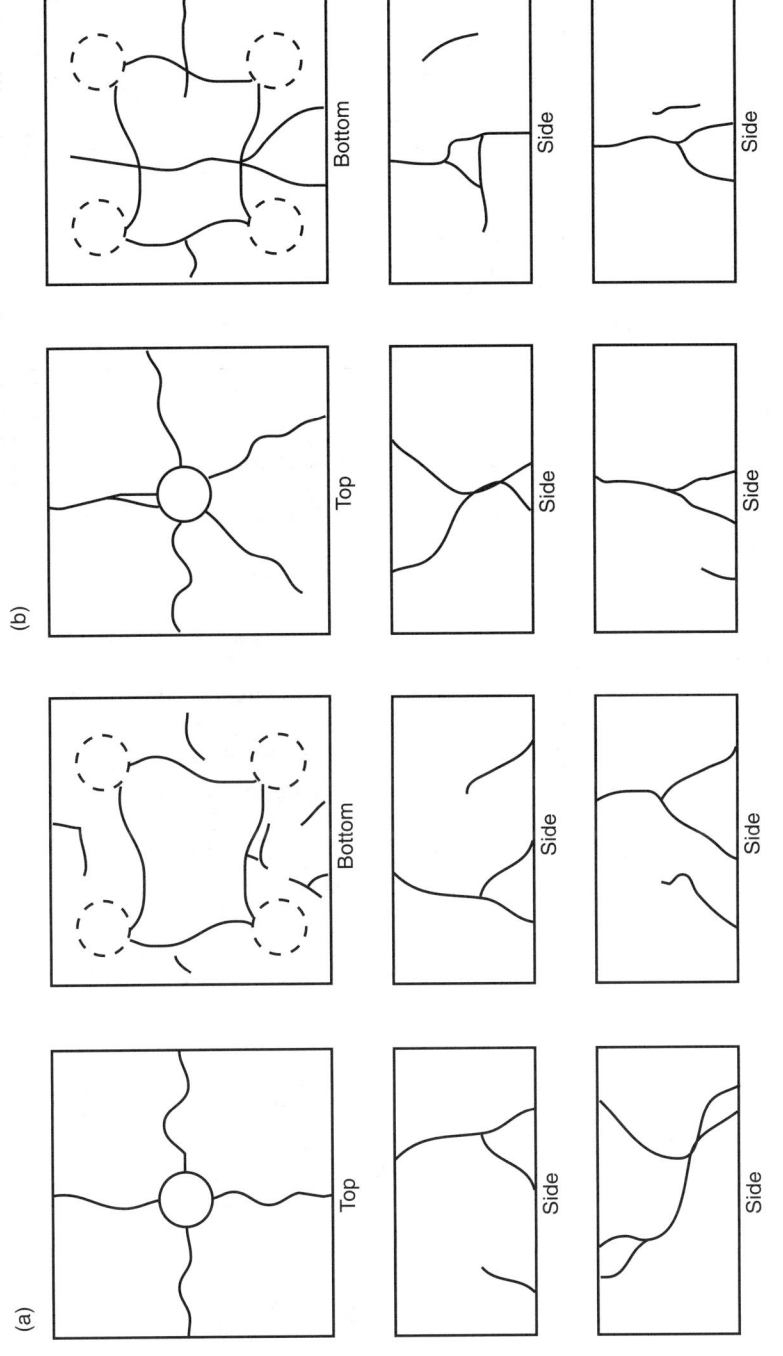

Figure 10.116 Typical crack patterns for specimen 3. (a) Crack pattern in 1/6 scale. (b) Crack pattern in 1/2.5 scale.

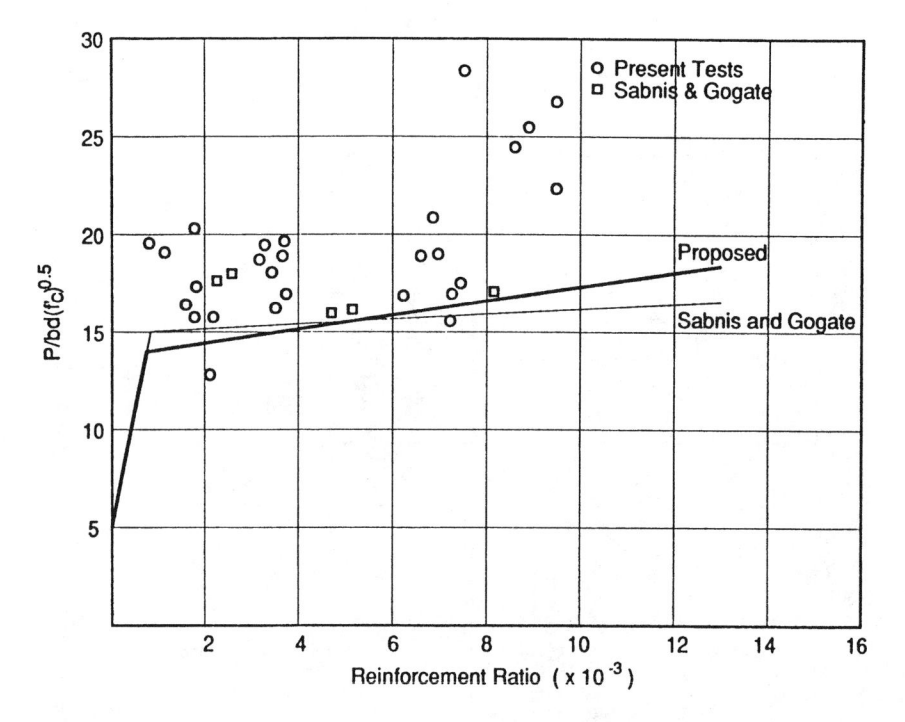

Figure 10.117 Ultimate shear strength vs. flexural reinforcement ratio.

The effect of amounts of steel on the load capacity of the specimens is shown in Figure 10.117. Clearly, there is no significant increase in capacity even with wide steel variation. Stress in steel also increased rapidly after the cracking took place and later remained almost constant, indicating yielding. The steel percentage does not play much of a role until the concrete is cracked, based on the minimum steel ratio of $200/f_y$. These results also indicate that the amount of flexural steel ratio did not increase the load capacity of the pile cap as long as the shrinkage and temperature reinforcements were provided as per the ACI Code.

In Table 10.20, results of calculations of loads are shown for test (failure) loads along with those calculated from the available methods/practice and a proposed method. Since all methods predict the loads on the low side, a lower bound was proposed using these results. The ACI and CRSI methods use the deep beam design; however, ACI assumes the constant shear strength at $8bd + \sqrt{f'_c}$, while CRSI uses a parabolic variation of strength up to $32bd + \sqrt{f'_c}$ as a/d reduces below a value of 2. On the other hand, the truss analogy method assumes the pile cap as a space truss, in which the compression members are made of concrete, while steel takes care of tension (Figure 10.118). Previous research Ndukwe (1982) indicates that this approach is better and recommended that a width of compression members are to be calculated as a "strut." In the proposed method, this approach is further modified to take into account the additional test data with larger variation of amount of steel. Figure 10.119 shows the relationship between shear strength of pile cap and the concrete strength. The proposed relation between the shear strength and f'_c in psi units may be represented as

$$R = 100 - \sqrt{f'_c}/100 \quad \text{for } f'_c < 5000 \tag{10.6}$$

$$= 0.3 \quad \text{for } f'_c > 5000 \tag{10.7}$$

Based on their work, Dagher and Sabnis (1989) concluded the following:

Table 10.20 Comparison of Experimental and Calculated Loads by Various Methods shown as Ratios with respect to Experimental Load

Pile Cap No.	Experimental Failure, kN	ACI	CRSI	Sabnis & Gogate	Truss Analogy	Yan	Blevot	Proposed Method
RD1	275.8	1.68	1.55	1.78	2.48	3.12	2.54	1.21
RD2	275.8	1.62	1.50	1.68	2.39	3.06	2.48	1.21
RD3	306.9	1.77	1.64	1.82	2.61	3.37	2.73	1.35
RD4	291.3	1.63	1.50	1.64	2.29	2.55	2.07	1.28
RD5	231.3	1.17	1.08	1.12	1.67	1.92	1.56	1.02
RD6	261.1	1.74	1.61	1.96	2.54	3.29	2.67	1.15
RD7	287.8	1.62	1.75	1.86	2.55	3.42	2.78	1.26
RD8	266.9	1.74	1.89	2.12	2.12	1.68	1.37	1.16
RD9	271.3	1.62	1.75	1.86	1.98	1.61	1.31	1.18
RD10	311.4	1.92	2.09	2.27	2.35	1.89	1.54	1.18
RD11	329.2	1.80	1.95	1.96	2.21	1.85	1.51	1.35
RD12	333.6	1.95	2.12	2.22	2.39	1.96	1.59	1.45
RD13	324.7	2.03	1.87	2.10	2.29	1.89	1.53	1.41
RD14	320.3	1.80	1.66	1.97	2.03	1.63	1.33	1.39
RD15	300.2	1.72	1.59	1.78	1.52	1.11	0.90	1.33
RD16	289.1	1.56	1.44	1.55	1.40	1.03	0.83	1.28
RD17	310.0	1.85	1.71	1.97	1.64	1.17	0.95	1.37
RD18	444.4	3.03	2.80	2.33	2.46	1.74	1.42	1.93
RD19	320.3	2.10	1.94	2.36	2.12	1.21	0.99	1.42
RD20	310.0	1.79	1.65	1.86	1.84	1.09	0.88	1.37
RD21	378.1	2.13	1.97	2.08	1.77	1.31	1.07	1.64
RD22	387.0	1.84	1.70	1.83	1.52	0.75	0.61	1.68
RD23	313.6	2.26	2.09	2.70	1.94	1.07	0.87	1.39
RD24	331.4	2.21	2.04	2.52	1.92	1.08	0.87	1.47
RD25	440.4	1.84	1.82	1.76	1.82	0.71	0.58	1.91
RD26	444.4	1.91	1.88	1.88	1.88	0.73	0.59	1.93
RD27	440.4	2.72	2.66	2.81	2.66	1.00	0.82	1.91
RD28	318.9	2.32	3.41	2.78	3.41	4.24	3.44	1.40
RD29	293.6	2.00	2.51	2.27	2.51	2.10	1.71	1.29
RD30	313.6	1.96	2.05	2.14	2.05	1.20	0.98	1.39
Average		1.88	1.97	2.06	2.14	1.86	1.49	1.36

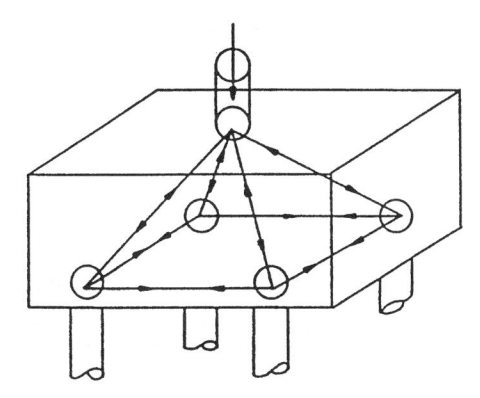

Figure 10.118 Truss analogy for a four pile cap.

1. It was determined that the flexural reinforcement must meet the minimum requirement of the ACI Code for shrinkage and temperature steel. The increase in this ratio will not significantly improve the load capacity of thick pile caps.

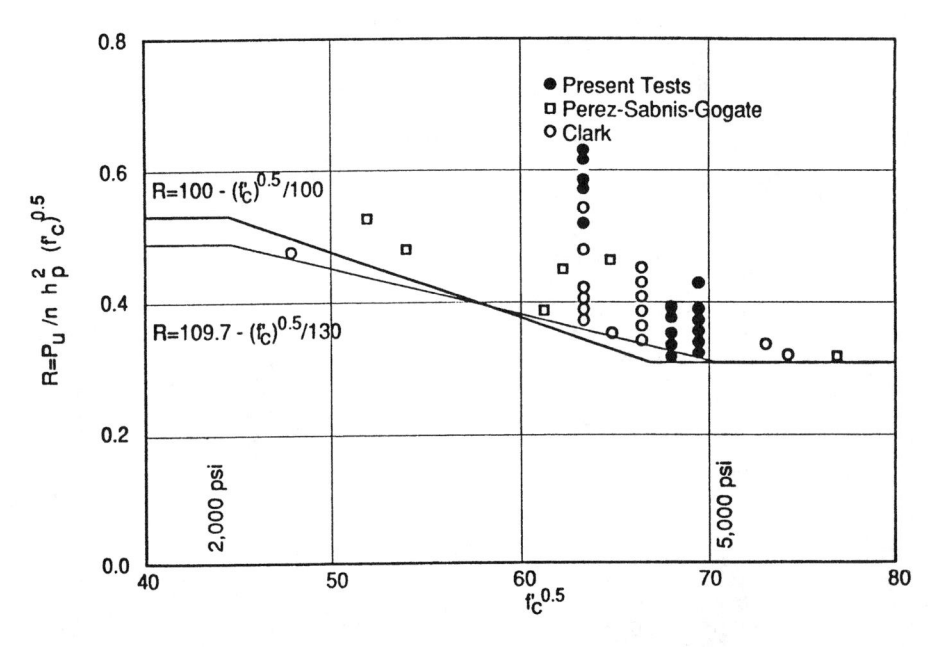

Figure 10.119 Nondimensional strength parameter "R" vs. compressive strength f'_c.

2. The ultimate shear strength is not constant at $8 + \sqrt{f'_c}$, once the span-to-depth ratio falls below 2 as the ACI Code specifies. These tests showed that shear strength is much higher and indicated the need to revise the ACI Code provisions to justify properly the behavior of thick pile caps.
3. The concrete compressive strength is related to the ultimate shear strength of pile caps through a nondimensional parameter. The increase results in the increase in the ultimate load capacity of a pile cap.
4. It was found that the span–depth ratio has an effect on the ultimate shear strength of a pile cap. Based on this research, the ultimate shear strength increased with decrease of the shear/span ratio (a/d).
5. Both scales of 2.5 and 6 were useful in investigating the behavior of pile caps; however, the small scale is preferred in view of the economy and the ease of handling of specimens.

10.3.6 Case Study H, Externally/Internally Prestressed Concrete Composite Bridge System

This case study involved the testing of four ⅓-scale bridge models as shown in Figure 1.23. The construction methodology was the same for each model with only minor variations of specific materials used (Grace and Sayed, 1996a,b; 1997a,b; 1998). The procedure is illustrated with the construction of Model DT-2 which was constructed in five stages. In the first stage, the two double-Tee (DT) girders were cast and internally post-tensioned. The prestressing forces in the internal strands were designed to support the deadweight of the bridge, the deck slab added later, and some live loads. This prestressing placed the cross-section of the girders in compression, as shown in Figure 10.120a. During the second stage, the DT girders were post-tensioned transversely through the tendon deviators and the cross beams. Only 50% of the transverse prestressing forces were applied. This linked the DT girders transversely without inducing significant moment. In the third stage, the reinforced deck was added over the entire bridge. The stirrups were projected from the DT girder webs, tendon deviators, and cross beams into the deck slab. Also, Skadur 32 epoxy was sprayed on the top surface of the flanges 1 h prior to pouring the deck slab. This combination of

MODEL APPLICATIONS AND CASE STUDIES 569

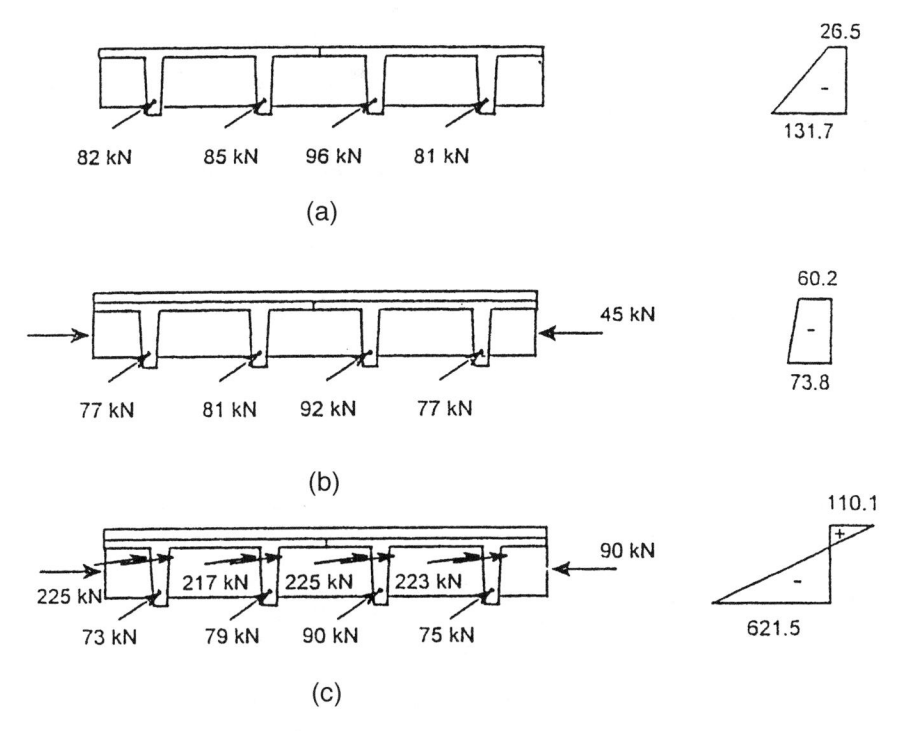

Figure 10.120 Prestressing forces and midspan strain during construction. (a) Internal prestressing (first stage). (b) Addition of deck slab (second and third stages). (c) External prestressing (fourth and fifth stages).

stirrups and epoxy ensured proper transfer of the horizontal shear forces and eliminated slippage between the deck slab and the girders. The strain distribution after the addition of the deck slab is shown in Figure 10.120b. Note that no tensile strains were allowed and hence no cracking at the bottom of the girders. In stage four (Figure 1.25), externally draped strands were post-tensioned. Figure 10.120c shows that considerable compressive strains and minimal tensile strains were experienced at the bottom and at the top of the girders, respectively. The tensile strain in the top of the girders was lower than the tensile strength of the concrete and as a result no cracks developed in the flanges. The stresses developed in the midspan were designed to counteract the live load stresses. Therefore, the bottom of the webs experienced no cracks under full live loading. In the last construction stage, the remaining 50% of the transverse post-tensioning forces was applied. This ensured that the deck slab was prestressed in both directions. Only the internal prestressing strands in the longitudinal direction were grouted after construction was completed.

Bridge model DT-1 was reinforced with glass fiber–reinforced polymer (GFRP) reinforcing bars and mild steel stirrups. It was prestressed using carbon fiber composite cable (CFCC) provided by Tokyo Rope (Grace and Sayed, 1966). Bridge model DT-2 was reinforced with 8-mm "Leadline" carbon fiber–reinforced polymer (CFRP) bars provided by Mitsubishi Chemical Corporation longitudinally and transversely. Multiple 8-mm CFRP bars were used for external post-tensioning. Stirrups were made from mild steel and 10-mm-diameter CFRP bars were used for internal and transverse post-tensioning. Bridge model DT-30 was reinforced in the same manner as bridge model DT-2, but used CFRP stirrups provided by Mitsubishi.

The four bridge models were subjected to sinusoidal fatigue loading oscillating between 8.9 and 53.4 kN up to 7 million cycles (Figure 1.24). Strains, prestressing forces, and deflections were monitored during fatigue testing. Strains at the top of the deck and the bottom of the webs at midspan for the first three models are shown in Figure 10.121. It can be seen that no significant changes occurred in either the top deck strains or the bottom web strains in any of the three models

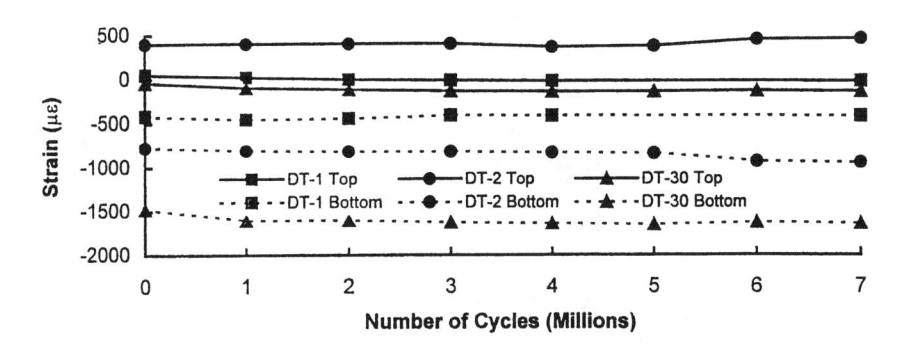

Figure 10.121 Change in top and bottom midspan strains due to fatigue loading.

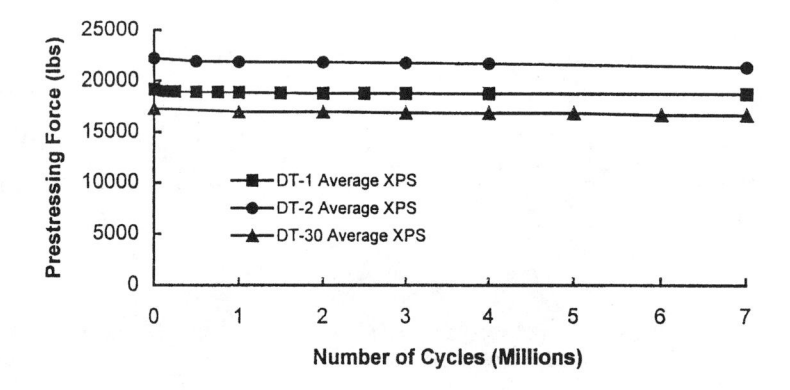

Figure 10.122 Change in average external prestressing forces due to fatigue loading.

up to 7 million cycles. Similarly, there were no significant changes to the external prestressing forces for the same number of fatigue cycles as shown in Figure 10.122. Measured reductions in the average prestressing force in 7 million cycles were 2.3, 4.2, and 3.7% for bridge models DT-1, DT-2, and DT-30, respectively.

To measure changes in stiffness of the bridge models, load–deflection tests were performed after the fatigue loading. Figure 10.123 shows the load–deflection responses of the first three bridge models at the beginning and conclusion (7 million cycles) of fatigue testing. It can be seen from this plot that the changes in response were not significant with minimum degradation in the stiffness of the bridge models due to fatigue loading.

After fatigue testing, the bridge models were loaded to failure under simulated AASHTO HS20-44 truck loading (Figure 10.124). The load–deflection curves of all four model bridges are shown in Figure 10.125. Bridge model DT-1 achieved a maximum deflection of 112 mm and a maximum load of 82.3 kN. The reasons for this lower than expected value were the dislocation of the internal prestressing (determined after testing) and the use of unbonded tendons which did not distribute the prestressing forces throughout the span. Bridge model DT-1 failed by crushing of the deck slab followed by rupture in one of the internal prestressing CFCC strands.

Bridge model DT-2 which was twice the width of DT-1 reached a maximum midspan deflection of 312.4 mm and a maximum load of 444.8 kN. DT-2 had bonded strands which helped to distribute the prestressing forces to the concrete along the bridge span. This model also failed by crushing of concrete in the deck followed by rupture in the internal prestressing bars.

Bridge model DT-30, which had a 30° skew, performed similarly to model DT-2. It reached a maximum midspan deflection of 254 mm and a maximum load of 524.9 kN. The failure was initiated by crushing of the deck slab concrete immediately followed by rupture in the internal prestressing bars.

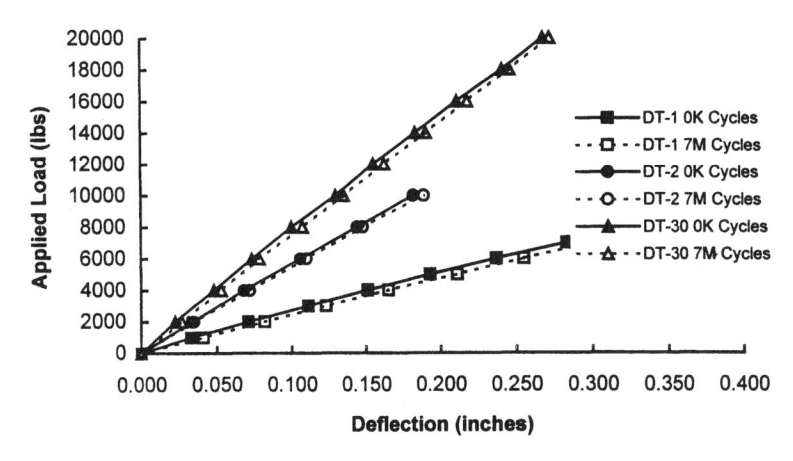

Figure 10.123 Change in load–deflection response due to fatigue loading.

Figure 10.124 Model bridge DT-30 during ultimate load testing. (Courtesy Prof. Nabil Grace, Lawrence Technological University, Southfield, MI.)

Bridge model DT-15, which had a 15° skew and was composed of three double-Tee sections, sustained a maximum midspan deflection of 221 mm and a maximum load of 502 kN.

The final cracking patterns of the first three bridge models are shown in Figure 10.126. The cracking experienced by specimen DT-1 (Figure 10.126a) was concentrated near the midspan because of the unbonded tendons and the tendon dislocation found after the test. Cracking in bridge model DT-2 was more extensive (Figure 10.126b), since the bonding of the internal strands allowed for more even distribution of prestressing forces and hence crack formation. The cracking behavior of bridge model DT-30 (Figure 10.126c) was similar to that of DT-2. Extensive cracking in bridge models DT-2 and DT-30, in contrast to the minimal cracking in DT-1, shows that the proposed bridge system can be designed to consume significant amounts of energy through secondary cracking. The mechanism for achieving this is the use of bonded internal strands in the longitudinal direction and unbonded strands in the transverse direction.

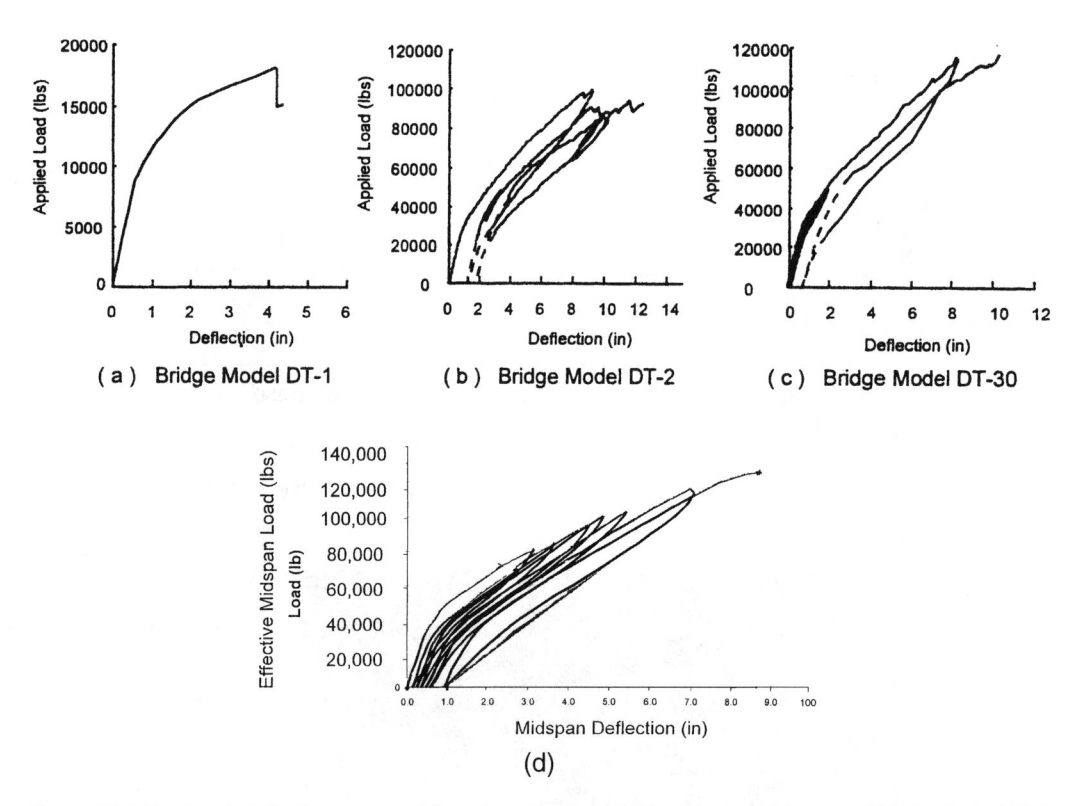

Figure 10.125 Load–deflection curves of four ⅓-scale model bridges tested at Lawrence Technological University. (a) Bridge model DT-1. (b) Bridge model DT-2. (c) Bridge model DT-30. (d) Bridge model DT-15.

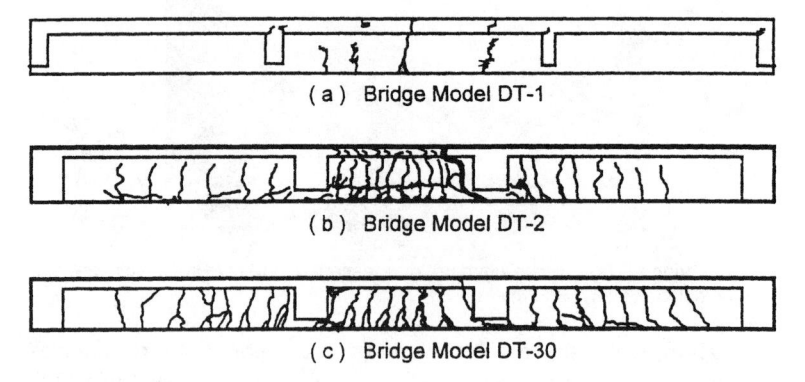

Figure 10.126 Postfailure cracks in DT bridge models.

10.4 SUMMARY

The results of several classes of structural model studies under static or quasi-static loading have been presented. Among these structures, buildings and bridges as well as special structures have been included. Materials of construction include reinforced and prestressed concrete, reinforced masonry, wood and steel structures. Detailed discussion of six case studies that were introduced in Chapter 1 is given in an attempt to convey to the reader a feeling of the breadth and power of the physical modeling technique.

MODEL APPLICATIONS AND CASE STUDIES 573

PROBLEMS

10.1 a. Design a 1/12-scale model for a singly reinforced concrete beam, using prototype and model properties given below:

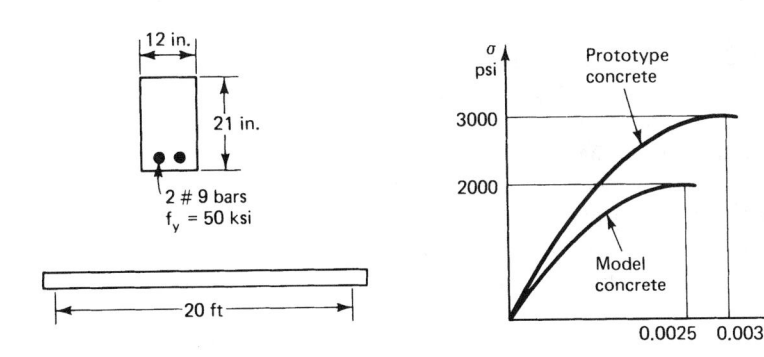

[Figure: Beam cross-section 12 in. wide × 21 in. deep with 2 #9 bars, $f_y = 50$ ksi; span 20 ft. Stress-strain curves showing prototype concrete peaking near 3000 psi at $\epsilon = 0.003$ and model concrete peaking near 2000 psi at $\epsilon = 0.0025$.]

b. Neglecting dead-load effects, and using elementary ultimate-strength calculation methods, determine the maximum midspan concentrated load each beam can carry. Verify your model design by scaling the computed model capacity up to the prototype and compare this with the calculated prototype capacity.

c. If the dead-load effect is to be included, how much additional uniform load must be applied to the model?

d. What are the requirements on the model concrete if shear capacity is to be modeled correctly?

e. Assuming you are using two bars for the model steel, what is the bond stress at failure under part b? How are the bond stresses in model and prototype related?

f. If the prototype beam also had two No. 9 bars as compression steel, would the same modeling of steel as used in part a be valid? Discuss.

10.2 You are asked to design a 1/10-scale model reinforced concrete beam made from gypsum-based model concrete with the uniaxial compressive stress–strain curve shown below. Prototype material properties are also given.

a. What other requirements would you place on model concrete properties, and why?

b. What are the advantages and disadvantages of using gypsum-based model concrete as compared with cement-based microconcrete?

c. Specify the required yield point of the model steel.

d. What are the major problems met in modeling the postcracking and ultimate-load behavior of reinforced concrete structures at this scale?

e. The model beam deflects 25 mm at midspan just prior to failure. What is the corresponding prototype deflection?

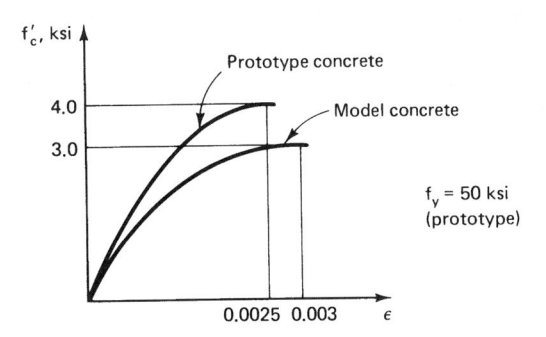

[Figure: Stress-strain curves with f_c' in ksi. Prototype concrete reaches 4.0 ksi at $\epsilon = 0.003$; model concrete reaches 3.0 ksi at $\epsilon = 0.0025$. $f_y = 50$ ksi (prototype).]

10.3 Design a reinforced concrete beam, either with single or double reinforcing, to fail in flexure or shear. Construct a model of the beam design and test it to failure, determining:
   a. Load–deflection curve
   b. Ultimate load capacity, including mode of failure
   Compare the test results with the theoretical predictions and discuss briefly.

   Ultracal (gypsum-based) model concrete with proportions of 1 part Ultracal, 1 part sand, and 0.31 parts water will produce a mix of about 21 MPa in 24 h. Rather than becoming involved with sealing the beam surfaces, it is suggested that you arrange the casting and testing of the beam, within a 24- to 27-h time period, such as casting on Monday and testing Tuesday afternoon.

   You may test the model steel in tension to determine its yield point.

10.4 The behavior of a prestressed concrete box girder bridge, made by post-tensioning 6-m-long segments together, is to be studied by model analysis in the laboratory. The tentative prototype structure design is

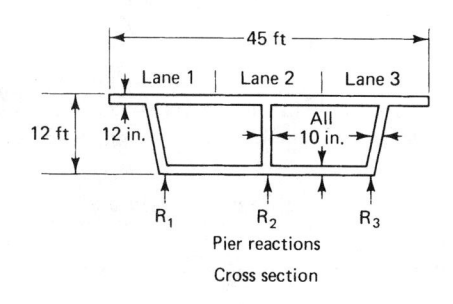

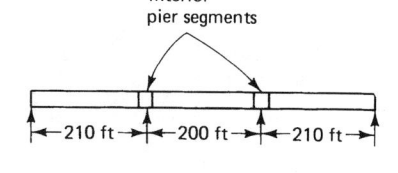

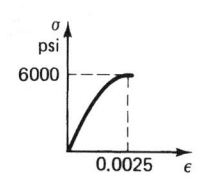

a) f'_c = 6000 psi at 28 days
   wt = 145 lb/ft^3
b) Reinforcing steel is 60 ksi yield strength, $\frac{3}{4}$ in. ϕ
c) Prestress steel is $1\frac{1}{2}$ in. diam. strand with E = 25,000,000 psi and f_y = 140 ksi
   Working (design) stress = 120 ksi

Each span is composed of similar segments. The construction sequence is to place a segment at each interior pier and then add segments in each direction alternately, in a cantilever mode, with continuous prestressing. The prestressing elements are placed in pipe conduits cast into the concrete. Outer portions of each end span are constructed on falsework.

Critical *loading conditions* that are to be studied experimentally are
a. Deadweight effects during the construction process.
b. Design live load, represented by a uniformly distributed surface load of 5 kN/m² on the entire top deck of the bridge.
c. Partial live load of 5 kN/m² on one side lane only (either lane 1 or 3) to produce maximum torsional effects.
d. Ultimate load capacity with a uniform loading on the full deck of the entire bridge.
e. Slab-punching effects produced by a heavy wheel load of 80 kN over an area of 0.2 m² (500 × 375 mm contact area of tire).

Quantities to be determined include:
f. Concrete stresses at critical sections.
g. Displacements.

MODEL APPLICATIONS AND CASE STUDIES

    h. Changes in force in the post-tensioning prestress elements.
    i. Reactions of the bridge on its supporting piers.
    j. Failure capacity and failure mode under full uniform load.
    k. Slab-punching strength from concentrated wheel load.

10.5 A footbridge similar to the tubular truss bridge shown below is to be investigated for a possible instability failure under a uniformly distributed live load. The spans of the structure are as shown below; cross-sectional properties are also given. A36 steel, 21 MPa concrete, and 276 MPa reinforcing steel are to be used in building the structure.

Assuming that the elastic buckling load is to be investigated experimentally on a model basis, describe in detail the model study you would undertake, including basic configuration (need the full structure be tested?), size of model, materials to be used, scaling relations, loading, method of applying load, instrumentation, and the actual method for determining the critical buckling load level. If resistance strain gages are to be used, specify their location and the types of gages to be employed, as well as any special circuitry you wish to use. Give justification for any distortion of true similitude. Comment on any unusual problems you feel this modeling will present, and discuss the confidence level you expect to achieve in the results.

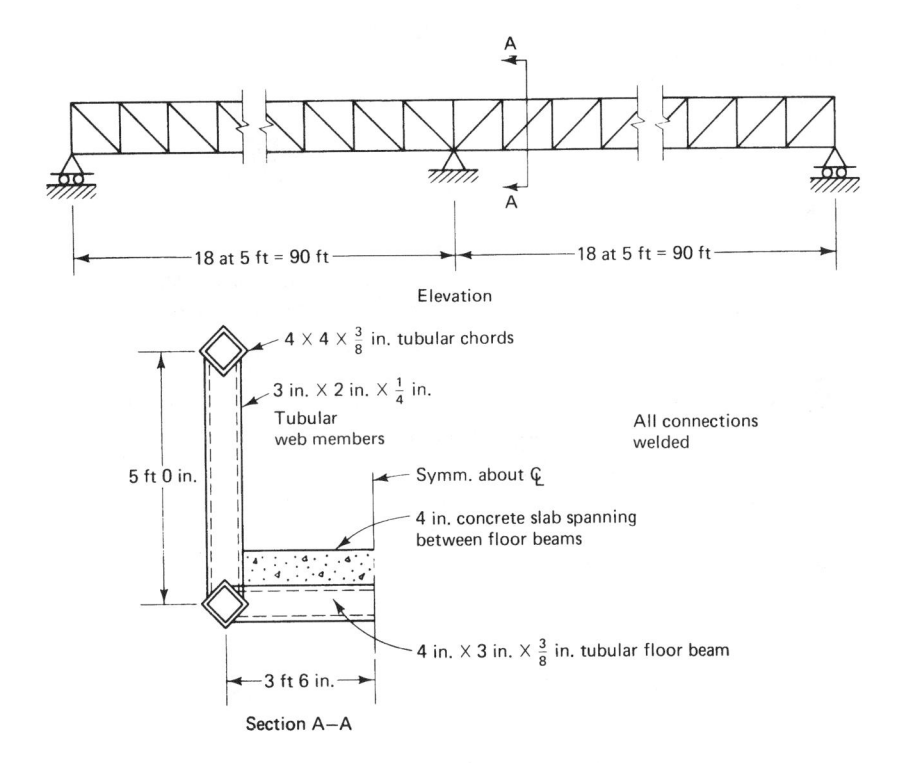

Finally, make a time and cost estimate based on the following rates:

Engineer (yourself)	$x/day
Mechanician	$(2/3 x)/day + 75% overhead
Technician	$(1/2 x)/day

Materials need not be estimated unless you wish to. Assume all testing equipment is available in the laboratory.

10.6 An existing underground storage tank (for liquids) is to be investigated for its strength to resist a high overpressure (blast loading) to be *applied* on the covering soil, at ground level. Since a full-scale test is not possible, it is proposed to study the strength of the structure with a model. The prototype tank is made of *reinforced concrete* and is detailed as shown below. The full reinforcing scheme is not shown here because a complete detailing of the model reinforcement is not warranted.

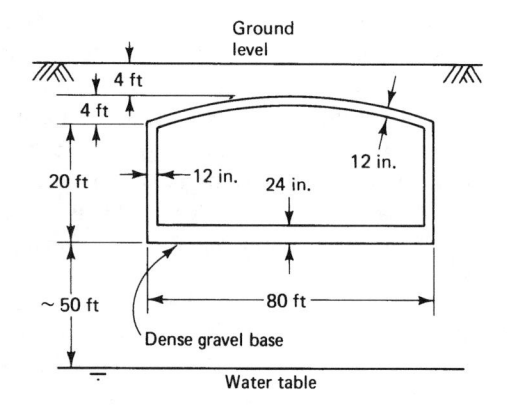

Prototype materials:

Concrete — f'_c = 4600 psi, with a strain at peak stress of 0.0032
E = 3,800,000 psi.

Reinforcing steel — yield strength = 50 ksi

Note: The backfill around the walls of the tank is of questionable quality; hence the tank is to be checked for vertical load only (no lateral passive or active pressures).

The structural engineer will be satisfied if the tank strength under *static* overpressure (that is, not dynamic) is known. The engineer wants to know the failure mode, which may be either strength controlled or stability controlled and the failure capacity, and would also like to know critical stresses at critical sections just prior to failure.

Outline a program for doing the model analysis for this situation. The program should include:

a. Decision on basic type of model, and choice of scale factor.
b. Choice of model materials. Two model concretes are available: one (E_1) = 23,000 MPa with a strength of 24 MPa and a strain at peak stress of 0.0025, and a second (E_2) = 29,000 MPa with a strength of 38 MPa and a strain at peak stress of 0.0035. Model reinforcing is available in whatever strength and diameter you need.
c. Fabrication techniques and a discussion of any special problems.
d. Model loading method, describing specifically how you will apply the load and measure it. Give scaling relationships to relate model loads to the prototype.
e. The necessary instrumentation to measure stresses, to detect any possible instability modes, and to determine the general deformations of the model. Types of strain gages and their lengths should be specified, along with the types of bridge circuits (single-arm, double-arm, or full-bridge) and how temperature compensation will be achieved.
f. Desired and expected level of accuracy in your results.
g. An estimate of the total person-days needed to complete the work, in terms of (1) engineer (yourself) and (2) technician (for construction, instrumentation, and testing).

10.7 A composite steel girder, concrete slab bridge, curved in plan, has been designed by an analytical approach that involves both elastic analysis and ultimate-strength concepts. The tentative prototype structure is shown below:

MODEL APPLICATIONS AND CASE STUDIES

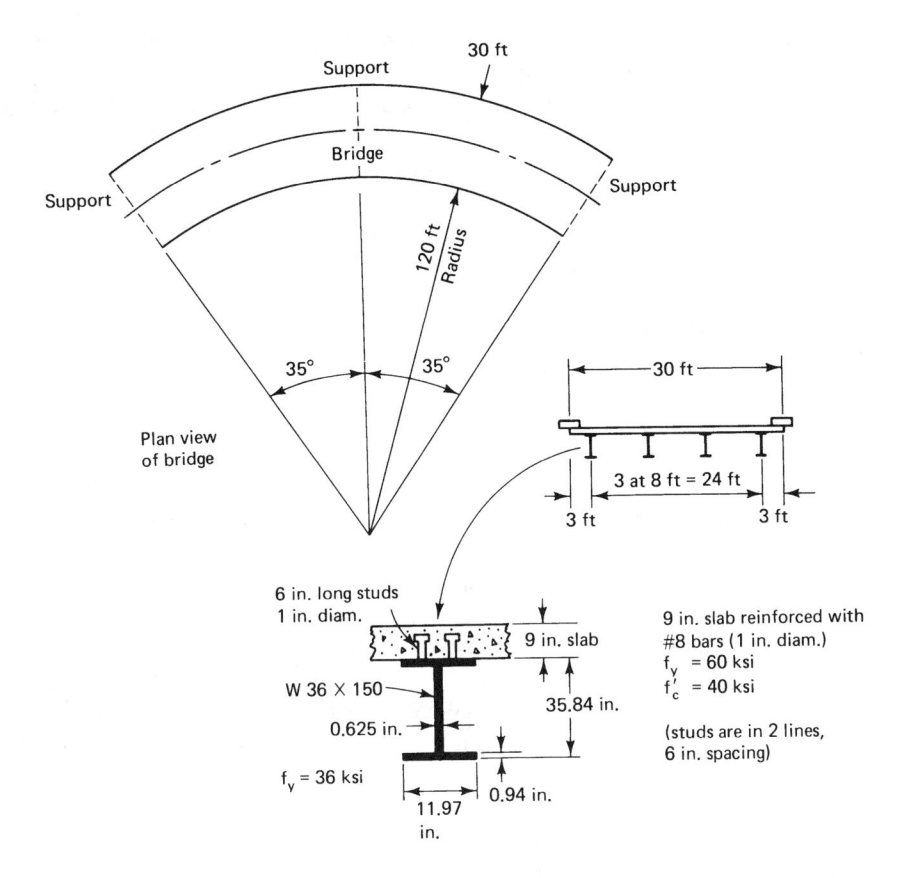

Plan view of bridge: 120 ft Radius, 35° + 35°, Supports at ends and bridge centerline, 30 ft width.

Cross section: 30 ft total; 3 at 8 ft = 24 ft; 3 ft overhangs on each side.

Girder detail:
- 6 in. long studs, 1 in. diam.
- 9 in. slab
- W 36 × 150
- 0.625 in. web
- f_y = 36 ksi
- 11.97 in. flange width
- 0.94 in. flange thickness
- 35.84 in. depth
- 9 in. slab reinforced with #8 bars (1 in. diam.)
- f_y = 60 ksi
- f'_c = 40 ksi
- (studs are in 2 lines, 6 in. spacing)

The shear connection between beam and slab is provided by metal studs welded to the upper flange of the beam. Such a connection can guarantee flexural strain compatibility between the two materials for loads appreciably beyond the design load. However, with inadequate, insufficient, or overstressed shear connectors, slippage of the slab relative to the beam will occur, thereby reducing the stiffness of the system and increasing stresses because the two elements then tend to act individually rather than as an integral unit. Thus, the potential failure modes include yielding of the steel beams (bending, shear, torsion effects), concrete crushing from flexural compressive stress, localized concrete failure from bearing stresses produced by shear studs, flexural or shear yielding of the studs, and combinations of the above. Lateral motions of the curved girders could also be a problem.

In constructing composite structures, the formwork for the concrete slab is normally supported by the steel beams; thus the deadweight of the slab is carried by the beams alone, and only live-load stresses are resisted by the composite, integral section.

It is proposed to use *physical model analysis* to check on the overall validity of the analysis and to establish the behavior of the structure at overloads leading up to failure of the bridge. Questions to be resolved include:

a. Accuracy of the analytical representation of the stiffnesses of the various elements (in particular, the curved beams).

b. Dead-load deformations produced by the weight of the fresh concrete acting on the steel girders.

c. Girder reactions and maximum stresses (bending, shear, torsion) produced by dead load.

d. Live-load influence lines for displacements, reactions, and peak stresses (within the range of behavior for normal live loads).

e. Ultimate load capacity (in shear) of the shear transfer line between slab and girder (shear connectors).

f. Failure mode and ultimate load capacity of the bridge subjected to a uniformly distributed loading that is meant to simulate maximum traffic.

Outline a program for doing the model analysis of this structure. The program should include:

g. Decisions on basic type of model (or models, if you choose to study one or more of the problems with a special partial model of the structure).

h. Choice of scale factor(s).

i. Choice of model material(s).

j. Fabrication techniques.

k. Loading arrangements, including deadweight effects.

l. Instrumentation. Concentrate on specifying instrumentation at one section of the model and then indicate how many sections would have similar instrumentation. Types of strain gages (foil, epoxy-backed, or wire, paper-backed) and their gage lengths must be given. Locations of gages must be shown clearly, and the strain gage circuitry used should be defined.

Give brief justifications for your decisions. If a distorted model is used, discuss the effects of the distortion. Discuss the level of accuracy you desire and the level you actually expect to achieve. Comment on any unusual problems you feel the modeling will present. Finally, make a rough estimate of the time required for completing the project. Time should be in two categories: engineer (yourself) and mechanic/technician.

10.8 A reinforced concrete beam–column is to be modeled for ultimate load capacity. The model is to be loaded to failure by increasing P and M in a fixed ratio. Prototype material properties and model concrete properties are given below.

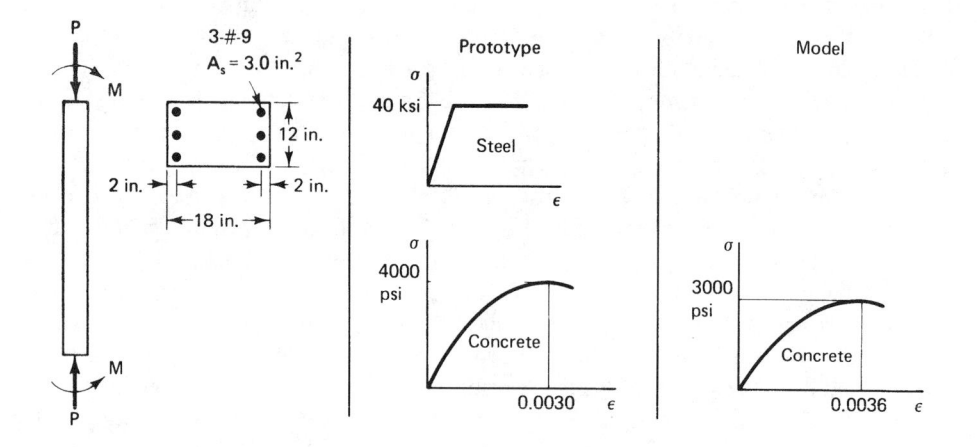

a. What is the minimum size model you would recommend for this prototype? Discuss your reasons for your choice.

b. Assuming a ⅛-scale model, establish:

  i. Required yield strength of the model steel;

  ii. Required areas of model steel;

  iii. Prediction factors for converting values of moment and axial load in the model to prototype values.

c. We cannot achieve true similarity with this model because the strain scale factor is not unity. Discuss the possible effects of this distortion on the model results. Will it tend to make the measured ultimate *P* and *M* higher or lower than they should be?
   d. Recommend a size of compressive cylinder for determining f'_c for the model material. Does the strength vary as a function of cylinder size?
   e. Sketch a suitable loading arrangement to achieve the conditions of combined axial load and equal end moments.
   f. Describe the instrumentation you would use for this study.

REFERENCES

AASHTO (1989). *Specifications of Highway Bridges,* 14th ed., American Association of State Highway and Transportation Officials, Washington, D.C.

Abboud, B. E. (1987). The Use of Small-Scale Direct Models for Concrete Block Masonry Assemblages and Slender Reinforced Walls under Out-of-Plane Loads, Ph.D. thesis, Civil and Architectural Engineering Department, Drexel University, Philadelphia.

Abboud, B. E., Hamid, A. A., and Harris, H. G. (1990). Small-scale modeling of concrete block masonry structures, *ACI Struct. J.,* 87(2), Mar.-Apr.

Abrams, D. P. and Paulson, T. J. (1990). Perceptions and observations of seismic response for reinforced masonry building structures, in *Proceedings of the Fifth North American Masonry Conference,* University of Illinois, June, 13 pp.

Abrams, D. P. and Paulson, T. J. (1991). Modeling concrete masonry building structures at one-quarter scale, *Struct. J. Am. Concr. Inst.,* Title No. 88-S50, 88(4), July-August, 475–485.

Anvar, S. A., Arya, S. K., and Hegemier, G. A. (1983). Behavior of Floor-to-Wall Connections in Concrete Masonry Structures, Report No. UCSD/AMES/TR-83/001, Department of Applied Mechanics and Engineering Sciences, University of California, San Diego, La Jolla.

Appa Rao, T. V. S. (1975). Behavior of concrete nuclear containment structures up to ultimate failure with special reference to MAPP-1 Containment, in *Symposium on Structural Mechanics on Reactor Technology,* B.A.R.C., Bombay, India.

Baglietto, E., Casirati, M., Castoldi, A., Demiranda, F., and Sammartino, R. (1976). Mathematical and Structural Models of Zarate-Brazo Largo Bridges, Report No. 85, ISMES — Instituto Sperimentale Modelli e Strutture, Bergamo, Italy, September, 46 pp.

Banerjee, A. C. (1973). Design of Pile Caps, Central Building Research Institute, Roorkee (India), March.

Billington, D. P. and Harris, H. G. (1981). Test methods for concrete shell buckling, ACI Special Publication SP-67, *Concrete Shell Buckling,* E. P. Popov and S. J. Medwadowski, Eds., American Concrete Institute, Detroit, 187–231.

Blevot, J. and Fremy, R. (1967). Sur pieus, *Annales, Institute Technique du Batiment et des Travaux Public* (Paris), Vol. 20, February.

Bouma, A. L. et al. (1962). Investigations on models of eleven cylindrical shells made of reinforced and prestressed concrete, in *Proceedings, Symposium on Shell Research,* Delft, 1961, Wiley-Interscience, New York, 70–101.

Breen, J. E., Ed., (1975. Research Workshop on Progressive Collapse of Building Structures, Summary, The University of Texas, Austin, Nov. 18–20.

Cheung, K. C. (1974). PCRV Design and Verification, Report No. GA-A12821 (GA-LTR-8), General Atomic Co., San Diego, CA.

Clark, J. L. (1973). Behavior and Design of Pile Caps with Four Piles, Technical Report No. 42.489, Cement and Concrete Association, Wexham Springs, Slough.

Corum, J. M., White, R. N., and Smith, J. E. (1969). Mortar models of prestressed concrete reactor vessels, ASCE *J. Struct. Div.,* 95(ST2), February, Proc. Paper 6419, 229–248.

CRSI (1986). Pile Caps for Individual Columns, in *CRSI Handbook,* Concrete Reinforcing Steel Institute, Chicago, IL, Ch. 13.

Dow, B. N. and Harris, H. G. (1978). Use of Small Scale Direct Models to Predict the Response of Horizontal Joints in Large Panel Precast Concrete Buildings, Structural Models Laboratory Report No. M78-3, Department of Civil Engineering, Drexel University, Philadelphia, June, 128 pp.

Fintel, M., Schultz, D. M., and Iqbal, M. (1976). Report 2: Philosophy of Structural Response to Normal and Abnormal Loads, Design and Construction of Large-Panel Concrete Structures, Office of Policy Development and Research, Department of Housing and Urban Development, Washington, D.C., March.

Fumagalli, E. and Verdelli, G. (1976). Research on PCPV for BWR — Physical Model as Design Tool — Main Results, Report No. 86, ISMES, Istituto Sperimentale Modelli e Strutture, Bergamo, Italy, September.

Grace, N. F. and Sayed, G. A. (1996a). Double tee and CFRP/GFRP bridge system, *Concr. Int.*, American Concrete Institute, 18,(2), 39–44.

Grace, N. F. and Sayed, G. A. (1996b). Feasibility of CFRP/GFRP prestressed concrete demonstration bridge in the USA, Fourth National Workshop on Bridge Research in Progress, National Center for Earthquake Engineering Research, Buffalo, NY.

Grace, N. F. and Sayed, G. A. (1997a). Ductility of prestressed concrete bridges using internal/external CFRP strands, Seventh International Conference & Exhibition, Structural Faults + Repair '97, Edinburgh, Scotland.

Grace, N. F. and Sayed, G. A. (1997b). Behavior of externally/internally prestressed concrete composite bridge system, *Third International Symposium on Non-Metallic (FRP) Reinforcement for Concrete Structures,* Sapporo, Japan, Oct. 14–16.

Grace, N. F. and Sayed, G. A. (1998). Ductility of prestressed concrete bridges usinginternal/external CFRP strands, *Concr. Int.*, ACI, June.

Guedelhoefer, O. C., Moreno, A., and Janney, J. R. (1972). Structural models of hangars for large aircraft, *Proceedings of the Symposium of the ACI Canadian Chapter on Models in Structural Design,* Montreal, 71–109.

Hanson, N. W. (1975). Interim report, Task 4 — Experimental Cantilever Test — Phase 1, *Design and Construction of Large Panel Concrete Structures,* prepared for the Office of Policy Development and Research, Department of Housing and Urban Development, by Portland Cement Association, Skokie, IL, December.

Harris, H. G. (1967). The Inelastic Analysis of Concrete Cylindrical Shells and Its Verification Using Small-Scale Models, Ph.D. thesis, Cornell University, Ithaca, NY.

Harris, H. G. and White R. N. (1972). Inelastic behavior of reinforced concrete cylindrical shells, *ASCE J. Struct. Div.,* 98(ST7), July, Proc. Paper 9074, 1633–1653.

Harris, H. G. and Muskivitch, J. C. (1977). Report 1: Study of joints and sub-assemblies — validation of the small-scale direct modeling techniques, in *Nature and Mechanism of Progressive Collapse in Industrialized Buildings,* Office of Policy Development and Research, Department of Housing and Urban Development, Washington, D.C., October, 165 pp., also Department of Civil Engineering, Drexel University, Philadelphia.

Harris, H. G. and Muskivitch, J. C. (1980). Models of precast concrete large panel buildings, *ASCE J. Struct. Div.,* 106(ST2), February, Proc. Paper 15218, 545–565.

Harris, H. G. and Oh, K. H. (1990). Seismic behavior of floor-to-wall horizontal joints between block masonry walls and precast concrete hollow core slabs using ⅓-scale direct models, in *Proceedings, The Fourth U.S. National Conference on Earthquake Engineering,* Vol. 2, Palm Springs, CA, May, 777–786.

Harris, H. G., Oh, K. H., and Hamid, A. A. (1992). Development of new interlocking mortarless block masonry units for efficient building systems, in *Proceedings of the 6th Canadian Masonry Symposium,* Saskatoon, Saskatchewan, Canada, Vol. 2, 723–734.

Harris, H. G., Oh, K. H., and Hamid, A. A. (1993). Development of New Interlocking Blocks to Improve Earthquake Resistance of Masonry Construction, Report to the National Science Foundation, Department of Civil and Architectural Engineering, Drexel University, Philadelphia, 146 pp.

Harris, H. G., Somboonsong, W., and Ko, F. K. (1997). A new ductile hybrid fiber reinforced polymer (FRP) reinforcement for concrete structures, in *Proceedings of the 1997 International Conference on Engineering Materials,* 8–11 June, Ottawa, Canada, Vol. I, 593–604.

Harris, H. G., Somboonsong, W., Ko, F. K., and Huesgen, R. (1998a). A second generation ductile hybrid fiber reinforced polymer (FRP) for concrete structures, in *Proc. Second International Conf. on Composites in Infrastructure,* Tucson, Jan. 15–17, Department of Civil Engineering and Engineering Mechanics, University of Arizona, Tucson.

Harris, H. G., Somboonsong, W., and Ko, F. K. (1998b). A new ductile hybrid fiber reinforced polymer (FRP) reinforcing bar for concrete structures, *ASCE J. Composites Construction,* 2(1), February, 28–37.

Hillerborg, A. (1975). *Strip Method of Design,* View Point Publications, London.

Hobbs, N. B. and Stein, P. (1957). An investigation into the stress distribution in pile caps with some notes on design, *Proceedings of the Institution of Civil Engineers*, Vol. 7, July, pp. 599–628.

Hoffman, N. S. (1988). Evaluation of Creosote Penetration in Pressure Treated Pennsylvania Area Red and White Oak, Unpublished Report, Department of Civil and Architectural Engineering, Drexel University, Philadelphia, July.

Hoffman, N. S. (1990). Behavior of Transversely Prestressed Wood Bridge Decks under Simulated Short-Term and Long-Term Traffic Loading, M.Sc. thesis, Department of Civil and Architectural Engineering, Drexel University, Philadelphia, March.

Ko, F. K., Somboonsong, W., and Harris, H. G. (1997). Fiber architecture based design of ductile composite rebars for concrete structures, in *Proceedings of the International Conference on Composite Materials*, Scott, M. L., Ed., July 14–17, Gold Coast, Australia, Vol. VI, *Composite Structures*, VI-723-VI-730.

Larbi, A. (1989). Behavior of Block Masonry Shear Walls under In-Plane Monotonic and Reversed Cyclic Loads Using ⅓-Scale Direct Models, M.Sc. thesis, Department of Civil and Architectural Engineering, Drexel University, Philadelphia, March.

Larbi, A. and Harris, H. G. (1990). Seismic performance of reinforced block masonry shear walls using ⅓-scale direct models, in *Proceedings of the Fifth North American Masonry Conference,* University of Illinois at Urbana-Champaign, June 3–6, Vol. 1, 321–332.

Larbi, A. and Harris, H. G. (1990a). Seismic performance of low aspect ratio reinforced block masonry shear walls, in *Proc. 4th U. S. Nat. Conf. Earthquake Engineering*, May 20–24, Palm Springs, CA, Vol. 2, 799–808.

Lee, S. J. and Lu, L. W. (1984). Studies on full-scale composite beam–column components, paper presented at the ASCE Convention, San Francisco, October.

Leyendecker, E. V. and Burnett, E. F. P. (1976). The incidence of abnormal loading in residential buildings, NBS Building Science Series 89, U.S. Department of Commerce, National Bureau of Standards, Dec.

Lim, J. Y. and McLean, D. I. (1991). Scale model studies of moment-reducing hinge details in bridge columns, *ACI Struct. J.*, 88(4), July-August.

Litle, W. A. and Foster, D. C. (1966). Fabrication Techniques for Small-Scale Steel Models, Department of Civil Engineering, Massachusetts Institute of Technology, Cambridge, October.

Johal, L. S. and Hanson, N. W. (1978). Horizontal joints tests, Supplemental Report B,*Design and Construction of Large Panel Concrete Structures,* prepared for the Office of Policy Development and Research, Department of Housing and Urban Development, by Portland Cement Association, Skokie, IL, April.

Mills, R. S., Krawinkler, H., and Gere, J. M. (1979). Model Tests on Earthquake Simulators Development and Implementation of Experimental Procedures, Report no. 39, the John A. Blume Earthquake Engineering Center, Department of Civil Engineering, Stanford University, June, 272 pp.

Muskivitch, J. C. and Harris, H. G. (1979a). Report 2: Behavior of precast concrete large panel buildings under simulated progressive collapse conditions, in *Nature and Mechanism of Progressive Collapse in Industrialized Buildings,* Office of Policy Development and Research, Department of Housing and Urban Development, Washington, D.C., January, 207 pp.; also Department of Civil Engineering, Drexel University, Philadelphia.

Muskivitch, J. C. and Harris, H. G. (1979b). Behavior of large panel precast concrete buildings under simulated progressive collapse conditions, in *Proceedings, International Symposium, Behavior of Building Systems and Building Components,* Vanderbilt University, March 8–9, Nashville, TN.

Muskivitch, J. C. and Harris, H. G. (1981). Report No. 3: Analytical investigation of the internal load distribution and the mechanism of progressive collapse in precast concrete large panel buildings, in *Nature and Mechanism of Progressive Collapse in Industrialized Buildings,* Office of Policy Development and Research, Department of Housing and Urban Development, Washington, D.C., June; also Department of Civil Engineering, Drexel University, Philadelphia, 248 pp.

Naaman, A. E. and Jeong, S. M. (1995). Structural ductility of concrete beams prestressed with FRP tendons, in *Non-metallic (FRP) Reinforcement for Concrete Structures,* L. Taewere, Ed., RILEM, E & FN Spon, London.

Ndukwe, A., (1982). Comparison of Methods of Analysis and Design of Thick Reinforced Concrete Pile Caps, Research Project Report, Howard University, Washington, D. C.

Oh, K. H. (1988). Monotonic and cyclic behavior of floor-to-wall intersections in masonry structures under in-plane shear loading — using ⅓-scale direct modeling technique, Master's thesis, Department of Civil and Architectural Engineering, Drexel University, Philadelphia, December.

Oh, K. H. (1994). Development and investigation of failure mechanism of interlocking mortarless block masonry systems, Ph.D. thesis, Department of Civil and Architectural Engineering, Drexel University, Philadelphia.

Oh, K. H. and Harris, H. G. (1990). Seismic behavior of floor-to-wall horizontal joints of masonry buildings using ⅓-scale direct models, in *Proceedings of the Fifth North American Masonry Conference,* University of Illinois at Urbana-Champaign, June 3–6, Vol. 1, 81–92.

Oh, K. H., Harris, H. G., and Hamid, A. A. (1993). New interlocking and mortarless block masonry units for earthquake resistance structures, in *Proceedings, the 6th North American Masonry Conference,* Drexel University, June 6–9, Philadelphia, Vol. 2, 821–836.

OHBDC (1983). *Ontario Highway Bridge Design Code,* 2nd ed., Highway Engineering Division, Ontario Ministry of Transportation and Communications, Downsview, Ontario, 357.

Ohio DOT, (1947). Investigation of the Strength of the Connection between a Concrete Cap and the Embedded End of a Steel H-Pile, Research Report No. 1, Ohio State Department of Highways.

Oliva, M. G. and Dimakis, S. R. (1988). Behavior of Post-Tentioned Wood Bridge Decks: Full-Scale Testing, Analytical Correlation, Design Guidelines, Structures and Materials Test Laboratory, 87-1, University of Wisconsin, May, 150 pp.

Paul, S. L. et al. (1969). Strength and behavior of prestressed concrete vessels for nuclear reactors, Civil Engineering Series, Structural Research Series No. 346, University of Illinois, Urbana, July.

Paulson, T. J., and Abrams, D. P. (1990). Correlation between static and dynamic response of model masonry structures, *Earthquake Spectra,* 6(3), August, 573–592.

Perdikaris, P. C. and Beim, S. R. (1988). RC bridge decks under pulsating and moving load, *ASCE, J. Struct. Eng.,* 114(3), March, 591–607.

Perdikaris, P. C. and Petrou, M. (1991). Code predictions vs. small-scale bridge deck model test measurements, in Transportation Research Record No. 1290, TRB, Vol. 1, *Bridges & Structures,* March 10–13, 179–187.

Perdikaris, P. C., Beim, S. R., and Bousias, S. (1989). Slab continuity effect of ultimate and fatigue strength of R/C bridge deck models, *ACI Struct. J.,* 86(4), July-August, 483–491.

Petrou, M. F. (1991). Behavior of concrete bridge deck models subjected to concentrated load-Ontario vs. AASHTO, M.Sc. thesis, Case Western Reserve University, Cleveland, OH, May.

Petrou, M. F., Perdikaris, P. C., and Wang, A. (1994). Fatigue behavior of con-composite reinforced concrete bridge deck models, in Transportation Research Record No. 1460, *Bridges and Structures,* Dec.

Petrou, M. F., Perdikaris, P. C., and Duan, M. (1996). Static behavior of non-composite concrete bridge decks under concentrated loads, *J. Bridge Eng.,* ASCE.

Porter, M. L. (1987). Sequential phase displacement (SPD) procedure for TCCMAR Testing, 3rd Meeting of the Joint Technical Coordinating Committee on Masonry Research, U.S.–Japan Coordinated Program for Masonry Building Research, Tomamu, Japan, Oct. 15–17.

Sabnis, G. M. and Dagher, R. (1989). Investigation of Reinforced Concrete Pile Caps, Proceedings of One-Day Conference on Life of Structures, Brighton, England.

Sabnis, G. M. and Gogate, A. B. (1980). Investigation of thick pile caps, *ACI J.,* 77(1), Jan.-Feb., 18–24.

Sagur, S. (1987). The Effects of Central Openings on the Ultimate Load in Simply Supported Reinforced Concrete Slabs, Structural Models Laboratory Report, Department of Civil and Architectural Engineering, Drexel University, Philadelphia.

Sagur, S. (1994). The Use of Small-Scale Modeling to Study the Behavior of Simply-Supported Reinforced Concrete Slabs with Openings under Uniform Loading, Ph.D. thesis, Department of Civil and Architectural Engineering, Drexel University, Philadelphia.

Schultz, D. M., Burnett, E. F. P., and Fintel, M. (1977). Report 4: A design approach to general structural integrity, in *Design and Construction of Large Panel Concrete Structures,* Office of Policy Development and Research, Department of Housing and Urban Development, Washington, D.C., October.

Shing, P. B., Noland, J. L., Spaeh, H., and Klamerus, E. (1986). Response of Reinforced Masonry Story Height Walls to Fully Reversed In-Plane Loads, Department of Civil Engineering, University of Colorado, Boulder, September.

Shing, P. B., Noland, J. L., Spaeh, H., and Klamerus, E. (1987). Inelastic behavior of masonry wall panels under in-plane cyclic loads, The 4th North American Masonry Conference, Los Angeles, August.

Shing, P. B., Klamerus, E. W., and Schuller, M. P., (1988). Behavior of single-story reinforced masonry shear walls under in-plane cyclic lateral loads, report of the Fourth Meeting of the U.S.–Japan Joint Technical Committee on Masonry Research, San Diego, CA, October 17–19.

Somboonsong, W., Ko, F. K., and Harris, H. G. (1998). Ductile hybrid fiber reinforced plastic (FRP) rebar for concrete structures: design methodology, *ACI Mater. J.*, 95(6), Nov.–Dec.

Swartz, S. E., Mikhail, M. L., and Guralnick, S. A. (1969). Buckling of folded plate structures, *Exp. Mech.*, 9(6), 269–274.

Takeda, T. et al. (1973). Pressure tests of PCRV Models, Ohbayashi-Gumi Report OTN. TY. 48100, December, presented at the 7th FIP/PCI International Conference, New York, May.

Tsien, H. S. (1942). A theory for buckling of thin shells, *J. Aeronaut. Sci.*, 9, 373.

Uang, C. M. and Bertero, V. V. (1986). Earthquake Simulation Tests and Associated Studies of a 0.3–Scale Model of a Six-Story Concentrically Braced Steel Structure, Rept. No. UBC/EERC-86/10, Earthquake Engineering Research Center, University of California, Berkeley, Dec.

Vandepitte, D. and Rathe, J. (1971). An experimental investigation of the buckling load of spherical concrete shells, subjected to uniform radial pressure, RILEM International Symposium—Experimental Analysis of Instability Problems on Reduced and Full Scale Models, Buenos Aires, Argentina, September 13–18.

Von Karman, T. and Tsien, H. S. (1939). The buckling of spherical shells by external pressure, *J. Aeronaut. Sci.*, 7, 43–50.

Vreedenburgh, C. G. J. (1966). *Heron*, Jaargang 14, No. 2, Delft, The Netherlands.

Wallace, J. B. and Krawinkler, H. (1985). Report No. 75: Small-Scale Model Experimentation of Steel Assemblies, U.S.–Japan Research Program, The John A. Blume Earthquake Engineering Center, Department of Civil Engineering, Stanford University.

Weymeis, G. (1977). Report on the investigation of instability of concrete domes conducted from 1967–1976 at the University of Ghent in Belgium (in Dutch), Rijks Universiteit Laboratorium voor Modelonderzoek, Grote Steenweg Noord 12, Swijnaarde, Belgium.

White, R. N. (1972). Structural Behavior Laboratory—Equipment and Experiments, Report No. 346, Department of Structural Engineering, Cornell University, Ithaca, NY.

White, R. N. (1975). Reinforced concrete hyperbolic paraboloid shells, *ASCE J. Struct. Div.*, September, pp. 1961–1982.

Whittaker, A. S., Uang, C. M., and Bertero, V. V. (1987). Earthquake Simulation Tests and Associated Studies of a 0.3–Scale Model of a Six-Story Eccentrically Braced Steel Structure, Rept. No. UCB/EERC-87/02, Earthquake Engineering Research Center, University of California, Berkeley, July.

Whittle, R. T. and Beattie, D. (1972). Standard pile caps I and II, *Concrete*, 6(1), Jan., pp. 34–36; 6(2), Feb., pp. 29–30.

Yan, H. T. (1954). The Design of Pile Caps, in *Civil Engineering and Public Works Review*, Vol. 49, May–June.

CHAPTER 11

Structural Models for Dynamic Loads

CONTENTS

- 11.1 Introduction 586
- 11.2 Similitude Requirements..587
- 11.3 Materials for Dynamic Models..588
  - 11.3.1 Dynamic Properties of Steel..588
  - 11.3.2 Dynamic Properties of Concrete..590
- 11.4 Loading Systems for Dynamic Model Testing..593
  - 11.4.1 Vibration and Resonant Testing...593
  - 11.4.2 Wind Tunnel Testing..597
  - 11.4.3 Shock Tubes and Blast Chambers 598
  - 11.4.4 Shaking Tables 600
  - 11.4.5 Drop Hammers and Impact Pendulums 601
  - 11.4.6 Centrifuges 604
- 11.5 Examples of Dynamic Models 604
  - 11.5.1 Natural Modes and Frequencies 604
    - 11.5.1.1 Buildings 604
    - 11.5.1.2 Bridges..605
    - 11.5.1.3 Special Structures...606
  - 11.5.2 Aeroelastic Model Studies of Buildings and Structures 611
    - 11.5.2.1 Wind Effects on Buildings...611
    - 11.5.2.2 Lions' Gate Bridge 611
  - 11.5.3 Blast Effects on Protective Structures 612
  - 11.5.4 Earthquake Simulation of Reinforced Concrete Structures 614
    - 11.5.4.1 Reinforced Concrete Frames..614
    - 11.5.4.2 Reinforced Concrete Bridges...618
    - 11.5.4.3 Precast Concrete Shear Walls 622
  - 11.5.5 Earthquake Simulation of Steel Buildings 625
  - 11.5.6 Earthquake Simulation of Masonry Buildings 628
    - 11.5.6.1 General 628
    - 11.5.6.2 Unreinforced Masonry 628
    - 11.5.6.3 Reinforced Concrete Masonry 639
  - 11.5.7 Impact Loading 641
  - 11.5.8 Soil–Structure Interaction Studies Using a Centrifuge 642
    - 11.5.8.1 Introduction 642
    - 11.5.8.2 Model Structure and Soil Foundation..646
    - 11.5.8.3 Experimental Results 646

11.6	Case Studies	649
	11.6.1 Wind Tunnel Tests of the Toronto City Hall	649
	11.6.1.1 The Problem	649
	11.6.1.2 Test Program	650
	11.6.1.3 Testing Techniques	650
	1. Steady Pressure Measurements	650
	2. Unsteady Pressure Measurements	650
	3. Flow Visualization Tests	651
	4. Conclusions	651
	11.6.2 Case Study B, Shaking Table Tests on R/C Frame-Wall Structures	652
	11.6.2.1 Initial Mechanical Characteristics	652
	11.6.2.2 Earthquake Simulator Test Program	652
	11.6.2.3 Test Results and Discussion	653
	11.6.2.4 Correlation of ⅕-Scale Model and Full-Scale Model	655
	1. Hysteretic Behavior	655
	2. Overall Lateral Load–Deformation	656
	3. Crack Pattern	657
	4. Failure Mechanism	658
	5. Conclusions	658
	11.6.3 Case Study D, Shaking Table Tests on Lightly Reinforced Concrete Buildings	660
	11.6.3.1 Cornell University ⅛-Scale Model	663
	1. Model Structure	663
	2. Test Results	664
	11.6.3.2 SUNY/Buffalo ⅓-Scale Model	669
11.7	Summary	673
Problems		673
References		675

11.1 INTRODUCTION

Time-dependent or *dynamic* loadings, because of their complex nature and effect on structures, have enabled the experimental technique of using small-scale models to compete on an equal basis with the more traditional analytical methods. Dynamic loadings of interest to the structural engineer range from wind- or traffic-induced elastic vibrations to blast and impact loadings that can cause considerable structural damage. Of special interest are the problems of wind and earthquake loading which, because of their widespread nature and potential destruction of life and property, have assumed a greater importance in the last few years. This is due to the fact that many regions of the world with exposure to turbulent high-magnitude winds or to high seismic activity have become more and more urbanized.

Dynamic modeling of structures plays an important experimental role in problems dealing with education, research, and design. Simple laboratory experiments are very useful in demonstrating basic concepts of vibration to undergraduate and graduate students. Some of these techniques will be illustrated in Chapter 12. In the area of structural research, the small-scale dynamic model has proved to be a powerful tool in extending knowledge and understanding of structural behavior to many complex dynamic situations where analytical techniques are inadequate. Analytical models of dynamic behavior are developed and fine-tuned by comparison to test results obtained from model testing. Of equal importance has been the ability of a carefully constructed model to play

STRUCTURAL MODELS FOR DYNAMIC LOADS

a significant role in the design process of many structures in which the main loading is dynamic in nature. Significant advances to the art and science of physical modeling of concrete and other types of structures have been made in the post World War II years when many structural problems dealing with dynamic loads had to be resolved.

As computers have become more powerful and sophisticated, the tools available to the structural engineer have multiplied. Bigger and more-encompassing computer models of complex structural systems with nonlinear capabilities can now be studied. Realistic verification and checking of the available analytical procedures developed for dynamically loaded structural systems, however, must still be carried out so that one can better understand structural behavior. Because of the large size of civil structures and the high cost and general lack of large testing facilities, only small component testing at full scale is usually available to the engineer. Limitations on size of available dynamic testing facilities dictate the use of small-scale models for check out and design purposes, especially if large segments of the structure or the whole structure must be included in the study.

11.2 SIMILITUDE REQUIREMENTS

If all that is known about a particular problem is that it involves a structural system and hence the basic quantities of force, mass, length, and time, the general model theory given in Chapter 2 gives a basic relationship which must always exist between model and prototype. Following the procedure of Chapter 2, the π term that is formed may be taken as

$$G(Ft^2/ml) = 0 \tag{11.1}$$

If this single π factor were to be made the same for the model and the prototype, complete dynamic similarity would be obtained:

$$F_m t_m^2 / m_m l_m = F_p t_p^2 / m_p l_p$$

Using the scale factor definitions,

$$F_p/F_m = S_F \quad t_p/t_m = S_t \quad l_p/l_m = S_l$$

the model–prototype relation then becomes

$$S_F = S_l S_m / S_t^2 \tag{11.2}$$

According to Focken (1953), Equation 11.2 can be regarded as an analytical expression of the principle of *dynamic similarity*.

Equation 11.2 can be applied directly to dynamic problems.

Example 11.1

If only gravity forces are important for a particular dynamic problem, what conclusions can be reached about the relationships between model and prototype?

Solution — In this case, $S_F = S_m$. Therefore, Equation 11.2 becomes

$$S_l = S_t^2 \tag{11.3}$$

Thus, a 1/9-scale model of a prototype structure will vibrate with 1/3 the period of the prototype and it will vibrate three times faster than the prototype.

Example 11.2

If in a dynamic structural problem, only elastic forces are of importance (usual case of many vibrating structures) what are the model to prototype relations imposed by Equation 11.2?

Solution — Assuming that model and prototype are of the same material, $S_m = S_l^3$. Equation 11.2 then becomes

$$S_F = S_l^4 / S_t^2$$

If model and prototype are made of the same elastic material, since strain is dimensionless, $S_F = S_l^2$, and the above expression becomes

$$S_l = S_t \qquad (11.4)$$

Thus, a ¼-scale model will have ¼ the prototype period. It will vibrate four times faster than the prototype.

Similitude requirements for dynamically loaded structures are given in Chapter 2. Scale factors are derived for the most common dynamic actions that affect structural behavior. Scale factors for elastic vibrations are summarized in Table 2.8. Similitude requirements for fluidelastic models such as long-span bridges lead to scale factors that are given in Table 2.9. Scale factors for structures loaded with blast loading are summarized in Table 2.10. The summary of scale factors for three possible types of earthquake load simulation are summarized in Table 2.11. Scale factors for testing structural/soil models in a centrifuge facility are presented in Table 2.12.

Similitude requirements for damping forces in concrete models were developed by Farrar et al. (1994). They reported that although damping is typically considered a material property, the similitude analysis shows that both material and system geometry must be considered when scaling the damping forces.

11.3 MATERIALS FOR DYNAMIC MODELS

11.3.1 Dynamic Properties of Steel

Experimental evidence indicates that the strength behavior of most materials depends on the rate of strain, especially at high strain rates during testing. In steel these changes are shown in Figure 11.1 for the case of ASTM A36 structural steel. The effects of increasing rate of strain can be summarized as follows:

1. The yield stress increases to some dynamic value (σ_{yd}).
2. The yield point strain (ε_y) increases.
3. The modulus of elasticity (E) remains constant.
4. The strain at which strain hardening begins (ε_{st}) increases.
5. The ultimate strength increases slightly.

The most important effect that will influence the design of steel structures to resist dynamic loads is the increase in the yield stress. In Figure 11.2 (Norris et al., 1959) the percentage increase in yield stress is given as a function of the rate of strain for two steels of different static yield stress. It is evident from the figure that the increase in the dynamic yield point is greater for steels with lower static yield strength, as is shown by the higher slopes of curve A. Figure 11.3 shows the

STRUCTURAL MODELS FOR DYNAMIC LOADS

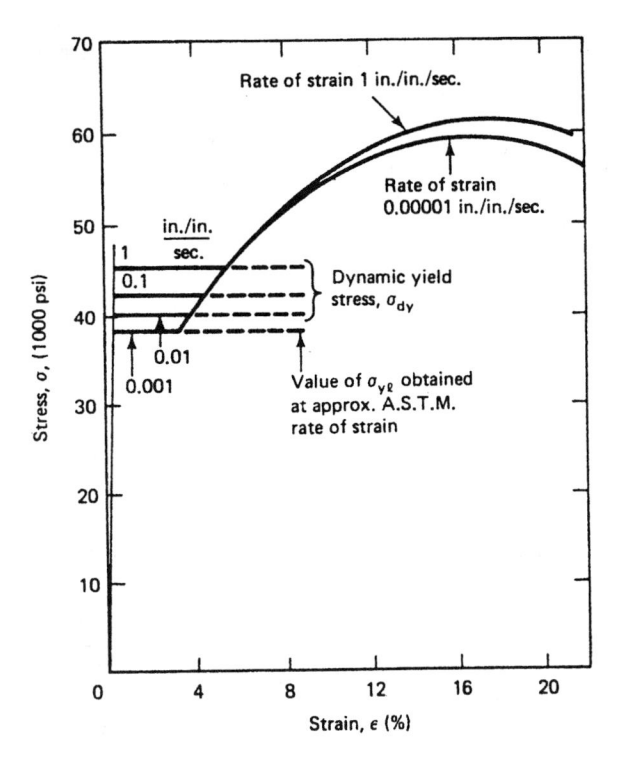

Figure 11.1 Effect of rate of strain on stress–strain curve for structural steel. (After Norris et al., 1959.)

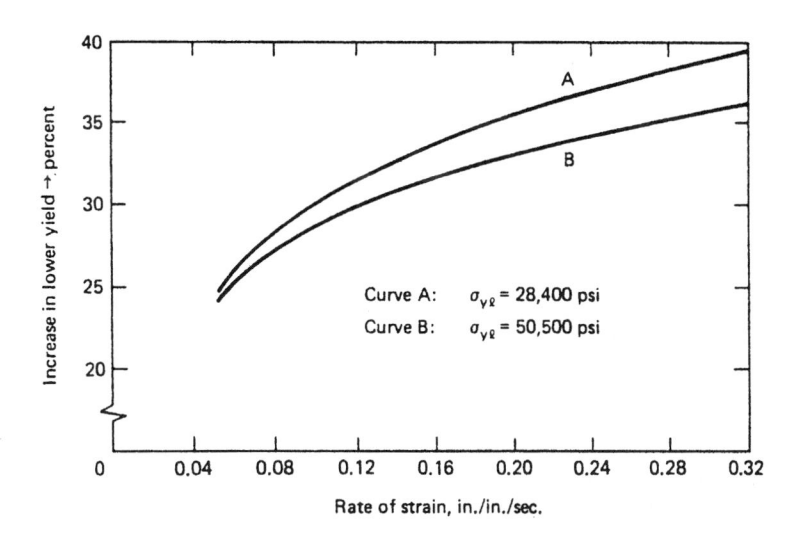

Figure 11.2 Increase of lower yield point of steel with strain rate. (After Norris et al., 1959.)

dynamic yield stress as a function of the time required to reach that value of stress (σ_{yd}) for ASTM A7 steel. From this curve, values of design yield stress could be found if the time to reach yield stress in a particular structure is known.

Useful data on the effect of strain rate on smooth and knurled steel wire suitable for model reinforcement has been obtained by Staffier and Sozen (1975) and is shown in Figure 11.4. Additional strain rate effects on steel tension specimens has been reported by Krawinkler et al. (1978) and Mills et al. (1979), with similar increases in yield strength with strain rate.

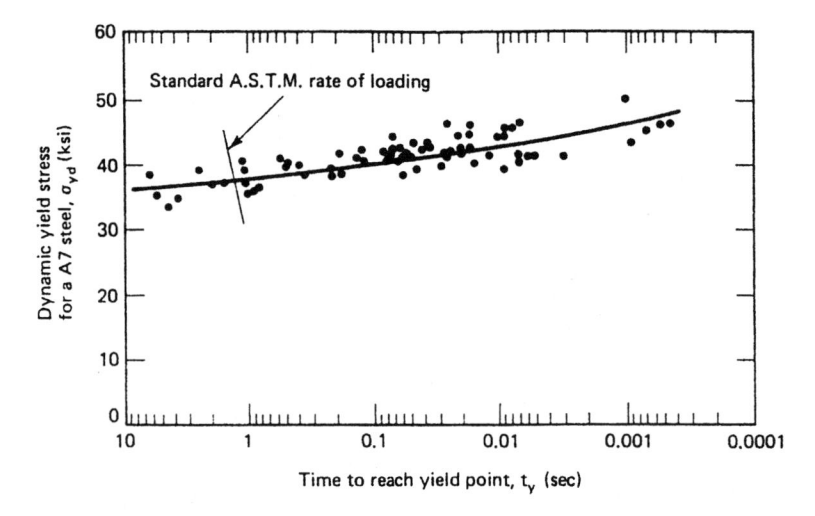

Figure 11.3 Effect of rate of strain on yield stress. (After Norris et al., 1959.)

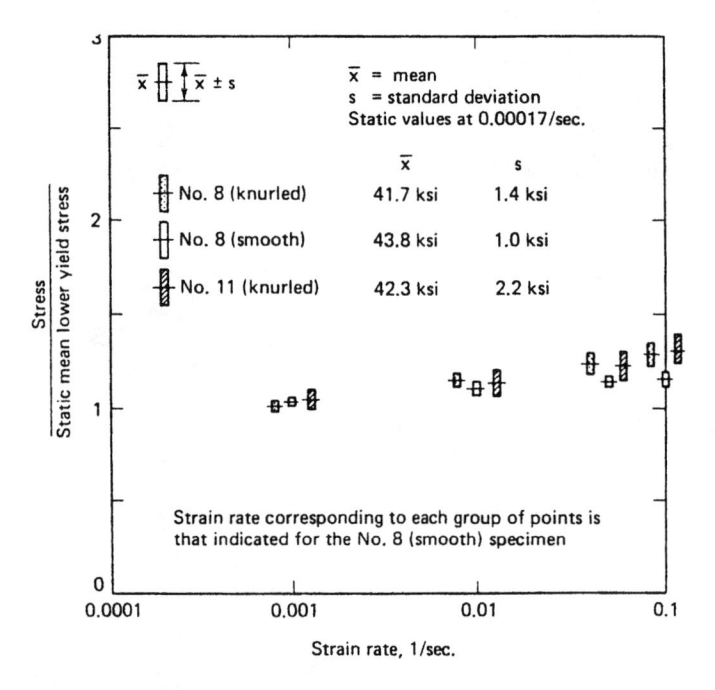

Figure 11.4 Ratios of standard deviation and mean lower yield stress to static mean lower yield stress at different strain rates for No. 8 and No. 11 gage black annealed wire. (After Staffier, S. R. and Sozen, M. A., 1975.)

11.3.2 Dynamic Properties of Concrete

For modeling of concrete structures subjected to inelastic deformations it is not possible to use anything but concretelike materials such as cement mortar or "microconcrete" and gypsum mortar. The static properties of model concrete and model reinforcement have been studied more extensively [Johnson, 1962; Little and Paparoni, 1966; Harris et al., 1966; 1970; Sabnis and White, 1969) than the dynamic properties. Strain rate effects on model concrete have been reported in connection with blast and impact loading effects (Ferrito, 1982), and cyclic reversed loading tests, albeit quasi-static,

STRUCTURAL MODELS FOR DYNAMIC LOADS 591

Table 11.1 Strain Rate Effects on Unconfined Compressive Strength 50 × 100 mm Cylinders

Average Strain Rate, mm/mm/s	Head Speed of Instron Machine, mm/min	f'_c = 11.9 MPa No. Specimens	f'_{cd}/f'_c	f'_c = 12.8 MPa No. Specimens	f'_{cd}/f'_c	f'_c = 15.7 MPa No. Specimens	f'_{cd}/f'_c	Average f'_{cd}/f'_c
1×10^{-5}	0.5	5	1.040	7	1.023	2	0.968	1.021
1×10^{-4}	5	4	1.109	7	1.116	3	1.153	1.122
1×10^{-3}	50	4	1.206	3	1.199	3	1.160	1.187
1×10^{-2}	500	3	1.367	0	—	3	1.303	1.335

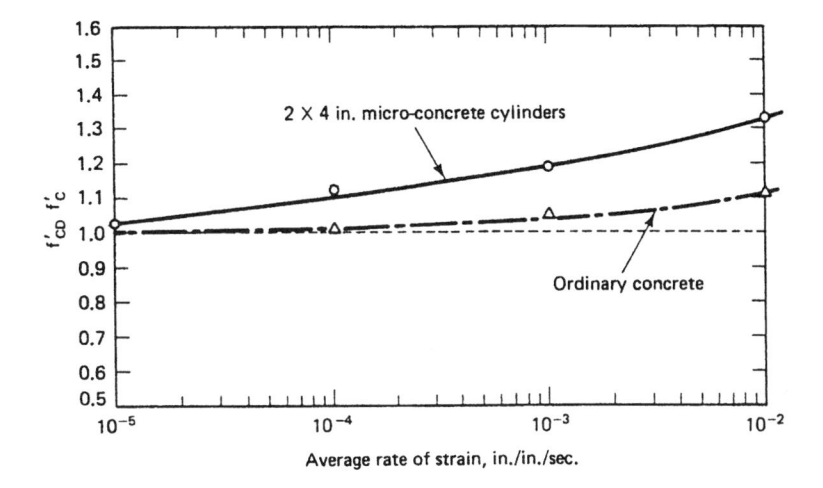

Figure 11.5 Effect of increased strain rate on the unconfined compressive strength.

have been conducted in connection with earthquake loading (Chowdhury and White, 1977; Kim et al., 1988).

A series of 50 × 100 mm cylinders of microconcrete with a mix of water:cement:sand of 0.9:1:4.5 were tested at increased strain rates in unconfined compression (Harris et al., 1963). The strain was measured using strain gages placed on opposing generators. A total of 44 specimens were tested at four rates of strain ranging from 10^{-5} to 10^{-2} (1/s). The results are shown in Table 11.1 as the ratio of dynamic unconfined compressive strength to the static value. The average values of f'_{cd}/f'_c vs. the average rate of strain are plotted in Figure 11.5. A comparison with similar results of ordinary concrete shows a higher rate of increase of f'_{cd}/f'_c for the microconcrete over the range tested. This is partly due to the size effect when testing smaller specimens of microconcrete, as discussed in Chapter 9.

Results from a series of microconcrete model cylinders 50 × 100 mm tested at high strain rates are given in Figure 11.6 (Krawinkler and Moncarz, 1982). Strain values were obtained by averaging the continuous readings of four 50 mm extensometers applied at 90° around the circumference. The results clearly show the expected increase in stiffness and strength and decrease in ε_u when the stain rate of loading is increased. The effect of strain rate on compressive strength from this series of microconcrete specimens is summarized in Figure 11.7.

A more extensive study of dynamic tests on model concrete has been reported by Ferritto (1982). Dynamic tests were conducted on microconcrete and gypsum concrete both having a maximum aggregate passing the No. 4 sieve. A description of the test matrix is given in Table 11.2. Cylinders 38 × 76 mm were tested in a pneumatic-hydraulic dynamic testing machine developed at the Naval Civil Engineering Laboratory, Port Hueneme, CA. Typical stress–strain curves of the microconcrete and a prototype concrete tested on the same machine are shown in Figure 11.8. The

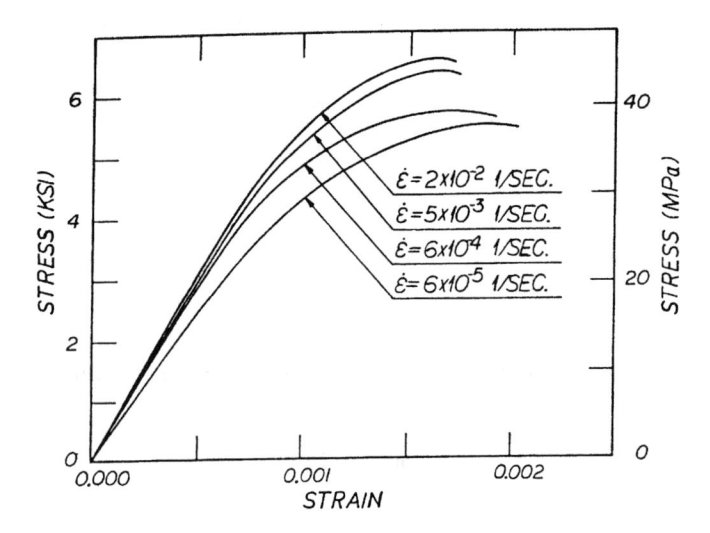

Figure 11.6 Stress–strain curves for microconcrete at different strain rates.

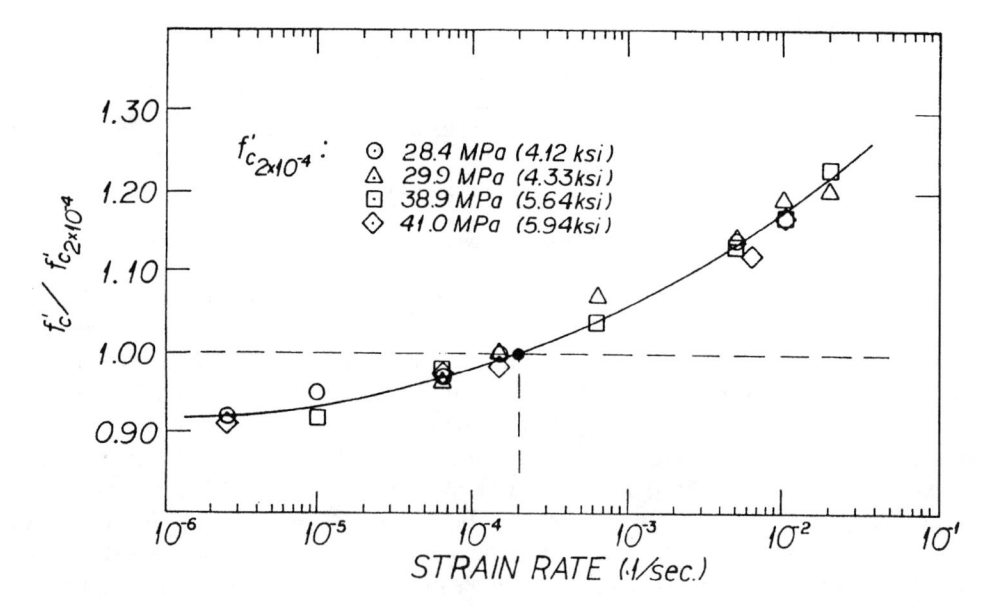

Figure 11.7 Effect of strain rates on compressive strength of microconcrete.

dynamic compressive strength increase factor ($f'_{c\,\text{dynamic}}/f'_{c\,\text{static}}$) as a function of strain rate for microconcrete with a No. 4 maximum aggregate gives good correlation with prototype values (Figure 11.9). Prototype concrete test results by Watstein and Boresi (1952) for two widely different concrete strengths are also shown in Figure 11.9. The increase in the ratio of dynamic to static modulus of elasticity with increasing strain rate gives reasonably good correlation with prototype concrete results (Figure 11.10). Considering its dynamic characteristics, this size microconcrete is well suited for use in direct models involving dynamic effects without special adjustment by a dynamic scale factor. The ratio of dynamic to static modulus of elasticity is also plotted vs. the ratio of dynamic strength increase factor in Figure 11.11.

Experimental modal analyses were performed by Farrar et al. (1994) to examine similitude of the dynamic parameters (resonant frequencies, mode shapes, and modal damping) of reinforced

STRUCTURAL MODELS FOR DYNAMIC LOADS

Table 11.2 Model Concrete Specimens

Type of Cylinder	Number of Cylinders	Size of Cylinder, mm	Cement	Maximum Aggregate Size	Age at Test (day)
Solid	16	38.1 × 76.2	Type III Portland	No. 4	18
Solid	16	38.1 × 76.2	Ultracal 30 gypsum	No. 4	2

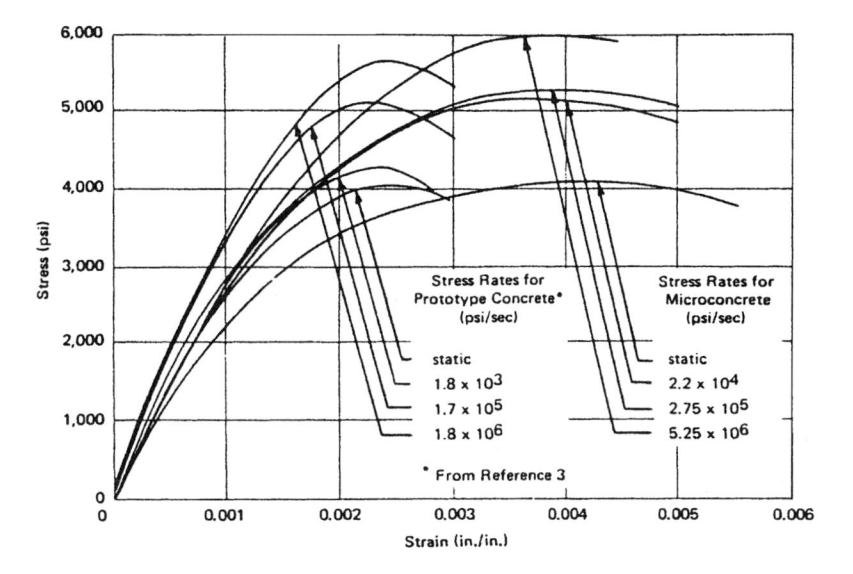

Figure 11.8 Comparison of stress–strain curves for various rates of loading.

concrete replica models. Results of the experiments show that the modal frequencies and the mode shapes of a prototype structure can be accurately predicted from tests on ⅓-scale model structures. Variations in equivalent viscous damping ratios identified on models and prototype were greater than variations for other measured dynamic parameters. However, all damping ratios were less than 2% of critical, and the observed variations would not significantly alter the dynamic response of the structures tested.

The dynamic strength increase factor for the gypsum concrete cylinders is somewhat less than that expected by comparing it to the prototype curves, Figure 11.9. The ratio of modulus of elasticity for increasing strain rates does not follow the same pattern as that of the prototype concrete (Figures 11.10 and 11.11). This may be a significant factor in the design and analysis of a model and may limit the use of gypsum concrete in dynamic model studies.

11.4 LOADING SYSTEMS FOR DYNAMIC MODEL TESTING

11.4.1 Vibration and Resonant Testing

Elastic vibration studies can be performed on full-scale structures using a variety of techniques (Hudson, 1967) to excite the natural modes and frequencies of the structure. These methods tend to be expensive, and alternative procedures using scaled models are therefore widely utilized.

Although free-vibration measurements are sometimes easily accomplished by pulling on the structure, then releasing quickly, and measuring the free motions of the structure, most vibration tests in the laboratory are performed by forcing the structure to vibrate in one of its natural modes.

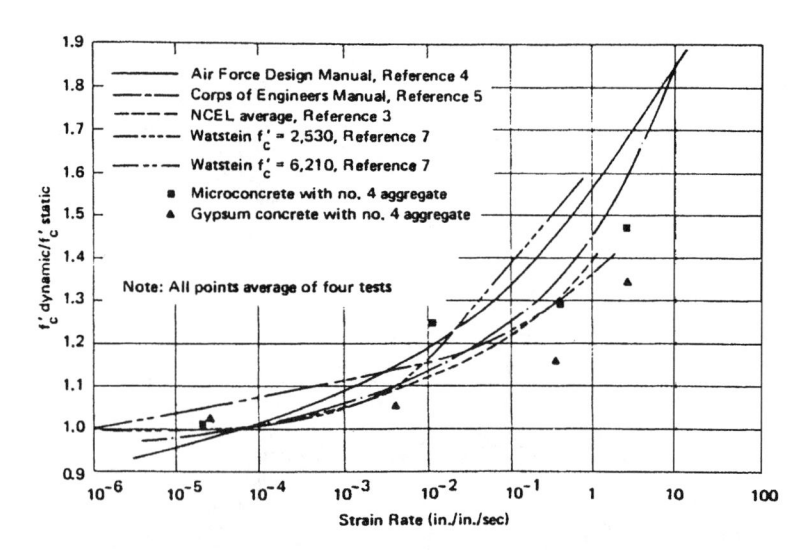

Figure 11.9 Ratio of compressive strengths vs. strain rate. (After Ferritto, J., 1982)

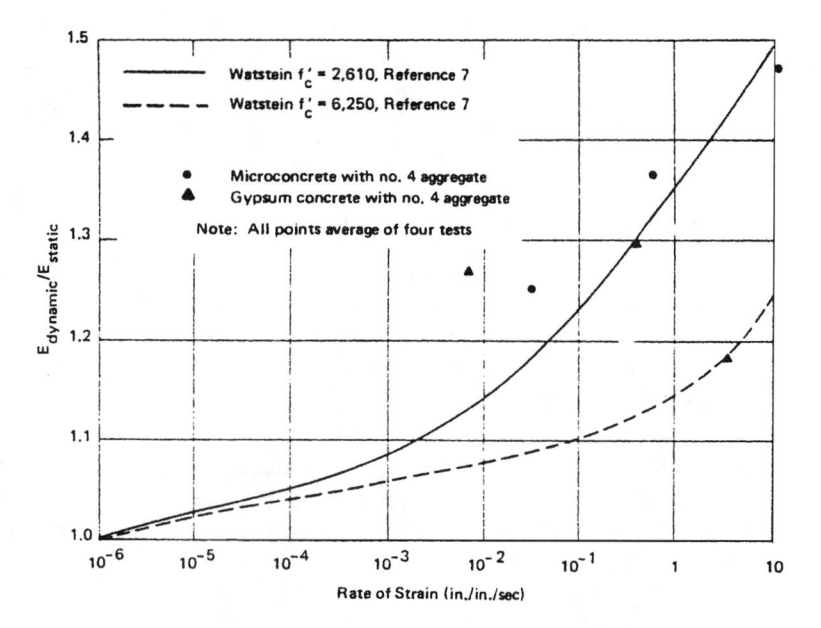

Figure 11.10 Strain rate vs. ratio of modulus of elasticity. (After Ferritto, J., 1982)

This is accomplished by the use of mechanical or electromagnetic oscillators or by placing the model on a shaking table. A setup for studying the natural modes and frequencies of a cantilever model plate is illustrated in Figure 11.12 (additional examples are presented in Chapter 12). Using a small-capacity electromagnetic shaker, the plate can be made to oscillate sinusoidally or in other types of programmed periodic motions. The plate structure is forced to oscillate with the shaker through a small coupling rod to which a load cell is attached (Figure 11.12). By changing the frequency of the forcing signal, the plate is forced to vibrate at different frequencies. The frequencies that correspond to the natural frequencies of the plate will result in zero applied force through the load cell. This condition can be observed on the oscilloscope and becomes therefore the point at which the natural mode shape and frequency are determined.

STRUCTURAL MODELS FOR DYNAMIC LOADS   595

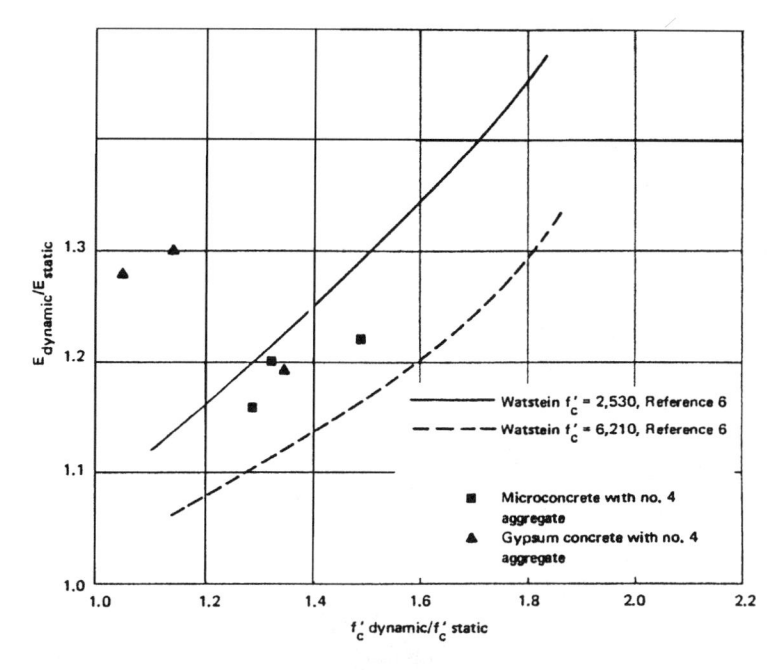

Figure 11.11 Ratio of dynamic strength increase factor vs. ratio of modulus of elasticity. (After Ferritto, J., 1982)

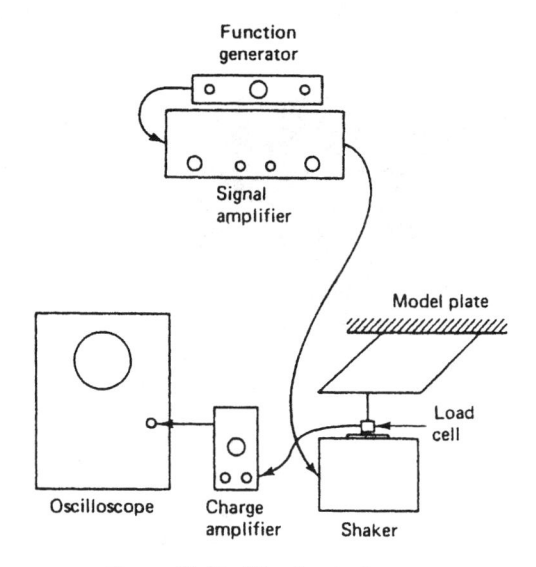

Figure 11.12 Vibration test setup.

Example 11.3

Design a small-scale demonstration model of a four-story shear building to be used in studying the dynamic characteristics (natural modes of vibration, mode shapes, and damping) of a multi-degree-of-freedom system. The model is to be able to be tested on a small shaking table and be instrumented with linear variable differential transformers (LVDTs) and accelerometers.

Solution — Since this is a demonstration model, the behavior will be limited to the linearly elastic range so that it can be used repeatedly. Considering the range of elastic materials, any number of possibilities exist, however, because of availability and ease of construction, a welded steel model at approximately ¹⁄₁₂-scale is chosen. The model dimensions are shown below. It consists of 178×330 mm steel plates 9.5 mm thick welded to four round steel columns 9.5 mm in diameter. A thicker steel base plate provides attachment to the shaking table and the necessary rigidity.

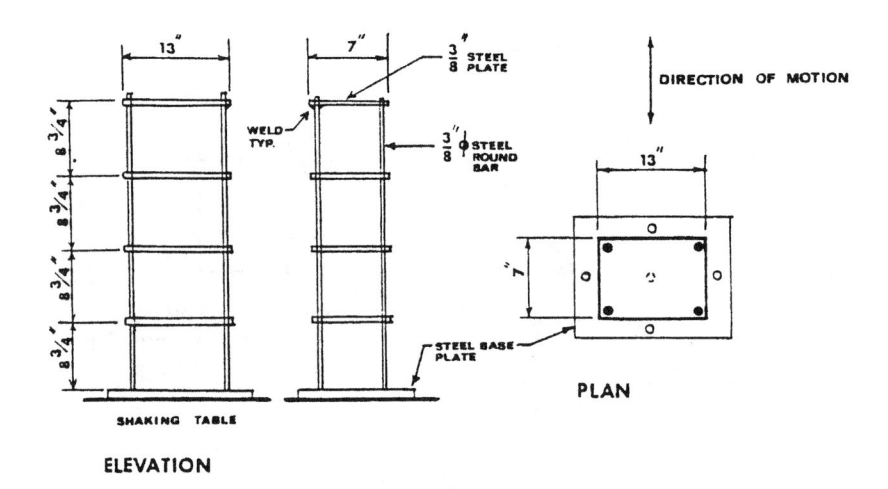

Testing — The loading for the model to determine the mode shapes and frequencies is to subject it to sinusoidal base motion. The excitation signal is produced using a function generator. Instrumentation for the model, which is bolted to the shaking table below, consists of five accelerometers attached to the centerline of each floor in the direction of motion and the base. In addition, 4 LVDTs are provided at each floor as shown in the picture. Note the rigid reference to which the LVDTs are attached.

STRUCTURAL MODELS FOR DYNAMIC LOADS

To determine the natural frequencies, the model is excited at the base with a low-frequency (starting at 2 Hz) low-amplitude sinusoidal motion. The frequency of the motion is then systematically increased until a relative maximum response is observed. An oscilloscope, monitoring the top-floor acceleration, is used to observe the response. This first relative maximum is the first mode frequency. Once an approximate value for the frequency is known, a closer value for the frequency can be determined by increasing the frequency of the base motion in smaller steps while observing the response.

To determine the mode shapes, the model is allowed to vibrate in the specific mode being analyzed. The acceleration traces at each floor level are recorded. In the first mode, the traces consist of four in-phase recordings. The relative magnitudes only are important here; therefore, all values are normalized to the top-floor magnitude.

Comparison with Analysis — A comparison of the experimental results with analytical predictions can be made as shown in the figure below. The frequencies measured in the model test are lower than the frequencies computed. This result is expected since damping was neglected in the analysis. Also, the additional mass at each floor level caused by the instrumentation was neglected.

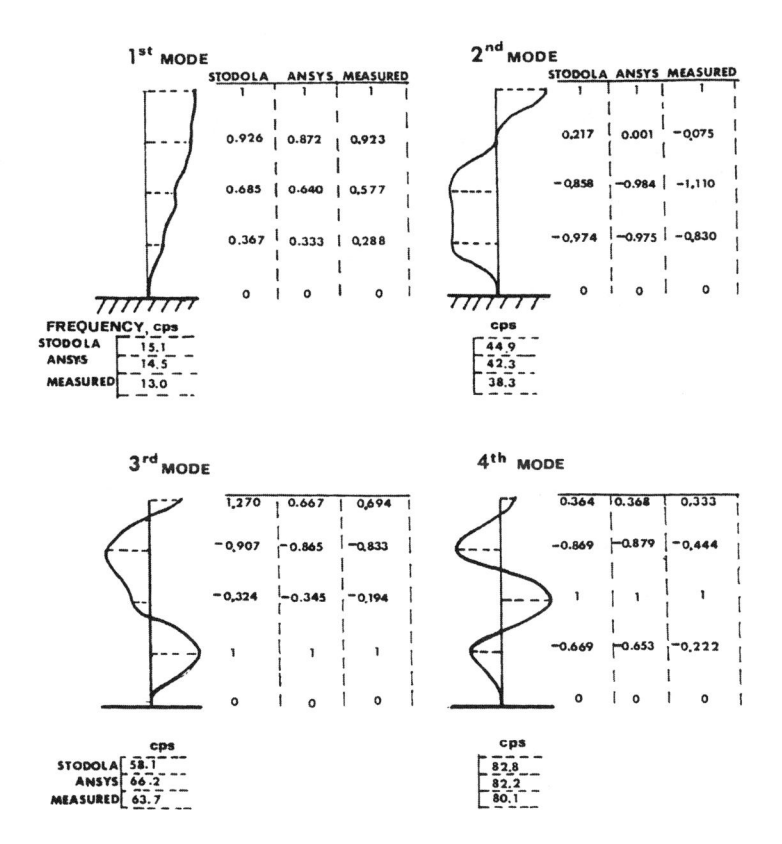

	1st MODE			2nd MODE		
	STODOLA	ANSYS	MEASURED	STODOLA	ANSYS	MEASURED
	0.926	0.872	0.923	0.217	0.001	-0.075
	0.685	0.640	0.577	-0.858	-0.984	-1.110
	0.367	0.333	0.288	-0.974	-0.975	-0.830
	0	0	0	0	0	0

FREQUENCY, cps
STODOLA 15.1 / ANSYS 14.5 / MEASURED 13.0

cps
44.9 / 42.3 / 38.3

	3rd MODE			4th MODE		
	1.270	0.667	0.694	0.364	0.368	0.333
	-0.907	-0.865	-0.833	-0.869	-0.879	-0.444
	-0.324	-0.345	-0.194	1	1	1
	1	1	1	-0.669	-0.653	-0.222
	0	0	0	0	0	0

cps
STODOLA 58.1 / ANSYS 66.2 / MEASURED 63.7

cps
82.8 / 82.2 / 80.1

11.4.2 Wind Tunnel Testing

Fluidelastic studies of structural models are performed in a variety of wind tunnels. At present, such facilities are available with the capability of simulating the characteristics of the atmospheric boundary layer which is essential for determining wind effects on buildings and other structures. These facilities can be divided into three basic types:

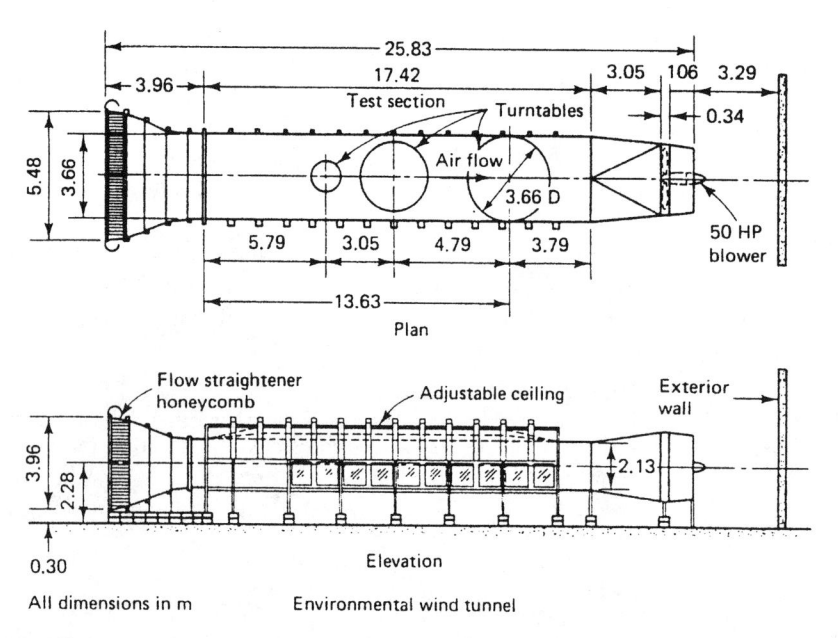

Figure 11.13 Wind tunnel for physical modeling of flow around buildings — Fluid Dynamics and Diffusion Laboratory, Colorado State University. (After Cermak, J. E., 1977.)

1. Long wind tunnels (of the order of 30 m) in which the boundary layer develops naturally over a rough floor. Examples of such tunnels are the environmental wind tunnel of Colorado State University (Figure 11.13) and the University of Western Ontario Boundary Layer Wind Tunnel (Figure 11.14) whose boundary layer development is shown schematically in Figure 11.15.
2. Wind tunnels with passive devices such as grids, fences, or spires which are used to generate a thick boundary layer. The flow then passes over a short section of roughness arrays. An example of such a tunnel is that of the National Aeronautical Establishment, Ottawa, Canada.
3. Wind tunnels with active devices such as jets or machine-driven shutters or flaps. Many tunnels used for aeronautical research fall into this category.

Wind effects on tall buildings require the determination of the shears, moments, and deformations along the height caused by the integrated effect of the wind pressures as well as the wind effect on glass panels and cladding. By introducing instrumented small-scale models of the structure and its surroundings into the test section of the wind tunnel, all of the above quantities can be determined experimentally.

An example of a rigid model of a building mounted in the wind tunnel is shown in Figure 11.16. Further examples of the testing techniques used in wind tunnel studies are presented by (Davenport and Isyumov, 1968, Cermak, 1977, Simiu and Scanlan, 1978, and Sachs, 1978).

11.4.3 Shock Tubes and Blast Chambers

Laboratory studies of external blast loading effects have been conducted successfully in "shock tube" facilities and specially constructed blast chambers. A shock tube consists of a straight, usually uniform, section separated into a high-pressure and a low-pressure portion by means of a diaphragm (Figure 11.17a). By quickly opening the diaphragm — by bursting, a compression wave is first propagated into the low-pressure region, followed by a rarefaction wave into the high-pressure region. The compression wave rapidly propagates into a shock wave as it progresses down the tube,

STRUCTURAL MODELS FOR DYNAMIC LOADS 599

Figure 11.14 Upstream view of boundary layer wind tunnel with model of rectangular building in foreground. (After Davenport, A. G. and Isyumov, N., 1968. Courtesy of Prof. A. G. Davenport, Director, Boundary Layer Wind Tunnel Laboratory, The University of Western Ontario, London, Ontario, Canada.)

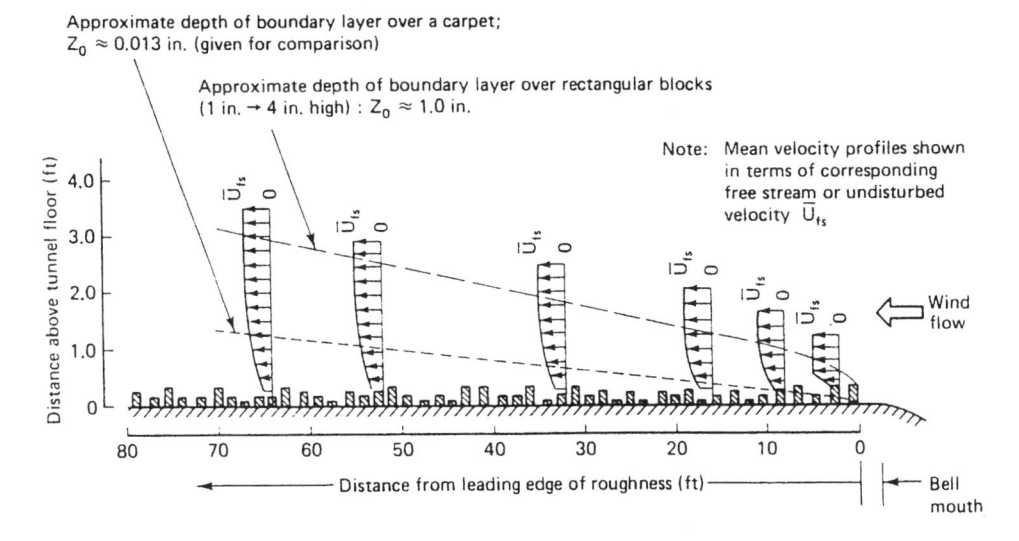

Figure 11.15 Development of boundary layer over typical surfaces. (After Davenport, A. G. and Isyumov, N., 1968.)

as illustrated in Figure 11.17. A scale model placed in the shock tube would therefore be subjected to the shock wave for a period of time prior to the eventual alteration of the shock by the interaction of the rarefaction with the closed high-pressure end of the tube and the compression wave with the open end of the tube. Many of these shock tubes are in military or private use.

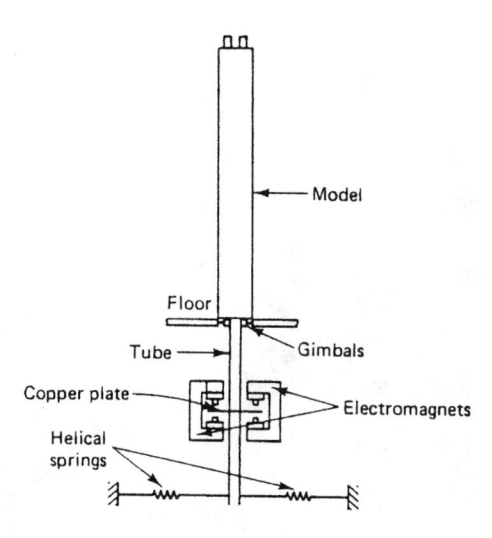

Figure 11.16 Rig for linear-mode models. (After Scruton, 1968.)

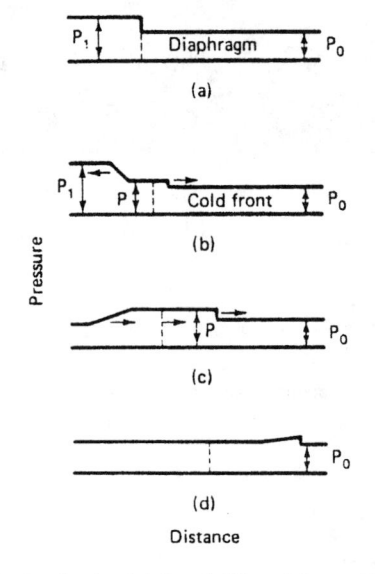

Figure 11.17 Pressure distribution in shock tube at different times. (After Norris, C. H. et al., 1959.)

Special concrete bunkerlike facilities have also been used to simulate blast effects on small-scale models (Norval and Cohen, 1970). A typical facility constructed at Drexel University consists of a 2.6 × 3 m boxlike bunker having 0.6-m-thick concrete walls. A blast deflector and masonry stack direct the blast and detonation products to the roof, where they are safely dispersed. The model is loaded by a blast wave from a scaled charge exploded inside the chamber. Instrumentation consists of pressure, deflection, strain, and acceleration pickups, and a quartz glass window provides access to the chamber for instrumentation cables or high-speed photography. Charges of up to 0.45 kg of TNT have been detonated without difficulty in this particular facility.

11.4.4 Shaking Tables

Shaking tables come in a variety of sizes and capabilities. Some are hydraulically actuated, and others, such as the 2 × 3 m table at ISMES, Italy (Castellani et al., 1976), are actuated by

STRUCTURAL MODELS FOR DYNAMIC LOADS 601

Figure 11.18 Layout of the test. (After Castellani, A. et al., 1976. Courtesy of ISMES, Bergamo, Italy.)

two synchronized electromagnetic shakers as shown in Figure 11.18. A classification of various shaking tables (Table 11.3) indicates that the smaller facilities are more suited to testing scale models. The model to be tested under simulated earthquake motions is usually bolted to the moving table as shown in Figure 11.18, which illustrates the testing of a model dam and its surroundings. The moving table is supported with various techniques indicated in the last column of Table 11.3.

A schematic diagram of the shaking table at the University of Illinois, Urbana-Champaign (Abrams et al., 1990), is shown in Figure 11.19. The experimental data from such earthquake simulator tests can be observed or plotted during the test and recorded in digital form for later processing.

11.4.5 Drop Hammers and Impact Pendulums

Structural testing under short duration dynamic (i.e., shock and impact) loads can be performed using a variety of devices: drop hammers, impact pendulums, and a variety of pneumatic guns shooting projectiles directly onto the specimen. Facilities that can perform impact testing of large structural models include an 8.9 m, 29 kN drop hammer device with a 7 m drop height which is housed in the Structural Engineering Laboratory of the Pennsylvania State University, State College, PA (Figure 11.20). An intermediate capacity impact pendulum facility consisting of a 4.25-m-high steel frame capable of swinging weights of up to 7.5 kN through an arc with a vertical drop height of about 3.65 m is also housed in the same laboratory. These systems are supported by various very high speed data collection systems (up to 1 MHz) for strain, displacement, force, etc., including a 12,000 frames per second SP200 camera and a copper vapor laser system.

To test small-scale reinforced concrete structural components, a large variety of impacting devices have been used. A setup used by Nilsson and Sahlin (1982) to study the nonlinear wave propagation behavior of reinforced concrete slabs and to correlate with theoretical results is shown in Figure 11.21. The use of high-precision impact tests to develop more precise analytical tools has been advocated by Krauthammer (1997).

Table 11.3 Classification of Various Shaking Tables*

Location (1)	Dimensions, m (2)	Payload Limit, kN (3)	a_{max}, g Horizontal (4)	Vertical	d_{max}, ±mm Horizontal (5)	Vertical	f_{max}, Hz (6)	Type of Support (7)
Small (<3 m)								
Stanford University	1.6 × 1.6	22.2	5	—	63.5	—	50	Roller bearings
University of Calgary	1.3 × 1.3	9	20	—	76.2	—	—	—
ISMES, Italy	3 × 2	1.3	100	—	—	—	800	Oil film
Drexel University	1.2 × 1.8	8.9	3.6	—	6.4	—	2000	Roller bearings
Medium (3–9 m)								
University of California, Berkeley	6 × 6	444.8	1.5	1.0	127.0	50.8	15	Air pressure
Corps of Engineers (triaxial)	3.7 × 3.7	587	2	1	±3	±6	60	—
SUNY/Buffalo	3.7 × 3.7	444.8	1	1	±6"	±3"	60	Actuators
Cornell University	1.5 × 1.9	89	5	—	±3"	—	100	Slip table
University of Illinois	3.6 × 3.6	44.5	7	—	101.6	—	100	Flexible support
Corps of Engineers	3.6 × 3.6	53.4	34	60	55.9	45.7	200	—
Wyle Lab, Huntsville, Ala.	5.5 × 3.5	42.3	8	8	76.2	76.2	500	—
Large (>9 m)								
National Research Center, Japan	15 × 15	4,448	0.6	1.0	30.5	—	16	—
Berkeley (proposed)	30 × 30	17,792	0.6	0.2	152.4	76.2	—	—

After Krawinkler et al. (1978).

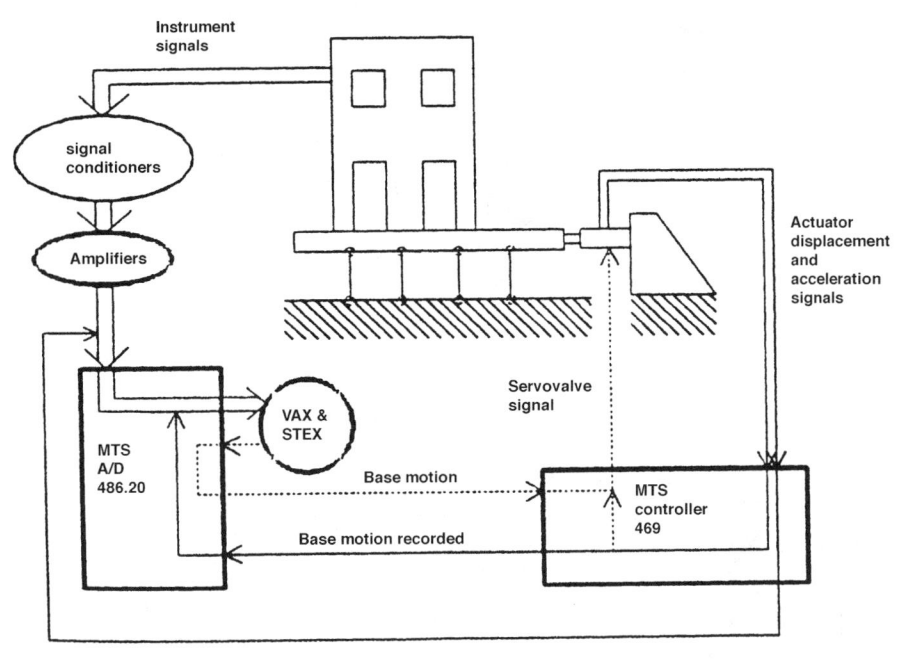

STRUCTURAL MODELS FOR DYNAMIC LOADS 603

Figure 11.19 Operation diagram of the University of Illinois, Urbana-Champaign. (After Abrams, P. et al., 1990.)

Figure 11.20 Pennsylvania State University drop hammer. (a) Hammer and frame. (b) Testing. (Courtesy of Dr. Theodor Krauthammer.)

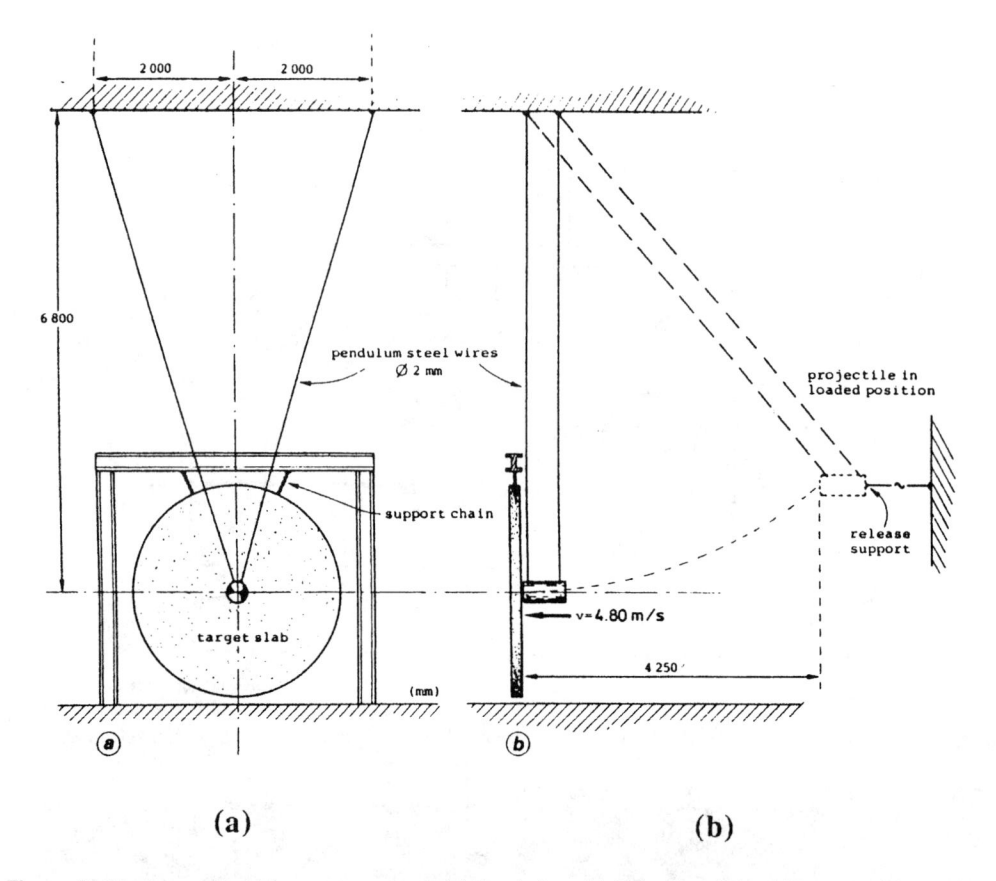

Figure 11.21 Impact pendulum arrangement. (a) Front view. (b) Side view. (After Nillson, L. et al., 1982.)

11.4.6 Centrifuges

In many massive structures such as dams and tunnels and many other structures that are founded and interact with soils, body forces are all-important to their engineering behavior. It is obvious from a dimensional analysis that in situations where gravity effects are important the model behavior would not replicate the full-scale prototype unless the model is tested under an increased body force field. While there are some methods of generating an increased body force field (added nonstructural mass, embedded pulling forces, etc.), a spinning centrifuge is the most convenient tool to achieve this requirement for testing of models of soils and structures. These fall into two main categories: the beam type and the drum type. An example of the former is shown in Figure 11.22 and an example of the latter is shown in Figure 11.23.

11.5 EXAMPLES OF DYNAMIC MODELS

11.5.1 Natural Modes and Frequencies

11.5.1.1 Buildings

The determination of natural vibration modes and frequencies of tall buildings is a necessary first step to an analysis involving any dynamic loads. It is during this phase of analysis or testing in which the dynamic characteristics (mode shapes and frequencies) of the structure are determined. The effect of dynamic loads, such as wind or earthquake, would be a tendency to excite the fundamental modes of the structure.

STRUCTURAL MODELS FOR DYNAMIC LOADS 605

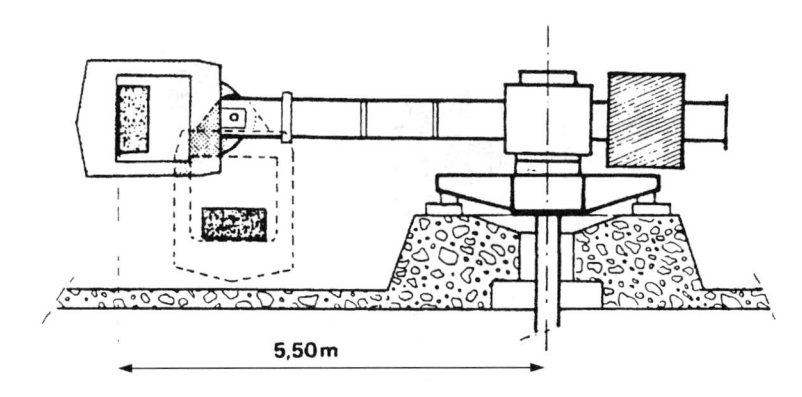

Figure 11.22 Acoutronic 680 Rotating beam centrifuge (Nicolas-Font, J., 1988).

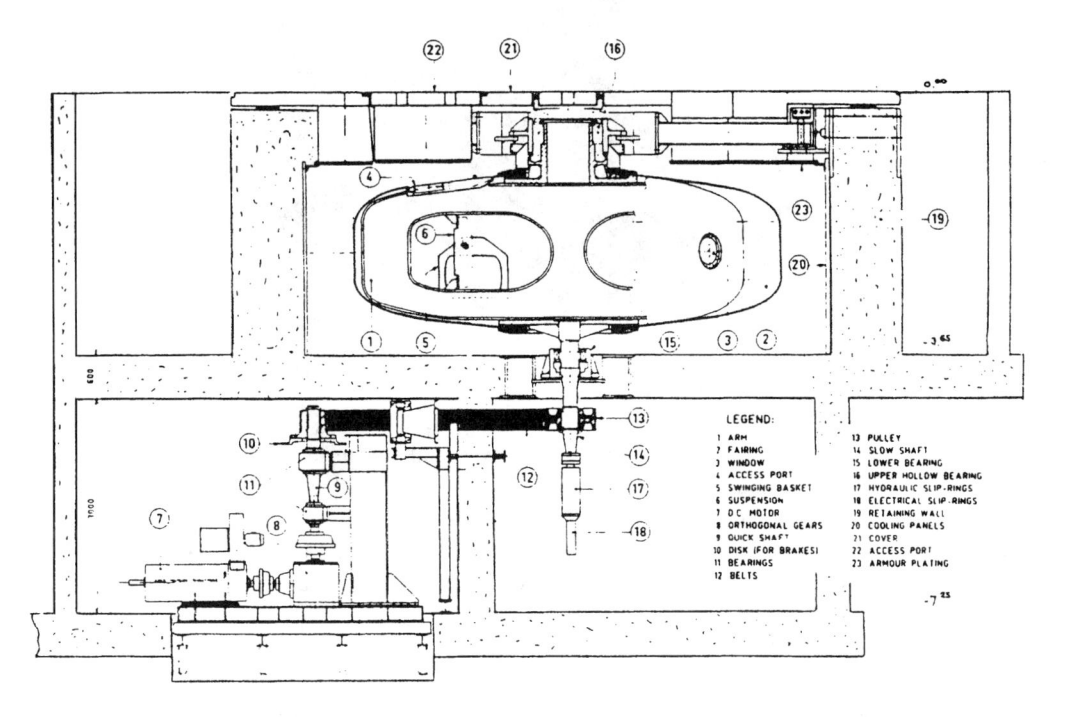

Figure 11.23 ISMES Geotechnical Centrifuge (IGC). Drum centrifuge (Baldi, G. et al., 1988).

Sometimes the complexity of the structure and its mass distribution require that experimental verification of the analytical procedures be carried out. Scale model testing can easily be used to carry out this step. An example of such a study (Castoldi and Casirati, 1976) is the determination of the natural vibration characteristics of the Parque Central high-rise building in Caracas, Venezuela. This study, which preceded the seismic testing, was carried out at ISMES, Bergamo, Italy. An elastic $1/40$-scale model was made using an epoxy resin as shown in Figure 11.24. The first four mode shapes and frequencies are shown in Figure 11.25, where it is noted that the torsional stiffness of the structure influences the vibration modes.

11.5.1.2 Bridges

An extensive static and dynamic model study (Baglietto et al., 1976) was conducted at ISMES for the design of the Zarate-Brazo Largo Bridges, Buenos Aires. These identical twin bridges are

Figure 11.24 Parque Central building model. (After Castoldi, A. and Casirate, M., 1976. Courtesy of ISMES, Bergamo, Italy.)

braced steel box and cable stayed, carrying both highway and eccentrically placed railroad traffic. This particular design arrangement was one of the main reasons of concern, which prompted a $1/33.3$-scale model study as the design work was progressing and not as a final check or afterthought to the design process. A view of the complete model is shown in Figure 10.43, with all simulated masses added and ready for vibration testing. A schematic of the method used in vibrating the model by means of four electromagnetic shakers is shown in Figure 11.26. The data control and recording equipment are also shown in this figure. Some typical results of the lowest resonant frequencies and corresponding mode shapes are shown in Figure 11.27 for the case with the deck loaded in the middle with the mass corresponding to two locomotives. A comparison of the analytically predicted natural frequencies with those obtained from the model vibration tests is made in Table 11.4. A close agreement of the results is indicated for both the dead-load and live-load cases.

Several extensive model studies (Williams and Godden, 1976; Godden and Aslam, 1978) of the dynamic response of bridges to earthquake loading have been conducted at the University of California at Berkeley. The models used in these studies were subjected to harmonic input as well as seismic inputs on the 6×6 m shaking table in the Earthquake Simulator Laboratory. The description of a $1/30$-scale ultimate strength model and its seismic behavior is given in Section 11.5.4.2.

11.5.1.3 Special Structures

Many types of structures other than those described in Figure 2.9 undergo dynamic loading during their useful operation. One class of such structures is aerospace vehicles. A vibration

STRUCTURAL MODELS FOR DYNAMIC LOADS 607

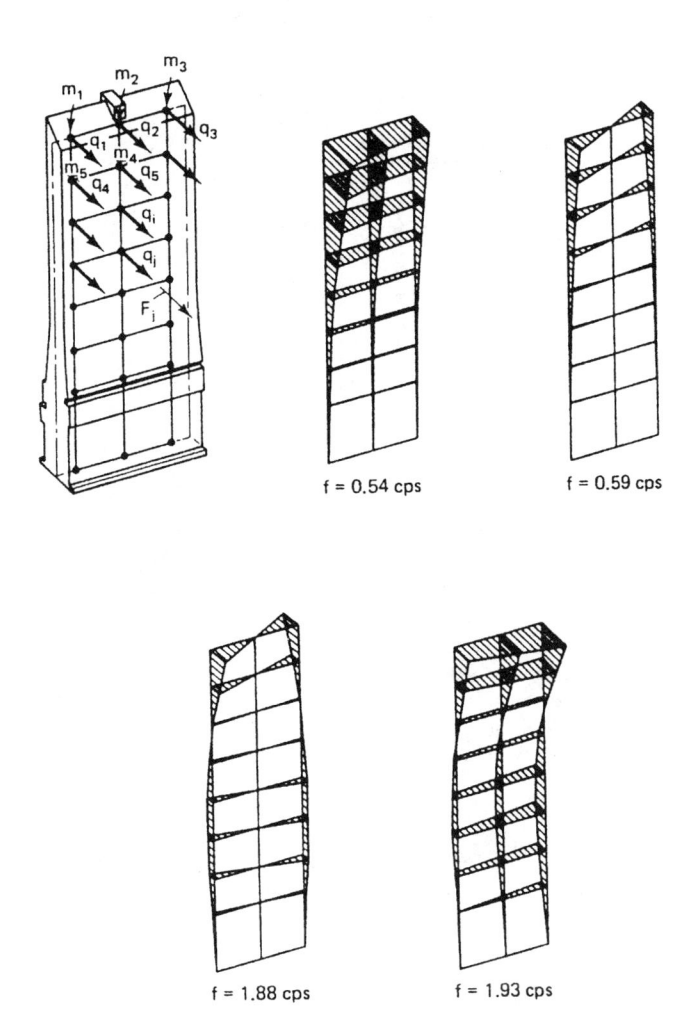

Figure 11.25 Parqe Central building. Stiffness matrix determination and vibration modes for transverse excitation. (After Castoldi, A. and Casirate, M., 1976).

evaluation is a first step in their design for dynamic loads. Dynamic models play a very important role in the design and analysis of these usually complex structures. As an example, the $1/10$-scale model of a portion of the Apollo simplified shell structure consisting of the conical lunar adapter (SLA), the short cylindrical instrumentation unit (IU), and the stiffened cylindrical shell of the S-IVB rocket forward skirt is shown in Figure 11.28. This model, simulating all the important axial and bending stiffness parameters, was vibrated through the four lunar module (housed within the conical adapter) attachment points by means of a small electrodynamic vibrator, seen in Figure 11.28. The lunar module was simulated by means of a cruciform structure and additional mass so that the inertial forces at the attachment points, interacting with the conical shell, were dynamically modeled. In addition, the mass and center of gravity of the command module on top of the conical shell was properly simulated. Mode shape surveys of the simplified shell model were made using a handheld accelerometer.

A comparison of the model and prototype lowest-mode test results (Harris, 1968) is shown in Figure 11.29 in an exploded view of the conical adapter. The conical shell vibrates in basically one circumferential wave everywhere except in the vicinity of the four lunar module attachment points at the adapter circumferential stiffening ring, where it vibrates essentially in five waves.

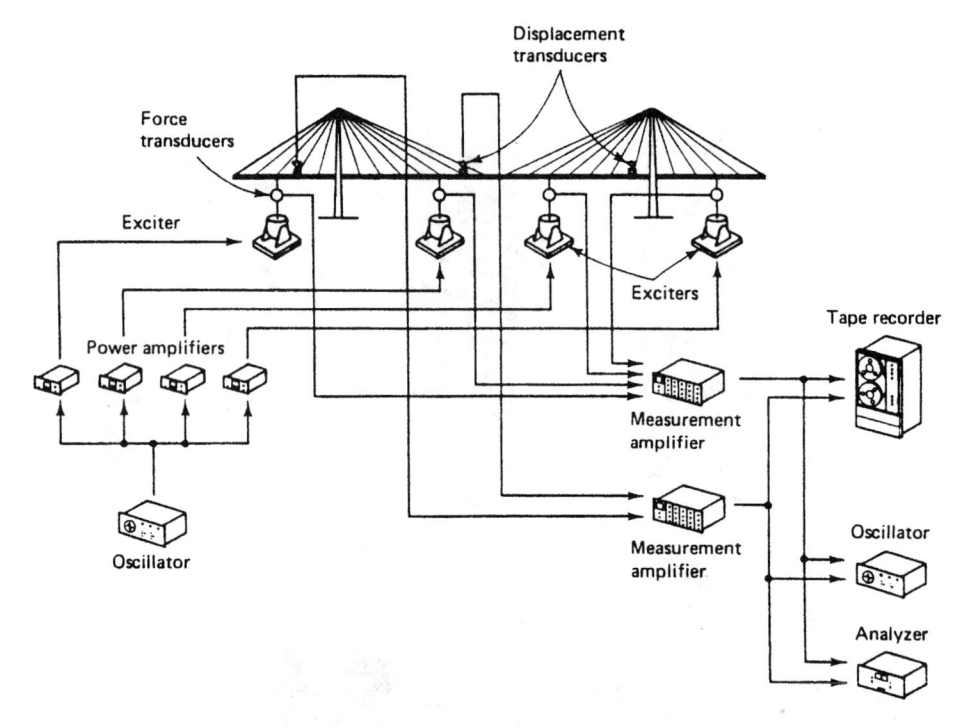

Figure 11.26 Block diagram of excitation and measurement system to determine vibration modes by means of concentrated forces. (After Baglietto, E. et al., 1976.)

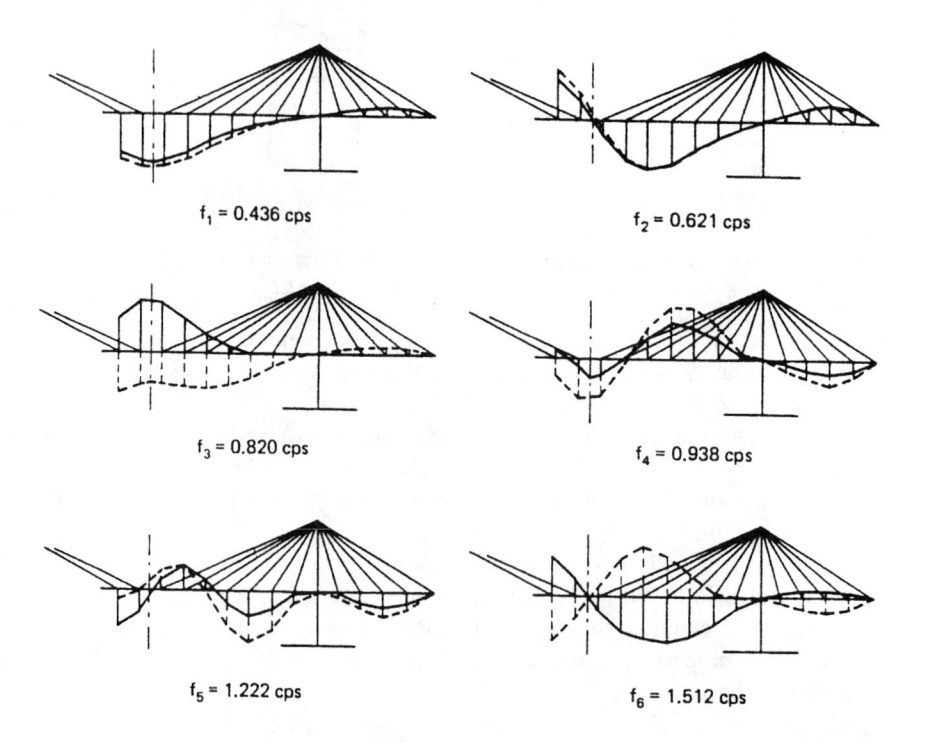

$f_1 = 0.436$ cps

$f_2 = 0.621$ cps

$f_3 = 0.820$ cps

$f_4 = 0.938$ cps

$f_5 = 1.222$ cps

$f_6 = 1.512$ cps

Figure 11.27 Deck vertical displacements at resonance frequencies. (After Baglietto, E. et al., 1976.) (— Railway side; ---- Highway side).

STRUCTURAL MODELS FOR DYNAMIC LOADS

Table 11.4 Comparison of Model Results with Analytical Predictions

		Frequency	
Mode	Shape	Dynamic Calculated, Hz	Structural Model, Hz
Dead loads, frequencies			
1	Symmetrical bending	0.44	0.472
2	Antisymmetrical bending	0.58	0.587
3	Symmetrical torsional	0.81	—
4	Symmetrical bending	0.88	0.945
5	Antisymmetrical bending	1.01	—
6	Antisymmetrical flexotorsional	1.09	—
Bridge with railway load in the central 330 m			
1	Symmetrical bending	0.42	0.436
2	Antisymmetrical bending	0.55	0.621
3	Symmetrical torsional	0.71	0.820
4	Symmetrical bending	0.82	0.938
5	Antisymmetrical flexotorsional	0.94	—
6	Antisymmetrical bending	1.04	1.222

Figure 11.28 Model of simplified Apollo shell shown driven by a 30 lb maximum force shaker at the +Z apex fitting. (After Harris, H. G., 1968.)

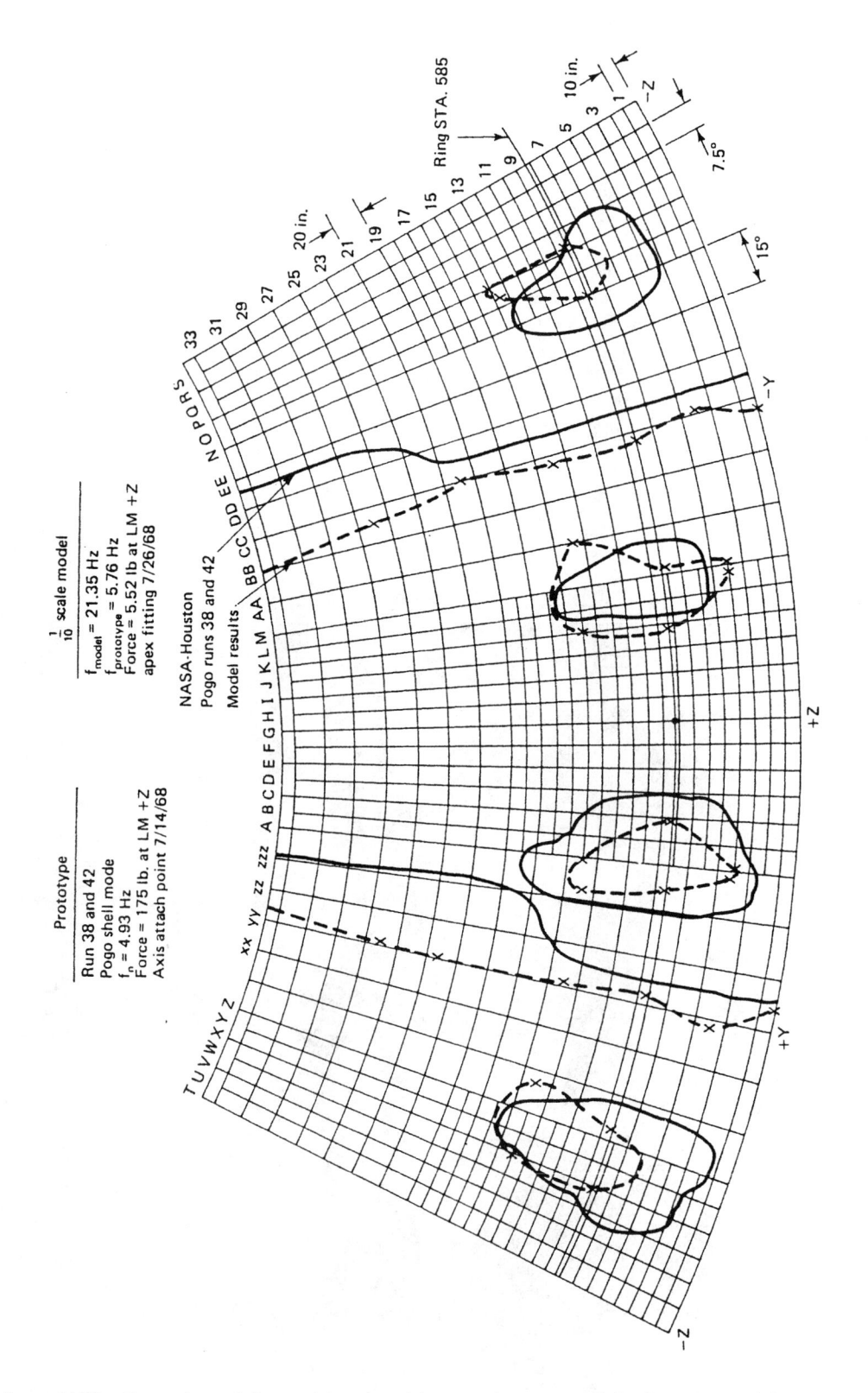

Figure 11.29 Comparison of the model and prototype mode surveys at the lowest natural frequency on developed shell. (After Harris, H. G., 1968.)

STRUCTURAL MODELS FOR DYNAMIC LOADS

11.5.2 Aeroelastic Model Studies of Buildings and Structures

11.5.2.1 Wind Effects on Buildings

For studies of wind effects on buildings, the model analyst is usually interested in both the mean deflection and sway of the structure as a whole and the magnitudes of the induced local pressures. The first effect is related to the design of the structural framing, and the second to the design of the skin (cladding and glass). Studies of aeroelastic and of rigid models of structures in representative wind flow can be performed to yield the information required to design both the structure and the cladding. A rigid model for such wind studies is shown in Figure 11.14 and an aeroelastic semirigid model, spring mounted at the base, is shown schematically in Figure 11.16.

An illustration of the main effects of fluctuating pressure distributions on the exterior of a model of a rectangular tall building tested by Davenport and Isyumov (1968) is shown in Figure 11.30. The model was situated in a scaled replica environment of a large city. In Figure 11.30 the oscillograph records of pressures at front, back, and side of the building at two different elevations are plotted as functions of the reduced time. These results show that the pressure distributions are similar to full-scale measurements on tall buildings. According to the authors:

> The complexity of fluctuating pressures, which in this case is further aggravated by the presence of surrounding buildings, emphasizes the difficulties associated in arriving at loads from a knowledge of velocity and stresses the advantages gained from obtaining the required structural response parameters from aeroelastic models.

A distinct advantage of the modeling of wind effects on structures is the ability to study the interference effects of groups of buildings and obstructions in the terrain in the immediate vicinity of a proposed structure. Such an investigation was made at the National Physical Laboratory, Teddington, England (Figure 11.31), in an attempt to estimate the amount of sway in typhoon winds of a proposed grouping of tower blocks. The buildings (Figure 11.31) were of basically octagonal section and had a height-width ratio of 4.25. The effect of turbulence is shown in Figure 11.32 for a single tower tested in isolation, which shows a significant increase of excitation of the turbulent flow over that of smooth flow. Two towers in line with the wind direction and spaced at 1.78 D apart (Figure 11.33) show larger amplitudes for the downstream tower and somewhat larger amplitudes in turbulent than in smooth winds. The dependence of the response of the leeward tower (tower B, Figure 11.33) on the spacing ratio S/D, is shown in Figure 11.34 for smooth flow. The optimum spacing was found to be 2.75 D for smooth airflow and 2 D for turbulent airflow. A conclusion drawn from this study is that because of the very rough surface of the towers, turbulence had only a small effect on the vortex excitation when there was mutual interference between a pair of towers.

11.5.2.2 Lions' Gate Bridge

Typical of aeroelastic model studies are the extensive experiments (Irwin and Schuyler, 1977) that were performed on a $1/110$-scale full aeroelastic model of the Lions' Gate Bridge, Vancouver, British Columbia. The general configuration and a proposed modification to the then 41-year-old bridge, using cantilever sidewalks to extend the three-lane roadway, are shown in Figure 11.35. The full aeroelastic model that was tested in simulated turbulent flow and also in smooth flow using the National Aeronautical Establishment of Canada 9×9 m wind tunnel is shown in Figure 11.36. Of prime concern in the program was the determination of the aerodynamic characteristics of the proposed changes as well as those of the present bridge. Two sectional models at $1/24$ scale (Irwin and Wardlaw, 1976), one of the present configuration and the other of the revised design, as well as a $1/110$ sectional model of the new version were tested in smooth flow under this general

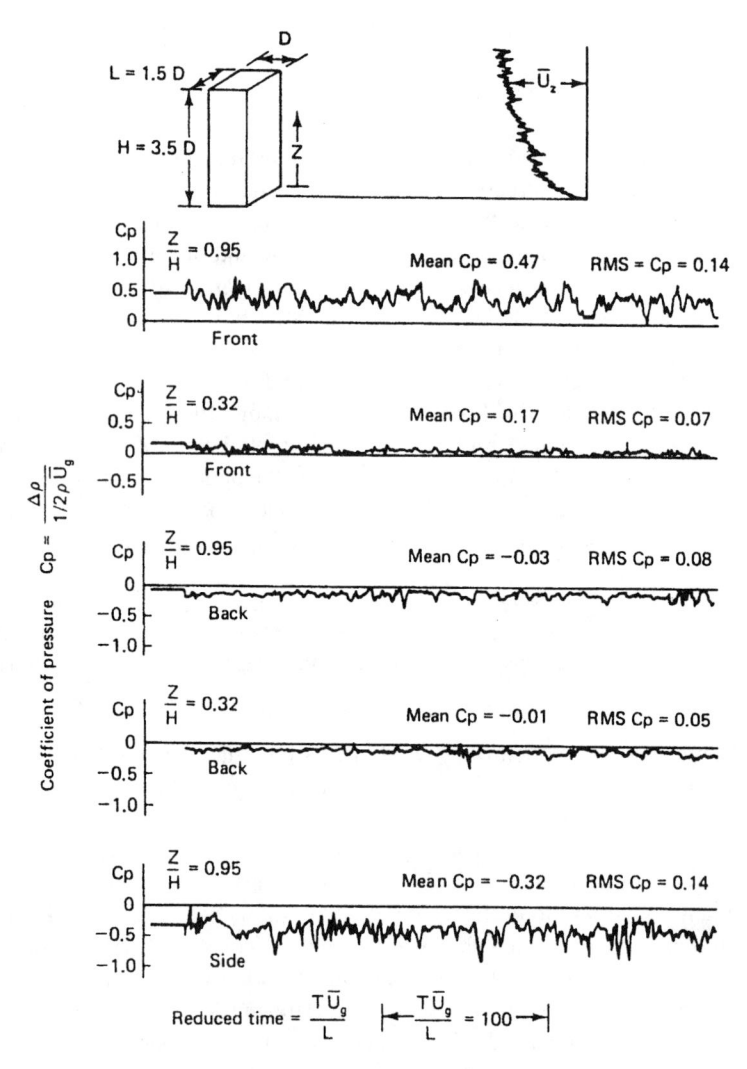

Figure 11.30 Exterior pressures on a rectangular building in boundary layer flow representative of wind over a large city. (After Davenport, A. G. and Isyumov, N., 1968.)

investigation. In smooth flow, the full-scale model was in good agreement with the behavior of the $1/110$-scale and $1/24$-scale sectional models. The presence of turbulence had a large effect on the critical velocity for flutter, raising it to a value above the velocity range tested. The results of these investigations imply that conventional methods of determining the stability of bridges using sectional models in smooth flow may lead to conservative (low) estimates of the critical velocity. Comparison of the experimental and analytical results for the vertical modes of vibration is shown in Figure 11.37. Excellent agreement is found between the two. Comparisons of the more complex torsional/lateral vibration modes are shown in Figure 11.38.

11.5.3 Blast Effects on Protective Structures

The recent number of bomb explosions on civilian targets in this country and around the world with devastating results on life and property has focused on the vulnerability of buildings not designed for any external or internal blast loading effects. Much remains to be done in learning to design civilian structures against blast loading and prevent a partial or total progressive collapse

STRUCTURAL MODELS FOR DYNAMIC LOADS 613

Figure 11.31 A grouping of model tower blocks of octagonal section. (After Scruton, 1968. Courtesy of University of Toronto Press, Toronto, Canada and the National Physical Laboratory, Teddington, England.)

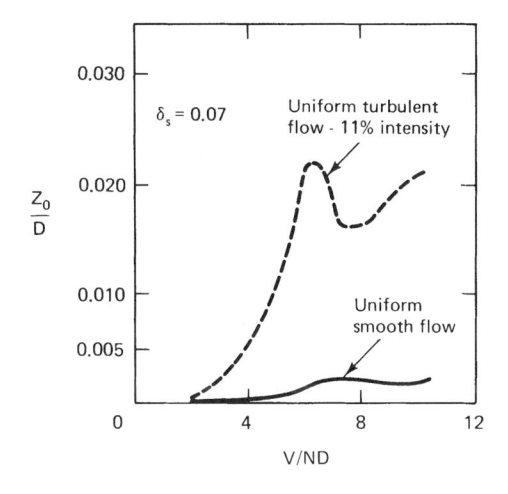

Figure 11.32 Variation of cross-wind amplitude with wind speed for an isolated tower block of octagonal section H/D = 4.25 immersed in smooth and turbulent flow. (After Scruton, 1968.)

of the structure. Small-scale physical models can play a very important role in this type of research. In fact, it was during the attempt to design protective structures under the specter of nuclear blast effects that much of the pioneering work of dynamic modeling was carried out. The relevance of that research in light of today's needs is obvious and for that reason the following example is retained in the second edition.

In order to evaluate the effects of high blast loading on structures, a series of spherical shells at $1/25$ scale were constructed at MIT (Smith et al., 1963) and then field-tested under blast wave loading conditions at overpressures ranging from 0.14 to 0.56 MPa. These were replica models of full-scale reinforced concrete domes tested in an aboveground nuclear explosion. Details of the model shell construction and the method of anchoring the edge ring to the base support are shown in Figure 11.39.

The wire meshes used for reinforcement were fabricated on a wooden form (Figure 11.40) of the same geometry as the domes, using a soldering technique. The form, ready for casting and using a rotating scribe to control the thickness of the shell, is shown in Figure 11.41. Field testing

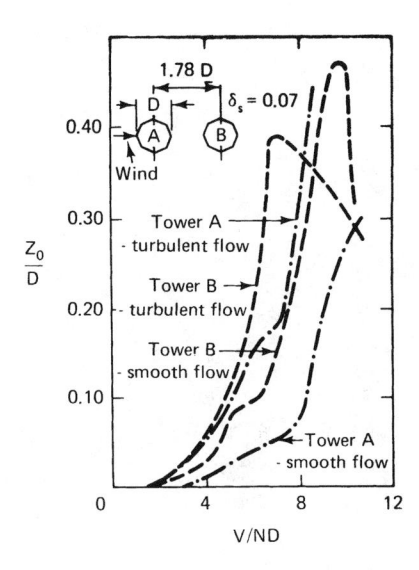

Figure 11.33 Amplitudes of oscillation (crosswind) of the towers of a twin-tower block configuration (H/D = 4.25) in smooth and turbulent airflow. (After Scruton, 1968.)

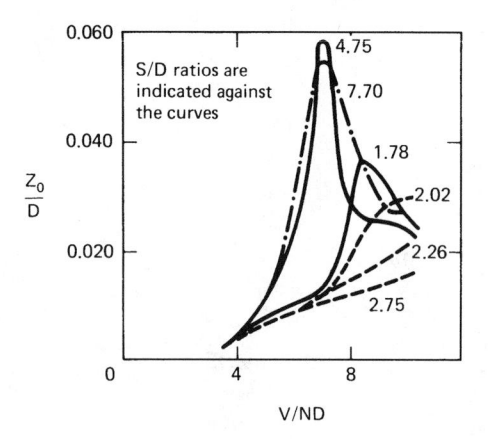

Figure 11.34 Variation of amplitude of leeward tower of a twin tower block configuration with the spacing (S) of the towers smooth airflow. (After Scruton, 1968.)

using a charge of 90.8 Mg of TNT was accomplished by arranging the 12 models at increasing distances from ground zero (Figure 11.42) so that the overpressures experienced would be as shown in Table 11.5, from a low value of 0.14 MPa to a high of 0.56 MPa. Typical before and after photographs of a dome model subjected to an overpressure of approximately 0.28 MPa during the test are shown in Figures 11.43 and 11.44, respectively. The mode of failure of the model domes was very similar to that experienced by the prototype domes.

11.5.4 Earthquake Simulation of Reinforced Concrete Structures

11.5.4.1 Reinforced Concrete Frames

A two-story reinforced concrete frame, representing a segment of a small office building, was studied on the 6 × 6 m shaking table of the Earthquake Engineering Research Center, University of California, Berkeley (Hidalgo and Clough, 1974; Clough and Bertero, 1977). This was a relatively

STRUCTURAL MODELS FOR DYNAMIC LOADS

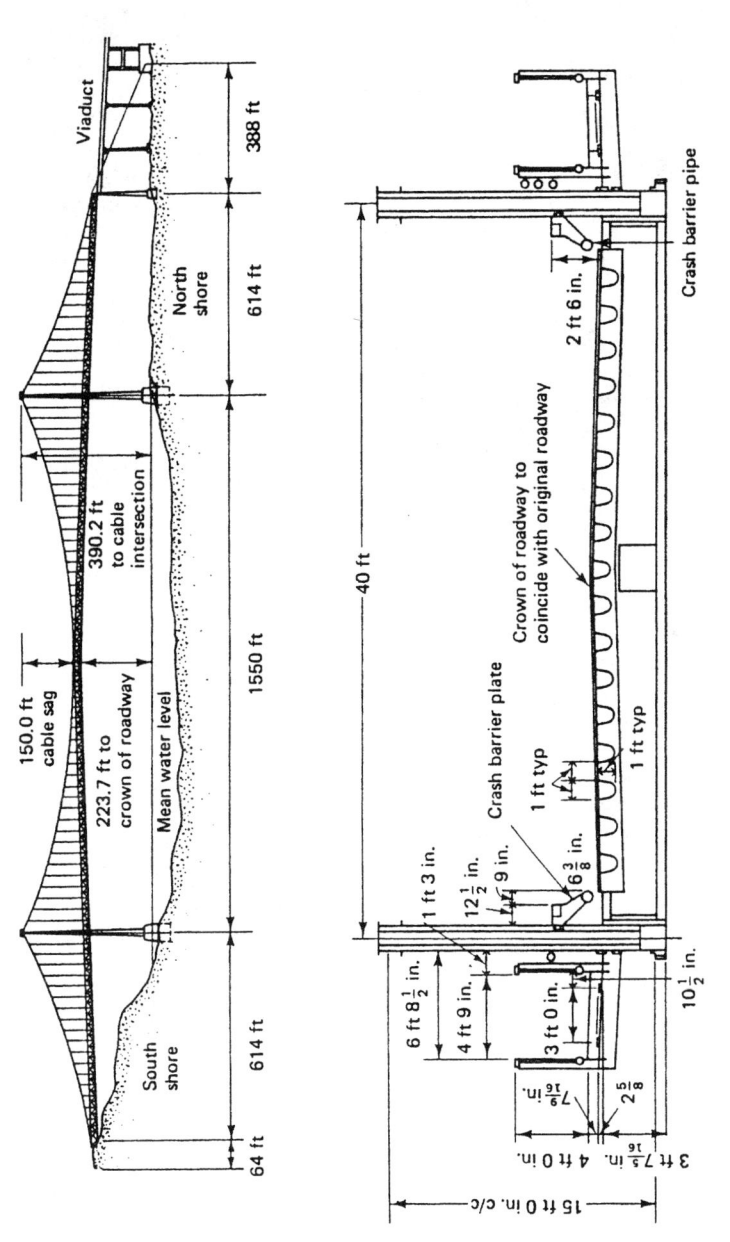

Figure 11.35 Lions' Gate Bridge with modified cross section. (After Irwin, H. P. A. H. and Schuyler, G. D., 1977.)

Figure 11.36 Aeroelastic model of the Lions' Gate Bridge in the 30 × 30 ft wind tunnel. (After Irwin, H. P. A. H. and Schuyler, G. D., 1977. Courtesy National Rsearch Council Canada, Ottawa, Canada.)

Mode shape — computed, ○ measured on model, (△ on other side span with sign change for anti-sym. modes)	Computed prototype n_r, Hz	Measured model $\dfrac{n_r}{\sqrt{110}}$, Hz	Model ζ_r
Tower — Mode 1 V	0.196	0.205 (run 243)	0.013
Mode 2 V	0.206	0.227 (run 250)	0.015
Mode 3 V	0.271	0.300 (run 252)	0.031
Mode 4 V	0.360	0.361 (run 254)	0.011
Mode 5 V	0.432	0.428 (run 250)	0.009
Mode 6 V	0.574	0.604 (run 243)	0.004

Figure 11.37 Vertical modes of model compared with computed prototype modes. (After Irwin, H. P. A. H. and Schuyler, G. D., 1977.)

STRUCTURAL MODELS FOR DYNAMIC LOADS

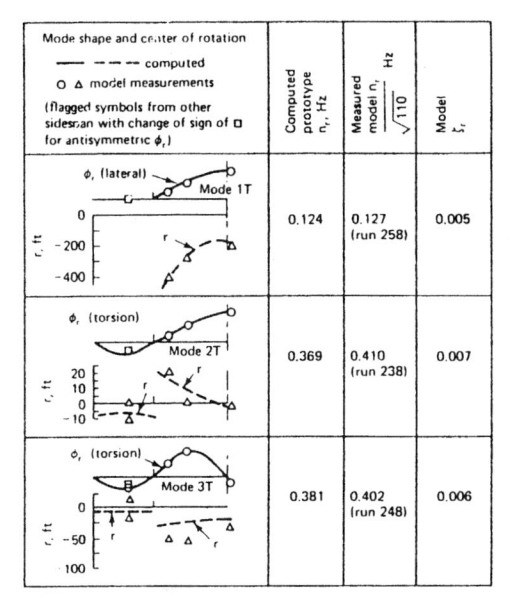

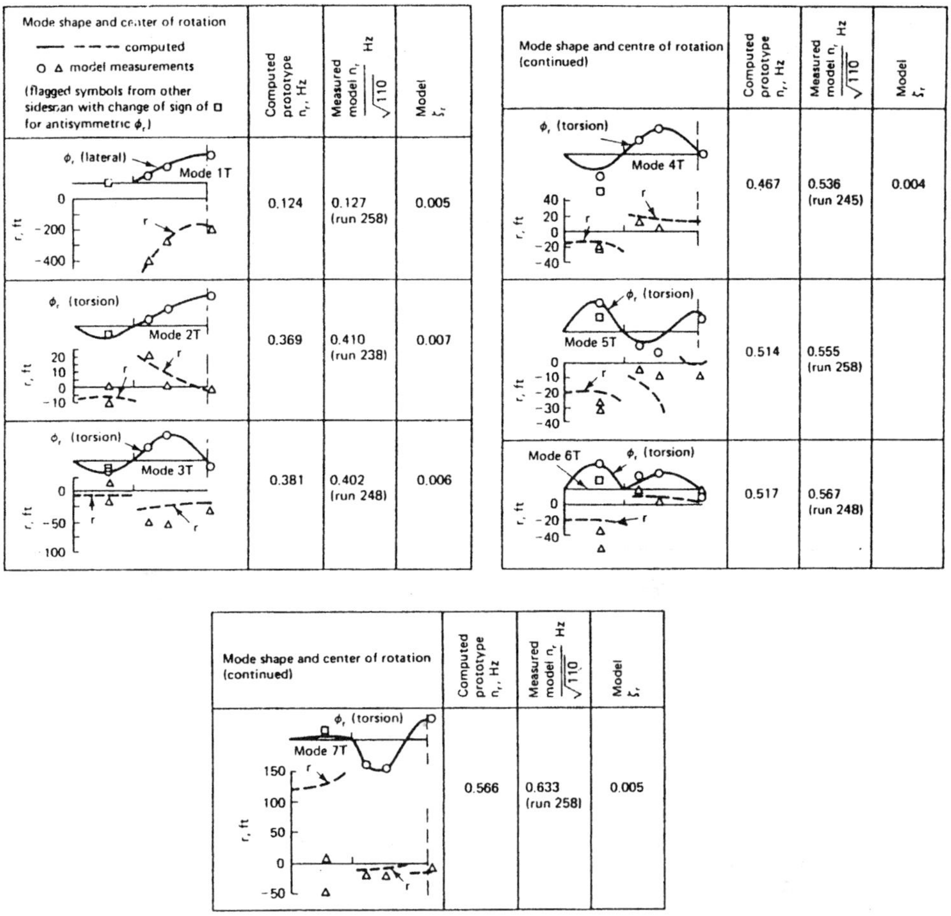

Mode shape and center of rotation ——— computed ○ △ model measurements (flagged symbols from other sidesran with change of sign of □ for antisymmetric ϕ_i)	Computed prototype n_i, Hz	Measured model n_i, $\frac{\text{Hz}}{\sqrt{110}}$	Model ζ_i
ϕ_i (lateral), Mode 1T	0.124	0.127 (run 258)	0.005
ϕ_i (torsion), Mode 2T	0.369	0.410 (run 238)	0.007
ϕ_i (torsion), Mode 3T	0.381	0.402 (run 248)	0.006

Mode shape and centre of rotation (continued)	Computed prototype n_i, Hz	Measured model n_i, $\frac{\text{Hz}}{\sqrt{110}}$	Model ζ_i
ϕ_i (torsion), Mode 4T	0.467	0.536 (run 245)	0.004
ϕ_i (torsion), Mode 5T	0.514	0.555 (run 258)	
Mode 6T, ϕ_i (torsion)	0.517	0.567 (run 248)	

Mode shape and center of rotation (continued)	Computed prototype n_i, Hz	Measured model n_i, $\frac{\text{Hz}}{\sqrt{110}}$	Model ζ_i
ϕ_i (torsion), Mode 7T	0.566	0.633 (run 258)	0.005

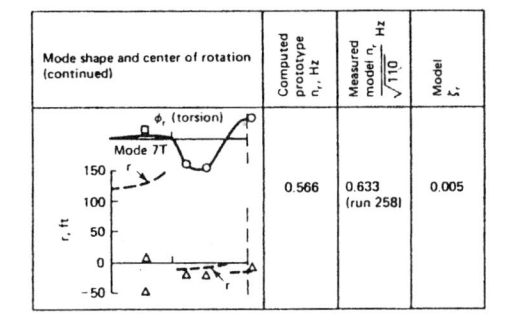

Figure 11.38 Torsional/lateral modes of model compared with computed modes. (After Irwin, H. P. A. H. and Schuyler, G. D., 1977.)

large-scale model ($7/10$) chosen such that normal reinforcing and fabrication procedures could be used. Additional concrete blocks were added for ballast, and lateral bracing was provided to constrain the building against any lateral and torsional motions, as shown in Figure 11.45.

Instrumentation installed to record the response of the test structure included accelerometers and displacement gages at each story level and on the shaking table, strain gages on the reinforcing bars at the column bases, and relative rotation-measuring devices at the ends of columns and girders. In addition, the midcolumn moments, shears, and axial forces were measured directly by means of transducers.

The input earthquake was one horizontal component of the Taft, CA, earthquake of July 1952 with no vertical components. The excitation was applied with successively increasing intensities, starting with a peak acceleration of $0.07g$ in the first run and reaching a maximum run having a peak acceleration of $0.44g$. The first run produced only elastic response, but the maximum intensity test caused considerable damage. Typical test records are shown in Figure 11.46. They depict the input table displacements and the resulting average column shears measured during the maximum test.

After each test, the free vibration frequency and damping of the test structure were determined by suddenly releasing a 6850 kN horizontal force applied at the first-story level. The successive changes of these properties, plotted in Figure 11.47, demonstrate the extent of damage inferred by

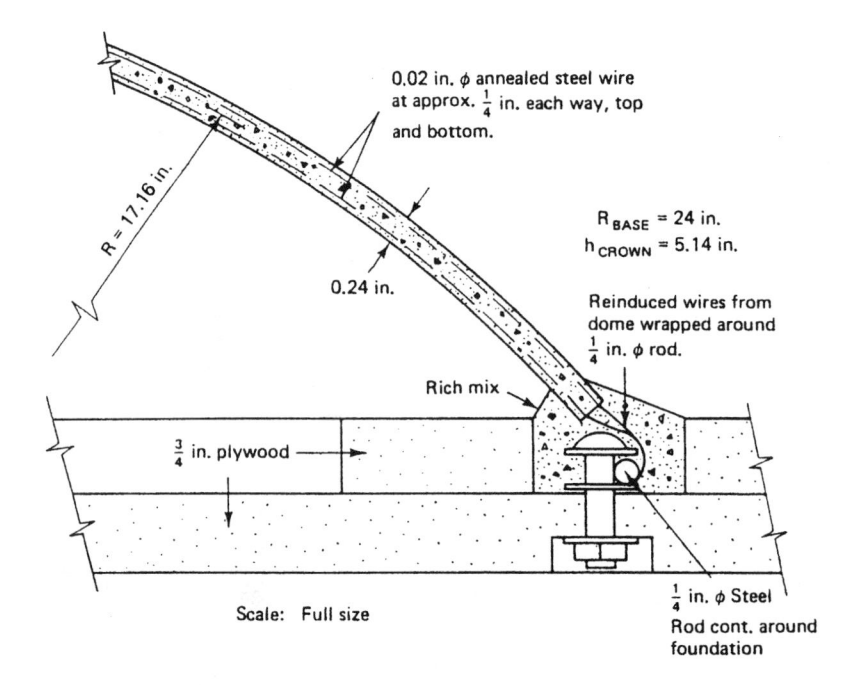

Figure 11.39 Part section through dome model showing base ring. (After Smith, H. D. et al., 1963.)

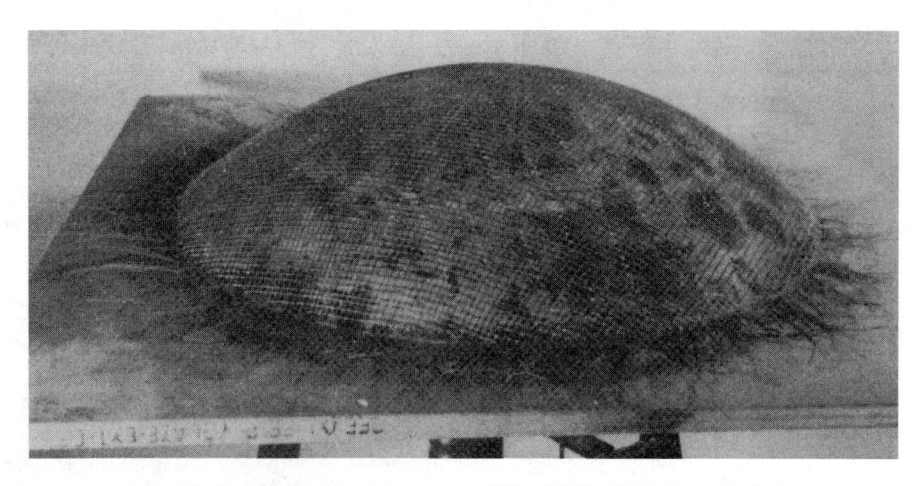

Figure 11.40 Model reinforcement. (After Smith, H. D. et al., 1963.)

the drop in natural frequencies and the increase in damping done to the structure in each test run. A comparison of the top-story displacement measured during the most intense test run with a nonlinear response analysis is shown in Figure 11.48.

11.5.4.2 Reinforced Concrete Bridges

The design, construction, and seismic testing of a model of a long, curved, reinforced concrete overcrossing structure was conducted at the University of California, Berkeley (Williams and Godden, 1976), to investigate the general dynamic behavior of such structures and to provide experimental data for comparison with theoretical solutions. A schematic drawing of the 1/30-scale model, which represents only a portion of the total structure, is shown in Figure 11.49. The multicell

STRUCTURAL MODELS FOR DYNAMIC LOADS

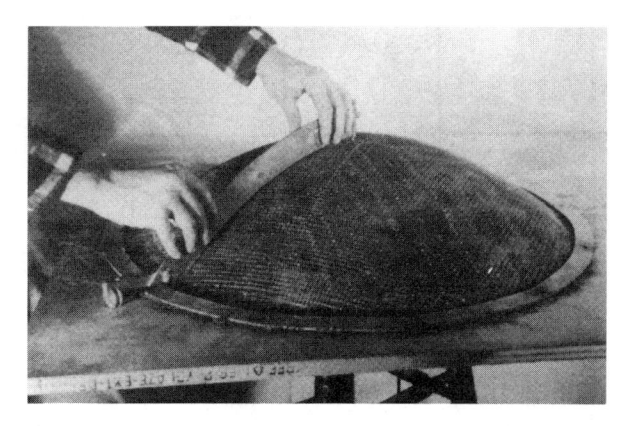

Figure 11.41 Casting technique. (After Smith, H. D. et al., 1963.)

Figure 11.42 View of model installations looking away from ground zero. (After Smith, H. D. et al., 1963.)

Table 11.5 Summary of Model Distance–Pressure Values

Expected Overpressure, MPa	Dome Number	Distance from GZ, m	Overpressure Probably Experienced, MPa
0.14	8	121.9	0.15
	9		
0.21	12	100.6	0.22
	7		
0.24	11	94.5	0.26
	6		
0.28	10	88.4	0.30
	5		
0.42	3	76.2	0.42
0.48	4	70.1	0.51
	2		
0.56	1	67	0.57

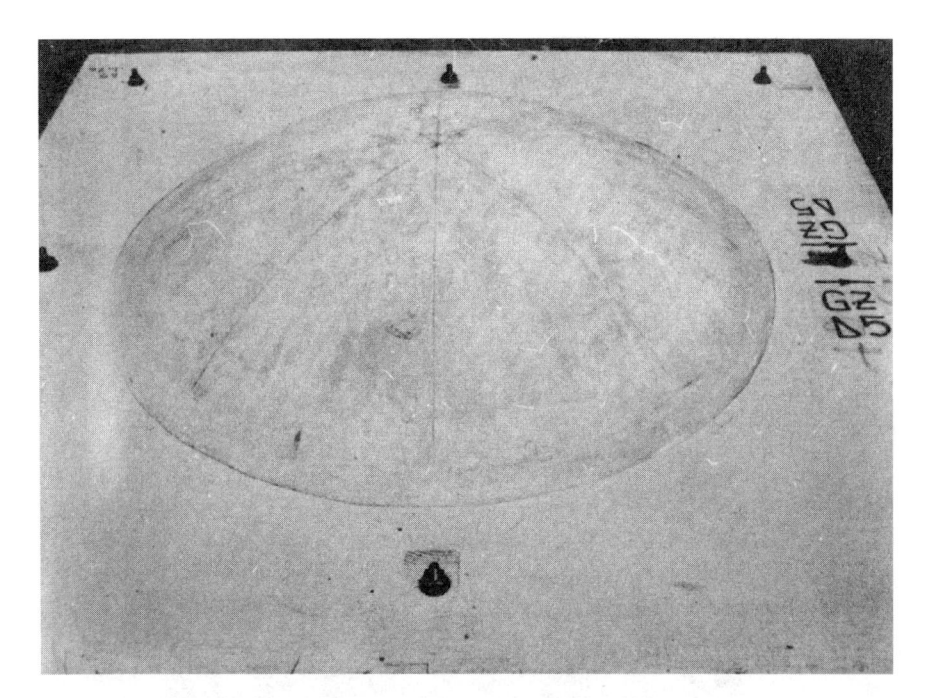

Figure 11.43 40-psi dome preshot. (After Smith, H. D. et al., 1963.)

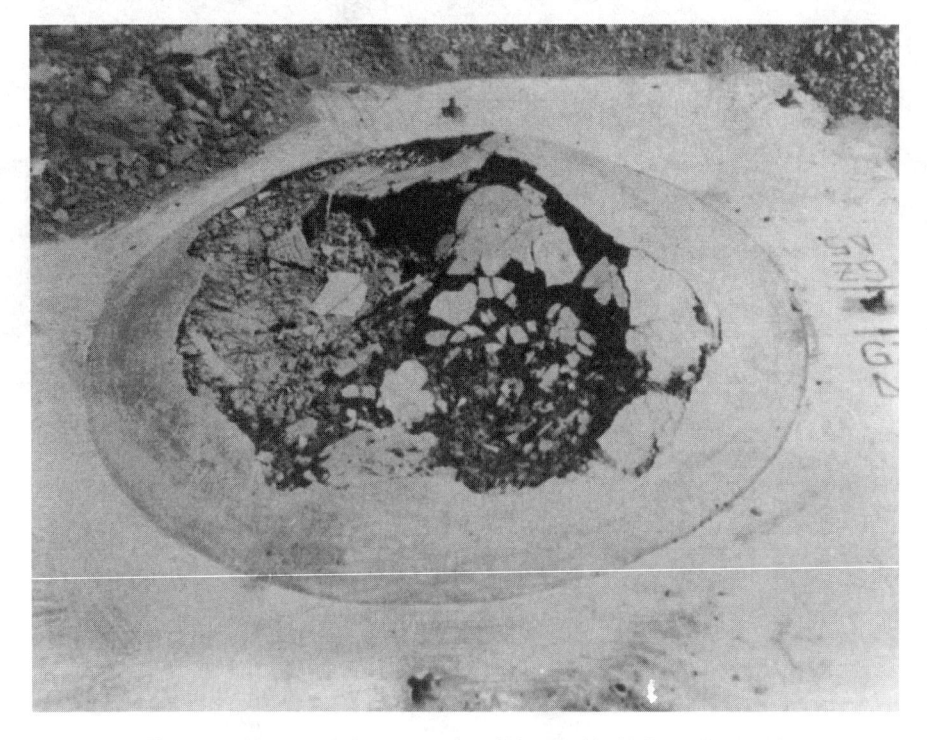

Figure 11.44 40-psi dome postshot. (After Smith, H. D. et al., 1963.)

box girder section of the prototype structure was modeled using an equivalent rectangular solid section to simplify the model construction. Mild steel deformed bars were used for the model longitudinal reinforcement, and plain mild steel wires for temperature effects, shrinkage, and shear

STRUCTURAL MODELS FOR DYNAMIC LOADS 621

Figure 11.45 Two-story R/C frame on shaking table. (After Clough, R. W. and Bertero, V. V., 1977. Courtesy of Prof. R. W. Clough, University of California, Berkeley, Earthquake Engineering Research Center.)

reinforcement (Figure 11.50). The model concrete utilized was a specially designed high-strength 55 MPa low-shrinkage mortar. Special attention was given to the fabrication of the two expansion joints (Figure 11.49). An equivalent-strength section was used for the columns with the longitudinal reinforcement butt-welded to steel plates at both ends of the model column (Figure 11.51). Since the model and prototype materials had similar densities, external weights were placed on the model to preserve the effective mass density required by dynamic similitude (Table 2.11). Care was taken to ensure that the added lead ingots used did not change appreciably the section stiffness or the system damping of the structure.

A large number of preseismic tests were conducted on the model components and on the assembled model to determine its most important dynamic characteristics. The seismic-type tests were conducted on the Berkeley 6×6 m shaking table by subjecting the model to prescribed table motions of increasing intensity. The measured dynamic response included displacements in the two horizontal directions at the expansion joints as well as the gap motion of the joint measured at both the inside and outside edges of the deck.

The model bridge was designed as a small representative structure on which trends of behavior could be studied and correlation studies with theory could be made. No attempt was made to model a specific prototype. For this reason, the mechanisms of failure are more significant than the maximum ground accelerations needed to produce these. Although the joints were supposedly overdesigned, they did suffer major damage but no catastrophic collapse. Typical results of the symmetric seismic response to an artificially generated table motion with a peak horizontal component of $0.48g$ on the model are shown in Figure 11.52.

The extensive seismic testing of the bridge model indicated that the most critical locations of the structure where damage occurs and hence where ductility is required are the expansion joints and the bases of the long columns. The joints are very susceptible to damage caused by multiple impacting in both the torsional and translational modes of response.

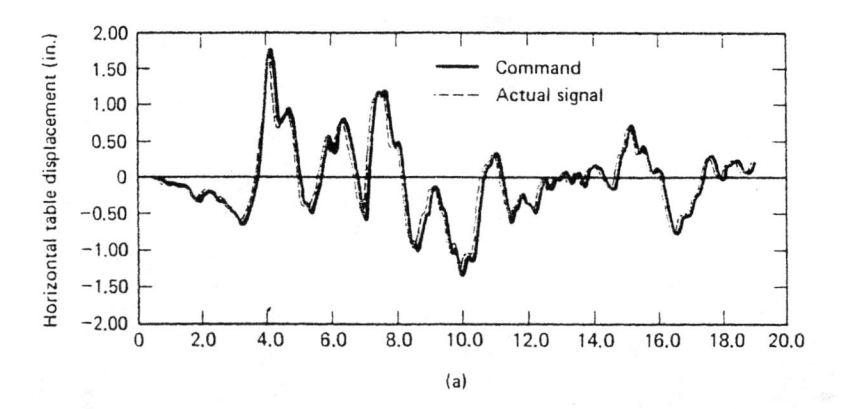

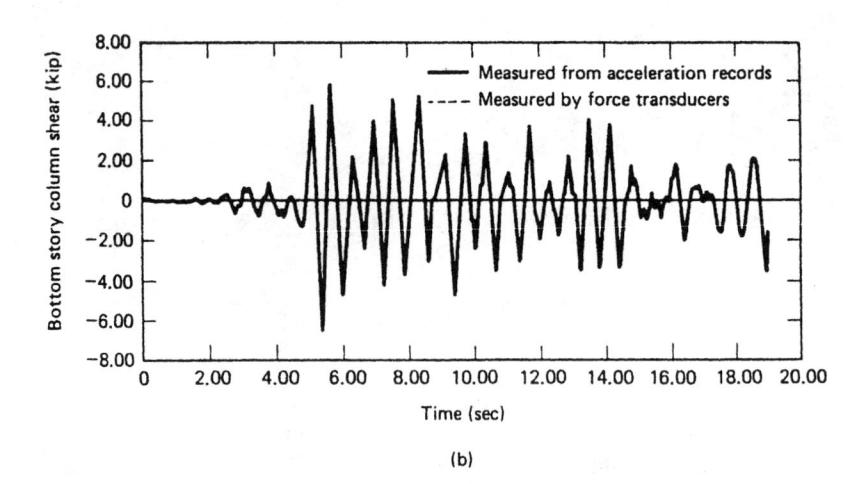

Figure 11.46 Shaking table motion and building response. (a) Table displacements: command vs. actual. (b) First-story column shear force. (After Clough, R. W. and Bertero, V. V., 1977.)

11.5.4.3 Precast Concrete Shear Walls

The main lateral load-supporting system of precast concrete large panel buildings designed originally for gravity loads with ties to mitigate possible progressive collapse mechanisms (Figures 10.29 and 10.37) are stacked precast walls with horizontal joints at every floor level. These are shown in Figure 11.53 for the typical double-loaded corridor arrangement with cross wall layout. An investigation to study the dynamic behavior of precast concrete shear walls was undertaken on a small-size shaking table (Caccese and Harris, 1982; 1984a; 1984b; 1987). Four $1/32$-scale microconcrete models of five-story simple shear walls were tested to failure under increasing levels of the following scaled earthquake records: El Centro 1940 N-S, Taft 1952 N21E, and Pacoima Dam 1971 S16E components. All joint details and tie systems were faithfully reproduced in the models. Artificial mass simulation consisting of various sizes of lead bars attached to floor slabs and walls was used to simulate the inertia loads on the models properly (Figure 11.54).

Acceleration and displacement response of the small-scale models was processed to quantify the maximum values recorded during each simulated earthquake test. Amplitude vs. time signatures, visual observation of cracking patterns, and post-test observations with the aid of movie records were used to study the rocking and slip response of the models. The latter two phenomena had been predicted by mathematical analyses performed at MIT and the University of California, Berkeley but had never before been verified experimentally until this study.

STRUCTURAL MODELS FOR DYNAMIC LOADS 623

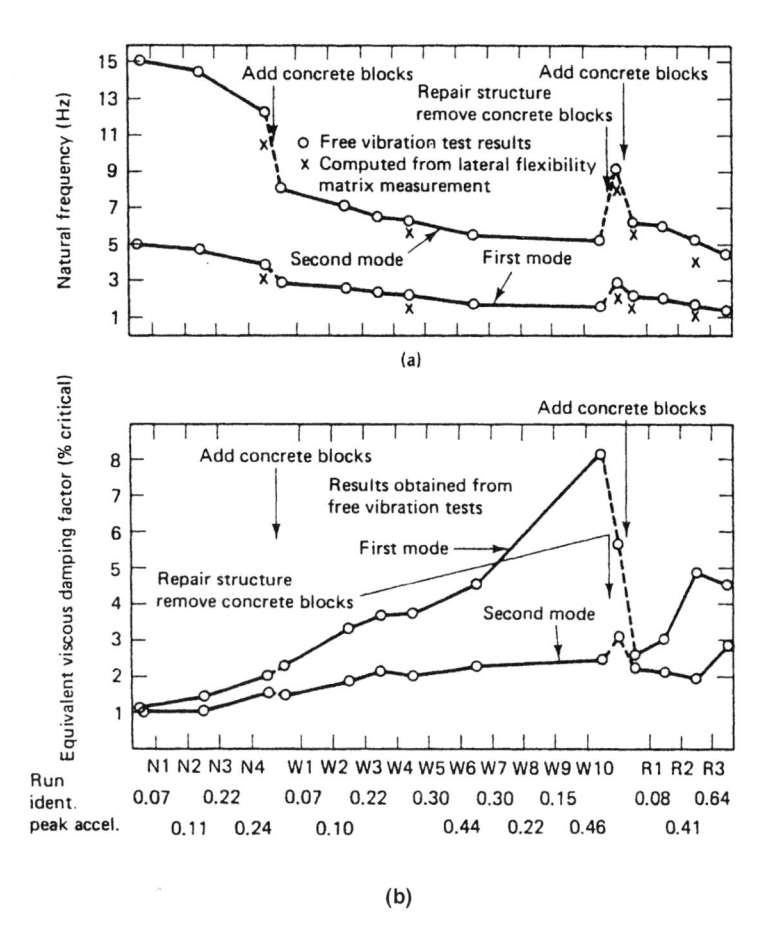

Figure 11.47 Changes of vibration properties during test sequence. (a) Frequencies vs. test number. (b) Damping ratio vs. test number. (After Clough, R. W. and Bertero, V. V., 1977.)

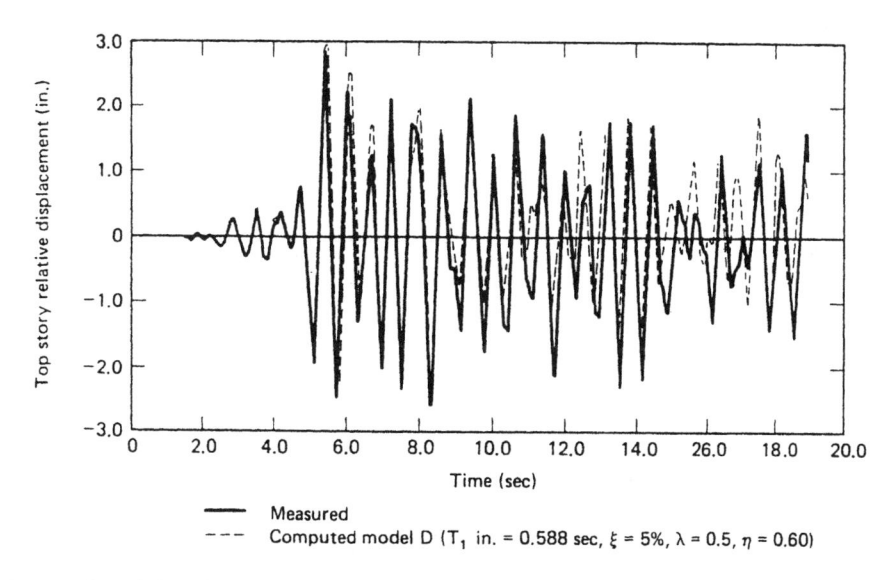

Figure 11.48 Correlation of top story displacement: measured vs. computed. (After Clough, R. W. and Bertero, V. V., 1977.)

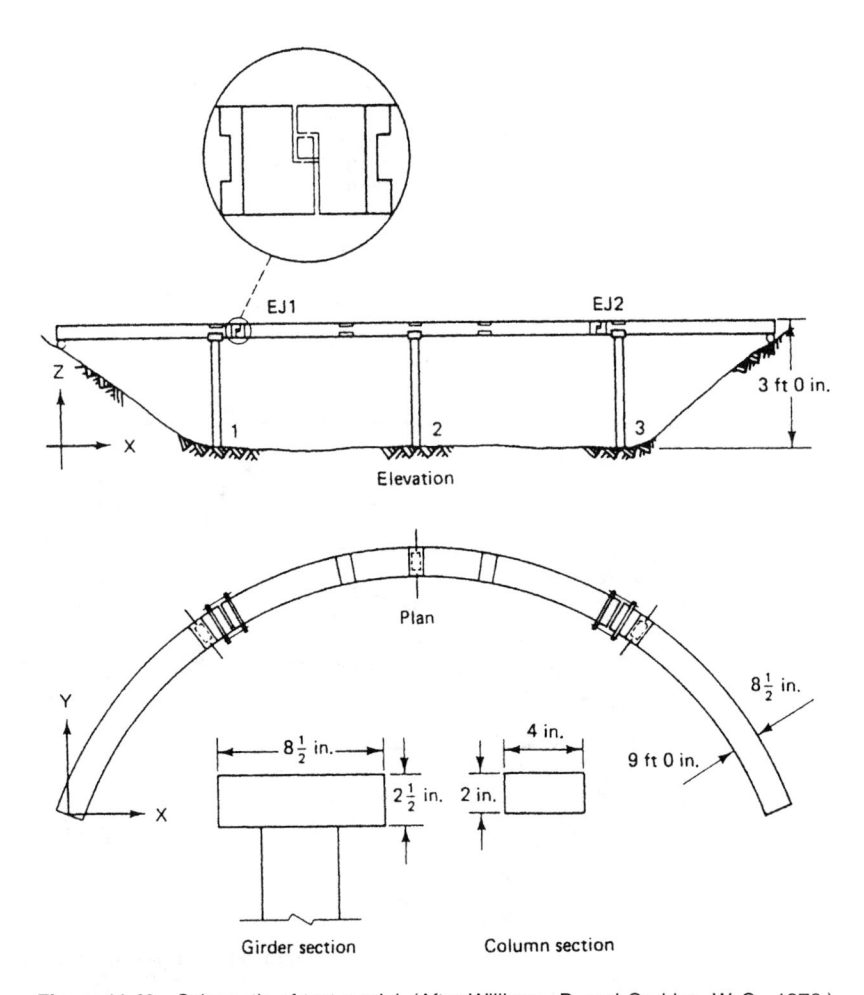

Figure 11.49 Schematic of test model. (After Williams, D. and Godden, W. G., 1976.)

The rocking displacements were recorded in each test run using LVDTs. These were defined as the relative vertical displacements between two wall panel ends, or between the first level wall and its footing in the case of rocking at the base level. The rocking rotation is the relative rotation, i.e., the relative angle change between two panels or between the first-level panel and the footing in the case of rocking at the base level. Typical rocking behavior is illustrated in Figure 11.55. A negative bias is seen in the front and rear displacement signatures of the shear wall. The horizontal joint is stiffer in compression; therefore, displacements in the vertical direction when the joint goes into compression will be reduced. The joint stiffness decreases when the joint goes into tension, thus the observed bias seen in Figure 11.55a and b. Comparison of the computed rocking rotation curve and the level 5 displacement of the same test run shows an almost identical duplication of the two wave forms (Figure 11.55c and d). This phenomenon indicates that the vertical rocking movement is related to the lateral movement.

The maximum rocking displacements vs. maximum input motion for the three models tested are shown in Figure 11.56. This figure shows the dramatic increase in rocking action observed as the input motion increased. Model 1-C, however, showed a slower increase in rocking response after the 0.4g maximum base acceleration level applied. More damage was imposed on this model due to intermediate applications of the Taft and Pacoima Dam records.

Damage as observed from the final cracking patterns was mostly concentrated in the upper levels as illustrated in Model 1-B (Figure 11.57). Longitudinal cracking was seen along the entire

STRUCTURAL MODELS FOR DYNAMIC LOADS 625

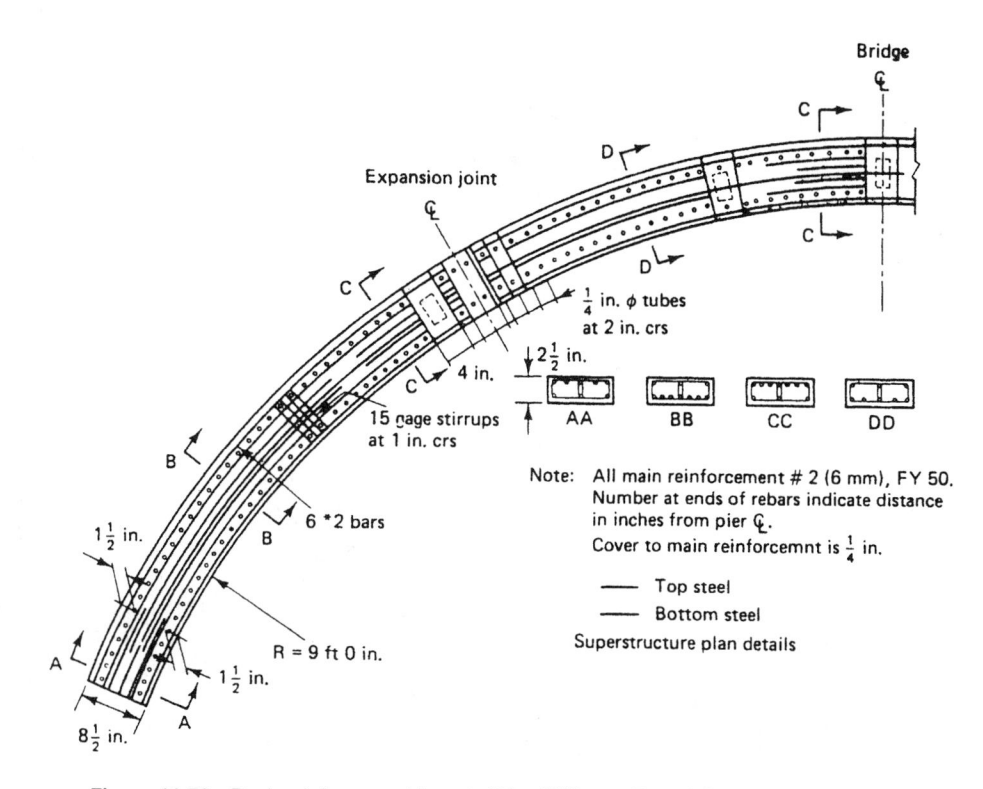

Figure 11.50 Deck reinforcement layout. (After Williams, D. and Godden, W. G., 1976.)

length of the base of the fifth and third walls and partially along the base of the fourth wall. The roof joint, the level 4 joint, and the level 2 joint were noticeably damaged. Also, the roof (level 5) damage was severe. Plan view 5T-5T shows the crack that existed along the entire length of this joint and the separation of the floor panels. The damage pattern of this model was controlled by large slippage at the roof level, a combination of global slip and rocking effects at level 4, and a combination of slip and rocking effects at level 2 (Figure 11.57).

The results of the simulated earthquake study show that the shear wall structure behaves in a nonlinear inelastic manner at even very low base acceleration levels. For low values of base acceleration (as low as 0.1 g) an elastic analysis will underpredict the response of the simple shear wall model. The inelastic mechanisms observed in both the small scale model and computer analysis consist of the combined action of the rocking and slip mechanisms. Rocking was most prevalent at the base and level 1 joints, whereas slip response was most prevalent at the level 5 and level 4 joints.

11.5.5 Earthquake Simulation of Steel Buildings

A three-story single-bay steel frame structure previously tested on the shaking table at the University of California, Berkeley was used as a prototype for a ⅙-scale model study at Stanford University (Mills et al., 1979). The Berkeley structure (Figure 11.58) was actually a reduced-scale model (S_l = 2.5) of a portion of a hypothetical UBC-designed (Uniform Building Code) steel framework (Clough and Tang, 1975; Tang, 1975). Thus, the Stanford model (Figure 11.59) had dimensions approximately ¹⁄₁₅th of actual size steel building frames. The dimensions of the Stanford model are shown in Figure 5.18. The main objective of this investigation was the development of a replica model to simulate all aspects of the prototype steel structural system which may contribute to its earthquake response characteristics. For this reason, artificial mass simulation, which consists of the addition of structurally uncoupled masses to augment the density of the model structure, was used to satisfy similitude requirements (Table 2.11) as shown in Figure 11.59.

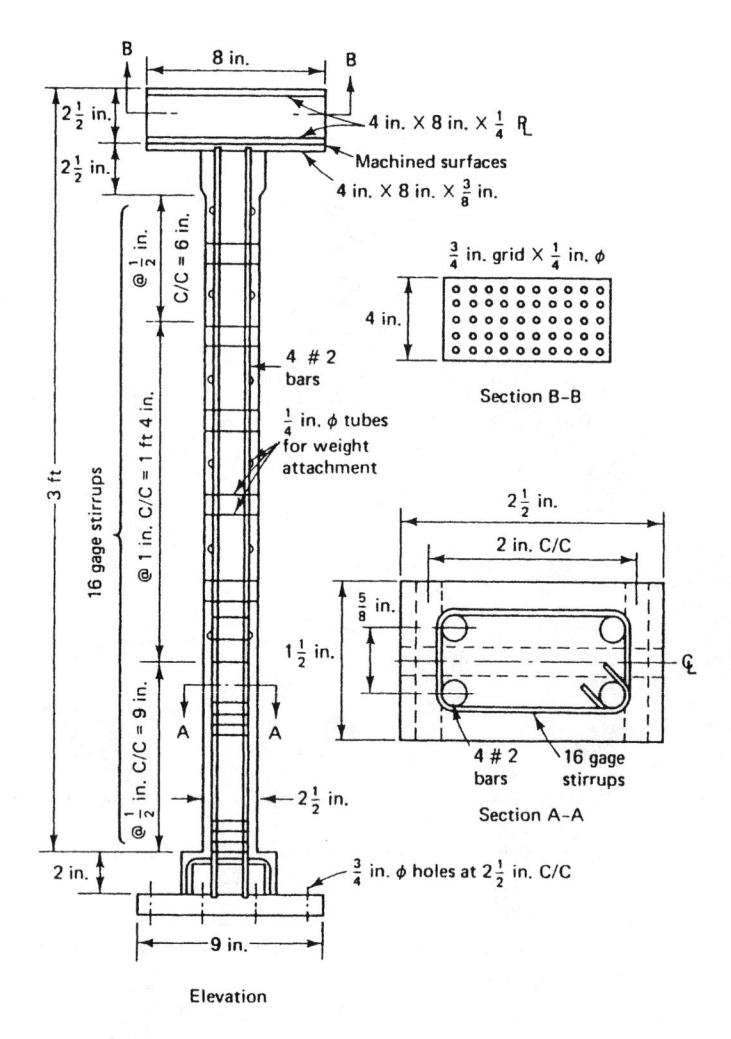

Figure 11.51 Model column details. (After Williams, D. and Godden, W. G., 1976.)

Many earthquake simulator tests of ranging intensity were performed on the prototype and model test structures to enable a complete evaluation of the elastic and inelastic response characteristics. Typical test result comparisons from this study are shown in Figures 5.60 and 5.61. Figure 11.60 gives the comparison of the floor displacements of the ⅙-scale model and prototype for the El Centro 1940 North-South record at an intensity setting of 100% of the prototype full-intensity earthquake. The comparisons of story drift for the same earthquake record is shown in Figure 11.61. Comparison of the joint panel deformations at 130% prototype full intensity of the same earthquake record is shown in Figure 11.62.

Several specific conclusions can be drawn from the results of the ⅙-scale model steel structure with regard to the accuracy of the prototype simulation. The nature of inelastic response is duplicated by the small-scale dynamic model by yielding of the joint panel zones in shear. Thus, the critical elements for model and prototype are identical, with the yielding of these zones producing similar response characteristics for the two test structures. In general, the correlation of the small-scale and prototype test results of these two well documented studies was very good to excellent. Any differences in dynamic behavior were found to be attributable to the three main sources of error: differences in the initial state of stress of the two test structures which influences mainly the elastic behavior, larger than desired welds in the small-scale model, and differences in the earthquake simulation as a result of using two separate shaking table simulators.

STRUCTURAL MODELS FOR DYNAMIC LOADS

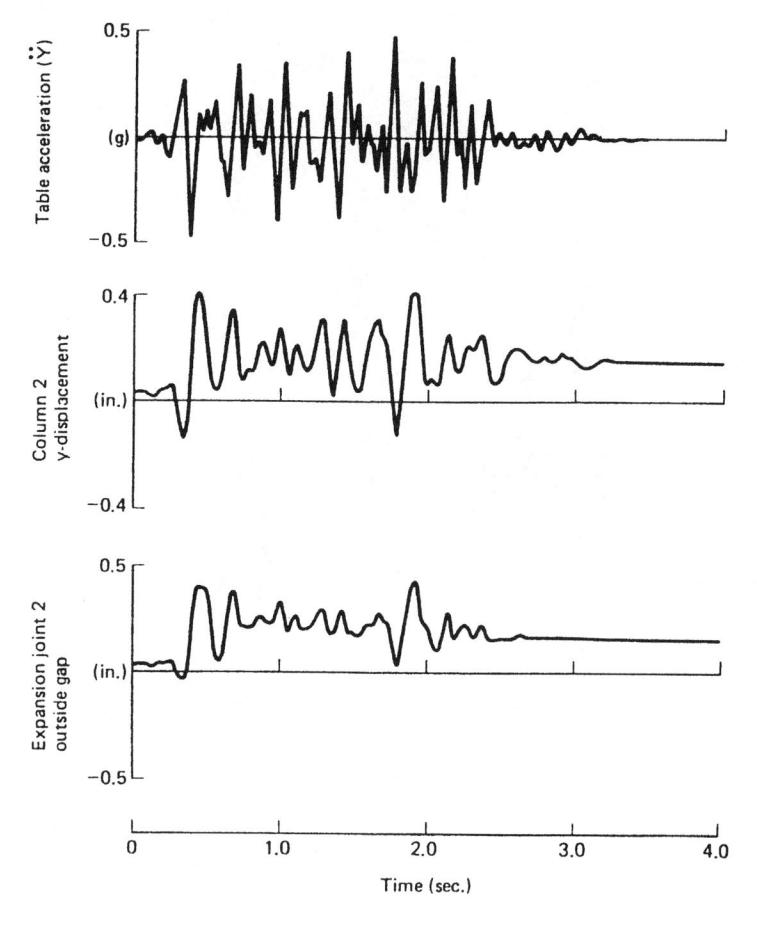

Figure 11.52 Response histories for test Y5A. (After Williams, D. and Godden, W. G., 1976.)

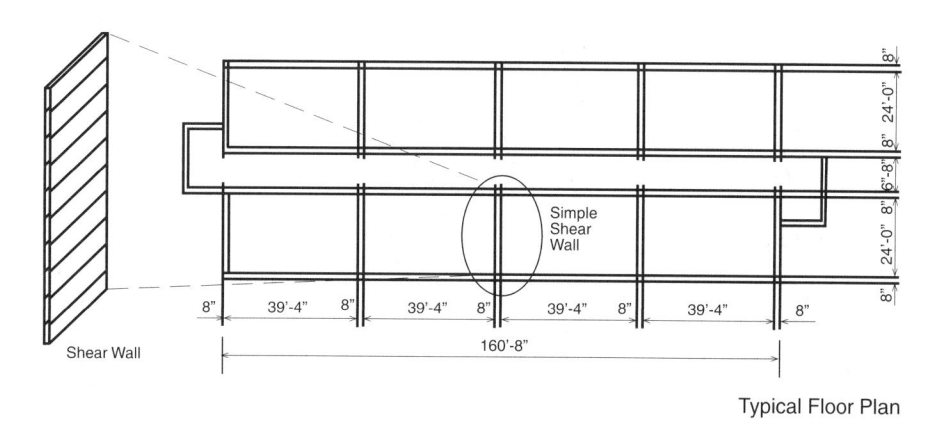

Figure 11.53 Typical cross wall building.

Figure 11.54 Dynamic model on the Drexel University shaking table.

11.5.6 Earthquake Simulation of Masonry Buildings

11.5.6.1 General

Dynamic tests of model masonry building structures using shaking table facilities have been reported in various parts of the world. They have been prompted, usually after major earthquakes, in an attempt to understand masonry behavior for the purpose of developing analytical models, evaluating existing codes, and developing strategies of reinforcing and strengthening (seismic upgrading) of such systems. A number of dynamic studies have been conducted in Ljubljana, Slovenia (formerly Yugoslavia) led by Professor Tomazevic (Tomazevic, 1987; Tomazevic and Weiss, 1990; 1994; Tomazevic et al., 1990; 1993a, b, c; Tomazevic and Velechovsky, 1992; Tomazevic and Lutman, 1996). Other European shaking table studies include Pomonis et al. (1992), Jurukovski et al. (1992), Modena et al. (1992), Limongelli and Pezzolli (1994), and Magenes and Calvi (1994). A considerable amount of dynamic testing of masonry structures has been conducted in the People's Republic of China (Zhu, 1986; Zhu et al., 1986; and Xia et al., 1990). In South America work on dynamic testing of model masonry has been reported by Bariola et al. (1990) and San Bartolome et al. (1992). The testing of masonry structures on shaking table facilities in the U.S. started with the full-size masonry tests of single-story brick and block houses reported by Clough et al. (1979), Manos et al. (1983), Gülkan et al. (1990), and Clough (1990). Paulson and Abrams (1990) tested two ¼-scale reinforced concrete masonry buildings on a shaking table to study dynamic response characteristics and compare them to static test results. Costley and Abrams (1996) studied the dynamic response of unreinforced reduced-size masonry buildings with flexible diaphragms.

11.5.6.2 Unreinforced Masonry

An example of studies on strengthening of existing historic masonry buildings in Europe is the seismic upgrading of old brick masonry urban houses (Tomazevic et al., 1996). Analysis of

STRUCTURAL MODELS FOR DYNAMIC LOADS

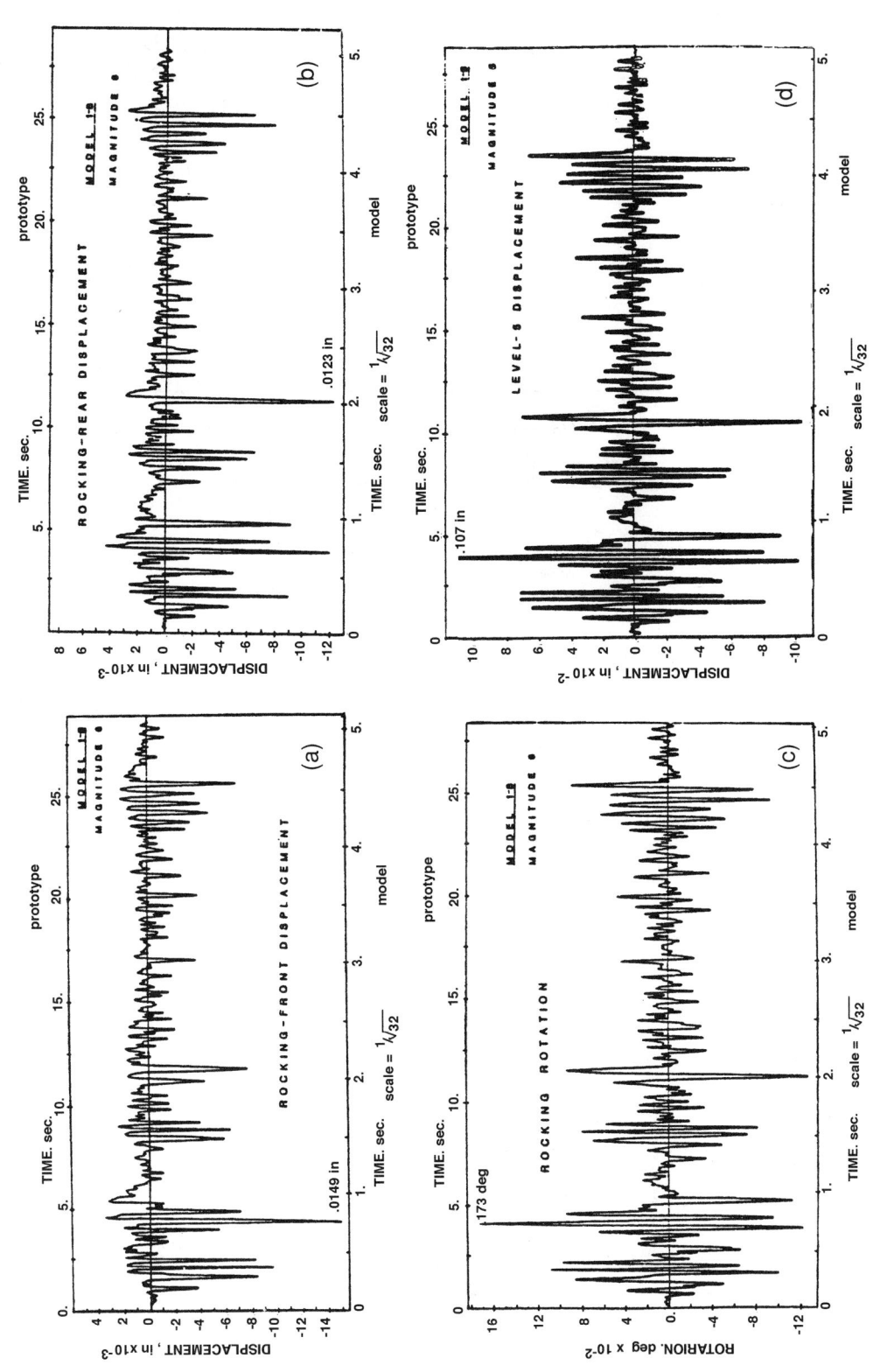

Figure 11.55 Typical set of rocking curves.

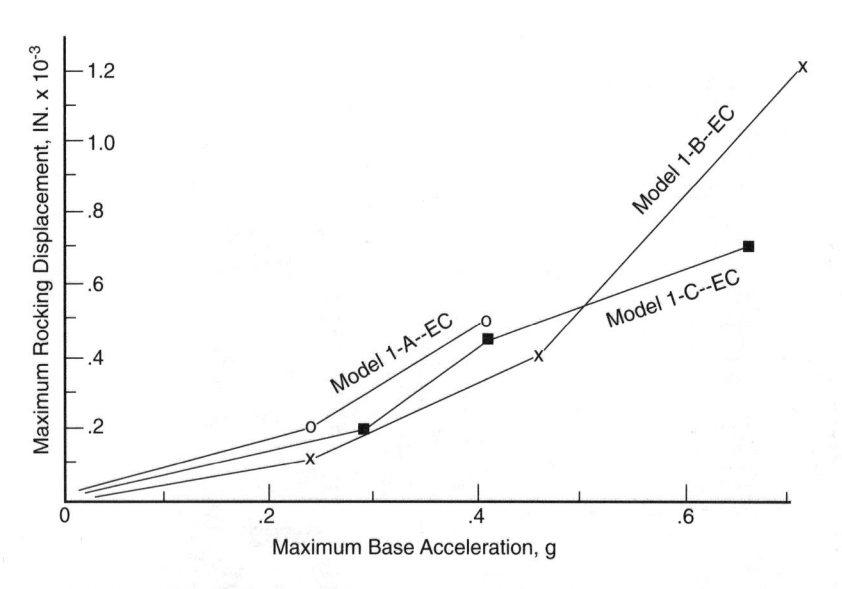

Figure 11.56 Maximum rocking response vs. maximum base acceleration.

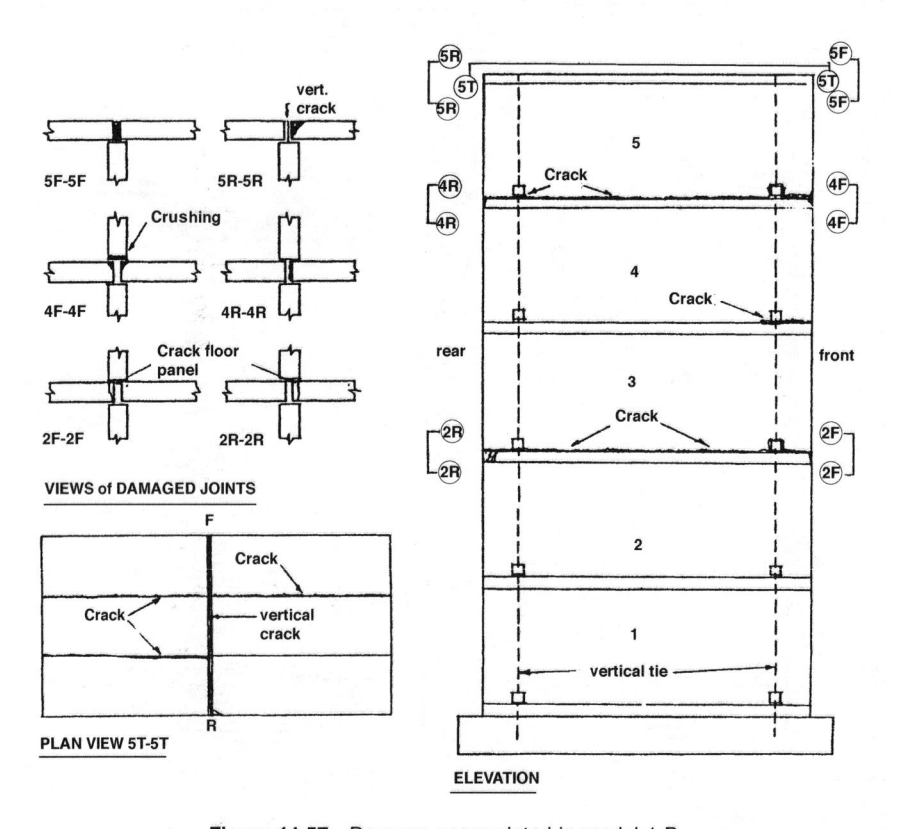

Figure 11.57 Damage accumulated in model 1-B.

STRUCTURAL MODELS FOR DYNAMIC LOADS

Figure 11.58 Prototype steel frame structure on University of California, Berkeley shaking table.

Figure 11.59 One-sixth scale model on Stanford University shaking table.

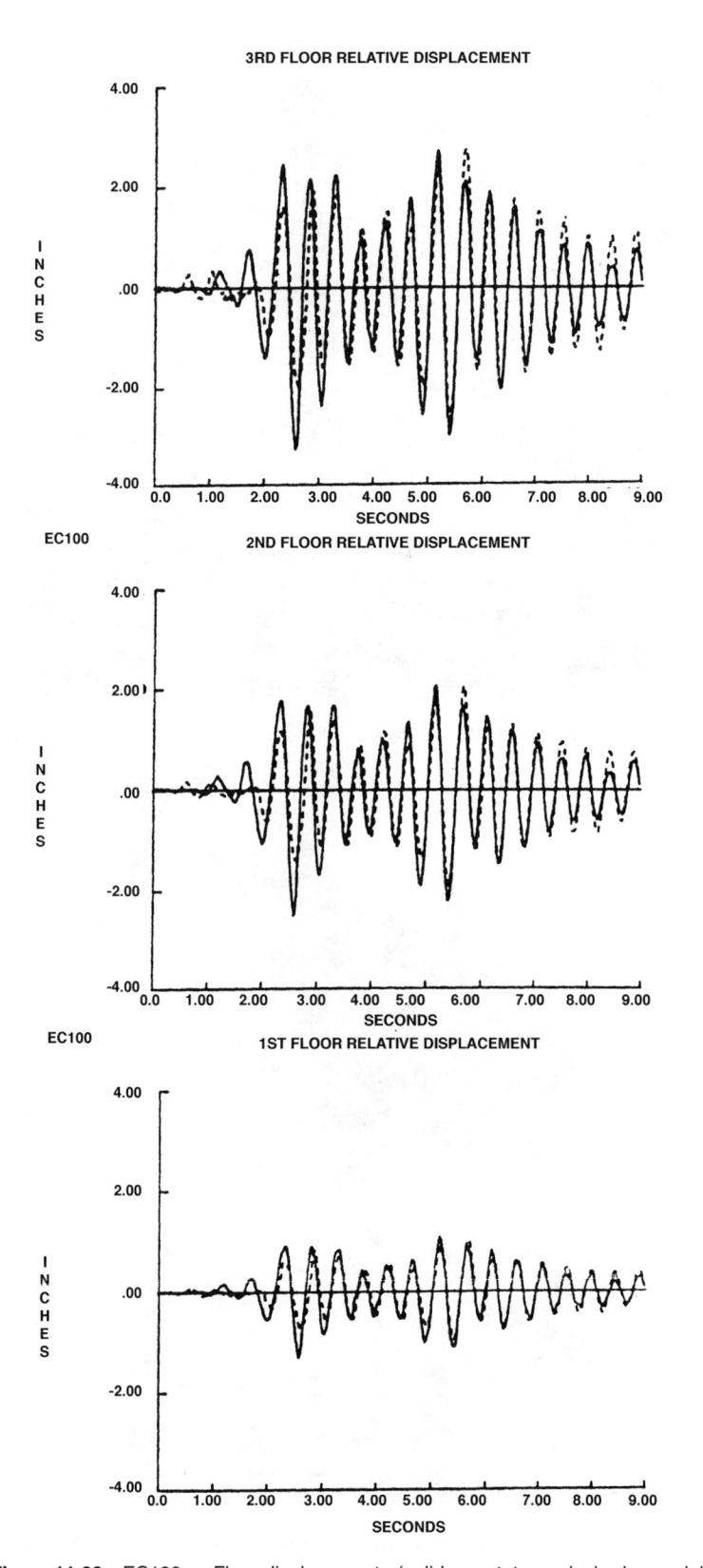

Figure 11.60 EC100 — Floor displacements (solid = prototype, dashed = model).

STRUCTURAL MODELS FOR DYNAMIC LOADS

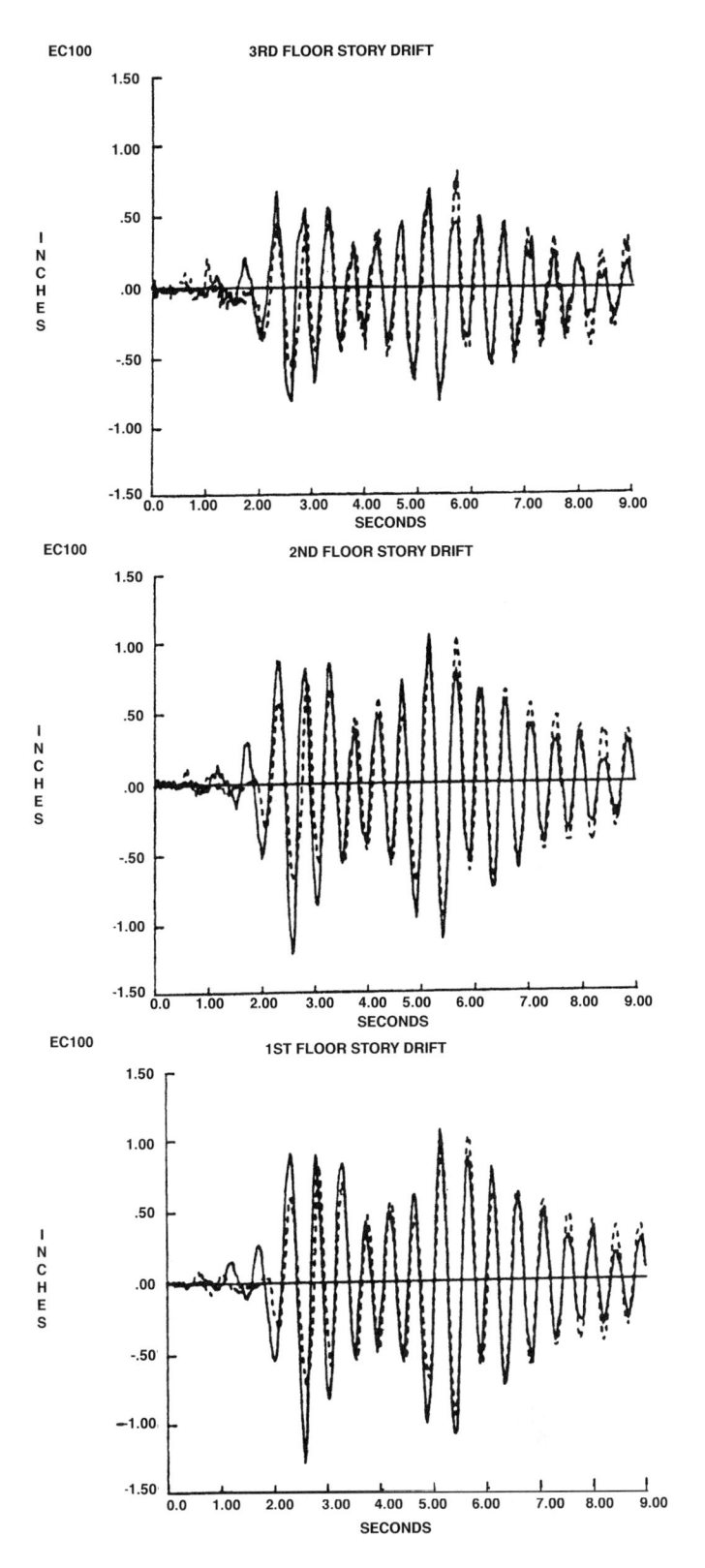

Figure 11.61 EC100 — Story drift (solid = prototype, dashed = model).

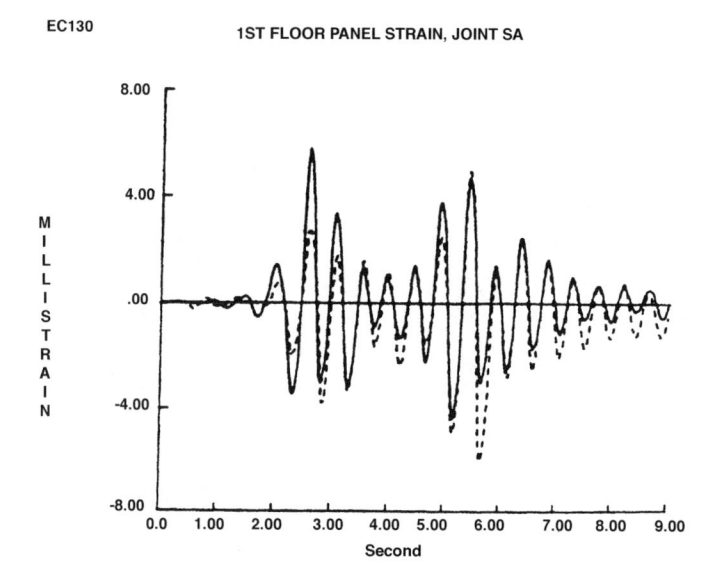

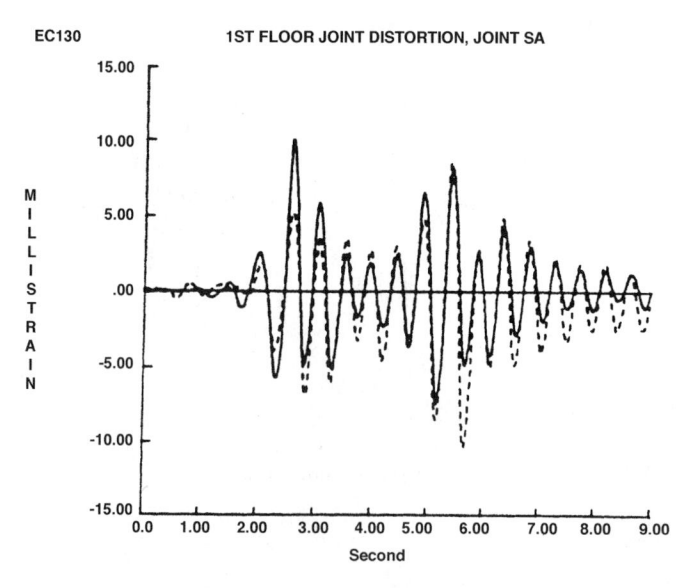

Figure 11.62 EC130 — Joint panel deformations (solid = prototype, dashed = model).

the damage to historic buildings due to earthquakes indicates that besides quality of materials and construction and distribution of structural walls in the plan, the wall-to-floor connections significantly influence the seismic resistance. In order to ensure integrity of unreinforced masonry structures during earthquakes, flexible wooden floors are often replaced by reinforced concrete slabs, anchored to supporting walls or the walls are tied with steel ties and wooden floors anchored to the walls and/or braced with diagonal ties. To investigate these strengthening techniques, four ¼-scale two-story brick masonry building models were tested on a shaking table.

The structural characteristics of all four models are given in Table 11.6. Model A was the reference structure and Model B used a reinforced concrete (R/C) slab to replace the wooden floor. All four models had the dimensions shown in Figures 11.63 and 11.64. Construction details of Models C and D in elevation and plan and the direction of motion are shown in Figures 11.63 and 11.64. Note the location of the wall ties at the wall–floor connection.

STRUCTURAL MODELS FOR DYNAMIC LOADS

Table 11.6 Structural Characteristics of Tested Models

Model Designation	Type of Floors 1st Floor	Type of Floors 2nd Floor	Steel Ties Along Walls	Steel Ties Diagonal	Steel Ties Prestressed
Model A	Wooden	Wooden	No	No	No
Model B	r.c. slab	r.c. slab	Bond-beams	—	—
Model C	Wooden	Wooden	Yes	No	Yes
Model D	Wooden	Wooden	Yes	Yes	No

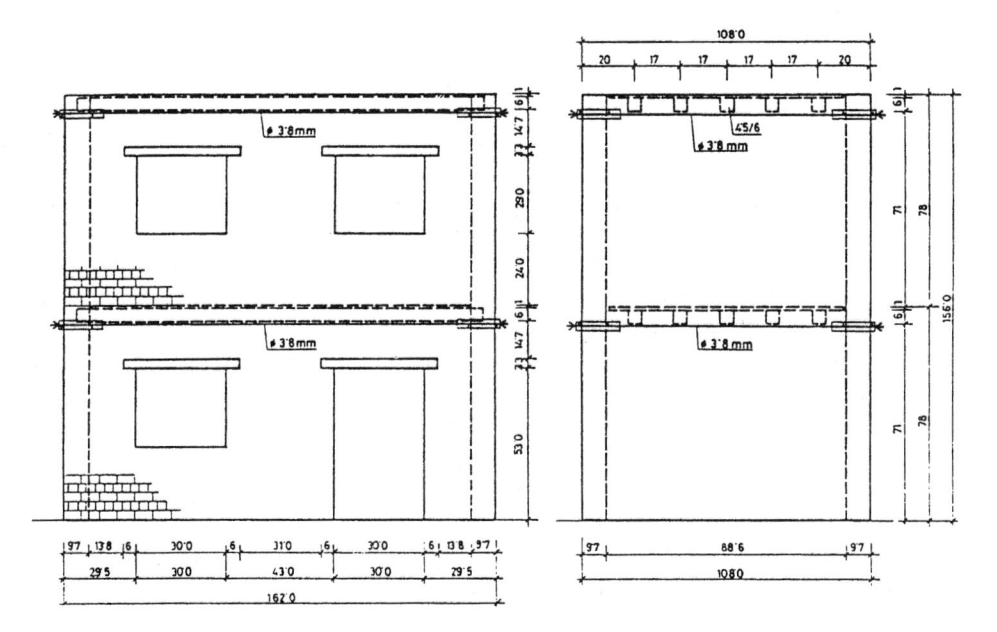

Figure 11.63 Vertical section of Model C with steel ties.

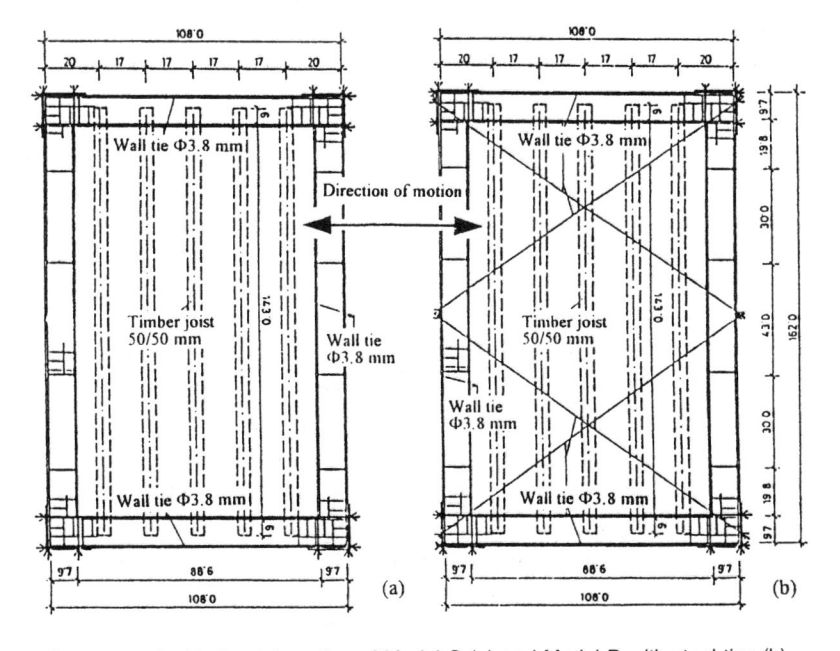

Figure 11.64 Horizontal section of Model C (a) and Model D with steel ties (b).

A casting technique was used to produce the 63 × 30 × 30 mm model bricks used to construct the models. Crushed brick aggregate, lime, cement (in proportion of 9:2:0.75), and water were cast in special steel forms. A cement:lime:sand mortar in the proportion of 0.4:1:1.1 was used in laying the model brick.

Additional mass was added at the floor level to preserve similitude. A modified ground acceleration record of the Montenegro earthquake of 1979 with peak ground acceleration of $0.43g$ was used to drive the shaking table with a gradually increasing intensity. All models were instrumented with accelerometers and displacement transducers. The changes in strain in the longitudinal steel ties were measured with strain gages.

The propagation of damage and the failure mechanism of each model was closely monitored. Cracking patterns of the two models with ties (Models C and D) were found to be essentially the same. Propagation of damage of all models is shown in Figure 11.65. It was observed from these tests that the wooden floors of Model A did not prevent separation of the walls, out-of-plane vibration, and disintegration of the upper story (Figure 11.66a). Models C and D with identical structure but with walls tied with steel ties retained integrity up until their final collapse. Both prestressed (Model C) and simply placed ties (Model D) efficiently prevented separation of the walls and excessive out-of-plane vibration. Figure 11.66b shows Model C after the test. Owing to retained integrity of the structure, lateral load resistance and deformability, as well as energy dissipation, capacity was significantly improved.

Model B with the reinforced concrete slabs replacing the wooden floors behaved differently. Propagation of damage from the first to the second floor was hindered by floor strength and rigidity. However, damage to corners in the case of Model B, in part caused by rocking motion, was also due to sliding of the slab on the top of the walls that resulted in pushing out of transverse walls and corners. This points to the need of good connection between replacement R/C slabs and existing masonry walls along the length of the slab if this method of seismic strengthening is used. The test results indicate that in the case where the masonry walls are adequately tied and the ties are prestressed, the replacement of wooden floors with R/C slabs is not the necessary condition to ensure adequate seismic behavior. The steel ties provided in the models tested can accomplish the same effect.

Dynamic tests on two two-story reduced-scale unreinforced masonry models with flexible diaphragms were conducted at the University of Illinios, Urbana-Champaign (Costley and Abrams, 1996). Each two-story building had four walls, two perforated in-plane, shear walls and two solid, out-of-plane walls. The two test structures had different window and door opening layouts in their main earthquake load–resisting walls. Test structure S1 (viewed from the window wall in Figure 11.67a) is shown on the earthquake simulator. Test structure S2 (viewed from the combination door and window wall) is also shown on the earthquake simulator in Figure 11.67b. In these test structures, model bricks were cut from standard pavers and the construction was carried out by qualified masons. The flexible diaphragms were designed with steel components to have the same properties as the usual wooden floor prototypes and had natural frequencies well below those of an equivalent rigid diaphragm structure.

The two test structures were subjected to increasing magnitudes of a modified version of the 1985 Nahanni earthquake record. Based on the results of a total of nine test runs on the two models, the following conclusions were drawn by the investigators:

1. Diaphragm and wall amplifications of base accelerations compared well with results measured on full-size buildings during actual earthquakes. Prior to cracking, both walls and diaphragms amplified base accelerations at a constant level while, after cracking, little or no amplification existed.
2. Flexible diaphragms amplified wall displacements prior to cracking in the walls. After cracking, diaphragm displacements relative to the walls were greatly diminished. Interstory drifts above the cracks also decreased after cracking.
3. Lateral forces were distributed equally between the two floor levels, not by the inverted triangular distribution normally assumed for rigid diaphragms.

STRUCTURAL MODELS FOR DYNAMIC LOADS

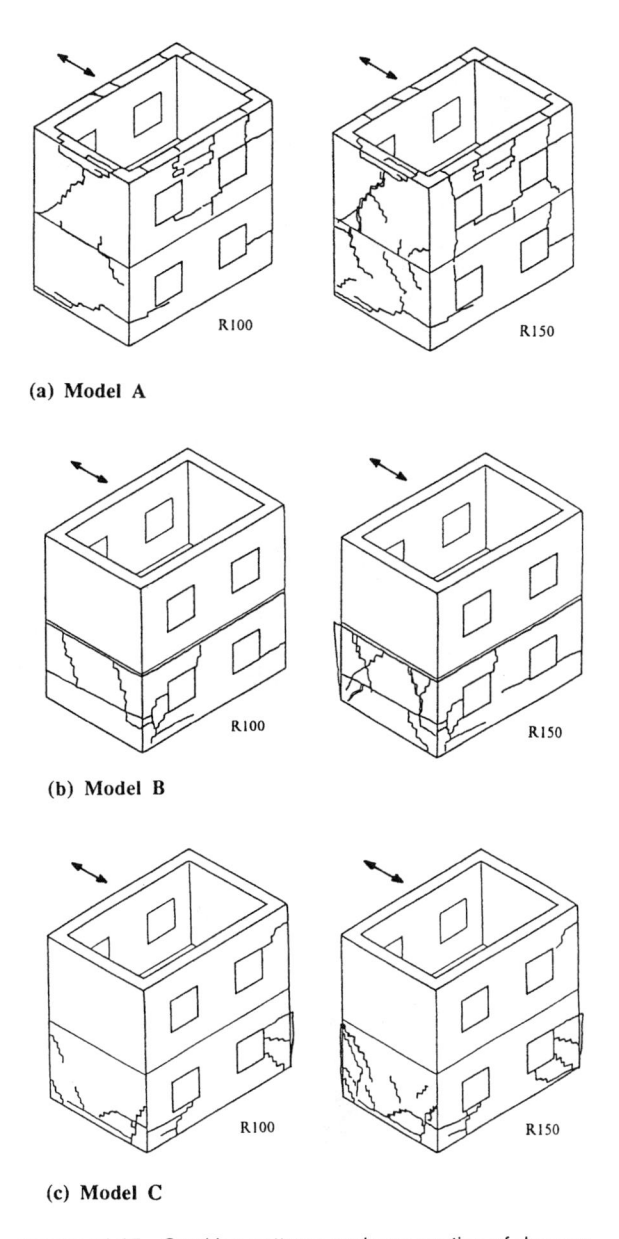

(a) Model A

(b) Model B

(c) Model C

Figure 11.65 Cracking patterns and propagation of damage.

4. Low masonry tensile strength resulted in horizontal cracks across the bases and tops of most of the piers.
5. First-story cracking drifts were approximately 0.1%.
6. Substantial strength and deformation capacity existed after cracking. This ductility resulted from pier rocking in the first story.
7. After cracking, up to 80% of first-story displacements were attributable to rocking.
8. Postcracking force–displacement curves were bilinear in shape which is indicative of rocking.
9. Natural frequencies decreased as structural damage, in the form of cracking, increased. Frequency measurements were dependent on the amplitude of the test. Calculated natural frequencies were much higher than measured frequencies, indicating a stiffer analytical model.

(a) (b)

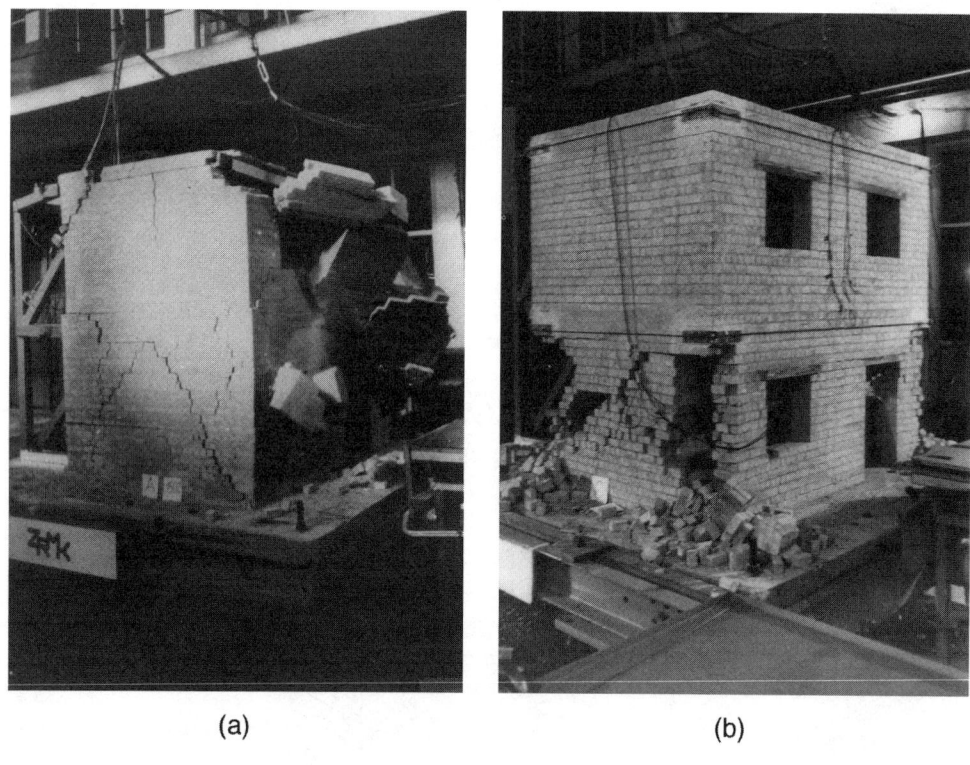

Figure 11.66 Photographs of the failure modes of Model A (a) and Model C (b). (Courtesy of Prof. M. Tomazevic, NBCEI, Ljubljana, Slovenia.)

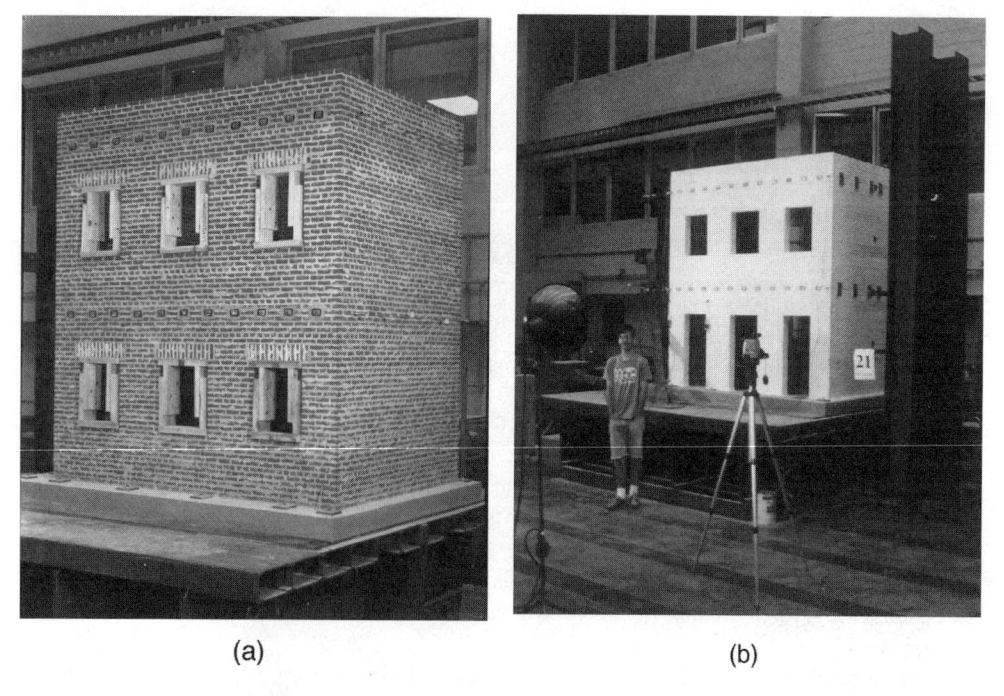

(a) (b)

Figure 11.67 Unreinforced masonry models S1 (a) and S2 (b) on the University of Illinois, Urbana-Champaign shaking table. (Courtesy of Prof. D. Abrams, University of Illinois, Urbana-Champaign.)

STRUCTURAL MODELS FOR DYNAMIC LOADS

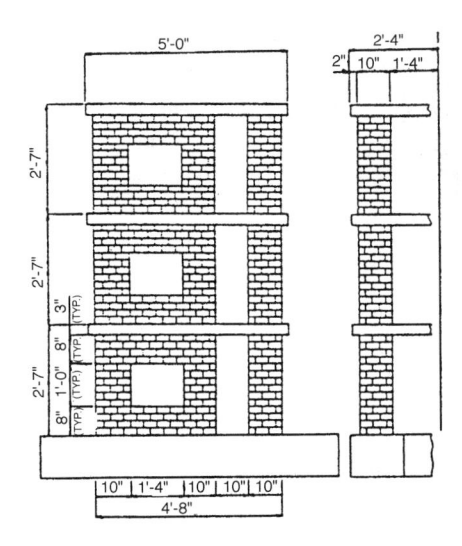

Figure 11.68 Test structure RM3 configuration.

11.5.6.3 Reinforced Concrete Masonry

Two ¼-scale three-story reinforced concrete masonry models were tested on the University of Illinois, Urbana-Champaign shaking table (Abrams et al., 1990; Abrams and Paulson, 1990; Paulson and Abrams, 1990). One model, RM1, was identical to the static test specimen shown in Figures 10.22 and 10.23. A second dynamically tested model, RM3, had an asymmetrical pattern of openings for comparison purposes (Figure 11.68). The flange width was also varied for the two dynamic models. The amount of vertical reinforcement was equal to 0.16% times the gross area. Horizontal reinforcement was nominally 0.094% of the gross area for model RM1. For model RM3, amounts of horizontal reinforcement were nearly double for the piers so that the vertical reinforcement in each pier could develop its full tensile strength.

The simulated earthquake events to which the two models were subjected were based on the El Centro 1940 NS component record. The time scale was compressed by a factor of 2.5 which was approximately the ratio of periods for a hypothetical full-scale masonry structure and the ¼-scale model structures. The amplitude of the base accelerations were varied for each test run to result in response (1) before cracking, (2) after cracking and before yielding of reinforcement, (3) at yield of reinforcement, and (4) at the ultimate limit state. Free vibration measurements were made before and after each earthquake simulation.

Typical measured base moment vs. top-level deflection curves (Figure 11.69) for the two dynamic models show the nonlinear nature of the response. However, it was found that the vibration of the models was dominated by a single frequency. The decrease in frequency with each test run is shown is Figure 11.70 which plots the apparent first-mode frequency vs. the peak value of the lateral drift (top-level displacement divided by height) for each particular run. The solid lines in Figure 11.70 are computed from Fourier spectra of top-level accelerations.

It was found from the testing that measured deflected shapes were nearly constant despite the amplitude of motion. This observation can justify the use of a single degree of freedom model for the nonlinear dynamic response analysis. Such an analysis is compared to the last test run for model RM1 in Figure 11.71. Whereas the complete history is not replicated, the simple analytical approach does estimate the amplitude and number of the large cycles very well.

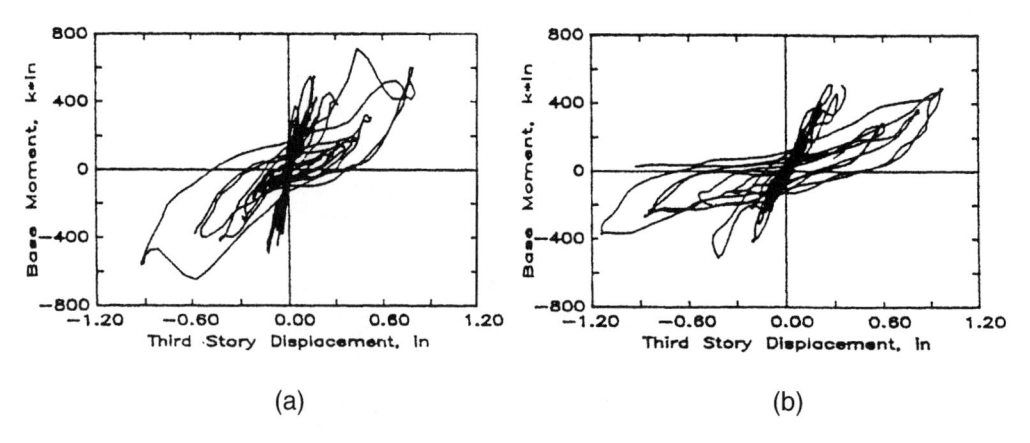

Figure 11.69 Measured base moment vs. top level deflection. (a) Structure RM1. (b) Structure RM3.

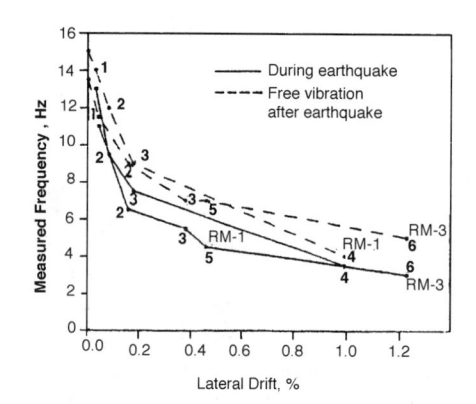

Figure 11.70 Decrease in frequency with test run.

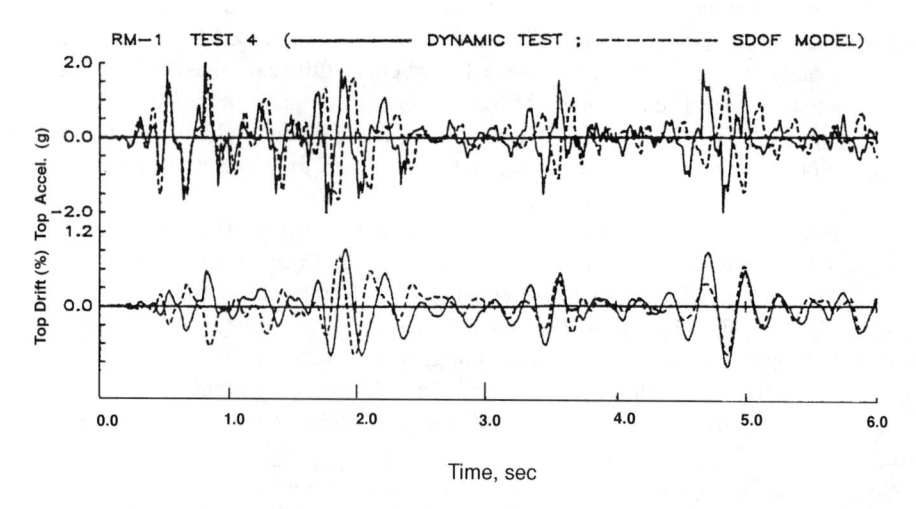

Figure 11.71 Measured and calculated response histories, model RM1.

STRUCTURAL MODELS FOR DYNAMIC LOADS

From their study, the authors conclude:

1. Although lateral force distributions did fluctuate substantially during shaking, The centroid of lateral force vacillated about a location equal to two-thirds of the height.
2. Lateral accelerations across a story were nearly constant.
3. Pier behavior was influenced by reversals in axial force. Story shear was not distributed to each pier symmetrically though the configuration was symmetrical.
4. Static base shear strength could be calculated accurately with simple pier models, however, dynamic strength exceeded static strength by 55%.
5. Story shear was not distributed in proportion to the stiffness of piers because of sliding along bed joints. As a result, available shear strength could not be utilized and the ultimate limit state of one specimen was attributable to diagonal tension failure of an exterior pier.
6. A capacity design approach appeared to be feasible for one structure.
7. Linear analytical models such as pier models, frame models and finite element models can be grossly over stiff even for uncracked response.

11.5.7 Impact Loading

Impulsive loads generated by loading machines based on electronic-hydraulic servo-mechanism principles or quick release of gas pressure in a piston have been used to simulate the effects of air blast loads. These techniques have been successfully used to test both full-scale and model structural components. Many such machines were built in the 1950s and early 1960s at universities and government and independent research institutions to study dynamic loading effects on structures. Illustrative of these dynamic loading machines is the device built at MIT (Massachusetts Institute of Technology) shown in Figure 11.72 for imparting impulsive loads to small-scale model structural components (Harris et al., 1962). The loading yoke mechanism is activated by a quick release of pressurized gas and impacts the test structure through a fast moving rod. Single-bay one-story $1/15$-scale steel frames were tested in this loading device while supported vertically under the loading ram. Dimensions, typical impulse load time histories and test results are shown in Figure 11.73.

Impact loading of model specimens can also be achieved by firing a missile by means of air pressure The missile strikes a stationary pressure bar that is in contact with the test model. Such devices have been used to conduct dynamic tests on a variety of concrete specimens (Garas, 1981; Harris, 1982).

Precision impact testing of structural components for development, verification, and validation of numerical analysis techniques has been advocated by researchers in modern fortification technology (Krauthammer, 1997; Krauthammer et al., 1993, 1996). The ability to obtain a wide variety of load functions using a drop hammer (see Figure 11.20) is illustrated by dropping a 26.75 kN hammer from different heights on a segment of a steel rail (Figure 11.74). Impact interface and support conditions were varied and the load pulses were measured with a load cell attached to the impacting face of the hammer. Details of six load cases are summarized in Table 11.7 and in Figure 11.75.

Power spectral density vs. frequency of the different loading cases show significant differences (Figure 11.76). Tests 1, 3, and 4 show dominant behavior in the lower frequency range of up to 40 Hz. In Test 2, however, a very large response has been noted in the frequency range after 40 Hz. In Tests 5 and 6, it is noted that the dominant behavior is in the frequency range of below 40 Hz, and that increasing the drop height affects the behavior strongly.

Clearly, one can study in great detail the characteristics of both the applied load and the structural response, thus deriving well-defined relationships between cause and effect. Such data would be very valuable for the validation and verification of computer codes.

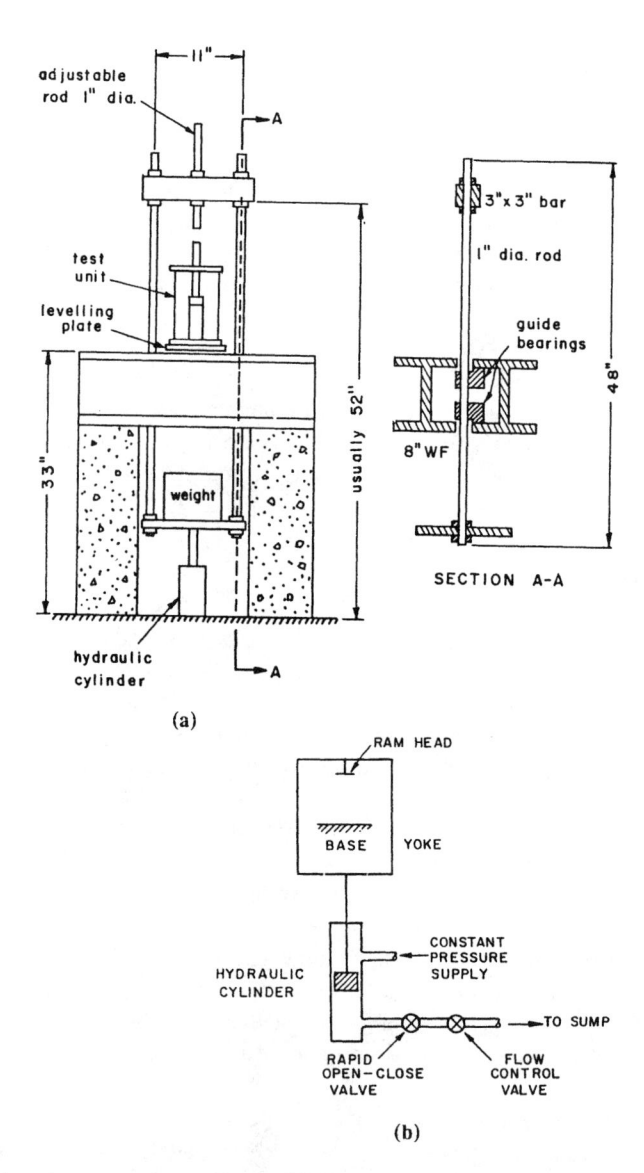

Figure 11.72 MIT impulse generating dynamic loading machine for small-scale model testing. (a) Impulse loading machine. (b) Schematic of controls.

11.5.8 Soil–Structure Interaction Studies Using a Centrifuge

11.5.8.1 Introduction

An interesting application of model testing under dynamic loading conditions in a centrifuge has been reported by Lassoudiere and Perol (1988). It involves the design of three-legged self-elevating drilling platforms operating in water depths of 90 to 100 m while exploring for gas or oil deposits. Most of these platforms have independent legs which are founded on soil through special footings. They are sensitive to the foundation stiffness and damping characteristics. Sea-wave loading is very different from earthquake or machine vibration loading both in terms of frequency (around 0.1 Hz) and duration. Strain levels in the soil around the foundation are in the

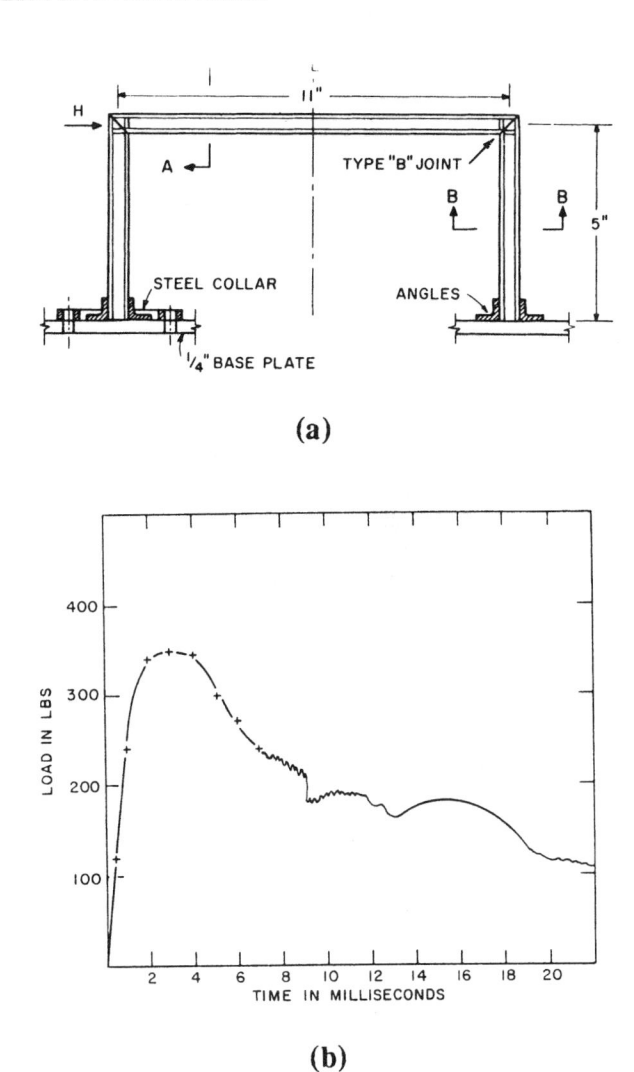

STRUCTURAL MODELS FOR DYNAMIC LOADS 643

Figure 11.73 One-fifteenth scale model steel frame impulsive test. (a) Dimensions. (b) Load pulse experienced by frame. (c) Model after the dynamic test.

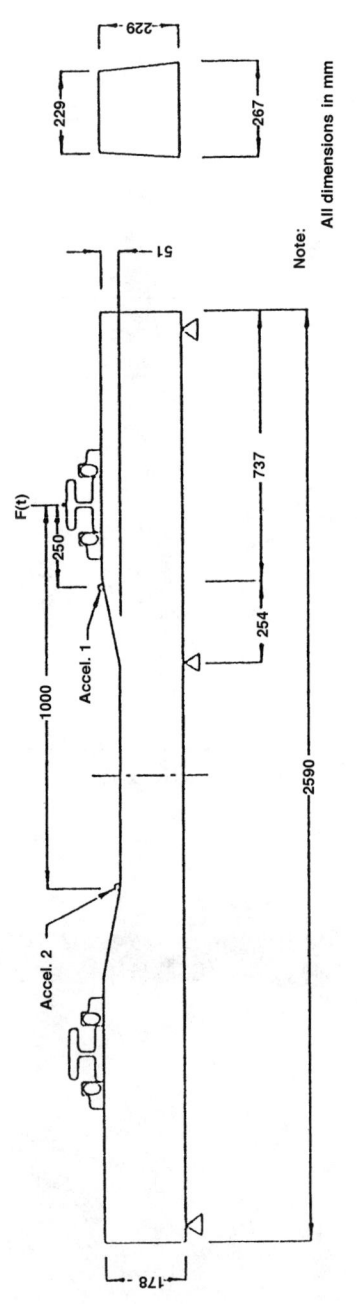

Figure 11.74 Test configuration.

STRUCTURAL MODELS FOR DYNAMIC LOADS

Table 11.7 Test Cases

Case	Drop Height, mm	Impact On	Supports	Actuator
1	150	Steel rail	Steel	Off
2	150	Steel rail	Rubber on steel	Off
3	150	Rubber pad over steel rail	Rubber on steel	Off
4	150	Steel rail	Rubber on steel	On
5	300	Steel rail	Steel	Off
6	600	Steel rail	Steel	Off

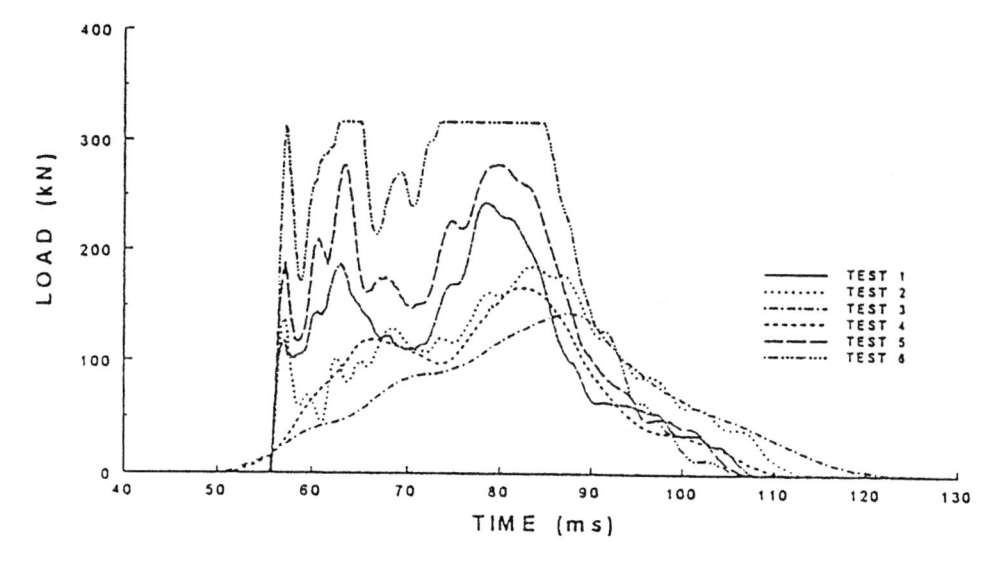

Figure 11.75 Load–time histories.

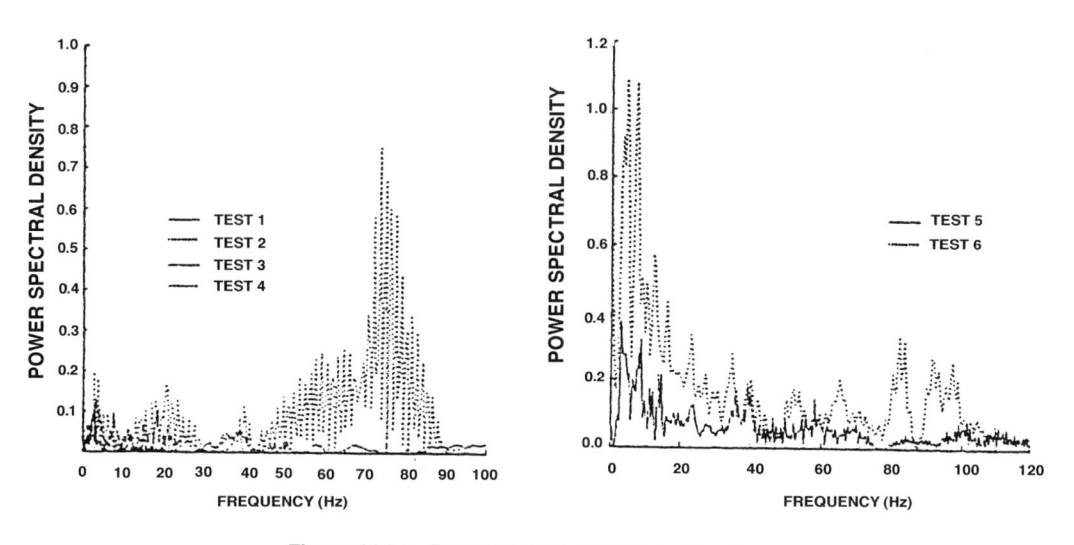

Figure 11.76 Power spectral density vs. frequency.

range of 0.1 to 5%. The reported study was undertaken to provide experimental data on soil–structure interaction in order to better understand the behavior of such undersea structures and to compare and calibrate numerical simulations. The tests were carried out on the centrifuge of CESTA (Centre d'Etude Scientifique et Technique d'Aquitaine, Le Barp, France).

Tall slender structures such as the self-elevating drilling platforms show vibration modes that can reach high energy wave periods of 8 to 10 s. Analysis of the structural response in such waves leads to large dynamic displacements incompatible with the structural integrity of the system. Therefore, in order to assess the self-elevating concept for future development, it is necessary to investigate the dynamic behavior of the structure submitted to cyclic loading in the vicinity of its natural period, and pay particular attention to the foundation response both in terms of stiffness and plastic dissipation near resonance.

11.5.8.2 Model Structure and Soil Foundation

Similitude was maintained using the scale factors given in Table 2.12. The $1/100$-scale model chosen for the study was one third of the drilling platform actual structure. The effect of the other two missing legs was simulated by adding horizontal stiffness at the deck level. The model dimensions are shown in Figure 11.77. The special footing of this type of platform was tested in two configurations: with and without embedment. The model was shaken at the deck level (+0.50 m) by a horizontal force supplied by an electromagnetic shaker. The connecting rod allowed for vertical displacement due to soil settlement. The $1/100$-scale model and the shaker mounted on the swinging basket of the centrifuge are shown in Figure 11.78.

The soil foundation consisted of Fontainebleau sand which has been extensively tested and used in centrifuge tests (Luong, 1980; 1986). The characteristics of the sand were a grain size of less than 0.5 mm, fine sand ($D_{50} = 0.2$ mm), and a dry bulk density $\gamma_d = 16$ kN/m^3, i.e., a relative density $D_r = 65\%$. The sand was placed in a tank $1.0 \times 0.8 \times 0.4$ m in which saturation was achieved by water circulation. The weight of the saturated sand pack reached 6 kN.

11.5.8.3 Experimental Results

The centrifuge process was carried out in steps at 25, 50, 75, and 90g. This enabled accurate measurement of the sand pack settlement under its own weight and determination of the relative settlement of the foundation (Figure 11.79). The deformed shape of the model at resonance obtained from integration of measured accelerations is shown in Figure 11.80. This shape of the first mode is in agreement with expected model behavior. The instrumentation provided continuous information about the loads applied to the foundation and its displacements and rotations. Figure 11.81 shows the cyclic settlement vs. frequency curve. An increase in the settlement can be seen during resonance, particularly in the nonembedded case. The final cyclic settlement remains limited however (7 cm with no embedment and 3.5 cm with a 3-cm embedment). The rocking stiffness determined as the ratio between the amplitude of the cyclic moment applied to the foundation and the amplitude of the rotation of the foundation is shown in Figure 11.82. The stiffness is nearly linear in the embedded case, with a mean value of approximately 370 Nm/degree (270 MNm/degree at full scale), while some nonlinearity occurs in the nonembedded case. The evolution of the energy dissipated inside hysteresis loops (Figure 11.83) is very similar in both cases and nearly linear. The smaller rotation of the foundation in the embedded case may explain the lower dissipation. As seen in Figure 11.83, the growth of the dissipation with the amplitude of the moment is not quadratic (dotted line) as would be expected in a linear viscoelastic model.

The experimental program has shown that, as long as the structural model is simple enough, it was possible to use the centrifuge to provide calibration information about the foundation behavior of an offshore structure submitted to low-frequency (sea-wave loading) near resonance. Information

STRUCTURAL MODELS FOR DYNAMIC LOADS 647

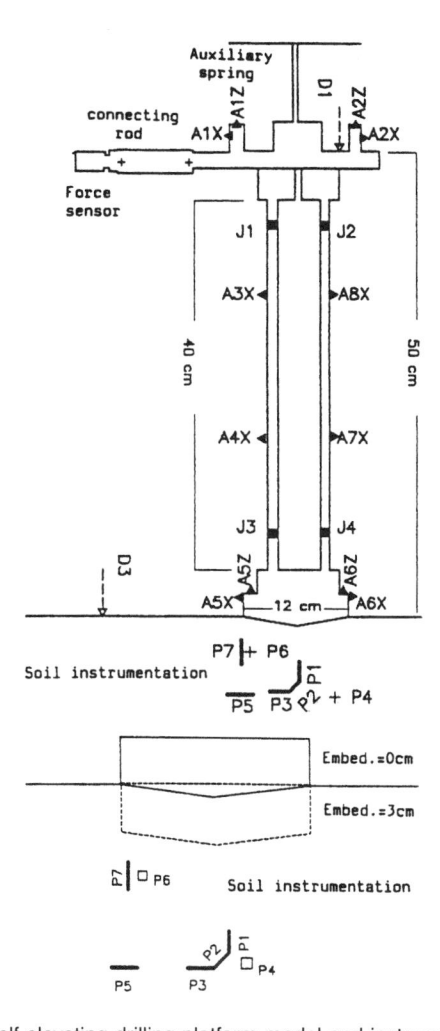

Figure 11.77 Self-elevating drilling platform model and instrumentation position.

Figure 11.78 Shaking machine, sand tank, and model mounted on the swinging basket of the centrifuge.

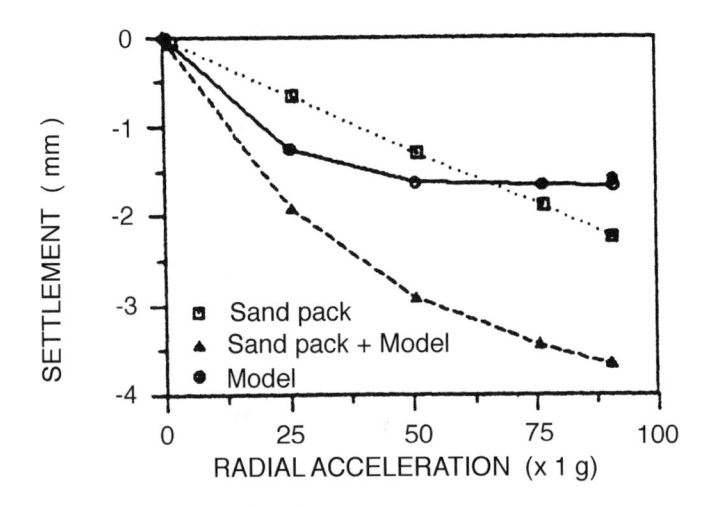

Figure 11.79 Settlement curves from 1*g* to 90*g*, before harmonic loading (embedment = 0 cm).

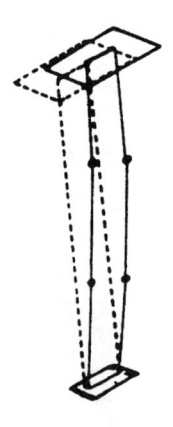

Figure 11.80 The deformed shape of the $1/100$-scale model at resonance.

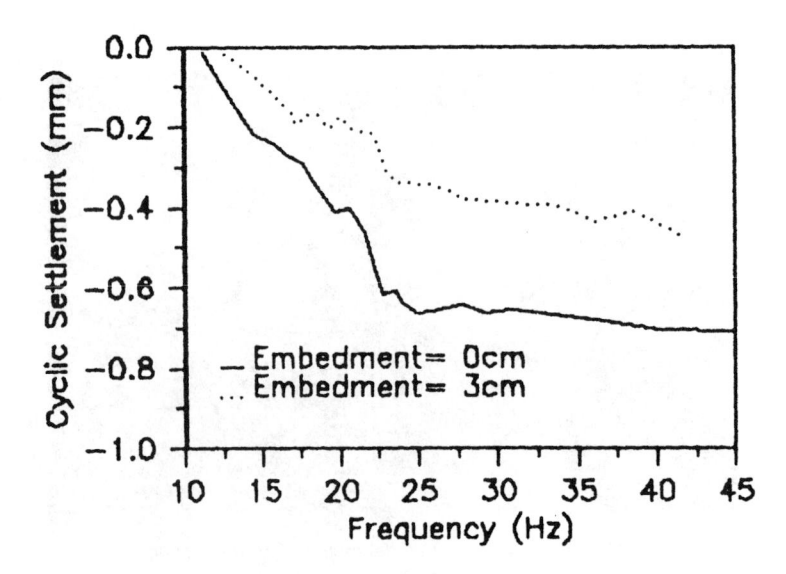

Figure 11.81 Foundation settlement during cyclic loading.

STRUCTURAL MODELS FOR DYNAMIC LOADS

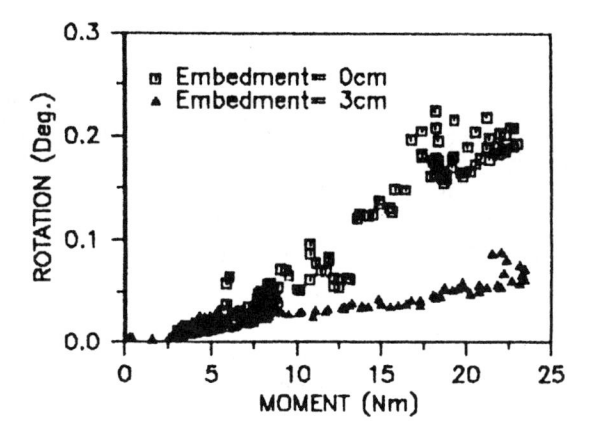

Figure 11.82 Evolution of the cyclic amplitude of foundation rotation vs. the amplitude of the moment applied to the foundation.

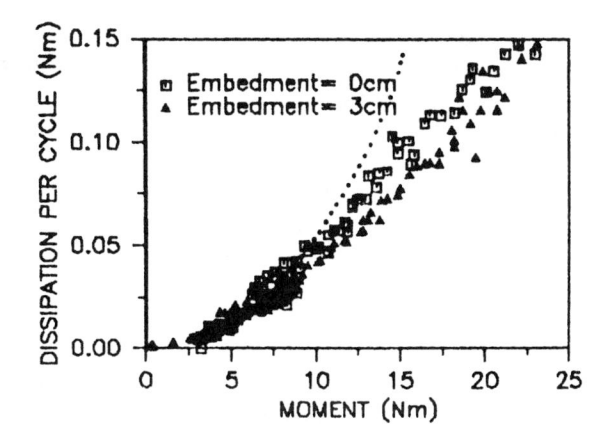

Figure 11.83 Evolution of the cyclic energy dissipation in the rocking movement of the foundation vs. amplitude of the moment applied to the foundation.

obtained concerning the foundation behavior (settlement, rotation, stiffness, energy dissipation, and soil pressures) is useful to compare and calibrate various methods of estimation of the mechanical characteristics of circular shallow foundation subjected to low-frequency high-amplitude sea-wave loading.

It was shown that structural resonance may induce some nonlinearities in the foundation response, both in terms of stiffness and irreversible settlement. These nonlinearities still remain limited, especially in the embedded case.

11.6 CASE STUDIES

11.6.1 Wind Tunnel Tests of the Toronto City Hall

11.6.1.1 The Problem

Aeroelastic response of the two crescent-shaped towers, one approximately 88.4 and the other 68.6 m high, of the new Toronto City Hall depends on the prevailing wind direction, and it may exhibit the flow characteristics of a diffuser, a nozzle, a semicylinder, or a complete cylinder. The pressure on the shell surface can vary considerably with the wind direction and can be steady or unsteady (that is, oscillating) for some specific wind directions. The unusual shape of the structure renders it weak against torsional loads. This along with the tower height led the engineers to suspect that the standard wind design pressures specified by the City of Toronto Building Code would not be applicable, and it was decided to conduct wind tunnel tests on a small-scale model of the structure at the Institute of Aerophysics of the University of Toronto.

Although the scale of the models was much smaller than desirable (no larger wind tunnels were immediately available), it was felt that despite the uncertainty associated with the scale effect it was preferable to base the design on these results rather than on the building code. The nature of the pressure distributions obtained gave the engineers sufficient confidence to make some changes in the design of the shell-like structure. To verify the aerodynamic characteristics on which to base the structural design, a second model, corresponding to the revised design, was also built and tested.

11.6.1.2 Test Program

Both models were tested in the 914 × 1067 mm test specimen of the University of Toronto subsonic wind tunnel. The original $1/276$-scale model, built using mahogany, was mounted on a circular plywood base plate, beveled and graduated at the outer edge (Figure 11.84). Three horizontal rows of 20 static pressure taps each, spaced evenly along the arc of the tower, were installed on both the inside and outside walls of the two towers. A plastic strip, 25 mm wide, consisting of 20 pressure tubes was mounted horizontally and flush with the wall at the desired level for the full width of the tower. Each tap was formed by drilling a hole from outside. The upstream ends of pressure holes were plugged with pins, while the downstream ends were connected by a strip of tubes to a multiple manometer board. The second model ($1/296$-scale) was equipped with only two pressure tap levels since tests on the original model showed little variation of pressure with height.

11.6.1.3 Testing Techniques

1. Steady Pressure Measurements — Although the maximum speed possible in the tunnel was in excess of 61 m/s, the very large drag of the model resulted in a maximum wind speed of 53.3 m/s. The objective was to obtain as high a Reynolds number as possible. A typical test consisted of rotating the model slowly in the airstream through a predetermined 60° sector while observing the pressure readings. A similar procedure was repeated for the remaining five 60° sectors. The wind direction that caused maximum suction on the outside of the tower was chosen for each 60° sector. This direction was maintained constant for all pressure readings on a given tower within each 60° sector. For a number of wind directions, pressure readings were taken in the two corners of each tower at a height of 38 mm from the base.

2. Unsteady Pressure Measurements — Unsteady pressures resulting from periodic vortex shedding from the building were measured at four pressure taps on the outside wall located 100 mm below the top edge. Vertical holes 100 mm deep and 15.9 mm in diameter were drilled into the tower at required positions. Each hole was equipped with a microphone registering pressure through a small (horizontal) drill hole. To ensure undisturbed airflow over the wall, the outside opening of the drill hole was closed with a tape that was subsequently punctured with a pin. The microphone was connected to a pen recorder, a sound-level meter, and an oscilloscope for immediate visual observation.

Seven pen recordings were made for two stations for two wind directions with the wind speed varying from 22.9 to 48.8 m/s. Sound-level readings in decibels were simultaneously taken on the sound-level meter. The vortex shedding frequency was calculated from the number of times the pen recording trace crossed the referenced line divided by twice the elapsed time.

STRUCTURAL MODELS FOR DYNAMIC LOADS 651

Figure 11.84 Solid mahogany wind model at 1/276 scale of the Toronto City Hall in the subsonic wind tunnel. (After Dau, 1961. Courtesy of the University of Toronto Institute for Aerospace Studies, Downsview, Ontario, Canada.)

3. Flow Visualization Tests — Flow visualization tests were conducted in a small smoke tunnel with a 25.4 × 305 mm test section. This precluded testing of a model properly scaled for height; instead a third 1/960-scale, two-dimensional model was used. The test consisted of taking still photographs of the streamline pattern around the model at 15° intervals in the wind direction, the wind speed being about 1.8 m/s (Figure 11.85). In addition, movie pictures were taken of the streamline patterns, while the model was rotated through 360° to observe the unsteady flow associated with vortex shedding at certain wind directions.

4. Conclusions — This investigation shows that the steady pressure can be predicted with reasonable accuracy from wind tunnel model studies, since viscous effects are negligible except in the regions of separated flow. In such cases, the pressure coefficients for the prototype are likely to be higher than those for the model.

Dau (1961) observed that the steady pressure distribution on the outside (convex) wall resembled that on a circular cylinder, with some suction peaks attaining a value of over twice the wind dynamic pressure. The resulting pressure distribution on the outer face causes torsional loads on the tower, while the pressure distribution on the inside (concave) contributes mainly to bending loads.

The wind tunnel test results were converted into design pressures using an assumed wind velocity distribution varying from 180 km/h at the tower top to 96.6 km/h at the bottom. The external design wind pressures were as high as 1.5 kN/m^2, and suctions as high as 3.5 kN/m^2. These high values, along with the unusual pressure distributions found from these tests, produced torsional and bending loads far in excess of those expected from standard design assumptions, thus proving the important value of the model study.

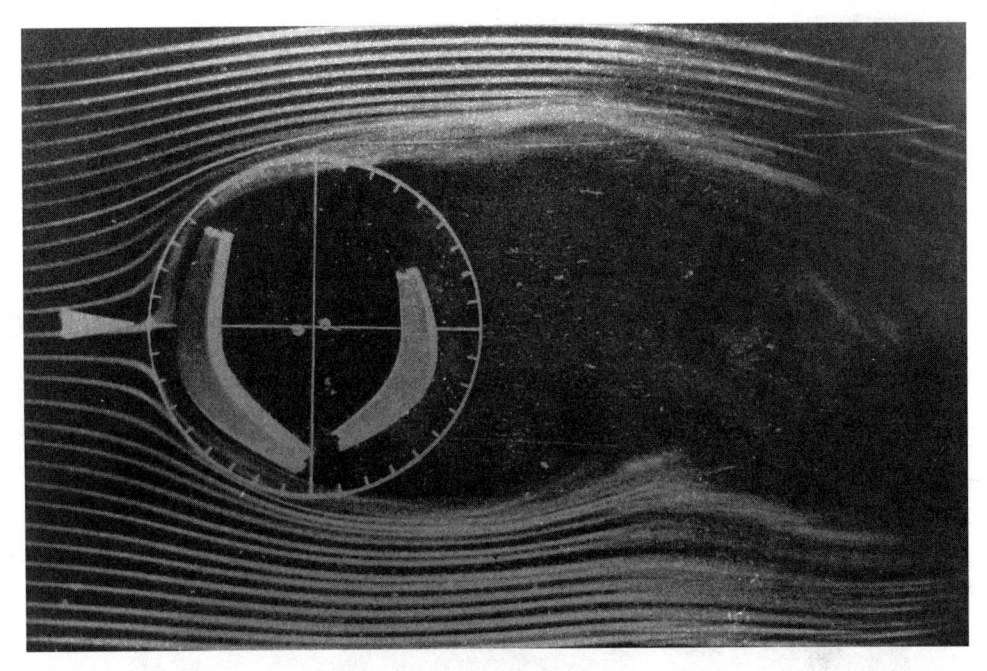

Figure 11.85 Unsteady flow around a distorted two-dimensional 1/960-scale model showing vortex shedding. (After Dau, 1961. Courtesy of the University of Toronto Institute for Aerospace Studies, Downsview, Ontario, Canada.)

Because of the large difference in the model and prototype scale, it was not possible to formulate any conclusions regarding the amplitude and frequency of unsteady pressures caused by vortex shedding on the prototype structure except that vortex shedding may occur for two wind directions.

11.6.2 Case Study B, Shaking Table Tests on R/C Frame-Wall Structures

11.6.2.1 Initial Mechanical Characteristics

The 1/5-scale model study of a seven-story frame-wall structure conducted on the University of California, Berkeley shaking table represents one of the largest physical model undertakings both in scope of work and also in physical size. The flexibility characteristics of the model (Figure 1.7) prior to and after the ballast had been applied are illustrated in Figure 11.86. As can be seen, the lateral stiffness of the structure increased significantly when the ballast load was added. Since the lead ballast gave rise to an increase in gravity load of more than 400%, the compressive axial force in the columns and particularly in the walls also increased significantly. This resulted in an increase in the average stiffness of the structure of approximately 40%. The lateral flexibility characteristics of the 1/5-scale model approached those of the full-scale structure as shown in Figure 11.86. The analytical flexibility characteristics were reasonably close to measured values, after similitude in the gravity stress level was satisfied.

11.6.2.2 Earthquake Simulator Test Program

The main earthquake simulation tests on the 1/5-scale model were conducted using as input to the shaking table the displacement time histories corresponding to two acceleration records designated as Miyaki-Oki (MO) and Taft (T). These records (normalized to a peak acceleration of $1.0g$ and with adjusted time scales, i.e., compressed by $\sqrt{5}$) are modified versions of recorded ground motions and are shown in Figure 11.87. The displacement time histories obtained by integration

STRUCTURAL MODELS FOR DYNAMIC LOADS 653

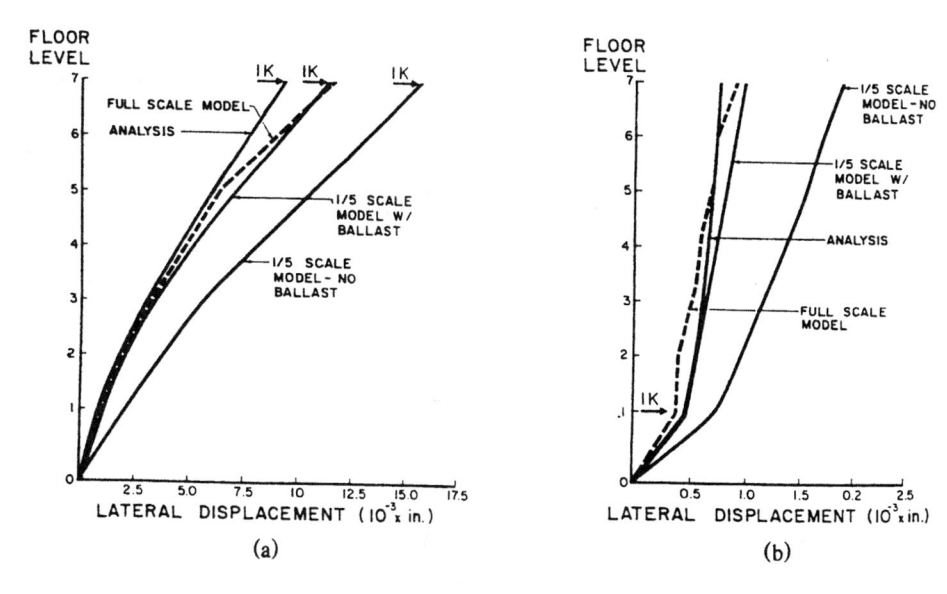

Figure 11.86 Displacement profiles for ⅕-scale model and full-scale structure (1 in. = 25.4 mm). (a) Loaded at seventh floor level. (b) Loaded at first floor level. (After Bertero, V. V. et al., 1984.)

of these modified acceleration records were used as input to the University of California, Berkeley earthquake simulator after scaling the time by a factor of $1/\sqrt{5}$ (Table 2.11). The Fourier amplitude spectra of these two records, shown in Figure 11.88, indicate that the T input had a broader frequency content (wider frequency range) with damage potential, while the MO input was considerably less intense over the complete range of frequency. However, the MO record appears to possess damage potential concentrated at certain frequencies in the range 1.5 to 3.0 Hz.

Table 11.8 lists all the dynamic tests to which the ⅕-scale model was subjected. It lists all the various levels of ground motion as well as the measured frequency and damping characteristic tests conducted. The earthquake simulator tests were classified into three series. The first series was intended to be diagnostic, i.e., low amplitude tests, conducted to check the operation of the earthquake simulator, data acquisition system, and instrumentation and to generate the initial uncracked serviceability limit state responses. Inadvertently, however, some cracking was noted in test 33. The second series, which consisted of base motions of increasing intensity, was designed to induce successive stages of damageability and collapse limit state responses, resulting in a complete flexural failure at the base of the main wall and extensive yielding throughout all frame elements.

Before the third series, the model was repaired and retrofit by strengthening and stiffening the bottom 165 mm of the wall. The model was then subjected to a series of particularly intense base motions to study the effectiveness of the repair as well as the collapse limit state response characteristics. As indicated in Table 11.8, the model was subjected to a total of 62 tests. Some were harmonic motion or free vibration tests conducted to determine changes in the frequency and damping characteristics of the model.

11.6.2.3 Test Results and Discussion

Only typical examples of the voluminous test results are discussed here and further reference should be made to Bertero et al. (1984). Time histories of base shear of the model during the MO 9.7, MO 24.7, MO 28.3, and T 40.3 tests are shown in Figure 11.89. These curves were evaluated from the measured translational accelerations and the lumped masses at each floor level. Since approximately 80% of the mass resulted from the ballast load, concentrated at each floor level of

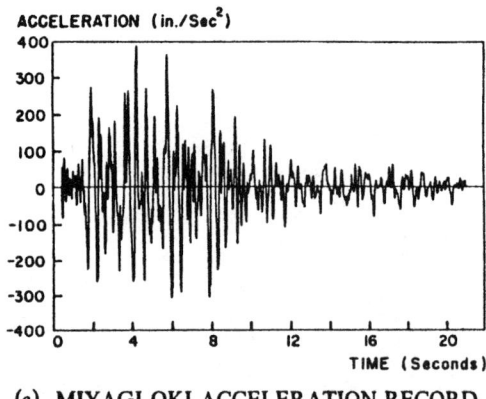

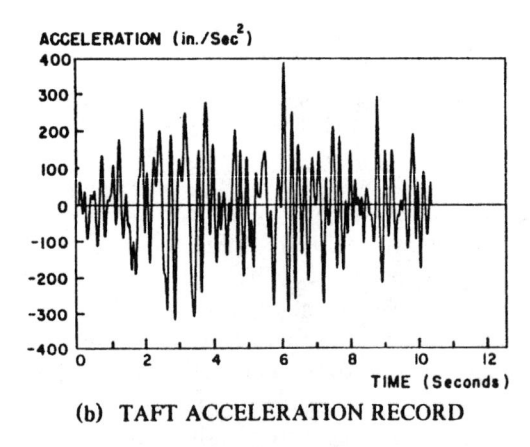

Figure 11.87 Acceleration records, normalized to peak acceleration of 1.0 g, used as source excitations in the test program (1 in. = 25.4 mm). (After Bertero, V. V., et al., 1984.)

the structure, the lumped mass idealization was considered justified. The displacement, shear, and overturning moment response was found to be in phase during each of the four excitations listed above, i.e., the zero crossings and peaks in displacements, shear, and moment responses occur nearly simultaneously.

The maximum base shear and overturning moment vs. the maximum interstory drift envelopes were determined for Tests 7 and 9 from Series 1 (serviceability limit state response), Tests 45, 46, 48, and 50 from Series 2 (successive damageability and collapse limit state responses), and Test 62 (collapse limit state) and are shown in Figure 11.90. These selected tests are the ground motion records MO 5.0, MO 9.7, MO 14.7, MO 28.3, T 40.3, and T 46.3, respectively (Table 11.8). The maximum base shear and overturning moment vs. the maximum interstory drift envelopes in Figure 11.90 indicate substantial changes in the stiffness characteristics after the MO 9.7, MO 24.7, MO 28.3, and T 40.3 tests. The effects of these tests on the structure, observed in the envelope of Figure 11.90, are reminiscent of the cracking, yielding, deformation hardening, and ultimate capacity of a reinforced concrete flexural element or of a basic flexural subassemblage.

The model structure was observed to have a high rate of deformation hardening during testing. A correct identification of the resistance mechanisms leading to this observed deformation hardening was considered to be extremely important. The outrigging provided to the wall by the frames oriented within and transverse to the plane of the wall was identified as a major cause of

STRUCTURAL MODELS FOR DYNAMIC LOADS 655

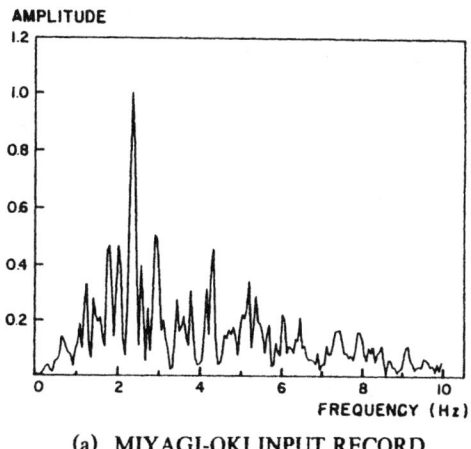

(a) MIYAGI-OKI INPUT RECORD

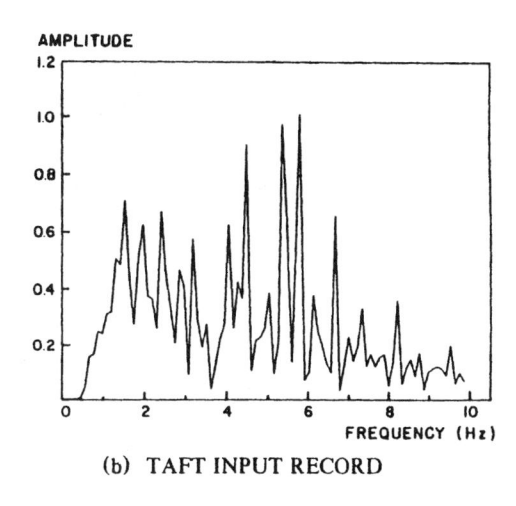

(b) TAFT INPUT RECORD

Figure 11.88 Fourier amplitude spectra of acceleration records. (After Bertero, V. V., et al., 1984.)

the deformation hardening. The wall tended to rotate with respect to its base as an almost rigid body after a plastic hinge had formed at its base, i.e., after all main reinforcement of the edge member had yielded. The outriggering action of the frames on the wall (illustrated in Figures 11.91a through c) and the deformation hardening of the wall at its base would be expected to have restrained the rigid body rotations of the wall. The axial compressive force at the base of the wall was measured to have increased from approximately 12 to 28% of its balanced axial force level (Figure 11.91d). The flexural stiffness of the complete diaphragm system contributed significantly to the stiffness of the outriggering system. The outriggering system was particularly beneficial during the collapse limit state responses to the third series of tests, after the complete flexural failure of the wall at its base. The restraint and axial compression provided by the outriggering system were so effective that the shear resistance of the wall remained practically constant even after the wall had failed in flexure.

11.6.2.4 Correlation of ⅕-Scale Model and Full-Scale Model

1. Hysteretic Behavior — The full-scale model and the ⅕-scale model exhibited excellent hysteretic behavior. Although the individual hysteresis loops do not match well since the models were

Table 11.8 Excitation Program for ⅕-Scale Model

| Initial Dynamic Characteristics ||| f = 4.75 Hz, ξ^a = 2.0% |
Series	Test No.	Input Signal	Peak Amplitude, %g
Diagnostic Tests	1–6	MO[b]	0.3–2.6
	7	MO	5.0[c]
	8	MO	7.6
	9	MO	9.7[c]
	10–13	T[d]	4.0–6.3
	14, 15	H (4Hz)[e]	1.5–2.0
	16, 31	H (3.25–4 Hz)	6.7
	32	MO	3.6
	33	MO	9.0
	34, 35[f]	FV[g] (22.24 kN pull)	f = 3.67 Hz, ξ = 3.5%
Damageability and collapse limit state responses	36–42	FV (22.4 kN pull)	f = 3.41 Hz, ξ = 3.7%
	43	MO	8.2
	44	MO	10.8
	45	MO	14.7[c]
	46	MO	24.7[c]
	47	FV (pulse)	f = 2.63 Hz, ξ = 6.87%
	48	MO	28.3[c]
	49	FV (pulse)	f = 2.50 Hz, ξ = 7.50%
	50	T	40.3[c]
	51, 52[h]	FV (pulse)	f = 2.33 Hz, ξ = 7.70%
Post-retrofitting responses	53–55	FV (pulse)	f = 2.33 Hz, ξ = 6.40%
	56	MO	8.6
	57	T	30.3
	58	FV (pulse)	f = 1.96 Hz, ξ = 6.20%
	59	T	48.4
	60	MO	32.9
	61	FV (pulse)	f = 1.96 Hz, ξ = 8.30%
	62	T	46.3[c]

[a] ξ = damping ratio.
[b] Miyagi–Oki record.
[c] The responses from these tests are evaluated in this paper.
[d] Taft record.
[e] Harmonic vibration.
[f] After this test the model was removed from the table and repaired by epoxy injection.
[g] Free vibration.
[h] The main wall was retrofit at the base after this test, with the structure remaining on the table.

subjected to different excitation histories, the shape of these loops (particularly for the most severe excitations) are quite similar (Figures 11.92 and 11.93). Although both structures show some pinching in their hysteretic behavior, the amount was significantly smaller than that which is observed from the experimental behavior of isolated shear walls. The main reason for this better behavior appears to be the three-dimensional interaction of space frame and wall, particularly the outriggering action of the surrounding ductile space frame on the wall after flexural yielding and even after the failure of its reinforcement. The energy dissipated by the ⅕-scale model considerably surpassed that of the full-scale model, again primarily due to the method of testing: dynamic vs. pseudodynamic (rate of straining) and particularly due to loading history.

2. Overall Lateral Load–Deformation — The correlation at the serviceability limit states was good (Figure 11.94). After roof drift indexes that induced yielding, the ⅕-scale model exhibited a relatively larger lateral strength (resistance) for similar lateral displacements. The maximum base shear for the ⅕-scale model (51%W) was 40% greater than that of the full-scale model (36.5%W) when loaded with an inverted triangular distribution of lateral force. This significant difference was

STRUCTURAL MODELS FOR DYNAMIC LOADS

(a) MIYAGI-OKI 9.7

(b) MIYAGI-OKI 24.7

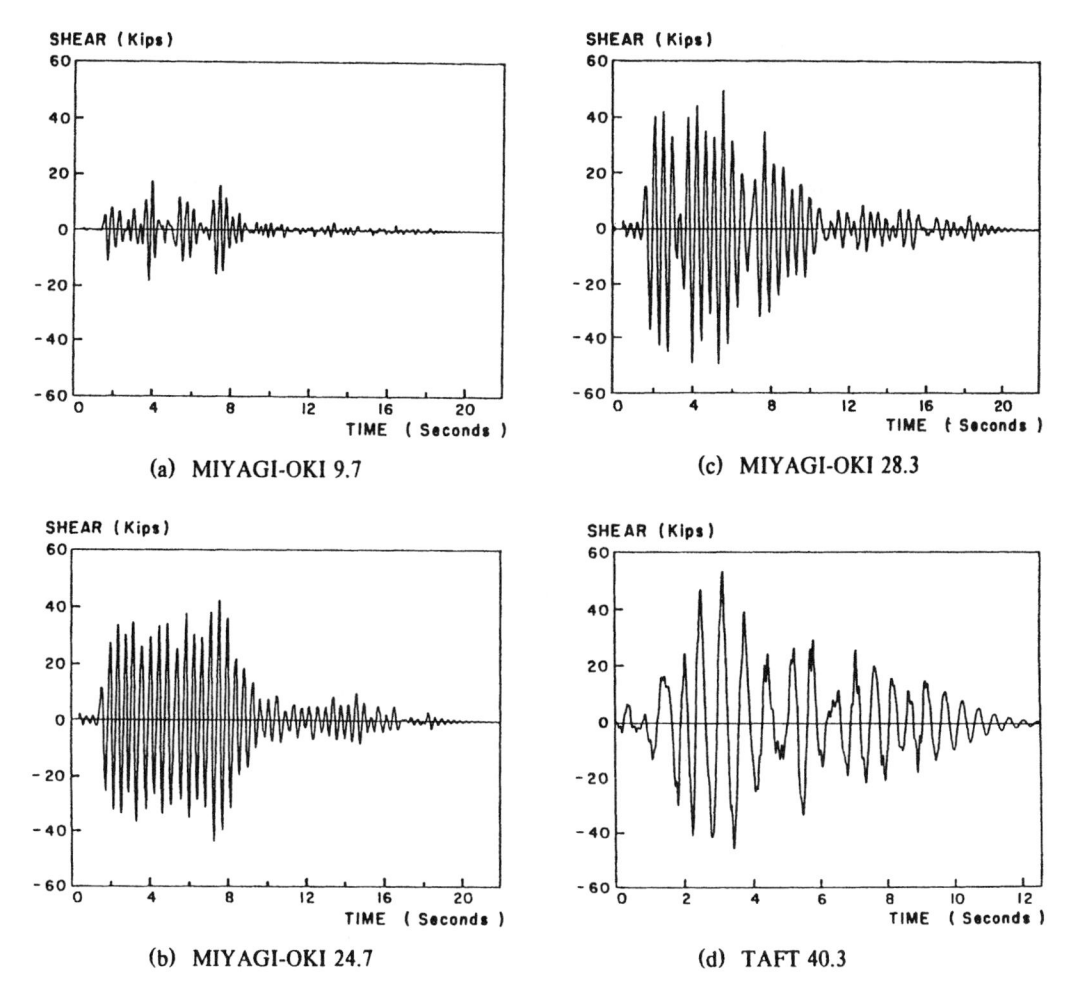

(c) MIYAGI-OKI 28.3

(d) TAFT 40.3

Figure 11.89 Base shear time histories of ⅕-scale model (1 kip = 4.45 kN). (After Bertero, V. V., et al., 1984.)

due to the methods of testing used, particularly the different distributions of total lateral force along the height of the structure. A somewhat better correlation is obtained when the envelope of overturning moment to roof drift is compared. When the full-scale structure was loaded under a uniform distribution of lateral load, its maximum strength was practically the same as the scaled maximum strength obtained in tests of the ⅕-scale model. The roof drift index at which the ⅕-scale and full-scale models reached their maximum lateral strength correlate very well: 1.4% for the ⅕-scale model and 1.5% for the full-scale model (Figure 11.94) for triangular loading and 1.35% in the case of uniform lateral loading.

3. Crack Pattern — The overall crack pattern at the critical sections was similar, but the number and spacing of cracks differed (Figure 11.95). The total number of cracks was smaller in the ⅕-scale model. Reasons for this lack of correlation are (1) the higher tensile strength of the microconcrete; (2) the considerably higher strain rate induced in the ⅕-scale model; (3) the higher gradient along the length of the critical regions of the members and through the critical sections of the members of the ⅕-scale model. The main cracks of the floor slabs along the edges of the main beams were similar. The cracks in the slabs of the full-scale model were, however, considerably larger in number due to the manner in which force was applied.

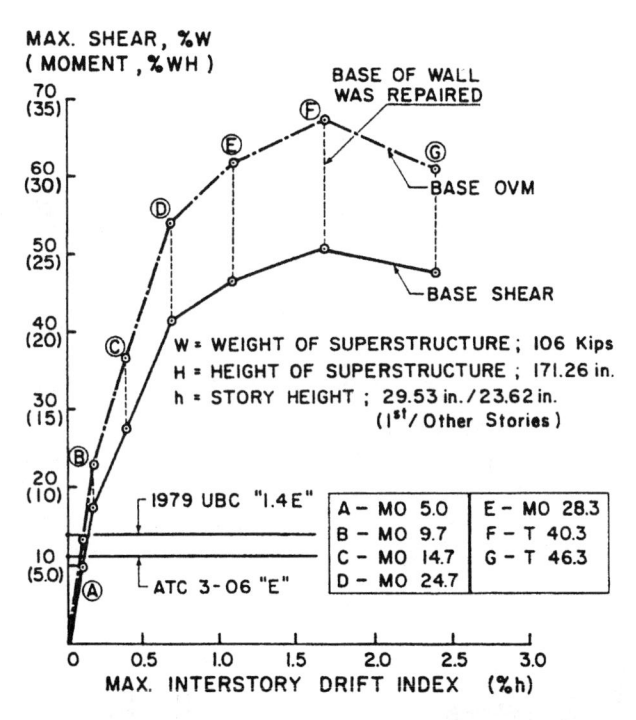

Figure 11.90 Maximum base shear and overturning moment–maximum interstory drift index envelopes from tests of ⅕-scale model (1 in. = 25.4 mm; 1 kip = 4.45 kN). (After Bertero, V. V., et al., 1984.)

4. Failure Mechanism — Failure of the full-scale model was due to crushing and spalling of the concrete at the wall edge member and particularly crushing in the wall panel which led to a final sliding shear failure of the wall at the first story with buckling and fracture of some of the reinforcement (Figure 11.96c). The ⅕-scale model failed due to buckling and/or due to tensile fracture of the wall main reinforcement after crushing and spalling of the concrete cover of the wall edge members at the base of the wall. These differences in failure mechanism are attributable to differences in crack pattern and to the fact that at failure the nominal shear stresses were practically the same in the full-scale and ⅕-scale model, while actual compressive strength and therefore shear resistance of the microconcrete was significantly higher in the ⅕-scale model. Furthermore, a close examination of the state of damage in the ⅕-scale model wall panel after failure revealed that there was some slight crushing and spalling as illustrated in Figure 11.96a, indicating that it was very close to a shear-compression type of wall panel failure such as that observed in the full-scale model. It can be concluded that the failure mechanism observed in the ⅕-scale model was very close to the margin separating the observed flexural failure from the shear-compression wall panel failure observed in the full-scale model under a uniform distribution of seismic force.

5. Conclusions — Despite significant differences in the methods of testing, time history of the applied excitations, and mechanical characteristics of the concrete materials, the correlation of behavior at the serviceability limit states and of the maximum lateral shear resistance and roof drift index was excellent. The results obtained in the tests conducted on the shaking table of the ⅕-scale model are of great importance especially since these results provide a better idea of the dynamic response of the structure to earthquake ground motion than the pseudodynamic tests as conducted on the full-scale structure. The advantages of testing reduced scale models on earthquake simulators over pseudodynamic testing of the full-scale model can be summarized as follows:

STRUCTURAL MODELS FOR DYNAMIC LOADS 659

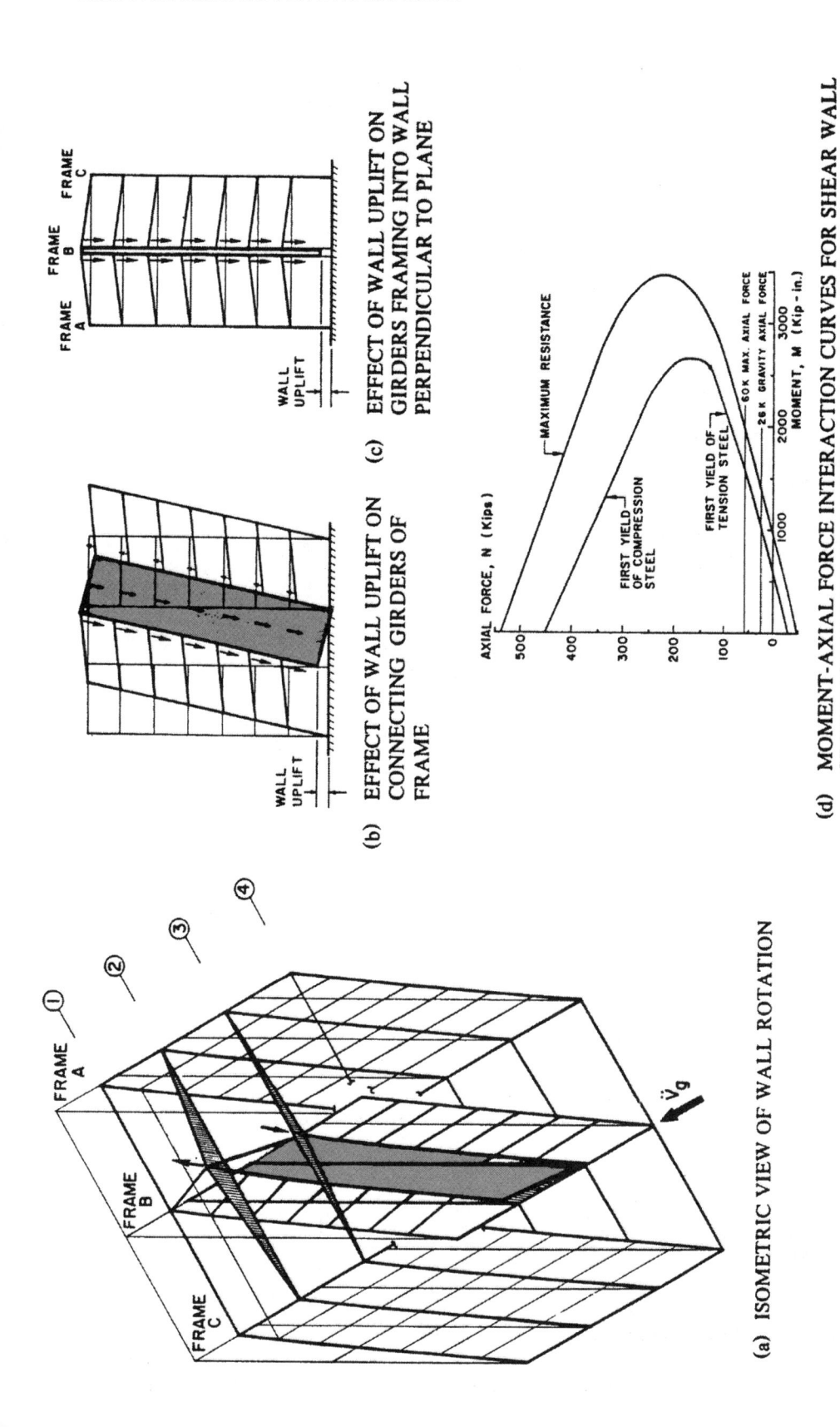

Figure 11.91 Effect of outriggering action of frames on wall on moment–axial force interaction of the wall section at its base (1 kip-in. = 0.113 kNm; 1 kip = 4.45 kN). (After Bertero, V. V., et al., 1984.)

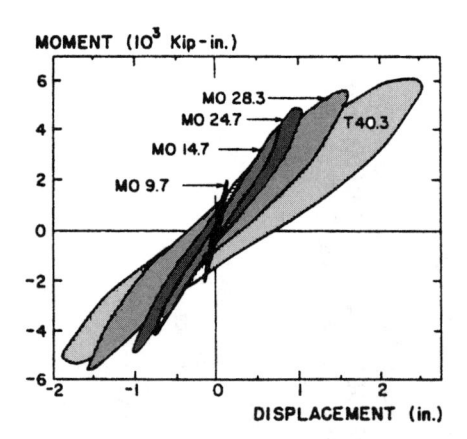

Figure 11.92 Base overturning moment vs. top-floor relative displacement hysteresis for critical durations of MO 9.7, MO 24.7, MO 28.3, and T 40.3 responses of ⅕-scale model. (After Bertero, V. V., et al., 1984.)

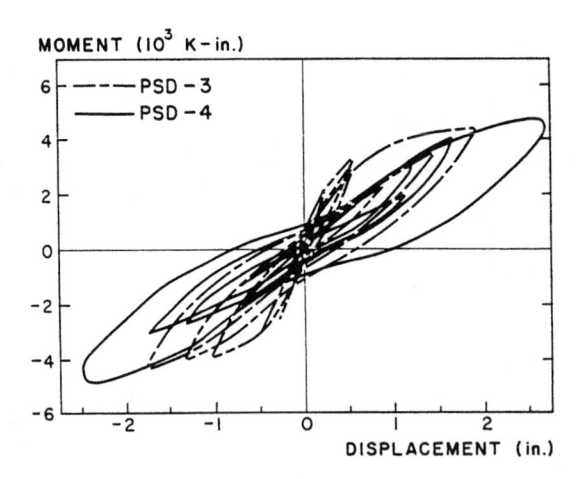

Figure 11.93 Base overturning moment–roof displacement hysteretic response of full-scale model converted to ⅕ scale. (After Bertero, V. V., et al., 1984.)

1. Reduced scale testing is more economical.
2. The structure is subjected to more realistic simulation of earthquake excitation and responses are therefore more realistic.

From a comparison of results from tests on the ⅕-scale model with those from tests on the full-scale model, and from analytical studies, it can be concluded that the shaking table tests of the ⅕-scale model provided reliable results from which the seismic behavior of the bare building structure could be predicted for the ground excitations used.

11.6.3 Case Study D, Shaking Table Tests on Lightly Reinforced Concrete Buildings

This case study involves the earthquake behavior of reinforced concrete buildings designed to resist purely gravity loads without regard to lateral loads (wind or earthquake forces). Two similar structures of a reinforced concrete building (Figures 1.12 and 1.13) designed using gravity load design (GLD) methods were studied. One, a ⅛-scale model, using small-scale modeling techniques,

STRUCTURAL MODELS FOR DYNAMIC LOADS 661

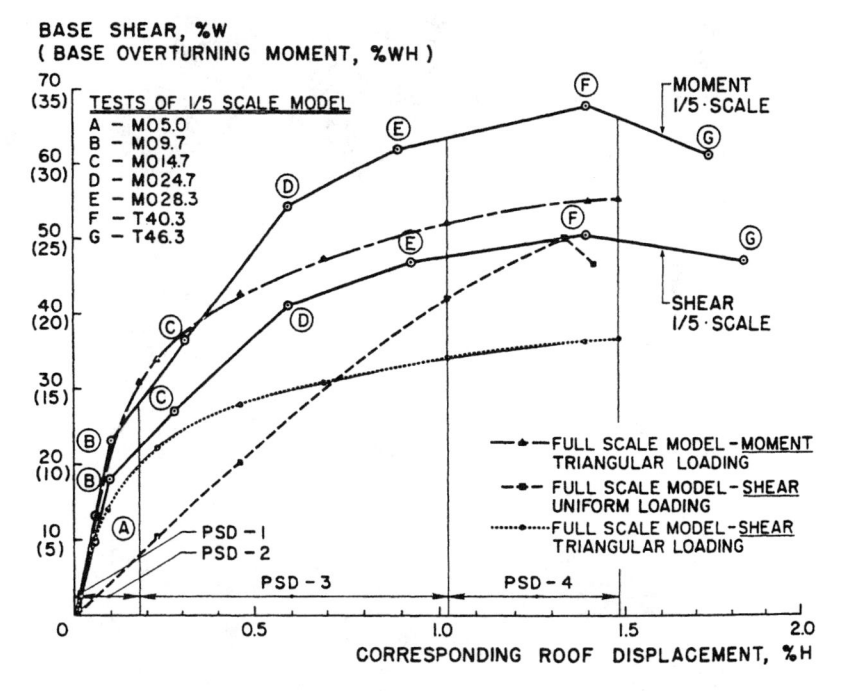

Figure 11.94 Comparison of envelope responses attained for the 1/5-scale and full-scale models. (After Bertero, V. V., et al., 1984.)

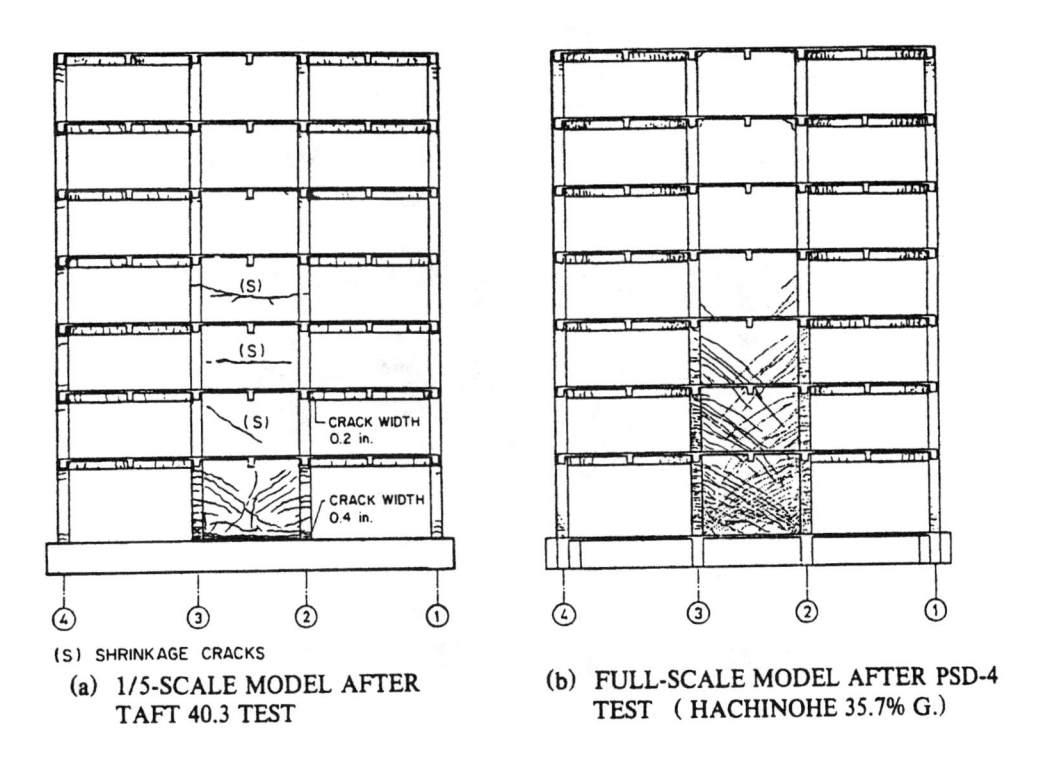

(a) 1/5-SCALE MODEL AFTER TAFT 40.3 TEST

(b) FULL-SCALE MODEL AFTER PSD-4 TEST (HACHINOHE 35.7% G.)

(S) SHRINKAGE CRACKS

Figure 11.95 Crack patterns in 1/5-scale and full-scale models after 1/4% roof drift (1 in. = 25.4 mm). (After Bertero, V. V., et al., 1984.)

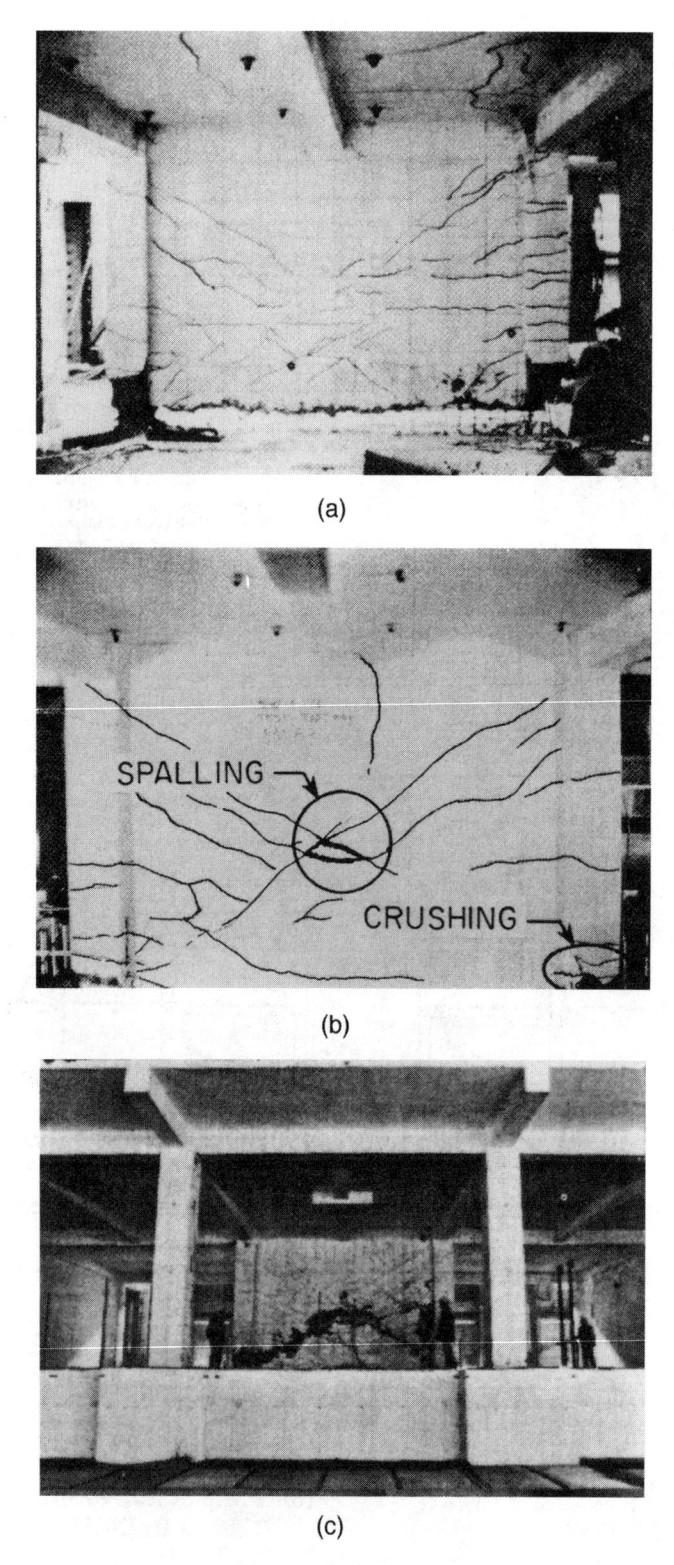

Figure 11.96 Photographs illustrating failure mechanisms. (a) Close-up of wall damage pattern at failure of $\frac{1}{5}$-scale model after removal of damaged concrete. (b) Close-up of wall damage. (Courtesy of Prof. Emin Aktan, Drexel University.) (c) Full-scale model. (After Bertero, V. V., et al., 1984.)

STRUCTURAL MODELS FOR DYNAMIC LOADS 663

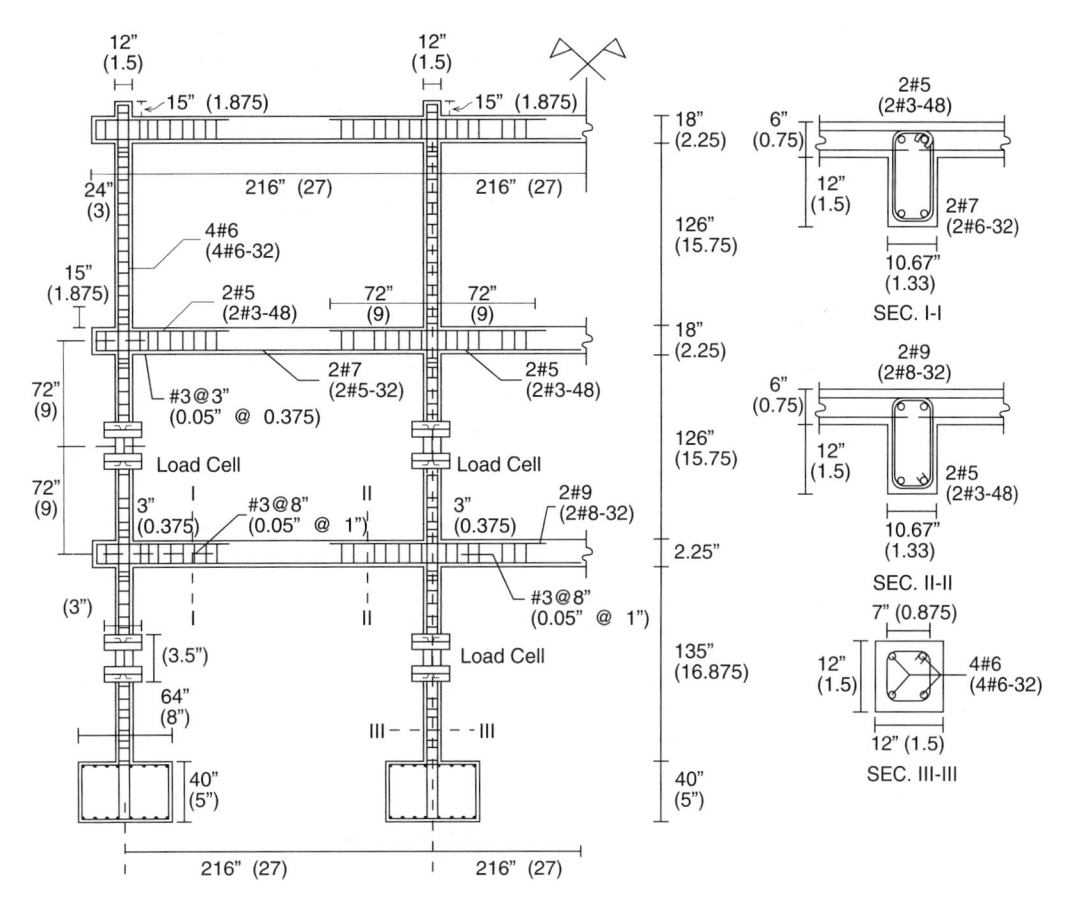

Figure 11.97 Reinforcement details of three-story prototype (model) building. (After El-Attar, A. G. et al., 1997.)

was tested on the Cornell University shaking table (El-Attar et al., 1991; 1997) and the other a larger, ⅓-scale model was tested on the shaking table of the State University of New York (SUNY) at Buffalo (Bracci et al., 1992, 1992a). The main findings of these two studies are summarized here. This research was supported by the National Center for Earthquake Engineering Research, with funding from the National Science Foundation, State of New York, and industrial sponsors.

11.6.3.1 Cornell University ⅛-Scale Model

1. Model Structure — The ⅛-scale model on the Cornell University shaking table is shown in Figure 1.14. The test structure was a three-story, one-bay by three-bay (in the shaking direction) office building designed for dead load plus 2.4 kN/m² live load and built with no in-fill walls. The reinforcement details are shown in Figure 11.97. To determine the most appropriate model reinforcement under earthquake loading, tests were conducted on two prototype and four ⅙-scale cantilever beam specimens (Kim et al., 1988). The best correlation was obtained with threaded steel bars as model reinforcement and these were used in the ⅛-scale model.

Self-weight similitude was achieved by adding lead blocks, with special attention given to the attachment technique to avoid any possible stiffening of the slab and beams by the attached blocks. Internal force transducers (load cells) were placed at midheight of the first- and second-story columns. The load cells measured axial force, bending moment, and shearing force in two perpendicular directions with a maximum interference of 3% (El-Attar et al., 1991). As shown in

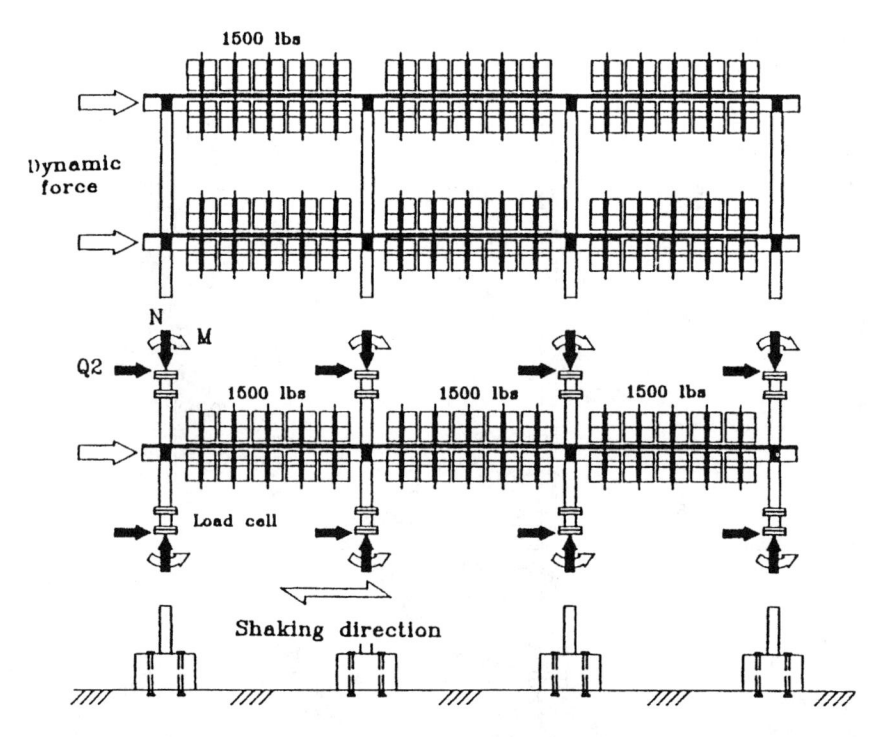

Figure 11.98 Forces acting on the first story during seismic tests. (After El-Attar, A. G. et al., 1991.)

Figure 11.98, the load cells give a direct reading of forces on the first story during the seismic tests, thus adding considerable knowledge of the dynamic behavior of the model.

The ⅛-scale model was tested using the time-compressed Taft 1952 S69E earthquake record with peak ground acceleration set at increasingly higher values. Each seismic test was preceded and followed by a static test and a free vibration test to determine the change in the structural properties (stiffness, natural frequency, and damping ratio).

2. Test Results — Top-story displacements and base shears recorded during test Taft 0.18g are given in Figure 11.99. As can be seen, all three stories are moving in phase, indicating the domination of the first mode. A summary of the seismic test results for the ⅛-scale model is given in Table 11.9 which indicates large stiffness degradation and reduction in fundamental frequency (18% reduction in fundamental frequency corresponding to 50% reduction in stiffness) during the Taft 0.18g test. After the Taft 0.35g run, the model fundamental frequency decreased to 1.65 Hz (25% reduction), indicating a 78% reduction in stiffness. Story shears and mode shapes recorded at the instant of maximum base shear are shown in Figure 11.100 for the three stable runs. It can be seen that in spite of considerable reduction in the model stiffness, the mode shapes remained essentially unchanged for all three complete runs.

Cracking. Cracks were detected at the top and bottom regions of the first- and second-story columns, within a distance equal to the column depth of 38 mm (plastic hinging region), after test Taft 0.18g. These cracks were localized and did not spread over the column height even after the Taft 0.35g run. No damage was observed in beams, joint regions, or splice termination areas, indicating that the nonseismic reinforcement details were not a critical source of damage to this particular GLD structure. Figure 11.101 shows the cracking pattern recorded at the end of test Taft 0.35g.

P-Δ Effect. The large flexibility of the tested frame structure without in-fill walls resulted in a pronounced *P-Δ* effect. At the instant of maximum story drift, during test Taft 0.18g, the sum of the column shears recorded by the internal force transducers installed at their midheights was 27% higher than the base shear computed from the story accelerations. This discrepancy was due to the fact that

STRUCTURAL MODELS FOR DYNAMIC LOADS

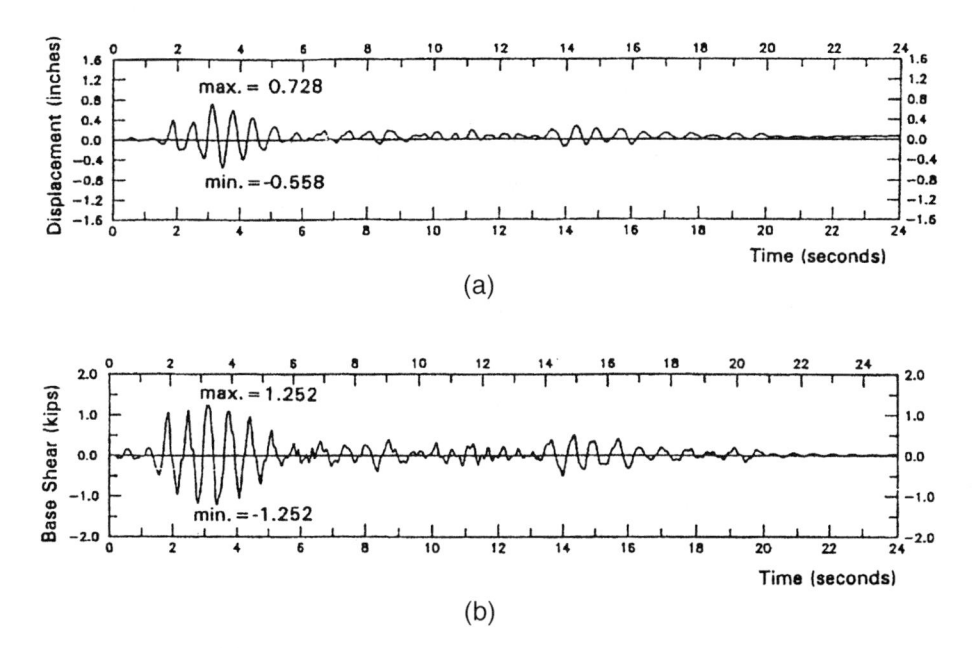

Figure 11.99 Three-story model response, run Taft 0.18 g: (a) third floor displacement; (b) base shear. (After El-Attar, A. G. et al., 1997.)

Table 11.9 Summary of Three-Story Model Seismic Test Results

Test	Top Story Drift, %	Maximum Base Shear, kN	Fundamental Frequency, Hz	First Mode Damping Ratio, % of ζ_{cr}
Taft 0.05g	0.19	1.50	2.20	1.30
Taft 0.18g	2.02	5.57	1.80	2.74
Taft 0.35g	2.84	6.16	1.65	2.76
Taft 0.80g	—	6.36	—	—

the load cells could capture the vertical load component developed in the columns as produced by the side sway of the structure, while the accelerometers on the story floors could only record their horizontal acceleration (Figure 11.102). This phenomenon was observed in all subsequent runs.

Correlation with Analytical Model. The correlation of the test results were made with the Program IDARC (Inelastic Damage Analysis of Reinforced Concrete Structures) developed at SUNY/Buffalo. Typical time history results of the calculated base shear for the three-story model is plotted against the measured base shear during test Taft 0.35g in Figure 11.103a. It can be seen that the calculated base shear correlated reasonably well with the measured shear and that both records were in phase during the entire run, indicating that the change in the mode period due to stiffness degradation was reasonably reflected in the analysis.

Failure Mechanism. The model collapsed during the Taft 0.80g test (after 7 s) in a soft story mechanism in the first-story columns. Failure was initiated at one of the interior columns, followed by failure of the remaining first-story columns. This failure mode (Figure 11.104) is consistent with the fact that interior columns are subjected to a higher axial force and consequently are less ductile than exterior columns.

Summary and Conclusions. Tests on the ⅛-scale model building indicates that GLD reinforced concrete structures will experience significant reduction in lateral stiffness after the first few cycles of a moderate earthquake. This reduction is mainly due to (1) the lack of ductility caused by poor

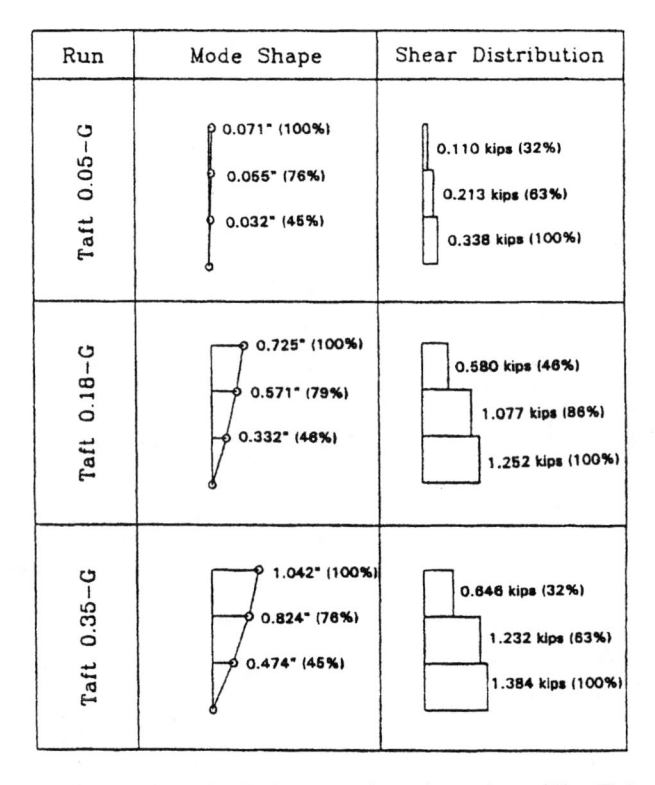

Run	Mode Shape	Shear Distribution
Taft 0.05-G	0.071" (100%) / 0.055" (76%) / 0.032" (45%)	0.110 kips (32%) / 0.213 kips (63%) / 0.338 kips (100%)
Taft 0.18-G	0.725" (100%) / 0.571" (79%) / 0.332" (46%)	0.580 kips (46%) / 1.077 kips (86%) / 1.252 kips (100%)
Taft 0.35-G	1.042" (100%) / 0.824" (76%) / 0.474" (45%)	0.646 kips (32%) / 1.232 kips (63%) / 1.384 kips (100%)

Figure 11.100 Mode shapes and shear distribution at maximum base shear. (After El-Attar, A. G. et al., 1997.)

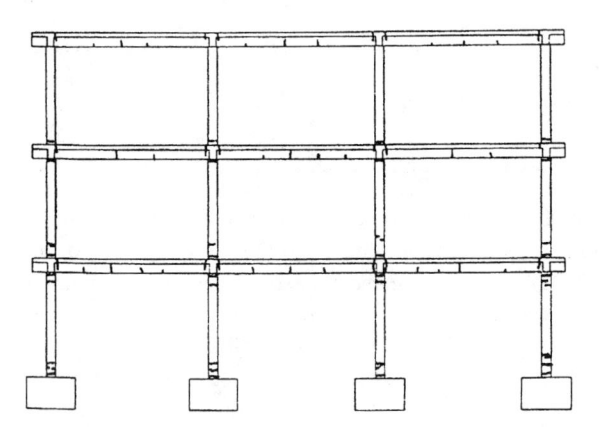

Figure 11.101 Cracking pattern of three-story model after run Taft 0.35 g. (After El-Attar, A. G. et al., 1997.)

confinement, especially in the joint regions, and (2) incipient pullout of the discontinuous positive beam reinforcement (Figure 11.105). The flexibility of the structure, tested without in-fill walls, resulted in pronounced P-Δ effect (up to 27% of the measured base shear).

The first mode of vibration dominated in all seismic tests. The mode shape was almost constant for all tests despite the significant reduction in the stiffness of the model (increase in their fundamental period).

It was also found that the share of the total story shear of each column is directly related to its axial force level. Interior columns in the three-story building resisted about twice the shear force acting on the exterior columns.

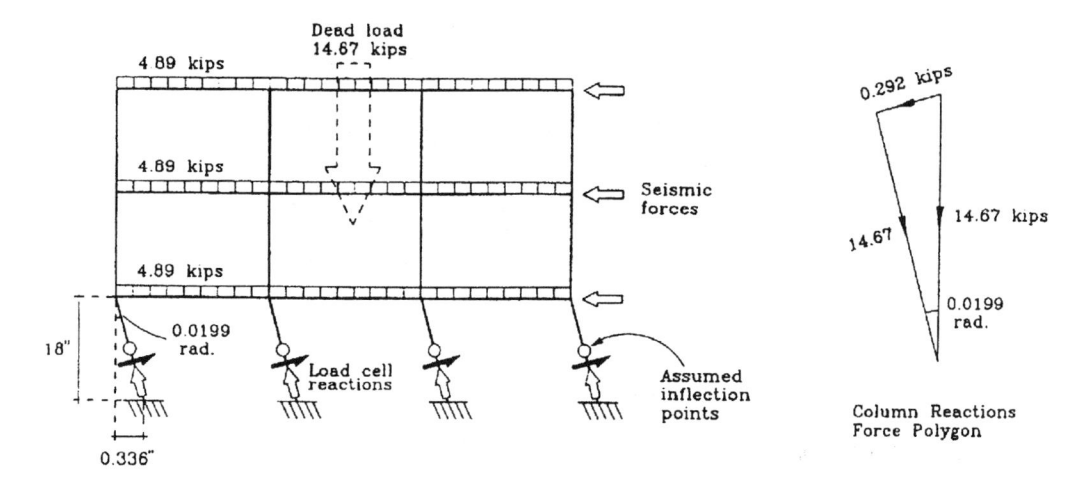

Figure 11.102 Simplified computation of P-Δ effect in first-story columns of three-story model during run Taft 0.18 g. (After El-Attar, A. G. et al., 1997.)

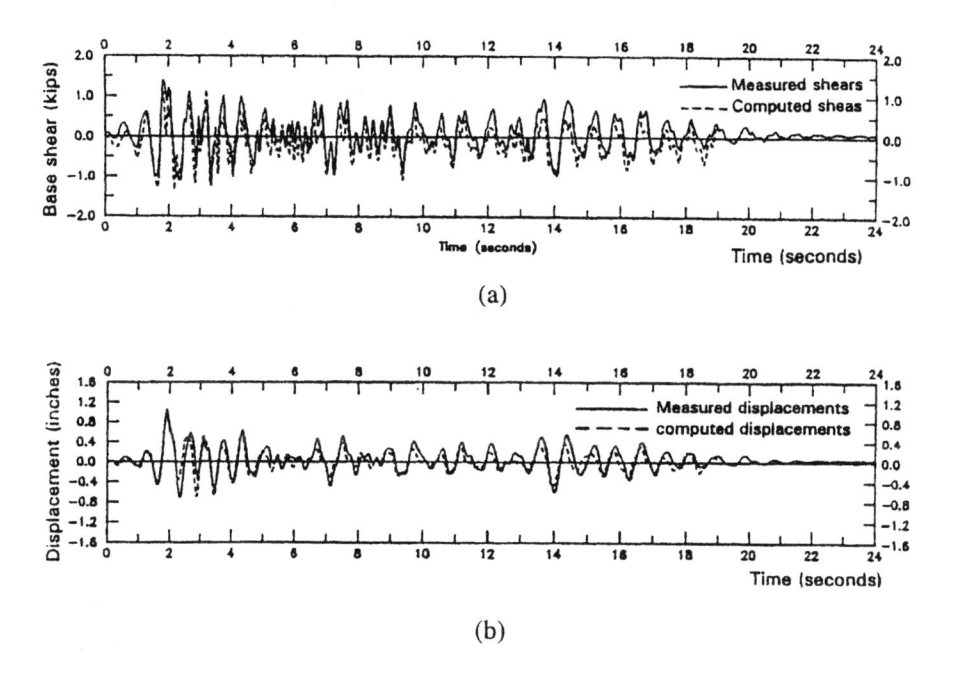

Figure 11.103 Computed vs. measured response, run Taft 0.35 g: (a) base shear; (b) third-story displacement. (After El-Attar, A. G. et al., 1997.)

Most of the deformation, damage, and energy dissipation occurred in the first-story columns. No significant damage was observed in the beams, at the splice locations, or in the joint panels. At failure, plastic hinges developed in the columns (strong beam–weak column behavior), producing a soft-story failure mode (see Figure 11.104)

Based on the experimental and analytical results the authors made the following conclusions:

1. Although the reinforcement details typically used in GLD reinforced concrete structures, may form a potential source of damage, they are probably not sufficient in themselves to develop a complete failure mechanism. Available experimental evidence indicates that the lack of sufficient strength of columns as compared to beams in typical GLD buildings

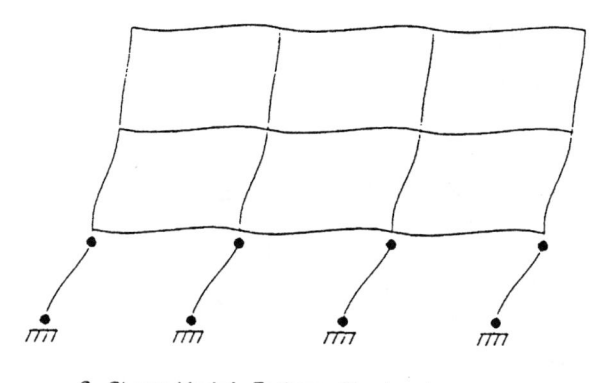

3-Story Model Failure Mechanism

Figure 11.104 Failure mechanisms of the model structures. (After El-Attar, A. G. et al., 1997.)

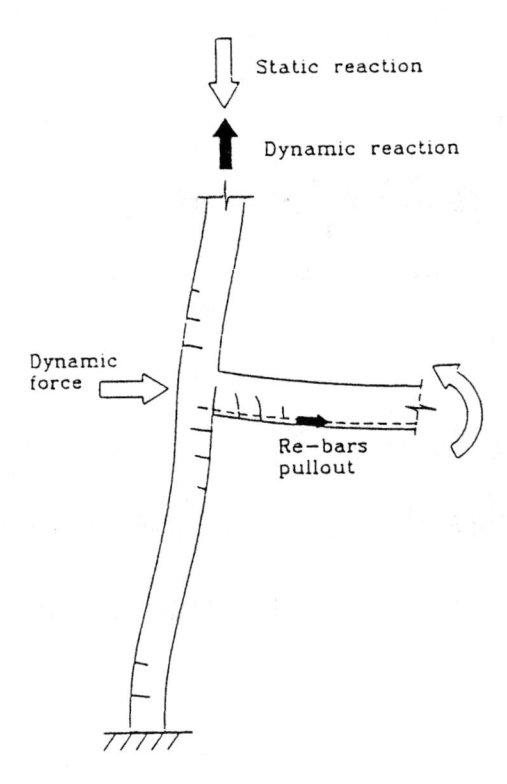

Figure 11.105 Rebars pullout at exterior beam–column joints. (After El-Attar, A. G. et al., 1997.)

usually leads to a premature soft-story mechanism before these details are subjected to significant demand.
2. GLD reinforced concrete buildings with no in-fill walls may experience large lateral deformations during a moderate earthquake, with deformations accentuated by substantial P-Δ effects.
3. Lateral stiffness of columns is a function of their axial force level.
4. Accounting for the slab contribution to the beam negative moment flexural strength is a vital step in the assessment of the performance of GLD reinforced concrete frames since it has the potential of altering the relatively ductile strong column–weak beam mechanism

Table 11.10 Shaking Table Testing Sequence for the Model

Test No.	Test Label	Test Description	Purpose
1	WHN_A	Uncompensated White Noise, PGA 0.024g	Table calibration
2	WHN_B	Compensated White Noise, PGA 0.024g	Identification
3	TFT_05	Taft N21E, PGA 0.05g	**Minor earthquake,** elastic response
4	WHN_C	Compensated White Noise, PGA 0.024g	Identification
5	WHN_D	Compensated White Noise, PGA 0.024g	Identification
6	TFT_20	Taft N21E, PGA 0.20g	**Moderate earthquake,** inelastic response
7	WHN_E	Compensated White Noise, PGA 0.024g	Identification
8	TFT_30	Taft N21E, PGA 0.30g	**Severe earthquake,** inelastic response
9	WHN_F	Compensated White Noise, PGA 0.024g	Identification

to a soft-story mechanism. This will be true even when a reduction factor is applied to positive beam flexural strength to account for the effect of possible pullout of the discontinuous positive moment reinforcement in the beams (Figure 11.105).

11.6.3.2 SUNY/Buffalo ⅓-Scale Model

The ⅓-scale model of the same GLD reinforced concrete building (Figures 1.12 and 1.13) that was tested on the SUNY/Buffalo Earthquake Simulation Laboratory shaking table forms the second part of this case study. The materials selection, model construction, and testing are described by Bracci et al. (1992, 1992a). The Taft 1952 N21E ground acceleration component was chosen as an appropriate base acceleration for this model. A time compression of the measured record with scaled peak accelerations of 0.05g, 0.20g, and 0.30g was used with a timescale factor of $1/\sqrt{3}$. The chosen three earthquakes are representative of minor–moderate, moderate–severe, and severe ground motions, respectively, in terms of ensuing structural damage. A summary of the shaking table testing sequence for the ⅓-scale model is shown in Table 11.10. The maximum response from the shaking table testing is shown in Table 11.11. Typical results such as the magnified overlay portion of the story displacements and story shears, respectively (Figure 11.106), during the moderate shaking indicate that the model stories are moving in phase. Figure 11.107 shows typical response comparison between experimental and analytical results for the severe earthquake shaking. As can be seen, the correlation is good for the third-story displacements and base shears.

The observed structural damage after the severe shaking test (Taft 0.30g) is shown in Figure 11.108. It consists of the following:

1. Cracking in the splice zone near the location of the transverse hoop reinforcement in the lower interior and exterior columns.
2. Cracking in the upper columns at the underside of the first- and second-story longitudinal beams and near the transverse hoop reinforcement.
3. Cracking of columns at construction joints fully around the columns and in the beam–column joints.
4. Vertical cracking in the web of the exterior longitudinal beams near the location of the transverse beam reinforcement.

Table 11.11 Maximum Response from Shaking Table Testing

Test	Story	Max. Story Displacement, mm	Max. Interstory Drift, %	Max. Story Shear, kN	Peak Story Acceleration, g
Minor shaking	Third	7.62	0.23	15.12	0.12
Taft N21E	Second	5.59	0.24	18.68	0.09
PGA 0.05g	First	3.56	0.28	23.57 (6.5%)	0.09
Moderate shaking	Third	33.53	0.54	24.91	0.20
Taft N21E	Second	28.96	1.07	41.37	0.20
PGA 0.20g	First	16.26	1.33	54.71 (15.2%)	0.25
Severe shaking	Third	59.69	0.89	31.58	0.25
Taft N21E	Second	52.07	2.24	51.60	0.22
PGA 0.30g	First	24.64	2.03	55.16 (15.3%)	0.29

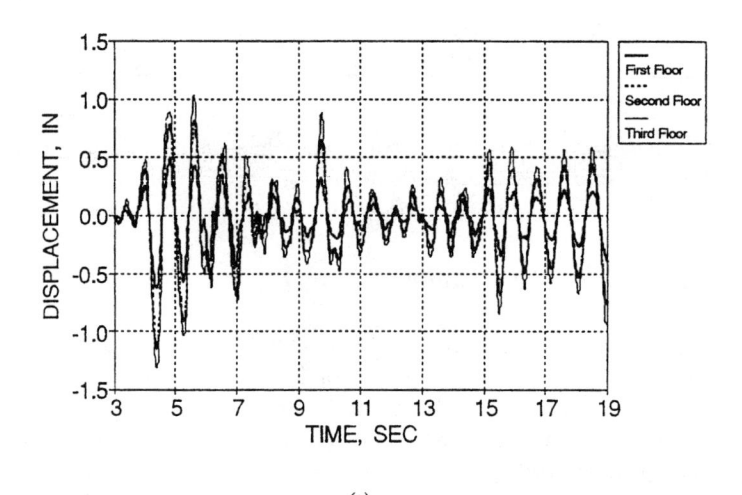

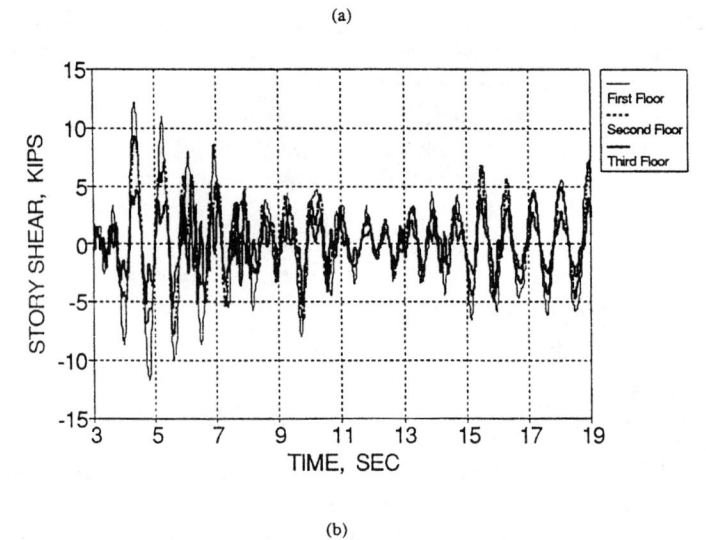

Figure 11.106 Overlaid global response time history segments for moderate shaking test (Taft 0.20 g). (a) Story displacements. (b) Story shear forces. (After Bracci, J. M. et al., 1992.)

STRUCTURAL MODELS FOR DYNAMIC LOADS

(a)

(b)

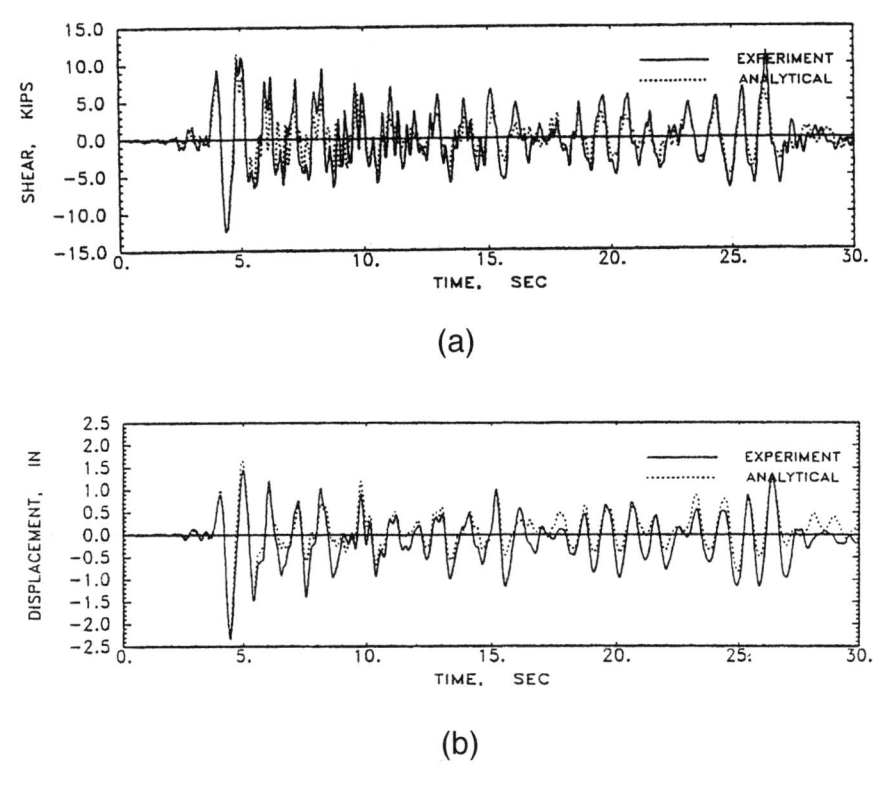

Figure 11.107 Base shear (a) and top-story displacement (b) for the severe shaking test (Taft N21E 0.30 g). (After Bracci, J. M. et al., 1992.)

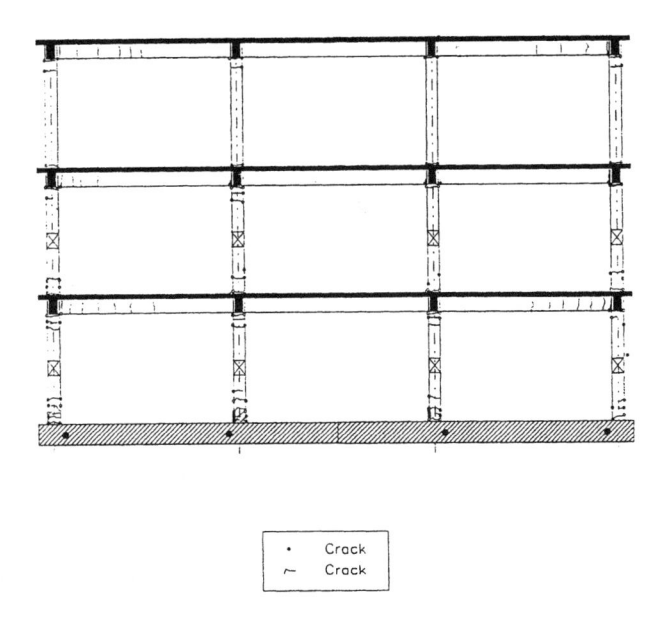

Figure 11.108 Observed structural damage from severe shaking test (Taft 0.30 g). (After Bracci, J. M. et al., 1992.)

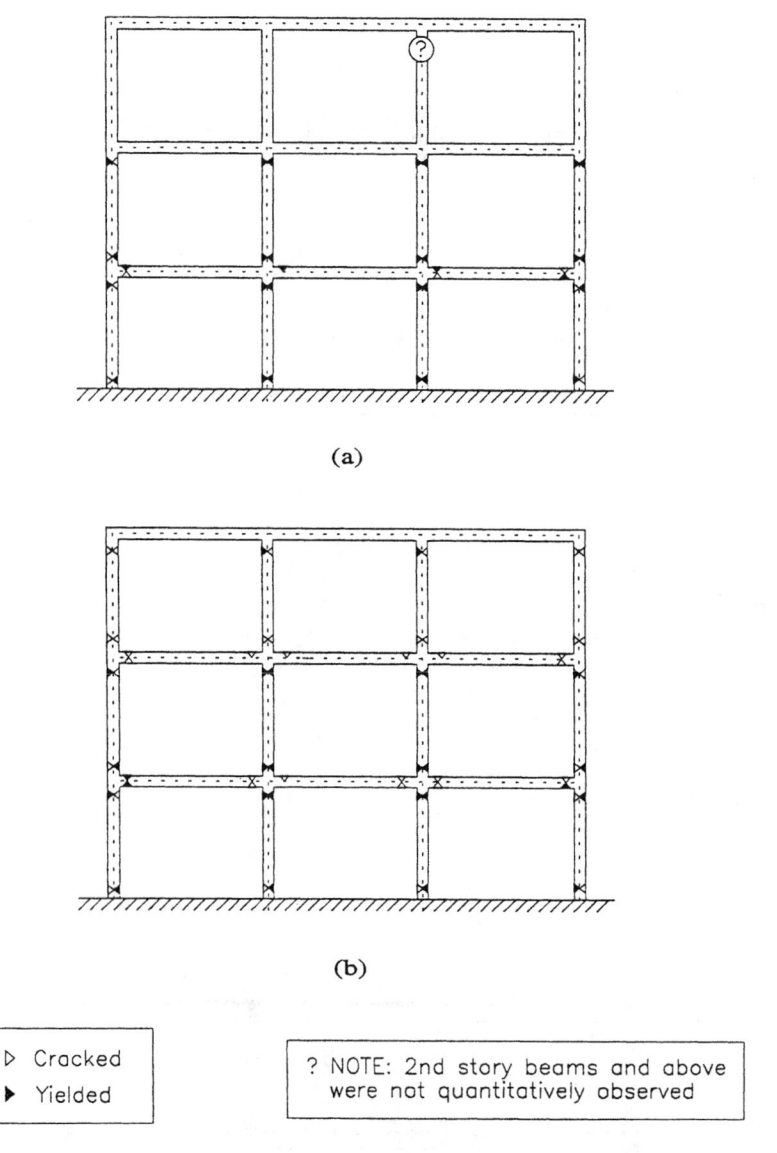

Figure 11.109 Comparison of damage states after severe shaking test (Taft 0.30 g). (a) Experimental. (b) Analytical. (After Bracci, J. M. et al., 1992.)

5. Large vertical cracking at the column face in the second-story exterior beams. Since large responses were observed at the second-story, pullout of the discontinuous rebars probably occurred.
6. Some slab cracking was observed along the transverse beams.

The damaged state of the ⅓-scale model after severe shaking is also compared to analytical predictions in Figure 11.109. It can be seen that yielding has occurred in the first- and second-story columns. In addition, some yielding has occurred in the beams. It is evident that a column sidesway or soft-story mechanism is developing similar to the one observed in the Cornell ⅛-scale model.

11.7 SUMMARY

A brief summary of some of the various dynamic models used in civil engineering practice has been presented. Types of loading effects on structures that have been considered include vibrations of elastic structures, wind effects, blast loading, impact, centrifuge loading, and earthquake loading. Material requirements for dynamic testing of model structures are given, and various types of loading facilities for testing such structures are described. Several case studies illustrating examples of dynamic models from a variety of structural design applications are presented. Dynamic models of structures are shown to be powerful tools in situations where the analytical techniques needed are lacking and there is strong need of test results for either calibration of analytical models or where practical designs must be generated.

PROBLEMS

11.1 Use the π theorem to develop an expression for the natural frequency of a freely vibrating fixed-ended beam, of prismatic cross section.

11.2 It is desired to build and test a Plexiglas model of a large cast steel flywheel having a heavy rim and radial spokes. Establish the similitude conditions. For a $\frac{1}{20}$-scale model, what are the stress and velocity scales for a prototype that has an angular velocity of 50 rad/s?

11.3 The natural frequency of a steel tuning fork is 200 vibrations/s. What is the natural frequency of a $\frac{1}{3}$-scale aluminum model of the tuning fork? The unit weight of aluminum is 36% that of steel, and its E value is $\frac{1}{3}$ that of steel. (A tuning fork is a freely vibrating elastic system.)

11.4 The idealized dynamic loading on a girder bridge is given by a forcing function that varies periodically with time. The exciting forces are fully specified by the frequency n and the maximum value of force F_0; for the prototype bridge, $n = 3$ Hz (cycles per second) and $F_0 = 10,000$ lb (44.5 kN). Determine similitude and scaling relations for a model bridge with a geometric scale factor $S_L = 30$. Both bridges are made of steel. Is it necessary to add additional mass (weights) to the bridge to satisfy similitude? Explain carefully.

11.5 For the girder of Problem 11.4, develop an expression for the maximum stress due to a concentrated load P at midspan, the beam self-weight, and a uniform temperature rise of ΔT. Assume elastic action only.

11.6 The torque on an airplane propeller depends only on the diameter d of the propeller, its angular velocity ω, the velocity of advance v, and the mass density m, and viscosity μ of the air. Using dimensional analysis, find an expression for the torque, and show that if the effect of viscosity can be neglected, then the torque is proportional to air density. (Courtesy of W. Godden, University of California, Berkeley.)

11.7 In order to study the performance of a high-speed train on its track, a $\frac{1}{10}$-scale model is made of the complete system. The track is both curved and banked (transverse slope) in places. The model train is true scale, and its effective density is found to be twice that of the prototype. The model is to be used to measure the forces exerted by the train on the track and to study the tendency for the train to overturn on corners. Specify the following model ratios required to simulate prototype behavior:

Track: **a.** Horizontal radius of curvature, r
              b. Transverse slope, ϕ
              c. Coefficient of friction, μ
Train: **d.** Velocity, v
              e. Acceleration, a

Forces: **f.** Centrifugal force on the track, F_c at points of curvature
g. Axial forces on track due to acceleration, F_a

Will this model correctly simulate the tendency for the train to overturn? (Courtesy of W. Godden, University of California, Berkeley.)

11.8 The structure of a tension roof system is composed of a series of freely hanging cables. Details of the prototype cables are as follows:

Span:	100 ft (30 m)
Diameter:	1 in. (25 mm)
Elastic modulus:	30×10^6 psi (200,000 MPa)
Specific weight:	500 lb/ft³ (8015 kg/m³)

It is required to study the aerodynamic properties of the roof in a steady wind. For this purpose it is proposed to use a 10-ft (3-m) span model in a wind tunnel. The model will be made of aluminum wires [$E = 10 \times 10^6$ psi (68,950 MPa), specific weight = 167 lb/ft³ (2677 kg/m³)]. It is known from prior experience on structures of this type that drag forces are not important.

a. What is the required diameter of the model wires?
b. Does the self-weight of the model wire have to be artificially increased? Explain.
c. Flutter in the model is observed at a wind velocity of 10 mph (16 km/h). At what wind velocity would you expect the same phenomenon in the prototype?
d. This flutter has a measured frequency of 5 cps in the model. At what frequency will the prototype vibrate? (Courtesy of W. Godden, University of California, Berkeley.)

11.9 It is required to study the response of a 400 ft (122 m) high concrete dam [properties of concrete: $E = 3.5 \times 10^6$ psi (24,133 MPa), specific weight = 150 pcf (2405 kg/m³)]. For this purpose a 4 ft (1.2 m) high model is made in gypsum plaster (material properties: $E = 3.5 \times 10^5$ psi (24,133 MPa), specific weight 100 pcf (1603 kg/m³). Strain gages are mounted on the surface of the model.

a. If it is required that the measured strains on the model are to be the same magnitude as those on the prototype, what is the required pressure in psi at the base of the model?
b. Then, what is the ratio of top displacements of model and prototype?
c. The freestanding model (that is, without back pressure) is put on a shaking table to determine its natural frequencies, and it is found that the first natural frequency of the model is 20 cps. What is the predicted first natural frequency of the prototype?
d. It is required to predict the dynamic stresses in the prototype as the ground shakes in harmonic motion at 2 cps at a maximum acceleration of $0.5g$. In order to have the same strains in model and prototype, at what frequency would you shake the model, and at what acceleration? (Courtesy of W. Godden, University of California, Berkeley.)

11.10 A beam 6 in. (152 mm) wide and 12 in. (0.3 m) deep, made of timber with $E = 1.2 \times 10^6$ psi (8274 MPa), is simply supported on an 8-ft (2.4-m) span and is to carry a moving vertical concentrated load. A model of the beam is constructed with a length scale of 20, but the beam is represented by a ⅛ in. (3.2 mm) diameter steel rod.

a. Neglecting shear effects, establish prediction equations for deflection, and for compressive stress on the top of the beam.
b. Establish prediction equations for shearing deflection and horizontal shearing stresses at the neutral surface.

11.11 A highly irregular (in plan and elevation) six-story structural steel framework is used in the construction of a chemical process plant. The proximity of this plant to a seismically active fault system makes it necessary to study the proposed structure and the large masses that it carries (attached piping and equipment) dynamically. A similitude analysis was made and it was decided to build and test a small-scale steel model on a medium-size shaking table available. How would you fabricate such a model so that an ultimate strength test could eventually be made? How would you instrument the model to capture its most important symmetric and antisymmetric dynamic behavior?

REFERENCES

Abrams, P., Paulson, T., and Colunga, A. (1990). Aspects of response for masonry building structures, in *Proceedings of Fourth U.S. National Conference on Earthquake Engineering,* Vol. 2, May 20–24, 57–66.

Baglietto, E., Casirati, M., Castoldi, A., Demiranda, F., and Sammartino, R. (1976). Mathematical and Structural Models of Zarate-Brazo Largo Bridges, Report No. 85, ISMES-Instituto Sperimentale Modelli e Strutture, Bergamo, Italy, September, 46 pp.

Baldi, G., Maggioni, W., and Belloni, G. (1988). The ISMES geotechnical centrifuge, Corté, J. F., Ed., Centrifuge 88, Proc. Int. Conf. Geotechnical Centrifuge Modelling, Paris, April 25–27, A. A. Balkema, Rotterdam, Brookfield, VT.

Bariola, J., Ginocchio, J., and Quinn, D. (1990). Out-of-plane seismic response of brick walls, in *Proceedings Fifth North American Masonry Conference,* Vol. I, 491.

Bertero, V. V., Aktan, A. E., Charney, F. A., and Sause, R. (1984). U.S.–Japan Cooperative Earthquake Research Program: Earthquake Simulation Tests and Associated Studies of a ⅕th-Scale Model of a 7-Story Reinforced Concrete Test Structure, Earthquake Engineering Research Center, Report no. UCB/EERC-84/05, June.

Borges, J. F. and Pereira, J. (1970). Dynamic model studies for designing concrete structures, paper SP-24-10, *Models for Concrete Structures,* ACI SP-24, American Concrete Institute, Detroit, MI.

Bracci, J. M., Reinhorn, A. M., and Mander, J. B. (1992a). Seismic Resistance of Reinforced Concrete Frame Structures Designed only for Gravity Loads: Part III—Experimental Performance and Analytical Study of a Structural Model, Technical Report, NCEER-92-0029, December 1.

Bracci, J. M., Reinhorn, A. M., and J. B. Mander (1992). Seismic Resistance of Reinforced Concrete Frame Structures Designed Only for Gravity Loads: Part I — Design and Properties of a One-Third Scale Model Structure, Technical Report, NCEER-92-0027, December.

Caccese, V. and Harris, H. G. (1982). Report 1 Description and Operation of the Drexel University Structural Dynamics Laboratory, Seismic Behavior of Precast Concrete Large Panel Buildings Using a Small Shaking Table, Structural Dynamics Laboratory Report No. D82-01, Department of Civil Engineering, Drexel University, Philadelphia, June.

Caccese, V. and Harris, H. G. (1984a). Seismic Behavior of Precast Concrete Large Panel Buildings Using a Small Shaking Table, Report 2 Small-Scale Tests of Simple Precast Concrete Shear Wall Models under Earthquake Loading, Drexel University, Philadelphia, December, 253 pp.

Caccese, V. and Harris, H. G. (1984b). Seismic Behavior of Precast Concrete Large Panel Buildings Using a Small Shaking Table, Report 3, Correlation of Experimental and Analytical Results, Structural Dynamics Laboratory Report No. D85-01, Department of Civil Engineering, Drexel University, Philadelphia, June, 218 pp.

Caccese, V. and Harris, H. G. (1987). Seismic resistance of precast concrete shear walls, *Earthquake Eng. Struct. Dyn.,* 15, 661–677.

Castellani, A., Castoldi, A., and Ionita, M. (1976). Numerical Analysis Compared to Model Analysis for a Dam Subject to Earthquakes, Report No. 83, ISMES-Istituto Sperimentale Modelli e Strutture, Bergamo, Italy, September, reprint for the *Proceedings of the Fifth International Conference on Experimental Stress Analysis,* Udine, Italy.

Castoldi, A. and Casirate, M. (1976). Experimental Techniques for the Dynamic Analysis of Complex Structures, Report No. 74, ISMES-Istituto Sperimentale Modelli e Strutture, Bergamo, Italy, February.

Cermak, J. E., (1977). Wind-tunnel testing of structures, *J. Eng. Mech. Div. ASCE,* 103(EM6), December, 1125–1140.

Chowdhury, A. H. and White, R. N. (1977). Materials and modeling techniques for reinforced concrete frames, *Proc. Am. Concr. Inst.,* 74(11), November, 546–551.

Clough, R. W. and Bertero, V. V. (1977). Laboratory model testing for earthquake loading, *ASCE J. Eng. Mech. Div.,* 103(EM6), December, Proc. Paper 13444, 1105–1124.

Clough, R. W., Gülkan, P., Manos, G. C., and Mayes, R. L. (1990). Seismic testing of single-story masonry houses: Part 2, J. Struct. Eng. , 116(1), Jan., 257–274.

Clough, R. W. and Tang, D. T. (1975). Earthquake Simulator Study of a Steel Frame Structure, Vol. I: Experimental Results, EERC Report No. 75-6, Berkeley, CA.

Clough, R., Mayes, R., and Gulkan, P. (1979). Shaking Table Study of Single-Story Masonry Houses, Vol. 3: Summary, Conclusions, and Recommendations, UCB/EERC-79/25, September, 96 pp.

Costley, A. C. and Abrams, D. P. (1996). Dynamic Response of Unreinforced Masonry Buildings with Flexible Diaphragms, Technical Report NCEER-96-0001, University of Illinois at Urbana-Champaign, Department of Civil Engineering, October.

Dau, K. (1961). Wind Tunnel Tests of the Toronto City Hall, UTIA Technical Note No. 50, Institute of Aerophysics, University of Toronto, Toronto, Canada.

Davenport, A. G. and Isyumov, N. (1968). The application of the boundary layer wind tunnel to the prediction of wind loading, in *Proceedings, International Research Seminar on Wind Effects on Building and Structures,* Vol. 1, University of Toronto Press, Toronto, 201–230.

El-Attar, A. G., White, R. N., Gergely, P., and Bond, T. K. (1991). Shake Table Tests of a ⅛-Scale Three-Story Lightly Reinforced Concrete Building, *Technical Report,* NCEER-91-0017, NCEER.

El-Attar, A. G., White, R. N., and Gergely, P. (1997). Behavior of gravity load designed reinforced concrete buildings subjected to earthquakes, *ACI Struct. J.,* 94(2), March–April.

Farrar, C. R., Baker, W. E., and Dove, R. C. (1994). Dynamic parameter similitude for concrete models, ACI Struct. J., 91(1), January–February, pp. 90–99.

Ferritto, J. (1982). Dynamic tests of model concrete, in *Dynamic Modeling of Concrete Structures,* Harris, H. G., Ed., ACI SP-73, American Concrete Institute, Detroit, pp. 23-33.

Focken, C. M. (1953). *Dimensional Methods and Their Applications,* Edward Arnold & Co., London.

Garas, F. K. (1981). Dynamic modelling of structures, Joint I. Struct. E./B.R.E. International Seminar, Garston, Watford, England, November 19, 20.

Godden, W. G. and Aslam, M. (1978). Dynamic model structures of the Ruck-a Chucky Bridge, preprint of the paper presented at the ASCE Spring Convention, Pittsburgh, April.

Gülkan, P., Clough, R. W., Mayes, R. L., and Manos, G. C. (1990). Seismic testing of single-story masonry houses: Part 1, J. Struct. Eng., 116(1), Jan., 235–256.

Harris, H. G. (1968). Simplified Apollo Shell One-Tenth Scale Model, Lunar Module Report LED-520-50, Grumman Aerospace Corporation, Bethpage, NY, October 1.

Harris, H. G. (1982). *Dynamic Modeling of Concrete Structures,* ACI Publication SP-73, Detroit.

Harris, H. G., Pahl, P. J., and Sharma, S. D. (1962). Dynamic Studies of Structures by Means of Models, Research Report R63-23, Department of Civil Engineering, MIT, Cambridge, September.

Harris, H. G., Schwindt, R., Taher, I., and Werner, S. (1963). Techniques and Materials in the Modeling of Reinforced Concrete Structures Under Dynamic Loads, Report R63-54, Department of Civil Engineering, Massachusetts Institute of Technology, Cambridge, December; also NCEL-NBY-3228, U.S. Naval Civil Engineering Laboratory, Port Hueneme, CA.

Harris, H. G., Sabnis, G. M., and White, R. N. (1966). Small Scale Direct Models of Reinforced and Prestressed Concrete Structures, Report no. 326, Department of Structural Engineering, Cornell University, Ithaca, NY, September, 362 pp.

Harris, H. G., Sabnis, G. M., and White, R. N. (1970). Reinforcement for small-scale direct models of concrete structures, paper No. SP-24-6, *Models for Concrete Structures,* ACI SP-24, American Concrete Institute, Detroit, MI, 141–158.

Hidalgo, P. and Clough, R. W. (1974). Earthquake Simulator Study of a Reinforced Concrete Frame, Report No. EERC-74-13, Earthquake Engineering Research Center, University of California, Berkeley, December.

Hudson, D. E. (1967). Scale model principles, in *Shock and Vibration Handbook,* Harris, C. M. and Crede, C. E., Eds., McGraw-Hill, New York, Chap. 27.

Irwin, H. P. A. H. and Wardlow, R. L. (1976). Sectional Model Experiments on Lions' Gate Bridge, Vancouver, Laboratory Technical Report No. LTR-LA-205, National Aeronautical Establishment, Ottawa, October.

Irwin, H. P. A. H. and Schuyler, G. D. (1977). Experiments on a Full Aeroelastic Model of Lions' Gate Bridge in Smooth and Turbulent Flow, Laboratory Technical Report No. LTR-LA206, National Aeronautical Establishment, Ottawa, October 18.

Johnson, R. P. (1962). Strength tests on scaled-down concretes suitable for models, with a note on mix design, *Mag. Concr. Res.,* 14(40), March.

Jurukovski, D., Kristevska, L., Allessi, R., Diotallevi, P, Merli, M, and Zarri, F (1992). Shaking table tests of three four-storey brick models: original and strengthened by RC core and by RC jackets, *Proceedings of the Tenth World Conference on Earthquake Engineering,* vol. 5, pp. 2795–2800.

Kim, W., El-Attar, A., and White, R. N. (1988). Small-Scale Modeling Techniques for Reinforced Concrete Structures Subjected to Seismic Loads, National Center for Earthquake Engineering Research, Technical Report NCEER-88-0041, November 22.

Krauthammer, T. (1997). Recent structural dynamics research with precision impact tests, in *Proc. 8th Int. Symp. on Interaction of the Effects of Munitions with Structures,* April 21–27, McLean, VA.

Krauthammer, T., Shanaa, H. M., and Assadi, A (1993). Response of structural concrete elements to severe impulsive loads, *Comput. Struct.* Vol. 53(1), 119–130.

Krauthammer, T., Jenssen, A., and Langseth, M. (1996). NDCS workshop of precision testing in support of computer code validation and verification, in *Proc. Specialty Symposium on Structrues Response to Impact and Blast,* Tel Aviv, Israel, 6–10 October.

Krawinkler, J. and Moncarz, P. D. (1982). Similitude requirements for dynamic models, in *Dynamic Modeling of Concrete Structures,* ACI, Publication SP-73, 1–22.

Krawinkler, J., Mills, R. S., Moncarz, P. D. et al. (1978). Scale modeling and testing of structures for reproducing response to earthquake excitation, The John A. Blume Earthquake Engineering Center, Department of Civil Engineering, Stanford University, Stanford, May.

Lassoudiere, F. and Perol, C. (1988). Centrifuge study of soil–structure dynamic interaction for a jack-up platform submitted to sea-wave loading, *Centrifuge 88, Proceedings of the International Conference on Geotechnical Centrifuge Modelling,* Paris, 25-227, April, Balkema, Rotterdam, ISBN.

Limongelli, M. P. and Pezzolli, P. (1994). Analysis of the seismic response of masonry buildings excited by a shaking table, *Eur. Earthquake Eng.,* 7 (2): 18–30, Bologna: Patron Editore.

Litle, W. A. and Paparone, M. (1966). Size effect in small-scale models of reinforced concrete beams, *Proc. Am. Concr. Inst.,* 63 (11) November, 1191–1204.

Luong, M. P. (1980). Phenomènes cycliques dans les sols pulvérulents, *Revue Françoise de Geotechnique,* No. 10, pp. 39–53.

Luong, M. P. (1986). Centrifugal models of piles and pile groups under lateral harmonic excitations and seismic actions, in Proc. Conf. on Numerical Methods in Offshore Piling, Paris, Techniped, pp. 511–525.

Magenes, Gu. and Calvi, G, (1994). Shaking table tests on brick masonry walls, in *Proceedings, Tenth European Conference on Earthquake Engineering,* (Vienna).

Manos, G., Clough, R., and Mayes, R. (1983). Shaking Table Study of Single-Story Masonry Houses: Dynamic Performance under Three Component Seismic Input and Recommendations, UCB/EERC-83-11, July, 156.

Mills, R. S., Krawinkler, H., and Gere, J. M. (1979). Model Tests on Earthquake Simulators Development and Implementation of Experimental Procedures, Report No. 39, The John A. Blume Earthquake Engineering Center, Department of Civil Engineering, Stanford University, Stanford, CA, June, 272 pp.

Modena, C., LaMendola, P., and Terrusi, A. (1992). Shaking table study of a reinforced masonry building model, in *Proceedings of the Tenth World Conference on Earthquake Engineering,* Vol. 6, 3523–3526.

Nicolas-Font, J. (1988). Design of geotechnical centrifuges, Corté, J. F., Ed., Centrifuge 88, Proc. Int. Conf. Geotechnical Centrifuge Modelling, Paris, April 25–27, A. A. Balkema, Rotterdam, Brookfield, VT.

Nilsson, L. (1979). A constitutive modelling, finite element analysis, and experimental study of nonlinear wave propagation, in *Impact Loading on Concrete Structures,* Department of Structural Mechanics, Chalmers University of Technology, Publication 79-1, Goteborg, Sweden.

Nilsson, L. and Sahlin, S. (1982). Impact of a steel rod on a reinforced concrete slab, in *Dynamic Modeling of Concrete Structures,* Harris, H. G., Ed., ACI SP-73, American Concrete Institute, Detroit, pp. 165–187.

Norris, C. H., Hansen, R. J., Holley, M. J., Jr., Biggs, J. M., Namyet, S., and Minami, J. K. (1959). *Structural Design for Dynamic Loads,* McGraw–Hill, New York.

Paulson, T. and Abrams, D., (1990). Measured Inelastic Response of Reinforced Masonry Building Structures to Earthquake Motions, SRS No. 555, UILU-ENG-90-2013, October, 290 pp.

Pomonis, A., Spence, R., Coburn, A., and Taylor, C. (1992). Shaking table tests on strong motion damagingness upon unreinforced masonry, in *Proceedings of the Tenth World Conference on Earthquake Engineering,* Vol. 6, 3533–3538.

Sabnis, G. M. and White, R. N. (1969). Behavior of reinforced concrete frames under cyclic loads using small-scale models, *J. Am. Concr. Inst.,* 66(9), September, 703–715.

Sachs, P. (1978). Wind tunnel techniques, in *Wind Forces in Engineering,* 2nd ed., Pergamon Press, New York, Chap. 5, p. 400.

Scruton, C. (1968). Aerodynamics of structures, in *Proc. Int. Res. Seminar Wind Effects on Buildings and Structures*, Vol. 1, University of Toronto Press, pp. 115–165.

Shaw, W. A. (1962). Static and Dynamic Behavior of Portal-Frame Knee Connections, U.S. Naval Civil Engineering Laboratory, Port Hueneme, CA, May.

Simiu, E. and Scanlan, R. H. (1978). The wind tunnel as a design tool, in *Wind Effects on Structures: An Introduction to Wind Engineering,* John Wiley & Sons, New York, Chap. 9, 458.

Simiu, E. and Scanlan, R. H. (1996). *Wind Effects on Structures: Fundamentals and Applications to Design,* Wiley-Interscience, John Wiley & Sons, New York.

Smith, H. D., Clark, R. W., and Mayor, R. P. (1963). Evaluation of Model Techniques for the Investigation of Structural Response to Blast Loads, Report R63-16, Department of Civil Engineering, Massachusetts Institute of Technology, Cambridge, February.

Staffier, S. R. and Sozen, M. A. (1975). Effect of Strain Rate on Yield Stress of Model Reinforcement, Civil Engineering Studies, Structural Research Series No. 415, University of Illinois, Urbana, February.

Tang, D. T. (1975). Earthquake Simulator Study of a Steel Frame Structure: Vol. 2, Analytical Results, UCB/EERC 75-36, Earthquake Engineering Research Center, University of California, Berkeley, September.

Tomazevic, M. (1987). Dynamic modelling of masonry buildings: Storey mechanism model as a simple alternative, *Earthquake Eng. Struct. Dyn.,* 15(6), August, 731–749.

Tomazevic, M. and Lutman M. (1996). Seismic behavior of masonry walls: modeling of hysteretic rules, *J. Struct. Eng.,* Vol. 122(9), September.

Tomazevic, M. and Velechovsky, T. (1992). Some aspects of testing small-scale masonry building models on simple earthquake simulators, *Earthquake Eng. Struct. Dyn.,* 21(11), November, 945–963, John Wiley & Sons, Chichester.

Tomazevic, M. and Weiss, P. (1990). A rational, experimentally based method for the verification of earthquake resistance of masonry buildings, in *Proceedings of Fourth U.S. National Conference on Earthquake Engineering,* May 20–24, Vol. 2, 349–358.

Tomazevic, M. and Weiss, P. (1994). Seismic behavior of plain — and reinforced — masonry buildings, *J. Struct. Eng.,* 120(2), February, ASCE, 323–338.

Tomazevic, M., Lutman, M., and Weiss, P. (1993a). The seismic resistance of historical urban buildings and the interventions in their floor systems: an experimental study, *Masonry Soc. J.,* 12(1), August, 77–86.

Tomazevic, M., Lutman, M., and Weiss, P. (1993c). The seismic resistance of historical urban buildings and the interventions in their floor systems: an experimental study, *North Am. Masonry Conf.,* 2, June, 669–680.

Tomazevic, M., Lutman, M., and Weiss, P. (1996). Seismic upgrading of old brick-masonry urban houses: tying of walls with steel ties, *Earthquake Spectra,* 12 (3), August, 599–622.

Tomazevic, M., Modena, C., Velechovsky, T., and Weiss, P. (1990). The influence of structural layout and reinforcement on the seismic behavior of masonry buildings: an experimental study, *Masonry Soc. J.,* 9(1), August, 26–50.

Tomazevic, M., Weiss, P., Lutman, M., and Petkovi, L. (1993b). Influence of Floors and Connection of Walls on Seismic Resistance of Old Brick Masonry Houses: Part One: Shaking Table Tests on Model Houses A and B, ZRMK/OI-93-04, pp. 119.

Watstein, D. and Boresi, A. P. (1952). The Effect of Loading Rate on the Compressive Strength and Elastic Properties of Plain Concrete, National Bureau of Standards, Report 1523, Washington, D.C., March.

Williams, D. and Godden, W. G. (1976). 'Experimental Model Studies on the Seismic Response of High Curved Overcrossings, Report No. EERC 76-18, College of Engineering, University of California, Berkeley, June.

Xia, J., Wei, Z., Huang, Q., and Gao, L. (1990). Shaking table test of multi-story masonry building with coal-slag-gas-concrete blocks, in *Proceedings, Fifth North American Masonry Conference,* Vol. I, pp. 165–176.

Zhu, B. (1986). A review of a seismic test for masonry structures in China, in *Proceedings of US-PRC Joint Workshop on Seismic Resistance of Masonry Structures,* III-5-1–II-5-12.

Zhu, B., Wu, M., and Zhou, D. (1986). Shaking table study of a five-story unreinforced block masonry model building strengthened with reinforced concrete columns and tie bars, in *Proceedings of US-PRC Joint Workshop on Seismic Resistance of Masonry Structures,* IV-11-1–IV-11-11.

CHAPTER 12

Educational Models for Civil and Architectural Engineering

CONTENTS

12.1	Introduction	680
12.2	Historical Perspective	681
12.3	Linearly Elastic Structural Behavior	681
	12.3.1 General	681
	12.3.2 Classroom Demonstration Models	681
	12.3.2.1 Bending, Torsion, and Buckling	682
	12.3.2.2 Shear Center of Thin Open Section Beams	683
	12.3.2.3 Frame Action	683
	12.3.2.4 Composite Behavior	683
	12.3.2.5 Folded Plate Action	684
	12.3.3 Laboratory Demonstration Models	685
	12.3.3.1 Simple Beams and Trusses	685
	12.3.3.2 Continuous Beams	687
	12.3.3.3 Gable Frames	690
	12.3.3.4 Euler Buckling Load	690
	12.3.4 Architectural Engineering Models	692
	12.3.4.1 Diaphragm Action	692
	12.3.4.2 Infilled Frames	693
12.4	Nonlinear and Inelastic Structural Behavior	694
	12.4.1 Ultimate-Strength Models of R/C Components	694
	12.4.1.1 Underreinforced/Overreinforced Concrete Beams	694
	12.4.1.2 Eccentrically Loaded Concrete Columns	698
	12.4.1.3 Two-Span R/C Beams	701
	12.4.1.4 Effective Width of Reinforced Concrete T-Beams	705
	12.4.1.5 Reinforced Concrete Arches	705
	12.4.2 Yielding and Inelastic Buckling of Steel Members	710
	12.4.2.1 Hinge Formation and Collapse Modes	710
	12.4.2.2 Inelastic Buckling	711
12.5	Structural Dynamics Concepts	712
	12.5.1 Basic Laboratory Instrumentation	712
	12.5.1.1 PM-25 Exciter	713
	12.5.1.2 The 2125 MB Power Amplifier	716
	12.5.1.3 Exact Electronics Model 502 Function Generator	716

		12.5.1.4	Evaluation of the Combined Drexel System 716
		12.5.1.5	Other Educational Vibration Systems 718
	12.5.2	Vibrations of Lumped-Mass Systems 718
		12.5.2.1	Single-Degree-of-Freedom Systems 720
		12.5.2.2	Multi-Degree-of-Freedom Systems 721
	12.5.3	Dynamics of Shear Buildings 722
		12.5.3.1	Four-Story Building Model 723
		12.5.3.2	Four-Story Building Model with Base Isolation 723
			1. Introduction 723
			2. Base Isolation System 723
			3. Test Results 723
12.6	Experimentation and the New Engineering Curriculum 725
	12.6.1	Historical Background 725
	12.6.2	New Engineering Curriculum 726
12.7	Case Studies and Student Projects 729
	12.7.1	Building Components and Systems 729
		12.7.1.1	Double-Tapered Glulam Beams 729
		12.7.1.2	Single-Fink Roof Truss 730
		12.7.1.3	Precast Building Elements 732
	12.7.2	Bridge Structures 736
		12.7.2.1	Concrete Segmental Bridge 736
		12.7.2.2	Cable-Stayed Bridge 737
	12.7.3	Special Structures 741
		12.7.3.1	Elevated Pedestrian Walkway 741
		12.7.3.2	Microwave Antenna Structure 742
		12.7.3.3	The Drexel Geodesic Tri-Span 747
12.8	Summary 751
Problems 752
References 753

12.1 INTRODUCTION

Educational models have long been used to convey basic and more complex structural engineering concepts to students. Carefully thought-out physical models have been used in courses in physics and mechanics for many decades. The logical extension of these ideas into the basic courses taken by the structural and architectural engineering student is addressed in this chapter. Small-scale physical models of a wide variety and complexity are discussed in what follows. These fall into three major categories of structural behavior: linearly elastic, nonlinear inelastic, and dynamic. Examples of all these types of structural behavior are discussed in a laboratory format to enable the student of structural and architectural engineering to understand basic concepts and to visualize the behavior of realistic structural systems.

This new chapter in the second edition attempts to help the instructor fix basic structural concepts in the mind of the student in the limited time that is allocated. The material contained in Chapter 12 together with the appropriate supporting sections of the chapters on materials (Chapters 3 through 5), model fabrication (Chapter 6), and instrumentation and loading (Chapters 7 and 8, respectively) can form the basis of a structural analysis and design course for undergraduate civil and architectural engineering students. With proper emphasis on the more complex nonlinear structural concepts, the material can be easily extended to the first-year graduate level for these same students. The material presented helps students or young engineers understand the behavior of structures by experiencing it in the classroom or the laboratory firsthand before their own eyes. In all of the examples chosen for discussion, the criteria used to qualify an example for demonstration are the following: Does it pass the "3 Ss test"? Is it *simple* to construct? Is it *simple* to demonstrate? Is it *simple* to understand?

12.2 HISTORICAL PERSPECTIVE

One of the first references of loaded models, particularly for classroom demonstration, was made by Rathbun (1934); he used wood blocks strung on wires to form an arch. Application of a concentrated load shows arch action clearly, while under distributed load the arch adjusts itself. Demonstration models were also made from heavy flexible rubber pads of structures such as beams, frames, bents, and arches. These show qualitative distributions of stress and deformation under large deformations.

Godden (1963) has reported an extensive study of demonstration models for teaching structural mechanics covering elastic theory, plastic theory, and photoelastic models. The carefully designed and constructed models developed could easily form the basis of a course in structural analysis. Shepherd (1964) describes the various model apparatus used at the Model Structures Laboratory, University of Canterbury, Christchurch, New Zealand. The 18 models shown in the paper were used in laboratory classes as part of a civil engineering undergraduate course, and as suggested by the author they could equally effectively be incorporated in any engineering or architectural science course. At the University of Sydney, Australia, substantial emphasis has been given to experimental techniques with appropriate laboratory facilities (Cowan, 1961). Laboratories using elastic models for teaching structural analysis of indeterminate systems existed in many universities in the U.S. some as early as the 1920s. Such facilities were found at such universities as MIT, Princeton, Lehigh, Carnegie-Mellon, Cornell, Johns Hopkins, University of California, Berkeley, College of the City of New York, and Drexel, among others.

12.3 LINEARLY ELASTIC STRUCTURAL BEHAVIOR

12.3.1 General

Linearly elastic behavior of structures forms the core of many analysis methods. In teaching structural analysis and engineering mechanics principles to undergraduate students, the physical model can be a powerful tool whose many advantages can be readily used in the classroom or in the laboratory. The next sections describe the techniques used for teaching basic structural principles to civil and architectural engineering students. The examples go progressively from simple to more complex, supplementing the engineering curriculum with visual insights and introducing experimental techniques.

12.3.2 Classroom Demonstration Models

Classroom demonstration models are the simplest form of physical scaled models. As they are to be used in the classroom during the lecture, they must be portable and very simple to operate. In addition they must be equally simple to comprehend otherwise the whole point of using a "visual aid" is missed. Models for classroom demonstration can be made of inexpensive materials such as paper, balsa wood, or plastic. Because these models must be kept as simple as possible, generally they have little or no instrumentation and demonstrate their point by exaggerated deformation.

Positive Moment

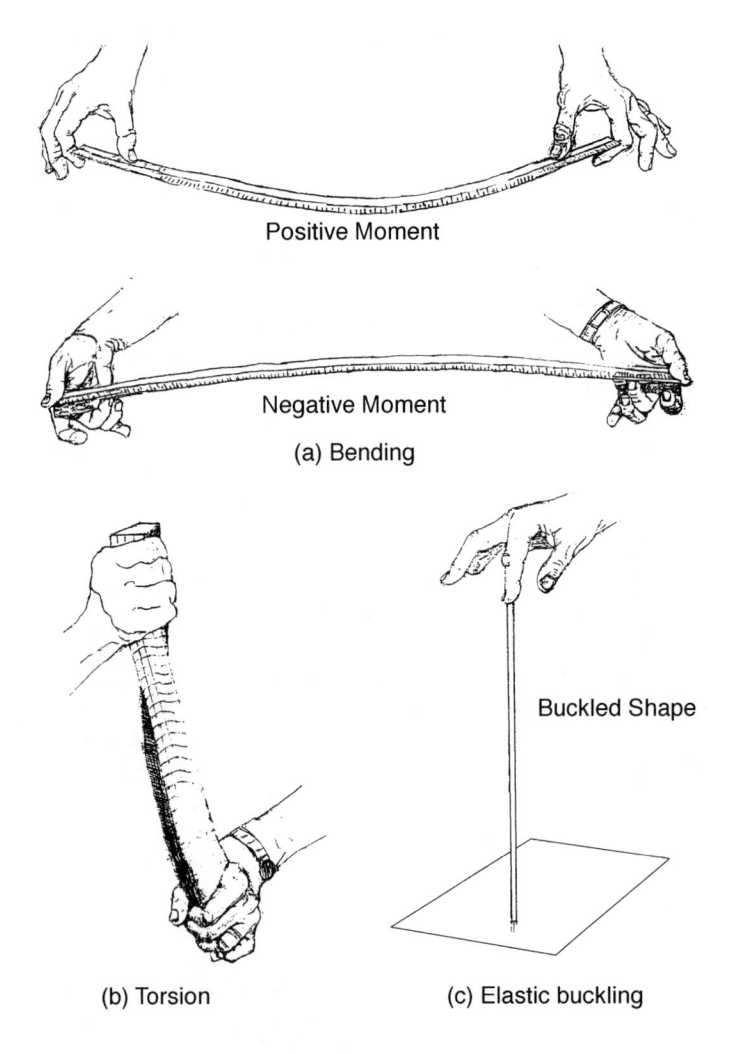

Negative Moment

(a) Bending

(b) Torsion

Buckled Shape

(c) Elastic buckling

Figure 12.1 (Courtesy of Helen T. Harris.)

12.3.2.1 Bending, Torsion, and Buckling

Bending of prismatic members is easily demonstrated to the class using finger loading as illustrated in Figure 12.1a. Two end couples provided by the fingers flex a thin wooden or plastic beam in positive or negative bending. More substantial models cut from blocks of rubber can demonstrate the concept of "plane section remain plane" and the distortion to the section due to the Poisson's effect.

Similarly, a square or rectangular prismatic member as opposed to a round section made out of rubber can illustrate very well the distribution of shearing strains caused by torsional unrestrained warping as shown in Figure 12.1b. Equally spaced lines forming a square grid on the surface distorts due to the existence of shearing strains and changes the right angle of the grid.

Elastic buckling can be demonstrated using a slender prismatic section made out of metal, plastic, or wood resting one end on a smooth surface and pressed by the fingers on the other end as shown in Figure 12.1c. If the same member were resting on a small platform balance, the actual bucking load could be determined.

EDUCATIONAL MODELS FOR CIVIL AND ARCHITECTURAL ENGINEERING 683

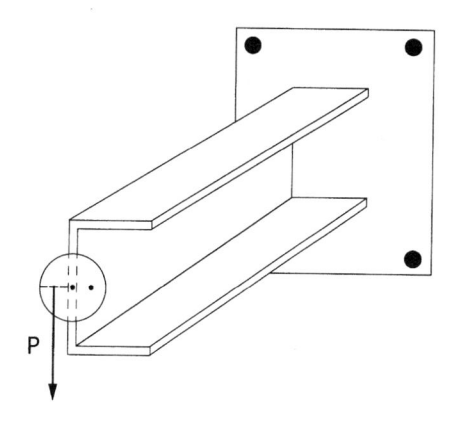

Figure 12.2 Shear center of thin open sections.

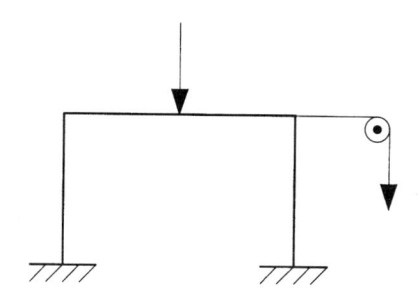

Figure 12.3 Frame action under lateral and gravity loading.

12.3.2.2 Shear Center of Thin Open Section Beams

To demonstrate the concept of the *shear center* of thin open section beams or the point through which the loads must pass in order that only bending will take place, without any torsion, small-scale plastic models can be used. The advantage of using plastics such as PVC or Plexiglas is that large elastic displacements can be generated with relatively small loads. Such a demonstration device illustrating how to locate the shear center in a channel section is shown in Figure 12.2. Small loads can be hung from a small plastic target glued to the end section at various locations as shown in Figure 12.2 until the section bends without twisting.

12.3.2.3 Frame Action

The bending of frame structures under vertical and horizontal loads can be readily demonstrated using a board-mounted spline-type model (Figure 12.3) made from any metal or plastic. Brass usually works well for these models because it can be easily soldered. Another material that works well for making realistic frame models is the machined PVC sections in the form of I, C, and L sections available in hobby and toy stores. The effect of support deformations and moment distribution can be demonstrated with such models by building joints that can be fixed or released at will during the demonstration.

12.3.2.4 Composite Behavior

Composite structural action can be demonstrated in the classroom with the three small-scale models shown in Figure 12.4. Model 1 (Figure 12.4b) consists of a set of three small balsa wood

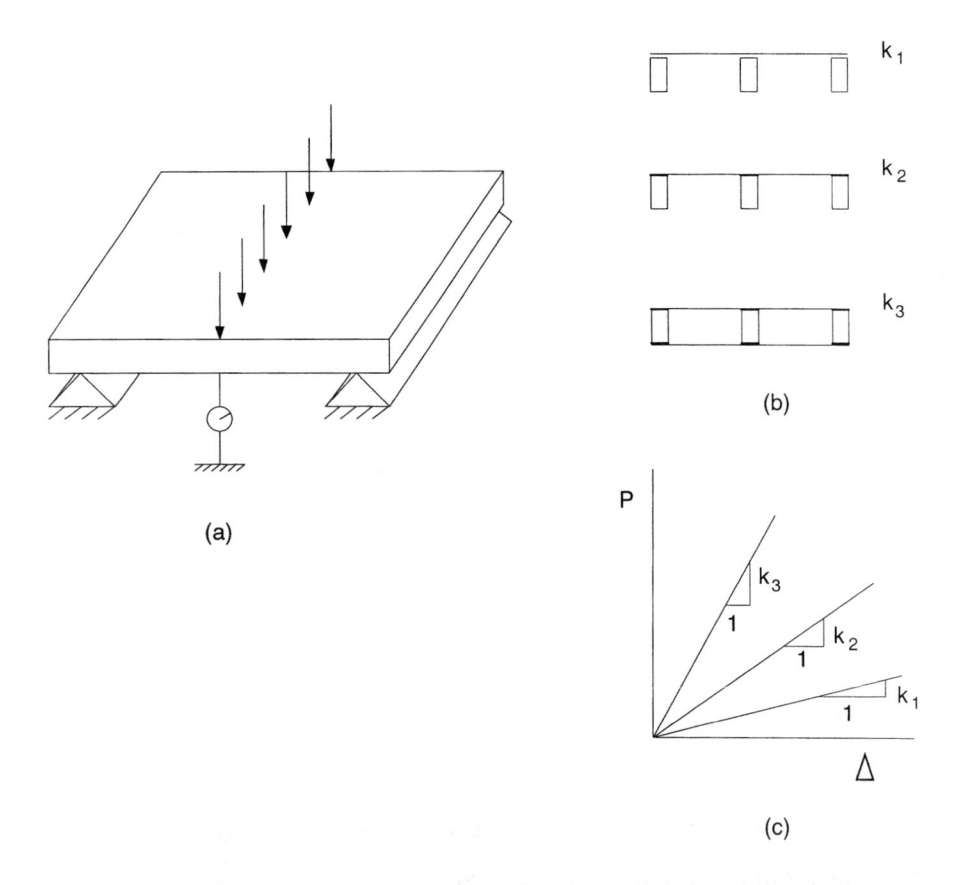

Figure 12.4 Composite action demonstration.

beams to which a slab made of thin cardboard rests unattached. The beams and the slab deflect separately under load (Figure 12.4 a) and the deflection under small graduated weights is measured with a dial gage. Model 2 and Model 3 are identical to Model 1 except that they have their slabs glued to the beams. In addition, Model 3 has a bottom slab glued to the bottom of the beams (Figure 12.4b). Similar load–deflection results are obtained for the Model 2 and Model 3 and the results compared as shown in Figure 12.4c. Only two to three points need be measured for each model since the load deflection behavior is linear. If an electronic deflection-measuring sensor such as an LVDT and its associated signal conditioning interface and software (Chapter 7) is used, the results can be displayed on the screen of a laptop computer and the load can be applied by hand pressure. It will be obvious from the three load–deflection curves obtained that there is a tremendous advantage gained (i.e., substantial increase in stiffness) in making the slab and the beams of the small-scale model work by composite fashion, i.e., acting together to carry the load.

12.3.2.5 Folded Plate Action

Of all the small-scale demonstration models that can be used in the classroom, the ones that illustrate the enormous load-carrying advantages that can be gained by changes in structural form are the most impressive. This point is well illustrated by comparing the load-carrying capacity of a plain sheet of writing or duplicating paper 216 × 279.4 mm supported at its longer ends. The sheet of paper has zero load-carrying capacity because it has no flexural resistance and will probably slide to the floor. Now take that same sheet of paper and fold it into longitudinal folds of equal width to form it into a folded plate. The pleated paper has considerably more stiffness; however,

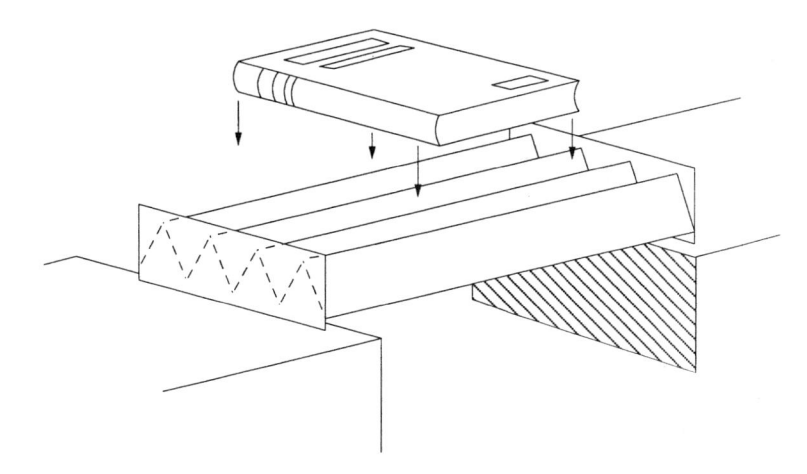

Figure 12.5 Folded plate action.

if loaded it tends to open up. It needs a perpendicular structure or "diaphragm" to prevent the pleats from changing shape. To achieve this, cut two equal strips of paper from another sheet and tape them with cellophane tape to the pleated folded plate thus forming two end diaphragms. The folded plate model of very thin slabs has remarkable stiffness and if supported on the end diaphragms can carry the heaviest textbook that you can find in the classroom laid slowly on the top of its ridges. This is certain to impress any student.

12.3.3 Laboratory Demonstration Models

Laboratory demonstration models are used in quite a different manner from classroom demonstration models. They are especially designed for measurement and are thus accurately constructed and instrumented. The instrumentation can vary from simple to complex depending on the type of model and the quantities that have to be measured. These models are primarily intended for students to use to measure quantities such as loads, deflection, strain, or natural frequencies and mode shapes. Comparisons are usually made with theoretical predictions that student's are already familiar with, thus enabling them to become more aware of the physical significance by observing the structural behavior. In more complex structural situations, the student may not yet be aware of the theoretical concepts but can obtain significant insight into structural performance by observing structural behavior firsthand. In addition, the student becomes familiar with structural testing and experimental techniques.

All laboratory demonstration models are tested on specially fabricated strong tables simulating the strong floor systems used to test full-scale structures and described in Chapter 8. These type of steel tables were originally reported by White (1972) and have been duplicated in some form or another at many university structural models laboratories including the one at Drexel University, where most of the examples shown in this chapter were tested.

12.3.3.1 *Simple Beams and Trusses*

Statically determinate beams, although simple to analyze, can be effectively used to demonstrate structural behavior to the novice. Deformations and strains can be measured by students and compared to theoretical predictions that they can perform. Gaining this important laboratory experience, students can then apply the modeling technique with confidence to more complex structural behavior situations which they may or may not have the theoretical tools to analyze the test structure fully.

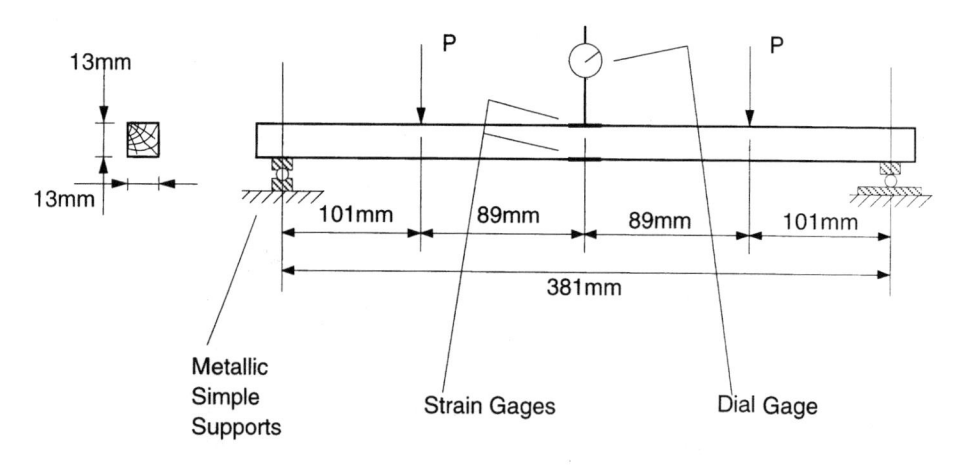

Figure 12.6 Simple beam under two-point loading.

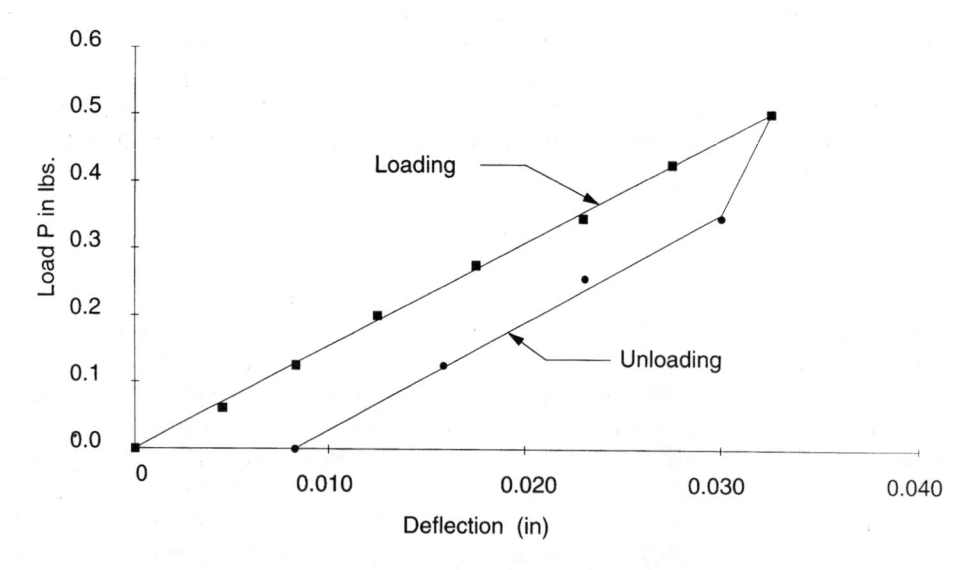

Figure 12.7 Load–deflection curves.

The first laboratory experiment that the student is asked to perform in an introductory structural analysis course with an experimental component is to determine the load–deflection and load–strain curves of a model wooden beam (Figure 12.6) that has been instrumented with strain gages on the top and bottom critical section fibers. From the experimental data observed by the student, Young's modulus and the bending stress–strain relationship are also determined. Typical load–deflection results from such an experiment are shown in Figure 12.7. Some delay in the unloading portion of the curve is due to the viscoelastic properties of the material. The bending stress is determined using the measured moment and dimensions of the model beam from the relation:

$$\sigma = \pm Mc/I \tag{12.1}$$

where: M = moment
c = distance from centroid to extreme fibers where strain gage is located, and
I = moment of inertia of beam section about its bending axis

The Young's modulus, E, is determined from the equation for the measured midspan deflection, δ_c, which follows:

$$E = Pa/24\, \delta_c I\, (3L^2 - 4a^2) \qquad (12.2)$$

where: P = half the total load on the beam
L = span length of the beam
a = distance from support to load point
δ_c = vertical deflection at midspan
I = moment of inertia of beam section about its bending axis

The basic theory of analysis of statically determinate structures is all that is needed to understand this experiment fully.

The simple balsa wood model bridge truss (like the one shown in Figure 3.20a) can be investigated under simulated moving loads. Deflections and bar forces can be determined experimentally and compared to theoretical predictions using a computer code for pin-jointed trusses such as the ones provided in the structural analysis course on statically determinate structures or by hand computation using the virtual work method (Equation 12.3).

$$1 \cdot \Delta = \sum uSL/AE \qquad (12.3)$$

where: 1 = external virtual unit load acting on the truss joint in the stated direction of Δ
u = internal virtual force in a truss member caused by the external virtual unit load
Δ = external joint displacement caused by the real loads on the truss
S = internal force in a truss member caused by the real loads
L = length of member
A = cross-sectional area of member
E = modulus of elasticity of member

The basic assumptions of truss analysis can thus be confirmed and appreciated by the student. Magnitude of specific bar forces can be verified by the student if the member is provided with strain gages (usually two symmetrically placed gages at midlength) and a portable strain indicator such as the one discussed in Chapter 7.

The accuracy of both the beam and the truss experiments can be made to be within 10% of the theoretical predictions when care is taken to use materials with uniform properties, accurate fabrication, and accurately determined geometric properties. Both these experiments are easy to execute by the student, taking less than an hour to perform, and can form an excellent basis of understanding of the real structural implications of the theory learned in their first structural analysis course.

12.3.3.2 Continuous Beams

The next level of challenge for the student is the model study of statically indeterminate structures covered in a second undergraduate or first-year graduate course in structures. These are structures whose internal forces cannot be determined from static considerations alone. Multiple span beams and fixed-ended frames fall into this category.

Continuous beams can be investigated under moving loads using any linearly elastic material such as wood, plastic, or metal. A Plexiglas I-shaped beam having the dimensions shown in Figure 12.8a is used in this experiment. Use the Müller–Breslau principle (Chapter 3) to determine the influence line for the reaction at B at a spacing interval of 50 mm. To accomplish this, reverse

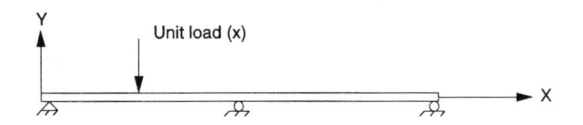

(a) Unit moving load on a two span beam

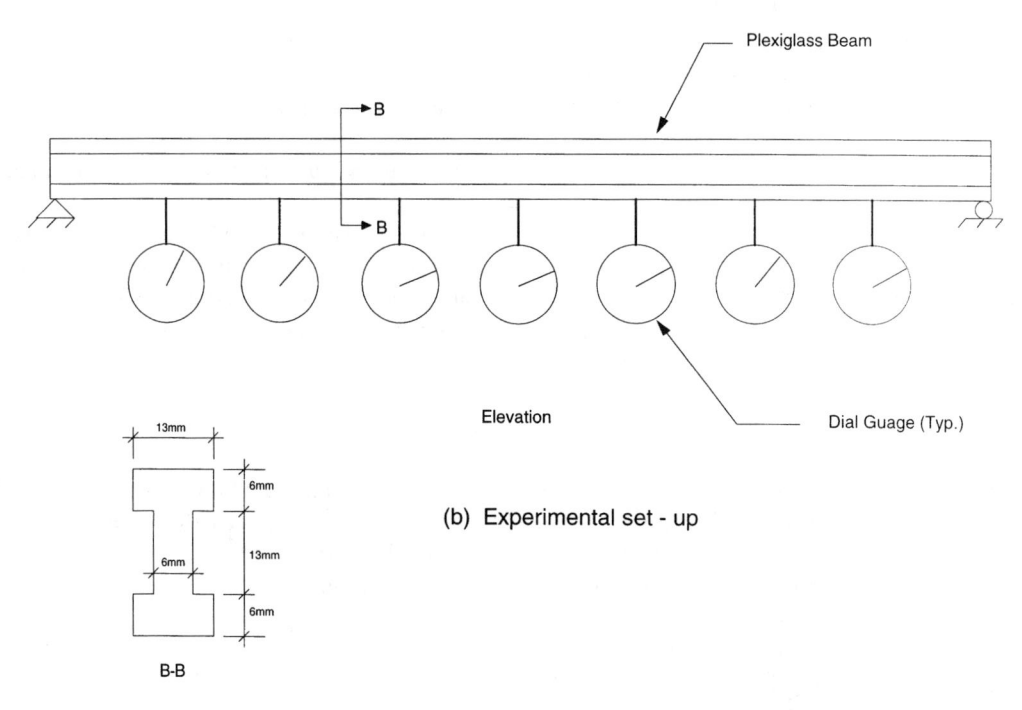

(b) Experimental set - up

Figure 12.8 Influence lines of statically indeterminate structures using the Müller–Breslau principle.

the reaction and then change the sign of the plotted ordinates. Figure 12.8b shows the test setup. Typical experimental results can be compared to the theoretical influence line ordinates obtained using the conjugate beam method:

$$R_b = (3L^2x - x^3)/2L^3, \quad 0 < x < L \tag{12.4}$$

By using symmetry, the rest of the influence line can be plotted. A comparison of the experimental and theoretical curves is shown in Figure 12.9a.

Use the same procedure to find the influence line for the reaction at point A. Use this influence line to obtain the influence line for the moment at B, M_b. Note that M_b (the negative moment over the interior support) is symmetric about B for equal spans. Check the experimental results against the theoretical value obtained using the conjugate beam method:

$$R_a = (4L^3 - 5L^2x + x^3)/4L^3, \quad 0 < x < L \tag{12.5}$$

and

$$M_b = R_a L - (1)(L - x)$$

$$= -x(L^2 - x^2)/4L^2, \quad 0 < x < L \tag{12.6}$$

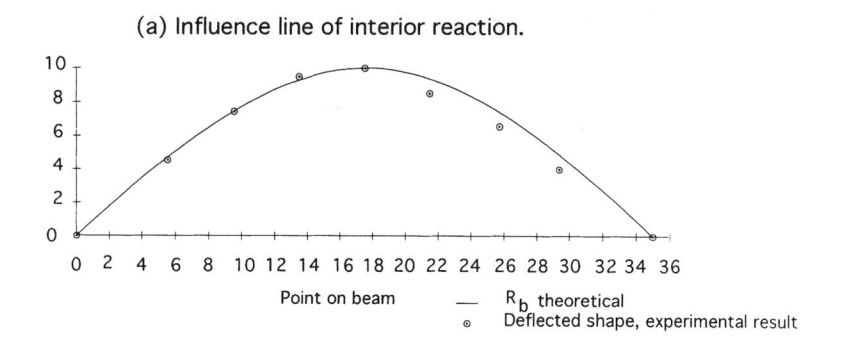

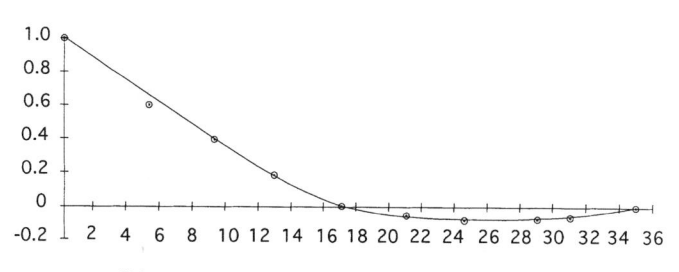

EDUCATIONAL MODELS FOR CIVIL AND ARCHITECTURAL ENGINEERING

(a) Influence line of interior reaction.

(b) Influence line for end reaction.

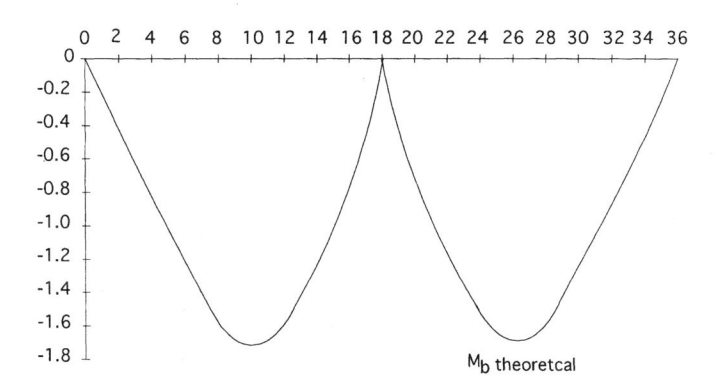

(c) Influence line for moment at interior support.

Figure 12.9 Influence lines for two-span continuous beam.

Typical results for the influence line of the reaction at A and the moment at B are given in Figure 12.9b and c, respectively.

This experiment is of intermediate difficulty and can be done in a 2-h period. The accuracy is within 10% of the theoretical predictions using the equations given. It helps the student fix ideas of influence lines of redundant structures and to better understand the Müller–Breslau principle.

An application of the experimentally derived influence lines obtained above can then be asked of the student. Suppose that the Plexiglas beam with a Young's modulus of 3.45 GPa represents a steel wide-flange section to a scale of $S_L = 48$. What will be the maximum negative moment for the H 20-16 series truck load in the prototype beam? Assume that the transverse section of the beam is such that all of the H 20-16 load acts on the beam.

(a)

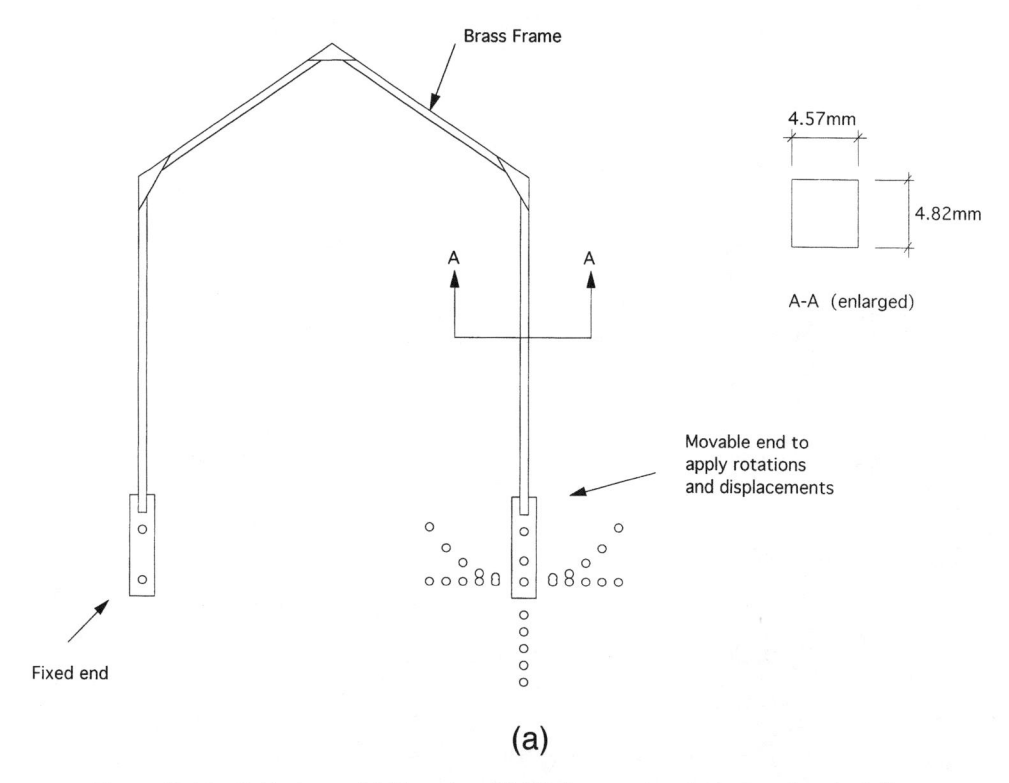

Figure 12.10 Gable frame. (a) Plan view. (b) Displacement under imposed end rotation.

12.3.3.3 Gable Frames

The Müller–Breslau principle can be used to find the redundant reactions for any loading of the gable frame shown in Figure 12.10 which is indeterminate to the third degree. The frame is made of brass and has the dimensions shown in Figure 12.10a. It is mounted on a smooth wooden board topped with a sheet of Plexiglas for frictionless movement. By giving end displacements in the direction of the three redundant forces (horizontal and vertical forces and bending moment at the right support), the influence lines of the frame can be drawn on graph paper with reference the original undeflected shape (Figure 12.10b) and then digitized for analysis. Values of the redundant forces can be computed for given horizontal and vertical loading (or, for example, wind loading) using the Müller–Breslau principle (Chapter 3).

This experiment is of intermediate difficulty and can be done in a 2-h period. The accuracy is within 10% of the theoretical predictions. It helps the student fix ideas of influence lines of redundant structures and to better understand the Müller–Breslau principle. The last two experiments using indirect models to obtain influence lines of redundant structures fit in well with a course in structural analysis of indeterminate structures.

12.3.3.4 Euler Buckling Load

Demonstration of the Euler buckling load can be easily studied by the student using very simple models. The equipment needed are a prismatic straight slender column made of any elastic material, small platform scale, ruler, and string. The column is first deflected by a central load when supported in a horizontal position as shown in Figure 12.11a. The ruler is pressed downward until some

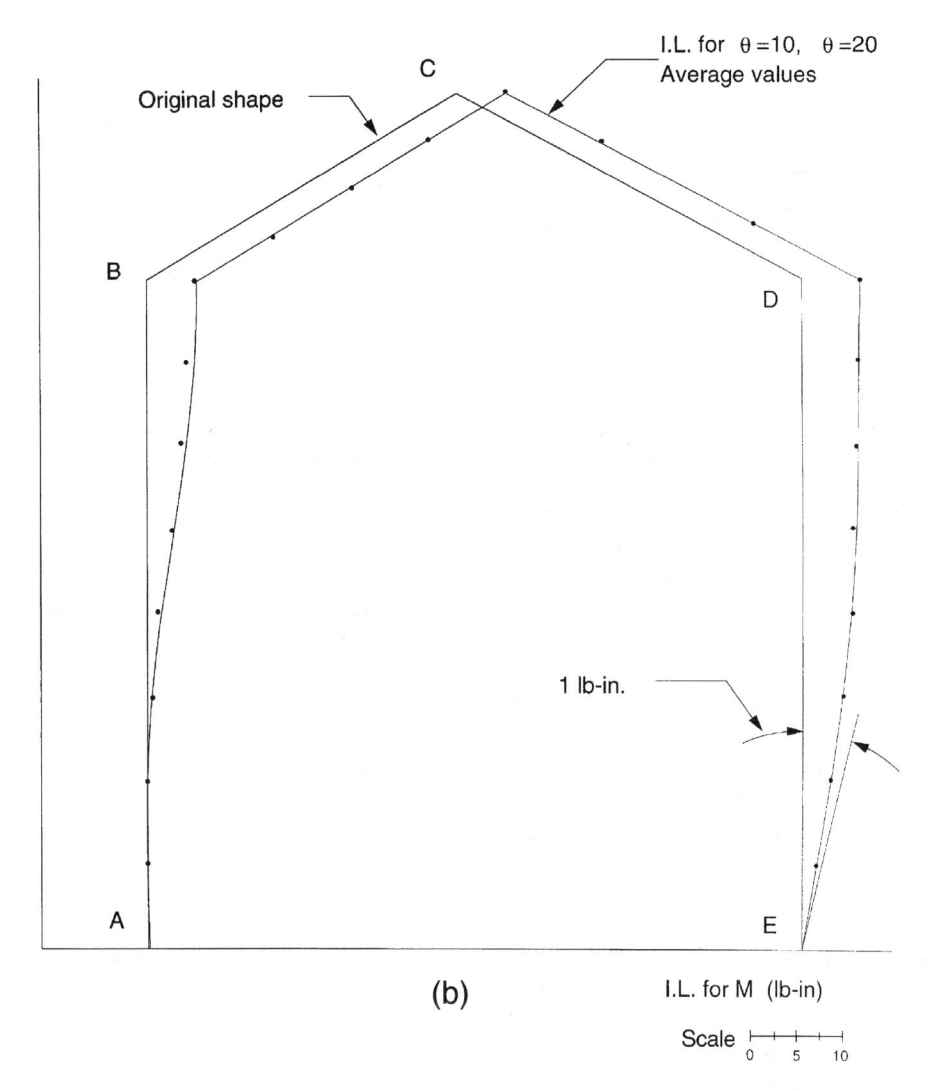

Figure 12.10 (continued)

definite deflection, Δ, is produced. The corresponding reaction, R, at the end of the beam is simultaneously read using the small platform scale. The centrally applied deflecting force, Q, on the beam is therefore equal to $2R$. The relation between deflection, Δ, and load Q is given by

$$\Delta = QL^3/48EI \tag{12.7}$$

Therefore, $EI = RL^3/24\Delta$. Substituting this for EI in the Euler buckling formula, the theoretical buckling load becomes

$$P_c = \pi^2 EI/L^2 = (\pi^2/L^2)(RL^3/24\Delta) = \pi^2 RL/24\Delta \tag{12.8}$$

Substitution of the values of R, I, and Δ into Equation 12.7 gives the value of the theoretical buckling load. Next the column is placed vertically with one end resting on the platform scale (Figure 12.11b).

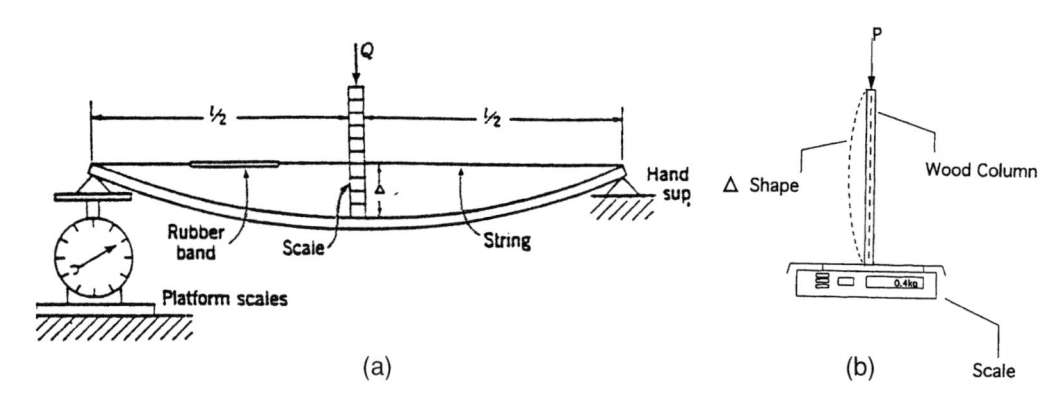

Figure 12.11 Simple elastic buckling experiment. (a) Column under end load and lateral load. (Adapted from Fairman, S. and Cutshall, C. S., 1953.) (b) Simply supported column on platform scale.

The upper end is pressed down carefully until the column deflects slightly, and the load is read on the scale. The computed value of P_c should be similar to the value read on the scale.

This experiment can be performed by students who have taken statics and strength of materials. It is easy to perform and takes approximately 1 h of laboratory time. The accuracy achievable is usually less than 15%. Care must be exercised in applying the required pin-ended conditions.

12.3.4 Architectural Engineering Models

Architectural engineering models range from simple demonstration models that give basically qualitative results to sophisticated scaled models of structural components or structural systems that can provide accurate quantitative test results. Examples of both types are illustrated below.

12.3.4.1 Diaphragm Action

Diaphragms are the structural elements that distribute the lateral loads from wind or earthquake origin to all the vertical load-carrying members. Efficient diaphragm action requires that horizontal elements of multistory structures act in a rigid manner to be able to distribute the horizontal load. Typical floor layouts of high-rise buildings can range from the cross or partition wall layout to complete skeleton or exterior wall and central core. The action of the floor diaphragm at each floor level is to distribute the horizontal loads to the lateral load-carrying elements.

To demonstrate this structural action, three three-story models are tested in this experiment under horizontal load (Figure 12.12a). The models are made of four bass wood square columns in a square layout representing a building frame at approximately ¹⁄₁₅ scale. The columns are fixed to a wooden base and horizontal load is applied by hanging small weights to string over pulleys. Three types of diaphragms made of different thickness cardboard are studied to determine the effectiveness of the diaphragm. Diaphragm effectiveness is measured by its ability to limit horizontal drift under lateral symmetric and unsymmetric load. Typical load–deflection curves from the model tests (Figure 12.12b) indicate the ability of the more rigid diaphragms to limit lateral displacements and to distribute the horizontal load to all four floor columns.

This experiment is easy to perform within a 2-h period. An accuracy of less than 10% can be achieved with theoretical predictions made from simple hand computations or finite element analysis (FEM). Any of the available planar (symmetric loads) or three-dimensional frame programs (unsymmetric loads) can be used for analysis. The input to the FEM program consists of the geometry, material properties of the model, and the magnitude and location of the loading. Separate tests

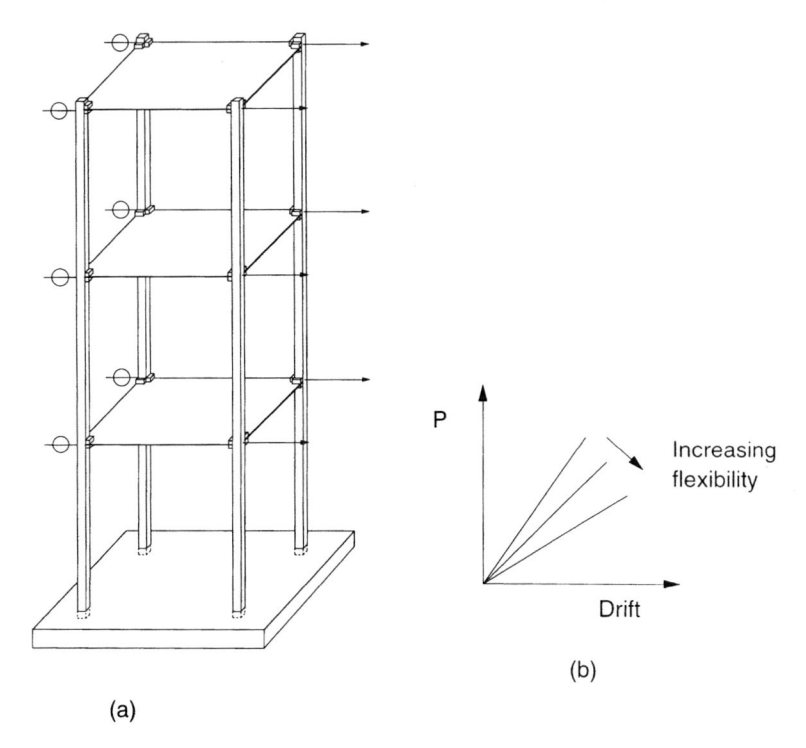

Figure 12.12 Experiment to study diaphragm action.

must be performed to obtain the properties of the bass wood columns and paper diaphragms. To appreciate the implications of the structural behavior fully, the student should have studied indeterminate structural analysis, although a qualitative feel of the structural behavior can be obtained after an introductory analysis course in statically determinate structures.

12.3.4.2 Infilled Frames

The use of structural walls and infilled frames in multistory buildings to resist lateral loads originating from wind or earthquakes has long been recognized. In the case of structural walls, a large portion of the lateral load, if not all, and the horizontal shear force resulting from the load are often assigned to these elements; hence they have been called shear walls.

Infilled masonry (Figure 12.13a) is another method of lateral stiffening of frame structural systems. An experiment that measures this stiffening effect is carried out using the three-story bass wood model used above to study diaphragm action. Cardboard panels are attached at each floor level of the frame to stiffen it in the direction of loading (Figure 12.13b). Since the elastic behavior of the masonry infill is being studied, this is permissible. Load and deflection measurements are made in the same way as in the previous experiment. Comparison of the horizontal load–deflection characteristics of the bare and the infill masonry–stiffened model is made.

This experiment is of intermediate difficulty and can be performed by students who have taken an analysis of indeterminate structures course and possibly an introduction to FEM techniques. It takes approximately 2 h of laboratory time. The accuracy achievable is usually less than 15%. Comparison of the experimental results are made with a simplified approximate structural analysis to get "ball park" results and with a more accurate FEM. Care must be exercised in applying the required stiffening panels and obtaining the material properties of the structural components.

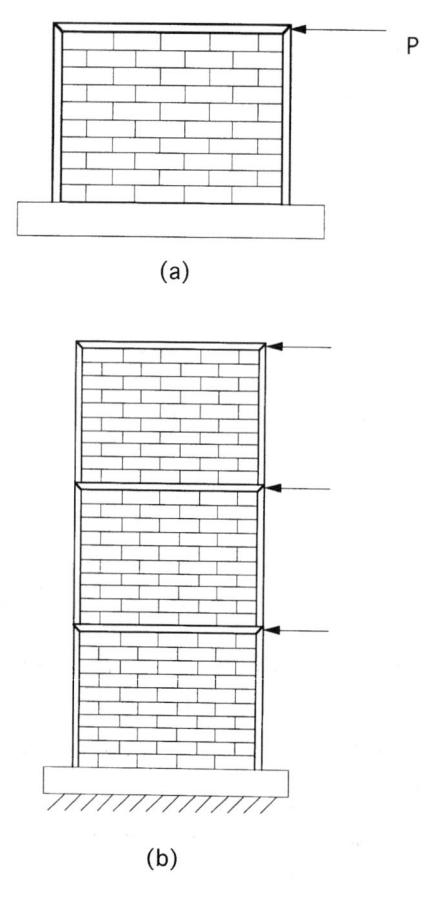

Figure 12.13 Masonry infilled frame models.

12.4 NONLINEAR AND INELASTIC STRUCTURAL BEHAVIOR

12.4.1 Ultimate-Strength Models of R/C Components

Nonlinear inelastic behavior of reinforced concrete structures is easily demonstrated to students using small-scale wire-reinforced microconcrete models. A large variety of structures can be studied using the techniques described earlier (Chapters 4 through 7). Only a few of the most important behavior patterns are illustrated in the examples chosen. It is required that an undergraduate course in the design of reinforced concrete structures has been taken by the student prior to attempting any of the following experiments.

12.4.1.1 Underreinforced/Overreinforced Concrete Beams

The flexural behavior of reinforced concrete beams is studied in an extended laboratory exercise over a 4-week period where the students, usually working in groups of three or four, help in the design of two approximately 1/10-scale underreinforced and overreinforced beams (Figure 12.14). They then take the leading role in the fabrication, casting, instrumentation, determination of the properties of all materials used in the models, beam testing, and data reduction culminating in a report documenting their collective and individual experience (Baker, 1995; Grove, 1995; Ronan, 1995; Wisser, 1995). They are assisted in their effort by the instructor and a laboratory-experienced

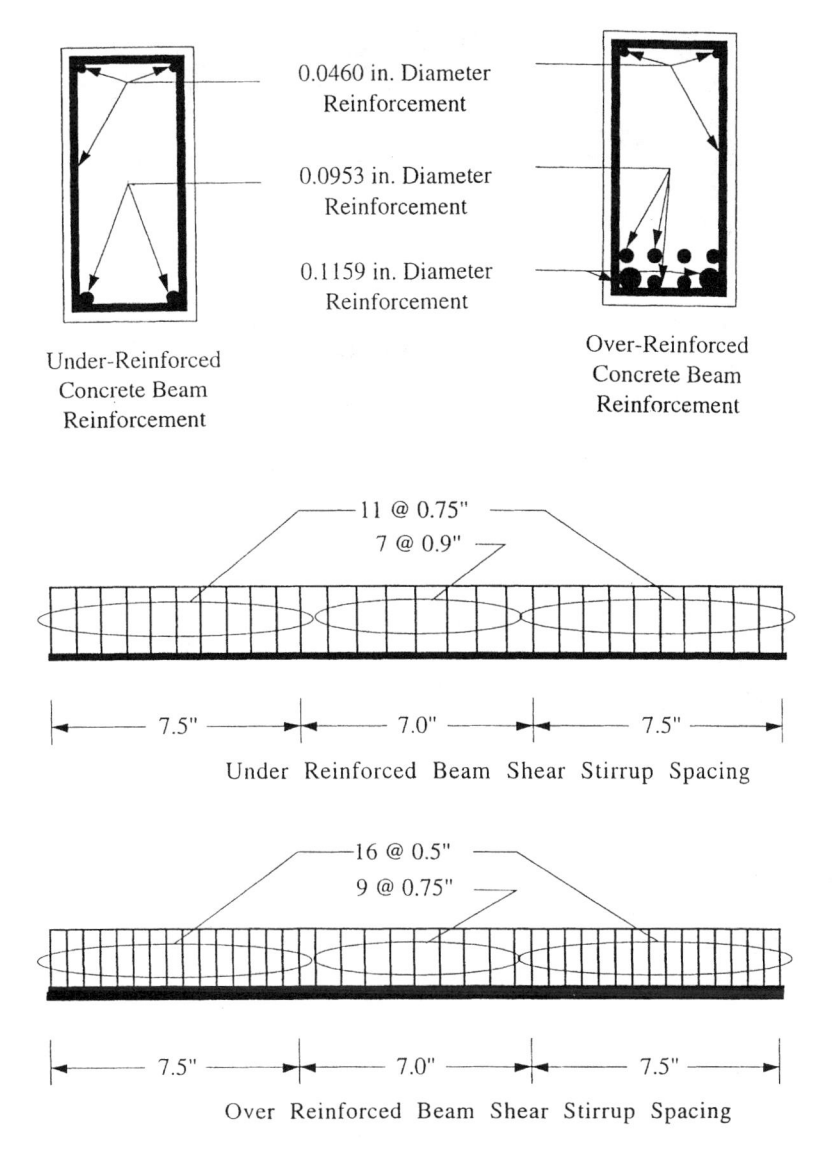

Figure 12.14 Model underreinforced and overreinforced beams.

teaching assistant. In addition to reporting and analyzing the test results, they must answer a set of questions on their observations covering such topics as experimental accuracy, adequacy of the model materials used, and the implications of the test results. Each student in the laboratory course makes a theoretical prediction of the expected ultimate strength of each beam and the winner who comes closest to the test value receives a prize.

The students help in the fabrication of the beam reinforcement and the casting and preparation of all specimens (Figure 12.15). Material test specimens are cast together with the model beams in order to study the compressive and tensile strength as well as the modulus of elasticity of the model concrete. The stress–strain behavior of all model reinforcement is also determined by the students (Figure 12.16). In addition to the standard 50×100 mm cylinders, smaller 25×50 mm cylinder specimens of the model concrete are cast together with the beams to determine the compressive and tensile strength. Unreinforced plain model concrete beams are also cast to determine the modulus of rupture and the modulus of elasticity. These beams are tested under two-point loads placed at the

Figure 12.15 Cast model beams and material control specimens.

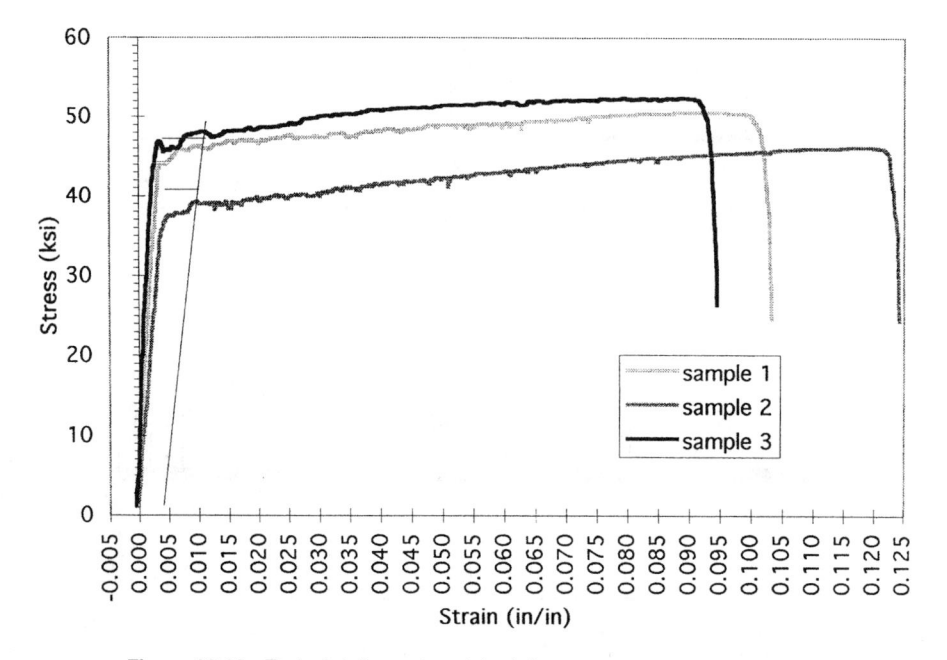

Figure 12.16 Typical deformed model reinforcement stress–strain curves.

third span and the deflections are measured with dial gages as shown in Figure 12.17. The modulus of elasticity of the model concrete is determined from the plain concrete beam elastic load–deflection curves. The model material strength values are used by the students to refine the theoretical calculations (which were based on nominal strength values) included in their report. The number of specimens cast and their sizes permit the students to study the effect of size and age of the model concrete in compression and tension by testing at 7, 14, and 28 days. Strength–age curves for the compressive strength are shown in Figure 12.18.

Testing of the model beams is planned at 28 days in such a way so as to allow load–deflection and moment–curvature relationships to be determined. The test setup is shown in Figure 12.19 where different types of instrumentation are used. The purpose of this is to give the student an appreciation of the different techniques used in obtaining deflection data, for cross correlation of the results, and for the safety of the more expensive LVDTs which can be disconnected before the ultimate load is reached to prevent damage to them. A Tinius-Olsen 44.5 kN miniature, bench-type

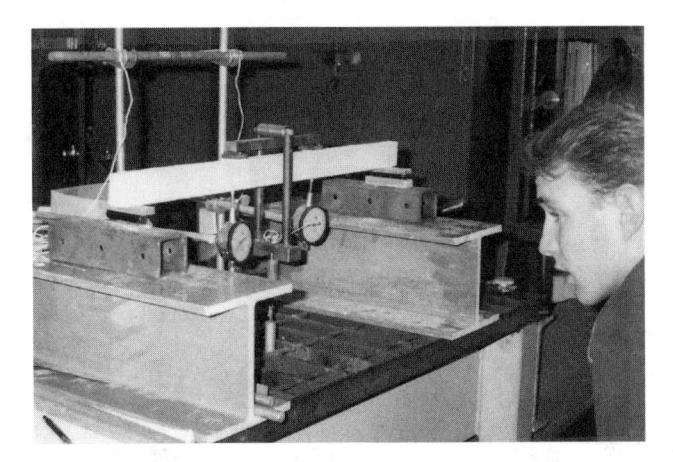

EDUCATIONAL MODELS FOR CIVIL AND ARCHITECTURAL ENGINEERING   697

Figure 12.17 Plain model concrete beam test.

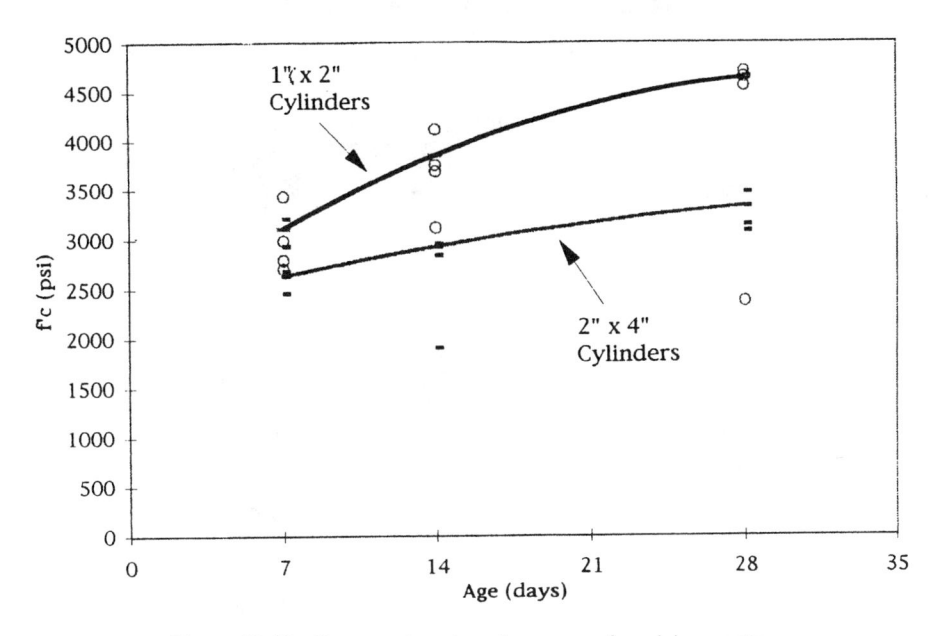

Figure 12.18 Compressive strength vs. age of model concrete.

universal testing machine controlled by an IBM PC (Figure 12.19) is used to obtain the material strength properties and to test the model beams. The students record all of the information and observe the behavior of the beams. Figure 12.20 shows the underreinforced concrete model beam after extensive cracking and yielding of the reinforcement.

The failed model beams together with some of the test specimens are shown in Figure 12.21a. The results of the load–deflection behavior of the underreinforced and overreinforced model beams are shown in Figure 12.21b. A comparison of the moment–curvature relations of the two beams, contrasting the ductile behavior of the underreinforced beam and the brittle behavior of the overreinforced beam, is given in Figure 12.21c. The concept of ductile and brittle failure is very well illustrated by these two small-scale model beams.

This experiment falls in a more demanding category than the elastic models discussed above. It requires a considerable amount of material preparation and planning that is facilitated by a properly equipped models laboratory. It requires four 2-h periods to do the casting, instrumentation,

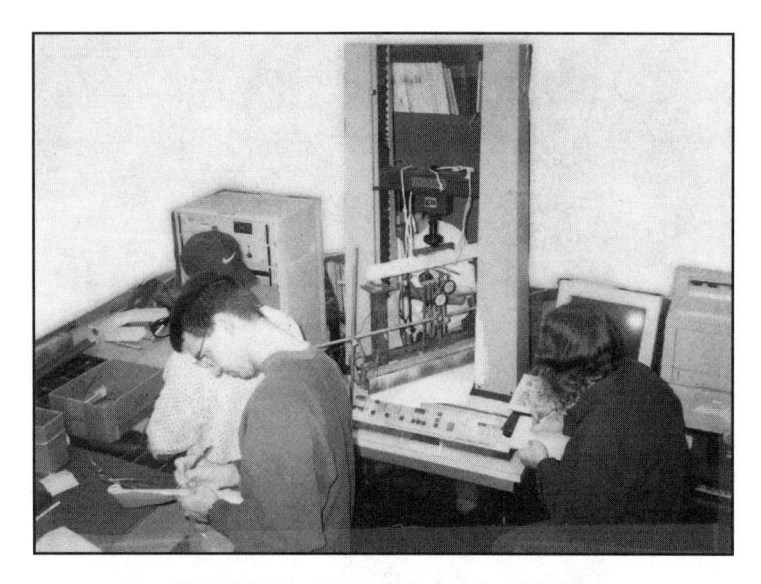

Figure 12.19 Model beam test in progress.

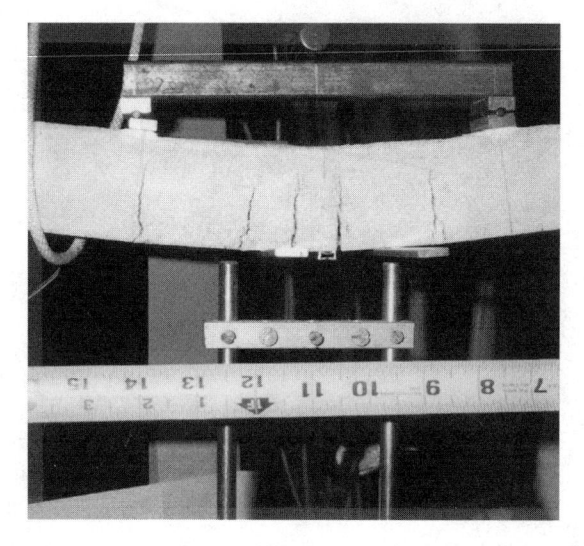

Figure 12.20 Underreinforced model beam during testing.

and testing. However, some prior help of a technician in the materials preparation (especially the model reinforcement) is necessary. An accuracy of less than 15% is achieved in this experiment.

12.4.1.2 *Eccentrically Loaded Concrete Columns*

As structural members, reinforced concrete columns are primarily intended to act in direct compression. In most instances, however, columns will be subjected to bending in addition to compression. A variety of causes can induce bending in columns and for this reason ACI 318-95 requires that all compression members be designed for a specified minimum eccentricity. The presence of bending in a column is an influencing factor in the behavior of the column when loaded, and can significantly affect the mode of failure of the column. Thus, the behavior of columns in compression and simultaneous bending is an important phenomenon for structural engineers to understand. The experiment demonstrating this behavior involves the design, fabrication, and testing

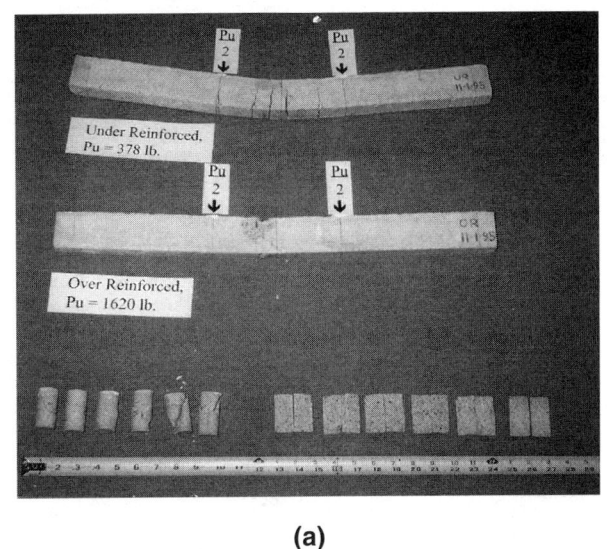

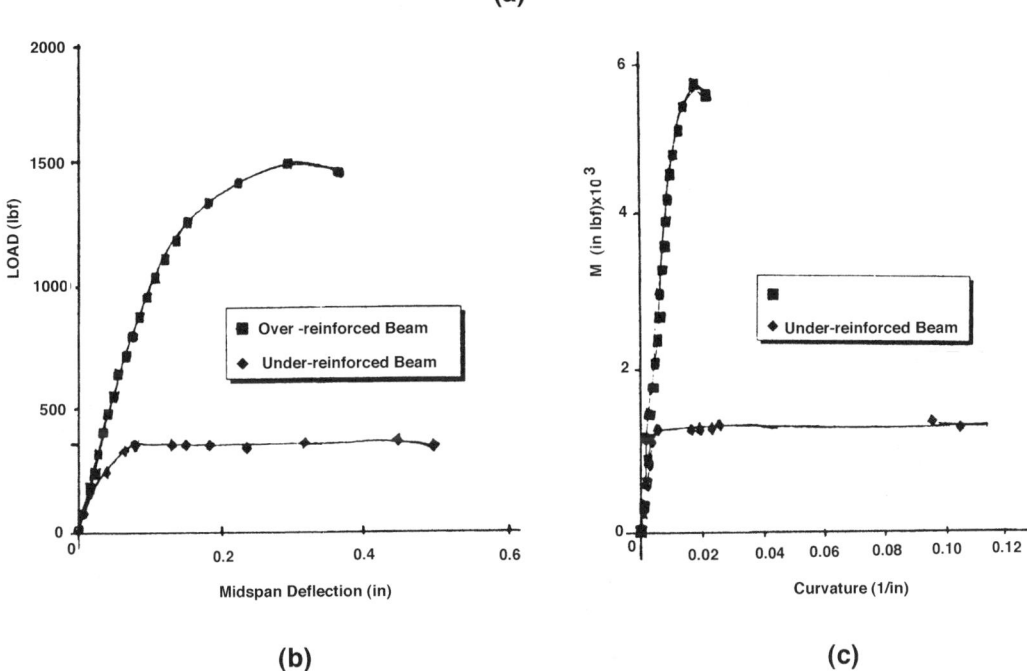

Figure 12.21 Under- and overreinforced model concrete beams. (a) Tested models of underreinforced and overreinforced beams. (b) Load–deflection behavior. (c) Moment–curvature behavior.

of a series of ¹⁄₁₀-scale short columns with varying amount of eccentricity of the applied loads (Felton, 1979). The dimensions chosen (Figure 12.22) have been chosen to eliminate any slenderness effects.

The model column reinforcement consisted of laboratory-deformed annealed wire with four 1.6-mm-diameter bars placed in both the tension and compression faces of the column. The ties consisted of 1.22-mm-diameter wire at 20 mm spacing (Figure 12.23). The model columns were cast in pairs in a Plexiglas mold (also shown in Figure 12.23). The loading and support system is shown in Figure 12.24. Loading was applied through knife-edges at a predetermined eccentricity (e = 0, 12.7, 19, and 25.4 mm). Additional corbel stiffening is provided to ensure that failure takes place in the column and not in the corbel or the connection.

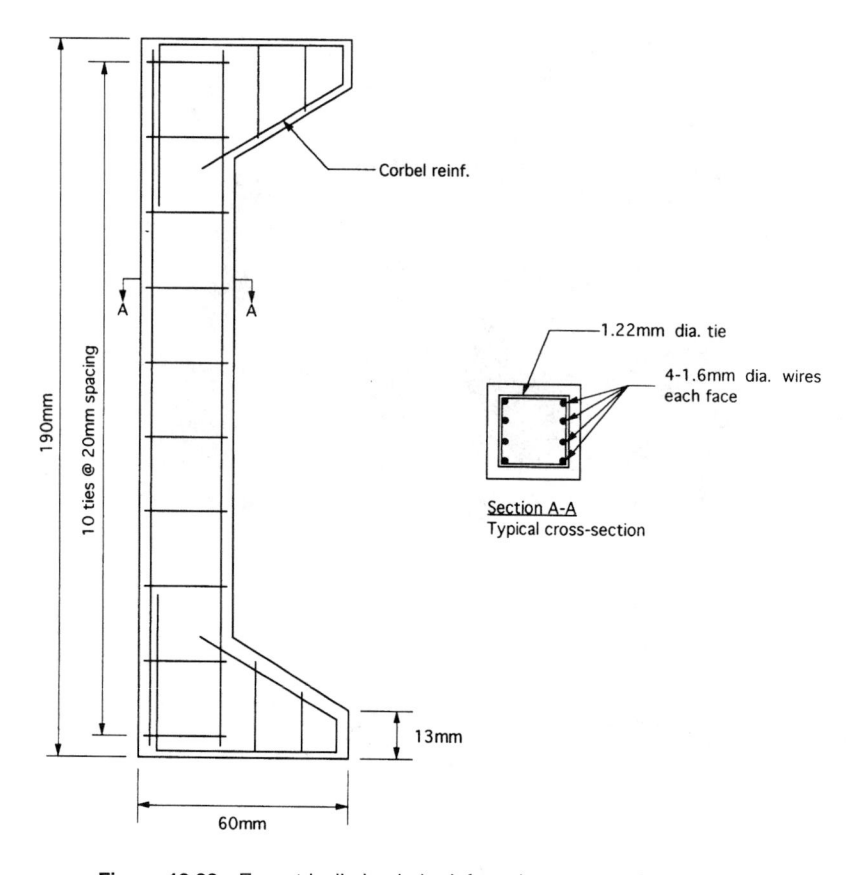

Figure 12.22 Eccentrically loaded reinforced concrete column model.

Figure 12.23 Fabricating techniques for eccentrically loaded columns.

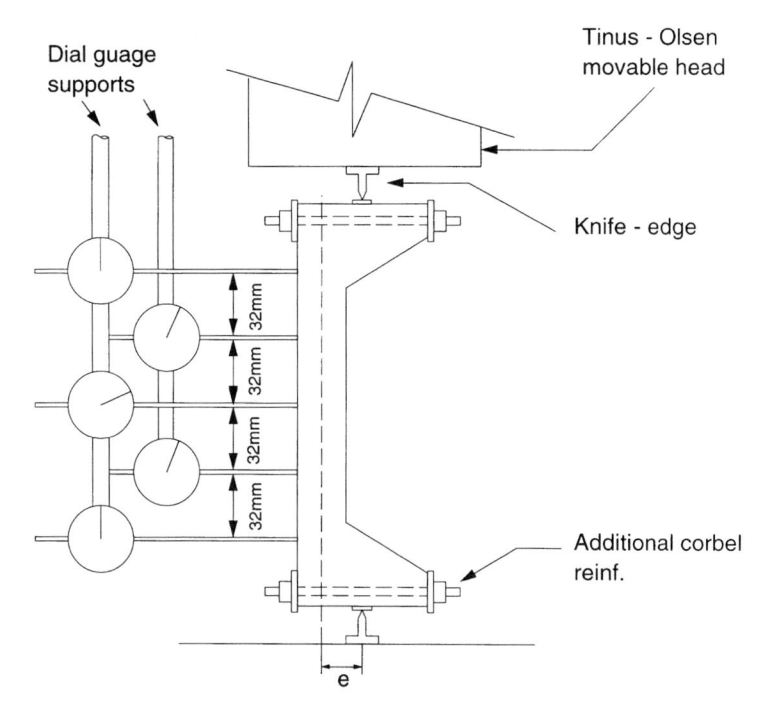

Figure 12.24 Test setup for model column.

Five dial gages are placed along one side of the column at 32 mm spacing to measure the deflections and curvature of the column during testing (Figure 12.24). Thus, the relationship between moment and curvature could be determined using the finite difference method. Typical results of column 4 with the largest eccentricity (25.4 mm) are shown in Figure 12.25. The theoretical interaction diagram between the ultimate axial load and the ultimate moment of the column together with the test results of the four columns tested are shown in Figure 12.26. Table 12.1 summarizes the results of this experiment and includes results from a similar test series (Harris et al., 1966) as well as extrapolations to the prototype scale. As can be seen from the results, the model experiment captures very well the behavior of eccentrically loaded concrete columns and gives the student considerable insight into their ultimate strength characteristics.

This experiment falls in the same difficulty category as the experiment above. It requires a considerable amount of material preparation and planning that is facilitated by a properly equipped models laboratory. It requires three 2-h periods to do the casting, instrumentation, and testing. However, some prior help of a technician in the materials preparation (especially the model reinforcement) is necessary. An accuracy of less than 15% is achieved in this experiment.

12.4.1.3 Two-Span R/C Beams

In addition to the course in reinforced concrete, a course in structural analysis of statically indeterminate structures is needed for this experiment. The two-span reinforced concrete beam is a good example of a single redundant structure where the effects of nonlinear behavior and redistribution of internal forces can be demonstrated to the student. A ¹⁄₁₀-scale model of a two-equal-span continuous beam under two point loads, P each, is shown in Figure 12.27 (Weisel, 1975). The center support can be treated as the redundant to be measured during testing and it is thus monitored with a small laboratory-fabricated load cell (Figure 12.27).

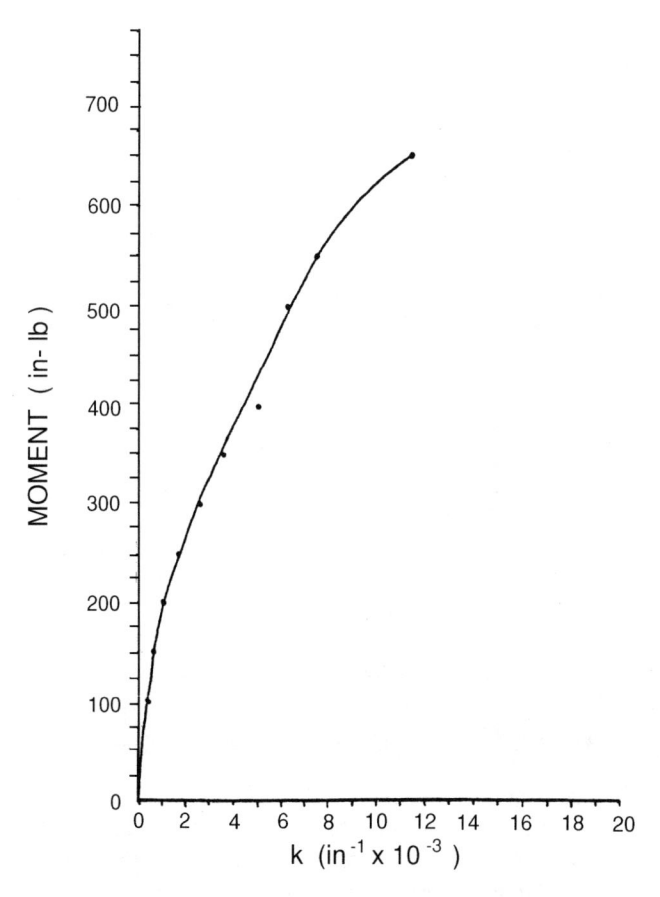

Figure 12.25 Moment–curvature diagram for column #4.

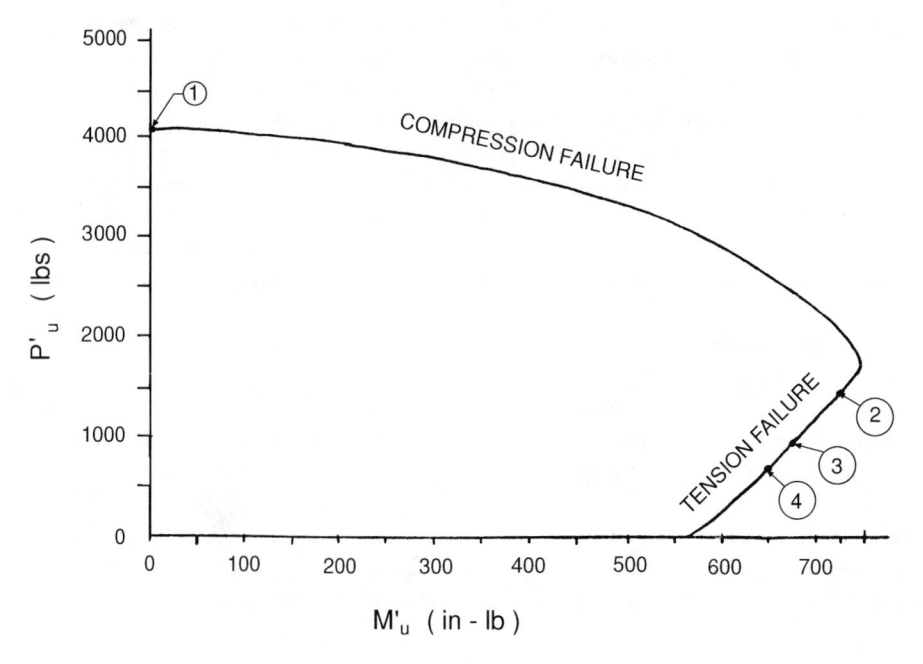

Figure 12.26 Column interaction curve.

EDUCATIONAL MODELS FOR CIVIL AND ARCHITECTURAL ENGINEERING

Table 12.1 Summary of Results

						Model			Prototype	
Column No.	e, mm	f'_c, MPa	f_{split}, MPa	P_{ult}[a], kN	P_{ult}[b] Theoretical, kN	P_{ult}[c] Cornell Study, kN	Comments on Failure	e, mm	P_{ult} Predicted, kN	P_{ult} Theoretical, kN
1	0.0	18.9	3.4	18.24	15.51	17.61	1. Compressive failure 2. Outer shell spalled	0.0	1824	1552
2	12.7	18.9	3.4	6.45	5.69	Not available	1. Tensile failure 2. Cracks in tensile face	127.0	645	569
3	19.1	19.5	3.7	4.00	3.62	4.55	1. Tensile failure 2. Numerous cracks in tensile face	190.5	400	362
4	25.4	19.5	3.7	2.89	2.45	Not available	1. Tensile failure 2. Pronounced cracking 3. Compressive face spall	254.0	289	245

[a] Actual test results.
[b] Theoretical predictions from equations, scaled to appropriate scale by similitude analysis.
[c] Results of a similar study at Cornell University. Note the following differences:
  $f'_c = 21.99$ MPa $f_y = 289.80$ MPa
  $f'_t = 3.69$ MPa   tie-spacing = 19.1 mm

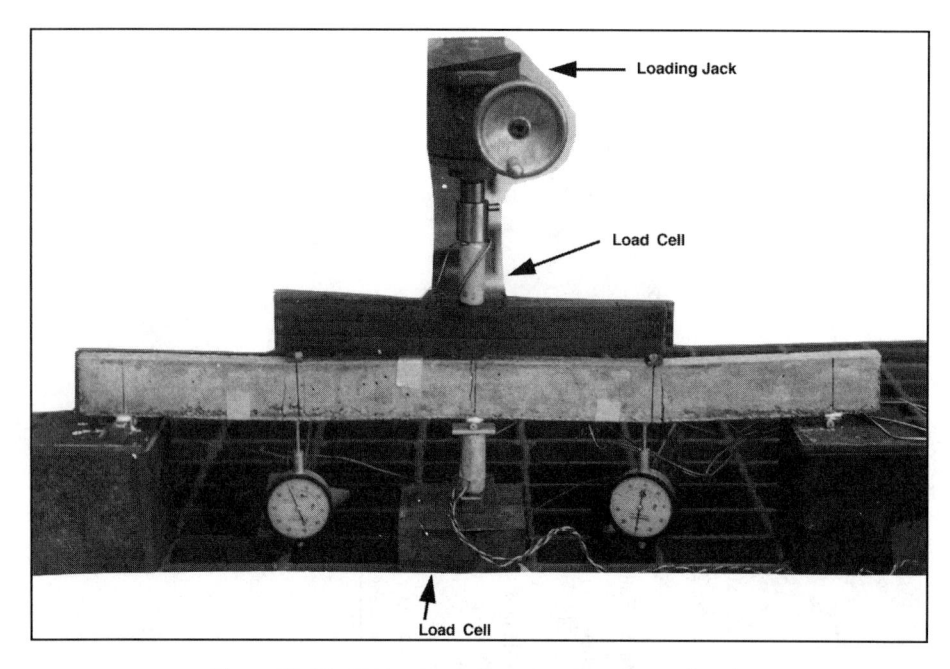

Figure 12.27 Test setup for two-span beam experiment.

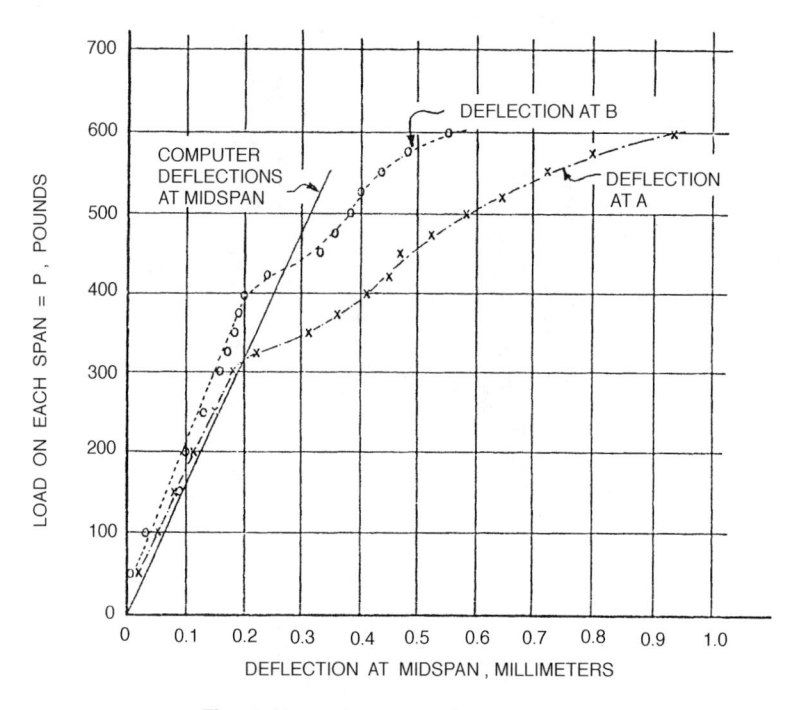

Figure 12.28 Load vs. deflection curves.

Elastic analysis of the statically indeterminate beam shows that the negative moment over the interior support is 2.06P and the positive moment in each span at 125 mm from the end supports is 1.72P. Test results of midspan deflections show good agreement with elastic analysis of the uncracked beam (Figure 12.28). After cracking at the interior support, there is redistribution of the internal forces with the interior support reaction continuously deviating from its initial linear value.

This experiment falls in the same difficulty category as the experiment above. It requires a considerable amount of material preparation and planning that is facilitated by a properly equipped models laboratory. It requires three 2-h periods to do the casting, instrumentation, and testing. However, some prior help of a technician in the materials preparation (especially the model reinforcement) is necessary. An accuracy of less than 15% is achieved in this experiment.

12.4.1.4 Effective Width of Reinforced Concrete T-Beams

In cast-in-place concrete floor systems part of the slab will act with the upper part of the beam to resist longitudinal compression. The resulting beam cross section is T-shaped rather than rectangular. The slab forms the beam flange, while the part of the beam projecting below the slab forms what is called the web or stem. It is evident that if the flange is but little wider than the stem width, the entire flange can be considered effective in resisting compression. For floor systems with widely spaced beams, however, it may be equally obvious that elements of the flange midway between the beam stems are less highly stressed in longitudinal compression than those elements directly over the stem.

When the T-beam is subjected to positive bending moments, part of the slab will act as the flange of the beam resisting the longitudinal compression balancing the tensile force in the reinforcement in the web. When the spacing between the beams is large, it is evident that simple bending theory does not strictly apply because the longitudinal compressive stress in the flange will vary with the distance from the web, the flange being more stressed over the web than in the extremities. This variation in flange compressive stress occurs because of shear deformations in the flange (shear lag), which reduce the longitudinal compressive strain with distance from the web.

An experiment to study this phenomenon (Sleiman, 1984) uses the small-scale modeling technique with strength models to study the problem of effective flange width of the T-beam shown in Figure 12.29. Three configurations of the flange width are studied with the same dimensions and reinforcement of the T-beam. All three models are $1/12$ scale and are made of the same materials. The reinforcement for the models consisted of 2-mm-diameter laboratory-deformed wire for the main bars and 0.79 mm wire for the stirrups. All three models are cast simultaneously in Plexiglas forms using a small variable frequency and amplitude vibrating table.

Instrumentation consists of dial gages along the length and strain gages in the flange at midspan. The test setup is illustrated in Figure 12.30. The moment–curvature curves (Figure 12.31) show a ductile behavior with stiffness increasing with increasing flange width. Strain distributions, measured at different load levels until the ultimate load, show strains approximately 10% less at the end of the flange than at the stem for the widest T-beam configuration (Figure 12.32). The final crack patterns are similar in all three models and the crack spacing is well distributed and extensive in both the beams and the flanges as shown in Figure 12.33.

This experiment falls in the same difficulty category as the experiment above. It requires a considerable amount of material preparation and planning that is facilitated by a properly equipped models laboratory. It requires three 2-h periods to do the casting, instrumentation, and testing. However, some prior help of a technician in the materials preparation (especially the model reinforcement) is necessary. An accuracy of less than 15% is achieved in this experiment.

12.4.1.5 Reinforced Concrete Arches

Reinforced concrete arches form very efficient load-carrying structural systems and are studied in this simplified experiment. The advantage of the arch over a straight beam is that, given identical cross sections, the arch could carry a greater load. This increase in capacity is due to the arch action, which is the action of the horizontal restraining force upon the structure, decreasing the bending moment caused by the loading of the arch and thereby enabling greater loads to be carried.

The configuration chosen for this experiment is the centrally loaded pin-connected circular arch whose model geometry at $1/10$ scale is shown in Figure 12.34 (Bowders 1979). This configuration

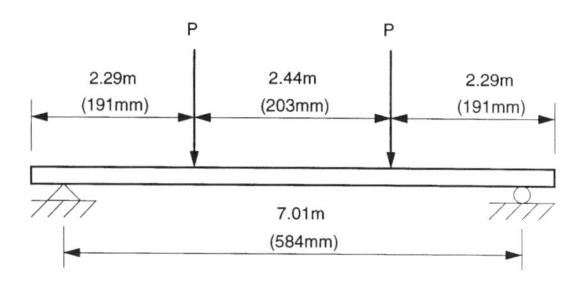

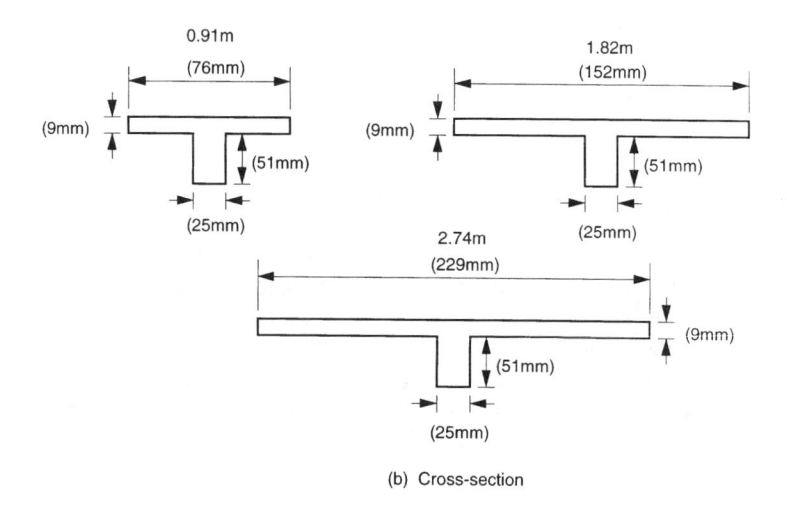

Figure 12.29 Dimensions of T-beam prototype (models).

(a) Elevation - same for all three beams.

(b) Cross-section

Figure 12.30 Loading and instrumentation of the model T-beam.

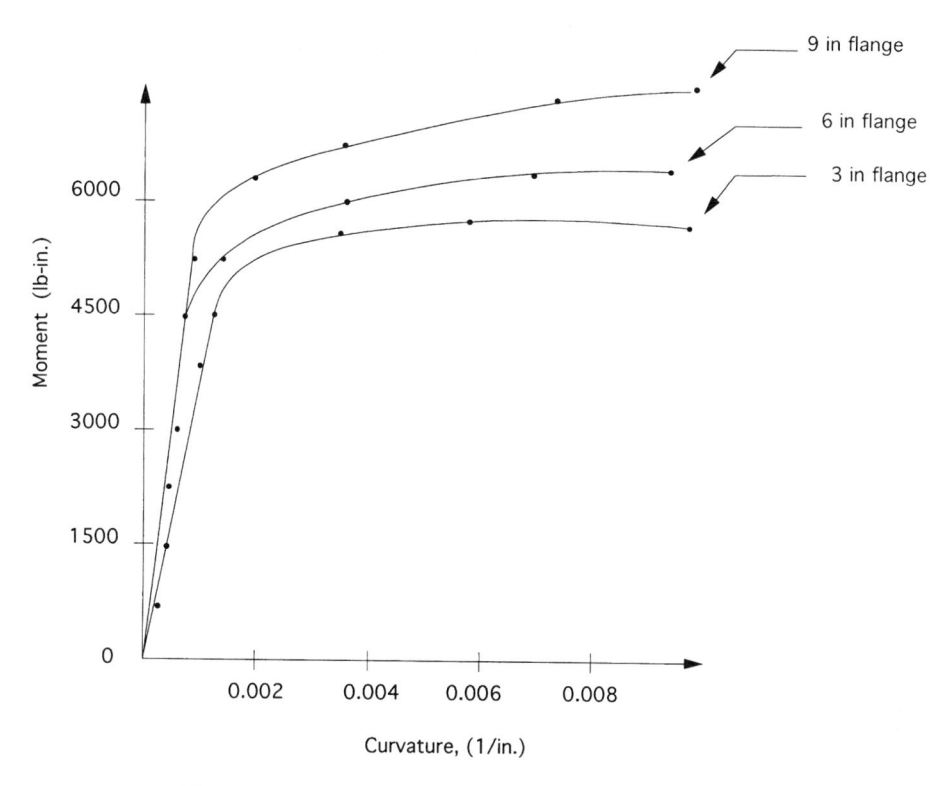

Figure 12.31 Moment–curvature curves for all three T-beams.

is indeterminate to the first degree and relies on rigid unyielding abutments for its strength. The section chosen (Figure 12.34) is doubly reinforced using four 1.22-mm-diameter deformed wires having a yield strength of 340.8 MPa. The average model concrete compressive strength is 30.4 MPa and the average split cylinder tensile strength is 3.9 MPa using both 25×50 and 50×100 mm cylinders to measure the strength. The model arch is cast horizontally in a Plexiglas mold.

During testing, five vertical and two horizontal displacements are measured, as shown in Figure 12.35. Testing proceeds incrementally using the test setup and lateral support system shown in Figure 12.35. The first crack appears in the positive moment region at the crown forming a three-hinge arch configuration and finally spreads to the negative moment regions between the supports and the crown as the failure load is reached. Typical load vs. crown deflection behavior for the model arch is shown in Figure 12.36.

The horizontal thrust H is found to be related to the centrally applied load 2W (Figure 12.34a) by the expression (Pippard and Baker, 1943)

$$H = (4 \cos \theta - 1 - 2\theta \sin 2\theta - 3 \cos 2\theta) \, W/(4\theta - 3 \sin 2\theta + 2\theta \cos 2\theta) \quad (12.9)$$

which for the geometry of the arch chosen (Figure 12.34a), $\theta = 45.1° = 0.787$ rad, becomes

$$H = 1.80 \, W \quad (12.10)$$

Now using statics, the maximum moment at midspan M_{CL} is related to the reactions V_B and H_B by the expression

$$M_{CL} = -W(L/2) + H_B(f) \quad (12.11)$$

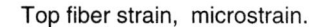

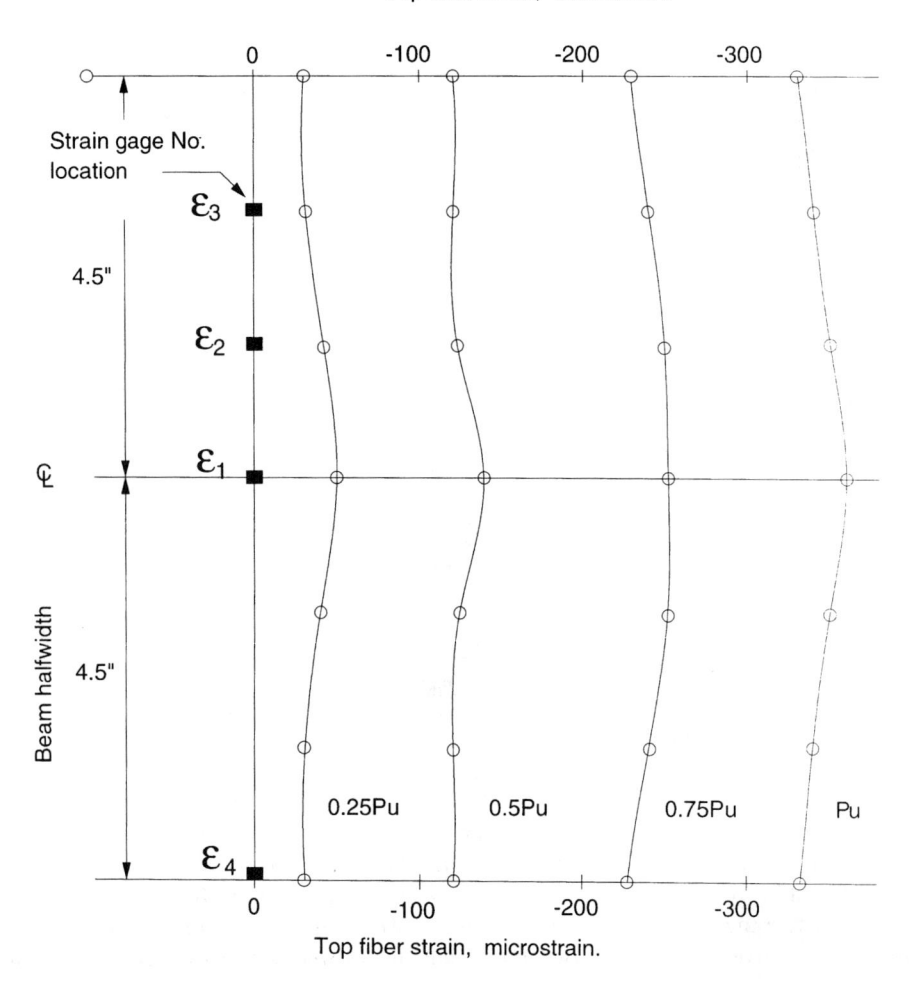

Figure 12.32 Strain distribution in model T-beam at different load levels.

where: f = arch rise
L = arch span

The ultimate strength of the reinforced concrete arch rib can be computed from the equation

$$M_{ult} = 0.33 f'_c \, b \, d^2 + f_y A'_s \, D \tag{12.12}$$

where: M_{ult} = ultimate bending moment
f'_c = concrete compressive strength
b = arch width
d = distance from top of arch to tension reinforcing steel
f_y = yield strength of steel reinforcement
A'_s = area of steel in compression
D = distance between compression and tension steel

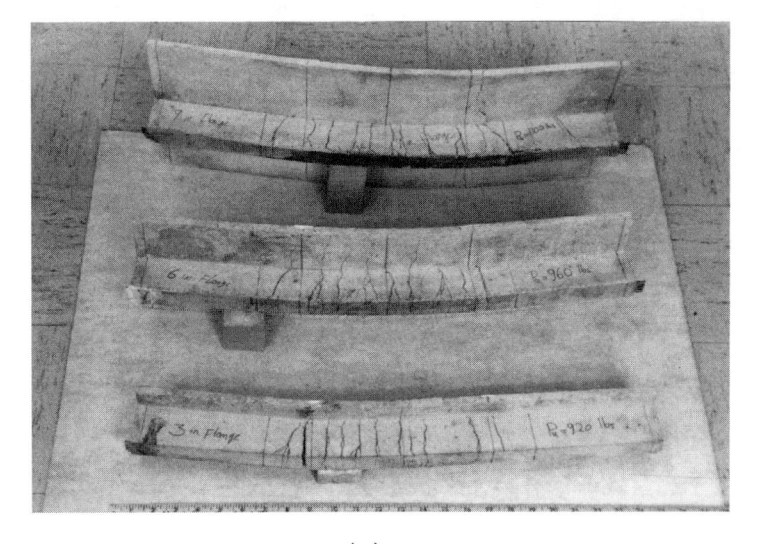

(a)

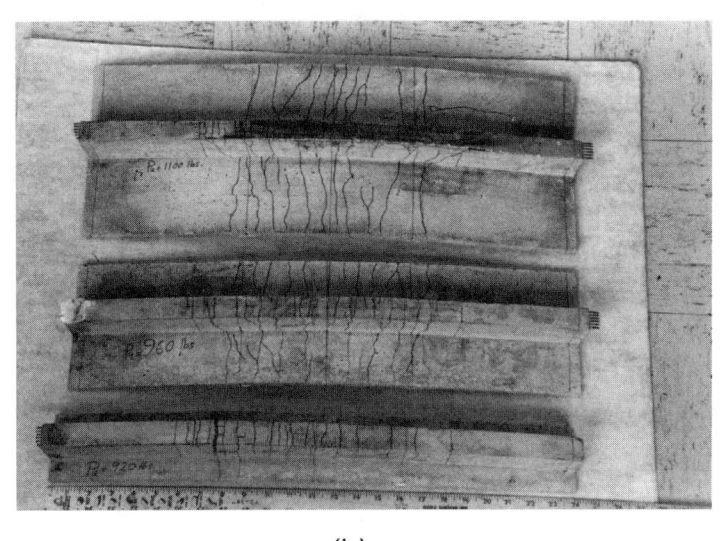

(b)

Figure 12.33 Final cracking patterns of model T-beams. (a) Web cracking. (b) Flange cracking.

The theoretical load can be computed from the above equations. For the geometry and measured material properties of the arch model, predicted load capacity of the arch is 1.68 kN. The ultimate load obtained during the test (Figure 12.36) is 1.56 kN.

This experiment falls in the same difficulty category as the experiment above. It requires a considerable amount of material preparation and planning that is facilitated by a properly equipped models laboratory. It requires three 2-hour periods to do the casting, instrumentation, and testing. However, some prior help of the technician in the materials preparation (especially the model reinforcement) is necessary. An accuracy of less than 15% is achieved in this experiment.

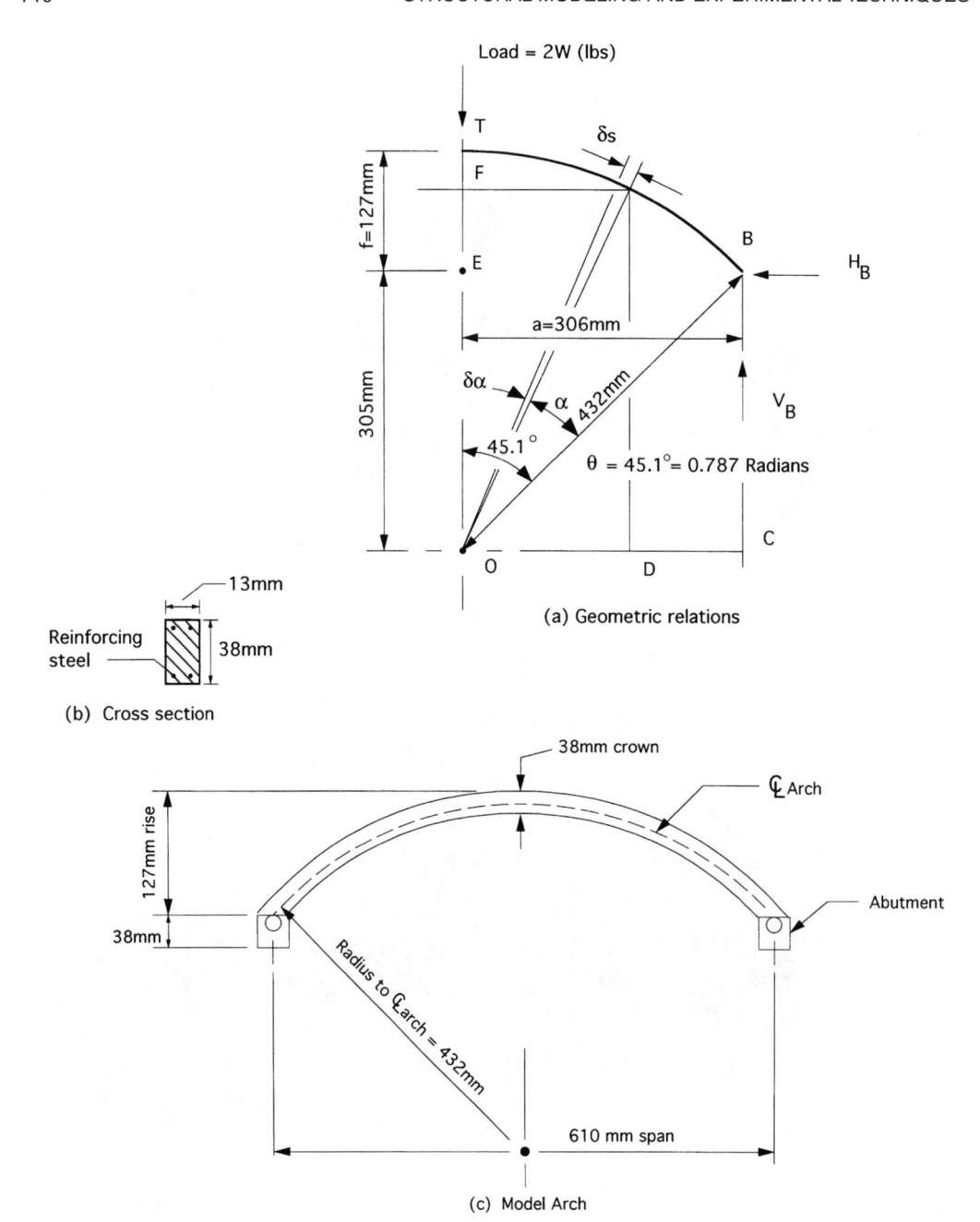

Figure 12.34 Dimensions of model arch.

12.4.2 Yielding and Inelastic Buckling of Steel Members

12.4.2.1 Hinge Formation and Collapse Modes

The demonstration of collapse modes of beams and frames under the assumption of the simple plastic theory was developed by Godden (1963). Using machined short rigid blocks, kept together

Figure 12.35 Test setup for model arch.

by means of tension of rubber tendons passing through them, a large variety of beams and frames could be constructed and loaded in any desired form of vertical and horizontal loads. The beam block geometry and the method of producing hinging action is illustrated in Figure 12.37. The simple plastic theory as applied to steel beams and steel frame structures assumes that such structures will collapse as a mechanism by developing a number of "plastic hinges." The calculation of collapse load by this theory thus depends on finding the correct mechanism of failure, that is, of finding the correct number and exact location of the hinges for any given condition of loading.

The technique developed by Godden allows the demonstration of collapse mechanisms in steel frame structures from which the collapse load of the structure can be computed. This is illustrated in Figures 12.38 and 12.39 for a beam and frame structure, respectively.

12.4.2.2 Inelastic Buckling

Buckling of structural steel beams and frames is an important structural consideration because of the high-strength thin sections that are used. To illustrate this phenomenon, a 1/10-scale model of a W18 × 96 wide-flange beam test is undertaken in this experiment. The model is machined from a solid piece of rectangular A36 bar stock in such a manner to minimize the tendency of warping of thin flanges. The model wide-flange beam is tested without lateral support, as shown in Figure 12.40. Under a load close to the ultimate, the beam continues to deflect in the inelastic range (Figure 12.40) and carry additional load until local buckling in the compression flange occurs. The load–deflection characteristics are shown in Figure 12.41 (Doukakis, 1980).

In this experiment close comparisons of the flexural behavior can be made with theoretical predictions since both strains and deflection measurements are continuously monitored up to failure.

This experiment is intermediate in difficulty and can be carried out in two 2-h sessions. It is required in this experiment that the model fabrication be carried out by an experienced machinist. The accuracy of the experimental results are within 10% of theoretical predictions. It is necessary to monitor the properties of the model materials in this experiment closely because they are sensitive to fabrication stresses. The student carrying out this experiment should have taken a course in design of steel structures.

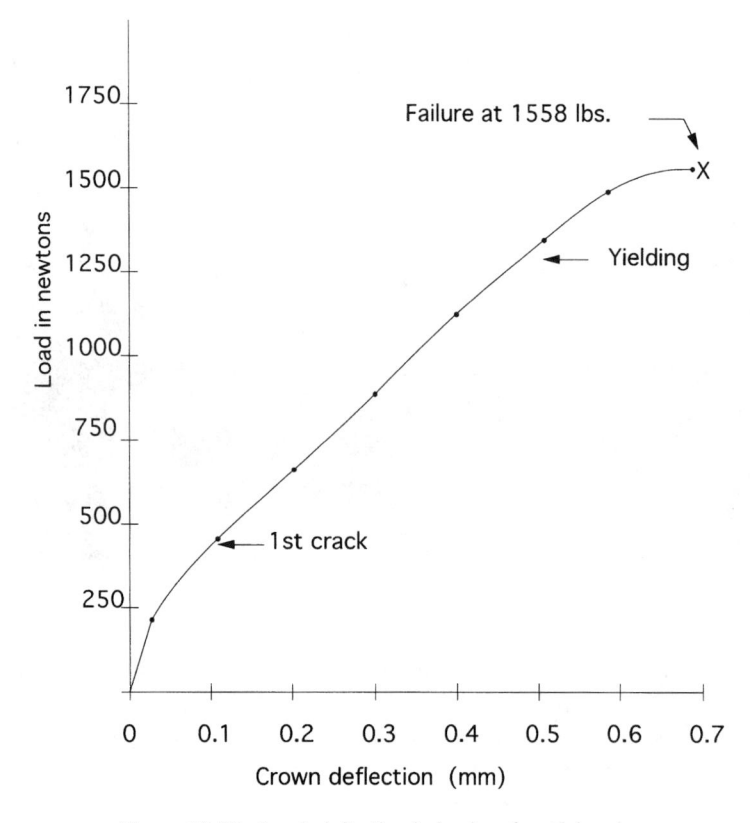

Figure 12.36 Load–deflection behavior of model arch.

12.5 STRUCTURAL DYNAMICS CONCEPTS

Educational models reach their highest potential when they can be used to demonstrate complex engineering phenomena to students who may not have taken advanced courses in the particular topic. Such an area where physical models can reach their full potential and power is in problems dealing with the dynamics of structures. The student making these experiments should have take courses in dynamics, statics, and strength of materials. Several examples are discussed below which illustrate how structural dynamics concepts can be taught to such students using small-scale physical models (Ergunay, 1993).

12.5.1 Basic Laboratory Instrumentation

The dynamic test facility in the Drexel University Structural Models Laboratory consists of an MB Vibramatic System (MB Electronics, New Haven, CT) which is specially designed for vibration testing of small-scale models or full-scale vibration testing of small machine components and subassemblies. The MB Vibramatic System has three main components: (1) PM-25 Exciter, (2) 2125 MB Power Amplifier, and (3) Exact Electronics, Model 502 Function Generator. In addition, the system is furnished with a Soltec 530 Dual-Beam Ocsilloscope to monitor the input motion waveform, magnitude, and input frequency, and a General Radio Co., Type 1531-A Strobotac to examine the structural response visually and determine the resonant frequencies of the test structure. The complete system is shown in Figures 12.42 and 12.43.

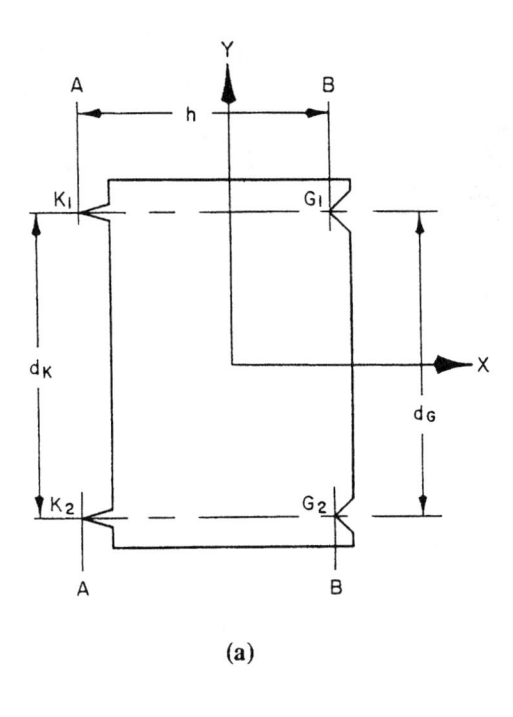

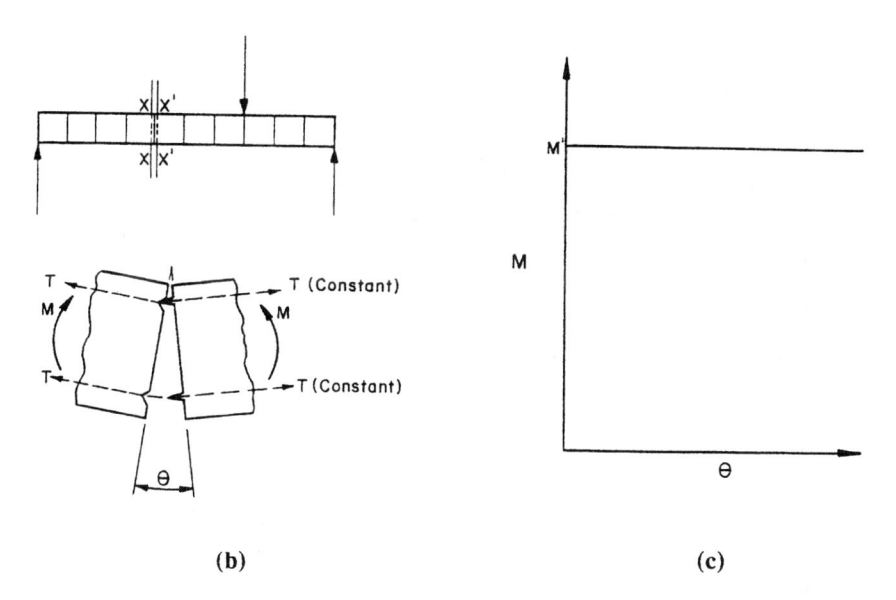

Figure 12.37 Demonstrating yielding in beams and frames. (a) Beam block. (b) Moment rotation. (c) Moment–rotation characteristics at any hinge in a beam composed of pretensioned blocks. (After Godden, W. G., 1963.)

12.5.1.1 PM-25 Exciter

The PM-25 Exciter has two main components: the exciter and the slip table. The exciter produces a mechanical motion/force by using electrical current flowing through a magnetic field. This magnetic field is created by a permanent magnet. The mechanical motion is transferred to the slip

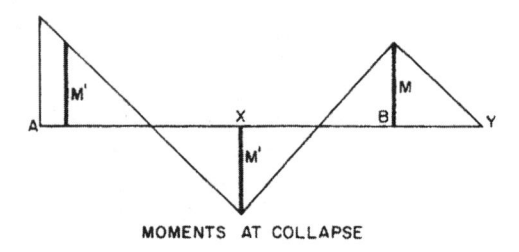

Figure 12.38 Collapse mode of beam due to critical values of W_X and W_Y (beam fixed at A on roller at B). (After Godden, W. G., 1963.)

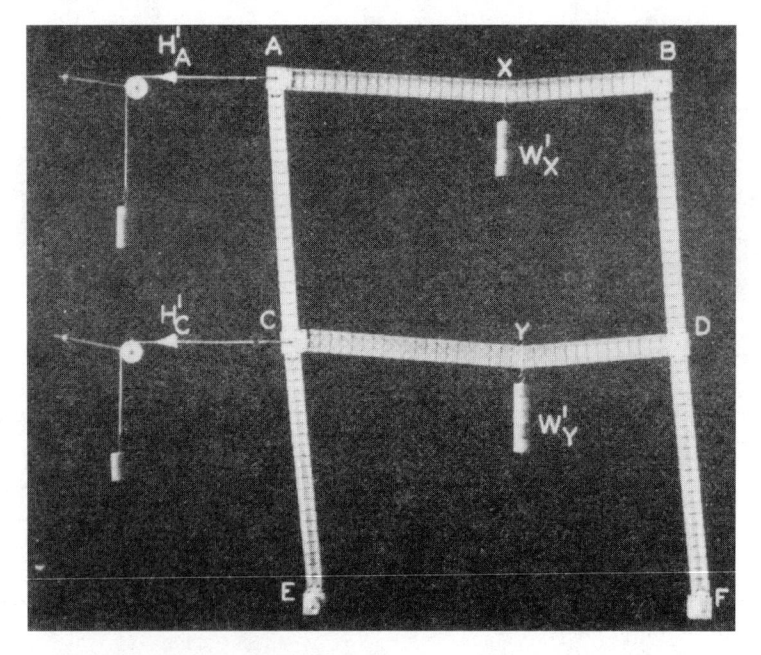

Figure 12.39 Collapse model of a two-story frame due to critical values of simultaneous horizontal and vertical loads. (After Godden, W. G., 1963.)

table by the driver unit. The slip table consists of a granite block and a magnesium slip plate (the testing area) to which the motion is actually transferred. The dimensions of the slip plate is 101.6 × 101.6 × 12.7 mm. The base can be set either to support horizontal or vertical motion depending on the type of test.

Figure 12.40 The test setup at approximately 5000 lb. Note deflections as well as angle of right roller assembly.

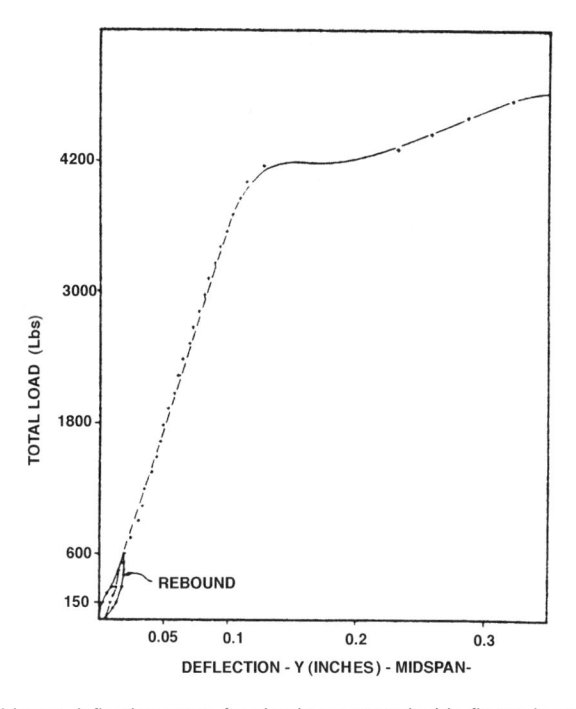

Figure 12.41 Load–midspan deflection curve for simply supported wide flange beam under two-point load.

The maximum capacity of the exciter is 111.2 N (25 lb) and the maximum acceleration that can be applied to the model depends on the output force provided by the system and the weight of the object. The following formula is recommended by the manufacturer to determine the maximum table acceleration dependent on the table load.

$$A_{max} = F_{rf}(W_{me} + W_o) \qquad (12.13)$$

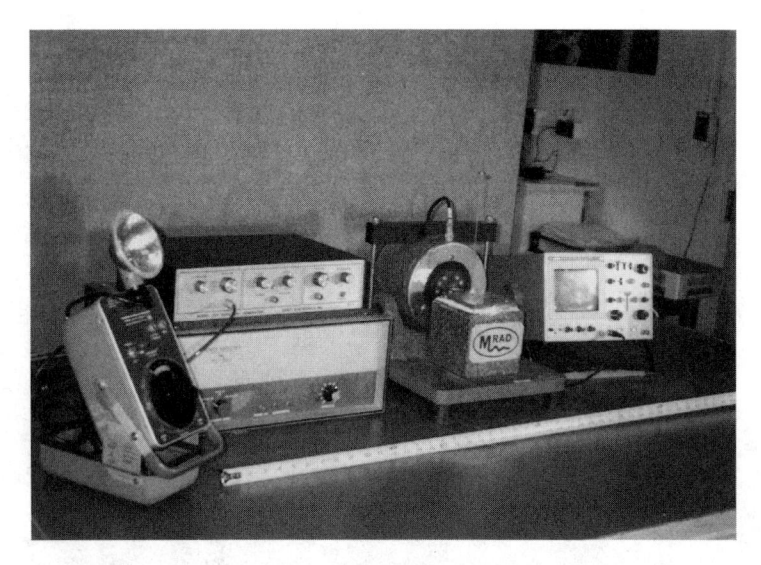

Figure 12.42 Drexel University dynamic system for model testing.

where: A_{max} = maximum allowable acceleration in g units
F_r = transmitted force
W_{me} = weight of the moving element assembly (22.2 N)
W_o = weight of the test object connected to the table

By using Equation 12.13 an upper limit for the maximum acceleration that can be applied to the table can be obtained as $41.5g$. This value corresponds to the case where no object is connected to the table ($W_o = 0$) and the force is at maximum ($F_r = 111.2$ N).

12.5.1.2 The 2125 MB Power Amplifier

The MB Vibramatic System uses a 125 VA transistorized power amplifier to drive the PM-25 vibration exciter. The amplifier consists of five basic sections: the preamplifier, driver stage, overcurrent protection stage, output stage, and unregulated power supply, and has a frequency range of 5 to 20,000 Hz. When it is coupled with the exciter, the frequency range narrows down to 5 to 10,000 Hz due to the lower operation frequency limit of the exciter. Performance limits for the 2125 MB Power Amplifier are given in Table 12.2. The combined system performance for the exciter and the power amplifier is shown in Figure 12.44.

12.5.1.3 Exact Electronics Model 502 Function Generator

The function generator is an integral part of the test facility. It can provide sinusoidal, square (both symmetrical and nonsymmetrical), and triangular waveforms at a frequency range of 0.001 Hz to 1 MHz, at a frequency stability rate between 0.05 to 0.6%. Since the operating frequencies of the other system components are limited to 5 to 10,000 Hz, the combined system frequency range is governed by these limits. The output deviation of the function generator ranges from 3 to 6%, which is a negligible rate for all practical purposes.

12.5.1.4 Evaluation of the Combined Drexel System

The main components of the combined system: the exciter, the power amplifier, and the function generator, are adequate for testing small-scale models for visual demonstration purposes. Although

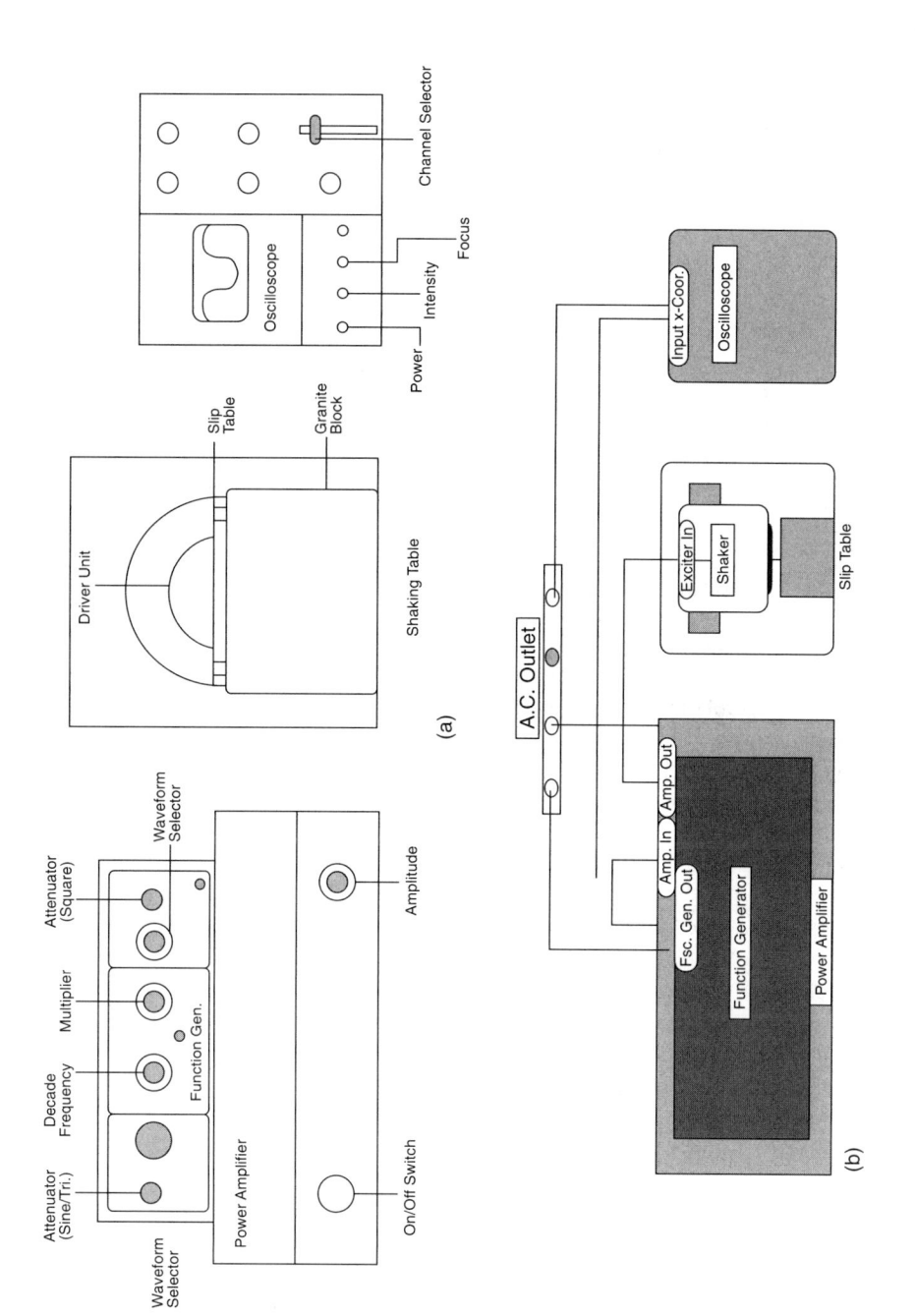

EDUCATIONAL MODELS FOR CIVIL AND ARCHITECTURAL ENGINEERING

Figure 12.43 System configuration. (a) Elevation. (b) Plan view.

Table 12.2 Performance Limits for 2125 MB Power Amplifier

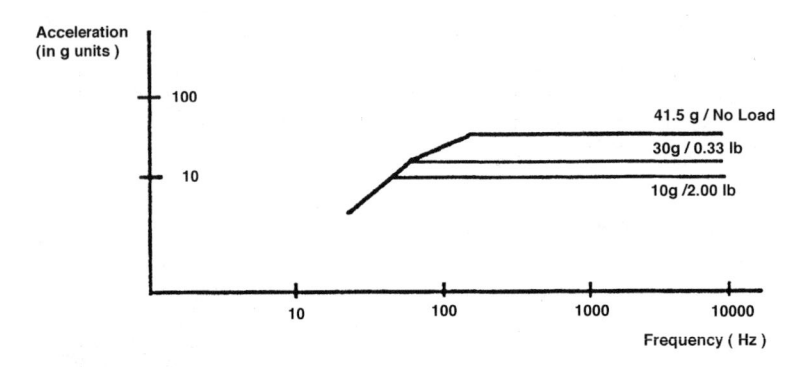

Power output	125 VA, 5–10,000 Hz
Frequency range	5–20,000 Hz
Input impedance	10,000 W
Frequency response	5–20,000 Hz
Distorsion	Less than 1% 5–3000 Hz
	Less than 3% 3000–20,000 Hz
Input power	110/220 V, 50/60 Hz
Dimensions	43.18 × 17.15 × 37.47 cm

Figure 12.44 Combined system performance curves for PM-25 Exciter and 2125 MB power amplifier.

there are obvious size limitations to the system, it can be adequately used, with proper small scaling techniques, to demonstrate a wide range of dynamic responses.

The oscilloscope and Strobotac are the secondary components that have no direct effect on the combined system performance. The oscilloscope constantly monitors and displays the input motion which is necessary for visual demonstration. The Strobotac helps to identify the higher modes of vibration and to determine the mode shapes accurately.

12.5.1.5 Other Educational Vibration Systems

A vibration system developed at the Mechanical Engineering Department, University of Maine (Martin, 1991) is used for educational and research purposes. This system consists of an MB Model EA1250 Exciter connected to a small shaking table mounted on Thomson ball bushing bearings (Figure 12.45). The performance limits of the small shaking table are given in Figure 12.46. A dedicated data acquisition system and especially written software enable the user of the vibration system to study a range of dynamic problems for education and research.

12.5.2 Vibrations of Lumped-Mass Systems

In the dynamic analysis of structures, lumped-mass systems are often used as a discretization of more-complicated continuous (distributed mass) systems. By lumping the mass at discrete points, an infinite-degree-of-freedom system with continuously distributed mass can be reduced to a finite-degree-of-freedom system with lumped masses. Lumped-mass discretization is mostly suitable for systems having a large mass concentration at some specific points, such as at floor levels of building frames.

Problems of mechanical vibrations are usually formulated by using the theory of differential equations to represent the time-dependent nature of the vibration phenomenon adequately and therefore to require an extensive background in mathematics. Furthermore, a basic understanding

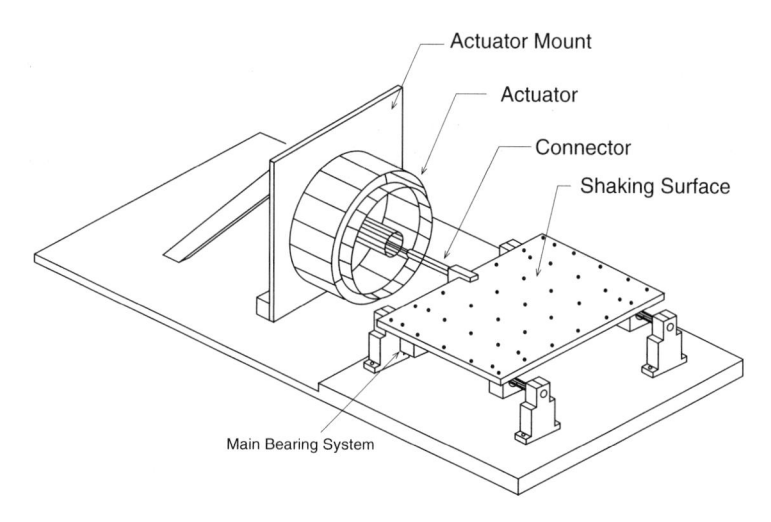

EDUCATIONAL MODELS FOR CIVIL AND ARCHITECTURAL ENGINEERING 719

Figure 12.45 Small linear shaking table. (University of Maine, Orono. Dana K. Martin, 1991.)

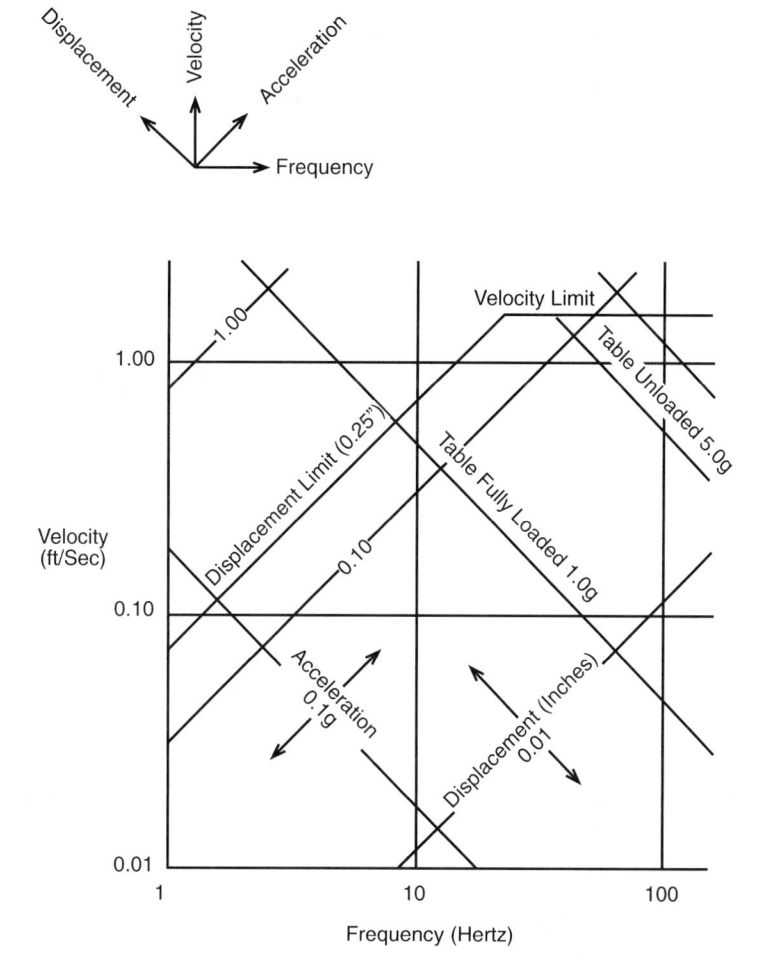

Figure 12.46 Performance limits of small shaking table, University of Maine. (After Martin, 1991.)

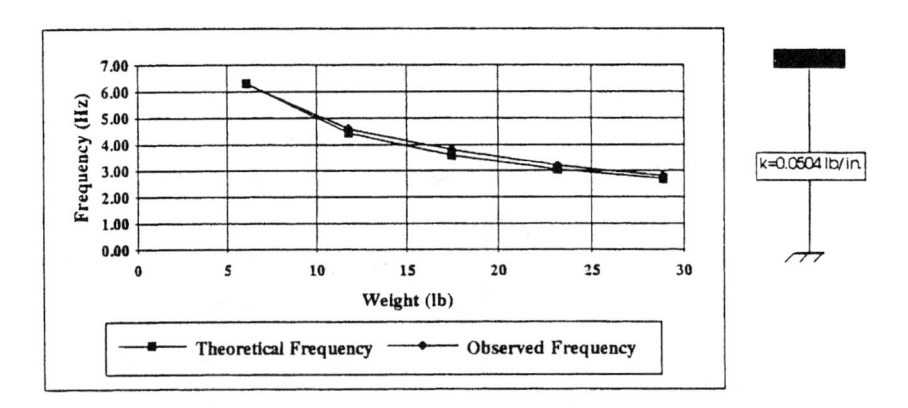

Figure 12.47 Weight–frequency curves for SDOF system (k = 0.0504 lb/in.).

Table 12.3 Instrument Settings for the SDOF Tests

Power amplifier amplitude	4 (approximately 44.48 N)
Function generator	
Attenuation	Max./2
Wave form	Sinusoidal
Signal	Symmetrical
Oscilloscope	
Time/div.	0.2 for $f \leq 5.0$ Hz
	0.1 for 5.0 Hz $< f < 7.0$ Hz
	50 ms for 7.0 Hz $< f < 12$ Hz
	20 ms for 12 Hz $< f < 15$ Hz
	10 ms for $f > 15$ Hz
Volts/div.	2.0

of the dynamic behavior of the system under consideration is necessary not only to model and formulate the system but also to determine the validity of the formulation and the solution. Determination of the natural frequencies and modes of vibration of a structural system is the first prerequisite for any dynamic analysis of the structure under any time-dependent loading such as forced vibrations, wind, or earthquake loading.

12.5.2.1 Single-Degree-of-Freedom Systems

Spring–mass systems are the simplest lumped-mass oscillatory systems, and they are often used in determining the main features of the behavior of more-complicated engineering systems under dynamic loading. The mathematical idealization and the physical model of a single-degree-of-freedom (SDOF) system is shown in Figure 12.47. In the analysis of spring–mass systems, the weight of the spring supporting the mass is usually neglected. Therefore, the variables involved in the analysis are the mass of the body and the stiffness of the spring.

In order to demonstrate the behavior of SDOF systems, a series of spring–mass systems with different mass and stiffness values are modeled using 4.24-mm-diameter brass bars with a unit weight of 1.68×10^{-5} kg/mm and cylindrical steel masses weighing 0.00572 kg each. The masses were attached to the bar through a 4.24-mm hole at their center, using a small setscrew. The brass bar is soldered to a 1-mm-thick brass plate that attaches directly to the slip table. With the settings shown in Table 12.3, the frequency and displacement amplitude of the small shaking table is varied slowly until resonance of the SDOF system is attained. A range of masses and stiffness can be tested in this manner and relationships among natural frequency, stiffness, and mass for SDOF systems can be explored by the student. The weight–frequency curve for a SDOF system is shown in Figure 12.47.

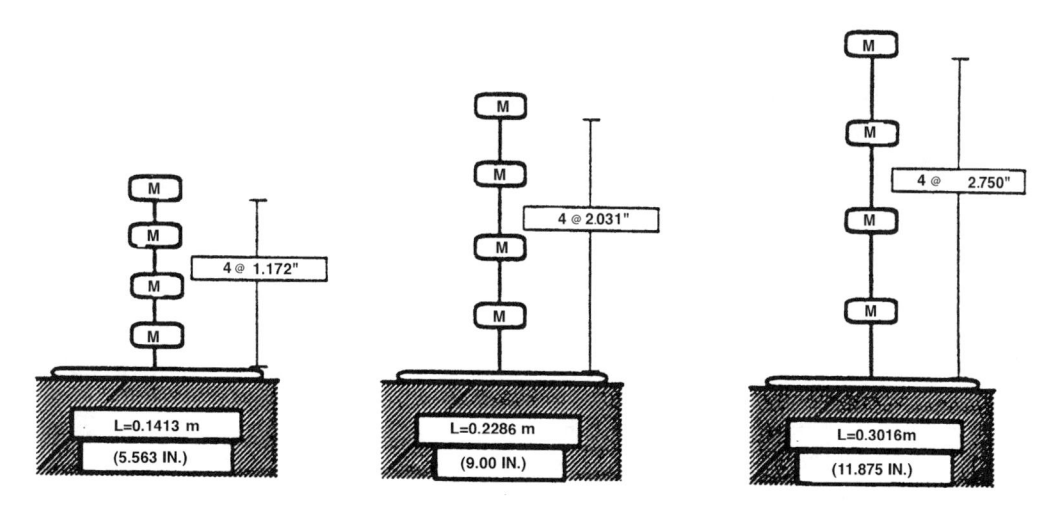

EDUCATIONAL MODELS FOR CIVIL AND ARCHITECTURAL ENGINEERING 721

Figure 12.48 Four degrees of freedom spring–mass model.

Table 12.4 Instrument Settings for MDOF Tests

Power amplifier amplitude	2–4 (corresponds approximately to 22.24–44.48 N)
Function generator	
Attenuation	Max./3
Wave form	Sinusoidal
Signal	Symmetrical
Oscilloscope	
Time/div.	0.1 for $f \leq 5.0$ Hz
	0.2 for 5.0 Hz $< f <$ 10.0 Hz
	50 ms for 7.0 Hz $< f <$ 12 Hz
	20 ms for 12 Hz $< f <$ 15 Hz
	10 ms for 15 Hz $< f <$ 20 Hz
	5 ms for 20 Hz $< f <$ 100 Hz
	2 ms for $f >$ 100 Hz
Volts/div.	2.0

12.5.2.2 Multi-Degree-of-Freedom Systems

The behavior of multi-degree-of-freedom (MDOF) systems are modeled by attaching additional cylindrical steel masses weighing 0.00608 kg to the stick models which were used in the demonstration of the behavior of SDOF systems. Models of three different lengths (141.3, 228.6, and 301.6 mm) are tested with up to four degrees of freedom to demonstrate the effect of change of stiffness (Figure 12.48).

The test procedure, similar to the one used to test SDOF systems, is used with the instrument settings shown in Table 12.4. The model is bolted on the slip table of the exciter from its base plate. The frequency of excitation is started from zero and increased by 0.3 Hz intervals at each step, until the maximum response is observed. The structure is excited at least 10 s at each frequency level to allow proper response. The frequency increase is dropped to 0.1 Hz intervals in the vicinity of the resonant frequency to improve the accuracy of the measurement. The frequency at the maximum response was noted as the natural frequency of the system. When this frequency is passed, a sudden decrease in the response is observed. After the first mode frequency is obtained, the input frequency intervals are increased to 0.3 Hz and the process is repeated until all frequencies (one for each degree of freedom) are obtained.

Table 12.5 Test Results of Three-Degree-of-Freedom Lumped-Mass Systems

System Number	Total Length of Model, cm	Average Stiffness per DOF, N/cm	Mode	Theoretical Frequency, Hz	Experimental Frequency, Hz	% Difference
1	14.13	8.38	1st	26.71	22.00	17.63
			2nd	73.85	73.00	1.15
			3rd	104.64	110.00	−5.12
2	22.86	1.93	1st	12.76	9.50	25.55
			2nd	35.45	36.00	−1.55
			3rd	50.62	55.00	−8.65
3	30.163	0.83	1st	8.36	6.00	28.32
			2nd	23.29	22.00	5.54
			3rd	33.35	32.00	4.05

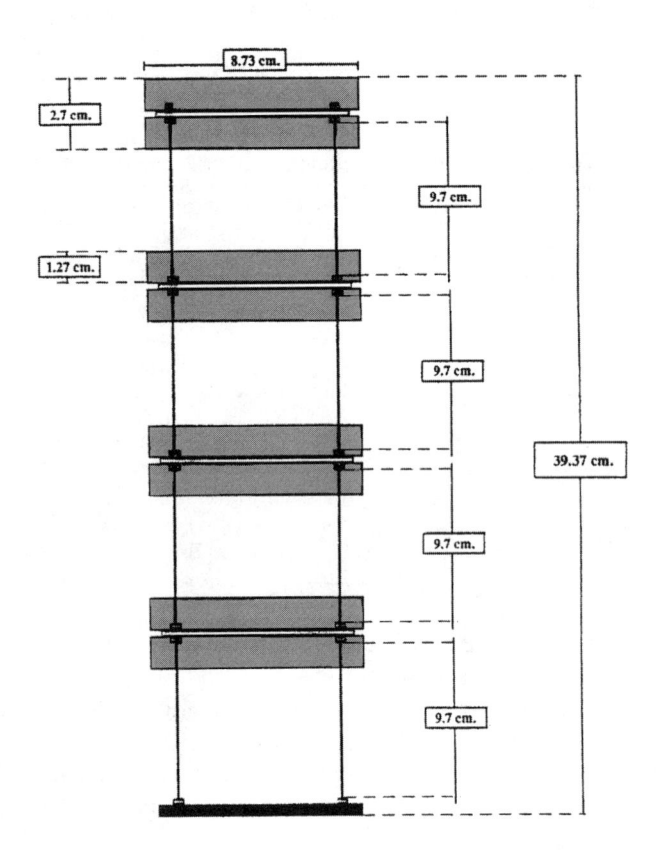

Figure 12.49 Four-story shear building model.

Typical test results compare well with theoretical predictions obtained from measured stiffness properties.

12.5.3 Dynamics of Shear Buildings

Dynamic analysis of multistory building frames is usually performed by idealizing the structural system as an MDOF lumped-mass system since the weight of the columns is much smaller than the total weight at the floor levels for most cases (Figure 12.49).

12.5.3.1 Four-Story Building Model

In order to demonstrate the dynamic behavior of building frames, a four-story shear building model is tested in this experiment. The model shown in Figure 12.49 is made up of brass plates and 1.59-mm-diameter piano wires as columns which are connected to the brass plates and the base plate with mechanical connectors. To augment the weight of the floor and roof levels, two rectangular steel bars are bolted to the brass plates. The height of the model is 393.7 mm with a clear story height of 97 mm. The weight at each degree of freedom is taken as 6.868 N of which 6.071 N is the weight of the brass plate and the added steel weights, and 0.777 N is the weight of the four columns lumped at the floor level.

The test procedure is the same as for the MDOF systems described earlier except that some changes in the instrument settings are warranted due to the increased weight and height of the model. Also, the time elapsed between two frequencies was increased to approximately 15 to 20 s, since it takes longer for the model to respond to the excitation. It should be noted that the range of amplitude of the input motion was decreased to avoid any permanent deformations in the columns. The instrument settings for this experiment are shown in Table 12.4. The four mode shapes and corresponding natural frequencies measured are shown in Figure 12.50. The experimental results compare very well to theoretical predictions as shown in Figure 12.51.

12.5.3.2 Four-Story Building Model with Base Isolation

1. Introduction — Seismic isolation is a relatively new technology to reduce the effects of earthquakes on structures. In most of its applications, the seismic isolation system is mounted beneath the structure and is referred to as a *base isolation* system. Basically, the base isolation system increases the flexibility of the structure and therefore increases its natural period. The dominant periods of typical earthquake accelerations are in the range of 0.1 to 1.0 s. and the maximum acceleration usually occurs in the range 0.2 to 0.6 s. Therefore, if the period of the structure is increased beyond these limits, the near-resonance response can be avoided and the structural response can be reduced. This experiment attempts to introduce the student to some of the concepts of base isolation. It uses the same four-story shear building model described above but modified to behave like a base-isolated system.

2. Base Isolation System — A base isolation unit was designed to fit the four-story building model. It consists of a cylindrical rubber element placed between two 3.97 mm stoppers which have the same hole diameter as the solid brass rods used for the columns of the building (see Figure 12.52a). The stoppers are held firmly in place with small setscrews. Inside the flexible rubber elements, a short piece of string, tightened and anchored to the setscrews, provides precompression to the rubber elements. The precompression provides additional stiffness to the columns and counters the overturning moment. The four base isolation units are mounted between two brass plates as shown in Figure 12.52b. This makes it possible to use the same model for testing with or without base isolation or to attach the base isolation units to other models so that the student can observe differences in behavior.

3. Test Results — The base isolation system used in this model is mounted on the base of the structure at the ground-floor level, acting like an additional floor. To represent the behavior of the model adequately, the stiffness characteristics to be used in a dynamic analysis are easily determined experimentally. In order to accomplish this, the flexibility matrix of the base-isolated model is determined by measuring the deflections at each story on graph paper under unit horizontal load as shown in Figure 12.53. The horizontal load is applied successively to each floor (Figure 12.54), thus generating the corresponding columns of the flexibility matrix.

The test procedure and instrument settings are the same for this experiment as those of the nonisolated case. Only small differences are required to account for the increased flexibility of the

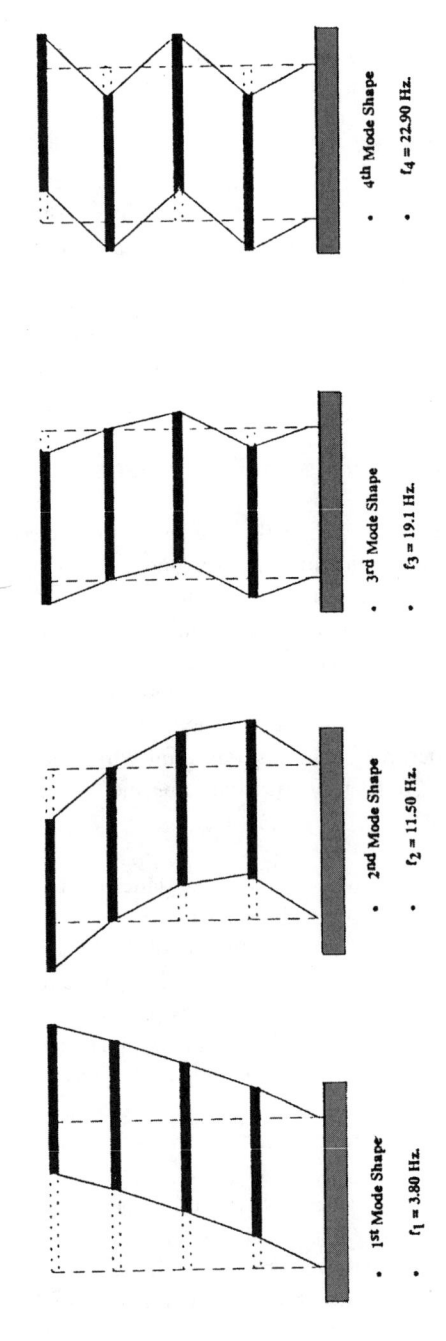

Figure 12.50 Vibration test results for the four-story building frame model.

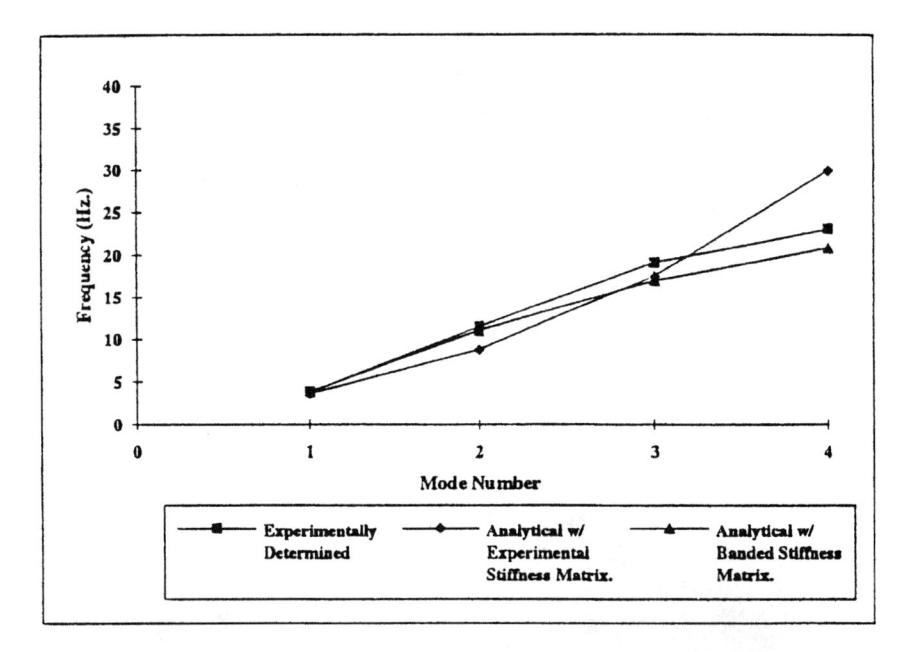

Figure 12.51 Graphical comparison of the experimentally determined frequencies with the analytical results.

base-isolated model. The measured natural frequencies and mode shapes are compared to the theoretical predictions in Figure 12.55. As can be seen, the combined behavior of the four-degree-of-freedom shear building with the base isolation unit agrees very well with the theory. It is observed that the base isolation increases the natural period as compared to the nonisolated four-story shear building model as shown in Figure 12.56.

The accuracy of the dynamic experiments are within 10% of the theoretical predictions. All of the various dynamic experiments described can be done in a 2-h period. The difficulty varies from easy to difficult when account is taken of the understanding of the theory needed for predictions of dynamic behavior. If predictions are not needed, and the models exist in good testing form, the actual carrying out of the dynamic experiments is relatively easy and straightforward.

12.6 EXPERIMENTATION AND THE NEW ENGINEERING CURRICULUM

12.6.1 Historical Background

In 1988, Drexel University began a 5-year experimental project entitled E4, "An Enhanced Educational Experience for Engineering Students." This multi-million-dollar program was jointly sponsored by the university and the National Science Foundation based on a proposal submitted by Dr. Robert Quinn, professor of electrical and computer engineering and Dr. Eli Fromm, then associate dean of the Drexel University College of Engineering and professor of electrical and computer engineering. The goal of E4 was to establish a new paradigm for undergraduate engineering education. New environments and experiences provided students with an early exposure to the engineering profession as a dynamic enterprise involving design, experimentation, problem solving, and communication.

In 1990, the program received the first ABET Award for Educational Innovation. Thanks to the combined efforts of over 60 faculty and 800 students led by Dr. Quinn, project director, with additional support from the Commonwealth of Pennsylvania, the General Electric Foundation, the Ben Franklin Partnership, Hewlett-Packard, Tektronix, and DuPont, E4 served as a model for the

(a)

Figure 12.52 Four-story building model with base isolation system. (a) Model. (b) Detail of the base isolation system.

Drexel Engineering Curriculum. This new curriculum provides an early introduction to the central body of knowledge that forms the fabric of engineering, modern experimental methods, the computer as a flexible and powerful professional tool, the importance of personal communication skills, and the imperative for continuous, vigorous, life-long learning.

12.6.2 New Engineering Curriculum

The new engineering curriculum, under development and implementation at Drexel University (1996) and other universities with strong undergraduate engineering programs, relies on teaching basic engineering concepts by giving the student hands-on experiences in the laboratory (Carr et al., 1989; 1995; Quinn, 1993a, b; 1994a, b, c). Mastery of experimental techniques is expected in the first and second years. The students become familiar with how data are acquired, processed, and analyzed as well as how basic experimental techniques, devices, and methods are used in a wide variety of engineering disciplines. The experimental experience is not lost in upper-level engineering courses but is integrated with the theory. This means that experimental techniques must become more widespread to reach a much wider segment of the engineering community.

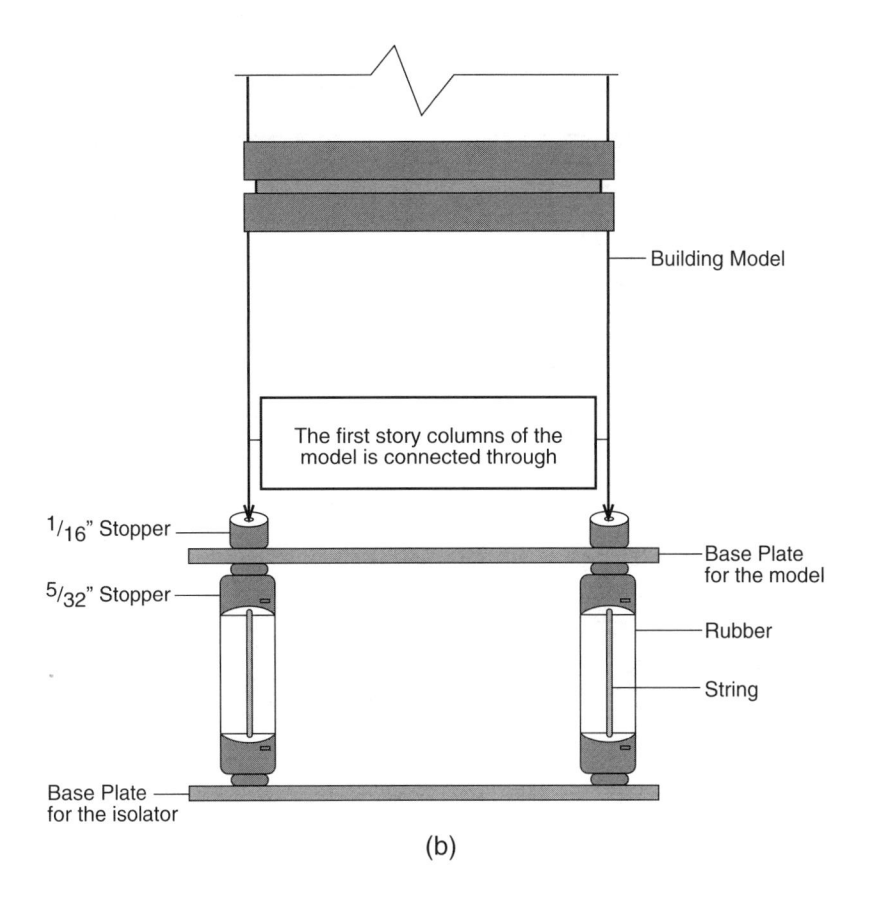

Figure 12.52 (continued)

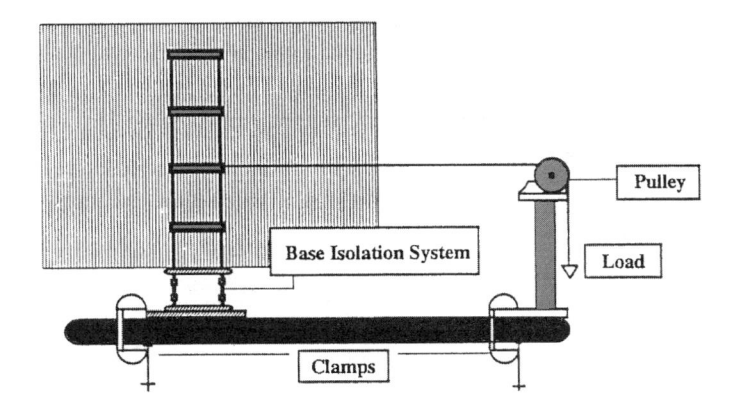

Figure 12.53 Test setup to determine the stiffness properties of the base isolation system.

It should be pointed out that in the new engineering curriculum experimentation is an integral part of the engineering student's professional development. The incoming freshmen start engineering courses from day one. Emphasis is given to experimental methods in engineering because of their wide use in analysis, design, development, and manufacturing. Special attention is given to the interpretation and effective presentation of experimental results in written and oral form. The computer is used effectively as a research and design tool. The students engage in professional

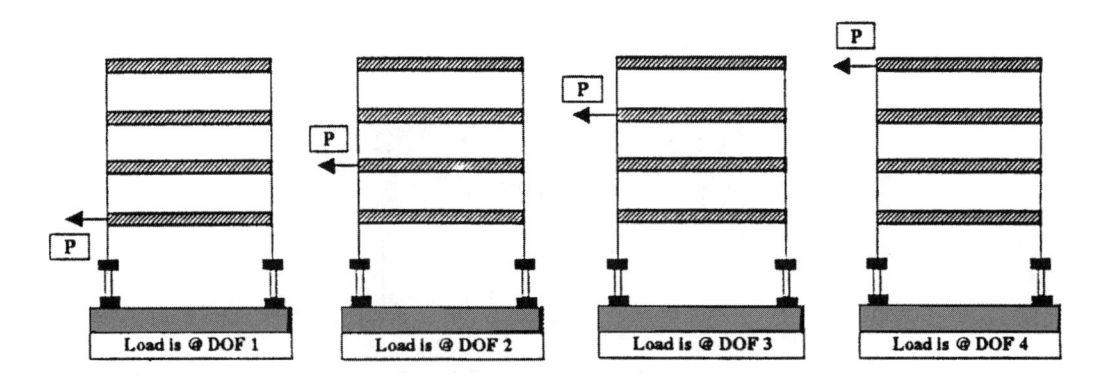

Figure 12.54 Determination of the flexibility matrix.

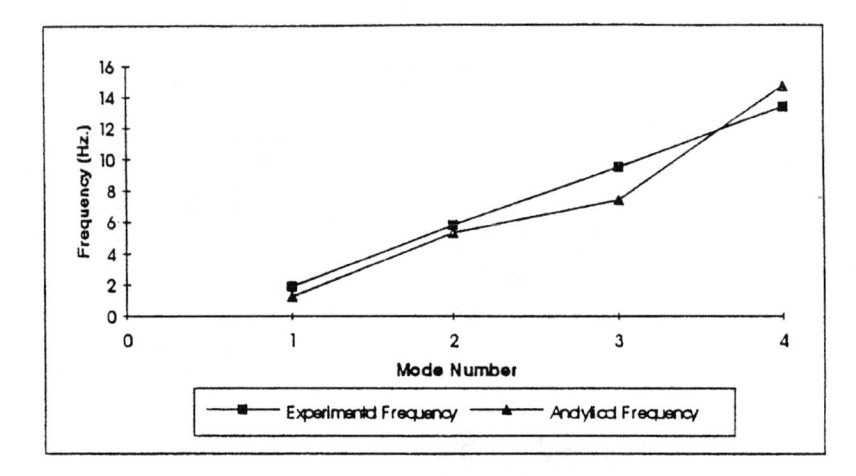

Figure 12.55 Comparison of experimental frequencies and analytical frequencies using the experimentally detrmined flexibility matrix for the base isolation model.

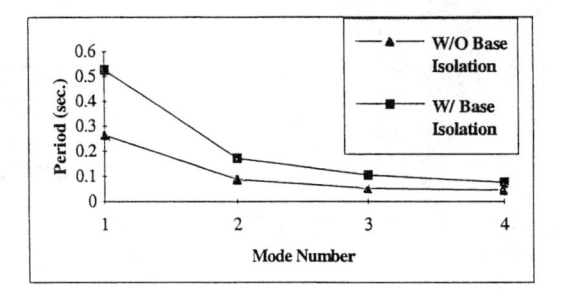

Figure 12.56 Experimental results and comparison with nonisolated case.

Table 12.6 Model Glulam Beams

Model	Length, mm	Depth, mm End	Depth, mm Centerline	Instrumentation Deflection	Instrumentation Strains
BM1A	762	25.4	88.9	Yes	Yes
BM1B	762	25.4	57.2	Yes	No
BM2A	762	25.4	57.2	Yes	Yes
BM2B	762	25.4	88.9	Yes	No

design projects solving real-world problems. They learn interactively, through teamwork, gaining life-long learning skills.

Regardless of discipline, practicing engineers perform a wide variety of experiments throughout their careers. Mastering the techniques of experimentation is therefore very essential. During the first and second years, students conduct a 3-h laboratory each week in the Engineering Test, Design, and Simulation Laboratory. This state-of-the-art facility provides students with opportunities to exercise their imagination, satisfy their curiosity, and experience the joy of engineering. Through the use of computers and computer-controlled laboratory instruments, students are able to explore how experimentation is used in engineering applications. They become familiar with how data are acquired, processed, and analyzed as well as how basic experimental techniques, devices, and methods are used in a wide variety of engineering disciplines.

During the upper years of the Drexel Engineering Curriculum, the laboratory component is shared among several courses rather than being an integrated course, as was the case in the first and second years. These courses teach the fundamentals of structural and architectural engineering and form the backbone of the young structural engineer's academic training.

12.7 CASE STUDIES AND STUDENT PROJECTS

The examples described in this section are chosen to illustrate the large variety of individual or group student projects that can be used in a structural models laboratory to enhance the education of civil and architectural engineering students. All of the examples shown are from the Structural Models Laboratory of Drexel University. The examples chosen are mostly one-term (10-week-long) projects with the exception of the last one, which was a three-term (one-academic-year) project and took one additional month for student volunteers to build. The difficulty of each varies and ranges from what was earlier termed intermediate to difficult. The accuracy of these projects, where theoretical results are available, were usually within 15 to 20%, which is acceptable for student-made models.

12.7.1 Building Components and Systems

12.7.1.1 Double-Tapered Glulam Beams

Results of the tapered glulam beam study using 1/12-scale models (Kopatz, 1984) described in Section 3.7.3 are given below. Details of the construction and modeling considerations for wooden structures are given in Chapters 3 and 6. Since the object of this study was to model the behavior of prototype glulam beams with available test results (Gutkowski et al., 1982), comparisons of the model and prototype behavior are made. In addition, a detailed finite-element analysis is performed using the SAP-IV code and the theoretical results are also compared. A total of four 1/12-scale models, consisting of two separate geometric configurations, were tested in this series (see Table 12.6).

Tests were performed using two symmetric point loads applied to the top of the beam as shown in Figure 12.57. This loading allowed for a constant-moment region through the center portion of the beam. Deflections and strains were measured at the locations shown in Figure 12.57. Two gages were mounted at the supports and five gages were place symmetrically along the beam. Three of the gages were placed in the critical central region. Only two of the models were instrumented with electrical strain gages (Figure 12.58) to monitor the longitudinal and transverse stresses. The test setup is shown in Figures 3.23d and 12.58. Typical load–deflection curves for the model glulam beams show a nonlinear behavior. The deflection profile for model beam BM1A is shown in Figure 12.59 at different load levels up to failure. The theoretical predictions from a finite-element idealization derived using the computer code SAP-IV are also shown in this figure. As can be seen, the agreement of the experimental and theoretical deflections is very good.

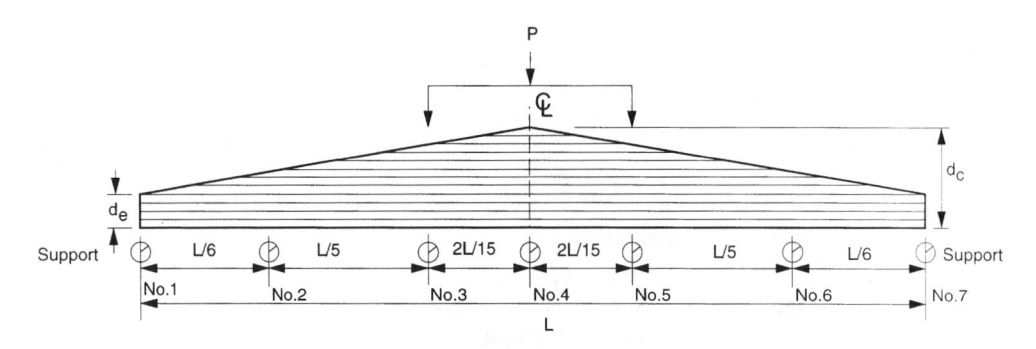

Figure 12.57 Instrumentation for double-tapered glulam beams.

Figure 12.58 Close-up of instrumentation in glulam model.

Failure of the prototype was caused by transverse stress which had a peak value at a point 686 mm above the bottom. The corresponding point on the model is 57.2 mm, and this is where the failure was expected to take place. On model beams BM1A and BM2B, transverse shear failure occurred at a point of 54 mm from the bottom.

12.7.1.2 Single-Fink Roof Truss

This experiment consisted of testing a ¹⁄₁₀-scale model of a Single-Fink truss (Figure 3.22) used in a garage roof structure (Heckler, 1980). The model was fabricated using the same materials and construction techniques as the prototype truss (see Chapters 3 and 6). The stress–strain characteristics for Eastern white pine used in the model and prototype structures are shown in Figure 12.60. It can be seen from the graph that there is a definite linearly elastic region with an elastic modulus of approximately 1×10^6 psi (6.895 MPa). Material properties for Eastern white pine are given in Table 12.7.

The truss model was instrumented with dial gages as shown in Figure 12.61. The top chord was prevented from moving laterally by providing support at the nodes. Loading was provided by hanging weights through strings attached to the top chord nodes as shown in Figure 3.22d. The model truss is shown during testing in Figure 12.62. A special supporting structure was made for each of the two supports (Figure 12.63). Loading of the three top chord nodes was carried out in 2.2 N increments on each of the three nodes up to a load of 28.9 N. At this load, the load–deflection

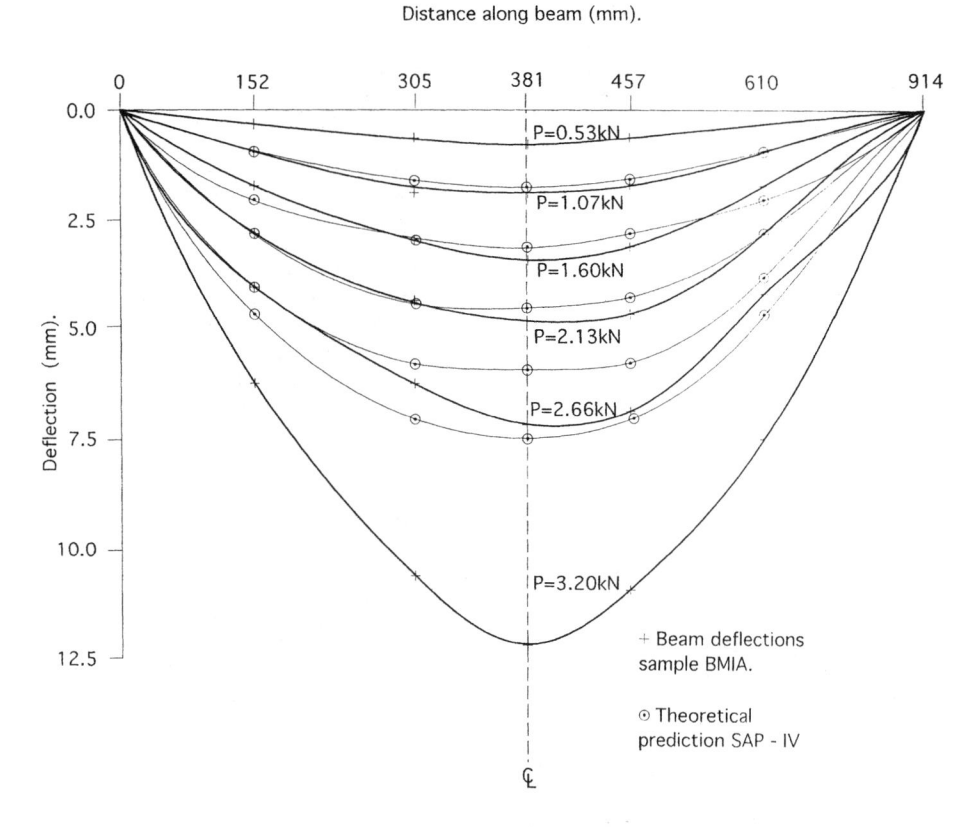

Figure 12.59 Deflected shape of Model BMIA with increasing load.

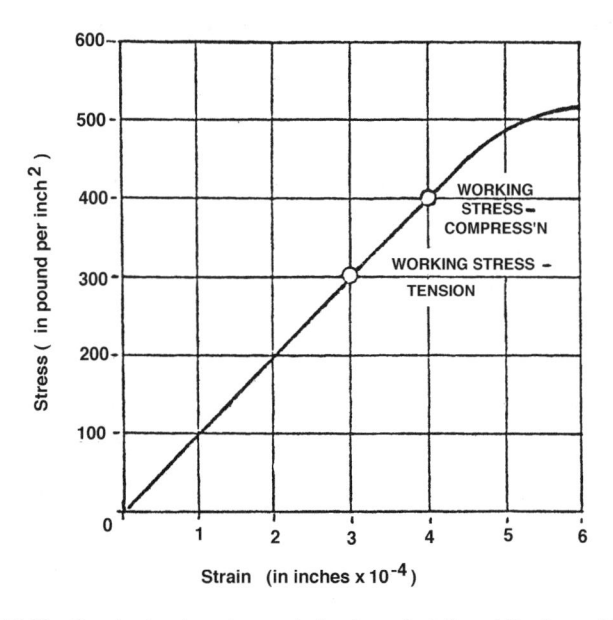

Figure 12.60 Graph showing stress–strain characteristics of Eastern white pine.

Table 12.7 Material Properties of Eastern White Pine

Specimen Number	Specimen Dimensions, mm × mm 3.18 × 5.56	3.18 × 11.91	Cross-Sectional Area, mm²	Failure Load, kN	Failure Stress, MPa	Comments
1	√		17.64	1601	82.95	Straight, hard grain
2	√		17.64	734	38.02	Jaw failure, soft grain
3	√		17.64	756	39.17	Soft grain
4	√		17.64	787	41.39	Soft grain
5	√		17.64	1027	53.90	Straight grain
6		√	37.80	1237	29.89	Oblique grain failure, soft grain
7		√	37.80	1526	36.88	Oblique grain failure
8		√	37.80	1312	31.72	Oblique grain failure
9		√	37.80	1312	31.72	Oblique grain failure
10		√	37.80	2647	63.98	Oblique grain failure, straight, tight grain

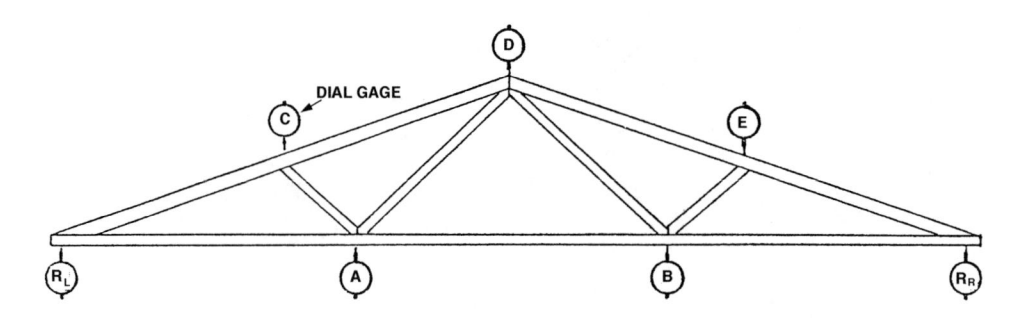

Figure 12.61 Showing positions and labeling of the dial gages.

behavior became markedly nonlinear and, as the intent was not to test to failure, the truss model was unloaded. Load–deflection characteristics for top and bottom chord nodes are shown in Figure 12.64. Theoretical predictions of the model truss deflections were also made using the virtual work method (Equation 12.3). The accuracy of this model was within 15% of the theoretical predictions.

12.7.1.3 Precast Building Elements

Industrialized building construction, with its promise of better quality and efficiency at reduced cost, is of great interest in building construction and other structural systems. Many such possibilities exist in all major construction materials. Two examples studied by students at Drexel University involving precast concrete elements are described here. The first study considers the earthquake resistance of large-panel building construction. Profiting from an ongoing research project on large-panel buildings at the time (see Section 10.2.1.3), the senior design group working on this experiment had access to fabricating molds and techniques for constructing model precast concrete elements. Using the developed procedures, they constructed a ³⁄₃₂-scale one-bay, six-story segment of a building (Figure 12.65) consisting of precast shear walls having the appropriate vertical, horizontal, and peripheral ties.

Since the object of this study was the evaluation of the dynamic characteristics of the building, a static determination of the flexibility matrix of the precast shear wall model was made by applying horizontal load to each story in turn and measuring the horizontal deflections as shown in Figure 12.65. Having determined the experimental stiffness characteristics of the model, the natural modes and frequencies were predicted from a dynamic analysis. The measured modes and frequencies

EDUCATIONAL MODELS FOR CIVIL AND ARCHITECTURAL ENGINEERING 733

Figure 12.62 Model truss during testing showing loading method.

Figure 12.63 Close-up of support system and instrumentation.

were then determined on the instrumented model (Figure 12.66) with added mass provided by attaching steel plates to each floor to preserve dynamic similitude (see Chapter 2). This was a difficult project for a group of senior civil engineering students but carried out over a three-term period it gave them time to study structural dynamics and to learn dynamic testing techniques.

The second study of precast concrete components involves the construction of a 1/24-scale segmented dome model of a circular concrete shell with a 24.38 m radius and enclosed half angle of 40° (Gambone et al., 1978)). The segmented dome consisting of 30 equal pie-shaped segments is shown under construction in Figure 12.67. The finished assembled dome model is shown in Figure 12.68.

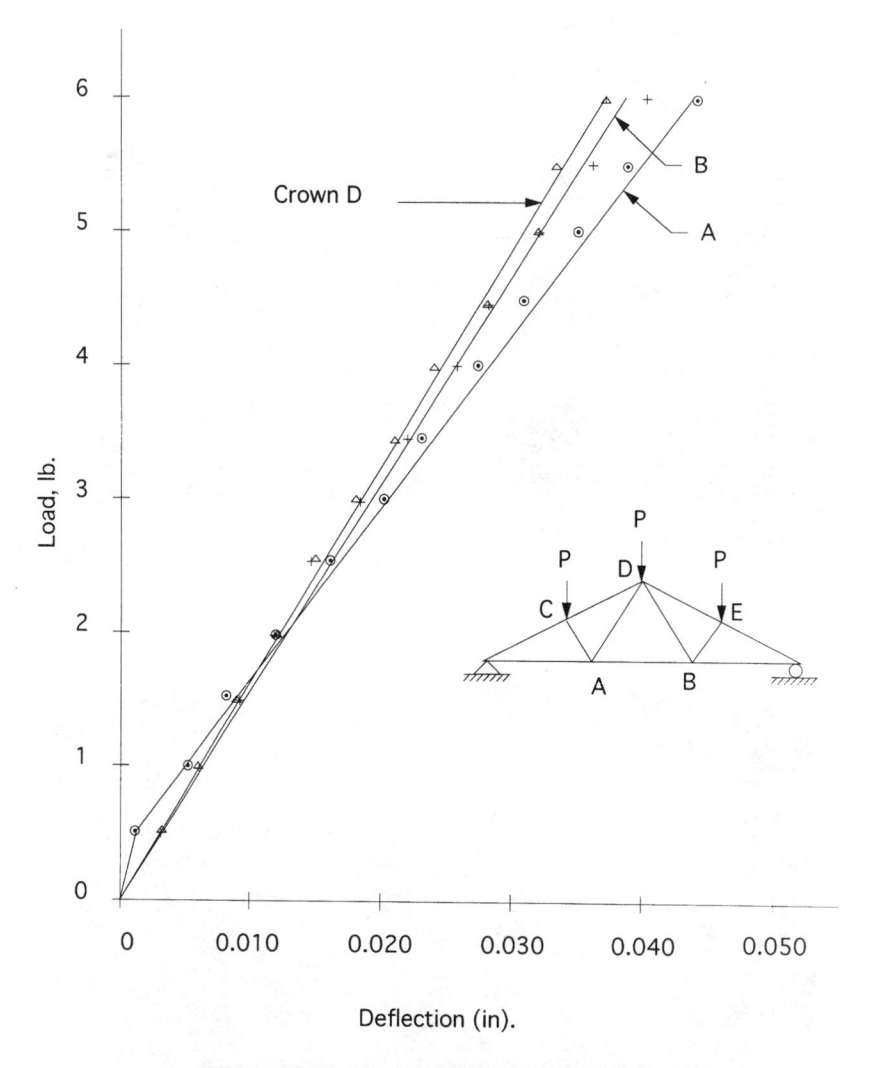

Figure 12.64 Load–deflection of model truss.

Figure 12.65 Determination of flexibility matrix for precast shear wall model.

EDUCATIONAL MODELS FOR CIVIL AND ARCHITECTURAL ENGINEERING 735

Figure 12.66 Model vibration tests on Drexel University shaking table.

Figure 12.67 Segmented precast concrete dome model.

Figure 12.68 Finished precast dome model.

Figure 12.69 Three-span section of 1/160-scale model of concrete segmental bridge.

12.7.2 Bridge Structures

12.7.2.1 Concrete Segmental Bridge

The prototype of this model study was a concrete segmental bridge, 533 m long and 32.3 m wide. The bridge consists of nine spans, seven of 66.7 m and two approach spans of 33.3 m. The cross section of the bridge is made up of two single-cell segments tied together with transverse post-tensioning. The purpose of the model test (Gatti, 1982) was to determine the dead and live load deflections, influence lines, and the transverse interaction of the two cells. After considerations of materials and test facilities, a 1/160-scale model of a three-span section of the bridge made of Plexiglas was chosen (Figure 12.69).

Deflections were measured using dial gages as shown in Figure 12.69. For the dead and live load tests, the dial gages were placed along the centerline of each cell at midspan, quarter points,

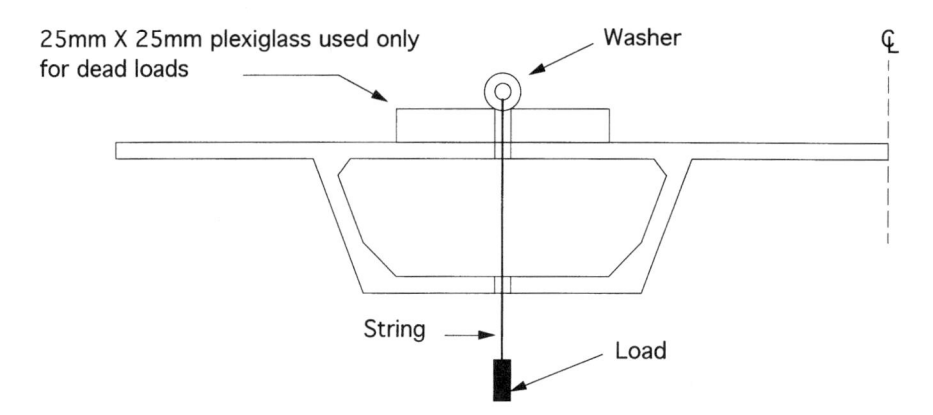

Figure 12.70 Load application scheme.

and the supports. For the transverse loading, dial gages were placed across the center of the span, on the cell centerline at the quarter points, and at the supports. In order to apply the dead and live load, 12 small holes were drilled in each span (6 along the centerline of each cell). A string tied to a small washer was threaded through each hole and the loads were hung from the end of the string (Figure 12.70). To apply the loads for the transverse load test, nine equally spaced holes were drilled across the section at the center of the span. Since the bridge model is symmetric, the transverse loading was only done on the center span and one end span.

The loads for the dead load test were applied in increments of 2.2 N up to 11 N per string. The load–deflections were linear and the deflections were measured quickly to avoid any creep effects. Three different live load combinations were added to the dead load. The center span was loaded with each of the two end spans and then the two end spans were loaded by themselves. A comparison of the results (Table 12.8) of dead load alone and dead load and live load together showed that the live load contribution was no more than 25% of the total. This indicates that the dead load is a major factor in the design of the bridge. The deflection profiles for the dead and live load combinations are shown in Figure 12.71.

For the transverse load test, a 8.9 N point load was used and placed at nine different locations across the model bridge. Deflections were measured over the entire model for each loading. From the deflection curves across the section, the interaction between the cells can be determined for each loading. Using the rest of the dial gages, one could get an idea of the shape of the influence surface of the bridge deck for each of the loadings. It was noted from the deflection measurements that the amount of interaction between the cells varies with the position of the load. When the load is at the edge of the cross section, there is not much contribution from the other cell (Figure 12.72). As the load moves inside the first web, there is much more interaction between the two cells and the deflections become more evenly distributed.

This bridge model was fabricated by an experienced technician. The testing was carried out by one senior-level student over a one-term (10-week) period.

12.7.2.2 Cable-Stayed Bridge

This study involves a 1/400-scale model of cable-stayed bridge (Socoloski, 1979). The prototype has six lanes and is supported by two box girders. It is of the double-plane harp cable arrangement type (Figure 12.73). The 1/400-scale model consisted of steel hollow tube girders and stainless steel cables. The main quantities that need to be measured are displacements along the panel points of the main girder, tensile strains in the cables, and extreme fiber strains at the girder midspan and other critical sections. Influence lines for cable forces, girder displacements, and girder bending moments are thus computed from these data.

LIVE LOAD DEFLECTIONS (left and right spans loaded)

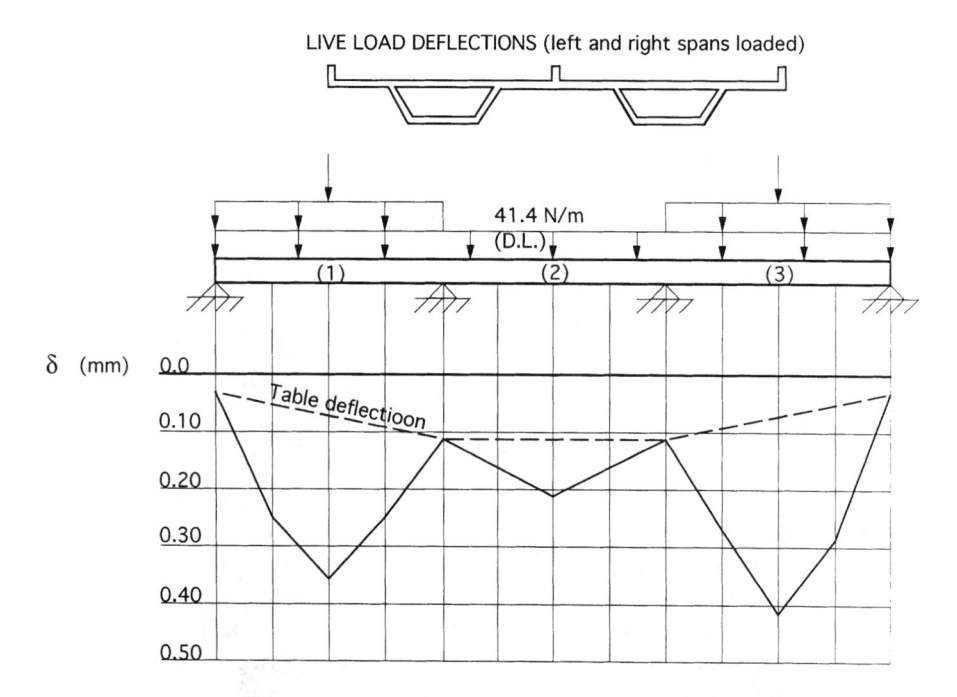

Figure 12.71 Deflections of dead load plus live load at left and right spans.

Table 12.8 Dead Load and Live Load Deflections

	Dead Load			
Span	Model Deflection, δ_m with Table Deflection, mm	Model Deflection, δ_m without Table Deflection, mm	Prototype Deflection, δ_p, mm	Deflection/Span Ratio δ/L
1	0.635	0.495	79.25	1/841
2	0.533	0.305	48.77	1/1025
3	0.838	0.699	111.76	1/597

		Live Load					
		Span 1		Span 2		Span 3	
Loading	Spans Loaded	δ_m, mm	δ_p, mm	δ_m, mm	δ_p, mm	δ_m, mm	δ_p, mm
1	Left Center	0.711	113.79	0.406	65.02	0.686	109.73
2	Center Right	0.584	93.47	0.406	65.02	0.813	132.08
3	Left Right	0.737	117.86	0.254	40.64	0.889	142.24

The geometric configuration and dimensions of the bridge model are shown in Figure 12.74. The main girder consists of two 1.52-m-long tubular steel sections. The cross section is 12.7 mm square and 1.59 mm thick. The span is continuous, simply supported at four points with one support acting as a pinned connection while the others simulate a roller connection, being able to resist only vertical motion. Figure 12.75 is a view of the pinned end connection, able to resist both vertical

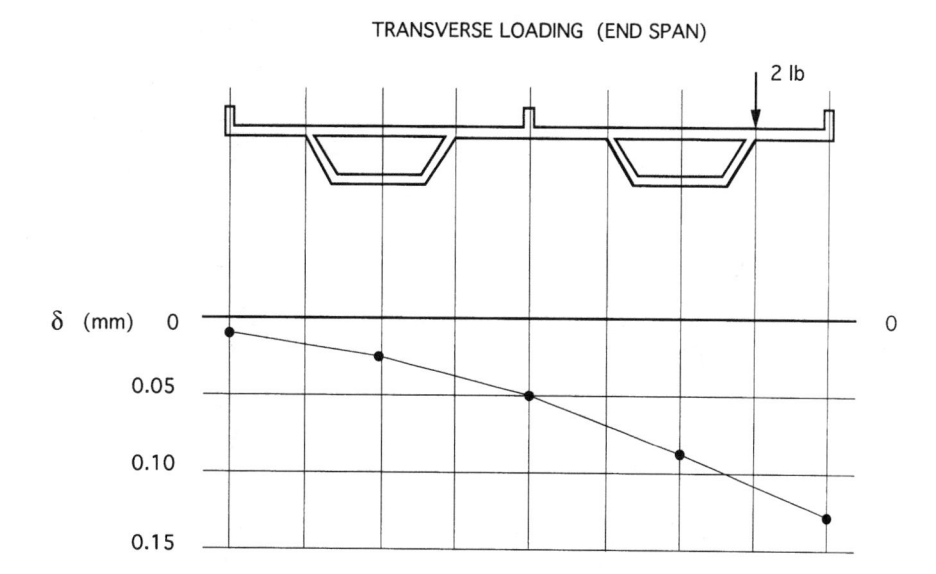

EDUCATIONAL MODELS FOR CIVIL AND ARCHITECTURAL ENGINEERING

Figure 12.72 Deflections under transverse loading on edge of span.

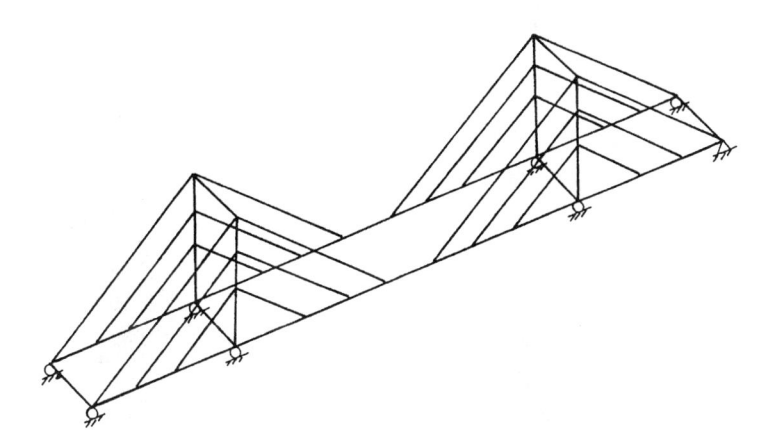

Figure 12.73 Isometric view of the double-plane cable-stayed bridge with harp system of cables.

and horizontal forces. The upright tower columns are 3.18-mm-thick steel bars, 12.7 mm wide at the bottom and 11.1 mm wide at the top. The two tower columns are spanned by three transverse girders which carry the cables through small notches (Figure 12.75). The tower support detail was designed to resist upward movements. The cable and tower connections and the tower support are shown in Figure 12.75.

The cables used in the model were 1.59-mm-diameter, seven-strand, seven-wire stainless steel cable having the properties shown in Figure 5.49. In order to provide known tension in the cables, the small cable turnbuckles shown in Figure 12.75 were instrumented with two small strain gages and thus became load cells that measured the cable tension. The use of another turnbuckle on the same cable allowed the tensioning force to be applied without disturbing the load cell. The calibration of the small load cells shows a linear relationship between the cable tension force and the strain output (Figure 12.76).

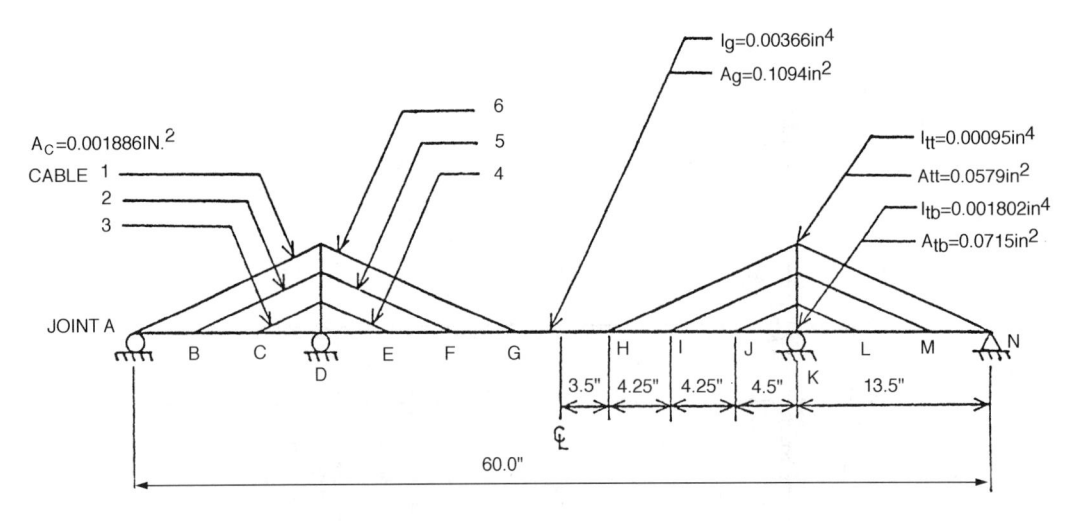

Figure 12.74 Model sectional and dimensional properties.

Figure 12.75 Cable and tower connections.

Dial gages measured the deflection at all panel points on the model and at the girder midspan as shown in Figure 12.77. The load–deflection characteristics of the model bridge are linear, as shown in Figure 12.78. Test results consisted of the determination of cable influence lines, bending moments, and deflections along the three spans.

This two-term project is challenging for a senior-level student but much can be learned about a complex structural form and experimental techniques. The accuracy of the results were within 20% of values determined from a larger model test. The model and hardware were fabricated by an experienced machinist.

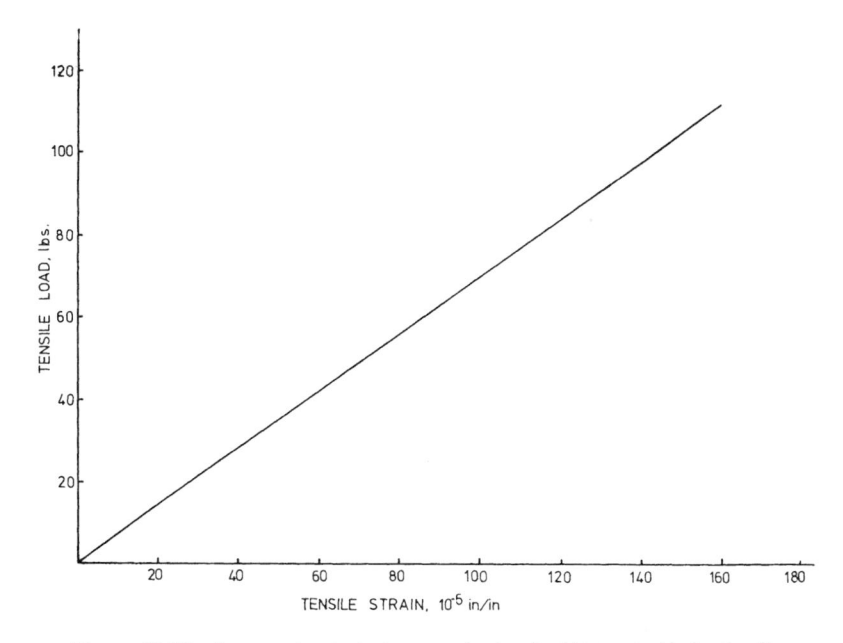

EDUCATIONAL MODELS FOR CIVIL AND ARCHITECTURAL ENGINEERING 741

Figure 12.76 Average load–strain curve for turnbuckle assembly load cell.

Figure 12.77 Test setup for model cable-stayed bridge.

12.7.3 Special Structures

12.7.3.1 Elevated Pedestrian Walkway

This experiment (Simonetti, 1982) involves the testing of a walkway between buildings at a height of approximately 61 m that is made of two steel girders, as shown in Figure 3.21a. A $\frac{1}{12}$-scale-model made of balsa wood (Figure 3.21b) was used to investigate the structural behavior. Lateral stability was of primary concern, and thus linearly elastic materials were used for the model. The walkway model was tested under two loading conditions: concentrated load at midspan (Test 1) and simulated uniform loading consisting of concentrated loads at midspan and quarter points (Test 2).

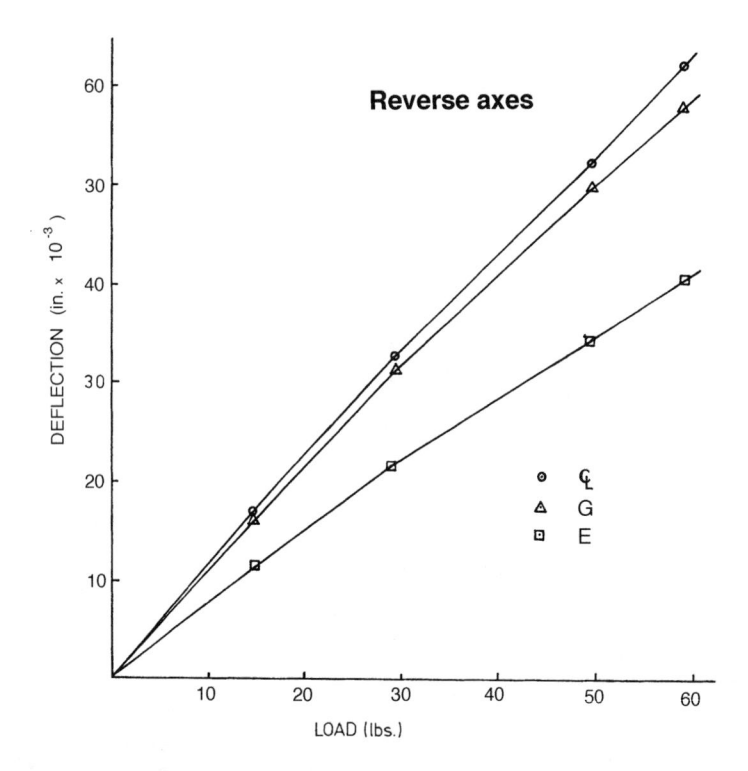

Figure 12.78 Incremental loads vs. deflection.

Dimensions of the two ¹⁄₁₂-scale beams needed to form the model walkway are shown in Figure 12.79. After attaching the lateral bracing to the top flange, the model was set up on a rigid steel testing table (Figures 3.21b and 12.80). As can be seen in Figure 12.80, end restraint was provided at the two end supports. Vertical and lateral deflections were measured in both tests as shown in Figure 12.81.

Under the concentrated load (Test 1), the load–deflection curves at midspan in both beams were essentially linear, as shown in Figure 12.82. Web buckling occurred at a load of 133.4 N, as noted in Figure 12.82 and the "jump" in the load–lateral deflection curve (Figure 12.83). The deflected shape of the compression flanges in both beams increased steadily with load, and at web buckling it took on shape.

Testing the model walkway under simulated uniform load (Test 2) presented a more severe loading condition. Web buckling was easily seen to occur between stiffeners (Figure 12.84). The load–deflection characteristics at midspan are also essentially linear. Five load-unload cycles were performed on this model to allow for the addition of stiffeners in the web. During the fifth cycle at a load of 173.5 N, the walkway experienced a failure when a fracture occurred in the buckled flange.

This study was a one-term project and the model was fabricated by a first year graduate student. The accuracy of the test results were within 15% when compared to theoretical buckling predictions.

12.7.3.2 Microwave Antenna Structure

This case study (Kopatz, 1982) involves the design and testing of a ¹⁄₉-scale balsa wood model of a prototype ³⁰⁄₂₀ GHz antenna operating on a new microwave theory. A ¹⁄₃-scale model of this prototype was designed using welded steel tubular sections. This formed the basis for the ¹⁄₉-scale balsa wood model. The final design of the tower structure of the ¹⁄₃-scale model consists of two structural frame pieces and two positioners (Figure 3.20b). The upper, cantilever section of this

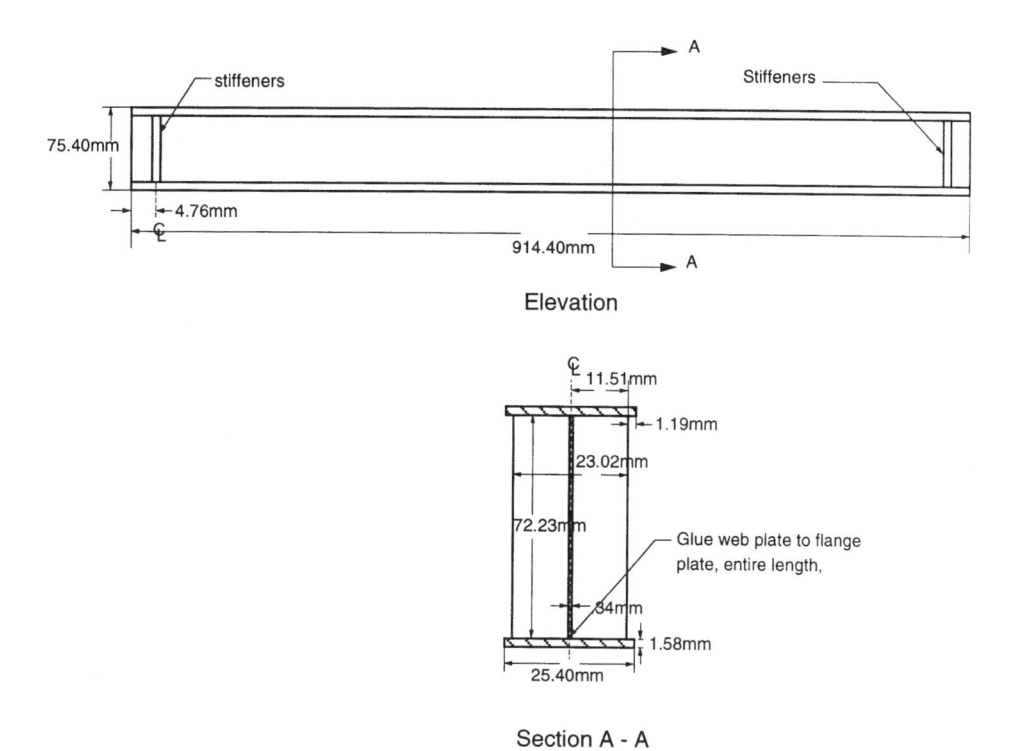

EDUCATIONAL MODELS FOR CIVIL AND ARCHITECTURAL ENGINEERING 743

Figure 12.79 Balsa wood 1/12-scale model plate girder.

Figure 12.80 Configuration of two plate girders forming supporting elevated walkway.

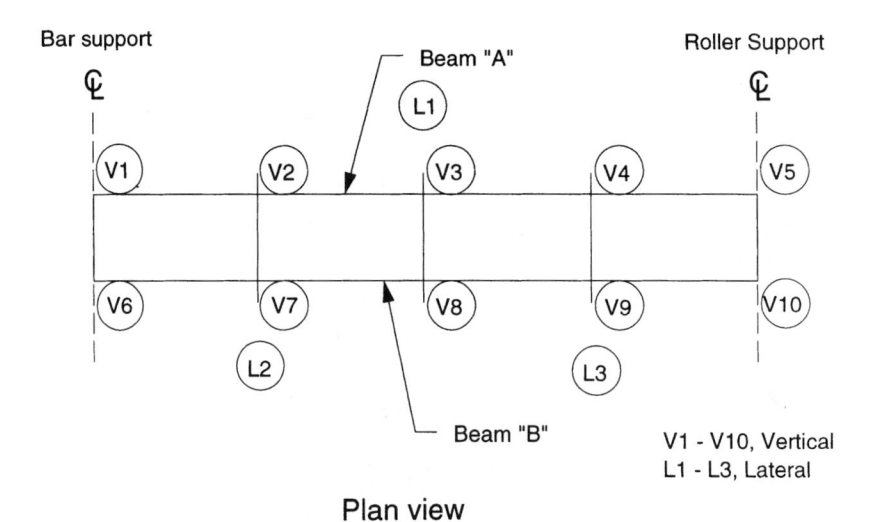

Figure 12.81 Location of instrumentation.

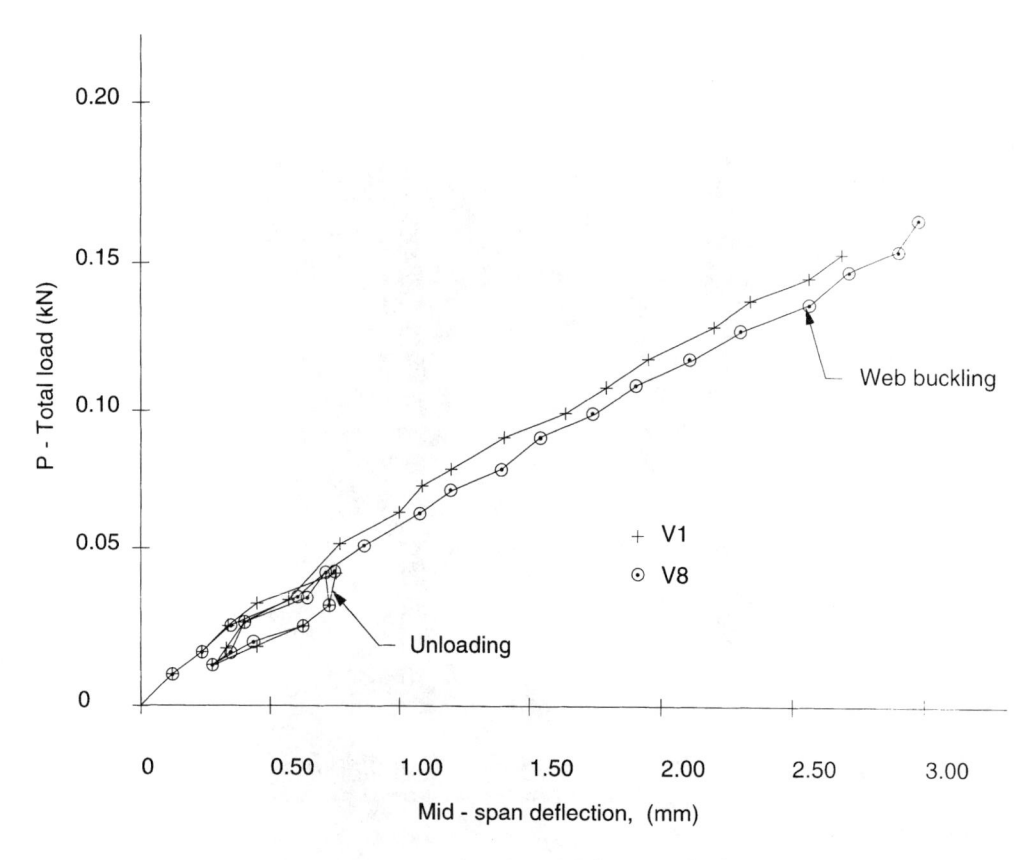

Figure 12.82 Load vs. lateral deflection behavior.

EDUCATIONAL MODELS FOR CIVIL AND ARCHITECTURAL ENGINEERING 745

Figure 12.83 Web buckling in both girders.

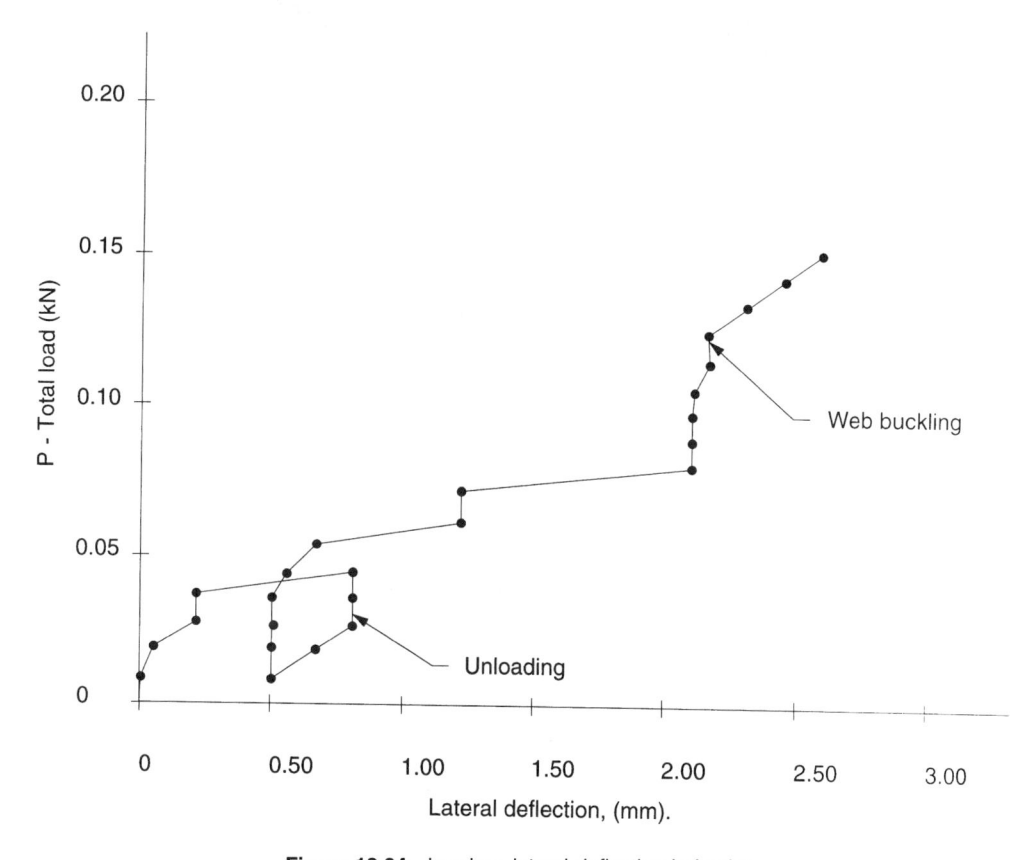

Figure 12.84 Load vs. lateral deflection behavior.

structure was selected for the ⅑-scale model analysis. This section is the most critical to the function of the antenna; yet it is the most difficult in which to limit deflection. It has a high degree of indeterminacy in an attempt to limit the differential displacement between the two reflectors. The

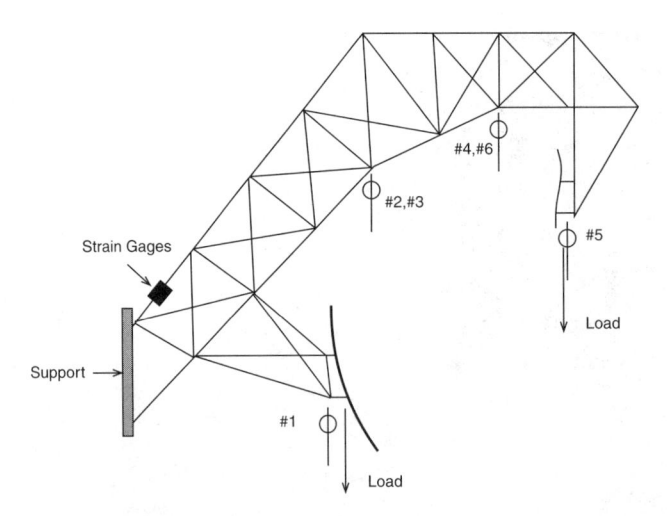

Figure 12.85 Location of instrumentation.

Figure 12.86 Microwave antenna model during testing.

differential displacement at ⅓ scale was set at 0.635 mm when rotated through 180° which for the ⅑-scale model meant a limit of 0.212 mm.

Instrumentation for the ⅑-scale model test consisted of deflection measurements and strain measurements, as shown in Figure 12.85. The model during testing is shown in Figure 12.86. A close-up of the bars with strain gages is shown in Figure 12.87. Loads were applied incrementally to the two reflectors (Figure 12.88) in the ratio of 3:1 to account for their weight difference. A linear load–deflection behavior is indicated in Figure 12.89 at gage #5 (Figure 12.85).

EDUCATIONAL MODELS FOR CIVIL AND ARCHITECTURAL ENGINEERING 747

Figure 12.87 Measuring strain in key truss members.

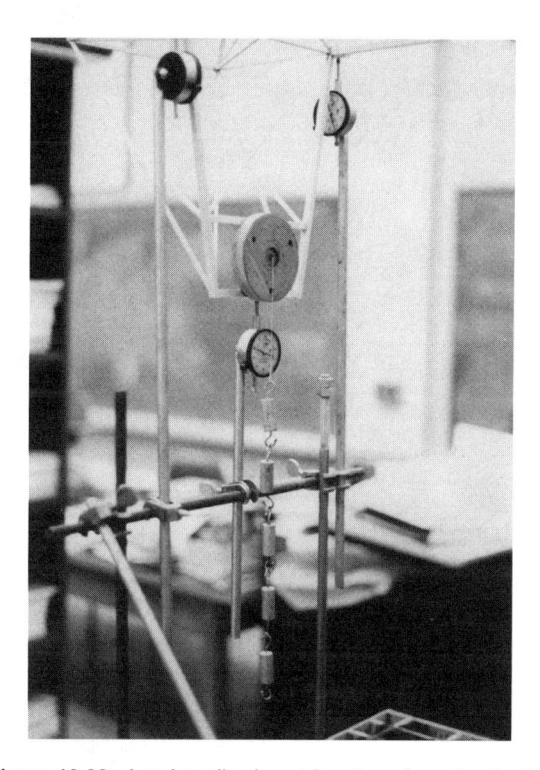

Figure 12.88 Load application at location of small reflector.

12.7.3.3 The Drexel Geodesic Tri-Span

The Drexel Geodesic Tri-Span (see Figure 12.90) was an open-air monumental structure, conceived, designed, and constructed by a group of senior civil engineering students as a bicentennial project (over a three-term period) on the university campus (Marriott et al., 1977; Anonymous, 1978). It was fabricated from bolted rectangular timbers which form planar triangular elements (the triangle is the symbol of Drexel University) having an average of 4.27 m on each side and

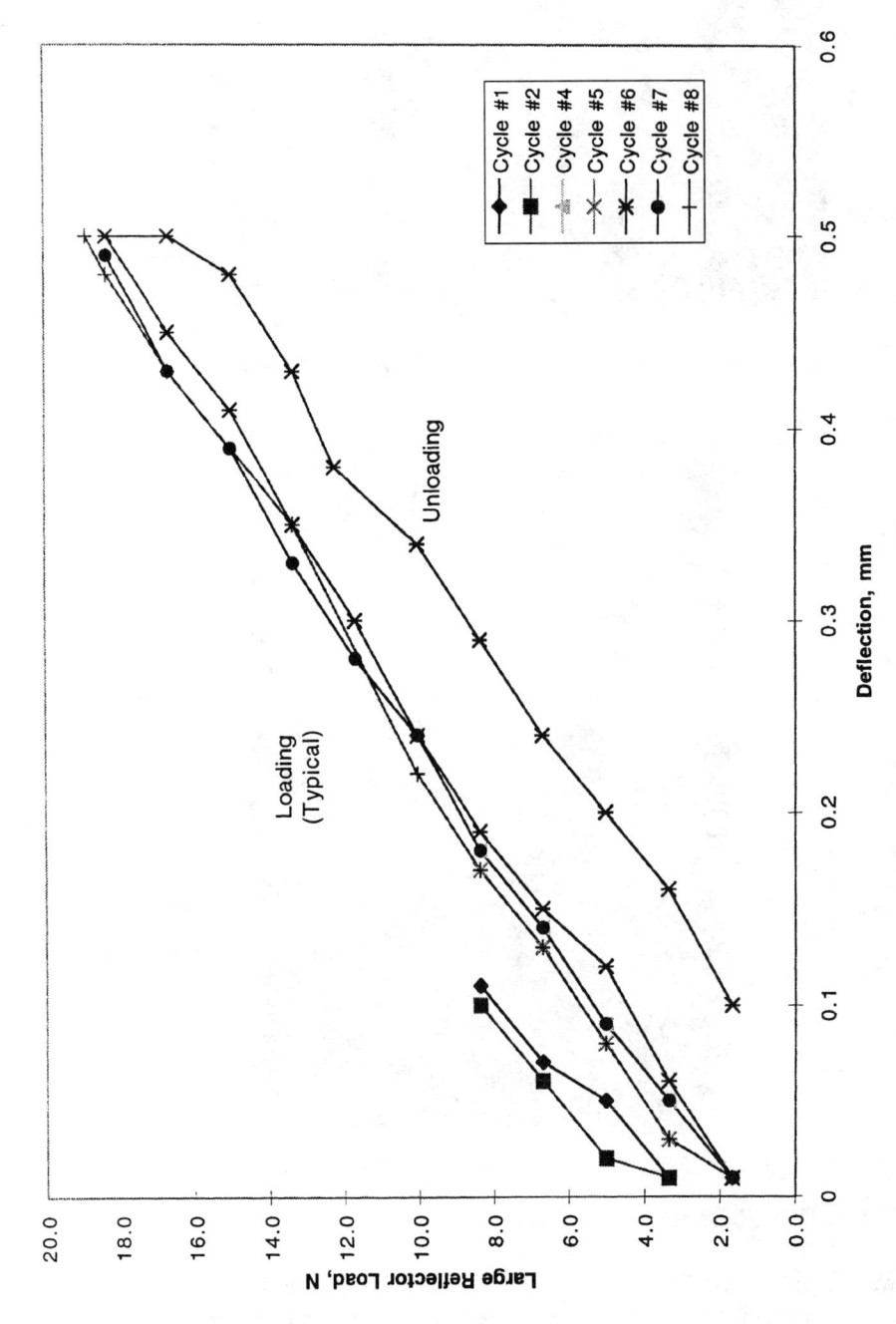

Figure 12.89 Load–deflection characteristics of antenna model at gage #5 (small reflector).

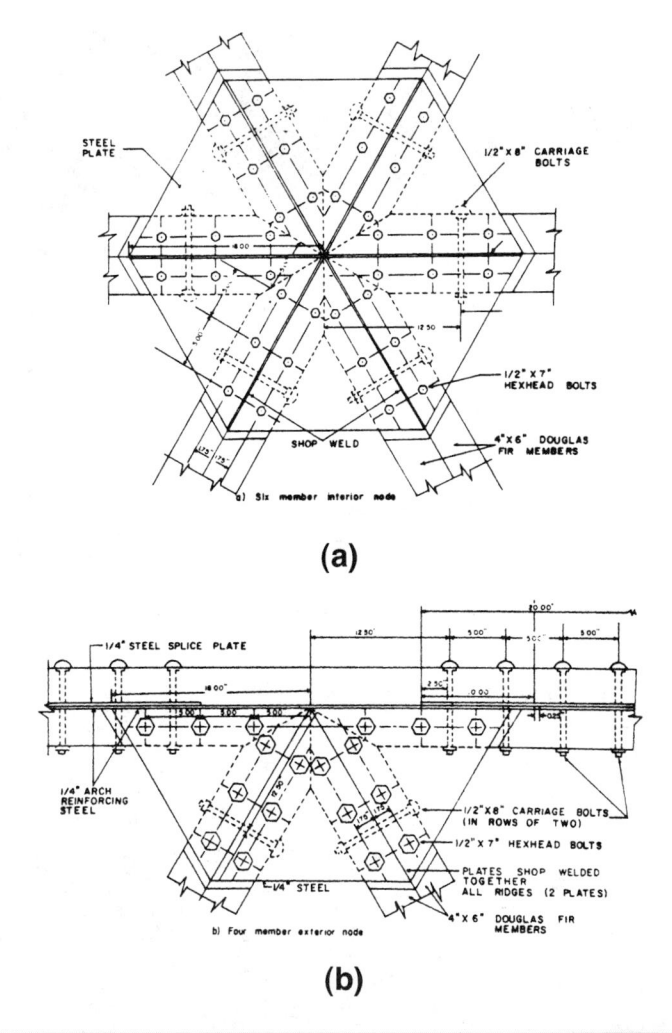

Figure 12.90 The Drexel University Geodesic Tri-Span (a) interior joint; (b) exterior joint; (c) finished structure..

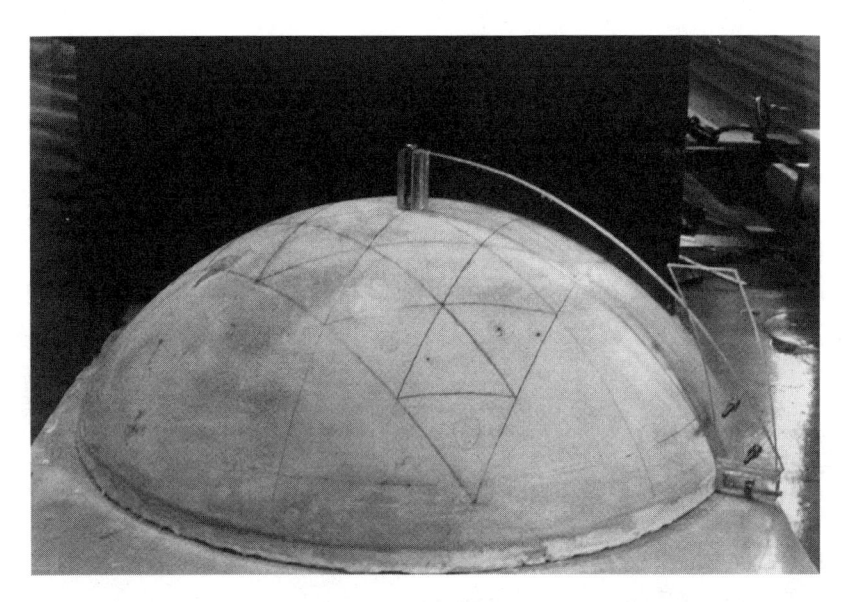

750   STRUCTURAL MODELING AND EXPERIMENTAL TECHNIQUES

Figure 12.91   Spherical surface on which $1/45$-scale model was built.

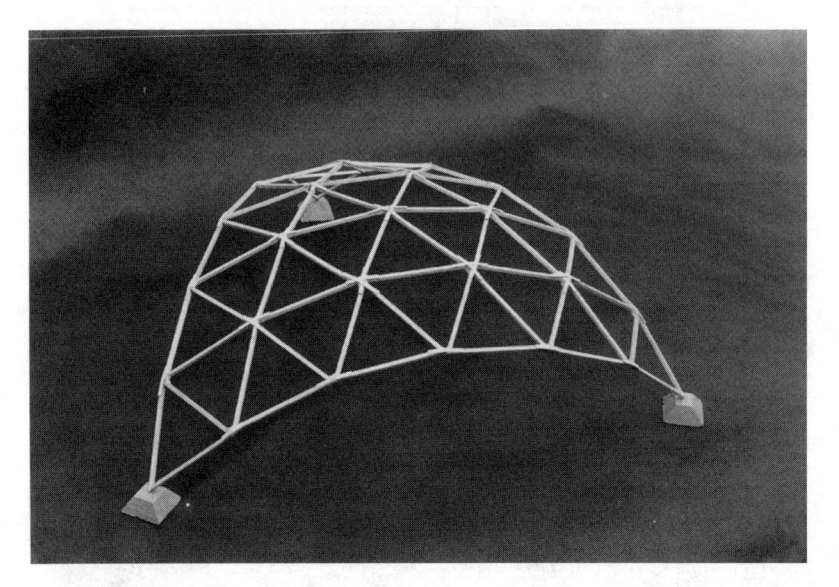

Figure 12.92   One-forty-fifth-scale model made from balsa wood.

whose vertices lie on a spherical surface with a radius of 15.32 m. The Geodesic Tri-Span has a triangular plan form and is supported on three podia forming a side span of 21.95 m. It rises 6.71 m at the crown, straddling an existing campus walkway over a city subway/surface trolley station.

A key part of the design of the structure was the construction and testing of a $1/45$-scale balsa wood model to obtain the load–deflection characteristics and the method of actual construction. The balsa wood pieces forming the various triangles were laid on a spherical surface of radius 0.34 m made by revolving a template on a built-up mold of wood and styrofoam and having a Hydrostone gypsum finish (Figure 12.91). Triangular elements were then glued together at the joints, using small strips of paper to simulate the steel plates (Figure 12.92). The testing of the model (see Figure 12.93) under simulated dead and live loading indicated that concentrated loads had large

EDUCATIONAL MODELS FOR CIVIL AND ARCHITECTURAL ENGINEERING

Figure 12.93 Test setup for the 1/45-scale model.

bending influences on the three-sided triangular dome supported at only three points. Obviously, it was these results that necessitated stiffening the three 21.95-m supporting arches. The reduced-scale model allowed the student designers to experience the structural behavior and also to visualize the overall physical structure. Testing the model convinced the students, who were inexperienced in the practical problems of construction, that their modified structure could in fact function effectively.

The connections between the various members framing at the nodes proved to be the critical item in the design of the Geodesic Tri-Span. The three supporting arches had considerable bending because of the lack of continuity that would normally be given by the portion of the full geodesic dome framework that is missing. The potential bending under unsymmetrical live load which may occur due to people climbing the dome put special requirements on the moment capability at the joints. The final design developed from considerations given to several types of steel connectors that could transfer moment from one wooden member to the others framing at the joint. Efforts to obtain commercially made joints from stamped sheet-metal proved futile. In order to minimize the anticipated problems, a preliminary design evaluation using flat strips at top and bottom of the beams and crisscrossing at each node was considered. But with this type of detail, a three-tier joint would be required because four to six members are framed at each node. Subsequently, the plates both on the inner and outer surface of the wood members would cross at different elevations. Such an arrangement was abandoned as architecturally undesirable. The selected joint design consisted of welded plates on the inner and outer surface of the dome which were made from segmented triangular pieces, shown in Figure 12.90, and bolted to the wooden members framing into the joint.

This was a remarkably successful student project by any comparison and epitomizes the power of the small-scale modeling technique. At the time this project was conceived, the analytical tools needed were not available to the designers. Thus, they resorted to a physical model approach which they could understand and which made the project possible.

12.8 SUMMARY

In this chapter small-scale model examples are discussed as they pertain to civil and architectural engineering education. A large variety of classroom and laboratory demonstration models are examined ranging in their characteristics from linearly elastic to nonlinear inelastic behavior.

Examples are given of architectural engineering demonstration models to supplement the student's early firsthand experience with structural components and structural building systems. Experiments are described that can be conducted by students without having advanced mathematical background and yet the results of which can give them the basic physical understanding needed to visualize structural behavior. The concepts of structural vibrations as they pertain to buildings and other structures are presented by model testing without the associated advanced mathematics. A large number of case studies mostly from the experience of the senior author are presented. These clearly show the wide range of structural problems that can be successfully used by senior-level undergraduate and first-year graduate students. The many examples described herein demonstrate the power of the physical modeling technique in educating young civil and architectural engineers.

PROBLEMS

12.1 Find the shear center of the irregular-shaped, thin open section shown below by making a simple cantilever beam and applying the shear load at various locations until only bending is measured without any torsional effects.

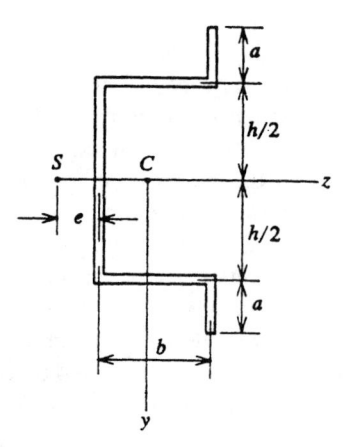

12.2 Set up a simple balsa wood test to determine the influence line for the diagonal bar in the second panel of the truss shown below. Assume that small paperbacked strain gages are available to you and that they can be directly applied to the member for this purpose.

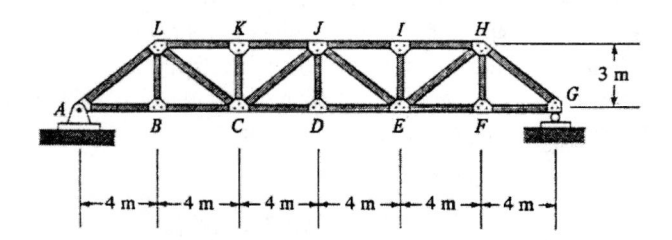

12.3 An existing four-story shear building model for the Drexel Models Laboratory shaking table needs to be stiffened at each level so that the three natural frequencies can be doubled from what they are at present. Describe techniques that will accomplish this requirement. Design the strengthening system for your best choice.

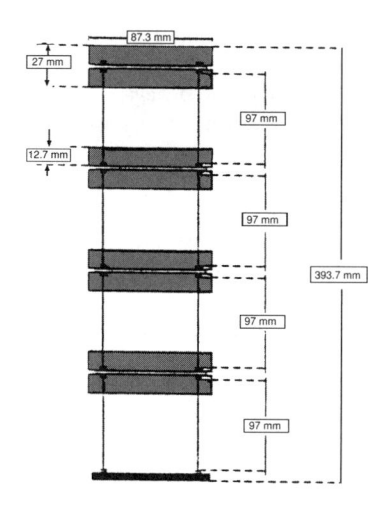

12.4 Use a rectangular section beam made from a very flexible material to demonstrate that "plane sections before bending remain plane after bending" by drawing a straight grid on its outer surface. What can you say about the Poisson's effect on the various fibers of the model beam? If you had strain gages available, how else could you prove this hypothesis attributed to Navier?

12.5 An infilled shear wall model is to be tested in a universal testing machine that has the available loading capacity. If the panel has an aspect ratio (width/height) of 1, i.e., it is square, how can you test it with no additional supports? Suppose the panel was rectangular with an aspect ratio of 2, how would you test it now? Design the supporting system.

REFERENCES

Anonymous, (1978). Drexel students build space frame as senior design, *Civil Eng. ASCE*, New York, September, 118–120.

Anonymous, (1996). *The Drexel Engineering Curriculum*, Drexel University, College of Engineering, LeBow Engineering Building, Philadelphia.

Baker, G. S. (1995). Ultimate Strength Behavior of Reinforced Concrete Beams, unpublished report for graduate course: *Model Analysis of Structures,* Department of Civil and Architectural Engineering, Drexel University, Philadelphia, December.

Bowders, J. J. (1979). Model Analysis of a Two-Pinned Reinforced Concrete Arch, unpublished report for graduate course: *Model Analysis of Structures,* Department of Civil and Architectural Engineering, Drexel University, Philadelphia, December.

Caccese, V. (1979). The Vierendeel Girder a Model Study, unpublished report for graduate course: *Model Analysis of Structures,* Department of Civil and Architectural Engineering, Drexel University, Philadelphia, December.

Carr, R., Venkataraman, T. S., Thomas, D. H., Smith, A., Tanyel, M., Fromm, E., and Quinn, R. G. (1989). An experiment to enhance the educational experience of engineering students, *J. Eng. Educ.*, April, 424–429.

Carr, R., Venkataraman, T. S., Thomas, D. H., Smith, A., Tanyel, M., and Quinn, R. (1995). The mathematical and scientific foundations of engineering, *J. Eng. Edu.*, 84(2), April.

Cowan, H. J. (1960). The use of models in the design of architectural structures, *Civil Eng.* (London), 55(646 and 647), May and June.

Cowan, H. J. (1961). The Architectural Science Laboratory of the University of Sidney, *Archit. Sci. Rev.*, 4(2), July.

Cowan, H. J. (1973). What Can We Do with Structural Models?, Models Laboratory Report MR11, Department of Architectural Science, University of Sydney, New South Wales, Australia.

Doukakis, C. (1980). A study of the structural behavior of a model wide flange beam under two-point load, *Drexel Tech. J.,* 43(1), Drexel University, Philadelphia, Fall.

Ergunay, T. (1993). Demonstrating the Basic Dynamic Behavior in the Laboratory Using Small-Scale Models, M.Sc. thesis, Department of Civil and Architectural Engineering, Drexel University, Philadelphia, September.

Fairman, S. and Cutshall, C. S. (1953). *Mechanics of Materials,* John Wiley & Sons, New York.

Felton, D. L., (1979). An Investigation of Eccentrically Loaded Short Columns Using Structural Modeling Techniques, unpublished report for graduate course: *Model Analysis of Structures,* Department of Civil and Architectural Engineering, Drexel University, Philadelphia, Fall.

Gambone, G., Georganas, N. et al. (1978). Design of a Large Span Reinforced Concrete Shell Made of Prefabricated Elements, unpublished senior design report, Department of Civil and Architectural Engineering, Drexel University, Philadelphia, May.

Gatti, D. L. (1982). Elastic Model of a Concrete Segmental Bridge, unpublished report for graduate course: *Model Analysis of Structures,* Department of Civil and Architectural Engineering, Drexel University, Philadelphia, May.

Godden, W. G. (1960). A new structural models laboratory at the Queen's University in Belfast, *Civil Eng.* (London), 55(647), June.

Godden, W. G. (1961). Demonstration Models for Teaching Structural Mechanics, National Science Foundation Report, University of Illinois, Urbana, August.

Godden, W. G. (1963). Demonstration Models for Teaching Structural Mechanics, Engineering Experiment Station Circular No. 78, University of Illinois, Urbana, Vol. 60, May.

Grove, J. (1995). Reinforced Concrete Beam Models: Lab. Study No. 4, Casting Procedure; Lab. Study No. 5, Material Strength Determination; Lab. Study No. 6, Instrumentation; and Lab. Study No. 7, Testing of R/C Concrete Beams to Failure, unpublished report for graduate course: *Model Analysis of Structures,* Department of Civil and Architectural Engineering, Drexel University, Philadelphia, December.

Gutkowski, R. M., Dewey, G. R., and Goodman, J. R. (1982). Full-scale tests on single-tapered glulam beams, *J. Struct. Div.,* ASCE, Vol. 108 (ST 10), October.

Heaney, A. C., The Versatile Structural Model Kit, University of New South Wales, Sydney, Australia.

Heckler, G. F. (1980). The Analysis, Fabrication and Testing of a Single Fink Truss, a Modified Single-Fink Truss and a Raised-Collar Truss and Similitude Study, unpublished report for graduate course: *Model Analysis of Structures,* Department of Civil and Architectural Engineering, Drexel University, Philadelphia, January.

Kopatz, K. W. (1982). Deflection Investigation of the $^{30}/_{20}$ GHz Antenna Using a Wooden Model, unpublished report for graduate course: *Model Analysis of Structures,* Department of Civil and Architectural Engineering, Drexel University, Philadelphia, May 5.

Kopatz, K. W. (1984). Model Analysis of Double Tapered Glulam Wooden Beams. term project, unpublished report for graduate course: *Experimental Analysis of Nonlinear Structures,* Department of Civil and Architectural Engineering, Drexel University, Philadelphia, Sept. 26.

Marriott, R., Markowski, R., Miller, R., DiBartolo, R., Asztalos, G., Hilferty, T., Kelly, T., and Cori, R. (1977). The Drexel geodesic tri-span—a bicentennial monumental structure, *Drexel Tech. J.,* Drexel University, Philadelphia, 8–14, Fall.

Pippard, A. J. S. and Baker, J. F. (1943). *The Analysis of Engineering Structures,* Edward Arnold, London.

Quinn, R. (1993a). Drexel's E4 Program: a different professional experience for engineering students and faculty, *J. Eng. Educ.,* 82(4), October.

Quinn, R., (1993b). The E4 Introductory Engineering Test, Design, and Simulation Laboratory, *J. Eng. Educ.,* 82(4), October.

Quinn, R. (1994a). The fundamentals of engineering: an introduction to the art of engineering, *J. Eng. Educ.,* 82(2), October.

Quinn, R., (1994b). The mathematical and scientific foundations for an integrative engineering curriculum, *J. Eng. Educ.,* August.

Quinn, R., Ed. (1994c). *Resource Manual for Freshmen Engineers,* Harcourt Brace, Orlando, FL.

Ronan, T. P. (1995). Models Lab #4,#5,#6,&#7 CIVE 501 Section 001 Under and Over Reinforced Concrete Beam Analysis, unpublished report for graduate course: *Model Analysis of Structures,* Department of Civil and Architectural Engineering, Drexel University, Philadelphia, December.

Sagur, S. M. (1987). The Effects of Central Openings on the Ultimate Load in Simply Supported R/C Slabs, unpublished report for graduate course: *Model Analysis of Structures,* Department of Civil and Architectural Engineering, Drexel University, Philadelphia, May.

Shepherd, R. (1964). Some models of use in the teaching of structural mechanics, *J. Eng. Educ.*, University of Canterbury, Christchurch, New Zealand, 54(10), June.

Simonetti, R. K. (1982). Model Analysis of a Steel Walkway System Using Balsa Wood Models, unpublished report for graduate course: *Model Analysis of Structures,* Department of Civil and Architectural Engineering, Drexel University, Philadelphia, April 25.

Sleiman, A. I., (1984). A Study of the Effective Flange Width of a Reinforced Concrete T-beam Using Small-Scale Models, unpublished report for graduate course: *Experimental Analysis of Nonlinear Structures,* Department of Civil and Architectural Engineering, Drexel University, Philadelphia, Sept. 26.

Socoloski, P. P. (1979). Model analysis of the cable-stayed bridge, unpublished senior seminar report, Department of Civil and Architectural Engineering, Drexel University, Philadephia.

Weisel, N. E. (1975). Reinforced Concrete Direct Model of a Two Span Continuous Beam, unpublished report for graduate course: *Model Analysis of Structures,* Department of Civil and Architectural Engineering, Drexel University, Philadelphia, August.

White, R. N. (1972). Structural Behavior Laboratory—Equipment and Experiments, Report No. 346, Department of Structural Engineering, Cornell University, Ithaca, NY.

Wisser, C. S., (1995). CIVE 501 Model Analysis of Structures, Laboratory Report #4–#7, Ultimate Strength Behavior of Reinforced Concrete Beams, unpublished report, Department of Civil and Architectural Engineering, Drexel University, Philadelphia, December.

APPENDIX A

Dimensional Dependence and Independence

CONTENTS

A.1 The Form of Dimensions 757
A.2 Method I: The Numeric Method 759
A.3 Method II: The Functional Method 761
A.4 Illustrative Examples 763
    A.4.1 Method II Solution 763
    A.4.2 Method I Solution 764
References 766

In Chapter 2 the conditions for dimensional dependence or independence between a group of physical quantities were stated without proof. The basis for these statements will be presented here. First, it will be necessary to discuss the functional form of the dimensions of physical quantities. Then, two different methods by which the results of the dimensional conditions can be obtained will be discussed.

A.1 THE FORM OF DIMENSIONS

Suppose that one is considering some n mechanical quantities. It is clear that the measures of these quantities can be established through the three *fundamental* dimensional units: force, length, and time. From a dimensional point of view then, these n quantities are known to be functions of force, length, and time.*

$$X_1 \doteq D_1(F, L, T)$$
$$X_2 \doteq D_2(F, L, T)$$
$$\cdots$$
$$X_n \doteq D_n(F, L, T)$$
(A.1)

---

* The symbol $\doteq$ will be used where dimensional equivalence is meant, and the symbol = will be left for those equations where numerical equivalence is also maintained.

It is of interest to ask whether there are any restrictions that should be placed upon the functions $D_1, D_2, \ldots, D_n$. Of basic importance in the concepts of dimensions as they have been formulated is that the relative magnitude of the numbers expressing the magnitudes of any two physical quantities remains unchanged when the size of the dimensional units is changed (e.g., the statement that John's car is traveling twice as fast as Joe's car does not depend upon whether speed is measured in feet per second, miles per hour, or centimeters per second). There cannot be any dispute over this fact: human beings simply established their measuring system on this premise. The important thing is to note that this fact places a definite restriction on the nature of the functional relationships $D_1, D_2, \ldots, D_n$. In fact, $D_1, D_2, \ldots, D_n$ *is restricted to be in the form of products of powers of the dimensional units.*

Proof

$$\text{John's speed} = D(aF, bL, cT)$$

$$\text{Joe's speed} = D(pF, qL, rT)$$

where a, b, c, and p, q, r are numbers that indicate the magnitude of the F, L, T dimensional units.

Now,

$$\frac{D(aF, bL, cT)}{D(pF, qL, rT)}$$

has absolute significance regardless of the size of the dimensional units; thus if the size of the force unit is changed to be $1/x$ as large, the length unit $1/y$ as large, and the time unit $1/z$ as large, then it must still be true that

$$\frac{D(aF, bL, cT)}{D(pF, qL, rT)} = \frac{D(axF, byL, czT)}{D(pxF, qyL, rzT)}$$

or

$$D(axF, byL, czT) = D(pxF, qyL, rzT)\frac{D(aF, bL, cT)}{D(pF, qL, rT)}$$

Now after applying the chain rule of partial differentiation with respect to x and remembering that F, L, and T represent dimensional units and are carried along only for completeness,

$$aF\frac{\partial D(axF, byL, czT)}{\partial axF} = pF\frac{\partial D(pxF, qyL, rzT)}{\partial pxF}\frac{D(aF, bL, cT)}{D(pF, qL, rT)}$$

Letting aF, bL, and cT vary while holding pF, qL, rT fixed, one gets when $x = y = z = 1$.

$$\frac{aF\dfrac{\partial D(aF, bL, cT)}{\partial aF}}{D(aF, bL, cT)} = \frac{pF\dfrac{\partial D(pF, qL, rT)}{\partial pF}}{D(pF, qL, rT)} \quad \text{constant}$$

or

$$\frac{\partial D(aF, bL, cT)}{D(aF, bL, cT)} = k_1\frac{\partial aF}{aF}$$

DIMENSIONAL DEPENDENCE AND INDEPENDENCE

This is a very special partial differential equation that can be treated as an ordinary differential equation, if it is kept in mind, of course, that the arbitrary constant of the ordinary differential equation becomes an arbitrary function for the partial differential equation. Thus, the integration yields

$$\ln D(aF, bL, cT) = k_1 \ln aF + \ln G(bL, cT)$$

or

$$D(aF, bL, cT) = G(bL, cT)(aF)^{k_1}$$

This process could be repeated, except that one would differentiate partially with respect to y and then z. The final result then becomes

$$D(aF, bL, cT) = \text{constant } (aF)^{k_1}(bL)^{k_2}(cT)^{k_3}$$

or

$$D(aF, bL, cT) = \text{constant } (F)^{k_1}(L)^{k_2}(T)^{k_3} \tag{A.2}$$

In this way it is seen that the functional forms of the dimensions of any physical quantity is necessarily a product. Naturally such a conclusion depends directly upon the way in which dimensions were first defined and used, and so it may be more satisfying just to note that the dimensions of all physical quantities do occur in the form of a single product.

The concepts dealing with dimensional dependence and independence can now be approached and rigorously deduced from either of two points of view. The first approach, which has been used since the early 1900s (Buckingham, 1914; Bridgman, 1922; Langhaar, 1951), leads to the conclusion that a set of physical quantities are dimensionally independent (or dependent) if the determinant formed from the powers of the fundamental units does not vanish (or does vanish). The second approach, suggested by Pahl (1962), considers the fact that if a set of functional relationships are independent (or dependent) with respect to certain arguments, then the Jacobian of these relationships with respect to the arguments does not vanish (or does vanish). Both approaches lead to the identical result, but since they are basically different it may be of interest to consider both.

A.2 METHOD I: THE NUMERIC METHOD

From the results of Section A.1 the dimensional form of mechanical physical quantities is known to be:

$$\begin{aligned} X_1 &\doteq F^{a_1} L^{b_1} T^{c_1} \\ X_2 &\doteq F^{a_2} L^{b_2} T^{c_2} \\ &\cdots \\ X_n &\doteq F^{a_n} L^{b_n} T^{c_n} \end{aligned} \tag{A.3}$$

A necessary and sufficient condition that r physical quantities be dependent (or independent) with respect to dimensions is that the determinant formed from the exponents of the "fundamental" dimensions of the quantities should vanish (or should not vanish).

The proof of the necessary portion can be effected in the following way. Assume that some three mechanical quantities, say, X_1, X_3, and X_6, are dimensionally dependent. That is, assume that there exists some constants k_1, k_2, and k_3 (not all zero) for which

$$X_1^{k_1} X_3^{k_2} X_6^{k_3} = 1 \tag{A.4}$$

It follows that

$$\left(F^{a_1} L^{b_1} T^{c_1}\right)^{k_1} \left(F^{a_3} L^{b_3} T^{c_3}\right)^{k_2} \left(F^{a_6} L^{b_6} T^{c_6}\right)^{k_3} \doteq 1$$

$$F^{(a_1 k_1 + a_3 k_2 + a_6 k_3)} L^{(b_1 k_1 + b_3 k_2 + b_6 k_3)} T^{(c_1 k_1 + c_3 k_2 + c_6 k_3)} = 1 \tag{A.5}$$

Now in order for Equation A.5 to hold, the exponents of F, L, and T must all equal zero. Thus, it is necessary that

$$a_1 k_1 + a_3 k_2 + a_6 k_3 = 0$$
$$b_1 k_1 + b_3 k_2 + b_6 k_3 = 0 \tag{A.6}$$
$$c_1 k_1 + c_3 k_2 + c_6 k_3 = 0$$

In order for a nontrivial solution of Equations A.6 to exist, it is necessary that the determinant of the coefficient matrix vanish. That is,

$$\begin{vmatrix} a_1 & a_3 & a_6 \\ b_1 & b_3 & b_6 \\ c_1 & c_3 & c_6 \end{vmatrix} = 0 \tag{A.7}$$

On the other hand, if this determinant did not vanish, then the trivial solution $k_1 = k_2 = k_3 = 0$ would be the only solution and there would be a contradiction to the initial assumption. The sufficiency condition is obtained by reversing steps.

One might wonder whether or not it would be possible to have a situation wherein the number of dimensionally independent quantities exceeded the number of fundamental dimensions. If one had four quantities and only three dimensions, Equations A.6 would contain four unknowns in only three equations. Such a situation would lead to the possibility that three of the unknowns could only be expressed in terms of the fourth, and infinitely many solutions exist. In fact, it is just this sort of reasoning which can be applied to a group of n physical variables. With n variables and only r dimensions, there would be an $(n - r)$-fold infinity of solutions to the set of Equations A-6. In other words, there would be $(n - r)$ infinity sets of the k terms that would satisfy the associated Equation A.4, thus making the product dimensionless. Accordingly, it would never be possible to find more independent quantities than there would be dimensions.

Since only r solutions can be dimensionally independent, it is of interest to know how the remaining $(n - r)$ quantities depend upon the dimensionally independent ones. This dependence can be obtained in the following way. First, Equation A.4 can be written to include all n terms.

DIMENSIONAL DEPENDENCE AND INDEPENDENCE

$$X_1^{k_1} X_2^{k_2} \ldots X_r^{k_r} \ldots X_n^{k_n} \doteq 1 \tag{A.8}$$

Equation A.6 now takes the form

$$a_1 k_1 + a_2 k_2 + \ldots + a_r k_r + \ldots + a_n k_n = 0$$
$$b_1 k_1 + b_2 k_2 + \ldots + b_r k_r + \ldots + b_n k_n = 0$$
$$\ldots$$
$$r_1 k_1 + r_2 k_2 + \ldots + r_r k_r + \ldots + r_n k_n = 0$$

This set of r equations in n unknowns can only be solved for r of the k values in terms of the remaining $(n - r)$ arbitrary k values. Thus

$$a_1 k_1 + a_2 k_2 + \ldots + a_r k_r = -a_{r+1} k_{r+1} - \ldots - a_n k_n$$
$$b_1 k_1 + b_2 k_2 + \ldots + b_r k_r = -b_{r+1} k_{r+1} - \ldots - b_n k_n \tag{A.9}$$
$$\ldots$$
$$r_1 k_1 + r_2 k_2 + \ldots + r_r k_r = -r_{r+1} k_{r+1} - \ldots - r_n k_n$$

where it is assumed that the first r numbered quantities are dimensionally independent (the quantities can always be renumbered to make this be so). Now, a particularly easy way to determine the dimensionally dependent relationship would be to set all but one of the $(n - r)$ arbitrary values in Equations A.9 equal to zero and then let that remaining one equal 1. Values of k_1 through k_n could then be obtained. Substituting all of the values back into Equation A.8, a dimensionless product is obtained. Another dimensionless product can be obtained by starting gain with Equations A.9, and letting the value of some other one of the $n - r$ arbitrary k values be 1. In the end, $(n - r)$ such dimensionless products can be obtained. These dimensionless products have often been referred to as *numerics*. It is for this reason that the title of this section is Method I: The Numeric Method.

It will not be proved here, but it is a fact that there are only $(n - r)$ independent dimensionless products that can be formed from a set of variables. These $(n - r)$ independent products are by no means unique, but they do correspond indirectly to the fact that Equations A.9 have, in general, an $(n - r)$-fold infinity of solutions. One such set of independent dimensionless products is obtained by the procedure outlined in the previous paragraph. Such a special set of products is known as a *complete set* and could be used in the application of Buckingham's theorem.

A.3 METHOD II: THE FUNCTIONAL METHOD

From the results of Section A.1 the dimensional form of physical quantities is known to be

$$X_1 \doteq F^{a_1} L^{b_1} T^{c_1}$$
$$X_2 \doteq F^{a_2} L^{b_2} T^{c_2}$$
$$\ldots$$
$$X_n \doteq F^{a_n} L^{b_n} T^{c_n}$$

It is shown in the calculus that *whenever r functions of r variables are functionally dependent, the Jacobian of the functions with respect to these variables vanishes identically. Conversely, if the Jacobian does not vanish, the r functions are independent.* The Jacobian is defined as

$$\frac{\partial(f_1, f_2, \ldots, f_r)}{\partial(m_1, m_2, \ldots, m_r)} = \begin{vmatrix} \frac{\partial f_1}{\partial m_1} & \frac{\partial f_1}{\partial m_2} & \cdots & \frac{\partial f_1}{\partial m_r} \\ \frac{\partial f_2}{\partial m_1} & \frac{\partial f_2}{\partial m_2} & \cdots & \frac{\partial f_2}{\partial m_r} \\ \vdots & \vdots & & \vdots \\ \frac{\partial f_r}{\partial m_1} & \frac{\partial f_r}{\partial m_2} & \cdots & \frac{\partial f_r}{\partial m_r} \end{vmatrix} \quad \text{(A.10)}$$

In the case of dimensional relationships, one attempts to find any r of the functions that the Jacobian of those r functions with respect to the r fundamental dimensions does not vanish. Suppose that it is found that

$$\frac{\partial(X_1, X_3, X_6)}{\partial(F, L, T)} = \begin{vmatrix} \frac{\partial X_1}{\partial F} & \frac{\partial X_1}{\partial L} & \frac{\partial X_1}{\partial T} \\ \frac{\partial X_3}{\partial F} & \frac{\partial X_3}{\partial L} & \frac{\partial X_3}{\partial T} \\ \frac{\partial X_6}{\partial F} & \frac{\partial X_6}{\partial L} & \frac{\partial X_6}{\partial T} \end{vmatrix} \neq 0$$

Then these three quantities are dimensionally independent. It is readily seen that the Jacobian determinant bears a strong resemblance to the determinant of the exponents that arose in the discussion of Method I. In fact, the two determinants are identical except that the Jacobian determinant may contain F, L, and T as well as numbers. That is, where the Method I determinant contained a 2, the Method II determinant might contain $2FT/L$.

Once a set of dimensionally independent quantities has been found, it is then necessary to obtain a complete set of dimensionless products. A very convenient way of obtaining a complete set (and, in fact, the set which is obtained is exactly the one which is obtained by the suggested procedure of Method I) requires knowledge of another theorem of calculus. Namely, if it is given that

$$X_1 \doteq D_1(F, L, T)$$

$$X_3 \doteq D_3(F, L, T)$$

$$X_6 \doteq D_6(F, L, T)$$

$$\frac{\partial(X_1, X_3, X_6)}{\partial(F, L, T)} \neq 0$$

then there exist *unique* functions

$$F \doteq f_1(X_1, X_3, X_6)$$

$$L \doteq f_2(X_1, X_3, X_6) \quad \text{(A.11)}$$

$$T \doteq f_3(X_1, X_3, X_6)$$

DIMENSIONAL DEPENDENCE AND INDEPENDENCE

With Equations A.11 one can return to Equations A.3 and substitute into the equations for X_2, X_4, X_5, X_7, X_8,...,X_n and thus determine the dimensions of X_2, X_4, X_5, X_7, X_8,..., X_n in terms of X_1, X_3, and X_6. These $(n-r)$ expressions can be transformed into dimensionless products. They will, in fact, be a complete set of dimensionless products.

A.4 ILLUSTRATIVE EXAMPLES

To illustrate how each method can be applied, suppose that one is considering the following set of physical variables:

Quantity	Units
X_1 Length	L
X_2 Force	F
X_3 Mass	$FL^{-1}T^2$
X_4 Stress	FL^{-2}
X_5 Strain	1
X_6 Acceleration	LT^{-2}
X_7 Displacement	L
X_8 Poisson's ratio	1
X_9 Modulus of elasticity	FL^{-2}

A.4.1 Method II Solution

Step 1. Search for three quantities for which the Jacobian of the three-dimensional functions with respect to F, L, and T does not vanish.

Try X_1, X_3, and X_7; then

$$\frac{\partial(X_1, X_3, X_7)}{\partial(F, L, T)} = \begin{vmatrix} \frac{\partial X_1}{\partial F} & \frac{\partial X_1}{\partial L} & \frac{\partial X_1}{\partial T} \\ \frac{\partial X_3}{\partial F} & \frac{\partial X_3}{\partial L} & \frac{\partial X_3}{\partial T} \\ \frac{\partial X_7}{\partial F} & \frac{\partial X_7}{\partial L} & \frac{\partial X_7}{\partial T} \end{vmatrix} = \begin{vmatrix} 0 & 1 & 0 \\ T^2 & -\frac{FT^2}{L^2} & \frac{2FT}{L} \\ 0 & 1 & 0 \end{vmatrix} = 0$$

so that X_1, X_3, and X_7 are dimensionally dependent. Clearly X_1 and X_7 cannot both be chosen.

Try X_1, X_3, X_5; then

$$\frac{\partial(X_1, X_3, X_5)}{\partial(F, L, T)} = \begin{vmatrix} \frac{\partial X_1}{\partial F} & \frac{\partial X_1}{\partial L} & \frac{\partial X_1}{\partial T} \\ \frac{\partial X_3}{\partial F} & \frac{\partial X_3}{\partial L} & \frac{\partial X_3}{\partial T} \\ \frac{\partial X_5}{\partial F} & \frac{\partial X_5}{\partial L} & \frac{\partial X_5}{\partial T} \end{vmatrix} = \begin{vmatrix} 0 & 1 & 0 \\ T^2 & -\frac{FT^2}{L^2} & \frac{2FT}{L} \\ 0 & 0 & 0 \end{vmatrix} = 0$$

so that X_1, X_3, and X_5 are dimensionally dependent. At first glance this result is surprising; however, it is seen that $X_5 \doteq (X_1)^0 (X_3)^0$ (or that the dimensions of X_5 equal the dimensions of X_1 to the power 0 times the dimensions of X_3 to the power of 0). Such an occurrence implies linear dependence just as strongly as if the exponents had been 1 and 2. In a certain respect this is a degenerate case, but it does point out that a dimensionless quantity can never be dimensionally independent from another group of quantities. In a similar manner one can prove that X_1, X_2, and X_4 are linearly dependent, etc.

Try X_1, X_6, and X_9; then

$$\frac{\partial(X_1, X_6, X_9)}{\partial(F, L, T)} = \begin{vmatrix} 0 & 1 & 0 \\ 0 & \dfrac{1}{T^2} & -\dfrac{2L}{T^3} \\ \dfrac{1}{L^2} & -\dfrac{2F}{L^3} & 0 \end{vmatrix} = -\frac{1}{L^2}\frac{2L}{T^3} = -\frac{2}{LT^3} \neq 0$$

so that X_1, X_6, and X_9 are dimensionally independent quantities.

Step 2. Since X_1, X_6, and X_9 are dimensionally independent, the functional relationships can be inverted and unique solutions for F, L, and T in terms of X_1, X_6, and X_9 can be obtained.

$$L \doteq X_1$$

$$T \doteq \sqrt{\frac{X_1}{X_6}} \qquad\qquad (A.12)$$

$$F \doteq X_1^2 X_9$$

Step 3. These in turn can be substituted into the remaining dimensional relationships to yield

$$X_2 \doteq F \quad \doteq X_1^2 X_9$$

$$X_3 \doteq FL^{-1}T^2 \doteq X_1^2 X_9 \frac{1}{X_1} \frac{X_1}{X_6} \doteq \frac{X_1^2 X_9}{X_6}$$

$$X_4 \doteq FL^{-2} \doteq X_1^2 X_9 \frac{1}{X_1^2} \doteq X_9 \qquad (A.13)$$

$$X_5 \doteq 1 \quad \doteq 1$$

$$X_7 \doteq L \quad \doteq X_1$$

$$X_8 \doteq 1 \quad \doteq 1$$

Therefore, the complete set of dimensionless products is

$$\frac{X_2}{X_1^2 X_9} \doteq \frac{X_3 X_6}{X_1^2 X_9} \doteq \frac{X_4}{X_9} \doteq X_5 \doteq \frac{X_7}{X_1} \doteq X_8 \doteq 1 \qquad (A.14)$$

A.4.2 Method I Solution

Step 1. Search for three quantities for which the determinant formed from the exponents of the functional dimensions does not vanish. Try X_1, X_3, and X_7; then

DIMENSIONAL DEPENDENCE AND INDEPENDENCE

$$\Delta = \begin{vmatrix} a_1 & b_1 & c_1 \\ a_3 & b_3 & c_3 \\ a_7 & b_7 & c_7 \end{vmatrix} = \begin{vmatrix} 0 & 1 & 0 \\ 1 & -1 & 2 \\ 0 & 1 & 0 \end{vmatrix} = 0$$

Try X_1, X_3, and X_5; then

$$\Delta = \begin{vmatrix} a_1 & b_1 & c_1 \\ a_3 & b_3 & c_3 \\ a_5 & b_5 & c_5 \end{vmatrix} = \begin{vmatrix} 0 & 1 & 0 \\ 1 & -1 & 2 \\ 0 & 1 & 0 \end{vmatrix} = 0$$

Try X_1, X_6, and X_9; then

$$\Delta = \begin{vmatrix} a_1 & b_1 & c_1 \\ a_6 & b_6 & c_6 \\ a_9 & b_9 & c_9 \end{vmatrix} = \begin{vmatrix} 0 & 1 & 0 \\ 0 & 1 & -2 \\ 0 & -2 & 0 \end{vmatrix} = -2 \neq 0 \quad (A.15)$$

so that X_1, X_6, and X_9 are dimensionally independent quantities.

Step 2. It must be possible to determine the dimensional relationship of all remaining quantities in terms of the dimensions of X_1, X_6, and X_9. This can be done in the same manner as was done in the Method II solution. Another procedure that can be made to lead to the identical results follows.

Any product π of the nine physical quantities has the form:

$$\pi = X_1^{k_1}, X_2^{k_2}, X_3^{k_3}, X_4^{k_4}, X_5^{k_5}, X_6^{k_6}, X_7^{k_7}, X_8^{k_8}, X_9^{k_9}$$

Regardless of the values of the constants $k_1, k_2, \ldots, k_9$, the corresponding dimension of π is

$$\pi \doteq [L]^{k_1}[F]^{k_2}[FL^{-1}T^2]^{k_3}[FL^{-2}]^{k_4}[1]^{k_5}[LT^{-2}]^{k_6}[L]^{k_7}[1]^{k_8}[FL^{-2}]^{k_9}$$

which can be rearranged to

$$\pi \doteq F^{(k_2+k_3+k_4+k_9)} L^{(k_1-k_3-k_4+k_6+k_7-k_9)} T^{(2k_3-2k_6)}$$

If it is now demanded that π be dimensionless, then each of the exponents of F, L, and T must vanish.

$$\begin{aligned} k_2 + k_3 + k_4 + k_9 &= 0 \\ k_1 - k_3 - 2k_4 + k_6 + k_7 - 2k_9 &= 0 \\ 2k_3 - 2k_6 &= 0 \end{aligned} \quad (A.16)$$

Here are three equations in nine unknowns (of course, k_3 and k_8 can just be considered to have zero coefficients). The result in Equation A.15 guarantees that this set of equations has a sixfold $(9-3=6)$ infinity of solutions; i.e., we can choose six of the k values arbitrarily and then solve for the remaining three. Suppose we choose $k_2 = 1$ and $k_3 = k_4 = k_5 = k_7 = k_8 = 0$ and then solve for k_1, k_6, and k_9. Equations A.16 reduce to

$$k_9 = -1$$

$$k_1 + k_6 - 2k_9 = 0$$

$$-2k_6 = 0$$

which have the unique solutions of $k_1 = -2, k_6 = 0, k_9 = -1$. Thus

$$\pi \doteq \frac{X_2}{X_1^2 X_9} \doteq 1$$

the same as the first of Equations A.14. The remaining five expressions in Equations A.14 can be obtained in exactly the same manner, by letting $k_3 = 1$, then $k_4 = 1$, etc., while holding the other k_2, k_3, k_4, k_5, k_7, and k_8 equal to 0.

It should be pointed out that Equations A.14 do not represent anything unique. There is an infinite number of ways that Equations A.14 could be written, corresponding to the fact that the six arbitrary k constants can be chosen in an infinity of ways. The particular choice of k constants that has been made above will prove to be the most useful.

REFERENCES

Bridgman, P. W. (1922). *Dimensional Analysis,* Yale University Press, New Haven, CT.
Buckingham, E. (1914). On physically similar systems, *Phys. Rev.,* London, 4(345).
Langhaar, H. L. (1951). *Dimensional Analysis and Theory of Models,* John Wiley & Sons, New York.
Pahl, P. J. (1962). A General Theory of Physical Models and Its Application to Structures of Significant Mass, MIT Publication T62-6, Department of Civil Engineering, Cambridge.

APPENDIX B

A Note on the Use of SI Units in Structural Engineering

CONTENTS

B.1 Geometry 768
B.2 Densities, Gravity Loads, Weights 768
B.3 Force, Moment, Stress, and Other Stress Resultants 768
B.4 Miscellaneous (Angles, Temperature, Energy, Power) 768
B.5 SI System Standard Practice 769

The International System of Units (Systeme International d'Unites), commonly called SI, is being adopted around the world as a uniform measurement system. Since usage of SI in engineering and scientific circles is proceeding rather rapidly, it will soon be essential that the modern civil engineer be experienced in using the SI system. For this reason, every effort has been made to make the second edition conform to the SI system of units.

The units adopted here for structural engineering work have been arrived at by careful study of SI references and using ASCE, AISC, AISI, and other technical recommendations. The SI user is urged to study the available references in the literature (e.g., ASTM, ASCE Standards) for a more general and complete treatment of this important topic.

The SI system differs from the MKS (meter-kilogram-second) system in the units of force and stress. In the MKS system, the unit of force is the kilogram force (kgf), while in SI it is a newton, which is explained below. Stress in the old metric system was expressed in units such as kilogram force per square centimeter (kgf/cm^2), while in SI it is newtons per square meter (N/m^2), known as pascal (Pa).

The basic and derived SI units for various categories of measurement are discussed in the following sections. A summary of pertinent SI units and conversion factors from U.S. Customary units is given at the end of the discussion. SI values for typical easily recognized quantities are also listed.

---

* This appendix is based on a similar one in White, R. N., Gergely, P., and Sexsmith, R. G., *Structural Engineering,* John Wiley & Sons, New York, 1976.

B.1 GEOMETRY

The length units, the meter (m) and the millimeter (mm), will be used for geometric quantities. The use of millimeter units for section modulus and moment of inertia (second moment of area) does involve large numbers for the majority of common structural shapes; this problem is met by listing steel section properties as (section modulus 10^3 mm^3) and (inertia 10^6 mm^4).

B.2 DENSITIES, GRAVITY LOADS, WEIGHTS

The standard of mass in SI is the kilogram (kg), equal to the mass of the international prototype of the kilogram (about 2.2 lb mass). This use of kg must not be confused with the old metric force called a kg or kgf.

Gravity loads exert forces on structures, and the conversion from mass to force becomes essential. In SI, the force unit is a newton (N), about ⅕ lb. It is the force required to accelerate 1 kilogram mass by 1 meter per second squared; a kilogram of mass exerts 9.80665 N on its support point. Load expressed in both kilogram and newton units will thus differ by the factor 9.80665 (9.8 for practical purposes).

B.3 FORCE, MOMENT, STRESS, AND OTHER STRESS RESULTANTS

The preceding section defines the newton (N) as the basic SI measure of force. The kilonewton (1000 N or kN) is about ⅕ of a kip and will be used widely in structural design. These force units are combined with meters to express loadings, bending and twisting moments, and other quantities involving length and force.

The basic stress unit in SI is the newton per meter squared (N/m^2), called the pascal (Pa). This is a very small unit (1 psi = 6895 Pa) that becomes practical only when used with a large prefix (k or M). The most convenient SI stress unit for structures is 1 000 000 Pa, the megapascal, or MPa, which is identical to 1 MN/m^2 (or 1 N/mm^2), and approximates ⅐ ksi. Since 1 ksi = 6.895 MPa, the modulus of steel as an example will be 200,000 MPa (or N/mm^2) in SI units.

Surface loadings and soil pressures have the units of pressure or stress, but the common usage will dictate their expression in kilonewtons per meter squared (kN/m^2). Surface loads in particular are well expressed in kilonewtons per meter squared because their effects must be converted into kilonewtons during structural analysis.

Moment is expressed in meter-newtons (m·N) or meter-kilonewtons (m·kN). These units are convenient since a meter-newton is close in value to a foot-pound (ft·lb) and a meter-kilonewton is close to a foot-kip.

B.4 MISCELLANEOUS (ANGLES, TEMPERATURE, ENERGY, POWER)

Plane angles are still measured in radians (rad) and solid angles in steradians (sr) in the SI system. Temperature in SI should be in degrees Celsius (C in old centigrade scale), but temperature in Kelvin (K) is also permissible, and the two are interchangeable for temperature gradients since 1°C = 1 K. Energy is expressed in joules (J): a joule is a newton-meter (N·m), and work is in watts (W), which is 1 joule per second (J/s).

B.5 SI SYSTEM STANDARD PRACTICE

There are several simple rules to be observed in using the SI system:

1. Preferred prefixes are to be selected from the following table, in which each prefix is a multiple of 1000:

Prefix	Symbol	Multiplication Factor
giga	G	10^9
mega	M	10^6
kilo	k	10^3
—	unit	1
milli	m	10^{-3}
micro	μ	10^{-6}
nano	n	10^{-9}

2. Use prefixes in the numerator only, except for kilogram, the base unit of mass. Thus, the stress unit of newton per millimeter squared (N/mm^2) is not recommended, rather meganewton per meter squared (MN/m^2).
3. Separate digits in groups of three, counting from the decimal sign. Do not separate with commas since the comma is used for the decimal point in many countries. Examples:

   $$1{,}234.57 = 1\ 234.57$$
   $$0.58729 = 0.587\ 29$$
   $$4789 = 4\ 789 \text{ or } 4789$$

4. Abbreviations of compound units, such as for moment, are written with a centered dot to indicate multiplication, such as m·kN. If the unit is spelled out, a hyphen or space is used (meter-kilonewton). Abbreviations of compound units that are divided are always written with a slash; thus kilogram per meter is abbreviated kg/m.

U.S. Customary Units, SI Units, and Conversion Factors (for converting U.S. to SI) in Structural Engineering

Property	U.S. Customary	×	Conversion Factor	=	SI Units
Overall geometry					
Spans	ft	×	0.3048[a]	=	m
Displacement	in.	×	25.4[a]	=	mm
Surface area	ft^2	×	0.0929	=	m^2
Volume	ft^3	×	0.0283	=	m^3
	yd^3	×	0.765	=	m^3
Structural properties					
Cross-sectional dimensions	in.	×	25.4[a]	=	mm
Area	in.^{2}	×	645.2	=	mm^2
Section modulus, volume	in.^{3}	×	16.39	=	10^3 mm^{3}[b]
Moment of inertia (second moment of inertia)	in.^{4}	×	0.4162	=	10^6 mm^{4}[b]
Material properties					
Density	lb/in.^{3}	×	27 680	=	kg/m^3
	lb/ft^3	×	16.03	=	kg/m^3
Modulus and stress values	psi	×	0.006895	=	MPa
	ksi	×	6.895	=	MPa

[a] Exact.
[b] AISC uses this style in SI units.

Loadings

		Mass Units				Force Units		
Concentrated loads	kip	× 0.4536	=	Mg[a]	kip	× 4.448	=	kN
Self-weight (density)	lb/ft^3	× 16.03	=	kg/m^3	lb/ft^3	× 0.1571	=	kN/m^3
Line loads (linear density)	k/ft	× 1488	=	kg/m	k/ft	× 14.59	=	kN/m
Surface loads	lb/ft^2	× 4.882	=	kg/m^2	lb/ft^2	× 0.0479	=	kN/m^2
	k/ft^2	× 4882	=	kg/m^2	kg/m^2 ×	47.9	=	kN/m^2

Stresses, moments

Stress	psi	× 6895	=	Pa
	ksi	× 6.895	=	MPa (MN/m^2 or N/mm^2)
Moment, torque	ft•lb (or lb•ft)	× 1.356	=	m•N (or N•m)
	ft•k (or k•ft)	× 1.356	=	m•kN (or kN•m)

Miscellaneous

Velocity	fps	× 0.3048	=	m/s
Energy	ft•lb force	× 1.356	=	N•m = J
Temperature	$t_C^\circ = (t_F^\circ - 32)(5/9)$			
	$t_k = t_F^\circ + 273.15$			
Linear expansion coefficient	°F^{-1}	× 1.8	=	°C^{-1} or K^{-1}

Typical Values

Property	U.S. Customary	Approximate SI
Water density	62.4 lb/ft^3	1000 kg/m^3 = 1 t/m^3
Concrete density	150 lb/ft^3	2400 kg/m^3
Steel modulus E	29,000,000 psi	200 000 Mpa
Concrete modulus E	3,500,000 psi	24 000 MPa
Allowable steel stress	25 ksi	170 MPa
Design live load	100 psi	5kN/m^2

Index

A

AASHTO
    Code, 539
    decks, 538
Acceleration, 78
    records, 654, 655
    top-floor, 597
Accelerometers, 595, 617
Acetone, 292
ACI, see American Concrete Institute
Acoustic emission, 353
Acrylic plastics, 88, 291
Actuator, hand-operated, 503
Adequate model, 56
Adhesives, 286, 328, 439
Aeroelastic models, 4, 616
Aerospace vehicles, 606
Aggregate(s)
    content, effect of, 138
    grading curves for, 164
    –gypsum ratio, 167, 169
    size, 421
Air bag pressure loading system, 393
Air-cured specimen, 140
Airplane
    parts, 324
    propeller, torque on, 673
AISC, see American Institute of Steel Construction
Aluminum, 111
American Association of State Highway and Transportation Officials Code, 20
American Concrete Institute (ACI), 131
American Institute of Steel Construction (AISC), 211
American Society for Testing and Materials (ASTM), 89, 767
    requirements, 66
    standards, 91, 199
    structural steel, 588
    type N masonry mortar, 181
Analog–digital converter, 501
Analytical predictions, 457, 597
Anchorage systems, 253
Annealing
    furnaces, 329
    processes, 250

Antenna model, load–deflection characteristics of, 748
Apollo
    shell, model of simplified, 609
    simplified shell structure, 607
Appalachian Trail, footbridges for, 126
Apparent modulus, 96
Arch
    load–deflection behavior of model, 712
    model, 710
    reinforced concrete, 705
    width, 708
Architectural engineering, 466, 680, 692, 752
Artificial mass simulation, 223
ASCE Standards, 767
ASTM, see American Society for Testing and Materials

B

Balsam fir, 116
Balsa wood, 298
    advantage of using, 108
    application of to model studies, 110
    buckling study model, 297
    joints, gluing of, 297
    models, 109, 113, 296, 750
    modulus of elasticity of, 108
    properties, 110
    shapes available, 105
    test, 752
Bar
    forces, 687
    stock, milling of, 225
Base
    isolation system, 723, 726, 727
    shear, maximum, 654
Basswood, 116
Beam(s)
    average deflection of edge, 514
    behavior of, 458
    bending, effect of load discretization on, 390
    block geometry, 711
    bonding strength of grouted bond, 562
    collapse mode of, 714
    –column
        assemblies, results of tested, 470

effects, 63
    reinforced concrete, 578
continuous, 687
cracking patterns of failed, 533
double-tapered glulam, 729
ductility indexes, 526, 534
failed model, 697
filler, 222
flange, 715
flexural bond strength for grouted, 559
floor, 467
materials, mechanical properties of, test, 529
model, 227
    glulam, 728
    underreinforced and overreinforced, 695
moment–curvature behavior of reinforced, 532
multiple wires in, 307
overreinforced, 432, 433
postcracking behavior of FRP-reinforced, 524
prismatic, 447
reinforcement
    in model, 454
    stress–strain relations for, 242
series, nonreinforced, 148
shear connection between slab and, 577
simple, 685
test(s), 219, 223, 426
    model, 698
    results, 524, 532
two-span
    continuous, 689
    R/C, 701
University of Texas bond, 262
yielding in, 713
Bed joint shear, 194
Beggs deformeter, 122
Bending, 682
    moment, ultimate, 708
    strains, 379, 381, 686
Birch veneer plywood, 115
Black ironwood, 106
Blast
    attacks, threats of terroristic, 488
    chambers, 598
    effects, on protective structures, 612
    loading, effects of high, 613
Block(s)
    geometric accuracy of interlocking, 557
    -making machine, Drexel University, 309, 310
    masonry, mortarless, 556
    -molding machine, 312
    specimens, WHD, 176
    splitting strength determination, model, 170
    unit, WHD, 557
Blunders, 439, 448
Boat hulls, 324
Boiling water reactors, 510
Bond
    beam
        test, symmetrical, 263
        University of Texas, 262
    similitude, 266

Boundary layer, development of over typical surfaces, 599
Braced frame specimen, 470
Bragg grating, fiber-optic, 370
Bridge(s)
    cable-stayed, 737
    decks, reinforced concrete, 537
    double-plane cable-stayed, 739
    load–deflection
        characteristics of model, 740
        curves of model, 572
    long-span, 14
    model(s), 35, 280
        extensive cracking in, 571
        external post-tensioning of, 36
        fatigue load test of, 36, 570
        postfailure cracks in DT, 572
        seismic testing of, 621
    piers, earthquake-resistant, 502
    prestressed wooden, 24, 542
    reinforced concrete, 618
    structures, 497
    system
        modified DT concrete girder, 33
        prestressed concrete composite, 31, 568
    truss, 687
    vibration model, 59, 60
    wooden, 26
Brittle coatings, 352
Brittleness, 420
Buckingham's pi theorem, 49, 53
Buckling, 682
    behavior
        paper and wood plate girder demonstrating, 120
        of thin shells, 144
    capacity, predicted, 390
    inelastic, 711
    load, 37
        elastic, 575
        Euler, 690
        theoretical, 691
    model, 401, 402
    program, experimental, 515
    studies, 396, 400
    tests, 126
Building(s)
    code, 650
    components, 729
    concrete large-panel, 486
    cross wall, 627
    dynamics of shear, 722
    earthquake
        -resistant masonry, 473
        simulation of masonry, 628
    elements, precast, 732
    frame, lateral load system for, 404
    GLD reinforced concrete, 668
    model, four-story, 723
    precast concrete large panel, 622
    shaking table tests on concrete, 660
    structures, 466
    wind effects on, 611

INDEX

Bundled strength theory, 415
Butt joints, failure of, 555

C

Cable tension force, 739
Calcite filler, 94
Calendars, 104
Calibration procedures, 457
Cantilever
    beam, elastic stiffness of, 51
    wall component assembly, 488
Capacitance gage, 326
Capillary welding, 292
Carbon fiber-reinforced polymer (CFRP), 31, 32, 569
Carlson stress meter, 358, 359
Cassegrainian antenna, 110
Casting technique, 619
Catalpa, 106
Cellulose
    acetates, 90
    nitrates, 90
Cement(s)
    –aggregate ratio, 132
    choosing 328
    heat-cured epoxy resin, 356
    particles, hydration of, 131
    Portland, 174
    rapid-hardening, 132
Cement and Concrete Association, 265
Centrifugal force, 81
Centrifuge(s), 604
    soil–structure interaction studies using, 642
    swinging basket of, 647
    testing, modeling for, 76
CFRP, see Carbon fiber-reinforced polymer
Circuit analysis, 332, 33
Civil engineering, 466, 767
    structures, 73
    undergraduate course, 681
Classroom demonstration models, 681
CMD, see Cross machine direction
Coarse aggregate, 131
Code(s)
    equations, 436
    provisions, 30, 560
    size factors in, 437
Collapse modes, 710
Column
    interaction curve, 702
    models, 215
    reinforcement, 234, 241
    specimens, reinforcement for, 504
    stirrups, 23
    strength formula, 79
Compaction density, 421
Complete set, 761
Composite action demonstration, 684
Compressibility time, 78
Compression
    load cells, 354
    tests, 218, 431
Compressive strength, 557
    effect of size on, 424
    effect of strain rate on, 141, 591
    relation of tensile strength to, 154
Computer
    codes, available to engineer, 86
    -controlled systems, 407
    as multipurpose laboratory tool, 367
Concrete(s)
    arches, reinforced, 705
    beam(s)
        design of reinforced, 574
        size sensitivity of web-reinforced, 436
        underreinforced/overreinforced, 694
    bridge(s)
        decks, 537
        model, prestressed, 386
        reinforced, 618
    building(s)
        GLD reinforced, 668
        lightly reinforced, 21
        shaking table tests on, 660
        three-story lightly reinforced, 23
    bunkerlike facilities, 600
    casting mode, 314
    column(s)
        eccentrically loaded, 698
        model, eccentrically loaded reinforced, 700
    comparison of compressive stress–strain curves for, 137
    components, precast, 733
    compressive strength of, 143
    cracks in, 352
    creep recovery of, 136
    crushing, 526
    deck, monitoring strain of, 376
    density, 770
    dynamic properties of, 590
    engineering properties of, 131
    frames, reinforced, 614
    gypsum-based model, 573
    large-panel buildings, 486
    long-term properties of, 429
    masonry
        models, fabrication of, 309
        test structure, reinforced, 484
        units, 171
    matrix strength, 421
    mechanical properties of, 155
    methods for testing, 148
    model(s)
        accuracy in, 450
        forms for casting reinforced, 302
        FRP reinforcement for, 255
        plywood forms for casting reinforced, 304
        reinforced, 63, 64, 266, 299, 505
    modulus, 770
    moisture content, 363
    pavements, 204
    piers, reinforced, 498
    prototypes, 153

reinforcement, 236
shell, model reinforced, 453
size effects in, 420, 431
slab(s)
    demonstration model of, 202
    with penetrations, 517
strength of, 418
stresses, 66
stress–strain
    behavior of, 133
    characteristics, 135
    curve, 156
structures
    GLD reinforced, 667
    inelastic response of reinforced, 53
    integration of fiber-optic sensors into, 373
T-beams, effective width of reinforced, 705
tensile strength of, 142, 426
Conjugate beam method, 688
Constant strain method, 99
Construction procedure models, 5
Consulting engineers, 498
Continuous beams, 687, 689
Conversion factors, 769
Corbel stiffening, 699
Cork, 106
Crack(s)
angle, 539
behavior, 520
detection methods, 351, 353
distribution of, 261
formation, 564
at interior support, 704
invisible, 268
motion, transverse, 543
pattern(s), 637, 657
    of CoH slabs, 525
    of deck regions, 538
    differences in, 658
    of failed beams, 533
    of failed specimen, 480
    final, 511, 512, 624
    in model frame, 269
    of model T-beams, 709
    observed in joints, 495
    path of for RCH slabs, 526
    of slab, 517
    for specimen, 565
    reproducing, 270
    under stationary pulsating load, 540
similitude, 267, 431, 432
surfaces, reversing shear movement of, 541
widths, 271
Crazing, 291
Creep, 321
characteristics, 87, 362, 438
curves, 95
strains, 125, 136
tests, 99, 430
Critical volume, 158
Cross machine direction (CMD), 119
Cube-root scaling law, 74

Curing, 140, 157, 168, 423
Cyclic loading, foundation settlement during, 648
Cyclic tests, displacement history for, 480

D

Damage
    pattern, of model, 625
    propagation of, 637
Damping ratio, 656
DAQ, see Data acquisition
Data
    acquisition (DAQ), 329, 365, 366
    dead-load, 536
    entry, 366
    observation of trends in, 457
    recording, types of, 364
    reduction, 377
    tensile test, 523
DCDT, see Direct current differential transformer
Dead load (DL), 387, 488
    data, 536
    deformations, 577
    effects, 573
Deadweight(s)
    effects, 315
    measurement of loads using, 456
    similitude condition, 406
    stresses, 404, 547
    suspended, 385, 387
Deck
    displacement, components of, 501
    elements, welding of, 501
    models, splice pattern for wood, 545
    panel(s)
        models, noncontinuous, 539
        simply supported, 538
    ratio, 569
    vertical displacements, 608
Deflection
    behavior, load vs. lateral, 744
    curves, load vs., 704
    measurements, 321
    -measuring sensor, 684
    predictions, full-scale, 556
Deformation
    hardening, 654
    requirements, 236
Delta gage configuration, 346
Delta rosette, 340, 378
Design
    applications, applicability of models in, 34
    code(s)
        recommendations, evaluation of, 474
        size effects and, 435
    live load, 770
    methods, 2
    and operating conditions, 57
    requirements, relaxation of, 60
    stress, 355

INDEX

Diagonal tension tests, 203, 562
Dial gage, 349
Diaphragm action, 692
Diffusion effect, due to drying, 420
Dimensional accuracy, 451
Dimensional analysis, 42, 45, 52
Dimensional dependence and independence, 757–766
    examples, 763–766
    form of dimensions, 757–759
    functional method, 761–763
    numeric method, 759–761
Dimensional units, fundamental, 757
Dimensionless products, 49
Direct current differential transformer (DCDT), 350
Direct models, 204, 383, 488
Direct tension test, 146
Discrete loading, 391, 400
Displacement
    control loading history, 482
    deformeter, 123
    dial gages, 494
    gages, 617
    history, for cyclic tests, 480
    measurements, 348
    model, 122
    top story, 623
    transducers, 18, 487
Distortion, types of, 59
DL, see Dead load
Dome(s)
    failure, 517, 520
    model
        finished precast, 736
        segmented precast concrete, 735
    postshot, 620
    preshot, 620
Double-Tee (DT) girders, 568
Douglas fir, 106, 116
Drilling platforms, 646, 647
Drop hammers, 601, 603
Drying, 420, 422
Drystack masonry systems, 556
DT girders, see Double-Tee girders
Ductile hybrid fiber reinforced polymer, 256
Ductility indexes, for beams, 526, 527
Dummy gages, 364
Dynamic loads, structural models for, 585–678
    case studies, 649–672
        shaking table tests on lightly reinforced concrete buildings, 660–672
        shaking table tests on R/C frame-wall structures, 652–660
        wind tunnel tests of Toronto City Hall, 649–652
    examples of dynamic models, 604–649
        aeroelastic model studies of buildings and structures, 611–612
        blast effects on protective structures, 612–614
        earthquake simulation of masonry buildings, 628–641
        earthquake simulation of reinforced concrete structures, 614–625
        earthquake simulation of steel buildings, 625–627
        impact loading, 641–642
        natural modes and frequencies, 604–610
        soil–structure interaction studies using centrifuge, 642–649
    loading systems for dynamic model testing, 593–604
        drop hammers and impact pendulums, 601–604
        shaking tables, 600–601
        shock tubes and blast chambers, 598–600
        vibration and resonant testing, 593–597
        wind tunnel testing, 597–598
    materials for dynamic models, 588–593
        dynamic properties of concrete, 590–593
        dynamic properties of steel, 588–590
    similitude requirements, 587–588
Dynamic models, 4
    examples of, 604
    materials for, 588
Dynamic response, 321

E

Earthquake(s)
    damage to historic buildings due to, 634
    events, simulated, 639
    forces, 8
    input, 617
    loading, 467, 586, 720
    modeling, 76
    prototype full-intensity, 626
    response, of structures, 77
    -resistant bridge piers, 502
    -resistant building, reinforced concrete frame-wall, 15
    simulation
        of masonry buildings, 628
        of steel buildings, 625
    simulator(s), 20, 222
        test program, 626, 652
        University of California, Berkeley, 653
Earthquake Engineering Research Center, 614
Earthquake Simulator Laboratory, 606
Eastern white pine, stress–strain characteristics of, 731
Ebony, 106
Educational models, for civil and architectural engineering, 679–755
    case studies and student projects, 729–751
        bridge structures, 736–741
        building components and systems, 729–735
        special structures, 741–751
    experimentation and new engineering curriculum, 725–729
        historical background, 725–726
        new engineering curriculum, 726–729
    historical perspective, 681
    linearly elastic structural behavior, 681–694
        architectural engineering models, 692–694
        classroom demonstration models, 681–685

laboratory demonstration models, 685–692
nonlinear and inelastic structural behavior,
    694–712
  ultimate-strength models of R/C
      components, 694–710
  yielding and inelastic buckling of steel
      members, 710–711
structural dynamics concepts, 712–725
  basic laboratory instrumentation, 712–718
  dynamics of shear buildings, 722–725
  vibrations of lumped-mass systems,
      718–722
Elastic buckling, 105, 682
Elastic displacement, maximum, 47
Elastic models, 85–128, 534
  determination of influence lines and influence
      surfaces using indirect models, 121–123
  effects of loading rate, temperature, and
      environment, 100–103
    coefficients of thermal expansion, 101–102
    effects of temperature and related thermal
        problems, 101
    influence of relative humidity on elastic
        properties, 103
    influence of strain rate on mechanical
        properties of plastics, 100–101
    softening and demolding temperatures, 103
    thermal conductivity, 102–103
  materials for, 87–88
  plastics, 88–96
    mechanical properties of polyester resin
        combined with calcite filler, 94–96
    tension, compression, and flexural
        characteristics of, 89–92
    thermoplastics and thermosetting plastics,
        88–89
    viscoelastic behavior of, 92–94
  special problems related to plastic models, 103–104
    influence of calendaring process on
        modulus of elasticity, 104
    modeling of creep in prototype systems,
        103–104
    Poisson's ratio considerations, 104
    thickness variations in commercial shapes,
        104
  time effects in plastics, 96–100
    determination of time-dependent modulus
        of elasticity and Poisson's ratio, 97–98
    loading techniques to account for time-
        dependent effects, 98–100
  wood and paper products, 104–120
    balsa wood, 105–114
    modeling of structural lumber, 114–115
    small-scale modeling of glue-laminated
        structures, 115–120
Elastic plate, free transverse vibrations of, 54
Elastic properties, influence of relative humidity on, 103
Elastic response, 3
Electrical resistance gages, disadvantages of, 325
Electromagnetic shapers, 601
Embedded strain gages, 358

Embedded stress meters, 357
End-web bonding, 192, 194
Energy dissipation, 506, 667
Engineering
  curriculum
    new, 725
    physical modeling and new, 8
  phenomena, measurements of, 442, 443
  programs, undergraduate, 726
Epoxy resins, 90, 91, 111, 356
Equation
  dimensionally homogeneous, 44
  prediction, 57
Equivalent formulation, 48
Error
  propagation, 444, 447
  types of, 439, 448
Euler buckling equation, 37
Experimental design engineer, 54
Experimental error, 439
Experimental stress analysis, 34

F

Fabrication
  accuracy, 451
  phase, planning of, 10
Failure
  mechanism, 87, 658, 665
  mode, 3, 176, 471, 513
Fastening techniques, 283
Fatigue
  deck behavior, 540
  loading, 570, 571
  testing, 33, 570
FBGS, see Fiber Bragg grating sensor
FEM, see Finite element analysis
FE model, see Finite-element model
FFPI, see Fiber Farby–Perot interferometer
Fiber Bragg grating sensor (FBGS), 368
Fiber Farby–Perot interferometer (FFPI), 368, 369
Fiber-optic sensor(s)
  earliest application of, 376
  elliptic-core two-mode, 371
  integration of into concrete structures, 373
  intracore Bragg grating, 370
  polarimetric, 371
Fiber-reinforced polymer (FRP), 14, 130, 210
  beams, moment–curvature relationships of, 524
  ductile, 522
  reinforcement
    for concrete models, 255
    nonductile, 256
  stress–strain characteristics of, 257
  systems, state-of-the-art linearly elastic, 528
Filler beams, 222
Fine aggregate, 131
Finite element analysis (FEM), 692, 693
Finite-element (FE) model, 559

Fink roof truss geometry, 117
First-order similarity, 60
Flange cracking, 709
Flexibility matrix, determination of, 728
Flexural bond, 192
Flexural failure modes, 484
Floor
 beams, 467
 displacements, 632
 loading system, string, 385
 slabs, composite, 473
 suspension mechanisms, 495
 system deflection pattern, 495
Flow visualization tests, 651
Fluidelastic models, 73
Folded plate action, 684
Force
 measurement, 353
 –deformation relationships, 320
 –displacement curves, postcracking, 637
Foundation
 link between super-structure and, 560
 settlement, during cyclic loading, 648
 stiffness, 642
Fourier amplitude spectra, 653
Fourier number, 69
Fracture mechanics, 414, 418
Frame
 action, 683
 model
  four-story building, 724
  steel-braced, 229
 test, 403
 -wall structures, 15
FRP, see Fiber-reinforced polymer
Full bridge, 337
Full-field strain measurements, 351
Function generator, 596, 716

G

Gable frames, 690
Gage(s), see also specific types
 capacitance, 326
 circuitry, resistance strain, 332
 configurations, 338
 embedding, 320
 factor
  error, 439
  value of, 334
 weldable, 326
Galileo's statement, 80
Gaseous explosions, in domestic surroundings, 75
Gaussian density function, 441
Geodesic Tri-Span, Drexel, 747, 749
Geometric distortion, 61
Geometric scale, 9
GFRP, see Glass fiber-reinforced polymer
Girder(s)
 balsa wood model plate, 743
 –column web joint detail, 222, 224
 reactions, 577
 web buckling in, 745
Glass fiber-reinforced polymer (GFRP), 569
GLD, see Gravity load design
Glue laminated (glulam)
 beams, 298
  double-tapered, 730
  model, 118, 728, 730
 structures, 115
Gluing techniques, 283
Glulam, see Glue laminated
Grain growth, 250
Graphical construction, 380
Gravitational acceleration, 77
Gravity deadweight stresses, 407
Gravity
 forces, 587
 forming, of shells, 293
 load, 403, 768
  design (GLD), 660, 667, 668
 simulation, 58
 stresses, simulation of, 395
Gross behavior observations, 457
Grout(ing)
 control specimens, 193
 effect of on prism compressive strength, 191
 model, 185, 186
 prisms, failure mode of, 193
 specimens, 198
 strength, 187
Gypsum
 Hydrostone, 174, 750
 mixes, 202
 mortar, 165
  mixes, 136
  size effects in, 430
  variation of modulus of rupture with size for, 428

H

Half bridge, 338
Hangar structures, 529
Hardwood species, North American, 26
Hickory, 106
High-speed train, 673
Highway bridges, 8, 9
Hinge
 details, 504
 formation, 710
Homemade rosette, 348
Hooke's law, 43, 92
Hoover Dam, 6
Hopkinson's law, 74, 76
Horizontal joints, 476, 488
Hydraulic material, 131
Hydraulic tension jacks, 491

Hydrostone gypsum, 174, 750

I

Impact
    loading, 641
    pendulums, 601
Independent random variable, 446
Indirect model, 3, 87, 122
Inductance strain gage, 327
Inelastic models, materials for concrete and concrete masonry structures, 129–208
    behavior in indirect tension and shear, 148–153
        correlation of tensile splitting strength to flexural strength, 152–153
        results of model split cylinder tests, 151–152
        tensile splitting strength, 149–151
        tensile splitting strength vs. age, 152
    design mixes for model concrete, 153–158
        choice of model material scale, 153–155
        important parameters influencing mechanical properties of concrete, 155–158
        properties of prototype to be modeled, 155
    engineering properties of concrete, 131–133
    flexural behavior of prototype and model concrete, 146–148
        influence of strain gradient, 148
        observed variations in modulus of rupture with changes in dimensions, 147–148
        rate of loading, 148
        specimen dimensions and properties, 147
        stress–strain curves, 147
    gypsum mortars, 165–170
        curing and sealing procedures, 168
        mechanical properties, 169–170
    model concrete mixes used by various investigators, 159–165
    modeling of concrete masonry structures, 170–188
        model grout, 185–188
        model masonry units, 171–181
        model mortars, 181–185
        prototype masonry units, 170–171
    prototype and model concretes, 130–131
    strength of model block masonry assemblages, 188–202
        axial compression, 189–192
        bed joint shear, 194–196
        diagonal tension strength, 202
        flexural bond, 192–194
        in-plane tensile strength, 196–199
        out-of-plane flexural tensile strength, 199–201
    tensile strength of concrete, 142–146
    unconfined compressive strength and stress-strain relationship, 133–142
        comparison of prototype and model concrete stress-strain characteristics, 135–136
        creep and creep recovery of concrete, 136–138
        effect of aggregate content, 138–139
        effect of strain rate, 139
        model concrete, 134–135
        moisture loss effects, 139–140
        prototype concrete, 133–134
        statistical variability in compressive strength, 142
        strength–age relations and curing, 140–142
Inelastic models, structural steel and reinforcing bars, 209–278
    bond characteristics of model steel, 259–266
    bond similitude, 266–267
    cracking similitude and deformation similitude in reinforced concrete elements, 267–272
    FRP reinforcement for concrete models, 255–259
        ductile hybrid fiber reinforced polymer, 256–259
        nonductile FRP reinforcement, 256
    model prestressing reinforcement and techniques, 252–255
        anchorage systems, 253–255
        model prestressing reinforcement, 252–253
    model reinforcement selection, 251–252
    reinforcement for small-scale concrete models, 230–252
        black annealed wire as model reinforcement, 233
        commercially deformed wire as model reinforcement, 234–239
        custom-ordered model wire, 233–234
        heat treatment of model reinforcement, 250–251
        laboratory wire-deforming machines, 239–249
        model reinforcement used by various investigators, 230–232
        wire reinforcement for small-scale models, 232–233
    steel, 210–214
        prestressing steels, 211–214
        reinforcing steel bars, 211
        structural steels, 211
    structural steel models, 214–229
        steel beams, 218–221
        steel columns, 215–218
        steel frames, 221–229
Infilled frames, 693
Influence
    diagrams, 3
    line, plotting of, 688
Instrument(s)
    error, 439
    manufacturers of, 329
Instrumentation, 319–382
    creep and shrinkage characteristics and moisture measurements, 362–363
    data acquisition and reduction, 364–367
        types of data recording, 364–365
        various data acquisition systems, 365–367
    displacement measurements, 348–351

INDEX   779

        linear resistance potentiometers, 350
        linear variable differential transformer,
           349–350
        mechanical dial gages, 349
    fiber optics and smart structures, 367–377
        criteria and selection of fiber-optic strain
           sensors, 372–373
        integration of fiber-optic sensors into
           concrete structures, 373–377
        types of, 368–372
    full-field strain measurements and crack detection
        methods, 351–353
        brittle coatings, 352
        other crack detection methods, 353
        photoelastic coatings, 352–353
    quantities to be measured, 320–321
    strain measurements, 322–351
        electrical strain gages, 323–332
        mechanical strain gages, 322–323
        resistance strain gage circuitry and
           applications, 332–348
    stress and force measurement, 353–361
        embedded stress meters and plugs, 357–360
        load cells, 354–356
        other measuring devices, 360–361
    temperature measurements, 361–362
Interference effects, 5
Internal blast effects, 75
Ironbark, 106
Iron
    –carbon system, of steels, 210
    castings, malleable, 417

J

Jacobian, definition of, 762
Jet aircraft, testing, 325
Joint(s)
    crack patterns observed in, 495
    failure of butt, 555
    gap, dry, 563
    mortar, 200
    openings, 489
    replacement, 23
    shear values, of model and prototype tests, 195
    stiffness, 624

L

Laboratory
    deformation technique, 240
    demonstration models, 685
    instrumentation, basic, 712
Lacquer, 352, 423
Laminar flow, 78
Large panel (LP)
    buildings, 486, 622
    construction, 284

Laser
    beam from He-Ne, 372
    system, copper vapor, 601
Lathe turning, 283
Lead wire resistance, temperature effect on, 337
LEFM, see Linear elastic fracture mechanics
Lifting equipment, 501
Light weight, 170
Lignum-vitae, 106
Lindberg open-tube furnace, 251
Linear elastic fracture mechanics (LEFM), 419
Linear potentiometers, 6
Linear variable differential transformers (LVDT), 6, 202,
    349, 595
Lions' Gate Bridge, 611, 615, 616
Liquid density, 78
Live oak, 106
Load(s), see also Loading
    application scheme, 737
    balancing of, 393
    cells, 329, 356, 663
        measurement of loads using, 456
        types of, 354
    deflections, dead and live, 738
    discrete vs. distributed, 389
    discretization, effect of on stress distribution in
        parabolic arch, 391
    indicator washer, 361
    modeling, 74
    rate, 87
    reaction systems, 384
    similitude requirements, 63
    spacing, effects of, 400
    tests
        front truss lateral, 536
        live, 28
    types of, 384
Loading
    beams, 491
    cases, 548
    comparison of stress–strain curves for, 593
    conditions, critical, 574
    devices, 385, 404
    equipment, 11
    to failure, 550
    history, 505
    impact, 641
    mistake in, 440
    pattern, maximum moment, 549
    platens, 422
    rate, 100, 148
    sea-wave, 646
    sequence, lateral, 503
Loading systems, laboratory techniques and, 383–409
    discrete vs. distributed loads, 389–390
    loading for shell models, 390–400
        discrete load systems, 395–400
        effects of load spacings, 400
        vacuum and pressure loadings, 392–394
    loading techniques for buckling studies and for
        structure subject to sway, 400–404
        shell instability, 401–402

structures undergoing sway, 403–404
miscellaneous loading devices, 404–407
    thermal loads, 406
    self-weight effects, 406–407
types of loads and load systems, 384–389
    loading devices, 384–386
    load reaction systems, 384
    pressure and vacuum loading systems, 386–389
Load and Resistance Factor Design (LRFD), 215
Locust, 106
Long-leaf pine, 106
Long-span bridges, 14
Longitudinal tie system, 492
LP, see Large-panel
LRFD, see Load and Resistance Factor Design
Lucite, 289
Lumber, modeling of structural, 114
Lumped-mass systems, vibrations of, 718
Lunar module, 607
LVDT, see Linear variable differential transformers

M

Machine
    direction (MD), 119
    parts, 324
Machining operations, 281
Mach–Zehnder interferometer (MZI), 368, 372
Mahogany, 106
Manila envelope paper models, 299
Maple, 106
Masonry
    anisotropic characteristics of, 201
    assemblies, strength characterizations of, 175
    blocks, mechanical properties of model concrete, 180
    buildings, 478, 482
    cement–aggregate ratio, 182
    failures, 196
    grouted drystacked interlocking, 559
    infilled, 693
    model(s)
        dynamic tests on unreinforced, 636
        grouting of, 310
        small-scale, 311
        unreinforced, 638
    mortars
        aggregate gradation curves for, 189
        size effects in, 434
    reduced-scale, 486
    reinforced concrete, 539
    scale factors for, 68
    shear walls, 475
    specimens, grouted conventional, 558
    -stiffened model, 693
    structure(s)
        fabrication technique for model, 313
        hypothetical full-scale, 639
        modeling of, 67, 170

system(s)
    drystack, 556
    interlocking, 28
    modified H-block, 29
test structures, static and dynamic response for, 488
units
    dry stack interlocking block, 12
    model, 171
unreinforced, 628
wall
    grouted model, 205
    panel, 479
Mass
    density, 17, 72
    simulation, artificial, 625
Material
    ductility, 256
    properties, 460
    requirements, 17
Materials systems and models, size effects, accuracy, and reliability in, 411–463
    accuracy in concrete models, 450–457
        accuracy in interpretation of test results, 456–457
        accuracy in testing and measurements, 455–456
        dimensional and fabrication accuracy, 451–454
        material properties, 454–455
    definition of size effect, 414
    errors in structural model studies, 437–439
    factors influencing size effects, 414–415
    influence of cost and time on accuracy of models, 458
    overall reliability of model results, 457–458
    propagation of random errors, 444–450
    size effects and design codes, 435–437
    size effects in metals and reinforcements, 433–434
    size effects in masonry mortars, 434–435
    size effects in plain concrete, 420–431
        evaluation of experimental research, 423–426
        evaluation of experimental work on tensile strength, 429
        experimental factors influencing size effects, 420–423
        experimental research on size effects, 423
        size effects in gypsum mortar, 430–431
        size effects in long-term properties of concrete, 429–430
        tensile and flexural strength, 426–428
    size effects in reinforced and prestressed concrete, 431–433
        bond characteristics, 431–432
        cracking similitude, 432
        ultimate strength, 432–433
    statistics of measurements, 441–444
    theoretical studies of size effects, 415–420
        classical theory of bundled strength, 415–416
        evaluation of theoretical studies, 420
        fracture mechanics approach, 418–420

other theoretical studies, 418
weakest link theory, 416–417
types of errors, 439–441
blunders, 439–440
random errors, 440
systematic errors, 440–441
Maxwell–Betti reciprocal theorem, 121, 122
MD, see Machine direction
MDOF, see Multi-degree-of-freedom
Measurements
accuracy in, 455
statistics of, 441
Measuring
devices, 360
system, establishment of, 758
Mechanical fastening, 283
Metal(s)
cutting of, 281
size effects in, 433
Methyl methacryenlates, 90
Microconcrete, 161
cement-based, 573
mix(es)
details, 160
study to improve, 161
model, 16
during construction, 536
effects of shrinkage or creep in, 454
elevation view of, 535
rate of moisture loss in, 142
shells made of, 515
tensile strength of, 167
Microducer, 360, 361
Microstrain, 460
Microwave antenna
construction of prototype, 297
model, 746
structure, 742
Milling, 217, 282
Missile
firing of by means of air pressure, 641
surfaces, testing, 325
Missouri corkwood, 106
Model(s), see also specific types
architectural engineering, 692
bar deformations, 247
beam section, 252
block masonry assemblages, strength of, 188
bridges, 33
for classroom demonstration, 681
column(s), 228
cross-section of model, 216
test setup for, 701
concrete(s), 130, 132, 134
casting, 314
dynamic tests on, 591
gypsum products used in, 165
mixes, 159, 160
–reinforcement interface, 265
static properties of, 590
confidence, 413, 437
cylinders, strength–age curves of, 144

deformation profiles, 512
dimensions, 244, 544
elements, tolerances of, 226
engineer, 320, 412
failures, comparison of, 516
girder, 228
grout, 185, 186
influence of cost and time on accuracy of, 458
installations, view of, 619
laboratories, 13
length, 78
load applied to three-dimensional, 494
masonry specimens, 199
material(s), 469
distorted, 62
scale, 153
strength of, 424, 426
mortar(s), 181
aggregate used for, 182
sand, 192
strength–volume relations for, 435
prestressing reinforcement, 252
reinforcement, 210, 243, 618
commercially deformed wire as, 234
deformed, 245
geometric properties of, 249
heat treatment of, 50
selection, 251
steps needed for producing, 231
reliability of, 401
response of, 11
results
comparison of, 609
reliability of, 457
scaled, 412
shear
test specimen, 197
wall, 479
slab(s)
test matrix of, 521
thickness measurements, 452
specimen, basic dimensions of, 471
spline-type, 683
steel frameworks, 221
strain, measurement of, 448
studies
errors in structural, 437
reliability in, 459
tabletop, 513
test(s), 7, 321, 509
application of under dynamic loading, 642
Drexel University dynamic system, 716
possible results of, 449
specimens, 468
two-mold drape forming of, 294
ultimate-strength, 459, 694
wall(s)
load–deflection characteristics of, 477
tests, 475
wire, customer-ordered, 233
Model applications, case studies and, 465–583
case studies, 529–572

externally/internally prestressed concrete composite bridge system, 568–572
  interlocking mortarless block masonry, 556–560
  pile caps, 560–568
  prestressed wooden bridges, 542–556
  reinforced concrete bridge decks, 537–542
  TWA hangar structures, 529–537
 modeling applications, 466–528
  bridge structures, 497–507
  building structures, 466–497
  special structures, 507–529
Model fabrication techniques, 279–317
 basic cutting, shaping, and machining operations, 281–283
  cutting of metal, plastic, wood, and paper products, 281
  drilling and milling, 282–283
  lathe turning and boring, 283
  shaping and machining operations, 281–282
 construction of plastic models, 288–293
  capillary welding, 292
  casting of plastic models, 296
  drape or gravity forming and drape molding of shell models, 293–294
  fabrication considerations, 288–292
  fabrication errors in thermal forming, 295
  spin forming of metal shells, 296
  thermal forming processes, 292–293
  vacuum forming, 294–295
 construction of structural steel models, 287–288
  silver soldering, 287
  tungsten inert gas welding, 288
 construction of wood and paper models, 296–299
  balsa wood models, 296–298
  cartridge or manila envelope paper models, 299
  glue laminated beams, 298–299
  structural wood models, 298
 fabrication of concrete masonry models, 309–312
  building model masonry components and assemblies, 309–311
  Drexel University/NCMA block-making machine, 309
  new Drexel model block-making machine, 309
 fabrication of concrete models, 299–308
  forms for casting reinforced concrete models, 302
  prestressed concrete models, 302–308
  reinforced concrete models, 299–302
 fastening and gluing techniques, 283–287
  epoxy resins, 286–287
  glues and adhesives, 286
  mechanical fastening, 283–284
  soldering, 284–285
  spot welding, 285–286
Modulus of elasticity, 50, 341, 438, 517
 effect of aggregate on, 141
 ratio of dynamic to static, 592
Modulus of rupture tests, 169, 431
Moisture
 gages, electrical resistance, 363
 loss
  effects, 139
  rate of in microconcrete, 142
 measurements, 362
Mold box, 311
Moment
 –curvature relationships, theoretical, 523
 deformeter, 6
 loading, 552
 rotation, 713
Mortar(s)
 beam dimensions, details of, 452
 –block interfaces, 198
 cylinders, model, 184
 gypsum, 165
 joint, 200
 mix, model, 185
 model, 181, 509, 515
 proportions, 188
 sands, 166, 192
 strength–volume relations for model, 435
 tensile strength of, 130
Mortarless blocks, 175, 556
Müller–Breslau principle, 121, 688, 690
Multi-degree-of-freedom (MDOF), 721, 723
Multispan girder, 46
Multistory buildings, elastic models of, 86
MZI, see Mach–Zehnder interferometer

N

National Center for Earthquake Engineering Research, (NCEER), 21, 663
National Concrete Masonry Association (NCMA), 173
Natural vibration modes, 604
NCEER, see National Center for Earthquake Engineering Research
NCMA, see National Concrete Masonry Association
Nickel–iron alloy, 328
Normal probability density function, 441
Normal weight, 170
Nuclear containment vessels, testing, 325
Nuclear reactor vessel, 8
Numerics, 761
Nusselt's number, 69

O

Offshore structures, 8
Ontario Highway Bridge Design Code, 20, 27
Oscillators, electromagnetic, 594
Oscilloscope, 597, 718

INDEX 783

P

Panel zone shear strength, 473
Paper
    models
        examples of, 120
        manila envelope, 299
    products, 104
        cutting of, 281
        used for structural models, 119
Partial differentiation, chain rule of, 758
Particle friction, 78
PCA, see Portland Cement Association
PCRV, see Prestressed concrete reactor vessel
PCs, 367
Pea gravel, 10
Pedestrian walkway, elevated, 741
Perfect universal strain gage, 322
Permeability, 78
Photoelastic coatings, 352, 353
Photomechanical models, 5
Physical modeling, introduction to in structural
            engineering, 1–39
    accuracy of structural models, 12–13
    advantages and limitations of model analysis,
            11–12
    choice of geometric scale, 9–10
    definitions and classifications of structural models,
            2–5
        models classifications, 3–5
        physical models in other engineering
            disciplines, 5
    historical perspective on modeling, 6
    modeling case studies, 13–34
        externally/internally prestressed concrete
            composite bridge system, 31–34
        interlocking mortarless block masonry,
            27–30
        lightly reinforced concrete buildings, 21–24
        pile foundations, 30–31
        prestressed wooden bridges, 24–27
        R/C frame-wall structures, 15–20
        reinforced concrete bridge decks, 20–21
        TWA hangar structures, 14–15
    modeling process, 10–11
    model laboratories, 13
    physical modeling and new engineering
            curriculum, 8–9
    structural models and codes of practice, 7–8
Pi terms, formation of, 50
Pi theorem, 80
Pier behavior, 641
Piezoelectric principle, 327
Pile
    cap, 560, 561
        model, 33
        truss analogy for, 567
    foundations, 30
Plane angles, measurement of, 768
Plasma arc process, 246
Plaster of Paris, tensile strength of, 417

Plaster ratio, aggregate–gypsum, 169
Plastic(s)
    acrylic, 291
    cutting of, 281
    flexural characteristics of, 89
    influences of strain rate on mechanical properties
            of, 100
    models
        casting of, 296
        problems related to, 103
    tests on, 91
    thermal conductivity of, 102, 294
    thermosetting, 98
    time-dependent behavior of, 97
    time effects in, 96
Plate
    middle surface, out-of-plane displacement of, 54
    ultimate strength of, 512
Plexiglas, 289, 451, 454
    beam, 689
    forms, for casting reinforced concrete models, 303
    model(s), 81, 673
        measuring strains on, 380
        tests on, 102
    mold, 699
    stress–strain curve for, 93
Plywood forms, 302
Poisson's ratio, 50, 341, 438
    considerations, 104
    determination of, 97
    discrepancy, 62
    mechanical properties of, 94
Polyester resins, 90
Polyethylenes, 90
Polymerization, 101
Polymethyl methacrylate, 293
Polyurethane, 423
Polyvinylchloride (PVC), 88, 90, 293, 454
    insulated leads, 362
    plastic model, 58
    stress–strain curve for, 93
Portal frame, 269
Portland cement, 131, 186, 417
Portland Cement Association (PCA), 489, 497
Postbuckling response, 472
Postcracking deflections, 261
Potentiometers, 365
Power
    amplifier, 716
    spectral density, frequency vs., 645
Preseismic tests, 621
Pressure loading, 388, 391, 392
Pressurized gas, use of directly against model, 386
Prestressed concrete reactor vessel (PCRV), 507
    physical models of for boiling water reactors, 510
    scale models, 508
Prestressing
    frame, 308
    system, 305
Pretensioning technique, 302
Prism(s)
    compression tests, 194

configuration of fabricated, 560
mode of failure for model, 195
prototype, 189
test(s)
    parameters derived from grouted, 561
    specimens, 191

Product
    development, 466, 522
    dimensionless, 760

Prototype
    bar, stress–strain curves of, 245
    beam, 321, 689
        load–deflection curves for, 268
        section, 252
    behavior, similarity between model and, 196
    bending, 60
    block masonry, 198
    building, beam–column joint details of, 24
    column, 228
    concrete, 63, 133, 203
        hypothetical, 158
        test results, 592
    data, correlating model data with, 434
    design of, 507
    dimensions, 244, 544, 706
    floor slabs, 227
    full-intensity earthquake, 626
    girder, 228
    gusset plates, 117
    joint openings, 493
    loads, representing, 384
    masonry
        concrete blocks, 183
        units, 170
    materials, 503
    mode surveys, 610
    multiple, 12
    piers, 500
    plate, 55
    prisms, 189
    properties of, 155
    reinforcement, 455
        geometric properties of, 249
        models, 248
    slabs, 270
    space truss members, 112
    steel frame structure, 631
    strains, 271, 274
    stresses, systematic error in predicted, 460
    structure(s)
        geometry of, 118
        modeling studies on, 2
    systems, modeling of creep in, 103
    test(s)
        cylinder size, 423
        data, 495
        joint shear values of, 195
        results, 200
        truss, 110
    wall strength properties, 480
Pseudodynamic tests, 658

Pullout tests
    average bond stress vs. L/D ratio from, 264
    concentric, 263
Punching shear failure, 21, 539, 564
PVC, see Polyvinylchloride

Q

Quarter bridge, 338

R

Radiation scaling, 69
Random errors, 444, 448
Random strength, 420
Reactor system, gas-cooled, 510
Recording equipment, 10
Recrystallization, 250
Rectangular rosette, 340, 380
Red oak, 106
References axes, 342
Reinforcement(s)
    bars, commercially available, 239
    bending of, 299
    –concrete interface, 130
    increasing horizontal, 476
    layout, 485
    placement, accuracy of, 300
    ratio, ultimate shear strength, 566
    size effects in, 433
    welding of, 300
Repeatability checks, 457
Replica model, 3
Residual stresses, 215
Resonant testing, 593
Rigid models, 4
Ring loading device, 393
Rockets, testing, 325
Rocking
    curves, 629
    response, base acceleration vs., 630
Rod gages, functioning as mechanical gages, 547
Roof
    displacements, base shear vs. roof, 474
    drift, 661
    structure(s)
        deflection analysis of, 314
        shell-and-dome, 121
    system, structure of tension, 674
    truss, Single-Fink, 730
Rosette(s)
    assembly, 358
    configuration, 347
    delta, 340, 344, 378
    homemade, 348
    Murphy's method for plotting strain, 348
    rectangular, 340, 380

INDEX

strain, 345
three-element, 339
types of, 343
Rotation-measuring devices, 617
Rubber, synthetic, 90

S

Sand
    –gravel ratio, 163
    tank, 647
Sanding sealer, 115
Scaffolding, 534
Scale
    factors, 9, 57, 579
    model testing, 605
Screeding strips, 536
SDOF, see Single-degree-of freedom
Sealing
    compounds, commercial, 168
    procedures, 168
Sea-wave loading, 646
Seismic isolation, 723
Seismic test(s), 664
    mode of vibration dominated in, 666
    results, three-story, 665
Self-temperature-compensated (STC) gages, 335
Self-weight effects, 406
Semicontinuous recording systems, 364
Sequential phased displacement (SPD), 475
Settlement curves, 648
Shaking machine, 647
Shaking table, 223, 225, 600
    classification of, 602
    Cornell University, 663
    Drexel Models Laboratory, 752
    Drexel University, 735
    NCEER, 22, 24
    performance limits of, 719
    small linear, 719
    Stanford University, 631
    studies, European, 628
    SUNY/Buffalo Earthquake Simulation Laboratory, 669
    testing
        maximum response from, 670
        sequence, 669
    U.C. Berkeley, 18
    University of Illinois, Urbana-Champaign, 639
Shaking test
    comparison of damage states after, 672
    severe, 671
Shape models, 4
Shear
    buildings, dynamics of, 722
    center, 683
    loading, 551, 552
    strength, 202
    test specimen, model, 197

Shell(s)
    geometry, 518, 520
    instability, 401
    model(s), 396
        dimensions of, 513
        drape molding of, 293
        suspended dead loads on, 395
    roof model, 402
    stiffness, 513
    structures, 512
    test apparatus, 394
    vacuum forming of doubly curved, 295
Shock tube(s), 598
    pressure distribution in, 600
    scale model placed in, 599
Shrinkage, 321, 429
    characteristics, 362
    deformations, 222
    effects of in microconcrete model, 454
SI units, use of in structural engineering, 767–770
    densities, gravity loads, weights, 768
    force, moment, and other stress resultants, 768
    geometry, 768
    miscellaneous, 768
    SI system standard practice, 769–770
Sieve analysis, 155
Silver soldering, 225, 287
Similarity, first-order, 60
Similitude
    criteria, 498
    requirements, 10, 42, 43, 62
        distortion, 65
        for elastic vibrations, 72
        load, 63
        for static elastic modeling, 86
    self-weight, 663
Single-degree-of freedom (SDOF), 720
Single-Fink root truss, 730
Site welding, 498
Size
    dependence, 435
    effect(s)
        experimental research on, 423
        factors influencing, 414
        theoretical studies of, 415
Slab(s)
    action, two-way, 539
    cracking, 517, 672
    failure mode of, 527
    multiple wires in, 307
    pressure systems for dynamic loading of, 389
    punching shear investigations of, 433
    -punching strength, 575
    shear connection between beam and, 577
    strip method of designing, 521
    text matrix of model, 521
    underreinforced, 432
Small-scale model(s), 321, 454
    acceleration and displacement response of, 622
    fabrication techniques for, 312
    power of, 751

size effects in, 472
Smart structures, 367
Soft-metal models, 280
Soil
    density, 78
    foundation, 646
    models, 604
    settlement, vertical displacement due to, 646
    –structure interaction, 642
Soldering, 284
    low-temperature, 306
    silver, 287
    use of, 300
Space truss, 566
SPD, see Sequential phased displacement
Specimen(s)
    basic dimensions of model, 471
    braced frame, 470
    crack pattern of failed, 480
    differential drying of, 430
    masonry, 558
    sizes, Weibull's plots for, 417
    steel tension, 589
Spin forming, of metal shells, 296
Spline-type model, 683
Split cylinder
    specimens, 150
    tests, 149, 151, 431
Spot welding, 285
Spray paint, 423
Spring balance, 99
Spring-mass systems, 720
Stability model, for folded-plate structure, 402
Stacking techniques, 29
Static deck behavior, 538
Static elastic modeling, 62
Static tests, 487, 500
STC gages, see Self-temperature-compensated gages
Steel(s)
    beams, 218
    bond
        characteristics of model, 259
        between concrete and, 263
    building(s)
        earthquake simulation of, 625
        model studies, U.S.–Japan, 226
    cars, reinforcing, 211
    columns, 215, 216
    –concrete interface, 66, 230
    deformed bars, 620
    dynamic properties of, 588
    frame(s), 221
        impulsive test, 643
        structure, prototype, 631
        tests, prototype, 225
    girders, 541
    hot-rolled, 219
    load–deflection behavior of, 531
    materials suitable for modeling, 272
    member, inelastic buckling of, 710
    mild, 111, 213, 490
    model(s)
        construction of structural, 287
        material simulating, 213
        structural, 214
    modulus, 770
    moment–curvature predictions of beams reinforced with, 530
    orthotropic, 20
    prestressing, 211
    ratio, 563
        flexural, 566
        isotropic, 538
    reinforcement
        isotropic, 20
        ratio, 529
        replacement of, 255
    ring beam support, 516
    rods, cold-rolled threaded, 230
    specimen, hot-rolled, 220
    stress, allowable, 770
    stressing rods, 551
    structure, plan view of prototype, 467
    tension specimens, 589
    variation of reinforcing, 561
    -to-wood ratio, 546
    yielding, 267
    Young's modulus of, 109
Stiff testing table system, for small models, 386
Stirrup fabrication, 301
Story
    drift, 633, 664
    shear, 641
Strain
    cell, 99
    data, recording of during test, 512
    distribution, 426
    field, 339
    gage(s), 327, 343
        adhesives, 345
        circuits, temperature compensation in, 335
        electrical resistance, 323, 324, 456
        embedded, 358, 359
        incomplete bonding of, 450
        instrumentation, 380
        major types of, 322
        selection chart, 327
        series selection chart, 330–331
        vibrating wire, 360
    increment factor, 357
    indicator equipment, 102, 273
    measurements, 3, 351, 564
    Mohr's circle for, 347, 378, 379
    rate, 139, 422
    readings, 340
    relieving, 250
    sensing filaments, 328
    sensors, embedment of fiber-optic, 373
Strength
    –age
        relations, 140
        tests, on compression cylinders, 184
    mode, 3
    summation theory, 418

Stress(es)
    calculation of, 340
    conditions, anticipated, 453
    meters, embedded, 357
    necessary instrumentation to measure, 576
    plug method, 357
    relaxation, long-term effects of, 545
    -strain relationship, 133
    strains converted into, 342
Stressteel
    bars, 507
    rods, 510
Stress–strain
    curves
        for cable, 253, 254
        experimental, 147
        nondimensional, 136
        for prestressing strands, 212
    laws, 61
Strip method, of designing slabs, 521
Strobotac, 718
Structural engineer, 576, 587
Structural model(s), 412
    accuracy of, 12
    definition of, 2
    new category of, 466
Structural problems, types of, 60
Structural steel, effect of rate of strain on stress–strain curve for, 589
Student projects, 729
Super-structure, link between foundation and, 560
Surface theory, 418
Suspension bridge oscillations, 73
Sway, structures undergoing, 403
Sycamore, 106
Symmetry checks, 457
Systematic errors, 440, 448, 450

T

T-beams
    cracking patterns of model, 709
    dimensions of, 706
    moment–curvature curves for, 707
Temperature measurements, 361
Tensile splitting
    strength, 149, 152
    testing, 162, 173
Tensile strain, effect of strain distribution on, 427
Tensile strength
    experimental work on, 429
    grout splitting, 201
    in-plane, 196
    properties, determination of, 107
    split cylinder, 707
Tensile test data, 523
Tension
    load cells, 354
    tests, 218
Terrain roughness, 73

Terroristic blast attacks, threats of, 488
Test
    beams, ductility indexes for, 527
    results, accuracy in interpretation of, 456
    structures
        concrete masonry, 484
        structural configuration of, 485
Testing
    accuracy in, 454
    machine, 403, 422
    table, frame fastened to models, 401
    techniques, 650
Theory, of structural models, 41–83
    dimensional analysis, 45–55
        additional; considerations in using dimensional analysis, 53–55
        Buckingham's pi theorem, 49
        dimensional independence and formation of pi terms, 50–52
        uses of dimensional analysis, 52–53
    dimensions and dimensional homogeneity, 42–45
    similitude requirements, 62–76
        models of masonry structures, 67
        reinforced concrete models, 63–67
        structures subject to dynamic loadings, 70–76
        structures subject to thermal loadings, 67–69
    structural models, 56–62
        distorted models, 61–62
        models with complete similarity, 56–57
        models with first-order similarity, 60
        technological difficulties associated with complete similarity, 58–60
Thermal conductivity, of plastics, 102
Thermal expansions, coefficients of, 101
Thermal forming
    fabrication errors in, 295
    processes, 292
Thermal loadings, structures subjected to, 67
Thermal modeling, scale factors for, 68
Thermal structural modeling, 62
Thermistor, 362
Thermocouples, 365
Thermoplastics, 88, 292
Three-dimensional model, 490, 494
TIG, see Tungsten inert gas
Tinius-Olsen extensometer, 257
Tinius-Olsen testing machine, 563
Titanium, 111
Torque
    –twist similitude, postcracking, 270
    wrench calibration, 307
Torsion test, 98
Tower block, isolated, 613
Transducer gaging, 332
Transverse stresses, 117
Trans World Airlines Maintenance Hangar Facility, 14
Truck loading, 555
True model, 56
Truss
    bracing, 499

during testing, 733
load–deflection of model, 734
member, measuring strain in, 747
model, 730
simple, 685
theory, 561
Tungsten inert gas (TIG), 220, 288, 290

U

Ultimate-strength
    model, 548
    test, load configuration for, 550
Ultracal model concrete mixes, 168
Undersea structures, 8
Unit compression strength, 185
Universal gage, 355
Universal load cells, 354
Universal temperature compensating gage, 336

V

Vacuum
    forming, 294
    loading, 387, 388, 391, 392
Venting area, 75
Vertical casting, 147π
Vertical tie system, 492
Vibrating wire gage, 359
Vibration(s)
    forced, 720
    modes, natural, 604
    problems, of elastic structures, 70
    properties, during test sequence, 623
    systems, educational, 718
    test(s)
        on Drexel University shaking table, 735
        results, for building frame model, 724
    problems of mechanical, 718
Viscoelastic material, 92
Void ratio, 78
Voltmeter, digital, 364

W

Walkway
    configuration of plate girders forming supporting, 743
    testing model, 742
Wall(s)
    amplifications, 636
    assembly
        deflection curves for first-floor interior, 496
        end deflections, 496
    component assembly, cantilever, 488
    damage pattern, 662
    displacements, amplified by flexible diaphragms, 636
    effect, 421
    failure, 491
    –floor connection, 634
    load–deflection characteristics of, 477
    model
        infilled shear, 753
        precast shear, 732, 734
        two-story, 491
    in out-of-plane bending, 473
    panel(s)
        masonry, 479
        removable, 493
    section, moment–axial interaction of, 659
    strength properties, prototype, 480
Water
    aggregate ratio, 159
    –cement ratio, 132, 136
    density, 770
Weakest link concept, 416, 417
Web
    buckling, 742, 745
    cracking, 709
Weibull
    plots, for specimen sizes, 417
    statistical theory, 418
Weld, prevention of oxidation of, 288
Weldable gages, 326
Welding
    capillary, 292
    of deck elements, 501
    of reinforcement, 200
    site, 498
    spot, 285
Weldwood® resorcinol glue, 126
WHD block, 180
Wheatstone bridge, 335, 337, 360, 378
Wheel-load setup, 22
Whiffle tree loading system, 395, 407
White oak, 106
White pine, 106
Whittemore gage, 323
Wind
    direction, 649
    effects model, 4, 531, 532
    loads, 86
    model, solid mahogany, 651
    pressures, distribution of external, 531
    profile, artificially generated, 534
    tunnel
        boundary layer, 599
        environmental, 598
        modeling, 7
        tests, of Toronto City Hall, 649
        University of Toronto subsonic, 650
Wire(s)
    -deforming machines, 239, 242, 247
    circuits, 323
    commercially deformed, 234
    customer-ordered model, 233
    deformed model, 262

deformeter
    Cornell University, 239
    Drexel University, 246
examples of deformed, 248
laboratory-deformed annealed, 699
meshes, used for reinforcement, 613
properties, 237
reinforcement, for small-scale models, 232
-soldering gun, 284
steel, chemical composition of, 235
temperature variation in lead, 336

Wood(s)
    cutting of, 281
    deck
        deadweight stresses in, 547
        models, splice pattern for, 545
    products, 104
    species, weights of common, 106
    strength properties of, 114, 116
    structure modeling studies, 124
    tensile strength of, 417
    viscoelastic properties of, 546

Wooden forms, plastic-coated, 451
Wooden models, examples of, 115
Workmanship-related errors, 309
Workstations, 367

Z

Zarate-Brazo Largo Highway-Railway Bridge, 497